Plant Biology

Plant Biology

Second Edition

Linda E. Graham
University of Wisconsin

James M. Graham
University of Wisconsin

Lee W. Wilcox
University of Wisconsin

PEARSON

Prentice Hall

Upper Saddle River, NJ 07458

Library of Congress Cataloging-in-Publication Data

Graham, Linda E.
   Plant biology / Linda E. Graham, James M. Graham, Lee W. Wilcox—2nd ed.
      p. cm.
   Includes index.
   ISBN 0-13-146906-1
      1. Botany—Textbooks.   I. Graham, James M.   II. Wilcox, Lee Warren.   III. Title.

QK47.G68 2006
580—dc22

2005043043

Publisher: Sheri L. Snavely
Project Manager: Karen Horton
Production Editor: Donna King
Executive Managing Editor: Kathleen Schiaparelli
Assistant Managing Editor: Beth Sweeten
Development Editor: Erin Mulligan
Editor in Chief, Development: Carol Trueheart
Managing Editor, Science Media: Nicole M. Jackson
Senior Media Editor: Patrick Shriner
Marketing Manager: Andrew Gilfillan
Manufacturing Buyer: Alan Fischer
Composition: Progressive Publishing Alternatives
Assistant Managing Editor, Science Supplements: Becca Richter
Director of Creative Services: Paul Belfanti
Art Director: John Christiana
Interior Design: Maureen Eide
Senior Managing Editor, Audio and Visual Assets: Patricia Burns
AV Production Manager: Ronda Whitson
AV Production Editor: Xiaohong Zhu
Art Production Coordinator: Dan Missildine
Artwork: Lee W. Wilcox
Director, Image Resource Center: Melinda Reo
Manager, Rights and Permissions: Zina Arabia
Interior Image Specialist: Beth Brenzel
Cover Image Specialist: Karen Senatar
Image Permission Coordinator: Debbie Latronica
Photo Researcher: Yvonne Gerin
Editorial Assistant: Lisa Tarabokjia
Cover Design: John Christiana/Maureen Eide
Cover Photo: Lee W. Wilcox

© 2006, 2003 by Pearson Education, Inc.
Pearson Prentice Hall
Pearson Education, Inc.
Upper Saddle River, NJ 07458

Printed in the United States of America

10 9 8 7 6 5 4 3 2

ISBN 0-13-146906-1

Pearson Education LTD., *London*
Pearson Education Australia PTY, Limited, *Sydney*
Pearson Education Singapore, Pte. Ltd
Pearson Education North Asia Ltd, *Hong Kong*
Pearson Education Canada, Ltd., *Toronto*
Pearson Educación de Mexico, S.A. de C.V.
Pearson Education—Japan, *Tokyo*
Pearson Education Malaysia, Pte. Ltd

*We dedicate this book to our
families, friends, students, and teachers.*

About the Authors

Linda E. Graham is Professor of Botany and Environmental Studies at the University of Wisconsin–Madison. She received her Ph.D. in Botany from the University of Michigan, Ann Arbor. Dr. Graham has taught a nonmajors plant biology course each year for more than 20 years. She has a strong desire to inspire students to learn about plants as a way of understanding and appreciating nature. Dr. Graham's teaching focuses on biological topics that every informed citizen should understand in order to make responsible decisions about both the environment and personal well-being. She also teaches courses on the biology of algae and bryophytes, contributes to an introductory biology course for majors, and has taught marine botany on a remote tropical island. Dr. Graham's research explores the evolutionary origin of land-adapted plants, focusing on their cell and molecular biology as well as ecological interactions. Dr. Graham's research and teaching are connected—both inspired by a desire to help preserve the life-sustaining properties of the natural world. Dr. Graham is the co-author of *Algae*, a majors textbook on algal biology, as well as the author of *Origin of Land Plants*. Dr. Graham is a past President of the Botanical Society of America and is a Fellow of the American Association for the Advancement of Science.

James M. Graham received his Ph.D. in Biological Science from the University of Michigan, Ann Arbor. He is an Honorary Fellow at the University of Wisconsin–Madison, where he conducts research in the area of microbial ecology. Dr. Graham contributed a chapter on phytoplankton ecology to the textbook *Algae*, by L. Graham and L. Wilcox. He has also taught a number of courses, including ecology, biology of algae, introductory biology for majors, and introductory botany for non-science majors.

Lee W. Wilcox received his Ph.D. in Botany from the University of Wisconsin-Madison. His research interests include symbiosis, algal evolution, and plant and algal cell biology. Dr. Wilcox designed the art programs for both *Algae* and *Plant Biology* and has provided many original photographs to both texts. He has also contributed scientific illustrations to a variety of other scientific articles and book chapters. During his experience as a graduate teaching assistant and, later, as a guest lecturer in a variety of courses, he became acutely aware of the need to illustrate subject matter clearly and with an eye toward aesthetics in order for students to best appreciate the material.

Brief Contents

Contents

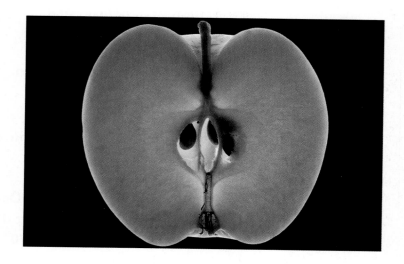

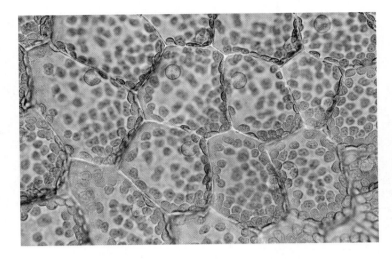

5 Photosynthesis and Respiration   80

9 Stems and Materials Transport 150

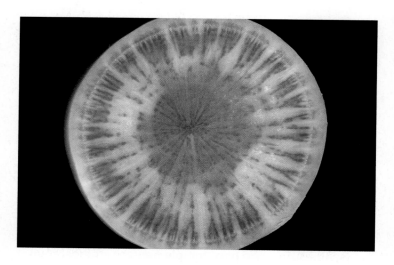

10 Roots and Plant Nutrition 170

PART III Plant Reproduction, Genetics, and Evolution

14   Genetics and the Laws of Inheritance   244

15   Genetics Engineering   264

16 Biological Evolution   278

PART IV   Diversity of Plants, Prokaryotes, Protists, and Fungi

17   Naming and Organizing Plants and Microbes   298

18   Prokaryotes and the Origin of Life on Earth   314

19 Protists and the Origin of Eukaryotic Cells   334

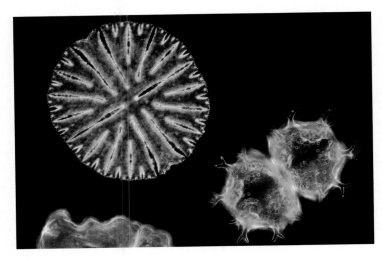

20 Fungi and Lichens   358

21 Seedless Plants   376

22 Gymnosperms, the First Seed Plants   400

23 Angiosperm Diversity and Reproduction 418

24 Flowering Plant and Animal Coevolution 440

PART V Ecology and Plant Adaptations to the Environment

25 Principles of Ecology and the Biosphere 460

26 Arid Terrestrial Ecosystems   482

27 Moist Terrestrial Ecosystems   512

28 Aquatic Ecosystems   544

29 Human Impacts and Sustainability 572

Preface

To the Student

Throughout your lifetime you will make many decisions that will influence the ecological and social integrity of our world, as well as your personal life. You will vote, support political candidates, and serve on juries. You may lead major corporations or government agencies, or teach thousands of students. You will decide whether or not to use herbal supplements, consume genetically modified foods, favor organically grown products, harvest wild plants or mushrooms, recycle wastes, purchase energy-efficient vehicles and appliances, select natural fiber clothing, plant gardens or woodlands, or support conservation organizations. A basic knowledge of biology—biological literacy—will provide invaluable background as you make these important decisions.

The study of plant biology is an exciting way to learn many fundamental and useful concepts in biology. Who can deny the appeal of beautiful flowers or breathtaking forested landscapes? While exploring the fascinating world of plants in your course, you will learn that plant biology is more than learning the names of plants and their parts. Plant biology also considers how and why plants are so important in the world, explaining many practical applications and issues appearing in the news media. This textbook is designed to aid your discovery by focusing on the biological concepts that every educated citizen should know in order to make well-informed decisions that will affect us all. We hope that you will enjoy your studies of plants as much as we have.

To the Instructor

Our increasingly technological society requires citizens who have been educated in basic science process and biology, because these serve as a basis for making rational decisions in many critical areas. Increasing the level of scientific and biological literacy among college students is certainly one of the most important instructional contributions we make, but achieving a high level of student involvement can be a challenge. Plants offer a particularly inviting way to impart biological and scientific literacy because plants are intrinsically attractive. Students are often very interested in plants as sources of useful materials and as components of familiar and exotic ecosystems. In writing *Plant Biology*, our principal goal has been to foster biological and scientific literacy. We work toward this by focusing on the function of plants in the world, linking important concepts to student interests and issues that appear in the news media, and providing many attractive, relevant, and compelling examples. We endeavor to promote student awareness of humans' evolutionary and ecological relationships with other life on Earth, in an effort to cultivate environmental ethics.

Special Features

Each chapter opens with a short list of key concepts that will be addressed in the chapter and an introductory story. To stimulate student interest, each chapter opening story links the key concepts to a familiar facet of everyday experience or human interest. For example, Chapter 16—Biological Evolution—opens with a description of Charles Darwin's concerns as a college student and how those influenced his epochal journey on HMS *Beagle*. The text includes many other examples of plant scientists and their discoveries, because these examples illuminate the process of science, and provide the means for students to search library databases and websites for additional information. Key concepts and opening stories illustrate why it is important for educated people to know something about the topics covered in this book. Essential chapter terms set in bold type and their glossary definitions provide students with a biological vocabulary that will remain useful long after they have completed your course.

Each chapter ends with Highlights that summarize the main concepts and information presented in each section. Review questions and questions that encourage students to apply the material are also provided at the end of each chapter. Suggested answers are provided at the end of the book.

Each chapter also includes one or more essays that provide examples of how humans apply basic biological information, or that extend learning. "The Botany of Beer, "Heavy-Metal Plants," "Supermarket Botany," "Molecular Detectives: DNA Fingerprinting Solves Crimes," "Killer Algae," and "Restoring a Lost Forest." are examples of essay topics that will engage and motivate your students. Several essays address many students' interest in cultural aspects of plant uses. These include "Ethnobotany: Seeds of Culture," "Native American Uses of Temperate Forest Plants, " and the new essay "Aspen Aspirin." Some essays are designed to extend learning, particularly for students who plan to continue their studies of biology or plant sciences. Examples include "Psi-chological Stress" (a new essay on plant cell water relations), "Growth Rings: Mirrors Into the Past," "The Guardian Spirit" (a new essay on molecular aspects of meiosis), and "PCR: The Gene Copier."

Throughout this book, plant structural and functional topics are integrated, rather than being treated separately. For example, plant water relations and water transport are such important topics that they are integrated into multiple chapters, including those focusing on cells, roots, stems, and leaves, as well as moist and arid ecosystems. Plant mineral nutrition is likewise addressed in multiple chapters, integrated with discussions of crop domestication, prokaryote and fungal symbioses, root mineral absorption, insectivorous plants, and degradation of aquatic ecosystems. In this way, particularly important concepts are revisited in multiple contexts, which aids student retention of knowledge.

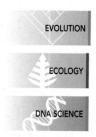

Three central themes—global ecology, evolution, and DNA science—that are both essential to biological literacy and relevant to everyday issues form this text's focal points. A chapter is devoted to each of these important topics, and in addition, they are integrated throughout the text, marked by icons to aid student recognition.

Art Program

The art program has been carefully crafted to attract and keep student interest and aid learning. Hundreds of original photos and illustrations of exceptionally high quality were developed by one of us (Wilcox) as the text was written. The art is thus tightly integrated with the text. Illustrations were designed to increase comprehension (even among students having little science or biology background) by stepping students through complex processes and highlighting relevant details.

Features New to the Second Edition

Features new to the second edition of *Plant Biology* include reorganization of content, in response to reviewer recommendations. For example, we have moved chapters on prokaryotes and protists to a new diversity section that also includes expanded coverage of seedless plants, seed evolution, gymnosperms, and plant life cycles. We have developed a new chapter, *Plant Behavior*, with expanded coverage of plant responses to physical and biological stimuli. Our *Biological Evolution* chapter includes new material illustrating rapid selection and adaptive response during evolution of bacterial antibiotic resistance. This new material provides very compelling evidence for the fact of evolution and its practical application to human affairs. As recommended by reviewers, we have also added new material on plant embryology, marine ecosystems, wood anatomy, and plant sex chromosomes. The ecosystem material has been reorganized into three coherent chapters that retain their capstone pedagogical function. We have added many additional examples of human uses of plants, enhanced many diagrams and added many new illustrations. We have also responded to calls for expanding the final chapter on sustainability. If you notice additional ways in which this text could be more effective, we are happy to receive your recommendations.

Acknowledgments

We continue to be very grateful for the indispensable assistance of our UW-Madison and Botanical Society of America colleagues with plant materials, photographs and art, and scholarly information used in this second edition, as well as contributions to the first edition of Plant Biology. Our special thanks to the faculty of the Department of Botany for continued intellectual stimulation, especially Robert Kowal, who provided a constant flow of articles describing important new botanical discoveries, and was a fount of information about plant taxonomy and systematics. Ever-helpful departmental office staff included Loraine Pilgrim, who handled a veritable river of mailings to and from Prentice Hall. Department academic staff Diane Derouen, Michael Clayton, Claudia Lipke, and Kandis Elliot continued to aid us in finding wonderful plant or fungal materials or illustrating them, as did Mohammed Fayyaz and his Greenhouse and Garden staff, and the staff of the Wisconsin State Herbarium. Biology Library staff lent their expertise in information retrieval. We also thank Don Schmidt, director of the greenhouses at the Department of Biological Sciences at Illinois State University.

Ellin Doyle added new end-of-chapter questions to those provided by Gary Wedemayer for the first edition. Ellin Doyle, Sharon Eversman, Sara Hoot, Andrea Lloyd, Diane Marshall, and Almuth Tschunko provided outstanding accuracy checking.

We also thank our family and friends for their encouragement and patience during first and second edition projects, in particular Michael and Melissa Graham, Martha Cook, and Norma Wilcox.

The efforts and encouragement of the Prentice Hall staff and those contracted by Prentice Hall are much appreciated. In developing this second edition, we acknowledge Sheri L. Snavely, Publisher; Karen Horton, Project Manager; Patrick Shriner, Senior Media Editor; Erin Mulligan, Development Editor; Donna King, Production Editor and Yvonne Gerin our photo researcher. They did a wonderful job building on the first edition guided by Teresa Chung, Barbara Muller, Travis Moses-Westphal, Jami Darby, Adam Velthaus, Colleen Lee, Xiaohona Zhu, and Chris Thillen.

During the preparation of this second edition content accuracy remained a priority, and we further endeavored to reorganize material according to reviewer recommendations. We are very appreciative of all the thoughtful and creative ideas suggested by the persons listed below.

Reviewers

Steve Ailstock, *Anne Arundel Community College*
Douglas Allan, *Santa Monica College*
Bonnie Amos, *Angelo State University*
Kathleen Archer, *Trinity College*
Tasneem Ashraf, *Cochise College*
John Averett, *Georgia Southern University*
Ellen Baker, *Santa Monica College*
Terese Barta, *University of Wisconsin Stevens Point*
Robert Bell, *University of Wisconsin Stevens Point*
Lisa Bellows, *North Central Texas College*
James Bidlack, *University of Central Oklahoma*
Brenda Blackwelder, *Central Piedmont Community College*
Meredith Blackwell, *Louisiana State University*
Karen Bledsoe, *Oregon State University*
Richard Bounds, *Mount Olive College*
James Brenneman, *Evansville University*
James Caponetti, *University of Tennessee*
Shanna Carney, *Colorado State University*
Curtis Clark, *California Polytechnic State University Pomona*
Danice Costes, *Troy State University*
Lawrence Davenport, *Samford University*
James Dawson, *Pittsburg State University*

Karen DeFries, *Erie Community College*

Roger del Moral, *University of Washington*

Pamela Diggle, *University of Colorado Boulder*

Ben Dolbeare, *Lincoln Land Community College*

Shannon Donovan, *University of Rhode Island*

R. Joel Duff, *University of Akron*

William Eisinger, *Santa Clara University*

Tom Evans, *University of Delaware*

Gerald Farr, *Southwest Texas State University*

Joseph Faryniarz, *Naugatuck Valley Community College*

David Francko, *Miami University of Ohio*

Nabarun Ghosh, *West Texas A&M University*

Simon Gilroy, *Pennsylvania State University*

David Gorchov, *Miami University of Ohio*

Michael Grant, *University of Colorado Boulder*

Sue Habeck, *Tacoma Community College*

Joel Hagan, *Radford University*

Everett Hansen, *Oregon State University*

Laszlo Hanzely, *Northern Illinois University*

Bob Harms, *St. Louis Community College*

Jill Haukos, *South Plains College*

David Herrin, *University of Texas Austin*

Scott Holaday, *Texas Tech University*

Tim Holtsford, *University of Missouri*

Elisabeth Hooper, *Truman State University*

Patricia Ireland, *San Jacinto College*

Thomas Jacobs, *University of Illinois Urbana-Champagne*

William Jensen, *Ohio State University*

Elaine Joyal, *Arizona State University*

Iris Keeling, *South Plains College*

Geoffrey Kennedy, *University of Wisconsin Milwaukee*

Joanne Kilpatrick, *Auburn University Montgomery*

Martina Koniger, *Wellesley College*

David Lemke, *Texas State University*

Marion Lobstein, *Northern Virginia Community College, Manassas*

Barry Logan, *Bowdoin College*

A. Christina Longbrake, *Washington & Jefferson College*

William Maple, *Bard College*

Michael Marcovitz, *Midland Lutheran College*

Will McClatchey, *University of Hawaii*

Roy McGowan, *Brooklyn College*

Dale McNeal, *University of the Pacific*

Susan Meiers, *Western Illinois University*

Larry Mellichamp, *University of North Carolina at Charlotte*

Richard Merritt, *Houston Community College*

Andrew Methven, *Eastern Illinois University*

Timothy Metz, *Campbell University*

Christopher Miller, *Saint Leo University*

Kathy Ann Miller, *University of the Puget Sound*

Priscilla Millen, *Leeward Community College*

Edgar Moctezuma, *University of Maryland*

Beth Morgan, *University of Illinois*

Lytton Musselman, *Old Dominion University*

Dawn Neuman, *University of Nevada Las Vegas*

John Olsen, *Rhodes College*

Laura Olsen, *University of Michigan*

Julie Palmer, *University of Texas Austin*

Bruce Parfitt, *University of Michigan Flint*

Lee Parker, *California Polytechnic State University San Luis Obispo*

Alan Pepper, *Texas A&M University*

Beverly Perry, *Houston Community College*

Carolyn Peters, *Spoon River College*

Jerry Pickering, *Indiana University of Pennsylvania*

Calvin Porter, *Texas Tech University*

Kumkum Prabhakar, *Nassau Community College*

Michael Renfroe, *James Madison University*

Stanley Rice, *Southeastern Oklahoma State University*

Jennifer Richards, *Florida International University*

Robert Robbins, *Utah Valley State College*

William Rogers, *Rice University*

Thomas Rosburg, *Drake University*

Manfred Ruddat, *University of Chicago*

Robert Rupp, *Ohio State University Agricultural Technical Institute*

Cindy L. Sagers, *University of Arkansas*

Michael Savka, *Rochester Institute of Technology*

Dan Scherier, *Northeastern University*

Edward Schilling, *University of Tennessee*

Bruce Serlin, *DePauw University*

Brian Shmaefsky, *Kingwood College*

Robert Shoemaker, *Towson University*

Barbara Schumacher, *San Jacinto College*

Beryl Simpson, *University of Texas Austin*

Richard Sims, *Jones County Junior College*

Don Smith, *University of North Texas*

James Smith, *Boise State University*

William Smith, *Wake Forest University*

Nancy Smith-Huerta, *Miami University of Ohio*

Teresa Snyder-Leiby, *State University of New York New Paltz*

F. Lee St. John, *Ohio State University Newark*

Lucy St. Omer, *San Jose State University*

John Stanton, *Monroe Community College*

Jon Stucky, *North Carolina State University*

Marshall Sunberg, *Emporia State University*

Stan Szarek, *Arizona State University*

Stephen Timme, *Pittsburg State University*

M. L. Trivett, *Ohio University*

Almuth H. Tschunko, *Marietta College*

Lowell Urbatsch, *Lousiana State University*

Rani Vajravelu, *University of Central Florida*

Staria Vanderpool, *Arkansas State University*

Mary Alice Webb, *Purdue University*

Cherie Wetzel, *City College of San Francisco*

James Willard, *Cleveland State University*

Robert Winget, *Brigham Young University Hawaii*

Wayne Wofford, *Union University*

Kathleen Wood, *University of Mary Hardin Baylor*

Todd Yetter, *Cumberland College*

Rebecca Zamora, *South Plains College*

Michael Zavada, *Providence College*

Plant Biology

Introduction to Plant Biology

Young sunflower head

- Plants are mostly photosynthetic organisms that are adapted to life on land.

- Modern groups of plants are bryophytes (such as mosses), lycophytes (such as club mosses), pteridophytes (such as ferns), gymnosperms (such as conifers), and flowering plants.

- Bacteria, protists (such as algae), and fungi are central to the lives of plants and thus are studied along with them.

- Plants and other living things are known by their scientific names.

- Scientific methods are used to learn about plants and associated organisms.

- Plants help maintain Earth's atmosphere and climate and are essential to the lives of humans and other animals.

Among the most wonderful plants in the world is the titan arum (*Amorphophallus titanum*), a native of the rain forests of Sumatra, in Indonesia. The titan arum is famous because it produces the largest known cluster of flowers (inflorescence), which may be 9 feet tall and 3–4 feet in diameter. When it blooms, which is but rarely, thousands of people flock to see the botanical extravaganza. The titan arum also produces a giant leaf that grows as tall as 20 feet and up to 15 feet wide. Titan arum can be grown in greenhouses from a huge tuber—an underground stem—that can weigh over 170 pounds. It is truly a titanic plant.

Plant biologists also appreciate the many ways that the titan arum dramatically displays the important roles of plants in nature. Like many other flowering plants, its flowers attract insect pollinators, which carry pollen from one plant to another. But to attract pollinators, the titan arum produces an incredibly powerful odor that smells like rotting fish to people. The odor is so strong and bad that titan arum is also known as the "corpse flower." If the flowers are pollinated, bright red fruits develop that are attractive to birds, which digest the fruit flesh and spread the seeds.

The titan arum thus illustrates plant evolution—like all other life-forms, the titan arum has become adapted to its environment in ways that maximize survival and reproduction. The titan arum's lifestyle is also a striking instance of the many ways in which Earth's organisms interact not only with their physical environments but also with each other—their ecology. The use of DNA science indicates that titan arum is a member of a plant group having human uses. Calla lilies and anthuriums, often used as cut flowers, are relatives of the titan arum, as is the common houseplant *Philodendron*.

In this book, many other examples of fascinating plants, their associations with other organisms, and relevance to human affairs will be described. Like the titan arum, other plant examples will illustrate basic biological concepts in evolution, ecology, and DNA science—three major themes of this book.

Titan arum

1.1 | What are plants?

Most people feel quite confident in their ability to recognize plants as rooted organisms that usually live on land and have green leaves (Figure 1.1). Most people also link plants with photosynthesis, their main ecological role. **Photosynthesis** is the production of organic food from inorganic molecules (carbon dioxide and water), with the use of light energy. But defining plants by these criteria does not always work. Even though most plants are photosynthetic, many plants are not. Mistletoes, sandalwood, and some morning glories are among the 3,900 species of plants known as parasites, because they obtain organic food from other organisms. Many of these plants lack green leaves and are not rooted in place (Essay 1.1, "Devilish Dodder").

Distinguishing plants from other organisms can also present a challenge. For example, some people think of **fungi** (such as mushrooms) as plantlike organisms because they appear to be nonmobile and have some other plantlike features (Figure 1.2a). For example, the cells (the basic units of living things) of fungi are enclosed by organic walls, as are those of plants. But plant cell walls are composed of **cellulose**, a material composed of linked sugar units, whereas fungi have walls composed of a different

FIGURE 1.1 Woodland ferns Most plants are recognizable as leafy green rooted trees, shrubs, or nonwoody herbs, such as these ferns.

sugar-rich substance, known as chitin. Fungi also lack photosynthesis and other distinctive features of plants. Many **bacteria** are photosynthetic, but these microscopic organisms have a much simpler structure than do plants (Figure 1.2b). **Protists**—organisms that are not bacteria, archaea, fungi, plants, or

---

ESSAY 1.1 DEVILISH DODDER

Dodder truly challenges our concepts of plant life (Figure E1.1). Unlike most plants, dodder has almost no chlorophyll—the green pigment that allows plants to convert sunlight energy to organic food molecules by photosynthesis. In addition, mature dodder is not rooted in the ground, as are most plants. Also known as strangleweed, devil's guts, goldthread, and by its scientific name *Cuscuta*, dodder could be the star of a botanical horror movie called *Attack of the String Monster*. That's because dodder twines its stringy, leafless, yellow or orange stems like tentacles around vulnerable green plants, then uses special feeding organs to suck water, minerals, and food from them. In fact, the reason for dodder's vampirelike behavior is its lack of chlorophyll and photosynthesis. Dodder is unable to produce its own food, so it must seize resources from other plants. Its lack of roots is explained by the fact that the plant obtains its water and minerals from plant victims, rather than from the soil.

Dodder is a widespread pest of crop fields and gardens, dreaded by farmers and gardeners alike. Dodder infestation causes serious losses of citrus, tomatoes, beets, potatoes, and other fruit and vegetable crops, animal forage crops such as alfalfa and clover, and ornamental flower crops, including roses. Dodder can also transmit microbes that cause plant diseases. After it has become established, dodder can be controlled only by hand-pulling and then incinerating every piece of stringy stem, because any fragments that are left will regenerate new dodder plants! Each dodder plant can produce masses of tiny cream-colored flowers and more than 16,000 seeds, which may remain alive in the soil for more than 60 years waiting for the opportunity to germinate and attack more green plants.

Dodder could be the star of a botanical horror movie called Attack of the String Monster.

Given these traits, you might wonder why dodder has not taken over the world! Fortunately, many plants—including corn and soybean—are genetically resistant. If dodder seedlings germinate near immune plants, they cannot gain access to the food, water, and minerals needed for growth, so they die. Despite its bad reputation, dodder has a positive side. Some of the 150 known dodder species help control the growth of other noxious weeds, such as kudzu, which blankets other plants with smothering growths in warm climates. Other dodder species are food plants for butterflies.

E1.1 Orange-colored dodder growing on a goldenrod plant

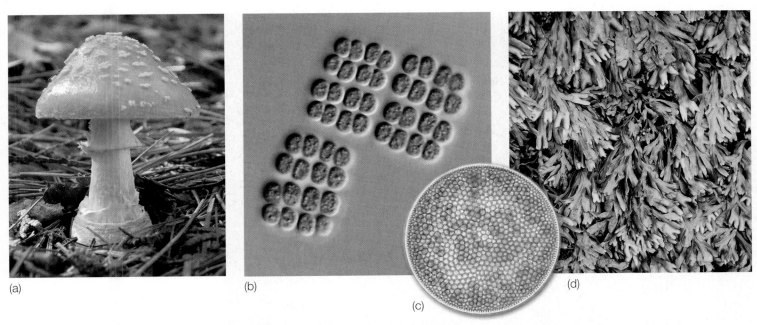

(a)   (b)   (c)   (d)

FIGURE 1.2   Other groups important to plant biology Among the organisms often studied with plants are (a) fungi, including the reproductive stage known as mushrooms; (b) photosynthetic bacteria; (c) single-celled algae, such as this diatom; and (d) seaweeds, such as rockweeds. However, none of these organisms fulfills all of the criteria of plants as defined in this text: photosynthetic, multicellular, and adapted in many ways to life on land.

animals—also include many photosynthetic forms, known as **algae** (Figure 1.2c,d). Even though most algae are photosynthetic, their structure and reproduction are usually simpler than those of plants, and algae primarily live in water. In contrast, adaptation to life on land is a characteristic of most plants. Together, relatively complex structure, cellulose cell walls, photosynthesis, and adaptation to land can be used to define and recognize plants.

In this book, **plants** are defined as organisms that (1) are composed of many cells, (2) have cellulose-rich cell walls, (3) have chlorophyll and are photosynthetic (or, if nonphotosynthetic, originated from photosynthetic ancestors), and (4) are adapted in many ways to life on land (or, if aquatic, are descended from land-adapted plants). Plants can be adapted to land conditions in many ways, but one example is a multicellular **embryo**—a young stage that develops from a fertilized egg within a mother plant's tissues. Because all groups of land-adapted plants possess such embryos, plants are often known as **embryophytes**.

Modern groups of embryophytes include (1) mosses and other simple plants known as **bryophytes**, (2) club mosses—**lycophytes**, (3) ferns and fern relatives—**pteridophytes**, (4) seed-bearing **gymnosperms** such as pine trees, and (5) flowering plants—**angiosperms** (Figure 1.3; Table 1.1). In the past, several other major plant groups existed, but they have become extinct and are now known only from their fossils.

1.2 | Bacteria, fungi, and algae are important to plant life

Botany, the study of plants, includes the study of bacteria, fungi, and protists—especially photosynthetic algae. Bacteria cause plant diseases, and some photosynthetic bacteria were

ancestral to plant **plastids**, the cellular location of photosynthesis. In nature, 80–90% of plants live in close association with fungi that help them obtain essential nutrients from soil. Beneficial fungal partners are credited with helping the first plants to become established on land, where early soils were poor. Other fungi are known as **pathogens** because they infect plants, producing disease. Plants evolved from microscopic green algae, but some other protists are serious plant pathogens. Plants also affect the growth of algae; fertilizers used to grow plant crops often wash into waterways, contributing to harmful algae blooms. Wetland plants absorb some of these fertilizer nutrients, thereby helping prevent harmful algal blooms in lakes and oceans. Thus, microbes (microscopic organisms that include bacteria, fungi, and protists) are closely tied to the ecology and history of plants (Table 1.2). All of the groups of plants, bacteria, fungi, and protists that are covered in this book play important roles in Earth's ecology; their relevance to human concerns is often reflected in media headlines (Figure 1.4).

The use of DNA and other characteristics to study relationships among organisms has made it increasingly apparent that all of Earth's organisms are related, descended from an ancient common ancestor. A "family tree of life," representing biologists' current understanding of organisms' relationships (Figure 1.5), reveals humans' relationship to plants, fungi, protists, bacteria, and other animals. Understanding how Earth's organisms have diversified through time—a process known as *evolution*—is a fundamental aspect of biology, because it explains why organisms are so diverse and function in so many different and interdependent ways. Throughout this book, an icon depicting a simple relationship tree will alert you to concepts and processes related to the evolution of plants and associated organisms.

EVOLUTION

FIGURE 1.3   **Embryophyte plant groups** These groups include (a) bryophytes, such as mosses; (b) lycophytes, such as this club moss; (c) ferns (pteridophytes); (d) seed-producing gymnosperms, including conifers; and (e) flowering plants, such as this wood lily.

TABLE 1.1	Major Groups of Embryophytes—Land-Adapted Plants			
Group	**Body Form**	**Reproduction**	**Ecological Roles**	**Human Uses**
Bryophytes	Under 1 meter (m) tall; no true roots, stems, or leaves (lack lignin-containing conducting tissues)	Single-celled spores; eggs and flagellate sperm (which allow cells to swim through water)	Carbon storage as peat; pioneers of bare soil or rocks; earliest land plants	Soil conditioners; decoration
Lycophytes	Low-growing or hanging from trees; true stems and roots with lignin-coated water-conducting tissues; leaves small with single unbranched vein	Single-celled spores; eggs and flagellate sperm	Formed coal deposits; modern-day ground cover	Ornamentals
Pteridophytes	Some are tree-sized; true stems; most have roots and true leaves with branching veins	Single-celled spores; eggs and flagellate sperm	Formed coal deposits; modern-day ground cover	Ornamentals
Gymnosperms	Most are trees; have true roots, stems, and leaves; woody	Seeds in cones; eggs and sperm (sperm of some, but not all, groups have flagella)	Early seed plants; dominant at high altitude and latitudes; food for seed-eating animals; carbon dioxide stored for long periods as wood	Timber and paper production; landscaping; seeds of some used as human food; source of products such as turpentine
Angiosperms	Some are woody trees, vines, or shrubs; some are nonwoody herbs	Flowers; seeds in fruits; eggs and nonflagellate sperm	Grow in a wide variety of habitats; animal foods; great species diversity	Major sources of human food, medicines, beverages, fibers, and building materials

TABLE 1.2   Other Major Groups Important to Plant Life and Often Studied with Them

Group	Body Types	Ecological Roles	Human Uses
Bacteria	Single cells, colonial, or multicellular; microscopic	Decomposers; some are plant associates; some cause diseases; some are photosynthetic; some produce toxins; earliest of Earth's organisms	Pharmaceutical and food manufacture
Algae	Single cells or multicelled; many are microscopic in size; usually with photosynthetic pigments	Photosynthesis; oxygen producers; food producers in aquatic habitats; some produce toxins	Food and industrial products
Fungi	One-celled or masses of filaments; no photosynthetic tissues	Decomposers; some cause diseases; some associate with algae to form lichens; some are poisonous	Food and food products; brewing

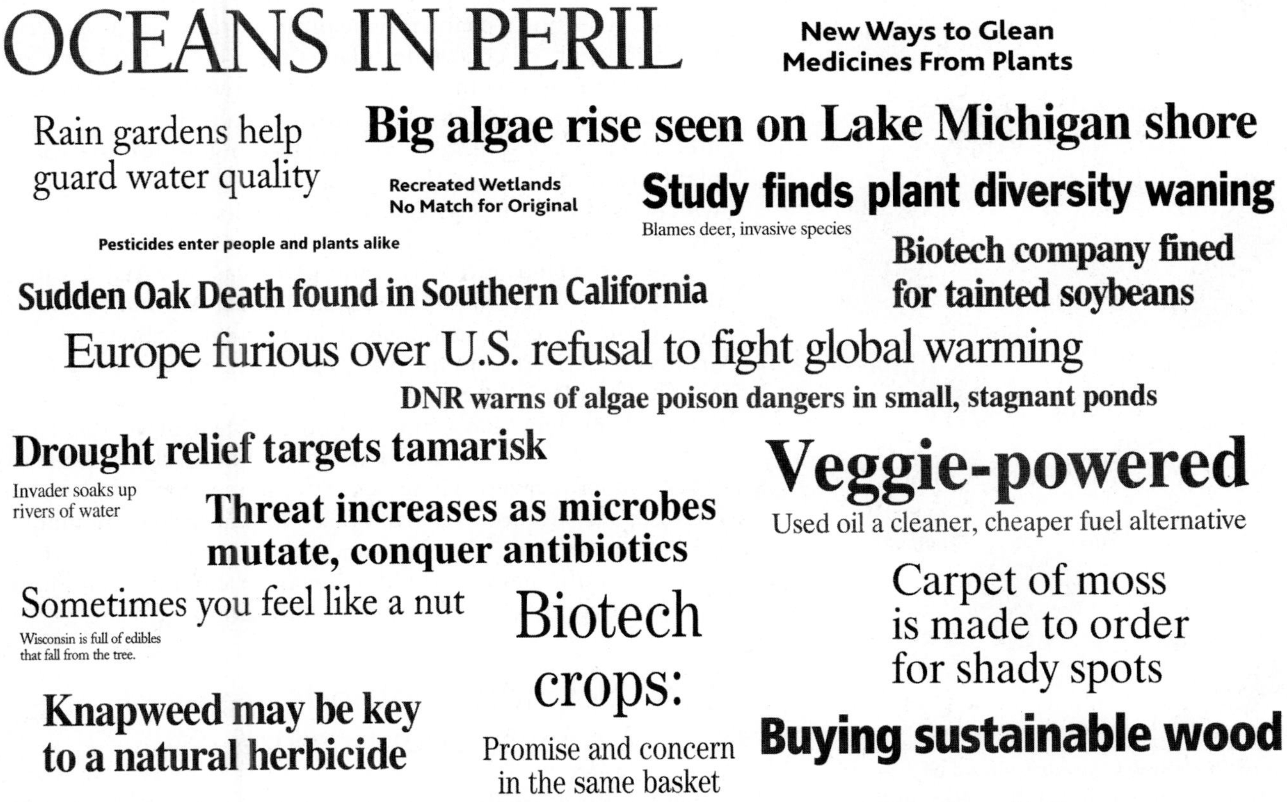

OCEANS IN PERIL

New Ways to Glean Medicines From Plants

Rain gardens help guard water quality

Big algae rise seen on Lake Michigan shore

Recreated Wetlands No Match for Original

Study finds plant diversity waning

Blames deer, invasive species

Pesticides enter people and plants alike

Sudden Oak Death found in Southern California

Biotech company fined for tainted soybeans

Europe furious over U.S. refusal to fight global warming

DNR warns of algae poison dangers in small, stagnant ponds

Drought relief targets tamarisk

Invader soaks up rivers of water

Veggie-powered

Threat increases as microbes mutate, conquer antibiotics

Used oil a cleaner, cheaper fuel alternative

Sometimes you feel like a nut

Wisconsin is full of edibles that fall from the tree.

Biotech crops:

Carpet of moss is made to order for shady spots

Knapweed may be key to a natural herbicide

Promise and concern in the same basket

Buying sustainable wood

FIGURE 1.4   **News media headlines** Media often feature plants and microbes, genetically modified crops, or environmental concerns that are related to plant biology.

1.3 | Plants and other organisms have scientific names

In order to study plants and other organisms, it is necessary to name them. Through the ages, humans have given organisms **common names** of local relevance, which often reflect appearance or usefulness. The problem with common names is that they often differ from place to place and from one language to another, and sometimes the same common name refers to more than one type of plant. For example, very different, unrelated plants are commonly named "snakeroot" because they have similarly twisted roots that reminded people of snakes. Confusion can result in mistakes, with serious consequences in commerce, agriculture, and health. Mistakes in communication can be avoided by the use of scientific names.

Scientific names, created by the biologists who first formally described organisms, ensure the accuracy and specificity needed for communication in commerce, agriculture, medicine, gardening, and other practical applications. Scientific names include two parts: a generic name whose first letter is always capitalized, and a species name (specific epithet). Both parts of a scientific name are properly underlined or italicized. The

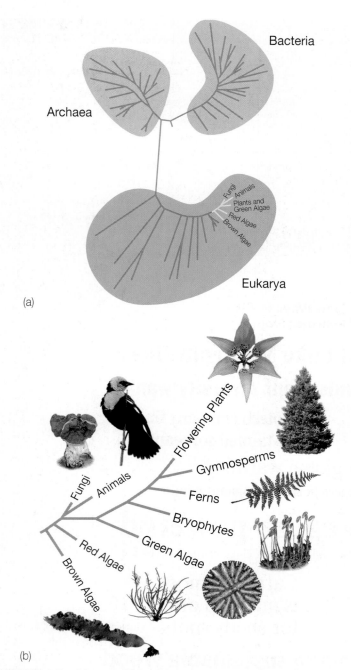

(a)

(b)

FIGURE 1.5 A tree of life shows current views of the relationships among plants, algae, fungi, bacteria, and other organisms (a) The three major groups of living things are Archaea, Bacteria, and Eukarya. Plants, algae (and other protists), fungi, and animals are classified within the Eukarya. Note that the organisms we are most familiar with—plants and animals—occupy only a tiny tip of one of the branches, indicating the tremendous diversity of life on Earth. (b) A closer view of the portion of the tree shaded in yellow in (a), which contains the major land plant groups as well as fungi, animals, and several groups of algae.

titan arum's scientific name, *Amorphophallus titanum*, is an example. Scientific names are generally based on Greek or Latin terms that often describe the organism. The use of classical languages is important because these languages are no longer changing and are understandable to biologists all over the world.

Plants used by humans to produce beverages provide excellent examples of the usefulness of scientific names. Coffee, tea, chocolate, and other beverages derived from plants have long been associated with human societies. For example, coffee has been an integral part of human life from ancient times in the deserts of Arabia and the coffee houses of Europe, where writers and thinkers met to discuss issues of the day, to the many coffee shops of today's city streets. In Polynesian societies, ceremonial drinking of kava-kava is an essential part of village meetings that gather for the purpose of making decisions and resolving conflicts. Coca, coco, cacao, castor, coffee, kola, kava—seven types of plants (Figure 1.6) that can be used to produce seven different beverages, but notice they have confusingly similar common names.

Unlike common names, the scientific names for these plants are not easy to confuse. *Erythroxylum coca* is the scientific name for the coca plant, from which both beverage flavoring and cocaine are produced. *Erythroxylum* means "red wood," reflecting a characteristic of these trees. *Cocos nucifera* is the coconut, or coco palm plant that grows abundantly on beaches throughout the tropics. Coconut milk is a common ingredient of tropical drinks, and coco palm has many other uses. *Theobroma cacao* is the scientific name for the cacao plant from which are derived chocolate and cocoa butter, widely used in cosmetics and food products. *Theobroma* literally means "food of the gods," a description with which chocolate lovers would agree. The source of castor oil, a laxative drink, is *Ricinus communis*. *Coffea arabica* is a major source of coffee, and cola flavors used in popular carbonated soft drinks are derived from the plant whose scientific name is *Cola nitida*. The plant *Piper methysticum* is the source of the Polynesian drink kava-kava and a food supplement of the same name is sold in industrialized countries for its claimed stress-reduction properties. These scientific names are not easy to confuse, they do not vary from place to place, they are unique to each plant, and they are understood around the world. Using scientific names provides a way to communicate more precisely than does the use of common names.

1.4 | Scientific methods are used to learn about nature

What is science? Many people view science as a collection of facts to be learned about some subject, but science is much more. One good description of science comes from Arthur Strahler in *Understanding Science* (1992): "Science gathers, processes, classifies, analyzes and stores information on anything and everything observable in the universe." As this description suggests, science is **empirical**; that is, it is based on the scientific methods of observation, measurement, and experiment (Figure 1.7). An **experiment** is a manipulation designed to test the validity of an idea. Scientific experiments should always include unmanipulated **controls** for comparison and many **replicates** to decrease the probability that an observation is based on error. Science is not immune to error, but the process has built-in correction procedures.

Scientists aim to describe and explain real phenomena by making scientific statements. To be scientific, a statement must

FIGURE 1.6 Common versus scientific plant names Coca (a), coco palm (b), cacao (c), castor (d), coffee (e), kola (f), and kava (g) plants are economically important plants that have similar-sounding common names, but distinctive scientific names.

be testable, and the tests must be reproducible. For example, in 1996, NASA scientists claimed to have found fossils of simple bacteria-like life-forms in a meteorite that originated from Mars. Their work, widely reported in the news media, implied that life might have arisen elsewhere as well as on Earth. Subsequently, other scientists tested the possibilities that the meteorite evidence of life might have originated in nonliving systems and found evidence that it could. These results did not disprove the existence of life on Mars, but did show that the question has not yet been resolved. The controversy surrounding the Martian meteorite illustrates that science works well, by raising questions and indicating further investigations that need to be performed in order to answer them. Individual scientists are human and can be biased in their statements. But if the investigations are reproducible, another scientist—who may or may not share the same bias—can repeat them and possibly draw different conclusions. Scientific objectivity arises out of the collective activities of the community of scientists.

Science has an error-correction capacity

Every scientific statement has associated with it a probability of being wrong, but science also includes an error-correction process. The scientific process of observation, measurement, and experimentation reduces the probability of a scientific statement being wrong. For example, in the 1950s—based on the best information available at the time—it was possible to make scientific statements such as "Mars might have primitive plant life" or "Venus may have oceans of hot water." Subsequent observations by spacecraft did not support either statement. Does this mean that science doesn't work? No, because science does not accumulate absolute truths as an end product—something that is a frequent source of confusion and criticism. Science accumulates a store of scientific statements for which the probability of being wrong has been significantly reduced. One way that scientists can achieve some measure of fame is by providing evidence that previous scientific statements are wrong or require significant change. This is a strong motivation for scientists to make new discoveries that are as accurate as possible. Because scientific views of how nature works change when new results appear, textbooks such as this one are typically revised every few years to reflect the newer concepts.

The scientific process is characterized by skepticism, the critical judgment of scientific work. Skepticism is built into science through the peer review processes of grant funding and publication. Scientists are expected to publish their results as

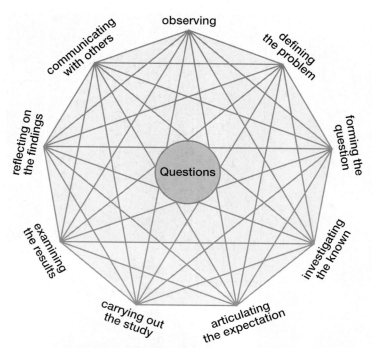

FIGURE 1.7 Processes involved in the scientific method
Scientists engage in these activities via diverse pathways and repeat activities as often as needed. This model of scientific work replaces an older, linear model (define hypothesis; test it; accept, reject, or modify hypothesis), which is now regarded as less representative of the way that scientists actually work.

articles in scholarly journals, of which there are thousands. Before an article submitted to these journals can be published, it must be examined and criticized by peers—qualified scientists who are knowledgeable about the work's topic and techniques. Reviewers may recommend acceptance with minor or major revisions, or they may recommend rejection. Reviewers may be biased in their review, so two or more reviewers usually evaluate a paper. If the article is rejected, its authors may have to perform more work and report additional results before the results can become part of the scientific record.

Scientific work is often expensive, so scientists often must seek outside funds from granting agencies. Plant scientists, for example, might obtain funding for their research work from the National Science Foundation (NSF), U.S. Department of Agriculture (USDA), and the U.S. Environmental Protection Agency (EPA). Grants are sums of money given to scientists (or, more accurately, to the institutions that employ them) to carry out specific research programs. Funds are always limited, and competition is intense for those available. Scientists must submit a grant proposal, which is a detailed plan of the proposed work that explains why and how the work will be done and justifies the funds requested. Peer scientists review these proposals and report on their quality to the granting agency. To avoid bias, the granting agency uses many reviewers. Acting as a peer reviewer of journal articles and grant proposals, typically without financial reward, is a major way that scientists provide service to their communities. Peer review of papers and grants are two important ways that skepticism is built into the scientific process.

The process of science may be inductive or deductive

The scientific process of making observations, measurements, and experiments and then evaluating the results can be approached from two directions, through **inductive reasoning** or **deductive reasoning**. In induction, scientists take measurements, make observations, analyze the data from these measurements and observations, and after a great many such investigations formulate a generalization called a theory. In modern society, the word *theory* is often used to mean a guess or an unsupported idea, when scientists would use the term "hypothesis." But in science, "theory" has a different meaning. A scientific **theory** is a broad explanation for diverse phenomena and is based on a large amount of information. For example, the Cell Theory explains that all of Earth's organisms are composed of cells, and all modern cells arise from parental cells (Chapter 4). Establishing the Cell Theory required many observations of plants, animals, protists, fungi, archaea, and bacteria.

By contrast, deductive reasoning begins with a **hypothesis**, an educated preliminary explanation based on past observations. To be useful, a hypothesis must be testable using available techniques. The hypothesis is then used to select the type of data that must be collected in order to test it. Testing hypotheses may consist of experiments or sets of observations. Such tests may support the hypothesis in question or not. If a hypothesis is not supported by the data, it must be altered and retested, or rejected. A set of related hypotheses that have been widely supported by data may eventually achieve the status of a theory. Thus a hypothesis is a tentative explanation based on some prior information that requires additional support for wide acceptance, whereas a scientific theory is a widely accepted generalization supported by a large body of data and experimentation.

1.5 | Plants and associated organisms play essential roles in maintaining Earth's environment

Plants, fungi, protists, archaea, and bacteria maintain the chemistry of Earth's atmosphere and our planet's climate at levels required for human and other life. Plants, fungi, protists, archaea, and bacteria also produce food that supports other life-forms. Thus, global ecology, and its relevance to all life, is a main theme of this book. Icons representing a leaf will occur throughout this book as a way to draw your attention to concepts and processes that are related to the roles of plants and microbes in Earth's ecology.

Plants, algae, fungi, archaea, and bacteria help maintain Earth's atmospheric chemistry and climate

Plants produce about half of the oxygen in the Earth's atmosphere. Algae and certain bacteria (cyanobacteria) produce the other half or more of atmospheric **oxygen** (O_2), without which people and most other organisms would die. Oxygen is a by-product of photosynthesis that escapes from plants (and photosynthetic microbes) into the atmosphere. Most of Earth's organisms require oxygen in order to harvest the chemical energy in organic food. Food energy supplies the needs of metabolism—all of the chemical reactions that occur in life. Less familiar, perhaps, is the fact that some of the oxygen produced by plants, algae, and bacteria is converted by solar radiation in the upper atmosphere into the ozone. Modern life on Earth could not exist without an ozone shield, which protects against harmful ultraviolet (UV) radiation. This is true even though ozone at ground level is harmful to living organisms.

Plants also perform another rarely recognized ecological service. They help lower the level of the gas **carbon dioxide** (CO_2) in the atmosphere. During photosynthesis, plants convert CO_2 to organic compounds that are stored in organisms, soil, or fossil deposits (not located in the atmosphere). Plants also harbor certain bacteria that need the oxygen produced by plant photosynthesis to consume atmospheric **methane** produced by archaea. These processes are important because carbon dioxide and methane, along with water vapor, are "greenhouse" gases. **Greenhouse gases** have the capacity to absorb heat—known as infrared radiation—preventing it from being lost to space. Such heat absorption by atmospheric gases warms the Earth. This is analogous to the glass walls of a greenhouse, which allow the entry, but not the exit, of thermal radiation so the interior warms. Such atmospheric warming is known as the "greenhouse effect" (Figure 1.8). Without this warming, Earth would be too cold for life as we know it. But excess amounts of greenhouse gases are related to **global warming**, a topic much in the news (Essay 1.2, "Global Warming: Too Much of a Good Thing"). Plants and associated bacteria are important in understanding and controlling global warming.

Many plants use some of the products of photosynthesis to construct lignin, a carbon-rich compound that does not readily decay. **Lignin** is a major constituent of wood, contributing to wood's strength and decay-resistant properties. Ancient plants with large amounts of lignin did not completely decay and were compressed into coal deposits. Over hundreds of millions of years, ocean-dwelling algae have likewise generated huge deposits of oil and natural gas, which along with coal are called **fossil fuels**. Although fossil fuels supply much of today's human energy needs, they also generate greenhouse gases when burned. If ancient plants and algae had not been so effective at removing carbon dioxide from the atmosphere, storing it in soils or fossil fuel deposits for a long time, Earth's climate would likely be too hot for most life.

Many bacteria and fungi are decomposers. They break down organic compounds, releasing the CO_2 and other minerals needed for photosynthesis. Without such decomposition, Earth's plants, algae, and cyanobacteria would not be able to make organic food upon which humans and other organisms depend.

Plants benefit from close associations with other organisms

In nature, plants do not live in isolation. They are intimately linked to the lives of bacteria, protists, fungi, and animals in a variety of ecological associations: food webs, symbioses, and

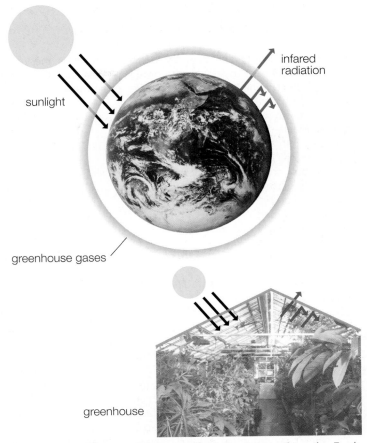

FIGURE 1.8 The greenhouse effect Sunlight striking the Earth or the plants and other contents of a greenhouse generates heat from infrared radiation (IR), some of which is radiated toward space. Greenhouse gases, like the glass panels of a greenhouse, allow most components of sunlight to escape, but not IR, and so the atmosphere heats up.

coevolutionary relationships. Often, plants gain real benefits from these alliances.

Food webs All of the organisms on Earth are enmeshed in complex feeding relationships known as food webs, in which plants, algae, and cyanobacteria are the main food producers (Figure 1.9). Small soil animals consume fungi and soil microorganisms, such as protists and bacteria. Many larger animals are **herbivores**, feeding on plant parts including leaves, fruits, seeds, flower nectar, and pollen. Even carnivores, which feed upon herbivores, indirectly depend on plants. In oceans, ponds, lakes, and streams, algae are food for aquatic animals because of their nutritional value; this association is the basis of commercial fisheries. As noted earlier, many fungi and bacteria obtain their energy by decomposing dead plants and animals and animal wastes, thereby cleaning the environment and recycling mineral nutrients from dead to living organisms.

Beneficial symbioses Plants and microbes are also involved in many kinds of close ecological associations that are described as symbioses. The word **symbiosis** means "living together"; it implies a relationship between two or more partners that is even more intimate than is typical of food-web associations. Symbiotic

ESSAY 1.2 GLOBAL WARMING: TOO MUCH OF A GOOD THING

Atmospheric gases warm Earth by retarding the loss of heat to space—the greenhouse effect. This is a good thing, because otherwise Earth would be like its neighbor planet Mars—too cold to support human life. But rapid, significant increase in global temperature, known as global warming, may be detrimental to humans and Earth's other life-forms, which may not be able to cope.

The existence of modern global warming—considered controversial even a few years ago—has now been convincingly demonstrated in the opinion of most climate scientists. The consensus, based on very large numbers of measurements, is that Earth's surface temperature has increased by 0.6°C during the past century. This may not seem like a very large increase, but its effects can be observed as the retreat of mountain summit glaciers and alarming rises in sea level experienced by South Sea island nations. Increased temperature caused the death of more than 90% of shallow corals on most Indian Ocean reefs during 1998. Temperate-zone lakes throughout the world are freezing significantly later in the fall and thawing sooner in spring than in the past. Alaska is warming particularly fast; winters there have recently warmed by 2°–3°C. As a result, sea-ice melting has increased and soils that usually remain hard and frozen (permafrost) have become soft, posing problems for wildlife, trees, and buildings. Alaskan spruce forests are being ravaged by an insect (spruce bark beetle) that is able to reproduce faster in the warmer summers than it did in the cooler past.

Global warming has been linked to increases in greenhouse gases resulting from human activities—the burning of forests and fossil fuels. Greenhouse gases, which trap heat in Earth's atmosphere, include water vapor, methane, and carbon dioxide (CO_2). Measurements of carbon dioxide in air bubbles trapped in ancient ice suggest that atmospheric levels were stable at 280 parts per million (ppm) for thousands of years and then rose at a rapid

Climate experts also warn that warming may melt sufficient polar ice to raise sea levels, drowning coastal cities and breaking down existing agricultural systems.

rate after the year 1800 (the dawn of the Industrial Age). Records kept by climate scientists since 1957 show that carbon dioxide levels in the atmosphere have risen from 315 to over 375 ppm (Figure E1.2). Atmospheric CO_2 is currently higher than at any time in the last 420,000 years and is expected to double by the end of this century.

What is the outlook for the future? Climate researchers use models and data obtained from around the world to make predictions. Climate models point to a 3°C temperature rise (with the range being 2.6°–4.0°C) by the end of this century. Some scientists predict that continued global warming will result in the destruction of some life-forms that are unable to adjust rapidly enough. Climate experts also warn that warming may melt sufficient polar ice to raise sea levels, drowning coastal cities and breaking down existing agricultural systems. Some experts think that warming also influences weather patterns, triggering violent

storms that kill and displace people, destroy crops, and increase the reproductive rate of crop pests, including insects, weeds, and disease microbes. Evidence for the latter includes a survey of the fossils left by plants and insects that lived during a warming period more than 50 million years ago. Temperature increase (deduced from geological data) was correlated with a rise in the diversity of pests and the extent of their damage to plants. Currently, 35–42% of growing or stored crops are lost to pests, resulting in an annual loss of $244 billion worldwide, and future climate warming is expected to increase such losses.

How can we reduce the threat of global warming? Local and global actions that reduce fossil fuel combustion and conserve forests may help. Forests play an important role in keeping atmospheric carbon dioxide low. Almost 90% of all organic carbon in living things is stored in trees. Half of all carbon that occurs in terrestrial ecosystems is stored in forests (trees and soil). When a forest is burned, organic carbon that has been stored there for 50–300 years is released within a few hours to the atmosphere as CO_2. Plants play an integral role in our understanding of and ability to control global warming.

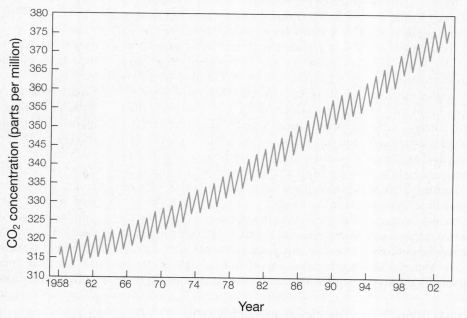

E1.2 Graph showing atmospheric carbon dioxide increases since the late 1950s as measured in Hawaii An annual fluctuation in levels causes the zigzag appearance.

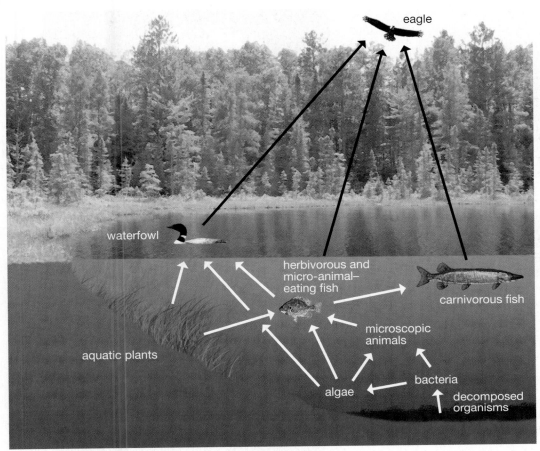

eagle

waterfowl

herbivorous and micro-animal–eating fish

carnivorous fish

aquatic plants

microscopic animals

bacteria

algae

decomposed organisms

FIGURE 1.9 **Food web** This example of a simple aquatic food web illustrates the importance of plant and algal photosynthesis in providing food for other organisms.

relationships can be beneficial to one or both partners, or to neither. If both partners benefit, the relationship is referred to as a beneficial symbiosis or **mutualism.** A familiar example of a mutualistic symbiosis is the teaming of photosynthetic algae and/or bacteria with a fungus to generate **lichens.** Lichens (Figure 1.10) are common on soil, rocks, walls of old buildings and tombstones, and tree bark; they grow better in these environments than can either microbial partner alone. The photosynthetic partner provides the fungus with needed sugar and oxygen and the fungus contributes water, minerals, carbon dioxide, and protection from excess light to algal and bacterial partners (Chapter 20).

Many single-celled green photosynthetic algae often live within the body of *Hydra*, a simple animal that is common in freshwaters and biology classrooms (Figure 1.11). In this alliance, the animal receives algae-produced organic food and oxygen—the by-product of photosynthesis. In return, the algae obtain carbon dioxide and minerals, which are produced as wastes by the animal. Here again, both partners benefit from the symbiosis. Many ocean corals contain single-celled algae that perform similar services. Associations of photosynthetic organisms (such as algae) with nonphotosynthetic fungi and animals are small-scale examples of the way that all of Earth's nonphotosynthetic organisms depend on plants, algae, and cyanobacteria.

Plants also benefit from mutualistic symbioses. For example, certain soil bacteria growing within roots of legumes

FIGURE 1.10 **Lichens are examples of symbiotic relationships among fungi, algae, and photosynthetic bacteria** Both the orange and gray-green patches on this rock are lichens.

and some other plants are an invaluable source of nitrogen-containing minerals that plants need in large amounts but that are often present in low levels in soil. The plants return the favor with photosynthetic products that can be used as food by bacterial partners (Chapter 10). Certain fungi are very commonly

FIGURE 1.11 *Hydra* This small aquatic animal is green because of the numerous green algal cells living symbiotically within its tissues.

FIGURE 1.12 A butterfly feeding on nectar produced by the milkweed plant (*Asclepias*) While feeding, butterflies transport pollen to or from the milkweed plant. The butterfly's long, tubular tongue is adapted for probing flowers for nectar, while the flower size and shape allows access to these insects, demonstrating the phenomenon of coevolution.

associated closely with plant roots, providing them with water and minerals in exchange for the plant's photosynthetic products (Chapter 20). Such fungal partnerships evolved very early in the history of plants, probably shortly after the first plants colonized land, and are thus regarded as a major reason that plants have succeeded so well on land since then (Chapter 21).

Coevolution Have you ever wondered why there are so many different kinds of flowers, and why many flowers are so attractive? The answer is that many flowering plant species have come to rely so closely on particular animal species that over time both have undergone changes in structure or behavior that link them even more closely. In other words, the evolutionary pathway of flowering plants has often depended on animals, and vice versa. This phenomenon is known as **coevolution**—plants and animals evolving together.

Flower structure, for example, has evolved in many different ways in response to the many types of animals that pollinate flowers (Figure 1.12). Many kinds of insects, birds, bats, and some mammals transport pollen from one plant to others of the same species. In return, the animals, known as pollinators, usually receive a food reward such as sugary nectar. Plants and animals involved in such pollination associations often become highly specialized in ways that improve pollination efficiency. Pollination coevolutionary relationships explain why there are so many different kinds of flowers; why flowers often have such attractive colors, forms, and fragrances; and why some animals are so diverse (Chapter 24). Pollination interactions are of practical importance to humans as well; many fruit crops depend on the pollination services of bees.

Various birds, bats, and mammals prefer to eat the fruits of particular plants. By so doing, they serve these plants as seed-dispersal agents, allowing them to colonize new habitats. Plant fruits are often sized and colored in ways that attract specific animal dispersers and provide the most appropriate kind of food reward (Figure 1.13). Many plants have come to

FIGURE 1.13 Fruits Many flowering plants encourage particular animals to disperse their seeds by offering a food reward in the form of colorful fruit. Here, a cedar waxwing is eating fruits of a hawthorn (*Crataegus*) plant. This interaction is another example of coevolution.

depend upon the presence of particular animal pollinators or dispersal agents, and vice versa. The disappearance of a food plant upon which an animal relies often means the animal may not be able to survive either. In pollination and dispersal relationships plant and animal species can greatly influence each other's survival. Humans have an important stake in learning as much as possible about how the plants, crops, and other organisms we depend upon are affected by activities that degrade habitats and thereby result in extinction. Humans have an interest in protecting Earth's ecological integrity because it is essential to our survival, as well as the survival of plants and other organisms.

1.6 | Plants, protists, fungi, and bacteria are important in human affairs

A list of the ways in which people use plants and microbes today or in the past would be very long indeed. Human agriculture, including production of food and fiber crops (Figure 1.14), livestock pasturage, and forestry are all based on plants. The baking, brewing, pharmaceutical, and other industries depend on plants and the fungi known as yeasts. Human ritual and ceremony (Figure 1.15) commonly employ decorative plants (and sometimes psychoactive plants and fungi).

Human nutrition is inextricably intertwined with plants. We know from research widely reported by the media that our diets should consist primarily of fruits, vegetables, and grains and that we gain demonstrable health benefits from grapes, red wine, (in moderation) soy products, chocolate, and other foods. Millions of people around the world are employed in food production, storage, and processing; marketing and advertising; transportation; research and biotechnology; engineering, con-

FIGURE 1.15   Leaves and a flower bud of the aquatic plant *Lotus nucifera* Lotus is regarded as a sacred plant in some cultures.

struction, and manufacturing; and commodities trading and economic analysis. Production of food supplements, including vitamins, amino acids, herbal extracts, and other products of plants and microbes, is a multimillion-dollar industry. As much as a quarter of all current prescription drugs contain ingredients derived from plants or microbes, and new plant or microbe-derived medicinal compounds are discovered every year. For example, curcumin, a major component of the spice tumeric (from the plant *Curcuma domestica* or *C. longa*) is nontoxic to humans and has potential use in treatment of cystic fibrosis and other diseases.

Lawns, gardens (Figure 1.16), parks, and nature preserves seem essential to the human psyche, representing our connection to nature and the universe. The horticultural industry is a

FIGURE 1.14   A cornfield test plot representing the importance of plants in modern agriculture

FIGURE 1.16   Gardens provide essential food and recreation for many people

major employer, supplying bedding plants and ornamental trees, shrubs, and grasses for landscaping, as well as a wide variety of houseplants. In addition to the plants themselves, this industry provides plant-derived gardening supplies, including peat moss, bark, and soil treatments.

The use of plant essences and derivatives in cosmetics and perfumes supports another major industry. We prize fabrics made of the plant fibers cotton and linen. Silk is a product of silkworms, which require the mulberry plant as food. For millennia, plants have inspired textile, tapestry, ceramic, jewelry, interior, and architectural design. Artists through the ages have used plants as a mainstay of their compositions (Figure 1.17). Wood serves sculptors as a medium and musicians as the principal material of many instruments and is an essential resource for the building and furniture industries, conferring beauty as well as strength and endurance. Additional examples of humans' long and varied associations with plants can be found in Chapter 2 and throughout this text.

In recent decades humans have developed **biotechnology**, which includes genetic engineering techniques for modifying the genetic material—**DNA**—of plants and microbes (and animals) so that these organisms become even more useful. But biotechnology is sometimes controversial and becomes the subject of widespread media attention (see Figure 1.4). Because of its importance in today's world, a more detailed consideration of plant and bacterial genetic engineering is presented in Chapter 15. In addition, DNA science is a major theme of this text. Throughout this book you will see an icon representing part of the structure of DNA. The icon is designed to draw your attention to concepts and processes related to DNA science, genetics, and biotechnology.

DNA SCIENCE

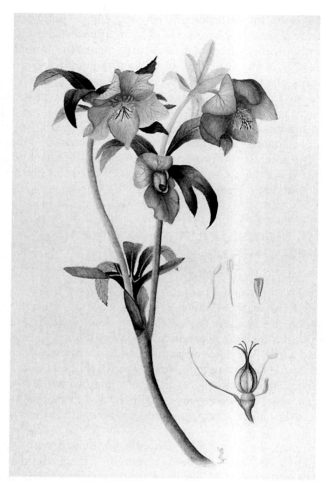

FIGURE 1.17 Artists have long used plants as subjects This painting by artist and botanical illustrator Mary Bauschelt shows the flowers of *Helleborus*.

HIGHLIGHTS

1.1 Plants, as defined here, are multicellular photosynthetic organisms with cellulose-rich cell walls; they adapted in many ways to life on land or, if aquatic, had ancestors with these features.

1.2 Bacteria, fungi, and protists (particularly photosynthetic algae) are central to the lives or history of plants and therefore are studied along with them. All of Earth's life-forms—archaea, bacteria, protists, plants, fungi, and animals—are related.

1.3 Unique, two-part scientific names are given to all living things. The use of scientific names reduces confusion and fosters accurate communication in science, commerce, agriculture, and medicine.

1.4 Scientific methods are used to learn about nature, including plants and associated organisms. Science has an error-correction capacity based on skepticism and peer review.

Scientific results are based on measurements, observations, or experiments involving replicates and controls. Results may be obtained through deductive or inductive approaches. A hypothesis is a tentative explanation that requires testing, whereas a scientific theory is based on extensive data, explains multiple phenomena, and is widely accepted among scientists.

1.5 People use plants and microbes as food and food supplements; in the food processing, baking, and brewing industries; as medicines and building materials; as subjects in art; and in many other ways.

1.6 Plants and microbes play essential roles in maintaining Earth's climate, atmosphere, and organisms. Plants and other organisms are often involved in beneficial relationships. Flowering plants and many animals have influenced each other's evolution.

REVIEW QUESTIONS

1. What is a plant as defined in this book? List its four criteria and briefly (one sentence) discuss each.

2. Besides plants, several other major groups of organisms inhabit this planet, including (a) animals, (b) fungi, and (c) bacteria. Briefly discuss in what ways each group is similar to plants, but most importantly, why each group does not qualify as plant life.

3. Why are plants sometimes referred to as embryophytes?

4. What are the major living groups of land-adapted plants?

5. Explain the ecological service provided by plants with regard to atmospheric levels of oxygen and carbon dioxide.

6. What is a "greenhouse" gas, and what is its relevance to current events in the late 20th/early 21st centuries?

7. What is a food web, and how does it differ from a symbiosis?

8. What is the scientific method? Briefly discuss the nature of science, how it gathers information, how it corrects past errors and misinterpretations, and so forth.

9. How do inductive and deductive reasoning in the scientific method differ?

APPLYING CONCEPTS

1. Chapter 1 discusses a type of symbiosis known as mutualism, in which both partners derive benefit from the relationship. Other types of symbiosis occur, such as commensalism, wherein one partner benefits and the other neither benefits nor is harmed; and parasitism, wherein one member derives benefit at the expense of another. Imagine a walk through a city, park, or campus near you; describe examples of each type of symbiosis you might encounter.

2. Imagine that you are walking through your house or apartment. Describe 10 different ways in which plants or plant products are being used in your own home right now.

CHAPTER

2 | Plants and People

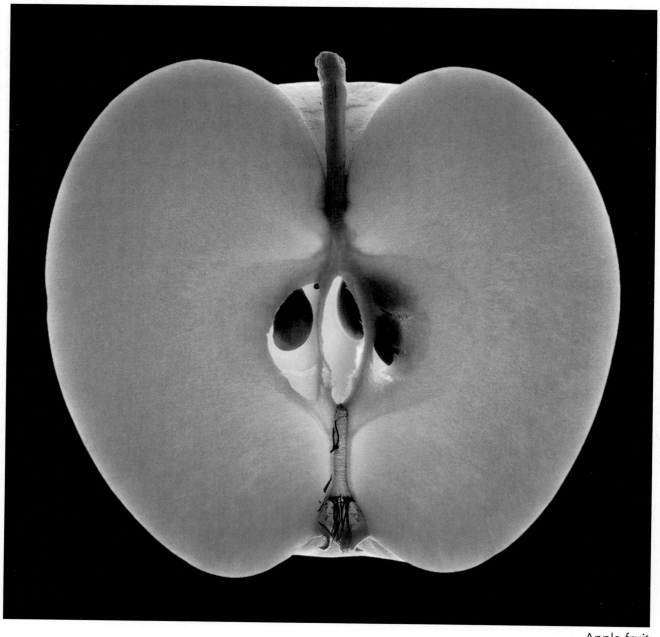

Apple fruit

- Scientific study of indigenous cultures has revealed new knowledge about food, medicinal, and other useful plants. Traditional farming techniques may help solve some of today's food supply problems.

- The study of agriculture's origins provides clues to the locations and identities of modern crop plants' wild ancestors, which harbor genes useful in modern food production.

- Ecological adaptations are responsible for the useful features of major crops and other plants utilized by people.

- Plants produce natural products (secondary chemical compounds) that aid in protection against attack by disease organisms or herbivores or provide other benefits. These compounds explain humans' uses of plants as medicine and drugs and in other applications.

Our group of botanists and marine biologists was temporarily marooned on a remote island in the Bahamas without normally available air transport when one member of our group suddenly required medical attention. She had gone tuna fishing with a local crew and, distracted by the excitement of landing a trophy-sized fish, acquired an extremely painful and extensive sunburn. We were nowhere near a clinic where we could seek medical advice or a pharmacy or store where we could obtain a soothing lotion. Then, one of us remembered seeing *Aloe* plants growing in a nearby garden. We cut several of the sword-shaped leaves and smeared the sap over our colleague's sunburn, whereupon she experienced immediate relief from pain. Additional applications prevented itching and infection of her damaged skin. We were grateful for aloe's healing properties, which allowed us to return to our studies of tropical plants and coral reefs.

Aloe—an asparagus relative that is native to Africa—has been known since ancient times as the "burn plant" or "medicine plant" for its efficacy in treating burns, poison ivy rashes, fungal infections, and other skin problems. It is planted in home gardens in warm areas around the world as a source of treatment for skin conditions. Modern science has shown that the thick, mucilaginous sap of cut leaves contains many compounds that are beneficial to skin. For this reason, modern skin lotions often contain aloe extract and plantations have been established to supply the market demand.

Aloe is but one example of an ancient folk-remedy plant that is economically valued today. Plants have long played an essential role in human medicine; they also provide us with food, beverages, spices, cosmetics, fibers, building materials, and many other products. Market demand for exotic spices influenced the course of human history, stimulating voyages of discovery in the 15th and 16th centuries. Successful cultivation of olive plants for oil production fueled the rise of Athens and, with it, Greek culture and the birth of Western civilization. This chapter will explore ancient and modern human uses of plants, as well as the materials produced by plants. It will provide insights important in developing strategies for discovering additional useful plants and in understanding why plants have been so useful to people. ■

Aloe plant

2.1 | Ethnobotany and economic botany focus on human uses of plants

For most people on Earth today, and for most humans in the past, material culture has been based primarily on plants (Figure 2.1). The scientific study of past human uses of plants or present-day plant uses by traditional societies is known as **ethnobotany**. The root *ethno* refers to the study of people, and *botany* is the study of plants, otherwise known as plant science. **Economic botany** is concerned with the use of plants by modern industrialized societies. These two areas of plant science are closely related in that ethnobotanical work reveals much indispensable information to developed societies, including the recognition of previously unknown medicinal properties of plants. Economic botany often involves investigations into the history of cultivation of modern crop plants to identify their wild relatives. These wild relatives harbor genetic material that is useful in developing crops that are more resistant to disease, insect attack, or drought.

Ethnobotanical studies have revealed that indigenous people—those who follow traditional, nonindustrial lifestyles in areas they have occupied for many generations—often use trial-and-error methods closely resembling the processes of modern science. These methods have led to the discovery of the useful properties of thousands of plants. People may first try a small amount of plant material or extract to estimate potential harm or benefit, then observe multiple incidents of use to make conclusions about its general utility. Development of agricultural crops and farming methods involved similar stages of trial, observation of effect, and modification. Using such methods, humans have discovered how to use more than 3,000 kinds of plants and how to cultivate hundreds. Folk traditions may also function to protect plants and their environments from overuse, improving food security. Modern industrialized societies owe enormous debts of gratitude to past and present indigenous cultures that over millennia have done the hard work of preserving habitats as well as developing and preserving knowledge of plant uses. Ethnobotanists try to bridge the gap between folk wisdom and modern science, bringing to industrialized societies traditional knowledge that may otherwise have been lost (Essay 2.1, "Ethnobotany: Seeds of Culture").

2.2 | The origin of agriculture was key to development of civilizations

Humans who first learned to cultivate food plants lived thousands of years ago, before the invention of writing. Thus, we have no written records of their monumental achievement. Yet it is possible to know a great deal about such ancient events.

How can we know something about the origins of agriculture?

Archeologists and anthropologists, together with ethnobotanists, have used many sources of information to learn about early human farming methods and crops. Such sources include ancient organic remains of plants such as seeds (Figure 2.2), grains, wood, pollen, and plant fibers that were accidentally preserved by being dried, charred in fires, or buried in lake sediments. Also useful are mineral crystals—phytoliths, or "plant stones," which are produced in plant cells but do not decay and are distinctive for particular plants (Figure 2.3)—and coprolites, fossilized human feces that retain undigested bits of food plants. Scientists who study early agriculture also gather information from artifacts that depict plants or that bear impressions of plant materials. Ancient documents aid in tracing the later progress and spread of agriculture. Botanists interested in tracing the origins of modern crops often travel to exotic locales to find wild relatives, then perform genetic crosses or compare the genetic material (DNA) of plants to understand relationships between modern crops and their wild relatives. Indigenous people are often quite knowledgeable about the locations of such wild relatives and traditional cultivation techniques and thus can be indispensable sources of information. A few modern cultures lack agriculture and depend on foraging for food supplies;

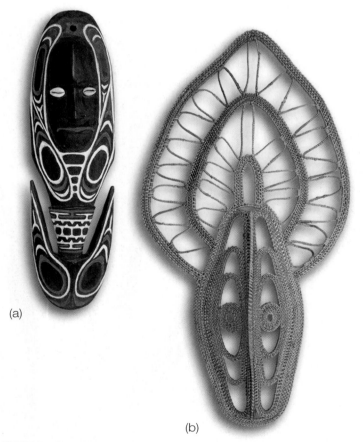

(a)

(b)

FIGURE 2.1 Plants as components of material culture Artifacts produced by aboriginal peoples of northern New Guinea include (a) this carved and decorated wooden hook used to hang food out of the reach of animals, beautiful and utilitarian baskets, and (b) masks woven from plant fibers that decorate cultivated yams for ceremonial purposes. To these people, yams are valuable not only for their high food value and long-keeping properties but also because they symbolize male pride and are thought to have spirits that communicate underground.

ESSAY 2.1   ETHNOBOTANY: SEEDS OF CULTURE

Ethnobotanists work with indigenous societies and with materials from archeological sites to piece together the history of plant uses by human cultures. Their training combines the study of botany with anthropology, archeology, folklore, linguistics, psychology, chemistry, medicine, pharmacology, and ecology. To accomplish their work, ethnobotanists must also know something of diplomacy, economics, and political systems.

An ethnobotanical study begins with identification of an interesting problem, such as the uses of plants by a culture that has previously been inaccessible or incompletely studied or a search for new medicines. Ethnobotanists must first obtain the approval of the government of the country in which the study will be done; plant materials and artifacts cannot generally be removed from the country of origin without permission. The ethnobotanist must meet and establish rapport with tribal elders and healers. It is usually necessary to negotiate some kind of return of knowledge or other reward to the local population or national government. Information about plant uses is best obtained directly, requiring mastery of

Most Antanosy people know medicinal plant remedies for aches and pains, colds, and skin infections.

local dialects and culture. Ethnobotanists assume the roles of participant-observer, teacher, and friend; they learn how the people identify, collect, prepare, and use plant materials. Ethnobotanists may take samples for chemical analysis and economic applications. They may also try to help develop ways to sustain cultural lifestyles, as these are often threatened by outside interests.

Linda Lyon (Figure E2.1A) of Frostburg State University in Maryland is an ethnobotanist. She works with the Antanosy people of southeastern Madagascar to learn how they use forest plants for food, shelter, and medicine. Most Antanosy peo-

ple know medicinal plant remedies for aches and pains, colds, and skin infections. However, because some of the plants are located in remote areas, the people buy plant remedies at the local market from a "pharmacist," (Figure E2.1B) who has collected the plants from nature and knows their uses. They also rely on advice from their shaman for more serious health and spiritual problems. Linda Lyon also determined that the Antanosy people harvest about 10 types of grasses and similar plants from local wetlands for producing baskets, mats, and hats for personal use or to sell but that these useful plants are not always found in plentiful supply. Her work will help perpetuate this culture's knowledge and natural habitat. It will also add to our store of knowledge of tropical plants and their uses.

E2.1A   Linda Lyon, an ethnobotanist

E2.1B   Traditional pharmacist in Madagascar

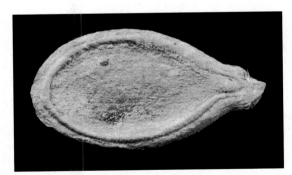

FIGURE 2.2   Plant evidence of domesticated crops This squash seed from a cave in Oaxaca, Mexico, has been dated to 8,910 years before present. The ridged edge and other features of this seed are characteristic of *Cucurbita pepo*, the same species as modern pumpkins and summer squash. Several other ancient seeds and pieces of squash rinds were found at the same site; they represent the oldest-known evidence of plant domestication in America.

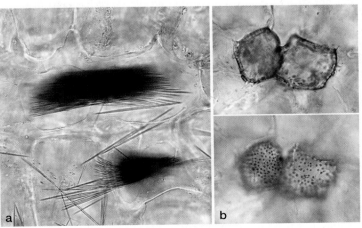

FIGURE 2.3   Phytoliths In (a), thin crystals called raphides are shown, photographed in cells of a banana peel viewed with a microscope. In (b), thick-walled cells ("stone cells") from a pepper are seen with the microscope focused in two different planes.

these groups model aspects of early human hunter-gatherer societies and help anthropologists understand this phase of human societal evolution. Taken together, these varied sources of information provide a picture of when, why, where, and how agriculture began and its role in nurturing human civilizations.

When and why did agriculture originate?

Ancient human societies obtained food by hunting, fishing, and gathering wild plants until approximately 10,000 years ago—the age of the earliest evidence to date for deliberate plant cultivation. The shift from foraging to agriculture as a primary mode of food acquisition occurred gradually and apparently independently in several different regions around the world at nearly the same time. Many experts think that this coincidence is related to a worldwide climate change that led to warmer conditions at the end of the most recent ice age. Several types of evidence indicate that atmospheric carbon dioxide levels also increased significantly (from 180 to 280 parts per million) at this time, a condition that would have increased the rate of plant photosynthesis (production of food using water, carbon dioxide, and sunlight energy). Scientists think that these environmental changes required human populations to adopt different ways of obtaining their food and favored the growth of grasses, legumes, and other plants that were nutritious and otherwise suitable for domestication.

Domestication is an evolutionary process, guided by humans, that results in crop plants having traits that differ in some useful way from those of wild relatives. Early farmers accomplished this by selecting plants having useful traits as the sources of seed for planting the next generation and continuing to perform this selection process over many years. As crop plants changed through time in ways that were beneficial to humans, agriculture become more efficient, allowing human populations to increase.

Subsequent development of methods for storage of food from year to year reduced the risk that food supplies could suddenly and unexpectedly fail—an ever-present danger for the forager. Because agricultural food production does not require that everyone be involved, some people were freed to practice other pursuits, including the arts, architecture, philosophy, science, medicine, law, and commerce. Major human civilizations, past and present, were built on the foundation of a productive agricultural system.

Where were plants first domesticated?

Agriculture originated independently in at least six geographic locations: the Near East, China, Mesoamerica, South America, the Eastern United States, and New Guinea. Agriculture may also have arisen separately in Africa, or might have been imported from the Near East. Plant remains found in archeological sites in the **Fertile Crescent** region of the Near East suggest that agriculture first began at this site about 11,000 years ago. People in this region domesticated the **cereals** wheat and barley and the **legumes** peas, chickpeas (garbanzos), and lentils. Cereals are plants belonging to the grass family, which produce fruits known as grains. Legumes are plants of the pea family,

which are named for their pod-shaped fruits, known as legumes. Comparison of the genetic material of modern wheat with that of its wild relatives has revealed that people living near the Karacadag Mountains in the southeastern part of modern Turkey started the wheat domestication process about 10,000 years ago. Other modern crops that were first domesticated in various parts of the Near East include olives, grapes, figs, dates, pistachios, almonds, and pomegranates.

Agricultural knowledge spread rapidly from the Fertile Crescent, reaching southern Europe by about 8,000 years ago, the Balkans and Central Europe by 7,500 years ago, and Scandinavia by 4,500 years ago (Figure 2.4). Europeans expanded the number of domesticated plants to include such valued crops as oats, apples, pears, plums, and cherries. Some language experts, using techniques borrowed from evolutionary biologists, find evidence that a Proto-Indo-European language originated in large, Neolithic settlements in the same Near Eastern region associated with the origin of agriculture. Linguists think that this ancestral language eventually gave rise to nearly 150 distinct Eurasian languages (including English), which spread and diverged alongside agricultural knowledge.

Plants were also domesticated at several locations in the Far East, including sites along the Yellow and Yangtse River Valleys in China. Plants first cultivated there include the modern cereal crops rice and millet, soybean (a legume), and hemp, whose fibers have been used in making cloth, paper, and rope for the past 5,000 years. Phytoliths and distinctive starch particles found stuck to the surfaces of stone tools indicate that bananas and taro (a starchy root crop from the plant *Colocasia esculenta*) were cultivated in New Guinea nearly 7,000 years ago. Coconut and sugar cane were also domesticated in tropical Southeast Asia.

Before the arrival of Europeans in the 15th century, Native Americans had domesticated more than 100 species of plants, including more than half the world's major modern crops. In Amazonia, farmers converted large areas of river-edge forest into fruit and nut orchards and fields of maize (corn) and

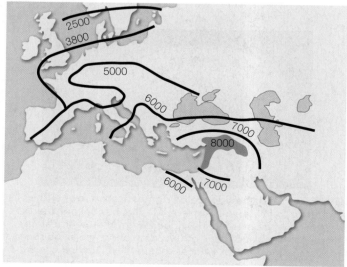

FIGURE 2.4 The spread of agriculture This map depicts the spread of agriculture from an origin in the Near East. Numbers are years B.C.

carbohydrate-rich manioc (cassava), a root crop also known as cassava. Amazonian farmers transported rich, dark muck from wetlands to their upland farms, in the process creating terra preta (black earth), which is still present. Terra preta is much more fertile than unmanipulated soils of the same region. The presence of such rich soil has led some anthropologists to think that early Amazonian agriculture may have supported much larger populations of people prior to European contact than has previously been realized. Although modern Amazonian rain forest is often regarded as an untouched wilderness, there is considerable evidence that ancient human populations significantly modified this landscape by their agricultural activities. Additional crops first grown by people living in present-day North, Central, or South America include sunflower, squash (first domesticated about 9,000 years ago; see Figure 2.2), four types of chili peppers, tomatoes, and potatoes. The protein-rich legumes peanut, common bean, and lima bean were also domesticated in the Americas (Figure 2.5).

Ecological adaptations are responsible for the useful features of cereals and legumes

A comparison of crops domesticated throughout the world (Table 2.1) reveals that most early agricultural societies cultivated both cereal and legume crops. Cereals and legumes afforded nutritional and storage advantages that are the results of adaptation to their environments by their wild ancestors.

Cereals—named for the Roman goddess Ceres—that were domesticated early in human agricultural history include wheat, barley, rice, and corn (maize). As members of the grass family, cereals have in many ways become well adapted to grassland environments (Chapter 26), which are typically drier than regions that support forests. Food production in cereals and other grasses is more efficient than in most other plants in dry conditions. Grass leaves are structured to maximize the harvesting of sunlight, and their grains are particularly rich in carbohydrates and proteins, which are needed to support the growth of seedlings. Humans learned to take advantage of the high growth rate of cereals and the nutritional content of their grains well before the origins of agriculture.

Because wheat is richer in protein than most food plants, including barley, rice, and corn (maize), it is used extensively today in the production of bread, pasta, and other food products. The average protein content of wheat is about 13%. The whole grain is most nutritious. It contains iron and B-vitamins that are lost during refining and are re-added to white flour. Corn (maize) is a particularly versatile crop that can be used directly as human food (sweet corn and popcorn); to feed livestock (dent or field corn); to manufacture corn oil, corn meal, corn flour, corn syrup, and corn starch—all of which are widely used in cooking and in processed foods; and to produce alcohol for use as a fuel and in beverages such as bourbon.

A major disadvantage of cereals is that they are low in the **amino acids** lysine and tryptophan, which are required in the

FIGURE 2.5 Foods domesticated in Central and South America Ancient Americans domesticated (a) zucchini, (b) corn (cereal), (c) chilies, (d) beans (legume), (e) squash, (f) peanuts (legume), and (g) potato.

TABLE 2.1 Some Crops Associated with Major Centers of Plant Domestication

Center	Uses	Common Name	Scientific Name
Near East	Cereal foods	Wheat	*Triticum monococcum, T. turgidum*
		Rye	*Secale cereale*
		Oats	*Avena sativa*
	Legume food	Lentil	*Lens culinaris*
	Legume forage	Alfalfa	*Medicago sativa*
	Fruit	Fig	*Ficus carica*
	Fiber	Flax	*Linum usitatissimum*
China	Cereal food	Rice	*Oryza sativa*
	Legume food	Soybean	*Glycine max*
	Fruits	Mulberry	*Morus alba*
		Orange	*Citrus sinensis*
	Fiber	Hemp	*Cannabis sativa*
New Guinea	Fruit	Banana	*Musa spp.*
	Other food	Taro	*Colocasia esculenta*
Mexico and Central America	Cereal food	Corn	*Zea mays*
	Legume food	Common bean	*Phaseolus vulgaris*
	Fruits	Avocado	*Persea americana*
		Squash	*Cucurbita pepo*
		Cacao	*Theobroma cacao*
		Red pepper	*Capsicum sp.*
	Other	Sweet potato	*Ipomoea batatas*
	Fiber	Cotton	*Gossypium hirsutum*
Central Andes	Legume food	Peanut	*Arachis hypogaea*
	Fruit	Pineapple	*Ananas sp.*
	Other food	White potato	*Solanum tuberosum*
		Manioc (cassava)	*Manihot esculentum (esculenta)*
	Other	Rubber	*Hevea brasiliensis*

human diet. Vegetarians know that legumes can supply these missing amino acids and that their meals should contain both legumes and cereals. Early agriculturists had also learned this principle.

Today, food crops described as legumes include soybean, pinto bean, green bean, broad bean, garbanzo bean, peanut, green pea, lentil, and lima bean, as well as alfalfa, clover, lespedeza, and bird's-foot trefoil, which are used as livestock fodder. Soybean—probably the world's most widely planted legume—is also used in the production of such industrial products as paints, plastics, oils, and adhesives. The part of legume plants most frequently eaten by people is the seed, although we consume the entire fruit of green beans and pea pods. Livestock fodder includes most of the legume plant. Legume-derived foods are richer in protein than any other plant products.

Legume crops are also valued because they enrich the soil where they are planted by adding nitrogen fertilizer. Farmers can rotate legumes with other crops that tend to deplete soil nitrogen levels in order to replenish soil nitrogen. This is of particular importance to farmers around the world who cannot afford to purchase and apply commercial nitrogen fertilizers. The nutritional value of legumes, their high protein content, and their soil-enriching capacity arise from their ability to harness the biochemical properties of soil microbes. Legume roots attract certain soil bacteria that are capable of producing nitrogen fertilizer from the nitrogen gas in air, a process that plants alone cannot perform. The bacteria multiply inside legume roots, nourished by food supplied by the plants, and, in return, provide needed nitrogen fertilizer to legume roots and the soils in which they grow. This partnership is valuable to the legume plant because it allows it to produce more seeds, increasing its reproductive success.

Because they are dry at maturity—an ecological advantage for plants of arid grasslands—legume seeds and cereal fruits are less likely to spoil during storage. The ability of early agricultural peoples to store cereal and legume crops for long periods and transport them over long distances was a tremendous advantage. The dry fruits and seed protein richness of cereals and legumes are examples of the general principle that ecological adaptations are responsible for the useful features of plants.

How were cereals and legumes domesticated?

No one is quite sure how people first thought of deliberately planting the seeds of food plants. One possibility is a few seeds remained in the unused parts of gathered food plants that were dumped in local trash heaps. These seeds later germinated, forming a dense growth that was more easily harvested than wild plants. Some enterprising person may have then tried scattering seed in a convenient location and soon realized how much easier it was to grow food plants nearby rather than to forage for them. In forested areas of the world, agriculture might have begun with human care of long-lived wild food plants in their natural environment. Repeated, tiring visits to prune, weed, and reap the harvest may have generated the idea to plant crops in plots closer to home.

When modern crops are compared to their wild relatives, differences are obvious. The grains of modern wheat are noticeably larger than those of wild relatives still growing in the grasslands of the Near East. Grains of wheat that accidentally fell into fires and became preserved as charcoal found at archeological sites show an increase in size over time. In addition, the fruit-bearing spike of modern wheat crops does not fall apart or shatter, as does that of the wild relatives. Shattering is advantageous to wild cereals because it helps spread the progeny of the parent plants as far apart as possible, which reduces competition for water, light, and minerals. But shattering is disadvantageous to humans because it makes wheat more difficult to harvest. Scientists think that occasional mutations in wild wheat and barley led to some plants having spikes that did not easily shatter. Such variants, being more easily harvested by humans, would have been preferentially selected as sources of seed for planting new fields. Over many generations the nonshattering types would come to dominate the crop. Similarly, humans repeatedly chose the largest cereal grains for seed, thereby contributing to a gradual increase in grain size.

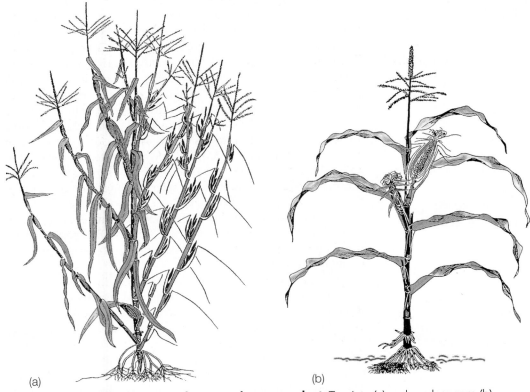

(a)　　　　(b)

FIGURE 2.6 Teosinte compared to a modern corn plant Teosinte (a) and modern corn (b). The dense, compact modern ear of corn represents an ear-bearing branch of teosinte that has been considerably shortened.

The process by which humans choose desirable plant traits and then preferentially use the seed from plants having those traits as seed for planting is known as **artificial selection.** This process, performed for generations, has resulted in crops that differ from their wild relatives, sometimes dramatically so. In the cases of wheat and barley, the genetic changes required for the human-induced increase in fruit size and nonshattering spike were relatively slight, explaining the evidence that these useful changes in plant structure occurred relatively quickly, within 30 to 200 years. Wild pea and lentil have fruits that open rapidly, an adaptation that is beneficial for seed dispersal; but cultivated peas and lentils have fruits that open much more slowly, facilitating harvest. This difference also results from a small genetic change and thus is likely to have occurred rapidly. Squash seeds nearly 9,000 years old (see Figure 2.2) and pieces of rind obtained from ancient storage pits in Mexican caves also show evidence of human domestication or deliberate selection of plant traits. The seeds found in human habitations are larger than those of wild squash, the shape of the fruit is different, and fruit color was bright orange, in contrast to the green- and white-striped coloration of wild squashes.

Probably the most dramatic alteration that humans have achieved in food plants has occurred with corn (maize). Genetic (DNA) evidence has shown that the closest wild relative of modern corn is a grass known as teosinte, which today occurs in southern Mexico (Figure 2.6). The cobs of teosinte and modern corn are very different (Figure 2.7). Teosinte cobs bear only six to ten grains (also known as kernels or fruits), which are covered with a very hard, inedible casing and lack a husk formed of leaves. In contrast, modern corn cobs contain many more grains, which are soft and edible, and they are enclosed by a husk, which increases ease of harvest and storage. Teosinte is much better adapted for survival in the wild than cultivated corn. Teosinte cobs shatter, scattering the one-seeded fruits effectively, and seed dispersal is not hindered by a husk. Modern corn has

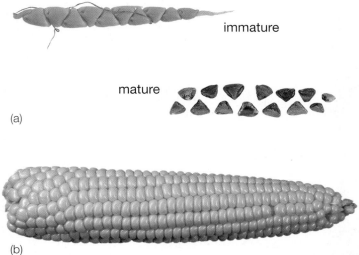

(a)

immature

mature

(b)

FIGURE 2.7 Domesticated corn cobs differ from those of ancestral grasses Teosinte (a) has only a few, inedible grains per cob, in contrast to modern corn (b), which has many, edible grains per cob. Human (artificial) selection is responsible for these differences.

been so altered by human selection that it could not survive for long in the wild.

Some experts think that hunter-gatherers first began to grow plants that were ancestral to corn as a sweet snack; they may have chewed the green ears and stalk pith much as Mexican children do today. Then ancient Americans may have noticed a teosinte plant in which a mutation had occurred that resulted in softer, edible kernel cases. Recognizing its food potential, people began to cultivate the mutant teosinte for its grains. At least four other genetic changes occurred during the transition from teosinte to modern corn, explaining the dramatic difference in their appearance. Study of DNA extracted from cobs at least 4,300 years old, found in caves, has revealed that humans had already begun to alter the genetic structure of maize thousands of years ago. The many generations of Native Americans who transformed corn into its modern form accomplished an amazing feat of plant modification.

2.3 | Food plant genetic resources and traditional agricultural knowledge need to be preserved

Conservation of wild relatives of modern crop plants and their habitats, as well as traditional varieties of crop plants from around the world, is now recognized as essential for the development of new crops and new varieties of older crops by plant breeding methods. Preservation of genetic material from crop plants and their wild relatives is essential to modern strategies for improving crops. Obtaining and preserving information about indigenous agricultural methods is likewise important.

Modern agriculture typically involves planting large areas with a single type of crop (Figure 2.8), a practice known as **monoculture.** Farmers in a particular region will often use the same hybrid seed variety, typically bred for increased food production. The advantage of monoculture is relative ease of planting and harvesting. But a negative aspect of monoculture was dramatically demonstrated in the 1970s, when in some areas of the United States as much as 50% of the corn crop was lost and the total U.S. harvest was reduced by more than 15%. This disaster was caused by a fungal infection called southern corn leaf blight. Crop scientists (agronomists) knew that the widely planted hybrid was more vulnerable than other corn varieties to this disease, but no one expected such a rapid and extensive spread of the fungus. The pest's spread was favored by the absence of genetic variation in the large populations of hybrid corn.

Agricultural experts now recognize that low genetic variation in modern crops is a widespread and serious problem. In response, crop scientists have collected and stored samples of crop varieties from around the world as sources of genetic material for plant breeding. They also collect wild relatives of modern crop plants, which have proven to be excellent sources of genes for pest and drought resistance. The presence of insect pests and microbial diseases, as well as climate variation in nature, favors the survival of pest- or drought-resistant plants. This process is known as **natural**

ECOLOGY

FIGURE 2.8 **Monocultured crops** This cornfield (a) and soybean field (b) in Central Illinois are familiar examples of modern cereal and legume crops that are grown in large populations of uniform genetic composition.

selection. People take advantage of this natural process by collecting potentially useful plants from nature and storing them in gene banks, seed banks, or germplasm repositories (Essay 2.2, "Chocolate: It Does a Body—and Ecosystems—Good"). Here plants can be maintained for long periods and made available to plant breeders as needed.

Unfortunately, loss of the natural habitats of the wild relatives of modern crop plants is a worldwide problem. Natural areas are often lost to agricultural development, which involves destruction of the native vegetation. Farmers may replace traditional crops of greater genetic variation with commercial crops of lower genetic variation. Also, farmers sometimes destroy wild relatives of commercial crops to prevent interbreeding. Preservation of natural habitats that harbor wild relatives of modern crops and traditional crop varieties has become an essential measure for improving food security for the world's population.

Another modern agricultural problem is the loss of knowledge about traditional methods for growing crops, especially techniques developed in tropical areas. Monoculture methods, which can work well in temperate areas once occupied by extensive grasslands and consequently climatically suited for cultivation of cereal crops, often do not work well in the moist tropics. This difference arises from a variety of climate and biological factors, including more rapid reproduction of pests and weeds in the tropics (Chapter 27). Furthermore, tropical farmers typically cannot afford to use chemicals for crop protection. Over the centuries, traditional farmers in the tropics have developed a variety of creative agricultural techniques that are well adapted to their local environments. One example is **chinampas** cultivation (Figure 2.9), invented by the ancient inhabitants of the Tehuacan Valley of Mexico 2,000 years ago in response to the challenge of farming swampy land. These farmers created raised, linear soil beds where they grew a mixture of corn, beans, squash, and other crops. This **polyculture** method is particularly well suited to the tropics because it reduces the ability of pests to find and entirely destroy any one crop. Crop beds were interspersed by water channels that provided a constant water supply to crops while also facilitating transport of the harvest using boats. The method was so successful that it formed the economic basis for the Aztec civiliza-

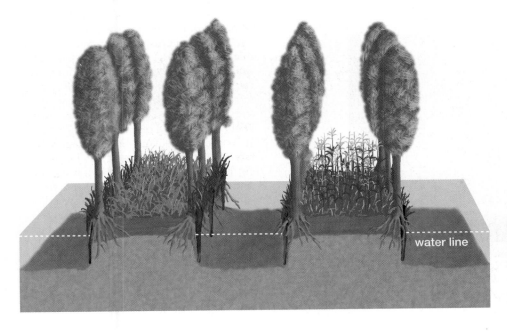

water line

FIGURE 2.9 **Diagram of chinampas agriculture** Crops of several types are grown together on raised linear beds interspersed with water channels. This cultivation method reduces losses from insects and diseases and at the same time provides for irrigation.

ESSAY 2.2   CHOCOLATE: IT DOES A BODY—AND THE ECOSYSTEM—GOOD!

For such a delicious treat, chocolate is actually quite a healthful food. Medical research has revealed that chocolate is a rich source of the organic compounds known as polyphenols—more precisely, a type of flavonoid known as procyanidins. Also found in foods such as cranberries, peanuts, tea, apples, and blueberries, procyanidins are beneficial antioxidants that prevent cell damage by oxygen. Flavonoids in chocolate and other plant foods lower the risk of cardiovascular disease. As little as 100 ml of cocoa can cause an increase in artery diameter that doubles blood flow rate for 6–8 hours. Flavonoids also affect blood clotting, reducing the potential for formation of harmful blood clots. Chocolate is so rich in the beneficial compounds that an effective dose (10–60 mg/day) can easily be achieved by consuming a small amount of chocolate. Of course, chocolate candy typically also contains saturated

The goals of chocolate farming include not only reliable and efficient production but also preservation of native species and the ecological functions they provide and a source of livelihood for local people.

fats and sugar, which are not so healthful. Because much of the fat has been removed during production of cocoa powder, hot cocoa can be healthful, if you avoid topping it with whipped cream!

Chocolate is processed from the seeds of the cacao tree, *Theobroma* (Figure E2.2), whose scientific name means "food of the gods." First domesticated in the Amazon, *Theobroma* has been cultivated in tropical areas for at least 2,000 years. More than 5 million small farmers worldwide make a living growing cacao trees, which can produce a crop for up to 100 years. Because *Theobroma* evolved in shady tropical forests, chocolate production is most sustainable when other trees are nearby to provide shade. Farmers often grow cacao together with other tropical tree crops—coconut, cashew, durian, and persimmon, for example. Growing multiple crops together, a farming technique known as polyculture, is beneficial because the crop aggregations

are similar to natural tropical forests. Thus, many native tropical organisms that provide important ecological functions are able to survive amongst the crops, decreasing the potential for extinction. Cocoa cultivation often occurs near major regions of high biological diversity where extinction is a concern. If grown as an unshaded monoculture (a field containing only one type of crop), associated organisms are less likely to survive, and their ecological benefits may also be lost. When grown in unshaded monocultures, chocolate production typically increases in the short term but, after a few years, the trees become so stressed by excess light and drought that they become vulnerable to disease microorganisms. When chocolate crops fail, farmers lose their cash-crop incomes, and the price of chocolate products increases. The goals of chocolate farming include not only reliable and efficient production but also preservation of native species and the ecological functions they provide and a source of livelihood for local people.

Though *Theobroma* is native to the Amazonian basin, 70% of the commercial chocolate crop is now grown in Africa, such as in Cameroon. Increasingly, *Theobroma* is being cultivated in Asia, including deforested areas of Vietnam. This is beneficial because the African and Asian crops are geographically separated from diseases that evolved with chocolate in its native habitat. Even so, diseases are still a major problem. One of the most serious chocolate diseases is witches' broom, caused by a fungus (*Crinipellis perniciosa*), that exudes compounds that disrupt natural plant growth. Infected *Theobroma* trees have abnormal brushy growths ("brooms") that are unproductive. Black pod disease, which destroys the chocolate fruit, is caused by the parasitic protist *Phytophthora*, whose name means "plant destroyer." Plant explorers have found wild *Theobroma* having disease-resistant genes in Amazonian habitats. Such natural habitats are disappearing rapidly as burgeoning human populations transform natural lands to farms and other uses. Seeds from wild plants with useful genes are stored in U.S. National Germplasm Repositories and international seed banks. Plant breeders can then use them to produce new, disease-resistant crop varieties.

E2.2   *Theobroma* fruit with seeds

tion (conquered by Cortéz in 1519). Modern Mexican farmers continue to use the chinampas technique in areas having persistently wet soil; it has the great advantage of low reliance on fossil fuels and chemical application. Many other locally adaptive agricultural techniques are being discovered through the study of indigenous people; these are expected to be helpful in solving present and future food production needs.

Some experts are concerned that the diversity of the food base in industrialized countries has become too low. Of the several hundred plants that have been domesticated, only 12 crops—primarily wheat, rice, corn, and potatoes—form the basis of most of the world's agriculture. A comparison of the diets of people in industrialized countries reveals that they eat fewer types of plants than those in nonindustrialized countries. Some experts think that reliance on just a few crop plants is a dangerous dependency.

2.4 | Natural plant products are useful to humans as medicine and in other ways

Crop plants are valued primarily because they are sources of carbohydrates, lipids (fats and oils), and proteins—which are primary compounds—as well as minerals and vitamins. These plant materials are important as major components of plant structure or in plant metabolism (the chemical reactions that occur in living things). They are essential components of all living things and are required in the human diet, explaining the importance of plants as food. People also use plants as sources of medicines, herbal dietary supplements, and psychoactive drugs (described in this chapter), as well as poisons, stimulating beverages, spices, perfumes, oils, waxes, gums, resins, fibers, and dyes (discussed elsewhere in this book). These uses are based on the presence in plants of organic compounds known as **natural products** or **secondary compounds** that help protect or strengthen plants or aid in reproduction. Together, plants produce about 100,000 different kinds of secondary compounds. The chemical makeup of secondary compounds differs from that of primary compounds (see Chapter 3).

The roles of secondary compounds in plants include defense against herbivorous animals and disease-causing microbes. Because plants cannot run away, they depend on chemicals for defense; consequently, they have become experts at chemical warfare. Plants are surrounded in nature by voracious herbivores, including swarms of ravenous insects, and deadly microbes—yet still they flourish. This is a testament to their ability to defend themselves chemically. Well-defended plants live to produce more descendants; thus, natural selection explains the presence of secondary compounds in plants. Plants also cannot move to accomplish mating or migration, as animals can, so they often use animals to transport reproductive cells or seeds. Secondary compounds are involved in producing flower and fruit colors and fragrances that entice animals to serve as dispersal agents (see Chapter 24).

ECOLOGY

Plants contain mixtures of many types of secondary compounds and may differ considerably in the amount of secondary chemicals that they contain. As a result, plants differ in their usefulness as beverages, spices, or medicines, as well as in their psychoactivity—influence on the animal nervous system, or toxicity. To deal with their many enemies, plants require many different types of chemical defenses. Plants growing in tropical rain forests, where the diversity of insect pests and disease microbes is very high, typically have more defensive chemicals than plants living where the defensive challenges are fewer. For this reason, new medicinal compounds are often sought in tropical plants, and indigenous healers in the tropics are often the focus of ethnobotanists' attention.

Plants are sources of medicine and dietary supplements

Among the common medications used in developed countries, more than one-quarter include plant-derived ingredients. One example is ipecac, used to induce vomiting in poisoning cases. Ipecac is derived from the plant *Cephaelis ipecacuanha*, native to South America. Cancer drugs originating from plants include podophyllin produced by the North American mayapple *(Podophyllum peltatum)* (Chapter 27), leukemia drugs from the Madagascar periwinkle *(Cantharanthus roseus)*, and taxol from the Pacific yew *(Taxis brevifolia)* (Chapter 7).

In other parts of the world, people often rely primarily on medicines derived from plants collected directly from nature. *Aloe barbadensis*' widespread use as a skin treatment was described at the beginning of this chapter, and willow bark has long been chewed for pain relief, as it contains an active compound similar to commercial aspirin. The poppy *Papaver somniferum*, native to Eurasia, has been used since ancient times to relieve pain. Poppy contains more than 26 active compounds, including the habit-forming morphine, but also non-habit-forming codeine—used in cough medicines—and papaverine—used in the treatment of intestinal spasms.

People have long used herbs as dietary supplements and as a source of perceived health benefits. In recent years, preparations from such plants as St. John's Wort, ginseng, garlic, *Ginkgo biloba*, goldenseal, saw palmetto, and *Echinacea* (Figure 2.10) have become very popular in industrialized countries. Typically, these plants are supplied in dried form, as in teas, or as extracts packaged in capsules; the active ingredients are not isolated. The market for such products—termed herbs, nutriceuticals, or phytochemicals—is worth billions of dollars annually in the United States alone. In 1999, U.S. consumers spent $400 million for herbal preparations of St. John's Wort. Companies in the United States are free to claim heath benefits for herbal products without regulation by any government agency. However, the fact that plants typically contain mixtures of compounds that can influence human health is a source of concern to the medical establishment in industrialized countries, and the extent to which such products should be regulated is a contentious issue. It is difficult to use established scientific testing methods to determine the safety and efficacy of these products because they each contain many

FIGURE 2.10 **Plants used in herbal preparations** (a) Purple coneflower *(Echinacea purpurea)* roots and underground stems were used by Native Americans to treat wounds, and extracts are now marketed as treatment for colds and other infections. (b) *Ginkgo biloba* (the maidenhair tree) is a nonflowering seed plant (gymnosperm) whose dried leaves and seeds are used in herbal preparations that purport to improve memory. Extracts of whole St. John's Wort (c) *(Hypericum perforatum)* plants are sold as preparations that are supposed to relieve depression. *Echinacea* and *Ginkgo* are also valued in gardens and streetside plantings.

types of secondary compounds that could vary in relative proportions from one harvest to another. Herbal products could affect individual people or their various health conditions differently or interact with prescription medicines in unexpected ways. For example, various herbal preparations interfere with the metabolism of prescription drugs in the human liver. *Ginkgo* inhibits a liver compound that normally breaks down drugs and thereby causes blood concentrations of drugs to increase to undesirable levels. St. John's Wort increases the activity of a different liver compound, causing undesirable decreases in contraceptive drugs or cyclosporine, a drug used to prevent the rejection of organ transplants. *Echinacea* decreases the action of some liver components but increases others, thus influencing different drugs in different ways. Researchers report very great variations among patient responses to herbal supplements and also great variation among different herbal products in active-ingredient content.

It is important for human health that decisions about herbal supplements are based on scientific evidence. Experts recommend that responsible governmental regulation and certification would encourage the supplement industry to invest more in high-quality nutrition research. Such research is expected to generate discoveries as important as recent findings that folic acid prevents birth defects and that taking antioxidant vitamins plus zinc slows the progression of retinal degeneration to blindness. In the United States, the National Institutes of Health (NIH) Center for Complementary and Alternative Medicine is a federal agency that supports scientific research on the medical efficacy of herbal treatments. The NIH is a unit of the Public Health Service, which in turn is part of the U.S. Department of Health and Human Services. The

NIH also supports an Office of Dietary Supplements, enabled by the 1994 Dietary Supplement Health and Education Act. This office is concerned with the development of reliable methods for analyzing chemical components of herbal supplements.

Plants are sources of psychoactive drugs

Some plants contain psychoactive compounds that help protect against animal herbivores by altering attackers' nervous system function. When herbivores consume these compounds, they may experience behavior changes that decrease feeding or increase vulnerability to their own predators. Since the human nervous system is similar to that of other animals, psychoactive plant compounds also affect humans. Depending on their effects on humans, psychoactive plant compounds are classified as stimulants, depressants, hallucinogens, or narcotics. Psychoactive compounds affect humans in a dose-dependent way; low doses may be stimulatory, but high doses can be dangerous—even lethal. Some psychoactive plant compounds are habit-forming (addictive) in humans, whereas others are not. Stimulants include cocaine, derived from the *Coca* plant; caffeine, derived from *Coffea*; and nicotine, derived from tobacco *(Nicotiana)*. Plants that have been used for their hallucinogenic (psychedelic) properties include *Cannabis* (marijuana), native to Asia; belladonna *(Atropa belladonna),* native to Europe; as well as *Datura* and the peyote cactus (Figure 2.11), both native to the U.S. desert southwest and other world regions.

The National Institute on Drug Abuse is a U.S. Health and Human Services Department unit that deals with habit-forming plant compounds. This institute publishes bulletins on the specific health impacts of drugs of abuse, including the addictive

ECOLOGY

FIGURE 2.11   The peyote cactus, *Lophophora williamsii*

substances nicotine, marijuana, heroin, and cocaine, which are all derived from plants. Nicotine is only one of more than 4,000 components of tobacco smoke, but it is the major one that acts on the brain. Although marijuana has medical uses in the treatment of cancer and AIDS patients, it also has deleterious brain effects and causes impairment of the heart and lung. Heroin, processed from morphine that occurs in poppy plants, acts in many places in the brain and nervous system. One effect is to depress breathing, sometimes to the point of death. Cocaine—though sometimes used medicinally—interferes with normal chemical communication among brain cells. Long-term cocaine use can generate full-blown paranoid psychosis, heart attack, breathing failure, and stroke. A knowledge of the beneficial and harmful effects of these and other plant secondary compounds is essential to maintaining good personal health and is vital for making sound societal policy decisions.

HIGHLIGHTS

2.1 Ethnobotany is the study of past and present traditional uses of plants. Economic botany is the science of plant applications in modern industrial societies.

2.2 Agriculture arose approximately 10,000 years ago in several regions of the world, including the Fertile Crescent in the Near East; China; New Guinea in the Far East; and North, Central, and South America. Cereals and legumes were among the earliest domesticated crops, providing important advantages in nutritional quality and ease of storage. Ecological adaptations explain why cereals and legumes are so useful. The increase in agriculture's food-production efficiency facilitated the rise of major human civilizations.

2.3 The wild ancestors of today's crop plants are valuable sources of genetic material for use in breeding plants with increased resistance to pests and diseases but are increasingly endangered by habitat destruction. Traditional agricultural methods have also proven useful in solving modern crop production problems and need to be preserved.

2.4 Plants are useful as sources of medicines, fiber, building and industrial materials, beverages, and spices, and because they contain natural products, also known as secondary compounds. Secondary compounds endow plants with defensive capabilities or play essential roles in plant reproduction.

REVIEW QUESTIONS

1. List some of the sources of information used by ethnobotanists, archeologists, and anthropologists to deduce the types of plants used and domesticated by ancient peoples.

2. Describe three major regions in which plants were first domesticated and the particular legumes and cereals characteristic of each.

3. What are the special advantages of cereals and legumes as crop plants?

4. How did the origin of agriculture contribute to the rise of major human civilizations?

5. What is artificial selection? Provide an example of the role of artificial selection in the domestication of a crop plant.

6. How does monoculture differ from polyculture, and what are the advantages and disadvantages of each?

7. Why do plants, and particularly those of tropical regions, produce so many types of secondary compounds?

8. Why might physicians be concerned about people's casual use of herbal preparations without obtaining medical advice?

9. Which plants produce compounds that are psychoactive in humans, and how are these compounds useful to the plants?

APPLYING CONCEPTS

1. If you were an archeologist who had discovered a previously unknown ancient human habitation, how could you decide from the plant remains uncovered in your excavation and a survey of the surrounding vegetation whether the ancient people who lived there merely collected native plants for use or farmed crop plants?

2. If you were lost in a remote area with little prospect for immediate rescue and needed to eat wild plants, how might you (as a mammal) deduce what kinds of plant foods might be least poisonous and most palatable?

3. If you were an agricultural advisor in a tropical area with nearby wetlands, how would you advise local farmers to improve soil quality?

4. If you were an agronomist (crop scientist), which plants would you advise a farmer to grow to feed farm animals if the soil were low in fixed nitrogen and the farmer could not afford commercial fertilizers?

3 | Molecules of Life

Sunflower (*Helianthus annuus*)

Sunflower fields are a dramatic sight—row upon row of giant stems, each bearing huge flower heads that not only glow with Sun colors but actually turn to follow the Sun's path from dawn to dusk every day. The Inca of Peru were so impressed that their worship included the sunflower as a Sun symbol. The Victorian Esthetic Movement, a reaction against aspects of the Industrial Age, needed a symbol of natural beauty, and sunflowers filled the bill. Victorians carved sunflowers into furniture decoration, cast them onto building facades, and forged them onto decorative ironwork. Sunflowers remain a popular decorative motif today.

The sunflower's scientific name, *Helianthus,* literally means "sun flower," and the 67 or so species in this genus are native to North American grasslands. Pioneers made fabric from sunflowers' tough stems and yellow dye from their flower petals. Today, sunflowers are grown in many places around the world for the floral and garden industries, for use as folk remedies, and for their seeds—used as birdseed, human snack foods, and as a source of food oil. One species, *Helianthus tuberosa* (the Jerusalem artichoke), is grown for its edible underground food-storage tubers. Amerindians first domesticated sunflowers about 3,000 years ago. Wild plant ancestors having highly branched stems and relatively small flower heads were transformed into 3-meter-tall, single-stalked plants that produce as many as 1,000 seeds per head.

Sunflower oil

Each giant sunflower head is actually composed of hundreds of tiny flowers, arranged in an artistic spiral. The flowers closest to the edge of the head produce the flamboyant petals. The hard-shelled, single-seeded fruits (which most people mistakenly call seeds) arise from the innermost *disk* flowers. The varieties used as snack foods have familiar striped fruits, but other varieties produce black fruits that are the source of oil used to make margarine. Sunflower oil is among the healthiest oils consumed by people and can also be used as massage oil and as fuel in farm machinery.

If you wonder why sunflower oil is better for you than some other oils, and why it can be useful as a fuel, you will find the answers in this chapter, which focuses on molecules of life. The structure of molecules is key to understanding how they function in plants and in us. This chapter will emphasize the critical role of water in living things, the main kinds of molecules that make up plants and other organisms, their functions in plants, and their human uses. ∎

3.1 | All physical matter is made up of elements composed of distinct atoms

All physical matter—living and nonliving—is composed of chemical elements. Understanding the properties of different elements in nature, and how elements are combined to form molecules, helps explain how life works. An **element** is a substance composed of only one kind of atom. An **atom**, composed of the subatomic particles known as protons, neutrons, and electrons, is the smallest possible unit of an element that has all the chemical and physical properties of that element. An element cannot be broken down into other substances. Iron, for example, is an element; but water is not, because water can be broken down into the elements hydrogen and oxygen.

More than 90 elements—ranging from hydrogen (the lightest) to uranium (the heaviest)—occur in nature. Each element has an abbreviated name, called a *chemical symbol*. The symbol is usually the first one or two letters of the element's name in English or Latin. Thus the symbol for hydrogen is H, for oxygen is O, and for nitrogen is N. But the symbol for iron is Fe and that for gold is Au because the Latin names for these elements are *ferrum* and *aurum*, respectively.

Plants contain and require more of some elements than others

Of all the elements found in nature, just nine account for 99.95% of the dry weight of plants. Carbon, hydrogen, oxygen, nitrogen, phosphorus, sulfur, calcium, potassium, and magnesium are called **macronutrients** because plants require larger quantities of these elements than others. Carbon is particularly important because its presence defines **organic compounds**, of which all living things are composed. Other elements in plants are considered **trace elements** or **micronutrients** because they are required in very small amounts. Table 3.1 lists macronutrients and micronutrients found in plants, as well as their chemical symbols and their average percentages of plant dry weight. If farm or garden soil is tested and found to be deficient in an element required by plants for growth, that element must be added to the soil to ensure good crop production. Farm and houseplant fertilizers contain macronutrients and micronutrients in proportions experimentally determined to be best for particular plants. Additional information on plant nutrients is presented in Chapter 10.

Atoms are made up of three subatomic particles: protons, neutrons, and electrons

Every element is composed of atoms that have a characteristic number of protons, neutrons, and electrons. Protons and neutrons each have considerable mass and are located in the center of atoms in a region called the *atomic nucleus*. Electrons, which have extremely small mass, are located outside the atomic nucleus in discrete orbits, like minuscule planets orbiting a tiny sun at various distances (Figure 3.1). Protons bear a positive electric charge $(+)$ and electrons carry a negative

TABLE 3.1 Essential Elements for Plants: Sources and Functions

Element (Symbol)	Atomic Number	Percent of Plant Dry Weight
Macronutrients		
Hydrogen (H)	1	6
Carbon (C)	6	45
Nitrogen (N)	7	1.5
Oxygen (O)	8	45
Magnesium (Mg)	12	0.2
Phosphorus (P)	15	0.2
Sulfur (S)	16	0.1
Potassium (K)	19	1.0
Calcium (Ca)	20	0.5
Micronutrients		
Boron (B)	5	0.002
Chlorine (Cl)	17	0.01
Manganese (Mn)	25	0.005
Iron (Fe)	26	0.01
Copper (Cu)	29	0.0006
Zinc (Zn)	30	0.002
Molybdenum (Mo)	42	0.00001

charge $(-)$. Neutrons, as their name suggests, carry no electric charge. The electrical attraction between the positively charged nucleus and the negatively charged electrons maintains the electrons in their orbits. Figure 3.1 shows diagrams of atoms of the six most common elements in plants. Notice that more than one electron can orbit a nucleus at the same distance. Electrons that orbit at the same distance are said to occupy the same shell. The first (innermost) shell may have up to two orbiting electrons, the second shell up to eight, and the third shell up to eighteen. The number of electrons an element has in its outermost shell largely determines its chemical properties.

Electrons in the same orbital shell possess the same energy level, but electrons in different shells possess different levels of energy. The farther an electron is from its nucleus, the higher its level of potential energy. Potential energy is energy of position. If an electron absorbs energy in the right amount, it can move from the first orbital shell to the second shell or from the second shell to the third, but not to any position in between shells (Figure 3.2). When an electron absorbs energy and jumps to a higher shell, it enters into an **excited state**. The movement of an

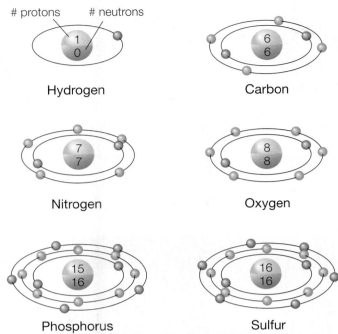

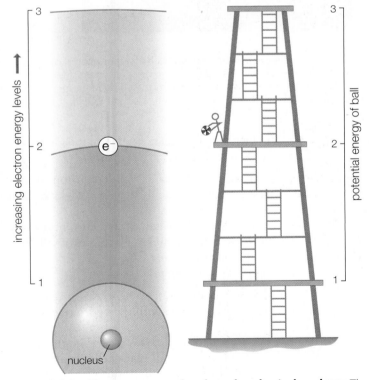

FIGURE 3.1 The six most common elements in plants These include hydrogen, carbon, nitrogen, oxygen, phosphorus, and sulfur. Electrons in the first, second, and third shells are colored red, green, and purple, respectively.

FIGURE 3.2 Electron energy levels and a physical analogy The left-hand portion shows a nucleus with three levels of electron energy. The right-hand portion shows a physical analogy. The higher the ball is carried up the tower, the greater its potential energy in Earth's gravity. The ball, like the electrons, can have only one of three distinct energy levels, as indicated by the platforms where the human figure can stop.

electron to a higher shell is the mechanism by which an atom absorbs energy from its surroundings. If that electron then moves back to a lower energy shell, energy may be released as heat or light or transferred to an electron around another atom.

Every element has an atomic number and a mass number

How can we distinguish elements? Each of the more than 90 naturally occurring elements has an assigned **atomic number**, equal to the number of protons in its atomic nucleus. The lightest element, hydrogen, has one proton in its atomic nucleus and therefore bears atomic number 1. Carbon has six protons in its atomic nucleus and atomic number 6. The heaviest element, uranium, has atomic number 92. All atoms of a given element have the same atomic number. Every element also has an **atomic mass number**, which is the sum of the number of protons and neutrons in the atomic nucleus. Protons and neutrons have about the same mass.

The atomic numbers are constant, but the atomic mass of an element may vary by adding extra neutrons. We can see how this works in the case of hydrogen (Figure 3.3), a gas that can be produced from plant materials and burned cleanly in state-of-the-art car engines. Ordinary hydrogen has one proton in its atomic nucleus. With the addition of another proton, it would no longer be hydrogen but helium, the element with atomic number 2 and the gas used in party balloons. If, instead, a neutron were added to the proton in ordinary hydrogen, the result would be not a new element, but a heavier form of hydrogen, one with a different atomic mass number, called an **isotope**. The isotope of hydrogen with one proton and one neutron is deuterium. Isotopes of an element have the same atomic number but different atomic mass numbers. The third isotope of hydrogen, tritium (see Figure 3.3), is radioactive and has many uses as a tracer molecule in plant and medical research. For example, scientists have used tritium to demonstrate that the hydrogen (H) of water (H_2O) is incorporated into organic compounds within plant tissue. After providing plants with tritium-containing water (which was taken up and used in photosynthesis in the same way as nonradioactive water), the experimenters then used a radioactivity meter to locate newly made radioactive organic compounds.

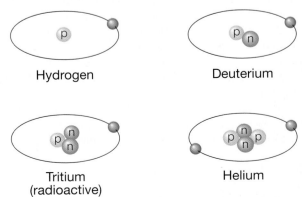

FIGURE 3.3 Diagrams of atoms of three isotopes of hydrogen (hydrogen, deuterium, and tritium) and helium

3.2 | Several types of bonds link atoms to form molecules

Atoms generally do not occur singly in nature. Rather, two or more atoms of the same or different elements combine into **molecules** by forming chemical bonds. For example, molecules of hydrogen (H_2), oxygen (O_2), and nitrogen (N_2) each consist of two atoms. A molecule of carbon dioxide (CO_2) has one atom of carbon and two atoms of oxygen. Water (H_2O) has two atoms of hydrogen and one of oxygen. Carbon dioxide and water are also compounds. A **compound** is a chemical substance that contains two or more different elements in definite proportions. There are three basic types of **chemical bonds**: **ionic**, **covalent**, and **hydrogen**.

Ionic bonds form when atoms gain or lose electrons

Atoms are normally electrically neutral and have the same number of protons and electrons. If an atom gains one or more electrons, it becomes negatively charged. If it loses one or more electrons, it becomes positively charged because it now has more (+)-charged protons than (−)-charged electrons. Whether an atom loses or gains electrons, it is a charged atom, called an **ion**. In the example of table salt, sodium chloride (NaCl), an atom of sodium loses an electron to an atom of chlorine (Figure 3.4a). The sodium atom becomes a (+)-charged ion (Na^+) and the chlorine atom becomes a (−)-charged ion (Cl^-). Because the two ions have opposite charges, they attract each other and join by an ionic bond.

Generally, ionic bonds form between atoms that have lost and gained electrons between them. Salts are crystalline structures (a solid, regular array of positive and negative ions) that are held together by ionic bonds (Figure 3.4b). It is important to realize that table salt (sodium chloride) is but one of many kinds of salts. Ionic compounds, including salts, tend to dissolve readily in water. The next time you cook spaghetti, notice how easily table salt dissolves into the water. Likewise, other types of salts made up of needed mineral elements dissolve in wet soils, making the ionic form of those elements readily available for uptake by plant roots.

Acids and bases contain ionic bonds

Acids and bases are common substances that also contain ionic bonds. Soda pop is a weak solution of carbonic acid—carbon dioxide dissolved in water—which makes the pop fizzy. The vinegar used in salad dressings is a weak solution of acetic acid. Hydrochloric acid is a strong acid sometimes used in drain cleaners. Household ammonia, which is pure ammonia (NH_3) dissolved in water, is an example of a strong base. Plants contain many kinds of acids and bases. For example, citric acid in plants helps give oranges and lemons their tangy flavors. Wood ashes contain bases that in past times were combined with fats to make soap.

An **acid** is any substance that releases hydrogen ions (H^+ or protons) in water. Hydrochloric acid dissolves completely into hydrogen ions and chloride ions:

$$HCl \longrightarrow H^+ + Cl^-$$

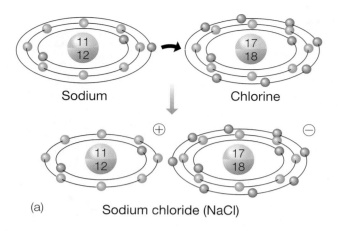

Sodium   Chlorine

(a)   Sodium chloride (NaCl)

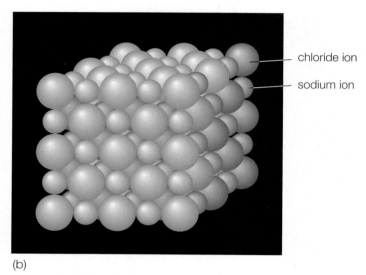

— chloride ion

— sodium ion

(b)

FIGURE 3.4   Diagram of ionic bonds between atoms of sodium and chlorine (a) The electron in the outermost shell of sodium is donated to the outer shell of the chlorine atom. (b) Diagram showing the crystalline structure of salt (NaCl).

The acetic acid in vinegar also releases hydrogen ions in water:

$$CH_3COOH \longrightarrow H^+ + CH_3COO^-$$

Acetic acid is an **organic** acid because of its carbon atoms; hydrochloric acid is an **inorganic** acid because it lacks carbon atoms.

When a **base** dissolves in water, it releases negatively charged hydroxide ions (OH^-), making a basic or alkaline solution. Ammonia dissolves in water as

$$NH_4OH \longrightarrow NH_4^+ + OH^-$$

When acids and bases are mixed, they neutralize each other because the hydrogen ions combine with the hydroxide ions to form water.

The balance between acids and bases is very important to the survival and growth of plants. All plant cells require a stable, specific, internal level of acidity. In nature, plants such as cranberries may thrive in an acid bog whereas other plants require an alkaline soil. A convenient way to express the relative acidity of a solution is by a number on the **pH scale**. Pure water is neutral at pH 7. Anything less than pH 7 is acidic, and anything greater than pH 7 is alkaline. Lemon juice has a pH of 2, vinegar a pH of 3, tomato juice a pH of 4, human blood is a little above a pH of 7, ocean water has a pH of about 8, and alkaline lakes may have a pH greater than 10. The pH of liquids or soil water can be measured with a meter or a paper indicator strip. If soil pH is not optimal for a particular farm crop or garden plant, the pH can be adjusted to get better yields. A basic soil can be made more acid by adding acidic chemicals like aluminum sulfate or *Sphagnum* moss from bogs.

In covalent bonds, two or more atoms share electrons

Covalent bonds occur when two or more atoms of the same or different elements share electrons. Covalent bonds are therefore stronger than ionic bonds. You would not be able to break covalent bonds by stirring substances into water, as is the case for salts (which have ionic bonds). Hydrogen has one proton and one electron but its outer electron shell could contain two electrons, an arrangement that is more stable. A hydrogen atom can gain another electron in its outer shell by sharing its electron with another hydrogen atom (Figure 3.5a). Both atoms together share two electrons. The result is a single, strong covalent bond. Molecular hydrogen, therefore, consists of two atoms, written as H_2. Now consider an oxygen atom with 8 protons and 8 electrons. Oxygen has 6 electrons in its outer shell, which could hold a maximum of 8. An oxygen atom can gain these two electrons by sharing two of its electrons with another oxygen atom, which shares two of its electrons in return (Figure 3.5b). The oxygen atoms form a double covalent bond because they share a total of 4 electrons in forming molecular oxygen, or O_2.

Nitrogen has 7 protons and 7 electrons and therefore only 5 electrons in its outer shell. As in hydrogen and oxygen, nitrogen fills its outer shell by sharing electrons with another nitrogen atom. A total of 6 electrons are shared in N_2, which has a triple covalent bond between the two atoms (Figure 3.5c). This triple bond is extremely strong and consequently, nitrogen gas is nonreactive—it does not react with other molecules. Plants need a lot of nitrogen, but they are unable to use N_2 because they cannot break the tough triple bond. Legumes and some other plants have solved this problem by forming partnerships with microbes that are able to break apart N_2 and make N-containing compounds that are usable by plants.

Covalent bonds join together the elements hydrogen, oxygen, nitrogen, and carbon to form organic molecules. Each of these elements can form only a specific number of covalent bonds. Hydrogen can form one, oxygen two, nitrogen three, and carbon four. The fact that carbon can bond at four positions means that carbon can form many different kinds of compounds and explains its importance to life.

Hydrogen bonds are weak attractions between molecules

When electrons are shared between two atoms of the same element (as in the gases H_2, O_2, and N_2), the electrons are shared equally. What happens when atoms of different elements share electrons? If the atoms differ in their ability to attract electrons, the electrons will be shared unequally, and a **polar covalent bond** will result. The best example of polar covalent bonds is the ordinary water molecule (H_2O). Water consists of one atom of oxygen and two of hydrogen (Figure 3.6a). The oxygen atom in water attracts the electrons in the two hydrogens toward its nucleus and thereby acquires a slight (−) charge. The two hydrogens each have a slight (+) charge because their electrons have been drawn

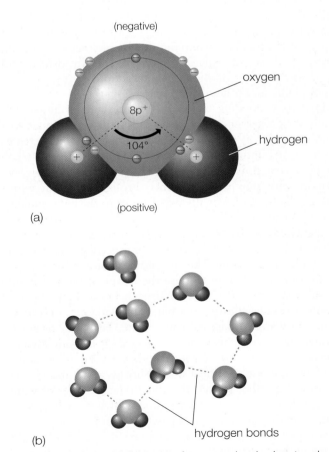

(a)

FIGURE 3.6   Water (a) Diagram of water molecule showing the 104° angle between H–O–H. (b) Hydrogen bonds among water molecules.

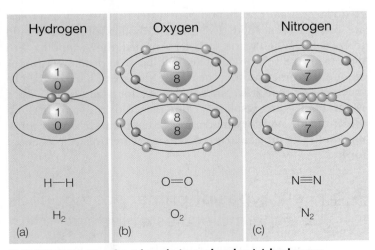

FIGURE 3.5   Covalent bonds in molecular (a) hydrogen, (b) oxygen, and (c) nitrogen These molecules have single, double, and triple bonds, respectively.

farther away from the protons of their nuclei. Water molecules therefore have distinct (−) and (+) poles, and water is known as a **polar compound**. There are other kinds of polar compounds in addition to water, and these include polar organic compounds. The (−)-charged pole attracts the (+)-charged poles of other polar molecules (Figure 3.6b). These attractions, which are weaker than covalent bonds, are called **hydrogen bonds**. Hydrogen bonding allows polar organic substances to dissolve in water and also plays a significant role in the structure and function of many organic molecules, including DNA.

3.3 | Water has unique properties because it forms hydrogen bonds

Water molecules are **polar molecules** that form hydrogen bonds between the slightly (+)-charged hydrogen atoms and the slightly (−)-charged oxygen atoms of adjacent molecules. These hydrogen bonds allow water to act in ways that are almost unique among compounds. If water behaved like structurally similar compounds such as hydrogen sulfide (H_2S) or ammonia (NH_3), it would be a gas at normal temperatures. Instead, at normal temperatures on Earth, water can exist as a solid, as a liquid, or as a gas, which is important for the existence of life. Hydrogen bonds give water unique life-giving properties.

Many substances dissolve readily in water. It is a highly effective solvent because of its polar nature. The weak charges on the hydrogen and oxygen atoms allow water to easily dissolve salts, polar organic molecules, and many gases such as oxygen, nitrogen, and carbon dioxide by interacting with slight charges on these substances. Therefore, it is an ideal medium for the vast number of chemical interactions that occur in the cells of plants and other organisms.

Hydrogen bonding makes water molecules cohesive—they literally stick together. Although a single hydrogen bond is weak, the collective effect of billions of such bonds in water creates a high level of bonding among molecules. Because of this cohesion, plants can draw water up from the soil through their roots, transport very thin water streams through the stem without breaking the flow, and finally expel water to the environment as a gas from leaf surfaces. The cohesion of water is strong enough that giant redwoods (*Sequoia sempervirens*) can raise water up to 110 m (350 ft). As a direct result of this cohesion, water also has a high **specific heat**. The specific heat of a substance is the amount of heat required to raise the temperature of a gram of that substance by 1.0°C. The hydrogen bonds that impart the cohesion to water also require that a large quantity of heat must be put into water to raise its temperature. The high specific heat of water means organisms are protected from wide swings in temperature by the water of their bodies. Lakes and oceans will store large quantities of heat during hot weather and give it up during cool weather so that the climate near bodies of water is moderated. An especially large amount of heat is necessary to change water from a liquid to a gas. Therefore, when plants emit water vapor from their leaves, they expel heat, cooling themselves and their surrounding environment.

Water has one further unusual property that is very important for life on Earth. Most liquids become more dense as they

FIGURE 3.7 Ice floats because of water's molecular structure The lattice of water molecules in the ice on this frozen Wisconsin lake makes ice less dense than liquid water, where the water molecules are more closely spaced.

cool, and when they freeze, their solid form is more dense than their liquid form at any temperature. This is not true of water. As liquid water cools, it becomes more dense down to about 4°C (39°F). Below 4°C, water becomes less dense. When it freezes and turns into ice at 0°C, the ice is less dense than the slightly warmer water around it. Consequently, water ice floats in liquid water (Figure 3.7). Although this may seem rather obvious, suppose it were not true. Suppose water behaved like most other liquids and its solid form sank. Then, lakes and even the oceans would freeze from the bottom up. There would be no large organisms in bodies of water because few can tolerate being frozen solid. Even worse, bodies of water would freeze solid. When warm weather returned, they would have to melt from the top down. This melting would take enormous amounts of time and solar heat. The climate of Earth would be much colder than it is. Since water ice floats and forms easily, water bodies are insulated from the cold and remain fluid under the ice, and they moderate the temperature of surrounding land. The molecular properties of water make it essential for life.

3.4 | Four types of primary compounds are the molecules of life

With this survey of atoms, molecules, and the properties of water behind us, we are now ready to consider organic molecules. Organic molecules are defined by the presence of the element carbon. There are four basic types of organic molecules (known as *primary compounds*) out of which living organisms are made: carbohydrates, lipids, proteins, and nucleic acids (Table 3.2). In each of these four groups, sets of smaller, simpler organic molecules are the building blocks of the larger, more complex organic molecules that carry out structural and metabolic functions in plants. The association between organic molecules, water, and life is so strong that the search for life beyond Earth is a search for water and organic molecules (Essay 3.1, "Molecules: Keys to the Search for Extraterrestrial Life").

Carbohydrates include sugars, starches, and cellulose

Carbohydrates function as energy molecules and structural components of plants and are also important in animal nutrition. Who can ignore carbohydrates? Sugar (from sugar cane or sugar beet plants) sweetens our morning coffee and breakfast cereal. The fruit on that cereal tastes sweet because of stored sugars. The cereal is largely starch plus a lesser amount of protein derived from ground grains like wheat or rice. Starch is prominent in potatoes, rice, and pasta products, as well as in the flour in all baked goods. Cellulose from raw fruits and vegetables contributes dietary fiber. It is also the main component of the paper on which this book is printed and of the cotton fibers in many of the clothes we wear.

Carbohydrates are organic molecules composed of carbon, hydrogen, and oxygen in the ratio 1C:2H:1O or $(CH_2O)_n$, where n may be any number from three to several thousand. The name *carbohydrate* means "carbon-water," since, in general, each carbon is attached to a hydrogen and a hydroxyl (—OH) group. There are three types of carbohydrates: **monosaccharides** (single sugars), **disaccharides** (two sugars), and **polysaccharides** (many sugars). Monosaccharides are the basic building blocks of carbohydrates; the other two types of carbohydrates are built by linking together monosaccharides.

Monosaccharides are **simple sugars** with three to six carbon atoms. Glucose is the most common monosaccharide in plants and the major carbohydrate circulating to your cells in your bloodstream. It is a 6-carbon sugar with the basic formula $C_6H_{12}O_6$. This formula shows only the numbers of each element (Figure 3.8c). Figure 3.8 also shows two other ways to represent simple sugars: Figure 3.8a shows the arrangement of atoms in three simple sugars, and Figure 3.8b shows how these sugars would appear if dissolved in water, where they assume a ring shape.

In the process of photosynthesis, plants synthesize glucose from carbon dioxide and water, releasing oxygen as a by-product (Chapter 5). Fructose, or fruit sugar, is another 6-carbon monosaccharide with the same formula as glucose. Figure 3.8 shows that fructose has a different three-dimensional structure from glucose. Fructose from corn syrup is widely used in processed foods

TABLE 3.2 The Major Groups of Molecules of Life

Type	Subtypes	Examples	Functions
Carbohydrates	Monosaccharides	Glucose, Fructose	Energy molecules
	Disaccharides	Sucrose	Short-term storage, main sugar transported in plants
	Polysaccharides	Starch	Energy storage
		Cellulose	Main structural component of plants
Lipids	Fats, oils	Corn oil	Energy storage
	Waxes	Waxes on plant surfaces	Reduce water loss, deter insects and pathogens
	Phospholipids	Phosphatidylcholine	Cell membranes
	Steroids	Digitalin	Deter insects
Proteins	Storage proteins	Zeatin in corn, wheat gluten	Energy storage
	Enzymes	Sucrase, rubisco	Regulate biochemical reactions
Nucleic Acids	Long-chain nucleic acids	Deoxyribonucleic acid (DNA)	Storage of genetic information
		Ribonucleic acid (RNA)	Translation of genetic information
	Single nucleotides	Adenosine triphosphate (ATP)	Energy carrier
		Cyclic adenosine monophosphate (cAMP)	Intracellular messenger

ESSAY 3.1 MOLECULES: KEYS TO THE SEARCH FOR EXTRATERRESTRIAL LIFE

We live on a small, rocky planet that possesses abundant water and teems with life-forms, at least one of which is curious about the world and universe around it. Are we alone on this little world in an incredibly vast universe? Over the past 30 years, many scientists have begun to look seriously for evidence of life beyond Earth. Much of that search has followed a chemical approach. Every organism on Earth is composed of liquid water and organic molecules. Organic molecules, called the molecules of life, are carbon-containing molecules. Based on the only example of life we know, the search for life beyond Earth consists of looking for past or present signs of water and organic molecules. This strategy is not quite as Earth-centric as it may appear. No other element is as versatile as carbon in terms of the variety and complexity of molecules it can form, and water is an ideal medium in which organic molecules can dissolve and interact.

... the search for life beyond Earth consists of looking for past or present signs of water and organic molecules.

How successful has the search been so far? Astronomers using large radio telescopes have detected the signals of simple organic molecules in the vast clouds of gas and dust particles that lie between stars in interstellar space (Figure E3.1A). Within our own solar system, the planet Mars has two small moons which are thought to be captured asteroids covered partially in organic molecules. Many other asteroids lying within the main belt of asteroids between Mars and Jupiter appear to be covered in organic matter.

In 1986, the European Space Agency's *Giotto* spacecraft flew directly through the cloud of dust surrounding Halley's Comet and discovered that its nucleus may be as much as 25% organic matter (Figure E3.1B)! Halley's Comet came from the vast, cold regions of space beyond the orbit of Neptune; its abundance of organic matter suggests that many other icy bodies in this remote region may be rich in organic matter. Organic chemistry has been very active throughout our solar system and out into interstellar space.

Does liquid water occur elsewhere in our solar system? If so, at least some kind of life might exist there. After three decades of sending space probes to Mars, scientists now know that water once flowed freely over the Martian surface in rivers and lakes and that an ocean once existed in its northern hemisphere. That water is now frozen in polar ice caps and permafrost. The *Galileo* spacecraft has discovered a present-day ocean of liquid water on Europa, a moon of the planet Jupiter. This fluid ocean lies under a crust of ice on the surface. Scientists do not yet know if either Mars or Europa harbors any past or present life forms.

E3.1A Great Orion Nebula

E3.1B Halley's Comet

because it is sweeter than table sugar. Glucose and fructose are energy molecules; cells break them down to provide energy for metabolic processes. Plants use solar energy to transform the raw materials carbon dioxide and water into these energy molecules.

Disaccharides are made by linking together two monosaccharides. In plants, disaccharides are used for short-term energy storage. The most common plant disaccharide is sucrose, or table sugar, which is obtained commercially from sugarcane (*Saccharum officinarum*) or sugar beets (*Beta vulgaris*). Sucrose is formed from glucose and fructose by a process called a **dehydration** (literally, "removing water") **synthesis** or dehydration reaction (Figure 3.9). Hydrogen is removed from fructose and a hydroxyl group (OH) from glucose, leaving the two simple sugars joined by single covalent bonds to oxygen. In the process, a water molecule is formed from the hydrogen and hydroxyl group, from which the term *dehydration* is derived. Sucrose is the principal sugar trans-

ported throughout plants. The maple syrup derived from the sap of maple trees is one example of how we take advantage of sucrose transport in plants.

Polysaccharides are formed by dehydration reactions just like those involved in joining disaccharides except that large numbers of monosaccharides become linked. Thus, monosaccharides are the building blocks of polysaccharides. Unlike simple sugars, polysaccharides are not water soluble. Plants retain tiny granules of the polysaccharide starch in their cells for long-term energy storage. Starch is an especially abundant polysaccharide in grains (corn, rice, and wheat) and root vegetables such as yams and potatoes (Figures 3.10a and b). Under light and electron microscopy, potato tubers can be seen to consist of cells crammed with starch granules (Figure 3.10c). Starch (Figure 3.10d) is synthesized from glucose by dehydration reactions. Molecules of starch may contain up to 1,000 glucose molecules in a complex, sometimes branched polysaccharide

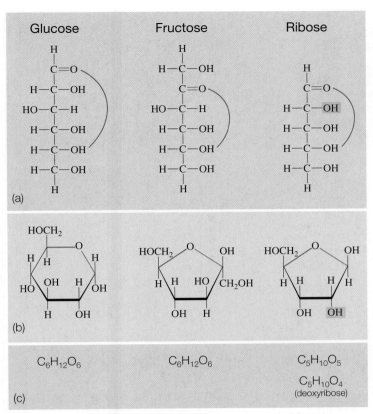

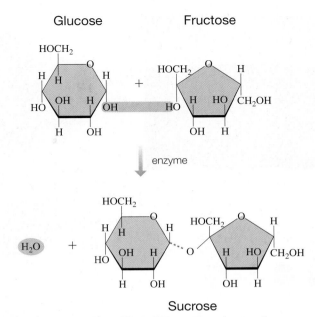

FIGURE 3.9 Diagram of dehydration synthesis of sucrose from glucose and fructose With the aid of an enzyme, the two monosaccharides are joined (red-colored bond) and a water molecule is formed.

FIGURE 3.8 Diagrams of the molecular structure of some sugars that occur in plants (a) The carbon atoms are shown in a linear array. (b) Sugar molecules assume a ring shape when dissolved in water. The red line in (a) equates to the bond shown in red in (b), and the double bond shown in (a) equates to the single blue bond in (b). The hydroxyl group in ribose—of RNA—(highlighted) is replaced with hydrogen in deoxyribose—the sugar found in DNA. (c) Molecular formulas for these sugar molecules. Note that although glucose and fructose have identical molecular formulas, they have different structures.

chain (Figure 3.10e). When energy is required, the glucose molecules are split off the starch molecule and metabolized. Plant cells that have used most of their carbohydrate stores have few or no starch granules.

Plants also make polysaccharides for use as structural components. The most abundant plant structural polysaccharide is cellulose. Cellulose makes up most of the material in plant cell walls and about half of that found in the trunks of trees. Like starch, the cellulose molecule is made up of many glucose molecules, but in this case they are linked differently, such that alternate glucose molecules are flipped over in the long linear chains of cellulose (Figure 3.11). This structural arrangement makes cellulose difficult to break down, and only certain bacteria, fungi, and animals, such as cows and termites with their special gut microbes, can digest it. Although humans cannot digest cellulose, it is still important in the human diet as dietary fiber, which aids in bowel function and may reduce the risk of colon cancer. This is one reason why you should eat grains and other fruits and vegetables.

Cellulose is especially important in the production of paper and cotton cloth. In paper production, wood is chipped and processed to liberate cellulose fibers, which are poured onto screens to drain. The fibers are then matted together by pressing them between rollers into thin sheets. Cotton fibers are about 90% cellulose. The fibers—hairs attached to the seeds of the cotton plant, *Gossypium*—are separated from the seeds in a process called ginning and spun into yarn for making cotton cloth.

Other useful indigestible polysaccharides include gums such as locust bean gum, used in beers, and pectins, which form the gels in fruit preserves. The soluble fiber in the gum from oat bran is thought to lower blood cholesterol.

Lipids include fats, oils, waxes, phospholipids, and steroids

Lipids are important to plants as energy stores and in the formation of protective surfaces on cells and plant leaves, stems, and roots. Lipids are also important to humans in many ways. Solid fats occur in meat and meat products. Vegetable oils like olive, corn, and peanut oil are used in cooking and salad dressings. Waxes protect the finish on our cars, polish our floors and furniture, and provide a soft light for candle-lit dinners. We use plant-made steroids as medicines to combat a variety of illnesses. Phospholipids, a type of lipid that contains phosphorus, are vital to life. Without them there would be no cells, the basic units of life.

Lipids are a more diverse group of organic compounds than carbohydrates. They generally have an oily texture and are insoluble in water. There are three types of lipids: (1) fats, oils, and waxes; (2) phospholipids; and (3) steroids. Lipids are primarily composed of hydrogen and carbon with only small

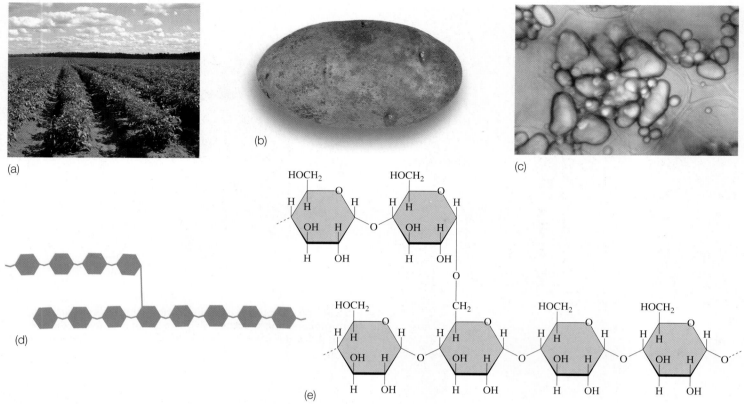

FIGURE 3.10 **Starch as an energy storage molecule in plants** (a) Potato crops provide a source of starch in the human diet. (b) Potato tubers are underground structures that store starch. (c) Starch granules occur abundantly within cells of potato tubers. (d) Diagram of part of a starch molecule, showing branching between adjacent strands. (e) Structural diagram showing how glucose molecules are linked in starch.

amounts of oxygen. Phospholipids, however, also contain phosphorus and sometimes nitrogen. The three groups have different structures and functions.

Fats and oils are easily distinguished. At room temperatures, fats are solid and oils are liquids. Both serve as energy-storage molecules; a gram of fat or oil contains twice as much energy as a gram of carbohydrate. Fats and oils are made from two building blocks: **glycerol** molecules and **fatty acids** (Figure 3.12). Glycerol is a 3-carbon molecule with three hydroxyl groups along one side. Fatty acids are long, unbranched chains of $-CH_2$ groups with an acidic carboxyl ($-COOH$) group on the end. The abundance of carbon-hydrogen bonds in fatty acids makes them energy rich; the bonds store sunlight energy during photosynthesis. The fat or oil is formed by three dehydration syntheses, which remove three H atoms from glycerol and an $-OH$ group from each of three fatty acids. Fats and oils are insoluble in water because they are nonpolar molecules—that is, they do not have ($+$) or ($-$) charges on their atoms. Nonpolar molecules cannot interact with polar water molecules and cannot dissolve in it. Because they do not dissolve in water, fats and oils are called **hydrophobic**, that is, "water fearing."

At the molecular level, fats and oils are distinguished by their fatty acids. If all the carbon atoms have the maximum number of hydrogens, the fatty acid is saturated, as in the top

fatty acid in Figure 3.12. If the fatty acid has one double bond between adjacent carbons, it is monounsaturated (the third fatty acid in Figure 3.12). A fatty acid is polyunsaturated if it has two or more double bonds in its chain, as in the second fatty acid in Figure 3.12. In fats, the fatty acids are mostly saturated, but in oils they are mainly unsaturated. Table 3.3 lists different vegetable oils in terms of the saturation of their fatty acids. Coconut and palm oil are fluid but have few unsaturated fatty acids. Peanut and olive oils contain significant amounts of monounsaturated fatty acids (Figure 3.13a–c); corn and sunflower oils contain more polyunsaturated fatty acids (Figure 3.13d–f).

Human consumption of monounsaturated vegetable oils tends to decrease blood cholesterol levels and therefore reduces the probability of arteriosclerosis, a disease in which fat builds up in the arteries. Polyunsaturated oils also reduce cholesterol levels, but they may also reduce the number of certain protective molecules that prevent cholesterol buildup in the linings of arteries. A diet including monounsaturated oils is the best choice for reducing blood cholesterol levels.

Waxes are similar to fats and oils, but in waxes the glycerol molecule is replaced by a long-chain carbon molecule with many hydroxyl groups along it and all the fatty acids are saturated. Waxes provide a waterproof coating over the surfaces of stems

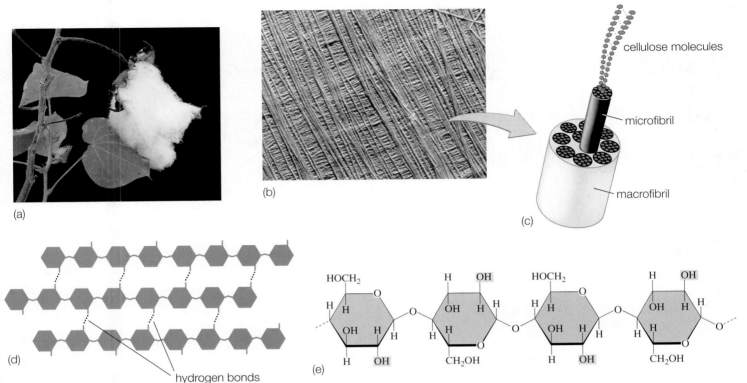

cellulose molecules

microfibril

macrofibril

(a)

(b)

(c)

(d)

hydrogen bonds

(e)

FIGURE 3.11 **Cellulose is a carbohydrate that provides plant structure** A large proportion of the biomass of plants is cellulose. (a) Cotton fibers come from cellulose-rich hair cells attached to cotton seeds. (b) Microfibrils (microscopic strands) of cellulose can be seen in plant cell walls with a high-magnification microscope. (c) Diagram showing that cellulose molecules bundle together to form microfibrils. These, in turn, are grouped to form larger macrofibrils. (d) Cellulose molecules are formed from chains of glucose molecules held together by hydrogen bonds. (e) Note that, unlike in starch, every other glucose molecule is flipped (compare to Figure 3.10e). Hydroxyl groups (highlighted) are involved in the hydrogen bonding shown in (d).

and leaves, which reduces water loss and protects the plant from injury by pathogens and insects. Waxes are especially thick on desert and chaparral plants (Chapter 26).

Phospholipids are structurally similar to fats and oils (Figure 3.14a). They have a glycerol backbone and long fatty-acid chains, which may be saturated or unsaturated. Phospholipids, however, contain a phosphate group ($-PO_4$) bonded to the glycerol molecule and an additional polar group attached to the phosphate. The phosphate and polar group are **hydrophilic,** which means, "water loving," and therefore soluble in water, whereas the fatty-acid chains are nonpolar and hydrophobic ("water fearing") and insoluble in water. The main function of phospholipids is structural. When placed in water, phospholipids assemble themselves into a double layer (Figure 3.14b), which is the basic component of cell membranes. The hydrophilic functional groups are located on the outside, facing the watery environment, or inside, facing cellular fluids. The hydrophobic fatty acids are inside the bilayer away from water. The entire membrane is very flexible. Phospholipids are one reason why phosphorus is such an important element in plant and animal nutrition. Without phosphorus, cell membranes could not function.

Steroids are structurally different from all other lipids. They are composed of four carbon rings with various functional side groups (Figure 3.15). The example shown is digitalin, a steroid extracted from the seeds of the purple foxglove (*Digitalis purpurea*). The functional side groups in digitalin are two sugar molecules. By making herbivores ill, digitalin deters herbivores like deer or rabbits from eating the plant leaves. Today, in the United States alone several million people rely on digitalin and related steroids from foxglove for treatment of congestive heart failure. These plant steroids increase the force of heart contraction while slowing the heart rate, a process that improves circulation, reduces fluid swelling (edema) in the lungs, and increases kidney output. The active drugs are still extracted from dried foxglove seeds or leaves, where the drug is most concentrated before flowering.

Proteins are large molecules composed of amino acids

Plant cells use proteins to store materials, build structures, perform movements, and transport materials, but the most important function of proteins is as biochemical catalysts that

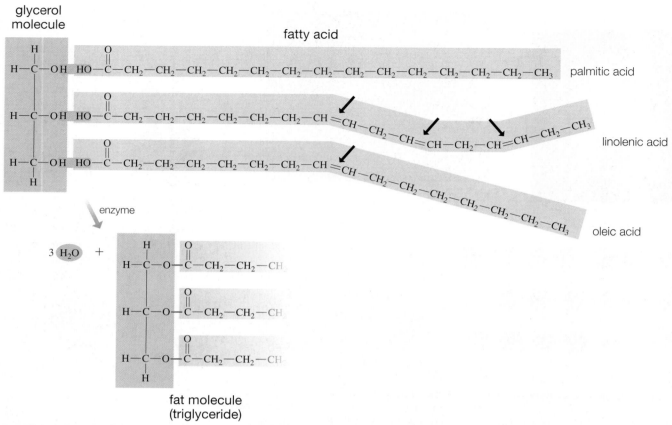

FIGURE 3.12 **Diagram showing the formation of a fat molecule (triglyceride), which consists of glycerol plus three fatty acids** As in sucrose synthesis, dehydration reactions are involved; here, three water molecules are liberated. Three types of fatty acids are shown: one saturated (palmitic), one polyunsaturated (linolenic), and one monounsaturated (oleic). Black arrows point to the double bonds in the fatty acids; note that the fatty acids bend or kink where double bonds occur. Such bending increases the tendency of a fat to be fluid.

facilitate essential chemical reactions. A **catalyst** is any substance that speeds up the rate of a chemical reaction without being used up itself. Certain proteins called **enzymes** act as biochemical catalysts and regulate the rates at which every biochemical process occurs in living plant cells. Enzymes are therefore absolutely essential for the survival of plants. Proteins are also important in human nutrition. The seeds of cereals and legumes, such as corn, rice, wheat, and beans, are especially rich in storage proteins, proteins stored in the seeds as food for the developing plant. Because protein is so essential to good nutrition, the amino acid subunits of protein are widely sold in health-food stores and added to health foods and beverages. Proteins derived from bacteria and fungi are used in industries such as waste treatment, dairy processing, and detergent manufacture.

Proteins are complex organic molecules that are composed of carbon, hydrogen, oxygen, nitrogen, and sulfur. Plants need nitrogen and sulfur from the soil because of their role in building proteins. Proteins are assembled from hundreds of smaller, simpler organic molecules called **amino acids**. Every amino acid consists of a central carbon atom to which are bonded a hydrogen atom, an amino group containing nitrogen ($-NH_2$) an acidic carboxyl group ($-COOH$), and a side chain, represented by the letter R. There are 20 different amino acids in proteins, and each has a unique side chain, which determines the properties of that particular amino acid. Some amino acids are hydrophobic, others are hydrophilic. They may be acidic or basic. Several examples of amino acids are shown in Figure 3.16.

Amino acids are linked together by peptide bonds into polypeptides (Figure 3.17). The hydroxyl group ($-OH$) from one amino acid is combined with a hydrogen from the amino end of another amino acid to remove water and form a covalent **peptide bond** between carbon and nitrogen. This is another example of a dehydration synthesis or reaction since it removes a molecule of water. As successive peptide bonds are formed, the peptide chain becomes longer and longer. Polypeptides typically have hundreds and even thousands of amino acids. The properties of polypeptides are determined by the properties of the amino acids in the chain and their positions. Thus if a polypeptide is made primarily of hydrophilic amino acids, it will be soluble in water.

Biochemists recognize four levels of organization in protein structure. The primary structure is the sequence of amino acids in the polypeptide chains (Figure 3.18a). The secondary

TABLE 3.3 Sources of Saturated, Monounsaturated, and Polyunsaturated Fatty Acids

Type of Fatty Acid	Sources
Saturated (all $CH_2 \text{—} CH_2$)	Coconut oil
	Chocolate
	Palm oil
Monounsaturated (one $CH \text{=} CH$)	Avocados
	Olive oil
	Peanut oil
	Canola oil
	Peanuts
	Almonds
Polyunsaturated (two or more $CH \text{=} CH$)	Sesame oil
	Soybean oil
	Corn oil
	Sunflower oil
	Safflower oil

structure arises from hydrogen bonding between amino acids in the chains (Figure 3.18b). Stretches within a polypeptide may coil into a spiral structure called an α (*alpha*) *helix*, which are important in anchoring proteins within cell membranes for most effective function. Alternatively, sections of polypeptides may link up through hydrogen bonds to form a β (beta) pleated-sheet structure (Figure 3.18c), which is common in structural proteins and provides elasticity. An example of an elastic structural protein is the keratin in your skin, nails, and hair. Most proteins also fold into complex three-dimensional tertiary structures that help determine their function (Figure 3.18d). Tertiary structures are held together by covalent bonds formed between specific amino acids (cysteines) located in different parts of the protein (see Figure 3.16). Cysteine units located opposite each other in the polypeptide chain form disulfide bridges (—S—S—), and these strong covalent bonds stabilize the tertiary structure of the protein. When a person gets a perm, it is these covalent bonds that are first broken then reformed during chemical processing of the hair protein.

In water, proteins will fold to expose hydrophilic amino acid R groups to the water and shield hydrophobic R groups from contact with water. This folding is the reason that the kinds of amino acids in a polypeptide chain and their positions are so important. Because a protein in water folds to expose its hydrophilic amino acid R groups, a change in the order of amino acids—such that a hydrophobic amino acid is exposed to water—could cause the chain to fold into a different tertiary structure, possibly altering its function. Mutations are changes

FIGURE 3.13 Plant-derived oils are important in human nutrition (a) Peanut field, (b) peanuts and (c) peanut oil, (d) corn field, (e) corn, and (f) corn oil.

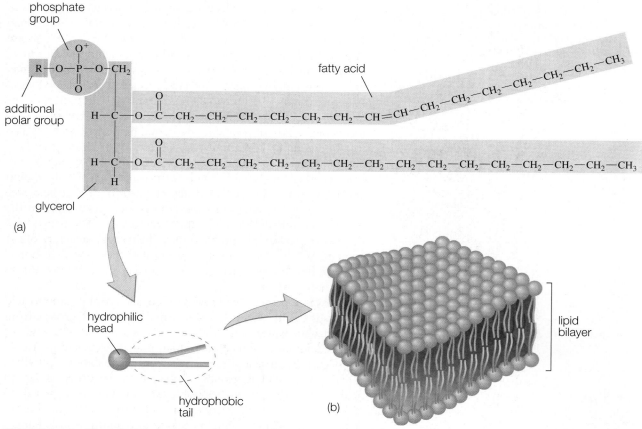

FIGURE 3.14 **Phospholipids** (a) Diagram of a phospholipid, which consists of a hydrophilic head and hydrophobic tail, consisting of two fatty-acid chains. (b) Diagram of a phospholipid bilayer in a cell membrane.

in DNA that cause alterations in protein sequence that may change protein function in organisms (Chapter 6).

Several polypeptides may join together to create a fourth level of organization, a quaternary structure. Hemoglobin, the oxygen-carrying protein in human blood cells has a quaternary structure consisting of two pairs of similar polypeptides held together by hydrogen bonds (Figure 3.18e). Each polypeptide chain holds an iron-containing molecule called a *heme*, which binds to, transports, and releases oxygen. Hemoglobin is a good example of a protein that needs trace amounts of a mineral (iron) in order to perform its function. Such trace elements are called *cofactors* and their presence is essential for many proteins to function. Other elements that act as cofactors include magnesium (Mg), sodium (Na), copper (Cu), and potassium (K).

Previously we mentioned only briefly some of the functions of proteins. Now that we have an understanding of protein structures, we can better understand their functions.

Storage proteins are important in human nutrition

Plants make all of the amino acids they need, but humans and other animals must obtain some of their amino acids through the protein in their diet. Amino acids that humans must obtain

in their diet are termed *essential* amino acids. The other amino acids are called *nonessential*, but only in the sense that we can make them ourselves from other amino acids.

Because some amino acids are essential, the protein and amino acid content of plants is very important to animal nutrition. Although the seeds of cereal crops such as wheat and corn are high in protein, the proteins in both grains are low in the essential amino acids lysine and tryptophan. Legumes such as peas, beans, and soybeans—from which the tofu used in East Asian and vegetarian cooking is derived—are also excellent sources of proteins, averaging about 25% protein by weight. Unlike cereal grains, legume proteins are rich in lysine and tryptophan but deficient in the amino acid methionine. Vegetarians know that a complete protein diet can be obtained by combining two or more complementary plant proteins such as those in rice and beans. Table 3.4 lists essential and nonessential amino acids for adult humans.

In today's world, chronic hunger and malnutrition are constant problems. Estimates suggest that 17 to 40% of the world's population is undernourished or malnourished. About 20 million deaths per year (mostly among children) result from these conditions. Two serious conditions stem from protein deficiency: kwashiorkor and marasmus. Kwashiorkor arises

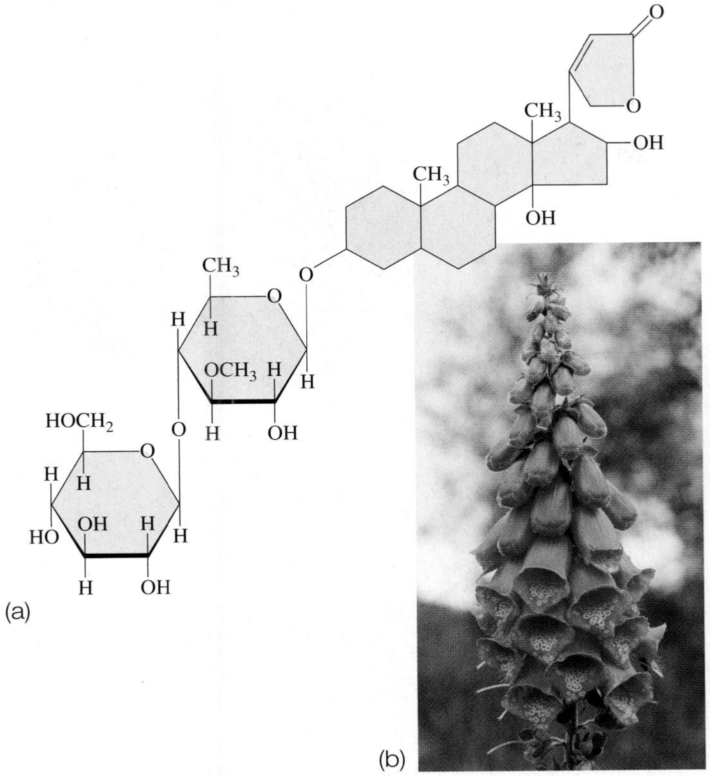

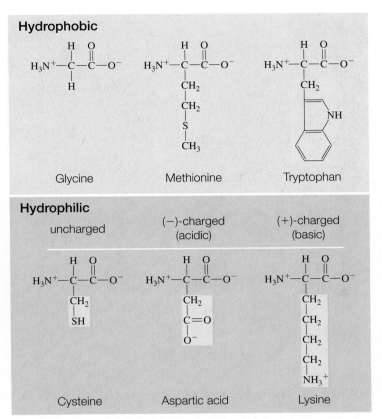

(a)

(b)

FIGURE 3.15 A plant steroid (a) Structure of the digitalin molecule. (b) Photo of purple foxglove, *Digitalis purpurea*, from which the medicine digitalin is obtained.

Hydrophobic

Glycine   Methionine   Tryptophan

Hydrophilic

uncharged   (−)-charged (acidic)   (+)-charged (basic)

Cysteine   Aspartic acid   Lysine

FIGURE 3.16 Amino acids Diagrams of selected amino acids, three hydrophobic and three hydrophilic. Notice that some parts of these amino acids are the same, but the side chains, or R groups (highlighted), are specific to particular amino acids and determine differences in their chemical properties. For example, the sulfur-containing part of cysteine allows this region to form covalent bonds with cysteines elsewhere in the same polypeptide, which helps proteins to fold properly.

when there is sufficient food but not enough protein in the diet. Symptoms include puffy skin and a swollen belly, reddish-orange hair, dermatitis, and listlessness. In marasmus, victims suffer from starvation and appear shriveled because the body digests heart and skeletal muscle to provide energy. Both conditions can be reversed, but mild mental retardation may be a permanent consequence.

Enzymes are proteins that act as biological catalysts

The most important function carried out by proteins is their role as **enzymes**, catalysts that regulate biochemical reactions in cells. In an ordinary chemical reaction, the initial molecules are referred to as the substrates or reactants, the final molecules are the products, and a highly reactive intermediate compound forms briefly during the reaction (Figure 3.19a). A chemical reaction involves the rearrangement of atoms in the substrate molecules. If the atoms in the substrate molecules share electrons in covalent bonds, those bonds are stable and represent a certain potential energy. Breaking and reforming those covalent bonds into a new arrangement requires an input of energy, called the **activation energy**. In nonliving systems, the activation energy may come from an increase in temperature. The higher the temperature, the more energy of motion in the molecules (kinetic energy) and the faster the reaction will proceed. But the cells of living organisms cannot

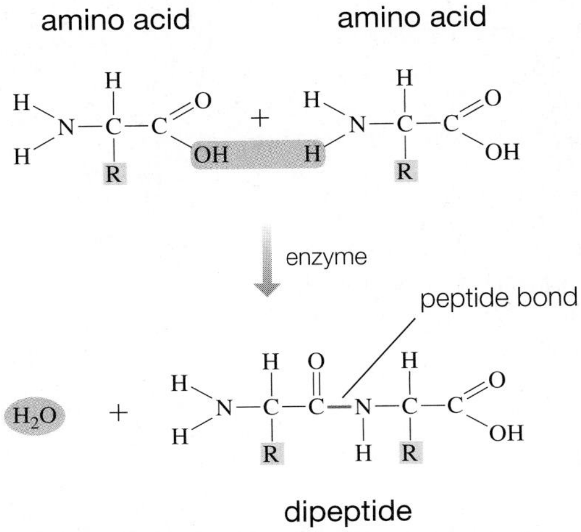

amino acid   amino acid

enzyme

peptide bond

H_2O

dipeptide

FIGURE 3.17 Diagram showing the formation of a peptide bond In a dehydration reaction, a water molecule is liberated during the formation of a carbon-nitrogen covalent bond (shown in red) between adjacent amino acids.

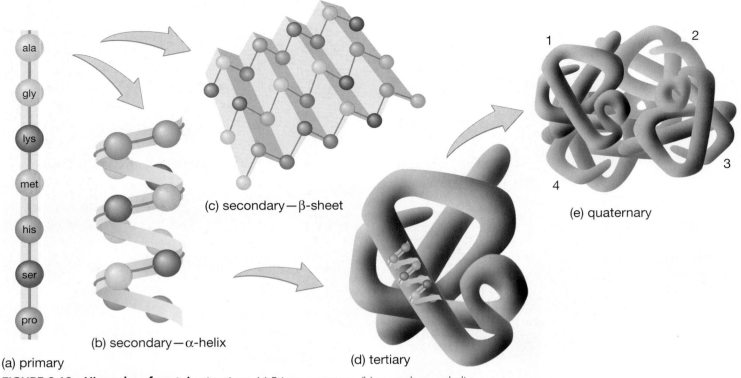

FIGURE 3.18 Hierarchy of protein structure (a) Primary structure, (b) secondary—α helix, (c) secondary—β sheet, (d) tertiary structure, (e) quaternary protein structure, of which hemoglobin is an example. Notice that this protein, which occurs in humans, is composed of four polypeptides, two of one type (reds) and two of another (blues).

TABLE 3.4	Essential and Nonessential Amino Acids For Adult Humans	
Essential	**Nonessential**
Histidine	Alanine
Isoleucine	Asparagine
Leucine	Aspartic acid
Lysine	Arginine
Methionine	Cysteine
Phenylalanine	Glutamic acid
Threonine	Glutamine
Tryptophan	Glycine
Valine	Proline
	Serine
	Tyrosine

speed up chemical reactions by raising temperatures without destroying themselves. Cells use catalysts to speed up reactions. A catalyst is any substance that speeds up the rate of a chemical reaction but is not used up in the process. It interacts closely with the substrate or substrates and speeds up reactions by lowering the activation energy (Figure 3.19b). In living cells these biological catalysts are called *enzymes*, which are large protein molecules with complex three-dimensional shapes. Enzymes are usually highly specific and operate on only one substrate. The three-dimensional shape of an enzyme determines its specificity. Enzymes are folded in such a way that they have a groove or pocket on their surface called the **active site**. The substrate of the enzyme fits very precisely into this active site, where the reaction catalyzed by the enzyme takes place.

A simple example is the enzyme sucrase, which catalyzes only the splitting of sucrose into glucose and fructose (Figure 3.20). Since sucrose is the compound on which sucrase operates, sucrose is said to be the substrate for sucrase. (Enzymes are usually named by adding the suffix *-ase* to part of the name of their substrate.) Sucrase binds a sucrose molecule into its active site, where water is added to the bond between the two monosaccharides. When the bond is broken, the two

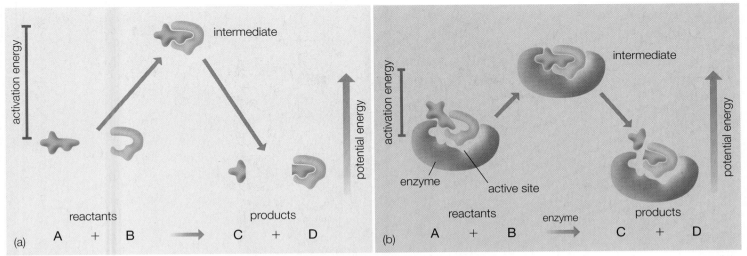

FIGURE 3.19 **A chemical reaction and an enzyme-catalyzed biochemical reaction** (a) Reactants and products are both characterized by a particular level of potential energy. In this example, the potential energy of the products is less than that of the reactants, and energy is released in the course of the reaction. For the reaction to proceed, the reactants must increase their potential energy by an amount called the **activation energy.** In a chemical reaction, a highly reactive substance called an **intermediate** forms and then reacts further during the conversion of reactants to products. (b) In an enzyme-catalyzed reaction, the reactants (substrates) first form a complex with the enzyme. The activation energy is much less than in a reaction without a catalyst.

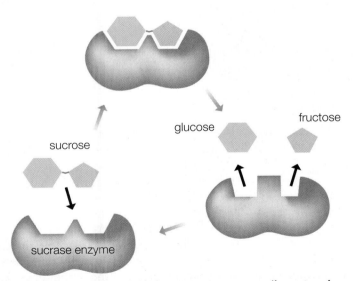

FIGURE 3.20 **Diagram of the enzyme sucrase (invertase) breaking sucrose into glucose and fructose** The protein has folded into a specific shape that determines the chemical function of this enzyme. Folding produces an active site into which sucrose just fits, but when sucrose is bound, the protein's shape changes, causing sucrose to split into its component simple sugars. Because glucose and fructose do not fit into the active site as well as sucrose, they are released, and the enzyme can bind and break another sucrose molecule. Because enzymes can do the same job over and over, only small amounts of each type are needed in cells.

sugars are released, and sucrase is ready to accept another sucrose molecule.

As mentioned earlier, humans and most other animals cannot digest cellulose. This is because we do not possess a cellulase enzyme to break it down. Only certain bacteria, fungi, and specific microbes—found in the rumens (stomachs) of cattle and sheep or the guts of termites and snails—have a cellulase to digest cellulose.

Some enzymes can operate simply as large proteins. Other enzymes, however, may require one or more nonprotein components called **cofactors** for their normal function. Cofactors may be metal ions such as magnesium (Mg) or an organic molecule. Cofactors that are organic molecules are called **coenzymes.** Vitamins, such as niacin and thiamine, are an important part of many coenzymes, which is why vitamins are so important in the nutrition of all organisms. Any living cell must have thousands of different enzymes to catalyze all the different reactions in the processes that make up its metabolism. Metabolism will be discussed in Chapter 5.

Nucleic acids such as DNA and RNA are composed of nucleotides

The instructions for all of the structures and functions of living organisms are encoded in and translated by nucleic acids. There are two types of nucleic acids: deoxyribonucleic acid (DNA)

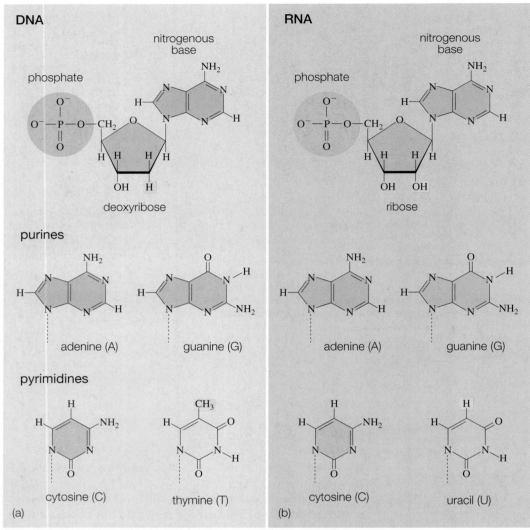

FIGURE 3.21 **Nucleotides and nitrogenous bases in DNA and RNA** (a) In DNA a nucleotide consists of a nitrogen-rich base (adenine in this case), a deoxyribose sugar, and a phosphate group. The four nitrogenous bases in DNA nucleotides are the pyrimidines cytosine and thymine and the purines adenine and guanine. (b) In RNA a nucleotide consists of a nitrogen-rich base (adenine is again shown here), a ribose sugar, and a phosphate group. The four nitrogenous bases that may be in RNA nucleotides are the pyrimidines cytosine and uracil and the purines adenine and guanine. The positions where ribose differs from deoxyribose and thymine from uracil are highlighted in yellow.

and ribonucleic acid (RNA). DNA is the largest molecule in the cell and contains genetic information organized into units called *genes*. RNA copies the information in DNA and translates it into proteins (Chapter 6).

Nucleic acids are large organic molecules composed of the elements carbon, hydrogen, oxygen, phosphorus, and nitrogen. The presence of nitrogen and phosphorus in nucleic acids is another reason why nitrogen and phosphorus are important plant nutrients. Nucleic acids are long chains of smaller subunits called **nucleotides**. Each nucleotide consists of three parts: a sugar, a phosphate group, and a base that is rich in nitrogen (Figure 3.21). The nitrogen-rich bases can be either a

double-ringed compound called a *purine* or a single-ringed compound called a *pyrimidine*. The specific purines in nucleotides are adenine (A) and guanine (G), and the pyrimidines are cytosine (C), thymine (T), and uracil (U). The nucleotides that compose DNA and RNA differ in two important ways. In DNA, the sugar is deoxyribose and the nitrogenous bases are adenine, cytosine, guanine, and thymine. In RNA, the sugar is ribose and the nitrogenous bases are adenine, cytosine, guanine, and uracil. The phosphate group, consisting of one atom of phosphorus covalently bonded to 4 atoms of oxygen, is the same in both DNA and RNA.

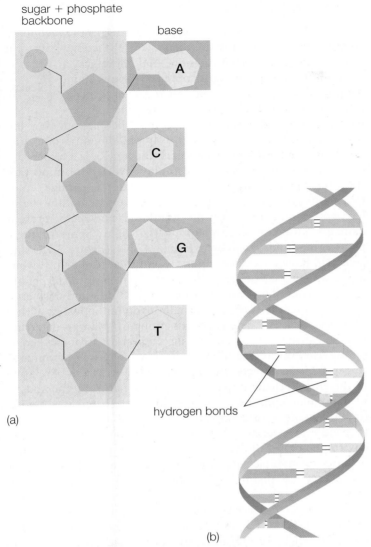

sugar + phosphate backbone

base

(a)

hydrogen bonds

(b)

FIGURE 3.22 **Portion of a nucleic acid of DNA and a double helix of DNA** (a) In this part of a nucleic acid, four nucleotides are shown joined by covalent bonds between the phosphate groups and deoxyribose sugars. (b) In DNA two nucleic acid strands spiral around each other in a double helix that is held together by hydrogen bonds between adenine and thymine and between cytosine and guanine.

sequence for the other strand can be deduced because adenine always bonds to thymine and guanine to cytosine. DNA molecules differ in length and base sequence, but all DNA molecules have a double-helix structure. RNA molecules are more diverse in structure and function than DNA. Most RNA molecules are single-stranded nucleic acid chains. The various RNA molecules are all involved in the synthesis of proteins in cells. Protein synthesis is considered in Chapter 6.

A few nucleotides have crucial roles in the cell apart from forming nucleic acids. All of the biochemical reactions that make up cellular metabolism require energy. Most of that energy is supplied by adenosine triphosphate (ATP), a nucleotide derivative that is the main energy carrier in all cells. ATP consists of the sugar ribose, the nitrogenous base adenine, and a chain of three phosphate groups, which carry energy in their covalent bonds. ATP releases energy when the enzyme ATPase removes its terminal phosphate group by adding

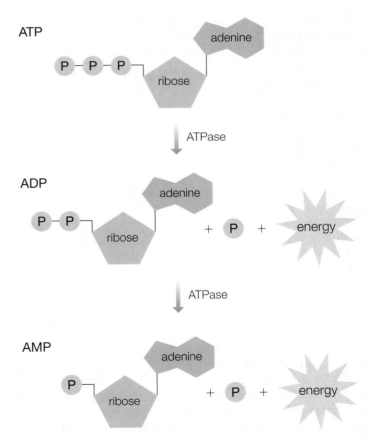

FIGURE 3.23 **Adenosine triphosphate (ATP)** When the terminal phosphate group on ATP is cleaved by an ATPase, energy is released and adenosine diphosphate (ADP) is formed. If the second phosphate group is also removed from ADP, energy is also released and adenosine monophosphate (AMP) is formed. (The reactions shown are reversible in that ATP can be regenerated from ADP and AMP by using energy from the breakdown of glucose in the cell.)

The nucleotides form the long chains of nucleic acids through dehydration syntheses between the phosphorus groups and the sugars (Figure 3.22). These syntheses are the same type of water-removing reactions that occur in the syntheses of sugars, lipids, and proteins. In DNA, two strands of nucleic acid wind about each other in a **double helix**, like a twisted ladder. The sidepieces of the DNA ladder are the covalently bonded sugars and phosphates, and the steps are the bases bonded to each other by hydrogen bonds. Guanine bonds to cytosine through three hydrogen bonds, and adenine bonds to thymine through two hydrogen bonds. If the base sequence is known for one nucleic acid strand, the base

water (hydrolysis) and produces a molecule of adenosine diphosphate (ADP) (Figure 3.23). A second phosphate group may also be removed, producing adenosine monophosphate and a similar amount of energy. In most chemical reactions within the cell, the phosphate is transferred to another molecule that is said to be *phosphorylated* and energized to undergo a chemical reaction. As an example, consider the formation of a molecule of sucrose from glucose and fructose (see Figure 3.20). The ATP/ADP system carries energy from energy-releasing reactions to energy-requiring reactions in cellular metabolism.

3.5 | Plants produce a wide range of secondary compounds

Secondary compounds are molecules produced by plants, algae, or fungi that are not found in all species. In contrast, primary compounds—carbohydrates, lipids, proteins, nucleic acids—are present in all organisms. Whereas primary compounds play central roles in metabolism, the thousands of types of secondary compounds are used for defense, specialized structures, or reproduction. Plants lack the supporting bony skeleton typical of animals. Individual plants cannot actively move to escape from predators or avoid sunburn, and they lack the mobile immune system cells that ingest pathogenic (disease-causing) microbes, as are present in animals. Neither can plants actively seek out mates or migrate to new habitats as can most animals. Secondary compounds help plants overcome these constraints. Some secondary compounds are so strong that they provide plants with structural support, which explains why trees can grow so tall. Other secondary compounds are distasteful or poisonous to animals that might feed on plants, preventing plants from being eaten. Yet other secondary compounds absorb UV radiation in sunlight (thereby preventing "sunburn") or prevent microbe attack. The beautiful flower pigments that aid in reproduction by attracting insects and other animals to carry pollen between plants, and fruit pigments that attract animals to transport plant seeds are secondary compounds (Chapter 24). The presence of particular secondary compounds explains the many ways in which people can use plants for building materials, fibers, decoration, spices, stimulating beverages, and medicines (Chapter 2); their importance to humans, and their roles in plant ecology.

Plants are master chemists, capable of producing a wide range of secondary compounds, the major types being terpenes and terpenoids, phenolics and flavonoids, and alkaloids. The complexity and diversity of plant chemical compounds illustrate the importance of chemistry to plants as a way of solving defense and other problems. Plant secondary compounds are chemically complex and diverse. This complexity explains why humans have not been able to produce many of these compounds industrially and thus must obtain them from plants.

Terpenes and terpenoids repel insects

The 25,000 different kinds of terpene and terpenoid secondary compounds vary greatly in their size and complexity, but they are all composed of the same building block, a molecule related to the gas isoprene. Various numbers of isoprene units can be put together to make different constructions. The names terpene and terpenoid are probably unfamiliar to most people, but they impact humans in many ways. For example, the important drug taxol, used in treatment of cancers (Chapter 7), is a terpene.

ECOLOGY

Terpenes and terpenoids also play a variety of important roles in plants. Carotenes, which give the orange color to carrots and to autumn leaves, participate in the light-harvesting reactions of photosynthesis and help prevent damage caused by excess amounts of light. Some of these plant compounds, such as extracts of rose (Figure 3.24) and lavender, have pleasant fragrances and therefore help flowering plants attract insect pollinators. Terpenes and terpenoid extracts from plants are also used in perfumes and cosmetics and in aromatherapy, a multibillion dollar industry. Other terpenes have a defensive function in plants—deterring disease-causing organisms or insect pests. Citronella, commonly used in candles or skin oils, is widely used by humans to repel biting insects. Pyrethrum, extracted from a relative of the chrysanthemum plant, is a potent natural terpene used in a popular commercial insecticide. Pyrethrum is superior to many manufactured insecticides because it is not very toxic to people or pets and does not persist for very long in the environment. Sunflowers and sagebrush produce terpenes in hairs on their leaves. The compounds are released when the plants are disturbed by animals and have a repellent effect.

Complex terpenoids include rubber, turpentine, rosin, and amber-hardened resin exuded from the trunks of some trees. The sticky resins on the trunks and needles of pines and firs are known to repel insects such as bark beetles. Scientists have been able to extract genetic material (DNA) from ancient organisms that became trapped and preserved in amber (but, so far, no dinosaur DNA has been obtained—despite what the makers of the fantasy movie *Jurassic Park* would have us believe).

Phenolics have antiseptic properties and flavonoids color many flowers and fruits

Many people are familiar with the antiseptic properties of phenol solutions and with the fact that flavonoids—phenolic compounds consumed in red wines, grapes, blueberries, and other fruits—are often cited in the media for their health benefits. You may have seen media stories about lignans, which are phenolic compounds especially abundant in green vegetables such as green beans and grains such as rye. Lignans are newsworthy because they appear to reduce the incidence of

Terpenes

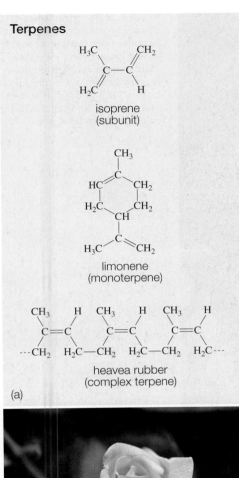

FIGURE 3.24   Terpenes (a) Chemical structures of isoprene, the building block of terpenes, and two terpenes: limonene, used as a solvent, and heavea rubber, a natural rubber compound. (b) Rose extracts (containing terpenes) are used in perfumes.

prostate and breast cancer. Plants produce an estimated 8,000 different phenolic compounds, and about 4,500 of these are flavonoids.

Like terpenes, phenolics have a building-block structure that can be combined in various ways to make many derivative compounds. The structure can confer the ability to absorb ultraviolet radiation from the Sun, thus protecting plants from genetic damage. Some phenolics give a bitter taste to plant parts, deterring herbivore feeding, and others have antimicrobial activity and prevent infection. Protective phenolics of celery, parsnips, figs, and parsley cause insects to die if they are exposed to light after feeding on these plants and can also cause contact skin rashes in people.

Tannins, which are toxic to many herbivorous animals and stop microbial growth, are also phenolic compounds. They are often produced in leaves and unripe fruit, where they prevent consumption before the seeds are mature. People have long used tannins derived from plants such as oak to preserve leather. Tannins also confer astringent flavors to tea and red wine. Additional examples of phenolics include the flavor compounds in cinnamon, nutmeg, ginger, clove, chilies, and vanilla (Figure 3.25). Natural vanilla is derived from the pod fruits of an orchid, *Vanilla planifolia,* which is native to Mexico. The Aztecs added vanilla to a drink that also included chocolate and chilies.

The burning sensation one gets when eating chilies is due to capsaicin, a compound that is related to vanilla. Capsaicin repels mammals that would otherwise destroy seeds while consuming the fruits. Mammals, including humans, have receptor molecules in their mucous membranes that bind capsaicin, triggering nerve impulses that the brain interprets as pain. However birds, which are useful to the plants as seed-dispersal agents, are impervious to capsaicin's painful effects, though the compound does act as a laxative, which aids in elimination of seeds. Capsaicin's repellence to mammals but not to birds illustrates the amazing ability of plants to target their chemical defenses. Medical researchers who are studying the mammalian capsaicin receptors think the compound will prove to be a general key to understanding and controlling pain in humans.

The phenolic polymer known as **lignin** gives wood its strength and also confers resistance to microbial attack. Among microbes (both bacterial and fungal) that function as decomposers, only a few can break down lignin. This is why wood rots slowly and makes a good building material. Lignin is also unpalatable to many animals.

Flavonoids include the beautiful **anthocyanin** pigments that attract animal pollinators or dispersal agents to flowers and fruits. The orange, pink, or red color of many flowers and fruits is due to the presence of flavonoid pigments known as *pelargonidins* (named after the geranium *Pelargonium*). Flavonoid pigments called *cyanidins* confer magenta and red colors, and delphinidins (named for *Delphinium*) produce purple and blue colors (Figure 3.26). Quercitin is a flavonoid found in onion and garlic that is used as a food supplement for its antioxidant and antihistamine effects. The health benefits attributed to extracts of *Ginkgo biloba* arise from flavonoids, as do extracts of grape seeds and green tea, which are sold as food supplements. Flavonoids extracted from peels of tangerines and oranges are potent cholesterol-lowering compounds that have fewer side effects than prescription drugs.

ECOLOGY

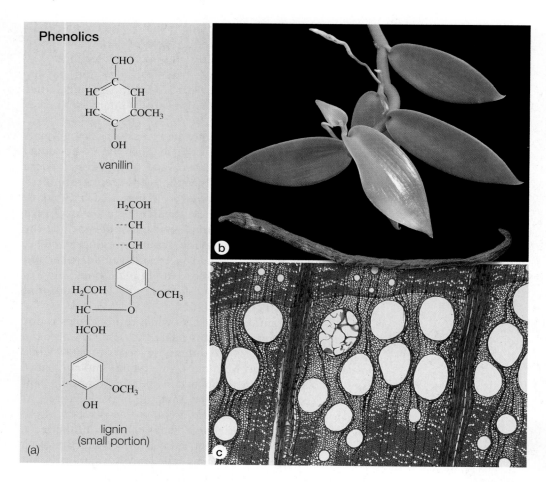

Phenolics

vanillin

lignin
(small portion)

(a)

(b)

(c)

FIGURE 3.25 Phenolics (a) Structures of vanillin (a flavoring agent) and lignin (an important component of wood). (b) Natural vanilla extract, which is valued more highly than artificial vanilla for culinary purposes, is extracted from the dried fruits ("vanilla beans," one of which is shown here) of the orchid *Vanilla*. (c) A stained slice of oak wood, which largely consists of lignin.

Many alkaloids are widely used as medicines

Various alkaloids may be familiar, such as the nicotine of tobacco and the caffeine of coffee. Alkaloids are widely used in medicine and are significant for the defensive roles they play in plants. The alkaloids are very diverse; there are over 12,000 known types, and some plants produce many types. For example, the rosy periwinkle *Catharanthus roseus*, known for production of drugs active against leukemia (Chapter 7), produces more than 100 different alkaloids.

All alkaloids contain some nitrogen, which distinguishes them from the other kinds of secondary compounds that we have described. Alkaloids are very effective in defending plants from hungry animals. All are toxic at high doses and some have potent effects on the nervous systems of animals. Caffeine, an active agent in tea and coffee, theobromine in chocolate, and nicotine in tobacco stimulate the human nervous system at low doses (Figure 3.27). In the case of strychnine and coniine (the active ingredient of poison hemlock, a drink used for executions in ancient times) even small amounts cause

death to humans. The Greek philosopher Socrates is probably the most famous person to be executed by drinking an extract of the poison hemlock, but he is almost certainly not this plant's only victim. Indigenous peoples make arrow-tip poisons from plants that contain the active alkaloid tubocurarine. This alkaloid is also used in modern medicine as a muscle relaxant before surgery.

Other alkaloids such as quinine, derived from the bark of *Cinchona*, a small Andean evergreen tree, are also used for medicinal purposes. Quinine helps control malaria, caused by a bloodstream-borne protist (Chapter 19), which affects more than 100 million people per year. Andeans, including the Inca, used quinine long before the arrival of Europeans. Although synthetic forms of quinine are now available, some types of malaria are resistant to them. So far, however, the disease-causing microbe has not become resistant to quinine from the natural source. Natural quinine contains 36 alkaloids, 4 of which are active against malaria. The widely used sedative reserpine, used to treat hypertension, is an alkaloid extracted

Flavonoids

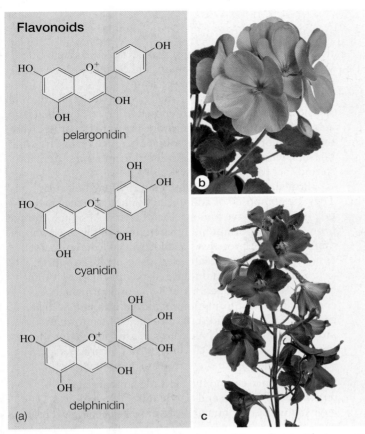

pelargonidin

cyanidin

delphinidin

(a)

b

c

FIGURE 3.26 Flavonoids (a) Structures of three flavonoid pigments: pelargonidin (found in the cultivated geranium *Pelargonium*), cyanidin (found in roses, for example), and delphinidin (found in *Delphinium*). (b) Cultivated geranium. (c) Larkspur (*Delphinium*).

from *Rauvolfia serpentina*. Indian healers used this plant, commonly known as snakeroot, for at least 3,000 years. Snakeroot contains many alkaloids, including several other medicinally effective compounds. Although reserpine has been synthesized, it is cheaper to extract it from plants grown throughout Southeast Asia.

Morphine and cocaine are legally controlled, habit-forming alkaloids that nevertheless have medicinal applications (see Chapter 2). Cocaine is extracted from leaves of the coca tree, *Erythroxylum coca* (see Figure 1.6), native to the Andes. The leaves contain 13 other secondary compounds. Andean workers have long used intact coca to depress hunger, relieve pain, and increase stamina; but they do not become addicted or experience harmful effects, in contrast to users of purified cocaine. Experts suspect that one or more of the other leaf compounds may modulate the effects of cocaine when it is consumed in the form of intact leaves. The widely used medical painkillers procaine and novocaine are synthetic compounds whose chemical design is based on the chemical structure of cocaine.

Plants have used chemistry to solve many difficult problems, including their need for structural support, sunburn protection, and defense against disease microbes and voracious animals. We all have attractive green chemical factories in our neighborhoods. The next time you admire a beautiful piece of wood or a brilliantly colored flower, smell the luscious fragrance of a garden, or taste exquisitely spiced foods, think of the many ways that humans enjoy the products of plant chemical industries.

Alkaloids

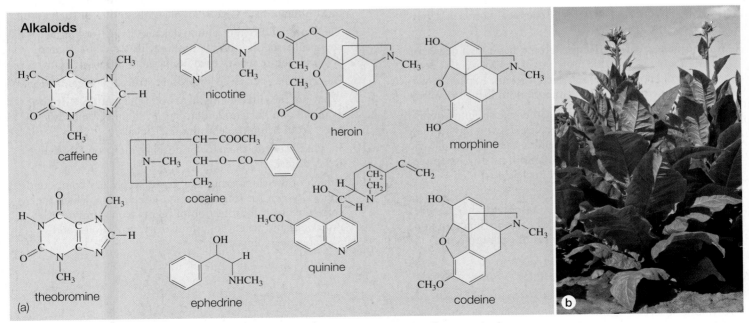

caffeine

nicotine

heroin

morphine

theobromine

cocaine

ephedrine

quinine

codeine

(a)

b

FIGURE 3.27 Alkaloids (a) Structures of several alkaloids, which have notable effects on the human body. (b) Cultivated tobacco (*Nicotiana*), after which the alkaloid nicotine is named.

HIGHLIGHTS

3.1 All physical matter is composed of chemical elements. Nine elements—hydrogen, oxygen, carbon, nitrogen, phosphorous, sulfur, calcium, potassium, and magnesium—make up most of the dry weight of plants.

Every element is composed of distinct types of atoms. Atoms are made up of protons, neutrons, and electrons.

3.2 Atoms of the same or different elements may combine through chemical bonds to form molecules. There are three basic types of chemical bonds: ionic, covalent, and hydrogen. Ionic bonds form when atoms gain or lose electrons and become ions. In covalent bonds, two or more atoms share electrons. If the sharing of electrons is not even, a polar covalent bond results, and the molecule will carry a slight charge. Water is one example of a polar molecule. Attractions between slightly (−)-charged oxygen and slightly (+)-charged hydrogens on other water molecules give rise to hydrogen bonds.

3.3 Because it is polar, water has unique properties and can exist as a solid, liquid, or gas at normal environmental temperatures.

3.4 Carbon forms the basis of organic molecules, of which there are four basic types: carbohydrates, lipids, proteins, and nucleic acids.

Carbohydrates include sugars, starches, and cellulose. Simple sugars, or monosaccharides, such as glucose and fructose are energy molecules. The disaccharide sucrose is

the main sugar transported in plants. The polysaccharide starch is used for energy storage, and cellulose is one of the major structural components of plants.

Lipids include fats, oils, and waxes; phospholipids; and steroids. Fats and oils are composed of glycerol and fatty acids. Waxes provide a waterproof protective coating over the surface of plants. Phospholipids arrange themselves into bilayers, which form the basis of cell membranes. In plants, steroids act to deter herbivores.

Proteins are the most abundant organic molecules. They are composed of amino acids linked together by peptide bonds and have four structural levels. Although storage proteins are important in human nutrition, the most important function of proteins is as biochemical catalysts called *enzymes*. Enzymes regulate every aspect of metabolism.

Nucleic acids contain the genetic information for all the structures and functions of living organisms. DNA contains genetic information organized into units called genes. RNA copies and translates the DNA information into protein synthesis. Nucleic acids are long chains of nucleotides, each of which consists of a 5-carbon sugar, a phosphate group, and a nitrogen-containing base.

3.5 Plant secondary compounds include terpenes and terpenoids, phenolics and flavonoids, and alkaloids. They have many uses for humans and have key ecological roles.

REVIEW QUESTIONS

1. Diagram the structure of an oxygen atom. Show how two oxygen atoms combine to form molecular oxygen. What kind of bond joins the oxygen atoms?

2. Describe the three basic types of chemical bonds and give examples.

3. Water is an extremely effective solvent, has a high degree of cohesion and high specific heat, and as solid ice is less dense than the cold liquid water in which it forms. Explain how each of these properties of water is important to living organisms.

4. What chemical reaction links together the subunits of all four basic types of organic molecules? How many of these reactions must take place to assemble each of the following molecules: (a) a disaccharide, (b) an oil, (c) a polypeptide with four amino acids, and (d) a nucleic acid nine nucleotides long?

5. Monosaccharides, disaccharides, and polysaccharides are all carbohydrates. Give an example of each, its function in a plant, and one or more human uses of each molecule.

6. Oils may contain saturated, monounsaturated, or polyunsaturated fatty acids. What does each term mean? Which type of oil is currently believed to be the healthiest for a human diet? Give a couple of examples of such an oil.

7. There are 20 different amino acids in living organisms. What chemical structures do all amino acids have in common? What structure makes each amino acid different?

8. Describe the four levels of protein structural organization.

9. What are the three components of a nucleotide? Give one example of a nucleotide that has a role in cell metabolism apart from being present in a nucleic acid.

10. In what ways do the nucleotides that make up DNA and RNA differ?

11. Which of the three groups of major secondary compounds is best known for fragrances and insecticides, which for spice flavors and protection of plants from ultraviolet radiation, and which for stimulatory, poisonous, and psychoactive compounds?

APPLYING CONCEPTS

1. A robotic spacecraft lands on Europa, one of the large moons of Jupiter, in the not-too-distant future. It scoops up a sample of surface ices and delivers it to an automated chemical laboratory. Along with water, the chemical lab reports the following formulas back to Earth:

(a) $C_4H_8O_4$

(b) $C_5H_{10}O_5$

(c) $C_3H_5(OH)_3$

(d) $H_2NCH(CH_2SH)COOH$

What will you tell the news media that the robot has found on Europa?

2. Starch and cellulose are both polysaccharides, but the bonds that bind their subunits together are different. How does this difference relate to the functions these two polysaccharides perform in plants?

3. The active site is the region in the tertiary structure of an enzyme where the chemical reaction takes place.

A mutation could give rise to a change in the amino acid sequence of the enzyme. Under what circumstances might a mutation occur but not affect the function of the enzyme?

4. Based on what you have learned about organic molecules in this chapter, make some recommendations about the components of a healthy human diet.

4 | Cells

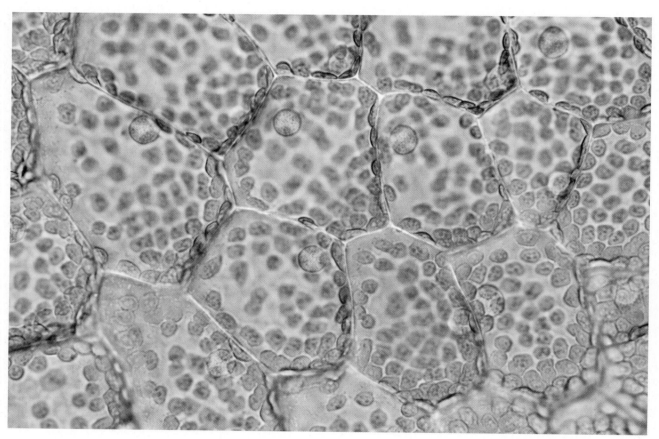

Fern gametophyte cells

English writer Jonathan Swift's famous satire and literature classic *Gulliver's Travels* includes a description of the imaginary land of Lilliput and its tiny inhabitants. In nature, there is a real lilliputian world of miniature beings, but it remained invisible to humans until the 1600s, when the first microscopes were invented. When people used these new instruments to examine samples of pond water, myriads of tiny organisms of many types were revealed. English physicist Robert Hooke used his handmade microscope to examine thin slices of cork from tree bark and saw rows and stacks of small cubical compartments. These cavities reminded Hooke of the monastery bedchambers—cells—occupied by monks of his time, and he coined the term *cell* in 1665.

Using increasingly powerful microscopes, later scientists examined many types of specimens, discovering a multitude of new types of microorganisms and cells. In 1838, the German plant scientist Matthias Schleiden was the first to conclude that all plants were composed of cells; in the next year, Theodor Schwann contributed the discovery that all animals were likewise constructed of cells. In 1858, Rudolf Virchow wrote, "Where a cell exists, there must have been a preexisting cell, just as the animal arises only from an animal and the plant only from a plant." Together, Schleiden's, Schwann's, and Virchow's observations make up the Cell Theory, which states that cells form the basic structure of all life and that all cells arise from parental cells. Further studies revealed that all cells have some features in common but that they differ in ways that reflect diverse functions.

These facts may seem obvious to us now, but the Cell Theory ranks among the major scientific achievements of all time. The Cell Theory is the basis of the many advances in biology that affect daily human life, including controversial issues such as cloning and cultivation of human stem cells, which appear often in the news media. For this reason, every educated citizen should have a basic knowledge of cells. In this chapter, we focus on the structure and function of plant cell components and their origins; but many of the concepts described apply to all cells, including those of the human body. ■

Algal cells

4.1 Organisms are composed of one to many microscopic cells

Many organisms are composed of just one tiny cell; they have **unicellular bodies**. These include most bacteria and archaea, many protists, and some fungi. Because of their unicellular construction, most microbes are so small that they can be seen only with the use of a microscope (Figure 4.1a). Organisms that are large enough to see easily—most seaweeds and fungi, plants, and animals—are composed of many minute cells and hence have **multicellular bodies**. Multicellular organisms often have one or more types of **tissue**—coherent groups of cells having similar structure and function (Figure 4.1b). The tiny chambers observed by Robert Hooke are the dead cells that make up cork tissues in the outer, protective layers of tree bark (Figure 4.1c). Different types of tissues are grouped to form **organs**, such as the stems, roots, and leaves of plants.

Some animal egg cells and certain seaweeds consist of a single very large cell that is visible without using a microscope. But most cells are exceedingly small—ranging in size from about 1 micrometer (1 μm; one-millionth of a meter) to 100 micrometers. Why are cells so tiny? Cells are small because this size maximizes the area of their surfaces through which materials are exchanged. As cells increase in size, the ratio of their surface area to cell volume decreases dramatically. Most cells are not large because in a large size they would be unable to import materials and export wastes rapidly enough to support their needs.

Because most cells are too small to see with the unaided eye, microscopes are essential tools for cell biologists, scientists who specialize in **cell biology**—the study of cells.

4.2 Microscopes are used to study cells

Imagine being able to observe objects that are less than a millionth the size of a person! Microscopes provide this power, opening up a world of amazing organisms and structures that most people have never seen. Cell biologists use several types of microscopes, each providing particular types of information and each useful in recording images of cells and their components at a range of magnifications.

Light microscopes use glass lenses and visible light to enlarge images

You may be familiar with **compound light microscopes** because they are widely used in biology classes, as well as in hospital, crime, and industrial laboratories (Figure 4.2b). Light microscopes are so called because they use glass lenses to enlarge images of specimens through which visible light has been transmitted (Figure 4.2a). If specimens are too thick to transmit light, they are sliced either by hand or with an instrument known as a microtome. Specimens are often stained with dyes to increase the visibility of cellular components (Figure 4.2c). Video cameras can be attached to light microscopes to record dynamic events in living cells (Figure 4.2d). **Stereo dissecting microscopes** (Figure 4.3a) are light microscopes used for study of specimens that are too large or thick to observe with a compound light microscope (Figure 4.3b). They are useful for manipulating and dissecting specimens, hence their name. Another specialized type of light microscope is the fluorescence microscope. Fluorescent materials emit light (glow) after being exposed to light. Biologists can detect these glowing materials with the use of fluorescence microscopes. Fluorescence microscopes are particularly useful for studying plant cells, because chlorophyll (Figure 4.4a) and other plant materials (Figure 4.4b) are naturally fluorescent. In addition, cells can be treated with fluorescent dyes that bind to specific cellular components, making it easier to locate them. For example, some dyes specifically bind to DNA, allowing cell biologists to locate DNA in cells and measure its amount (Figure 4.4c). Confocal laser scanning microscopes are fluorescence microscopes in which a laser provides highly organized beams of light. The laser scans the specimen at a precisely determined depth, which results in particularly clear images of cell interiors (Figure 4.4d).

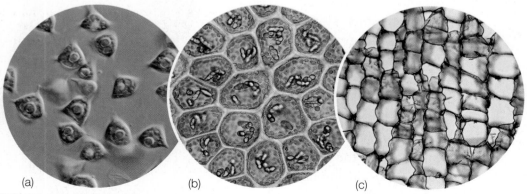

(a)　　　　(b)　　　　(c)

FIGURE 4.1   Cells and tissues (a) Several cells of the unicellular photosynthetic green algal protist *Chlorotetraedron*. (b) Plant tissue composed of numerous small adherent green cells, each surrounded by a pale wall. (c) A section through a cork bottle cap, showing the individual cells in cork tissue, whose dark-walled cells are dead at maturity. This is similar to the view Robert Hooke observed when he coined the term *cell*. Commercial cork is obtained from the bark of the cork oak, *Quercus suber*.

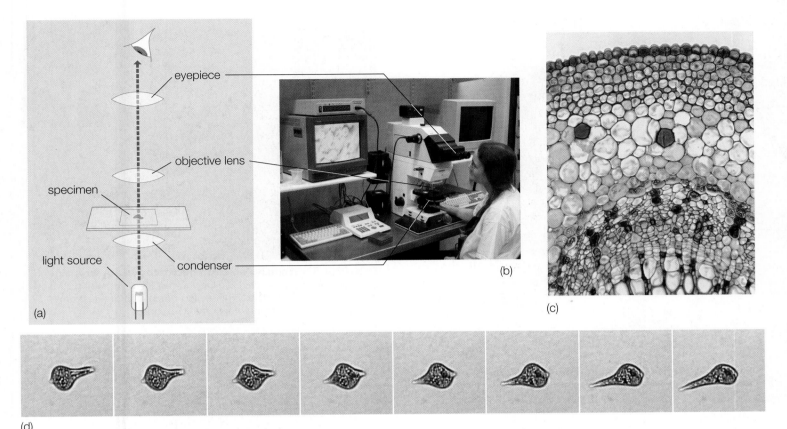

FIGURE 4.2 Compound light microscopy (a) Diagram showing the major components of a compound light microscope. A light source produces a light beam that passes through a thin specimen. The condenser lens helps concentrate this beam. Different combinations of objective and eyepiece lenses can be used to vary the magnification (and resolving power). (b) Scientist using a modern research-grade compound light microscope. (c) A stained plant specimen (a young stem of the basswood, *Tilia americana*). (d) Several frames of a video sequence photographed with a compound light microscope, showing the unusual movement of the protist *Euglena gracilis*.

Electron microscopes use magnetic lenses to focus beams of electrons

Electron microscopes have about 100 times the magnifying power of light microscopes because they use electrons and magnetic lenses to form images. Compared to light microscopes, electron microscopes are better able to distinguish very tiny objects lying close together because electrons have much shorter wavelengths than visible light—the shorter the radiation wavelength, the greater its power to distinguish small objects. Specimens to be examined with electron microscopes are usually preserved in a chemical solution that prevents changes in cell structure, a process that kills the cells.

Scanning electron microscopes (SEMs), Figure 4.5a, scan a beam of electrons across the surface of a specimen that has been coated with a very thin layer of gold, palladium, or platinum. Electrons that have bounced off the specimen surface form the image. The viewer sees an enlarged picture of elevations and depressions in the specimen surface (Figure 4.5b).

Transmission electron microsopes (TEMs), Figure 4.6a, are used to examine extremely thin slices (called *sections*) of specimens made by use of an ultramicrotome. This is possible only after specimens are infiltrated with liquid plastic

that hardens, providing stability. After the sections are cut, they are stained with heavy metals such as lead and uranium salts; these heavy metals bind to some cell materials better than others. Uranium stains DNA and RNA, and lead stains lipids and proteins, increasing the contrast between these molecules and their background. When sections are placed within the microscope, the electron beam is transmitted through them in places where the heavy metals have not been bound, but it is deflected wherever the heavy metals are located. Images of the cell interior are formed by transmitted electrons (Figure 4.6b, c).

These various types of microscopes provide different and complementary ways of "seeing" cells. For example, the transmission electron microscope has revealed that cell interiors of plants, animals, fungi, and protists are remarkably complex and similar in many ways, containing structures that do not occur within cells of bacteria and archaea. The TEM also reveals that plant cells share some distinctive features. These differences are more difficult or impossible to observe with light microscopes or SEMs. On the other hand, it is often easier to observe how cells are organized into tissues with light microscopes, and the SEM is the best way to detect some features of cell surfaces.

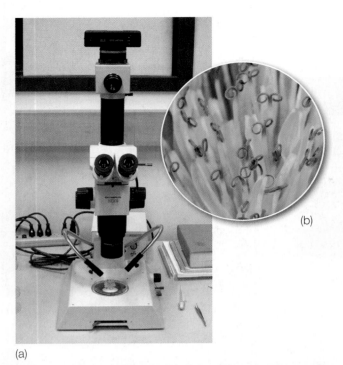

(a)

FIGURE 4.3   Stereo dissecting light microscopy (a) A stereo dissecting microscope. (b) Photograph of a specimen (close-up of a dandelion flower head) taken with a stereo microscope.

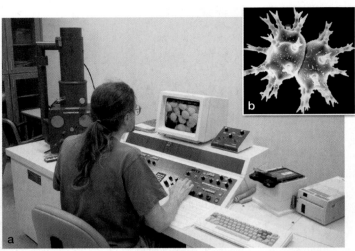

FIGURE 4.5   Scanning electron microscopy (a) A scanning electron microscope, which is especially useful in visualizing surface details. (b) Specimen (the single-celled green alga *Staurastrum*) visualized with a scanning electron microscope.

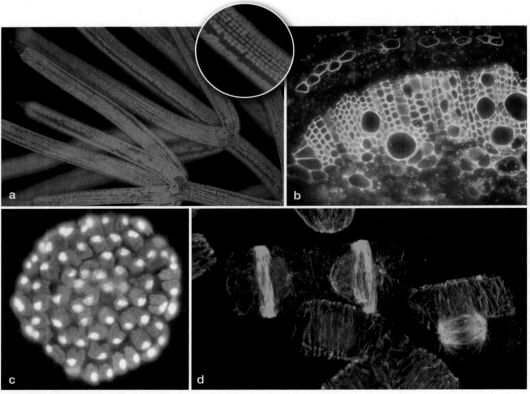

FIGURE 4.4   Fluorescence microscopy (a) Autofluorescence (self-fluorescence) of the chloroplasts in the green alga *Nitella*. The many small chloroplasts are arranged in very regular rows in these large cells (inset). (b) A section through a tomato plant stem. The lignin-coated water-conducting cells emit a yellow-green fluorescence. (c) Cells (of the green algal protist *Coleochaete*) stained with a fluorescent dye that binds to DNA, so that the nuclei (white spots) are better seen. As in (a) and (b), the chloroplasts are auto-fluorescing a red color. (d) An image of plant cells made with the use of a confocal microscope to localize particular cell components. In this image, nuclei are red and filaments known as microtubules appear green.

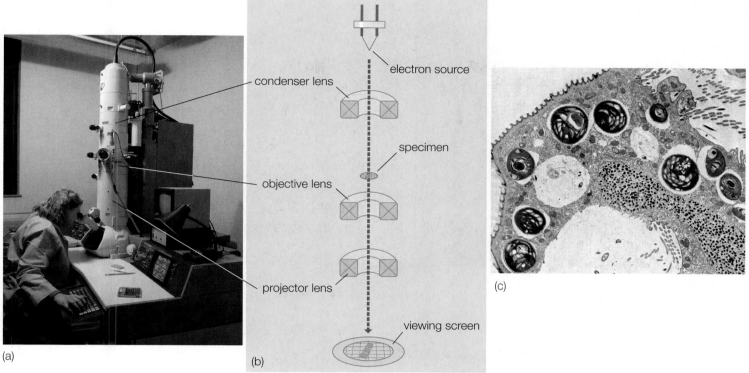

FIGURE 4.6   Transmission electron microscopy (a) A transmission electron microscope. (b) Simplified diagram of a transmission electron microscope. In general terms, this type of electron microscope is quite similar to a compound light microscope, except that it is inverted. A beam of electrons is focused onto a thin specimen by an electromagnetic condenser lens. The objective and projector lenses determine the magnification. Unlike compound light microscopes, which allow different lenses to be rotated into the light path, the current running through the electromagnetic lenses can be varied so that a considerable range of magnifications can be obtained. (c) Photograph of a thin section viewed with a transmission electron microscope (the specimen is a protozoan that contains algal cells).

4.3 | Two major types of cells are eukaryotic cells and prokaryotic cells

Plants, animals, fungi, and protists are known as eukaryotes, because they are composed of complex eukaryotic cells. In contrast, bacteria and archaea are known as prokaryotes, because they are composed of simpler prokaryotic cells. **Eukaryotic cells** are defined as cells that contain a nucleus and other membrane-enclosed structures collectively known as **organelles** (little organs). Mitochondria and plastids are examples of organelles. Plant cells are eukaryotic and possess nuclei, mitochondria, plastids, and other structures (Figure 4.7). In contrast, nuclei and other organelles are absent from prokaryotes (Figure 4.8).

Cell membranes, cytoplasm, and ribosomes occur in all cells

All eukaryotic cells and prokaryotic cells have at least three components in common: an outer cell membrane, also known as a plasmalemma or plasma membrane; internal cytoplasm; and numerous ribosomes. The **cell membrane**, which defines the cytoplasm's outer limit, is composed of phospholipids,

proteins, and other materials. The **cytoplasm** is a watery solution that includes ribosomes and other cell structures. **Ribosomes**, composed of proteins and RNA, are the cell's protein synthesis machinery.

The cell membrane functions in communication and transport of materials

How do cells perceive stimuli from the environment and influence other cells? How can cells allow entry of needed materials while preventing entry of harmful ones? How can cells get rid of wastes? The cell membrane plays a key role in all of these essential processes.

All cells possess a cell membrane having embedded proteins. The cell membrane is described as selectively permeable because it allows free passage of some materials, but not others. Very small, uncharged molecules—water; gases such as oxygen, carbon dioxide, and nitrogen; and some other small molecules—can easily pass through the cell membrane. In contrast, cell membranes do not allow passage of larger molecules and ions (molecules with an electrical charge) unless specific membrane transporter proteins are present (Figure 4.9). Some cell-membrane proteins transport nutrient molecules into cells or wastes out of cells, while others receive communication signals from the environment.

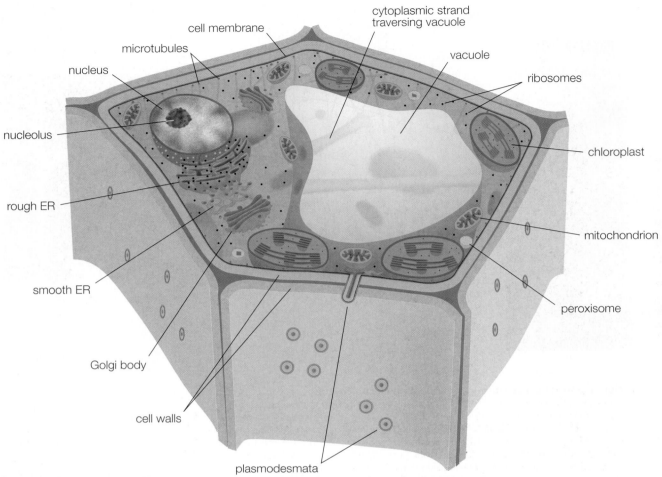

FIGURE 4.7 Diagram of a plant cell The major features found in plant cells are shown. Many of the same structures are also found in other types of eukaryotic cells.

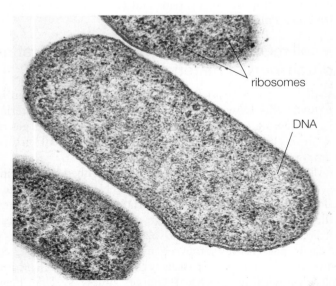

FIGURE 4.8 A prokaryotic cell as viewed with the transmission electron microscope Note that this cell (of *Escherichia coli*) lacks the internal complexity of plant or other eukaryotic cells. The small dots that fill this cell are ribosomes, and these ribosomes are smaller than those of eukaryotic cells. Thin strands of DNA are also apparent.

Some cell-membrane proteins perceive environmental information Receptor proteins in the cell membrane bind to chemical messengers in their environments. Receptors with messengers bound to them transmit signals to the cytoplasm, within which a response occurs. Cell-membrane receptors allow cells to sense their chemical environments and respond appropriately. For example, receptors in membranes of cells in one part of a plant may perceive hormone molecules emitted from other plant parts. In plants, hormones influence development, flowering, fruit ripening, and many other processes (Chapter 12). Communication systems involving hormones and cell-membrane receptor proteins also occur in humans and other animals.

Some cell-membrane proteins transport materials into or out of cells Molecules travel into or out of cells through passageways known as **transport proteins** embedded within the cell membrane. Much like electronic systems that regulate human entry to buildings or rooms based on keycards, codes, or fingerprint or retinal scans, cell-membrane transport proteins allow needed materials or information to move inside cells but restrict entry of harmful or unneeded substances. Transport proteins are so important to cells that they compose

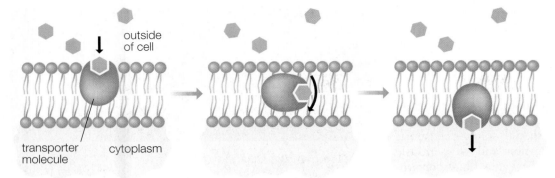

FIGURE 4.9 Cell-membrane carrier protein Each carrier protein shuttles a specific type of large molecule across the cell membrane. A molecule (green) binds to the carrier, which changes its shape in response, eventually releasing the molecule on the other side of the membrane.

50–75% of membranes. They are known as carrier, channel, or pump proteins.

For example, plant root cells must use channels that transport ions to obtain minerals such as phosphate from the soil. Recall that plants need phosphate ions to construct many essential cell molecules, including phospholipids, ATP, and DNA (Chapter 3). Nitrate and ammonium—ions that plants use to produce amino acids and proteins—are also obtained via ion transport proteins in the root-cell membrane. Other cell-membrane proteins import organic food molecules such as sugars or amino acids. For example, insectivorous plants, which grow on soils low in nitrogen minerals, catch and digest insects as a source of nitrogen (Chapter 11). The insect-digesting parts of these plants have separate proteins in their cell membranes that import ammonia, amino acids, and small proteins.

Selective membrane permeability is the basis for osmosis If you water your houseplants with saltwater rather than freshwater, they will probably die. Why? The answer requires an understanding of osmosis, a process that influences many aspects of cell and plant behavior.

Osmosis is the movement of water across the cell membrane according to the relative concentration of dissolved substances in the watery solutions on the insides and outsides of cells (Figure 4.10). Dissolved materials, known as **solutes**, include salts (ions), sugars, and other low-molecular-weight molecules that do not pass easily through cell membranes. Osmosis occurs because water behaves like other substances in tending to move from an area of high concentration to a region of lower concentration, a process known as **diffusion**.

If the concentration of solutes inside and outside a cell is the same, the cell is osmotically balanced—its net water content does not change. Over a period of time, the same number of water molecules moves into the cell as moves out (Figure 4.10a). In this case, the surrounding watery medium is described as **isotonic** to the cells. In contrast, if the solute concentration is lower outside a cell than inside, the outside solution is **hypotonic** to the cell. More water will enter the cell than can leave, possibly causing it to burst (Figure 4.10b). Cells exposed to water low in solutes—such as freshwater ponds or rain—commonly encounter this problem. Most prokaryotes, many protists, fungi, and plants

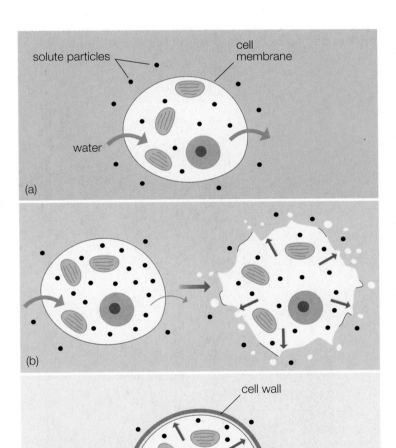

FIGURE 4.10 Solute balance (a) Isotonic solutions have the same amount of solutes as cells, so that the amount of water entering and leaving cells (blue arrows) is equal. (b) Hypotonic solutions have lower concentrations of solutes than are present inside the cell. In this case, more water diffuses in, possibly to the point where the cell bursts. (c) Cell walls prevent cells from bursting when the cytoplasm expands outward (red arrows) under conditions where solute concentrations are higher inside than outside the cell (when cells are in hypotonic solutions).

have acquired cell walls that limit the extent to which cell volume can increase by taking in water (Figure 4.10c). Some freshwater protists have **contractile vacuoles**—membrane sacs within cells that accumulate excess water and periodically expel it from the cell (Figure 4.11).

The cells of animals do not require walls to prevent bursting, because they are bathed in body fluids whose salt content balances that of cells. This is why salt is a necessary component of animal diets, why humans use special solute-rich sports drinks to replace fluids lost by perspiration, and why seriously dehydrated people are treated with intravenous fluids containing salts and sugars rather than plain water.

If the concentration of solutes outside cells is higher than that inside, water will leave the cell, causing the cytoplasm to shrink—a process known as **plasmolysis** (Figure 4.12). In this case, the outside solution is described as **hypertonic** to the cell. A hypertonic medium such as salty water can damage plant cells to the point of death by causing dehydration—loss of

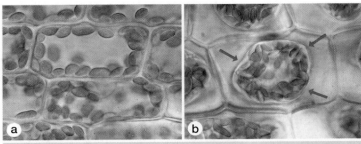

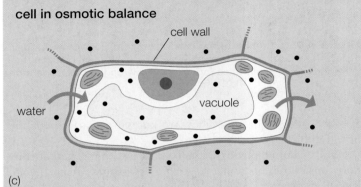

cell in osmotic balance

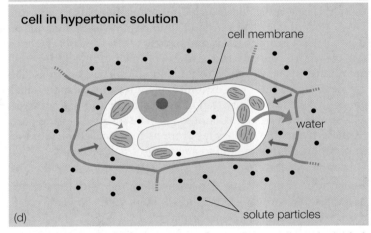

cell in hypertonic solution

FIGURE 4.12 Plasmolysis (a) Photograph of plant cells (in the leaf of *Elodea*) in osmotic balance with the environment. (b) When dehydrated by placing the leaf into a hypertonic solution, the cell membrane (arrows) pulls away from the cell wall. This phenomenon is shown diagramatically in (c) and (d). The cell in osmotic balance with the environment (c) has an equal amount of water (blue arrows) diffusing into and out of the cell. The cell in a hypertonic solution (d) incurs a net loss of water and thus the cell membrane pulls away from the cell wall (red arrows).

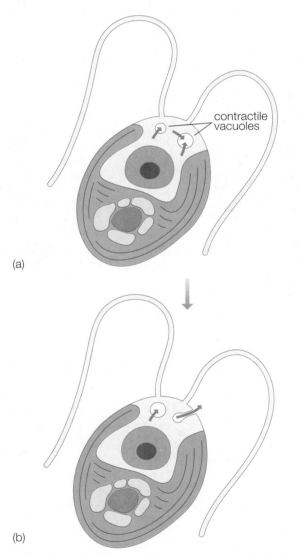

FIGURE 4.11 Contractile vacuoles Some protists deal with environments having low solute concentrations by using contractile vacuoles. Water accumulates in membranous sacs (a) and is periodically emptied to the outside of the cell (b).

water. In areas where salt is used as a road de-icer, roadside plants can be damaged from salty spray or soil (Figure 4.13). This damage results from the osmotic loss of plant water to the salty external environment. Osmotic dehydration of plant cells will also occur if you water your houseplants with salty water. Plant biologists describe the responses of plants and their cells to salty solutions (and other types of water stress) in terms of a concept known as water potential (Essay 4.1, "Plant 'Psi'chological Stress").

Some plants are adapted to salty environments Some plants are able to grow in salty habitats such as deserts and

FIGURE 4.13 Salt-damaged plants The dramatic red-brown needles on these pine trees are the result of osmotic damage, caused by spray containing salt (used to de-ice roadways) that was thrown by passing automobiles.

Eukaryotic cells share some features that prokaryotes lack

In addition to a nucleus, eukaryotic cells of plants (see Figure 4.7), animals, fungi, and protists share several features that distinguish them from prokaryotic cells. All eukaryotic cells have a nucleus, an endomembrane (meaning "internal membrane") system consisting of endoplasmic reticulum (ER) and Golgi bodies, a cytoskeleton (cell skeleton) composed of protein fibers, and several types of motor proteins that cause movements within the cytoplasm.

Most eukaryotic cells also possess mitochondria (singular, *mitochondrion*), important sites of chemical energy transfer, and peroxisomes—organelles that help protect the cytoplasm from destructive molecules and that are important in plant photosynthesis. Because of their widespread occurrence in modern eukaryotes, biologists deduce that these features were also present in early eukaryotic cells.

Nuclei contain most of the eukaryotic cell's genetic information

The **nucleus** is the major site of genetic information—DNA—storage in eukaryotic cells (Figure 4.16). Though most eukaryotic cells contain a single nucleus, some organisms possess eukaryotic cells having more than one nucleus.

A **nuclear envelope** composed of two membranes encloses the interior of the nucleus—the **nucleoplasm**—during most of a cell's lifetime. Nuclei also contain **nucleoli**, bodies in which cytoplasmic ribosomes are constructed. During division of plant and animal cells, the nuclear envelope breaks down so that the nucleoplasm mixes with the cytoplasm. The nuclear envelope re-forms at the end of the division process (see Chapter 7). Pores in the nuclear envelope—**nuclear pores**—are gateways through which the nucleoplasm communicates chemically with the rest of the cell.

Eukaryotic nuclear DNA is combined with proteins to form units known as **chromosomes**. Species of organisms have characteristic numbers of chromosomes; in plants, the number varies from just a few to over a thousand chromosomes crowded into a nucleus. Chromosomes can be stained with colored dyes and are individually visible with the light microscope in dividing cells (Figure 4.17). Most of the time, however, the threadlike chromosomes are not individually distinguishable; they are clustered into a mass known as **chromatin** (colored material).

The endomembrane system constructs and transports cell materials

How do cells construct the proteins, carbohydrates, and lipids needed to grow and reproduce, and how are these materials transported to correct locations within cells? The answer to both of these questions is the endomembrane system. **Endomembranes** are phospholipid membranes with embedded proteins that are quite similar to the cell membrane but occur within the cytoplasm. The **endomembrane system** is the sum of

coastal salt marshes. Such plants are known as **halophytes** (salt-loving plants). Halophytes have a variety of adaptations that help prevent osmotic damage to their cells. Many accumulate inorganic salts such as NaCl (table salt) in their vacuoles and organic solutes in their cytoplasm; this balances solute concentrations inside and outside cells. Other halophytes excrete salt. The coastal mangrove, for example (Figure 4.14), secretes salts onto its leaf surfaces. Other halophytes isolate excess salts into special tissues. The desert shrub *Atriplex*, for instance, transports excess salt into surface bladders that then break off or burst harmlessly. Halophytes have potential as crops that can be grown on salty soils or watered with saline waters, which are common in many parts of the world and harmful to most plants.

ECOLOGY

Endocytosis and exocytosis also transport materials across cell membranes What if cells need to transport particles or other materials for which membrane transporter proteins won't work? Eukaryotic (but not prokaryotic) cells can accomplish such transport by exocytosis and endocytosis (Figure 4.15). **Exocytosis** is the transport of materials out of cells by enclosing them in a sphere of membrane, known as a vesicle. Since they are made of similar material, vesicles easily fuse with the cell membrane, thereby dumping their contents outside the cell. Plant root cells use exocytosis to secrete large amounts of polysaccharide mucilage that helps lubricate roots as they grow through soil. **Endocytosis** is a process that brings materials into cells by enclosing them in a sphere of cell membrane, then pinching off a vesicle within the cytoplasm. An important example of endocytosis in plants is the entry of nitrogen-fixing bacteria into legume root cells (Chapter 10). One form of endocytosis is **phagotrophy**, which means "particle-feeding." Many protists use phagotrophy to ingest food such as bacteria and other small particles. Phagocytosis is evolutionarily important as the mechanism by which eukaryotic cells acquired some of their organelles—mitochondria and plastids (Chapter 19).

ESSAY 4.1 PLANT "PSI"CHOLOGICAL STRESS

Water stress occurs when plants contain a less than optimal amount of water. Most plants experience water stress at least some of the time, and plants that live in arid, saline, or cold habitats often suffer water stress.

Plant biologists describe the degree of water stress experienced by a plant or plant cell by the phrase "total water potential." The total water potential is symbolized by the Greek letter psi, with subscript w (for water)—ψ_w—expressed in Pascals (Pa). (A Pascal is a unit of pressure equal to 1 Newton per square meter.) Total water potential is affected by several factors, some of which do not affect the total very much, but two factors do have a large impact. One of these important factors is the number of particles dissolved in water inside and outside of cells. Such particles are known as *solutes*, and their quantity is symbolized by ψ_s, where the subscript s stands for solutes. The other important factor is the physical force exerted on water, a quantity known as *pressure potential*. This quantity is symbolized by ψ_p, where p stands for pressure. The expression that includes both effects is $\psi_w = \psi_s + \psi_p$. The total water potential, ψ_w, can be used to predict the movement of liquid water into or out of a plant cell. This is because water will diffuse across cell membranes from a region of high total water potential to low total water potential (from high ψ_w to low ψ_w). Water potential describes the osmotically driven flow of water across cell membranes—not only the outer cell membrane (plasmalemma) but also membrane-bound compartments within cells. In plant cells, these will include mitochondria, plastids, and the vacuole.

If a plant cell is placed into pure water, the change in ψ_w will cause water to enter the cell until the cell is swollen and the cell membrane presses up against the cell wall. This is cell turgor pressure, which will increase until the water potential is equal on both sides of the cell membrane and net water transport across the membrane

Most plants experience water stress at least some of the time, . . .

is zero (Figure E4.1A). Conversely, if a plant cell is placed into a concentrated salt solution, water will flow out. Such a cell loses turgor and becomes plasmolyzed as the cell membrane pulls away from the cell wall (Figure E4.1B). If the solute concentration inside a nonturgid (plasmolyzed) cell becomes greater than the solute concentration of the outside medium, water potential will cause water to move into the cell, and turgor may be restored as the cell expands.

Relative water content (RWC) is a quantity that is often measured and reported with plant total water potential. To determine this quantity, the fresh weight (unmanipulated weight) of plants or samples taken from them is measured, then the turgid weight (the maximum amount of water that the plant cells can take up) is measured. To measure turgid weight, plant tissues are floated on water in an enclosed, lighted chamber until the weight of the tissues remains constant. Then the plant or sample is dried and

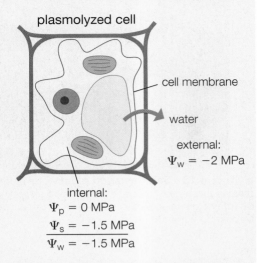

E4.1B Example water potential calculations for a plasmolyzed plant cell

internal:
$$\Psi_p = 0 \text{ MPa}$$
$$\underline{\Psi_s = -1.5 \text{ MPa}}$$
$$\Psi_w = -1.5 \text{ MPa}$$

Ψ_w internal > Ψ_w external so water leaves cell

weighed to obtain dry weight. The $\text{RWC} = [(\text{fresh weight} - \text{dry weight})/ \text{turgid weight} - \text{dry weight})] \times 100$. When a plant's RWC drops below a critical value, tissue death results. Though plants vary greatly in this critical value, many die when the RWC drops to a value less than 50%.

As an adaptation to avoid death by water loss from cells, plants that are resistant to dry, cold, or saline habitats are able to adjust the solute concentration of their cytoplasm. This is known as *osmotic adjustment*. The additional solutes decrease the cells' ψ_w, drawing water into the cell from the outside and preventing plasmolysis. Solutes commonly used by plants for osmotic adjustment include the amino acid proline, monomeric sugars such as glucose and fructose, and sugar alcohols such as mannitol. Plants may also adjust to drought by increasing the number of cell-membrane proteins that form pores which conduct water—such proteins are known as *aquaporins*. The presence of aquaporins allows water to move into cells more rapidly than by diffusion alone. Aquaporins help plant cells quickly recover turgor when water becomes available.

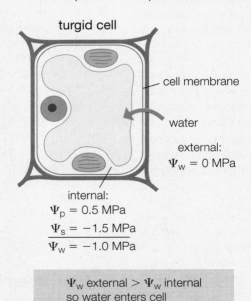

E4.1A Example water potential calculations for a nonplasmolyzed plant cell

internal:
$$\Psi_p = 0.5 \text{ MPa}$$
$$\underline{\Psi_s = -1.5 \text{ MPa}}$$
$$\Psi_w = -1.0 \text{ MPa}$$

Ψ_w external > Ψ_w internal so water enters cell

all of the endomembranes in a cell's cytoplasm. The nuclear envelope is often connected to the endomembrane system and is structurally similar. In fact, both the nuclear envelope and the endomembrane system are thought to have evolved by inward growth of the cell membrane in the ancestors of the first eukaryotes (Chapter 19).

The two major components of the endomembrane system are the **endoplasmic reticulum** (Figure 4.18a) and the **Golgi**

FIGURE 4.14 Salt-adapted plants The black mangrove adapts to its salty environment by excreting excess salt from the surfaces of its leaves.

apparatus (Figure 4.18b). Though each has specific functions, the endoplasmic reticulum and the Golgi apparatus often work together.

The endoplasmic reticulum

The endoplasmic reticulum—abbreviated as **ER**—is a network of flattened membrane sacs or tubes and the small, round, membrane-coated vesicles that arise from them. *Endoplasmic* means "inside the cytoplasm," and *reticulum* means "network." ER occurs in two forms—**smooth ER** and **rough ER**—that are connected, though each has distinctive functions. The smooth ER constructs fatty acids and phospholipids, which are needed to produce membranes. Smooth ER is also the site where harmful toxins are broken down into less dangerous molecules that can be excreted from cells.

The surface of rough ER is bumpy because it is covered with ribosomes. Ribosomes are the small, globular structures—composed of protein and RNA—that generate proteins from amino acids (Chapter 6). Ribosomes also occur free in the cytoplasm and within mitochondria and plastids. The ribosomes associated with rough ER produce proteins that will be secreted from the cell, and the ER is the pathway by which these proteins begin their outward journey.

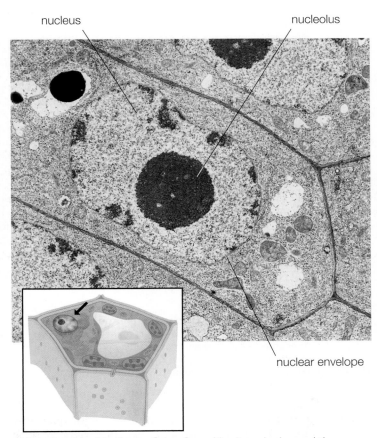

FIGURE 4.16 Nucleus of a eukaryote A nucleolus and the double-membrane nuclear envelope are shown.

Proteins produced by rough ER have particular amino acid sequences at their ends that enable them to enter the ER tubes through protein gateways, much as you would use a ticket at the entrance to a subway train. Once inside the ER, chains of sugars may be added to proteins to form **glycoproteins**. The sugar addition may help proteins to fold properly. The rough ER network then conveys glycoproteins to the vicinity of Golgi bodies for further chemical processing. Glycoproteins are transported from the ER within tiny, round vesicles. These are pinched off by the action of special proteins that coat the outside of the ER membrane, bending it into small spheres (Figure 4.19).

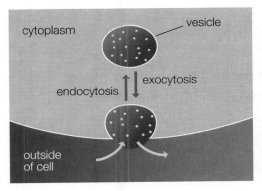

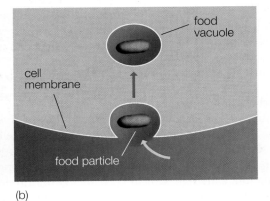

(a)   (b)

FIGURE 4.15 Exocytosis, endocytosis, and phagocytosis (a) Substances can move out of the cell in vesicles that fuse with the cell membrane (exocytosis). Substances can move into a cell by reversing the process (endocytosis). (b) Phagocytosis is a type of endocytosis that allows small cells and other particles to enter the cell.

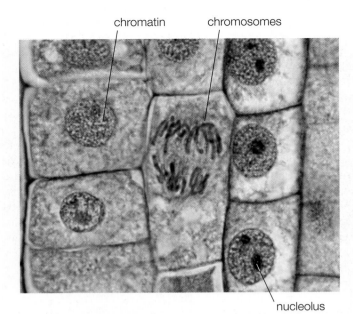

chromatin   chromosomes

nucleolus

FIGURE 4.17 Chromosomes, chromatin, and nucleoli In this specimen of an onion root, the DNA has been stained purple. The large cell in the center is dividing and has distinct chromosomes. In cells not undergoing cell division, the DNA is more diffuse (distinct chromosomes are not visible) and is termed *chromatin*.

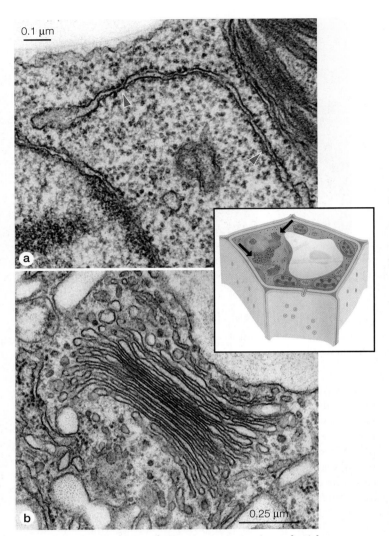

0.1 μm

a

b   0.25 μm

FIGURE 4.18 Endomembrane system as viewed with a transmission electron microscope (a) Endoplasmic reticulum (ER) (arrowheads). Shown here is rough ER, so called because ribosomes are found on its surfaces. (b) Golgi body.

The Golgi apparatus The function of the **Golgi apparatus** is to produce, modify, and distribute cell materials to their appropriate locations. The Golgi apparatus of a cell consists of one or more **Golgi bodies**, each a stacked array of pancakelike membrane sacs, known as **cisternae**. Typically, cisternae are flattened in the center and inflated at their edges (see Figure 4.18b). One of the outermost cisternae, known as the "forming face" or the "cis face," has membrane proteins that recognize, bind to, and fuse with ER vesicles, in this way receiving glycoproteins from the ER. Distribution of Golgi products occurs in vesicles produced at the "secreting face," also known as the "trans" face.

Each of the cisternae in a Golgi body contains different sets of enzymes. Materials move directionally through the stack of cisternae, in the process undergoing an orderly sequence of chemical modifications, much as a machine is put together on an assembly line. Golgi enzymes trim or modify the sugar chains that were added to proteins in the ER; these chemical changes tailor proteins for specific cell functions. Plant Golgi bodies also manufacture noncellulose polysaccharides—**pectins** and **hemicelluloses**—that will help form the cell wall (Section 4.4). Secretion of these polysaccharides is particularly active during formation of new cell walls after cell division and during cell enlargement, when the wall must increase in surface area.

How do proteins and cell-wall carbohydrates move from the Golgi bodies to their final destinations? The products of Golgi bodies are released in distribution vesicles that pinch off the edges of cisternae, particularly the secreting face. Distribution vesicles are coated with proteins that bind to specific receptor molecules at their final destination. Different vesicle coat proteins are used to target vesicles to various cell locations.

Vesicles loaded with cell-wall polysaccharides, for example, have membrane coats that bind only to the cell membrane. When the vesicles arrive at the appropriate location, they fuse, releasing their contents by exocytosis (see Figure 4.19). Vesicle fusion is also a mechanism for depositing membrane proteins into the cell membrane, allowing it to grow. Golgi-produced vesicles can also travel to and fuse with cell vacuoles, discharging their contents inside. Vesicle coat proteins play an essential role by ensuring that the many types of cell materials produced by the endomembrane system end up in their appropriate locations.

The cytoskeleton and associated motor proteins generate cell movements

The insides of living cells are truly motion-filled places, and some types of cells are able to move in their environments. These cell motions result from activities of the cytoskeleton and associated motor proteins. Although we often think of plants as motionless compared to animals, movement of cell structures occurs in plant cells in the same way as for animal

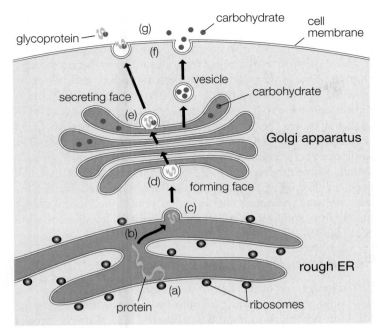

FIGURE 4.19 Endomembrane system Proteins to be secreted from cells are synthesized by ribosomes associated with (rough) ER (a). The newly synthesized protein moves through the rough ER (b) until it is packaged into small vesicles (c). These vesicles move to a Golgi body (all Golgi bodies in a cell are collectively known as the Golgi apparatus), where they fuse with the forming face. (d) The protein makes its way through the stacks of membranes making up the Golgi body, undergoing various modifications en route. Carbohydrates produced in the Golgi body also make their way through the stacks. At the secreting face, the proteins, carbohydrates, or molecules containing both (glycoproteins) are packaged into vesicles (e) that make their way to and fuse with the cell membrane (f), where their contents are released to the cell's exterior (g). This process also adds more surface area to the cell membrane.

and other cells. One example of motion within plant cells, which can be easily observed with a light microscope, is the rapid cycling of cytoplasm in cells of the very thin leaves of the water plant *Elodea*. This movement is known as **cytoplasmic streaming**.

Cytoskeletal components

The **cytoskelton** consists of three major types of long, thin, protein fibers: hollow tubes known as **microtubules**, thinner **microfilaments**, and filaments of intermediate size—**intermediate filaments**. The walls of microtubules are constructed of many molecules of the protein tubulin, microfilaments are made of two intertwined strands of the protein actin, and intermediate filaments are formed of other types of protein subunits. These fibers provide structural support in cells. For example, intermediate filaments located just inside the nuclear envelope help maintain nuclear shape. Cytoskeletal fibers also function something like railroad tracks, along which the motor proteins move attached organelles and vesicles. Each type of fiber has particular uses in cells, much as ropes of various types are made for different purposes.

Types of motor proteins

Motor proteins interact with microtubule or microfilament tracks to produce cellular move-

ments. There are three types of motor proteins: kinesin, dynein, and myosin. They share the ability to convert chemical energy into movement but have distinctive motions that are each useful in cells. Myosin moves rapidly, kinesin moves more slowly but steadily, and dynein has a sliding motion. Cytoskeletal fibers are each associated with particular motor proteins. Kinesin and dynein associate with microtubules, whereas myosin interacts with microfilaments.

The myosin-microfilament system is familiar to many people because it is the major component of animal skeletal muscle. Movement of myosin along microfilaments is the basis for the contraction ability of muscles. Although plants lack muscle tissue, their cells do contain mini-muscles, in the form of microfilaments and associated myosin. The microfilament/myosin assembly is responsible for the cytoplasmic-streaming phenomenon described earlier. It is also essential to the process by which plant and other eukaryotic cells divide.

Active movement of chromosomes during nuclear division of eukaryotic cells operates by microtubule-motor systems (Chapter 7). Microtubules also serve as tracks along which membrane-bound organelles move. By this system, mitochondria, Golgi bodies, and vesicles linked to kinesin are able to travel from one place to another within cells (Figure 4.20).

Flagella Some eukaryotic cells have one or more **flagella** or **cilia**, elongate extensions from the cells that can bend and flex, allowing cells to swim. Flagella are typically longer than cilia, but are otherwise similar. Flagella are common among algae and other protists, occur on some fungal cells, and are produced by the sperm cells of animals as well as sperm of some plants. Flagellar motions allow sperm to actively travel toward egg cells in order to accomplish fertilization. Flagellar action is due to bending motions resulting from the operation of dynein-microtubule systems. Eukaryotic flagella typically contain a pair of single microtubules, surrounded by a cylinder of nine

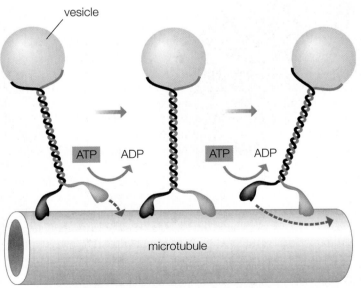

FIGURE 4.20 The motor protein kinesin moves vesicles along microtubule tracks Chemical energy in the form of ATP is used to perform the work of cell motions such as this.

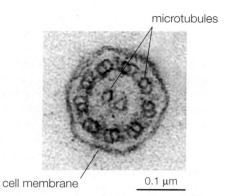

microtubules

cell membrane   0.1 μm

FIGURE 4.21   Cross section of a flagellum showing microtubules The microtubules in nearly all flagella and cilia have this so-called 9 + 2 arrangement (nine pairs surrounding two central microtubules).

paired microtubules linked by dynein and other proteins and encased by the cell membrane (Figure 4.21). Prokaryotic flagella lack such microtubule arrays.

Mitochondria are major cellular sites of chemical energy transformations

Mitochondria constitute the main power production structures of the eukaryotic cell. Each cell may contain one or dozens of mitochondria, each enclosed by an envelope composed of an outer membrane and a highly folded inner membrane (Figure 4.22). These inner membrane folds, known as **cristae**, provide a large surface area for the attachment of enzymes and other molecules essential to efficient energy metabolism

EVOLUTION

(Chapter 5). Within mitochondria is a watery **matrix** containing soluble enzymes that are also essential to energy metabolism. The mitochondrial matrix also contains DNA and ribosomes similar to those of prokaryotic cells. Another similarity between mitochondria and prokaryotes is that both reproduce themselves by pinching in half (binary fission—see Chapter 7). However, most prokaryotes are capable of independent existence, whereas mitochondria cannot persist outside of their host cells. Also, most prokaryotes have cell walls, whereas mitochondria do not. The similarities between mitochondria and bacterial cells are not accidental. Mitochondria descended from ancient bacteria that had been engulfed by a phagotrophic cell but not digested. Rather, the bacteria were maintained because they contributed useful energy metabolism and were gradually transformed into mitochondria (Chapter 19).

Peroxisomes contain protective enzymes

Peroxisomes are organelles that contain sets of enzymes but have only one surrounding membrane. A characteristic enzyme found in peroxisomes is catalase (Figure 4.23). Catalase breaks down hydrogen peroxide, a product of cell metabolism that is very harmful. For example, people use solutions of hydrogen peroxide to cleanse wounds because it kills harmful bacterial cells (the solutions are weak enough that human tissues are not harmed). Peroxisomes of green plant tissues help prevent loss of photosynthetic products from plant cells when photorespiration occurs (Chapter 5). Related organelles, known as glyoxysomes occur in cells of peanut and sunflower seeds, which have high oil content. **Glyoxysomes** transform storage lipids into sugar, which is used as a source of energy during seed germination.

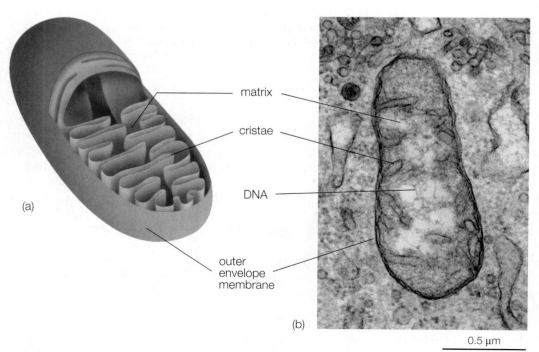

matrix

cristae

DNA

outer envelope membrane

(a)

(b)

0.5 μm

FIGURE 4.22   Mitochondrion (a) Diagrammatic view of a mitochondrion. (b) Transmission electron microscope view of a mitochondrion. The mitochondrion, like the chloroplast, is bounded by a double-membrane envelope. The inner membrane is infolded to form sheets or tubes of membranes (depending on the organism), termed *cristae*. The nonmembranous center of the mitochondrion is termed the *matrix*.

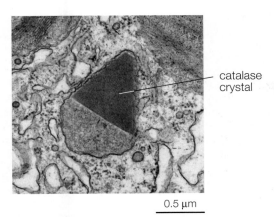

catalase crystal

FIGURE 4.23 Peroxisome Transmission electron microscope view. The large crystal within the peroxisome is composed of the enzyme catalase.

0.5 µm

4.4 | Plant cells have the general features of eukaryotic cells and additional components

Plant cells are particularly complex because they have all of the components generally found in eukaryotic cells (Section 4.3) as well as several additional features. Plant cells are distinguished from most other eukaryotic cells by a **cellulose-rich cell wall** that lies outside the cell membrane; **plasmodesmata**, cytoplasmic channels that penetrate cell walls; a **large, central vacuole** consisting of watery fluid surrounded by a membrane; and one or more photosynthetic organelles, the **plastids** (see Figure 4.7 and Table 4.1). Some green algae share these cell features with plants, revealing a close evolutionary relationship between plants and green algae.

The presence of photosynthetic plastids in plant cells, but not in those of animals or fungi, explains why plants are autotrophic (produce their own food) and animals and fungi, which are heterotrophic, must obtain food from other sources. Plant cell walls limit cell movement, which explains why plants do not actively move in the same way as animals. The sperm of some plants are an exception—they are released from the walls of sperm-producing cells, allowing the sperm to swim.

Cellulose-rich plant cell walls provide support and protection

Stress relief is a major role played by plant cell walls. As noted earlier, the walls of plant cells provide protection from osmotic stress, and they help plants resist attack by disease-causing microbes. Walls also influence cell shape and contribute to plant body strength. Plant cell walls are rich in the polysaccharide cellulose but also contain noncellulose polysaccharides, glycoproteins, and other materials essential to cell-wall function.

Cellulose The major structural component of plant cell walls is long cables of the carbohydrate polymer **cellulose**,

held together by other carbohydrates and protein. Within cellulose cables are tiny fibrils (microfibrils) that are extensively hydrogen-bonded (Chapter 3). This makes cellulose very strong, explaining its function in plant cell walls. Cellulose constitutes at least 15%, and often more than 30%, of the dry mass of plant cells. Because cellulose is so important to plants and plants are so abundant, cellulose is the single most common biological material on Earth. The abundance and strength of cellulose have long made it very useful to people for making paper, cloth, and many other products. The cell membranes of plants, some algae, and certain bacteria contain integral membrane proteins—enzymes known as cellulose synthases—that spin out cellulose chains at the cell surface. Cellulose is made from the raw material glucose, a simple sugar provided by the cytoplasm. During the early evolution of plant cells, the ability to produce a protective cellulose wall may have been obtained from bacteria. Several types of bacteria are able to produce cellulose, and it has been proposed that the ancient protistan ancestors of plants obtained the genes for cellulose synthesis from ingested bacteria.

Noncellulose components of plant cell walls In addition to cellulose, plant cell walls contain other polysaccharides, such as hemicelluloses and pectin, and proteins. **Hemicellulose** is a polymer of glucose and other sugars that is important in binding cellulose fibrils together. Xyloglucan, a polymer containing the sugars xylose and glucose, is characteristic of land plant cell walls, but not walls of closely related green algae (Chapter 21). Thus, xyloglucan may be an adaptation to the terrestrial environment. **Pectin** is another typical plant cell wall polysaccharide. Pectin binds with calcium ions and water to form a jellylike matrix that fills spaces between cellulose fibrils as well as between cells (Figure 4.24). People use the gelling qualities of pectin extracted from plant cell walls in making jelly, candy, and other food products.

Primary and secondary plant cell walls Plant cells that are still growing have only a single wall layer, known as the primary cell wall (Figure 4.25), which retains its ability to stretch as the cell expands. Cell enlargement involves controlled loosening of wall components and uptake of water by the cell vacuole. Increased water pressure provides the force needed to stretch the cell and its wall. After plant cells stop enlarging, a secondary cell wall (Figure 4.25), also composed of cellulose microfibrils and cohesive polysaccharides, is deposited on the inside of the primary wall, whereupon the cell wall becomes unable to extend further.

Both the primary and secondary walls are easily permeable to water and substances dissolved in it. However, some mature plant cells—for example, those of water-conducting tissues, including wood—acquire a partial layer of lignin. Lignin is a phenolic polymer (Chapter 3) that confers rigidity, resistance to microbial attack, and waterproofing to cell walls. The wall portions that have been coated with lignin are no longer permeable to water. Long series of such cells function as water conduits analogous to the pipes of a building.

TABLE 4.1 Cell Components

Structure	Description	Function
1. Structures Generally Found in Eukaryotic Cells		
Cell membrane (plasma membrane)	Lipid bilayer proteins	Regulates passage of materials into and out of cells; receives chemical signals from environment
Nucleus	Membranous nuclear envelope with pores; encloses DNA	Cell's control center; DNA storage
Ribosomes	Protein + RNA	Protein synthesis
Endoplasmic reticulum (ER)	A network of membrane sheets and tubes	Protein modification and transport
Golgi bodies	Stacks of flattened membranous sacs	Protein processing, cell-wall polysaccharide synthesis; materials packaging and export
Mitochondria	Organelle enclosed by a double membrane envelope; inner membrane highly folded to form cristae	Chemical energy transfer
Cytoskeleton	Microfilaments of actin, microtubules of tubulin, intermediate filaments	Cell support and shape, movement of cells and structures within cells
Motor proteins	Myosin, kinesin, dynein	Movement of cells and structures within cells
Peroxisomes	Enzymes enclosed by a membrane	Prevents chemical damage; processes products of photosynthesis
2. Additional Structures Characteristic of Plant Cells		
Cell wall	Cellulose fibrils + pectin + hemicelluloses	Resistance to osmotic stress; structural support and protection
Central vacuole	Membrane-bound, water-filled sac	Water, ion, and pigment storage; supports cell by turgor pressure, cell enlargement
Chloroplasts	Organelle enclosed by a double membrane envelope; chlorophyll bound to internal membranous thylakoids	Photosynthesis
Plasmodesmata	Tubular connections between cytoplasm of adjacent cells	Communication between cells

Lignin-walled, water-conducting cells are discussed more completely in Chapter 9.

Plant cells are connected by plasmodesmata

Adjacent cells of plants and of many algae, even though separated by walls, are still able to communicate with one another. This is accomplished by means of **plasmodesmata,** channels running through cell walls that connect the cytoplasm of neighboring cells (Figure 4.26). Nearly all cells of plants are interconnected by plasmodesmata. Most plasmodesmata are produced when new cell walls are formed at cell division, a process described more completely in Chapter 7. Plasmodesmata can also develop later, when portions of the cell walls of adjacent cells are digested to form channels.

Plant plasmodesmata are not merely open channels, but are lined by the cell membrane and contain a central strand of endoplasmic reticulum. The opening between the endoplasmic reticulum and the cell membrane contains proteins that are thought to regulate the passage of materials from one cell to another. Ions and molecules as large as nucleic acids can be transported between cells via plasmodesmata. Thus, these channels are very important in plant development and in roots. When plants are wounded, plasmodesmata at the damaged site become plugged with polysaccharide cement known as **callose.** This helps prevent the loss of cytoplasm from adjacent cells as well as infection by pathogenic microbes. Plasmodesmata function like building doors that can be opened, closed, locked, or even permanently blocked, controlling access.

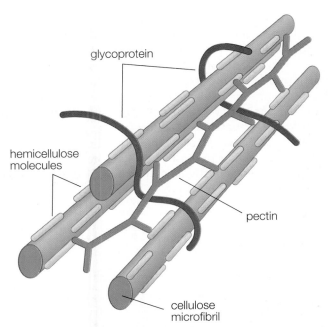

glycoprotein

hemicellulose
molecules

pectin

cellulose
microfibril

FIGURE 4.24 **Plant cell-wall components** Diagram showing the arrangement of cellulose fibrils, gluelike hemicellulose and pectin, and glycoproteins.

Plastids are sites of photosynthesis and other functions

EVOLUTION

Leaves and other green parts of plants contain organelles called **chloroplasts,** which in turn contain **chlorophyll**—the pigment essential to photosynthesis. Plastids are also important as sites of amino acid and fatty acid production. The more general term **plastid** is often used interchangeably with chloroplasts as well as for related organelles that lack chlorophyll. Plant plastids originated when an ancient plastid-less protist ingested a cyanobacterial cell via phagocytosis, but didn't digest it (cyanobacteria are

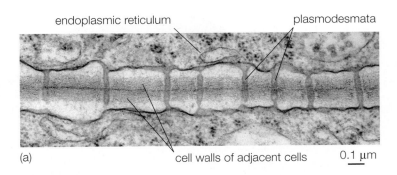

endoplasmic reticulum   plasmodesmata

(a)
cell walls of adjacent cells   0.1 μm

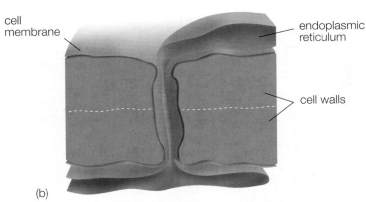

cell membrane

endoplasmic reticulum

cell walls

(b)

FIGURE 4.26 **Plasmodesmata** (a) Transmission electron microscopic view. (b) A diagram showing that endoplasmic reticulum extends through plasmodesmata, from one cell to an adjacent cell.

chlorophyll-containing bacteria that carry out plantlike photosynthesis—see Chapter 18). Rather, the cyanobacterial prisoner was maintained within its host cell as a source of organic compounds produced by photosynthesis. Eventually, the cyanobacterium was transformed into a plastid (Chapter 19). Plants are descended from an ancestral protist that had acquired green plastids in this way.

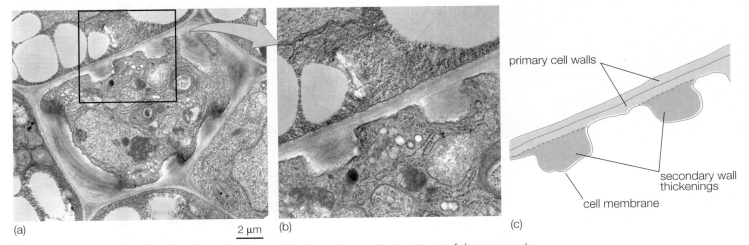

(a)   2 μm (b)

primary cell walls

secondary wall thickenings

cell membrane

(c)

FIGURE 4.25 **Plant cell walls** Shown here is a cell in the water-conducting tissue of the commonly studied plant *Arabidopsis thaliana*. In this type of cell, shown at low magnification in (a), the primary cell wall is deposited around the cell's entire periphery. Secondary cell-wall material is deposited later, but not over the entire surface of the primary wall. The area shown in (b) corresponds to the box in (a). In (c), the primary and secondary walls are shown diagrammatically.

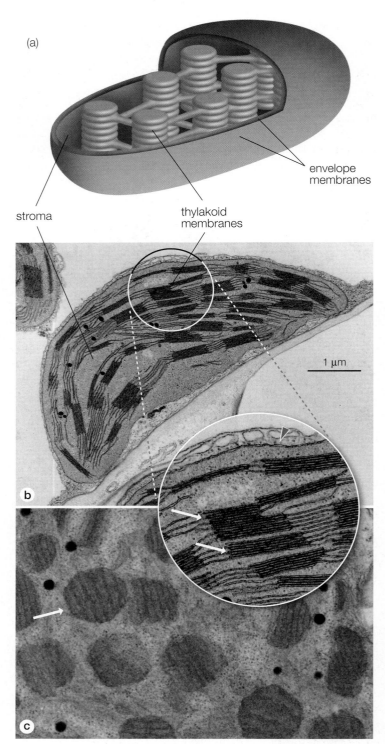

FIGURE 4.27 **Chloroplast** (a) Diagrammatic representation of a chloroplast. (b) Transmission electron microscope view of a chloroplast (from corn) showing the stroma and thylakoid membranes, some of which are stacked like coins. The inset shows the area indicated by the circle at higher magnification. The double-membrane chloroplast envelope is indicated by the arrowhead. Arrows point to two of the thylakoid stacks. (c) In this section, a chloroplast was cut such that the coinlike stacks of membranes are viewed approximately end-on (arrow).

Plant and algal cells contain from one to many plastids per cell. Plant cells often contain 40–50 plastids, enclosed by an envelope composed of two membranes. Pigmented plastids contain many internal membranes known as **thylakoids,** which are arranged in stacks (Figure 4.27). Chlorophyll and other molecules important to photosynthesis are bound to the thylakoids. The portion of the chloroplast that is not occupied by thylakoids is known as the **stroma.** The stroma is also the location of various enzymes required for photosynthesis. DNA and ribosomes similar to those of prokaryotes also occur in the plastid stroma. Plastids divide by pinching in two, in the same way as prokaryotic cells.

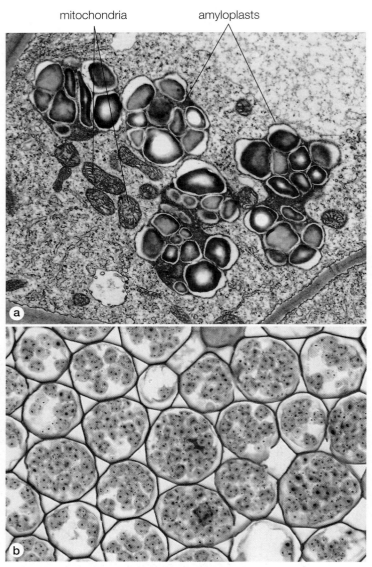

FIGURE 4.28 **Amyloplasts in root cells** (a) Transmission electron microscope view. The amyloplasts contain numerous starch grains but lack the extensive thylakoid membranes found in chloroplasts. (b) Light microscopic view of root tissue with starch grains—stained blue-purple—contained in amyloplasts.

Starch is formed in plastids of plants and green algae

In plants and green algae, the carbohydrate products of photosynthesis are stored in the stroma of plastids as conspicuous starch grains. **Amyloplasts** are nongreen, starch-rich plastids, containing few thylakoids, that occur in roots and other starch-storing tissues of plants (Figure 4.28). This starch can be transformed into sugars for energy production or other purposes when needed, particularly when photosynthesis may not proceed at optimal rates. One example is sugar maple trees, which transform the starch stored in their roots and stems into a sugar solution that is pumped to stem tips and young leaves to fuel growth in early spring, when temperatures may not be warm enough for photosynthesis. By tapping the trees for maple syrup and sugar production, humans intercept some of this flow. Starch is an important component of human nutrition, and several crop plants—including potato, taro, and manioc—are grown specifically for their high starch content.

Chromoplasts are nongreen, pigmented plastids of flowers and fruits

Chromoplasts are plastids that are colored bright orange, yellow, or red by lipid-soluble carotenoid pigments (Figure 4.29). Chromoplasts confer color to ripe fruits, autumn leaves, and flower petals. Plant carotenoids are important to human nutrition; consumed in fruits and vegetables, they generate vitamin A, which is essential for skin and eye health.

Vacuoles play several important roles in plant cells

Why do plants wilt? The cell vacuole holds the answer. A large, central, water-balloon-like **vacuole** (Figure 4.30) is typical of mature plant cells and may occupy 90% of the cell volume. Although nonplant cells may contain vacuoles, these are generally much smaller than those of plant cells. A major role of the plant vacuole is to maintain water pressure, also known as **turgor pressure,** which helps maintain plant structure. If plant cells become so dehydrated that the vacuole loses its water, the plant will wilt. As soon as you detect wilting of your house or garden plants, you should water them to restore vacuolar water pressure. With few exceptions, prolonged dehydration will lead to the death of plants. Loss of vacuole water in the dry conditions of a refrigerator is also what makes celery and lettuce become limp; soaking in cool water may restore their lost crunch.

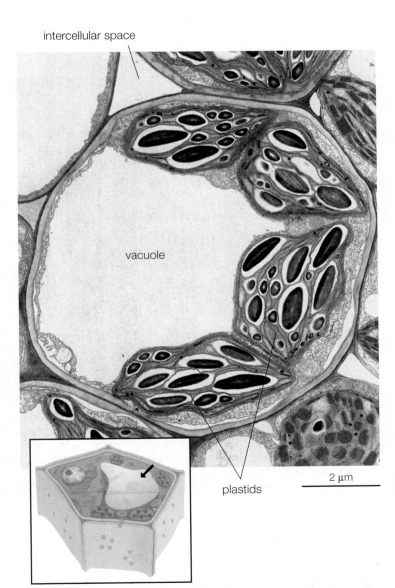

FIGURE 4.30   **Plant cell with large, watery vacuole** This cell is one that lies next to a vein (food- and water-conducting tissue) in a corn (maize) leaf. Chloroplasts and other organelles lie next to the cell wall, with the vacuole occupying much of the cell's center.

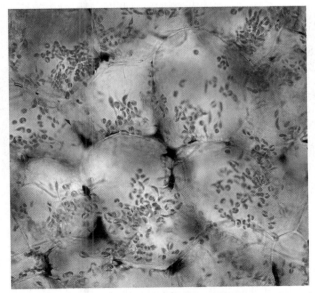

FIGURE 4.29   **Chromoplasts in a pepper** These numerous tiny orange plastids impart the bright color to the pepper fruit.

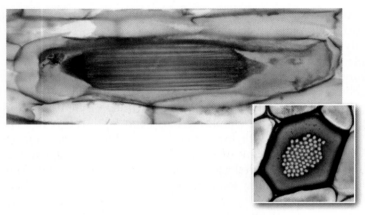

FIGURE 4.31 **Crystals in plant cell vacuole** These crystals are known as raphides, which are long, needlelike, and usually occur in bundles. They are seen here in a section through their longest dimension and across their midpoint (inset).

The plant cell vacuole is enclosed by a single membrane known as the *tonoplast*, which has specific proteins embedded in it that allow accumulation of ions and other substances. Vacuoles can serve as storage compartments for acids in the case of citrus fruits; water-soluble red, purple, or blue pigments in the case of red beets and various flowers; or sugar in the case of sugar beet and sugarcane. Vacuoles can also store compounds used to defend the plant against herbivores, such as bitter-tasting tannins, irritating crystals (Figure 4.31), and heavy metals (see Essay 4.2, "Heavy-Metal Plants").

ESSAY 4.2 HEAVY-METAL PLANTS

A number of plants such as the mustard (*Brassica*) or alpine pennycress (*Thlaspi caerulescens*, Figure E4.2) are able to grow on soils rich in heavy metals that are toxic to most plants. Such soils often occur near abandoned mines or military bases; naturally nickel-rich serpentine soils are common in the tropics and subtropics. Some resistant plants are able to survive by taking up very large amounts of metals such as zinc, cadmium, lead, copper, mercury, selenium, chromium, or nickel, binding the metals with small peptides and then storing them in cell vacuoles where they cause no harm. This phenomenon is known as *metal hyperaccumulation*, and the plants that exhibit it are called *hyperaccumulators*. Hyperaccumulation is an evolutionary response to metal-rich soils of particular geological regions.

Environmental scientists have become interested in using metal-accumulating plants to clean up contaminated soils . . .

Hyperaccumulators import heavy metals by the use of cell-membrane transport proteins similar to those used by all plants to acquire mineral nutrients at roots or store mineral ions in vacuoles. Hyperaccumulators differ from other plants in that some of these transport proteins have been altered so that they can carry heavy metals. Using these altered transport proteins, *Thlaspi caerulescens* can accumulate up to 3% dry weight of zinc without experiencing toxicity. Brake fern (*Pteris vittata*) hyperaccumulates arsenic at levels more than 100 times those found in soils. In some cases, the stored metals are useful to the hyperaccumulators as a defense against herbivores, who reject their metallic taste. Environmental scientists have become interested in using metal-accumulating plants to clean up contaminated soils in

E4.2 *Thlaspi caerulescens*, a hyperaccumulator

which heavy metals are the major form of toxic waste or to treat metal-rich sewage sludge. Metal-rich plants can be harvested for metal extraction or buried in toxic storage areas, where they occupy much less mass than do contaminated soils. Using plants to help clean up wastes is known as *phytoremediation*.

HIGHLIGHTS

4.1 All life-forms are composed of cells having a cell membrane and enclosed cytoplasm that includes ribosomes—the protein-producing structures in cells.

4.2 Microscopes of various types are essential tools of cell biology because cells are typically too small to be seen with the unaided eye. Light microscopes use light and glass lenses to form magnified images. Electron microscopes use electrons and magnetic lenses to form even more greatly magnified images of very thin slides of chemically preserved cells. Small size maximizes cells' ability to take in nutrients and rid themselves of wastes.

4.3 Eukaryotic cells are distinguished by a nucleus, an endomembrane system composed of endoplasmic reticulum (ER) and Golgi bodies, and a protein cytoskeleton that provides support and interacts with motor proteins to produce cell motions. Most eukaryotic cells also possess one or more mitochondria, the power plants of cells. Mitochondria originated from bacterial cells that were ingested by an ancient protistan ancestor of modern eukaryotes.

4.4 Plant cells are distinguished by cellulose-rich cell walls, large watery vacuoles, and plastids. Chloroplasts contain chlorophyll bound to thylakoids and are the sites of photosynthesis and starch storage. Amyloplasts are plastids that lack chlorophyll but rather function as starch storage organelles. Chromoplasts are plastids that confer color to fruits and flowers. Plastids originated from photosynthetic cyanobacteria that were ingested by an ancient protistan ancestor of plants.

REVIEW QUESTIONS

1. Describe several ways in which prokaryotic cells are similar to eukaryotic cells, and several ways in which they differ.

2. What is an integral membrane protein? Give examples of integral membrane proteins found in the plasma membranes of eukaryotic cells.

3. Explain the concepts of hypotonicity, isotonicity, and hypertonicity. Which type of environment is most likely to cause cell plasmolysis, and why? Describe some ways that animal and plant cells have evolved to cope with hypotonic and hypertonic environments.

4. Distinguish endocytosis from exocytosis, and provide examples of each in the plant or protistan worlds.

5. The nucleus is a vital feature of eukaryotic cells. Discuss the roles of the following structures in the nucleus: (a) nuclear envelope, (b) nucleoplasm, (c) nuclear pores, (d) nucleoli, and (e) chromosomes.

6. What is the endomembrane system of a eukaryotic cell? What are its two major components, and what are the functions of those two components?

7. Briefly distinguish or describe the following types of microscopes: (a) light microscope, (b) compound light microscope, (c) stereo dissecting microscope, (d) fluorescence microscope, (e) confocal laser scanning microscope, (f) transmission electron microscope, and (g) scanning electron microscope.

8. What are halophytes? How have they adapted to growing in a high-salt environment?

9. What is the cytoskeleton, what are its major components, and what motor proteins associate with them? Give an example of a plant activity involving each type of motor protein.

10. What are the functions of the following chemical and structural components of the plant cell wall: (a) cellulose, (b) hemicellulose, (c) pectin, (d) plasmodesmata, (e) callose, and (f) lignin?

11. Describe the plastids that occur in plant cells, including their typical colors and functions.

APPLYING CONCEPTS

1. For each of the following scenarios, discuss whether it would be most appropriate to use a light microscope (LM), scanning electron microscope (SEM), or a transmission electron microscope (TEM) based on the goals provided: (a) You wish to observe the types and placement of cells and tissues in the stem of a new plant species you have just discovered. (b) You wish to observe the process of cell division from start to finish in an algal cell. (c) You wish to stain the cytoskeletal fibers in a cell and examine the entire network all at once. (d) You wish to observe fine details of the mitochondria as well as the motor proteins on the cytoplasmic microtubules. (e) You wish to observe the tiny scales covering the surface of a unicellular alga. (f) You wish to observe the individual atoms of hexokinase protein, an enzyme associated with glucose metabolism.

2. Is the primary cell wall laid down inside or outside the secondary cell wall? Why is the secondary wall not produced until the cell has stopped elongating? Where is the cell membrane relative to these walls?

3. Speculate on the cellular consequences to an animal or plant that possesses a defective motor protein. In answering this, suggest specific cell processes that would be altered if (a) kinesin, (b) dynein, or (c) myosin were nonfunctional. Would you expect all three of these to be lethal mutations?

4. If life were to be discovered on Mars or Europa, why do you think it is unlikely to consist of large unicells the size and shape of a beach ball?

5 | Photosynthesis and Respiration

- Life on Earth is solar powered.

- Photosynthesis is the process by which cells trap the energy in sunlight as chemical bonds in simple sugars.

- Plant cells and those of other photosynthetic organisms construct the complex carbohydrates, lipids, nucleic acids, and proteins of life from simple sugars.

- Respiration is a series of chemical reactions that release energy from chemical bonds in the organic molecules which were produced by photosynthesis.

- When oxygen is present, most cells break down simple sugars completely into carbon dioxide and water.

- If oxygen is absent, cells such as yeasts can harvest some of the energy in simple sugars by converting them to alcohol; this process is fermentation.

- Fermentation is the basis of bread making, cheese production, and brewing.

id you realize that your morning cereal and toast, your lunchtime sandwich, and your dinner rolls all derive from a grass plant—wheat? Wheat is the most widely grown agricultural crop in the world; it provides a major percentage of the food consumed by the global human population. Wheat—like rice and corn—is a grass that bears small, inconspicuous flowers that give rise to hard, dry fruits called *grains*. The grains can be stored easily or ground into flours to make breads and breakfast cereals.

Domesticated wheat had its origins in the Near East more than 9,000 years ago. When wild wheat formed a hybrid with another type of wild grass, known as goat grass, it gave rise to a new form of wheat called *emmer*. Emmer wheat *(Triticum turgidum)* yields semolina flour, from which pasta is made. Emmer wheat formed yet another hybrid with goat grass and gave rise to bread wheat *(Triticum aestivum)* around 8,000 years ago. Compared to other species of wheat, bread wheat has a higher protein content, particularly a protein called *gluten*. If bread-wheat flour is mixed with water, the gluten allows the dough to stretch. When a rising agent such as yeast or baking soda is added to bread dough, carbon dioxide gas is produced and the bubbles stretch the dough, forming leavened bread. Bread often represents more than half of all the daily food supply for many peoples.

How is it that a single species of plant can provide so much of the daily human food supply? Wheat is able to carry out this vital function because of two important processes in its metabolism. This chapter examines those two important metabolic processes—photosynthesis and respiration. **Photosynthesis** is the process by which wheat and other photosynthetic organisms capture solar energy and use it to produce carbohydrates. **Respiration** is the process by which organisms release energy from carbohydrates and other organic molecules. That energy is then used to carry out the multitude of processes that make up metabolism. Metabolism is the sum of all the chemical reactions that take place within an organism. Because photosynthesis and respiration are both complex biochemical processes, this chapter begins with an overview of both processes and of the chemical reactions that occur in them. ■

5.1 Photosynthesis and respiration are the processes by which living organisms capture solar energy and release it to sustain life on Earth

As a planet, the Earth is surrounded by a thin layer of living organisms present in soils and on the surfaces of rocks, in the air that forms the atmosphere, and in the waters of lakes and oceans. This layer of life is called the *biosphere*. Life is an energy-intensive process that demands a constant input of energy. Photosynthesis is the main route by which that energy enters the biosphere of Earth. Plants and other photosynthetic organisms use photosynthesis to assemble high-energy organic compounds like carbohydrates from the low-energy inorganic compounds carbon dioxide and water, using the energy in solar radiation. The process can be summarized in an equation:

$$6CO_2 + 12H_2O \rightarrow C_6H_{12}O_6 + 6O_2 + 6H_2O$$

In this equation carbon dioxide (CO_2) and water (H_2O) are the reactants or substrates, and oxygen (O_2), water, and simple sugars ($C_6H_{12}O_6$) are the products (Chapter 3). The energy of solar radiation powers the reaction and is captured in the covalent bonds of simple sugars. The equation describes what happens but does not explain how it happens. Virtually all organisms on Earth are directly or indirectly dependent on this process for their energy.

The capture of solar energy in the chemical bonds of simple sugars is only part of the story. To sustain and power life on Earth, that captured energy has to be released and used in cellular metabolism. Respiration is the process by which most organisms release the energy in the covalent bonds of simple sugars. As in photosynthesis, the process of respiration can be summarized in an equation:

$$C_6H_{12}O_6 + 6O_2 \rightarrow 6CO_2 + 6H_2O$$

Again, the equation indicates what happens but not how it happens. If you burn a cube of sugar, you release carbon dioxide and water, but all the energy is lost as heat. In respiration, however, a significant part of the released energy is transferred to molecules of ATP (Chapter 3). In the form of ATP, the energy currency of the cell, the energy originally captured from solar radiation is available to carry out the many biochemical reactions that make up cellular metabolism. Figure 5.1 summarizes the overall process of energy flow from solar energy through photosynthesis and respiration to the production of ATP to power cellular metabolism. Simple sugars may be made into disaccharides or polysaccharides, or converted into any of the other primary or secondary compounds found in plants. Before we consider the details of how photosynthesis and respiration capture and release the energy that sustains life on Earth, we will present some of the chemical reactions that occur in them and how those reactions are organized.

5.2 Metabolism includes many kinds of chemical reactions organized into series called pathways

In every cell of your body, and in every plant cell, hundreds of chemical reactions are taking place at any single moment. The **metabolism** of a cell is the sum of the chemical reactions and processes occurring in that cell. The chemical reactions that make up cellular metabolism in general, and photosynthesis

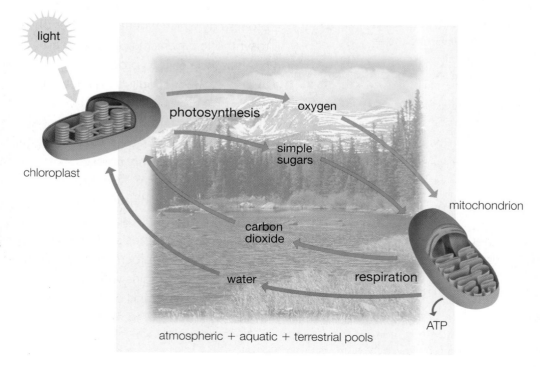

FIGURE 5.1 The path of energy flow on Earth Photosynthesis traps solar energy in the form of simple sugars and releases oxygen. Respiration breaks down those simple sugars in the presence of oxygen to produce the ATP that drives cellular metabolism. The products of respiration—carbon dioxide and water—enter global pools in the oceans and atmosphere where they may be reused by photosynthesis.

and respiration in particular, are directed by biological catalysts called *enzymes*. Enzymes are large protein molecules, and, being catalysts, speed up the rate of a chemical reaction without being used up in the process. The individual chemical reactions in metabolism may be of various kinds. Some enzyme-mediated reactions may require an input of energy in order to proceed; other reactions may release energy. Many of the chemical reactions of cellular metabolism involve the movement of an electron from one atom to another. Such reactions are called *oxidation–reduction reactions*, and they play a large role in photosynthesis and respiration. Chemical reactions in cellular metabolism do not occur in isolation but are linked together in series. Such linked series of reactions are called *pathways*.

Chemical reactions may be exergonic or endergonic

Chemical reactions in metabolic pathways can be grouped according to whether or not they release or require energy. If the potential energy of the substrates is greater than that of the products, energy is released and the process is said to be **exergonic** (energy-out). A reaction that is exergonic can occur spontaneously without an input of energy from outside the system. Respiration is an exergonic process. Conversely, if the potential energy of the substrates is less than that of the products, the reaction is **endergonic** (energy-in) and requires an input of energy from outside the reaction for the reaction to proceed. Many of the chemical reactions in photosynthesis and other biosynthetic reactions are endergonic processes. Where does a cell obtain the energy to drive endergonic reactions?

Cells often drive endergonic reactions by coupling them to exergonic reactions that provide a surplus of energy. Because of this surplus of energy, the entire coupled reaction is exergonic and proceeds spontaneously. The energy, however, does not pass directly from the exergonic to the endergonic reaction. In most coupled reactions, **adenosine triphosphate** or **ATP** (Chapter 3) carries energy from the exergonic to the endergonic reaction (Figure 5.2).

ATP is the main energy carrier of cells. It has three phosphate groups (PO_4), and the covalent bonds that link the last two phosphates contain appreciable potential energy. If the third phosphate bond is broken, energy is released, and ATP becomes adenosine diphosphate (ADP). If the second phosphate bond is also broken, a comparable amount of energy is released, and ADP becomes adenosine monophosphate (AMP). The energy released by an exergonic reaction can be captured in one or more phosphate bonds of ATP. This energy can then be transferred to an endergonic reaction, which will then proceed spontaneously. In most reactions, the terminal phosphate group is not just removed but is transferred with its energy to another molecule. The molecule receiving the phosphate group is phosphorylated and energized to take part in another reaction.

Oxidation–reduction reactions are highly important in cell metabolism

In many chemical reactions, electrons (often symbolized as e^-) may shift from one energy level to another and move from one

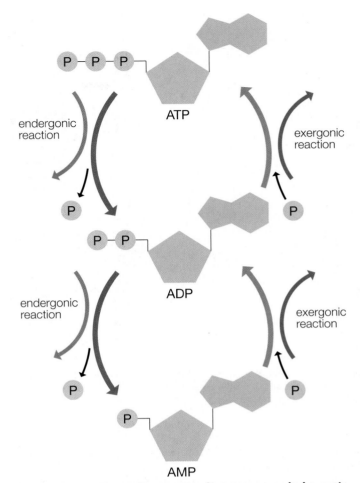

FIGURE 5.2 ATP transfers energy between coupled reactions
In cellular metabolism, endergonic reactions (blue arrows) are coupled to exergonic reactions (red arrows) that provide a surplus of energy. The net reaction is therefore exergonic and proceeds spontaneously. ATP and ADP transfer energy between these coupled reactions through the energy in the covalent bonds of their phosphate groups.

atom to another. Such reactions are called **oxidation–reduction reactions**. When an atom or molecule loses an electron, it is oxidized. A **reduction** is a gain of an electron. Oxidation and reduction reactions occur in pairs; an electron lost by an oxidized atom moves to another atom that is reduced (Figure 5.3a). In the metabolic reactions of living cells, the electron often moves to another atom accompanied by a proton (H^+) so that a hydrogen atom is transferred together with its potential energy ($1e^- + 1H^+ = 1$ hydrogen atom). In respiration, for example, hydrogen is transferred from the simple sugar glucose to oxygen. The oxygen atoms in molecular oxygen (O_2) are reduced to water while the carbon atoms in glucose are oxidized to carbon dioxide (Figure 5.3b). The energy level of the electrons is lowered during the transfer, and energy is released. Oxidation of glucose is exergonic. In contrast, during photosynthesis electrons and protons are transferred from water (which is oxidized to oxygen) to carbon dioxide, which is reduced first to a simple 3-carbon sugar and then to glucose. The electrons are moved to a higher energy level, and an input

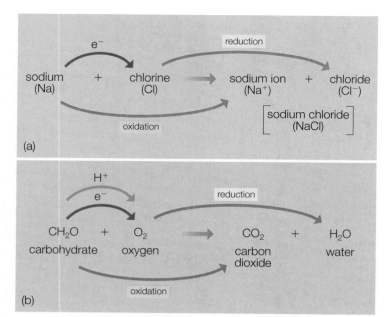

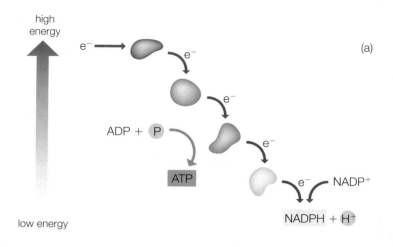

FIGURE 5.3 Oxidation and reduction reactions (a) Oxidation and reduction in common table salt, sodium chloride (NaCl). Oxidation and reduction reactions always occur together. When one atom (in this case, sodium) loses an electron, it is oxidized. At the same time, another atom (here, chlorine) must gain an electron and be reduced. (b) In many oxidation–reduction reactions, a hydrogen ion or proton (H^+) is transferred along with the movement of electrons. In the equation for respiration, a simple sugar is oxidized to carbon dioxide, and molecular oxygen is reduced to form water.

of energy from solar radiation is required. The reduction of carbon dioxide to sugar is an endergonic process.

Electrons do not move directly from glucose to oxygen during respiration, nor do they move directly from water to carbon dioxide during photosynthesis. The electrons and their accompanying protons instead transfer to electron acceptor molecules together with some of their energy. Two very important electron acceptor molecules are **nicotinamide adenine dinucleotide** (NAD^+) and **nicotinamide adenine dinucleotide phosphate** ($NADP^+$). $NADP^+$ is NAD^+ with a phosphate group (PO_4) added. $NADP^+$ plays a major role in photosynthesis, and NAD^+ has an important role in respiration.

Electron acceptor molecules may be grouped into a series called an **electron transport chain** (Figure 5.4). Like an assembly line, an electron transport chain receives electrons at one end and transfers them through a series of acceptor molecules until the electron reaches the end of the chain. As each acceptor molecule receives an electron, it is reduced. When it passes that electron on to the next link in the chain, it becomes oxidized. An electron transport chain is thus a series of oxidation–reduction reactions. At each link in the chain, the energy level of the electron is lowered. Part of that energy is dissipated as heat, and part is used to perform metabolic work. Electron transport chains play crucial roles in both photosynthesis and respiration.

In metabolism, chemical reactions are organized into pathways

In cells, chemical reactions are linked together in series like an assembly line. The product of one enzyme-catalyzed reaction

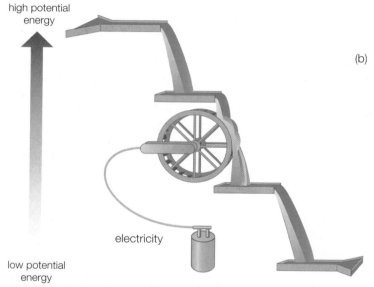

FIGURE 5.4 An electron transport chain and a physical analogy (a) As electrons move down the transport chain, their energy level is lowered in discrete steps by a series of electron acceptor molecules (colored shapes). The electron acceptor molecules go through oxidation–reduction reactions. Part of the released energy is dissipated as heat, and part is recovered as ATP to perform useful work. (b) Similarly, as water flows down a series of chutes, part of its potential energy is dissipated as heat and part is recovered by the spinning of the waterwheel and conversion to electricity.

becomes the substrate for the next enzyme-catalyzed reaction (Figures 5.5 and 5.6a). Such an ordered series of reactions forms a metabolic pathway. All metabolic pathways are not, however, simple linear assembly lines. In some pathways a branching point may occur, from which the product of a reaction may proceed along either of two pathways (Figure 5.6b). Cycles are another type of pathway in which the starting molecule in the cycle is always regenerated in the last step of the cycle (Figure 5.6c). Cycles may take in simple molecules and produce larger, more complex molecules; or they may begin with a large, complex molecule and break it into simpler molecules. Photosynthesis and respiration both include cycles.

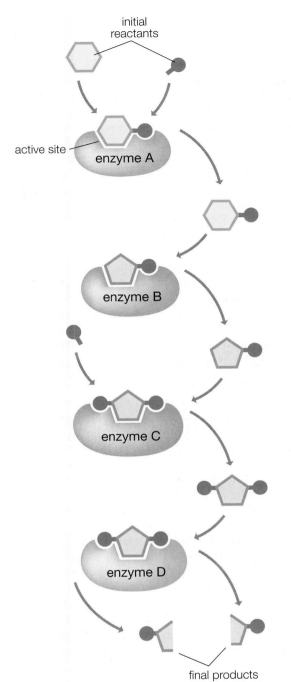

FIGURE 5.5 An ordered series of chemical reactions The product of one reaction is the substrate for the next reaction. In this example, a different enzyme catalyzes each of the steps.

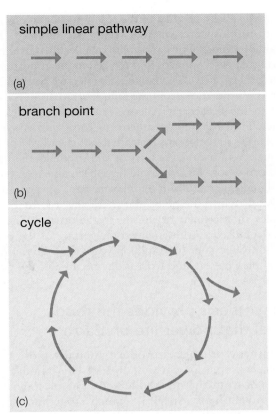

FIGURE 5.6 Three basic types of metabolic pathways (a) A simple linear pathway of chemical reactions. (b) A branch point in an otherwise linear pathway. (c) A cycle pathway. Cycles always regenerate their starting compound.

5.3 | Photosynthesis harvests solar energy to sustain life on Earth

Photosynthesis is the process by which high-energy organic compounds such as carbohydrates are assembled from two low-energy inorganic compounds—carbon dioxide and water—using the energy in solar radiation. Photosynthesis is the main route by which energy enters the biosphere of Earth. Virtually all organisms on Earth are directly or indirectly dependent upon this flow of energy. Photosynthesis not only sustains life on Earth now but it also transformed the Earth in the past such that complex life became possible.

Photosynthesis changed the early Earth so that multicellular life became possible

The composition of the atmosphere of the early Earth was very different from the atmosphere of Earth today. The present-day atmosphere is 78% nitrogen gas (N_2), 21% oxygen (O_2), and about 1% argon, water vapor (H_2O), ozone (O_3), and carbon dioxide (CO_2). The first atmosphere was essentially volcanic in origin, and scientists think it contained a number of gases such as molecular hydrogen (H_2) water vapor, ammonia (NH_3), nitrogen gas, methane (CH_4), and carbon dioxide (CO_2). Methane and carbon dioxide were present at much higher levels than at present. Scientists agree that oxygen was absent. The modern atmosphere supports complex lifeforms, including humans, that could not have existed in the atmosphere of the early Earth. How did this dramatic change in the atmosphere occur?

Photosynthetic microorganisms, swarming in the seas of the early Earth some 3.5 billion years ago, took up the abundant carbon dioxide and used the solar energy from the Sun to reduce that carbon dioxide into simple carbohydrates. Solar energy was stored in the chemical bonds of the carbohydrates, and oxygen

EVOLUTION

was released as a by-product of the reaction. That oxygen drifted up into the atmosphere, where it slowly accumulated. The concentration of atmospheric carbon dioxide declined over eons and that of oxygen rose. Some of that oxygen in the atmosphere combined to form ozone (O_3). Ozone is an effective shield against ultraviolet (UV) radiation from the Sun. UV radiation is extremely harmful to life because its energy can break complex organic molecules. The early photosynthetic microorganisms would have occupied levels in the oceans beneath a sufficient layer of water to screen out UV radiation. As oxygen and ozone levels rose, however, these microorganisms would have been able to enter surface waters. Finally, some 600 million years ago, oxygen and ozone had accumulated to the point that there was a virtual explosion of complex multicellular organisms in the oceans (Chapter 19). Complex multicellular life invaded the surface of the land some 150 million years later (450 million years ago). Humans and all other complex life on land exist today because of the process of photosynthesis.

Photosynthesis provides the food and fuel that power life on Earth

Photosynthetic organisms produce annually more than 250 billion metric tons of carbohydrates. This annual production derives from the metabolic activities of some 500,000 photosynthetic species, which together support— either directly or indirectly—the remaining 3 million known species of organisms on Earth. Cereal crops such as wheat, rice, corn, and barley provide through their photosynthesis about 60% of all human annual food consumption.

ECOLOGY

Beyond the provision of food, photosynthesis also supplies a wide range of products that affect every aspect of our lives. Many textiles, such as cotton and linen, are derived from plant fibers. Wood, of course, is widely used for construction, furnishings, and the manufacture of paper and paper products. The United States uses around 3 billion cubic feet of wood per year to make paper. The products of photosynthesis also include spices, perfumes, drugs and cosmetics, waxes, oils, and rubber. Photosynthesis provides the bulk of the fuels used by humans. Most human populations still depend on wood for cooking and for heating their homes. All the fossil fuels, such as coal, oil, and natural gas, used to power the modern world represent solar energy trapped by photosynthetic organisms in

the past. Thus, the modern world depends on past photosynthesis for its power, but fossil fuels are in limited supply and nonrenewable.

When fossil fuels are burned for the generation of power, their carbon enters the atmosphere as carbon dioxide. Since the Industrial Revolution, the atmospheric content of carbon dioxide has risen from about 0.028% to the present 0.037%. Because carbon dioxide in the atmosphere traps solar heat, global temperatures have been rising along with the increase in carbon dioxide. If all the carbon dioxide from the burning of fossil fuels had remained in the atmosphere, global temperatures would be higher than they are. A significant part of that carbon dioxide has been removed by increased photosynthesis and stored in oceans, soils, and peatlands. Photosynthesis may be buffering the atmosphere from the full impact of human consumption.

The modern world runs on the past and present products of photosynthesis. How does the process of photosynthesis, which is so vital to the world as we know it, capture solar energy in the chemical bonds of carbohydrates?

The interaction between light and pigments is crucial to the capture of solar energy

During photosynthesis, the energy that is captured is that of visible light—light we can see with our own eyes. As Isaac Newton showed more than 300 years ago, visible light is made up of a spectrum of colors ranging from violet at one end to red at the opposite end. In the 19th century, British physicist James Clerk Maxwell demonstrated that visible light was a small part of a much larger spectrum of radiation called the *electromagnetic spectrum* (Figure 5.7). The radiation within the electromagnetic spectrum has the properties of both particles (called photons) and waves, like the waves on a pond surface. All radiation has associated with it a characteristic wavelength, which can be thought of as the distance between the crest of one wave and the crest of the next wave on a pond. The wavelengths of gamma rays, X-rays, UV radiation, and visible light are all very short and measured in nanometers (nm), or billionths of a meter. Visible light ranges from about 400 nm (violet) to 700 nm (red). At the opposite end of the spectrum, the infrared rays of heat have wavelengths in the micrometer (μm) to millimeter (mm) range, and radio waves have wavelengths measured in meters.

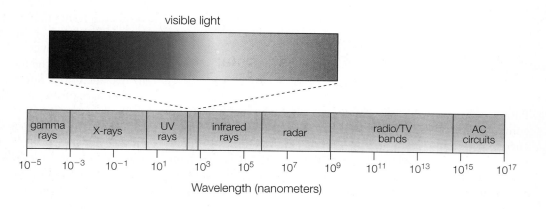

FIGURE 5.7 The electromagnetic spectrum Visible light is only a small portion of the electromagnetic spectrum; its wavelength is situated between UV rays and infrared rays (heat).

The longer the wavelength, the lower the energy carried by that photon or wave of energy. Within the range of visible light, violet has the highest energy and the shortest wavelength, whereas red has the longest wavelength and the lowest energy. Radiation of wavelengths shorter than violet light, such as ultraviolet (UV) and X-rays, has sufficient energy that it can break chemical bonds and damage complex organic molecules. Radiation with wavelengths in the infrared is absorbed by water and increases the motion or heat of organic molecules.

Photosynthesis uses visible light because visible light has the right amount of energy to excite electrons in organic molecules. The lowest energy state of an electron is called its *ground state* (Figure 5.8a). When an electron absorbs energy from visible light, it jumps to a higher orbital position around the nucleus of its atom (Figure 5.8b) and is said to be in an **excited state**. When an electron is excited, there are three possible routes that energy can follow: (1) The electron may return directly to a lower energy state, and the energy will be emitted as heat or light (Figure 5.8c); (2) the electron returns to a lower energy level, but the energy is passed to another molecule, causing it to become excited (Figure 5.8d); or (3) the excited electron may itself be transferred to another molecule called an *electron acceptor molecule* (Figure 5.8e). When an organic molecule absorbs visible light, electrons move into higher energy levels. What kinds of molecules absorb light?

A pigment is any substance that absorbs light. Most pigments absorb only certain wavelengths of visible light and reflect those they do not absorb. **Chlorophyll *a*** is the most abundant light-absorbing pigment in cyanobacteria, algae, and in the chloroplasts of plants (Figure 5.9). Chlorophyll *a* reflects green light, which is why plant leaves usually look green. The pattern of light absorption by a pigment is called its **absorption spectrum**. The absorption spectrum of chlorophyll *a* shows that it mainly absorbs in the red and the blue-violet regions of the visible-light spectrum (Figure 5.10). When chlorophyll *a* absorbs visible light, an electron is excited. That excited electron may leave the chlorophyll molecule and transfer to an electron acceptor molecule.

Chlorophyll *b* and carotenoids are examples of **accessory pigments**. Because accessory pigments have absorption spectra different from chlorophyll *a*, they increase the range of visible light available for photosynthesis (see Figure 5.10). They transfer that absorbed energy to chlorophyll *a*. Both chlorophyll *a* and accessory pigments absorb energy from visible light to power photosynthesis.

Photosynthesis occurs in the chloroplasts of algae and plants

In plants and algae, the process of photosynthesis occurs within organelles called *chloroplasts* (Figure 5.11). The interior of chloroplasts is crisscrossed by a network of membranes in the form of thin, flat, platelike vesicles called **thylakoids**. The thylakoids are embedded in a viscous fluid called the **stroma**. The thylakoids and the stroma are the sites of different parts of the process of photosynthesis. In the thylakoids, the energy in visible light from the Sun is converted

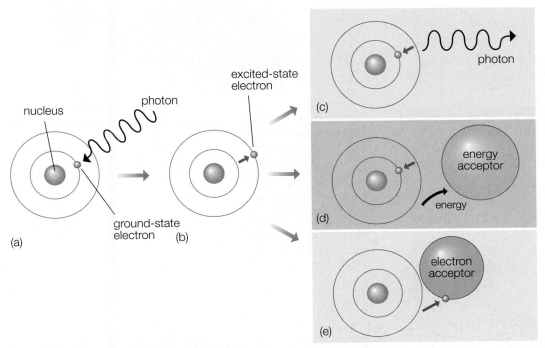

FIGURE 5.8 The excitation of an electron by a photon of visible light (a) The electron is in a ground state, but a photon of visible light is about to be absorbed. (b) The excited electron jumps to a higher orbital position around the nucleus of its atom. The absorbed energy can follow one of three routes: (c) return to the ground state and emit the energy as a photon of light, (d) return to a ground state but transfer the energy to excite an electron in another atom, or (e) transfer the excited electron to another atom in an acceptor molecule.

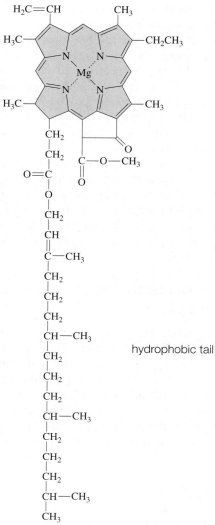

FIGURE 5.9 The structure of chlorophyll *a* The colored region absorbs light energy from the Sun and the hydrophobic tail anchors the molecule in the thylakoid membrane.

hydrophobic tail

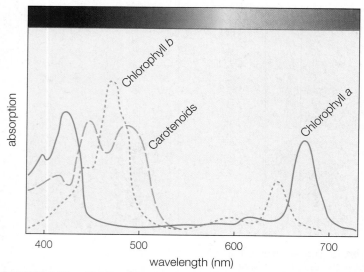

FIGURE 5.10 Absorption spectrum of chlorophyll *a*, chlorophyll *b*, and carotenoids Chlorophyll *b* and carotenoids, which can pass absorbed light energy on to chlorophyll *a*, extend the range of wavelengths that are usable for photosynthesis.

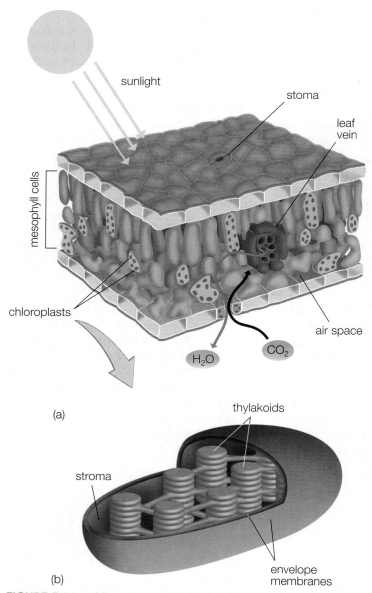

FIGURE 5.11 Chloroplasts within leaf cells are the primary sites of photosynthesis (a) If a cross section of a leaf is examined, the photosynthetic mesophyll cells occupy the leaf center, bound on the upper and lower surfaces by a layer of cells that typically lack chloroplasts. The three main ingredients necessary for photosynthesis—light, water, and carbon dioxide—are labeled. Veins conduct food and water into and out of the leaf (food-conducting tissue is shown in blue, and water-conducting in red). Stomata (singular, stoma) are more common on the underside of the leaf and provide a path for carbon dioxide to enter the leaf (and water to leave). The photosynthetic cells are surrounded by air space, which allows for gas exchange. (b) Chloroplast diagram showing thylakoid membranes and stroma.

into chemical energy in the form of adenosine triphosphate (ATP) and the electron acceptor molecule nicotinamide adenine dinucleotide phosphate (NADPH) in its reduced form. In the stroma, that energy is used to reduce carbon dioxide to a simple sugar.

In plants, the chloroplasts are primarily located in special cells called **mesophyll cells,** located inside leaves (see Figure 5.11).

Water, carbon dioxide, and sunlight are all brought together in the mesophyll cells of leaves for photosynthesis to proceed. Leaves typically have large surface areas to intercept light. Water passes up from the roots through the vascular system to the veins in the leaf, from which it diffuses out to the mesophyll cells. Carbon dioxide enters the interior of the leaf through the **stomata** (singular, stoma), the tiny pores on leaves through which gases are exchanged with the outside environment. Carbon dioxide moves through the open spaces among the mesophyll cells and dissolves in the water on their outer surfaces. It then diffuses into the mesophyll cells and enters the chloroplasts.

Photosynthesis converts light energy into chemical energy stored in sugars

Solar panels on the roof of a house convert solar energy directly into usable electricity. Leaves on plants convert solar energy into usable chemical energy in the form of chemical bonds of organic molecules by the process of photosynthesis. In photosynthesis, carbon dioxide (CO_2) and water (H_2O) are the substrates or reactants, and oxygen (O_2), water and simple sugars ($C_6H_{12}O_6$) are the products. As presented earlier, the overall process can be summarized by the equation

$$6CO_2 + 12H_2O \rightarrow C_6H_{12}O_6 + 6O_2 + 6H_2O$$

This equation indicates what happens in photosynthesis but does not explain how the process occurs. The process actually consists of two groups of reactions: light reactions and carbon-fixation reactions.

The light reactions capture solar energy   The light reactions occur in the chloroplast thylakoids. When chlorophyll *a* and accessory pigments absorb visible-light energy, electrons in these molecules become excited or energized. That energy in excited electrons is used in one of three processes. (1) Some of the energy in excited electrons is used to split water molecules. The oxygen from those split water molecules is released as molecular oxygen (O_2), most of which enters the atmosphere. (2) Some of that energy is captured in the energy carrier adenosine triphosphate (ATP). (3) The electrons (e^-) and protons (H^+) from water combine with the electron acceptor nicotinamide adenine dinucleotide phosphate ($NADP^+$) to form NADPH, which is an energy carrier for those excited electrons. In summary, the light reactions use solar energy to split water to release electrons (e^-), protons (H^+), and oxygen. The electrons and protons are used to make ATP and NADPH. As an aid in understanding the light reactions of photosynthesis, refer to the mechanical analogy presented in Figure 5.12.

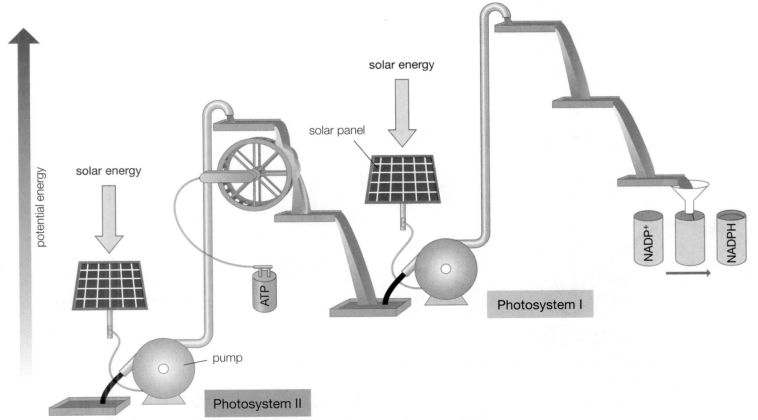

FIGURE 5.12   Mechanical analogy to the light reactions in the chloroplast thylakoids The two light reactions are represented as water pumps. Solar energy strikes arrays of solar cells and powers the pumps. Water is treated as analogous to electrons. The pumps raise water to higher levels of potential energy. As the water flows down the first series of chutes and waterwheel, part of that potential energy charges up a battery to generate ATP. Water flowing down the second series of chutes has its potential energy trapped in containers labeled NADPH.

How do the light reactions actually carry out these three processes? Two discrete groupings of proteins and pigment molecules are involved, photosystem I and photosystem II (named I and II in the order of their discovery). Both photosystem types are embedded in the thylakoid membranes. Each contains two components: an **antenna complex** consisting of accessory pigment molecules, which collect light energy and pass it to the second component, the **reaction center**. Each reaction center contains two special molecules of chlorophyll *a*, which are the only ones that can pass an electron to an electron acceptor molecule and therefore have electrons removed. An electron transport chain links together the two photosystems. The system ends with a final electron transport chain, which reduces $NADP^+$ to NADPH.

The process begins in photosystem II when an accessory pigment in the antenna complex absorbs a photon of light, transferring the absorbed energy from molecule to molecule until it is passed to one of the special chlorophyll *a* molecules called P_{680} in the reaction center (Figure 5.13). The P in P_{680} stands for pigment and the number indicates the wavelength of light for best absorption. When a molecule of P_{680} is excited, the energized electron leaves the P_{680} chlorophyll *a* molecule, joins with a primary acceptor molecule, and enters the first electron transport chain. The electron lost from the chlorophyll *a* molecule is replaced by splitting a water molecule into electrons (e^-), protons (H^+), and oxygen, a process termed *photolysis* (from *photo*, "light," and *lysis*, "breaking"). The excited electron passes down the first electron transport chain. Part of its energy is dissipated as heat, but a significant part is captured in molecules of ATP. At the end of the first electron transport chain, the electron transfers to the special chlorophyll *a* molecule P_{700} in the reaction center of photosystem I. There the electron replaces one that was previously excited to a higher energy state (by light energy absorption) and entered the

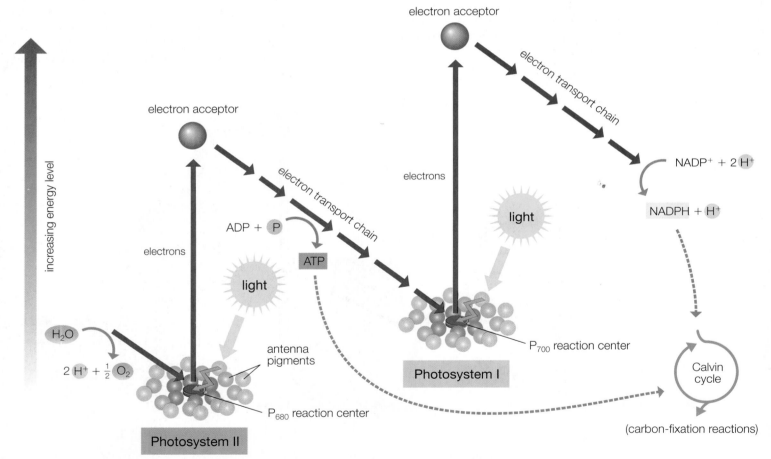

FIGURE 5.13 Schematic diagram of the energy levels of electrons during the light reactions
Light energy absorbed by pigment molecules in the antenna complex is transferred to the special chlorophyll a molecules in the reaction center of photosystem II, thereby exciting electrons to a higher energy level. The excited electrons are replaced by electrons that are generated when water is split in photolysis. The splitting of water also yields protons and oxygen. The excited electrons are passed to the first molecule in a series of electron receptors that make up the first electron transport chain, with some of the energy being used to form ATP from ADP and a phosphate group. The electrons are passed along to the reaction center of photosystem I, where light once again is used to excite them. The reexcited electrons are passed along a second electron transport chain, finally being accepted by $NADP^+$ to form NADPH and protons. Both the ATP and NADPH formed in the light reactions can be used in the carbon-fixation reactions. This zigzag pattern of electron energy levels is called the Z scheme.

second electron transport chain. At the end of the second electron transport chain, this electron joins with a proton (H^+) and $NADP^+$ to form the energy carrier NADPH.

Figure 5.13 is a schematic diagram (called the Z scheme) of the energy levels of the electrons as they flow through photosystem II and photosystem I of the light reactions. The physical location of photosystem II, photosystem I, and their associated electron transport chains are the thylakoids of the chloroplasts (Figure 5.14). The photolysis of water occurs on the inside of the thylakoid in the space called the *thylakoid lumen*, near photosystem II. Photolysis causes a buildup of protons (H^+) inside the lumen. The electrons freed by splitting water replace those leaving the P_{680} chlorophyll *a* molecules. The electrons that have been excited to leave P_{680} in the reaction center enter the first electron transport chain at a primary electron acceptor then pass through a series of molecules called *plastoquinones* (indicated as PQ in Figure 5.14), which are soluble in lipids and move about freely within the thylakoid membrane. The electrons then pass through a cytochrome complex before arriving

finally at a copper-containing protein called *plastocyanin*. The cytochrome complex contains several different cytochrome molecules, each of which contains a ring structure and an iron atom that functions as an electron carrier. From plastocyanin, the electrons enter photosystem I, where they replace those energized in the P_{700} molecules in the reaction center. Energized electrons in the P_{700} chlorophyll *a* molecules pass into the second electron transport chain at another primary electron acceptor. Electrons then flow through a series of iron-sulfur proteins (indicated by Fd in Figure 5.14) until they reach an enzyme complex ($NADP^+$ reductase), where they are combined with protons and $NADP^+$ to form NADPH.

In summary, electrons derived from the splitting of water flow through the reaction center of photosystem II, down the first electron transport chain, enter the reaction center of photosystem I, and finally flow down the second electron transport chain, where they are united with $NADP^+$ and protons to form NADPH. This process is referred to as **noncyclic electron flow** because the flow of electrons is in one direction. For every

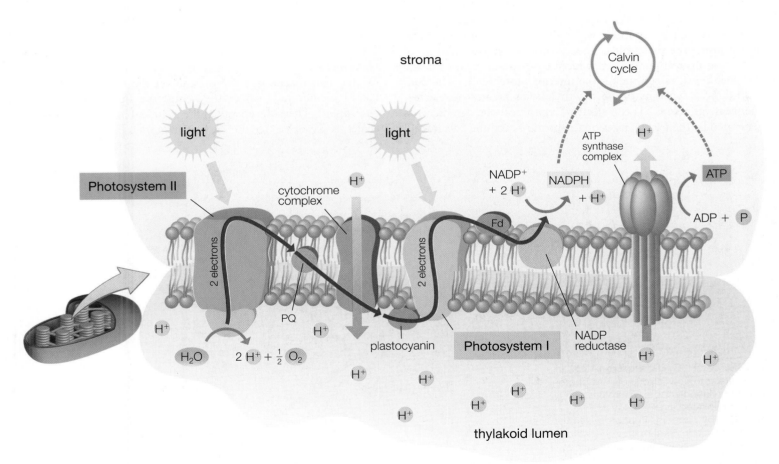

FIGURE 5.14   The location of the molecules of the light reactions in the thylakoid membrane
Electrons move from water through photosystem II to the primary electron acceptor through the first electron transport chain, which consists of plastoquinones (PQ), the cytochrome complex, and plastocyanin, to photosystem I. The electrons replace those energized from the photosystem I reaction center. Energized electrons from photosystem I flow down the second electron transport chain through iron-sulfur proteins (Fd) to end at $NADP^+$ reductase. The splitting of water occurs inside the thylakoid lumen, where protons (H^+) accumulate. Protons are also pumped from the stroma across the thylakoid membrane into the lumen by the flow of electrons down the first electron transport chain. The resulting electrochemical gradient powers the synthesis of ATP as protons flow down through the ATP synthase complex to reenter the stroma.

8 photons absorbed in the reaction centers, 4 electrons move through noncyclic electron flow and yield a total of 3 ATP and 2 NADPH (1 NADPH for every 2 electrons).

Photosystem I is also capable of a **cyclic electron flow** (Figure 5.15). In this process, an excited electron from the reaction center of photosystem I moves down the second electron transport chain for a few steps, then hops over to the electron transport chain between photosystems I and II, where ATP is generated. The electron returns to the reaction center in photosystem I. No water is split and no oxygen released. Cyclic electron flow increases the amount of ATP produced but does not make NADPH. It provides additional ATP to drive the carbon-fixation reactions in the next steps of photosynthesis.

Chemiosmosis and photophosphorylation produce ATP molecules in the chloroplast

Both photosystems are embedded in the thylakoid membranes of the chloroplasts. As excited electrons pass down the electron transport chains, part of their energy is used to pump protons across the thylakoid membranes from the stroma into the space (the lumen) inside the thylakoids. The enzymes that split water into electrons, protons, and oxygen lie on the inside of these thylakoid membranes, and the protons formed by splitting water are released into this same lumen inside the thylakoids. Thus protons from the splitting of water and proton pumping produce a buildup of protons inside the thylakoids, which makes the interior more acidic. The concentration of protons inside the thylakoids creates an electrochemical gradient. The "electro-" part of the gradient is generated by the (+) charges on the protons, and the "chemical"

part arises from the concentration of hydrogen ions (H^+) that make the lumen more acidic than the outside stroma of the chloroplasts. The resulting electrochemical gradient across the thylakoid membranes acts like a battery and represents potential energy. That potential energy can drive the production of ATP if a channel allows the protons to flow back to the stroma.

The **ATP synthase complex** provides that channel (see Figure 5.14). Protons flow back to the stroma through the channel inside the ATP synthase complex by a mechanism called *chemiosmosis* ("chemi" refers to "chemical" and osmosis means a "thrust"). In the same way that electric current flowing along wires from a battery can light a lamp, protons moving down the channel in the ATP synthase complex assemble ATP from ADP and phosphate. The tail of the ATP synthase complex anchors the unit in the thylakoid membrane, and the head contains the ATP synthase enzymes. This type of phosphorylation of ADP to ATP is called **photophosphorylation** because the source of the energy for the phosphorylation comes from the Sun. Current studies indicate 1 ATP is formed for each 4 protons moving down the ATP synthase complex. The ATP synthase complex is the key structure generating ATP from the energy captured from sunlight and makes the subsequent carbon-fixation steps possible.

The carbon-fixation reactions reduce carbon dioxide to simple sugars

The molecules of ATP and NADPH formed by the light reactions are produced on the outside of the thylakoid membranes, where they enter the fluid stroma of the chloroplast. There the energy in the chemical bonds of these molecules is available to the carbon-fixation reactions to reduce

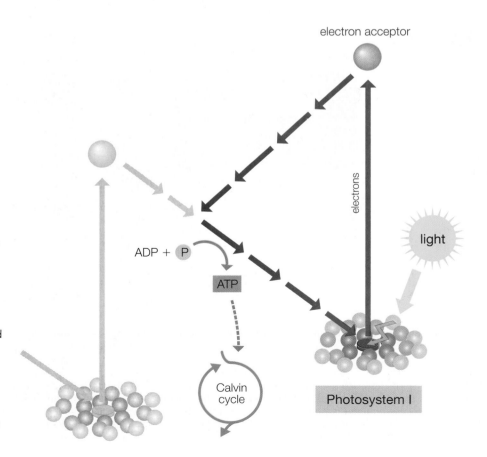

FIGURE 5.15 Cyclic electron flow In cyclic electron flow (also known as photophosphorylation), photosystem II is not involved. Instead, light energy is absorbed by pigments in photosystem I, exciting reaction center electrons, which begin to move down the second electron transport chain (see Figure 5.13). Before they can make their way down the second chain, they are shunted to the first electron transport chain, where some of their energy is used to form ATP, which can be used in the carbon-fixation reactions (the Calvin cycle). Unlike the scheme shown in the previous figure, no NADPH is formed and water is not split. Cyclic electron flow is a means of generating additional ATP.

electron acceptor

electrons

light

ADP + P

ATP

Calvin cycle

Photosystem I

carbon dioxide to simple sugars. In most plants and photosynthetic protists, the series of reactions by which carbon dioxide is reduced is called the *Calvin cycle*, after its discoverer Melvin Calvin, who won the Nobel Prize for this work in 1961. The Calvin cycle is the only known pathway for making carbohydrate from carbon dioxide. The chemical reactions and enzymes of the Calvin cycle are all located in the chloroplast stroma.

The chemical reactions in the Calvin cycle are fairly complex, but the essential points can be summarized briefly. As in all cyclic metabolic pathways, the Calvin cycle begins and ends with the same chemical, a 5-carbon sugar called ribulose 1,5-bisphosphate, or RuBP (Figure 5.16). A single molecule of carbon dioxide is attached or fixed to it through the operation of the enzyme ribulose 1,5-bisphosphate carboxylase/oxygenase, which is commonly called **rubisco**. The final terms indicate that the enzyme rubisco can act as either a carboxylase, and combine with carbon dioxide, or as an oxygenase, and combine with oxygen. When it combines carbon dioxide with RuBP, the resulting 6-carbon sugar is immediately split into two 3-carbon molecules of 3-phosphoglycerate, or PGA. Because the first detectable product of the Calvin cycle is the 3-carbon molecule PGA, this metabolic pathway is called the **C_3 pathway**. The C_3 pathway is the most widespread carbon-fixation pathway, and about 85% of known plant species use it exclusively. The remaining 15% of plants have an additional pathway for assimilating and fixing carbon dioxide. Common

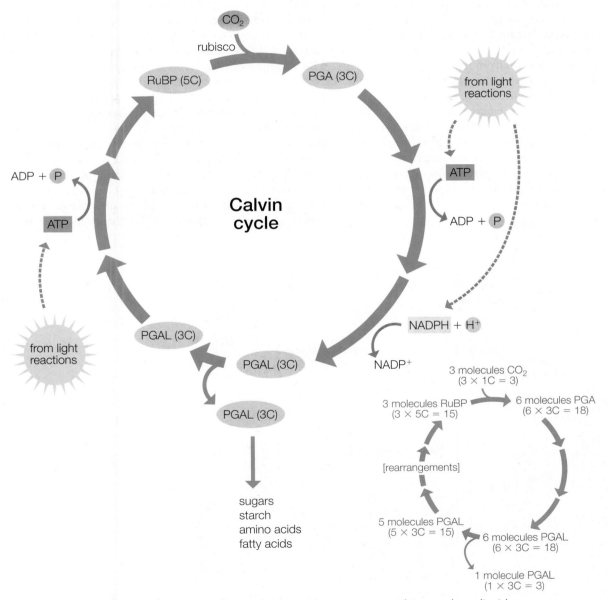

FIGURE 5.16 **Calvin cycle** In the Calvin cycle, the rubisco enzyme combines carbon dioxide with the 5-carbon RuBP to yield two 3-carbon molecules (PGA). ATP and NADPH from the light reactions are used to convert PGA to the 3-carbon molecule PGAL, some of which is used by the cell to make sugars and other carbon-containing compounds. With energy from ATP, the remaining PGAL is converted back to the 5-carbon RuBP, completing the cycle. The inset shows that three turns of the cycle are required to produce one molecule of PGAL (see text).

cereal grasses such as wheat, rye, and oats as well as potatoes and soybeans use the C₃ pathway and are therefore referred to as **C₃ plants.**

Rubisco is the most abundant enzyme in the world and one of the most important. All plants and photosynthetic protists use some form of rubisco to fix carbon dioxide into simple sugars. Without rubisco, carbon dioxide could not be incorporated into sugars, and there would be no photosynthesis to produce food.

Photosynthesis is usually shown as producing the 6-carbon sugar glucose. Actually, however, photosynthesis produces very little glucose. The first product of carbon fixation is the 3-carbon molecule PGA. After a series of chemical reactions in the Calvin cycle, PGA is reduced to the 3-carbon molecule phosphoglyceraldehyde, or PGAL. This reduction reaction is endergonic and requires energy from ATP and the reducing power of NADPH, both products of the light reactions (see Figure 5.16). PGAL is the true product of photosynthesis. Most of the PGAL is rearranged to regenerate RuBP, but some PGAL emerges from the cycle to be made into sugar or starch. For every three turns of the Calvin cycle, one molecule of PGAL emerges from the cycle. After six turns of the cycle, two molecules of PGAL emerge and can be made into one 6-carbon molecule of a simple sugar like glucose or fructose (see Figure 5.16 inset). If the PGAL produced in the Calvin cycle is retained in the chloroplast, it is converted to starch. Starch is the main storage product in plants and green algae. If PGAL molecules are exported from the chloroplast out into the cell, the PGAL is made into sucrose. Sucrose is the principal transport sugar in plants.

To summarize, the carbon-fixation reactions in the chloroplast stroma utilize the ATP and NADPH produced by the light reactions to reduce carbon dioxide to simple 3-carbon molecules of PGAL. The enzyme rubisco carries out the essential carbon-fixation step. PGAL molecules are combined into sucrose and starch for the transport and storage of energy in the form of chemical bonds.

Photorespiration makes the pathway inefficient

Plants that use the C₃ pathway have a built-in metabolic inefficiency in their photosynthetic process. Plants acquire carbon dioxide from the atmosphere through their stomata. They also lose water through the stomata when they are open. When C₃ plants are subject to stress during hot dry weather, they close their stomata to reduce water loss. As a result, the concentration of carbon dioxide in the interior of leaves drops. The enzyme rubisco then combines oxygen, rather than carbon dioxide, with the 5-carbon sugar RuBP. This results in only one molecule of PGA being formed (instead of the two PGA that are formed when rubisco combines with carbon dioxide), along with another organic molecule, which passes down a salvage pathway that consumes oxygen and releases carbon dioxide (Figure 5.17). This process is called **photorespiration** because it occurs in sunlight, consumes oxygen, and releases carbon dioxide. In C₃ plants, photorespiration increases with light intensity and temperature; thus C₃ plants are less efficient in hot summer weather.

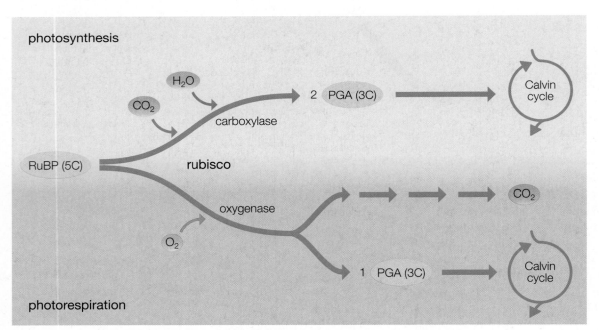

FIGURE 5.17 Photorespiration When the rubisco enzyme behaves as was described for the Calvin cycle (i.e., as a carboxylase), RuBP is combined with CO_2, giving rise to two PGA molecules, which continue along in the Calvin cycle (top pathway). When CO_2 levels are low, however, rubisco can act as an oxygenase (bottom pathway), combining RuBP with oxygen, which yields only one PGA molecule and, following a series of reactions, CO_2. By causing the loss of carbon that might otherwise have been used to construct sugars or other molecules, photorespiration can considerably reduce the efficiency of photosynthesis.

Photorespiration may return as much as half of all fixed carbon to free carbon dioxide.

How could such a wasteful system develop? Rubisco may have evolved when there was little oxygen and much greater amounts of carbon dioxide in the atmosphere of Earth (Chapter 18). Because the present-day atmosphere contains 21% O_2 and only 0.037% CO_2, photorespiration occurs in C_3 plants because carbon dioxide may be available in only trace amounts. The recent trend of increasing levels of atmospheric carbon dioxide has actually increased the efficiency of C_3 plants grown as crops. But other plants have evolved mechanisms to overcome photorespiration.

EVOLUTION

C_4 plants and CAM plants have mechanisms to reduce photorespiration

C_3 plants rely on the Calvin cycle alone for the fixation of carbon dioxide into simple sugars. Consequently, if the level of carbon dioxide around their mesophyll cells drops to low levels, rubisco reacts with oxygen rather than carbon dioxide and photorespiration occurs. About 15% of known plant species have evolved alternative mechanisms to ensure that the rubisco in their photosynthetic cells is not exposed to low levels of carbon dioxide.

C_4 plants reduce photorespiration by preconcentrating CO_2
Did you ever wonder why crabgrass is so rampant in your lawn, or why crops such as corn and sugarcane are so productive? To a large degree these plants are successful in warm weather because they have an additional photosynthetic pathway, called the **C_4 pathway**. In these plants the first product of carbon fixation is a compound called *oxaloacetate*. Since oxaloacetate has four carbons rather than three, this pathway is called the C_4 pathway, and plants that use it are **C_4 plants**.

ECOLOGY

In the mesophyll cells of the leaves of C_4 plants, there is no Calvin cycle and carbon dioxide combines with a 3-carbon compound called *phosphoenolpyruvate (PEP)* to form oxaloacetate in a reaction catalyzed by the enzyme PEP carboxylase (Figure 5.18). Oxaloacetate is then very rapidly reduced to malate. The unusual feature of C_4 plants is that the mesophyll cells do not carry out the remaining steps of photosynthesis, because there is no Calvin cycle. They transport the malate to special cells called *bundle-sheath cells*, which surround the veins of the leaf. There the malate is converted into carbon dioxide and the 3-carbon pyruvate. The carbon dioxide enters the Calvin cycle, and rubisco fixes it into PGA, just as in C_3 plants. Carbon dioxide is fixed twice, once in the mesophyll cells and again in the bundle-sheath cells. The pyruvate returns to the mesophyll cells, where it is converted back to PEP using ATP. C_4 plants spatially separate the C_4 pathway and the Calvin cycle: The C_4 pathway operates in the mesophyll cells, and the Calvin cycle in bundle-sheath cells.

C_4 plants greatly reduce photorespiration by using this C_4 pathway. The export of malate and release of carbon dioxide in the bundle-sheath cells concentrates carbon dioxide there to levels 10 times greater than in the atmosphere. With these high levels of carbon dioxide available to rubisco, photorespiration is essentially eliminated. As a result, under high light and temperature conditions C_4 grasses such as corn and sugarcane can

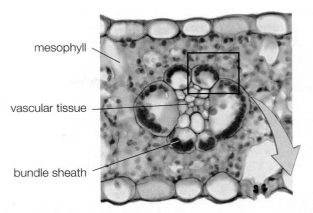

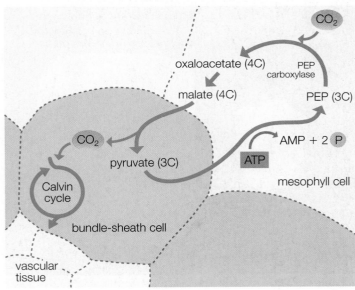

FIGURE 5.18 C_4 pathway The initial fixation of carbon in the C_4 pathway is mediated by the enzyme PEP carboxylase, which combines CO_2 with the 3-carbon PEP to form 4-carbon oxaloacetate. This reaction takes place in the mesophyll cells in the leaf of C_4 plants (a section through a corn leaf is shown here). The oxaloacetate is converted to malate, which is transported to the specialized bundle-sheath cells. There it is split into pyruvate and CO_2, which is used in the Calvin cycle (fixed by rubisco). The 3-carbon pyruvate returns to the mesophyll cells, where it is converted to PEP with energy derived from ATP. PEP is then ready to begin the cycle again.

have higher net rates of photosynthesis than C_3 grasses such as wheat and rice.

CAM plants reduce photorespiration by fixing CO_2 at night
In deserts, plants such as cacti and agaves are frequently exposed to high temperatures and long periods of water stress. To reduce water loss, desert plants must close their stomata during the day. These conditions would lead to high levels of photorespiration in C_3 plants. Many desert plants, especially succulent plants with thick, fleshy stems or leaves (Chapter 26), have evolved a different mechanism than C_4 plants for reducing photorespiration. Because this mechanism was first recognized in plants called stonecrops in the family Crassulaceae, it is called **crassulacean acid metabolism (CAM)**. **CAM plants** use both the

ECOLOGY

C_4 pathway and the Calvin cycle, as do C_4 plants, but CAM plants separate the two pathways in time rather than in space (Figure 5.19). At night, when temperatures are lower, CAM plants open their stomata to allow air to enter. They fix carbon dioxide in their mesophyll cells by the C_4 pathway and store the resulting malate in large cell vacuoles as malic acid. As dawn approaches, the stomata close and malic acid leaves the vacuoles and enters the cell cytoplasm as malate. In the cytoplasm, malate is broken into carbon dioxide and pyruvate, and the carbon dioxide enters the chloroplasts, where it is fixed in the Calvin cycle. CAM plants thus operate the C_4 pathway at

FIGURE 5.20 **Examples of CAM plants** (a) Snake plant (*Sansevieria*) and (b) *Crassula rupestris*, a stonecrop.

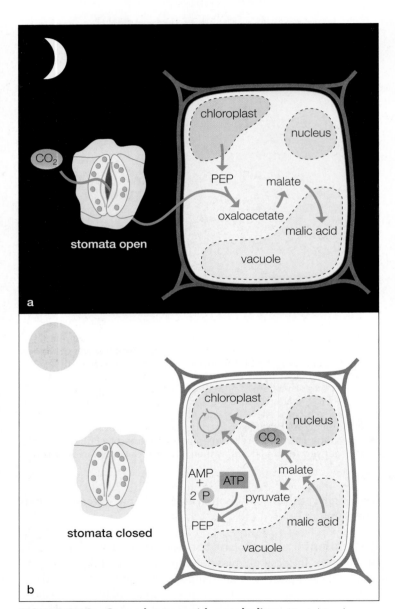

FIGURE 5.19 Crassulacean acid metabolism At night, when the stomata are open, CO_2 enters the plant and, through a series of steps, is converted to the 4-carbon malate, which is stored in the vacuole as malic acid. In the daytime, stomata close (conserving water for the plant) and the malic acid is removed from the vacuole and is converted to malate and a CO_2 molecule is split off, which can be fixed in the chloroplast by rubisco in the Calvin cycle. The 3-carbon pyruvate molecule is, through the input of energy, converted back to PEP.

night when their stomata are open and the Calvin cycle in the day when the stomata are closed. Cacti and agaves are all CAM plants, as are some familiar houseplants such as the snake plant (*Sansevieria*, Figure 5.20a) and the jade plant (*Crassula argentea*) and other *Crassula* species (Figure 5.20b).

If CAM and C_4 plants are so efficient at photosynthesis, why have they not displaced C_3 plants? C_3 plants are adapted to cooler, moister conditions and lower light levels than are C_4 plants. In most parts of the United States, lawns consist of C_3 grasses such as Kentucky bluegrass; but as hotter, drier weather arrives in summer, the C_4 plant crabgrass takes over many lawns. C_4 plants are better adapted to hot, dry weather than C_3 plants, but the carbon-concentrating pathway in C_4 plants does not come free. Operation of the carbon-concentrating pathway requires extra energy in the form of ATP (see Figure 5.18). If this extra energy cost is less than that of photorespiration, which is the case at high temperatures, then C_4 plants will be favored. CAM plants are even better suited to hot, dry climates, but succulent CAM plants generally cannot tolerate freezing temperatures because of the high water content of their cells. Their water-filled cells freeze readily, and the plants are damaged. Most succulent CAM plants are restricted to latitudes and elevations where they are not exposed to freezing temperatures. The carbon-concentrating pathway of CAM plants also costs energy to operate (see Figure 5.19). They grow more slowly than other plants, and C_3 and C_4 plants can displace them if moisture is plentiful. Thus, C_3, C_4, and CAM plants persist because each group is adapted to a different set of climatic conditions.

5.4 | Respiration and fermentation release energy for cellular metabolism

When athletes prepare for a strenuous exercise, they may "carbo load" their cells by eating pasta or other carbohydrate-rich foods. Similarly, plants load their seeds with starch or lipids so that those seedlings can make a fast start in spring growth. The chemical reactions that break down these organic compounds release energy and are called exergonic (energy-out) reactions. The energy released is captured in energy-rich chemical bonds of adenosine triphosphate (ATP) when a phosphate (PO_4) is attached to adenosine diphosphate (ADP). Plants and green algae store their food reserves as starch. Plants transport chemical energy to their cells as sucrose. To release the energy in these compounds, starch and sucrose are first broken down into glucose. Glucose then enters the chemical pathway known as **respiration.**

In everyday usage, respiration is often used to mean breathing—the rhythmic movement of air in and out of the lungs. Breathing is more correctly called *ventilation*. Respiration is the metabolic process by which glucose is broken down in the presence of oxygen into carbon dioxide and water and the energy released is captured in ATP. The process of respiration can be summarized by the equation

$$C_6H_{12}O_6 + 6O_2 \longrightarrow 6CO_2 + 6H_2O$$

This equation indicates the overall process—the substrates and products of respiration—but it does not indicate how the process takes place. When one molecule of glucose is oxidized to carbon dioxide and water, 36 to 38 molecules of ATP are assembled. These ATP molecules represent about 38% of the energy in the glucose molecule. The rest of the energy is dissipated as heat. If you burned a cube of sugar, you would still obtain water and carbon dioxide, but all of the energy would be dissipated as heat. Thus, in the presence of oxygen, respiration captures a significant part of the energy in glucose for use in cellular metabolism. How is this accomplished?

Respiration captures the energy in the chemical bonds of glucose by processing it through a series of enzyme-controlled reactions. These reactions take place in five distinct processes: (1) glycolysis, (2) acetyl coenzyme A formation, (3) the Krebs cycle, (4) the electron transport chain, and (5) chemiosmosis and oxidative phosphorylation (Figure 5.21). These processes occur in different parts of living cells.

Respiration occurs in the cytoplasm and mitochondria of cells

In respiration, glycolysis occurs in the fluid portion of the cytoplasm. The remaining groups of reactions all occur in the mitochondria (Chapter 4). A mitochondrion consists of an outer membrane and an inner membrane that is highly folded (Figure 5.22). The space between these membranes is called the *intermembrane space*. The folds of the inner membrane are known as **cristae**. Inside the inner membrane lies a fluid **matrix**. The formation of acetyl coenzyme A (acetyl CoA) and the Krebs cycle occur in the matrix of the mitochondria. The enzymes of the electron transport chain and oxidative phosphorylation lie in the membranes of the cristae.

Glycolysis is the splitting of glucose into two molecules of pyruvate

Glycolysis takes place in the cytoplasm of every living cell, whether photosynthetic or nonphotosynthetic, prokaryotic or

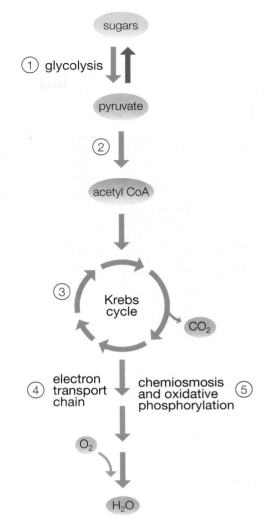

FIGURE 5.21 The five processes in respiration (1) Glycolysis, (2) the formation of acetyl coenzyme A, (3) the Krebs cycle, (4) the electron transport chain, and (5) chemiosmosis and oxidative phosphorylation.

eukaryotic. The term *glycolysis* comes from two Greek words, *glyco* meaning "sugar" and *lysis* meaning "breaking." Oxygen is not required for its operation.

In glycolysis, each 6-carbon-containing molecule of glucose is split into two 3-carbon-containing molecules of pyruvate. The process of glycolysis consists of a series of 10 steps, each of which is catalyzed by a particular enzyme (Figure 5.23). The first steps are spent rearranging the glucose molecule and preparing it for splitting into two 3-carbon molecules. The remaining steps convert this initial 3-carbon compound into pyruvate ($C_3H_3O_3$). There is a net gain in energy for the cell of 2 ATP per glucose molecule passed through glycolysis. These two ATP molecules represent only a little more than 2% of the energy available in the glucose molecule. Glycolysis also captures some energy by reducing two molecules of nicotinamide adenine dinucleotide (NAD^+) to NADH with electrons and protons removed from glucose. Most of the energy in the glucose is still contained in the two molecules of pyruvate.

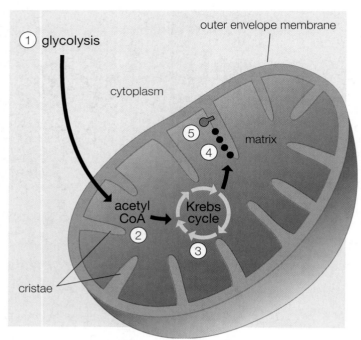

FIGURE 5.22 Locations of the five respiration processes
Glycolysis takes place in the cytoplasm. The Krebs cycle occurs in the matrix of mitochondria, and the electron transport chain, chemiosmosis, and oxidative phosphorylation take place on the inner mitochondrial membrane (which is infolded to form cristae).

Because oxygen is not required for glycolysis, the process is thought to be primitive and to have arisen before the atmosphere of Earth contained oxygen (Chapter 18). In EVOLUTION the absence of oxygen, under what is called **anaerobic** conditions, glycolysis will be followed by **fermentation**.

Fermentation extracts energy from organic compounds without oxygen

In yeasts and most plant cells, the process of fermentation breaks down pyruvate into ethyl alcohol and carbon dioxide (CO_2) and oxidizes NADH back to NAD^+. NAD^+ is then available to oxidize the 3-carbon sugars in glycolysis (see

Figure 5.23). Alcohol fermentation is the basis of the brewing, baking, and winery industries. In brewing and wine making, yeasts (fungi) ferment sugars to produce ethyl alcohol in the beverages. Yeasts cannot tolerate alcohol levels above about 12%, which is why wines have a maximum alcohol content no greater than that. In beer, the alcohol content may vary from 3 to 12%, but most range around 4 to 6% (refer to Essay 5.1, "The Botany of Beer").

In baking, the important by-product of fermentation is carbon dioxide, whose bubbling causes bread to rise (Figure 5.24a). The alcohol evaporates in the oven. A dramatic demonstration of carbon dioxide production can be made by placing yeasts in a warm sugar solution in a bottle capped with a balloon. The activity of the yeasts will produce enough carbon dioxide to expand the balloon (Figure 5.24b).

Many bacteria, fungi, protists, and animal cells carry out a fermentation process that results in the formation of lactate. Bacteria and fungi are utilized commercially to produce cheese, yogurt, sausage, pickles, sauerkraut, and soy sauce by lactate fermentation. In humans and other animals whose muscle cells are subjected to heavy exercise, the supply of oxygen to the cells may not keep up with the demands of exercise. In this situation, muscle cells switch to lactate fermentation and accumulate lactate, leading to muscle fatigue. The oxygen needed to break down the lactate into carbon dioxide and reoxidize NADH is the oxygen debt of the muscle tissue.

Although the energy yield from anaerobic glycolysis and fermentation is low, it is sufficient for many organisms. If oxygen is present, however, under **aerobic** conditions, the process of respiration can break down pyruvate molecules into carbon dioxide through the formation of acetyl CoA and the Krebs cycle, and much more usable energy can be released.

Pyruvate is split into CO_2 and an acetyl group attached to coenzyme A

Pyruvate molecules move from the cytoplasm into the matrix of the mitochondrion. There a molecule of carbon dioxide is split off from pyruvate (Figure 5.25). The remaining 2-carbon compound, called an *acetyl group*, is attached to a large molecule

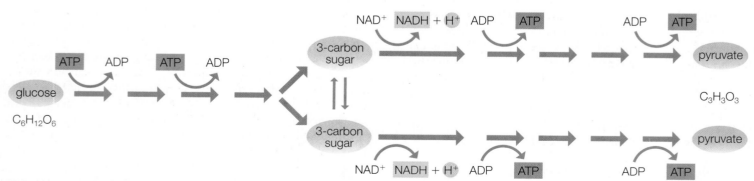

FIGURE 5.23 Glycolysis For each molecule of glucose, 4 ATP molecules are produced, but because 2 ATPs are also used up (in the first and third steps), the result is a net gain of only 2 ATP molecules. Two molecules of NAD^+ are also converted to NADH and protons for every glucose molecule. The end product of glycolysis is pyruvate.

ESSAY 5.1   THE BOTANY OF BEER

No one knows when humans first began to brew beers, but brewing was well established before recorded history. Malt and hops come from two species of plants. Grains of barley (*Hordeum vulgare*) provide the malt that contains the sugars and enzymes to begin the process (Figure E5.1A). On a commercial scale, malt is made from barley grains by washing the grains in flowing water for 8 to 10 hours in large tanks. After washing, the grains are left in standing water for 40 hours. The grains are then spread out in warm malting rooms until they sprout short roots. The barley

Beer brewing is a process of controlled fermentation involving three ingredients —water, hops, and malt.

grains are sprouted so that they will produce enzymes to convert starch in the grains to sugars. The grain is then kilned (heated) to 130–200°C (266–392°F) to kill the seedlings and preserve the enzymes. Malt is the germinated, heat-killed, and dried grain of barley.

To brew a beer, barley malt is mixed with water in a vat called a mashing tun, which is heated to 68–73°C (154–163°F) for 2 to 6 hours. Enzymes from the malt enter the liquid and break down the starch into sugars. After this mashing process, the liquid, called wort, is filtered through the bottom of the tun and sent to another vat, where it is boiled. The hops are added at this point, which is the actual brewing stage. Hops are flowers from a plant called *Humulus* (Figure E5.1B). Volatile oils in the flowers lend a pleasant taste and aroma to beers, and they also produce a clear beverage. Breweries usually add hops as dried, pressed pellets.

The wort is then cooled, filtered, and sent to fermentation tanks, where yeasts are added. The yeasts perform the glycolysis and fermentation that break down the sugars into ethyl alcohol. After 7 to 12 days the liquid—now called a green beer—

E5.1B   Hops

is moved to aging tanks for 2 to 3 weeks. Breweries in the United States pasteurize (partially sterilize) and filter their brews to remove yeast cells and add carbonation before bottling.

Light beers have become popular in recent years. Regular brews are high in calories because of carbohydrates left in the beers, which the yeasts cannot ferment. In light beers, enzymes are added to the fermentation tanks to break down these carbohydrates into sugars usable by yeasts. This produces a beer with fewer calories but higher alcohol content. Water is then added to reduce the alcohol percentage. Modern breweries control all of these operations through automation and computer monitoring.

E5.1A   Barley

called **coenzyme A (CoA)** to form acetyl CoA. Energy is captured as a molecule of the electron acceptor and reduced to NADH for each acetyl group formed. Because each glucose molecule makes two pyruvate molecules in glycolysis, two acetyl groups enter the Krebs cycle for each molecule of glucose broken down in respiration.

In the Krebs cycle, the acetyl group is broken into two carbon dioxide molecules

The breath you exhale is rich in carbon dioxide that is produced mostly by operation of the Krebs cycle in the mitochondria of your cells. Plant mitochondria also produce carbon dioxide by the Krebs cycle, which was named for Sir Hans Krebs, a British biochemist who worked out the reactions in the 1930s and later received a Nobel Prize. Acetyl CoA enters the Krebs cycle by transferring the acetyl group to oxaloacetate, a 4-carbon molecule, to yield a 6-carbon molecule of citrate (see Figure 5.25). The formation of citric acid is the reason the Krebs cycle is also

called the **citric acid cycle.** With each turn of the Krebs cycle, one acetyl group is oxidized to two carbon dioxide molecules. Some of the energy in the acetyl group is captured as one molecule of ATP, but most of the energy is transferred to nicotinamide adenine dinucleotide (NAD^+) and flavin adenine dinucleotide (FAD). These electron acceptor molecules receive the electrons and protons removed from the acetyl group and become reduced to NADH and $FADH_2$. Three molecules of NADH and one of $FADH_2$ are made for each acetyl group moving through the Krebs cycle. Oxaloacetate is regenerated at the end of each cycle (the predecessor of oxaloacetate in the Krebs cycle, incidentally, is malate, which is also involved in C_4 and crassulacean acid metabolism). Because two acetyl groups enter the Krebs cycle for each glucose molecule processed, 2 ATP, 6 NADH, and 2 $FADH_2$ are produced from each glucose molecule passing through the cycle.

By this point in the respiratory process, one molecule of glucose has been broken completely into carbon dioxide. Glycolysis has yielded 2 ATP and the Krebs cycle 2 ATP, for a total

FIGURE 5.24 Production of CO$_2$ in fermentation (a) Bread rises due to CO$_2$ produced by yeast expanding the dough. (b) Yeast placed in a sealed container with sugar will produce CO$_2$, causing an attached balloon to expand. The experiment shown here was run for 1 hour.

of 4 ATP. This amount of chemical energy is not sufficient to meet the energetic needs of most cells. Most of the energy present in the original glucose molecule is now contained in the high-energy electrons attached to the carrier molecules NADH and FADH$_2$, but these molecules are not directly useful to a cell for cellular work. An analogy may be helpful at this point. Cars need a supply of fuel to power them. Crude oil has a lot of energy in its chemical bonds, but it is not usable in a car engine. Crude oil has to be refined into gasoline to power a car. Cells also need "gas" (ATP) to power their metabolism, but by the end of the Krebs cycle most of the energy is still tied up in the cellular equivalent of crude oil—NADH and FADH$_2$. That energy is converted into ATP during the final two processes of respiration—chemiosmosis and oxidative phosphorylation. These final processes recover the bulk of the energy captured from the glucose molecule.

The electron transport chain generates a proton gradient across the inner mitochondrial membrane

The electron transport chain acts like the alternator in a car and charges the battery of the oxidative phosphorylation process. The chain consists of a series of electron carrier molecules, which are located in the inner mitochondrial membranes of the cristae (Figure 5.26). There are four multiple-subunit protein complexes, called Complexes I through IV. Complex I oxidizes the NADH generated in the Krebs cycle back to NAD$^+$ and reduces a molecule of ubiquinone. Ubiquinone molecules are hydrophobic electron acceptors that move about freely within the inner mitochondrial membrane. Complex II oxidizes FADH$_2$ from the Krebs cycle back to FAD and reduces ubiquinone. The reduced ubiquinone (called *ubiquinol*) carries electrons from Complex I and Complex II to Complex III. Complex III passes electrons from ubiquinol to cytochrome *c*, which is the only protein in the electron transport chain that is not tightly associated with one of the four complexes. Cytochrome *c* transfers one electron at a time from Complex III to Complex IV, where oxygen is reduced to water in the final step. For every four electrons transferred through cytochrome *c*, one molecule of oxygen is reduced to two molecules of water. Any substance that interferes with the ability of oxygen to accept electrons—as do many poisons, such as cyanide—will shut down electron transport and the production of energy in those cells.

In summary, the carrier molecules take high-energy electrons from NADH and FADH$_2$ and pass them down the chain in a series of oxidation–reduction reactions. NADH and FADH$_2$ are oxidized to NAD$^+$ and FAD, which can again accept electrons in the Krebs cycle. With each transfer in the chain, the energy level of the electrons is lowered. Some of that energy is dissipated as heat, and some is used to pump the protons (H$^+$) from NADH and FADH$_2$ across the inner mitochondrial membrane and into the intermembrane space between the inner and outer membranes of the mitochondrion. Complexes I, III, and IV pump protons across the inner membrane, but Complex II does not (see Figure 5.26). This pumping of protons puts the charge on the "battery" for the next step in the process.

As the electron transport chain pumps protons across the inner membrane, the intermembrane space becomes more acidic relative to the matrix. Because protons also carry an electric charge, the buildup of protons on one side of the inner membrane creates an electrochemical gradient, which is the charge on the battery. This electrochemical gradient across the mitochondrial cristae is basically the same as the electrochemical gradient across the thylakoids in chloroplasts. In both cases, that electrochemical gradient represents potential energy. That potential energy is put to work in the final process of respiration.

Chemiosmosis and oxidative phosphorylation generate ATP in the mitochondrion

The potential energy of a battery can be used by providing a channel in the form of a wire to connect the battery to a motor or lamp. In the mitochondrion, the potential energy of the proton gradient can be accessed through the **ATP synthase complex**. This complex provides a channel through which protons in the intermembrane space can flow back to the matrix. The flow of protons is called *chemiosmosis*, as it was in the chloroplast thylakoids. As the protons flow back, their energy is captured in chemical bonds formed by converting adenosine diphosphate (ADP) into ATP (see Figure 5.26). The process is called **oxidative phosphorylation** because oxygen is the final

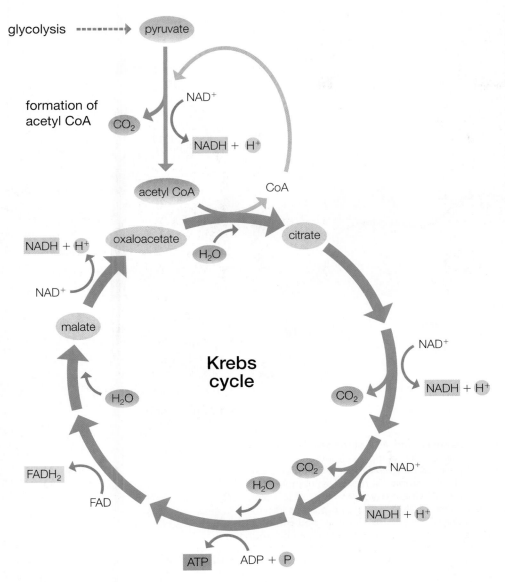

glycolysis ----------> pyruvate

formation of
acetyl CoA

NAD^+

CO_2

$NADH + H^+$

acetyl CoA

CoA

oxaloacetate

citrate

H_2O

$NADH + H^+$

NAD^+

malate

**Krebs
cycle**

NAD^+

$NADH + H^+$

H_2O

CO_2

CO_2

NAD^+

H_2O

$NADH + H^+$

$FADH_2$

FAD

ATP   ADP + P

FIGURE 5.25   **Acetyl CoA formation
and the Krebs cycle** In the first of these
two processes, the 3-carbon pyruvate from
glycolysis is converted to the 2-carbon
acetyl group attached to CoA (releasing
CO_2 and storing some energy in the form
of NADH). In the Krebs cycle, the acetyl
group is combined with oxaloacetate
(4-carbon) to form 6-carbon citrate, while
the CoA is released and can combine
with another pyruvate. Through a series
of steps, the citrate is converted to the
4-carbon malate, which is converted to
oxaloacetate, completing the cycle. The
two carbon atoms removed from citrate
are released as CO_2, and high-energy
compounds (ATP, NADH, and $FADH_2$)
are formed.

electron acceptor in the chain and energy is captured in a phosphorylation when a phosphate group (PO_4) is added to ADP to
form ATP.

The yield of ATP from electron transport and oxidative
phosphorylation is substantial. Up to 34 ATP can be generated
from one glucose molecule during these final processes. All of
the previous processes together produce only 4 ATP. Thus, the
maximum yield of ATP from complete aerobic respiration of
one glucose molecule is 38 ATP.

In conclusion, photosynthesis and respiration are complex
biochemical processes that involve many steps. In the chloroplast thylakoids, the light reactions of photosynthesis capture
solar energy in the form of molecules of ATP and NADPH.

The carbon-fixation reactions in the chloroplast stroma use
that ATP and NADPH to fix CO_2 into the 3-carbon sugar
PGAL. PGAL molecules can then be combined to produce
sugars such as glucose, sucrose, and starch. Respiration takes
simple sugars like glucose and breaks them down completely into carbon dioxide. The energy released from the breakdown of sugar is captured in the form of ATP to drive cellular
metabolism. Both photosynthesis and respiration use electron
transport chains to generate electrochemical gradients across
membranes. Both processes also then use that electrochemical
gradient to produce ATP from ADP and phosphate. The role
of the ATP synthase complex is crucial to both photosynthesis
and respiration.

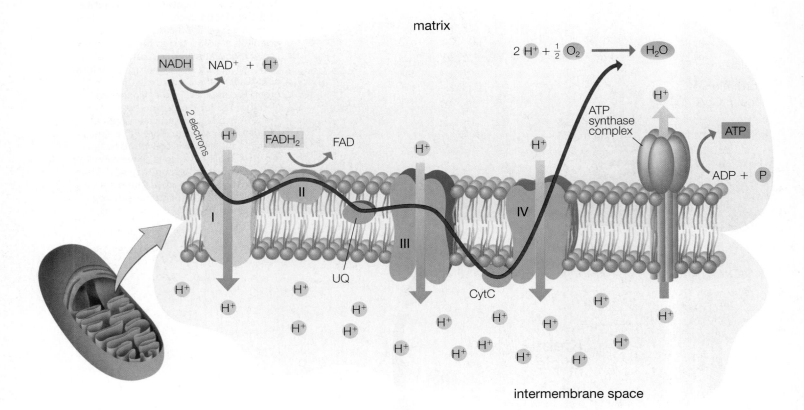

FIGURE 5.26 The electron transport chain, chemiosmosis, and ATP synthase The electron transport chain in the inner mitochondrial membrane consists of four multiple-subunit protein complexes referred to as Complexes I, II, III, and IV. Complex I oxidizes NADH, and Complex II oxidizes $FADH_2$, passing the electrons and protons to ubiquinone (UQ). Complex III reoxidizes ubiquinone and passes the electrons on to cytochrome *c* (CytC) and Complex IV, where oxygen is reduced to water. During the operation of the electron transport chain, Complexes I, III, and IV pump protons into the space between the inner and outer mitochondrial membranes. This creates a proton gradient in which the positively charged protons in the intermembrane space have potential energy that can be exploited by the ATP synthase complex to form ATP.

HIGHLIGHTS

5.1 On Earth the energy that sustains life flows from solar radiation through photosynthesis and respiration to the energy molecules of ATP to power cellular metabolism.

5.2 Cellular metabolism is the sum of the chemical reactions and pathways in a cell. Chemical reactions and pathways may be endergonic or exergonic, or involve oxidations and reductions. Electron transport chains are a series of oxidation–reduction reactions that play major roles in photosynthesis and respiration.

5.3 In the process of photosynthesis, solar energy is captured by chlorophyll *a* and used to reduce carbon dioxide to simple sugars. Photosynthesis consists of two sets of reactions: the light reactions in the thylakoids, which trap the energy of visible light as ATP and NADPH, and the

carbon-fixation reactions in the stroma, which reduce carbon dioxide to a simple 3-carbon sugar by passage through the Calvin cycle. Certain plants, such as CAM plants and C_4 plants, use different mechanisms to acquire carbon dioxide from the atmosphere, but still synthesize carbohydrates by the Calvin cycle.

5.4 Respiration is the process by which cells oxidize simple sugars to carbon dioxide to release energy and perform cellular work and metabolism. In the presence of oxygen, respiration is aerobic and proceeds through glycolysis, acetyl CoA formation, the Krebs cycle, and an electron transport chain and oxidative phosphorylation. When oxygen is absent, glycolysis is followed by fermentation.

REVIEW QUESTIONS

1. Distinguish between photosynthesis and respiration. Name several ways in which they are similar and several ways in which they are different.

2. Define the following terms, which are related to chemical reactions: (a) substrates, (b) products, (c) activation energy, (d) exergonic reactions, (e) endergonic reactions.

3. What is a catalyst? Define the following terms, which are related to catalysis: (a) enzymes, (b) active site, (c) cofactors, (d) coenzymes.

4. How does oxidation differ from reduction? Do either of these reactions require oxygen? Why do these reactions always occur together? What is the most common example of an oxidation–reduction reaction in biological systems?

5. What is a pigment? Examine the absorption spectrum of chlorophyll (see Figure 5.10), and discuss why chlorophyll *a* is a green pigment. Why is chlorophyll *a* the crucial pigment in the light reactions of photosynthesis? Name some examples of accessory pigments, and describe the role they play in the light reactions.

6. Distinguish between C_3, C_4, and CAM plants. What type of climate is each best adapted to? Which type of photosynthesis is used exclusively by most plant species?

7. In one paragraph each, summarize the light reactions and the dark reactions (Calvin cycle) of photosynthesis. What is the net product of each set of reactions?

8. How does cyclic electron flow differ from noncyclic electron flow?

9. What is photorespiration, and why does it appear to be a wasteful process?

10. Discuss the reactions of glycolysis and the citric acid cycle with regard to (a) starting reactants (substrates), (b) reaction products, (c) number of ATP molecules directly produced, (d) number of NADH and $FADH_2$ molecules produced, and (e) the net fate of the carbon atoms that were originally found in the glucose before glycolysis.

11. Briefly describe the processes of electron transport and oxidative phosphorylation. Which compound(s) donate electrons for electron transport? What is the terminal electron acceptor, and what does it become after accepting an electron pair? How is a proton gradient established across the inner mitochondrial membrane, and how is this gradient harnessed to make ATP?

APPLYING CONCEPTS

1. In this chapter, you were given the overall equation for photosynthesis:

$$6CO_2 + 12H_2O \longrightarrow C_6H_{12}O_6 + 6O_2 + 6H_2O$$

which can be reduced to

$$6CO_2 + 6H_2O \longrightarrow C_6H_{12}O_6 + 6O_2$$

This is the opposite of the overall equation for respiration:

$$C_6H_{12}O_6 + 6O_2 \longrightarrow 6CO_2 + 6H_2O$$

Thus it has been said that photosynthesis is the opposite of respiration. But is it really? Describe several of the similarities between photosynthesis and respiration, and list several ways in which these reactions are quite different from each other.

2. Many people believe that plants, being photosynthetic, do not need to obtain any energy at all from respiration. Give some examples of where and when plant cells do use respiration instead of photosynthesis to obtain the ATP energy they need.

3. The reactions of respiration—including glycolysis, the citric acid cycle, electron transport, and the creation and harvesting of a proton gradient to phosphorylate ADP to ATP—are accomplished in a series of many small steps. Why do you think biological systems use such an incremental approach rather than just completing the oxidation in a couple of large, quick steps?

6

DNA and RNA
Genetic Material and Protein Synthesis

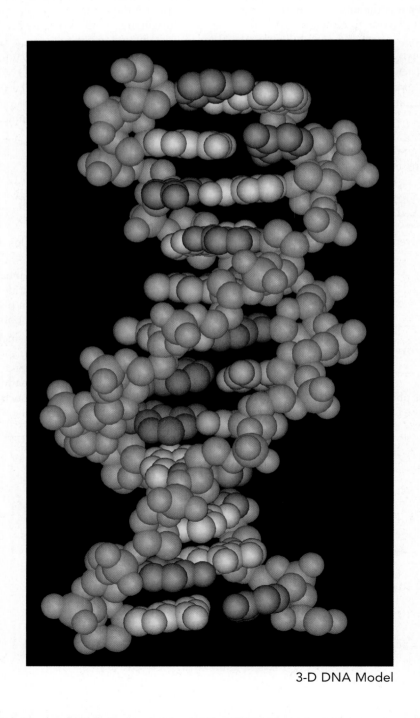

3-D DNA Model

• DNA contains the coded genetic information to construct every living organism.

• The information in DNA is translated into proteins, which control the structure and function of the cells in organisms.

• Differences in DNA account for the differences among organisms.

• Individuals have unique patterns of DNA, which can be used for identification in criminal and social justice cases.

W eeds are plants that grow where humans do not want them. But one small weed, *Arabidopsis thaliana*, has proven to be of enormous value. *Arabidopsis* is the first plant whose entire DNA sequence has been determined. *Arabidopsis* was chosen for intensive study because it has the smallest known total amount of DNA, or genome, of any plant. Small size, few chromosomes (5), short reproduction time, ease of use in cross-breeding experiments, and high seed production are additional traits of *Arabidopsis* that encourage plant geneticists to use it as an experimental organism. *Arabidopsis* will continue to be a gold mine of information on plant genetics for years to come—quite an important role for an inconspicuous little weed!

In the 1950s, DNA (deoxyribonucleic acid) was an obscure molecule known only to a handful of biochemists. Today, it is common knowledge that DNA is the molecule of heredity that contains the genetic information to construct every organism, from a bacterium to a giant redwood tree. Molecular biology is the science dedicated to the study of DNA, and, along with genetic engineering, it is playing an increasing role in our lives. DNA analysis has assumed an important role in criminal justice cases. In medicine, it has led to a molecular understanding of many genetic diseases and promises to lead to gene therapies. In agriculture, genetic engineering has led to crop improvements such as resistance to plant diseases.

In this chapter, we will present a short history of how the structure of DNA was discovered. In DNA, as in much of biology, structure is the key to function. We will discuss how the structure of DNA allows it to contain all the information that defines a bacterium, protist, plant, or animal; how that information is coded as groups of nucleotides; and how that information is replicated so that it can be transmitted from one generation to the next. We will show how the information in DNA is copied or transcribed into specific RNA molecules, which are then translated by other types of RNA molecules into the multitude of proteins and enzymes that direct the metabolism of organisms. Finally, we will learn how differences in DNA account for differences among organisms.

Arabidopsis plant

6.1 | DNA is a long molecule composed of subunits called nucleotides

Chapter 3 briefly introduced nucleic acids as one of the four basic types of organic molecules in living organisms. In prokaryotes, including bacteria, and in chloroplasts and mitochondria, DNA exists as a large, closed, circular molecule. In the nuclei of eukaryotes, DNA is present in **chromosomes** together with proteins called **histones**. The chromosomes carry the **genes**, the units of genetic information, which usually contain the coded information for the synthesis of proteins. Each chromosome contains a single DNA molecule in the form of a chain that is hundreds of thousands of **nucleotides** long. Each nucleotide contains three parts: a 5-carbon sugar called *deoxyribose*, a phosphate group, and one of four different nitrogen-containing bases: adenine, guanine, cytosine, or thymine (Figure 6.1). Adenine and guanine are similar in having a double-ring structure; they are called *purines*. Thymine and cytosine, with a single ring, are also structurally similar and are called *pyrimidines*. The nucleotides form the long chains of DNA by covalent bonds (Chapter 3) between the phosphate groups and the deoxyribose sugars (Figure 6.2). But what form do the chains of nucleotides assume in DNA?

In 1950, two-time Nobel laureate Linus Pauling, at Caltech (California Institute of Technology) in Pasadena, had shown that proteins often assume the form of a coiled spring or α helix (see Figure 3.18b) and that the coiled structure was maintained by hydrogen bonds. Pauling suggested that DNA might have a similar α-helix structure. At about the same time, Rosalind Franklin and Maurice Wilkins at Kings College in London carried out X-ray studies of DNA that strongly indicated some sort of helical structure. In addition, Erwin Chargaff at Columbia University collected data showing that in the DNA from a wide range of organisms, the percentage of adenine was about the same as thymine and the percentage of cytosine was the same as guanine. The significance of these data was not immediately apparent but suggested a relation between specific pairs of nucleotides.

6.2 | DNA contains two nucleotide strands that wind about each other in a double helix

Scientists believed the structure of DNA was the key to understanding its function. They hoped that understanding the function of DNA would lead to many applications and benefits in medicine and agriculture.

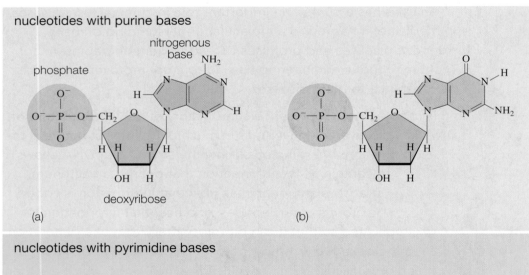

nucleotides with purine bases

(a) phosphate — nitrogenous base — deoxyribose

(b)

nucleotides with pyrimidine bases

(c)

(d)

FIGURE 6.1 The four nucleotides found in DNA The deoxyribose sugar and a phosphate group are common to all; the nucleotides differ only in the nitrogen-containing bases—adenine (a), guanine (b), thymine (c), and cytosine (d)—attached to the deoxyribose sugar.

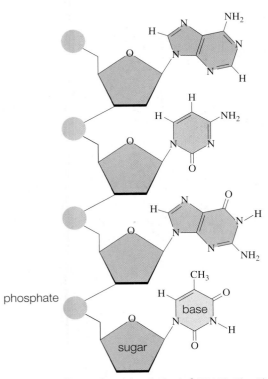

FIGURE 6.2   **A single nucleotide chain (of DNA)** The "backbone" of the chain is formed by sugar-phosphate chemical bonds.

Early in the 1950s, a young research fellow named James Watson went to the Cavendish Laboratory in Cambridge, England, to study molecular structure. There he met physicist Francis Crick. Together they began work to build a model of DNA that would agree with the existing data and explain how DNA could carry the genetic information of organisms. The proposed model would have to show how DNA could be highly variable in order to carry information, yet be capable of precise replication so that the information could be passed accurately from cell to cell and generation to generation.

Watson and Crick were able to synthesize a workable model of the structure of DNA from the data collected by other researchers. They proposed that DNA was not a single-stranded α-helix like many proteins, but, rather, a **double helix** of two strands twined about each other (Figure 6.3a). Imagine taking a ladder and twisting it into a helix. The two sides of the ladder are made up of the deoxyribose sugars and phosphate groups. The rungs of the ladder are the nitrogen-containing bases—adenine (A), thymine (T), cytosine (C), and guanine (G)—with two bases forming each rung of the ladder. The pairs of bases meet in the interior of the helix, where they are joined by hydrogen bonds. The nucleotides along any one strand of the double helix could be arranged in any order, such as ATCTGTA (using the abbreviations for the bases). Since

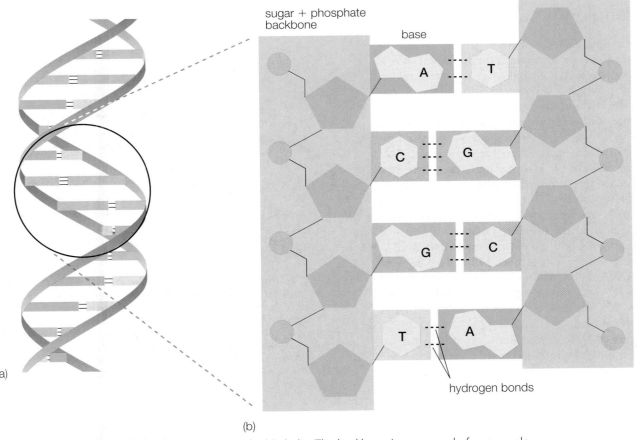

FIGURE 6.3   **Structure of DNA** (a) The DNA double helix. The backbone is composed of sugar molecules and phosphate groups. The horizontal rungs represent the nitrogenous bases. (b) Complementary base pairing in DNA. Adenine pairs with thymine through the formation of two hydrogen bonds (dashed lines). Three hydrogen bonds are formed between the bases guanine and cytosine.

a DNA molecule can be thousands of nucleotides long, it can be highly variable and potentially carry a large amount of information.

When Watson and Crick constructed the model of the matching DNA strand, which corresponded to the opposite side of the ladder and half of the rungs, they made an important discovery. Adenine would pair only with thymine, and guanine would pair only with cytosine. This pattern occurs because adenine can form two hydrogen bonds only with thymine, and cytosine can form three hydrogen bonds only with guanine (Figure 6.3b). Hydrogen bonds cannot form properly between adenine and either guanine or cytosine. The model Watson and Crick built explains the base composition of DNA determined by Erwin Chargaff. The percentage of adenine equals that of thymine, and the percentage of cytosine equals that of guanine. The two strands of nucleotides in DNA are not the same; rather, they are **complementary**—each strand contains a sequence of bases that is the complement of the other. Thus if one strand had the sequence ATCGCTA, the complementary strand must be TAGCGAT because A pairs only with T and C pairs only with G. This complementary pairing suggests a mechanism for DNA to copy itself.

6.3 | DNA replicates by separating its two strands and synthesizing two new complementary strands

The ability to manipulate DNA to cure a genetic disease or improve a crop plant depends on understanding how DNA is copied in cells—by a process that is analogous to photocopying important documents. Before a cell can undergo division into two daughter cells, the DNA of the parent cell must replicate to provide each daughter cell with a copy. The double-stranded molecule separates into two strands as a special enzyme destabilizes the double helix and opens it up for replication. This enzyme acts like the pull-tab on a zipper and unzips the two strands of DNA. Each single strand then acts as a template or pattern for the synthesis of a new complementary strand along its length (Figure 6.4a). Wherever an A is present in an original strand, a T will be placed in the opposite position on the new strand. If a G is present, a C will be placed opposite it, and so forth. When the process is complete, there will be two double-helix molecules of DNA. Each has one new strand and one of the original strands. This type of replication is called **semiconservative** because half of the DNA molecule is the original (conserved) molecule and half is new.

The basic process of DNA replication is simple in principle. Two parental strands separate and serve as templates for newly synthesized complementary strands. The actual mechanisms, however, are considerably more complex. Each step in DNA replication requires different enzymes, and many years of research have gone into working out the general outline of the process.

DNA replication does not begin at the end of the molecule. In prokaryotes like bacteria, the DNA is a continuous loop with no free ends. DNA replication always begins at a specific site

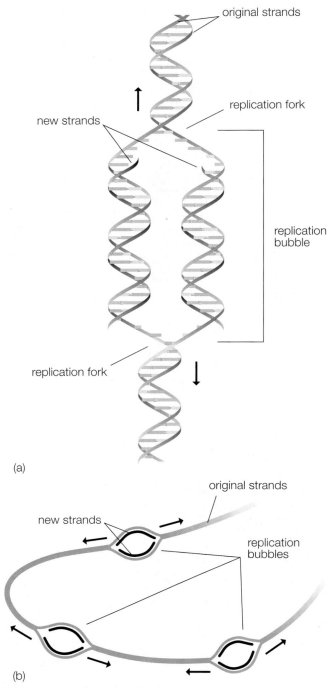

(a)

(b)

FIGURE 6.4 Replication of DNA (a) The original double-stranded molecule (red and purple) is split open, and new strands (in black) are synthesized such that the daughter molecules each contain an old strand in addition to a newly formed strand. The Y-shaped structures where the original strands unravel are termed the *replication forks*. The region where the new strands are forming is called the *replication bubble*. (b) Diagram of DNA of a eukaryote, showing three replication bubbles. The newly formed strands are shown in black.

called an **origin of replication**. Special initiator proteins and enzymes called **helicases** break the hydrogen bonds between the two strands at the origin of replication and open up the double helix for replication. At each point where the double strands are being separated, the DNA molecule forms a Y-shaped pattern called a *replication fork*. The region between the two replication forks, where new strands are being synthesized, is called a *replication bubble* (Figure 6.4a). Binding proteins keep the two strands separated while a complex of enzymes called **DNA polymerases** synthesizes the new complementary strands. DNA polymerases must recognize each base in the original strands and bring the nucleotide having the complementary base to them. Then the polymerases must bind the nucleotide to the growing strand. In prokaryotes, there is only one replication bubble and two replication forks. The two replication forks are opening in opposite directions along the DNA molecule away from the origin of replication. Replication proceeds around the circle of DNA until the two replication forks meet on the opposite side. Replication is very fast in prokaryotes, up to 500 nucleotides per second.

In eukaryotes, the process of replication is different from that of prokaryotes because there are multiple replication bubbles in the DNA of each chromosome (Figure 6.4b). The two replication forks in each bubble are opening in opposite directions along the DNA molecule away from their origin of replication. Despite these multiple origins of replication and replication bubbles, the replication rate of DNA is slower in eukaryotes than in prokaryotes; the rate in mammals, including humans, is about 50 nucleotides per second.

6.4 | Stability of genetic information depends on efficient mechanisms for DNA repair

Many human diseases can be traced to DNA damage and mistakes during replication. What would happen if an error were made during DNA replication and an incorrect base were inserted in a strand of DNA? What would happen if ultraviolet

(UV) radiation or chemicals damaged one or more bases in a strand of DNA? If the error were not corrected or the damage not repaired, the change in that DNA strand would be transmitted to the next generation of cells. Such a change in genetic information is called a **mutation**. Some rare mutations may be beneficial; these make up the raw material of new genetic information upon which natural selection and evolution operate. But most mutations occur randomly. In complex cells, random changes in genetic information are highly unlikely to improve the cell structure or metabolism. UV radiation can cause severe sunburn, which eventually heals, but years later the underlying DNA damage may result in a skin cancer in that individual. Thus many mutations are harmful, if not lethal, and a high rate of mutations could have disastrous consequences for complex organisms. Many scientists are therefore concerned about the decline in atmospheric ozone, because ozone shields the Earth from much of the UV radiation.

EVOLUTION

ECOLOGY

The survival of cells and organisms depends on efficient mechanisms for repair of errors in DNA replication and DNA damage. DNA polymerases are remarkably accurate in replicating DNA. Only about one error is made in every 1 billion base-pair replications. If an error in replication does occur, DNA polymerases can reverse themselves and remove the error. If this were not the case, any error in replication would be transmitted to the descendants of that cell. Cells possess many enzymes to repair damaged DNA. Since DNA is double-stranded, cells possess two copies of their genetic information. Damage done to DNA rarely occurs to both strands at the same time. The damaged DNA strand is recognized and removed by special enzymes called **DNA repair nucleases**. These enzymes cut out the damaged length of DNA by breaking the sugar-phosphate bonds that bind together the damaged nucleotides (Figure 6.5). DNA polymerase then uses the undamaged strand as a template to fill in the gap with new

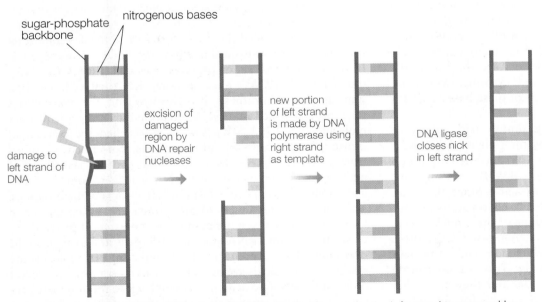

FIGURE 6.5 Generalized mechanism of DNA repair Damage to the left strand is removed by DNA repair nucleases. DNA polymerase then fills in the gap with new nucleotides. Finally, DNA ligase seals the remaining gap called a "nick" in the repaired strand.

nucleotides. A third enzyme called **DNA ligase** completes the repair process by sealing the remaining break in the damaged strand. Thus most mutations are not transmitted to the next generation of cells and the genetic information is maintained intact. This brings us to the question of exactly how DNA contains genetic information.

6.5 | Genetic information is coded in DNA as groups of three nucleotides

Genetic information is contained in DNA in the form of genes, which are the basic units of genetic information. Most, but not all, genes contain information for making proteins. Proteins are composed of 20 different kinds of amino acids, linked together by peptide bonds (Chapter 3). Each chain of amino acids is a polypeptide, and proteins are composed of one or more polypeptides. At any moment, living cells are producing hundreds of proteins. Photosynthesis, for example, consists of many chemical reactions, each of which is controlled by an enzyme (Chapter 5). Enzymes are proteins whose production is determined by genes. Thus photosynthesis can be seen as a trait of green plants determined by numerous genes in the plant DNA. More obvious traits such as flower color are also under gene control. Several enzymes that assemble a pigment may produce the red color in the petals of a flower. A mutation in one of these genes could change the color of the flower. Everything about the structure and function of an organism is controlled by the information in the genes that determine the production of proteins. How is that genetic information for cell structure and metabolism contained in the genes of the DNA molecule?

The genetic information for every protein is contained in genes in the form of a code consisting of sequences of nucleotide bases. A code is any system of communication in which arbitrary symbols are given definite meaning. For example, the letters on this page are arbitrary symbols that, when grouped into words, contain definite meaning. Similarly in the genetic code, the arbitrary letters are the nucleotide bases adenine, thymine, guanine, and cytosine, which are abbreviated as A, T, G, and C. Groups of these bases code for specific amino acids.

If one nucleotide coded for one amino acid, only four amino acids could be represented by the four nucleotides. If two nucleotides (such as AA, AT, or CG) specified an amino acid, then there could be a total of $4 \times 4 = 16$ combinations of nucleotides to code for 16 amino acids. Since there are 20 amino acids in proteins, a code based on pairs of nucleotide bases is not sufficient. If groups of three nucleotides, however, are used in a code (such as AAA or CGC), there could then be $4 \times 4 \times 4 = 64$ possible codes and more than enough to specify all 20 amino acids. Trios of nucleotide bases do in fact form the **genetic code**, and they are called **triplets**. The triplet AAA, for example, codes for the amino acid phenylalanine. Most triplets code for an amino acid, and some amino acids have more than one triplet. A few triplets do not code for any amino acid and instead signal the end of protein synthesis.

Subsequent research has shown that the genetic code is essentially the same in all organisms, from bacteria to plants and animals. This discovery was a striking confirmation of the common origin of all life on Earth. Because the genetic code is essentially the same in all organisms, genes from bacteria can function in plants, and plant and animal genes can be made to function in bacteria. The common origin of all life on Earth makes genetic engineering possible (see Chapter 15).

EVOLUTION

We now know how genetic information is coded in molecules of DNA. We can therefore address the question of how these instructions are translated into the proteins that govern all the structures and functions of cells.

6.6 | Protein synthesis involves three forms of RNA in the cytoplasm

RNA (ribonucleic acid) was long suspected of playing a major role in protein synthesis. Cells that are synthesizing large amounts of protein also have large amounts of RNA. DNA is confined to the nucleus of eukaryotes, but RNA is found in the cytoplasm, where protein synthesis occurs. Cells making protein contain large numbers of ribosomes, which are small particles containing protein and a large amount of RNA. Some viruses contain only RNA and protein (Chapter 17). The existence of RNA viruses implies that RNA must be capable of carrying information about protein structure.

RNA is a long, thin molecule of nucleic acid, as is DNA, but it differs from DNA in three important features (see Figure 3.21): (1) RNA has ribose as a 5-carbon sugar instead of deoxyribose; (2) Instead of thymine, RNA contains the nitrogenous base uracil (U), which pairs with adenine; (3) RNA exists as a single strand and does not form a helix as DNA does. However, hydrogen bonds can form between different complementary regions of some types of RNA molecules.

Research revealed that three forms of RNA play important roles in the processes leading from the coded information in DNA to the synthesis of proteins: messenger RNA (mRNA), transfer RNA (tRNA), and ribosomal RNA (rRNA). The first step in protein synthesis involves copying the instructions for a protein coded in a DNA molecule into a complementary molecule of mRNA. This first step is called the process of **transcription** because the information is transcribed or copied from DNA to mRNA using the same code of sequences of nucleotides. The molecule of mRNA detaches from the DNA and moves out to the cytoplasm, where it joins with ribosomes. The ribosomes contain rRNA and proteins. They are the actual location of protein synthesis. In the second step of protein synthesis, **translation**, tRNA molecules bring the amino acids to the ribosomes. The coded instructions on the mRNA molecule are read, the amino acids are brought to the ribosome in the correct order, and the amino acids are assembled into the complete protein. This second step is called translation because the information in the nucleic acid mRNA is translated from a sequence of nucleotides into the sequence of amino acids, which defines the

First position in codon	Second position in codon				Third position in codon
	U	C	A	G	
U	phenylalanine	serine	tyrosine	cysteine	U
	phenylalanine	serine	tyrosine	cysteine	C
	leucine	serine	stop (UAA)	stop (UGA)	A
	leucine	serine	stop (UAG)	tryptophan	G
C	leucine	proline	histidine	arginine	U
	leucine	proline	histidine	arginine	C
	leucine	proline	glutamine	arginine	A
	leucine	proline	glutamine	arginine	G
A	isoleucine	threonine	asparagine	serine	U
	isoleucine	threonine	asparagine	serine	C
	isoleucine	threonine	lysine	arginine	A
	methionine (AUG)	threonine	lysine	arginine	G
G	valine	alanine	aspartic acid	glycine	U
	valine	alanine	aspartic acid	glycine	C
	valine	alanine	glutamic acid	glycine	A
	valine	alanine	glutamic acid	glycine	G

FIGURE 6.6 **The genetic code** The genetic code is given in terms of the mRNA codons. Most of the codons specify amino acids, and most amino acids have more than one codon. Methionine, with codon AUG, is the signal for the start of protein synthesis. Any of the three stop codons will end protein synthesis. The start and stop codons (which are spelled out) serve as examples of how to read the code in the figure.

protein. We will examine each of the two steps in protein synthesis in somewhat more detail in the next sections.

Instructions for protein synthesis coded in DNA are first transcribed into a coded mRNA molecule

Scientists are now able to produce a number of medicines such as insulin and human growth hormone more cheaply because we can control some aspects of cell protein synthesis in bacterial systems on an industrial scale. The instructions for synthesis of a particular protein are coded in DNA as a series of triplets, such as AAA or CGC. That information is transcribed into a molecule of mRNA in a manner similar to DNA replication. The double helix is opened up, and an enzyme called **RNA polymerase** synthesizes a strand of mRNA along one of the two strands of DNA, which acts as a template. Complementary base pairs are added just as in DNA replication, except that wherever a nucleotide with adenine occurs in the DNA, a nucleotide with the base uracil is added instead of thymine. Thus the triplet AAA in DNA becomes the **codon** UUU in mRNA. The genetic code is in fact defined in terms of the mRNA codons. Scientists broke the code by creating artificial mRNA molecules, each with a single type of codon, and then determining the single amino acid in the resulting protein. An artificial mRNA containing only the codon UUU produces a protein with only the amino acid phenylalanine.

The entire genetic code in terms of mRNA bases is given in Figure 6.6. Amino acids are specified by 61 of the 64 possible codons. The other three codons are **stop** codes to terminate protein synthesis. Since 61 codons are used for 20 amino acids, many amino acids are coded by more than one codon. The genetic code is therefore said to be redundant. The important amino acid methionine has only one codon, AUG, but it is the **start** signal for protein synthesis.

RNA polymerase attaches to a special region on the DNA strand called the *promoter* that lies close to the coding region of the gene (Figure 6.7). Transcription proceeds along the DNA template strand for anywhere from 500 to 10,000 nucleotides, depending on the size of the protein. When RNA polymerase reaches the end of the *transcription unit*, transcription stops, and the mRNA molecule is released.

In a prokaryotic cell, the mRNA molecule is ready for protein synthesis even before transcription is complete. In a eukaryotic cell, however, the mRNA must undergo further

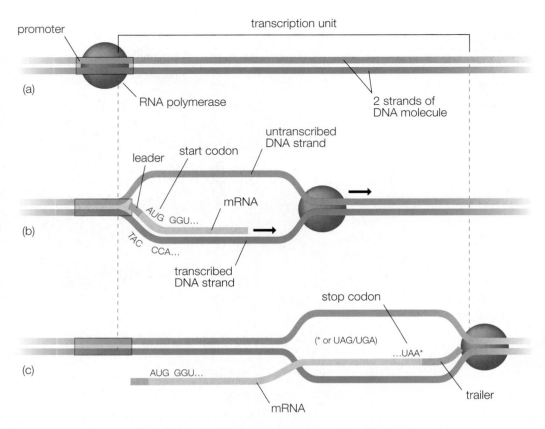

FIGURE 6.7 Transcription of mRNA from a template strand of DNA (a) RNA polymerase attaches at the promoter sequence and moves along the DNA molecule as the strands become unzipped. (b) As the RNA polymerase moves down the DNA strand, an mRNA molecule forms that is complementary to the transcribed DNA strand. The leader and trailer regions are transcribed but do not code for a part of the protein. (c) When RNA polymerase reaches the end of the transcription unit, transcription ceases and the mRNA molecule is liberated.

processing. In eukaryotes, the DNA sequence that codes for a particular protein (a gene) is often interrupted by sequences that do not code for any protein. These noncoding sequences are called **introns** (literally, "the piece within"), whereas the regions that code for a protein are called the **exons** (meaning "the outside segment"; Figure 6.8). In eukaryotes, the initial mRNA molecule is released inside the nucleus. Before it leaves the nucleus, the introns are snipped out and the exons spliced together. A cap and a tail of adenine nucleotides (poly-A tail) are also added to the mRNA before it leaves the nucleus. Prokaryotes typically do not have introns.

The coded information in mRNA is translated into a protein with the aid of ribosomes and tRNAs

Scientists can now produce desired proteins in the laboratory using ribosomes as protein factories. Whether from a prokaryote or a eukaryote, the ribosome binds to a ribosome-binding site at the beginning of the mRNA and then scans down to the initiation codon (AUG). A ribosome consists of two subunits, the small subunit and the large subunit. Both contain ribosomal RNA (rRNA) and proteins. Specific genes in DNA code for these rRNAs; they

are an example of genes that do not code for proteins. The small subunit contains the binding site for the mRNA molecule, and the large subunit contains three binding sites for the transfer RNA molecules (tRNAs), which bring in the amino acids.

The tRNA molecules are the actual translators in the process of protein synthesis. In every cell there can be more than 60 different tRNA molecules, one for each codon. The amino acid phenylalanine, for example, has two tRNAs, one for each of its two codons (see Figure 6.6). The genes for the tRNAs are also coded in DNA and are thus a second example of genes that do not code for proteins. Each tRNA molecule is a short, single strand of about 80 nucleotides that folds back on itself, forming several "stems" and "loops" (Figure 6.9). The base sequence and arrangement of the stems and loops is characteristic for each tRNA molecule.

Each tRNA has two important attachment sites: an **anticodon** on the central loop and the amino acid attachment site on the opposite end. The anticodon is simply a set of three nucleotides that recognizes and binds to the complementary codon on the mRNA molecule. If the codon on the mRNA were UAU, the anticodon on the corresponding tRNA would be AUA. On the opposite end of each tRNA molecule, the amino acid attachment site is always indicated by the three nucleotides CCA. When an amino acid is bound to its tRNA,

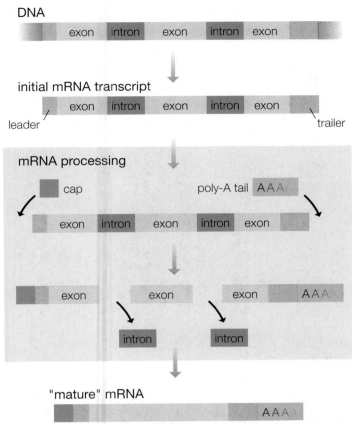

DNA

initial mRNA transcript

leader | trailer

mRNA processing

cap | poly-A tail AAAA

exon intron exon intron exon

exon | exon | exon AAAA

intron | intron

"mature" mRNA

AAAA

FIGURE 6.8 **The processing of a primary mRNA transcript in the eukaryotic nucleus** The initial transcript of RNA contains a leader, a trailer, and several introns, which do not code for protein. The introns are snipped out during mRNA processing; a "cap" is added at one end, and a "tail" composed of a stretch of adenine nucleotides (the "poly-A tail") is added at the other end. The cap and tail protect the completed (mature) mRNA from being degraded in the cytoplasm before it is translated.

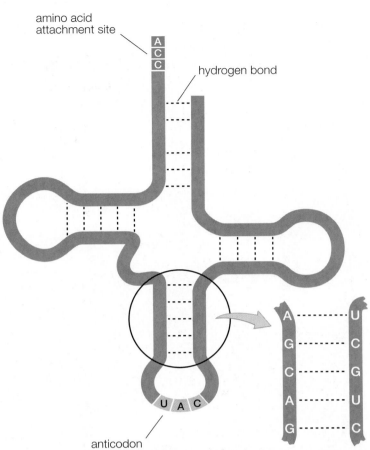

amino acid attachment site

hydrogen bond

anticodon

FIGURE 6.9 **Diagram of the structure of a tRNA molecule** The shape of the molecule is important and is achieved through hydrogen bonding (base pairing), which gives rise to stems and loops. Ribosomal RNA (rRNA) also has stems and loops that, together with proteins, give ribosomes their characteristic shape.

the resulting molecule is called an *aminoacyl-tRNA*. Enzymes called *aminoacyl-tRNA synthetases* match up each tRNA with the appropriate amino acid (Figure 6.10). Each of these important enzymes recognizes both the characteristic shape of a specific tRNA and the unique side chain that defines the amino acid to which the tRNA is to be attached. These synthetases therefore play an important role in translation of the genetic code contained in mRNA.

With this information about the different RNAs, we are now ready to consider the actual process of protein assembly at the ribosome. You will find the information in Figure 6.11 very helpful in following this discussion of protein synthesis. Protein synthesis consists of three stages: (1) initiation, (2) elongation, and (3) termination. In plant cells the initiation stage begins with the attachment of an initiator tRNA molecule to the small ribosomal subunit (Figure 6.11-1a). Methionine is the amino acid that is linked to the initiator tRNA. Next, the small subunit attaches to an mRNA molecule, and the anticodon of the initiator tRNA binds to the **start** codon (AUG) of the mRNA (step 1b). The large

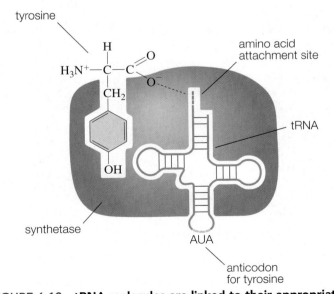

tyrosine

amino acid attachment site

tRNA

synthetase

AUA

anticodon for tyrosine

FIGURE 6.10 **tRNA molecules are linked to their appropriate amino acids by enzymes called aminoacyl-tRNA synthetases** Different synthetases recognize the shape of their corresponding tRNA molecules as well as the side chain of the appropriate amino acid (e.g., tyrosine); they attach the amino acid to the attachment site on the tRNA. The anticodon will pair with the complementary codon on an mRNA molecule.

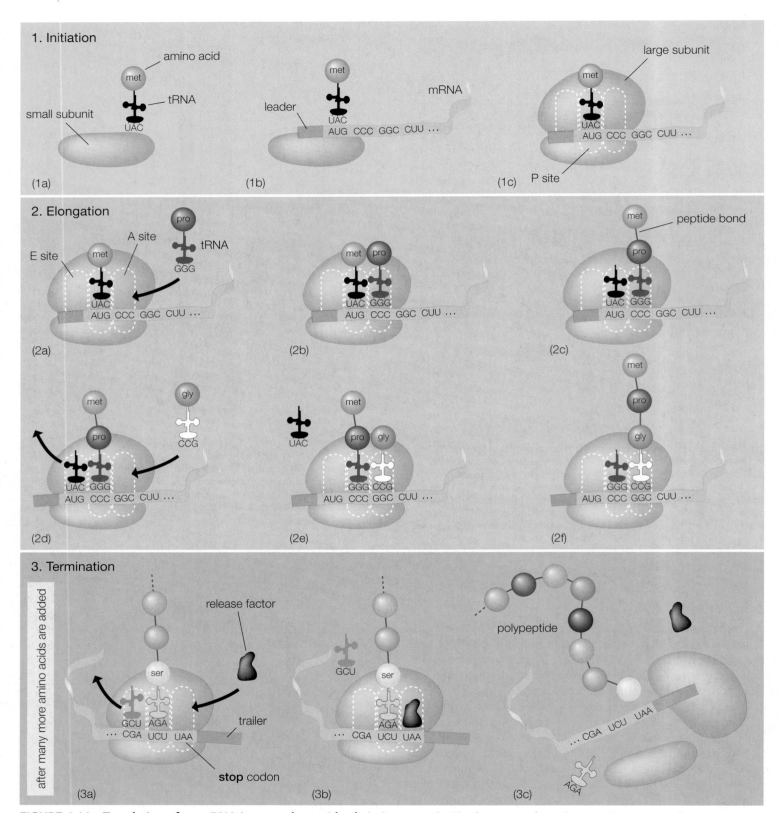

FIGURE 6.11 Translation of an mRNA into a polypeptide chain in a protein The first stage of translation is (1) initiation, where a tRNA carrying methionine attaches to a small ribosomal subunit and then to the start codon (AUG) of an mRNA molecule, after which the large subunit of the ribosome attaches to the small subunit. The second stage is (2) elongation, where tRNAs carrying their corresponding amino acids temporarily attach to the ribosome, where peptide bonds are formed between amino acids in the growing polypeptide as the ribosome moves along the mRNA molecule. (This process is explained in detail in the text, using this figure as an example.) The final stage is (3) termination, in which a release factor recognizes the stop codon on the mRNA molecule, liberating the newly synthesized polypeptide as well as the other components in the translational system. (met = methionine; pro = proline; gly = glycine; ser = serine)

subunit of the ribosome then attaches to the small subunit such that the methionine-tRNA lies in the P site of the large subunit (step 1c). The term *P site* refers to the fact that the polypeptide chain of the protein will be attached here.

The elongation phase now begins. A aminoacyl tRNA with the anticodon for the second codon on the mRNA now binds to the mRNA molecule and, with its amino acid, sits in the A site. In Figure 6.11-2b, the new amino acid is proline. The term *A site* means that this site receives the new aminoacyl-tRNAs. The close position of the tRNAs and particularly the tRNA at the P site, assist in the formation of a peptide bond between methionine and proline such that both amino acids are attached to the second tRNA in the A site (step 2c). The ribosome then shifts down the mRNA molecule by one codon, and several events occur simultaneously. The tRNA that brought in methionine moves to the E (exit) site and leaves the ribosome (step 2d). The tRNA with its attached peptide at the A site moves to the P site, and the A site is open to receive a new aminoacyl-tRNA.

The next amino acid (glycine in Figure 6.11) is now brought into the ribosome. The aminoacyl-tRNA binds to the codon on mRNA and lies in the A site (step 2e). A peptide bond is formed between proline and glycine, and the growing polypeptide chain is transferred to the tRNA in the A site (step 2f). Again the ribosome moves down one codon, a tRNA is released from the E site, the polypeptide and tRNA move to the P site, and the A site is open for another aminoacyl-tRNA. This process repeats for as many amino acids as there are codons in the mRNA molecule. Once the first codon is exposed to the cytoplasm, another ribosome can bind to it and begin translating the same mRNA molecule. When several ribosomes are translating the same mRNA molecule, they are referred to as a *polyribosome.*

The termination stage is reached at the point where one of the three codons for **stop** (UAG, UAA, or UGA) occurs on the mRNA molecule (step 3a). A cytoplasmic protein called a *releasing factor* (not a tRNA) binds to any **stop** codon at the A site (step 3b). The protein and the last tRNA are released, and the ribosomal subunits separate (Figure 6.11-3c).

This account of the process of protein synthesis generally applies to both prokaryotes and eukaryotes. We conclude with a brief discussion of DNA differences between organisms.

6.7 | Differences in DNA account for differences among organisms and even among individuals

Industries based on the cultivation of living organisms such as brewing, baking, and biotechnology must closely monitor their organisms, often with DNA-based methods, because a change in those organisms could be disastrous for production. For this reason, it is important to understand how the DNA of organisms differs.

TABLE 6.1	The Quantity of DNA in Various Prokaryotes and Eukaryotes in Terms of Number of Base Pairs (one kilobase [kb] is 1,000 base pairs)
Species (organism)	**Number of Base Pairs (kb)**
Prokaryotes with circular DNA	
Archaea (Archebacteria)	
Methanococcus jannaschii	1,700
Bacteria (Eubacteria)	
Mycoplasma genitalium	600
Synechocystis sp.	3,600
Escherichia coli	4,700
Eukaryotes with linear DNA	
Fungi	
Saccharomyces cerevisiae (yeast)	13,500
Plants	
Arabidopsis thaliana	100,000
Zea mays (corn)	4,500,000
Trillium	100,000,000
Animals	
Drosophila melanogaster (fruit fly)	165,000
Homo sapiens (humans)	3,000,000
Amphiuma sp. (salamander)	76,500,000

The amount of DNA varies enormously from organism to organism. The numbers in Table 6.1 are expressed in kilobases (thousands of base pairs). Since the average protein has about 300 to 400 amino acids, which require about 1,000 base pairs to code the genetic information, the numbers in the table indicate roughly how many structural genes an organism could contain if all the DNA coded for protein. Prokaryotes have the smallest amount of DNA. *Mycoplasma*, a bacterium, contains one of the smallest known amounts of DNA, enough for only about 760 proteins. Our common intestinal bacterium *E. coli* has 4.7 million base pairs, which code for about 4,000 proteins. Prokaryotes are generally very efficient in using their DNA.

Eukaryotes, in contrast, have much larger amounts of DNA, but much of it (90%) does not code for any proteins. About 20 to 40% of the DNA in eukaryotes consists of noncoding repeated sequences and has as yet no known function. Introns that interrupt sequences coding for proteins are another example of noncoding DNA. Among plants, *Arabidopsis* has a very small amount of DNA, which has made it the focus of a great deal of genetic research to identify and sequence all of its genes. What the common spring flower *Trillium* (Figure 6.12) needs with 100 billion base pairs of DNA is unknown. By comparison, human DNA has about 3 billion base pairs, as little as 1% of which may code for proteins.

DNA typically contains short sequences (most commonly four or six nucleotides long) called *recognition sequences*, which are symmetrical, meaning one strand is identical to the other when

FIGURE 6.12 *Trillium*, a plant with one of the largest known genomes

read in the opposite direction, such as CTGCAG/GACGTC (Figure 6.13). Due to genetic differences and mutations, every human (or other complex organism) has a unique pattern of these recognition sequences. If DNA is treated with special enzymes derived from bacteria, called *restriction endonucleases*, these recognition sequences will be cut by the enzymes and produce a series of DNA fragments (Figure 6.14). For example, the restriction

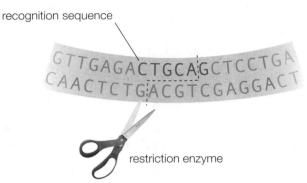

FIGURE 6.13 Symmetrical recognition sequences in strands of DNA The two sequences are identical when read in opposite directions. Restriction endonucleases will cut these sequences.

FIGURE 6.14 RFLPs—Restriction Fragment Length Polymorphisms
(a) Restriction endonucleases cut DNA at specific recognition sequences (also known as restriction sites) and (b) produce DNA fragments of various sizes. The DNA fragments can be separated by size using gel electrophoresis. A gel is a semisolid matrix containing a lattice of tiny channels through which molecules such as fragments of DNA can pass. Small samples of DNA are added to "wells" in the gel, usually with a colored dye (c). An electric current applied to the gel moves the fragments of DNA through the gel. DNA is negatively charged and thus it moves toward the positive pole. Smaller fragments travel farther than larger ones, and a pattern of DNA fragments is produced with the larger pieces at the top and the smaller toward the bottom (d). After staining or labeling, the resulting pattern of bands is called a DNA fingerprint.

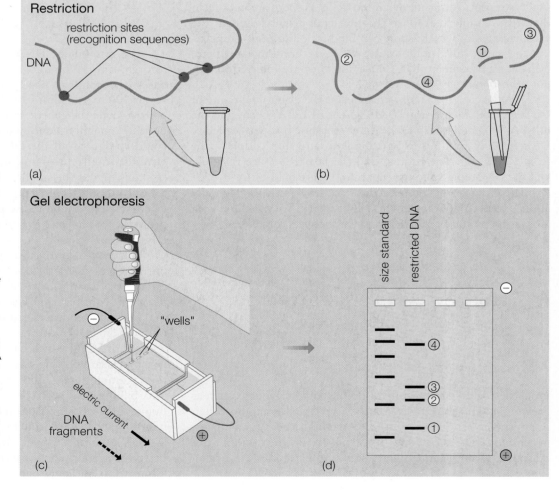

endonuclease *Pst I* cuts DNA where the sequence CTGCAG is present. These fragments can be separated out on a gel material (derived from red algae) by applying an electrical current in the process known as gel electrophoresis. Gel electrophoresis separates the fragments according to size; the smallest fragments migrate the farthest through the gel, and the largest migrate the least distance.

Since every human (or other animal or plant) has a unique arrangement of recognition sequences, each will have a unique pattern of these DNA fragments on a gel. The variability of these fragment sizes is called **restriction fragment length polymorphism**, or **RFLP** (pronounced "Riflip"), where *polymorphism* means "many forms." A gel showing RFLPs is also called a **DNA fingerprint**. For some uses of DNA fingerprinting, refer to Essay 6.1, "Molecular Detectives: DNA Fingerprinting Solves Crimes."

ESSAY 6.1   MOLECULAR DETECTIVES: DNA FINGERPRINTING SOLVES CRIMES

DNA fingerprinting (Figure E6.1) has had a major impact on the field of forensic science, which is the science of gathering evidence for criminal and social justice cases. DNA fingerprinting is routinely done in connection with a second technique called *polymerase chain reaction* (PCR; see Chapter 15). PCR allows the DNA present in minute samples such as a spot of blood, semen, or skin cells to be amplified until enough DNA is obtained to produce a DNA fingerprint. The technique was first used in a criminal case in England. In 1986, the bodies of two 15-year-old girls were found in the Leicestershire countryside. They had been raped and strangled. A DNA fingerprinting test, developed by geneticist Alex Jeffreys to look for genetic disease markers, was used to obtain a DNA fingerprint of the murderer from semen on the bodies. Then forensic investigators collected more than a thousand blood samples from men in three villages around the murder scene. The DNA fingerprint of a young baker matched that of the rapist and murderer. The suspect confessed.

Even DNA fingerprints from plants have helped solve murders. In Phoenix, Arizona, in 1992 a woman was found murdered near a palo verde tree (Chapter 26). The principal suspect denied he was at the scene, but his pickup truck contained

In Phoenix, Arizona, in 1992 a woman was found murdered near a palo verde tree.

palo verde seed pods. Detectives went to Dr. Timothy Helentjaris at the University of Arizona, who had done DNA fingerprints of many crop plants. He was given seed pods from 12 different palo verdes in the area of the murder, including the one nearest the body, plus 18 other samples from trees around the Phoenix area. Helentjaris did not know which was which. He obtained DNA fingerprints for each sample by PCR, and only two samples matched. The seed pods from the pickup truck matched the ones from the tree near the body. The suspect was found guilty of first-degree murder.

Mitochondrial DNA, which is the circular DNA found in the mitochondria of eukaryotic cells (Chapter 4), and PCR have been used to return kidnapped children to their families. In 1976, the civilian government of Isabel Perón was overthrown in Argentina and a military regime began a 7-year program of kidnapping and murder. At least 12,000 people were killed and 210 children kidnapped or born in prison before their mothers were executed. These children were given away or sold. A human rights organization called the Grandmothers of the Plaza of May was founded in Buenos Aires to track the missing children. In 1983, a new civilian government came to power, and it became possible to seek a return of these children. A group of grandmothers came to the United States seeking help; they needed a way to establish family

connections in children suspected of being kidnapped years earlier. Dr. Mary-Claire King provided that help through genetic tests and PCR on mitochondrial DNA, which children inherit from their mothers. While many children have been returned to their families, the work will continue for years as kidnapped children reach adulthood and seek out their families.

DNA science also aided in the identification of victims of the attack on the World Trade Center on September 11, 2001. Genetic profiles from remains were matched to DNA samples obtained from articles owned or worn by victims.

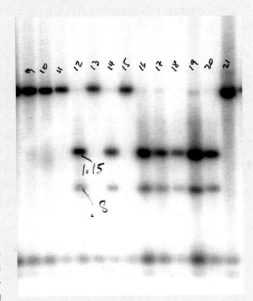

E6.1   DNA fingerprint

HIGHLIGHTS

6.1 DNA is a long, thin molecule composed of thousands of nucleotides. Each nucleotide contains the 5-carbon sugar deoxyribose, a phosphate group, and one of four different nitrogen-containing bases: adenine, guanine, cytosine, or thymine.

6.2 DNA exists in the form of two nucleic acid strands, which wind around each other in a double helix. The two strands are complementary to each other because adenine can pair only with thymine and guanine can pair only with cytosine.

6.3 DNA replicates by unwinding the two strands. Each strand acts as a template for synthesis of a new strand by complementary base pairing.

6.4 Mutations are changes in genetic information that can arise through errors in replication or damage from radiation or chemicals. Enzymes called *DNA repair nucleases* remove damaged DNA.

6.5 Genetic information is coded in DNA by sets of three nucleotides called *triplets*.

6.6 Three forms of RNA translate the information in DNA into proteins. Messenger RNA (mRNA) contains a transcript of the information coded in DNA for a protein. The mRNA joins with ribosomal RNA (rRNA) at the ribosomes, which are the protein factories of the cell. Transfer RNAs (tRNAs) bring the amino acids to the ribosome, where the codons in mRNA are translated into protein.

6.7 Species differ enormously in the quantity of DNA in their cells. Prokaryotes have the fewest number of DNA base pairs, and some eukaryotes have as many as 100 billion base pairs. Individuals have unique patterns of DNA segments, which can be identified in DNA fingerprinting.

REVIEW QUESTIONS

1. Describe the roles of the following structures/compounds with respect to DNA: (a) chromosomes, (b) histones, (c) genes, (d) nucleotides, (e) deoxyribose and phosphate, and (f) the nitrogenous bases adenine, guanine, cytosine, and thymine.

2. The bases adenine, cytosine, guanine, thymine, and uracil are found in DNA and/or RNA molecules. Which bases are found (a) in both DNA and RNA, (b) in DNA only, and (c) in RNA only? Which of these bases (d) contain nitrogen, (e) are purine compounds, or (f) are pyrimidine compounds?

3. Discuss the base-pairing rules in DNA and RNA. Are all bases normally paired in DNA? In RNA? If one strand of a DNA fragment had the base sequence AATATACCG, what would be the sequence of the complementary strand?

4. Briefly describe the semiconservative mode of DNA replication. What are the roles of origins of replication, DNA polymerases, helicases, and binding proteins in this process?

5. For many years in the latter part of the 20th century, there was a dictum in the field of biology: "One gene, one protein." That is, each gene encoded the sequence of one and only one protein. As important as protein encoding is, we now recognize that different segments of DNA may have functions other than just encoding amino acid sequences. Give several examples of such alternate functions of DNA.

6. Although DNA and RNA are both nucleic acids, they differ in several fundamental ways. List at least three properties they share and at least three ways they are different.

7. List the three types of RNA and discuss their functions within the cell.

8. Protein synthesis proceeds through three stages: (a) initiation, (b) elongation, and (c) termination. Briefly discuss each of these stages and their roles in the formation of a new protein molecule.

9. Different organisms possess different amounts of DNA and differing numbers of genes. Compare these differences between (a) prokaryotes and eukaryotes, (b) various prokaryotes, and (c) various eukaryotes.

10. Describe a ribosome, including the compounds that compose it, and describe the sites found on its surface that are important in protein translation.

11. One major difference between prokaryotic and eukaryotic genes is that in prokaryotes, the mRNA, once formed by transcription, is usually immediately ready for translation, but in eukaryotes the mRNA must typically be processed before it can be translated. Briefly describe this processing, discussing the role of introns and exons.

12. Briefly, in one paragraph, distinguish between the processes of replication, transcription, and translation.

APPLYING CONCEPTS

1. Consider a hypothetical fragment of a DNA molecule containing a very short gene (genes are typically much longer than this). Its base sequence is as follows:

   TTTACAAACCAGTCGCTCCCATCCAA

   Give the resulting base or amino acid sequences of the products after the following operations have occurred: (a) replication; (b) transcription of the hypothetical strand, into mRNA; and (c) referring to Figure 6.6, the genetic code, translation of the mRNA into a short polypeptide. When you are translating the sequence, keep in mind that translation begins at **start** codons and terminates at **stop** codons.

2. Continuing with the example from Question 1, consider the following single-base changes in the sequence of the original DNA fragment and predict their consequences with respect to both the resulting mRNA and polypeptide sequences. In the following sequences, changes have been highlighted in boldface:
   The original DNA sequence

   TTTACAAACCAGTCGCTCCCATCCAA

   (a) A single substitution,

   TTTACAAACCAGTCGCTCC**G**ATCCAA

   (b) A different substitution,

   TTTACAAACCAGTC**G**GCTCCCATCCAA

   (c) Another substitution,

   TTTA**G**AAACCAGTCGCTCCCATCCAA

   (d) Another substitution,

   TTTACAAACCAGTCGCTCCCAT**G**CAA

   (e) A single base (T) deletion,

   TTTACAAACCAGTCGC-CCCATCCAA

   (f) A single base (T) insertion,

   TTTACAAACCAGTCGCT**T**CCCATCCAA

3. Imagine an extraterrestrial life-form whose DNA uses the same four nitrogenous bases as terrestrial life but must encode for some number of amino acids other than 20. What would be the minimum codon size, in bases, if only 14 amino acids must be coded for? 50 amino acids? 70 amino acids?

4. Suppose that different groups of organisms defined codon usage in different ways, so that perhaps one group would interpret a UUU codon as encoding phenylalanine, while in another group UUU would encode another amino acid (e.g., tyrosine). Imagine that this phenomenon was widespread and that terrestrial life consisted of hundreds, even thousands, of groups of organisms, each using the same nitrogenous bases and the same 20 amino acids but interpreting the genetic code in their own unique way. What would be the implications of such a situation regarding the theory of the common origin of life on Earth?

5. Recognition sequences in DNA are usually symmetrical, meaning that the complementary strands have the same, though oppositely oriented, sequence. The recognition sequence for the PstI restriction enzyme provides an example:

   CTGCAG
   GACGTC

   Such a stretch of DNA, which reads the same forward or backward (though on opposite strands) is called a *palindrome*. Common words, phrases, and even sentences can also be palindromic. Examples of palindromic words are *pop*, *mom*, and *radar*. Much more elaborate examples can be found in palindrome compilations, such as Jon Agee's delightful book *Go hang a salami! I'm a lasagna hog!* (New York: Harper Collins, 1991). Here are some examples from Agee's book (ignore punctuation):

   "Neil, an alien"
   "Yo! Bozo boy!"
   "Pint a' catnip!"
   "Amy, must I jujitsu my ma?"
   "I madam, I made radio! So I dared! Am I mad? Am I?"

   Can you think of any additional palindromic words, phrases, or sentences?

Taxus brevifo[l]

KEY CONCEPTS

- All organisms are composed of one or more cells; all cells pass through a series of events called the cell cycle.

- The cell cycle and cell division are defining attributes of life.

- Prokaryotic organisms and cell organelles reproduce by binary fission.

- Eukaryotic organisms have separate processes of nuclear division and cytoplasmic division.

- Mutations in the genes that control nuclear division can lead to cancers in both plants and animals, including humans.

Not long ago infectious diseases caused by bacteria ravaged most human populations but, in modern societies, cancer, of which there are more than 200 recognized types, is one of the leading causes of death. In North America more than half a million people die of cancer each year. For thousands of years human cultures have been aware of cancer and tried to treat it with various plant extracts. In the 1950s, the National Cancer Institute began a program to screen plant extracts for potential anticancer activity. Fertile chicken eggs were infected with cultured cancer cells. Plant extracts were added to the fertile eggs, and the effects on the developing chicken embryo were monitored. Thousands of plants were tested. Hundreds were found to have some inhibitory effect on cancer growth. Many more thousands remain untested, and, with the present rate of plant species extinction, life-saving compounds may vanish before they can be discovered.

Two major success stories are the discovery of the anticancer compounds in the Madagascar periwinkle, *Catharanthus roseus*, and the Pacific yew tree, *Taxus brevifolia*. Both plants produce secondary plant compounds called *alkaloids* (Chapter 3). The Madagascar periwinkle is a tropical perennial herb used by traditional healers as a treatment for diabetes. It is known to contain more than 70 different alkaloids. Two alkaloids, known as vincristine and vinblastine, are used in chemotherapy. Vincristine is used in the treatment of childhood leukemia, where it produces a 99% remission rate and 50% survival after 3 years. Vinblastine is used in treating cancers of the lymphatic system, such as Hodgkin's disease, which now has a 40% survival rate.

The bark of the Pacific yew tree yielded an extract that contained the anticancer alkaloid taxol. At first the supply of taxol was a problem because the Pacific yew tree was rare. Other species of yew tree, such as the English yew, *Taxus baccata*, also were found to have taxol in their bark and leaves. Taxol has recently been synthesized in the laboratory. Taxol is especially promising for the treatment of advanced ovarian and breast cancers.

Cancers arise when cells become abnormal and divide rapidly in an uncontrolled manner. Vincristine, vinblastine, and taxol act against cancer by stopping this uncontrolled cell division. To understand how they stop uncontrolled cell division, it is necessary to understand the normal process of cell division. This chapter focuses on the normal processes of cell division in prokaryotes and plants, which produce daughter cells that are essentially identical to the parent cell. ■

Catharanthus

7.1 | Cell division and the cell cycle

All organisms are composed of cells. An organism may consist of a single cell (and thus be termed unicellular) or may contain many cells (multicellular). All cells go through a process called the **cell cycle.** The cell cycle is the orderly sequence of events that constitute the life history of a cell. In a typical cell cycle there is a relatively long interval during which the "parent" cell duplicates all its structural components including membranes, organelles such as mitochondria or chloroplasts (Chapter 4), and the genetic information in its DNA (Chapter 6). A relatively brief period then follows in which the parent cell divides these duplicated components between two "daughter" cells. This division of components is called **cell division.** The daughter cells, which result from normal noncancerous cell division, are essentially identical to the parent cell. Cell division and the cell cycle are among the defining attributes of life.

In prokaryotes (bacteria and archaea) and single-celled eukaryotic protists, cell division leads to an increase in the number of individuals in a population. In single-celled protists and prokaryotes, this type of cell division is called asexual reproduction or **binary fission.** Binary fission literally means a breaking into two parts. In multicellular eukaryotic organisms such as plants and animals, cell division leads to an increase in the size of the organism or to replacement of worn-out or damaged cells in the body. This type of cell division, in which the daughter cells are identical to the parent cell, is called **somatic**—or body cell—division in multicellular organisms.

When something goes wrong with the orderly process of the cell cycle and cell division, the consequences can be severe.

Scientists now understand that normal cells become cancerous when certain mutations or changes occur in their genetic material. These mutations may arise spontaneously or, more often, result from exposure to radiation or to chemicals called *carcinogens.* Certain viruses can also induce genetic changes that initiate cancerous growths or **tumors.** Plants, as well as animals, can develop tumors. The soil bacterium *Agrobacterium tumefaciens* attacks many commercial crops, including apples, grapes, cherries, and walnuts. The bacterium invades the plant through wounds and transfers part of its own DNA into the host plant's cells. The transferred bacterial DNA directs the plant cells to produce substances that cause them to divide rapidly in an uncontrolled manner. The result is a tumorous growth called a *crown gall* that is similar to cancer in animals (Figure 7.1). Severely infected plants die. *Agrobacterium* has proven to be a useful model system for studying cancer and a valuable tool in plant genetic engineering (Chapter 15).

7.2 | Division in prokaryotes, mitochondria, and plastids occurs by binary fission

Prokaryotes, both bacteria and archaea, are living cells that possess a cell cycle and reproduce by cell division, or binary fission. Prokaryotes have one circular chromosome containing a closed, double-stranded loop of DNA. The single circular chromosome typically contains only one copy of each gene. Because prokaryotes have only one copy of each gene, their genome is referred to as **haploid.**

DNA SCIENCE

In prokaryotes, the chromosomal DNA tends to aggregate in a specific location within the cell, where it can be seen with the electron microscope as a discrete region, the **nucleoid** (Figure 7.2). In some prokaryotes, the nucleoid is large enough to be seen with a light microscope if special stains are used. The common intestinal bacterium *Escherichia coli* has a single chromosome whose DNA contains 4,700,000 nucleotide base pairs. This number is expressed as 4,700 kb (1 kb or kilobase equals 1,000 base pairs). Since the average protein requires about 1,000 base pairs to code for its genetic information, *E. coli* would have 4,700 structural genes if all of its DNA

FIGURE 7.1   **Crown gall on a *Kalanchoë* plant**

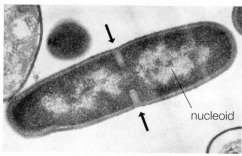

FIGURE 7.2   **Nucleoid region of a bacterium**   In this electron micrograph of a dividing bacterial cell, a nucleoid is present in each half and a septum (arrows) is forming between them.

coded for proteins. This genome size is large for a prokaryote but small in comparison to eukaryotes (refer to Table 6.1). If *E. coli* cells are gently broken open (lysed) and prepared for electron microscopy, the DNA spills out of the cell and can be seen (Figure 7.3). If all of this DNA were stretched out in a straight line, it would be about 1 millimeter (mm) long. But this DNA is packed into a cell with dimensions of about 0.5 micrometer (μm) by 1–2 μm long. Consequently the DNA in the bacterial chromosome is highly coiled and the coils are stabilized by proteins (Figure 7.4).

Prokaryotes reproduce by a process called *binary fission*. In binary fission, the bacterial chromosome is first duplicated to produce two daughter chromosomes (Figure 7.5). Each of the daughter chromosomes is attached to a separate inward fold in the cell membrane. The area between the two attachment points elongates until the cell is about twice as long

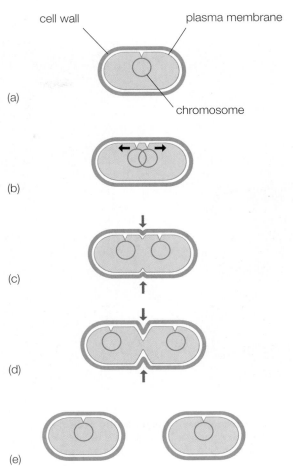

FIGURE 7.5   **Binary fission in bacteria** (a) A bacterial cell's chromosome is attached to the plasma membrane. (b) Following replication of the DNA, the two chromosomes (each attached to the plasma membrane) begin to move apart as the cell elongates (c) and (d). As the cell continues to elongate and the chromosomes move farther apart, a septum begins to form (red arrows) between what will become the daughter cells. (e) The two daughter cells, each with their own chromosome.

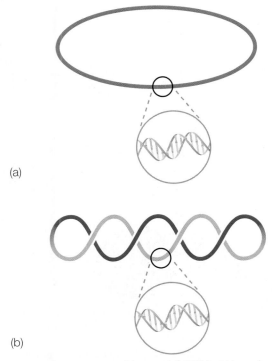

FIGURE 7.3   DNA of *E. coli* This cell was broken open and the DNA (colored orange in this false-colored electron micrograph) has spilled out.

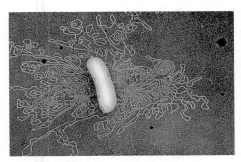

FIGURE 7.4   Arrangement of bacterial DNA Although a bacterial chromosome is a circular molecule of DNA and often diagrammed as a simple circle (a), the DNA is actually twisted back on itself in a "supercoiled" arrangement (b).

as the original parent cell. Cell elongation carries the daughter chromosomes apart. The cell wall and cell membrane then begin to grow inward, forming a partition, the **septum** (see Figure 7.2). Finally, the two daughter cells are pinched off as separate cells. Under optimal conditions of temperature and nutrients, *E. coli* can complete this fission cycle in about 20 minutes.

Organelles such as the mitochondria and chloroplasts found in eukaryotic cells are derived from prokaryotic cells, and like their ancestors these organelles maintain a prokaryotic structure, cell cycle, and division process. Mitochondria, which are the location of cellular respiration, and chloroplasts, the site of photosynthesis (Chapter 5), reproduce by binary fission. The DNA of mitochondria and chloroplasts occurs as one or more chromosomes, each containing one double-stranded molecule of DNA. As in prokaryotes, during binary fission the mitochondrial DNA is duplicated, and the daughter chromosomes are attached at different points to the inner mitochondrial membrane. Elongation of the mitochondrion separates the

EVOLUTION

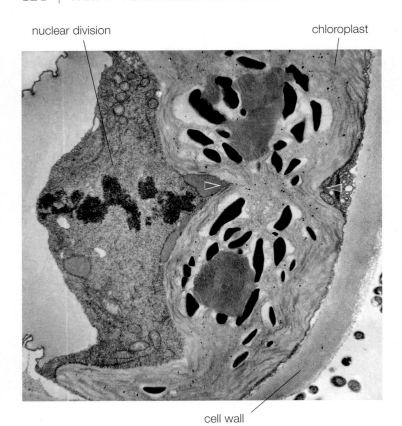

nuclear division   chloroplast

cell wall

**FIGURE 7.6 Chloroplast division in the green alga
*Coleochaete*** The chloroplast in this cell is dividing by binary
fission (arrowheads). Nuclear division is also taking place.

daughter chromosomes. The mitochondrial membranes then
constrict and separate the two daughter organelles. Plant cells
may have a few hundred to several thousand mitochondria
per cell.

The mechanism of division in chloroplasts is essentially the
same as described for prokaryotes and mitochondria. The final
stage of fission is marked by a deep constriction in the plastid
membrane (Figure 7.6). Plastids often divide all at the same time,
in synchrony with the cell division of their eukaryotic host cell.
Synchronous plastid division assures that both daughter cells
will receive some plastids when the eukaryotic cell divides. If the
plastids did not divide at the same time, there is a real chance
that one of the eukaryotic daughter cells might not receive any
plastids because there is no mechanism to divide the plastids
evenly between daughter cells.

7.3 | Eukaryotic cells have separate processes of nuclear and cytoplasmic division

Cell division in eukaryotic cells is much more complex than in
prokaryotes or organelles. Animal cells have two genomes in
two compartments to replicate, the mitochondrial and the nuclear genomes. Plant cells have three genomes—in the nucleus,

plastids, and mitochondria. We have presented the process by
which organelles are reproduced. Our focus in the rest of this
chapter is on the distinctly eukaryotic processes of nuclear and
cytoplasmic division. In **nuclear division**, the nucleus of the
eukaryotic cell duplicates all the genetic material in its chromosomes and then divides the duplicated chromosomes equally
between two daughter nuclei. Cytoplasmic division then follows nuclear division; it is the division of the cell cytoplasm—
all the cellular material outside of the nucleus—between the
two daughter cells.

The accurate distribution of copies of genetic material is
a much more complex process in eukaryotes than in prokaryotes because eukaryotic cells contain about a thousand times
more DNA than prokaryotic cells (compare data in Table
6.1). Eukaryotic DNA is organized into long, linear molecules
that form distinct chromosomes, rather than circular molecules, and it is associated with proteins called *histones* that
give structure to the DNA and chromosomes. The DNA is
wound around histone proteins to form spool-like units
called *nucleosomes*. The nucleosomes are then packed into
denser fibers, which are then looped into coils and condensed
into the chromosomes that we can see with the light microscope (Figure 7.7). This winding and coiling of DNA prevents

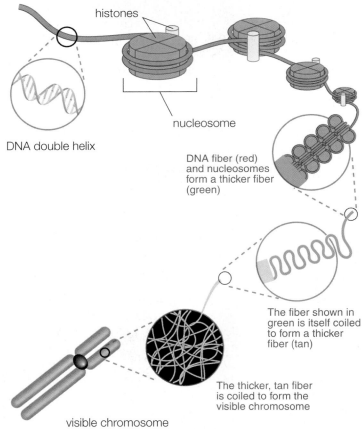

histones

nucleosome

DNA double helix

DNA fiber (red)
and nucleosomes
form a thicker fiber
(green)

The fiber shown in
green is itself coiled
to form a thicker
fiber (tan)

The thicker, tan fiber
is coiled to form the
visible chromosome

visible chromosome

FIGURE 7.7 Nucleosome and chromosome structure In eukaryotes, DNA wraps around proteins called *histones*, which make up
spool-like structures termed *nucleosomes*. The DNA and nucleosomes
are packed closely together to form a fiber. This fiber, in turn, is looped
and coiled into the condensed DNA of the visible chromosome.

tangling and breakage, but it also makes the process of replication of DNA and its separation into two daughter cells more complex.

The number of chromosomes is characteristic for each species, although some species may have the same number. Humans have 46 chromosomes, wheat, 42, and cabbage, 20. When any of these cells divide, each daughter cell must receive one copy of each chromosome. In contrast to prokaryotes, in which there is only one chromosome and the genome is haploid, in many eukaryotes (including plants and animals), each chromosome occurs as a member of a pair. Wheat has 42 chromosomes, which occur as 21 pairs, and cabbage has 20 chromosomes in 10 pairs. Eukaryotic cells in which each chromosome occurs as a member of a pair are **diploid**. The members of a pair of chromosomes are called **homologous chromosomes**. Chromosomes usually differ enough in length, staining pattern, and shape that the members of a homologous pair can be recognized at the microscopic level. Diploid organisms carry two versions of each chromosome and, therefore, two copies of each gene are present. (Even more copies of some genes that code for materials needed in large amounts—ribosomal RNA genes, for example—may be present, regardless of whether an organism is haploid or diploid.)

In addition to a large amount of genetic material, eukaryotes also possess a nuclear envelope. The nuclear envelope surrounds the chromosomes and separates them from the cytoplasm. It also protects the chromosomes from enzymes that might degrade them, by controlling exit and entry of molecules at the **nuclear pores**. At the time of cell division, the nuclear envelope forms a barrier to the separation of chromosomes between daughter cells. The challenge of a large coiled genome and nuclear envelope barrier is solved by a mechanism that is both elegant and precise. **Mitosis**, or nuclear division, is a series of steps by which a set of duplicated chromosomes is separated into two daughter nuclei.

Preparation for cell division occurs during interphase

In eukaryotic cells, the cell cycle consists of a repeating sequence of processes. A single cell cycle covers the interval from the beginning of one cell division to the beginning of the next. The length of time to complete a cell cycle depends on the type of cell and external conditions such as temperature, but it generally ranges from a few hours to several days. The eukaryotic cell cycle can be represented as a circular chart (Figure 7.8). The cell cycle is divided into a relatively long **interphase** and a brief division phase. The division or **M phase** includes both mitosis and the division of cytoplasm, or cytokinesis. Interphase precedes mitosis and cytokinesis. It is a period of active growth and synthesis of cellular components, including the replication of DNA and duplication of chromosomes. Interphase is divided into three phases, called the G_1, S, and G_2 phases. (The letter G stands for gap phase, and S stands for synthesis phase.)

The G_1 phase is a period of intense synthesis of molecules and structures   During the G_1 phase, or first

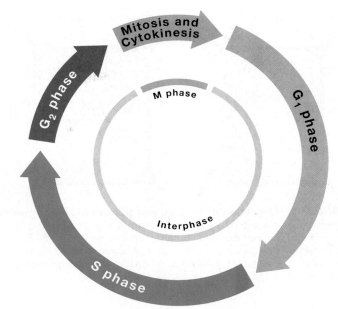

FIGURE 7.8   The cell cycle The cell spends most of the cell cycle in interphase, which consists of three phases. The first gap phase, G_1, is a period of intense synthesis of structures. In the S phase, DNA is replicated; and the second gap phase, G_2, completes preparation for cell division.

gap phase, the cell increases in size and synthesizes enzymes, ribosomes, and membrane systems. Organelles such as plastids and mitochondria reproduce during this phase. Microtubules, which are long, thin (about $0.025\ \mu m$ in diameter) tubules composed of the protein tubulin, and filaments of actin—one of the major proteins found in muscle—are assembled out of newly synthesized proteins. Microtubules move the chromosomes during mitosis. Membrane structures such as Golgi bodies, vesicles, and vacuoles are constructed from lipids and proteins.

In a process unique to plant cells, the nucleus migrates into the center of the cell, if it was initially at the periphery. The nucleus becomes fixed in the center of the cell by strands of cytoplasm (Figure 7.9). These strands merge to form a sheet that bisects the cell along the plane where it will later divide into two daughter cells. This sheet, the **phragmosome**, contains microtubules and actin filaments. At the end of the G_1 phase, a control system called a **checkpoint** may stop the cell cycle or signal the beginning of the S phase. The checkpoint might stop the cell cycle if insufficient materials, such as membranes or microtubules, had been synthesized to proceed or if the external environment had become adverse.

DNA replication occurs during the S phase   The main events of the **S phase**, or synthesis phase, are the replication of DNA and the synthesis of the associated histone proteins. At the end of the S phase, the chromosomes have been duplicated, but they cannot be seen under a light microscope because the DNA molecules are long and threadlike. As discussed in Chapter 6, DNA can replicate only when it is in

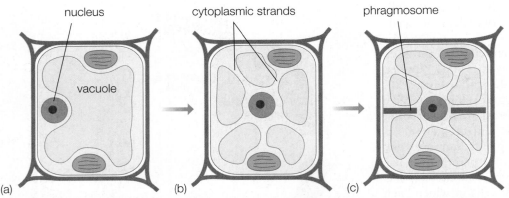

nucleus

cytoplasmic strands

phragmosome

vacuole

(a)　　　　(b)　　　　(c)

FIGURE 7.9 Migration of nucleus and phragmosome formation prior to cell division
(a) Plant cell nuclei often lie toward the periphery of nondividing cells, and a large vacuole is present in many cells. (b) Prior to cell division the nucleus moves to the center of the cell, where it is suspended by strands of cytoplasm. (c) A phragmosome of microtubules and actin filaments forms in the plane in which the cell is about to divide. The phragmosome forms a flat sheet, which is shown cut in half in this figure.

this long threadlike condition because only then can the two complementary chains be opened to permit replication by complementary base pairing. It is absolutely vital that the pairs of duplicated chromosomes, called *sister chromatids*, remain attached to each other. If they do not, there is no way the mechanism of mitosis can ensure that each daughter cell receives one, and only one, copy of each chromosome. Current research is directed toward understanding the mechanisms that bind duplicated chromosomes together until the correct moment in mitosis.

The G₂ phase completes preparations for cell division The G_2 phase, or second gap phase, completes the final preparations for cell division. Protein synthesis increases as the cell assembles the structures necessary for distributing a complete set of chromosomes to each daughter nucleus and dividing the cytoplasm. These structures include microtubules and actin filaments necessary for formation of the mitotic spindle and phragmosome. At the end of G_2, a second checkpoint can stop the cell cycle or proceed to mitosis. If DNA damage has occurred or replication is not complete, the checkpoint can halt the cycle until repairs or replication are finished. The G_2 phase ends and mitosis begins as the diffuse chromosomes of the interphase nucleus slowly condense into visible chromosomes.

Mitosis consists of four phases

Mitosis, or nuclear division, is a continuous sequence of events whose beginning is marked by the gradual condensation of diffuse chromosomes. When the chromosomes first become visible in the light microscope, they appear as very thin threads. This threadlike appearance is the origin of the word *mitosis*, because in Greek *mitos* means "thread." Most cellular metabolism, such as protein synthesis, ceases during the

brief time the cell undergoes division. Mitosis is typically divided into four phases: prophase, metaphase, anaphase, and telophase.

In prophase, chromosomes condense until they appear as sister chromatids The beginning of **prophase** (from the Greek *pro*, meaning "before") is marked by the appearance of threadlike strands of fine chromosomes (Figure 7.10), whose DNA was duplicated in the S phase of interphase. The DNA in these chromosomes may be several centimeters long, but as prophase continues, the chromosomes coil up and condense until they are only 5 to 10 μm long. If chromosomes did not condense in this manner, they would become tangled and break during the later phases of mitosis, and each daughter cell would not receive a complete genome.

As they condense and become more distinct, the duplicated chromosomes can be seen to consist of two **sister chromatids** with a constriction somewhere along their length called a **centromere** (Figure 7.11). The centromere is a special region on each chromosome with specific DNA sequences that binds the chromosome to the mitotic spindle.

During prophase, microtubules appear in a clear area around the nucleus. At first these microtubules are randomly oriented, but by late prophase they have lined up along the future spindle axis. This is the first appearance of the **mitotic spindle**, which is a structure composed of microtubules that aids in the separation and movement of the sister chromatids into the daughter nuclei. The equator of the mitotic spindle lies in the same plane as the phragmosome. The poles of the mitotic spindle lie at the farthest points away from the equator, and the microtubules converge at the poles (Figure 7.10). The alkaloids vincristine and vinblastine, which were described in the chapter introduction, stop the uncontrolled cell division of cancer by preventing cancerous cells from assembling the microtubules to make the mitotic spindle. Without the spindle, the sister chro-

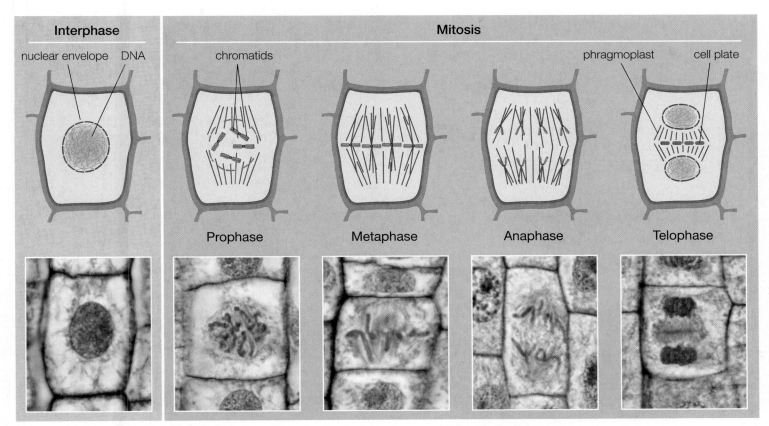

FIGURE 7.10 **The phases of mitosis** In interphase, the chromosomes are not visible with the light microscope and the nuclear envelope is intact. In prophase, the DNA condenses and chromosomes become apparent. The mitotic spindle begins to form and the nuclear envelope breaks down. Chromosomes align themselves in the center of the cell at metaphase. Sister chromatids separate in anaphase and move to the poles. In telophase, the sister chromatids have completed their poleward migration and the chromosomes become diffuse. The nuclear envelope re-forms. Cytokinesis begins during telophase with a phragmoplast and cell plate being visible. The photographs shown here are of mitosis in an onion root tip. Because of its simple structure, the growing root is an excellent place to observe cell division in plants.

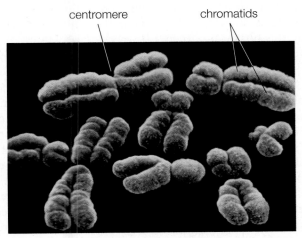

FIGURE 7.11 **Chromosomes viewed with the scanning electron microscope**

matids cannot move and cell division stops. Toward the end of prophase, the nuclear envelope breaks down.

In metaphase, chromosomes align on the equator of the mitotic spindle The mitotic spindle becomes fully formed in **metaphase** (from the Greek *meta*, meaning "after"). At the centromere, each chromatid develops a special protein complex called a **kinetochore** (Figure 7.12). The kinetochores lock onto the ends of microtubules at what is considered their plus (+) ends. The minus (−) ends of the microtubules are all located at the spindle poles. A typical kinetochore may bind to 30–50 microtubules. As soon as one kinetochore attaches to microtubules, the chromosome begins to move toward the pole from which those microtubules extend. Soon microtubules attach to the kinetochore on the sister chromatid, and the chromosome is drawn back toward the opposite pole (Figure 7.13). The resulting tug-of-war aligns each chromosome along the equator of the mitotic spindle, midway between the poles (Figure 7.10). Once all the chromosomes are aligned along the spindle equator, the

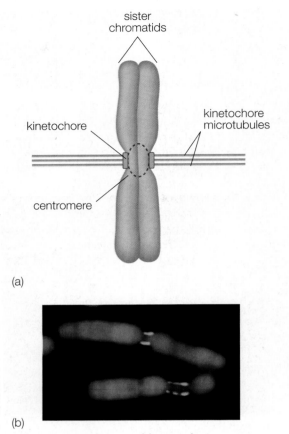

(a)

(b)

FIGURE 7.12 **Chromosome and kinetochores** (a) Diagrammatic view of a chromosome. The centromere is the constricted portion where kinetochores are located. (b) The disk-shaped kinetochores (green) as seen with the fluorescence light microscope. A fluorescent molecule is attached to an antibody that reacts with a protein found at the kinetochore, making it visible against the red fluorescent chromosome.

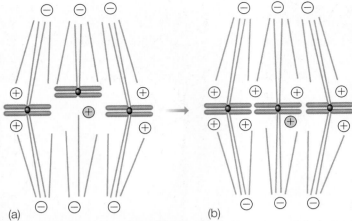

(a)   (b)

FIGURE 7.13 **Diagram showing "tug-of-war" that centers chromosomes in a cell at metaphase** Spindle microtubules attach to the kinetochores of chromosomes at their (+) ends, and their (−) ends lie at or near the poles. As microtubules attach to chromosomes (numerous microtubules attach to each kinetochore), the chromosomes are tugged toward one pole or the other, depending upon how many microtubules are attached to each of the two kinetochores. Once the numbers are approximately the same, equal force is exerted from each direction, and the chromosomes come to lie at the cell's equator. The center chromosome in (a) is attached to only the top pole and begins moving toward it until microtubules from the lower pole (blue circle) attach to the other kinetochore, moving the chromosome back toward the lower pole (b).

chromatids are ready to separate. Essay 7.1, "When to Stick and When to Split: Mechanisms of Chromosome Cohesion and Separation," describes some current research on the biochemical mechanisms by which sister chromatids are bound together through interphase and early phases of mitosis, and how they are separated at anaphase. The anticancer drug taxol, described in the chapter introduction, operates by stabilizing the spindle microtubules and preventing the spindle from breaking down into tubulin and actin. As long as the spindle is intact, the cell cannot complete division.

Sister chromatids separate in anaphase The shortest phase in the process of mitosis is **anaphase** (from the Greek *ana*, meaning "away"). All at once the sister chromatids separate at the centromeres and begin to move apart. With the centromeres and kinetochores leading, the arms of the chromatids appear to lag behind (Figure 7.10). After separation, the sister chromatids are now called daughter chromosomes.

Movement is rapid; anaphase can be complete in plant cells in as little as 2 minutes.

How the chromosomes move apart at anaphase is still somewhat uncertain. Figure 7.14 presents the currently accepted model of how chromosomes are moved toward the spindle poles. The **motor protein** called **dynein** (see Chapter 4) uses energy from ATP to generate the force that propels the kinetochore and chromosome toward the pole. As the kinetochore moves up the microtubule, the protein in the microtubule shortens. Anaphase ends when the chromosomes are at the poles, which themselves move farther apart during this phase.

In telophase, chromosomes become indistinct Telophase (from the Greek *telo*, meaning "end") is the final phase of mitosis, during which the two complete sets of chromosomes are fully separated. A new nuclear envelope forms around each set, and the spindle fibers disappear. The chromosomes uncoil and become less distinct as each daughter cell enters interphase. The actual length of interphase varies widely, especially the length of the G_1 phase. Some cells may remain in a G_1 phase for so long that they appear to have stopped proceeding through a cell cycle. Cells in such a quiet or resting state are said to be in a G_0 state. In telophase, mitosis ends and the first steps in cytokinesis begin.

FIGURE 7.14 **The motor protein model of how chromosomes might move to the poles** The motor protein dynein, with energy derived from ATP, "walks" the kinetochore to the pole along the kinetochore microtubules, which fall apart at the kinetochore (+) end. Dynein acts like a train engine moving along the track of the microtubule and pulling the chromosome behind it.

In cytokinesis, cytoplasm is divided between daughter cells

As explained, mitosis is nuclear division, in which sister chromatids are segregated into daughter cells. The process of mitosis is accomplished by a system of microtubules and actin filaments composing the mitotic spindle. Cytokinesis is the process of cytoplasmic division between daughter cells. Like mitosis, the process of cytokinesis is carried out by a system of microtubules and filaments composed of actin and myosin (the proteins that make up skeletal muscle), which in plants form a structure called the **phragmoplast.** Animal cells divide by pinching into two cells, but plant cells are constrained by a cell wall and must build a partition called a **cell plate** across the middle of the parent cell. How the microtubules and filaments of the phragmoplast function to construct a new cell wall is not understood with certainty at this time, but the basic steps can be observed with electron microscopy.

Beginning at the end of anaphase and continuing through telophase, cytokinesis starts with the formation of the phragmoplast between the two daughter nuclei (Figure 7.15a). At first, the phragmoplast microtubules are concentrated in the center of the cell directly between the two daughter nuclei. Partitioning of the cell begins with the appearance of the cell plate—a disk of material lying on the division plane among the phragmoplast microtubules. Secretory vesicles originating in the Golgi complex of the cell form the cell plate. These vesicles are apparently guided toward the plane of division by the phragmoplast microtubules. At the division plane, the vesicles fuse. Their membranes make up the cell membranes of the new daughter cells, and their contents form the new cell-wall material. Some of the plasmodesmata, the minute threads of cytoplasm that extend through pores in cell walls and connect the cytoplasm of adjacent cells, are formed at this time as bits of smooth endoplasmic reticulum are caught between the fusing vesicles. Plasmodesmata are discussed in more detail in Chapter 4.

The phragmoplast and cell plate develop first at the center of the cell and grow outward toward the periphery (Figure 7.15b). Thus the phragmoplast microtubules encircle the cell plate as it grows toward the cell wall. When the cell plate reaches the parent cell wall, it fuses with it. Each daughter cell then lays down a new primary cell wall around itself and continuous with the cell plate (Figure 7.15c). As the new daughter cells expand, the old parent cell wall stretches and breaks. The cell walls formed by the daughter cells now become the main cell wall. The fully separated daughter cells have now completed a cell cycle and enter into early interphase of the next cycle.

ESSAY 7.1 WHEN TO STICK AND WHEN TO SPLIT: MECHANISMS OF CHROMOSOME COHESION AND SEPARATION

In eukaryotic cells, chromosomes are duplicated during the S phase of the cell cycle, and they may remain attached to each other for a long time before they are separated in mitosis. Without this cohesion between sister chromatids, the orderly process of cell division in mitosis would not be possible. If sister chromatids drifted apart after duplication, there would be no way the cell, or mitotic spindle, could "know" which chromatids were sisters and should be segregated into daughter cells. Cell division would produce cells with odd numbers of chromosomes and missing parts of the genome. This condition in cells is called *aneuploidy*, meaning "odd number," and is associated with cancerous tumors.

In normal mitotic cells, the chromosomes are aligned along the equator in metaphase and are under considerable tension from the spindle fibers. In undamaged cells, the force exerted by the spindle fibers toward the poles is balanced by the cohesion between the sister chromatids. Anaphase appears to be triggered by the loss of cohesion between sister chromatids, beginning at the centromeres and proceeding out to the arms of the chromatids (Figure E7.1A). If a laser is used to cut one group of spindle fibers away from a pair of chromatids, both sister chromatids will move rapidly toward the opposite pole. Despite the importance of cohesion between chromatids, scientists have only recently begun to understand the mechanisms of binding and release.

Research on a number of plants, animals, and fungi has begun to clarify the mechanisms of sister chromatid cohesion and release at anaphase. Mutants in corn

> *Anaphase appears to be triggered by the loss of cohesion between sister chromatids . . .*

and fruit flies, in which sister chromatids fell apart prematurely, indicated that the substances involved in cohesion were special proteins. Studies with yeasts and animals identified a group of four proteins that were essential for sister chromatid cohesion. These proteins were given the appropriate name **cohesin**. The cohesin proteins bind the DNA molecules together during DNA replication (Figure E7.1B). Studies with the electron microscope have suggested that cohesin has a symmetrical structure consisting of a central loop and lateral, V-shaped hinges. Establishment of cohesion between sister chromatids in S phase is essential for proper chromatid separation in M phase. In yeasts, cohesin maintains cohesion between chromatids until anaphase.

Since sister chromatids are bound together by proteins, it was reasonable to hypothesize that breakdown of the cohesin proteins might cause separation. At first, a complex of proteins called

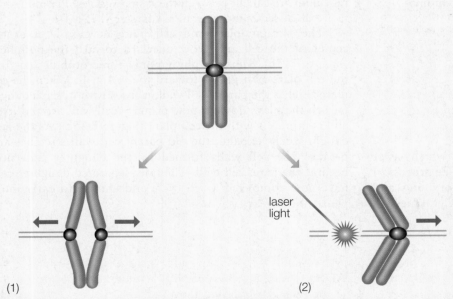

(1)   (2)

E7.1A **Experiment demonstrating the poleward force kinetochore microtubles exert on chromatids** (1) Normal anaphase, (2) Experimental laser treatment.

laser light

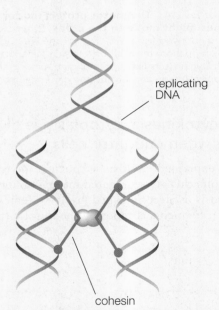

replicating DNA

cohesin

E7.1B Cohesin proteins bind DNA during replication

the **anaphase-promoting complex,** or **APC,** was thought to break down cohesin. But subsequent research showed that the role of the APC was indirect. The APC acts by destroying an inhibitor of the actual sister-chromatid separating mechanism. These inhibitor molecules are called **securins** because they control the separation of sister chromatids (Figure E7.1C). Securins have been identified in yeasts, fruit flies, and humans. They have an enzyme partner called **separase.** In yeasts and fungi, separase is essential for the separation of sister chromatids. The current hypothesis is that separases are the proteins that cut the cohesin proteins, but they are held in check by securins until anaphase. The APC acts to free the separases from the

securins. But what tells the APC to release the separases from the securins?

The answer to that question can be seen directly in dividing cells. Chromosomes that have arrived at the spindle equator must wait for any chromosomes still at the poles. Only when all chromosomes are at the equator does anaphase proceed. This mechanism, by which anaphase is held up until all chromosomes are aligned, is called the **spindle assembly checkpoint.** It is now known that in most eukaryotic cells kinetochores not attached to spindle fibers produce a signal protein, Mad2, that inhibits the APC and prevents the breakdown of securins. By blocking the destruction of securins, Mad2 prevents sister chromatid separation.

The Mad2 pathway is now thought to be essential for regulating mitosis in many organisms. Defects in this pathway caused by mutations may be linked to cancers. Human securin was first isolated from a tumor, where it was produced at a high level. Cells isolated from human colon cancers have defects in genes involved in the spindle assembly checkpoint. Most cancerous tumor cells show aneuploidy, and their odd numbers of chromosomes may be caused by defects in the Mad2 pathway. Someday an understanding of the basic mechanisms of mitotic cell division and the defects that can arise through mutations may lead to new methods for treating cancers. Separase and cohesin also play important roles in meiosis (see Essay 13.2)

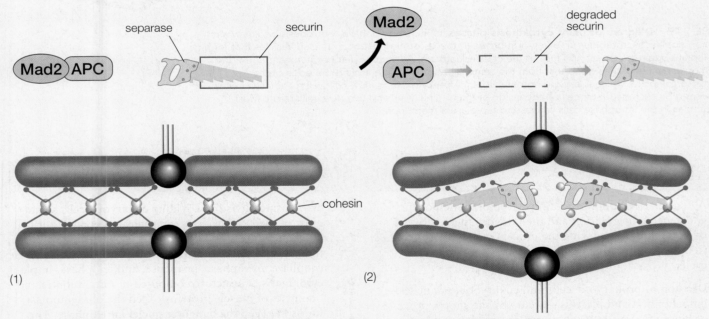

E7.1C   The APC-separin pathway separates sister chromatids (1) Prior to anaphase, the signal protein Mad2 keeps the APC from degrading securin, which acts like a sheath to keep separase from separating the sister chromatids. (2) With Mad2 levels lowered, APC degrades securin and liberates the separase enzymes to "saw" through the cohesin molecules that hold the sister chromatids together, beginning at the centromere region. Once the cohesion is lost between sister chromatids, they can each move to their respective pole.

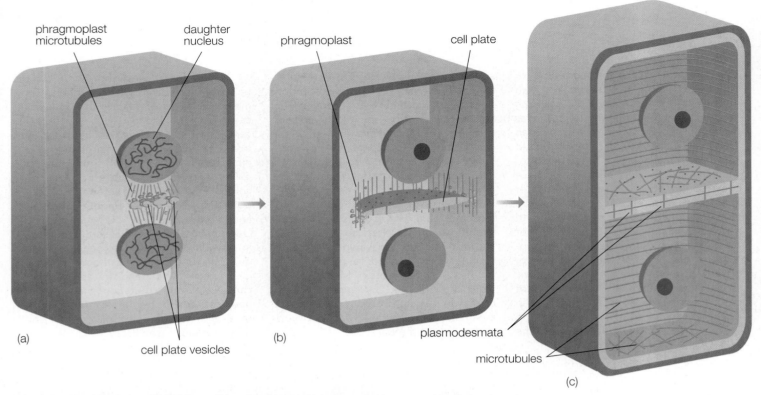

FIGURE 7.15   Diagram of plant cytokinesis phases (a) In late telophase, cytokinesis begins when the phragmoplast microtubules (in blue) form between the re-forming daughter nuclei. Vesicles that form the cell plate begin to appear and fuse at the cell's equator. (b) As the cell plate continues to form, the phragmoplast microtubules disappear from the center and form a circle around the edge of the developing cell plate, which is expanding outward. (c) After cytokinesis, new cell-wall material (in yellow) is deposited by each daughter cell on the inside of the original wall, and two new cells result. Most microtubules in the interphase cells lie close to the plasma membrane.

HIGHLIGHTS

7.1 A "parent" cell duplicates all its structural components and genetic information and then divides them between two daughter cells during the process of cell division. Cell division is part of a larger process called the cell cycle.

7.2 Division of prokaryotic cells and eukaryotic organelles—mitochondria and plastids—occurs by the simple process of binary fission. In prokaryotes, the single, circular chromosome is duplicated to produce two daughter chromosomes.

7.3 Cell division in eukaryotes occurs by separate processes of mitosis, or nuclear division, and cytokinesis, or

cytoplasmic division. Dividing eukaryotic cells spend most of their cell cycle in interphase, preparing for cell division. Mitosis and cytokinesis occur during M phase, the phase of cell division. Mitosis consists of four phases: prophase, metaphase, anaphase, and telophase. In plant cytokinesis, a structure composed of microtubules and filaments of muscle proteins called the phragmoplast forms between the daughter nuclei in telophase. The phragmoplast directs the formation of a cell plate between the daughter nuclei, completing the process of cell division.

REVIEW QUESTIONS

1. What is the cell cycle, and what cellular events characterize the two major phases of that cycle?

2. Discuss the differences in DNA intracellular location with respect to prokaryotes and eukaryotes. Where, and in what state, is the DNA found in each type of cell? Is the location of DNA in eukaryotic, DNA-containing organelles such as mitochondria and plastids more prokaryote-like or eukaryote-like?

3. Summarize the process of binary fission by which prokaryotic cells divide.

4. What is a centromere? What are sister chromatids, and how do they differ from chromosomes? When do chromatids become chromosomes in their own right?

5. How do mitosis and cytokinesis differ? Do these two processes occur in prokaryotes, eukaryotes, or both?

6. Distinguish between haploid and diploid cells. How many chromosomes do haploid and diploid cells possess? What are homologous chromosomes, and are they found in haploid cells, diploid cells, or both?

7. Briefly discuss the events in each of the following phases of the cell cycle: G_1 phase, S phase, G_2 phase, prophase, metaphase, anaphase, telophase, and cytokinesis. Which of these are part of interphase, and which are part of the division phase of the cell cycle?

8. What is a kinetochore, and what is its role in mitosis?

9. What is a phragmoplast, and what is its function in cell division? Are phragmoplasts found in prokaryotes? In eukaryotes? In animals? In plants?

APPLYING CONCEPTS

1. What is the basic accomplishment of mitosis with regard to the genetic material of the cell?

2. At one time in the past, some scientists suggested that spindle fibers might not be real structures but rather optical illusions caused by lines of stress formed during mitosis. Suggest several lines of research that allow you to investigate this idea further.

3. As discussed in the chapter opening, vinblastine is used to treat certain cancers of the lymphatic system. In light of the finding that this drug interferes with the assembly of microtubules, suggest a mechanism by which vinblastine might slow the growth of these cancerous tumors.

4. Generally in higher animals, anything other than diploidy in somatic cells is deleterious and usually lethal. This is not necessarily true in plants, where viable cells are known with 3, 4, 6, 8, 10, and even more sets of chromosomes present in the nucleus. Such a condition, with more than two sets of chromosomes present, is termed *polyploidy*. Suggest names for the polyploidy conditions just listed.

5. Using a high-quality light microscope to examine highly mitotic tissue such as the tip of an onion root, suggest a way to determine the portion of the cell cycle that these cells spend in interphase and in each phase of mitosis.

8 | Plant Structure, Growth, and Development

Opium poppy (*Papaver somniferum*)

Dolly the sheep won a place in history books and became a media star because she was among the first mammals to result from the cloning process. Biologists extracted a cell from the mammary gland of an adult sheep, then fused this cell (together with its nucleus) with an egg cell from which the nucleus had been removed, obtained from a different sheep. In the lab, this hybrid cell was provided with the nutrients necessary to undergo cell division, so that it grew into a multicelled embryo—an early stage of animal development. The embryo was implanted into the uterus of a surrogate sheep mother, where it generated many types of specialized cells, tissues, and organs required for normal sheep development. Dolly was a major scientific achievement because, for the first time, biologists had succeeded in treating an adult sheep's mammary cell, whose structure was highly specialized for milk production, so that it behaved like a young, unspecialized embryo cell. Using this new technique of taking specialized cells and coaxing them to behave like unspecialized embryonic cells, biologists can now clone mature animals that are known to have valuable traits, such as high-quality wool, meat, or milk.

Plant scientists, while appreciating Dolly's importance, have pointed out that techniques for cloning plants from specialized cells were discovered in the early 1950s. At Cornell University, F. C. Steward and his associates, working with carrots, developed similar techniques for working with plant cells almost 50 years before Dolly made front-page news. Single cells removed from an adult plant can be multiplied in the lab, then treated to generate many identical young plants. Today, plant cloning is a widespread, standard technique for producing ornamentals and other plant crops.

Plant and animal cloning requires an understanding of how adult bodies grow and develop from single cells, including the processes that generate specialized cells, tissues, and organs. In this chapter we describe principles of plant form and growth that are useful in agriculture, home gardening, and houseplant care. Knowledge of plant structure and development also helps us comprehend the great variation in plant form that we see in nature. ∎

8.1 | Plant structural variation is ecologically and economically important

The more than 250,000 different kinds of plants that now occur on Earth vary greatly in size and shape. Structural variation is important to plants' ecological roles and also relevant to human uses of plants. A comparison between duckweeds—among the smallest of plants—and redwoods—among the largest plants—provides a useful example.

Vast numbers of nearly microscopic duckweed plants (Figure 8.1a) blanket many pond surfaces, providing essential food for ducks and other animals. Their small size and the way their structure traps air enables duckweeds to float on the water's surface. Floating on the surface of still water, duckweed leaves are exposed to abundant sunlight and carbon dioxide from the air, while their roots have access to ample water and minerals. Duckweed plants are very abundant because their size and structure are well adapted to acquire the essential factors for plant growth. Giant redwoods (*Sequoia sempervirens*) and other cone-bearing trees in forests of western North America provide dramatic contrast in size and structure (Figure 8.1b). Giant redwood trees not only are among the largest of plants but also are among the oldest—living for thousands of years. Their great height enables redwood trees to successfully compete for sunlight in forests crowded with plants of many types. These magnificent trees are able to achieve large size because they produce abundant wood, which provides the necessary girth and strength to support their size and enables efficient water movement to the tall crown. In consequence, redwoods are important components of coastal forests and are valuable sources of wood products.

What explains the dramatic difference in size and structure of duckweeds and redwoods? Part of the answer is that the growth processes of redwood and duckweed are somewhat different. In plants, **growth** is a process by which the number and size of cells increase. Redwood trees feature a particular type of growth that greatly increases stem girth and generates wood. This type of growth is absent from duckweeds, whose stems remain slender and nonwoody. Redwoods, duckweeds, and other plants also vary in the number and arrangements of their organs, such as leaves, roots, and stems. These differences arise from variations in body development. **Development** is the process by which a few- or single-celled reproductive structure is transformed into a multicellular adult body by means of growth and cell specialization. For example, redwoods possess many more leaves than do duckweeds because the tree stem has developed many more branches on which leaves can arise. Different plant types also feature different amounts and types of tissues and cells having specialized structure and function. For example, wood is a type of tissue—composed of specialized water-conducting cells—that is absent from duckweeds. We will begin this chapter by surveying the types of organs, tissues, and cells that form the basis of plant structure, then we will consider how plants grow and develop.

8.2 | Plant bodies are composed of organs, tissues, and many types of cells

Plant bodies, like those of animals, are organized from smaller components—**organ systems** composed of one or more types of organs, **organs** composed of one or more types of tissues, and **tissues** composed of one or more types of cells (Figure 8.2). Here, we describe the major plant organ systems and organs, as well as examples of plant tissues and specialized cells. More extensive descriptions of plant tissues and cell types can be found in books on plant anatomy, plant morphology, and plant development—the sciences of plant form and structure.

Shoots, roots, leaves, flowers, fruits, and seeds are plant organs or organ systems

Plants typically have an upright, aboveground organ system known as the **shoot** and a belowground **root system** consisting of one or more main roots, which are often branched. The shoot

ECOLOGY

up to 135 meters

actual size

(a)   (b)

FIGURE 8.1 Small and large plants (a) The tiny duckweed plant (*Lemna*), with its small green body and clear roots, is contrasted with (b) a redwood tree (*Sequoia sempervirens*) growing in Northern California.

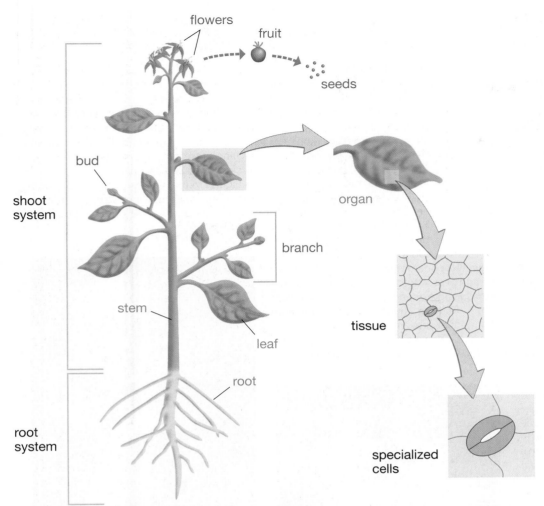

FIGURE 8.2   The plant body and its component organs, tissues, and specialized cells
Organs (blue labels) consist of tissues of one or more types of specialized cells. Examples of specialized cells include the two (dark-green-colored) cells that surround pores (stomata) in the surface tissue of leaves. Complex plant structures (made up of more than one organ) are labeled in red and include buds, flowers, and seeds. The aboveground portion of the plant is the shoot system and the belowground portion, the root system.

of flowering plants includes a **stem** with branches bearing **leaves, flowers, buds** for new flowers and branches, and **fruits** containing **seeds** (see Figure 8.2 and Essay 8.1, "Supermarket Botany"). Stems, roots, and leaves are the basic plant **organs.** Flowers, buds, fruits, and seeds are more complex in structure, being composed of more than one of the basic plant organs.

Flower and branch buds, for example, are very young shoots (organ systems) produced by most kinds of plants, including annual, biennial, and perennial flowering plants. Annuals live for only one growing season, and biennials flower in the second year of a two-year life span. The buds of perennial plants—those capable of living for more than two years—are different in that they can persist through stressful winter conditions, protected by tough specialized leaves known as bud scales (Figure 8.3). All buds contain a tiny stem, with several miniscule leaves in an arrested state of development, and a terminal growth point. In warm, moist conditions, buds open and begin to grow, forming new branches with maturing leaves, or flowers.

Flowers are also complex shoots, which bear leaf-derived organs known as sepals, petals, stamens, and carpels (Chapter 23). Fruits are complex because they develop from a portion of the carpel and contain seeds. Seeds are complex plant structures because they include dormant embryos, complete with young shoot and root, together with stored food—all enclosed in a protective seed coat derived from structures produced by the parent plant.

Plant organs are composed of tissues whose cells are linked by plasmodesmata

Plant tissues are groups of adherent, related cells that result from the mitotic division of precursor cells (Chapter 7). Tissues may be organized into tissue systems composed of several cell types having a cooperative function. There are more than a dozen kinds of nonreproductive plant tissues alone (Table 8.1), as well as tissues specialized for reproduction. Plant tissues differ in the activities of specific genes.

ESSAY 8.1 SUPERMARKET BOTANY

The produce section of your local food market is full of examples of plant organs and organ systems that people use as food sources. Many of these are easily recognizable. Potatoes are stems, indicated by the presence of buds (the "eyes") (Figure E8.1A). Pieces of potato that include at least one bud are able to grow into new potato plants. Leaf lettuce, kale, and spinach are recognizable examples of leaves; and most people realize that carrots are roots. Apples, grapes (Figure E8.1B), melons, and lemons are recognizable as fruits because they contain seeds.

However, other produce may be a bit more difficult to identify as a plant part once it is removed from the plant and packaged for sale. For example, tomato, squash, green beans, and bell peppers (Figure E8.1C) are commonly considered to be vegetables, but they are really fruits because they contain seeds. Peanuts, walnuts, filberts, pecans, and other nuts sold in the shell are also fruits, as are peppercorns. The parts of celery and rhubarb that we eat are actually leaf stalks (Figure E8.1D). Onions are really large buds having a small

Onions are really large buds having a small stem with numerous fleshy leaves.

stem with numerous fleshy leaves. Brussels sprouts are not seedlings (as are alfalfa and mung bean sprouts), but rather are large axillary buds (Figure E8.1E) plucked from the plant's stem. The part of kohlrabi that is eaten is a swollen stem together with bases of the leaves. The tops of broccoli stalks are masses of tiny unopened flowers. You may have noticed that if you leave broccoli in the refrigerator too long, the flower buds will begin to open, turning the broccoli yellowish and less attractive. Cauliflower heads are huge aggregations of meristems, which result from a human-induced genetic change that prevents shoots from elongating and flowers from forming (Figure E8.1F). Cabbage heads are an entire plant

shoot—a stem with apical meristem, very short internodes, and tightly folded leaves. In fact, cabbages, broccoli, cauliflower, Brussels sprouts, and kohlrabi are all closely related cultivars of *Brassica oleracea* that were modified by artificial selection from an ancestral form known as colewort by ancient agriculturists (these are all pictured in Figure 16.5). The next time you go food shopping, you may be able to impress fellow shoppers with your botanical knowledge.

E8.1E Brussels sprouts—axillary buds

E8.1C Pepper—a fruit

E8.1A Potato—a stem with buds

E8.1D Rhubarb—leaf petioles

E8.1F Cauliflower—a mass of shoot meristems

E8.1B Grapes—fruits

FIGURE 8.3 **Young shoots emerging from protective bud scales** (a) The terminal and two lateral buds of the buckeye (*Aesculus*). The buds are breaking, with leaves beginning to emerge (the opening bud scales are indicated by the arrowheads). (b) A flower shoot and young leaves that have begun to expand.

The cells of plant tissues are linked by narrow cytoplasmic connections, known as **plasmodesmata** (Figure 8.4). Tissues whose cells are joined by plasmodesmata are said to have **symplastic continuity**, meaning that their cytoplasms are continuous, allowing chemical communication. Systems of plasmodesmata, which connect the many cells of a particular tissue, are somewhat like a local area computer network that allows unrestricted communication among members of a limited group for special purposes. Proteins, messenger RNA molecules, and viruses are known to travel from cell to cell via plasmodesmata. Plasmodesmata are not just simple pores. Rather, they contain an inner tube of endoplasmic reticulum (ER) and a variety of proteins. The plasmodesmata ER and proteins regulate the passage of materials, much as filters do in an industrial pipeline.

Primary plasmodesmata form as new plant cell walls develop between sibling cells during cytokinesis (Chapters 4, 7). Secondary plasmodesmata can be generated when enzymes remove wall material between two cells that may or may not be siblings. For example, secondary plasmodesmata form between plant organs that are grafted together, integrating them. Grafting shoots of one plant onto root systems of another (akin to organ transplants in animals) is a common technique in agriculture. For example, wine grape production often involves grafting of shoots that produce desirable fruit onto rootstocks from plants whose roots are more resistant to disease.

Some plant cells with special functions are isolated from the rest of the plant body because they have few or no plasmodesmatal connections to other cells. Other cell junctions have many more plasmodesmata than the average, or very highly branched plasmodesmata, indicating a region of intense interaction among cells. The cell membranes of animal cells can be in very close contact and communicate by various means, but animal tissues lack cytoplasmic channels resembling plant plasmodesmata. Plants inherited plasmodesmata from ancestral green algae.

Plants grow by production of new tissues and cell enlargement

How do plants increase in height, length, or girth? Plant growth results from two basic processes: (1) production of new cells and tissues by meristems and (2) cell enlargement.

Primary apical meristems produce primary tissues

All plants have shoot and root tip growth points—known as primary **apical meristems**—that generate new primary tissues at root and shoot tips (Figure 8.5). Root apical meristems increase the length of roots (Chapter 10), and shoot apical meristems increase the height or length of shoots (Chapter 9). The genes *stm* and *clavata* are very important in determining the behavior of plant apical meristems.

Apical meristems of most plants consist of several layers of cells and generate three types of **tissue systems:** (1) the outermost **dermal ("skin") tissues** or **epidermis,** (2) the **vascular ("conducting system") tissues,** and (3) **ground tissues,** located between dermal and conducting tissues and in the center (the **pith**) of the stem (Figure 8.6). These tissue systems develop from precursor tissues known as protoderm, procambium, and ground meristem, respectively. Each of these tissue systems is represented in the various plant organs. A few cells embedded in both shoot and root meristems divide only infrequently and act as a reserve. When necessary, they replace aging or damaged components of the actively dividing portions of primary apical meristems.

The activity of the primary shoot apical meristem produces new stem tissue and young leaves, as well as

TABLE 8.1 Major Nonreproductive Plant Tissues and Specialized Cells (*Note:* Not all of the listed tissues and cells have been described in this chapter but are included for use in extending explorations of plant structure.)

Basic Plant Organs*	
Stems	
Roots	
Leaves	

Simple Primary Plant Tissues	Specialized Plant Cells
Parenchyma	Parenchyma cells
Collenchyma	Collenchyma cells
Sclerenchyma	Fibers, sclereids
Root endodermis	Endodermal cells
Pericycle	Pericycle cells

Complex Primary Plant Tissues	Specialized Plant Cells
Leaf mesophyll	Spongy mesophyll cells, palisade mesophyll cells
Xylem	Fibers, parenchyma cells, tracheids, vessel elements (primarily flowering plants), transfer cells
Phloem	Fibers, parenchyma cells, sclereids, sieve-tube members and companion cells (flowering plants) or sieve cells and albuminous cells (ferns, gymnosperms), transfer cells
Epidermis	Shoot and root epidermal cells (including root-hair cells and root-cap cells), trichomes, cells associated with pores (stomata): guard cells and subsidiary cells
Secretory tissues	Transfer cells, secretory cells

Complex Secondary Plant Tissues	Specialized Plant Cells
Secondary xylem	Fibers, parenchyma cells, tracheids, vessel members (primarily flowering plants)
Secondary phloem	Fibers, parenchyma cells, sclereids, sieve-tube members and companion cells (flowering plants) or sieve and albuminous cells (ferns, gymnosperms)
Periderm	Cork cells, cork cambium, parenchyma cells

*Each organ is composed of the three tissue systems—dermal, ground, and vascular.

axillary buds located in the angle (axil) between stem and leaf stalks (Figure 8.7). The region of a stem from which one or more leaves or branches emerge is a **node**. The stem regions between nodes are the **internodes** (see Figure 8.7). Plant height increases by addition of new nodes (via cell division activity at the meristem) and by elongation of internodes. Plants vary a great deal in the length of internodes. If internodes are short, branches and leaves will be more closely spaced than if internodes are longer. For example, the internodes of duckweed are very much smaller than those of redwood trees. In contrast to shoots, roots are not organized into nodes and internodes.

Secondary meristems produce wood and bark

Maturing and adult trees, shrubs, and woody vines and other plants that produce some wood possess two meristems in addition to the primary root and shoot meristems—the **vascular cambium** and the **cork cambium** (Figure 8.8). These **secondary meristems** produce **secondary tissues**. Wood and the inner bark are secondary conducting tissues produced by the vascular cambium. Outer bark includes both old secondary conducting tissues and secondary surface tissues that are generated by the cork cambium. Secondary meristems increase girth of stems and roots. Descriptions of secondary meristem activity and wood and bark structure can be found in Chapter 9.

Plants also grow by cell expansion

Both plant and animal cells can grow in size by increasing the amount of cytoplasm, by making more proteins, and replicating organelles. But plant cells can expand greatly by a mechanism that is not present in animals—uptake of water into the large central

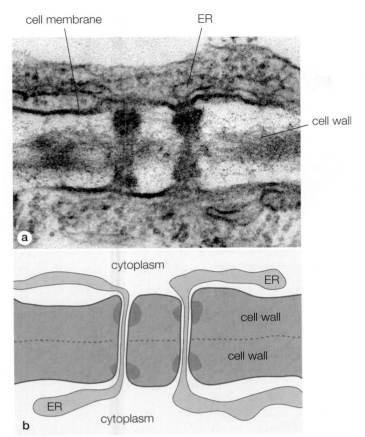

FIGURE 8.4 **Plant plasmodesmata** (a) Transmission electron micrograph showing plant plasmodesmata. The two plasmodesmata shown here are from *Sphagnum* peat moss. A small tubule of ER (endoplasmic reticulum) runs through each of the plasmodesmata. (b) An interpretive diagram of *Sphagnum* plasmodesmata.

FIGURE 8.5 **Plant primary meristems** Meristems (red) that lead to lengthening of a plant are primary meristems. They are found at the tips of shoots, in axillary buds (which grow into branch shoots), and the tips of roots.

vacuole, accompanied by cell-wall expansion (Figure 8.9). The mechanisms by which plant cells and vacuoles take up water are described in Chapter 4 (see Essay 4.1). Proteins unique to plants, known as expansins, unlock the linkages between cell-wall polysaccharides so that the cell wall can stretch, allowing cells to enlarge. Cell expansion by water uptake, together with meristematic activity, allows some plants to grow at astounding rates. Bamboo (Figure 8.10), for instance, can grow more than a meter per day, or 30 meters in less than 3 months. Plants' reliance on water as a means of cell enlargement, together with their use of water as an internal transport medium and as a reactant in photosynthesis, explain why water is such an essential resource for plants.

Plant tissues are composed of one to several cell types

Some plant tissues are composed of only one or two cell types, but others include several cell types. Tissues that contain just one or two cell types are known as simple tissues. **Parenchyma** is a simple tissue composed of only one type of cell, thin-walled cells of varying shape known as **parenchyma cells**. One function of these cells is starch storage, but parenchyma cells may have a variety of other roles. For example, parenchyma cells in fruits and stems of the opium poppy *(Papaver somniferum)* contain enzymes that begin the synthesis of morphine.

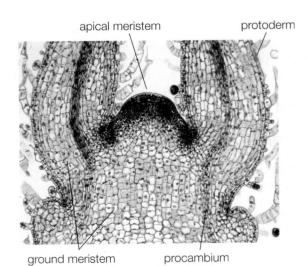

FIGURE 8.6 **Apical meristem of *Coleus*** Very early in the development of the plant shoot, cells destined to form the dermal, ground, and conducting tissue systems are distinguishable (referred to at this stage as protoderm, ground meristem, and procambium, respectively).

FIGURE 8.7 Axillary buds, nodes, and internodes on a plant stem (a, b) Axillary buds (arrows) and leaves are attached at each node of these plants, the buds situated in the angle (axil) formed between the stem and leaf. The internode is the region on the stem between adjacent nodes. (c) Internodes closer to the apical meristem are shorter because cells in that part of the plant are still elongating.

Collenchyma (from the Greek word *kolla*, meaning "glue") is another example of a tissue composed of only one cell type. Collenchyma is commonly found just beneath the epidermis of herbaceous stems or leaf stalks, the petioles (Figure 8.11). Those annoying strings we encounter when eating or cutting celery are composed largely of collenchyma. This tissue provides flexible support, functioning somewhat like an athlete's elastic knee brace, and is composed of **collenchyma cells** (see Figure 8.11) whose walls are unevenly thickened with the carbohydrate pectin. The thickened areas lend strength, while the intervening thinner wall areas allow cells to stretch during rapid growth. If collenchyma cell walls were thickened throughout, or contained rigid materials, these cells would not be flexible.

Sclerenchyma (from the Greek word *skleros*, meaning "hard") occurs in seed coats, nut shells, and the pits of peaches and other "stone fruits." This tissue also gives the flesh of pears its gritty texture. Sclerenchyma is composed of two types of cells whose walls are evenly thickened with the addition of the phenolic polymer lignin—(1) star- or stone-shaped **sclereids**, also known as "stone cells" (Figures 8.12, 8.13); and (2) long, thin **fibers** (Figure 8.14). The tough walls of these cells resist damage from mechanical stress and attack.

Examples of complex plant tissues include **xylem**, whose functions are to support the plant and conduct water and minerals, and **phloem** (Figure 8.15), which transports sugars and other organic compounds in a watery solution. Xylem includes several types of cells: sclerenchyma, parenchyma, and pipelinelike arrays of specialized water-conducting cells whose walls are reinforced with lignin. Phloem tissue is complex because it contains sclerenchyma, parenchyma, and arrays of specialized food-conducting cells.

Specialized cells arise by the process of differentiation

Plant tissues, like those of animals, vary because they contain different types and numbers of **specialized cells** having particular structures and functions. Specialized cells contain the same genetic material found in all cells of the same organism but differ from one another in the particular sets of genes that are expressed (i.e., in the kinds of enzymes and other proteins they produce). Plants produce many kinds of hormones, including auxin, cytokinin, gibberellic acid, abscisic acid, and ethylene, which influence growth and development. These internal chemical signals, as

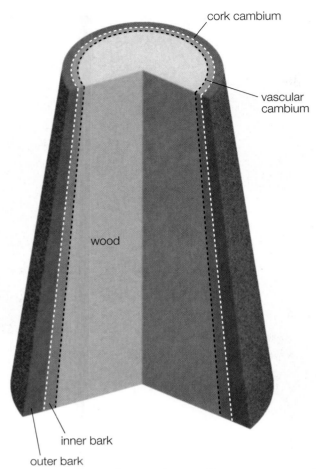

FIGURE 8.8 The secondary meristems There are two secondary meristems in woody plants—the vascular cambium (black dashed line) and cork cambium (white dashed line). These cylindrical (actually, slightly conically)-shaped meristems produce wood and bark and increase the girth of woody plants. The vascular cambium forms wood to the inside and inner bark tissues to the outside. The cork cambium contributes to the outer bark. (The increase in girth is exaggerated for purposes of the illustration.)

well as external signals (such as light) influence cell specialization, in addition to tissue and organ development, by inducing changes in gene expression (Chapter 12).

Differentiation, the process by which specialized cells take on distinctive structures and functions, differs somewhat in animals and plants. Animal development involves movement of embryonic cells from one place to another; this relocation influences their fates. In addition, the lineage of an animal cell often determines cell fate, much as inheritance of titles and property might be restricted to the oldest child in some societies. By contrast, most plant cells lack mobility, and experiments have revealed that other mechanisms are more important than lineage effects in plants. Unequal (asymmetric) cell division (Figure 8.16a) can result in nonuniform distribution of cytoplasmic components to newly formed plant cells, causing the two cells to develop differently. For example, the specialized cells on leaf surfaces that form pores, known as stomata (see Figure 8.2), begin their development with an asymmetric division (Figure 8.16b). Differences in cell positions may cause one cell to experience an environment that differs from the other (Figure 8.17), leading to distinct developmental pathways. In the example of leaf-pore development just cited, the smaller of the two cells resulting from asymmetric division secretes a protein that binds to nearby cells, preventing them from dividing to form stomata. In contrast, cells farther away from developing pores are not affected by that protein and may proceed with the unequal divisions needed to form more stomata. This helps to prevent the formation of too many pores on leaf surfaces and helps to maintain their even distribution.

Secretory cells and trichomes are additional examples of specialized plant cells. **Secretory cells** produce protective secondary compounds such as tannins, resins, or drugs. For example, in the opium poppy, secretory cells known as lactifers accumulate the secondary compounds morphine and codeine (see Chapter 2). **Trichomes** are spiky or hairlike projections on the surfaces of many leaves that offer protection from excessive light, ultraviolet radiation, extreme air temperature, or attack (see Figure 11.10). Broken trichomes of the stinging nettle (see Figure 25.9), for example, release a caustic substance that irritates the skin of

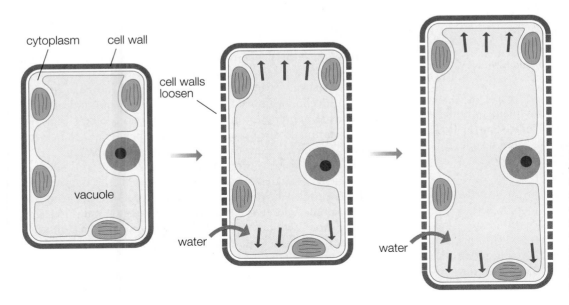

FIGURE 8.9 Growth of plant cells through water uptake As more and more water is taken up into the vacuole and the cell wall is loosened, the cell expands. The direction of cell expansion can be controlled by the orientation of cellulose microfibrils and where wall loosening takes place. Here, the cell is permitted to increase in length, which would be common in stems.

FIGURE 8.10 **Fast-growing bamboo in a Japanese garden**

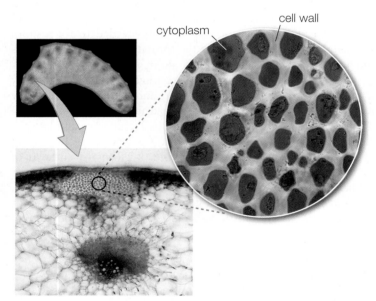

FIGURE 8.11 **Collenchyma** This simple tissue is shown at low and higher magnification in a celery stalk (which is part of the leaf).

mammals (including humans) and causes them to avoid these plants. Forty or so types of specialized cells occur in flowering plants (see Table 8.1; many of these cell types are described in Chapters 9–11).

8.3 | Plants develop from single cells or small pieces

Plant development is an amazing thing. A tiny redwood seed, for example—weighing a miniscule 0.003 g—can develop into a tree weighing more than 1,000 tons, a weight gain of more

FIGURE 8.12 **Sclereids in pear fruit** Pear's (inset) gritty texture results from the presence of sclereids, viewed at (a) low and (b) higher magnification.

than 250 billion times! Many differences in size, structure, and complexity, as well as the various ways in which plant bodies differ from those of animals, can be explained by a consideration of how plants develop.

Plants can develop from zygotes, spores, or excised pieces

Adult plant bodies, like those of animals, can develop from a single-celled **zygote**, which is the product of sexual fusion of **egg** and **sperm**. Zygotes of both plants and animals develop into multicellular **embryos** and adults by repeated mitotic cell divisions (Chapter 7). Young embryos of both plants and placental mammals are attached to and dependent on maternal tissues for nutrition, at least early in their development.

However, plants differ from animals in having not one, but two multicelled bodies that develop from single cells. One of these bodies is produced from a zygote and the other body is produced from a **spore**. A spore is a cell that can develop into a multicellular adult body without having first fused with another reproductive cell. A zygote is a cell that typically results from sexual fusion of two gametes. The two plant bodies are called the **sporophyte** and the **gametophyte**. Although these two life stages take different developmental pathways, they both start from a single cell. The

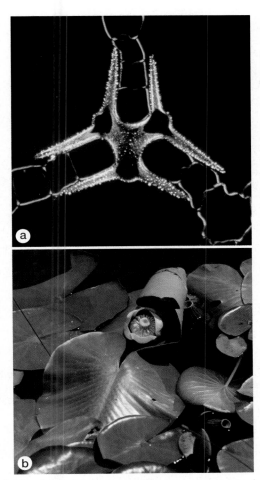

FIGURE 8.13 **Sclereids in water lily** (a) A branched sclereid as seen in polarized light, which makes crystalline materials glow brightly, from the water lily *Nuphar* (b).

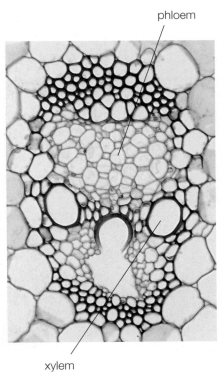

phloem

xylem

FIGURE 8.15 **Complex plant-conducting tissues—xylem and phloem** Conducting tissues in a young corn stem are shown here.

sporophyte develops from a zygote and is a spore-producing plant body. In contrast, the gametophyte develops from a spore and is a gamete-producing plant body. The two bodies are of different sizes and structure. Typically, these two bodies alternate during the plant's life cycle. All plants have these two distinct types of bodies, but bryophytes and ferns provide particularly good examples of how plant gametophyte and sporophyte bodies

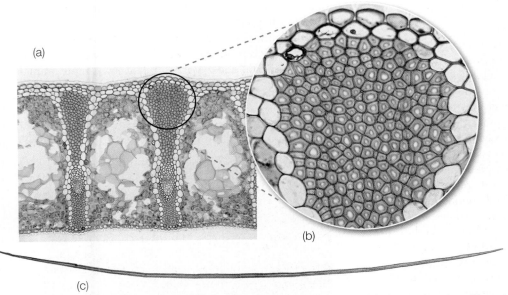

(a)

(b)

(c)

FIGURE 8.14 **Plant fibers** The fibers shown in cross section in (a) and (b) are in the plant *Phormium*, which is native to New Zealand and used for fiber production. (c) An isolated fiber, obtained by treating a plant in a way that breaks apart the tissues.

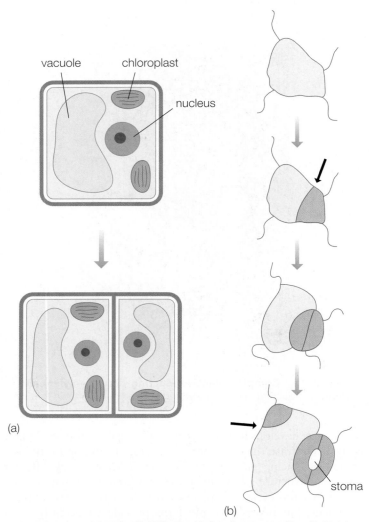

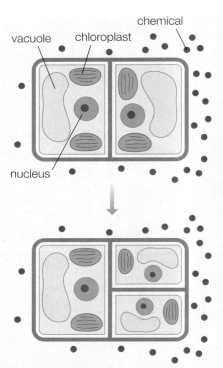

(a)

(b)

FIGURE 8.16   Plant cell differentiation by unequal cell division
(a) General diagram of unequal cell division. (b) Specific example of unequal cell division (black arrows) during development of epidermal pores (stomata). The cells in green develop into the specialized (guard) cells that flank the pore.

FIGURE 8.17   Plant cell differentiation may result from position effects One cell may experience a different environment from another. This could be increased exposure to a chemical factor or physical influence, for example.

several very different-looking stages of development during their lives—a process known as metamorphosis—but this process differs from sporophyte-gametophyte alternation in plants. The animal transformations (think of tadpoles and frogs) do not involve changes in chromosome numbers, whereas transition between plant sporophyte and gametophyte stages typically does involve changes in chromosome numbers resulting from zygote formation and a type of cell division known as meiosis. More information about the role of meiosis in plant reproduction can be found in Chapter 13.

Another difference between plants and animals is that plants more commonly reproduce asexually, that is, without the formation of eggs, sperm, and zygotes. Only a few simple animals are

differ. In mosses, the bryophyte sporophyte is small and always grows on its (usually) more conspicuous green gametophyte (Figure 8.18). The fern sporophyte is the familiar leafy stage often used as a houseplant; it produces spores on the undersides of leaves. The egg- and sperm-producing gametophytes of ferns are much smaller than the sporophyte, only about the size of your thumbnail or smaller (Figure 8.19). A microscope is needed to see the tiny gamete-producing structures of fern gametophytesc. Fern sporophytes have vascular tissues, including phloem and lignified xylem (as do seed plants), and produce roots, but fern gametophytes lack both roots and specialized conducting tissues.

Flowering plants also produce gametophyte stages, but these are only a few cells in size, are not green, and occur within the tissues of flowers on the leafy, green sporophyte stage. The gamete-producing life stages of flowering plants cannot grow independently in nature; they rely on food produced by the sporophyte stage in which they live. Flowering plant sporophytes develop from seedlings, which in turn develop from embryos contained in seeds (Chapter 23). Some animals go through

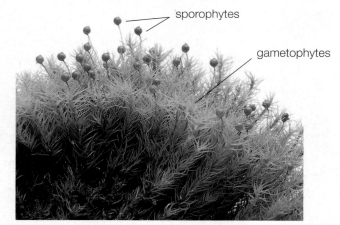

FIGURE 8.18   Bryophyte sporophytes are structurally distinct from the gametophytes that bear them

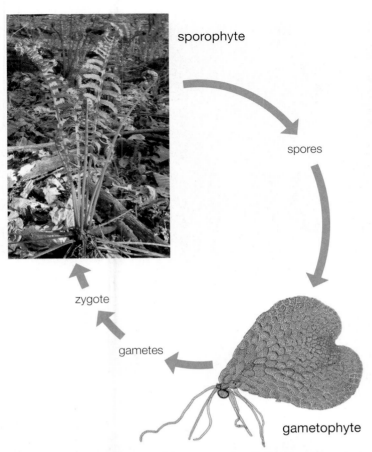

sporophyte

spores

zygote

gametes

gametophyte

FIGURE 8.19 **Ferns provide an example of distinct sporophyte and gametophyte plant bodies** The fern sporophyte produces single-celled spores, which develop into gametophytes. Fern gametophytes, which are much smaller than sporophytes (the two bodies are not shown to scale here), produce single-celled gametes that combine in pairs to form single-celled zygotes. Sporophytes develop from these zygotes.

a

b

FIGURE 8.20 **Asexual reproduction** (a) Plantlets of *Kalanchoë*. The small plantlets found on the leaf margins fall to the soil, where they can grow into new adult plants. (b) Runners of a species of *Potentilla*, a relative of strawberries. Note the plantlets spaced along the runners.

Embryos produced by somatic embryogenesis are much like those formed by sexual reproduction. Somatic embryogenesis was first discovered in carrots but has now been accomplished in a wide variety of plants, including other important crops.

FIGURE 8.21 **Asexual reproduction** Propagation of begonia plants (top) and English ivy (bottom) via cuttings.

able to undergo **asexual reproduction**, regenerating new adult bodies from small pieces of tissue or specialized "buds," whereas most types of plants can reproduce in this way. Like Dolly the sheep, organisms resulting from asexual reproduction are clones; they have the same genetic composition. The advantage of plant asexual reproduction is the rapid spread of organisms whose genetic composition is especially well adapted to a particular environment (see Chapter 13). An example of plant asexual reproduction can be observed in the common houseplant *Kalanchoë*, also known as the "maternity plant" because it produces many tiny plantlets along the edges of its leaves (Figure 8.20a). The plantlets eventually drop off, take root in the soil, and grow into adults genetically identical to their parent. The horizontal stems of grasses, strawberries, and other plants (Figure 8.20b) can generate clones of new plants, and the roots of some trees and shrubs produce "sucker shoots" that can generate new plants. Many types of plants can be propagated in greenhouses or nurseries from stem, root, or leaf pieces (Figure 8.21).

In the laboratory, nonreproductive cells taken from plants can be made to form embryos, then new adult plants. This is a cloning process known as **somatic embryogenesis**—production of embryos from body cells rather than by sperm and egg fusion.

Plant bodies have polarity, radial symmetry, and indeterminate growth

Development of embryos within seeds begins with an unequal cell division of the single-celled zygote. Such division separates a smaller portion of the zygote cytoplasm from a larger one, so that each contains distinctive messenger RNAs. Thus, from their very beginnings, plant embryos have distinct tops and bottoms, known as **apical-basal polarity**, as do animals. The smaller cell of the two gives rise to an embryo by continued division, while the larger cell serves as a kind of "umbilical cord." It links the developing embryo to the maternal food supply. These early events provide another example of the important role that unequal cell division plays in plant development. As plants mature, this early-developed polarity is retained, and is manifested by the upper shoot and lower root systems.

Unlike higher animals, which are outwardly bilaterally symmetrical (the same on right and left sides) and have a distinct front and back, most plants lack a distinct front and back. Their organs and organ systems, such as leaves, flowers, and flower parts, are often produced in attractive circular or spiral patterns around the stem axis (Figure 8.22). Such patterns can be described mathematically and are known as phyllotaxy. The plant body thus has **radial symmetry**, which is established very early in plant embryos. The genetic basis of plants' radial symmetry is as yet unclear, though some experts propose that it may begin with early development of a cap of protoderm. Recall that protoderm generates the plant's epidermis (skin). In these cells, expression of genes that control synthesis of protective surface materials establishes a distinction between inner and outer embryo tissues. This radially symmetrical distinction between the inside and the outside tissues persists in mature plants and their meristems. Though plant shoots and roots typically have radial symmetry, some plant parts can have bilateral symmetry. Leaves and many flowers provide examples (Chapters 11, 23, 24).

Another major difference between animal and plant bodies is in the number of organs they possess and the timing of organ production. In animals, organ number is determined very early in development, at the embryo stage, and does not change in the adult. You have only one heart, for example, and you cannot grow another, no matter how useful this ability might be. In contrast, adult plants typically have many more organs than do their embryos. For example, the embryos of flowering plants, located within seeds, have one or two tiny leaflike organs (cotyledons); but adult plants may produce many thousands of leaves in their lifetimes, with new crops of leaves appearing at the beginning of each growing season in perennial plants. Many plants can produce multiple stems and roots, and even those with only a single main stem or root are able to produce a large number of branches from them. The capacity to produce many organs throughout its lifetime equips a plant with the ability to adjust its leaf and root surface area for efficient collection of sunlight, water, and minerals. A tree sapling growing under the shady canopy in a forest of taller trees with larger root systems may remain small for years, unable to compete for resources until a storm or human harvesting removes taller trees. The sapling is then able to respond to

ECOLOGY

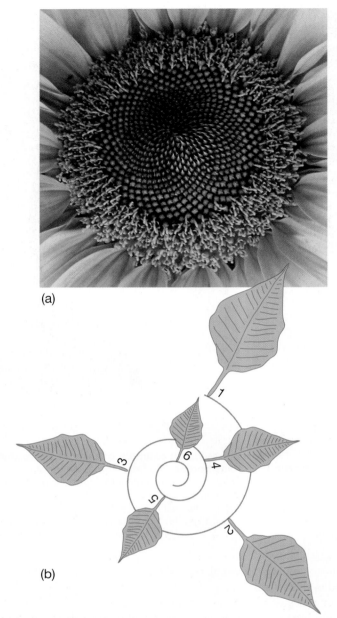

(a)

(b)

FIGURE 8.22 Phyllotaxy (a) The sunflower blossom is actually a cluster of many individual flowers that are arranged in an obvious spiral pattern on a very short, compact shoot. (b) When plant parts are located farther apart on a longer shoot, the spiral pattern is less obvious but can be seen when observed closely. Here, the points on the stem where leaves are attached on a *Poinsettia* plant are mapped on a spiral. The lowermost leaf is #1.

increased light and soil water and mineral availability by rapidly multiplying leaves and roots and increasing in size.

Likewise, the number of reproductive organs produced by a plant (flowers, fruits, and seeds, in the case of flowering plants) is not predetermined at the embryo stage, as it is in animals. Instead, the number of reproductive organs produced by a plant at any one time depends on its environment. Plants thus commonly have **indeterminate growth**, as contrasted with **determinate growth** of animals. Even so, plant growth is also under genetic control; duckweed plants, for example, do not produce as many leaves as most other plants, nor can they ever grow to the giant redwood's height.

HIGHLIGHTS

8.1 Plants have multicellular bodies that vary structurally in ways that adapt them to different environmental conditions. Plants grow by addition and enlargement of new cells. Plant development is the transformation of single reproductive cells or small pieces of plants into multicellular bodies containing many types of specialized cells, tissues, and organs.

8.2 Leaves, stems, and roots are basic plant organs. Buds, flowers, fruits and seeds are composed of multiple organs. Shoots and roots are organ systems. Plant growth and body form is a function of (1) the cell division activity of primary apical meristems and often also secondary meristems, (2) cell enlargement after division, and (3) the types of specialized cells that develop by differentiation. Meristems are localized regions of cell division activity. Shoot and root primary meristems increase plant height or length by producing primary plant tissues and also generate leaves and branches. There are three types of primary tissue systems that arise from primary meristems: dermal, vascular, and ground. Parenchyma, collenchyma, and sclerenchyma are examples of simple primary tissues containing one or two cell types. Secondary meristems increase plant girth by generating wood and bark; they produce complex conducting tissues such as wood, as well as cork. Sclereids, fibers, trichomes, and secretory cells such as lactifers are examples of the 40 or so types of specialized cells present in flowering plants. Cell specialization occurs by means of unequal cell divisions or exposure to influences present at particular locations. Differentiation of cell types reflects differential gene expression. Chemicals produced within the plant body (hormones) as well as external chemical and physical influences (such as light) influence plant development, including tissue and cell specialization.

8.3 Plant bodies develop from zygotes, spores, or asexual structures. Plants that reproduce sexually have two distinct types of reproductive bodies. Sporophytes develop from zygotes and produce spores. Gametophytes develop from spores and produce gametes, which mate to form zygotes. These structurally different bodies alternate during the sexual reproductive cycles of plants. In the laboratory, plant embryos can be produced from nonreproductive body cells without the previous formation of zygotes. Unlike most animals, many plants are able to reproduce asexually from small pieces of parental plants. Embryos and mature plant bodies have apical-basal polarity, a property shared with animals. Plant bodies bear appendages in radial or spiral patterns (phyllotaxy). Plant bodies thus have radial symmetry, in contrast to the bilaterally symmetrical bodies of most animals. However, leaves typically have bilateral symmetry, as do some flowers. Plants commonly have indeterminate growth, meaning that they can continue to add organs during their lifetimes. In contrast, most animals have determinate development, meaning that they cannot add new organs during development.

REVIEW QUESTIONS

1. What are the three basic plant organs, and how are these arranged into the aboveground and belowground plant systems? Give several examples of more complex plant organs that are admixtures of the three basic plant organs.

2. Distinguish between annual, biennial, and perennial plants.

3. What are plasmodesmata, and why are they more than simply pores between adjacent cells? What is symplastic continuity? Distinguish between primary and secondary plasmodesmata.

4. A plant may produce a variety of meristems over the course of its life. What is a meristem? Define *primary meristem*, *apical meristem*, and *secondary meristem*. Are the vascular and cork cambia meristems?

5. Plant and animal cells may both grow in size by increasing the amount of cytoplasm in the cell. Describe a different means of cell growth available to plant cells but not to animal cells.

6. Although plants and animals share many of the same biochemical and genetic processes, nonetheless some striking differences exist between the two groups. For instance, discuss the differences in life cycles between plants and animals, being sure to mention sporophytes and gametophytes. Do animals possess these stages in their life cycles? Also, discuss a common means of reproduction in plants in addition to sexual reproduction. Do animals have the ability to reproduce in this manner?

APPLYING CONCEPTS

1. The cells of plant tissues are linked by narrow cytoplasmic connections, known as plasmodesmata. Describe some of the consequences of these cytoplasmic connections, both positive and negative, for plants.

2. Plants produce two types of bodies known as the sporophyte and the gametophyte. Describe the differences between these two types of multicellular plant bodies and give examples from mosses, ferns, and flowering plants.

3. Some plants are capable of asexual reproduction. Compare this process to the laboratory process known as somatic embryogenesis. Are the two processes, one natural and the other human-directed, similar or dissimilar?

9

Stems and Materials Transport

Basswood forest

- Stems are plant organs that bear buds, leaves, flowers and fruits, or cones.

- Stems allow plants to increase their length, surface, and mass.

- Stems also conduct water, minerals, and organic compounds between leaves and roots.

- Stems of some plants function as food-storage or photosynthetic organs.

- Herbaceous plants have stems that lack large amounts of wood or bark.

- Stems of woody plants— shrubs, trees, and woody vines—are relatively thick and have a protective bark surface.

- Wood provides plants with large water-transport capacity, great strength, and decay resistance.

- Wood is used in many ways by people.

B ristlecone pine trees living in the mountains of Nevada and California are the world's oldest trees, reaching an astounding 5,000 years of age. In contrast, the oldest of the most common tree species in the eastern United States, sugar maples and red oaks, are 400 or fewer years of age. Eastern forest trees seem to more readily succumb to attacks by insects and pathogens and damage by ice and windstorms. But recently, scientists have found surprisingly old trees in eastern North America, where most native forests have been gone for more than 170 years. Many of these ancient trees have been hiding out in steep, swampy, or otherwise inaccessible places. Some were protected by conservation-minded landowners, by location on military bases, or in forest preserves. For example, loggers in North Carolina bypassed swamp bald cypresses that are now an amazing 1,700–2,000 years old. In Ontario, some ancient white cedars have proven to be over a thousand years old. Cedars of similar age have been reported from riverside ledges in eastern Tennessee. They survived because their steep cliff habitat was inaccessible to loggers. A nearly 700-year-old tupelo was found in a New Hampshire swamp. The oldest known pitch pine has been growing on a mountain ridge in upstate New York since 1617.

These old trees have provided evidence of past climate stress, such as periods of drought lasting up to 5 years. One extreme drought episode corresponds with the disappearance of the earliest English colony on Roanoke Island, VA, in the 1580s, possibly explaining the colony's failure. These trees provide evidence of past climatic events that could occur again, having devastating impacts on today's larger human populations with their greater water needs. The trees provide a warning that should be heeded in planning for sustainable water supplies.

Amateur tree enthusiasts discovered many of the very old trees. Such people combine their enthusiasm for hiking with a passion for searching out protected sites where old trees may have survived. Working with scientists, amateurs have been essential to locating ancient trees. But sometimes their search ends unhappily. Increasingly, only the stumps of ancient trees

are found, as logging on unprotected lands continues. The trunks of trees are examples of plant stems, the topic of this chapter. Here, we survey the structure and functions of stems, including the woody stems of trees. This information is useful in understanding both human uses of plant stems and forest ecology. ■

9.1 | Stems are fundamental plant organs having multiple functions

Vascular plants are those plants that have a conducting system composed of vascular tissue (xylem and phloem). Vascular plants include all land plants except bryophytes (Figure 9.1). During the evolution of vascular plants, stems appeared well before leaves and roots. Evidence for this is the fact that the earliest-known fossils of vascular plants have stems, but not roots or leaves. In fact, several types of stemmed, but leafless and rootless, plants lived hundreds of millions of years ago and left fossils (Figure 9.2). Plant stems often fossilized because their high lignin content helped preserve them from microbial decay. Fossil plant stems (Figure 9.3) have been very useful in tracing the evolutionary history of vascular plants and in understanding how vascular systems evolved. Such fossils, and increasing DNA evidence, suggest that leaves and roots evolved from ancestral stems (Chapter 21). Among modern vascular plants' organs, stems are the oldest and most fundamental. This is why stems are described prior to our focus on roots and leaves in Chapters 10 and 11.

EVOLUTION

Stems are indispensable organs for most plants. All other organs or organ systems—roots, buds, leaves, flowers and fruits, or cones—are attached to stems. Twigs bear distinctive scars where leaves and bud scales were formerly attached (Figure 9.4). Stems enable plants to increase their height or length, mass, and surface by the activity of apical meristems. Recall from reading Chapter 8 that shoot apical meristems are localized regions of cell division that generate new tissues at stem and branch tips. Apical growth gives plants increased access to sunlight. Plant stems are usually branched, which allows increase in mass and the amount of surface available for attachment of leaves and reproductive structures. The more leaves on a stem, the greater the amount of sunlight they can harvest in photosynthesis. The more flowers and fruits or cones on a stem, the greater a plant's ability to reproduce. Stems also store food materials, and sometimes serve as photosynthetic organs (Essay 9.1, "Weird and Wonderful Stems").

Stems transport water and minerals collected by roots from the soil to leaves, where these materials are needed for photosynthesis. Stems also conduct sugars produced in leaves to roots and any other places where sugars are needed to fuel metabolism and growth. Conduction occurs in specialized tissues—xylem and phloem. **Xylem** conducts water, minerals, and certain organic compounds. **Phloem** transports organic compounds in a watery solution. Xylem is defined by the presence of the tough, waterproofing compound **lignin** on walls of specialized water-conducting cells—**tracheids** and **vessel elements** (Figure 9.5). Lignin also strengthens the walls of fibers, which provide support to vascular tissues (see Figure 9.5). Recall that lignin is a very tough polymer constructed of phenolic secondary compounds (Chapter 3). Lignin confers strength, compression resistance, and waterproofing qualities to the vascular systems of plants. Thanks to lignin, the vascular systems of plants not only function as transport systems but also contribute to structural support. Vascular plants are said to have true stems, because xylem and phloem are present. Mosses have stemlike structures that bear leaflike appendages (Figure 9.6) and sometimes have conducting tissues resembling xylem and phloem. However, mosses and other bryophytes are regarded as nonvascular plants lacking true stems, leaves, and roots because lignin is not present in these early-evolved plants.

9.2 | The structure of conducting tissues helps explain their functions

How do plants manage to conduct a stream of water containing minerals or sugar through the entire length of tall corn plants—or giant sequoia trees, for that matter? Building an answer to this question first requires an understanding of how cells and tissues are arranged within stems. As is the general rule in biology, structure helps to explain how organisms function. Our integrated consideration of structure and function also explains the distinctive structure of vascular plants.

Conducting tissues in plants occur in vascular bundles

In **herbaceous** (nonwoody) stems and the young stems of woody plants, xylem and phloem tissues differentiate from precursor tissue (procambium) formed by the action of the apical meristem (Chapter 8). Mature conducting tissues formed in this way are known as **primary xylem** and **primary phloem.** These primary conducting tissues are located near each other within elongate **vascular bundles** (Figure 9.7). Vascular bundles are often surrounded by hard, supporting (sclerenchyma) tissue and are separated by varying degrees of nonconducting tissues. If you cut a thin cross section of a young corn stem, you will see vascular bundles scattered throughout the stem (Figure 9.8a). This pattern of vascular bundles is typical of grasses and other plants that are classified as monocots—plants having one seed leaf. In contrast, young stems of alfalfa and many other dicots—plants having two seed leaves—have vascular bundles arranged as a ring (Figure 9.8b). Other plants have vascular tissue that forms a more or less complete cylinder, with the individual vascular bundles separated by only a tiny bit of nonvascular tissue (Figure 9.8c). Conducting cells in xylem and phloem are arranged end to end to form pipelinelike arrays. The vascular tissue in a plant is interconnected—extending from the roots, through the main stems, and into branches, leaves, and other organs—much as the water pipes of a house link the plumbing in all rooms with the main water supply (Figure 9.9).

Living phloem tissues conduct organic compounds in a watery solution

Phloem tissues include pipeline components known as **sieve elements**, which may consist of sieve cells or sieve tube members (see Chapter 8) end to end (Figure 9.10). Sieve elements possess pore-containing end walls known as *sieve plates.* Their perforations develop by expansion of plasmodesmata—the intercellular channels that often penetrate walls of adjacent plant

FIGURE 9.1 **Vascular plants have true stems** (a) Lycophytes, (b) ferns, (c) gymnosperms, and (d) flowering plants have vascular systems whose water-conducting cells bear lignin coatings. The fern stem in (b) is the dark, horizontal structure that is seen extending between two clumps of young fronds (leaves). A few moss plants are growing on it.

cells. Sieve plate pores range from 1 to 15 micrometers (μm) in diameter and can easily be observed in slices of stem tissue viewed with a light microscope. Pores in the end walls of sieve elements allow phloem sap—a watery solution of sugars and other organic molecules—to move freely from one cell to the next (Figure 9.11). Phloem also includes **companion cells**, which are necessary for the functioning of sieve elements (see Figure 9.11). Phloem parenchyma and fibers, which provide support, are additional components of phloem tissue.

Phloem sieve elements are alive at maturity, possessing a cell membrane, some endoplasmic reticulum, plastids, and mitochondria. But the nucleus and some other cell components are degraded during sieve element development and are thus absent from mature cells. Removal of these structures appears to be necessary to allow phloem sap to flow through sieve elements. In order to function, sieve elements require the help of the adjacent companion cells, which have nuclei and provide materials to sieve elements via plasmodesmata. Companion cells and sieve elements are sister cells; they arise from an unequal division of the same precursor cell (Figure 9.12).

An example of the dependence of sieve elements upon companion cells can be observed when a plant is cut or wounded. In this event, P protein (phloem protein) masses along the sieve plate of sieve elements, forming a "slime plug" (see Figure 9.11). Such plugs function to reduce loss of phloem sap, much as clots reduce blood loss from

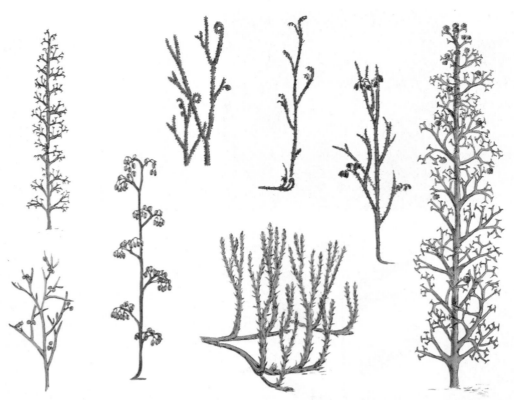

FIGURE 9.2   Many early vascular plants had true stems with xylem but lacked leaves and roots
These drawings are reconstructions of early vascular plants based on fragmentary fossils.

the vascular systems of animals. P protein also binds components of the walls of microbes and may thus also help prevent infection of plant wounds. Experiments have shown that sieve elements lack the ability to produce P protein messenger RNA; thus the mRNA and/or P protein itself are contributed by companion cells. Another wound response is deposition of the carbohydrate callose, which also helps to plug phloem sap leaks.

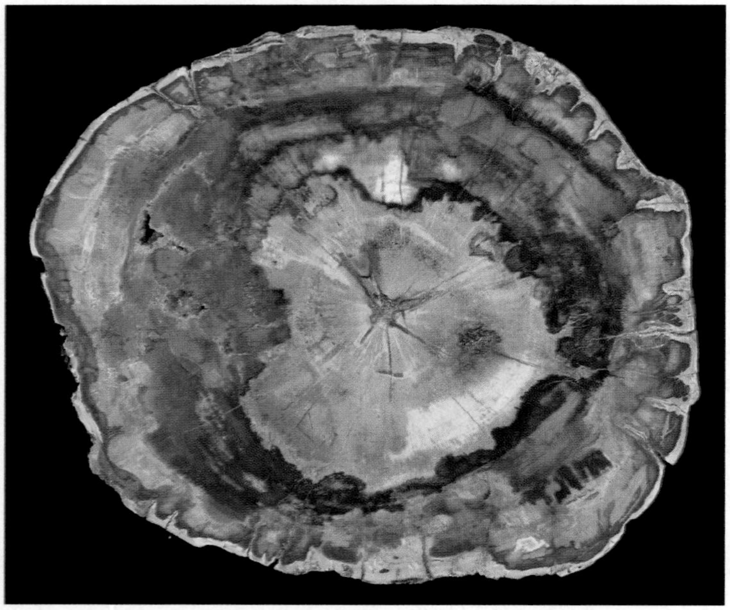

Phloem conducts sugars from their source to the sites of utilization

Stem phloem is continuous with phloem of roots, leaves, and other plant organs. Thus, phloem provides plants with a long-distance transport system. The direction of transport in phloem is from the source of organic molecules to sites—known as "sinks"—where organic molecules are utilized. Thus, the direction of movement of materials in the phloem is described as "source to sink." For example, sugars produced in leaves are transported to roots, which require organic food because their cells are deprived of the light necessary for photosynthesis. Sugars are also required by young leaves whose photosynthetic apparatus is not fully mature and by flowers, fruits, seeds, and food storage tubers, such as potatoes. The sugar transported to these sites by stem phloem is made by photosynthetic cells or is derived from the breakdown of starch stores.

How does sugar enter the phloem? In some plants, sugars are loaded from the cells producing them either directly into sieve elements or indirectly, via companion cells through plasmodesmata. This process is known as **symplastic loading** (Figure 9.13a). Recall from reading Chapter 8 that cells in plant tissues are often linked by plasmodesmata into a cytoplasmic continuum. This continuum, known as a *symplast*, enables sugars and some other materials to move readily from one cell to another, without having to cross cell membranes or cell walls.

FIGURE 9.3   A fossil stem This slice of a petrified (mineral impregnated) stem is from a conifer.

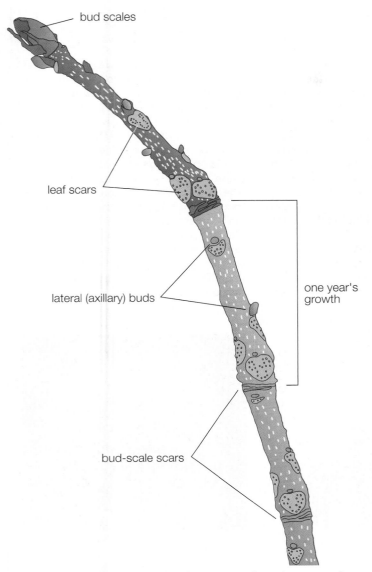

bud scales

leaf scars

one year's growth

lateral (axillary) buds

bud-scale scars

FIGURE 9.4   Twig surfaces possess scars where leaves and bud scales were formerly attached

Other plants have **apoplastic loading** of phloem, which occurs from intercellular spaces (Figure 9.13b). Apoplastic loading of sugars into living cells requires cell-membrane transport proteins (refer to Chapter 4). Energy in the form of ATP is often required to power the movement of sugars across cell membranes. Samples of phloem sap taken from sieve tubes show that sucrose is the most abundant transported sugar; but some other sugars, other organic compounds, and certain ions can also be loaded into phloem for long-distance transport.

The force that moves organic compounds within the phloem is known as **pressure flow** or **mass flow** and is based on osmosis and cell water potential (see Chapter 4). Recall that osmosis is water movement across membranes in response to solute concentration differences. Water readily flows from nearby xylem tissues into leaf sieve elements having high sugar concentrations, but their rigid walls prevent cell expansion, so hydrostatic pressure builds within sieve elements. This pressure causes sugar-rich phloem sap to move into the next element of a continuous sieve tube (Figure 9.14). Sugars move from sieve tubes into cells that

have sugar concentrations lower than that of the phloem sap, helping to maintain the flow of sugar from regions of high to low concentration. The flow of phloem sap will continue as long as there is a difference in the concentration of sugars along the sieve tube path.

Dead xylem tissues are structured to facilitate water transport

The structure of water transport cells in xylem tissues helps us to understand how water and dissolved minerals can move so rapidly through them. Xylem tissues of flowering plants usually include two types of elongate cells—tracheids and vessel elements. Nonflowering vascular plants and certain flowering plants primarily use tracheids for water transport. Both tracheids and vessel elements undergo programmed cytoplasmic death (Figure 9.15). During this process the cytoplasm disappears, which facilitates water flow through the empty cells, whose pipelike cell walls remain intact.

Tracheids are narrower than vessel elements and form long rows through which water can readily flow (Figure 9.16a). Their end walls are often slanted so that they fit closely together. Vessel elements have large perforations in their end walls and are stacked to form pipelinelike arrays known as **vessels** (Figure 9.16b). The sidewalls of tracheids and vessels are strengthened and waterproofed by lignin, a tough material applied from the insides of these cells. The lignin coating is not complete but, rather, occurs in spiral or ringlike patterns in primary xylem (see Figure 9.5a). Many small, round, nonlignified wall areas known as **pits** occur in the later-formed xylem (see Figure 9.16). Pits allow water movement into and out of water-conducting cells through their sidewalls. Pectins (see Chapter 4) help regulate water flow by swelling or contracting in response to the mineral content of water passing through cells' sidewalls.

Water and solutes move through stems as the result of transpiration

With few exceptions, plants obtain their water and minerals from the soil and move these materials via the root's xylem into the stem (Chapter 10). The stem xylem's main function is to transport water and minerals to other organs. Water moves in the xylem as the result of **transpiration**—the evaporation of water from plant surfaces mostly via epidermal pores, the stomata. A stream of water rises through the plant as each water molecule lost from a cell at the surface is replaced by another from inside the cell, which in turn exerts an attractive force on nearby water molecules, causing the water to rise. This process, known as *cohesion-tension*, is described more fully in Chapter 11.

In order to meet increased demand for sugar by growing shoots in the springtime, xylem also transports sugary solutions, which sometimes are tapped for making syrup. When days become warm but nights are still cold, trees convert starch stored in stem cells into sugars that flow upward through the xylem, supplying the needs of leaf and flower bud growth. Maple trees are particularly good sources of such xylem sap and are cultivated in large numbers in plantations known as sugar bushes, where they are tapped for the production of maple syrup

ESSAY 9.1 WEIRD AND WONDERFUL STEMS

Many kinds of stems are not easily recognized because they occur in unexpected places or forms, serving a variety of functions. However, a familiarity with unusual stems is useful in gardening, tending houseplants, and understanding how plants perform in nature.

The fact that grasses have horizontal stems explains why you can mow or cut grass without killing it, . . .

Take horizontal stems, for example. Sometimes mistaken for roots, these stems grow parallel to the ground—either above or below the soil surface—an adaptation that protects the delicate apical meristem from hungry animals or climate extremes. The fact that grasses have horizontal stems explains why you can mow or cut grass without killing it, and how grasses survive severe winter cold and summer heat. Plants that produce horizontal stems often spread rapidly; this is how a small patch of crabgrass can rapidly take over your lawn.

Strawberries have horizontal stems that grow along the ground surface and are known as *stolons*. Irises are examples of plants having horizontal stems that grow just below the surface; these stems are called *rhizomes* (Figure E9.1A). Many people are familiar with the fact that iris rhizomes must be dug up and divided after several years of growth, to reduce crowding and foster proper growth. *Gladiolus grandiflorus*, crocus, and cyclamen are popular garden or houseplants having a type of compact underground stem known as **corms** (Figure E9.1B). Corms are underground storage stems that hold food reserves for use by the plant during development and growth. In contrast, bulbs—which outwardly resemble corms—are large buds with thick, nongreen, food-storage leaves clustered on very short stems. Familiar plants that produce bulbs include onion (*Allium cepa*), daffodils, and *Narcissus*. If you cut a bulb in half, you can observe the leaves, which are not present within corms.

Tendrils of Virginia creeper (Figure E9.1C—left) and grape (Figure E9.1C—right), which coil around other plants, fence wires, or other supporting structures, are also examples of unusual stems (pea tendrils are the modified tips of leaves). Tendrils allow plants to reach upward without having to produce wood.

Stems can also serve as photosynthetic and water-storage organs in arid-zone plants. Cacti, which are native to American desert regions, and euphorbs—inhabitants of arid Africa—are known as succulents because their fleshy stems accumulate water when it is available and store it in large cells for drier times. The green stems are the primary organs of photosynthesis for these plants, because having large leaves would result in excess water loss (Figure E9.1D). Cacti do have leaves, but they mostly occur in the form of spines, which have a defensive function.

E9.1B Gladiolus corm

E9.1D Barrel cactus

E9.1A Iris rhizome

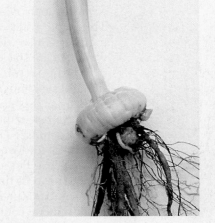

E9.1C Virginia creeper and grape tendrils

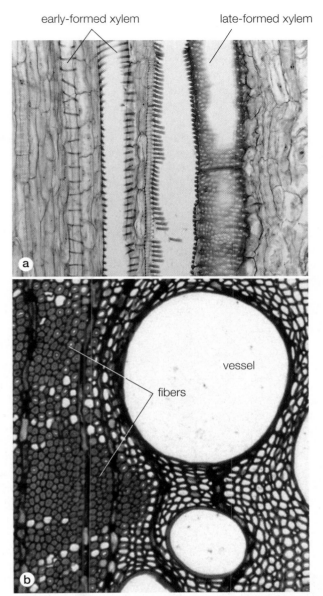

FIGURE 9.5 **Xylem tracheids, vessels, and fibers bear lignin coatings** The lignified walls take on a red stain in these preparations. (a) Vascular tissue in squash. Early-formed cells of the primary xylem usually have ring-shaped or spiral thickenings, which allow the cells to expand as the plant grows in length. When the stem has stopped elongating, more heavily lignified xylem cells with pits are produced. These cells are rigid and cannot lengthen. (b) Vessels and fibers of red oak wood, which is heavily lignified (explaining its structural strength).

(Figure 9.17). With care, the same sugar bush tree can be tapped year after year and yield about 150 liters (L) of sap per year without being harmed.

Wood and bark arise by the activity of secondary meristems

Many plants produce little or no wood or bark, remaining herbaceous throughout their lifetimes. In contrast, woody plants produce wood tissue and bark by the action of two meristems, the **vascular cambium** and the **cork cambium.**

FIGURE 9.6 **Mosses have leaf- and stemlike structures but lack lignified conducting tissue, and true stems or leaves**

These are localized regions of cell division that increase plant girth (see Figure 8.8). The girth increase of a tree trunk is a process known as **secondary growth.** The vascular cambium and cork cambium are **secondary meristems.** Recall that primary growth by activity of the primary apical meristem is a process that increases plant height or length. Primary growth is a property of all plants, because all plants have a primary meristem. In contrast, secondary growth occurs only in plants that have secondary meristems. Though many herbaceous plants produce a small amount of wood, **woody plants** are those in which secondary growth is extensive and whose surface is covered with bark. Woody plants include **trees,** having one or only a few main stems; **shrubs,** having several to many main stems; and **lianas,** which are woody vines (Figure 9.18).

The vascular cambium produces wood and inner bark

In woody plants, the vascular cambium forms by the coalescence of meristematic tissue situated between xylem and phloem in vascular bundles (the fascicular—"bundle"—cambium) with meristematic cells located between adjacent vascular bundles, known as the interfascicular ("between bundles") cambium (Figures 9.7, 9.19). The mature vascular cambium takes the form of a cylinder. The vascular cambium produces lignin-rich **secondary xylem** tissue to the inside—known as **wood**—and **secondary phloem** to the outside, which makes up the inner bark (Figure 9.20). (In grasses and other monocots, secondary growth involving a vascular cambium does not occur.)

There are two types of cells in the vascular cambium—vertically elongate **fusiform initials** and cuboidal **ray initials** (Figure 9.21). Division of fusiform initials generates secondary xylem—secondary tracheids and vessel members—toward the inside of the stem, and secondary phloem—such as sieve elements—toward the stem surface. More secondary xylem is produced than secondary phloem. Addition of a thick cylinder of wood requires that the circumference of the vascular

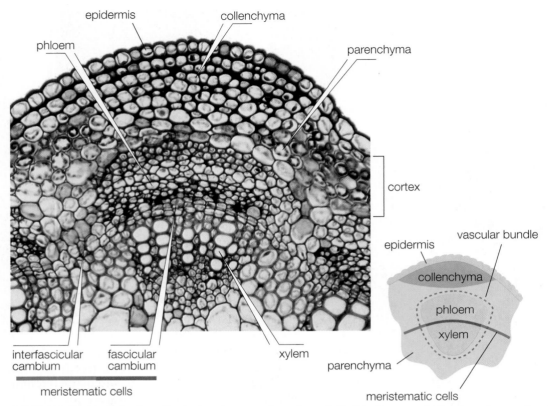

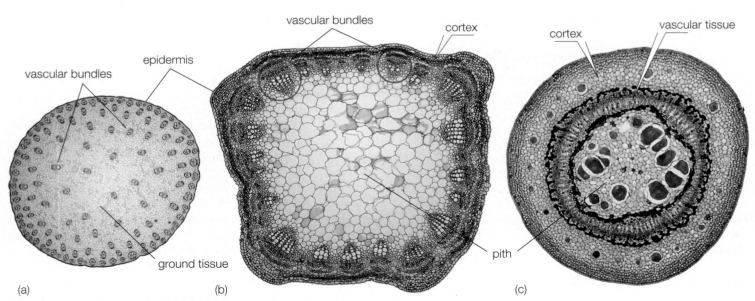

FIGURE 9.7 Vascular bundle In this close-up view of an alfalfa *(Medicago sativa)* stem, xylem and phloem of a single vascular bundle are seen along with the cortex, which is the region between the bundle and the epidermis. In the regions where the stem is ribbed, collenchyma cells beneath the epidermis provide support for the plant. The inner cortex (nearer the vascular bundles) usually consists of parenchyma cells, which carry out photosynthesis. Cells that retain meristematic activity are found between the xylem and phloem, the fascicular (bundle) cambium, and extending between the vascular bundles, the interfascicular cambium.

FIGURE 9.8 Cross sections of monocot and herbaceous dicot stems (a) Cross section of a monocot stem (corn—*Zea mays*). Monocots have a number of vascular bundles scattered throughout nonconducting stem tissue. (b) The alfalfa stem, like those of other herbaceous dicots, has discrete vascular bundles arranged in a ring. (c) In some plants like basswood *(Tilia)*, the vascular bundles are separated by only a very thin layer of cells and so appear as a cylinder.

FIGURE 9.9 Vascular system of a plant Within a plant, the vascular bundles are interconnected, forming an extensive network extending throughout the roots, stems, leaves, and organ systems. Only the larger bundles are shown here.

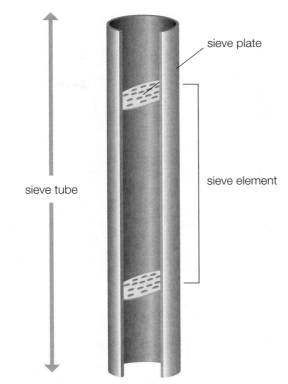

FIGURE 9.10 Sieve elements forming sieve tube Overlapping end walls (sieve plates) of adjacent sieve elements are perforated by large pores through which dissolved sugars and other molecules can travel. The chain of sieve elements is known as a *sieve tube*.

cambium must increase, necessitating the addition of new vascular cambium cells by cell division.

Ray initials produce ray parenchyma cells and ray tracheids that, together, form the **vascular rays** (Figure 9.22). Rays store starch, protein, and lipids. Rays also transport food laterally from the secondary phloem to the secondary xylem, and water and minerals laterally from the secondary xylem to the secondary phloem.

During each growing season, the vascular cambium produces new cylinders of secondary xylem, adding to the stem's accumulation of wood and forming growth rings (see Essay 9.2, "Growth Rings: Mirrors into the Past"). Xylem produced in previous years may remain functional in water transport, but much of the older wood becomes nonfunctional in water transport because conducting cells become clogged by protrusions of cell-wall material (known as *tyloses*), generated by neighboring parenchyma cells. The innermost wood of a tree trunk may be impregnated with decay-resistant chemical compounds. This is the heartwood (see Figure 9.22), whose deeper color and higher density is often valued for making furniture and other objects. Sawing wood in different planes (Figure 9.23) yields different

patterns relevant to lumber production, furniture building, and other human concerns.

In contrast to the xylem, usually only one year's production of phloem is active in food transport. The sieve elements that constitute much of phloem tissue typically live for only a year, after which they are crushed and die due to their lack of thick,

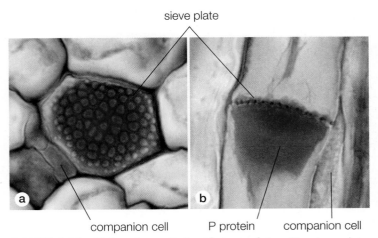

FIGURE 9.11 Light micrographs of sieve plates The sieve plate is seen in face view in (a) and side view in (b). The bright red material is P protein (phloem protein), which during processing for microscopic observation has formed a slime plug. The protein slime has been stained with a red dye. Companion cells occur adjacent to sieve elements.

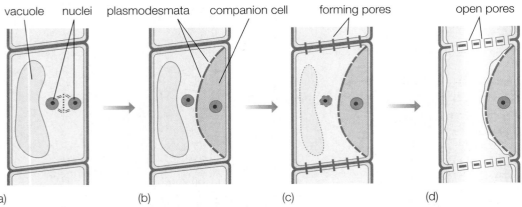

(a) (b) (c) (d)

FIGURE 9.12 Development of sieve element and companion cell after unequal cell division
(a) Unequal cell division leads to formation of the sieve element and a companion cell (b). (c) The nucleus of the sieve element degenerates, and the pores of the sieve plate begin to develop from expansion of plasmodesmata. (d) A mature sieve element lacks a nucleus and has sieve plates with large pores.

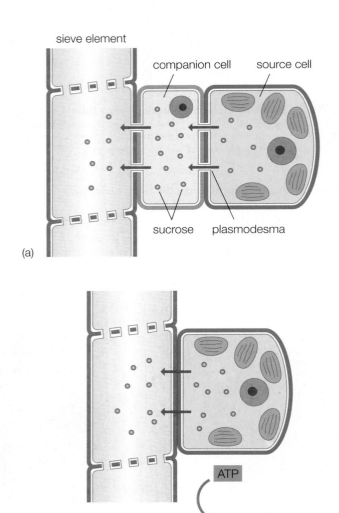

FIGURE 9.13 Symplastic and apoplastic phloem loading (a) In some plants, sugars are symplastically transported into the phloem from other cells via plasmodesmata. (b) Other plants transfer sugar into phloem from intercellular spaces. This requires sugars to move across the sieve tube element's cell membrane (not through plasmodesmata), a process that requires protein carrier molecules and energy in the form of ATP.

lignified walls. Thus, only a thin layer of phloem—the innermost bark—is responsible for most of the sugar conduction in a large tree. Because this thin phloem layer is located in the inner bark, a tree's food transport can be seriously damaged by abrasion of the bark. For example, people remove a circumferential strip of bark (girdling) in order to kill a tree. Young trees are vulnerable to bark damage, because this may have the same effect as girdling.

The cork cambium produces a protective covering for older woody stems

As a young woody stem begins to enlarge, the delicate epidermis eventually ruptures and its protective role is replaced by **cork**. Cork is produced to the outside of a secondary meristem called the **cork cambium**, which also produces some parenchyma cells to the inside. Together, the cork, cork cambium, and parenchyma cells make up the **periderm** (Figure 9.24a). Most woody plants produce a series of periderms during their lifetime. The first periderm usually originates just beneath the epidermis and may survive for many years in some species. Eventually, it may be unable to keep up with the stem's expanding girth and the plant forms a new periderm, which originates in the inner bark (living secondary phloem). As the plant grows older, a series of additional periderms may form in this way until remnants of several are visible in the **outer bark**, which consists of all the tissues outside of the innermost periderm (Figure 9.24b). The outer bark contains only dead tissues, whereas the inner bark is living. The periderm(s) can take on different appearances depending on the species and can therefore be useful in tree identification (Figure 9.25).

The walls of cork cells are well equipped to protect the stem from pathogens and other damage. Cork cell walls have layers of lignin and suberin. **Suberin** is composed of two materials: phenolic compounds, which help prevent microbial attack, and waxes, which retard water loss from the stem surface. Cork tissues also contain tannins—phenolic compounds that bind to pathogens' proteins, inactivating them. Cork tissues also exude gums and latexes that likely also have protective functions.

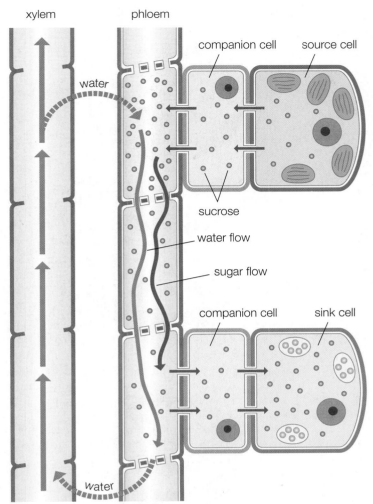

FIGURE 9.14 **Mass flow in phloem** As sugars are "loaded" into sieve elements (mostly from leaf cells, which are "sources"), the higher concentration of sugar in the sieve element compared to nearby xylem vessels causes water to move from the xylem to the phloem by osmosis (upper dashed arrow). Sugars are actively "unloaded" from the phloem at "sinks" such as root cells. This reduces the sugar concentration such that water can move into the xylem and be transported back upward. Since the sugar concentration in the sieve tubes is lower the farther away from the source, the amount of water moving from the xylem vessels to the sieve tubes (by osmosis) declines.

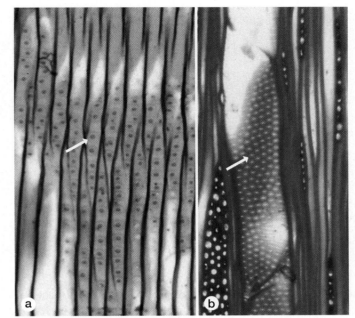

FIGURE 9.16 **Pits are nonlignified regions of the walls of water-conducting cells** Water can more easily move through the sidewalls of these cells at pits (arrows), which are not coated with lignin. In (a), pits are seen in pine wood tracheids. In (b), they are evident in the wall of a maple wood vessel member.

Lenticels, visible with the unaided eye as variously shaped, slightly raised patches (e.g., circles, ovals, or lines) on stem surfaces (Figure 9.26), are interruptions in the bark's cork layer that provide gas exchange for inner stem tissues. Lenticels may also occur on the surfaces of roots and fruits (for example, the small dots on apples are lenticels).

Some plants can grow tall without extensive wood

Wood is an extremely useful support material because it is relatively light and strong. But tree fern, cycad, and palm stems, which lack vascular cambia and extensive wood, can nevertheless grow to surprising heights (Figure 9.27). The primary meristems of these plants produce thickening tissues before

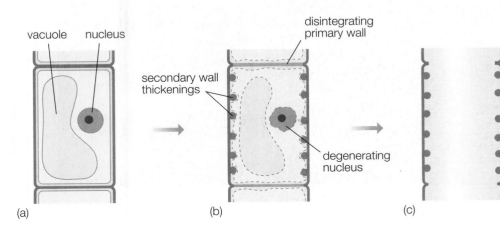

FIGURE 9.15 **Vessel development** (a) A cell destined to become a vessel member has a nucleus early on. (b) As the cell matures, secondary wall thickenings are deposited, and later the cytoplasm and nucleus begin to degenerate through programmed cell death. (c) At maturity, the cell is dead (i.e., there is no cytoplasm remaining). There are often large pores between the individual cells, through which water can easily pass.

FIGURE 9.17 A sugar bush A plantation of sugar maple trees that are tapped in spring for sugar-rich sap, from which maple sugar and syrup are made.

these stems increase in height, and overlapping leaf bases also provide some protection and support (Figure 9.28). Tree ferns, cycads, and palms branch only sparingly, if at all. Having few branches reduces drag forces exerted by wind and helps these plants avoid falling over in storms.

9.3 | Humans use stems in many ways

Stems of both woody and nonwoody plants have been useful to people in many ways, both in the past and the present. These include production of paper, cork, bamboo building materials, lumber, furniture, and other products.

Paper

Paper has been used to transmit information in written form for thousands of years since its invention in China. But an earlier stage along the evolutionary trail to paper was the Egyptian invention of papyrus, the material from which paper takes its name. Papyrus was made from the thick stems of the papyrus plant, *Cyperus papyrus*. Papyrus is an

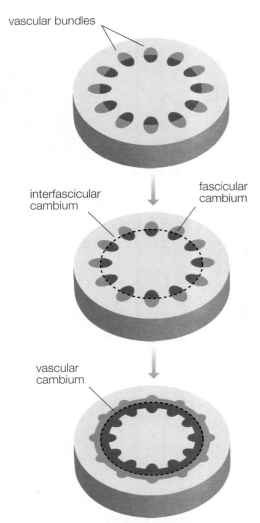

FIGURE 9.19 Vascular cambium formation The vascular cambium develops from meristematic cells between the xylem and phloem—the fascicular (bundle) cambium—together with meristematic cells lying between bundles—the interfascicular (interbundle) cambia. Together, they form a cylinder of meristematic cells. The relative contribution of fascicular versus interfasicular cambia in forming the vascular cambium varies from species to species, depending upon how widely the bundles are spaced.

FIGURE 9.18 Woody plants Included among woody plants are trees, shrubs, and vines (lianas). They are represented here by (a) an oak tree, (b) a dogwood, and (c) a bittersweet vine.

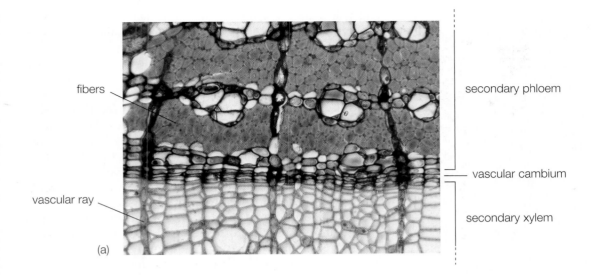

(a)

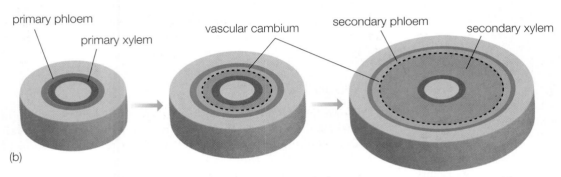

(b)

FIGURE 9.20 **Secondary xylem and phloem develop from vascular cambium** (a) In older portions of stems in woody plants such as basswood, the vascular cambium has begun to produce secondary xylem toward the inside of the stem and secondary phloem to the outside. Large numbers of fibers are visible here in the secondary phloem. (b) A diagram showing the expanding girth of a woody stem as secondary xylem and phloem are produced by the vascular cambium (dashed line). Note that the primary xylem and phloem are pushed apart as the secondary conducting tissues are produced.

herbaceous plant that grew abundantly along the banks of the Nile River (Figure 9.29a). Ancient Egyptians used papyrus to make rafts, sails, cloth, mats, and cord. To make paper, they peeled the outer layers of papyrus stems off, exposing the inner tissue (pith), whose cell walls were pliable but tough, thanks to cellulose. The pith was sliced into thin strips, then placed side by side in two overlapping layers. Then workers pounded the layers, releasing starchy glue that cemented the strips and layers together. The thin sheets that resulted were dried in the sun and polished to a smooth surface usable for writing (Figure 9.29b). Papyrus sheets were glued together to make long scrolls that could be rolled for storage. Some scrolls were 100 feet (30 meters) or more in length. Papyrus documents, and later records on paper made from other plants, have been invaluable in tracing the evolution of many human enterprises from the time of the Egyptians, Greeks, and Romans.

Today, most paper is made from wood pulp generated from trees grown in plantations. Genetic engineering has in recent years been used to tailor the characteristics of trees for more efficient production of paper products. Genetic engineers have succeeded in producing normally structured aspen trees whose growth is enhanced and whose wood contains 45% less lignin and 15% more cellulose than is normal. Production of more cellulose and less lignin is a particular advantage when wood is to be pulped for production of paper. During pulping, lignin by-products generate dioxin and other toxic compounds that can contaminate drinking water supplies. These toxic compounds are present in lower concentrations if the wood has reduced lignin content.

Cork

The cork oak, *Quercus suber,* generates a cork layer that is unusual in being several inches thick (Figure 9.30). This thick cork layer protects the oak from the frequent fires characteristic of its native Mediterranean habitat. Commercial cork is obtained by stripping the outer bark, beginning when the tree is 20–25 years old—a process that does not harm the cork oak. After the cork is stripped, new cork develops from underlying

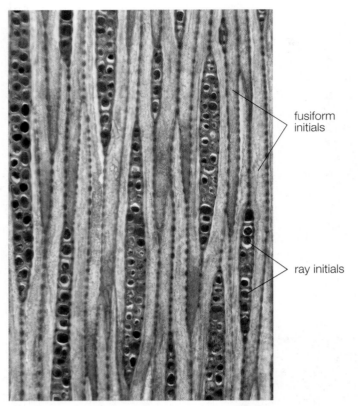

FIGURE 9.21   Fusiform and ray initials of the vascular cambium If the vascular cambium is observed in glancing (tangential) view (as if you were to make a longitudinal slice of the outermost portion of a garden hose), two types of cells can be seen: fusiform (spindle-shaped) initials and ray initials. Fusiform initials are elongated vertically, whereas ray initials are more or less equal in all dimensions.

periderm layers and within 10 years or so has produced enough cork for another harvest. Harvests can continue for the rest of the tree's life—about 150 years. Cork is filled with many small air spaces, explaining why this material is lightweight

FIGURE 9.22   Tree cross section showing heartwood The dark heartwood and lighter sapwood make up the secondary xylem. Note also the vascular rays, which appear as fine white lines radiating outward from the center of the stem.

and will float. No synthetic material has all of cork's natural properties. Cork is widely used as insulation and for flooring, wall coverings, shoe soles, and bottle stoppers.

Bamboo

The tall tropical grass known as bamboo (see Figure 8.10) has strong, resilient stems having many uses. For example, a million or so people live in bamboo housing. Recently, engineers associated with the U.K. Department of International Development have developed earthquake-proof houses based on the use of bamboo in construction. This strategy includes use of waterproofed bamboo to construct sheet roofing material and incorporation of bamboo in concrete walls, to provide reinforcement. Use of bamboo in buildings costs half as much as other basic constructions, allows assembly in as little as 3 weeks, and yields building predicted to last 50 years. Testing suggests that such bamboo-based houses could withstand earthquakes rating more than 7 on the Richter scale. Since many earthquake-related deaths result from collapse of traditional mud-brick buildings, many lives might be saved by the use of bamboo in reconstruction.

Wood

People have long valued the strength and beauty of wood harvested from trees for many purposes, as they do today. One example of ancient wood use is in the construction of observatories. A thousand years ago an astronomical observatory existed at Cahokia, a Native American city whose remains lie near the modern metropolis of St. Louis (U.S.A.). The observatory was an important part of life, allowing the people to track seasonal events that affected their lives, indicating times of planting and harvest. The Cahokia observatory consisted of a great circle—more than 400 feet in diameter—of trimmed red spruce poles arranged so that the sun aligned with specific poles on the first day of each season, much like the famous stones of Stonehenge near Salisbury, England. Because the Cahokians used tree trunks instead of stones, the Cahokia site is known as Woodhenge. Though the original poles disappeared long ago, the holes into which they were anchored remain, and new poles have been erected, restoring Woodhenge (Figure 9.31). Four thousand years ago, British farmers erected a similar observatory ring—21 feet in diameter—of oak posts surrounding a single large central post. In 1998, remains of these posts were discovered as a result of sea erosion of the soil that had buried the observatory for millennia. This observatory has been named Seahenge. In addition to observatory construction, ancient peoples used trees in many other ways—for constructing defensive stockades, flutelike musical instruments, tools, bowls, weapons, and canoes.

Modern humans continue to use the wood from many types of trees for construction material, fuel, and paper production. Modern wood industries represent more than 1% of the world's economy. Wood beams and columns support and decorate buildings. Chairs, tables, and other articles of furniture are constructed from wood because of its strength and beauty.

Almost everyone has seen growth rings in the wood of tree stumps, but did you know that tree rings tell us about climate conditions during long-past times, before people began keeping records? Scientists are interested in learning about past climates as one way of understanding the effects of climate change on plants and human societies.

The reason tree rings occur is that wood produced during the early part of the growing season has larger cells with thinner cell walls—and thus is less dense—than wood produced later in the growing season. The difference in cell size reflects warm, moist, spring growing conditions versus cooler conditions in fall, heralding the onset of winter. The change from less-dense "early" wood to more-dense "late" wood may be fairly gradual. But at the boundary between one year's late wood and the next year's early wood there is a pronounced difference, thereby making each year's annual growth rings stand out (Figure E9.2A). An annual ring is composed of one series of early to late wood. Counting the number of annual rings provides an estimate of a tree's age and is the basis for the science of dendrochronology ("tree time-keeping").

Dendrochronology was pioneered by astronomer Andrew Douglass, who needed a long-term record for his studies of climate. Douglass used local climate

. . . dendro-chronology has become a standard archeological technique for analyzing many types of wood remains . . .

records to establish that the amount of wood produced by a tree in a year's time—the width of a tree ring—closely correlated with the amount of rainfall. If light, temperature, and rainfall are favorable for tree growth and wood production, tree rings are wider than if climate is not so conducive to tree growth.

Since the vascular cambium, which generates each year's new cylinder of wood, is located just inside the inner bark, the outermost rings are younger than rings closer to a trunk's center. Thus, the width of rings produced in recent years for which climate records are known can be used to calibrate ring width for particular tree species. Also, trees of the same species growing in the same area produce rings of the same widths in response to the same degree of climate variation. This allows scientists to deduce climate of the remote past from very old trees, including the nearly 5,000-year-old bristlecone pines (Figure E9.2B) that still live in forested mountains of eastern Nevada and California (U.S.A.), or pieces of dead wood. Wood from living trees has been matched with dead wood of the same species to build a continuous record of tree rings and climate dating back more than 8,200 years.

Archeologist Clark Wissler, working in New Mexico (U.S.A.), first realized that dendrochronology could be used to date wood samples obtained from ancient ruins. Since then, dendrochronology has become a standard archeological technique for analyzing many types of wood remains: complete or partial artifacts, construction and carpentry debris, or fuel wood remains. In addition to information about age, wood analysis provides insight into a society's wood selection habits, evolution of woodworking technologies, environmental change influences on culture, and effects of humans on local forest ecosystems. For example, tree-ring analysis of wood used in ancient structures provided much of the evidence that climate change influenced the decline of Anasazi culture in the southwestern United States about a thousand years ago.

Other examples of climate change recorded in tree rings include deformed rings produced by Siberian pine trees during the years 536–537. These rings contain cells that were destroyed by unusual summer cold that froze the trees' sap. An unusually cool climate is also recorded in very compact tree rings produced in high-elevation European forests during the years 1625–1720. Slow tree growth produced particularly dense wood that was selected to make superior musical instruments, possibly including the famed Stradivarius violins.

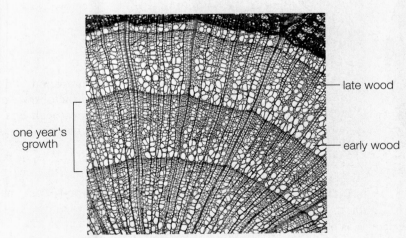

one year's growth

late wood

early wood

E9.2A Growth rings

E9.2B Bristlecone pine

FIGURE 9.23 **Cross, tangential, and radial sections of wood reveal distinctive patterns** Shown here are sections of red oak wood. (a) Cross-sectional view. Note the fibers as well as several large vessels in the early wood. (b) Tangential section (cut perpendicular to the vascular rays). Angiosperm wood typically has both narrow and wide rays in contrast to the narrow rays of conifer wood, which are usually only one cell wide. (c) Radial section, cut along the vascular rays, which have a bricklike appearance in this view. The structures shown here microscopically are responsible for the grain of wood, which varies considerably in appearance depending upon how the tree is sawed.

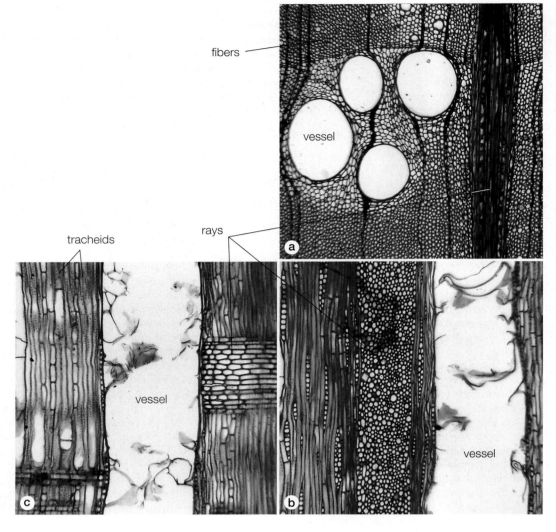

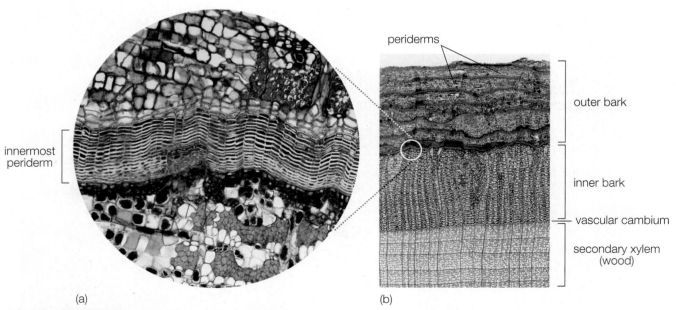

FIGURE 9.24 **Periderm** (a) A close-up view of a late-formed periderm in an old basswood stem (most of what is evident is cork; the other layers are less obvious). (b) A lower-magnification view showing the multiple periderms lying outside of the inner bark (which consists of secondary phloem). Secondary xylem (wood) lies internal to the vascular cambium.

FIGURE 9.25   Examples of bark (a) The smooth bark of a beech tree *(Fagus grandifolia).* (b) The warty or ridged bark of hackberry *(Celtis occidentalis).* (c) The smooth, papery bark of a birch *(Betula)* species. The horizontal lines on the surface of (c) are lenticels.

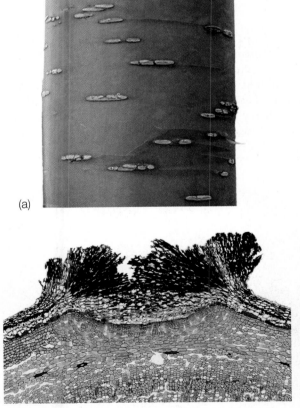

FIGURE 9.26   Lenticels (a) Lenticels on the Amur chokecherry *(Prunus maackii)* tree. (b) A section through a lenticel on an elderberry *(Sambucus canadensis)* stem. The numerous intercellular spaces between cells that fill the lenticel allow air to reach the deeper stem tissues.

FIGURE 9.27   Tall but nonwoody plants (a) Tree ferns, (b) cycads, and (c) palms.

FIGURE 9.28 **Persistent leaf bases of a palm help support and protect the stem**

FIGURE 9.30 **The cork oak (Quercus suber)** Trees from which bark has been stripped. The inset shows a tree with the typically thick bark.

Woods from western red cedar *(Thuja plicata)*, Douglas fir *(Pseudotsuga menziesii)*, redwood *(Sequoia sempervirens)*, and red and white oak *(Quercus rubra* and *Q. alba)* are valued for shipbuilding because of their durability in water. Basswood (linden) *(Tilia americana)*, yellow birch *(Betula alleghaniensis)*, black cherry *(Prunus serotina)*, and sugar maple *(Acer saccharum)* woods are valued for making musical instruments, because their wood structure contributes to beautiful tone.

(a) (b)

FIGURE 9.29 **Cyperus papyrus, the plant from which ancient Egyptians made the first paper** (a) Drawing of the plant. (b) Modern papyrus made by the ancient methods.

FIGURE 9.31 **Woodhenge** Native Americans who built mounds at Cahokia, IL, also constructed an astronomical observatory by erecting red spruce trunks in a giant circle. A modern restoration can be observed at the archeological site.

HIGHLIGHTS

9.1 Vascular plants are defined as plants having a vascular system whose water-conducting cells are reinforced with lignin, which provides strength and waterproofing. Modern groups of vascular plants include lycophytes, pteridophytes, gymnosperms, and flowering plants. Stems are the most ancient organs of vascular plants. Stems serve in organ attachment, conduction of materials from roots to other organs, food storage, and sometimes in photosynthesis.

9.2 Water-, mineral-, and sugar-conducting tissues occur in vascular bundles that are scattered in stems of grasses and related monocots but are arranged in a ring in dicots. Phloem tissues include living sieve elements, which conduct organic molecules in a watery sap. Phloem conducts sugars from their source to the sites of utilization. The force that moves materials in phloem is pressure flow, based on osmosis and cell water potential. Dead xylem tissues, which may include lignified tracheids and vessels, transport water and minerals. Cohesion of water molecules as transpiration occurs at plant surfaces is the force that draws water up through stems. Xylem also transports sugary sap upward from roots to shoots in spring. Wood and bark of woody plants arise by the activity of secondary meristems, which generate secondary tissues. Wood is secondary xylem, and inner bark is secondary phloem; both are generated by the vascular cambium. Annual rings of wood are useful in studying past climate change and archeology. In older woody plants, a protective covering is produced by the cork cambium. Cork cells are equipped in several ways to protect stem surfaces. Tree ferns, cycads, and palms, which lack extensive wood, can nevertheless grow tall.

9.3 Humans use plant stems to produce paper, cork, wood, and objects made from them. Genetic engineering is used to generate plants having desirable wood characteristics.

REVIEW QUESTIONS

1. Describe some of the benefits that stems provide to plants.

2. What materials are transported in the xylem and phloem, and how does the direction of flow differ in these two tissues? Which tissue is tapped in the springtime when making maple syrup?

3. Xylem and phloem are tissues specialized for the movement of substances about the plant body. Which groups of plants possess these tissues, and approximately when did they appear in the fossil record? Why are the conducting tissues of bryophytes and some algae not considered as true vascular tissue?

4. How does primary vascular tissue differ from secondary vascular tissue? What are vascular bundles, and how are they arranged within the (a) monocot and (b) herbaceous eudicot stem? How are secondary xylem and phloem arranged within the woody eudicot stem?

5. What is the shape of the vascular cambium (a) in a cross section of a stem and (b) in three dimensions? Describe how the cambium produces secondary xylem and secondary phloem.

6. What is the periderm, and what is its function? How is it formed? What are the roles of cork cells, suberin, lignin, tannin, gums, and latexes in the function of the periderm?

7. Compare the mechanisms used by xylem and phloem to accomplish transport within the plant body.

8. Discuss the process resulting in the formation of growth rings in woody stems.

9. Describe the science of dendrochronology.

10. Xylem and phloem each contain several different cell types that give them their functionality. Describe the major cell types in each, along with their function.

11. Classically, stems are considered as the upright axes connecting subterranean roots with leaves situated some height above ground level. Yet, there are many examples of valid stems that deviate from this theme. Describe several of these modified stems along with their function.

APPLYING CONCEPTS

1. The economic uses of plant stems are extremely varied. Besides the multitude of uses wood and paper are put to, name at least five other uses for plant stems in your household right now.

2. What would you expect to be the effect on growth rings of woody plants if there was little change in weather between the late wood of one season and the early wood of the following season? Can you think of a climate where this might occur regularly?

3. Imagine that you are examining a cross section through the trunk of a giant redwood tree. This section is a whopping 5 meters (m) in diameter and is on display where it covers an entire wall of a large room. Where are the very oldest cells (or cell remnants) in this trunk? Be sure to justify your reasoning by reflecting on how a woody stem adds new growth in height and in girth.

4. Imagine that in hanging a sign you pound a nail into the bark of a 10-m-tall tree. If the nail were initially 2 m above ground level, at approximately what height would you expect to find the nail when the tree reaches 20 m in height?

5. Imagine that you are a wood-boring insect, and you eat your way from the outside of a tree stem right down to the very core. Describe the tissues you will encounter, in order, on your gourmet journey.

10 | Roots and Plant Nutrition

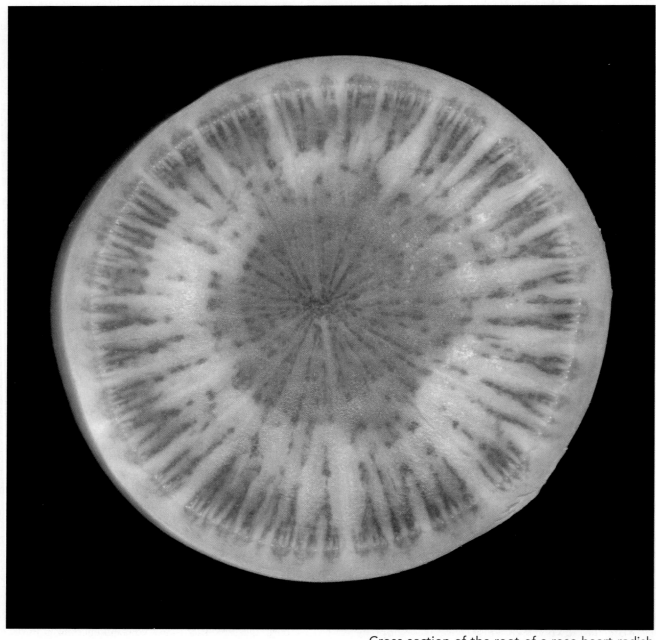

Cross section of the root of a rose-heart radish

- Roots play a variety of roles in plants: anchorage, water and mineral absorption, carbohydrate storage, synthesis of hormones and secondary compounds, stem support, and oxygen uptake.

- Root structure—including an apical meristem, root hairs, branch roots, and internal tissues—is closely related to root function.

- Plant roots are heterotrophic—they use oxygen to break down organic compounds supplied by the shoot to produce the energy-currency compound ATP needed for uptake of minerals from soil.

- Plant roots are often associated with beneficial fungi and bacteria that help them obtain needed nutrients.

If you were to take a cruise along tropical and subtropical coastlines, you would probably see vast seaside forests of mangrove trees and shrubs. Located between the sea and the land, mangroves serve essential ecological functions. Mangroves trap sediments and soil nutrients such as phosphorus and nitrogen brought by rivers and tides. Mangrove forests thereby protect offshore coral reefs. These reefs die when blanketed by sediment or smothered by seaweed, which grows too abundantly when ocean waters become polluted with unnaturally high nutrient levels. Mangroves also act as breakwaters, preventing coastal erosion by storms and waves. These forests are essential to the protection of some port cities. Birds such as the flamboyant scarlet ibis and others nest in the branches of mangroves, and these watery thickets also support large numbers of oysters, crabs, prawns, and fish harvested for food by many people. Together with tidal salt marshes, mangrove forests are estimated to provide nearly $2 billion worth of ecosystem benefits each year. However, mangrove forests are increasingly threatened by overgrazing, pollution, river diversions, shrimp pond construction, deforestation during timber extraction, and the rise in sea level associated with global warming.

One of the most striking aspects of mangrove trees is their huge, woody root systems. The roots of these plants constitute 50–60% of plant mass, and this root system is often conspicuous because much of it extends above the ground. Aboveground mangrove roots may appear as tangles of arches or arrays of upward-growing roots that may extend 2 meters (6.6 ft) or more above the waterline. Mangrove roots can be a formidable barrier to human passage but are key to the success of the entire mangrove ecosystem. Mangroves occupy permanently waterlogged soils that are too low in oxygen to allow the growth of most plants, whose roots would suffocate. But the aboveground roots of mangroves function like lungs, absorbing essential oxygen needed not only by submerged portions of the mangrove roots but also by animals inhabiting the stagnant waters.

In this chapter we will take a closer look at plant roots—components of plants that people usually don't think much about because they mostly grow underground. Yet roots are essential to plant survival, and successful agriculture and houseplant cultivation depend on a knowledge of root structure and function. A survey of the diverse forms and functions of roots will also help explain how plants, including mangroves, are able to grow in diverse and often stressful habitats. ■

Mangrove roots

10.1 | Roots play a variety of roles in plants

Whether you grow houseplants, garden vegetables, or flowers, whether you are interested in the uses of roots for food or as traditional herbal medicines, a knowledge of the many ways that roots function in plants can be useful.

Roots anchor plants and absorb water and minerals

When seeds germinate, the first plant organ to emerge is the **embryonic root—the radicle**—and a primary root is soon present on young plants (Figure 10.1). The rapidity of root development reflects the importance of roots in anchorage and in the absorption of water and minerals. It is essential that young plants become firmly attached to the soil substrate so they can begin absorbing water and minerals from the soil as soon as possible. Shoot development depends on enlargement of cells by water uptake. And photosynthesis requires water to serve as the necessary electron donor. Both of these processes are highly dependent on an early water and mineral supply. As plants develop, the primary root is replaced by a more extensive, branched root system. In a heavy clay or rocky soil that prevents root systems from developing properly, most plants will not thrive and may ultimately die. For this reason, farmers and gardeners sow seeds in soil that has been loosened by tilling machinery or a garden fork. Addition of organic material to heavy clay soils also helps break soil up enough that seedling root systems can develop properly.

ECOLOGY There are a few examples of plants that lack roots. These include bryophytes (in which roots never evolved); Spanish "moss," a flowering plant that grows on tree branches; the aquatic flowering plant bladderwort (*Utricularia*); the water fern *Salvinia*; and filmy ferns of moist tropical forests. Because these plants have thin leaves that directly absorb water and minerals from their very wet environments, roots are not required. But for most plants, roots are essential for anchorage, water and mineral absorption, and other functions.

Due to the need for a very large amount of root absorptive surface area, many plants can produce prodigious amounts of

FIGURE 10.2 **Grass plant showing extensive soil-binding root system**

root material. Amazingly, the roots of a single rye plant added together can reach a length of 500 kilometers (300 miles) and have a combined surface area of about 640 square meters! If you attempt to pull up a clump of crabgrass from your lawn, you can appreciate the extent of grass root systems and their astonishing ability to bind large clumps of soil (Figure 10.2). The soil-binding capacity of such extensive root systems explains why a cover of plants, especially grasses, is so important in preventing soil erosion by water and wind.

Some roots are useful as human food because they store carbohydrates

Biennial plants, those requiring two years to produce seed, store carbohydrates in fleshy **storage roots** during the first year of growth. In the second year, biennials use this food to help fuel the development of flowers, fruits, and seeds. Carrots, sugar beets, parsnips, and rutabagas are familiar biennial plants that are grown for food because they have food-storage roots (Figure 10.3; see Essay 10.1, "The Root of the Matter: Human Uses of Roots").

Roots are important sites of hormone and secondary compound production in plants

Roots produce the plant hormones cytokinins and gibberellins, which are transported in the xylem to the shoot, where they influence growth and development (Chapter 12). Roots may also be the main sites for producing protective secondary compounds. For example, the alkaloid nicotine is synthesized in the roots of tobacco plants and then transported to leaves, where it acts as a poison that helps prevent herbivore attack by causing nervous system malfunction (something to consider if you are thinking of taking up smoking). The roots of an African tree (*Bobgunnia madagascariensis*) have long been known to produce a yellow substance used by traditional healers to treat leprosy and syphilis, diseases

FIGURE 10.1 **Radish seedling showing primary root with root hairs**

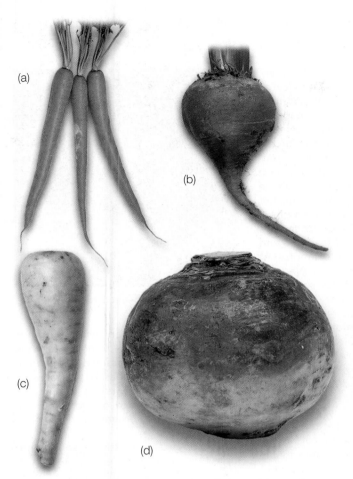

FIGURE 10.3 **Edible roots** (a) carrot, (b) beet, (c) parsnip, (d) rutabaga.

caused by bacterial infections. Recent scientific studies have shown that the yellow compound, identified as a terpene, does indeed kill bacteria as well as fungi and may have future applications in medicine and crop protection.

FIGURE 10.5 **Buttress roots in *Ficus***

The roots of some plants help support stems

If you look closely at the base of corn plants, you may notice **prop roots** growing from the stem into the soil. These specialized roots help the tall corn plants stay upright even though they lack woody tissue (Figure 10.4). Many tropical trees grow in soils so thin that they would easily fall over in high winds without the presence of impressive **buttress roots,** which can be as large as a person (Figure 10.5). The tropical banyan tree *(Ficus benghulensis)* sends down ropelike **aerial roots** from its branches, which anchor in the soil and become very thick, forming massive columns that support the heavy branches (Figure 10.6). Single banyan trees can cover several acres and produce hundreds of these trunklike support roots.

ECOLOGY

FIGURE 10.4 **Prop roots of corn**

FIGURE 10.6 **Banyan tree with aerial roots that support the branches**

ESSAY 10.1 THE ROOT OF THE MATTER: HUMAN USES OF ROOTS

Humans have long valued roots for their food value or as sources of other useful compounds. In premodern Europe, human-shaped roots of the mandrake plant, *Mandragora officinarum* (Figure E10.1A), were thought to indicate special powers, and extracts were used as a pain reliever and anesthetic. The Carthaginian general Hannibal is reputed to have prevented an uprising by his troops by adding sleep-inducing mandrake extracts to their wine. But today we know that in large doses, mandrake root is poisonous, so its consumption is legally restricted.

South American tribes have long extracted a compound known as rotenone from roots, seeds, and leaves of various tropical plants for use as a poison in fishing. Today, nearly 700 products containing rotenone have been marketed as a "natural" insecticide for use in organic gardens and on household plants, in flea powders, and for control of fire ants. Fish managers also add rotenone to lakes and reservoirs to

> *Mandrake and rotenone are examples of the general principle that plants produce secondary compounds for their own defense.*

kill pest fish species. Unfortunately, recent research with rats has implicated rotenone exposure in the development of Parkinson's disease in humans. Because Parkinson's disease is one of the most common brain degenerative diseases—affecting nearly 1 million people in the United States—further research is being conducted to evaluate rotenone's involvement. Mandrake and rotenone are examples of the general principle that plants produce secondary compounds for their own defense that may prove to be both useful and dangerous to people.

As foods, roots are low in protein and fat content, but those of several plants are excellent sources of dietary carbohydrates and vitamins. Carrots have been in cultivation for over 2,000 years, beginning in Europe. Carrots have been bred for their high production of beta-carotene, an orange-red substance that is the source of vitamin A, an essential human nutrient that helps prevent eye and skin disorders. Yams (also a good source of vitamin A), turnips, rutabagas, parsnips, and horseradish are other common modern root crops. The cassava (manioc) (*Manihot esculenta*) is a shrubby South American plant whose roots have long been cultivated for the food farina, now a staple for millions of people in the tropics. Cassava (Figure E10.1B) contains a poisonous latex that must be extracted before it is safe for human consumption. The starchy pellets used in making tapioca pudding and newly popular bubble tea are also derived from cassava. About 35% of the world's sugar comes from the roots of sugar beets—paler-fleshed relatives of common garden beets (*Beta vulgaris*)—that have been bred for high sugar content.

Roots are also used as flavorings. One of the main traditional flavoring

E10.1B Cassava (manioc) root

components of root beer, which traces its origin to 19th-century America, comes from sassafras (*Sassafras albidum*) roots. Roots of sarsaparilla (*Aralia nudicaulis*) and other plants are often used as well, in addition to extracts and oils from other plant organs (bark, leaves, seeds, etc.).

The physical properties of some roots make them useful in other ways. Owing to their tensile strength and flexibility, spruce roots are used by Native Americans in the construction of traditional birch-bark canoes (Figure E10.1C). Long, thin roots are peeled and split to form laces that are used to stitch together pieces of bark.

E10.1C Athabascan master boat builder with harvested spruce roots

E10.1A Mandrake plant

Pneumatophores help provide oxygen to underwater roots of some mangroves

Pneumatophores ("breath bearers") are specialized roots of some types of mangrove trees (Figure 10.7a). The pneumatophores, which grow upward into the air, absorb oxygen-rich air via surface openings—**lenticels.** Water does not readily enter lenticels when aerial roots are covered by water at high tide because these openings are lined with water-repellent materials. However, when the tide recedes, air is sucked into the roots. The pneumatophores contain tissue—known as aerenchyma—which has open channels filled with air (Figure 10.7b). Oxygen can diffuse through the aerenchyma's passageways to the submerged roots, which grow in an oxygen-poor environment. Submerged roots use the oxygen provided by pneumatophores to generate ATP, the cellular energy-currency molecule, which is needed for nutrient uptake by roots, a topic that is explored more extensively in Section 10.3 of this chapter. If sufficient oxygen is not present, mangrove roots can use fermentation (Chapter 5) to produce a small amount of ATP, which can help in the short term. But long-term survival depends on a reliable supply of oxygen from pneumatophores or other specialized aerial roots. Pneumatophores and other roots also provide mangroves with support in the unstable soils and turbulent tidal environments of coastal swamps.

Some plants produce other types of specialized roots

Some herbaceous plants, including dandelions and hyacinths, produce **contractile roots,** which are able to shorten or contract by collapsing their cells. This process pulls the shoot deeper into the ground where the soil is relatively warm, helping these plants survive changeable early spring weather.

ECOLOGY

Parasitic plants, such as dodder *(Cuscuta)* and mistletoe, obtain water, minerals, and organic food from host plants. Mistletoes infect a variety of plants, including mangroves and pines, by producing rootlike **organs** that penetrate the host's stem and tap into the host's vascular system (Figure 10.8).

(a)

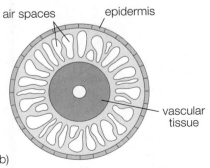

(b)

FIGURE 10.7 Pneumatophores (a) Roots of the black mangrove, *Avicennia.* (b) Cross-sectional diagram of a pneumatophore showing the spaces through which air is transported downward to underground portions of the roots.

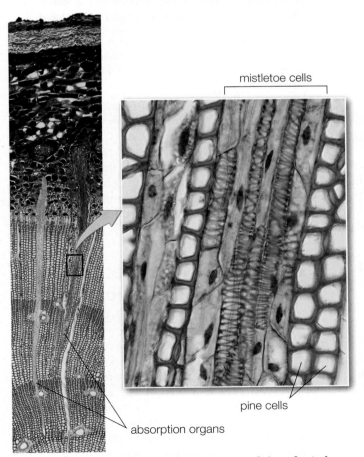

FIGURE 10.8 Rootlike absorption organ of dwarf mistletoe This portion of a cross-sectioned young pine stem shows two such organs, one of them highlighted. A portion of the other is shown at higher magnification, where it is surrounded by cells of the pine host's secondary xylem (wood).

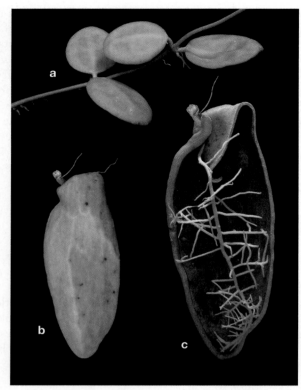

FIGURE 10.9 **The flower pot plant (Dischidia rafflesiana)** This epiphytic plant has aerial roots that grow into modified hollow leaves ("flower pots") that catch water and minerals. (a) This plant also produces leaves that are not hollow. (b) An intact hollow leaf. (c) A hollow leaf that has been cut open to show the roots inside.

FIGURE 10.10 **Taproot of dandelion** The dandelion is known as the "yellow-flowered earth nail" in the Yangtze River Valley of China because of the long taproot that anchors it in the soil.

Epiphytic plants, that grow non-parasitically on other plants, often have specialized roots. Epiphytic orchids have aerial roots that are photosynthetic (see Chapter 27). The "flower pot plant" *(Dischidia rafflesiana)* has aerial roots that emerge from the stem and grow downward into specially modified hollow leaves, which trap water and debris (Figure 10.9). Ants may live in the pots and contribute nutrients, including nitrogen, to the plant through their waste products. The roots absorb water and minerals from the pots.

10.2 | Taproots, fibrous roots, and feeder roots are major types of underground root systems

If you have weeded a garden or helped dig a stump or transplant trees, you have observed that plant roots can differ in their external form. These differences result from variations in the fate of a seedling's primary root. For example, in gymnosperms such as conifers, as well as in eudicot angiosperms, the primary root generates a **taproot system**—a single main root from which many branches emerge. Dandelions, familiar to many people who have pulled these weeds from their lawns, have a well-developed taproot system (Figure 10.10). In contrast, the primary root of grasses and other monocot flowering plants lives for only a short time and is replaced by a system of roots that develop from the plant's stem. Roots that develop

from the stem (or sometimes from the leaves) are known as **adventitious roots** (the word *adventitious* is derived from the Latin word *adventicius*, meaning "not belonging to"). In monocots, the roots originate from stem tissue. Many adventitious roots may contribute to the development of a root system in monocot plants, so no single root is most prominent. Rather, a **fibrous system** of many highly branched roots develops (Figure 10.11). Fibrous root systems are usually shallower in

FIGURE 10.11 **Fibrous root system of a grass plant**

FIGURE 10.12 Adventitious roots (a) A tomato plant was stimulated to form adventitious roots by turning it on its side with the lower stem lightly covered by soil. (b) A close-up view of adventitious roots emerging from the stem, photographed about one week later.

the soil than taproots. The prop roots of corn, mentioned earlier, are also adventitious roots.

Plants other than monocots also produce adventitious roots. For example, when planting young tomato plants (which are eudicots), people often strip off the lowermost leaves and bury the lowermost stem horizontally in the soil to foster the growth of adventitious roots from the stem (Figure 10.12). This step encourages rapid development of a more extensive root system, which gives the young plant a faster start.

Feeder roots, produced by both taproot and fibrous root systems, are the fine (less than 2 mm in diameter) peripheral roots that are most active in absorbing water and minerals from the soil. They often occur very near the soil surface (within 6 to 12 in. of the surface is common) but extend quite far from the plant. For example, the feeder roots of corn (*Zea mays*) extend about 1 m in all directions from the plant. Fine feeder roots have limited lifetimes and have to be continually replaced. Until recently, fine roots were thought to live for only about one year, but more recent studies have revealed that these roots can live for several years. One way that feeder root lifetimes have been studied is by photographing root growth from within clear tubes inserted into soil. Another method for determining the lifetimes of fine roots is measuring the amount of a naturally radioactive form of carbon in roots; this gives an

estimate of the time elapsed since carbon in the atmosphere was incorporated into the roots.

Knowledge of feeder roots is useful in landscaping and gardening. When transplanting, care is needed to excavate enough soil to include these shallow feeder roots and avoid tearing them. Otherwise, the plant may not survive the move, or the transplant's growth will be slowed while new feeder roots are produced. When digging a hole for a tree or shrub obtained from a nursery, make sure the hole is only deep enough so that the top of the root ball is at the same level as the surrounding soil surface. This precaution facilitates rapid growth of feeder roots.

10.3 | Root structure and function are intimately related

Roots of all types can do amazing things. They can grow through extremely dense soils, and they can sense gravity. Roots can defend themselves from pathogenic soil microbes as well as form close associations with beneficial microbes that aid in water and mineral absorption. Roots can branch repeatedly to generate huge surface areas for absorption of water and minerals. Some roots can become tough and woody. Roots can select minerals needed for plant metabolism and at the same time prevent harmful materials from entering the rest of the plant. How is all this possible? As we shall see, the key to understanding how roots grow and function is root structure. We will examine root structure at three increasingly finer levels: external structure, root tissues as seen with the use of a light microscope, and a close-up view of some important root-cell types.

External root structures include branch roots, root hairs, and the root tip

The external structure of a generalized, nonwoody root is shown in Figure 10.13. **Branch roots,** also known as lateral roots, are present, decreasing in age from the soil surface to the root tip. In other words, the youngest branch roots occur closest to the root tip. The feeder roots described earlier are young branch roots. Soil texture influences root branching. Plants that must grow through hard, dry soil have fewer branch roots than those growing in moist, loose soil. Branch roots and the main root axis are covered by an epidermis, which sometimes has a thin surface covering of cuticle but is generally exposed to the soil to facilitate water and mineral uptake.

A region closer to the root tip is fuzzy with countless **root hairs**—fingerlike extensions from some epidermal cells (see Figure 10.1). For most roots, these hairs are the main location of water and mineral absorption, and root hairs are a major site of uptake selectivity, the ability of plant roots to discriminate between useful and harmful soil minerals. Later, we will examine root-hair cells in more detail to discover how they accomplish this essential function.

At the cone-shaped root tip there is a **root apical meristem (RAM).** This is a region of meristematic cells, which divide rapidly, increasing the number of cells in the main portion of

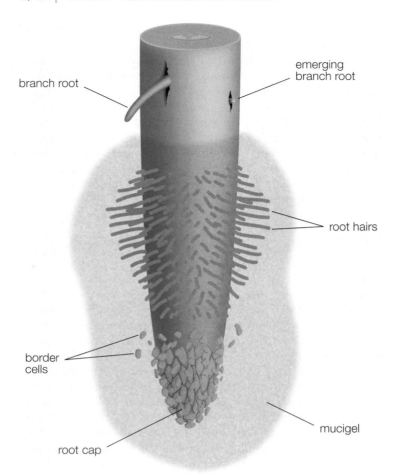

FIGURE 10.13 **External structure of a root** The root is surrounded by a zone of mucigel, polysaccharides produced by the root cells. This zone contains large numbers of bacteria and other microorganisms. The root cap protects the apical meristem as the root grows through the soil.

do not then just die but apparently help modify the external root environment in ways that prevent attack by microbes and tiny soil worms (nematodes). Root border cells have been described as biological "goalies," because they seem to ward off dangers to the root as it grows through the soil.

The tips of roots, including the root-hair region, are embedded in a blanket of **mucigel**, composed of gluey polysaccharides that are secreted from the Golgi apparatus of root-tip epidermal cells. It has been estimated that the roots of corn plants growing in 1 hectare (about 2.5 acres) secrete about 1,000 m^3 of mucigel, enough to fill the interior of a large house. Mucigel has several important roles. It lubricates the root, aiding in passage through the soil. Mucigel also helps in water and mineral absorption, prevents root drying, and provides an environment favorable to beneficial microbes.

An internal view of root tissues reveals how root cells grow and specialize

If a root is thinly sliced lengthwise or crosswise, the various root tissues and their relationships can be observed (Figure 10.14). The last few millimeters of a root tip consist of four major zones: (1) the previously mentioned root cap; (2) the root meristem, a zone of cell division; (3) the

the root. Protecting the RAM is a thimble-shaped **root cap**, whose cells are also generated by the apical meristem. Cells in the center of the root cap contain starch-rich plastids, **amyloplasts**, which are heavy enough to fall from one position to another as root position changes. Some experts think that amyloplasts operate as gravity sensors, signaling the downward growth path normal to most root cells. However, other experts have suggested alternative mechanisms for **gravitropism**—a root's growth response to gravity. Experiments designed to test some of these hypotheses have been conducted in space, where Earth's gravity field is weak. As yet, however, plant biologists do not completely understand how roots respond to gravity. They generally grow downward but may sometimes grow upward, as in the example of mangrove pneumatophores. Roots also grow toward a high concentration of water and minerals, a phenomenon experienced by homeowners who have had to replace or clean out leaky sewer pipes that have been clogged by invading tree roots.

About four to five days after root-cap cells are produced, they slough off from the root tip. Thus they must continually be replaced. These dispersed cells, known as **root border cells,**

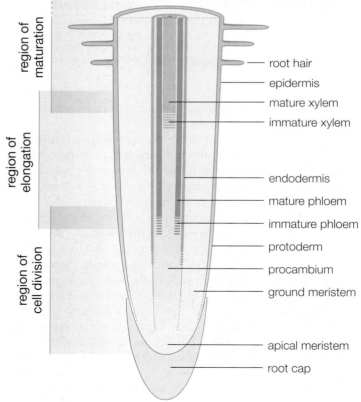

FIGURE 10.14 **Zones of root cap, meristematic cell division, cell elongation, and tissue maturation** The primary meristems—protoderm, procambium, and ground meristem—develop into the epidermis, vascular cylinder, and cortex, respectively.

elongation zone, where cell length increases dramatically; and (4) the zone of maturation, where tissue specialization (cell differentiation) occurs and where root hairs are present. Although root zones are not sharply delineated and their size can vary among roots of different plant types, they are characterized by specific types of gene expression. An expression map for more than 22,000 genes has been made for roots of the model plant *Arabidopsis thaliana*.

The root apical meristem produces primary tissues

The tissues produced by the root apical meristem are known as **primary tissues** because they are the first tissues produced by this plant organ (see Figure 10.14). Similar primary tissues are also produced by the shoot apical meristem (Chapters 8 and 9). Both root and shoot apical meristems first give rise to three meristematic tissues (primary meristems), known as the **protoderm**, **ground meristem**, and **procambium** (Figure 10.15).

Root cells enlarge and begin to specialize in the zone of elongation

The cells generated by the apical meristem are small, of more or less equal size in all dimensions, and look much alike. In the zone of elongation, these cells enlarge greatly—primarily in the vertical direction—through the absorption of water into the expandable cell vacuole (Chapters 4, 8). This is the main way in which root length increases. As the cell vacuole enlarges, it exerts pressure on primary-cell walls, which in turn respond by stretching. Cell enlargement also requires an increase in the amount of cytoplasm.

Cell specialization also begins in the elongation zone. Sieve tubes of phloem start differentiating from procambial cells earlier—at a lower point in the root—than other vascular tissues (see Figure 10.14). Because phloem brings organic materials from the shoot to root cells, its early appearance may reflect the root tip's high requirements for these materials in order to maintain high rates of cell division and to produce new cytoplasm and copious mucigel. Immature xylem develops next (from the procambium), followed by mature xylem elements.

Specialized cells and tissues are present above the zone of maturation

During tissue specialization, the procambium gives rise to the mature **vascular tissues** in the core of young roots, protoderm develops into the **epidermis** of young roots, and ground meristem produces a tissue known as the **cortex**, which occupies the region between the epidermis and vascular tissues.

Cortex Cells in the cortex usually lack chlorophyll but often store starch (Figure 10.16). Air spaces are usually very common in the cortex; these provide a way for oxygen-containing air to diffuse throughout the root. For reasons that we shall shortly describe, oxygen is just as necessary for the survival and function of plants and their roots as it is for animals.

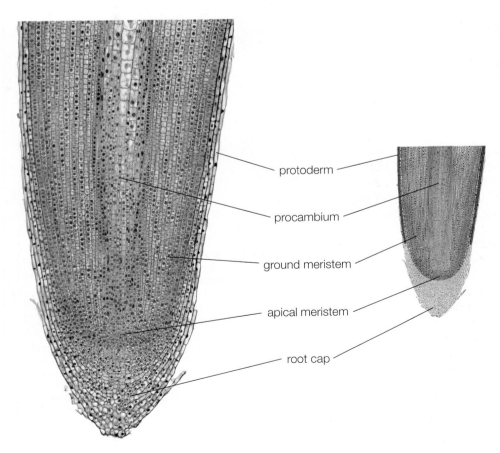

protoderm

procambium

ground meristem

apical meristem

root cap

FIGURE 10.15 Longitudinal section of onion *(Allium cepa)* root The apical meristem, root cap, and the primary meristems (protoderm, procambium, and ground meristem) are visible. The rapid growth and soft tissues make root tips excellent subjects for observing mitosis, either by sectioning the root or squashing it to separate the cells.

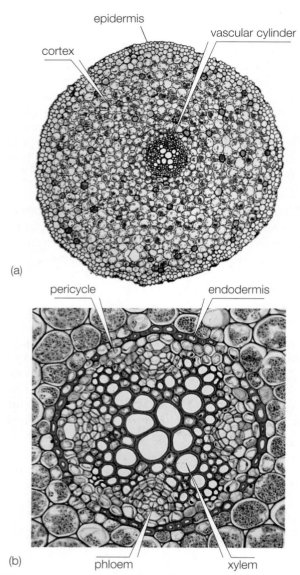

(a)

(b)

FIGURE 10.16 Cross sections of a buttercup (Ranunculus) root (a) A low-magnification overview showing the cortex, epidermis, and vascular cylinder. The purple structures in the cortical cells are plastid-contained starch grains. (b) A higher-magnification view of the vascular cylinder showing the primary xylem, which forms star-shaped lobes that alternate with primary phloem, and the pericycle. The endodermis lies to the outside of the pericycle.

Water and dissolved minerals can also move from the environment into roots and through spaces between cortex cells.

Endodermis The **endodermis** consists of cells of the innermost cortex layer, which lie very close together, forming a waterproof barrier between the cortex and the root core, where vascular tissues lie. Water and minerals that have entered roots via cell walls and spaces between root cells cannot cross the endodermis to gain access to vascular tissues. The reason is that endodermal cell walls of eudicots, unlike those of other cortex cells have a strip of **suberin,** a water-repellent material, that forms the **Casparian strip.** The ribbonlike suberin layer occurs on the tops and bottoms and two side walls of endodermal cells, but not on the other two side walls. In other words, the radial and trans-

verse endodermal cell walls have the suberin strip, but walls facing inward and outward do not (Figure 10.17). Water from the cortex can easily pass through walls with no suberin, and dissolved minerals can enter the cytoplasm of endodermal cells if their cell membranes contain the appropriate transporter proteins. Water and minerals for which membrane passage is allowed can also move from endodermal cytoplasm into intercellular spaces surrounding xylem (see Figure 10.17). In this way, endodermal cells act as a one-way filter for minerals. Those minerals for which carrier proteins are present can take the

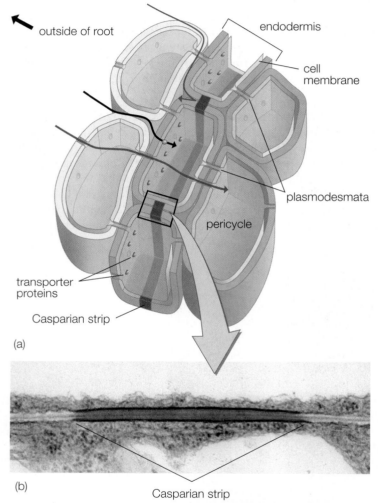

(a)

(b)   Casparian strip

FIGURE 10.17 Endodermis and Casparian strip (a) Cells of the endodermis are surrounded by a band of waterproof material known as the Casparian strip, which impedes the further inward flow of minerals dissolved in water that has moved through cortical cell walls and intercellular spaces. Therefore, minerals must pass through the cytoplasm of endodermal cells in order to reach the vascular cylinder. Purple arrow indicates minerals that pass through cell walls and intercellular spaces that are stopped by the Casparian strip. Other minerals pass through plasmodesmata into the endodermis (red arrow). The black arrow indicates minerals that move to the endodermis through cell walls and are actively transported across the endodermal cell membrane. (b) Transmission electron micrograph of a Casparian strip. Notice that the endodermal cell membrane is pressed tightly to the suberized portion of the cell wall; this prevents dissolved minerals from slipping into the space between the Casparian strip and cell membrane, effectively sealing the junction.

cytoplasmic route through the endodermis, thus bypassing the Casparian strip. But water and dissolved minerals that have not been permitted to transit through the endodermal cytoplasm are stopped by the suberin strip.

Suberin is composed of layers of waterproofing wax and antiseptic phenolic compounds (Chapter 3). Suberin also occurs in bark (Chapter 9) and in the abscission zone of leaves (Chapter 11), sites where having a waterproof, antiseptic barrier is advantageous in protecting plants from desiccation and disease-causing microbe attack. Roots of some plants have an outermost cortex layer whose cell walls are suberin-coated and act to prevent water loss and microbe attack.

Primary Vascular System The primary vascular system includes a core of xylem and phloem and a surrounding cylinder of tissue known as the **pericycle** (see Figure 10.16b). In some plants, alkaloids that accumulate in leaves are synthesized in pericycle cells and then transported to leaves via vascular tissues. But the pericycle has another important function—the production of branch roots. The pericycle retains meristematic activity and is the site from which new branch roots start to grow. Because branch roots start to form while they are embedded in root tissues, they must push their way through the cortex to the surface (Figure 10.18). In contrast to roots, stems and leaves lack pericycle, and shoot branches arise from axillary buds (Chapters 8, 9, 11). Otherwise, roots and shoots are similar in many ways, and roots probably evolved from shoots (Chapter 21).

The primary xylem of dicot roots has strands of primary phloem located between the xylem protuberances (see Figure 10.16b). Recall that xylem transports water and minerals upward to shoots and phloem conducts a watery solution of organic compounds from their source to the point of utilization—from "source to sink."

In plants that produce woody roots, cell division in the pericycle contributes to formation of a vascular cambium similar to that of stems, which generates secondary xylem and phloem (Figure 10.19). As in stems, secondary xylem forms

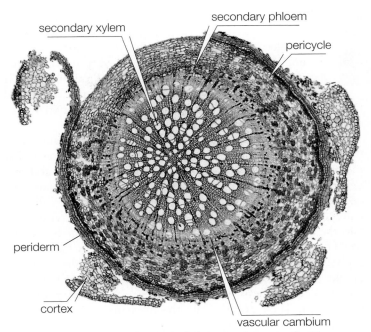

FIGURE 10.19 Cross section of the woody root of willow (*Salix*) after one year of growth The secondary xylem (wood) is formed internal to the vascular cambium, which also produces secondary phloem to the outside. The epidermis and cortex (remnants of which are still present) are eventually replaced by periderm(s), whose cork cells function to protect the root.

wood, and a cork cambium (Chapters 8 and 9) produces protective, suberin-coated cork tissue, which replaces the primary cortex and epidermis.

Epidermal Root Hairs Observing the root surface, the maturation zone—in which root epidermal cells reach their mature form—can be identified as the fuzzy root-hair region (see Figure 10.1). Microscopic observation of thin root slices reveals that some of the epidermal cells in this zone extend

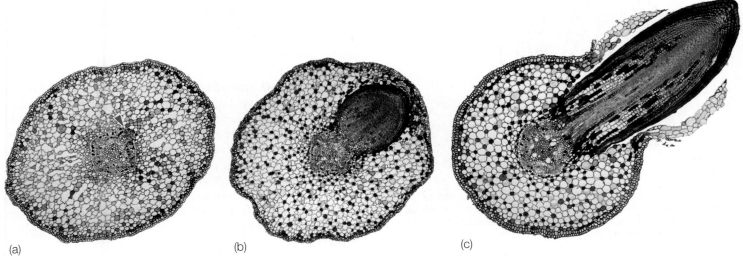

(a)   (b)   (c)

FIGURE 10.18 Branch root formation in willow (*Salix*) In (a), cells of the pericycle have just begun to divide (arrowhead), initiating the formation of a branch root. The forming branch root grows through the cortex (b), eventually bursting through the epidermis (c).

a fingerlike outgrowth into the soil; these are the root hairs. Root-hair length may reach 1.3 cm, about the width of your little finger; thus masses of root-hair cells are visible to the unaided eye. But root hairs are very narrow, only about 10 μm in diameter—about one-thousandth their length. This is an advantage in reaching the water and minerals present in soil pores that are mostly too narrow for even the smallest roots to enter.

Within a few hours, root hairs grow to a thousand times their original cell length. This remarkable growth requires that the outermost epidermal cell wall become very stretchy and that new cytoplasm be quickly produced to fill the hair. As root hairs develop, their cell membranes become richly embedded with proteins that selectively transport materials from the environment into the root-hair cytoplasm, in conjunction with the hydrolysis of ATP to ADP and phosphate. Energy supplied by the ATP hydrolysis generates a proton gradient that is used to drive mineral uptake (Figure 10.20). Recall that a proton gradient is used to power ATP synthesis in photosynthesis and respiration (see Chapter 5). These proteins are the mechanism by which root

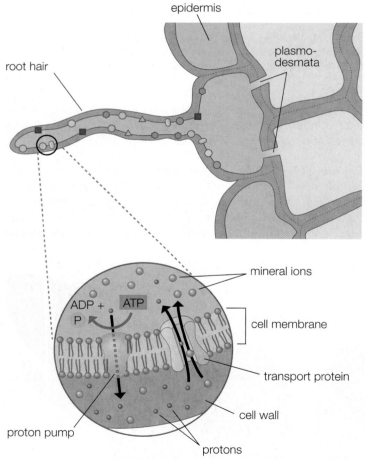

FIGURE 10.20 **Root hair with cell membrane transporters and plasmodesmata** Minerals move easily through the fibrous cell wall but can pass through the cell membrane only if appropriate transporter proteins are present. ATP is hydrolyzed to drive the uptake process, through generation of a proton gradient. As protons move back into the cell through the transporter protein, a mineral ion (for example, nitrate) is carried with it. Once in the epidermal cell cytoplasm, ions can move to adjacent cells through plasmodesmata.

hairs discriminate between materials that are required by plants and materials that are harmful. Transport proteins are shaped to exclude harmful materials, but they allow useful materials to bind to them and then move into root-hair cytoplasm. Once minerals have been transported into root-hair cytoplasm, they can pass from one cortex cell's cytoplasm to another—into endodermal cells—via the plasmodesmata, the intercellular passageways. A watery solution of minerals then passes into the dead xylem cells through nonlignified regions of their walls (Chapters 8 and 9).

Root hairs are so delicate that they are easily damaged by abrasion as roots grow through the soil; their average life span is only 4–5 days. As a result, root hairs do not occur on older root regions that lie above the zone of maturation. Roots must continually produce new root hairs from new protoderm tissues generated by the apical meristem. For some plants, the average rate of root-hair production has been estimated at more than 100 million per day!

Root mineral absorption is selective and requires energy

Root xylem obtains minerals and water in one of two ways: (1) Water and minerals are selectively taken up by root hairs and then transmitted from one cell's cytoplasm to others via plasmodesmata, avoiding contact with endodermal walls; or (2) water and minerals that penetrate root tissues within intercellular spaces and cell walls are selectively absorbed at the cell membrane of nonsuberized surfaces of endodermal cells and then released from endodermal cytoplasm on the other side of the endodermal barrier (Figure 10.21; also see Figure 10.17). These are the two mechanisms by which roots allow the entry of beneficial minerals and exclude harmful ones. Mineral passage from root hairs through the cortex and endodermis via plasmodesmata is known as **symplastic** transport. When minerals dissolved in water diffuse from the root's environment into epidermal cell walls, then through walls of cortical cells to the endodermis, such movement is known as **apoplastic** transport. Water-conducting cells of the xylem are dead and therefore incapable of discriminating between useful and harmful minerals. Any toxic minerals that might reach the root's xylem would be readily transmitted to leaves and other shoot organs, with dire consequences for the plant. The cell membranes of root hairs and endodermal cells provide the essential mineral-sorting function.

Root hairs and endodermal cells are sites of selective absorption As we have seen, roots play an essential role in screening the mineral content of water transmitted to the plant shoot. We have observed that the cells that perform this filtration process are the epidermal root hairs and cells of the endodermis. A closer look at mineral uptake by these cells explains how plant roots concentrate essential mineral ions—such as phosphate—from low concentrations in soils and how roots protect themselves from toxic minerals such as aluminum ions. It also explains why roots, like animals, require a supply of oxygen and organic food in the form of sugar.

Many metal ions, including iron, copper, manganese, and magnesium, are needed by plants for the proper functioning of enzymes and other complex molecules in plant cells (Table 10.1).

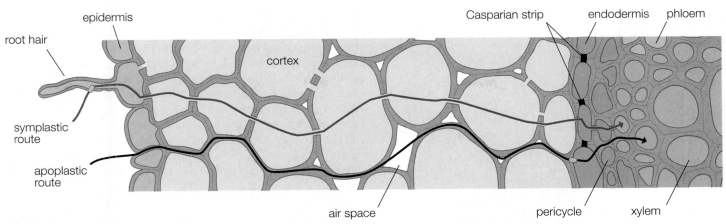

FIGURE 10.21   Minerals can move from cell to cell through plasmodesmata or via cell walls and intercellular spaces en route from root hair to vascular tissue In this diagram showing part of a root viewed in cross section, the red line represents mineral ions moving from root hairs of epidermal cells through plasmodesmata of cortical cells, the endodermis, and into the vascular cylinder (symplastic route). Minerals can also move through the cell walls and intercellular spaces (apoplastic route—black line), but the Casparian strip prevents them from moving into the vascular cylinder via this route, unless they are allowed to pass through the cytoplasm of the endodermal cells.

TABLE 10.1	Essential Elements for Plants: Sources and Functions			
Element (Symbol)	**Atomic Number**	**Percent of Plant Dry Weight**	**Source**	**Function(s)**
Macronutrients				
Hydrogen (H)	1	6.0	Water	All organic molecules
Carbon (C)	6	45.0	Air as CO_2	All organic molecules
Nitrogen (N)	7	1.5	Soil	Proteins, nucleotides
Oxygen (O)	8	45.0	Water and air	All organic molecules
Magnesium (Mg)	12	0.2	Soil	In chlorophyll
Phosphorus (P)	15	0.2	Soil	Nucleotides, phospholipids
Sulfur (S)	16	0.1	Soil	Proteins, vitamins
Potassium (K)	19	1.0	Soil	Ionic balance in cells
Calcium (Ca)	20	0.5	Soil	Cell-wall component
Micronutrients				
Boron (B)	5	0.002	Soil	Uncertain
Chlorine (Cl)	17	0.01	Soil	Ionic balance in cells
Manganese (Mn)	25	0.005	Soil	Enzyme cofactor
Iron (Fe)	26	0.01	Soil	Enzyme cofactor
Copper (Cu)	29	0.0006	Soil	Enzyme cofactor
Zinc (Zn)	30	0.002	Soil	Enzyme cofactor
Molybdenum (Mo)	42	0.00001	Soil	Enzyme cofactor

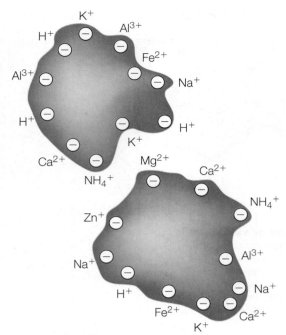

FIGURE 10.22 Positively charged mineral ions bound to negatively charged soil particles

These positively charged mineral ions are associated with small, negatively charged soil particles (Figure 10.22). Examples of the essential functions of minerals include magnesium in chlorophyll and iron in the electron carriers of photosynthesis and respiration (Chapter 5).

Aluminum ions, however, are both very abundant in soils and toxic to plants. Aluminum binds to organic acids, phosphates, and sulfates in proteins and nucleotides, causing disruption in membrane function. The first and most evident symptom of aluminum toxicity in plants is that roots stop elongating within 5 minutes of exposure. Aluminum toxicity is a major limitation to the ability to grow enough crops, particularly in the 30% of world soils that are acid. Such soils are particularly common in the tropics, where high human populations strain agricultural production capacity, as well as in northern Europe, Canada, and the northeastern United States, which receive acid precipitation (Chapter 27). Acid rain decreases soil pH—the logarithm of the reciprocal of hydrogen ion concentration in grams per liter. When soil pH is below 5, abundant hydrogen ions bind to negatively charged soil particles, thus releasing aluminum ions that were previously bound to soil. These ions are then dissolved in the soil water and available for absorption. As a protective response, plants can release organic acids from roots that bind to aluminum, and they can also sequester aluminum ions in their cell walls. But a major way that plants avoid aluminum ions is by not allowing the transport of this ion across their cell membranes and into their cytoplasm.

In contrast, plants require a great deal of the phosphate ion for construction of phospholipid membranes, production of ATP, and for the nucleic acids DNA and RNA. Plants simply cannot survive, grow, or reproduce without a supply of phosphate throughout their lifetime. Phosphate is one of the most important components of crop and houseplant fertilizers and is abundant in manure. But phosphate forms strong chemical bonds with iron and aluminum oxide minerals in soils, reducing its availability to plant roots. Plant roots can respond by releasing organic acids or hydrogen ions, which help dissolve bound phosphate, and by increasing the amount of root branching and the density and length of root hairs in order to maximize absorptive surface area. They also exude enzymes that release phosphate from soil organic compounds.

Importantly, plant root cell membranes contain transporter proteins whose shapes enable them to bind even small amounts of soil phosphate and actively ferry it into their cytoplasm, using energy supplied by the hydrolysis of ATP (see Figure 10.20). When soil phosphate is low, root cells increase the number of phosphate transporter proteins in their membranes. Upon releasing the phosphate to the cytoplasm, the transporters are able to ferry more phosphate from outside. The specificity and high efficiency of plant phosphate transporters is the result of a high degree of control over transporter protein amino acid sequences, encoded by DNA in the cell's nucleus. Separate root cell-membrane proteins are specialized to collect and import other essential plant nutrients, such as nitrogen compounds, magnesium, iron, and other needed minerals (Essay 10.2, "Food for Thought: Plant Mineral Nutrition"), even if these are of low concentration in the soil. Genetic engineering methods offer an opportunity to improve the efficiency of root mineral-import systems (Chapter 15).

DNA SCIENCE

Plant roots require organic food and oxygen and produce carbon dioxide Plant root cell-membrane transporters are amazingly efficient at mineral nutrient absorption, but energy input in the form of ATP is necessary. This is because plant roots typically must accumulate mineral ions from a dilute solution in the soil. If the concentration of needed ions in the soil is greater than that within cells, ions can readily diffuse into the cells until the inside and outside concentrations are the same (i.e., they would reach equilibrium), given that appropriate import channels composed of proteins are present in the cell membranes. Such **facilitated diffusion** would not require energy input. However, the concentration of needed mineral ions in soil is usually much lower than inside cells; in this case, work must be done to import minerals into cells. The source of energy to perform this work in cells is ATP. Mineral uptake that requires ATP is known as **active transport**. ATP is also required to power cell division at the root tip, synthesize new cell-wall materials as root cells elongate, and conduct other metabolic activities. The protein gradient established by proton pumps in the plasma membrane of root-hair cells is exploited to move mineral ions into plant roots. As a proton moves into the cell across the plasma membrane (along its concentration gradient), a mineral ion is ferried across with it (against its concentration gradient) (see Figure 10.20).

ESSAY 10.2 FOOD FOR THOUGHT: PLANT MINERAL NUTRITION

It is commonly stated that plants produce their own food, referring to organic compounds such as sugars generated from the raw materials CO_2 and water. However, it is also true that, in addition to organic compounds and water, plant growth requires more than a dozen minerals. The "plant food" that you dissolve in water and apply to your houseplants is composed of these minerals. Plants require ammonium or nitrate for making proteins and other nitrogen-containing compounds such as nucleotides. Protein synthesis also requires sulfate, and phosphate is needed to make DNA and RNA as well as ATP, phospholipids, and other phosphorus-containing compounds. Magnesium is essential for chlorophyll, potassium helps maintain the balance of plant cell ions, and calcium is needed in cell walls. Because these minerals are required in relatively large amounts, fixed nitrogen, sulfate, phosphate, magnesium, potassium, and calcium, together with carbon, oxygen, and hydrogen, are known as plant **macronutrients** (Table 10.1).

Lesser amounts of other minerals are also required, and these are the **micronutrients** boron, chlorine, manganese, iron, copper, zinc, nickel, and molybdenum (see Table 10.1). Several micronutrients are essential components of enzymes. For example, iron is required for molecules that perform electron transport during energy transformations in plant mitochondria (and those of other eukaryotes). Iron and molybdenum are required for construction of nitrogenase, the bacterial enzyme critical to nitrogen fixation in legume root nodules and other nitrogen-fixation associations between plants and microbes.

Minerals taken up by plants are returned to the soil when the plants and the animals that consumed them die and decompose.

In natural systems, minerals taken up by plants are returned to the soil when the plants and the animals that consumed them die and decompose. In agricultural systems, however, crops are removed and the minerals leave with them. As a result, soil fertility decreases over time, and farmers and gardeners must add minerals in the form of fertilizers. Organic fertilizers include compost and manure. Inorganic fertilizers are commercial mixtures of chemicals, both organic and inorganic. However, it is only the inorganic form that enters the plant. The numbers on inorganic fertilizer packages, such as 30-10-10, refer to the proportions of nitrogen, phosphorus, and potassium in the mix. Organic fertilizers work more slowly (because they must be broken down to inorganic compounds) but are longer lasting than inorganic fertilizers. Inorganic fertilizers not immediately absorbed by plant roots may run off into waterways or enter the groundwater, where they can cause pollution.

Some commercial farmers grow plants such as tomatoes and lettuce in greenhouses without soil, using a mineral nutrient solution (Figure E10.2). This technique is known as **hydroponics**. In arid lands, plants grown by hydroponics need less water than plants in outdoor irrigated fields. Hydroponics is thus an attractive option for desert countries. Hydroponics may cost more than conventional agriculture, but it can be cost-effective if fresh foods must otherwise be imported from long distances at great expense.

E10.2 Lettuce plants growing hydroponically

The most efficient mode of ATP production is aerobic respiration, in which oxygen is used to break down organic molecules to carbon dioxide (Chapter 5). Thus, production of sufficient ATP to import mineral ions on a constant basis requires roots to acquire both organic compounds (i.e., sugars) and oxygen. But roots, being underground, cannot photosynthesize. It is the photosynthetic shoot that provides sugars to roots via sieve tubes of the phloem, and oxygen comes from the atmosphere as a product of shoot photosynthesis. Because they require oxygen and organic food for metabolic processes, roots are classified as **heterotrophic**. Like heterotrophic animals, roots depend on green plant organs as a source of organic food and oxygen and release carbon dioxide.

ECOLOGY

Root-produced carbon dioxide dissolves in soil water to produce carbonic acid, which causes a series of soil-mineral chemical reactions known as weathering.

One result of weathering is that carbon dioxide is converted to dissolved bicarbonate ions, which wash into oceans and eventually become part of carbonate rocks. This is one process by which plants reduce the amount of carbon dioxide in the atmosphere. The other process is the formation of large amounts of decay-resistant organic compounds such as lignin and suberin, which form soil humus. Recall that carbon dioxide is a greenhouse gas that helps warm Earth's atmosphere, so reduction of atmospheric carbon dioxide is related to climate-cooling events. Ever since plant roots first evolved, weathering has dramatically influenced Earth's atmospheric chemistry and climate. For example, plants caused a dramatic drop in atmospheric carbon dioxide at the end of the coal age about 286 million years ago. This change in atmospheric chemistry resulted in climate cooling that appears to be related to the rise of seed plants (Chapter 22).

10.4 | Plant roots are associated with beneficial microbes

Plant roots can be remarkably efficient in finding and absorbing mineral nutrients, but the challenges of growth in soils of low mineral nutrient content often require the aid of beneficial microorganisms that form symbiotic relationships with plant roots. These include **mycorrhizal fungi** (Chapter 20) and the **nitrogen-fixing bacteria**, which live within roots of legumes and some other plants (Chapter 18). These microbes help plants obtain the large amounts of those minerals needed for growth, growth that would otherwise be limited if they were not present in sufficient quantity. In nature, plants that have beneficial microbial associates are better competitors, and crop plants with helpful microbes have higher yields than plants that lack them. Mycorrhizal fungi are particularly important in providing phosphate, although they supply other minerals and water as well. Almost all vascular plants have mycorrhizal fungal partners.

Nitrogen-fixing bacteria, which are uniquely able to convert gaseous nitrogen in the air into combined nitrogen, supply the nitrogen compounds required by plants to produce amino acids and proteins. Many plants probably benefit from loose associations between their roots and soil nitrogen-fixing bacteria. But legumes and some other plants have much closer associations with such bacteria, producing special root nodules whose tissues harbor nitrogen-fixing bacterial partners.

Legume-bacterial relationships begin with a chemical conversation. First, legume roots secrete flavonoids into the soil (Figure 10.23). Recall that flavonoids are plant secondary compounds that play many important roles in plants (Chapter 3) and are also beneficial to people (Chapter 2). Legume-root flavonoids serve as signals to soil nitrogen-fixing bacteria, which respond to the flavonoid signal by secreting small organic molecules into the soil. Legume-root epidermal cell membranes contain receptor molecules that recognize and bind molecules excreted from particular nitrogen-fixing bacteria. These bacterial compounds (called Nod factors because they signal the plant root to begin the process of nodule formation) cause the root hairs to curl around the bacterial cells which then become incorporated within special root cells. Root nodules contain both infected and uninfected cells, and mature nodules (Figure 10.24) possess vascular tissue that connects with the root vascular system (see Figure 10.23). The vascular system distributes

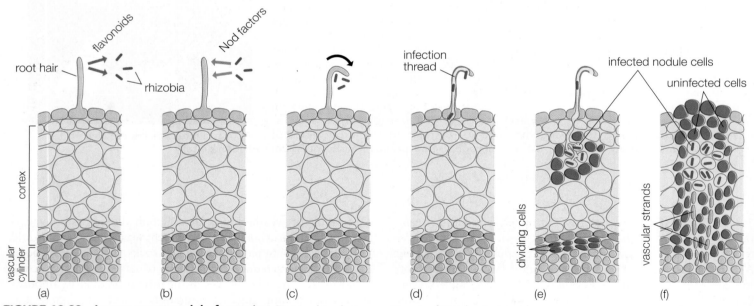

FIGURE 10.23 Legume root nodule formation Partnerships between nitrogen-fixing bacteria and legume roots begin with secretion from roots of flavonoid signals to soil bacteria (a). Particular nitrogen-fixing bacteria "smell" the chemical signal and respond by secreting chemicals known as Nod (for nodulation) factors (b). Nod factors diffuse through the soil and, when they reach legume roots, induce root hairs to curl around the selected bacteria (c), starting the process by which bacterial cells enter roots via infection threads (d). Legume roots produce lumpy nodules, shown developing in (d) and (e). Some nodule cells harbor numerous bacterial cells, while others do not. The conditions inside infected cells favor bacterial reproduction and nitrogen fixation. The bacteria use some of the nitrogen they fix. Additional nitrogen fixed within nodules travels via vascular tissues, providing a link to the main root. This nitrogen is transported throughout the legume host, including developing seeds, for protein production. This partnership explains why legumes are such rich sources of protein in human nutrition.

FIGURE 10.24 Mature root nodules on a legume

compounds containing nitrogen throughout the plant. Nod factors differ slightly among related nitrogen-fixing bacteria, and legume-root receptors recognize the Nod factors produced by specific bacterial partners, allowing particular pairs of legume and bacterial species to form. For example, the bacterium *Sinorhizobium meliloti* is allowed to infect alfalfa *(Medicago sativa)*, but not peas *(Pisum sativum)*. In contrast, *Rhizobium leguminosarum* infects peas, but not alfalfa. The chemical communication system consisting of specific Nod factors and particular legume receptors explains this partnership specificity.

An amazing recent discovery is that some of the genes controlling legume-bacterial communication are also involved in the establishment of mycorrhizal partnerships! Plant-fungal partnerships are very ancient (more than 400 million years old—see Chapters 20, 21), whereas legume-bacterial associations are much younger (perhaps 70 million years). The new data suggest that legumes modified an older system for establishing root-fungal partnerships to enable new root partnerships with nitrogen-fixing bacteria. This is an example of the evolutionary principle "descent with modification," which explains many traits of modern plants and their roots.

EVOLUTION

HIGHLIGHTS

10.1 Roots generally function in water and mineral absorption but can play a wide variety of additional roles. These include food storage, production of plant hormones and secondary compounds, structural support, and oxygen absorption.

10.2 Eudicots typically produce taproot systems having a single main root, whereas monocot plants have fibrous, adventitious root systems. Systems of shallow, feeder branch roots that play a major role in nutrient absorption can be produced by both taproots and fibrous roots.

10.3 External root structures include branch roots, root hairs, and root tip (but not nodes and internodes or buds, characteristic of stems). A root cap protects the delicate root tip and secretes mucilage that lubricates root growth through the soil. The root apical meristem produces primary tissues. Cells of these tissues enlarge and begin to specialize, forming a zone of elongation. Root cells complete their development in the zone of maturation. Important tissues present in mature and functioning roots include epidermis with epidermal root-hair cells, starch-storing cortex, suberin-wrapped endodermis, pericycle that generates branch roots, and internal vascular transport tissues (xylem and phloem). Minerals are selectively taken

into cells by transporters in cell membranes of root-hair and endodermal cells. Because soil concentrations of essential minerals are often too low to allow diffusion into root cells, active transport involving energy input (ATP) is often required. Roots require oxygen and sugar in order to undergo cell respiration, which generates ATP. Oxygen comes from aerated soils (or from the air, in the case of plants adapted to waterlogged soils). Sugar is provided by photosynthetic shoots via phloem transport. Minerals and water absorbed by root-hair cells or endodermal cells can move from cell to cell through plasmodesmata, ultimately reaching the xylem, which transports these materials to shoots.

10.4 Plant roots are often associated with beneficial microbes that help them obtain needed minerals. Nitrogen-fixing bacteria live within root nodules of legumes and may also be associated with roots of other plants. Mycorrhizal fungi that live on the surfaces or within the intercellular spaces of plant roots help most plants obtain phosphorus and other minerals. Similar genes are involved in chemical communication between legume roots and bacteria on the one hand and mycorrhizal fungi and roots on the other.

REVIEW QUESTIONS

1. What three primary meristems develop from the root apical meristem, and what types of tissues do each of them produce? Does the shoot apical meristem produce these meristems as well?

2. Briefly describe some functions of roots.

3. What are pneumatophores, and what kinds of plants produce them? Describe how their structure enables them to perform their aeration function. Why do roots require air (oxygen) in the first place?

4. Distinguish among taproots, adventitious roots, fibrous roots, and feeder roots.

5. Briefly describe the external structure of a typical root tip, starting from the region sporting the last branch roots and proceeding to the actual tip of the root.

6. In addition to the root cap, the terminal few millimeters of a root tip consists of three internal zones of activity. Briefly describe these three zones, explaining how they participate in the formation of mature primary root tissue.

7. Describe the two methods by which required minerals gain access to the xylem of a root. Discuss why both routes are classified as "selective," and describe the location of the membrane that functions as a selective barrier in each case. What prevents mineral ions from gaining entry to the root xylem by simple diffusion through the intercellular spaces between cortical cells?

8. Describe some of the mechanisms used by plants to (a) avoid aluminum toxicity when soil aluminum levels are high and (b) obtain sufficient phosphate ions when soil phosphate levels are low or soil phosphate is unavailable.

9. Distinguish between active transport and facilitated diffusion. Why must roots generally use the former process to take up minerals from the soil?

10. Provide several examples of ways that humans use roots.

11. Explain the chemical communication process involved in establishment of legume partnerships with nitrogen-fixing bacteria and why this is important.

APPLYING CONCEPTS

1. Imagine that you are a tiny root-boring beetle and you have just begun eating your way into the interior of a mature eudicot primary root. As a no-nonsense sort of insect, you bore directly from the periphery to the root center, passing through the primary phloem as you go. Beginning with the epidermis, describe the tissues you would encounter—in the order you would encounter them—during your culinary adventure.

2. Compare the *external* anatomies of a primary root and a primary stem, noting which structures or cell types they share and which are unique to one or the other.

3. Compare the *internal* anatomies of a primary root and a primary stem, noting which structures or cell types they share and which are unique to one or the other.

4. Imagine that you germinate a seed on a piece of wet filter paper and it sprouts a primary root that currently is about 2 cm in length. You use an India-ink pen to place black marks on the root. To accomplish this, you simply start at the root tip and mark the root at 1-mm intervals, for a total of about 20 marks. You now allow the root to grow to 5 cm in length. Which area(s) of the root will not have elongated, so that the marks there will still be spaced at 1-mm intervals, and which area(s) will have changed their length and hence changed the original 1-mm spacing between the India-ink marks?

5. Why do root hairs occur back in the zone of maturation of a root instead of closer to the tip?

6. Imagine that you and a friend are in the produce section of a grocery store. You are examining the Idaho potatoes and casually mention that potatoes are tubers and thus are actually stems, not roots. Your friend disagrees. What evidence could you provide to support your case?

11

Leaves: Photosynthesis and Transpiration

Young bur oak leaves (*Quercus macrocarpa*)

- Leaves, which are the major organs of photosynthesis in plants, come in a wide variety of sizes, shapes, and arrangements on stems.

- Plants evaporate huge quantities of water through their leaves.

- The senescence and fall of leaves is a normal part of plant development.

- Leaves perform many functions in addition to photosynthesis, such as attracting pollinators, storing water and food, defending the plant from animals, and capturing animal prey.

On a hot summer day in the Midwest, a walk along a concrete sidewalk can be an uncomfortable experience, especially if you are barefoot. How pleasant it is to pass under the shade of large maple trees lining the sidewalk. If the day is unusually hot, the drop in air temperature under the maples is clearly perceptible on your skin. Better yet, leave the concrete sidewalk and walk on grass. Even in full sun, the grass will feel cool on bare feet compared to the burning heat of the concrete. Plants are cool—literally. Why should this be the case? After all, the grass is exposed to the same hot sun as the concrete. But you seldom see anyone sunbathing while lying on a concrete sidewalk!

Plants are cool because they have the capacity to evaporate large quantities of water through their leaves. When liquid water in the leaves of grass absorbs large amounts of heat energy, the water transforms into a gaseous vapor. The heat is then carried away when the vapor leaves the leaves. Because of this evaporation, the grass is actually cooler than its surroundings, a fact plainly evident to your feet. A single acre of lawn grass will evaporate about 102,000 liters (27,000 gallons) of water during a summer week. The shade under a maple tree is not just cooler because the tree acts like an umbrella. A single mature maple tree may have about 100,000 leaves and can evaporate about 200 liters of water per hour on a hot, sunny day. Plants are cool because they evaporate huge quantities of water. Because they evaporate so much water, our environment is cooler on hot summer days.

This chapter will survey the form, composition, and function of leaves. It begins by describing some of the vast array of shapes of leaves and their arrangements on stems. It then introduces the structure of leaves, showing how that structure relates to their primary function as organs of photosynthesis and then presents the path of water through leaves. The discussion of photosynthetic leaves ends with a description of leaf loss in autumn. Examples of functions that leaves perform other than photosynthesis, including support, defense, water and food storage, and predation, are then described. The chapter concludes with a survey of the many human uses of plant leaves. ■

11.1 | As photosynthetic organs, leaves occur in a vast range of forms

Leaves come in a wide range of forms and sizes. Because many plants have their own characteristic leaf shape and structure, their form alone may identify the species. Leaf form is so variable that there is an extensive vocabulary of terms to describe shapes, margins or edges, vein patterns, and arrangement of leaf attachments to the stem. Here we will present only a minimal amount of that terminology.

In most eudicots, which includes flowering plants in the families of roses, maples, and daisies (Chapter 23), the typical foliage leaf consists of a wide, flat portion—the **blade**—and a stalk portion—the **petiole** (Figure 11.1). Many leaves have little leaflike structures called **stipules** at the base of the petiole where it joins the stem (see Figure 11.21). Where the leaf joins the stem, it forms an angle between the petiole and the stem called the **axil**. Buds often occur in these axils, and they are appropriately called **axillary buds**. (*Axil* comes from the Greek *axilla* or armpit, so axillary buds might be regarded as armpit buds!) The point on the stem where a leaf is attached is called the **node**, and the space along the stem between nodes is the **internode** (see Figure 11.1). Many leaves lack petioles and are termed **sessile**, meaning the blades are seated directly on the stem. In monocots such as lilies, tulips, onions, and grasses—and in a few eudicots—the base of the leaf is expanded out into a **sheath**, which wraps around the stem (Figure 11.2).

Most of the variation in leaves is in the form of the blade

Leaves may be **simple**, meaning they have only one blade, or **compound**, in which the blade consists of two or more

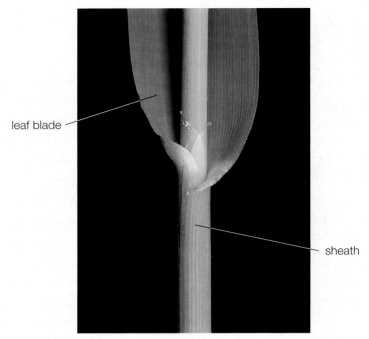

FIGURE 11.2 Monocot leaf Leaves of monocots have a sheath that wraps around the stem, as in the grass shown here.

leaflets. Simple leaves may have margins that are smooth, toothed, or deeply lobed (Figure 11.3). There are two types of compound leaves. In **pinnately** compound leaves, the leaflets are arranged along a central axis that is an extension of the petiole (Figure 11.4a). In **palmately** compound leaves, the leaflets all arise from a common point at the end of the petiole (Figure 11.4b). Leaflets can be distinguished from leaves because leaves have axillary buds and leaflets do not.

Leaves also possess a wide variety of patterns of **veins**, which contain the xylem and phloem, the conducting tissues of the plant. Monocots such as grasses, lilies, and iris have **parallel** veins (Figure 11.5). Magnolias and eudicots generally have **netted** veins that branch from a common point or main axis. In **pinnately netted** veins, major veins branch off along a central vein, as in the leaves of elm trees (Figure 11.6a). **Palmately netted** veins have several major veins branching from a common point, as in Norway maples (Figure 11.6b).

Leaves are arranged in distinct patterns on stems

Leaves can be arranged along the stem in a number of different orderly patterns. The most common arrangement is **alternate**, where there is one leaf at each node (Figure 11.7a). In plants with alternate leaves, the leaves typically form a spiral around the stem. Examples include oak and walnut trees. In maples and ash trees, however, there are two leaves across from each other at each node; this leaf arrangement is called **opposite**. In plants with opposite leaves, the leaves at one node are often rotated 90° with respect to those at the nodes above and below them (Figure 11.7b). Some plants have three or more leaves at each node—an arrangement termed **whorled** (Figure 11.7c).

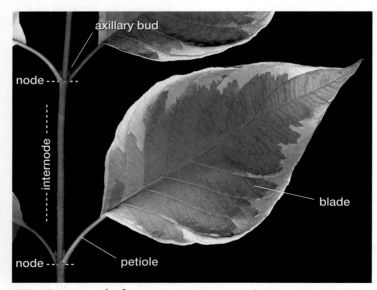

FIGURE 11.1 A leaf Leaves typically consist of an expanded blade, which is attached to the stem by the petiole. An axillary bud is located in the angle formed between the petiole and the stem. The point on the stem where a leaf (or leaves, as in the case here with variegated dogwood—*Cornus*) attaches to the stem is termed the node. The region between adjacent nodes is known as the internode.

(a)
(b)
(c)

FIGURE 11.3   Simple leaves (a) Dogwood (*Cornus*) leaves have smooth margins. (b) The leaf of the basswood tree (*Tilia americana*) has a toothed margin. (c) Bur oak (*Quercus macrocarpa*) leaves are lobed.

a

b

FIGURE 11.4   Compound leaves (a) The pinnately compound leaf of prickly-ash (*Zanthoxylum americanum*). (b) The palmately compound leaf of Virginia creeper (*Parthenocissus quinquefolia*).

FIGURE 11.5   Parallel veins A small portion of a day-lily leaf, showing the parallel veins.

FIGURE 11.6   Types of vein patterns (a) Pinnately veined elm *(Ulmus)* leaf (an arrangement like the divisions of a feather or *pinna*). (b) The palmately veined leaf of Norway maple *Acer platanoides* (an arrangement like fingers on the *palm* of your hand). (inset) A close-up view of the smaller veins of a maple leaf. Smaller-sized veins branch from the major veins, forming a network that extends throughout the leaf.

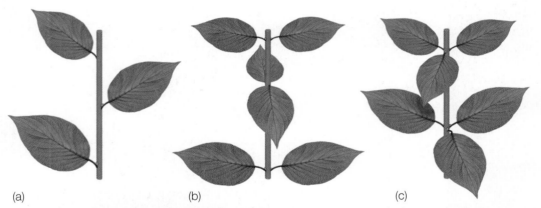

FIGURE 11.7   Arrangements of leaves on stems (a) Alternate—one leaf per node. (b) Opposite—two leaves attached at each node. (c) Whorled—three or more leaves per node.

An example is the catalpa tree. Different leaf arrangements may represent different ways to maximize light absorption and minimize self-shading.

11.2 | The major tissues of leaves are epidermis, mesophyll, xylem, and phloem

Leaves are complex structures composed of four main types of tissues—epidermis, mesophyll, xylem, and phloem (Figure 11.8). The **epidermis** is the tough outer cell layer of the leaf that helps to protect the inner tissues and provides structural support. **Mesophyll** tissue is the main photosynthetic tissue of

the leaf. Xylem and phloem are the conducting tissues, which together with other tissues form the **vascular bundles,** or veins, of the leaf.

The structure of leaves depends to a large extent on the environment in which the plant is growing, particularly the availability of water in that environment. Plants growing in very wet environments where they are wholly or partly submerged in water are called **hydrophytes.** Plants with leaves adapted to dry habitats are **xerophytes,** and plants that require a moderately moist environment are **mesophytes.** These categories are not exclusive, however, and many leaves show a combination of features associated with different habitats. Leaves adapted to different habitats show variations in their four main tissues associated with that environment.

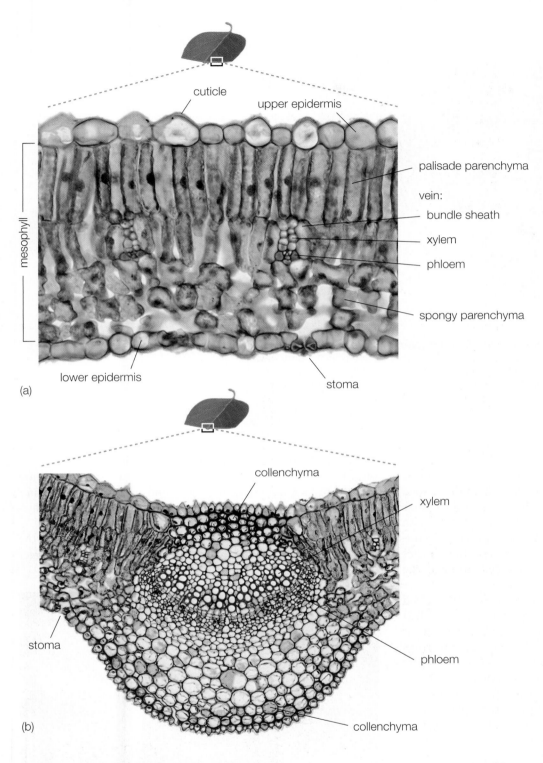

(a)

(b)

FIGURE 11.8 Anatomy of a lilac (*Syringa vulgaris*) leaf (a) This leaf, typical of those of mesophytic plants, consists of the epidermis (upper and lower); mesophyll, which consists of palisade and spongy parenchyma; and xylem and phloem, which are located in veins. Part of a blade is shown here in cross section, revealing two small veins. Note the bundle sheaths that surround the veins as well as the stoma on the lower surface. (b) Collenchyma cells provide support along the midrib of the leaf, where a greater amount of vascular tissue is present.

The epidermis provides structural support and retards water loss

The epidermis is composed of tough, compact cells that cover both the upper and lower surfaces of foliage leaves. In most plants, the epidermal cells lack chloroplasts, which are the organelles responsible for photosynthesis, and they are therefore translucent, permitting sunlight to pass through them to the photosynthetic mesophyll tissue (see Figure 11.15). Their tough, compact structure lends considerable strength to leaf form. Because leaves present a large surface area, there is a large potential for water loss through their surface. Epidermal cells, however, secrete a waxy layer—the **cuticle**—on their outer surfaces, which helps retard water loss (Figures 11.8 and 11.9). As you might expect, the thickness of the cuticle is greatest in plants growing in semiarid and arid climates, where the wax may be thick enough to scrape off with a knife. Plants in Mediterranean climates, where summers are hot and dry and winters cool and wet, often have very thick cuticles and may also have multiple layers

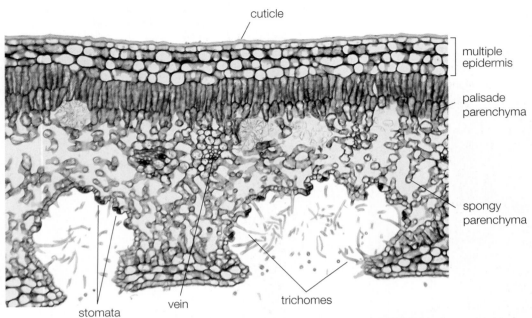

cuticle

multiple epidermis

palisade parenchyma

spongy parenchyma

trichomes

vein

stomata

FIGURE 11.9 **Anatomy of the xerophytic leaf of oleander (Nerium oleander)** In this leaf, which is adapted for dry conditions, there is a multiple epidermis and a very thick cuticle, both of which help to reduce water loss. Another adaptation in this plant is the presence of "stomatal crypts," which are depressions in the lower epidermis. These crypts, with their numerous hairs or trichomes, help to retain moisture near the stomata, which are located only within the crypts.

of epidermal cells to further reduce water loss (Figure 11.9). Leaves of mesophytes have a thicker cuticle on the upper surface, which is exposed to direct sunlight, than on the shaded lower surface. Among hydrophytes the thickness of the cuticle is related to the degree of submersion in water. Plants with emergent or floating leaves generally have well-developed cuticles. Submerged hydrophytes, however, have very thin cuticles and even have chloroplasts in the epidermal cells to bring the photosynthetic organelles as close as possible to sunlight.

Epidermal cells may be specialized to form hairs called **trichomes**. In some plants the hairs are so dense the leaf appears fuzzy (Figure 11.10). Trichomes serve a variety of functions. In arid regions leaves often have dense coverings of trichomes, and studies have shown that leaf hairiness increases reflectance of solar radiation, reduces leaf temperature, and decreases water loss. In tropical rain forests many epiphytic plants (plants that grow upon other plants) use their leaf trichomes to absorb water and minerals. Other types of trichomes deter plant eaters (herbivores) from consuming the leaves. On some plant species, hooked trichomes can impale insects. Other plants produce swollen glandular trichomes that contain oils that repel herbivores.

Epidermal tissue typically contains many thousands of minute pores called **stomata** (singular, **stoma**), which permit gas exchange between the outside environment and the leaf's interior. When the stomata are open, water vapor and oxygen move out from the leaf and carbon dioxide moves into the interior of the leaf. Each stoma consists of a pore surrounded by two **guard cells** that contain chloroplasts (Figure 11.11). When the

guard cells are filled with water—that is, when they are **turgid**—they bend so as to open the pore. If the guard cells are limp from lack of water, they close and prevent further water loss from the leaf. A more detailed discussion of this phenomenon is found in section 11.3.

In most mesophytic plants, the stomata occur only on the lower epidermis of leaves, where they are shielded from direct

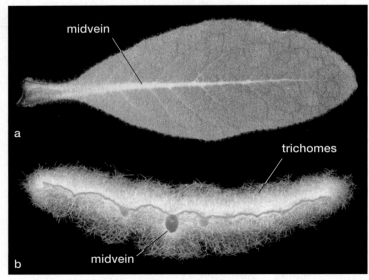

midvein

a

trichomes

midvein

b

FIGURE 11.10 **Trichomes** (a) The fuzzy leaf of the mullein plant (Verbascum thapsus). (b) A cross section of a young mullein leaf, showing the extensive trichomes on the upper and lower surfaces.

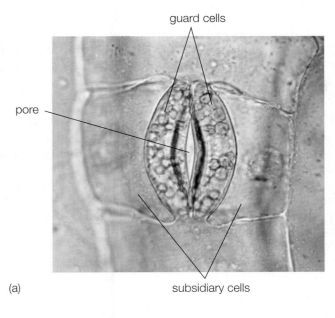

(a)

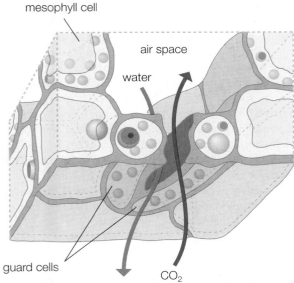

(b)

FIGURE 11.11 **Stoma** In most plants, the guard cells are the only cells of the epidermis with chloroplasts. (a) In some plants, such as *Zebrina*, shown here, there are a pair of subsidiary cells, which lie adjacent to the guard cells. (b) Cutaway diagram of stoma, looking upward toward underside of leaf.

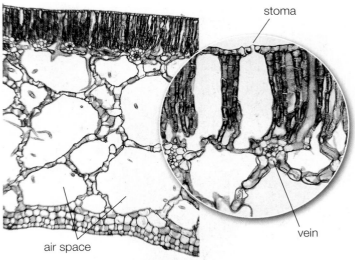

FIGURE 11.12 **Water-lily leaf** The water lily *Nymphaea* has leaves well suited to the aquatic environment in which it grows. The leaf has a great deal of air space, which helps keep it afloat and provides channels to conduct air into submerged parts of the plant. (inset) Stomata are found in the upper epidermis only, and xylem tissue is reduced compared to the leaves of mesophytes.

sunlight. In hydrophytes with floating leaves, however, stomata occur only on the upper epidermis (Figure 11.12), and submerged leaves may lack stomata entirely. Xerophytic leaves may contain large numbers of stomata, but such leaves may be shed when moisture becomes scarce. The leaves of eudicots have stomata scattered about the surface; but monocot leaves, like those in maize (corn) and other grasses, have stomata arranged in rows parallel to the leaf veins.

The upper epidermis of grass leaves has one other specialized cell type, the **bulliform cell**, from the Latin *bulla* meaning "bubble" (Figure 11.13). Bulliform cells are large cells arranged in longitudinal rows that act to fold and unfold the leaves. When bulliform cells are turgid with water, the leaf is unfolded to capture sunlight. When they are flaccid or limp from water loss, the leaf folds up to reduce further water loss.

Mesophyll is the photosynthetic tissue of leaves

The mesophyll is located between the upper and lower epidermis in the middle of the leaf (the Greek word *meso* means "middle" and *phyllos* means "leaf," see Figure 11.8). Mesophyll cells are **parenchyma** cells packed with chloroplasts and surrounded by a large volume of intercellular space, which facilitates the gas exchange necessary for efficient photosynthesis. The intercellular space is connected to the external atmosphere by the stomata.

Mesophyll tissue is often organized into two layers, a **palisade layer** and a **spongy layer**. The cells of the palisade layer (or palisade parenchyma) are columnar whereas those of the spongy layer (spongy parenchyma) are irregular in shape (Figures 11.8, 11.14a). Although the palisade layer may appear highly compact in leaf sections, the palisade cells are actually surrounded by intercellular spaces, and most of the photosynthesis takes place within them rather than in the spongy layer. The open arrangement of the cells in the spongy layer promotes rapid gas exchange, with carbon dioxide entering and oxygen and water vapor leaving through the stomata. Hydrophytes especially have a great deal of air space in their mesophyll tissue (see Figure 11.12). This open space not only helps keep the leaves afloat but also conducts air throughout the plant and down to the roots.

The palisade layer is usually located on the upper side of the leaf and the spongy layer on the lower side. But leaves in high light environments may have two or more layers of palisade cells on the upper side, and xerophytes may have a palisade layer on

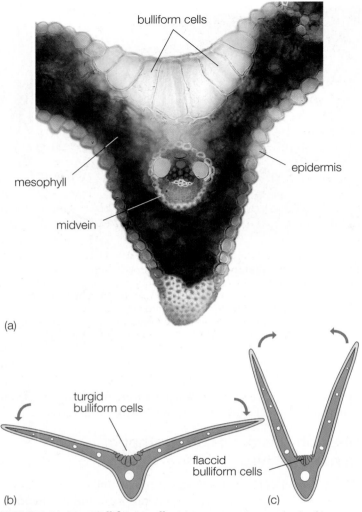

(a)

FIGURE 11.13 Bulliform cells (a) In many grasses, the leaf has a V-shape in cross section. A row of cells called bulliform cells (seen here in a live specimen) operate in opening and closing the V. (b) When the cells are turgid, the leaf is open. When flaccid, the leaf folds, thereby conserving water.

both the upper and lower sides of their leaves. Monocot leaves, such as those in maize and other grasses, have a simpler mesophyll tissue. All the mesophyll cells are similar in shape, and there are no distinct spongy and palisade layers (Figure 11.14b).

Xylem and phloem are the conducting tissues of leaf veins

The veins we can readily see in leaves contain the xylem and phloem. No mesophyll cells are ever very far from a vein because either parallel veins lie close together (monocots) or branching is extensive (eudicots); see Figures 11.5 and 11.6. Xylem conducts water and dissolved minerals to the leaf tissues. It is usually located on the upper side of each vein. Phloem, which conducts dissolved sugars and other photosynthates from the leaf tissues, is located on the lower side of the veins (Figure 11.8).

Veins are surrounded by a bundle sheath made up of parenchyma cells. The bundle sheaths extend out to the tips of the veins, so that any substances leaving or entering the veins must pass through the bundle-sheath cells. In many leaves the bundle sheaths are connected to the upper and lower epidermis by bundle-sheath extensions (Figure 11.15). Bundle-sheath extensions act like support columns and give extra mechanical support to the leaf structure. In some leaves they also conduct water from the xylem to the epidermis.

The leaves of certain grasses such as sugarcane and maize have a special type of bundle-sheath anatomy called **Kranz anatomy**. In Kranz anatomy, mesophyll cells form a ring around the bundle-sheath cells surrounding the vascular bundles of xylem and phloem (see Figure 11.14b). In cross section this arrangement of cells resembles a wreath, from which the German name (*Kranz*) is derived. It is associated with a special photosynthetic pathway called C_4 *photosynthesis*. The significance of C_4 photosynthesis and its relationship to Kranz anatomy was discussed in Chapter 5.

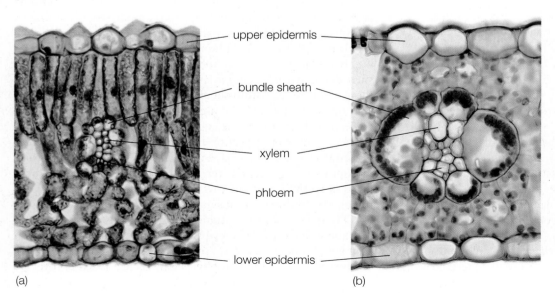

(a)   (b)

FIGURE 11.14 Comparison of eudicot and monocot leaves (a) An eudicot leaf (lilac) in cross section. Note the distinct palisade and spongy layers and the bundle sheath, as in Figure 11.8. (b) A monocot leaf (corn). Note that the mesophyll does not have distinct palisade and spongy parenchyma. Bundle-sheath cells are very conspicuous, and a layer of mesophyll cells is closely appressed to them.

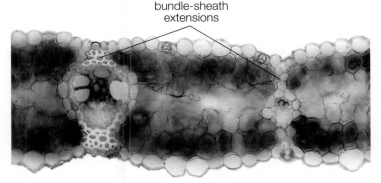

FIGURE 11.15   Bundle-sheath extensions In this section of a living grass leaf, bundle-sheath extensions are visible on the veins. Note that the epidermal cells are clear, which allows light to pass through to the green mesophyll tissue.

11.3 | Plants lose large quantities of water through transpiration

During their lifetimes, plants take up enormous quantities of water from the soil through their roots, and nearly all of that water evaporates as water vapor through the stomata. More than 90% of the total water evaporated by plants passes out through the stomata of leaves in the process of **transpiration**. The rest passes out through stems and leaf surfaces as cuticular transpiration, in which water passes through the epidermis and cuticle. A single tomato plant *(Solanum lycopersicum)* transpires about 125 liters (L) of water during its growing season. A single corn plant *(Zea mays)* will transpire about 200 L over its lifetime—about 100 times its own mass! Why do plants transpire so much water?

Plants transpire large amounts of water because they must carry out the process of photosynthesis, which is the main function of foliage leaves. Energy for photosynthesis comes from sunlight. A plant must create and spread out a large surface area to capture the sunlight that powers photosynthesis. This large surface area also allows for water loss by transpiration. But plants also need carbon dioxide, which is in fairly low amounts in the atmosphere compared to other gases. In order to enter the mesophyll cells in the interior of leaves, carbon dioxide must dissolve in water, and therefore the mesophyll cells must be moist. Because they are moist and in contact with the air, evaporation takes place. Thus plants transpire large quantities of water while taking up CO_2 for photosynthesis. Plants have not evolved any mechanism to take up CO_2 without losing some water, but they have adapted in ways to reduce water losses.

Stomatal movements control transpiration

Because more than 90% of water lost by transpiration is lost through the stomata, the opening and closing of stomata is important in the regulation of transpiration. In most plants, especially mesophytes where water is readily available, stomata are open during the day and closed at night. Because CO_2 is needed when sunlight is available, most plants open their stomata when transpiration is likely to be greatest, during the day.

Stomatal opening is controlled by an increase in the amount of water in the guard cells. Dissolved substances, especially potassium ions (K^+), are actively accumulated in the guard cells under the influence of daylight (Figure 11.16). This increase in dissolved substances causes a movement of water into the guard cells by **osmosis** (Chapter 4). As water builds up in the guard cells, it creates **turgor pressure**, which causes the guard cells to bend and open the stomatal pores. This bending of the guard cells is crucial to opening the pore and therefore to stomatal function. The guard cells bend under increasing turgor pressure because of two structural features of their cell walls. The guard cell walls contain radially oriented cellulose microfibrils that allow the cells to lengthen under increased turgor pressure but not to expand in width. In addition, the guard cell walls are attached to each other at their ends. This attachment prevents the guard cells from simply sliding past each other when they lengthen under increased turgor pressure. At night, the reverse process closes the stomata. Potassium and other dissolved solutes are actively pumped out of the guard cells. Water moves out of the cells by osmosis, turgor pressure drops, and the guard cells unbend and close the pore. Although most plants use this mechanism to open their stomata during the day and to close them at night, a number of environmental factors can override this pattern.

Environmental factors can affect stomatal movements

In plants that normally have their stomata open in the daytime, water stress can cause the stomata to close. Water stress causes the production of a plant hormone called **abscisic acid (ABA)**. ABA causes solutes to move out of the guard cells, resulting in a loss of turgor pressure and pore closure. Thus, water loss will overcome the normal pattern of stomata opening in the daytime.

An increase in the concentration of carbon dioxide within the mesophyll tissue may also cause stomata to close. Some plants in hot climates will close their stomata around midday, when water loss by transpiration exceeds water absorbed from roots and carbon dioxide has built up in the interiors of the leaves. High midday temperatures increase rates of respiration and therefore generate more CO_2 inside the leaves. The plants can close their stomata, reduce water loss, and maintain photosynthesis by using this excess CO_2 from respiration. In this situation, however, the plant cannot gain any fixed carbon from photosynthesis. The plant is simply recycling carbon dioxide from respiration back to photosynthesis. (For more about stomata and global atmospheric concentrations of CO_2, see Essay 11.1, "Plant Leaves Track CO_2 Levels in the Atmosphere.")

Many succulent or fleshy desert plants such as cacti and many epiphytes such as orchids and bromeliads open their stomata at night when water losses are low. These crassulacean acid metabolism (CAM) plants fix CO_2 at night as an organic acid. In the daytime they close their stomata and cleave the

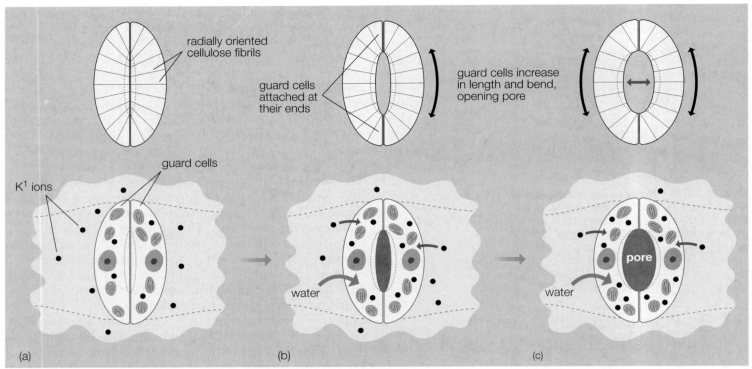

FIGURE 11.16 **Diagram of stomatal function** (a) Stomata close when potassium ion concentrations are equal in guard cells and other cells of the epidermis. (b) As potassium ions are pumped into guard cells from adjacent cells (red arrows), the higher solute concentration results in water moving into the guard cells by osmosis (blue arrow) and the pore begins to open. (c) As the guard cells become turgid, the pore fully opens and gases and water vapor pass through it. The accompanying diagrams for parts (a)–(c) indicate the structural basis for stomatal opening. Radially oriented cellulose microfibrils in the guard cell walls prevent cells from expanding in width as they become turgid (they are constrained to increase in length). Because the guard cells are attached to one another at their ends, they must bend as they increase in length, thereby opening the pore.

organic acid to release CO_2 for use in photosynthesis. CAM is discussed in Chapter 5.

Transpiration is an important component of the global process in which water cycles through the oceans, atmosphere, and land. Transpiration puts water vapor into the atmosphere, where it condenses into clouds and precipitation. Vegetation cover therefore has a significant impact on local climate and rainfall. Where the vegetation is removed, as in tropical deforestation or in desertification in grasslands, regional rainfall declines and droughts become more frequent.

11.4 | The cohesion-tension theory explains movement of water through plants

Water moves from the soil to the plant roots and up the conducting tissues into the veins of the leaves. It then passes from the veins into the cells of the mesophyll, where it wets their outer surfaces and evaporates into the air spaces of the interior of the leaves. From the leaves it moves out of the plant by transpiration through the stomata. But what makes it move at all?

There are only two possibilities for water movement—either the water is pushed up from the roots or it is pulled up by the

leaves. Root pressure does exist. When soil moisture is high and transpiration is low, water enters the roots and can be forced out the ends of veins in leaves to form droplets. This process is called **guttation,** and if soil moisture is high it may occur at night when transpiration is normally shut down (Figure 11.17). But measurements of root pressure show that it is too weak to force water up a tall plant, especially trees like giant redwoods; furthermore, it does not operate when soil moisture is moderate to low.

It is easy to demonstrate that water can move up a plant to the leaves without roots at all. Simply take a potted plant and immerse it in a sink full of water. Cut off the plant at the level of the soil while holding it under water. Place the cut stem into a beaker of water and remove it from the sink. Prop up the plant in sunlight and add some colored dye to the beaker of water. Soon the plant stem and leaves will take on the color of the dye (Figure 11.18). Water moves up the plant when no roots are present!

Water must therefore be pulled up by the leaves. The theory that explains this upward pull of water is called the **cohesion-tension theory.** Water has a high degree of cohesion due to hydrogen bonds linking together adjacent water molecules (Chapter 3). Cohesion is the force that allows water molecules to stick together even when a pulling force, or tension, is exerted on them. As water evaporates from the surfaces of mesophyll cells, the water at the surfaces is replaced by water from the interior of the mesophyll cells. When water is drawn out of these mesophyll

ECOLOGY

ESSAY 11.1 PLANT LEAVES TRACK CO₂ LEVELS IN THE ATMOSPHERE

Global warming and greenhouse gases are frequent subjects in the news. Over the past 200 years, human industrial activity, car exhaust, and tropical deforestation have raised global atmospheric concentrations of carbon dioxide gas and altered the climate of our planet. At the beginning of the 19th century, the atmospheric level of CO_2 was around 0.028%. The Industrial Revolution, which was the complex of social and economic changes that accompanied the mechanization of production, has raised that amount to 0.037%—a 28% increase—in just 200 years. Short-term experiments with plants exposed to elevated levels of CO_2 under field conditions have shown that plants develop thicker leaves as CO_2 levels increase. Soybeans, for example, develop a third palisade layer in their mesophyll tissue. Apparently, under elevated carbon dioxide levels leaves can obtain enough CO_2 internally to support additional cell layers.

To find out if plants had changed in any way since the beginning of the

... when dinosaurs roamed the Earth, the level of CO_2 in the atmosphere was higher than at present.

Industrial Revolution, Dr. F. I. Woodward of the Department of Botany at the University of Cambridge examined plant specimens stored in his university herbarium. He studied leaves from reproductive stems of eight species of common trees, such as maples, oaks, and beeches (Figure E11.1), collected between 1750 (before the Industrial Revolution began) and the present. He found that the density of stomata on the leaves, the number per unit of area counted, had declined by 40% over about 200 years! Plants were responding to increasing levels of CO_2 by reducing the number of stomata. Since CO_2 was becoming more abundant, fewer stomata were needed to take up CO_2 for photosynthesis, and water loss through transpiration should therefore be reduced.

The same technique of counting the density of stomata has been applied to fossil leaves to determine atmospheric CO_2 levels millions of years ago. Fossil leaves from more than 65 million years

E11.1 Herbarium specimen of a beech tree (*Fagus grandifolia*)

ago have few stomata. Thus, when dinosaurs roamed the Earth, the level of CO_2 in the atmosphere was higher than at present.

FIGURE 11.17 **Guttation** Under some conditions, root pressure can cause water droplets to form at the tips or leaf margins of some leaves, as in these barley seedlings.

FIGURE 11.18 **Dye uptake by a plant without roots** Blue dye was added to water in a test tube, into which a rootless jewelweed (*Impatiens*) plant was placed. After a relatively short period of time, the dye was observed moving upward in the vascular bundles of the stem (arrows).

cells, the concentration of solutes in the cells rises and more water is drawn into the cells by osmosis from adjacent cells. Surface evaporation creates a tension or water deficit that causes those mesophyll cells to draw more water by osmosis from adjacent cells. Those adjacent cells, in turn, draw water from other cells by osmosis until a vein is reached. The tension is then transmitted to the water in the xylem, which is drawn by osmosis through the xylem all the way up from the roots, which draw water from the soil. Because water molecules are so cohesive, evaporation of water in leaves thus draws water by osmosis all the way up from the roots, even in trees like giant redwoods. In fact, the cohesion of water is great enough to raise a column of water in xylem to 1,980 meters (6,500 feet). By comparison, the tallest trees are only about 115 meters tall (380 feet).

The cohesion-tension theory treats the movement of water in a plant as if the water were one continuous piece. All the pull occurs at the top end, and the column of water is drawn up from root to leaf as if it were a long string. Living cells in the vascular tissue are assumed to play no role in this movement of water. Although the cohesion-tension theory has been widely accepted for about 100 years, it is currently being challenged. A new hypothesis, proposed by Martin Canny at Australian National University, suggests that living vascular tissue provides a tissue pressure in addition to the pulling force assumed in the cohesion-tension theory. This tissue pressure helps maintain the flow of water and repairs any breaks that might occur in the water columns. Debate and research continue.

11.5 | Senescence and leaf fall are a normal part of plant development

Senescence, or aging, is a series of irreversible changes in plants that eventually leads to death. Senescence may occur at the level of individual cells, organs, or whole plants. At the cellular level, some of the cells that make up the xylem of plant vascular tissue undergo a genetically programmed cell death in order to assume their final function as water-conducting elements. Annual plants complete their entire life cycle in one year and normally age and die after flowering. At the plant organ level, the bright colors of flowers attract animal pollinators, after which the flower rapidly ages and dies. The flowers of the morning glory *(Ipomoea tricolor)*, for example, senesce after being open for only one day. But the most familiar example of plant organ senescence is the drop of leaves.

At some time or times during the year, plants will **abscise**, or drop their leaves and remain bare for some period. Desert plants with leaves may drop them during drought periods and leaf out in moist intervals. Certain tropical forests have distinct wet and dry seasons and drop all their leaves as the dry season arrives. Conifer trees or evergreens may drop some of their leaves (needles) at any season, but they often shed the largest amount in fall before winter snows begin. The leaves of evergreens may remain on the trees for several years. Those of the bristlecone pine *(Pinus longaeva)* have a life span of about 45 years. The most familiar example of leaf abscission,

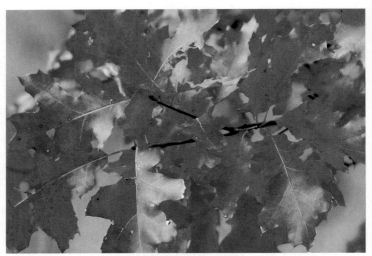

FIGURE 11.19 **A red oak *(Quercus rubra)* in fall**

however, is the pattern shown by temperate deciduous woody shrubs and trees.

Autumn leaf abscission is preceded by a period of senescence

Temperate deciduous trees and shrubs prepare for winter with massive leaf senescence preceding the actual leaf abscission. This genetically programmed aging and death of leaf organs is initiated by declining day length after midsummer. Leaf senescence is an orderly process of disassembly of leaf cellular components and recovery of nutrients. As autumn sets in, starches and proteins in the leaves are broken down into sugars and amino acids and returned to the stem. Essential minerals such as nitrogen and phosphorus are recovered. Chlorophyll is broken down, revealing the bright yellow xanthophylls and orange carotenoid accessory pigments. These and water-soluble pigments called *anthocyanins* are responsible for the wide range of colors of autumn leaves (Figure 11.19). Nutrients recovered from senescing tissues are recycled to surviving plant parts.

Leaf abscission has two main adaptive functions in temperate deciduous trees and shrubs. First, winter freezes lock up all the moisture in the upper soil layers and as snow at the surface; moisture is no longer available to the roots for movement to the leaves. Thus leaves can no longer carry out their function as photosynthetic organs or be provided with sufficient water for transpiration. Leaf abscission removes these nonfunctional organs and conserves water. A second reason for leaf abscission can be seen when sudden early snows hit a region before the trees and shrubs have lost their leaves. Heavy snow can build up on leaves and bend branches and trunks down to the ground, snapping off limbs and even killing some trees.

EVOLUTION

Leaf abscission is preceded by formation of an abscission zone

Leaf abscission is a complex series of events that end with the drop of the leaf. The plant hormone **ethylene** promotes leaf abscission. An **abscission zone** is formed at the base of each leaf

FIGURE 11.20 **Abscission zone** The abscission and protective layers are evident in this section through the abscission zone in a maple leaf.

FIGURE 11.21 **Tendrils on a pea plant (*Pisum sativum*)** These tendrils (arrows) occur at the tips of the compound leaf. Also note the large stipules.

petiole (Figure 11.20). In woody flowering plants, the abscission zone consists of two layers—an abscission, or separation, layer and a protective layer. The separation layer contains short cells with thin, weak walls. Beneath the separation layer toward the stem, the protective layer develops as a layer of cork cells impregnated with suberin, a waxy, waterproof substance that isolates the leaf from the stem before the leaf drops. Enzymes then act on the cell walls in the separation layer. Finally, only a few strands of vascular tissue hold the leaf on the stem. These strands may be broken by the weight of the leaf or by the wind or rain. The protective layer remains to seal the area and appears as a leaf scar on the stem.

11.6 | Leaves perform many functions in addition to photosynthesis

Not all leaves function as photosynthetic organs. Leaves may undergo extensive modifications in structure to perform other functions. These include acting as support structures, substitute flower petals, and traps for animal prey as well as providing winter cold protection, food and water storage, and defense against animals.

One common leaf modification is the formation of **tendrils**. A tendril is a threadlike structure that helps plants climb over other plants or objects and gain access to light or space. Most tendrils are formed from leaves, but a few are formed from stems, as in the grapevine. The garden pea (*Pisum sativum*) forms a tendril at the end of its pinnately compound leaf (Figure 11.21).

In temperate climates, the buds of woody plants are protected over the winter by modified leaves in the form of **bud scales** (Figure 11.22a). Wax-impregnated bud scales prevent desiccation and insulate the bud from the winter cold. They also contain growth inhibitors to prevent the buds from growing until spring. Thus bud scales are a modification to adapt plants to environments where winters are cold.

Many plants that appear to have large, showy flowers actually have small inconspicuous flowers surrounded by colorful modified leaves called **bracts**. The "flower" of the poinsettia is really a circle of brightly colored bracts surrounding small yellow flowers in the center (Figure 11.22b). In temperate climates, dogwood "flowers" consist of white to pink bracts around a cluster of yellow-green flowers. Bracts are leaves modified to assume the role of flower petals in attracting pollinators.

Some leaves are specialized for water or food storage

In arid environments, many plants have leaves modified for internal storage of water. Such plants are called *succulents* and have thick, fleshy leaves (Figure 11.22c). The century plants (*Agave*) of the deserts of the United States and the aloes (*Aloe*) of African deserts both have thick, succulent leaves that store water and carry out photosynthesis. In tropical environments, many species of plant grow as epiphytes (literally, "on plants") on the branches of other plants. Many such epiphytes have leaves arranged and modified to store rainwater in tanks (Figure 11.22d).

Many familiar garden plants have leaves modified to store nutrients. A cabbage head is really a short stem with many thick, overlapping leaves. Cabbage and kale are both varieties of *Brassica oleracea*. Lettuce (*Lactuca sativa*) and spinach (*Spinacia oleracea*) are also storage leaves, and the edible parts of celery

FIGURE 11.22 Modified leaves (a) Bud scales on a hickory tree *(Carya ovata)* that have opened, revealing the young shoot. (b) Poinsettia *(Euphorbia splendens)* bracts. (c) An *Agave* plant with succulent leaves. (d) The water-holding tank of a bromeliad (pineapple family member). (e) An onion *(Allium cepa)* bulb, which consists of many tightly packed leaves on a very short stem.

(Apium graveolens) and rhubarb *(Rheum rhabarbarum)* are the petioles of leaves. Onions *(Allium)* and plants such as lilies and tulips form **bulbs**, which are short, conical, underground stems surrounded by fleshy storage leaves (Figure 11.22e).

Leaves are modified for defense in some plants

Many desert plants have leaves that are reduced to form sharp, nonphotosynthetic spines. In the cacti of North and South America, spines act to deter plant-eating animals or herbivores (Figure 11.23a,b). The terms *spine*, *thorn*, and *prickle* are often used interchangeably, but technically a spine is a modified leaf and a thorn is a modified stem that arises from the axil of a leaf (Figure 11.23c). A prickle is neither a stem nor a leaf but a sharp outgrowth from the epidermis or cortex of a stem (Figure 11.23d). Rose thorns are really prickles. All three structures serve as defenses against herbivores.

Many plants protect their leaves from herbivores by placing various types of secondary compounds in them (Chapter 3). The secondary compounds may make the leaves taste bad or make the herbivore ill if it ingests enough of them. One of the most spectacular plant defense strategies involves *Acacia* plants, whose modified leaves house populations of ants that protect the plant. This association is described in detail in Chapter 24.

Leaves of some plants capture animal prey

Some of the most amazing modifications of leaves occur among carnivorous plants, which trap live animal prey. These plants grow in bogs in temperate and tropical regions where soils are deficient in minerals such as nitrogen. They acquire scarce minerals by attracting, trapping, and digesting various prey. Traps may be active or passive.

FIGURE 11.23   Spines, thorns, and prickles Spines represent modified leaves, thorns are specialized stems, and prickles are extensions of the epidermis or cortex. In (a), a spine from barberry (*Berberis thunbergii*) is seen with a young branch growing from its axil. (b) Cactus spines. In this case, numerous spines (leaves) are growing from shortened shoots that arise from the photosynthetic stem. (c) Thorns on honey locust (*Gleditsia triacanthos*). (d) Prickles on the stem of a rose species (*Rosa* sp.).

Carnivorous plants with passive traps have leaves modified into pitchers that trap rainwater (Figure 11.24). In North America, pitcher plants grow on boggy ground among mats of moss and produce pitchers grouped as erect clusters. Tropical Asian pitcher plants grow as vines and form their pitchers at the ends of leaves, whose petioles wrap around branches for support. The plants secrete digestive enzymes into the water and lure insect prey with bright colors, odors, and nectar—mimicking flowers.

When an insect lands on the pitcher, it falls in and is unable to climb out because of the slippery sides and downward-pointing trichomes. The insect drowns and is digested. Some tropical pitcher plants are quite large and can hold a liter or more of liquid. The pitfall traps in large pitchers may snare lizards and frogs.

Active traps are of three types. The Venus flytrap (*Dionaea muscipula*), a plant native to North and South Carolina, has special leaves that function like a steel leg trap (Figure 11.25). Each side of the leaf trap has several touch-sensitive hairs. If an insect touches one hair twice or two hairs in quick succession, the trap snaps shut in only half a second. After about one week, the insect is digested and the trap reopens.

The leaves of sundews (*Drosera*) have a coating of club-shaped hairs that bear beads of clear, sticky mucilage (Figure 11.26). The mucilage attracts insects, and the insects become tangled in it. The hairs bend inward and surround the prey as the leaf folds up. Enzymes in the mucilage digest the insect, and the glandular trichomes then absorb the nutrients released from the insect. Research on sundews in the United Kingdom has shown that two *Drosera* species derive an average of 50% of their nitrogen from insect prey.

The bladderworts (*Utricularia*) are carnivorous plants that grow submerged or floating in shallow, nutrient-deficient waters (Figure 11.27). Among the finely divided leaves of the plant lie a number of tiny bladders shaped like small, flattened pears attached to short stalks. The bladders are chambers containing water and the inner chamber walls are lined with specialized glands. The plant expends energy to pump water out of the chambers, generating a negative pressure and causing the walls of the bladders to pinch inward. The free end of the chamber contains a trapdoor surrounded by a number of sensory hairs. A set trap acts somewhat like a squeezed bulb on a medicine dropper. When a small aquatic animal bumps into the sensory hairs, the trapdoor snaps open inward and the negative pressure sucks in water and the unfortunate animal with it. The door snaps shut, and the trapped animal is then digested. The cells in the walls of the trap absorb the minerals and organics.

11.7 | Humans use leaves in many ways

In addition to the direct use of leaves as food, humans use leaves as seasonings in foods and as a source of scents in some perfumes. Leaves also serve as sources of waxes, medicines, psychoactive drugs, beverages, fibers, and dyes.

In botanical usage an **herb** is any nonwoody plant. The "herbs" used in seasoning foods consist of fresh or dried leaves from a variety of woody and nonwoody plants found in temperate regions of the world. The mint family (Lamiaceae) contains more herb species widely used as food flavorings than any other family of flowering plants. The mint family includes rosemary, thyme, marjoram, oregano, basil, sage, and mint. Rosemary (*Rosmarinus officinalis*) not only flavors foods but the oil responsible for its flavor also is used in some perfumes and hair conditioners. Thyme (*Thymus vulgaris*)

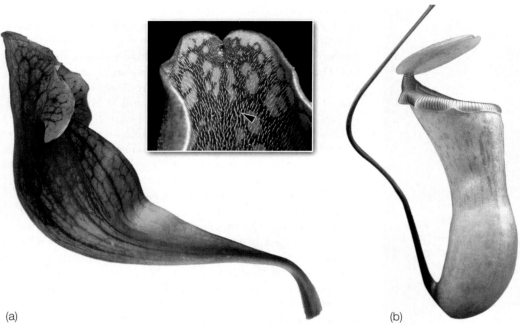

(a)   (b)

FIGURE 11.24 **Pitcher plants** (a) Leaf of a North American pitcher plant *(Sarracenia purpurea)*. The inset shows a close-up view of the leaf hood, which has many downward-pointing hairs (arrowhead). (b) A tropical pitcher plant *(Nepenthes)*.

b

FIGURE 11.25 **Venus flytrap *(Dionaea muscipula)*** The edges of the leaves of this carnivorous plant are modified to trap insects. The trap closes when insects brush against special hairs (arrows) on the inside of the trap.

contains a flavorful oil that has antiseptic properties and is used in mouthwashes and cough drops. Marjoram *(Origanum marjorana)* and oregano *(Origanum vulgare)* are closely related but different in flavor. Oregano is widely used in the United States because of the popularity of Italian food such as pizza. Basil *(Ocimum vulgare)* is another herb used in Italian cooking, especially with tomatoes in pasta sauces. Sage *(Sativa officinalis)* is an important part of the flavor of stuffing for Thanksgiving turkey. A number of species of mint *(Mentha)* are used in cooking and as a flavoring in mint jelly, candies, toothpastes, mouthwashes, and medicines.

The carrot family (Apiaceae) contains additional widely used herbs, including parsley, dill, and cilantro. Parsley *(Petroselinum crispum)* is widely used as a garnish on plates of restaurant food, but it is also a good flavoring herb whose leaves contain high levels of vitamins A and D. Dill *(Anethum graveolens)* is used in pickled foods, and cilantro *(Coriandrum sativum)* is a popular ingredient in Mexican foods. Two final examples of leaves used as food flavorings are tarragon *(Artemisia dracunculus)* in the family Asteraceae and bay leaf *(Laurus nobilis)* in the Lauraceae. Tarragon is popular in Europe as a flavoring in soups, salads, sauces, vegetables, and pasta dishes. Bay leaves are more often considered a spice than an herb because the leaves come from a tree. Other common herbs, such as cumin, fennel, and coriander, are derived from ground fruits rather than leaves.

The leaves of some plant species are used in the manufacture of perfumes. Various processes are used to extract fragrant oils and the resulting scents are referred to as notes. Perfume notes derived from leaves include oils of rose, thyme, ivy *(Hedera)*, cypress *(Cupressus)*, fig *(Ficus)*, tobacco *(Nicotiana)*,

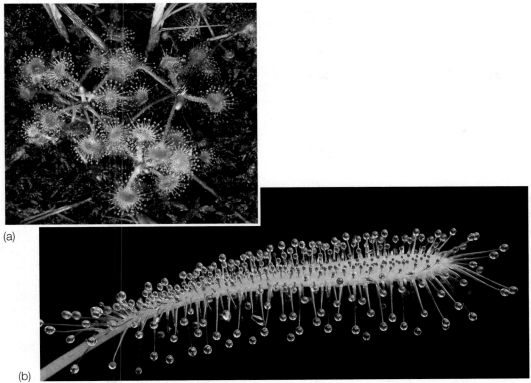

(a)

(b)

FIGURE 11.26 Sundews (Drosera) The trichomes on the leaves of this carnivorous plant bear beads of sticky mucilage that attract and trap insects. (a) A round-leafed sundew in its boggy habitat. (b) A narrow-leafed species at closer range.

FIGURE 11.27 Bladderwort (Utricularia) The traps on this carnivorous plant are the tiny bladders that draw in animal prey by suction. Numerous developing traps are visible on the apical portion of this submerged plant.

and tea *(Camellia sinensis)*. Many more notes are derived from flowers, fruits, aromatic woods, mosses, lichens, and ferns. A master perfumer combines notes to compose a perfume for commercial sale.

Waxes are a type of lipid (Chapter 3). All flowering plants make waxes to cover their exposed aerial surfaces and reduce water loss, but few plants produce enough wax for commercial harvest. One exception is the palm *Copernicia cerifera* from northeastern Brazil. Young leaves of this palm are collected and dried. After drying, the leaves are beaten to remove the particles of wax, and the particles are shipped to importing countries as carnauba wax. Carnauba wax is used in car waxes and shoe polishes because it is harder than beeswax or synthetic waxes.

The use of plant leaves in medicines and psychoactive drugs was discussed in Chapter 2. Often the difference between a plant secondary compound acting as a medicine, psychoactive drug, or poison is one of the dose consumed.

Tea brewed from the leaves of *Camellia sinensis* is one of the two most important and widely drunk beverages in the world. Tea plants are small evergreen trees that are grown commercially in plantations located in the mountains of warm regions where rainfall is good and constant, cool temperatures prevail. Tea is stimulating because of the alkaloid caffeine in the leaves, which is extracted into boiling water. The characteristic brown color and flavor of tea is due to tannins. Herbal teas do not contain leaves of *C. sinensis*, but many use blackberry leaves *(Rubus)* along with other plant parts.

Natural dyes and fibers can be obtained from the leaves of a number of plant species. Leaves of grape *(Vitis)* contain a yellow dye, peach *(Prunus persica)* a green dye, and indigo *(Indigofera tinctoria)* a blue dye. Natural dyes have been almost completely replaced by synthetic dyes derived from coal tars. Henna, an orange dye from the leaves of *Lawsonia inermis*, is still used as the base of many hair-color products. Most commercial leaf fibers are obtained from only two tropical genera, *Agave* and *Musa*. The fibers of *Agave* are used in ropes and twines. The outer mature leaves are cut at the base, hauled to a factory, and crushed between rollers to remove the sap and pulp. The fibers are then washed and dried. Manila hemp is derived from the leaves of *Musa textiles*, a relative of the banana plant. Manila hemp is used in tea bags, paper money, and Manila envelopes.

HIGHLIGHTS

11.1 Photosynthetic leaves come in many forms. A typical leaf consists of a broad blade, petiole, and an axillary bud at the petiole base. Leaves may be simple or compound, and compound leaves may be pinnate or palmate in form. Leaves contain veins, which may have a parallel arrangement, as in monocots, or a netted pattern, as in eudicots. Leaves are attached along stems at nodes. The arrangement of leaves along a stem may be alternate, opposite, or whorled.

11.2 The major tissues of leaves are the epidermis, mesophyll, xylem, and phloem. The cells of the epidermis form a tough, translucent covering over the surfaces of leaves, permitting sunlight to penetrate. A waxy cuticle outside the epidermis reduces water loss and minute pores called stomata allow for gas exchange. Mesophyll is the main photosynthetic tissue of the leaf. Xylem and phloem are the conducting tissues of the leaf. Xylem brings water and dissolved minerals up from the roots to the leaves, and phloem conducts sugars made in the leaves throughout the plant.

11.3 Plants take up enormous quantities of water from the soil, and most of that water evaporates through the stomata of the leaves in transpiration. In most plants, stomata are open during the day and closed at night. Stomata open and close because of changes in turgor pressure brought about by osmosis.

11.4 The cohesion-tension theory explains the movement of water from roots to leaves. Evaporation of water from mesophyll cells causes them to draw water by osmosis from adjacent cells, which in turn draw water from a vein. The vein transmits that pull to the root, which pulls water from the soil.

11.5 Leaf abscission is the end step in a process of genetically programmed leaf senescence. Before leaves fall off, sugars, amino acids, and minerals are withdrawn.

11.6 Leaves perform functions other than acting as photosynthetic organs. They may be modified as tendrils to support climbing plants, as bud scales to protect buds from winter cold, or as bracts that function as flower petals. In deserts, succulent leaves are modified for water storage. Many garden vegetables have leaves adapted for nutrient storage. Leaves may be modified as spines to defend plants from herbivores, and some plants have leaves modified to trap animal prey.

11.7 Humans use leaves for many purposes, including directly as food, as herbs for seasoning many foods, and as perfumes, waxes, medicines, psychoactive drugs, beverages, dyes, and fibers.

REVIEW QUESTIONS

1. Define the following words, which refer to the external anatomy of a leaf: (a) *blade*, (b) *petiole*, (c) *stipules*, (d) *axils* and *axillary buds*, (e) *node*, (f) *internode*, (g) *sessile*, and (h) *leaf sheath*.

2. What are leaflets? Distinguish between simple and compound leaves and, regarding the latter, between pinnately and palmately compound leaves.

3. Discuss the arrangement of leaves along stems by defining alternate, opposite, and whorled leaves.

4. Discuss some of the mechanisms or anatomical structures used by leaves to conserve water.

5. What are mesophytes, xerophytes, and hydrophytes, and how are the epidermises and cuticles of each modified to permit growth in their respective habitats?

6. Discuss the structure and function of stomata. What causes the guard cells to open and close? Under what circumstances would you expect to find the guard cells open?

7. Describe the stomatal distribution patterns found in the leaves of hydrophytes, mesophytes, and xerophytes, as well as among monocots and eudicots.

8. Why do plants abscise their leaves? Do all plants abscise their leaves at once? Do evergreen plants abscise their leaves at all? Describe the important tissue and physiological changes leading to leaf abscission.

9. Describe how transpiration and the cohesion-tension theory can explain the movement of water from the roots through the leaves to the atmosphere.

10. Describe some of the variety of functions that leaves may perform in addition to photosynthesis, giving examples.

APPLYING CONCEPTS

1. Imagine you are walking through the produce section of a well-stocked grocery store. Can you name at least five different vegetables that consist primarily of leaves?

2. On a trip to the grocery store with a friend, you casually mention that celery is basically a bunch of leaves from the celery plant. Your friend disagrees, arguing that celery stalks are stems. Who is correct, and how could you settle the question?

3. Imagine that you discover a new species of herbaceous plant. The flowers of this plant clearly place it with the eudicots. But the plant body has a peculiar, jointed appearance, and you are not sure whether the large photosynthetic structures are leaf blades or flattened stems. Name several anatomical criteria that would allow you to decide whether these structures are leaves or stems.

12

Plant Behavior

Regrowth of prairie plants after a fire

Consider the smell of smoke. It's pungent, aromatic, and somewhat alarming—warning of fire. So it is not surprising that humans and other animals can detect and respond to the presence of smoke; this ability is critical to our survival. But it may seem amazing that many plants can also "smell" smoke and respond to it. How and why do plants exhibit such behavior?

After a fire, landscapes rarely remain bare for long. Multitudes of seedlings rapidly arise from seeds that previously lay dormant in the soil. How do buried seeds "know" that new growing space has become available and respond by germinating? One part of the answer is that many types of seeds can detect the presence of smoke. More precisely, seeds detect butenolide, a component of smoke that results when cellulose is burned. Since cellulose is a major component of all plant cell walls, butenolide is present in all soils after vegetation burns. Butenolide is stable at high temperatures, explaining its persistence after a fire. This compound is also water-soluble, so butenolide can be absorbed into seeds and even low amounts detected, much like animal noses receive odor signals. Butenolide can be applied to seeds to promote their germination in the absence of fire. This is useful in horticulture, agriculture, and habitat restoration projects.

Light sensitivity and detection also helps to explain seed germination after fires. Plant seeds can "see" that after a fire, light availability has increased enough to support seedling growth. Plant "vision" is based on the presence in plant cells of several types of light-sensitive pigments, as is animal vision.

When these pigments are exposed to light, they switch on a series of chemical reactions that result in plant responses. By these and other processes, plants sense and respond to changes in day length and temperature (change of seasons), drought and flooding, wind and other types of touch, gravity, and attack by disease microorganisms or hungry animals. In other words, plants exhibit many types of behavior that help them to survive environmental stress, just as animals do. In this chapter you will learn more about plant behavior and how this information is useful in producing food and other crops. ■

Prairie plant seedling

12.1 | Plants sense and respond to external and internal signals

The environments of plants and other organisms include stimuli of many types (Figure 12.1). Abiotic signals are external and physical; these include light, temperature, pressure, and gravity. Other signals originate from other organisms or from within an organism's own body; these are biotic signals. Organisms of all types possess cells capable of recognizing and responding to internal and external environmental signals. Recall that the membranes of cells bear many types of proteins and that some of these have the role of receiving information from the environment (Chapter 4). Some of these proteins recognize and bind particular information molecules, such as hormones produced elsewhere in the body. Other membrane proteins, or molecules in the cytoplasm, undergo physical change (become activated) in response to light or other physical signals (Figure 12.2). When signals are received, they must be transmitted to cell components that either continue to pass the signal along or directly cause a cell/organism response. The process by which signals are transmitted from the reception point to the response site is known as **signal transduction**.

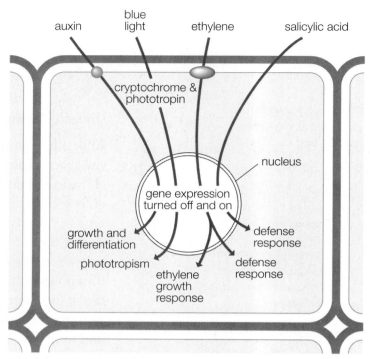

FIGURE 12.2 **Signal transduction** Plant cells respond to signals such as hormones produced within the plant body, as well as external signals, such as light. Receptor molecules in the cell membrane or cytoplasm receive signals. These receptors transmit messages via signal transduction pathways. Ultimately, the messages influence gene expression, causing changes in plant behavior, development, or chemistry.

Signal transduction begins when activated receptor molecules cause chemical changes to occur in nearby, cytoplasmic **messenger molecules**. Often, the message consists of the chemical addition of a phosphate group to specific messengers. These messenger molecules pass the chemical signal along to other messenger molecules in the signal transduction pathway. One activated receptor can send off many messenger molecules, leading to hundreds of specific molecular responses. In this way, the initial signal is greatly amplified as it moves along the pathway. Sometimes the chemical signals reach the cell nucleus, causing specific genes to be switched on or off. Other times, the signals cause changes in the flow of ions into cells, which result in particular responses, such as closing of leaf stomata (Chapter 11). In plants, signals often cause the release of calcium ions stored in the vacuole or endoplasmic reticulum. The released calcium binds to particular proteins that produce cell responses. There are several hundred kinds of **calcium-binding proteins** in plant cells, reflecting the great importance of calcium messengers in plants. Many aspects of signal transduction are similar in plants, animals, and microbes. For example, nitric oxide—NO—is a well-known chemical messenger in animal signal transduction. Recently, plant biologists have discovered that NO also functions as an important cellular signal in plants. For example, NO relays external signals as well as internal signals (hormones) to nuclear genes that control flowering. NO may integrate these two sources of information so that flowers are produced at the best time. In the next section, we explore hormones, chemical signals produced within a plant's body.

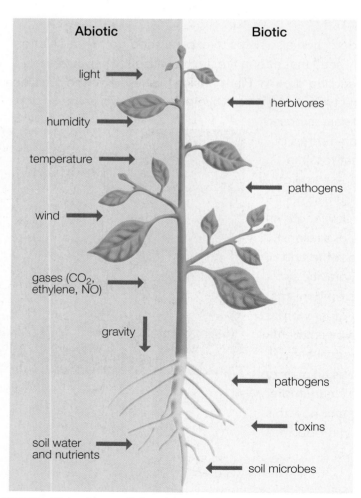

FIGURE 12.1 **Some of the abiotic and biotic signals received by plants**

12.2 | Hormones regulate plant growth and development

Most people know that animals produce hormones that influence development and body function. **Hormones** (from the Greek word meaning "to excite") are chemical compounds produced in one part of the body that exert their effects in another part of the body. For example, animal sex hormones produced by the brain influence the function of sex organs. Animal hormones are notable for being effective in tiny amounts. Like animals, plants produce a variety of chemical compounds that are known as hormones because they are effective in small amounts and move within the plant, influencing development and growth. Plant hormones are also known as **plant growth substances**.

There are several types of plant hormones

The plant hormones include a diverse collection of small organic molecules. Several types of **auxin** compounds, several types of **cytokinins**, many types of **gibberellins**, **ethylene**, **abscisic acid**, **brassinosteroids**, **systemin**, **salicylic acid**, and **jasmonic acid** are known plant hormones. There is also evidence that **sugars** function as hormones in plants. All of these compounds play diverse and essential roles in plant development. Differences in the chemical structures of plant hormones determine differences in their functions (Table 12.1). Several of these hormones were discovered only recently, and it is quite possible that additional plant hormones will be identified. For example, experiments suggest the existence of a hormone that promotes flowering (florigen), but the exact identity of this substance is still to be determined. Some bacteria, algae, and fungi produce several of the plant hormones and can thereby influence the growth and development of plants. The following survey of plant hormones and their effects reveals many ways in which these compounds are applied everyday in agriculture or gardening.

Auxins If you like to eat fruit, you have to appreciate auxins, because they are crucial to fruit production. The maturation of plant ovaries into fruit depends on production of auxin by embryos within developing seeds. Fruit can even be produced in the absence of pollination or fertilization by treating flowers with auxin. Orchard owners often spray their apple or pear trees with auxins in fall, in order to prevent the fruit from dropping prematurely. In addition to fruit production, there are many other ways in which auxins influence plant growth.

Auxins are produced from the amino acid tryptophan in young leaves, then transported to the shoot apex, where they induce formation of new leaves or flowers. Auxins induce cells to excrete acid into their cell walls, causing the walls to loosen so cells can expand. Cell elongation contributes to shoot growth, as described in Chapter 8. Shoot responses to changes in the direction of light or gravity are also mediated by auxin (see sections 12.3 and 12.4). At the young shoot tip, auxins also induce the development of new vascular tissue. As the vascular tissue matures, it creates a path for efficient auxin transport to lower regions of the plant. Auxins are transported within parenchyma cells of the phloem. Recall that series of phloem cells extend end to end from the shoot tip to the root tip (Chapters 8, 9). Phloem parenchyma cells import auxins by means of specific carrier proteins located only at the upper ends of these cells. Different proteins, located only at the lower ends of phloem parenchyma cells, transport auxin out, where it is accessible to the upper end of the next phloem cell (Figure 12.3). This explains auxin's downward movement in plants, influencing tissues as they pass through. The auxin produced by buds at the tops of plants inhibits growth of lateral buds (those located at the sides of shoot tips). This phenomenon is known as **apical dominance**. If the topmost, auxin-producing bud is removed, the lateral buds will grow, producing branches (Figure 12.4). This is because lateral buds are no longer inhibited. Professional growers and gardeners alike use this technique to increase the bushy appearance of plants.

In woody plants, auxins stimulate the secondary meristems that produce wood and bark. Auxins are also responsible for the downward growth of root tips and contribute to production of new adventitious roots on stems of cuttings. One form of auxin, 2,4-D (dichlorophenoxyacetic acid), is an herbicide commonly used to kill broadleaf weeds in lawns. This auxin doesn't kill the grass because grass plants (which are monocots) can inactivate the herbicide, whereas broad-leafed (dicot) weeds can't. A mixture of 2,4-D and a related compound (2,4,5-T) composed the infamous Agent Orange that was extensively sprayed by U.S. forces during the war in Vietnam to cause tree leaf drop (defoliation) and thus increase visibility. Later, it was discovered that during its manufacture, the 2,4,5-T used to make Agent Orange had become contaminated with dioxin, a highly potent carcinogen (cancer-causing agent). Dioxin, which is not a plant hormone or produced by plants, has since been recognized as the cause of major health problems in U.S. troops and Vietnamese who were exposed to it during the war.

Cytokinins Florists often spray cut flowers with cytokinins; because these hormones have an anti-aging effect, sprayed flowers will last longer. Cytokinins are also widely used (together with auxins) in plant laboratories to clone valuable plants for commercial production. Thousands of identical plants such as African violets and orchids, all having the same desirable characteristics, can be produced for market. In this process, known as **tissue culture** (Figure 12.5), pieces of stem, leaf, or root are removed from a plant. Their surfaces are sterilized to prevent growth of microbes and then placed in a dish containing mineral nutrients, vitamins, and sugar. If auxin and cytokinin are also supplied in the proportion of about 10 to 1, the plant cells will divide, forming a lumpy mass of white tissue known as **callus**. If the callus is then transferred to a new dish containing the same nutrients, but a higher proportion of auxin (greater than 10 to 1), the callus will form roots. Reducing the proportion of auxin to cytokinin causes the callus to turn green and develop shoots. By dividing a single callus into many pieces, then treating each with these hormones to induce roots and shoots, many hundreds of identical new plants can be produced.

TABLE 12.1 Plant Hormones: Structure and Effects

Plant Hormones	Chemical Structure	Functions
auxins	indoleacetic acid (IAA)	Apical bud dominance (retards growth of lateral buds immediately below); mediate growth response to light direction; induce development of vascular tissue; promote activity of secondary meristems; induce formation of roots on cuttings; inhibit leaf and fruit drop; stimulate fruit development; stimulate ethylene synthesis
cytokinins	zeatin	Promote cell division in shoot and root meristems; influence development of vascular tissues; delay leaf aging; promote development of shoots from undifferentiated tissue in lab culture
ethylene	ethylene	Promotes ripening of some fruits; promotes leaf and flower aging and leaf and fruit drop from plants; affects cell elongation and seed germination; helps plants perceive and respond to pathogen attack and mechanical stress
abscisic acid	abscisic acid	Promotes transport of food from leaves to developing seeds; promotes dormancy in seeds and buds of some plants; helps plants respond to water stress emergencies; regulates gas exchange at the surfaces of leaves
gibberellins	gibberellic acid (GA)	Stimulate both cell division and cell enlargement during shoot elongation; promote seed germination; stimulate flowering in some plants
brassinosteroids	brassinolide	Stimulate shoot elongation; reduce plant stress caused by heat, cold, drought, salt, and herbicide injury
salicylic acid	salicylic acid	Helps plants perceive pathogen attack

TABLE 12.1 Plant Hormones: Structure and Effects

Plant Hormones	Chemical Structure	Functions
systemin	systemin (amino acids)	Signals that wounding has occurred
jasmonic acid	jasmonic acid	Helps plants resist fungal infection and other stresses; induces plant production of protective secondary compounds (alkaloids)
sugars	glucose	Helps regulate amounts of chlorophyll and other photosynthetic components

Cytokinins are derivatives of adenine, a compound that you will recognize as a component of ATP and as one of the four bases found in DNA and RNA. Cytokinins are produced primarily in root tips (but also in shoots and seeds) as part of the pathway that

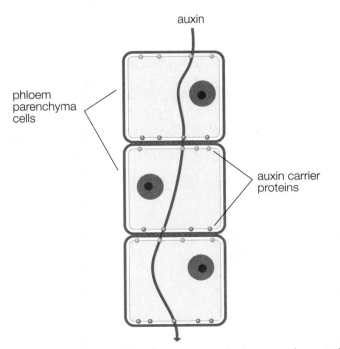

FIGURE 12.3 Directional auxin transport Auxin moves downward in plants through parenchyma cells found in the phloem. Cell-membrane carrier proteins (in purple) located in the lower portion of cells transport auxin out of the cells, where it can be taken into the adjacent cell by different carrier proteins (in blue) located in the top of the cell.

also leads to terpenes (see Chapter 3). Cytokinins are transported in the xylem to meristems, seeds, leaves, and fruit. These hormones stimulate cell division, a major component of plant growth, as described in Chapter 8. Cytokinins work by inducing cell-cycle transition to the M (mitosis) stage (described in Chapter 7). Additional cytokinin effects include the activation of secondary meristems and production of roots along the lower parts of stems (adventitious roots). Cytokinins are also involved in establishing partnerships between plant roots and beneficial bacteria—such as the association of *Rhizobium* and other nitrogen-fixing bacteria with legumes and other plants (Chapter 10). Cytokinins are also involved in association of plant roots with fungi to form beneficial mycorrhizae (Chapter 20).

Ethylene You can find fully ripened fruits of many types at your local market year-round because the supplier has treated the fruits with the ripening hormone ethylene. The effects of ethylene gas on plants were first discovered in the 1800s when people noticed that street-side trees exposed to leaking gas (which contained ethylene) unexpectedly lost their leaves. Plants produce ethylene within their bodies from the amino acid methionine. In addition to leaf drop and fruit ripening, ethylene influences many other aspects of plant development—cell specialization, sex determination, flower aging, defense against pathogens, and response to mechanical stress.

Some nuts dry and mature when ethylene causes rupture of cell membranes, resulting in tissue water loss. During ripening of some fruits, ethylene promotes the production of sugars that increase fruit sweetness and cell-wall softening by digestion of pectin. Ethylene also "switches on" expression of genes for synthesis of carotenoids, thus causing fruit color to change. These

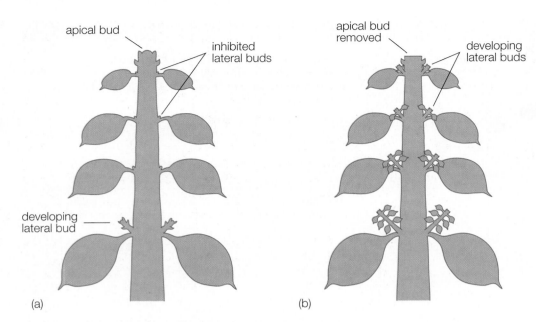

apical bud

inhibited lateral buds

developing lateral bud

apical bud removed

developing lateral buds

(a)   (b)

FIGURE 12.4 Apical dominance
(a) Apical buds produce auxins that inhibit the growth of nearby lateral buds. Lateral buds located lower on the plant may start to develop because the influence of auxin is reduced. (b) If the tops of plants are pinched off, the inhibition is removed, allowing lateral buds to produce branches.

changes increase fruit tastiness and edibility, adaptations that foster seed dispersal by animals (Chapter 24). An overripe banana or apple releases enough ethylene to hasten ripening of nearby fruit. This process is the scientific basis for the saying, "one bad apple spoils the barrel." For this reason, picked apples are often stored in an atmosphere that is low in ethylene. Ripe fruit is more vulnerable than unripe fruit to damage caused by shipping. Fruit can be genetically engineered so that ethylene synthesis is slowed, which delays the fruit ripening process long enough that fruit can be transported to market with less damage.

Emerging seedlings must muscle their way through crusty soil. Ethylene helps in this process by inducing what is called "the triple response": (1) inhibition of stem and root elongation; (2) swelling of stems, making them stronger; and (3) formation of a stem hook, which holds the seed leaves in a downward, protective "tuck" position (Figure 12.6). The tougher stem hook thus pushes through the soil, preventing damage to the more delicate apical meristem. The hook is formed when ethylene causes an imbalance of auxin across the stem axis, so that cells on one side of the stem elongate faster than cells on the other side.

Gibberellins
Grapes available in markets are larger than in the past, not only because of improvements in grape breeding

but also because growers spray growing grapes with gibberellins to dramatically increase their size. Gibberellins, also known as GA (an abbreviation for gibberellic acid), are produced via the terpene pathway in apical buds, roots, young leaves, and embryos of plants. These compounds promote stem and leaf elongation by stimulating cell division. GA also causes cell-wall loosening, allowing uptake of water and cell expansion, and stimulates the synthesis of cell-wall components. Many kinds of dwarf plants are short because they produce less gibberellin than taller varieties of the same species. If these dwarf varieties of plants are treated with gibberellin sprays, their stems will grow to normal heights. Dwarf wheat and rice crops are less vulnerable to storm damage than are taller varieties, and dwarf crops are also often more productive. Most of the genes that control dwarfing encode proteins that either regulate gibberellin synthesis or influence its signaling pathway.

In germinating seeds, gibberellins produced by embryos cause breakdown of starch and seed storage proteins into sugars and amino acids useful for seedling growth. The hormone is thus a "feed me" message sent by embryos to nutritive seed tissues (Chapter 26). Barley malt used in brewing beer is prepared by applying gibberellins to large numbers of barley seeds, causing them all to start germinating at the same time. The brewmaster

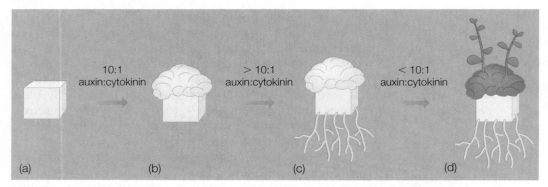

10:1 auxin:cytokinin

> 10:1 auxin:cytokinin

< 10:1 auxin:cytokinin

(a)   (b)   (c)   (d)

FIGURE 12.5 Diagram showing the use of hormones in plant tissue culture (a) A piece of plant material (e.g., the inside of a stem). (b) Colorless callus forms. (c) Roots are stimulated to form. (d) The callus greens up and shoots form.

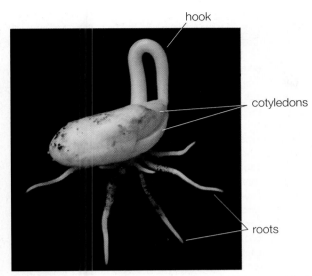

FIGURE 12.6 Bean seedling The hook (part of the stem) is the first portion of a young bean seedling to emerge from the soil. The seed leaves and upper part of the shoot are later pulled up through the soil, helping protect the apical meristem.

then bakes the germinated seeds at a high temperature to kill them, and in the process plant sugars are caramelized, generating distinctive beer flavors (see Essay 5.1 The Botany of Beer).

Several plant hormones play protective roles

Abscisic acid, brassinosteroids, salicylic acid, systemin, jasmonic acid, and sugars have protective effects in plants, helping them to cope with stress.

Abscisic acid, abbreviated as ABA, was named when botanists thought it played a role in leaf or fruit drop, also known as *abscission.* But plant biologists now know that abscisic acid plays a different role, acting to slow or stop plant metabolism when growth conditions are poor, functioning much like the brakes on a car. ABA may induce the onset of **dormancy** (from the Latin word *dormire,* meaning "to sleep"), a condition in which tissues protected within buds and seeds do not grow until specific environmental signals or cues occur. During the growing season, abscisic acid accumulates in the shoots of perennial plants, where it stimulates formation of tough, protective bud scales around apical meristems, preparing them to survive winter. This hormone likewise accumulates in seed coats of plants such as apple and cherry, preventing these seeds from germinating unless temperature and moisture conditions are suitable for growth. ABA also plays an important role in conserving plant water during drought and exposure to high salinity and cold. Water-stressed roots produce ABA via the terpene pathway. The ABA travels to shoots, where it helps to prevent water loss from leaf surfaces by inducing stomata to close (see Chapter 11).

Brassinosteroids, which are chemically related to animal sex hormones, are named after the plant *Brassica,* from which they were first identified. They are found in seeds, fruits, shoots, leaves, and flower buds of all types of plants. During plant development, brassinosteroids promote cell elongation by a mechanism different from auxins. When applied to crops, they help protect plants from heat, cold, salt, and herbicide injury. Brassinosteroids also retard leaf drop and promote xylem development.

Systemin, salicylic acid, jasmonic acid, and sugars mobilize plant defenses in response to viral, bacterial, and fungal infections. Plants can detect the presence of microbes by sensing compounds released from the attacker's cell wall, something like an animal detects an alarming odor. This induces the formation of defense hormones and a series of responses (see Section 12.6).

Systemin is a polypeptide that is formed in wounded leaves, traveling within 60–90 minutes to other leaves via the conducting tissues of the stem. Systemin thus operates something like a sentry to sound the alarm and rouse defenses. Its mechanism of action is like the inflammatory responses of animals to trauma. **Salicylic acid,** whose chemical structure is similar to aspirin, also has a role in defense and, additionally, retards flower aging. If you dissolve an aspirin tablet in water in a vase of cut flowers, they will last longer. Plants produce salicylic acid in the phenolic pathway. The presence of salicylic acid in plants is the basis for the ancient use of willows (*Salix* species) (Figure 12.7a), poplars (*Populus*),

FIGURE 12.7 Plants rich in salicylic acid that were traditionally used in pain relief (a) Willow (*Salix*), (b) wintergreen (*Gaultheria procumbens*).

ESSAY 12.1 ASPEN ASPIRIN

Quaking aspen (*Populus tremuloides*) (Figure E12.1), balsam poplar (*Populus balsamifera*), cottonwood (*Populus deltoides*), and several species of willow (*Salix*) are closely related trees or shrubs that are native to North America. Balsam poplar is common across Canada, and the other plants are common in temperate and arctic zones of North America. Native Americans and European settlers used leaves, inner bark, sap, roots, or fragrant spring buds of these trees as medicines. The bark or buds were added to hot water to make a tea (though the buds taste bitter unless pretreated to remove the bitter components). Alternatively, a salve or ointment was made by gently heating buds on a low fire with fats or oils. Beeswax could be added to achieve the desired consistency. These tree materials were used to relieve fever, sore eyes, sore throat, toothache, arthritis, headaches, colds, pain of bites and wounds, and other conditions that are often treated today with commercial aspirin. These plant materials were effec-

These plant materials were effective because they are rich in aspirin-like compounds.

tive because they are rich in aspirin-like compounds.

Many plants produce salicylates (salicylic acid or a related, sugar-containing compound—salicin). These terms reflect the scientific name of willow (*Salix*) and the fact that these molecules are ring-shaped (cyclic). Salicylates help plants detect and respond to local infection. Plants such as tobacco, cucumber, and potato that have been experimentally treated with salicylic acid show increased resistance to pathogen attack. Salicylic acid also controls heat production during flowering of the voodoo lily (*Sauromatum guttatum*), a process that aids in attracting pollinating insects. Salicin is one of several types of phenolic and terpenoid compounds that are abundant in poplar buds. Willow bark contains at least 12 different salicylate compounds, especially salicin, which sometimes reaches more than 11% of bark dry weight. The constituents vary by season and species.

When people consume willow bark extracts, gastric juices break down sali-

E12.1 Aspen (*Populus tremuloides*) leaves

cylates into sugars and salicyl alcohol, which is easily absorbed into the bloodstream. Blood and liver enzymes convert salicyl alcohol into salicylic acid—the compound responsible for the pain-relieving, fever-reducing, anti-inflammatory properties associated with aspirin. Plants were the original source of salicylic acid for aspirin production, but now this compound is produced industrially.

---

and wintergreen *(Gaultheria procumbens)* (Figure 12.7b) to relieve pain (see Essay 12.1, "Aspen Aspirin"). This compound works in animals by retarding production of the animal hormones known as prostaglandins. **Jasmonic acid** is chemically similar to mammalian prostaglandins (chemicals released during inflammation); both jasmonic acid and prostaglandins are produced from fatty acids. Jasmonic acid is very fragrant and is a component of jasmine oil, which is used in the perfume industry. In addition to playing a role in plant defense, this hormone inhibits seed and pollen germination, retards root growth, and induces fruit ripening and color changes.

Sugars have only recently been recognized to have hormonal function in plants. In addition to roles in defense, they regulate several photosynthetic genes and nitrogen use in the plant and interact with ethylene signal pathways. The balance between ethylene and sugar signals can influence seed germination, leaf development, and flowering. In the next sections, we will survey plant responses to external stimuli, such as light, touch, drying, flooding, and attack. Hormones often play important roles in these responses, as well as the roles we have just described.

12.3 | Plants use pigment-containing molecules to sense their light environments

As many as a billion plant seeds are buried in every hectare (about 2.5 acres) of soil; if these germinate beneath soil layers too deep for light to penetrate, or under a canopy of plants that does not allow sufficient light penetration to support seedlings, they will die. Seeds thus need a mechanism for metering light. Flowering plants also require a way to determine the time of day and season of the year so they can flower at the best time for pollination and seed production. Temperate perennial plants must have a method for predicting the onset of cold weather so they can produce dormant buds before winter arrives. Tree leaves need a way to assess their light environment in order to grow in patterns that prevent overlap and shading.

Amazingly, plants and their seeds are able to measure light and time periods by using light-responsive molecules that are composed of a pigment and protein. Plants use these molecules to determine if light levels are high enough to support photosynthesis of leaves or seedlings. Plants also measure night length, which is directly related to the length of days and to the season of the year.

The plant molecules that perceive light are **phytochromes**, which react to red and far-red light, and **cryptochrome**, **phototropin**, and **zeaxanthin**, which absorb blue and ultraviolet radiation. Light absorption by the pigment component of these molecules causes changes in the protein component, thereby starting a series of signal transmissions ending with a cell response (see Figure 12.2).

Cryptochrome helps seedlings determine if their light environment is bright enough to allow photosynthesis, or if they need to grow taller in order to obtain more light. Seedlings that grow in the dark contain a nuclear protein, known as COP1, which prevents expression of genes for chlorophyll and photosynthesis (Figure 12.8). Dark-grown seedlings continue to elongate in search of more light. Blue light alters cryptochrome so that it can bind to COP1. When COP1 is bound by cryptochrome, it stops inhibiting photosynthesis genes, so seedlings are able to produce leaves and chlorophyll. Phototropins are involved in plant growth responses to changes in light direction. Zeaxanthin plays a role in stomatal responses to light. In this chapter, we will focus on phytochromes because they play a variety of roles in plant biology.

DNA SCIENCE

Phytochrome controls seed and spore germination

Water-soaked lettuce seeds will not germinate in darkness, but they will germinate if exposed to as little as 1 minute of red light. However, if the red-light treatment is followed by a few minutes treatment with far-red light (light of a longer wavelength than red light), the lettuce seeds will not germinate. By means of this and other experiments, biologists have learned that plant "light sensor" molecules function like electric light switches—at any point in time they are either turned on or off (Figure 12.9). Phytochrome occurs in two states: a "switched-

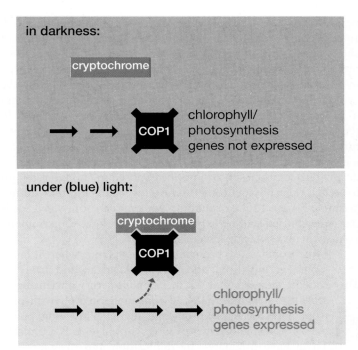

FIGURE 12.8 Seedlings use cryptochrome to sense their light environment In darkness, the protein COP1 prevents the expression of genes involved in chlorophyll production and photosynthesis (which would be wasteful under dark conditions), and seedlings lengthen in an attempt to reach light. When light is present, the blue-light receptor cryptochrome undergoes a change so that it can bind with COP1, which, once removed, allows for the expression of photosynthetic genes and causes the seedlings to produce chlorophyll and turn green.

off" state that can absorb red light (P_R) and thus be turned on, and a "switched-on" state that is capable of absorbing far-red light (P_{FR}), and thus being turned off. When lettuce seeds are exposed to red light, the light-absorbing portion of the phytochrome molecule causes changes in the protein component and switches the molecule to the "on" position. The longer the

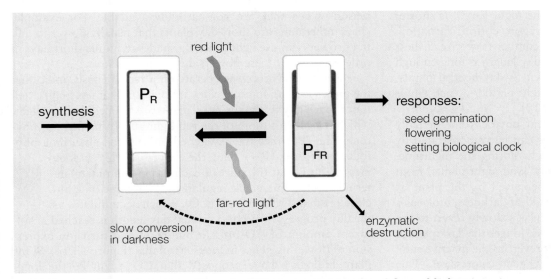

FIGURE 12.9 Phytochrome enables plants to respond to red and far-red light Phytochrome acts like a switch that controls many aspects of plant development (think of the switch used to turn electric lights on and off). Sunlight produces more P_{FR} than P_R.

period of exposure to red light, the more phytochrome molecules become activated. In some cases, such as in regulation of chloroplast position within cells, "switched-on" phytochrome molecules exert their influence within the cytoplasm. However, in other cases, activated phytochrome molecules travel from the cytoplasm to the nucleus, where they induce chemical changes in other molecules that turn gene expression "on" or "off." In the case of seeds exposed to red light, activated phytochrome turns on genes essential for ending seed dormancy and stimulating seed germination. In general, seeds kept in darkness or deeply buried in the soil will not be exposed to enough red light to switch their phytochrome to the active state and they do not germinate. If you till or spade up your garden soil, weed seeds that were buried will be exposed to enough light to germinate, eventually necessitating a weeding job.

Phytochrome is also present in seedless plants, where it influences spore germination, and in various algae and bacteria. This evidence suggests that phytochrome is a very ancient light-sensing system and that seed plants inherited phytochrome from ancestral spore-producing plants. Flowering plants possess four major types of phytochrome—A, B, C, and E. These have originated by duplication and divergence of ancestral genes. Phytochrome A is particularly sensitive in low light conditions; this phytochrome may have been especially useful to the earliest flowering plants, which are thought to have inhabited shady places.

EVOLUTION

Phytochrome helps control the timing of flowering and dormancy

Flowering plants can be classified according to the way their flowering corresponds to day length. Asters, strawberries, dahlias, poinsettias, potato, and soybean are classified as **short-day plants** because they flower only if the day length is shorter than a critical period (that is, when the night length is longer than a critical period). In contrast, lettuce, spinach, radish, beet, clover, gladiolus, and iris are **long-day plants** because they flower in spring or early summer, when the light period is longer than a critical length (that is, when the night length is shorter than a critical period). Roses, snapdragon, cotton, carnation, dandelion, sunflower, tomato, and cucumber flower regardless of the day or night length (as long as day length is long enough for plant growth) and are thus known as **day-neutral plants**. Plants' response to the length of the light period in each day is known as **photoperiodism** (Figure 12.10).

In the photoperiodic response, leaf phytochrome actually measures the length of the night. This can be demonstrated by exposing plants to a brief flash of light during the nighttime, which prevents short-day plants (really, long-night plants) from flowering, since the night length is perceived by the plant as being two short periods. During a period of darkness, molecules of phytochrome in the activated state (P_{FR}) slowly revert to the inactive form (P_R) (see Figure 12.9). The longer the dark period, the more phytochrome molecules will revert to the inactive state and leave the cell nucleus, reducing gene expression. This process is something like the slow movement of sand from the top to the bottom half of an hourglass. A flash of light during the night that contains red wavelengths reconverts phytochrome

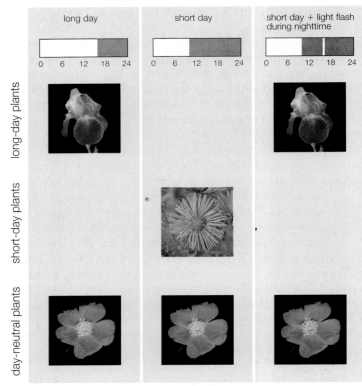

FIGURE 12.10 Photoperiodism Long-day plants flower when the night period is short, while short-day plants flower when the night is long. Day-neutral plants flower regardless of day length. Exposure to a brief flash of light during the middle of the night has the effect of shortening the length of the nighttime such that short-day plants no longer flower. A more fundamental description of this phenomenon is presented in the text.

to the active form (P_{FR}), analogous to turning the hourglass to its original position before all of the sand had passed through.

Day length effects on flowering affect the distribution of plants on Earth. Few short-day plants occur in the tropics because the nights are not long enough. Ornamental plant growers manipulate day length to produce flowers for market during seasons when they are not naturally available. For example, chrysanthemums are short-day plants that usually flower in fall, but growers can use light-blocking shades to mimic short days in order to produce these flowers during any season.

Many plant processes operate on a regular basis, indicating the presence of an internal clock mechanism that lets plants "tell time." Each day, as the sun rises, plants are flooded with red light, which transforms more of the red-light absorbing form of phytochrome (P_R) into the "switched-on" far-red-absorbing phytochrome form (P_{FR}) than the reverse. This sudden increase in P_{FR} at the start of each day resets plants' internal clocks. The clock regulates the amount of a nuclear protein, CONSTANS (CO), which accumulates as the day progresses. When a critical day length is reached, a sufficient amount of CO protein builds up and turns on expression of flowering genes in long-day plants (Figure 12.11). Some plant biologists have proposed that CO might be the long-sought flowering hormone, florigen. Interactions between the clock and phytochrome also regulate the daily closing of some flowers in the evening.

DNA SCIENCE

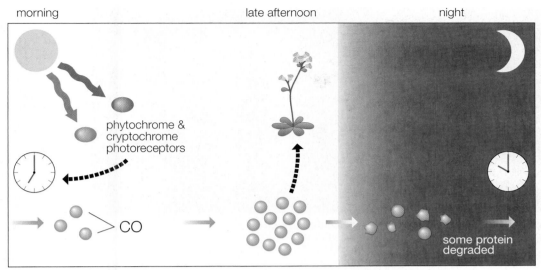

morning late afternoon night

phytochrome & cryptochrome photoreceptors

CO

some protein degraded

FIGURE 12.11 Day length and an internal clock work together to influence flowering in long-day plants A plant's internal clock is reset each day at sunrise, when photoreceptor molecules detect the sun's light signal. As the day proceeds, more of the protein CO is produced than is destroyed. When a critical level of CO is reached, long-day plants flower. At night, more CO is destroyed, so cell levels are lowest at sunrise.

Plants also use light-sensing systems to control shoot dormancy. For example, birch trees *(Betula)* respond to shorter day lengths by closing off the plasmodesmatal connections between cells of the apical meristems, thereby shutting down growth during winter. Buds become acclimated to cold, increasing their ability to survive through winter. When the day length increases in spring, shoot apical meristems are reactivated and resume growth.

Plant shoots can sense shading and grow into the light

You may also have noticed that houseplants placed in a window often grow toward the light (Figure 12.12). The growth of shoot tips toward light is known as **phototropism**. A tropism is

FIGURE 12.12 These radish seedlings illustrate phototropism by their growth toward the light

growth toward or away from an external stimulus; the prefix *photo* refers to light. The stem curvature results from a greater amount of cell elongation on the shaded side of the tip, because auxin has accumulated there. Auxin migrates to the shaded side under the influence of blue-light receptors.

Leaves within a tree's dense canopy detect shade as an increased proportion of far-red to red light. This reaction occurs because the chlorophyll in overlying leaves absorbs more red light than far-red light. So more of the phytochrome in the shaded leaves is in the inactivated state than is the case for leaves in the sun. Activated phytochrome inhibits the growth of shoot internodes, but phytochrome in the inactivated state does not. So shoots bearing shaded leaves can increase shoot internode length, which has the effect of extending leaves into the light. Shoots bearing leaves that are not shaded do not extend because leaf phytochrome is primarily in the activated state and thus inhibits shoot internode elongation. Blue-light sensors and the hormone ethylene form an additional plant mechanism used to sense shading.

12.4 | Plants respond to gravity and touch

If you turn a potted plant over on its side, after some time you will notice that the shoot has bent back upward (Figure 12.13). Auxins are responsible for the upward growth of shoots against the force of gravity. This is because plants display **gravitropism**— growth in response to the force of gravity. Roots also exhibit gravitropism, while growing downward. Plant biologists are still not sure exactly how plants detect Earth's gravitational force. Some think that settling of starch-heavy plastids (known as statoliths) to the lower sides of cells causes redistribution of calcium signals. Others suggest that the weight of the cytoplasm pulls on different

FIGURE 12.13   Gravitropism is illustrated by a tomato plant that has resumed upward growth after being placed on its side Upward growth started about 4 hours after the plant was turned sideways; this photo was taken 20 hours later.

time zero	< 1 sec	2 min	6 min	12 min

FIGURE 12.14   The touch response of the sensitive plant (Mimosa pudica) If touched, the leaves fold rapidly, then gradually return to their original positions.

parts of the cell wall, depending on the direction of gravitational force. It may be that both mechanisms help plants perceive the direction of gravitational force.

Plants exhibit growth responses to touch that reflect adaptation to the influences of wind, rain, or animal encounters. For example, in very windy places, shorter plants might be less likely to blow over than taller plants. Plant scientists have observed that rubbing plant stems can result in shorter plants. Even touching plants while measuring them can influence growth. Some plants are called "touch specialists" because they display rapid responses to touch. An example is the rapid leaf folding of the sensitive plant when touched (Figure 12.14). This may reflect an adaptation that reduces exposure of leaf stomata to the drying effects of wind. Other plant growth responses to touch include the coiling of tendrils around a supporting structure. Such growth responses are known as **thigmotropisms** (Figure 12.15). The young tips of vines often sway in circular patterns, a process known as **nutation**. These dance-like movements, visualized by time-lapse photography, allow vines to locate potential support structures and begin to grow around them.

Plants may also grow toward or away from water (*hydrotropism*), and the leaves or flowers of some plants track or avoid the sun (*heliotropism*). Growth toward or away from chemical influences is known as *chemotropism*.

FIGURE 12.15   Thigmotropism is illustrated by tendrils of a cucumber plant that have wrapped themselves around a nearby grass plant

12.5 | Plants can respond to flooding, heat, drought, and cold stress

The most harmful effect of flooding on plants is inhibition of their roots' ability to absorb minerals from soils. Roots can only take up mineral ions from the soil when oxygen is present (see Chapter 10). In very wet soil, plant roots lack sufficient oxygen for their metabolism, and ethylene builds up in the roots. This happens for two reasons: Low oxygen stimulates ethylene synthesis, and the ethylene gas cannot diffuse away from the root in waterlogged soil. Ethylene induces some root cortical cells to undergo controlled death. This generates continuous root channels that allow oxygen to diffuse from shoots to roots, functioning something like snorkels. Tissues that contain such channels are known as **aerenchyma** (see Chapter 10).

Plants respond to drought in a variety of ways (Chapter 26). **Stomatal closure**, controlled by the plant hormone abscisic acid (see Chapter 11) is a common response. But if this method of water conservation fails, what other options do plants have for surviving drought? Some plants respond to predictable periods of drying by timing their flowering so that seeds are produced prior to the dry period. Others survive as underground bulbs, tubers, or rhizomes (see Chapter 9). **Desiccation tolerance**, the ability to survive alternating dry and wet periods, is an option exercised by many bryophytes, some lycophytes (such as the "resurrection plant," *Selaginella lepidophylla*, Figure 12.16),

a

b

FIGURE 12.16 A resurrection plant *Selaginella lepidophylla* is one of several types of plants described as "resurrection plants" because they can recover from an extremely dry condition.

and ferns, and a variety of flowering plants that are adapted to arid habitats. These plants typically survive severe water loss for 6 months and some can survive in a dry state for 10 months of the year. Water shortage induces the production of abscisic acid, which induces plant changes that help them survive in a dry condition. Desiccation-tolerant plants typically exhibit high concentrations of disaccharide sugars, such as sucrose. These sugars bond to phospholipids, stabilizing cell membranes. Desiccation-associated proteins (such as dehydrins, which are also found in dehydrating seeds) also bind to membranes, protecting them.

Plants (and other organisms) cope with heat stress by increasing amounts of **heat shock proteins**. These proteins wrap around critical cell proteins such as enzymes, keeping them from coagulating in response to heat. This protection allows enzymes to retain their function despite the heat. Some plants exude the gas isoprene when they are stressed by too much light or heat. This is thought to help protect the photosynthetic system from damage. Plants that are adapted to seasonally cold climates can sense the onset of cold weather (though it is as yet unclear how) and respond to it in various ways (see Chapter 27). For example, cold-resistant plants may increase the content of solutes—dissolved materials such as ions and sugars—in their cells' cytoplasm. This helps prevent water movement out of cells when the formation of ice crystals in the intercellular spaces lowers extracellular water potential (see Chapter 4). The presence of these solutes lowers the temperature at which cell water freezes, just like adding antifreeze prevents automobile gas line and windshield washing fluids from freezing. Deciduous trees drop their leaves as a way to reduce water loss from leaf surfaces (see Chapter 11).

Vernalization is a series of changes occurring during a seasonal cold period that delays flowering until springtime. This adaptation prevents plants that live in seasonal climates from flowering at the wrong time. In barley, wheat, and other cereals native to regions with cold climates, winter wheat varieties are sown in the fall, because they require a long period of low-temperature exposure (vernalization) in order to flower. In contrast, spring wheats don't require the cold period and can be planted in spring. Recently, agricultural scientists have identified two vernalization genes, *VRN1* and *VRN2*, whose expression explains these variations. In winter wheat, the gene *VRN2* encodes a protein that represses flowering but is destroyed by cold. When the weather improves, the product of the *VRN1* gene increases, promoting spring flowering. Loss of function of *VRN2* genes by natural mutation has given rise to spring wheat varieties.

DNA SCIENCE

12.6 | Plants can defend themselves against attack

Animal cells defend themselves against attack by microscopic pathogens (disease agents such as viruses, bacteria, protists, and fungi) by engulfing and destroying the invaders within special immune system cells. Though plants lack this type of defensive system, they are well protected by other means. Plants use preformed barriers to infection (such as leaf cuticles) or toxins to prevent disease, or they recognize pathogens and prevent them from spreading. Plants respond within minutes to the presence of attacking microorgansims. Only a small proportion of pathogens that infect plants actually cause disease.

The **hypersensitive response (HR)** occurs when a plant has a dominant **resistance gene** (*R*) that encodes a specific membrane receptor protein and the pathogen has a complementary, dominant **avirulence gene** (*Avr*), which encodes a soluble protein that binds the plant receptor. When these two proteins bind, plants respond as if they were extra- (hyper) sensitive, by producing compounds that kill pathogens or retard their growth and by sealing off the infection site. Plants often increase production of hydrogen peroxide (H_2O_2). This is a very effective germ killer that people commonly use to prevent skin wounds from becoming infected. Infected plant cells also produce signal molecules: NO, salicylic acid, ethylene, and jasmonic acid. These stimulate cells to fortify their walls with protective callose and lignin, helping to prevent pathogen spread. Some plant cells may be sacrificed, undergoing controlled cell death as a means of containing the pathogen. Under these circumstances, the pathogen does not cause disease because the plant is able to contain the infection. As a result of infection or hypersensitive responses, plants may bear dead (necrotic) spots illustrating where the battles have occurred (Figure 12.17). But if one (or both) of these genes is recessive, disease may occur. Thus, agricultural scientists try to breed resistance genes into crop plants.

Systemic acquired resistance (SAR) is a response of the whole plant that results from previous infection or the hyper-

FIGURE 12.17 Necrotic spots reveal the locations where plants have contained attacks by pathogenic microorganisms

sensitive response. Like vaccination of an animal, SAR results in much-reduced disease symptoms but offers even broader protection against a variety of different pathogens. The mobile peptide systemin may spread information throughout the plant that pathogen attack or successful infection has occurred. Remote plant parts respond by taking preemptive, protective actions to avoid infection. Plants may produce defensive enzymes that break down the attackers' cell walls or tannins, which are toxic to microorganisms. A number of commercial products are designed to induce SAR without exposing plants to pathogens. When sprayed onto crop plants, these products stimulate plant "immunity," thereby preventing or reducing the effects of diseases.

HIGHLIGHTS

12.1 Plants are able to sense and respond to external abiotic signals such as gravity, light, temperature, wind, gases, and soil, water and nutrients. Plants also respond to stimuli from herbivores and pathogens and internal information in the form of hormones. Stimuli are received by cell-membrane or cytoplasmic receptor molecules. Signals are transmitted within cells by messenger molecules, such as calcium-binding proteins, in signal transduction pathways. Ultimately, signals influence gene activity and elicit responses such as changes in plant structure, development, or chemistry.

12.2 Plant hormones are small, organic molecules. Auxins stimulate cell elongation by fostering cell-wall loosening and foster fruit development. Cytokinins stimulate cell division. Ethylene influences fruit ripening and seedling emergence. Gibberellin influences cell division and expansion and seed germination. Protective hormones include abscisic acid, brassinosteroids, systemin, jasmonic acid, and sugars.

12.3 Phytochrome is an important light-sensing molecule in plants. Several blue-light receptors also aid in light perception by plants. Seed and spore germination depend

on the presence of phytochrome in its "switched-on" P_{FR} form. The timing of flowering is influenced by phytochrome's determination of day length and by an internal clock that is set at sunrise. Phytochrome also allows plants to detect shading and respond by growing into the light.

12.4 Plants respond to gravity, ensuring that shoots typically grow upward and roots downward. Plants respond to touch, including the influence of wind, and some are able to locate and grow around supporting structures.

12.5 Flooding, heat, drought, and cold stress can be detected by plants, which respond in protective ways. Vernalization is a series of changes that occur during a cold period that allows flowering to occur in spring.

12.6 Plants respond to attack from genetically compatible pathogenic microorganisms by mounting a hypersensitive response that limits disease progression. Once having been attacked, plants can become less susceptible to pathogens by taking preemptive protective actions.

REVIEW QUESTIONS

1. Briefly describe some major effects of the following plant hormones on the plant body. Give examples, where possible, of commercial applications of that hormone. (a) Auxins, (b) cytokinins, (c) ethylene, (d) gibberellins, (e) abscisic acid, (f) brassinosteroids, (g) salicylic acid, (h) systemin, (i) jasmonic acid, and (k) sugars.

2. What is phytochrome, and what are its chemical constituents? How is it turned "on" and "off," and how does the cell respond to these states? Give some examples of plant processes that are controlled by phytochrome.

3. What are short-day, long-day, and day-neutral plants? How do these plants sense the length of the day?

4. Describe the processes that occur when a plant "wakes up" at sunrise (think "alarm clock!"). How do these determine whether or not the plant will flower.

5. Explain why very few pathogen attacks actually result in plant disease.

APPLYING CONCEPTS

1. Tomato plants have been genetically engineered so that transcription of one of the genes required for ethylene production has been blocked. What effect would you anticipate this loss of endogenous ethylene to have on the tomato plant? How could this effect be easily reversed?

2. A man leaves his landscaped garden unattended for several years and returns to find that many of the shrubs have become spindly looking—the plants produce their leaves far apart on thin stems, with little branching. He tells his gardener to cut back all the spindly shrubs, making sure to remove all the apical buds in the process. How will this succeed in producing a fuller, denser-looking plant? What phenomenon is responsible for the paucity of branching?

3. Imagine that you are growing a short-day plant, a long-day plant, and a day-neutral plant together in an indoor greenhouse (which receives no sunlight). You set your artificial lights so that the plants receive 10 hours of light and 14 hours of darkness each day. Is it possible that all three plants will flower at the same time? What if you change the day:night schedule to 14 hours of light and 10 hours of darkness?

4. Poinsettias are familiar plants to most people. Sold in late November and December, their verdant foliage and crimson-red bracts brighten the holidays for many people. Yet such plants, maintained as houseplants, rarely bloom a second time. Knowing that poinsettias are short-day plants, why would you suspect that they rarely reflower, and how might you get them to flower again?

5. Many fruits, such as apples, oranges, and grapefruits, are sold in grocery stores in perforated bags—plastic bags with holes in them. What is the purpose of the holes in these bags?

CHAPTER 13

Reproduction, Meiosis, and Life Cycles

Goldenrod (*Solidago*)

- Many plants and other organisms reproduce sexually—by a process that involves gametes and zygotes. Others reproduce asexually—without the involvement of gametes or zygotes—or by both sexual and asexual means.

- Asexual reproduction allows organisms to reproduce and disperse rapidly, but sexual reproduction has better potential for generating genetic diversity.

- Meiosis is a cell division process that is essential to sexual reproduction.

- Plants typically have a sexual life cycle known as alternation of generations, with two types of multicellular bodies—one that produces gametes and another that produces spores by meiosis.

A mbrosia is a native American plant named for the delectable food eaten by mythical Greek gods in order to gain immortality. But *Ambrosia artemisiifolia* and *A. trifida*, commonly known as ragweeds because of the ragged shapes of their leaves, have a less favorable modern reputation. Ragweeds, which have inconspicuous green flowers, are a major cause of hay fever—allergic reactions to their pollen. Not only do ragweeds typically produce very large amounts of airborne pollen but the surfaces of their pollen grains are coated with proteins that cause sneezing, watery eyes, and breathing difficulties in susceptible people. Ragweeds bloom in late summer and fall, during the same period as goldenrods, which have more conspicuous, bright yellow flower clusters. When people see goldenrod flowers, they think of pollen. Consequently, goldenrods are often blamed for the allergy problems that are actually caused by ragweeds. Goldenrods produce relatively little pollen, which is transported by insects rather than wind, so they are not important sources of allergenic pollen. At some other times of the year grass and tree pollens cause allergies.

Despite its allergenic properties, pollen is needed for fruit production by flowering plants. Without pollen there would be no apples, cherries, and other fruit and seed crops because pollination is an essential stage in the sexual reproductive cycle. However, many plants and other organisms are able to reproduce without sex, that is, asexually. In this chapter we take a closer look at sexual and asexual reproduction in plants and certain protists and fungi of medical, economic, or ecological importance to humans.

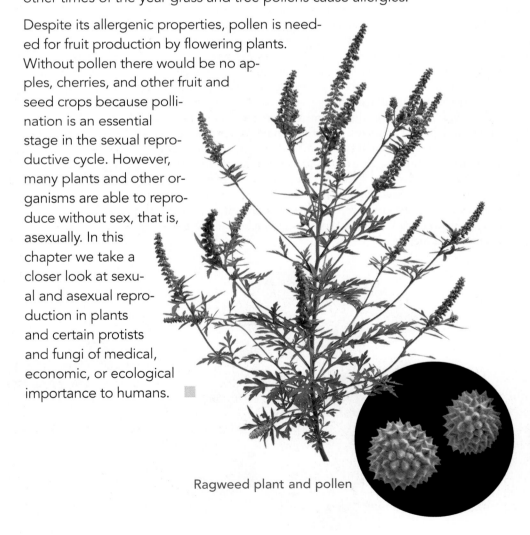

Ragweed plant and pollen

13.1 | Sexual and asexual reproduction confer different advantages

Many plants and other organisms depend on sexual reproduction, which is common and often conspicuous. Flowers, for example, are the sexual reproductive organ systems of angiosperms (Figure 13.1). However, some flowering plants and many other organisms reproduce only asexually, that is, without sex (Figure 13.2). Yet other organisms, including many plants, reproduce by both sexual and asexual means. Understanding the differences between sexual and asexual reproduction, and the benefits or disadvantages of each, is important in agriculture, medicine, and ecological studies.

Sexual reproduction accelerates adaptation

Sexual reproduction is the fusion of male and female gametes to form a zygote. **Gametes** are single cells, produced by adult organisms, which are specialized for mating; gametes often die if mating is not accomplished. **Sperm** are male gametes. In seedless plants, sperm have flagella, which enable the sperm to swim though water to the egg, much like animal sperm swim to egg cells. Seed plants also produce sperm, though these mostly lack flagella and are transmitted to eggs by tubular extensions that grow from germinating pollen grains. **Eggs** are female gametes, which are usually larger than sperm and lack flagella. All plants, from bryophytes to angiosperms, produce naked (wall-less) sperm and nonflagellate egg cells, as do animals. In contrast, many protists produce gametes that look similar to each other, and protist gametes sometimes have cell walls. **Zygotes** are the single cells that result from fusion of gametes. If conditions are favorable, zygotes may develop into adults, completing the reproductive cycle. Sexual reproductive cycles of eukaryotes also require **meiosis**—a nuclear division process that reduces the number of chromosomes by one-half (Figure 13.3). Sexual reproduction in eukaryotes is characterized by gametes, zygotes, and meiosis.

FIGURE 13.2 **Some flowering plants reproduce only by asexual means** Dandelions *(Taraxacum officinale)*, shown here, are a common example.

Sexual reproduction is expensive. Sex requires organisms to invest scarce resources and expend considerable energy to produce gametes, accomplish mating, and undergo meiosis. Elaborate mating behaviors of many animals and the large, showy flowers of many plants are familiar examples. So why did sex evolve? This is an intriguing question for biologists, who have found evidence that sexual reproduction is advantageous in several ways. Sexual reproduction combines the DNA of two different organisms (usually of the same species), creating new combinations

EVOLUTION

FIGURE 13.1 **Flowers usually function as sexual reproductive organs of angiosperms**

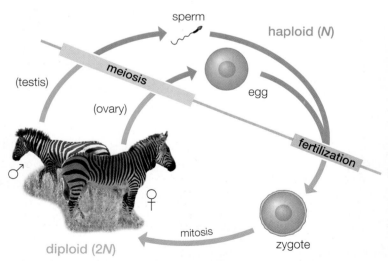

FIGURE 13.3 **The sexual reproductive cycle of animals** Animals, including humans, have a life cycle in which gametes (sperm and eggs) fuse to form zygotes. Gametes are produced as the result of meiosis, and have N chromosomes. Zygotes combine the chromosomes of egg and sperm and thus have 2N chromosomes. In humans, $N = 23$ and $2N = 46$ but other species have different chromosome numbers.

of valuable genetic traits. New genetic combinations give species the potential to colonize new types of environments or adapt to changing conditions (Figure 13.4). By comparison to asexual reproduction, sexual reproduction also speeds the loss of deleterious genes (a process known as purging) and allows potentially beneficial genes to escape the influences of inhibitory genes that previously held them in check.

As a likely consequence of these advantages, sexual reproduction is very common among eukaryotes. Sexual reproduction is widely present in animals, plants, fungi, and all but a few groups of protists. DNA evidence suggests that sex probably originated independently several times among protists and that animals and plants inherited sexual reproduction from separate protist ancestors. Consequently, many of the genes that control sexual reproduction in plants differ from those of animals.

DNA SCIENCE

Prokaryotes lack sexual reproduction involving fusion of two gametes to form zygotes, and meiosis is also absent from prokaryotes. But the advantages of genetic exchange are so great that even prokaryotes have evolved ways to transfer genes among themselves (Chapter 18).

Asexual reproduction can occur rapidly

Given the advantages of sexual reproduction just described, why do many organisms also or exclusively reproduce asexually? One answer is that asexual reproduction often requires fewer resources and less energy than sexual reproduction and therefore can be accomplished more quickly and efficiently. In addition, asexual organisms do not expose themselves to the risk of gametes not finding mates. The gametes of most organisms are so specialized for sexual function that they die if they do not undergo fertilization and thus represent wasted effort. Asexual reproductive structures, on the other hand, are able to develop into new organisms if the environment is suitable.

Another advantage of asexual reproduction is that it requires only one parent. Even if only one sex occurs in a population or only a single organism is present, progeny can be produced asexually. In contrast, species that reproduce only by sexual means depend on the presence of both sexes. Another advantage of asexual reproduction is that all of the progeny are genetically alike, so the chances are good that many will survive if their environment is stable, homogeneous, and similar to the one in which their parents were well adapted (see Figure 13.4). In addition, asexual reproduction may be more effective than sexual reproduction in harsh environments. In such habitats there may be too few animal pollinators to transmit pollen among flowering plants or too little liquid water to allow sperm of seedless plants to swim to eggs. These advantages explain why asexual reproduction has commonly evolved in nature.

Many organisms that reproduce only asexually evolved from sexually reproducing ancestors

There are many examples of protists, fungi, and plants whose closest relatives reproduce by both sexual and asexual means but are thought to have lost the ability to reproduce sexually through the evolutionary process. Such organisms now depend on asexual reproduction; despite this limitation, many are so abundant in nature that they are regarded as weeds.

EVOLUTION

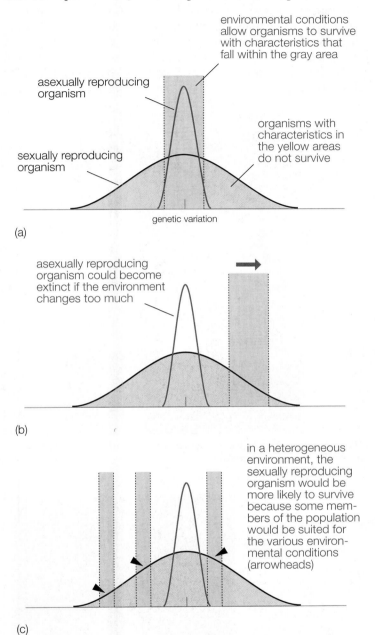

FIGURE 13.4   Advantages of sexual and asexual reproduction (a) When the environment is homogeneous and stable, a population of an asexually reproducing organism that is well adapted to that habitat is very successful and does not lose many individuals that are less well adapted; greater diversity is present among sexually reproducing organisms, and only the well-adapted survive. (b) In a changeable environment, the conditions may alter to such a degree that an asexually reproducing organism might not be able to survive because it is not genetically equipped to survive the new conditions. A population of a sexually reproducing organism, in contrast, may be able to survive because some members would have features that permit their survival. (c) In heterogeneous environments, once again, a sexually reproducing organism would be at an advantage, with genetic variability making survival under different conditions more likely.

A freshwater green alga, *Cladophora glomerata* (Figure 13.5a), provides an example from among the photosynthetic protists (algae). Most of *Cladophora*'s relatives grow along marine shorelines, where they reproduce both asexually and sexually, but the widespread freshwater species *C. glomerata* is known to reproduce only asexually (Figure 13.5b). This alga forms weedy growths worldwide in freshwater lakes and streams that have been polluted by excess amounts of phosphate. The phosphate acts as a fertilizer, allowing rampant reproduction by single-celled asexual structures known as zoospores. Zoospores have flagella and can thus move, as can animals (the Greek root *zoon* means "animal"). Zoospores, produced in abundance, are released from the parent cells, attach to rocks or other surfaces, and grow into large populations of algae (Figure 13.5c).

More than 17,000 species of fungi also appear to have lost the ability to reproduce sexually. These fungi typically reproduce by small, asexual spores known as *conidia* (Chapter 20), which are discharged from the parent fungal body and dispersed in air currents. When they land in a suitable moist place, conidia grow into the black or green molds often found on bread or citrus peel. If inhaled, conidia can cause allergic responses and breathing problems similar to pollen reactions. *Aspergillus* is a common mold that causes potentially fatal lung disease in humans; its airborne asexual conidia are important elements of its propensity to cause infections. On the other hand, *Penicillium* and close relatives are the sources of important antibiotics such as penicillin. Asexual conidia provide the means for industry to transport and maintain valuable strains of these organisms. DNA evidence shows that *Aspergillus, Penicillium,* and other asexually reproducing fungi are closely related to fungi that can reproduce sexually. This evidence suggests that the ability to reproduce sexually was lost by the ancestors of these important molds.

The water weed *Elodea canadensis* (Figure 13.6) is an angiosperm that, unlike closely related flowering plants, reproduces primarily by asexual means—growth of broken-off shoots. Its asexual reproduction has allowed *Elodea* to become very common in freshwater habitats. This plant is widely used in biology classes to demonstrate photosynthesis and plant cell structure because the leaves are very thin and thus can be readily examined with a microscope. *Elodea*'s asexual reproduction makes it easy for teachers to propagate this plant in aquaria for classroom use.

FIGURE 13.6 The common water plant *Elodea* reproduces primarily by asexual means

(a)

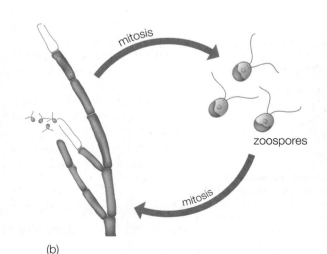

(b)

(c)

FIGURE 13.5 The asexually reproducing, weedy, freshwater green alga *Cladophora glomerata*
(a) Microscopic view showing the branched filaments attached to rocks and other available substrates.
(b) Asexual reproductive cycle of *Cladophora*. Cells in the filaments contain numerous nuclei. During the formation of zoospores, the cytoplasm is divided, so that some is associated with each nucleus. These small units of cytoplasm plus nucleus develop into the zoospores, which are released through a pore in the cell wall. Zoospores settle after a time and develop into new filaments through mitotic divisions. (c) A large amount of *Cladophora* growing near the shore of a lake rich in phosphate. Storms may cause large amounts of *Cladophora* to detach and wash up onto shore, where it decays.

Dandelion (*Taraxacum officinale*; see Figure 13.2) and related hawkweeds (*Hieracium* species; Figure 13.7) are but two examples from among hundreds of plants that reproduce exclusively by seeds that are formed asexually, a process known as apomixis—from the Greek *apo*, "away from" and *mixis*, "mixing" (Chapter 23). In contrast, sexual reproduction

is required for most plants to produce seeds. Apomictic plants can reproduce prolifically, explaining why they can be so abundant in lawns, fields, or roadsides.

Many organisms reproduce by both asexual and sexual means

The relative advantages of sexual and asexual reproduction are illustrated by a wide variety of organisms that are able to reproduce both sexually and asexually. This ability is often important in agriculture.

Phytophthora, whose name means "plant destroyer," is a common protist that is related to golden and brown algae (Chapter 19). This protist causes destruction of a wide variety of crop plants, including potato, avocado, tomato, papaya, apple, soybean, citrus, pineapple, eucalyptus, and tobacco. *Phytophthora* was responsible (together with destructive social conditions) for the Irish Potato Famine of 1846–47, which influenced the subsequent immigration of many Irish people to the United States. *Phytophthora* has a reproductive cycle that includes both asexual and sexual processes (Figure 13.8); both are important in its ability to cause devastating crop losses. *Phytophthora*'s asexual cycle begins with **sporangia**, structures that produce and enclose spores. **Spores**, single-celled reproductive structures that are dispersed from the parent, can result from both sexual and asexual processes. Sporangia are released from *Phytophthora* and carried to plant leaf surfaces by wind, after which they release zoospores that rapidly spread by swimming through films of water present when the weather is cool and rainy. The zoospores then germinate, producing a threadlike body that invades the leaf tissue and extracts food from it, in the process depleting the plant's resources for growth. Diseased plants are unable to produce normal numbers of tubers in the case of potatoes, or fruit, wood, or

FIGURE 13.7   Hawkweeds reproduce primarily by means of seeds that are asexually produced

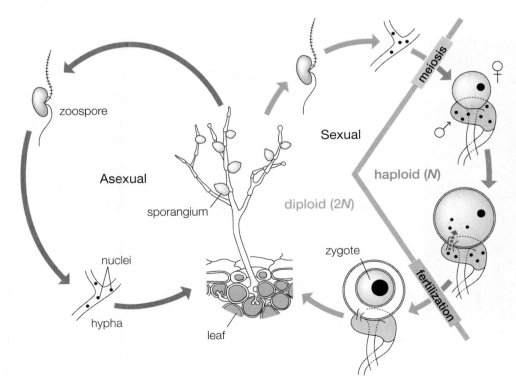

FIGURE 13.8 *Phytophthora*, **the plant destroyer** This microbe, which causes diseases in many plants, reproduces both sexually and asexually. The tubelike body (each tube is called a *hypha*) can obtain food for growth and reproduction from plant leaf tissues. The food is used to produce branches with many sporangia that break off and are dispersed to other plant surfaces. The sporangia-bearing structure shown here is emerging from a leaf stoma. In asexual reproduction, each sporangium releases many genetically similar zoospores that disperse in water films and grow into new *Phytophthora* bodies, thus spreading the organism. In sexual reproduction, the hyphae produced by zoospores generate male and female gametes, and fertilization generates zygotes that are genetically diverse.

other economically valuable commodities in other plants. *Phytophthora* uses materials obtained from the plant host to produce a new supply of sporangia, which disperse in the wind to other plants to continue the infection cycle.

In the sexual cycle of *Phytophthora* (see Figure 13.8), gametes mate to form a zygote. The zygote functions in two ways: first, as a resting stage that can resist harsh conditions and germinate when conditions improve; and second, as a source of new gene combinations. Genetic variation contributes to this pathogen's ability to increase its host range, that is, infect additional kinds of plants.

Among plants, there are many examples of species that reproduce by asexual and sexual means. Aspen trees (*Populus tremuloides,* Figure 13.9) often form large groves of hundreds of individuals that are genetically alike because they have been generated by asexual reproduction—in this case, by development of young sapling shoots from horizontally growing roots of parent trees. Yet aspens also produce seeds by sexual means.

Many grasses (including the lawn nuisance, crabgrass) produce huge spreading populations from underground, horizontal stems (rhizomes) as well as by seeds. New strawberry plants can arise from both horizontal "runners" or stolons (aboveground stems) and seeds. The common houseplant *Kalanchoë* has leaves edged with tiny plantlets (Figure 13.10) that can detach and take root in the soil below, another example of asexual reproduction. Yet this plant also produces flowers, fruits, and seeds by sexual means. In the case of dates and some other crop plants, agricultural propagation is often accomplished by asexual processes rather than sexual ones, even though the plants readily produce seeds (Essay 13.1, "The Perfect Date").

In summary, sexual and asexual reproduction are both useful to plants and other organisms, and an understanding of their relative roles is important to humans both in agriculture

FIGURE 13.10 *Kalanchoë* This plant reproduces both sexually and asexually, by means of plantlets produced at the edges of leaves. Small roots are evident on the plantlets.

and the study of ecology. A comparison of the cell division process mitosis, which is associated with asexual reproduction, to meiosis, a cell division process associated with sexual reproduction, provides additional insight into reproduction of plants and other organisms.

13.2 | Meiosis is essential to sexual reproduction

It's the rule: **Meiosis** is a cell division process that is associated with sexual reproduction, but not with asexual reproduction. Why is this true?

Meiosis prevents buildup of chromosomes as the result of sexual reproduction over many generations

Consider what happens when two gametes fuse during sexual mating, a process known as **syngamy** or fertilization. The cytoplasm of the two cells combines, and the nuclei of the two cells also meld together. As a consequence, the zygote formed from syngamy is a single cell having cytoplasmic components and nuclear chromosomes contributed by both parents. Mating usually occurs only between individual organisms of the same species that have the same number of chromosomes in their nuclei. Thus, fusing gametes usually possess an equal number of

FIGURE 13.9 A grove of quaking aspen (*Populus tremuloides*)
Because they can reproduce asexually, it is possible that all the trees in an aspen grove are genetically identical, in which case the grove could be considered to be one large organism.

ESSAY 13.1 THE PERFECT DATE

Dates have been cultivated for at least 5,000 years. They originated in North Africa and were subsequently planted throughout the Middle East and other parts of the world. Dates were very important to the lives of ancient people, offering an energy-rich food that could easily be dried for storage and transport. The high sugar content of dried dates inhibits the growth of decay microbes (in such a high-solute environment, microbes lose their cell water by osmosis). Today, dates are an important crop in the California desert, which has a climate similar to that of the Middle East. Most dates are sold as dried fruit, which is valued for its rich taste as well as high potassium and fiber content (Figure E13.1A).

Dates are the fruit of the date palm, *Phoenix dactylifera*, a flowering plant that can grow more than 100 feet tall (Figure

Sexual reproduction combines characteristics to yield the perfect date, and asexual reproduction provides a way for people to grow many of them.

E13.1B). Date palms produce numerous flowers on long, branched stalks that grow from the leafy crowns. Individual trees produce only female flowers or male flowers, but fruit production requires transfer of pollen from male flowers to female flowers. Thousands of years ago, people learned that they could increase fruit production by hanging male flower stalks high in the tops of female trees, to enhance pollen transfer. Now, growers use fans to waft pollen from male trees when the female flowers are receptive. Fruit is produced in bunches of about 1,000 dates, and a single tree can yield 400 pounds of fruit per year. Fruit production is highest when date palms have at least 100 days during which the temperature reaches 100°F or higher. But temperatures higher than 110°F can harm the trees. Sufficient water, usually provided by irrigation, is also required.

Inside each date fruit there is a single long, hard pit that includes the seed. You can easily grow a houseplant-sized date palm by planting a date pit. But commercial date farmers do not use seeds to produce groves of date palms because the fruits of the resulting plants would be highly variable in their characteristics. Dates can be red, brown, or yellow in color; spherical, oval, or elongate in shape; from 1 to 3 inches long; and variable in taste and storage features. Many combinations of these characteristics result from the gene mixing that occurs during meiosis and sexual reproduction. For the market, growers need groves of trees that produce uniform fruit having desired characteristics. They achieve this by using asexual reproduction, propagating offshoots that grow from the bases of trees whose fruit is best in quality. Over its lifetime, a single date palm can produce 12–15 offshoots, each genetically identical to the parent. Sexual reproduction combines characteristics to yield the perfect date, and asexual reproduction provides a way for people to grow many of them.

E13.1A Dried dates

E13.1B Date palm

chromosomes. The basic number of chromosomes present in gametes is known as N—the **haploid** chromosome number. The value of N varies greatly among organisms. In the case of one plant, $N = 2$, yet in some other plants N can be greater than 500. For humans, $N = 23$.

Because zygote nuclei contain all the chromosomes contributed by each gamete, a zygote's chromosome number is $2N$—the **diploid** chromosome number (46 for humans). If the zygote grows into an adult organism by means of mitosis, as is the case for animals and plants, each adult cell nucleus will contain $2N$ chromosomes. For most people, body cell nuclei contain 46 chromosomes, with the exception of gametes, which contain only 23 chromosomes.

Now imagine what would happen if adult organisms with a $2N$ chromosome number produced gametes also having $2N$ chromosomes. Further imagine that such $2N$ gametes fuse—the zygote formed from this mating would have $4N$ chromosomes! Many generations later, cell nuclei would include immense numbers of chromosomes—too many for cells to separate evenly during mitosis. Meiosis prevents the problem of chromosome buildup resulting from sexual reproduction by reducing the chromosome number in a cell by one-half at some point prior to gamete production. This means that gametes routinely have a chromosome number of N. Sexual mating of gametes with N chromosomes results in zygotes and adult organisms having the $2N$ chromosome number typical for the species.

Meiosis has another critical function in sexual reproduction—it shuffles and recombines the genetic information contained in the chromosomes contributed by each parent. This is the process that generates beneficial genetic variability in a sexually reproducing population of organisms. Genetic variation is the basis for organism adaptation and evolution, an essential attribute of living things.

Meiosis contributes to genetic variability

Meiosis works by separating pairs of homologous chromosomes. **Homologous chromosomes** are more similar to each other than they are to other chromosomes in a cell's nucleus because homologous chromosomes usually bear the same types of genes (though not always exactly the same DNA sequences). An organism's male and female parents each contribute an equal number of chromosomes to their progeny. More specifically, each parent contributes one chromosome of each homologous pair present in their offspring. For example, humans have 23 pairs of homologous chromosomes. The male parent has contributed 23 chromosomes (one of each homologous pair), and the female parent has contributed an equal number (and the other member of each homologous pair in the offspring). When an organism becomes sexually mature and some of its cells undergo meiosis, the 23 chromosomes originally contributed by its male parent are mixed with the 23 chromosomes contributed by the female parent. Thus, cells produced by meiosis contain various combinations of parental chromosomes. In addition, pieces of chromosomes that came from one parent can be exchanged with the homologous chromosomes that came from the other parent, contributing even more genetic variation. Understanding these processes in a bit more detail is very helpful in comprehending differences between sexual and asexual reproduction and how organisms evolve.

13.3 | Meiosis resembles mitosis in some respects, but differs in important ways

Understanding how meiosis compares with mitosis is essential to understanding how organisms reproduce. A review of the basic processes of mitosis (Chapter 7) is useful at this point. Recall that mitosis is the process by which eukaryotic cells produce two identical copies of a cell nucleus, and that mitosis (along with cytoplasmic division—cytokinesis) is the basis for population growth of single-celled eukaryotes and body development in multicellular organisms. Mitosis occurs in shoot and root tips and other meristematic tissues of plants (Figure 13.11a) and is thus associated with growth (Chapter 8). In contrast, meiosis occurs in flower parts or other tissues (of nonflowering plants) that are involved in sexual reproduction (Figure 13.11b).

Meiosis follows DNA replication and uses a spindle apparatus, as does mitosis

Meiosis and mitosis both occur after DNA replication, which produces **chromatids**—identical copies of each chromosome that remain attached by a centromere (Chapter 7; Figure 13.12Ia, IIa).

In both mitotic and meiotic prophase, chromosomes condense to the degree that they become visible with the use of a light microscope, the nuclear envelope often disappears, and a microtubular spindle forms (Figure 13.12Ib, IIb). The spindle serves as a structural framework for chromosomal movement in both mitosis and meiosis, and in metaphase of both meiosis and mitosis the chromosomes line up along the midpoint of the spindle (Figure 13.12Ic, IIc). These similarities, together with the fact that all eukaryotes exhibit mitosis while early diverging groups of eukaryotes lack sex and meiosis, suggest that meiosis evolved from mitosis.

EVOLUTION

Homologous chromosomes pair, then separate during meiosis I

How do meiosis and mitosis differ? As you now know, meiosis and mitosis occur in different plant tissues, and they are associated with sexual and asexual reproduction, respectively. At the cell level there are additional important differences.

First, meiosis (see Figure 13.12II) is a double division process involving two sequences of prophase → metaphase → anaphase → telophase → cytokinesis. In contrast, mitosis involves only one such sequence of nuclear division events (see Figure 13.12I). The first division sequence of meiosis is known as meiosis I (shaded green on the diagram), and the second is known as meiosis II (shaded brown). Meiosis I also differs from mitosis in that pairs of replicated homologous chromosomes line up along the middle of the spindle apparatus (Figure 13.12IIc) and then separate, each homologue moving to opposite spindle poles (Figure 13.12IId). In contrast, pairing and separation of homologous chromosomes does not occur during mitosis. While homologous chromosomes are paired in meiosis I, they can exchange segments, thereby creating new genetic combinations (Figure 13.12IIb, IIc). This chromosome segment exchange process is known as **crossing over** (see Figure 14.12); it is important in explaining inheritance phenomena (Chapter 14).

Meiosis I contributes in another very important way to the production of genetic variation. During metaphase I, chance alone determines how chromosomes align themselves in the division plane and, therefore, how many chromosomes derived from the male parent (or female parent) move to the same spindle pole during anaphase. Consider an organism having in its nuclei four chromosomes—two pairs of homologous chromosomes (Figure 13.13a). At anaphase of meiosis I, it is possible that both chromosomes derived from the male parent would move toward one pole and that both derived from the female parent would move toward the other pole (Figure 13.13b). But it is equally likely that one chromosome from the male parent could travel with one originally contributed by the female parent (Figure 13.13c). As a result (leaving aside for the moment the effects of crossing over), two different genetic types of meiotic progeny could be produced by a single cell undergoing meiosis, and four types could be generated by a population of meiotic cells from the same organism. If a plant cell has six chromosomes per nucleus, there are eight possible chromosome

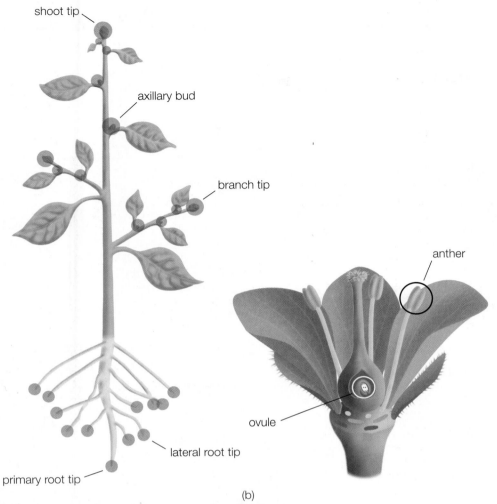

shoot tip

axillary bud

branch tip

anther

ovule

lateral root tip

primary root tip

(a)

(b)

FIGURE 13.11 Location of mitosis versus meiosis in plants (a) Most of the mitotic divisions in plants occur at growing points called primary *meristems* (red dots), many of which are situated at the tips of shoots and roots. Plants that produce some wood also possess secondary meristems. (b) Meiosis in plants occurs in reproductive structures and gives rise to spores. In flowering plants, for example, meiosis occurs in the male and female parts of flowers (circled).

combinations among its meiotic products (spores). For humans, approximately 8 million possible combinations of 23 chromosomes may occur in gametes, which are produced by meiosis. When the effects of crossing over are added, the potential amount of genetic variation arising from events in meiosis I can be enormous. Additional genetic variation arises from random mating during sexual reproduction. Any single person represents one of 64 trillion possible chromosome combinations (8 million squared). It is no wonder that individual siblings arising from the same pair of parents (whether plants or people) can look so different!

Another difference between meiosis and mitosis occurs during anaphase. At mitotic anaphase, the centromeres that join chromatids divide, so that each of the two progeny telophase nuclei receives one chromatid; once separated, chromatids are known as chromosomes (see Figure 13.12Ie). In contrast, centromeres do not divide during anaphase of

meiosis I; the chromatids remain joined and travel together to one pole or the other (see Figure 13.12IId). As a result, the number of chromosomes in nuclei formed at the end of meiosis I has been reduced. For this reason, meiosis I is also known as "reduction division."

Meiosis I is an elegant way for cells to reduce their chromosome number without disturbing the normal balance of chromosomes in a cell nucleus. This very precise "dance of the chromosomes" ensures that progeny cells will have one, and no more than one, of each type of chromosome typical of the species. As noted earlier, chromosome imbalances—too many chromosomes of a particular type or missing chromosomes—cause cell and organism malfunction, cancer, or death. If homologous chromosomes do not separate at anaphase of meiosis I, an event known as **nondisjunction** (because homologous chromosomes do not become disjunct), abnormal gametes are formed. Nondisjunction occurs in plants

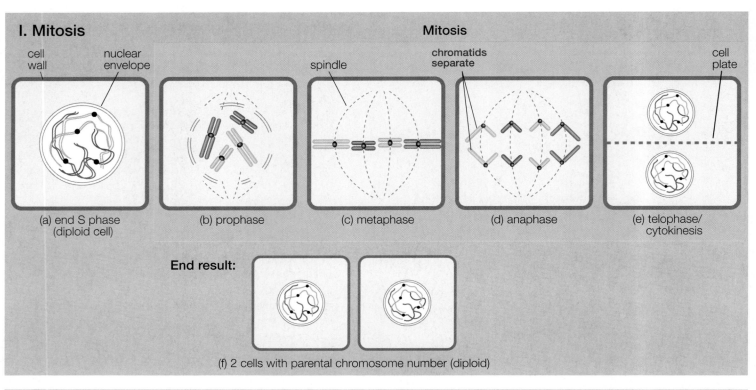

I. Mitosis

Mitosis

cell wall
nuclear envelope

(a) end S phase (diploid cell)

(b) prophase

spindle

(c) metaphase

chromatids separate

(d) anaphase

cell plate

(e) telophase/cytokinesis

End result:

(f) 2 cells with parental chromosome number (diploid)

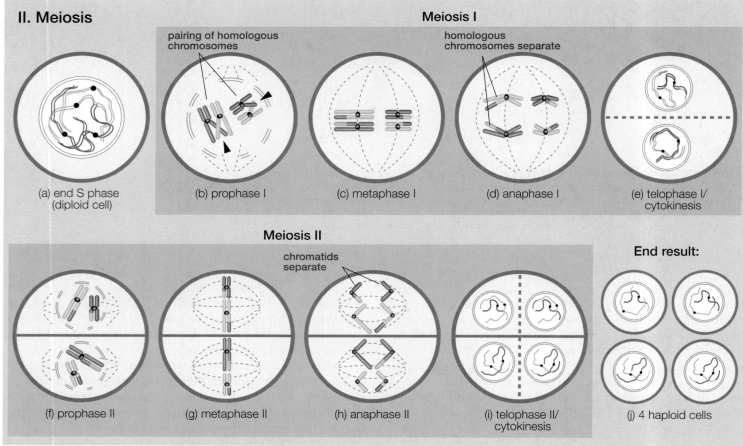

II. Meiosis

Meiosis I

pairing of homologous chromosomes

homologous chromosomes separate

(a) end S phase (diploid cell)

(b) prophase I

(c) metaphase I

(d) anaphase I

(e) telophase I/cytokinesis

Meiosis II

chromatids separate

End result:

(f) prophase II

(g) metaphase II

(h) anaphase II

(i) telophase II/cytokinesis

(j) 4 haploid cells

FIGURE 13.12 Comparison of mitosis and meiosis (Ia–f) Mitosis. (IIa–j) Meiosis. The dark- and light-red chromosomes are homologous (i.e., one was contributed by the male parent and the other by the female) as are the dark- and light-blue chromosomes. DNA replication occurs prior to both mitosis and meiosis. The major difference between mitosis and meiosis lies in the behavior of homologous chromosomes. In mitosis (mitosis proper shaded blue), homologous chromosomes do not pair during prophase and chromatids are separated. In meiosis, homologous chromosomes pair in prophase I (meiosis I, shaded green). Also, in meiosis, crossing over can occur between homologous chromosomes, such that portions of the chromosomes are exchanged (indicated by the arrowheads in IIb). Meiosis II (shaded brown) is like mitosis in that chromatids separate. Meiosis also differs from mitosis in that there are four progeny cells, each having only half the number of chromosomes of the parent, whereas mitosis yields only two cells, whose chromosome number is the same as that found in parental cells. In this example, the parent had four chromosomes and the meiotic progeny two.

syndrome, characterized by some degree of mental retardation and characteristic physical traits. The chances of nondisjunction increase with age, and so genetic testing and counseling are often recommended for older expectant mothers. An understanding of meiosis is very useful in comprehending the value of such genetic testing. Understanding the cellular and molecular basis for differences between mitosis and meiosis (Essay 13.2, "The Guardian Spirit") is essential to developing ways to prevent human miscarriage and birth defects that arise from mistakes in meiosis.

Chromatids are separated during meiosis II

What is the function of meiosis II, the second phase of meiosis? During anaphase of meiosis II, the two chromatids that together constitute each chromosome are separated. This process is identical to anaphase of mitosis except that the cells resulting from meiosis have only half the number of chromosomes that were present in cells before they entered meiosis. Cells that undergo mitosis do not experience any change in chromosome number, whereas the number of chromosomes in cells resulting from normal meiosis is exactly half that found in parental cells.

A final difference between meiosis and mitosis is the number of cells produced at the end of these division processes. Meiosis II generates four haploid cells, whereas mitosis generates two progeny cells whose chromosome numbers are the same and equal to that of the parental cell. If a parental cell is diploid, the mitotic progeny will also be diploid; if the parental cell is haploid, the mitotic progeny will also be haploid.

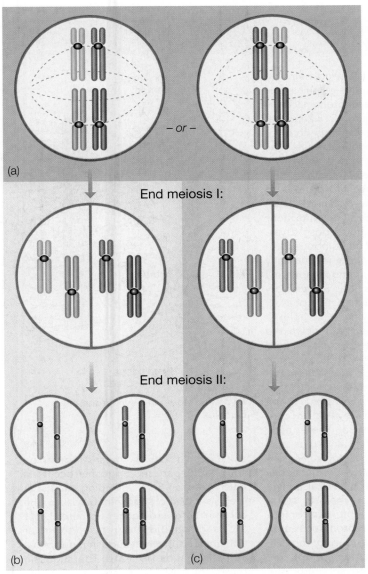

End meiosis I:

End meiosis II:

FIGURE 13.13 Segregation of chromosomes in meiosis The diploid cells in a plant, including those that undergo meiosis, have two sets of chromosomes, one (in dark red and dark blue) that came from the female parent and one (in light red and light blue) that came from the male parent. (a) In metaphase I of meiosis, the homologous chromosomes could line up such that the resulting cells get a set of chromosomes just like those in the parent plants (as shown in b) or, equally likely, they could be arranged such that each resulting cell would receive chromosomes of both parental types (as in c). In (c), two of the four haploid cells have a combination of chromosomes (and alleles of genes found on them) that were not present in the parent plants. This is one way in which new genetic combinations can be obtained through sexual reproduction. The effects of crossing over are not shown in this figure; additional new combinations would be generated through this process.

13.4 | Life cycles link one generation to the next

Asexual organisms typically have very simple life cycles. They produce genetically identical reproductive structures, such as the zoospores produced by aquatic algae and fungi, airborne conidia released by many terrestrial fungi, and plantlets of plants. Mitosis is the only division process involved in asexual reproduction (Figure 13.14a).

Sexually reproducing organisms have more complex life cycles involving haploid and diploid stages and specialized reproductive cells—gametes, zygotes, and sexual spores. There are three basic types of sexual life cycles—gametic, zygotic, and sporic—named for these three types of reproductive cells. Life cycles differ among organisms in the point at which meiosis occurs, the relative importance of haploid versus diploid stages, and the types of reproductive cells generated by meiosis (Figure 13.14b–d).

Gametic life cycles are typical of animals and some algae

The reproductive cycle of humans and other animals (see Figures 13.3 and 13.14b) is the life-cycle type most familiar to people. Meiosis occurs during the production of

and other eukaryotes. In humans, gametes normally have 23 chromosomes (each of which has been assigned a number). However, nondisjunction can result in gametes having more than 23 chromosomes. If a gamete with two number 21 chromosomes fuses with a gamete having a normal chromosome number, the child that results will likely suffer from Down

ESSAY 13.2 THE GUARDIAN SPIRIT

Errors in mitosis or meiosis can lead to missing or extra chromosomes (aneuploidy). In humans, mitosis-related aneuploidy often leads to cancer, and meiotic aneuploidy results in miscarriage or birth defects. In humans, errors in meiosis are responsible for more than 30% of miscarriages and are the most common cause of mental retardation. About 20% of human egg cells have abnormal chromosome numbers due to failure of meiosis to accurately distribute chromosomes during egg production. Mistakes also occur during plant meiosis. A detailed understanding of the cellular, biochemical, and molecular basis of meiosis will help in understanding the kinds of errors that may occur and aid in developing ways to prevent such problems.

Something apparently protects centromere cohesion during meiosis I. This protector has now been identified as a protein complex known as sugoshin (which means "guardian spirit" in Japanese).

Early in meiosis, chromosome ends—known as telomeres—cluster at a specific place on the nuclear envelope. Experts think that this process helps homologous chromosomes find and begin to pair with each other, a defining feature of meiosis. Recall that each homologous chromosome consists of two chromatids (the products of DNA replication). Chromatids are held together by protein aggregates known as cohesin, as also occurs during mitosis (see Essay 7.1). During homolog pairing, cohesin and other proteins produce a ladder-shaped synaptonemal complex (Figure E13.2A) that glues homologous chromosomes together while they condense and shorten. The synaptonemal complex also fosters the formation of chaismata—X-shaped DNA breaks and reconnections between homologs. Chiasmata hold homologous chromosomes together until they separate during anaphase I of meiosis. This is necessary because during metaphase I, an enzyme complex—separase—breaks down the synaptonemal complex (and also breaks down the cohesin that holds chromatid arms together) (Figure E13.2B). The *phs1* gene recently discovered in corn affects one of these early steps in homolog pairing and thus may aid in understanding its molecular control, which has been mysterious until now.

Until recently, experts were also mystified by the fact that in meiosis, separase does not sever the centromere connections between chromatids until meiosis II. In contrast, during mitosis, separase cuts chromatids completely apart, allowing them to separate (see Essay 7.1). Something apparently protects centromere cohesion during meiosis I. This protector has now been identified as a protein complex known

E13.2A Model of a synaptonemal complex

as sugoshin (which means "guardian spirit" in Japanese). Sugoshin occurs in diverse eukaryotes, including plants. A meiosis-specific sugoshin binds to the cohesin surrounding centromeres. This allows chromatids to remain together until meiosis II, when sugoshin degrades and chromatids separate. The sugoshin thus "guards" centromeric cohesin, hence its captivating name. The discovery of sugoshin's involvement in meiosis led to the observation that a different type of sugoshin helps stabilize

gametes, and gametes represent the only haploid stage in the life cycle. Other life-cycle stages—zygotes, embryos, juveniles, and adults—are diploid. The fact that all sexually reproducing modern animal groups have this life-cycle type suggests that it arose very early in the evolutionary history of animals. Gametic life cycles are not unique to animals or multicellular organisms, however. They also occur in several protist groups, including the unicellular silicon-walled algae known as diatoms as well as some seaweeds (Chapter 19).

An advantage in being primarily diploid is that organisms have two copies of all necessary chromosomes (the homolo-gous chromosomes) and thus possess some genetic redundancy. If a deleterious mutation occurs in a gene on one chromosome of a diploid organism, the organism will likely not suffer, because there is a "spare" normal gene on the homologous chromosome. The environment may change enough that the variant gene form becomes more useful than the "normal" gene. If an organism possesses such gene variants, known as **alleles,** it will probably grow and reproduce better than organisms lacking them in the new environment. The diploid advantage helps explain the great evolutionary success of animals (and diatoms, of which there may be more than 10 million living and fossil species).

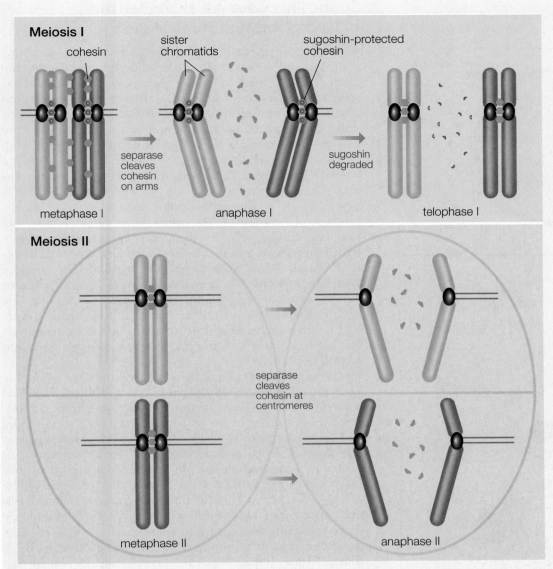

E13.2B Cohesin, separase, and sugoshin in meiosis

chromosomes during mitosis. Mutations in the gene that encodes the meiotic form of sugoshin can cause errors leading to birth defects and miscarriage, and mutations in the mitosis-related sugoshin may be involved in cancer. Knowledge of the "guardian spirit" proteins will aid in preventing or identifying mutations that lead to disease.

Zygotic life cycles are common among protists

Many protists (algae, protozoa, and other simple eukaryotes) are primarily haploid. The only diploid cells in their life cycle are zygotes, which arise by the fusion of gametes (Figure 13.14c). An advantage of this life-cycle type is that gametes can be produced quickly by mitosis. Meiosis occurs during division of the zygote. The thick-walled zygotes of many protists serve as survival stages during periods unfavorable for growth. When conditions improve, meiosis generates genetically diverse progeny.

Because haploid organisms have only one chromosome of each type rather than homologous pairs, they have only one allele of a unique gene. If the function of an essential but unique gene is disrupted by mutation, the organism may die because there is no normal gene to mask the effect of the mutation. The organism has no "spare genes," no genetic redundancy. Also, only altered genes that have little effect on the organism's survival are likely to be transmitted to the next generation. If the environment changes so that a more dramatic gene change would be advantageous, the mutation would have to reappear in the population. This process could take such a long time that the rate of evolutionary change would be slow.

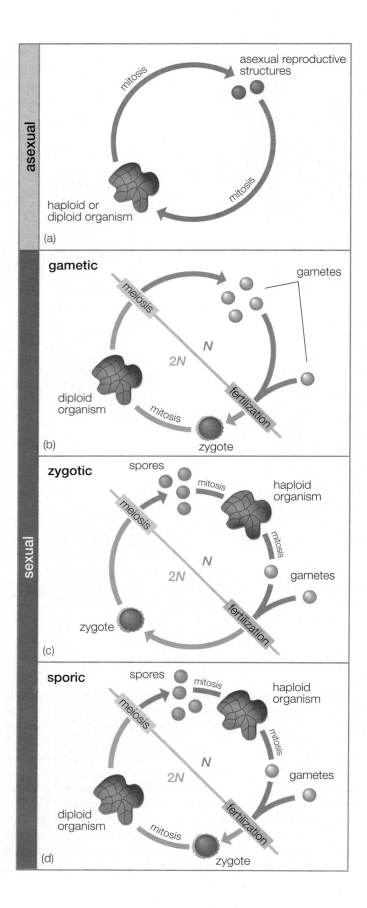

(a)

gametic

(b)

zygotic

(c)

sporic

(d)

Sporic life cycles are characteristic of land plants and some algae

All plants, from simple bryophytes to complex flowering plants, as well as some algae (including the ecologically and economically important brown algae known as kelps) have sporic life cycles—also known as **alternation of generations** (Figure 13.14d). One of the most important points to understand is that plant spores are the products of meiosis. Plant gametes are produced as the result of mitosis, rather than meiosis. This contrast is a major difference between plant and animal reproduction. Plants also differ from animals in that the plant life cycle involves two types of multicellular bodies—a diploid, spore-producing sporophyte and a haploid, gamete-producing gametophyte. The gametophyte generates eggs and sperm by mitosis (not meiosis), and the zygote that results from fertilization grows into the plant sporophyte by repeated mitotic divisions. When sporophytes are mature, they produce spores by meiosis. You may have observed spores as brown dust produced by tiny **sporangia,** which are arranged in visible clusters known as *sori* on the backs of fern leaves (Figure 13.15). Fern spores grow into thumbnail-sized gametophytes that produce gametes. Mating generates zygotes that grow into adult fern plants.

Ferns are good plants in which to study the plant life cycle because both the large sporophyte and thumbnail-sized gametophyte are visible to the unaided eye (Figure 13.16). Plant groups differ in the relative sizes of their sporophytes and gametophytes. Sporophytes are the larger, conspicuous stage of ferns, gymnosperms, and flowering plants; in contrast, the gametophytes of these plants are small and inconspicuous (Figure 13.17). Seed plants produce pollen grains—spores that contain tiny gametophytes, which produce male gametes. Given the genetic advantages of diploidy noted earlier, it is not surprising that the relative sizes of plant gametophytes and sporophytes have changed through evolutionary time (Figure 13.18). Bryophytes, in which the diploid sporophyte stage is small and dependent upon the gametophyte, are probably much like the earliest land-adapted plants. The later-appearing seed plants reveal the benefits of reducing the size of ecologically vulnerable haploid gametophyte stages. Small gametophytes can be enclosed by tough spore walls (in the case of pollen) or within female tissues of cones or flowers and thus protected from exposure to environmental stresses. In seed plants, sporophytes are the dominant life-cycle stage.

FIGURE 13.14   A comparison of life cycles (a) Asexual cycle, with no change in chromosome numbers. There are three types of sexual life cycles: (b) gametic, (c) zygotic, and (d) sporic. In gametic life cycles, gametes represent the only haploid stage. Organisms with zygotic life cycles are haploid, the only diploid stage being the zygote. In sporic life cycles (also known as alternation of generations), there are two types of individuals—haploid gametophytes and diploid sporophytes. Meiosis in this case leads to the production of spores, not gametes.

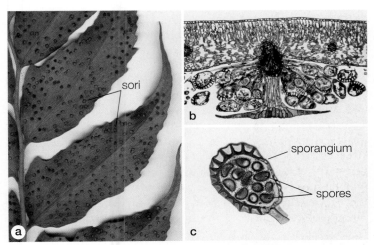

FIGURE 13.15   Fern sporangia occur in clusters known as sori on leaf undersides (a) The underside of a leaf (frond) of the fern *Cyrtomium* showing a number of sori (brown dots). (b) A section through an individual sorus, which contains numerous sporangia. (c) An individual sporangium, containing many spores, which are dispersed in air when the sporangium opens.

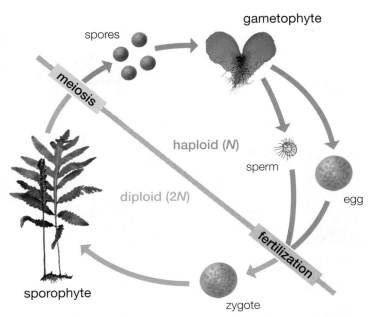

FIGURE 13.16   Life cycle of a fern In ferns, the typical plant life cycle with alternation of a diploid sporophyte with a haploid gameto- phyte is particularly obvious because both phases can be easily seen without the aid of a microscope. Meiosis yields spores, which grow into gametophytes. Gamete-bearing structures are located on the gametophytes. Sperm are released into water droplets, but eggs are retained on the gametophyte. Union of a sperm cell and egg yields a zygote, which develops into a new sporophyte. Fern fronds are the leaves of the sporophytes; in most ferns they are attached to a hori- zontal stem (rhizome). In the example shown here (the sensitive fern *Onoclea*), the brownish frond on the left is the fertile one; it contains the reproductive cells that will undergo meiosis to form spores. The other (vegetative) frond is strictly a photosynthetic organ.

The plant life cycle is difficult for many people to compre- hend because it is quite different from the human life cycle. Yet understanding the plant life cycle is important because we are ultimately dependent on plants. Devising agricultural improve- ments in plant reproduction requires knowledge of the plant life cycle. Knowledge of plant reproduction helps us under- stand why seed plants need to produce pollen, even though it may make us sneeze.

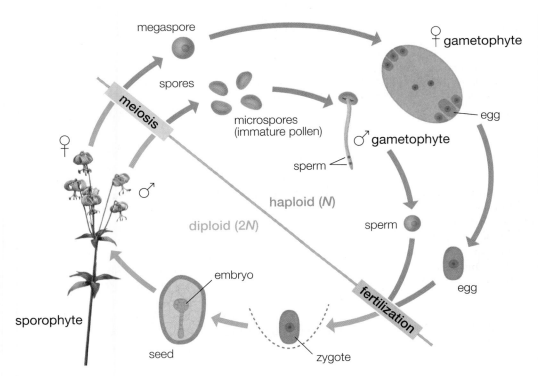

FIGURE 13.17   Seed plant life cycle In seed plants, including the flowering plants, there is an alternation of a large sporophyte generation with microscopic gametophytes (male and female). Meiosis gives rise to spores—either "megaspores" or "microspores." Microspores produce mature pollen grains after one or more mitotic divisions within the spore wall. After being transported to the female portion of a flower, pollen grains germi- nate to form a pollen tube, which trans- ports sperm nuclei to the egg. The germinated pollen grain with pollen tube is the male gametophyte. Although four megaspores are produced by meiosis, only one survives. It develops, via mitotic divisions, into a few-celled female game- tophyte, which contains the egg cell. Fertilization yields a diploid zygote, which divides by mitosis to form a small embryo encased in the seed. When the seed germinates, the diploid embryo grows into an adult sporophyte, completing the cycle. More detailed descriptions and illustrations of seed plant life cycles are found in Chapters 22 and 23.

sporophyte

bryophytes   ferns   gymnosperms   angiosperms

gametophyte

bryophytes   ferns   gymnosperms   angiosperms

FIGURE 13.18 Change in the relative size of plant sporophytes and gametophytes over evolutionary time In the green algal ancestors of plants, the only diploid stage in the life cycle was the zygote. In bryophytes, which represent the most ancient lineage of plants still living today, the diploid stage consists of a small sporophyte that is dependent upon the larger gametophyte. As other plant groups evolved, the trend has been toward reduction in the size of the gametophyte relative to that of the sporophyte. In ferns, the sporophyte is considerably larger than the fingernail-sized gametophyte. In seed plants (gymnosperms and angiosperms), the gametophytes—both male and female—are microscopic in size, whereas the sporophytes may be gigantic in some cases.

HIGHLIGHTS

13.1 Sexual reproduction (involving gametes and zygotes) and asexual reproduction (not involving gametes and zygotes) are both common in nature. Sexual reproduction provides genetic variation—the raw material for organism evolution—and thus accelerates adaptation. Asexual reproduction allows organisms to reproduce very rapidly and in harsh environments that do not favor sexual reproduction. Many organisms that reproduce only asexually evolved from ancestors that reproduced sexually. Asexual reproduction explains the weedy nature of some plants and algae and the widespread occurrence of some medically important molds. Many plants and other organisms reproduce both sexually and asexually.

13.2 Meiosis is a cell division process that reduces the number of chromosomes in cells by one-half. Meiosis is essential to

sexual reproduction because it prevents the buildup of chromosomes that would otherwise occur after many generations of sexual reproduction. Meiosis also contributes to genetic variation in populations of organisms by shuffling the chromosomes donated by parents. Meiosis additionally provides the opportunity for chromosomes to exchange segments, a process known as crossing over.

13.3 Meiosis shares some features with mitosis. This and other evidence suggests that meiosis arose multiple times from mitosis during the evolutionary origin of sexual reproduction in different groups of eukaryotes. Other features of meiosis are unique. For example, meiosis occurs in two stages, meiosis I and II. During meiosis I, homologous chromosomes pair, then separate, a process that does not occur in mitosis. Crossing over and chromosome shuffling also occur during meiosis I, but not mitosis. Nondisjunction (failure of homologous pairs to separate during meiosis I or II) can cause genetic disease in humans. Chromatids formed at DNA replication are separated during meiosis II.

13.4 Life cycles are the series of reproductive processes that link generations. There are three major types of life cycles, which differ in the point at which meiosis occurs and the types of reproductive cells generated by meiosis. The gametic life cycle, in which gametes arise by meiosis, characterizes

animals and some algae. A zygotic life cycle is typical of many protists; meiosis occurs during zygote germination. Sporic life cycles, in which sexual spores are generated by meiosis, occur in land-adapted plants and some algae, notably giant kelps. Plants and algae with sporic life cycles have two alternating multicellular bodies, the sporophyte (which produces sexual spores) and the gametophyte (which produces gametes). Thus, the plant life cycle is also known as alternation of generations. Understanding life cycles is important in agriculture and ecological studies.

REVIEW QUESTIONS

1. List several of the advantages and disadvantages that sexual reproduction confers on a species, and then do the same for asexual reproduction.

2. Why does meiosis always occur in the life cycle of any sexually reproducing organism; that is, why do sexual reproduction and meiosis always go together? What would happen if one occurred without the other?

3. Compare mitosis and meiosis, listing several ways in which they are similar and several ways in which they are different.

4. Briefly summarize (a) gametic, (b) zygotic, and (c) sporic life cycles, and list examples of organisms exhibiting each life-cycle type.

5. Define the following terms: (a) *sexual reproduction*, (b) *gametes*, (c) *sperm*, (d) *eggs*, and (e) *zygotes*.

6. One hypothesis states that sexual reproduction probably had multiple and independent origins among the protists of long ago. If this assertion is true, what is the consequence to plants and animals today, and what line of evidence supports it?

7. What do the terms *haploid* and *diploid* mean? Is diploidy the highest ploidy level known? How many chromosomes does a haploid organism possess?

8. What are homologous chromosomes? How many homologues are present in (a) a haploid (1N) cell? (b) A diploid (2N) cell? (c) A tetraploid (4N) cell?

9. What is crossing over, when does it occur, and what is its importance?

10. What is nondisjunction? Give an example of an abnormal condition in humans caused by nondisjunction.

APPLYING CONCEPTS

1. Do prokaryotes have homologous chromosomes? Discuss why or why not.

2. What is alternation of generations? *Obelia* is an invertebrate animal related to jellyfish. It has two diploid stages in its life cycle, one an attached stage called a *polyp* and the other a free-floating, jellyfish-like stage known as a *medusa*. Polyps give rise to medusans through asexual reproduction. Medusans, in turn, produce gametes (via meiosis) that fuse and develop into a new generation of polyps. What type of life cycle does *Obelia* show? Would you say that it has true alternation of generations?

3. In red algae, male gametophytes produce nonflagellated male gametes, and female gametophytes produce nonmotile female gametes. Male gametes must be passively carried by water currents to the female for fertilization to occur. Clearly, these male gametes are quite different from the flagellated sperm we see in so many other groups of organisms. If both gametes lack flagella and hence are nonmotile, on what basis are they recognized as male or female gametes?

4. Imagine that a single mold spore falls upon the pristine surface of nutrient agar in a petri dish. The spore germinates and begins to grow on the agar. As it grows, it encounters two types of growing conditions. Initially it finds conditions optimal for growth—abundant nutrients, water, and air as

well as plenty of agar real estate to spread out over. The mold reproduces continually and rapidly colonizes the entire dish. Soon, the good times are over. The dish becomes overcrowded, and nutrients, water, and fresh air are severely limited. At this point, the molds change their reproductive strategies to a second type of reproduction. Explain when and why it would be advantageous for the mold to reproduce asexually, and when sexual reproduction would be a better strategy. When would you expect to see the most evidence of sexual reproduction?

5. The sporophyte is the dominant plant body in both ferns and flowering plants, although both do produce gametophytes. Fern gametophytes are about the size of thumbnails and are nutritionally independent individuals. They can be found growing on the ground near the sporophytes. Fern gametophytes help nourish the very young sporophytes, after which time they disintegrate. The gametophytes of flowering plants, however, are much less conspicuous than those of the ferns. Where would you look for these small gametophytes?

6. In plants, although triploid individuals are often healthy, vigorous growers, they are also sterile. Suggest an explanation for this sterility in terms of the complications to meiosis that a triploid individual would experience.

14

Genetics and the Laws of Inheritance

Sweet pea (*Lathyrus odoratus*)

- Gregor Mendel, an Austrian monk who experimented with garden peas, worked out the basic patterns of inheritance of genetic traits in the 1860s. Mendel's work was rediscovered in the 1900s and expressed as laws of inheritance.

- Because plants can be grown and crossed easily and large numbers of offspring can be obtained, many fundamental gene interactions were discovered in plants, such as dominance and recessiveness.

- Genetic discoveries made in plants elucidated the inheritance of genetic traits in humans.

K nowledge of genetics and the laws of inheritance has had a profound impact on agriculture and human health. Early in the 1920s, plant breeders began to apply this knowledge to improve the quality and yield of many crop plants. Plant breeders developed high-yield varieties that required mechanization, fertilizers, pesticides and, in some cases, irrigation to achieve maximum productivity. Their efforts produced a modern, efficient agricultural system in the United States. In the 1940s, plant breeders began to apply these same genetic techniques to improve the agricultural systems in developing nations. A number of international agricultural research centers were established to conduct this research. By the late 1960s, these efforts had produced a "Green Revolution" in terms of rising crop yields in developing nations. The protein quality of such widely grown tropical grain crops as corn and sorghum was improved. Plant breeders also developed new high-yield dwarf varieties of wheat and rice that thrived in tropical countries. The wheat harvest in Mexico doubled over a period of 20 years as a result of this research. The research centers are now working to improve tropical crops such as cassava and sweet potato and to develop new disease-resistant varieties.

Human populations contain genes that can cause serious illness. When these genes occur frequently enough that some offspring suffer, the disease can be recognized as a genetic disorder. Many such genetic disorders are now known and their inheritance understood. In many cases, blood or enzyme tests can detect them before the disease symptoms develop. Modern DNA technology holds out the promise of eventual gene therapy to cure these disorders. Just 100 years ago, however, none of this information was known—not even the basic mechanisms by which a trait was passed from parent to offspring. How was this knowledge of inheritance discovered, with all that it implies for human health and the breeding of better crops in agriculture? It all began with an Austrian monk conducting experiments in a monastery garden with common pea plants. ■

14.1 | Gregor Mendel's experiments with garden peas revealed the pattern of inheritance of genetic traits

Before he became a monk in the monastery of St. Thomas in Brünn, Austria (now Brno in the Czech Republic), Gregor Mendel studied botany and mathematics at the University of Vienna. This training was crucial to the experiments on peas he later performed in the monastery garden. He described his results in the 1860s, but his work was ignored or misunderstood until it was rediscovered in the 1900s. To appreciate the originality of Mendel's work it is necessary to describe briefly some of the ideas about heredity that prevailed before Mendel's time.

FIGURE 14.1   White and purple pea flowers in garden pea, *Pisum sativum*

Early hypotheses assumed hereditary material blended in the offspring

One of the principal early concepts of heredity was that traits were transmitted directly from parents to their offspring. Each body part was assumed to transmit information from each parent into the developing embryo. As recently as 1868, Charles Darwin, who was unaware of Mendel's research at that time, suggested that cells excreted microscopic particles called *gemmules* that passed into the embryo and directed its development. Since both parents contributed to this direct transmission, the offspring would represent a blending of their characteristics. Parents with red hair and blond hair should produce children with reddish-blond hair. A pea plant with yellow peas and one with green peas should yield offspring with yellow-green seeds.

These ideas of blending inheritance created a problem. If the traits within each species blended at each mating, then all the members of any species ought to look alike after only a few generations. This is clearly not the case, since the individuals of most species differ considerably from each other and continue to differ from generation to generation. The answer to this problem had actually been determined long before the 1860s. In 1760, the German botanist Josef Koelreuter successfully crossed different varieties of tobacco to produce fertile **hybrids.** The hybrids were different from either parent strain. When the hybrids were crossed with each other, some offspring resembled the hybrids and some the original parental varieties. These experiments clearly contradicted the idea of blending inheritance. If traits blend, how could a trait that disappeared in the hybrid generation reappear in the next generation?

Others followed up on the work of Koelreuter. In the 1790s, the English farmer T. A. Knight crossed two true-breeding varieties of garden peas, one with purple flowers and one with white flowers (Figure 14.1). True-breeding varieties remain the same from generation to generation. A cross between two varieties that differ in only one trait is called a **monohybrid cross.** The progeny of the cross all had purple flowers. When this hybrid generation of purple-flowering peas were crossed, many of the offspring had purple flowers but some had white flowers. The

trait for white flowers had been latent or masked in the first generation and reemerged in the second generation. In current terminology, a geneticist would say that the traits of the parental varieties were **segregating,** or separating out, in the offspring of the hybrid generation. In these simple results lay the clues to the science of genetics, but the clues were not understood for another hundred years. Botany was then a qualitative, not a quantitative, science, and workers did not count the results of their experiments. With his training in mathematics, Gregor Mendel counted his results and thereby discovered the process of heredity.

Mendel's use of garden peas had many advantages

Gregor Mendel chose the common garden pea *(Pisum sativum)* for his experiments. A considerable body of knowledge already existed about garden peas and crosses among them. There were many varieties of true-breeding peas available in Mendel's time, and Mendel initially examined 32 different ones. Because of work done by previous investigators, he knew that hybrid peas could be produced and that traits would segregate among the offspring of these hybrids. Following the earlier studies of T. A. Knight, Mendel selected varieties that differed in seven easily recognizable traits, such as yellow versus green seed color or purple versus white flowers.

Peas have the additional advantage of being fast and easy to grow. Fast growth means they take a short time to produce many generations of seeds because more than one generation can be grown in a single year. Therefore, data can be gathered in a shorter time period than with a plant having only one generation per year.

Perhaps the greatest advantage of peas as experimental plants is that their reproduction can be controlled. Peas, like many flowers, have on the same flower both "male" sex organs, the stamens, and "female" sex organs, the carpels (Figure 14.2a) (the terms *male* and *female* are in quotes because the sporophytic stage of most plants is not assigned a gender—see Chapter 24). Unlike most flowers, however, gametes from a single pea flower can fuse to form healthy offspring. This process is called **self-fertilization** and occurs normally if the flower is not

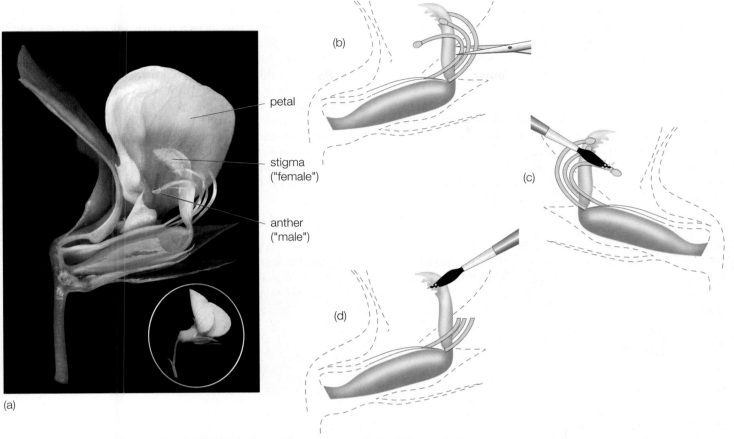

FIGURE 14.2 Cross-pollination of pea flowers (a) Section through a pea flower showing the "male" and "female" reproductive structures. Pollen is produced in the yellow anthers and is deposited on the stigma. A considerable amount of pollen is present on the stigma here. The inset shows an intact flower, where the reproductive structures are nearly completely enclosed by petals. Cross-pollination is carried out by (b) removing the anthers on one flower (to prevent self-pollination) and transferring pollen from a second flower with a different trait (c) to the stigma of the first flower (d).

disturbed by pollination. But it is easy to interrupt the normal process by removing the anthers before fertilization occurs (Figure 14.2b). The experimenter can then transfer pollen from another variety of pea that has different traits to the stigma (the receptive part of the carpel) of the flower whose anthers were removed (Figure 14.2c and d). This process (cross-pollination) results in **cross-fertilization**. Self-fertilization allows the experimenter to see what traits are present within a single individual. Cross-fertilization permits the experimenter to create hybrids between varieties with different traits.

Mendel's experimental design used both self-fertilization and cross-fertilization. Initially, he allowed his varieties to self-fertilize for several generations to verify that the traits he was studying were true-breeding. Peas with yellow seeds always produced yellow seeds, and those with green seeds always produced offspring having green seeds.

Mendel then performed cross-pollinations, resulting in cross-fertilizations between varieties with contrasting traits such as purple versus white flowers or yellow versus green seeds. In these experiments, he removed the anthers from flowers of peas with one trait and pollinated the carpels with pollen

from anthers of a variety with the contrasting trait. He pollinated plants with purple flowers with pollen from plants with white flowers, and he also performed the reverse cross by pollinating plants with white flowers with pollen from those with purple flowers. The offspring of these crosses were hybrids between the two varieties.

Lastly, Mendel allowed the hybrids to self-fertilize for several generations. This phase of his research permitted the alternative forms of the traits he was studying to segregate out into the offspring. In contrast to previous investigators, Mendel counted the numbers of offspring with the alternative traits (i.e., the number with yellow or green seeds, or purple or white flowers, in each generation). It was the relationship between these numbers that revealed the process of heredity.

Mendel's experiments focused on seven distinct traits in peas

Mendel studied seven traits in garden peas that were readily distinguishable from each other. Each trait was present in separate pea varieties in one of two easily recognizable forms, such as yellow versus green seed color and round versus wrinkled seed shape. Mendel's experiments with these seven traits through two generations of monohybrid crosses are summarized in Figure 14.3. We will discuss in detail his work with flower color and seed color, but all of the other traits could be described in the same way.

The F₁ generations revealed dominant and recessive traits

For his parental generations, Mendel used two varieties of true-breeding peas with contrasting forms for a single trait. He then performed a cross-pollination, which resulted in cross-fertilization. The resulting hybrid offspring are referred to as the F_1 generation, where F_1 refers to the **first filial** generation (*filius* is Latin for "son," and *filia* is Latin for "daughter"). In the cross between purple-flowered peas and white-flowered peas, all members of the F_1 generation were purple, not pale lavender as blending inheritance would predict. For each of the seven traits Mendel studied, one form was expressed in the F_1 generation and the other was hidden or latent (see Figures 14.4 and 14.5, later in the chapter). Mendel referred to the trait expressed in the F_1 pea plants as **dominant** and the unexpressed form of the trait as **recessive**. Thus, in the case of flower color, purple was the dominant form and white the recessive. Similarly, yellow seed was the dominant form and green seed the recessive form when considering the trait of seed color.

Trait	Parental varieties		F₁ generation (Dominant trait)	F₂ generation Dominant form	Recessive form	Ratio
Flower color	Purple	White	Purple	705 Purple	224 White	3.15:1
Seed color	Yellow	Green	Yellow	6022 Yellow	2001 Green	3.01:1
Seed shape	Round	Wrinkled	Round	5474 Round	1850 Wrinkled	2.96:1
Pod color	Green	Yellow	Green	428 Green	152 Yellow	2.82:1
Pod shape	Round	Constricted	Round	882 Round	299 Constricted	2.95:1
Plant height	Tall	Dwarf	Tall	787 Tall	277 Dwarf	2.84:1
Flower position	Axial	Top	Axial	651 Axial	207 Top	3.14:1

FIGURE 14.3 **Mendel's experiments with seven individual traits in garden peas**

The F$_2$ generations had dominant and recessive forms for each trait in a 3:1 ratio Mendel allowed the pea plants in his F$_1$ generation to self-pollinate and then planted the seeds to see which form of the traits would appear in the F$_2$ or **second filial generation.** For each trait, some of the F$_2$ individuals exhibited the recessive form of the trait. Mendel went a step further than his predecessors and counted the numbers of each form of each trait in the F$_2$ generations (see Figure 14.3). From the purple-flowered F$_1$ generation, he obtained 929 F$_2$ offspring, which included 705 (75.9%) with purple flowers and 224 (24.1%) with white flowers. Mendel obtained the same result with each of the other traits he examined. In the F$_2$ generations, about 3/4 of the offspring had the dominant form of the trait and 1/4 had the recessive form. Another way to say the same thing is that the ratio of the dominant form to the recessive form was 3:1. Note that in the studies with seed color and seed shape, where Mendel counted the greatest numbers, the ratios are the closest to the expected ratio of 3:1 for all the traits examined. With larger numbers of observations, chance events cancel out, and the observed ratio approaches the expected ratio more closely.

Mendel's model of the pattern of inheritance

From his experiments on these seven traits of garden peas, Mendel was able to construct a simple model of the process of inheritance. This model has five main points.

1. Parents do not transmit traits directly to their offspring but, rather, transmit information about those traits, which Mendel called *factors*. These factors (today called *genes*) act in the offspring to produce a form of the trait. Mendel assigned a simple set of symbols to these factors. In the case of flower color, the symbol *P* represented the purple flower factor. *P* is written in uppercase because purple is the dominant factor. The symbol for white flower, the recessive factor, was written as *p*, the same symbol written in lowercase. Similarly, the factor for yellow seed was written as *Y*, and that for green seed as *y*.

2. Each individual receives two factors, one from each parent, that may code for the same or different forms of the trait.

3. Not all factors for a particular trait are the same. Thus *Y* and *y* are factors (now called *alleles*) for the trait of seed color, but they are not the same. They can produce different forms of the trait, such as yellow or green seeds.

4. When two different factors for the same trait are present in the same individual, the factors neither blend nor alter each other in any way but remain discrete. Mendel described the factors as "uncontaminated."

5. The presence of a factor does not mean that the form of the trait coded by that factor will be expressed. Only the dominant factor is expressed in individuals with two different factors present. The recessive factor, although present, is latent and unexpressed.

Mendel used this model and his symbols to see if it would predict the results of his experiments. In the experiment with purple- and white-flowered peas, *PP* represents true-breeding purple-flowered plants and *pp* stands for true-breeding white-flowered plants (Figure 14.4a). A cross between two varieties is

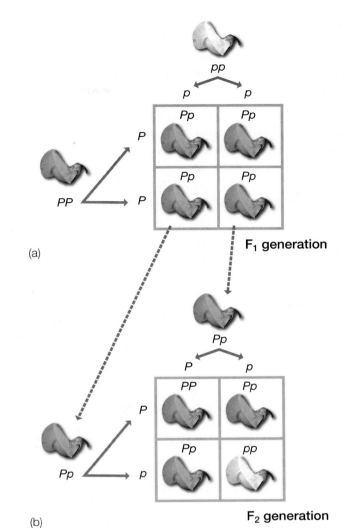

(a)

F$_1$ generation

(b)

F$_2$ generation

FIGURE 14.4 Mendel's experiment with purple and white flowered peas (a) True-breeding purple and white peas are crossed, yielding an F$_1$ generation with all purple flowers. (b) When plants from the F$_1$ generation are crossed, the resulting F$_2$ generation has purple and white-flowered plants in a 3:1 ratio. Mendel's "factors" are represented by letters (*P* or *p*). The phenotypes are represented by the photographs.

identified by the symbol × and Mendel's experimental cross as *PP* × *pp*. Purple-flowered peas can contribute only the dominant factor *P* for purple flowers. White-flowered peas can contribute only the factor *p* for white flowers. All members of the F$_1$ generation must therefore be *Pp* and purple. The combinations of factors (*Pp*) in the F$_1$ generation are arranged in a box diagram called a **Punnett square** after its originator, British geneticist Reginald Punnett. The factors contributed by the white-flowered peas (*pp*) are shown across the top of the Punnett square (see Figure 14.4), and those contributed by the purple-flowered peas (*PP*) are arranged along the left side.

If the F$_1$ generation is crossed either by self-fertilization or cross-fertilization, each F$_1$ individual can contribute either a *P* or a *p* factor to the F$_2$ generation. The factors for alternate forms of the trait segregate from each other in individuals that possess both factors. This is **Mendel's First Law of Heredity,** the **Law of Segregation.** The Punnett square shows that three types of F$_2$ offspring are possible: *PP* and *Pp* have the purple-flower trait, and *pp* has the white-flower trait (Figure 14.4b). The ratio of dominant to recessive traits is 3:1 (purple:white). The ratio of factor combinations is 1 *PP*:2 *Pp*:1 *pp*. Of the purple-flowered individuals in the F_2 generation, the model predicts that 1/3 will be true-breeding purple and 2/3 will not be true-breeding for purple but will reproduce the 3:1 ratio of purple to white flowers in later generations when self-fertilized. This is exactly what Mendel observed among F_2 purple-flowered plants. The same analysis works for each of the other seven traits, such as yellow versus green seeds (Figure 14.5).

The testcross revealed the true nature of dominant traits

As a further test of his model of inheritance, Mendel devised a special type of cross-fertilization procedure he called a **testcross.** Assume that you have a purple-flowered pea plant. How can you determine if the two factors in it are *PP* or *Pp*? In the case of a plant derived from a yellow seed, how do you determine if it is *YY* or *Yy*? Mendel realized that the best way to answer such questions was to cross the plant having the dominant trait with one containing the recessive trait (Figure 14.6). In the first example, a purple-flowered pea plant is crossed with a white-flowered pea plant, which has the factors *pp*. If the purple-flowered pea plant is

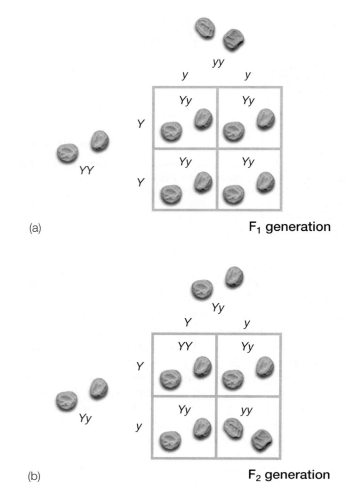

(a) F₁ generation

(b) F₂ generation

FIGURE 14.5 Mendel's experiment with yellow and green seed color (a) True-breeding plants with green and yellow seeds are crossed to yield an F₁ generation, all having yellow seeds. (b) The F₂ generation consists of 3 yellow:1 green-seeded plants.

PP, all the offspring of the cross will be *Pp* and purple-flowered (Figure 14.6a). But if the purple-flowered plant is *Pp*, half of the offspring will be purple-flowered and half will have white flowers. In the second example, if the pea plant grown from a yellow seed is *YY*, all of the offspring from a cross with plants having green seeds will be *Yy* and yellow (Figure 14.6b). Conversely, if the yellow seed was *Yy*, half of the offspring in a cross with plants having the recessive green trait will be green also. Thus his model predicts that if a plant with the dominant trait has one of each factor for that trait, the dominant and recessive traits would appear in a testcross in the ratio of 1:1; that ratio is just what he observed in the case of each of his seven traits.

14.2 | Mendel's model in terms of genes, alleles, and chromosomes

One remarkable aspect of Mendel's work is that he arrived at the correct mechanism of the inheritance of traits without any knowledge of the underlying cellular mechanisms that were responsible for the patterns he observed. In Mendel's time, chromosomes had not been described nor meiosis discovered (Chapter 13). Certainly there was no knowledge of genes, alleles, or DNA. Yet if one simply substitutes the word *allele* for Mendel's term *factor*, the previous descriptions of his studies become essentially modern. Mendel's factors for a trait, such as flower color, were the alleles of the gene for that trait.

The symbols Mendel used to represent his factors are the ones used today by geneticists to represent genes. Thus *P* is the symbol for the dominant gene for purple flower color in the garden pea, and *p* is the symbol for the recessive gene for white flower color. *P* and *p* are **alleles,** different forms of the gene for flower color. In molecular terms, a gene is a length of nucleotides within a much larger DNA molecule, and the alleles of a gene are slightly different versions of that nucleotide sequence, which can produce different versions of a particular trait.

Mendel hypothesized that every pea plant had two factors, but he would not have been able to say where they were located in a plant cell. We now know that his factors (alleles) are located on chromosomes in the nuclei of the plant cells. Pea plants are **diploid;** that is, they have two genes for each trait, and the nucleus of each cell has two copies of each type of chromosome, one from each parent (Figure 14.7). Pairs of chromosomes that have genes involved with the same traits are called **homologous chromosomes.** A gene coding for a particular trait is located at the same site or **locus** on each of the homologous chromosomes. Considering the gene for seed color in peas, both members of the pair of homologous chromosomes might have the same allele of the gene and be *YY* or *yy*. In this case, the plant would be **homozygous** for the gene for that trait. If the members of the homologous pair had different alleles for seed color (*Yy*), the individual plant would be **heterozygous** for seed color. Since one allele is often dominant over another allele, there is a difference between the observable traits of an organism and the underlying genetic makeup of that organism. In the case of pea seed color, a plant with yellow seeds may be either *YY* or *Yy*. The observable

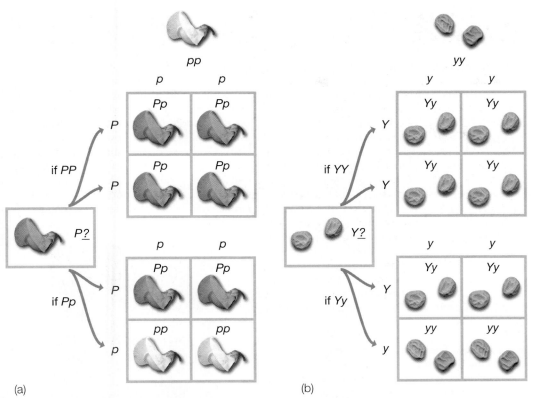

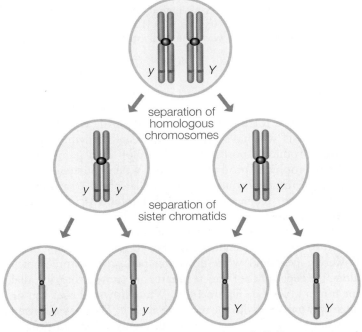

(a)   (b)

FIGURE 14.6   Testcross for purple flower color and yellow seed color A testcross is a cross where a plant with a dominant phenotype whose genotype is unknown is mated with one having the recessive phenotype. If the unknown genotype is homozygous dominant, all offspring plants will have the dominant trait. If heterozygous, the resulting progeny will consist of both dominant and recessive individuals in a 1:1 ratio. (a) A testcross with flower color and (b) one with seed color.

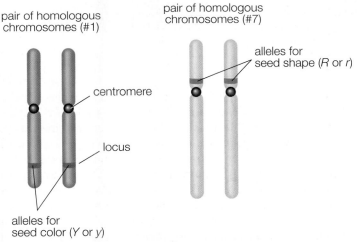

FIGURE 14.7   Two homologous pairs of pea chromosomes One pair of chromosomes (number 1) has the alleles for seed color (Y and y), and the other pair (number 7) has the alleles for seed shape (R and r).

trait is the **phenotype,** and the underlying genetic makeup is the **genotype.** In recessive traits such as green seeds or white flowers, the phenotype reflects the actual genotype.

Mendel also knew that his factors segregated independently in F_2 generations so that recessive traits reappeared. He did not know that the underlying mechanism of this segregation of factors was the process of meiosis (Figure 14.8). Pairs of

FIGURE 14.8   Meiosis and the segregation of alleles Meiosis is the mechanism whereby alleles of a particular gene (for seed color, in this case) are segregated. The homologous chromosomes, each bearing a different allele, are separated (segregated) during the first meiotic division. Each spore then receives one chromosome with one copy of the allele, and resulting gametophytes and gametes will have the same allele.

ESSAY 14.1   PSEUDOSCIENCE AND THE LYSENKO AFFAIR

Astrology, water dowsing, UFOs, psychic phenomena, and medical quackery are all contemporary examples of pseudosciences. They pretend to be scientific, but they do not use the scientific process. Pseudosciences begin with the assumption that their main hypothesis is absolutely true. Their science then consists of gathering material—usually personal testimonies—that supports their hypothesis. But when pseudosciences are put to rigorous scientific testing, their predictions prove to be no better than chance.

Pseudosciences are prevalent in part because the media find them profitable. It is easier to sell alien abductions than real scientific discoveries. Pseudosciences are widely accepted because the public in general is not good at separating the real from the unreal. Some pseudosciences are harmless, but if a government adopts one as national policy, it can be dangerous. Consider the Lysenko Affair in the former Soviet Union.

Trofim Lysenko was an agronomist of minimal scientific training but with real talent for character assassination and political demagoguery (Figure E14.1). He believed that heredity was a general internal property of living matter and that genes and chromosomes played no role in it. He claimed that plants could be altered in predetermined directions by the environment. Thus, cold environments did not lead to selection for cold-resistant plants, but to a direct change of heredity toward greater cold resistance. Lysenko and his supporters never offered any scientific proof of these assertions, but promoted them through political intimidation.

Pseudosciences are widely accepted because the public in general is not good at separating the real from the unreal.

Because Lysenko had strong political support from the Communist Party under Joseph Stalin and Nikita Khrushchev, his ideas became national policy from the 1930s to the 1960s. The Communist Party supported him because his ideas rejected Western science. If heredity could be changed by environment, then harsh environments should produce hardier, more productive plants.

Lysenko was able to brand serious Soviet geneticists as fascists, reactionaries, and enemies of the people. Ironically, geneticists were accused of practicing a pseudoscience! Genetics laboratories were abolished, all genetic literature was removed from libraries, courses were abolished, and faculty and scientists were dismissed. Many were imprisoned and some died while incarcerated. All genetic research in crop plants, animal husbandry, and human genetic disease ceased. Lysenko saddled Russia with an inefficient agricultural system for more than 30 years. He was removed only after Khrushchev was forced to resign in 1964—in part because of mounting failures in agriculture. Politicians can support a pseudoscientific agenda, but they cannot repeal the laws of nature.

E14.1   Trofim Lysenko

---

homologous chromosomes are separated during meiosis. If each member of a pair of homologous chromosomes contains a different allele for a particular trait, then the cells that result from meiosis will have different alleles. Recall that in plants meiosis produces spores, spores develop into gametophytes, and gametophytes produce gametes. When gametes fuse with other gametes, new combinations of alleles will arise and recessive traits can reemerge. Meiosis (Chapter 13) is the mechanism of Mendel's Law of Segregation, although meiosis was unknown at the time of Mendel's work.

If we return to Mendel's cross between true-breeding purple-flowering peas and white-flowering peas, the factors depicted along the edges of the Punnett squares are gametes bearing alleles for flower color (see Figure 14.4). White-flowered peas can produce only gametes with the recessive allele p. The purple-flowered peas produce gametes with the P allele only. All F_1 offspring are therefore heterozygous (Pp) for flower color and phenotypically purple. When the F_1 generation is crossed or self-fertilized, each purple-flowered plant can make gametes bearing either the P allele or the p allele. When these gametes fuse, they can produce three genotypes: PP, Pp, and pp. The ratio of phenotypes in the resulting F_2 generation will be 3 purple to every 1 white, which is very close to what Mendel actually observed (see Figure 14.3).

Gregor Mendel published his research on hybrids in 1866, but the significance of his results was not accepted, probably because they contradicted prevailing ideas. In 1900, three biologists—Carl Correns, Hugo de Vries, and Erich Tschermak—rediscovered the mechanisms of inheritance and Mendel's original paper. Hugo de Vries actually named the discovery "Mendel's Law of Segregation."

The science of genetics has made great advances over the decades since the 1900s. At the beginning of the new millennium, genetics is developing therapies for human genetic disorders, and genetic engineering is being applied to crop plants and livestock. But in the period from 1929 to 1965 in the then Soviet Union, it was possible for individuals to lose their jobs and be thrown in prison for believing in genetics! For more about this antigenetics movement, see Essay 14.1, "Pseudoscience and the Lysenko Affair." The remainder of this chapter addresses variations on Mendelian genetics discovered in the years following the 1900s and the relationships between genes and chromosomes.

14.3 | Variations on Mendelian genetics

In his experiments with peas, Mendel had selected a relatively simple genetic system. Each of the seven traits he examined was governed by a single gene with two alleles, one of which was dominant and the other recessive. Heterozygotes therefore showed the dominant trait. But this is not always the case.

In incomplete dominance, the heterozygote has an intermediate phenotype

In snapdragons a cross between homozygous red-flowered plants (RR) and homozygous white-flowered plants ($R'R'$) does not produce an F_1 generation of red-flowered hybrids (Figure 14.9). Instead, the F_1 generation consists of pink-flowered plants that are all (RR'). The alleles for flower color are all shown as capital letters because they are neither dominant nor recessive. This is not an example of blending inheritance. In the F_2 generation, the R and R' genes segregate and recombine to produce red-flowered plants, pink-flowered plants, and white-flowered plants in the ratio of 1 red:2 pink:1 white. If blending inheritance were occurring, all F_2 plants would have pink flowers.

Snapdragon color is a trait that exhibits **incomplete dominance**. The R allele codes for an enzyme involved in the formation of red pigment, the R' allele makes a defective enzyme, and the RR' genotype can produce only enough red pigment to make a pink flower. Two R genes are needed to produce a red flower, and two R' genes result in a white flower. In incomplete dominance, the phenotype is a direct reflection of the genotype, since each genotype produces a characteristic flower color. A similar process of incomplete dominance governs red and white flowers in a number of other plant species.

In pleiotropy, a single gene affects several traits

Many genes control only a single trait, but some single genes can affect multiple traits and thus profoundly alter the phenotype of the organism. This phenomenon is called *pleiotropy* (Greek *pleios*, meaning "more," and *trope*, meaning "turnings"). During his breeding experiments with peas, Mendel actually found an example of pleiotropy. In crosses between peas having purple flowers, brown seeds, and a dark spot on the axils of the leaves with a variety of peas having white flowers, light-colored seeds, and no spot on the axils of the leaves, the three traits for flower color, seed color, and leaf axil spot all were inherited as a single unit. The reason for this pattern of inheritance was that the three traits were controlled by a single gene with dominant and recessive alleles.

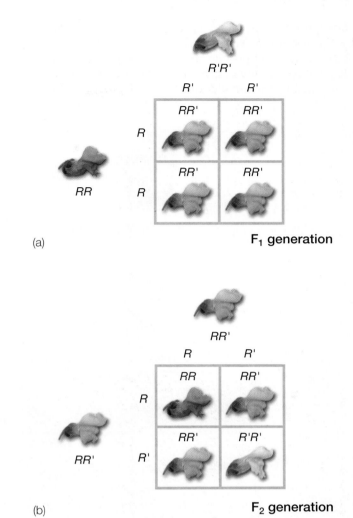

(a)   **F_1 generation**

(b)   **F_2 generation**

FIGURE 14.9 Incomplete dominance in snapdragons (a) If a snapdragon with red flowers (RR) is crossed with one with white flowers ($R'R'$), the resulting F_1 plants will have flowers with an intermediate, pink color (RR'). (b) When F_1 plants are crossed, the F_2 generation has plants in the ratio 1 red:2 pink:1 white.

In polygenic inheritance, several genes combine to affect a single trait

Many traits are the result of the sum of several genes, which are known as **polygenes**. Each individual gene in a group of polygenes makes a contribution to the phenotype of the organism. Traits controlled by polygenes show a wide range of variability. Plant fruit size and height are examples of traits controlled by polygenes, as is height in humans. The more genes that are involved, the more the range of variability approaches a continuous pattern, in which phenotypes can no longer be separated from each other.

The first experiments that documented polygenic inheritance were conducted by Swedish scientist H. Nilson-Ehle on color in wheat kernels. Kernel color is controlled by two genes, each with two alleles, with red kernel color dominant to white. Nilson-Ehle crossed two true-breeding varieties of wheat, one dark red with genotype $R_1R_1R_2R_2$ and the other white with genotype $r_1r_1r_2r_2$ (Figure 14.10a). All members of the F_1 generation were $R_1r_1R_2r_2$ and medium red. Every F_1 wheat plant could produce four types of gametes: R_1R_2, R_1r_2, r_1R_2, and r_1r_2. In the F_2 generation, the intensity of red color is determined by the number of R genes (Figure 14.10b). Four R

genes produce a dark-red kernel color. Three R genes make a medium-dark red, and two R genes a medium-red kernel. One R gene is light red, and the absence of any R gene is white. The R genes act in an additive manner to produce the red kernel color. If the numbers of each phenotype are plotted against the intensity of red kernel color, the distribution of phenotypes resembles a bell-shaped curve (Figure 14.10c). If any more genes had been involved in this wheat trait, the phenotypes might not have been distinguishable, and the bell curve would have appeared continuous.

The environment can alter the expression of the phenotype

Genes interact in complex ways to determine the phenotype of an organism, but the environment in which the organism lives can affect the expression of those genes in the phenotype. For example, a plant with the "best" genes cannot realize its full potential in a nutrient-poor soil. The common garden plant *Hydrangea macrophylla* normally produces large globose clusters of blue flowers, but the same plant may produce pink flowers instead (Figure 14.11). The color of the flowers is determined by soil acidity. Acid soils produce a blue floral phenotype, and more basic soils yield a pink phenotype.

Many aquatic plants produce two kinds of leaves—one kind for submerged leaves, another kind for leaves held above water. Submerged leaves are often long and thin or highly dissected, but aerial leaves are broad and flat. Thus the external environment affects the way the genes for leaf shape are expressed in these plants.

In some snapdragons temperature and light level can affect the phenotype of flower color. In these snapdragons when a true-breeding red-flowered plant is crossed to a true-breeding ivory-flowered plant, the phenotype of the F_1 generation (the heterozygotes) is red if the plants grow in bright light at cool temperatures but ivory if the plants grow in shade at warmer temperatures.

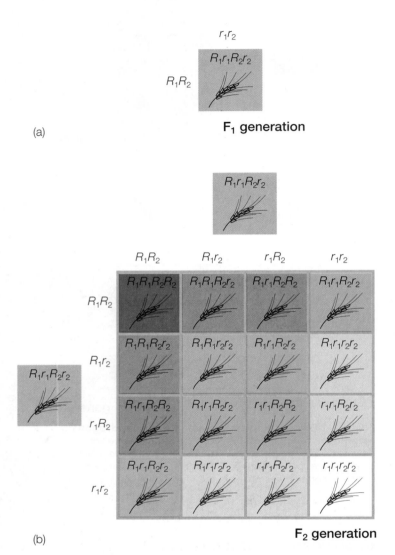

FIGURE 14.10 **Polygenic inheritance in wheat kernel color**
(a) The F_1 generation formed by crossing a plant having dark-red kernels ($R_1R_1R_2R_2$) with one having white kernels ($r_1r_1r_2r_2$) has medium-red kernels. (b) The F_2 generation consists of plants having kernels of five different colors depending on the number of R alleles. (c) The numbers of each phenotype appearing in the Punnett square are shown graphically. The color distribution resembles a bell-shaped curve.

FIGURE 14.11 **Effect of soil acidity on flower color in** *Hydrangea macrophylla* An acid soil yields blue flowers (a) and a neutral-to-basic soil produces pink flowers (b).

14.4 | Genes and chromosomes

Once they learned that the information about traits, or genes, was located on the chromosomes, scientists realized that there were many more genes than there were chromosomes, and therefore each chromosome must bear many genes. Genes located on the same chromosome may be **linked.** Genes linked together on the same chromosome will tend to be inherited together.

In sweet pea, the genes for flower color and pollen shape are on the same chromosome

The English geneticist William Bateson first discovered gene linkage in 1905 while studying the sweet pea *(Lathyrus odoratus).* Bateson crossed sweet peas that were homozygous recessive for red petals, *dd,* and round pollen grains, *ee,* with a second variety that was homozygous for the dominant purple flowers, *DD,* and long pollen grains, *EE.* Bateson used the F₁ generation, which was *DdEe,* to obtain an F₂ generation. If the traits were truly linked together, the F₂ generation should have consisted of plants with purple flowers and long pollen and plants with red flowers and round pollen. The F₂ generation, however, consisted of the plants listed in Table 14.1.

Most of the plants were the same as the original parental plants, but 42 plants were completely new combinations! This result occurs because sometimes genes on the same chromosome do not always stick together. In the first prophase of meiosis, homologous chromosomes come together and may exchange pieces of themselves in a process called **crossing over** (Figure 14.12). Crossing over is a fairly common process; at least one exchange will occur at any meiosis. When crossing over occurred in sweet peas between the genes for petal color and pollen shape, two new types of gametes resulted—those with the alleles *De* and *dE,* in addition to the parental types *DE* and *de.* Crossing over produces **genetic recombination**—the formation of new combinations of alleles by the exchange of segments between homologous chromosomes. By chance, some of the gametes in Bateson's experiment fused to yield plants with red petals and long pollen and purple petals but round pollen (see Table 14.1).

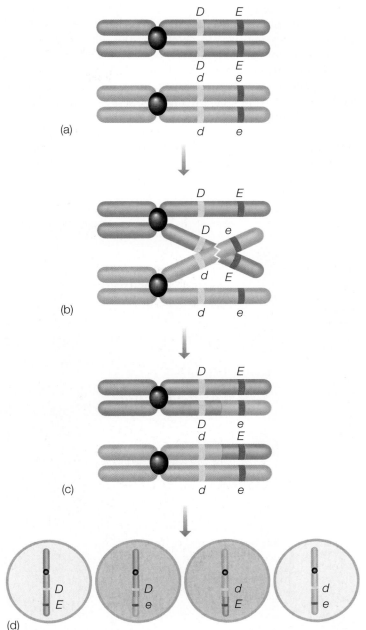

FIGURE 14.12   Crossing over in the sweet pea, *Lathyrus odoratus* (a) Genes controlling petal color (*D* or *d*) and pollen shape (*E* or *e*) are linked on the same chromosome. If no crossing over occurs, the generation resulting from a cross between a dominant with purple petals and long pollen and a recessive with red petals and round pollen would be 3 purple long to 1 red round, just as in a single-trait cross. In (b), a pair of chromosomes with the genes for petal color and pollen shape is shown crossing over. (c) After crossing over, the rearrangement is more apparent. (d) Four types of gametes result: *DE* and *de* (the parental types) and *De* and *dE* (the cross-over types, which are shaded).

Genetic maps show the order and position of genes on chromosomes

Genes that are close together show little crossing over because there is little room between them for the homologous chromosomes to intertwine and exchange segments. The

TABLE 14.1	The Generation in Bateson's Cross		
Number of Plants	**Petal Color**	**Pollen Shape**	**Type**
284	purple	long	parental
55	red	round	parental
21	purple	round	cross-over
21	red	long	cross-over

farther apart genes are on their chromosome, the more frequently crossing over occurs. Scientists therefore used the frequency of crossing over to determine the relative position of genes on chromosomes and to construct the first maps of genes on chromosomes. Modern DNA technology now allows geneticists to construct gene maps using less laborious methods. A partial genetic map of a pea chromosome is shown in Figure 14.13.

The entire genomes—the nucleotide sequences of all the DNA of an organism—have been determined for a number of organisms, including bacteria, yeast, the flowering plant *Arabidopsis*, and humans. In plants, gene maps can aid in the development of new varieties of crop plants. Knowledge of the location of genes involved in human genetic disorders may lead to new diagnostic procedures and eventually to gene therapies.

In a dihybrid cross, genes segregate independently if they are on separate chromosomes

Mendel's research with garden peas demonstrated that the alternative alleles for a single trait segregated independently during crosses. When the genes for two different traits are located on different, nonhomologous chromosomes, the genes for those traits will also segregate independently (Figure 14.14). A cross between individuals or plant varieties that differ in two traits is called a **dihybrid cross.**

As a change from our examples taken from pea plants, consider a dihybrid cross in the garden tomato. The genes for fruit color and stem hairiness lie on different pairs of homologous chromosomes. The allele for red fruit, *R*, is dominant to that for orange fruit, *r*. The allele for hairy stem, *H*, is dominant to that for smooth stem, *h*. The original parental varieties are homozygous dominant for red fruit and hairy stems, *RRHH*, and homozygous recessive for orange fruit and smooth stems, *rrhh* (Figure 14.15). The homozygous dominant plants can produce only gametes that are *RH*, and the recessive plants can make only gametes of the genotype *rh*. The F_1 generation consists entirely of plants of genotype *RrHh* with red fruits and hairy stems. Each plant in the F_1 generation can make four different types of male and female gametes—*RH, Rh, rH,* and *rh*—because each parent donates one gene for each trait. Representing all of these gametes requires a Punnett square with four cells on each side, for a total of 16 combinations. The resulting F_2 generation—which receives two genes for each trait, one from each parent in the F_1 generation—is shown in the 16 cells. The ratio of red to orange fruit is 3:1, and the ratio of hairy to smooth stems is 3:1, just as if we had considered each trait separately. The overall ratio of the phenotypes is 9:3:3:1, or 9 red hairy:3 red smooth:3 orange hairy:1 orange smooth. If we raised 1,600 F_2 offspring, we would expect about 900 red hairy, 300 red smooth, 300 orange hairy, and 100 orange smooth individuals.

Mendel performed the same type of dihybrid crosses with his pea plants, using such traits as yellow versus green seeds and round versus wrinkled seeds. These traits assorted themselves independently, and his discovery is called **Mendel's Second Law of Heredity** or the **Law of Independent Assortment.** Geneticist T. H. Morgan named the law for Mendel in 1913. Mendel's second law, however, holds if the traits are located on separate, nonhomologous chromosomes. Mendel knew nothing about chromosomes or linkage between genes on the same chromosome. Garden peas have seven pairs of homologous chromosomes, and not all of the seven traits Mendel studied are located

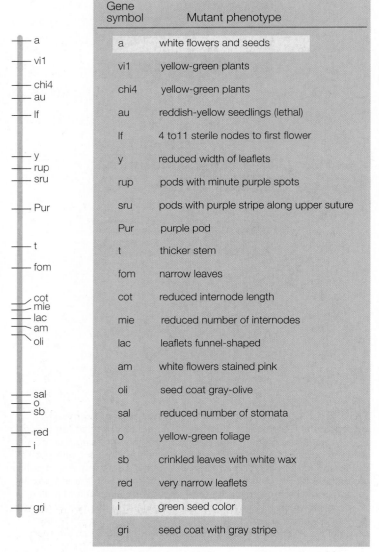

Gene symbol	Mutant phenotype
a	white flowers and seeds
vi1	yellow-green plants
chi4	yellow-green plants
au	reddish-yellow seedlings (lethal)
lf	4 to11 sterile nodes to first flower
y	reduced width of leaflets
rup	pods with minute purple spots
sru	pods with purple stripe along upper suture
Pur	purple pod
t	thicker stem
fom	narrow leaves
cot	reduced internode length
mie	reduced number of internodes
lac	leaflets funnel-shaped
am	white flowers stained pink
oli	seed coat gray-olive
sal	reduced number of stomata
o	yellow-green foliage
sb	crinkled leaves with white wax
red	very narrow leaflets
i	green seed color
gri	seed coat with gray stripe

FIGURE 14.13 Map of genes on chromosome 1 of pea (Pisum sativum) Only a small number of the genes known to occur on this chromosome are listed. The alleles for genes determining flower and seed color that were studied by Mendel (highlighted) are now designated *A* or *a* (flower color) and *I* or *i* (seed color). The mutant phenotypes of the genes are listed. Note that mutants whose symbols are lowercase are recessive and those starting with a capital letter are dominant.

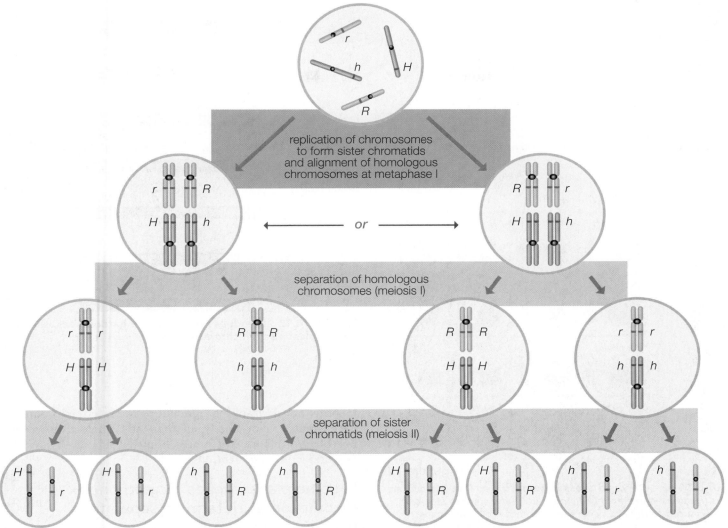

FIGURE 14.14 **Meiosis and the independent assortment of alleles** A tomato plant is heterozygous for two genes: fruit color (*Rr*) and stem hairiness (*Hh*). During meiosis the alleles assort independently because they are located on different chromosomes that separate independently of each other.
Meiosis may follow two different paths (depending on how the homologous chromosomes line up during
metaphase of meiosis I), leading to production of spores, gametophytes, and gametes with the alleles *rH*
and *Rh* (left side) or *RH* and *rh* (right side). The heterozygous tomato plant can thus form four different
kinds of gametes.

on separate pairs (Figure 14.16). Seed color and seed shape
clearly are located on different pairs, and they showed
independent assortment in Mendel's dihybrid crosses. But
pod shape and plant height are close together on the same pair,
and they would have shown tight linkage. Mendel apparently
did not study this pair of traits in a dihybrid cross, or at least
there is no record of such a study.

Two or more genes interact to produce a trait in epistasis

In the early 1900s, some commercial varieties of corn (maize)
had kernels with a purple pigment called anthocyanin in their
seed coats. Geneticist R. A. Emerson crossed two true-breeding

(homozygous) varieties of white corn and obtained F_1 hybrids
that had all purple kernels. When the F_1 purple hybrids were
crossed, the F_2 generation contained 56% purple kernel corn
plants and 44% white kernel plants.

Emerson deduced that two genes were involved in the production of purple anthocyanin pigment. Therefore, the F_1 generation cross had actually been a dihybrid cross. In a typical
dihybrid cross with independently assorting genes, there are
16 different combinations and a 9:3:3:1 ratio of phenotypes.
Emerson realized that 56% of 16 is 9 and 44% of 16 is 7; thus
his F_2 generation of corn plants had a modified dihybrid ratio
of 9:7 instead of 9:3:3:1 (Figure 14.17). Emerson had discovered the first example of **epistasis** (Greek, meaning a "stopping") in genetics. The two genes interact in such a way that

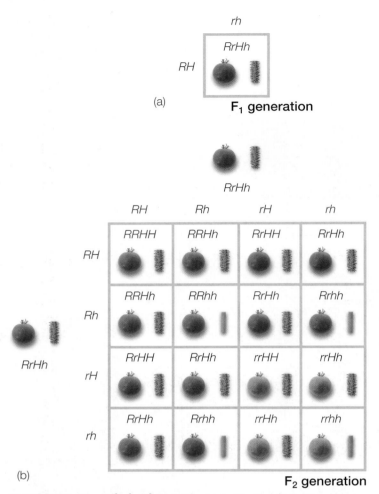

(a) **F₁ generation**

(b) **F₂ generation**

FIGURE 14.15 Dihybrid cross in tomato Red fruit (*R*) and hairy stems (*H*) are dominant over orange fruit (*r*) and smooth stems (*h*). Crossing a red-fruited, hairy-stemmed homozygous dominant plant (*RRHH*) with an orange-fruited, smooth-stemmed homozygous recessive plant (*rrhh*) yields an F₁ generation that is heterozygous for both traits (a). (b) The F₁ generation produces four types of gametes, and the F₂ generation consists of 16 combinations of alleles.

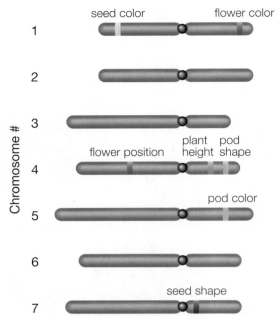

FIGURE 14.16 Mendel's seven traits in garden peas, shown on the plant's seven chromosomes

either gene is capable of stopping the phenotypic expression of the other.

To produce the purple pigment, corn must have a dominant gene for each of the two genes involved in pigment production. Anthocyanin is made in a two-step process (Figure 14.18). The dominant alleles for each gene code for functional enzymes; the recessive alleles code for nonfunctional enzymes. As long as a corn plant has at least one dominant allele for each gene, it can make purple pigment because it can make both enzymes in the biochemical pathway for anthocyanin production. But if it is homozygous recessive for either or both genes, the kernels will be white. A nonfunctional enzyme at either step stops the production of the end product because production is a sequential process.

Many genetic mechanisms have been discovered in the course of crossing varieties of plants to produce hybrids.

Hybrids are desirable in agriculture because they frequently have greater vigor and yield than any single true-breeding line of crop plants. For more about hybrids see Essay 14.2, "Hybrid Corn and Hybrid Orchids."

In some plants, genes located on sex chromosomes determine separate male and female organisms

About 94% of all flowering plants have only one type of individual, which produces flowers containing both "male" floral organs (the stamens) and "female" floral organs (the carpels). Such plants are termed *sexually monomorphic* because they have only one form. Some 6% of flowering plants (about 14,620 species) have two separate sexes and are called *dimorphic*. "Male" plants produce flowers with stamens but no carpels and "female" plants produce flowers with carpels but no stamens. Researchers interested in the mechanisms of sex determination in plants must use plants other than the widely used *Arabidopsis thaliana* because it is sexually monomorphic.

In the white campion, *Silene latifolia* (Figure 14.19), distinct **sex chromosomes**, designated X and Y, are responsible for sex determination. When sex chromosomes are present, the diploid sporophyte (including flowers) can be male or female. The male plant has both X and Y chromosomes, and the female plant has two X chromosomes. The Y chromosome bears the genes for the development of male plants and the formation of stamens. The X chromosomes carry the genes for female plant development and carpel formation. The X and Y chromosomes are morphologically distinct

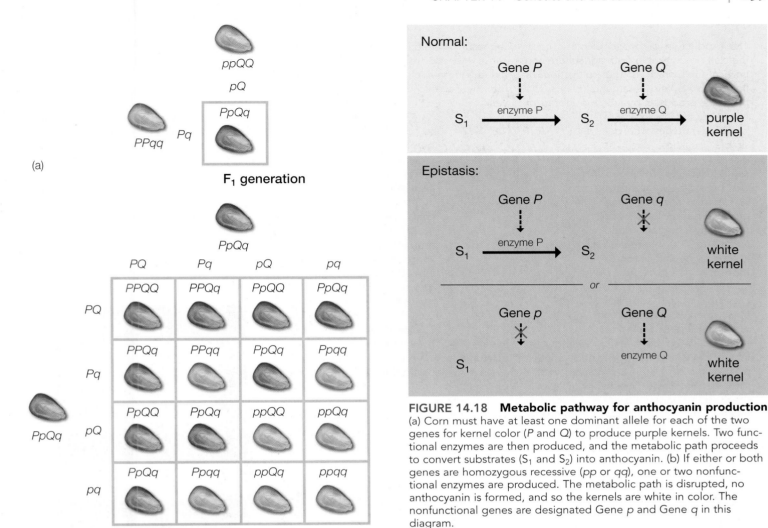

(a)

F₁ generation

(b)

F₂ generation

(c)

FIGURE 14.17 Epistasis in corn for the genes for anthocyanin production (a) The F₁ generation formed by crossing a homozygous dominant plant having purple kernels (*PPQQ*) with a homozygous recessive plant having white kernels (*ppqq*). (b) The F₂ generation has a 9 purple:7 white phenotypic ratio. (c) An ear of corn from an F₂ plant. You might try counting the number of purple and white kernels to determine their ratio.

FIGURE 14.18 Metabolic pathway for anthocyanin production (a) Corn must have at least one dominant allele for each of the two genes for kernel color (*P* and *Q*) to produce purple kernels. Two functional enzymes are then produced, and the metabolic path proceeds to convert substrates (S₁ and S₂) into anthocyanin. (b) If either or both genes are homozygous recessive (*pp* or *qq*), one or two nonfunctional enzymes are produced. The metabolic path is disrupted, no anthocyanin is formed, and so the kernels are white in color. The nonfunctional genes are designated Gene *p* and Gene *q* in this diagram.

(Figure 14.20). Essentially, the same system of sex determination occurs in humans. Human males are XY, females are XX, and the sex chromosomes are morphologically distinct. Chromosomes other than the sex chromosomes are called **autosomes**.

Recently, researchers in Hawaii discovered sex chromosomes in the papaya *(Caria papaya)*. Papaya has 17 pairs of autosomes and one pair of sex chromosomes. Male papaya plants are XY and female plants are XX. Unlike the white campion and humans, however, the sex chromosomes in papaya look like the autosomes. The sex chromosomes are

The production of hybrid plants is commercially important in both agriculture and **horticulture**, which is the cultivation of flowers, fruits, and ornamental plants. Hybrids are produced by crossing inbred or true-breeding varieties of known genotypes with desirable qualities, such as disease resistance or large seeds. Hybrids are important because they are larger, more vigorous, and produce better seed crops than any single inbred variety—a phenomenon known as **hybrid vigor** or **heterosis.**

Hybrid corn is made by crossing inbred varieties. One variety is detassled to prevent self-pollination and then pollinated by a second variety. In Figure E14.2A, two single-hybrid varieties (B × A and C × D) are produced in this manner. The single-cross hybrid C × D is then used to pollinate the B × A hybrid to make a double-cross hybrid for commercial planting. In crops grown from hybrid seed,

Hybrids are important because they are larger, more vigorous, and produce better seed crops than any single inbred variety ...

the resulting seed cannot be replanted because genetic recombination will have made undesirable new gene combinations. New hybrid seed must be purchased each year.

Orchid hybrids are bred to create new and more beautiful varieties for commercial sale. Hybrids are made between different genera and species to bring in new traits. *Cattleya* orchids are very popular because they are large and fragrant, making them a favorite for corsages (Figure E14.2B). Orchids in the genus *Laelia* (Figure E14.2C) can hybridize with *Cattleya* orchids to produce hybrid orchids called *Laeliocattleya* (Figure E14.2D). One procedure for making a hybrid large yellow *Cattleya* orchid is given in Genetics Problem 6 at the end of this chapter.

In the wild, orchids are pollinated mainly by insects. Their pollen is molded into waxy masses called *pollinia*. Hybrid

orchids are made by collecting these pollinia on a probe and transferring them to the "female" receptive organ or stigma on another species or genus. If the cross is successful, the flower wilts within two days, and the ovary develops into a seed capsule over the next 9 months. The tiny, dustlike seeds can then be collected, disinfected in antiseptic solution, and sown in flasks with nutrient agar. After about a year of growth in sterile flasks, the young seedlings are ready for transplant into flats filled with a mixture of fine bark and sand. Orchids require 4 to 5 years before they will flower. (This account of orchid culture is only a bare outline, but there are many books available on growing orchids.)

E14.2B *Cattleya*

E14.2C *Laelia*

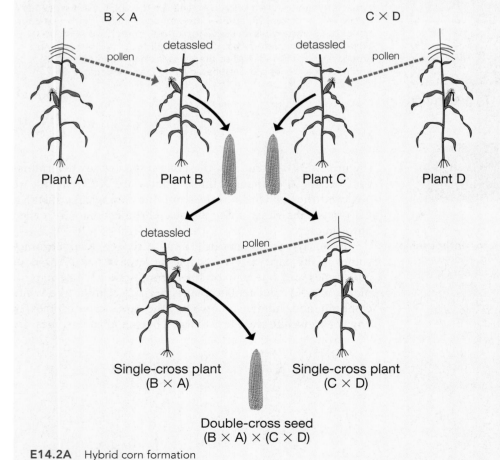

E14.2A Hybrid corn formation

E14.2D *Laeliocattleya*

functionally distinct because the Y chromosome carries the genes for male development and organs, and the X bears the female developmental genes. The researchers suggest that the sex chromosomes in papaya provide evidence supporting the hypothesis that morphologically distinct sex chromosomes evolved from autosomes that carried sex-determining genes.

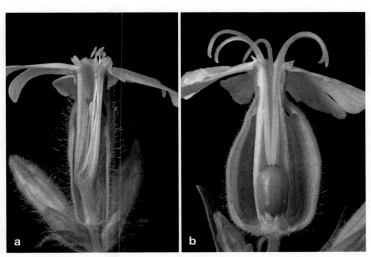

FIGURE 14.19 The two types of flowers in *Silene latifolia* (a) Male flower. (b) Female flower. Part of each flower has been cut away to reveal the interior reproductive organs.

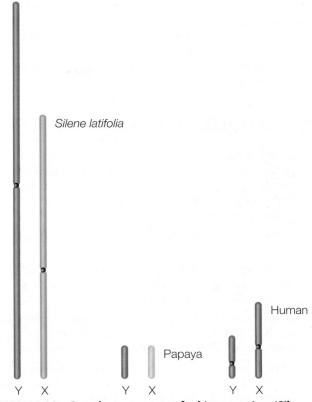

FIGURE 14.20 Sex chromosomes of white campion (*Silene latifolia*), papaya (*Caria papaya*), and human (*Homo sapiens*) The sex chromosomes of the white campion and of humans are of different sizes and can be distinguished from autosomes, but those of the papaya are similar in size. Note that the Y chromosome in white campion is larger than the X, unlike the situation in humans. (The location of centromeres is uncertain in papaya.)

HIGHLIGHTS

14.1 Early experiments showed that hereditary material did not blend in hybrids and parental traits reemerged in later generations. Mendel focused on seven traits in peas, each of which existed in two alternate forms in separate parental varieties. In each trait the F_1 generation showed only one form, the dominant one. In the F_2 generation, however, the ratio of dominant to recessive forms was 3:1. The factors of the trait segregate from each other independently, a fact that later became known as Mendel's Law of Segregation.

14.2 In contemporary terms, Mendel's factors are the genes that code for a trait. The alternate factors are the alleles of a gene. If the alleles are the same, the individual is homozygous for that trait; if they are different, the individual is heterozygous. The observable trait is the phenotype, and the underlying genetic makeup is the genotype. Meiosis is the cellular mechanism responsible for Mendel's Law of Segregation.

14.3 Flower color in certain plants is controlled by a pattern called *incomplete dominance*. In pleiotropy, a single gene may affect more than one phenotypic trait. Many genes combine in an additive manner to produce a single trait in polygenic inheritance. The environment can alter the way in which the genotype is expressed in the phenotype.

14.4 Genes located on the same chromosome are linked. Homologous chromosomes may break and exchange segments during meiosis in a process called *crossing over*. When two individual plants or varieties differ in two traits, a cross between them is called a *dihybrid* cross. If the two traits are located on different pairs of homologous chromosomes, the traits will assort themselves independently, a fact known as Mendel's Law of Independent Assortment. In epistasis, two or more genes interact such that either may mask the phenotypic expression of the other. Some plants have separate male and female sexes determined by sex chromosomes.

REVIEW QUESTIONS

1. Discuss the theory of blending inheritance, the prevailing theory in the 1800s when Gregor Mendel was working with garden peas. What were Charles Darwin's views on this theory? What is a major objection to this theory?

2. What are monohybrid and dihybrid crosses? Give examples of each.

3. In performing his famous genetics experiments, what advantages did Mendel gain by choosing garden peas over some other species of plants?

4. What is the first filial generation? What is the preceding generation called? If the plants of the first filial generation are self-pollinated, what would the subsequent generation be called?

5. Distinguish the following pairs of words: (a) *dominant* vs. *recessive traits*; (b) a *gene* vs. an *allele*; (c) *homozygous* vs. *heterozygous*; (d) *phenotype* vs. *genotype*; (e) *self-fertilization* vs. *cross-fertilization*.

6. Briefly describe the five main points of Mendel's model of inheritance.

7. Discuss Mendel's Laws of Heredity: (a) the Law of Segregation and (b) the Law of Independent Assortment. Who actually coined the phrases for each law?

8. Is Mendel's second law, the Law of Independent Assortment, always accurate? Do traits always assort independently?

9. What is the phenomenon of incomplete dominance? Using floral color in snapdragons, discuss the mechanism of incomplete dominance in molecular/enzymatic terms.

10. What is linkage? What is crossing over, and when does it occur? Does crossing over occur frequently between any two chromosomes, or only between homologous chromosomes? Did Mendel encounter linkage?

11. What is epistasis, and how does it differ from polygenic inheritance?

12. In the white campion and humans, sex is determined by sex chromosomes that are distinct from the autosomes. In what way are the sex chromosomes of white campion and humans different?

GENETICS PROBLEMS

Basic Problems

1. In garden peas, a cross is made between true-breeding yellow-seeded plants (Y) and true-breeding green-seeded plants (y). Yellow seed color is dominant to green seed color.

   (a) What is the genotype of the F_1 hybrid plants?

   (b) What kinds of gametes can the F_1 plants produce?

2. In a population of wildflowers, blue flower color (B) is dominant to white flower color (b).

   (a) If two heterozygous, blue-flowered plants are crossed, what fraction of the offspring will have blue flowers and what fraction white flowers? Show your work in a Punnett square.

   (b) What would be the best way to determine the genotype of a blue-flowered plant? Show what crosses you would make and their results.

3. In four o' clocks, true-breeding red-flowered plants are crossed with true-breeding white-flowered plants. All the F_1 plants bear pink flowers. When the F_1 plants were crossed, the F_2 generation consisted of 102 red-flowered plants, 206 pink-flowered plants, and 99 white-flowered plants.

   (a) What type of inheritance does this demonstrate?

   (b) If R and R' are the genes for flower color, show the genotypes of the three phenotypes of the F_2 generation.

   (c) Why is this not an example of blending inheritance?

4. In garden peas, tall plants (T) are dominant to short plants (t), and round pea pods (R) are dominant to constricted pea pods (r). A true-breeding, homozygous tall plant with round pods ($TTRR$) was crossed with a true-breeding short plant with constricted pods ($ttrr$). All F_1 plants were tall with round pods ($TtRr$). When the F_1 plants were crossed, the F_2 generation contained 75 tall/round and 24 short/constricted plants.

   (a) What can you conclude about the location of these genes on chromosomes?

   (b) When some of the F_1 plants were crossed to plants that were short with constricted pods, the following plants were obtained: 211 tall/round, 198 short/constricted, 23 tall/constricted, and 19 short/round. What is the most likely explanation for the two new phenotypes?

Advanced Problems

5. In garden peas, yellow seed color (Y) is dominant to green seed color (y), and round seed shape (R) is dominant to wrinkled seed shape (r). List the types of gametes and the ratio of phenotypes in the offspring that would occur in the following dihybrid crosses:

   (a) $YyRr \times YyRr$
   (b) $YyRr \times yyrr$

6. In orchids, a true-breeding *Cattleya* with large purple flowers is crossed to another true-breeding orchid with small yellow flowers. Yellow flower color (Y) is dominant to purple color (y). Flower size, however, shows incomplete dominance. Thus LL produces a large flower, LL' an intermediate flower, and $L'L'$ a small flower. All of the F_1 hybrid orchids are $YyLL'$ and have yellow intermediate-sized flowers. The orchid grower wants to get a true-breeding large yellow-flowered orchid.

   (a) What types of gametes do the F_1 hybrids produce?

   (b) In a Punnett square, show the genotypes and phenotypes that result from a cross of the F_1 hybrid orchids.

   (c) What fraction of the F_2 generation will be true-breeding for large yellow flowers?

7. In squash, true-breeding plants for disk-shaped fruits are crossed with true-breeding plants with elongate fruit. The F_1 plants all have disk-shaped fruits. When the F_1 plants are crossed, the F_2 plants have fruit in the ratio of 9 disk-shaped:6 spherical:1 elongate.

   (a) What genetic mechanism of inheritance is most likely demonstrated by these squash plants?

   (b) Using the symbols A or a and B or b for the genes, indicate the genotypes for disk-shaped fruit, spherical fruit, and elongate fruit.

CHAPTER 15 | Genetic Engineering

Genetically engineered corn

Corn—*Zea mays*—is among the world's most important food crops. But corn also has a very bright future in production of safer, cheaper medicines. This is possible because plant DNA can be altered by genetic engineering techniques, so that plants become able to synthesize human proteins such as growth hormone and many other compounds that can be extracted for medicinal use. Plants can thus be engineered to serve as pharmaceutical factories, offering many advantages over conventional industrial methods. Farms are more economical than industrial facilities—the Sun's energy powers photosynthesis, yielding the organic carbon and ATP needed for synthesis of complex pharmaceuticals. Agricultural technologies have already been developed for harvesting and processing plant products on large scales. The risks of contaminating plant-grown medicines with human pathogens and toxins are lower than with conventional industrial methods. Corn and other cereals offer the particular benefit that their seeds are rich in proteins that are stable during storage. Companies have already begun to plant fields of corn that has been genetically altered to produce medicines. Despite the benefits of this and other examples of plant genetic engineering, there are valid concerns that foreign genes might escape from engineered plants into food crops or wild relatives.

Genetic contamination of seed supplies for several major U.S. crops—including corn—has already occurred, according to a recent study by the Union of Concerned Scientists. Seed contamination can result if genetically altered seeds are accidentally mixed into unaltered seed supplies, or if crop plants grown for seed have received pollen from genetically altered plants. The genetic material of wild relatives of crop plants can also become contaminated with engineered genes, with potential ecological effects.

Educated people need to know something about genetic engineering in order to make informed decisions about genetically modified food crops. Citizens also need to become knowledgeable about products made by genetically modified microbes, efforts to genetically alter humans and other animals, and the possibility that weeds, pests, or disease microbes could acquire altered genes that make them harder to control. This chapter describes major genetic engineering techniques and discusses some of the benefits and pitfalls of gene alterations, with a focus on plants. ■

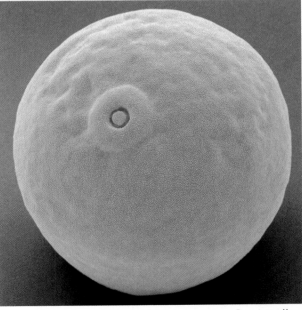

Corn pollen

15.1 | What is genetic engineering?

Genetic engineers use molecular tools that are naturally present in living organisms to alter the genetic composition of crops and other organisms for human benefit. Organisms that have undergone genetic engineering are also said to have been genetically altered, genetically modified (GM), or transformed. They are also known as GEOs (genetically engineered organisms) or transgenic organisms. Their DNA has been changed, usually only slightly, so that **GM organisms** produce proteins or other materials that would otherwise not be present but that are useful to people. Because the changes are made in the heritable material DNA, the new genetic traits are present not only in the engineered organisms themselves but also in their progeny.

Genetic engineering includes modification of bacterial DNA (see Section 15.3) so that it can encode the production of drugs useful in medicine. Examples of such drugs include interferon used to fight infections, human insulin needed by diabetics (Figure 15.1), and human growth factor—a medicine (previously obtained from cadavers at great expense) that is used to treat dwarfism. Large populations of GM bacteria can be grown in vast industrial tanks, where they cheaply produce large amounts of medicines and other compounds that are otherwise more difficult to obtain. Plants can also be genetically modified to enhance crop yield or quality both for food production and other applications (section 15.4). Gene therapy involves modification of human DNA to cure disease. Genetic engineering is a major aspect of **biotechnology**—the use of biological organisms and processes to solve human technological problems.

Genetic engineers use tools that are common in nature

Nature is a common source of inspiration for new engineering materials, processes, and designs. For example, a Swiss engineer, George de Mestral, was inspired to invent the now ubiquitous Velcro® fasteners by observing fruits with hooked spines that attached themselves to his pants and his dog's fur while on hiking trips. In nature, these hooks attach fruits to the fur of animals, thereby helping plants to disperse. Likewise, genetic engineers use molecular tools and procedures that evolved in nature as adaptations that help organisms to survive.

ECOLOGY

Some genetic engineering tools are derived from bacteria and viruses

Bacteria produce enzymes that slice up the nucleic acids of invading viruses. Genetic engineers put these enzymes, known as **restriction enzymes**, to work as molecular scissors to cut DNA in very precise ways. Organisms also produce enzymes to repair damaged DNA. Genetic engineers use these enzymes, called **ligases**, like glue to join pieces of DNA together.

Some viruses, including HIV (human immunodeficiency virus), which causes AIDS (acquired immunodeficiency syndrome), use RNA rather than DNA to encode their genes (Figure 15.2). Such viruses produce an enzyme that copies their RNA-encoded genes into DNA form. The enzyme is known as **reverse transcriptase** because it reverses the process by which DNA is normally transcribed into RNA (Chapter 6). This process (reverse transcription) allows the viral genes to insert themselves into a host organism's chromosomes, where they are replicated along with that organism's own DNA. (This ecological strategy is much like that of cowbirds, which leave their eggs in other birds' nests to hatch and be fed among the foster birds' own young.) Molecular biologists have co-opted viral reverse transcriptase to make stable double-stranded DNA copies of RNA extracted from cells of higher organisms. This is advantageous because the resulting DNA is less fragile than RNA and thus easier to study yet contains the same genetic information. Using this viral tool allows molecular geneticists to more easily learn how gene expression differs among the specialized cells, tissues, organs, and life stages of higher organisms. This information is indispensable to genetic engineers, who need to know which of many similar genes to alter in order to effect a desired change in an organism's structure or function.

ECOLOGY

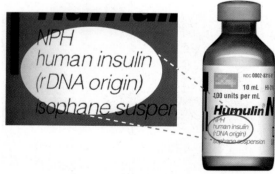

FIGURE 15.1 **Recombinant insulin is a product of genetic engineering** On the label, "rDNA" stands for recombinant DNA (meaning it has been genetically engineered).

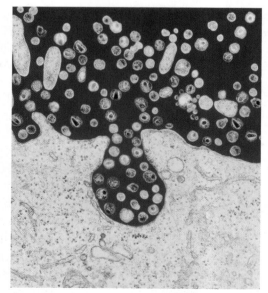

FIGURE 15.2 **HIV (human immunodeficiency virus) viewed with the transmission electron microscope** This is an RNA virus that uses reverse transcriptase to insert its genetic material into host cells. In this false-colored image, the HIV virus particles are colored pink-red.

Other examples of useful molecular tools that scientists have borrowed from nature include the DNA-synthesizing enzymes—**DNA polymerases**—of bacteria that live in hot springs like those in Yellowstone National Park (Figure 15.3). These "thermophilic" (heat-loving) bacteria have enzymes that remain stable under high heat, which allows them to replicate their DNA and grow at temperatures that would coagulate and destroy the function of most organisms' DNA-synthesizing enzymes. Molecular biologists have learned to use these heat-stable enzymes to make many identical copies of specific DNA pieces, in a process that operates something like a photocopy machine. This copy process, known as the **polymerase chain reaction** or **PCR**, has made it easier to study and manipulate genes by providing researchers a much larger sample of DNA having the same sequence (Essay 15.1, "PCR: The Gene Copier"). PCR has become an important tool in the investigation of crimes. Tiny amounts of DNA left by criminals at the scenes of their crimes are PCR-amplified into amounts large enough to allow comparison to the DNA of suspects (see Essay 6.1, "Molecular Detectives: DNA Fingerprinting Solves Crimes"). It is important for everyone to know something about how PCR works and how DNA sequences are obtained and compared, because DNA evidence has become widely used in criminal cases. It would be tragic if a defendant were wrongly convicted or released because judges, lawyers, or jury members were unable to comprehend DNA evidence.

Plants contribute other molecular tools

Plants and other organisms have pieces of mobile DNA—also known as **transposons**, transposable elements, or "jumping genes." Transposons are able to excise themselves from chromosomes and then settle themselves (or a copy of themselves) in a different chromosomal location. They play important roles in evolution by inactivating genes, altering gene regulation, assembling new genes, and adding new parts to existing genes. For example, transposons generated white grapes by disrupting the genes for color in purple-fruited ancestors. Transposons were first discovered by geneticist

FIGURE 15.4   Indian corn kernels often have colored spots or streaks that reveal the action of transposons

Barbara McClintock, who observed their effects on the colors of Indian corn kernels (Figure 15.4) and won the Nobel Prize for her discovery. Half of corn DNA (and much of human DNA) consists of transposons. Genetic engineers have learned how transposons control their entry and exit and use them to "tag" other genes, thereby disrupting the function of specific genes. Many gene sequences are known whose function, though potentially useful to genetic engineers, is unclear. Transposon tagging helps molecular biologists deduce gene function because they can observe the effect on the organism when normal gene behavior is lost.

Many plants are capable of "cloning" themselves from small pieces, a form of asexual reproduction that is widely used by gardeners and in commercial nurseries to propagate plants with desirable characteristics (Chapter 8). Plant scientists use this capability to regenerate many identical plants from single genetically engineered cells by means of a process known as *tissue culture* (Chapter 12). As we learn more about nature, genetic engineers acquire additional materials and tools that increase their capacity to make new and more effective constructions.

15.2 | Plant genetic engineering resembles crop breeding but is faster and more versatile

Genetic manipulation of plants and other organisms is sometimes viewed as a dramatic modern departure from older human technologies. Actually, genetic engineering is an extension of breeding techniques that humans have used since the dawn of agriculture to alter plant and animal characteristics to better suit human needs.

Humans have long altered the genetics of domesticated plants and animals

Archeological evidence indicates that humans have been genetically modifying their food crops (and livestock) since the dawn of agriculture, some 10,000 years ago (Chapter 2). Ancient agriculturalists learned that if they planted seeds from plants having the most desirable food or fiber characteristics, the next

FIGURE 15.3   Heat-stable enzymes (a) Commercial DNA polymerase enzyme. Such enzymes were originally discovered in bacteria found in thermal pools in Yellowstone National Park (b).

ESSAY 15.1 PCR: THE GENE COPIER

The polymerase chain reaction—known as PCR for short—is a method for making many copies of a particular gene or DNA sequence of interest. It is very rapid and can be done relatively cheaply and easily with a minimum of equipment. PCR has become indispensable in many modern biological research laboratories and in criminal investigations.

PCR is made possible because the hydrogen bonds holding together the two strands of the DNA double helix can be intentionally broken and reformed simply by changing the temperature. In addition, the enzyme known as DNA polymerase, which makes complementary copies of single-stranded DNA template molecules, is easy to use and readily available. Further, by using heat-stable polymerase—originally obtained from bacteria growing in extremely hot environments like thermal springs—that is able to survive repeated cycles of heating and cooling, it is not necessary to add new enzyme after each heating step. This makes PCR practical in that technicians can simply start the process and let it proceed on its own.

When a piece of DNA is heated to high temperatures—usually to around 95°C (a little over 200°F)—the hydrogen bonds that link the two strands become weaker, allowing them to separate. If the temperature is then lowered, new complementary DNA strands can form on each of the separated strands, provided that DNA polymerase and the four types of nucleotides are present. It is possible to specify which part of the DNA is copied by including two small DNA pieces (10–30 nucleotides long)

PCR has become indispensable in many modern biological research laboratories and in criminal investigations.

known as *oligonucleotide primers*, which have a sequence of choice. The primer sequences are chosen to be complementary to the two ends of the DNA fragment one wishes to copy. Primers are made by DNA-synthesizing machines, which add specific nucleotides to growing chains one at a time, like stringing beads onto a necklace.

The PCR process is outlined in Figure E15.1. Following the heating (step 1), the DNA, nucleotide, polymerase, and primer mixture is cooled (typically to around 50–55°C). The primers bind ("anneal") to the separated DNA strands (step 2), and the DNA polymerase can begin synthesizing DNA, forming new strands that are complementary to the remaining portion of the strands. This "extension" step, during which the complementary copies are made by DNA polymerase, is usually carried out at approximately 72°C (step 3). When replication of the DNA segment has been completed, which occurs within a matter of seconds to minutes, the mixture is again heated (to 95°C). This separates the just-formed DNA strands so that they too can serve as templates and be copied when the temperature is lowered again (steps 1–3). Repeated many times, this chain reaction yields many identical copies of the DNA region of interest. Thirty cycles will produce millions of gene copies within minutes or a few hours. PCR is typically carried out in special instruments (thermocyclers) that can be programmed to automatically raise and lower the temperature to a wide variety of set points for a variable number of cycles.

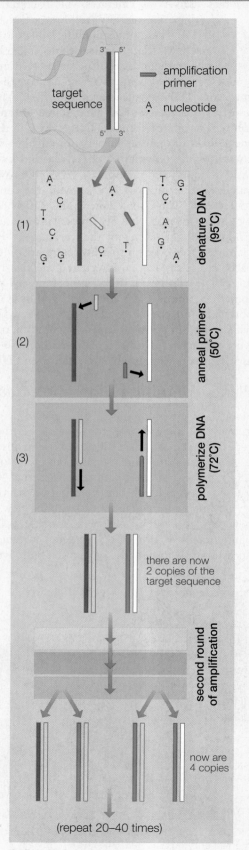

E15.1 Polymerase chain reaction

crop would have more of the desired features than if they used seed from random plants. This process, known as *artificial selection*, mirrors natural selection—the improved survival and reproduction of organisms having characteristics that are advantageous in their particular environments. Humans have also long successfully practiced breeding programs in which domesticated plants and animals having different desired features are cross-bred to produce offspring having more of the favored characteristics than their parents. Breeding programs are based on Mendelian genetics, in which new combinations of chromosomes are produced (Chapter 14). Desirable characteristics encoded by particular alleles and carried on specific chromosomes that are segregated at meiosis are combined with others at mating. The new and variable chromosome combinations of the progeny represent reshuffling of the genetic cards—chromosomes. Sometimes, a winning hand results—a chromosome combination that encodes an organism with particularly desirable traits. A champion racehorse may result, or a cow whose milk production sets a new record, or a new crop variety that is both more nutritious and resistant to pest attack. People have actually been engaged in a form of genetic engineering for a long time—using controlled matings to manipulate organisms' chromosome combinations.

Genetic engineering overcomes some drawbacks of traditional breeding methods

Although traditional crop and animal breeding programs have been very beneficial to humankind, they suffer some major disadvantages. Breeding programs demand long time periods—hundreds or thousands of crosses may be required to find just a few valuable types—and the genetic material that can be used is limited because only the DNA of the parents is available. Months to years are required for agricultural plants and animals to reach reproductive maturity so that mating crosses can be made. Further time is needed for the resulting progeny to grow to the point that their characteristics can be evaluated so that the best parents of the next generation can be selected. Many generations may be needed to produce the desired results. In contrast, changes can be made via genetic engineering much more quickly. New genes can be inserted into cell genomes within a matter of minutes, and individual cells that have been transformed can be identified (selected) in a matter of days. Within weeks, genetically modified plants can be grown to a size large enough to determine whether or not traits have been altered in desired ways. If so, such plants can be propagated. As a result, seeds of genetically engineered plants can be brought to market considerably faster than is usual for traditional breeding methods.

Some plants that could benefit from traditional breeding programs have not, because they reproduce primarily by asexual means (Chapter 13). An example is the banana, an important food crop that is susceptible to fungal and bacterial diseases. Cultivated bananas are seedless; they are propagated by removal and planting of shoots that emerge at the base of the trunk (Figure 15.5). Further, banana is polyploid (having more than two copies of all chromosomes), another feature that makes classical breeding more difficult. Genetic engineering offers the possi-

FIGURE 15.5 Cultivated banana plants are propagated asexually by transplanting basal shoots Two basal shoots arising from the main shoot (with bananas) are indicated by the arrowheads.

bility of increasing banana's resistance to disease pathogens, thus increasing crop yield.

Another disadvantage of traditional breeding programs is that they rely upon genes that are present in wild or domesticated populations of a particular organism or its close relatives. If the desired variants are absent, the trait cannot be added. This was the situation with rice—genetic engineers wished to incorporate genes for beta carotene, a nutrient absent from rice but needed for eye health and limb development. Engineering beta carotene into rice has the potential to prevent many of the hundreds of thousands of cases per year of blindness in children that are caused by beta carotene deficiency. However, no wild relatives or cultivated varieties of rice having beta-carotene in grains were available for use in traditional breeding. Genetic engineering allows the genes of a much wider range of organisms, including very distant relatives, to be incorporated into crop plants. For example, genes from bacteria and daffodils were successfully added to the genetic material of cultivated rice to make rice colored golden by beta carotene. The insect-repellent properties of genetically modified crops are derived from a bacterium. Genes added to bacteria for industrial production of interferon, insulin, and human growth factor came from humans. Genetic engineers are also able to synthesize and introduce into organisms completely new "designer genes" that do not (to our knowledge) exist in any living natural organisms. Genetic engineering allows people to overcome the constraints of time and gene availability that limit traditional breeding programs. Genetic engineering represents a quantum leap in our ability to tailor crop plants and livestock to people's needs.

15.3 | Bacteria can be genetically engineered to produce useful materials

Certain characteristics of bacteria make them ideal systems for **gene cloning**—generating many copies of genes and the products encoded by them. Bacteria have small and relatively simple cells that reproduce rapidly and can incorporate foreign DNA. In addition to the bacterial cell's main DNA (the genophore), bacteria often include one or more **plasmids**—small pieces of DNA that usually take the form of a closed circle (Figure 15.6). Plasmids are key to the ability of bacteria to incorporate foreign DNA into their cells. Plasmids can be constructed so that they contain foreign genes, that is, to form **recombinant DNA**. Plasmids often contain **antibiotic resistance genes,** or they can be modified to include these or other genes, which function as markers or labels later in the cloning process (Figure 15.7). Plasmids and viruses are known as **vectors,** because they are capable of moving DNA from one organism to another. Both plasmids and viruses are used by molecular biologists in genetic engineering.

There are four main stages in the cloning process (Figures 15.8–15.10). The first step is to prepare the vector DNA (a plasmid or virus) and the DNA that will be inserted into the vector. The second step consists of linking these two DNA segments together. In the third step, the recombinant vector bearing foreign genes is inserted into receptive bacterial cells (usually *Escherichia coli*, or *E. coli* for short). The final step consists of growing large populations of modified bacterial cells. These cells can be used to produce large amounts of foreign DNA for laboratory study. Alternatively, materials produced within the bacterial cells, by expression of the new genes, can be harvested for use.

Restriction enzymes are used to prepare DNA and vectors for cloning

Vectors are generally obtained through commercial sources and can quickly be prepared for cloning. The most critical factor is that the ends of the vector DNA and foreign DNA are compatible (i.e., they allow the two pieces of DNA to be joined successfully). This compatibility is accomplished through the use of restriction enzymes. Each type of restriction enzyme recognizes, binds to, and cuts DNA at a specific nucleotide base sequence, the recognition site. Recognition sites are usually four, six,

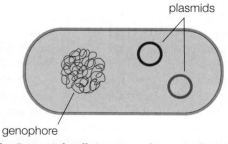

FIGURE 15.6 Bacterial cell, its genophore, and plasmids Plasmids are typically circular pieces of DNA that occur apart from the cell's main DNA (the genophore). The plasmids are enlarged relative to the genophore and cell in this diagram.

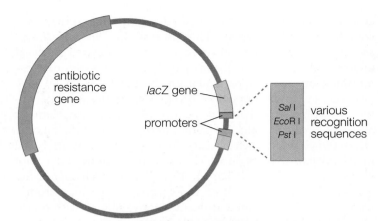

FIGURE 15.7 Simplified diagram of plasmid with antibiotic resistance gene marker A variety of components have been engineered into the plasmids that are typically used in cloning DNA, including antibiotic resistance genes. When the plasmid is present in a bacterial cell, such a gene confers antibiotic resistance to that cell, meaning that cells with no plasmids (which would not be of interest to geneticists) do not survive when grown in culture media containing a particular antibiotic. In addition to this gene, various other sequences are added to assist in cloning (usually as a battery of recognition sequences for restriction enzymes). Others aid in sequencing, and some (promoters) are involved in expressing cloned genes (making the protein that the gene codes for). The *lacZ* gene is used in a color detection system (depicted in Figure 15.10) to show whether or not foreign DNA has been incorporated.

or eight nucleotides in length. When DNA is cut by the same restriction enzyme, the ends are always the same, even if the DNA molecules come from two different organisms. For example, the enzyme known as *Eco*R I (pronounced "eco-R-one"—named for the bacterium *E. coli,* from which it is extracted) always cuts DNA at the sequence GAATTC.

When the vector DNA is cut with a restriction enzyme, the circular vector molecule opens up into a linear fragment of DNA (Figure 15.8a). Foreign DNA is treated with the same enzyme (Figure 15.8b), which ensures that the ends of the vector and foreign DNA molecules match and can join together through complementary base pairing. Some restriction enzymes cut DNA such that the ends are "blunt"; that is, both strands are cut at the same point, which results in no overhanging "sticky" ends.

Foreign DNA is "glued" into the vector

Once the two fragments of DNA have attached to each other through complementary base pairing, the sugar-phosphate "backbone" of the DNA molecule still needs to be joined together. This is accomplished through the use of an enzyme called DNA ligase (Figure 15.8c). Following this step, the foreign DNA is fully integrated into the vector (Figure 15.8d).

Depending on the aims of the researcher, only particular small fragments of foreign DNA might be cloned, or all genes of an organism can be cloned to construct a **genomic library**. To make a genomic library, an organism's entire complement of DNA is treated with the same restriction enzyme used to cut the vector. This generates many DNA fragments of different sizes—but all with ends that match those of the vector. The vectors used in making genomic libraries are able to incorporate large pieces

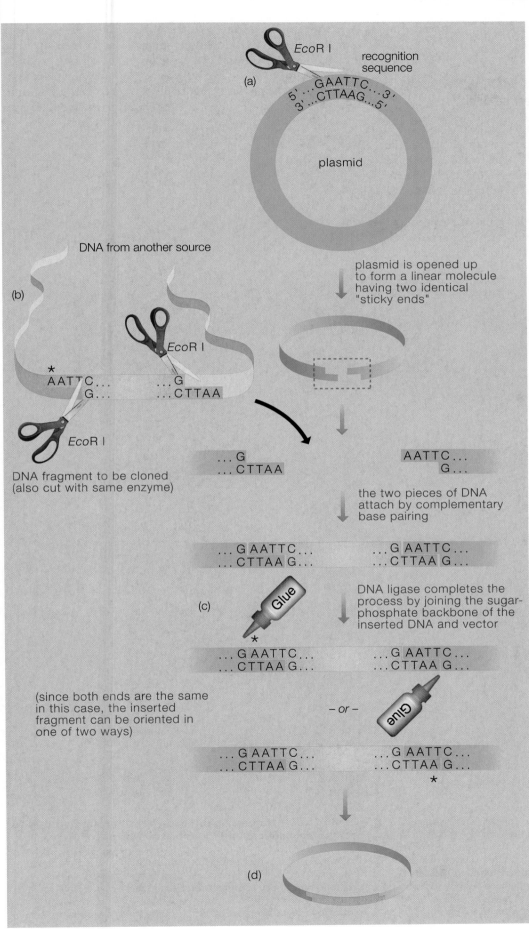

(a)

EcoR I

recognition
sequence

5'···GAATTC···3'
3'···CTTAAG···5'

plasmid

DNA from another source

(b)

EcoR I

*
AATTC...   ...G
    G... ...CTTAA

EcoR I

DNA fragment to be cloned
(also cut with same enzyme)

plasmid is opened up
to form a linear molecule
having two identical
"sticky ends"

...G   AATTC...
...CTTAA   G...

the two pieces of DNA
attach by complementary
base pairing

...G AATTC...   ...G AATTC...
...CTTAA G...   ...CTTAA G...

(c)

Glue

DNA ligase completes the
process by joining the sugar-
phosphate backbone of the
inserted DNA and vector

*
...G AATTC...   ...G AATTC...
...CTTAA G...   ...CTTAA G...

(since both ends are the same
in this case, the inserted
fragment can be oriented in
one of two ways)

– or –

Glue

...G AATTC...   ...G AATTC...
...CTTAA G...   ...CTTAA G...
                                    *

(d)

FIGURE 15.8 Restricting the plasmid and DNA to be cloned and joining the two together (a) The plasmid is restricted (cut) with a restriction enzyme (here, EcoR I). This makes the circular plasmid linear. (b) The DNA fragment to be cloned is cut with the same enzyme (or enzymes—two different enzymes could also be used), making its ends compatible with those of the plasmid. Under the proper conditions the complementary ("sticky") ends of the plasmid and cloned DNA piece will attach to each other through the hydrogen bonds formed during complementary base pairing. An enzyme, DNA ligase, repairs the breaks in the sugar-phosphate backbones of the DNA, completing the process (c). The cloned piece of DNA is now fully incorporated into the once-again circular plasmid (d), which can be introduced into a bacterial cell. If a single restriction enzyme is used, the inserted piece of foreign DNA can be oriented in one of two ways. If two enzymes are used (in "directional cloning"), the orientation would be known, which can be advantageous.

of foreign DNA, so that fewer vector molecules are needed to clone the entire genome.

Modified vectors are incorporated into bacterial cells, which are then grown to large populations

Plasmids carrying foreign DNA are mixed with bacterial cells that have been treated in such a way they will readily take up plasmids from the surrounding medium (Figure 15.9a). Bacterial cells that have taken up plasmids bearing foreign DNA have been **transformed**. Once inside the host bacterial cells, plasmids replicate along with the bacterial genophore and are transmitted to progeny cells. The proliferation of foreign DNA and plasmids in growing bacterial populations is the basis of gene cloning.

How do scientists distinguish bacterial cells that have been transformed from those that have not taken up plasmids? A single bacterial cell typically takes up only a single plasmid, and if spread out sufficiently on an agar plate containing growth medium (Figure 15.9b), each cell grows into a separate colony (Figure 15.10). The nutrient medium also

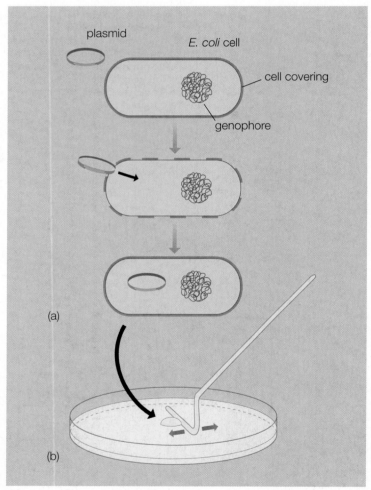

FIGURE 15.9 Insertion of plasmids into bacterial cells—transformation (a) Treatment of *E. coli* cells with particular chemicals renders them capable ("competent") of taking up plasmids by loosening their cell covering. Plasmids and competent *E. coli* cells are mixed and warmed for a short time (usually to 42°C) and then quickly placed on ice, which aids in moving the plasmids into the cells. The mixture containing transformed cells is poured out onto agar that has been supplemented with antibiotics and/or chemical markers that select for *E. coli* cells that contain plasmids with inserted pieces of foreign (cloned) DNA. (b) The droplet of transformed bacteria is spread out on the agar surface, usually with a glass rod that has been bent into an appropriate shape.

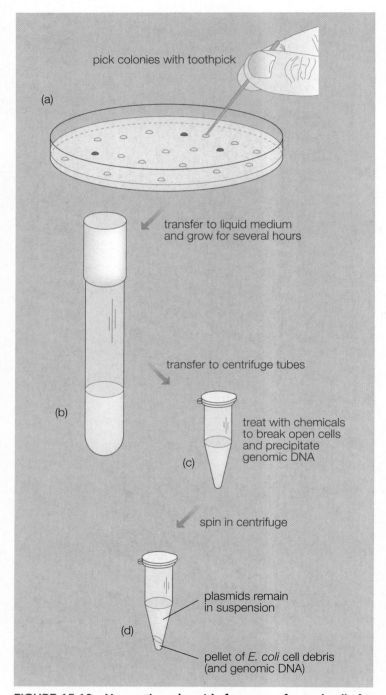

FIGURE 15.10 Harvesting plasmids from transformed cells for further manipulation (a) "Picking" colonies from agar plate with a toothpick. (b) Growing up the bacteria from separate colonies in liquid culture. (c) Treating the bacterial cells with chemicals to break open the bacteria and salts to precipitate the bacterial genomic DNA. (d) After being briefly spun in a centrifuge, the bacterial DNA and cellular debris forms a white pellet, while the plasmids remain in solution and can be removed from the liquid with a pipette.

contains an antibiotic or other chemical that can reveal the presence of markers that were incorporated into plasmids. Any cells that have taken up a plasmid will also possess a gene that confers resistance to the antibiotic or makes other markers visible. Cells that were not transformed will either die or remain unmarked in this selection process. Transformed bacteria can be removed and grown to large populations, and their DNA can then be removed for determining its sequence or for other types of analyses.

If the bacteria were exposed to a mixture of plasmids containing different pieces of foreign DNA, different colonies will contain different foreign DNA segments. These transformed bacterial colonies represent the "books" of a genomic library, which can be kept in storage and "read" at different times. Particular genes present in the cells of each colony can be identified by the cells' binding of single-stranded, labeled segments of known DNA. A cloned gene can be expressed using the bacterial transcription and translation systems to make the protein it encodes.

FIGURE 15.11 **A gene gun** This device is used to shoot DNA-coated gold particles into cells in order to transform them. Some of the foreign DNA will integrate into the cell's DNA and be expressed.

15.4 | Plants can be genetically engineered

Plant genetic engineering starts with single plant cells. If one can transform a plant cell with the gene of interest, an entire adult plant whose every cell contains the new gene can be generated from it. This occurs because during mitosis, the process by which a multicellular plant develops from a single cell, DNA replication faithfully copies all of a cell's DNA—including added genes—and each set of copied DNA is accurately provided to both of the cells resulting from each cell division (Chapter 7).

How can a new gene be inserted into a plant cell? One method is to use the soil bacterium *Agrobacterium tumefaciens,* which can transfer a plasmid bearing its own genes into the cells of higher plants (Chapter 18). Genetic engineers can incorporate cloned genes from plants, fungi, protists, or bacteria into modified *Agrobacterium* plasmids. They also engineer such plasmids to contain marker genes (see section 15.3) and a promoter derived from a virus. Then the plasmid is inserted into *A. tumifaciens,* which in turn inserts the new gene and its promoter into plant chromosomes. Promoter DNA starts messenger RNA production from the attached new gene, causing it to be expressed by the plant cell. *Agrobacterium* was used to transfer daffodil and bacterial genes into rice embryos to produce the famous beta-carotene-rich golden rice plants.

Viruses and electroporation are two other methods for inserting foreign genes into plant cells. Electroporation is a method by which an electric field is used to make temporary holes in plant cell membranes, through which DNA can enter the cells. In addition, genes can be introduced by using a gene gun (Figure 15.11), a pistol-like device that blasts gold or tungsten particles coated with foreign DNA into plant cells. Cells that have incorporated the foreign DNA can be identified by the presence of markers and then grown into adult plants by the use of hormone treatments (Chapter 12).

15.5 | Genetic engineering has produced valued new forms of crops

GM crops are widely planted throughout the world; in 2002, nearly 60 million hectares of land were planted with GM crops. New GM crops are continuously being developed because they are viewed as having valuable properties. These may include greater nutritional value; better pest or drought resistance; increased ability to obtain soil mineral nutrients; production of compounds useful in industry; and resistance to damage during harvest, transport, and storage. In the 1990s, a major epidemic of the papaya ringspot virus nearly destroyed the papaya industry in Hawaii, but growers were able to recover by using a new variety of papaya that had been engineered for viral resistance. There are many other examples of beneficial plant genetic engineering; some are described in the following subsections.

The roots of some GM crops can more effectively obtain soil phosphate

Phosphate levels in soils often limit agricultural productivity, and phosphate fertilizers are not affordable by many world farmers. In some soils, phosphate is present but tightly bound in mineral complexes too large for plant roots to absorb. Plant roots normally exude organic acids, such as citric acid, to release phosphate from these complexes. Citric acid secreted from roots has another benefit: It helps reduce plant absorption of aluminum ions, which are abundant in about 12% of farmland soils and harm plants by stunting root growth. Genetic engineers have succeeded in modifying some plants so that they release more citric acid from their roots. This added capability gives the GM plants an advantage not only by allowing them to absorb more of the

ECOLOGY

phosphate present in soils but also by providing more resistance to harmful aluminum ions. The genetic change increases the expression of the enzyme citrate synthase, merely causing the plant to more effectively perform a natural function.

Plants can be genetically modified to produce new types of starch

Starch, the storage polysaccharide produced within plant chloroplasts, is a major component of the diet of many people and is also used in industry. Starch obtained from plants such as potatoes also can be used to make biodegradable substitutes for nondegradable plastics, such as those used for packing materials or disposable diapers. The starch polymer comes in two major forms having different properties—linear amylose and the more highly branched amylopectin. Starches that are rich in amylose are best for making fried snacks because they produce a crispier product and retard oil penetration. In contrast, amylopectin-rich starches improve the quality of frozen foods and are better for use in the paper and adhesive industries, as well as in livestock feeds. Genetic engineering is used to transform the starch-producing systems of plants so that more starch is produced, or starches are genetically designed for particular industrial applications.

GM crops can produce antibody and vaccine proteins for use in human medicine

Antibodies are proteins produced within the bodies of humans and other animals that provide protection against pathogenic microbes and viruses. A number of antibodies have been approved for use as medicines in the treatment of disease. Plants do not normally produce antibodies, because they have evolved other chemical means of protecting themselves against pathogen attack (Chapters 2 and 12). Genetic engineers have nonetheless succeeded in transforming plants so that they are able to produce inexpensive "plantibodies" useful in human medicine. Applications include treatment of tooth decay, viral infections, and cancer. Several genes are involved in the synthesis of a single antibody, and it is difficult to transfer them all into a plant cell at the same time. However, plant biologists have been able to transform each gene into separate plants and then use conventional crosses to obtain plants having all of the necessary new genes. Vaccines consist of inactivated pathogens or their components (often proteins) that stimulate the animal body to develop immunity to disease-causing viruses and bacteria. Plants have been engineered to produce proteins for use in vaccines that stimulate immunity to viruses (including those that cause hepatitis B and rabies in humans) and bacteria (including toxic *E. coli* and the agent of cholera).

As noted in the chapter opening, there are several advantages in using GM plants for pharmaceutical production rather than animal or bacterial systems. Animal cell cultures can become contaminated with viruses that could infect patients, but plant cells do not support the growth of viruses that are pathogenic to animals. Plants can produce larger amounts of vaccine and antibody proteins than can bacteria, and plants secrete the proteins into extracellular fluids for easy harvest.

Finally, there is the prospect that vaccines and antibodies could be delivered to patients in the form of tasty foods.

15.6 | Genetically engineered crops pose some concerns

Many people seem to accept some of the basic goals of genetic engineering. These include genetic alterations of organisms in scientific research labs for the purpose of learning more about biology, the use of genetically modified bacteria to produce medicines, and even gene therapy (modifications made in human genes in attempts to cure genetic diseases). GM cotton crops (Figure 15.12) that have been modified to resist bollworm attack have been praised as a way to reduce negative health and environmental effects of pesticides. Plantations of closely packed "designer" trees that produce a higher proportion of wood than usual might reduce economic pressures to destroy ecologically valuable natural forests. In these cases, the public seems to perceive the rewards as being greater than the risks.

In contrast, genetic alteration of plants used in the production of human food has engendered substantially greater public concern. Critics of GM food crops have asked some valid questions for which answers are not yet clear—"Is genetic engineering of food crops really necessary? Does biotechnology draw resources away from other approaches to solving world food sufficiency problems? Has the introduction of GM crops into our food supply been too rapid? Are some kinds of crop genetic alterations more justifiable than others? Should foods be labeled if they contain GM crop products? Will human health or environmental problems arise from the production of GM crops?" Several of these questions merit further consideration.

Will genetic engineering help solve world food sufficiency problems?

Biotechnology advocates sometimes argue that GM crops will help feed the world's growing population and help prevent starvation. Many countries, states, and universities have invested large amounts of money and human resources in building

FIGURE 15.12 Genetically modified cotton

biotechnology programs to generate economically promising products. Yet to repay the funds invested in their invention, GM crops are usually patented, such that only patent holders—often large corporations—are allowed to produce seed. This means that GM seed is likely to cost more than what farmers in impoverished areas of the world can afford. Further, some analysts think that solving food supply problems caused by inadequate food distribution systems, diminishing water and fossil fuel supplies, climate change, and other economic, political, and environmental factors may be of equal or greater importance, yet lack the required levels of support. Supplying the financial resources necessary to slow human population growth is of primary importance in achieving worldwide food security but is resisted by large segments of society.

Should foods containing genetically modified crop products be labeled?

Food industries that use GM crops often resist food-labeling requirements, fearing that societal concerns might deter consumer purchases. They argue that GM crops are not really different from new crop varieties that have been produced by traditional breeding methods and thus should not be distinguished by labels. On the other hand, some human health advocates argue that people who must avoid food allergens need this information to make appropriate purchasing decisions. An example illuminates this issue.

Genetic engineers working for a major U.S. crop-seed-producing company transferred a gene from Brazil nuts into soybean, thereby improving the nutritional quality of the altered soybean. But the scientists discovered that the gene for an allergy-inducing compound (allergen) had also been incorporated into the GM soybean, making it potentially harmful for human consumption, and so the project was discontinued. The possibility that GM crops might cause harmful allergic responses in consumers will need to be addressed for all food products that contain GM components; but at present it is not easy to predict the allergenic potential of GM food products. Even if the probabilities of allergies and the numbers of persons likely to be affected are judged to be low, people prone to food allergies may desire the ability to identify and avoid labeled GM foods.

Might GM crops have harmful environmental effects?

Crops that have been engineered with genes to help them resist attack by microbes or insects (or herbicide applications) have also attracted critics, who fear that these crops might have harmful environmental effects. An example of such crops is corn that has been transformed with a bacterial gene that encodes a protein (Cry9C) capable of destroying the intestinal systems of insect larvae such as the European corn borer, killing it before it can cause much damage to the crop. This protein is one of a group of toxins known as Bt toxins (after *Bacillus thuringiensis*, the bacterial species that naturally produces the toxins and from which the *Bt* genes have been identified). Other examples include crops that have been engineered to resist particular herbicides, so that the

ECOLOGY

herbicides, which would kill nonengineered plants, can then be applied to fields to kill weedy competitors. Crops may also be engineered with viral genes, in efforts to generate resistance to viral diseases. On first inspection these alterations are appealing, but closer examination reveals some serious issues. Environmental scientists interested in the impact of genetically engineered organisms on nature have pointed out that the critical experiments needed to assess risks have not been done and that studies done so far have often produced conflicting results.

Evolution of resistance to pest control measures might offset the value of some GM crops One problem with the widespread use of crops that have been engineered for greater resistance to pests is that pest organisms can rapidly become tolerant to the engineered crops, much as they respond to chemical pesticides, with the result that resistant pest populations may rise. The mechanism by which pests respond to this pressure is natural selection (Chapter 16). Naturally occurring mutants constantly arise in populations, and sexual recombination constantly reshuffles chromosomes, producing organisms with new traits and combinations of characteristics. One insect, the diamondback moth, has already been observed to have become resistant to Bt toxin in nature. Ten other moth species, four species of flies, and two types of beetles have become resistant in laboratory studies. Other insect pests that are tolerant to Bt toxin as well as weeds that are resistant to the particular herbicides used with some GM crops can be expected to appear and flourish. Widespread evolution of pests resistant to control measures would wipe out the advantages of such GM crops.

EVOLUTION

GM crops might have harmful effects on nonpest species Other potential problems with GM crops include harmful effects on nontarget insects. For example, although some studies suggest various insects that consumed pollen from crops engineered with Bt toxin genes may not be affected, other studies suggest population declines in monarch and black swallowtail butterflies and honeybees. Bees, butterflies, and other insects are essential pollinators for most of the flowering plants—carrying pollen from one flower to another of the same species (Chapter 24). A decline in pollinator populations would have serious effects on the reproduction of many natural plants and fruit crops that depend on insect pollination.

ECOLOGY

GM crops might poison the natural enemies of crop pests In experiments, insects, such as lacewings and ladybird beetles, that fed on insects having consumed crops altered to contain Bt toxin suffered reduced survival and reproduction. These observations suggest the possibility of wider food-web consequences in nature. A decline in the natural enemies of crop pests would be a serious problem indeed.

GM crop plants might interbreed with wild relatives to form "superweeds" Evidence suggests that some transgenic crops can hybridize with close relatives, transmitting foreign genes into natural plant populations. Such gene flow has been measured at high levels

ECOLOGY

in sunflowers and strawberries (Figure 15.13), for example. This problem, known as *gene pollution*, occurs by dispersal of pollen from genetically engineered plants to wild relatives or by escape of GM seeds into nature. GM corn and potato pollen can travel more than 1 kilometer, and canola pollen is transported as far as 8 kilometers from fields. Spillage of seeds from trucks is a common source of GM plants along roadsides. Environmental scientists are concerned that this might lead to the formation of new weeds that would be difficult to control—so-called superweeds. Weeds are noncrop plants that grow profusely, outcompeting crops for water, light, and soil minerals. Seven of the world's top 13 crops—wheat, rice, soybean, sorghum, millet, beans, and sunflower—are known to have hybridized with wild relatives, thereby producing new types of weeds. Controlling weeds often requires the intensive use of herbicides. Because herbicide applications can be expensive and may have undesirable effects on native plants and other nontarget species, reducing chemical applications has been one goal of biotechnology. Involvement of GM crops in the formation of new weeds would be a strong argument against their use.

GM crops pose other environmental concerns

It is unclear whether toxic bacterial proteins (Bt toxins) in GM crops might accumulate in the environment after the decay of crop residues in the fields. Some data suggest that soil microbes can easily decompose Bt toxins, but conditions that foster or prevent rapid breakdown of environmental Bt toxins are not well understood. Crops engineered with viral genes might contribute to the formation of new viruses, which can evolve by recombination with related forms. Another potential problem is that DNA from the cauliflower mosaic virus is widely used in creating GM crops. The viral DNA is needed to signal the

FIGURE 15.13   **Strawberries are examples of crop plants in which transfer of genes from GM to nonengineered plants has been documented**

plant's DNA translation apparatus to start work, causing the foreign gene to be expressed. But at least one investigation has shown that cauliflower mosaic viral DNA can move from one location to another within a GM plant's DNA. This could have unintended and unpredictable genetic effects.

It is important for educated citizens to understand biotechnology issues well enough to judge the conclusions of scientific experiments that are performed to evaluate the safety of GM crops and other products of genetic engineering. When evaluating the pronouncements of experts or industrial representatives, people must consider how such opinions may have been influenced by the prospect for financial gain. The public good is not well served either by uncritical promotion of genetic technologies or by close-minded resistance to them.

HIGHLIGHTS

15.1   Genetic engineers use molecular tools borrowed from bacteria, viruses, plants, and other organisms. These molecular processes evolved in nature as adaptations that help organisms survive.

15.2   Plant genetic engineering is an extension of agricultural breeding technologies used by humans for thousands of years to improve crops and domesticated animals. However, genetic engineering has several advantages over conventional breeding techniques. Genetic engineering can yield faster results, be used to transfer genes among unrelated species, and be applied to asexually reproducing plants that are difficult to modify by conventional methods.

15.3   Genes can be cloned into bacteria. Bacterial enzymes can be used to cut DNA from other organisms into small pieces having ends that match up with the ends of bacterial plasmids or viruses that have been cut with the same enzyme. Ligase enzymes can be used to complete the incorporation of foreign DNA pieces into plasmids by "gluing" together the sugar-phosphate backbones of the DNA. The recombi-

nant plasmids can be inserted into bacterial cells, which then may express the foreign DNA. This process is used to produce pharmaceuticals such as recombinant insulin.

15.4   Crop plants can be genetically engineered to express new traits such as pest resistance and increased nutritional quality. Plant genetic engineers can incorporate genes of interest into plasmids from the bacterium *Agrobacterium*. Plant cells or tissues to be genetically modified can be exposed to *Agrobacterium*, which has the capacity to transfer its foreign gene-bearing plasmid into plant DNA. Viruses and gene guns can also be used to introduce foreign DNA into plant cells. The genetically modified plant cells can then be grown into adult GM plants.

15.5   Genetic engineering has generated valuable new forms of crop plants that are able to more readily obtain soil nutrients, resist disease microbes and other pests, and produce industrially and medically useful materials.

15.6   GM crops raise important issues regarding effects on human and environmental health.

REVIEW QUESTIONS

1. What is a GM organism, and what other names are used to refer to them? Give one actual example.

2. The following enzymes are useful tools to the genetic engineer: (a) restriction enzymes, (b) ligases, (c) reverse transcriptase, (d) DNA polymerase. What are the natural roles of these enzymes in the cells in which they naturally occur?

3. How and why do genetic engineers use the enzymes listed in Question 2?

4. What is a transposon? Who discovered transposons, and what is their role in evolution? What common item in the grocery store produce section illustrates the action of transposons? How do genetic engineers use them?

5. What are some of the drawbacks of traditional breeding methods, and how can they be overcome by genetic engineering?

6. Can genes as foreign as those encoding human antibodies be inserted and expressed in plants? What are some of the advantages of doing so?

7. Plasmids are an exceedingly important bioengineering tool. What are plasmids, what is their importance in nature, and how are they used by genetic engineers?

8. What makes bacteria such an ideal system for gene cloning?

9. Describe the steps required to clone a gene into a bacterial cell. What can be done with the cells once the gene has been cloned?

10. Transforming plant cells is a little different from transforming bacterial cells. Describe how a new gene might be inserted into a plant cell.

APPLYING CONCEPTS

1. How are transposons similar to viruses, and how do they differ?

2. Chapter 15 listed several concerns regarding GM organisms, especially crop plants. What were some of those concerns? Can you think of others?

3. Following is a list of common restriction enzymes and their restriction sequences: EcoR I, GAATTC; BamH I, GGATCC; HinD III, AAGCTT; Bgl I, AGATCT. (In the list, the sequence of only one strand is shown, and its end is on the left.) Examine the sequences carefully. Do you notice anything peculiar about them?

Galápagos Island Flora

KEY CONCEPTS

- Charles Darwin developed the theory of evolution by natural selection largely as the result of observations he made as a young college graduate on a voyage around the world.

- The theory of evolution addresses such questions as why there are so many species on Earth, how they have become adapted to different environments, and how they originated.

- Evolutionary theory helps explain the increasing resistance of bacteria to antibiotics as well as the resistance of weeds and insects to herbicides and pesticides.

The idea of evolution is so tightly linked to Charles Darwin that the two are almost inseparable in the public perception. Though usually depicted as an elderly man with a great flowing beard (which is how he looked near the end of his career in the late 1870s), Darwin made his most original insights and began the work that led to his fame when he was a young man of 22.

Like many modern students, Darwin was uncertain about his future career. The son of a country doctor, Darwin had begun to study medicine, but the subject failed to interest him. In letters home he complained about "long stupid lectures" on medical subjects, and he admitted he could not stand the sight of surgery. In desperation, his family sent him to Cambridge University to study theology in the hope he would become a country parson. He finally obtained a degree in theology, but he had little more interest in it than in medicine.

Like many students today, Darwin did best in subjects that interested him, and what interested him was natural history. Darwin liked to hike about the countryside, observing plants and animals, collecting specimens and classifying them. At Cambridge, Darwin met botanist John Henslow, who had similar interests, and Darwin became his constant companion on field trips. Then an opportunity arose that changed the direction of Darwin's life.

In 1831 the British government was preparing to send the ship *Beagle* on a 5-year survey expedition along the coasts of South America and around the world. A naturalist was needed to collect geological and biological specimens for the British Museum. Henslow recommended Darwin for the position. Darwin was uncertain about accepting it, but the captain of the *Beagle*, Robert Fitzroy, convinced Darwin to join the expedition. Darwin later wrote, "If it is desirable to see the world, what a rare and excellent opportunity this is. Perhaps I may have the same opportunity of drilling my mind that I threw away at Cambridge."

Darwin did more than drill his mind. His experiences aboard the *Beagle* led to the theory of evolution by natural selection, an idea so revolutionary that the eminent geneticist Theodosius Dobzhansky later wrote, "Nothing in biology makes sense except in the light of evolution." This theory is the subject of this chapter—how it came about, its credibility, and its synthesis with modern genetics and molecular biology. ■

Charles Darwin

16.1 | Pre-Darwinian science held that species were unchanging

Before Darwin developed the theory of evolution, natural science was greatly influenced by theology and the concept of special creation. According to this concept, all species were created simultaneously in their present forms and have remained unchanged to the present. Borrowing from ancient Greek philosophers, the concept of special creation further proposed that each species had an ideal form, of which any individual was only an approximation. Thus variation in a species was unimportant; only the ideal form was significant. Earth itself was thought to be about 6,000 years old, as determined by counting the generations in the Old Testament.

By the 18th century, however, new discoveries were beginning to challenge the idea of a young, static natural world. The science of geology, in particular, made considerable advances in the 18th and early 19th centuries. Geologists recognized that rocks occurred in distinct layers, and those layers often contained **fossils**—the Latin word for "dug up." It became apparent that fossils were the remains or impressions of plants or animals that had died long ago. Before these dead organisms decayed, ancient sediments had covered and preserved them. Over long periods of time, geological forces converted these sediments into rock. A new branch of science called **paleontology** arose and was dedicated to the study of fossils.

Study of rock layers and their fossils led to several conclusions. Certain fossils were always found in specific rock layers and never in others. Younger rock layers always lay on top of older rock layers, and the younger the rock layer, the more the fossils resembled present-day organisms. Older rock layers contained fossils of plants and animals that were **extinct**, meaning no members of those species were still living on Earth. In fact, there were more extinct species than living ones.

The richness of fossil species created a problem for the concept of special creation. Why were so many species created, if most of them were to become extinct? French paleontologist Georges Cuvier proposed that all species were created initially, but a series of catastrophes destroyed most of them. These catastrophes were responsible for the rock layers in which the fossils were found. If living species are survivors of all those catastrophes, however, some individuals of living species should have been preserved as fossils from the earliest catastrophes. Yet they were not. French geologist Louis Agassiz offered an alternative: A new creation occurred after each catastrophe, with modern species representing the product of a recent creation. But the fossil record required at least 50 separate catastrophes and new creations!

Some geologists, like James Hutton and Charles Lyell, took a different approach. They proposed that rock layers were the result of normal geological processes such as erosion, glaciation, sedimentation, and volcanism operating over long periods of time. This principle is called **uniformitarianism**, which means that processes observable today also operated in the past. If slow natural physical processes produced layers of rock thousands of feet thick, then Earth must be millions of years old, not just a few thousand. Today geologists estimate the age of the Earth at more than 4.5 billion years. Lyell and Hutton proposed an Earth old enough for evolution to operate.

16.2 | Some early biologists proposed that species could evolve

Before Darwin, a few naturalists had proposed that species might be capable of changing. The French naturalist Georges Louis LeClerc had suggested that the original creation might have produced only a few founder species, and modern species evolved by some natural process. Charles Darwin's own grandfather, Erasmus Darwin, had suggested that species might transmute into new species.

The first biologist to present a mechanism for evolution was Jean Baptiste Lamarck. He hypothesized that organisms can evolve through the inheritance of acquired characteristics. In his best-known example, Lamarck suggested that ancestral giraffes had short necks, which they stretched to feed on leaves at higher levels in trees (Figure 16.1). The giraffes that stretched the most acquired longer necks and passed them on to their offspring. Over time, this process produced the modern giraffe. Today such a hypothesis seems naive, but in the 18th century no one understood how inheritance worked. Although Lamarck's hypothesis was wrong, he introduced the idea that species could change by some natural process.

16.3 | During the voyage of the *Beagle*, Darwin made observations that revolutionized biology

Darwin set sail aboard the *Beagle* on December 27, 1831. With a degree in theology and given the prevailing ideas about species, Darwin had probably been schooled in special creation. What he saw on his historic voyage suggested an alternative.

First, the *Beagle* sailed along the coasts of South America, making frequent stops along the route (Figure 16.2). While others on the voyage mapped coastlines and harbors, Darwin studied the plants, animals, fossils, and geological formations of the coastal and inland areas. In Argentina, he found fossil bones of large, extinct animals. When he observed a snake with rudimentary hind limbs, he noted that it marked "the passage by which Nature joins the lizards to the snakes."

Darwin made his most significant observations on the Galápagos Islands, located 950 kilometers (km) (590 miles) west of Ecuador. Darwin observed that Galápagos plants and animals most closely resembled those in nearby South America, but entire groups of continental organisms were absent. If the island forms were the result of special creation, why should they resemble those in South America? Why should they not be something unique? Furthermore, the plants and animals varied among the different islands (Figure 16.3). The Galápagos Islands were inhabited by giant tortoises

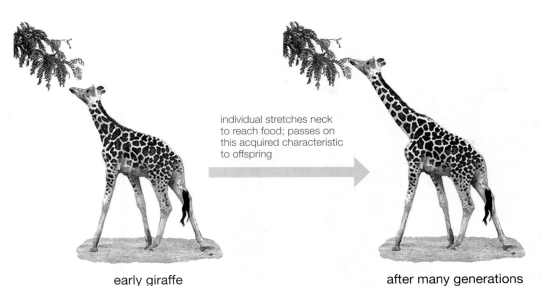

early giraffe

after many generations

individual stretches neck
to reach food; passes on
this acquired characteristic
to offspring

FIGURE 16.1 Lamarck's hypothesis of the inheritance of acquired characteristics Lamarck incorrectly proposed that giraffes acquired their long necks through stretching. Those giraffes passed their acquired longer necks on to their descendants. After many generations of stretching, modern giraffes had acquired their present long necks.

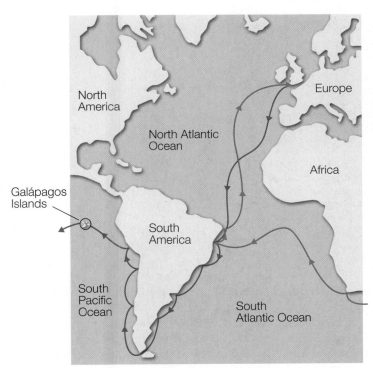

FIGURE 16.2 The voyage of the *Beagle* The *Beagle* spent about $3\frac{1}{2}$ years surveying the coasts of South America, during which time Darwin made several inland expeditions. About a month was spent in the Galápagos Islands, but that month resulted in the most significant observations for the science of biology. (Several excursions along the southeastern coast of South America are not plotted on the map.)

(*galápagos* is Spanish for "tortoise"), and tortoises on the various islands differed markedly in form. Tortoises and plants seemed to affect each other. On islands without tortoises, prickly pear cacti grew with their pads and fruits close to the ground; but on islands with tortoises, the cacti grew

tall trunks that lifted their pads and fruits above the reach of the tough-mouthed tortoises.

The organisms that most amazed and confounded Darwin were the finches. The many finches on the islands differed mainly in the size and shape of their beaks, apparently to utilize different food resources (Figure 16.4). There were parrot beaks, large beaks, small beaks, and straight beaks, all of which existed only on these islands. Darwin wrote in his journal, "One might really fancy that, from an original paucity of birds in this archipelago, one species had been taken and modified for different ends." For years to come, Darwin concentrated on the problem of the diversity of organisms on the Galápagos Islands.

16.4 | Over the next two decades, Darwin developed his theory of evolution by natural selection

Darwin returned to England in 1836. He proceeded to build his reputation as a scientist through the publication of an account of his voyage and several works on geology and natural history. But the problem of the origin of species always remained in his mind. He deduced from his observations in the Galápagos that evolution had occurred, but he did not know how species changed over time. What was the mechanism of evolution? He was aware that plant and animal breeders practice selective breeding of domestic species to produce new varieties for agriculture. This **artificial selection** can produce significant changes in domesticated species in a few generations. From this knowledge Darwin concluded that because all species vary in the characteristics of their individuals, selection of some individuals and elimination of others could be the key to evolutionary changes. In artificial selection, human choice is

FIGURE 16.3 Galápagos Island biota The Galápagos Islands consist of 13 main volcanic islands and numerous smaller ones located about 950 km west of Ecuador. (a) Giant prickly pear cactus (*Opuntia echios*) on Santa Cruz reaches a height of 12 m (39 feet). (b) Shrubby *Opuntia* on an island without tortoises, which eat cactus pads. (c) Giant tortoise from Isabela Island, where the vegetation is relatively lush. (d) Tortoise from the arid island of Española. The high arch at the front of the shell allows the tortoise to reach up for food.

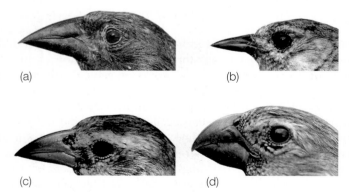

FIGURE 16.4 Darwin's finches Representative finches with different ecological roles include (a) ground finches, which mainly eat seeds; (b) the warbler finch, which eats insects off leaves; and tree finches, which mainly eat insects. Included among the tree finches is the woodpecker finch (c), a bird that uses cactus spines to pry insects out of rotten wood, and the vegetarian tree finch (d), which has a parrotlike bill and eats buds.

the selective mechanism. What was the mechanism of selection in nature?

Darwin had read an essay on human population by Thomas Malthus (1798), an English clergyman and economist. Malthus wrote, "It may safely be pronounced, therefore, that [human] population, when unchecked, goes on doubling itself every 25 years, or increases in a geometrical ratio." Human population is controlled by environmental factors such as wars, famines, and pestilences. Darwin surmised that a similar principle applied to all species in nature. All species reproduce in a manner that would overflow the environment in short order if growth went unchecked. But natural populations tend to remain constant over long periods. Consequently, in every generation many individuals must die without reproducing. Which individuals die in each generation is not always or usually a matter of chance, but depends on the characteristics of the individuals. Individuals with favorable characteristics are more apt to survive to reproduce, while others with less-favorable characteristics are more likely to die. Whether a characteristic is favorable or not is determined by the environment. Favorable characteristics, if inheritable, become more common in populations. These conclusions became the basis of the theory of evolution by natural selection. **Natural selection** is the differential survival and reproduction of individuals with different inheritable characteristics. In natural selection, nature is the selective mechanism.

For the next 20 years, Darwin gathered data to support his theory. He had nearly finished writing his book on the origin of species when, in 1858, he received a manuscript from a relatively unknown naturalist named Alfred Russel Wallace. Wallace, who had been collecting plants and animals in Malaysia and Indonesia, had conceived the idea of evolution through differential reproduction. He sent his manuscript to Darwin asking for critical comment. Darwin was justifiably upset! Fortunately, Darwin's colleagues arranged for a presentation of the theory of evolution by both men at a meeting of the Linnaean Society in London. Darwin's paper was presented first, and he was given first credit. Darwin quickly finished his book, and in 1859 published *The Origin of Species*.

16.5 | The theory of evolution by natural selection can be summarized as a series of observations and conclusions

In common usage, the word *theory* has the meaning of a "guess, hunch, or idea." In scientific usage, however, a theory is a principle or a conclusion that derives from and explains a considerable group of observations and/or experiments. Darwin's theory of evolution is not a guess but, rather, a tested principle that explains observations about species in nature. This theory or explanation can be summarized as a simple chain of logic.

1. **Every species is capable of producing more offspring than can survive to maturity.** Consider the common dandelion, *Taraxacum officinale*. Because a single plant can produce hundreds of offspring, in a short time dandelions could potentially cover every inch of land surface on Earth! Although they may sometimes seem to be doing this in lawns and parks, they have not overrun the Earth. The same argument applies to any species—of plants or animals.

2. **The sizes of natural populations tend to remain fairly constant over time.** A **population** of plants or animals includes all the members of a single species of plant or animal in a particular area. Over time, populations tend to vary in numbers about some fairly consistent level. The resources available in a particular area, such as food, water, and usable space, as well as factors such as disease and other populations of plant-eating herbivores or predators, determine this consistent population level.

3. **Therefore, there is competition for survival and reproduction.** Because the supply of resources can support only so many individuals of a species in a particular area, and because every species has the capacity to produce more offspring than can survive to reproduce, those individuals and their own offspring have to compete for limited resources, and many individuals do not survive.

4. **Individuals in a population vary in traits that affect their chances for survival and reproduction.** Some traits help individuals obtain food, evade predators, or survive extreme temperatures.

5. **Therefore, those individuals with the most advantageous traits are more likely to survive to produce the most offspring.** As Alfred Wallace expressed it, those individuals that survive "must be, on the whole, those which have some little superiority enabling them to escape each special form of death to which the great majority succumbed." This is the origin of the phrase "survival of the fittest." Natural selection is the process by which the environment selects individuals for survival and reproduction because they have traits that make them better adapted to that environment. Over generations, the frequency of favorable traits will increase, while the frequency of less-favorable traits will decrease. The species will become better adapted or may in time become a new species. **Evolution** is the change in the frequency of traits in a population over generations of time.

16.6 | The synthetic theory of evolution combines Darwinism with genetics and molecular biology

The logic of Darwin's theory of evolution and the mass of data he gathered to support it were highly persuasive. After reading Darwin's book, biologist Thomas Huxley remarked, "How extremely stupid not to have thought of that!" But there were some aspects of the theory of evolution Darwin could not explain fully, because the level of knowledge in related scientific areas was insufficient at the time.

Darwin could not have understood the source of the variation among individuals in a species. That understanding required knowledge of chromosomes, genes, and DNA, which was obtained in the 20th century. Nor could Darwin explain how traits were passed from one generation to the next. Although Darwin was a contemporary of Gregor Mendel, who discovered the basic process of inheritance (Chapter 14), Mendel's work was not recognized until the early 1900s. Today we know that variation in populations is primarily due to genetic mutations and recombination. Variable traits are expressions of the alleles of genes, which are segments of DNA located in chromosomes. Alleles are segregated into separate reproductive cells by meiosis. The reproductive cells then recombine in fertilization during sexual reproduction to form new individuals. Mutations and the genetics of populations will be considered in later sections of this chapter.

Darwin also could not offer a complete explanation for the origin of new species. The first widely accepted model for the

DNA SCIENCE

origin of new species was developed in 1942. Other models have since been proposed and verified, and new mechanisms of speciation are still being explored. Mechanisms of speciation are discussed in the final section of this chapter.

Almost all biologists accept the modern **synthetic theory of evolution**, which combines Darwinism with modern genetics and molecular biology, but they may have different opinions about mechanisms of speciation. Such controversy is all part of the scientific process (Chapter 1). Controversy among scientists on these points does not mean they doubt the concept of evolution.

16.7 | Many areas of science provide evidence for evolution

An overwhelming body of data supports the modern synthetic theory of evolution. In this section, we will survey some of the major lines of evidence.

Artificial selection demonstrates that species can be modified

Darwin thought that artificial selection made a strong argument for evolution by natural selection. He focused on two examples of artificial selection—the strawberry and the domestic pigeon. About 40 years before Darwin wrote *The Origin of Species*, strawberries had begun to be artificially selected in England, and already a large number of varieties had been produced. Also, the artificial selection of pigeons had produced varieties so different

that, if found in nature, they might be classified as distinct species in more than one genus. But all the varieties were known to be descended from the wild rock pigeon, and all were capable of interbreeding and producing fertile offspring!

The common grocery vegetables broccoli, Brussels sprouts, cabbage, cauliflower, kale, and kohlrabi are quite distinct in appearance, but all are the results of artificial selection of one species, *Brassica oleracea* (Figure 16.5). Kale is most similar to the original wild ancestor. If humans could produce all these different crop plants by artificial selection in a few thousand years, natural selection and other mechanisms of evolutionary change should be capable of producing the observed diversity of living organisms over hundreds of millions of years.

Comparative anatomy reveals many evolutionary relationships

Comparative anatomy involves the examination and comparison of the structural details of organs found in different organisms. Common sense suggests that organisms that are related to one another by descent from a common ancestor should be classified together. In an ideal evolutionary classification system, organisms would be grouped according to **homologous** structures (*homologous* comes from the Greek word *homologia*, meaning "agreement"). Homologous structures share a common origin but may differ in their current function. Bud scales, bracts, spines, and floral parts are homologous structures because they are all modifications of the same basic organ—foliage leaves (Chapter 11). Darwin recognized that the existence of

(a)
(b)
(c)
(d)
(e)
(f)

FIGURE 16.5 *Brassica oleracea* **and artificial selection** These common vegetables—broccoli (a), Brussels sprouts (b), cabbage (c), cauliflower (d), kale (e), and kohlrabi (f)—look very distinct, but all are derived from wild *Brassica oleracea*, following thousands of years of artificial selection by farmers.

homologous structures modified to perform different functions in related organisms was exactly the predicted outcome of evolution.

Structures that look alike and perform similar functions, however, are not always homologous. The tendril of a pea plant is derived from a leaf, but the tendril on a grape vine is a modified stem. Both pea and grape tendrils perform the same function of climbing support, but they are **analogous** organs (from the Greek *analogos*, meaning "proportionate"). Analogous organs perform similar functions but have different evolutionary origins. Some plants have no foliage leaves but instead have wide, flat stems or branches that perform photosynthesis (Figure 16.6). These flattened stems or branches and foliage leaves are analogous organs.

Analogous organs are of evolutionary interest because they reveal how different organs can be modified to perform similar functions in unrelated organisms. Natural selection acts to adapt organisms to their environment. If unrelated organisms are placed in similar environments, selection might then produce similar adaptive structures, a process called **convergent evolution**. In deserts, certain groups of plants have adapted by developing thick, fleshy stems for water storage and leaves modified into protective spines. In the deserts of North and South America, the cactus family has evolved these structures. In Africa and Asia, however, the spurge and milkweed families have these same features. Convergent evo-lution has produced strikingly similar-appearing plants in three separate families (Figure 16.7). All three families have even independently evolved CAM photosynthesis (Chapters 5 and 26), which allows them to keep their stomata closed during the day.

One form of convergent evolution is **mimicry**, in which one organism evolves a resemblance to another organism or inanimate object. Mimicry may hide an organism by providing it with **protective coloration**. In the deserts of southwestern Africa, plants in the genus *Lithops* resemble small stones. Hungry herbivorous animals cannot easily distinguish them among pebbles (Figure 16.8). Mimicry can also cause the flower of one species to resemble that of another. In western Europe and the Mediterranean region, a spectacular rose-colored orchid occurs together with blue bellflowers (Figure 16.9). The flowers of the two plants are structurally similar, and solitary bees pollinate both flowers. From the bellflowers the bees receive pollen as food, but from the orchids they receive nothing. To human eyes the flowers of the two plants appear in different colors, but the bees cannot see red; to them, the two flowers look the same. The bees are deceived into pollinating the mimetic orchid without receiving any reward in return!

Changes in proteins and DNA trace evolutionary changes

Organisms that are closely related and structurally similar can also be shown to be biochemically similar. The more closely related two species are, the more similar will be their proteins and DNA. Conversely, the longer two species have

ECOLOGY

FIGURE 16.6 Stems or branches can look and function like leaves Some plants have structures that resemble leaves but are actually modified stems or branches. They are analogous to leaves, meaning that they perform the same function (photosynthesis) but have different origins. Shown here are branched stems of *Phyllanthus* (a). Flowers are shown emerging from the margins of the flattened stems in (b).

FIGURE 16.7 Convergent evolution in desert plants Convergent evolution has produced strong structural similarities among desert plants that belong to different families. Members of the three families shown here—(a) milkweed, (b) cactus, and (c) euphorb—have evolved thick, fleshy stems and have leaves that have been greatly reduced, often to spines. They have also evolved similar biochemical features.

FIGURE 16.8 Stoneplants In the deserts of southwestern Africa, small plants with thick, fleshy stems have evolved protective coloration and form to resemble stones. These stoneplants (members of the genus *Lithops*) are well protected from plant-eating (herbivorous) animals by growing among small pebbles. (The inset maps the location of the stoneplants.)

FIGURE 16.9 Pollination by deceit—floral mimicry between orchid and bellflowers Most flowers reward their insect pollinators with food in the form of pollen or nectar. Some flowers have evolved ways to cheat their pollinators, however, by mimicking other flowers that provide food. These mimic flowers provide no food reward but, because the pollinator cannot differentiate them from the ones that do, they are also visited by the pollinator. The red orchid (*Cephalanthera rubra*) (a) provides no food, but bees pollinate it because it resembles the common bellflower (*Campanula*) (b), which does. (Bees, unlike humans, cannot see the red color of the orchid.)

been evolving separately, the more different they will be in terms of their proteins and DNA. It is therefore possible, by examining a set of proteins or the nucleotide sequence for a particular gene, to determine the pattern and extent of evolutionary change that has occurred within a group of plants (Chapter 17). The same techniques can trace long-distance dispersal events that occurred hundreds of thousands of years ago.

All of the native plants in the Hawaiian Islands, for example, arrived by long-distance dispersal from other islands or the continents around the Pacific Rim.

Most are derived from tropical regions, but recently an arctic connection was discovered. In North America, violets are small herbaceous plants but, in Hawaii, some are woody trees and shrubs. Scientists had assumed that Hawaiian violets came from South America, but that assumption was determined to be incorrect. Molecular analysis of DNA in Hawaiian, South American, and Alaskan violets carried out by Harvey Ballard and Kenneth Sytsma at the University of Wisconsin showed that the Hawaiian woody violets were most closely related to *Viola langsdorffii*, a bog and meadow species in Alaska (Figure 16.10). Because many Alaskan bird species winter in Hawaii (as do many humans), dispersal could have occurred via seeds on the feathers or feet of migrating birds.

Fossils provide a record of large-scale evolutionary changes

Fossils are the preserved remains (or impressions of the remains) of organisms that lived in the past (Figure 16.11). The rocks that contain them can be dated accurately by using radioactive isotopes. The kind of rock reveals much about the environment in which the fossil was formed, and the fossil itself can indicate a great deal about the ecology of the organism. By tracing specific fossils through different geological eras, the evolutionary history or **macroevolution** of that organism can be reconstructed. Macroevolution is change that results in new species over geological time.

For many years scientists have studied the fossil record. In Darwin's day, paleontology was a new science, the fossil record had not been extensively studied, and the incompleteness of that record received considerable attention. But today the evolutionary history of many organisms has been worked out in detail, including many plant groups and numerous animals. Even human evolution is now understood in considerable detail. Gaps and controversies still exist, because science is an ongoing process, not a finished task.

One recent controversy among evolutionists who study the fossil record is between the ideas of **gradualism** and **punctuated equilibrium**. Gradualism is the older Darwinian approach and maintains that evolution proceeds at a steady, gradual rate as changes slowly accumulate. All of the gradual steps in evolutionary change are not preserved in the fossil record. In 1972, Niles Eldredge and Steven Jay Gould proposed that the fossil record was not as incomplete as was thought. They were impressed by the fact that many of the fossil species they studied appeared suddenly in the record, lasted 5 or 10 million years with little change, and then disappeared suddenly—to be replaced by another related but distinctly different species. What if this is actually what happened? As a new species arises it exhibits many structural changes from its ancestors. Then it changes very little over a long period of time—an equilibrium. The equilibrium is then interrupted or punctuated by a new period of rapid speciation and rapid structural changes, perhaps due to rapid changes in the environment.

This punctuated equilibrium model of speciation has generated considerable controversy and reexamination of data,

FIGURE 16.10   Hawaiian woody violets originated from Alaskan tundra violets (a) *Viola langsdorffii*, an herbaceous bog and meadow violet from Alaska identified by DNA sequence analysis as sister to the Hawaiian violets. (b) *V. kauaensis*, an herbaceous bog violet from Kauai in the Hawaiian Islands. (c) *V. helenae*, a shrub on the banks of streams on Kauai. Refer to the map of the Hawaiian Islands in Figure 16.20. (d) *V. chamissoniana*, a small, dry-forest tree found on Kauai, Oahu, and Molokai.

FIGURE 16.11   Fossil plants (a) A fossilized stem of a horsetail plant. (b) A leaflet from a fern.

but no consensus has yet been reached. The controversy does not mean the theory of evolution is in doubt, only that scientists are still exploring how evolution operates. In the next section, we consider how evolution operates at the level of genes in populations.

16.8 | Evolution occurs when forces change allele frequencies in the gene pool of a population

Gregor Mendel discovered the principles of inheritance that describe the frequencies of gene combinations (genotypes) in the offspring of two parents. When Darwinian evolutionary theory was combined with Mendelian genetics, the science of **population genetics** arose as a new branch of biology. A *population* refers to all individuals of a single species living in a particular area. Every population has a gene pool, which is the sum of all the alleles of all the genes in that population. Consider the population of flowering plants shown in Figure 16.12. In the diploid plants shown, each individual plant has two alleles for the gene determining flower color. There are three alleles for flower color in this population; R produces a red pigment, R_1 a yellow pigment, and r makes no pigment. In this example, the gene pool for flower color consists of all the alleles—R, R_1, and r—in this population. Population genetics is the branch of science that examines genes and their allelic frequencies in entire populations.

Population geneticists are interested in gene pools and how and why they change over time. Individuals with favorable combinations of alleles tend to survive and leave more offspring than individuals with less-favorable combinations of alleles. The frequency of these favorable combinations of alleles will tend to increase in the next generation because individuals with less-favorable combinations of alleles will be less likely

FIGURE 16.12 The gene pool of a population of plants Only one gene (with three alleles) for flower color is shown. The alleles determine the following flower colors: *R* produces a red pigment (*RR* is therefore red), R_1 makes a yellow pigment (R_1R_1 is yellow), and *r* makes no pigment, yielding a white flower (*rr*). The heterozygous RR_1 are orange, *Rr* are pink, and R_1r are pale yellow. Each diploid plant has two alleles for flower color.

to survive and reproduce. Consequently, the frequency of these less-favorable alleles will decrease in the next generation. In the example shown in Figure 16.12, suppose that red flowers produce more offspring than white because pollinators visit red flowers more often than white flowers. Then, over time, the frequency of the *R* allele will increase while that of the *r* allele will decrease, and the gene pool will change over time. Such small-scale changes in the frequency of alleles from one generation to the next are referred to as **microevolution**. Evolution results

from the accumulation over time of such small-scale changes in the gene pools of populations. When the forces of mutation, nonrandom mating, genetic drift, migration, or natural selection act upon a population, allele frequencies in the gene pools will be changed from one generation to the next.

Mutation provides new variation to a gene pool

Any change in the hereditary material of an organism is a **mutation**, but to be heritable the mutation must affect reproductive cells. Mutations may occur at the gene level or at the chromosome level. In gene mutations an allele changes into another form. Chromosome mutations can involve pieces of chromosomes, single whole chromosomes, or entire sets of chromosomes.

A gene mutation normally involves only one or, at most, a few nucleotides in a strand of DNA. Such mutations occur randomly in nature due to errors in transcription or may be induced by agents such as chemicals or radiation. A gene mutation alters the normal nucleotide sequence, and that alteration may be beneficial or harmful, depending on how it changes the function of that gene. If it does not alter gene function, the change may be neutral.

DNA SCIENCE

Crossing over, in which two homologous chromosomes exchange equal segments during meiosis, is a normal and common process in meiosis (Chapter 13). Sometimes, however, crossing over may involve an unequal exchange of segments. In this event, a segment of one chromosome may be lost in a deletion (Figure 16.13). In other cases, one homologous chromosome

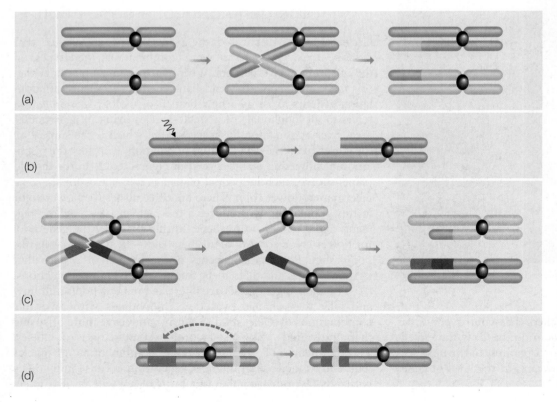

FIGURE 16.13 Chromosome mutations (a) Normal crossing over between homologous chromosomes, where equal segments are exchanged. (b) Deletion caused by radiation damage. (c) Duplication of a segment (shaded region) in one homologous chromosome and deletion of that segment in the other. (d) Movement of a gene from one location on a chromosome to another.

may acquire two copies of a segment, a duplication. Deletions and duplications are types of chromosome mutations and usually produce a change in the organism.

Normally, genes occur in a fixed location in a chromosome, but on occasion they may move to a new location in the same chromosome. These movable genes, popularly referred to as "jumping genes," were discovered in corn plants by Barbara McClintock, a discovery that earned her the Nobel Prize in 1984. Genes that have jumped into another gene will destroy this other gene's function.

Entire chromosomes may be lost or added to the basic set, or the entire set of chromosomes may be duplicated. If a single chromosome is lost or added to a set, the condition is referred to as *aneuploidy*. If the entire set is duplicated, the condition is called **polyploidy**. Polyploidy is highly significant in the development of new plant species. It will be discussed in a later section.

Nonrandom mating alters the frequency of alleles

If individuals choose their mates based on their phenotypes, that is, their outward physical appearance, the frequencies of alleles in the gene pool can change. Among plants, self-fertilization—as in Mendel's garden peas—is a form of non-random mating. Animals often choose mates based on their phenotype, which reflects their genotype. The appearance of male peafowl or peacocks is the end result of peahens preferring males with ever more elaborate tail feather displays as a sign of fitness.

In small populations, genetic drift can cause alleles to be lost

In small populations, random or chance events can cause sudden changes in allele frequencies. Consider a population of 100,000 individuals with an allele w present at a frequency of 1%. One percent of 200,000 alleles is 2,000 w alleles. The loss by chance of a few individuals with the w allele would have little effect on the gene pool. But suppose the population were only 100 individuals. One percent of 200 alleles is only 2 individuals with the w allele. A storm or predator could easily destroy both of these individuals and eliminate the w allele. The removal of an allele by chance alone in a small population is called **genetic drift**. Because of their small population sizes, endangered species are especially subject to genetic drift. Conservation of endangered species is made more difficult by this loss of genetic variation.

Genetic drift plays a significant role in a phenomenon called the **founder effect**. Assume there is a large population of plants on a continent (Figure 16.14). That population has a number of alleles for a gene for flower color. A storm arises and blows seeds from some of the plants across an ocean. A few seeds land on an island, where they grow. Those seeds contain only a small portion of the alleles in the gene pool of the continental population. The island population now has a different gene pool from that of the parent population. Many plant species in the Hawaiian Islands have reduced genetic diversity due to the founder effect.

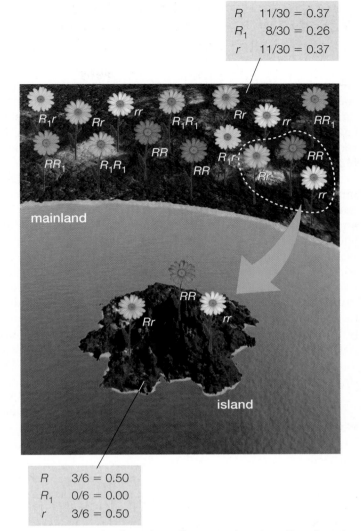

R	11/30 = 0.37
R_1	8/30 = 0.26
r	11/30 = 0.37

R	3/6 = 0.50
R_1	0/6 = 0.00
r	3/6 = 0.50

FIGURE 16.14 The founder effect The founder effect describes the result when a small number of seeds forms a new population or colony on an isolated area such as an island. The colonizing seeds may contain fewer alleles than are present in the parental population, and their gene pool will remain different even as the population increases.

Migration causes alleles to flow into or out of a population

Most species exist as a number of separate local populations spread out across an area called a **range**. Individuals of a species may migrate from one local population to another, and the seeds and pollen of plants may move between local populations. When migrating individuals reproduce with members of local populations, their genes join the local gene pool. Migration therefore represents a **gene flow** between populations. If the levels of migration and gene flow are high, they will keep the gene pools of the local populations similar. If migration and gene flow are minimal, however, natural selection and possibly genetic drift will make the gene pools of the local populations different. Thus migration counteracts the effects of natural selection and genetic drift.

Through natural selection, allele frequencies change such that populations become better adapted to their environment

Natural selection is the process by which individuals in a population with more favorable combinations of alleles and chromosomal arrangements will have higher survival and reproduction rates. The population as a whole becomes better adapted to its environment. Because the environment is always changing in some respect, natural selection and adaptation are a continual process.

The process of natural selection is not a random process but is highly selective. The selective agent is the environment, and the mechanism—differential survival and reproduction among individuals—will lead to different combinations of alleles. Natural selection is limited by the availability of alleles. Natural selection and evolution do not occur by chance. The source of the variation in a population, however, is mutation, and mutation is a random or chance event. Mutations are caused by physical agents such as transcription errors, radiation, chemicals, or chromosome breakage. When we say mutation is a random process, we mean that a mutation in any particular gene or chromosome occurs purely by chance. Radiation does not target specific genes! Mutation has driven the process of evolution since life originated on Earth.

There are three major types of natural selection based on the effects they have on a population over time—**directional selection**, **stabilizing selection**, and **disruptive selection**.

Directional selection   In directional selection, individuals with one trait at the extreme of its character range are favored over individuals with either the average or the opposite extreme for that trait. The evolution of the long necks of giraffes was likely due to directional selection. Giraffes with longer necks could reach more leaves than giraffes with shorter ones and therefore left more offspring. Martin Cody and Jacob Overton at the University of California, Los Angeles, reported a striking case of directional selection for dispersal ability among certain weedy plants in the daisy family (Asteraceae) growing on small islands near Vancouver Island, Canada (Figure 16.15). These weedy species produce wind-dispersed seeds that consist of a tiny seed and a fluffy parachute called a *pappus*. Cody and Overton found that when these weedy species colonized small islands, their ability to disperse seeds by wind declined rapidly. This decline was indicated by an increase in seed size and a decline in parachute size. Rapid directional selection against wind dispersal occurred because, on small islands, wind dispersal just scatters seeds into the ocean where they are lost. Evolution of herbicide resistance in weeds and antibiotic resistance in bacteria are other examples of directional selection.

Stabilizing selection   Stabilizing selection favors individuals with the average condition for a particular trait. In plants, stabilizing selection may be operating to determine the optimal number of flowers in a cluster (Figure 16.16). Flowering plants make flowers to produce seeds, but flowers also cost the plant in

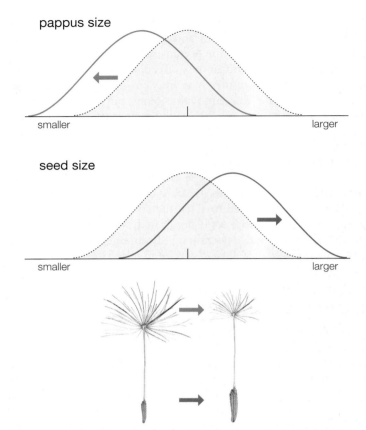

FIGURE 16.15   Directional selection An example of directional selection comes from a study of weedy members of the daisy family on islands off the coast of Vancouver Island in British Columbia, Canada. These weedy plants all have wind-dispersed seeds consisting of the tiny seed and a large mass of fluff called a *pappus*. The pappus acts as a parachute in wind dispersal. Plants on islands developed reduced dispersal ability by making larger seeds and smaller parachutes. On the islands, high dispersal ability meant being blown out to sea and therefore lost to the population. Natural selection acted rapidly in a directional manner to reduce dispersal within 5 to 10 years. The curves show that as one extreme of the normal distribution of the trait is favored (selected for), the distribution shifts in that direction.

terms of energy. Stabilizing selection may act to balance the cost of the number of flowers produced against reproductive success.

Disruptive selection   In disruptive selection (Figure 16.17), individuals in a population with the extremes of the range for a trait are favored over individuals with the average for that trait. A number of families of flowering plants contain species that have more than one form of flower. Darwin studied several such species and published his results in 1877. In species where there are two distinct floral forms, Darwin used the term *pin flower* for the form with a long female organ (called a *style*) and short male organs (*anthers*). Those with a short style and long anthers he termed *thrum flowers*. Pin and thrum flowers cannot fertilize themselves. They are self-incompatible. A pin flower requires pollen from a thrum flower to produce seed and, conversely, a thrum flower needs pollen from a pin flower. Thus the different flower forms promote exchange of pollen with other individual plants (outcrossing) and new genetic combinations. Disruptive

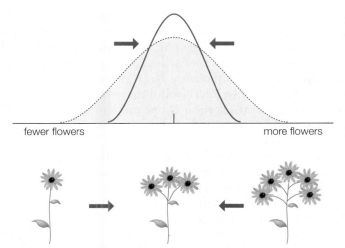

FIGURE 16.16 **Stabilizing selection** In stabilizing selection, the average condition is favored and so natural selection prunes the extremes, leading to a narrowing of the distribution. Stabilizing selection might act in controlling the number of flowers produced on a plant. Production of flowers entails a cost in energy to the plant, but the plant must produce flowers to reproduce. Too few flowers mean a reduced chance for reproduction. Too many mean too high a cost in energy to the plant. Stabilizing selection for flower numbers would balance reproductive success against the energy cost of producing flowers. An optimal number of flowers per cluster would result from selection operating against the extremes.

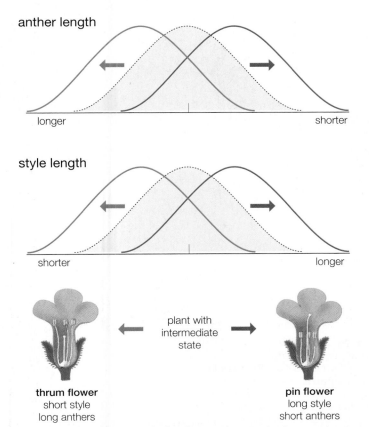

FIGURE 16.17 **Disruptive selection** In disruptive selection, the average condition is less successful than the extremes. Pin and thrum flowers evolved by disruptive selection from flowers with male and female organs of the same length. Pin flowers developed long styles and short anthers, whereas thrum flowers acquired long anthers and short styles.

selection is thought to have produced these two different flower forms out of a population with a single flower form with variable lengths of anthers and styles. The first step was the development of a gene for self-incompatibility. Disruptive selection then produced all the morphological differences between flower forms. For more on the evolution of different flower forms, refer to Chapter 24.

16.9 | New species originate through the development of reproductive isolation

In the remaining subsections of this chapter, we discuss the mechanisms by which natural selection brings about speciation, but first we will consider just what is a species.

The concept of species is based on genetic isolation

Species in Latin means "kind," and, in the simplest sense, species are distinct kinds of organisms. Historically, species were recognized based on their structure or morphology, where *morphology* refers to the form of organisms. This is the **morphological species concept**. But morphology alone is not sufficient to define species. Broccoli, kale, cabbage, and cauliflower are all members of the same species, *Brassica oleracea*; yet they are so distinct in form that if they were found wild in nature, they would likely be classified as separate species. With the development of population genetics as a science, a species came to be defined as a group of populations whose members interbreed with each other but either cannot or usually do not interbreed with members of other populations. This is the **biological species concept**, and it is based on reproductive or genetic isolation. Species are defined by the presence of reproductive barriers; that is, they do not reproduce or exchange gene pools with members of other species.

There are a number of problems with the biological species concept. One is that it applies only to species that reproduce sexually. Numerous microorganisms and plants can reproduce asexually. Many plants, like Mendel's garden peas, are self-fertilizing. Such species have to be defined on morphological or other characteristics. Another problem with the biological species concept arises when trying to use reproductive isolation as a criterion. Many species that do not interbreed in nature may do so if brought together in a greenhouse or laboratory. The biological species concept usually adds the provision that species do not commonly interbreed in nature.

If a population becomes isolated from other populations of the species, the gene pool is also divided. If the gene pools then become subject to different evolutionary forces, they will diverge along different evolutionary paths and become separate species. At present, there are considered to be two main patterns of speciation. The most common pattern is geographic isolation followed by differential evolutionary forces, a process known as **allopatric speciation**, where the word *allopatric* means "other homeland." Under special circumstances in animals, but rather commonly in plants, speciation can occur without geographic isolation in **sympatric speciation**, where *sympatric* means "same homeland." It is likely that these are not the only mechanisms of speciation.

Allopatric speciation requires geographic isolation

Darwin was never able to fully explain how new species could arise. The person who first proposed the mechanism for speciation was Harvard biologist Ernst Mayr in 1942. Mayr developed the biological species concept. He also proposed that speciation requires two factors—geographic isolation and genetic divergence. Geographic isolation is one means of preventing gene flow between populations. Geographic isolation can be achieved by many factors. Forests may be fragmented, glaciers advance and retreat on mountains, islands arise by volcanism, and large lakes may dry down and break up into smaller lakes (Figure 16.18). Long-distance dispersal can lead directly to geographic isolation. Oceanic islands like the Hawaiian Islands and the Galápagos are geographically

isolated from the continents, and the plants and animals that reached them were at once isolated from other populations. Mayr recognized that geographic isolation was not a condition sufficient to ensure speciation. During isolation, populations must acquire sufficiently large genetic differences that they will be unable to interbreed if they were to come in contact again. Natural selection is one mechanism by which the isolated populations could diverge and achieve speciation. Genetic drift is another.

Particularly striking examples of allopatric speciation occur when a plant species colonizes a geographically isolated island group. The islands present a wide range of new and essentially empty habitats for the colonizing plant species to fill. Consequently the plant species undergoes a sudden diversification into many new species, in a process called **adaptive radiation**. In the Hawaiian Islands, there is a group of native plants known as the silversword alliance, which consists of 28 species in 3 genera belonging to the sunflower family Asteraceae (Figure 16.19). These species show a tremendous amount of morphological variation. They range from small shrubs to tall trees and even vines, and they grow in a wide range of habitats from lava fields to forests and bogs. Despite this high degree of diversity in morphology and habitat, these species are all closely related and can interbreed. The

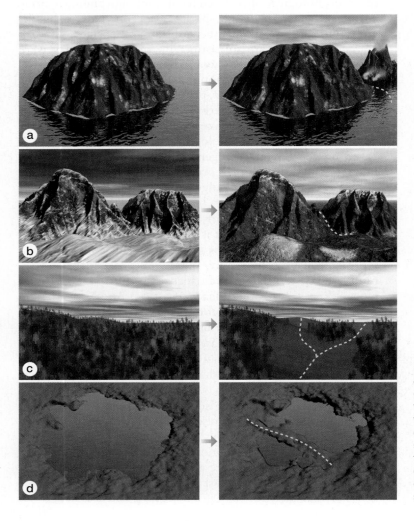

FIGURE 16.18 Barriers leading to allopatric speciation Four examples of geographic barriers that might lead to the formation of new species include (a) the formation of new islands; (b) the recession of glaciers to form isolated snow masses on separate mountain peaks; (c) the fragmentation of habitat, leading to isolated patches of vegetation; and (d) the drying down of a large body of water to form several smaller ones.

FIGURE 16.19 **Adaptive radiation of silverswords in the Hawaiian Islands** (a) Map of the Hawaiian Islands. Silverswords spread from Kauai to successive islands (arrow). (b) *Wilkesia gymnoxiphium*, (c) *Argyroxiphium sandwicense*, (d) *Dubautia ciliolata*.

silversword alliance appears to have evolved following the arrival of a single colonist from California on the island of Kauai, which was formed by volcanic action about 6 million years ago. Kauai contains the most specialized and distinctive members of the silverswords. The species found down the island chain are all the result of dispersal from Kauai. All the species show the result of genetic drift and founder effects, namely, loss of alleles and decreased variation in their gene pools.

Sympatric speciation can occur when polyploidy arises in plants

Sympatric speciation is especially common in plants because plants can form polyploids. Many organisms are diploid and have two complete sets of chromosomes, but many flowering plants—estimated at 47 to 70%—are polyploid and have four or more complete sets of chromosomes. Many important food crops such as wheat and potatoes are polyploids. Because of polyploidy, plants can give rise to new species in short time periods. Polyploidy can arise in plants by two processes.

In **autopolyploidy** (*auto* means "self"), the number of chromosomes doubles within individuals of a single species (Figure 16.20). Autopolyploidy usually arises by a failure of the chromosomes to separate during meiosis, producing diploid gametes. When these diploid gametes fuse in self-fertilization, they form tetraploid individuals with four complete sets of chromosomes that are a new species. The Dutch botanist Hugo de Vries was the first to observe autopolyploidy when a new species of evening primrose with double the normal number of chromosomes suddenly appeared among his plants.

More commonly, sympatric speciation occurs by **allopolyploidy** (*allo* means "other") in which two distinct species produce a fertile hybrid (see Figure 16.20). Normally, such a hybrid would be sterile because the chromosomes could not pair during

meiosis. But if the chromosome number doubles after fertilization due to a failure of the chromosomes to separate after duplication, the resulting plant will be fertile because its chromosomes can pair normally in meiosis. The resulting allopolyploid is a new species that cannot produce fertile offspring with either of the parental species but can breed with other similar allopolyploids.

One of the best-documented examples of new species arising by allopolyploidy can be seen in the genus of weedy plants, *Tragopogon* (goat's beard; Figure 16.21). Three species—*T. dubius*, *T. porrifolius*, and *T. pratensis*—were introduced and became naturalized in Washington State and Idaho. All three species were diploid, and each could cross with the others to form sterile hybrids. Then, in 1949, two new species of fertile tetraploid hybrids were discovered in Washington. One, named *T. mirus*, arose by allopolyploidy from *T. dubius* and *T porrifolius*; the other, *T. miscellus*, came from *T. dubius* and *T. pratensis*. The fertile polyploids have since spread over a number of western states, perhaps due to the superiority of the hybrids over their parents, a phenomenon called *hybrid vigor*.

Wheat plants of the genus *Triticum* constitute one of the most economically important polyploid groups. Common bread wheat is the most widely cultivated crop in the world. Sympatric speciation is therefore of immense economic importance. (For more about the development of wheat as a cereal crop, refer back to Chapter 5.)

Polyploidy is not the only mechanism by which sympatric speciation can occur. Recent studies suggest that new species may arise by changes in a few major genes that effect reproductive biology. Refer to Essay 16.1, "Big Beneficial Gene Changes Can Separate Species."

Microbial species can evolve in the lab

One long-running criticism of evolutionary theory is that no one has ever seen a species evolve. This criticism means that no one

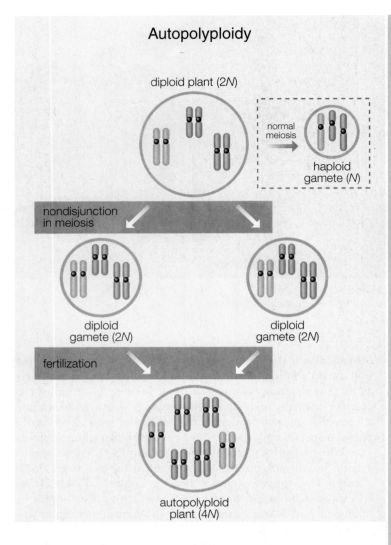

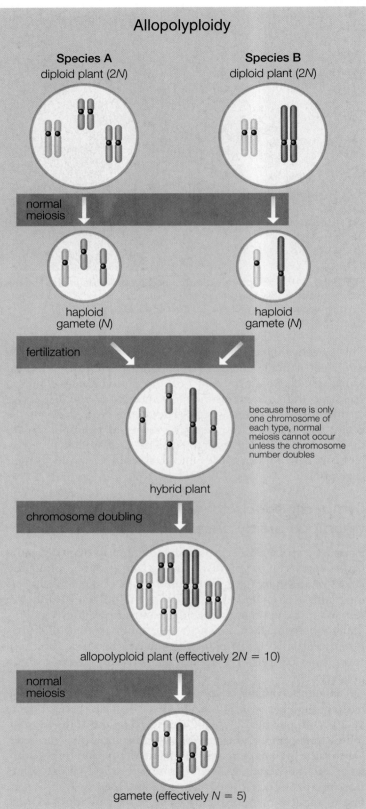

FIGURE 16.20 Autopolyploidy and allopolyploidy In autopolyploidy, an anomoly in meiosis (nondisjunction, i.e., nonseparation) leads to the formation of diploid gametes. These fuse to form an autopolyploid (tetraploid) plant. In allopolyploidy, two different species interbreed to form a hybrid individual. Such hybrids are typically sterile (their chromosomes lack corresponding homologous chromosomes, meaning they cannot successfully complete meiosis). If, however, the number of chromosomes is doubled (as occasionally occurs), the resulting allopolyploid plant can successfully reproduce through self-pollination or by breeding with a similar allopolyploid. (Recall that in plants, meiosis does not directly give rise to gametes but, rather, to spores.)

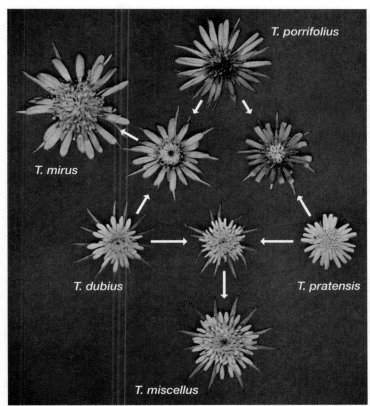

FIGURE 16.21 Hybrids in *Tragopogon* (goat's beard) The three diploid species of *Tragopogon* (*T. porrifolius*, *T. dubius*, and *T. pratensis*) are shown at the vertices of the triangle indicated by the white arrows. The arrowheads point to the sterile hybrids formed from the species at the vertices. Two of the sterile hybrids gave rise to fertile hybrids by allopolyploidy. *T. mirus* arose by allopolyploidy from *T. porrifolius* and *T. dubius*, and *T. miscellus* arose from *T. dubius* and *T. pratensis*.

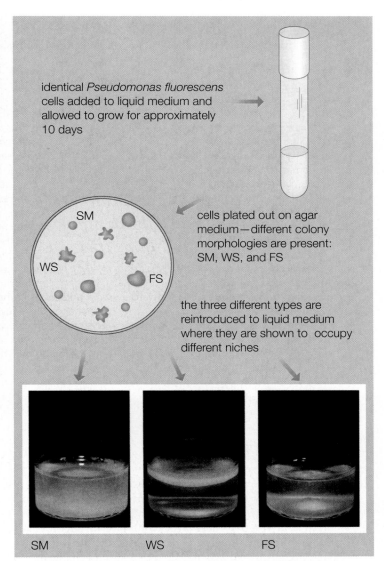

FIGURE 16.22 Rapid development of microbial diversity in vials of sugar water The common plant bacterium *Pseudomonas fluorescens* gives rise to three separate variants in stratified vials of nutrient medium within 10 days. The three variants are named for their colony appearance on agar plates: smooth morph (SM), wrinkly spreader (WS), and fuzzy spreader (FS). When grown in liquid medium, SM spreads throughout the medium, WS forms a film on the surface, and FS coats the bottom.

scientific observer has seen a new species arise in a time frame that is short compared to the life span of the observer. According to the standard evolutionary models, speciation should require many generations. For multicellular organisms, generation times are too long, compared to the life span of a research scientist, to follow evolution through hundreds, much less thousands, of generations. In the past decade, a number of investigators have solved this generation time problem by following evolution in microorganisms, where generation times are on the order of 3 to 4 hours or less, and genetic changes can be tracked by molecular techniques.

Paul Rainey and Michael Travisano at the University of Oxford used microbes to examine how organisms undergo adaptive radiations and how predictable they are. They seeded vials of sugar water with the common plant bacterium *Pseudomonas fluorescens*. The vials are not mixed so that they stratify into different environments. The surface layer has abundant oxygen, and the bottom layer has abundant nutrients but depleted oxygen. In almost every replicate vial *P. fluorescens* split into three variants. Rainey and Travisano named the three variants for the appearance of their colonies on agar plates (Figure 16.22). Smooth morph is the ancestral form that spreads through the liquid medium. Wrinkly

spreader grows on the upper surface as a mat of cells. Fuzzy spreader coats the bottom.

What happens in these little vials is analogous to what happens to a plant newly arrived on an island. The plant and the bacterium diversify into the available empty habitats. In the vials, the adaptive radiation leads each time to the same result, which is determined by the available environments. But the genetic paths taken by *P. fluorescens* to become wrinkly spreader are different in each case. The environment determines what the bacterium becomes by the process of evolution, but chance mutations determine how it arrives there. Studies such as these are transforming evolution into a replicative and predictable experimental science.

ESSAY 16.1 MAJOR BENEFICIAL GENE CHANGES CAN SEPARATE SPECIES

According to the standard model of allopatric speciation, two populations of a species are separated geographically, and speciation occurs by the slow accumulation of random mutations. But is this standard model of speciation correct? With modern molecular techniques, researchers can trace the actual mutations that separate species, and they are finding that a few big beneficial mutations mark major shifts in reproductive biology that can separate populations and drive them rapidly toward speciation.

Evolutionary biologist Doug Schemske and geneticist Toby Bradshaw at the University of Washington, Seattle, stud-

With modern molecular techniques, researchers can trace the actual mutations that separate species.

ied monkeyflowers in Yosemite National Park (1996). One species is pollinated by bees and another by hummingbirds (Figure E16.1). Genetic mapping revealed that the two species differed from each other in only a few genes, but changes in those few genes brought about changes in flower color, petal shape, and the amount of nectar produced. The resulting differences in floral display attracted different pollinators and kept the two species reproductively isolated. This work and several other recent studies show that changes in a few genes with large effects can produce sudden changes in basic biology and rapid adaptation. Evolution and speciation may occur faster than previously imagined.

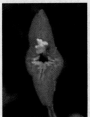

E16.1 Switch of pollinators in monkey flowers *Mimulus lewisii* (on left) is pollinated by bees and the bright-red *M. cardinalis* (on right) by hummingbirds (recall that bees cannot see red but birds can). Changes in three major genes switched this plant from bee-pollinated to hummingbird-pollinated in a sympatric speciation event. Separate pollinators now maintain their reproductive isolation.

ECOLOGY

HIGHLIGHTS

16.1 Until the early 19th century, the natural sciences believed that the Earth was young and that species were unchanging. The science of geology, however, established that rocks occurred in layers, that these layers contained distinct fossils, and that the older the layers, the less the fossils resembled living organisms. Fossils indicated that species changed. If normal geological processes formed these rock layers, then the Earth must be very old.

16.2 Several early naturalists proposed that species might be capable of changing, but they could not propose a reasonable mechanism of change.

16.3 During the voyage of the *Beagle*, Charles Darwin visited the Galápagos Islands, where he observed that single species seemed to have been modified to produce many different forms.

16.4 Darwin knew that artificial selection, as practiced by plant and animal breeders, could effect significant changes in domestic species in a few generations. Because the individuals of all species vary in their characteristics, selection of some individuals and elimination of others could be the key to evolution. An essay by Thomas Malthus on human population growth suggested a mechanism for natural selection.

16.5 The theory of evolution can be summarized in a series of statements. (1) Every species can produce more offspring than can survive. (2) Natural populations tend to remain fairly constant in size over time. (3) Thus there is com-

petition for survival. (4) Individuals vary in traits that affect their survival. (5) Therefore, individuals with the most advantageous traits survive to produce the most offspring.

16.6 The synthetic theory of evolution combines Darwinism with genetics and molecular biology. Variation in individuals is due primarily to mutations that produce new alleles and recombination that produces new combinations of alleles.

16.7 Numerous areas of scientific data support the synthetic theory of evolution, including artificial selection, comparative anatomy, changes in proteins and DNA, and the fossil record.

16.8 Every population possesses a gene pool that is the total of all the alleles of all the genes in that population. Evolution occurs when forces change the frequencies of alleles in a gene pool. These forces include mutation, nonrandom mating, genetic drift, migration, and natural selection.

16.9 New species arise through the development of reproductive isolation. In allopatric speciation, species become geographically isolated and undergo genetic divergence. In sympatric speciation, populations become reproductively isolated due to polyploidy or the sudden appearance of changes in major genes that effect a shift in reproductive biology. The evolution of microbial species can now be followed in the laboratory.

REVIEW QUESTIONS

1. Prior to Darwin's release of his theory of evolution, what were the prevailing beliefs for the age of the Earth and the origin of species? How did the science of the day account for the variation seen within a species? Did everyone believe that species were immutable and unchanging?

2. Summarize the chain of logic that Darwin put forth in support of the idea of natural selection. What is the significance of natural selection in regard to his theory of evolution?

3. How are artificial selection and natural selection similar, and how are they different?

4. Briefly list some lines of evidence that support evolution.

5. What is the difference between gradualism and punctuated equilibrium? Which one is currently believed to best account for the appearance of new species in the fossil record?

6. Define the following words: (a) *population*, (b) *gene pool*, (c) *population genetics*, (d) *microevolution*, (e) *genetic drift*, and (f) *gene flow*.

7. Describe several types each of gene mutations and chromosome mutations.

8. Distinguish between aneuploidy and polyploidy, and describe how each is produced.

9. What are the differences between (a) directional selection, (b) stabilizing selection, and (c) disruptive selection?

10. How does the morphological species concept differ from the biological species concept?

APPLYING CONCEPTS

1. A mutation is a change in the sequence of DNA in a cell. Are all mutations heritable?

2. In the late 1800s, an extremely powerful winter storm killed many sparrows. An ornithologist named Hermon Bumpus collected some of the dead birds and compared several of their traits with those of birds who had survived the tempest. His studies revealed a tendency for the dead birds to possess abnormal characteristics (characters at the extremes of the character ranges), whereas the surviving birds possessed traits clustering closer to the norm. What type of selection had Bumpus discovered? Would this type of selection be the basis for evolutionary change?

3. Why are only heritable mutations and phenotypes important to the process of evolution?

17

Naming and Organizing Plants and Microbes

The genus *Solanum* includes nightshade, potato, and tomato

- The use of scientific names for plants and other organisms helps avoid communication problems and errors in agriculture, commerce, medicine, and other areas of human activity.

- Scientific names are based on classical Greek or Latin root words and therefore are not politically offensive, and their meanings do not change through time.

- Taxonomic keys are used to distinguish and identify plants. Collections of expertly identified plant specimens are stored as reference "libraries" at museums, botanical gardens, universities, and other places.

- A system for organizing, or classifying, organisms is necessary because there are so many different kinds.

- Modern classification systems are based on relatedness, recognized by similar DNA and reproductive characteristics.

Beautiful but deadly, the plant *Aristolochia fangchi* was mistakenly administered to 10,000 Belgian people in herbal diet pills prescribed by weight-loss clinics in the 1990s. Dozens of these patients suffered kidney failure and cancer because chemical compounds in the *Aristolochia* had bound to their DNA, causing mutations. Many of the victims had to undergo surgery to remove their damaged organs. This medical error resulted from a slipup in plant identification. *Aristolochia*, which was previously known to damage kidneys and cause cancer in animals, had been mistaken by pharmacists for another plant, *Stephania tetrandra*, which had long been used by Chinese herbalists in weight-loss treatments without obvious harm.

Other species of *Aristolochia*—known by the common names *birthwort, heartwort,* or *snake root*—have been used as herbal remedies by ancient Egyptians and Native Americans, and in India, Africa, and Europe to treat a wide range of illnesses from skin diseases to infections. Ancient people believed that plant flowers, roots, or other parts were shaped in ways that indicated their medicinal uses—the medieval doctrine of signatures. *Aristolochia* has flowers that are curved, somewhat resembling a fetus or a snake. *Aristolochia* was thus widely used in childbirth and cases of snakebite, though today we know there is no scientific basis for such treatments.

Now that *Aristolochia* is known to be toxic to humans, import or use of this plant as a medicine has been legally restricted in Germany since 1981, in Belgium since 1992, but in the United States since only 2000. The dramatic misuse of *Aristolochia* that occurred in Belgium graphically illustrates why it is important for people to have accurate methods for naming and identifying plants that can be communicated across language, cultural, and national boundaries.

Scientific names were briefly introduced in Chapter 1. In this chapter, we explain how newly discovered organisms are given scientific names—the field known as **taxonomy**. The chapter also describes plant identification—an essential component of plant science and an enjoyable hobby for many people—as well as **systematics**, the methods by which biologists organize and classify plants and other organisms. Professional biologists who specialize in naming and classifying organisms are known as taxonomists or systematists. ■

Aristolochia

17.1 | Scientific names originated with Linnaeus, the father of biological taxonomy and systematics

The system of scientific names, still used today to describe living things, was first described in the book *Systema Naturae*, published in 1758 by Carl von Linné. Also known as Carolus Linnaeus (scholars of the time normally used the latinized form of their names), he was a professor of both medicine and botany at Uppsala University near Stockholm, Sweden. Linnaeus was also a pioneer ethnobotanist. At the age of 25, he visited the Sami people of Lapland, learning their language and their uses for plants in much the same way ethnobotanists do today.

Linnaeus' invention of scientific names was somewhat accidental. Linnaeus described all plants then known to Europeans in his book *Species Plantarum* ("Species of Plants") published in 1753. At that time it was proper procedure to give each plant or animal a Latin name up to 12 words long, and Linnaeus followed this tradition. But these names were so cumbersome to write, and so difficult to remember, that Linnaeus used a shorter, two-word name (binomial) for his own convenience. His shorthand idea became popular, and now a **binomial scientific name** is given to each of the millions of known plants, animals, and microbes. Viruses are exceptions to the rule; their names are composed of more than two words (Essay 17.1 "Viruses—Extreme Minimalists").

Originally, Latin was used to name organisms because it was the language of scholarship. Latin and classical Greek continue to be useful for scientific names because these languages are no longer changing, and they are not politically offensive to anyone. Though their use might at first seem pretentious—scientific names might appear strange to those who have not previously studied classical languages—Latin- and Greek-derived names are advantageous because they can be recognized anywhere in the world, by speakers of any language. Places where you would commonly encounter plant scientific names include gardening catalogs, tags on potted plants or trees for sale in a nursery, or labels on plants in display gardens (Figure 17.1).

Each kind of organism has a unique scientific name

Each species of organism has a unique scientific name. What does this mean? It does not mean that each individual has a different name (as is common for people), but, rather, that the whole population of organisms of the same type has the same scientific name. A population is all of the individuals of a particular species that occupy a defined area at the same time. For example, all of the more than 6 billion human beings on Earth have the same scientific name—*Homo sapiens*, meaning "wise human." In contrast, populations of very similar types of creatures that do not interbreed must be given different species or generic names, depending on the degree of difference. If the distinction is small, the two

organisms may be designated as separate species belonging to the same genus. If the difference is larger, the two organisms must be placed into different genera (*genera* is the plural form of *genus*). The animals most closely related to humans are chimpanzees and bonobos, but because these animals differ from humans in many ways, they are classified into the genus *Pan* (after the Greek god who was thought to control fertility of the land). Genera may contain from one to many species (Figures 17.2 and 17.3). Although only one species of *Homo* lives today (that's us), fossil evidence indicates that in the past several other *Homo* species existed and have since become extinct.

Scientific names are structured to provide useful information

The two parts of a binomial scientific name are first, the **generic name**, which is a noun, and second, the **specific epithet**, an adjective (Table 17.1). Together they form the name of the species. A **species** is a group of individual organisms that are similar because they are descended from a common ancestor. All of the members of a group of closely related species have the same generic name. In the scientific name for coffee, *Coffea* is the generic name and *arabica* is the specific epithet. The generic name can be used by itself but the specific epithet, used alone, is meaningless because a wide variety of very different organisms can have the same specific epithet. For example, many different plant genera contain a species named *vulgaris*, because this epithet means "common."

In addition to providing accurate communication, scientific names are often designed to convey information about an organism, such as its appearance, location, or usefulness. For example, in the scientific name for the cacao plant, *Theobroma* means *"food of the gods,"* implying that it tastes really, really good. In the scientific name for the coca plant, *Erythroxylum* means "red wood"—the wood of this plant has a red color. On occasion, scientific names may honor a scientist or some other real or mythical person. An example is the twinflower, *Linnaea borealis* (Figure 17.4), named after Linnaeus. Some generic names are also used as common names. Some familiar plant examples include *Vanilla, Hydrangea, Narcissus* (Figure 17.5), *Gladiolus, Hibiscus*, and *Begonia;* an animal example is *Gorilla*, another primate closely related to humans and chimps.

There is a correct way to write scientific names

It is important to write scientific names correctly so that they can be recognized. The first letter of the generic name is typically capitalized (viruses are exceptions), but the first letter of the specific epithet is usually not capitalized. Also, scientific names are typically underlined or italicized. When an organism's name is used many times or several species of the same name are listed, the generic name can be abbreviated. For example, among the 60 or so species of *Coffea, C. arabica, C. canephora*, and *C. liberica* are the major cultivated species.

ESSAY 17.1 VIRUSES—EXTREME MINIMALISTS

Viruses and viroids are infectious particles that are much smaller than the smallest bacterial cell and can be viewed only with the use of an electron microscope. Viruses consist primarily of small amounts of genetic material—either DNA or RNA, but not both—enclosed within a protective protein coat. Viroids lack the protein coat and consist of naked nucleic acid. Viruses lack typical cell structure and rely on host prokaryotic or eukaryotic cells for metabolic functions and resources needed for reproduction. A single virus can generate hundreds or thousands of new viruses within a single infected host cell (Figure E17.1A). Because viruses lack some of the properties of living things (cellular structure and metabolism), they are not

More than 400 types of viruses are known to cause more than a thousand plant diseases, which typically stunt plant growth and reduce crop yields.

alive, strictly speaking. But because they have such minimal structure, viruses can evolve very quickly—that is, rapidly change characteristics from one generation to another. Manufacturers of flu vaccines have to take evolutionary change in human flu viruses into account when they decide which types of vaccine to produce each year.

Studies of DNA sequences indicate that viruses did not have a single origin or common ancestor. Various groups of viruses probably originated at different times from cellular organisms by escape of small bits of DNA or RNA. This explains why viruses have limited host ranges, meaning that they can cause disease in only a limited range of related host species. Diverse viral origins also explain why viruses that attack bacteria do not attack eukaryotes (and vice versa), and why animal and plant viruses are usually quite different from each other. More than 400 types of viruses are known to cause more than a thousand plant diseases, which typically stunt plant growth and reduce crop yields. Symptoms of plant viral disease include leaf curling, ring-shaped spots, light green or yellow blotches, or streaks on leaves or other green parts (Figure E17.1B). The latter are known as mosaic diseases because of the mosaic-like mottling patterns. There are no chemical treatments for plant viral diseases, so crop scientists try to reduce infection by breeding plants that are resistant to viral disease or by controlling the spread of viruses from one plant to another. Plant viruses are commonly spread from infected plants to uninfected ones by insects that use piercing-sucking mouthparts to suck plant juices. Viruses can also enter plants via surface abrasions and can be transmitted in cuttings, seeds, pollen, or grafts. Streaked tulips (Figure E17.1C) are examples of an

E17.1B Plant infected by a virus

inherited viral infection. Viruses move from one plant cell to another via plasmodesmata, and they move throughout plants in the phloem sap.

Names of viruses are determined by the International Committee on the Taxonomy of Viruses (ICTV). Viral family and genus names are capitalized and italicized, but species names are not. The first component of a virus's name reflects the host, the next components are symptoms, and the final word is the genus. Tomato ringspot *Nepovirus*, clover yellow vein *Potyvirus*, and carrot red leaf *Luteovirus* are examples of virus names. Viruses are distinguished and classified based on their sizes, shapes, coverings, and genes.

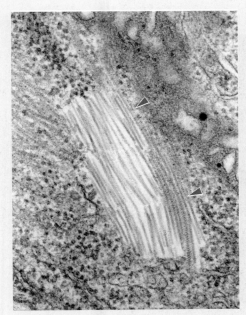

E17.1A Viruses within a plant cell
Tobacco mosaic viruses (elongate tubes—arrowheads) in a mesophyll cell of tobacco leaf

E17.1C Streaked tulips

FIGURE 17.1 Scientific names are included on labels of plants in botanical gardens and greenhouses

Plant names signify subspecies, varieties, or cultivars

The two-part scientific name may be followed by a subspecies name. A **subspecies** is a group of organisms that inhabits a geographical area or type of habitat different from other members of the same species. An example is corn, *Zea mays* subspecies *mays* (Figure 17.6c). Its ancestor—the tropical grass teosinte, which grows in southern Mexico—is *Zea mays* subspecies *parviglumis*.

Scientific names of commercial ornamental or garden plants are often followed by a variety name or cultivar name in the consumer's language. **Varieties** are members of the same species that have small, but consistent, differences from each other. Varieties may occur in nature but more commonly result from human cultivation. A **cultivar** is a variant that does not occur in nature and grows only in cultivation. Cultivar names are designed to attract customers. For example, *Orchid Lace*, *Rainbow Loveliness*, and *Sonata* are poetic names given to some of the cultivars of the ornamental flowering plant *Dianthus* (common name "pink").

Hybrids have distinctive scientific names

Hybrids result from crossbreeding between two species or genera. Plants appear to form hybrids more easily than animals, both in nature and in cultivation. Hybrids that maintain characteristics different from their parent species over many

FIGURE 17.2 Genera may include as few as one species The ginkgo (*Ginkgo biloba*) is the only living species in its genus. A tree is shown in (a) and the leaves in (b).

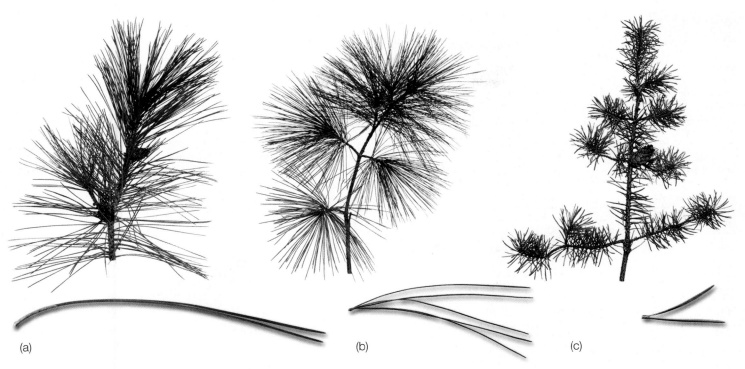

FIGURE 17.3 **Genera may include many species** The conifer genus *Pinus* (pine) includes many species, including (a) red pine (*P. resinosa*), (b) white pine (*P. strobus*), and (c) jack pine (*P. banksiana*).

generations (and that do not crossbreed with them) are given their own species names. One example is a desert sunflower, *Helianthus anomalus*, which arose from two other species, *H. annuus* and *H. petiolaris*. If a hybrid has resulted from crossing of two genera, the hybrid's generic name combines all or parts of the parents' names. For example, the hybrid orchid, *Laeliocattleya* (Figure 17.7) is made by crossing the parents *Laelia* (which provides bright colors) and *Cattleya* (which contributes attractive shape).

Naming new plants must follow an established procedure

New kinds of plants are continuously being discovered in natural environments. For example, in 1994, a new genus of conifer tree, more than 40 meters tall, was found in a national park not far from Sydney, Australia, and named *Wollemia nobilis* (Figure 17.8). Although *Wollemia* is rare—there are only about 40 plants in nature—it is so beautiful that the tree is being propagated for future markets. In addition, plant breeders constantly generate new types of cultivated plants for sale in garden stores. Newly discovered or artificially produced plant types require new names.

Naming a new plant involves following several required steps. Rules for naming plants are listed in the International Code of Botanical Nomenclature (there is a separate code for naming animals, bacteria, and viruses). These rules are established by professional taxonomists who meet at intervals to revise and refine them. The discoverer must check reference books listing plant names to make sure that the chosen name

has not been used before. He or she must also deposit a reference specimen of the plant in one or more herbaria. Last but not least, the discoverer must publish a description of the new plant in a scientific journal that is widely available to botanists around the world. The description must explain why the new plant is different from any other plant previously named and must include a Latin description of the new plant. Following these steps is a lot of work, but the reward is a measure of immortality because the discoverer's name (as the describing authority) forever follows the specific epithet when it is formally presented (Figure 17.6).

17.2 | Identifying and archiving plant specimens

Plant collections can be stored as herbarium specimens

Around the world there are many important collections of plant specimens; these collections are known as **herbaria** and are often associated with major universities, museums, or botanical gardens (Essay 17.2, "Botanical Gardens: Science and Art All in One!"). Herbaria, which often contain thousands of specimens from all over the world, serve as a kind of reference library. Specimens that have been expertly identified are widely exchanged among herbaria for study by plant scientists throughout the world. Plant collections are also sources of DNA used by evolutionary biologists to trace plant relationships. It is easier

DNA SCIENCE

TABLE 17.1	Plant-Specific Epithets Cited in This Chapter, with Their Meanings
Epithet	**Meaning**
annuus, annua, annuum	annual
anomalus, anomala, anomalum	unusual
arabicus, arabica, arabicum	of Arabia
aureus, aurea, aureum	golden yellow
borealis, boreale	northern
cacao	Aztec name for cocoa tree
canephorus, canephora, canephorum	like a basket bearer
coca	name used by South American Indians
communis, commune	growing in clumps
cryophilus, cryophila, cryophilum	cold-loving
deliciosus, deliciosa, deliciosum	delicious
edulis, edule	edible
esculentus, esculenta, esculentum	tasty
fluitans	floating on water
glomeratus, glomerata, glomeratum	aggregated
libericus, liberica, libericum	from Liberia, West Africa
mays	from Mexican name for Indian corn
methysticus, methystica, methysticum	intoxicating
nitidus, nitida, nitidum	glossy
nucifer, nucifera, nuciferum	nut-bearing
petiolaris, petiolare	having a leaf stalk (petiole)

FIGURE 17.4 Some plants are named to honor people *Linnaea borealis*, whose common name is twinflower, is named after Linnaeus, the father of taxonomy and systematics.

FIGURE 17.5 *Narcissus* Also known as the daffodil, this is an example of a species whose scientific name is also used as a common name.

to obtain DNA from a small sample taken from a stored herbarium specimen than to make a long and expensive collecting trip to a remote part of the world. People also use herbarium collections of plants to teach and learn about more kinds of plants than are available locally.

Herbarium specimens of plants commonly consist of dried plants that have been pressed and mounted onto high-quality paper sheets labeled with useful information about where the plant was collected, its name, and who identified it (Figure 17.9). In this form, plant specimens occupy little

(a)   (b)   (c)   (d)

FIGURE 17.6 Plants and an alga known and named by Linnaeus (a) The bryophyte *Riccia fluitans* L., (b) the fern *Polypodium aureum* L., (c) the seed plant *Zea mays* L., and (d) a green alga that was renamed *Cladophora glomerata* (L.) Kütz. by a later expert. The "L." following the binomial is short for Linnaeus. One can identify a renamed organism by the fact that the name (or abbreviation) of the person(s) who first named it appears in parentheses, followed by the name of the person(s) who renamed it.

FIGURE 17.7 Plant hybrids have distinctive names *Laeliocattleya* is a beautiful orchid hybrid developed from two different parent genera, *Laelia* and *Cattleya*.

FIGURE 17.8 New genera and species are constantly being discovered and named *Wollemia nobilis* is a very large conifer tree that was only recently discovered in Australia.

space, do not readily break apart, and are easily examined. Amazingly, dried, pressed plants retain many aspects of the appearance of the living plant. Plant collections often contain frozen plants and specimens in bottles of liquid preservative as well.

Amateur botanists often find that making personal collections of plant specimens is an enjoyable hobby. However, collectors should not remove rare plants from their habitats, as this can lead to local extinction. When new specimens are added to personal or professional herbaria, they must be identified. Learning to correctly identify plants is also essential for anyone who collects wild plants for food or herbal medicines, because some plants are poisonous and many plants contain toxic compounds. Ecologists and conservation biologists also need to identify plants in the natural habitats they study. How do amateurs and professionals identify plants?

ESSAY 17.2 BOTANICAL GARDENS: SCIENCE AND ART ALL IN ONE!

Botanical gardens are excellent places to see representatives of the major flowering plant families, as well as gymnosperms, ferns, and bryophytes. Botanical gardens are primarily collections of living plants, but they sometimes also serve as a repository of seeds and/or dried herbarium specimens. Many countries maintain one or more botanical gardens; together, they contain living examples of about one-third of the known plants of the world. Botanical gardens serve as a scientific resource for botanists who need specimens of plants that otherwise might be very difficult to obtain. They also conserve rare species that are in danger of extinction in their natural habitats. Botanical gardens are sources of material used by government agencies, seed companies, and the pharmaceutical and biotechnology industries. Botanists associated with botanical gardens travel to remote parts of the world, whose plants are poorly known, in order to collect specimens for study, work closely with local biologists,

Many countries maintain one or more botanical gardens; together, they contain living examples of about one-third of the known plants of the world.

and help train indigenous people to catalogue and conserve their own natural resources.

Botanical gardens are also visited by landscape architects and home gardeners looking for new ideas. Artists and photographers find gardens to be sources of creative inspiration, and botanical gardens are popular places for social events, such as weddings. The Royal Botanical Gardens at Kew, in the United Kingdom, is one of the most famous gardens in the world, with scenic vistas and large glass conservatories that protect many unusual and delicate plants (Figure E17.2A). The Missouri Botanical Garden in St. Louis features many plants native to the continental United States, as well as collections of tropical and desert plants maintained in greenhouses, including the famous Climatron (Figure E17.2B). In this geodesic dome, the climate is carefully controlled so that tropical plants of many types can be maintained. A Japanese garden, with traditional plantings of mosses, is another special feature of the Missouri Botanical

E17.2C   Desert Botanical Garden

Garden. The Desert Botanical Garden, located near Phoenix, Arizona, is a good place to see examples of desert plants, including many varieties of cacti (Figure E17.2C). There are also outside exhibits depicting ancient Native American uses of desert plants. The Fairchild Tropical Garden's collection of palm species and other outdoor exhibits of tropical plants are spectacular (Figure E17.2D). These gardens are located in Coral Gables, Florida.

E17.2A   Kew Gardens

E17.2B   Missouri Botanical Garden

E17.2D   Fairchild Tropical Garden

Resources for identifying plants include identification keys

Among the resources that can aid nonexperts in identifying plants (and other organisms) are **identification keys** (Figure 17.10), which are widely available on websites or in the nature section of bookstores. Identification keys illustrate plants and provide strategies for their identification. They often include

important information about plants, such as their toxicity or relative rarity. Dictionaries of terms used by botanists to describe the many plant parts, textures, and shapes that are used in identification keys are also helpful for beginners.

Professional biologists use **dichotomous keys** to identify organisms. The word *dichotomous* means "forking into two branches," so using these identification keys involves making a series of choices between two possibilities. The final choice

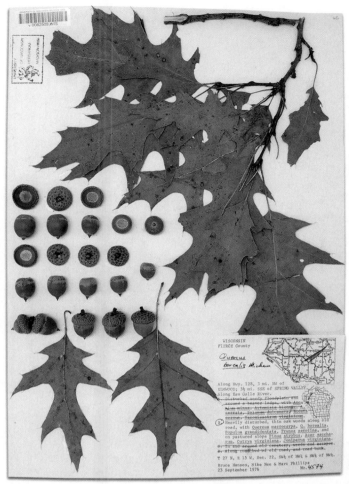

FIGURE 17.9 Herbarium specimen A pressed, dried plant specimen with informative label is a major type of herbarium specimen.

FIGURE 17.10   Some plant identification keys are designed for nonexperts

provides the organism's name (Figure 17.11). A good way to learn to use identification keys and to gain confidence in your ability to identify plants is to "key out" local plants whose names you already know, such as labeled plants in a display garden of native plants. If you fail to get the right answer the first time, you can backtrack to determine which of your choices were incorrect.

17.3 | Plants and other organisms are classified according to their relationships

There are more than 250,000 known species of plants, including flowering plants, gymnosperms, ferns, lycophytes, and bryophytes. Species of algae, fungi, and bacteria are thought to number in the millions. How can biologists bring order to this tremendous diversity? In the past, biologists have devised classification systems that grouped organisms according to their outward appearance, but modern systematists try to organize plants according to their genetic relationships.

The first person known to have tried to classify plants was Theophrastus, who lived some 2,300 years ago. He organized the 500 plants known by the ancient Greeks into three main categories—trees (having one main stem), shrubs (having multiple main stems), and herbs (nonwoody plants). Plants were then classified into smaller groups on the basis of leaf characteristics.

Today we know from DNA and other evidence that trees are not all closely related to each other. Some trees are more closely related to various shrubs or herbs than to other trees. There are also many examples of plants having similar leaves that are not at all closely related to each other (Figure 17.12), as well as many plants that are very closely related, but have quite different leaves (Figure 17.13). Why is this? The structure of plant stems, leaves, and roots is often more strongly influenced by the environment in which the plant has evolved than by its relationship to other plants. Unrelated plants growing in similar environments will often have similar leaves or stems because these are adapted in ways that help plants cope with drought, shade or other stresses.

Linnaeus advanced the science of classification by using floral characteristics to organize flowering plants. This method is a better way to group related plants together because reproductive characteristics (such as features of flowers and DNA) often reflect relationships more accurately than vegetative characteristics (features of stems, leaves, and roots). Reproductive characters (Figure 17.14) tend to be more stable to environmental difference because large changes in them could impair a plant's ability to mate and leave progeny.

Modern classification systems are based on **phylogenetic systematics**, a field of study in which biologists organize organisms according to their ancestry—**evolutionary relationships**. Modern systematics relies heavily on an approach known as **cladistics** or **cladistic analysis**, which

DNA SCIENCE

EVOLUTION

1. Unicellular or sometimes attached temporarily after cell division 2

1. Filamentous 5

    2. Cells elongate, sometimes arc-shaped . . . . *Closterium*

    2. Cells not elongate or arc-shaped 3

3. Cells incised only at the isthmus, apices of semicells without processes . . . . *Cosmarium*

3. Semicells variously incised 4

    4. Cells flat in end view . . . . *Micrasterias*

    4. End views of cells radiate . . . . *Staurastrum*

5. Cells with scarcely perceptible median constrictions . . . . *Hyalotheca*

5. Cells with obvious median constrictions . . . . *Desmidium*

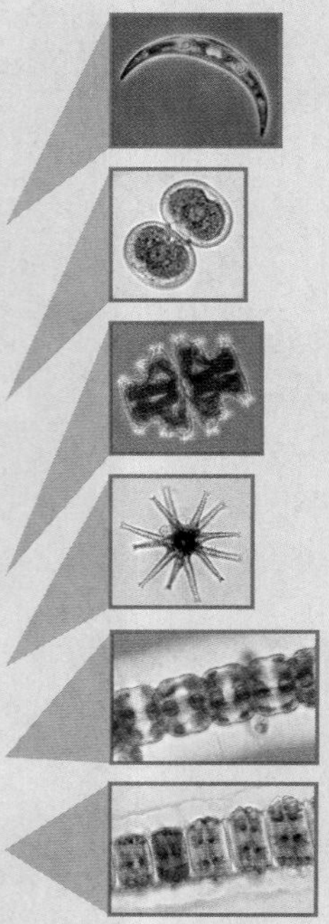

FIGURE 17.11 **Dichotomous keys** Professional biologists use dichotomous keys that present a series of two choices (arrows). Shown here is a portion of a key to a group of green algae, to which photographs of the organisms have been added.

a   b

FIGURE 17.12 **Unrelated plants can have similar leaves** *Ricinus* (castor bean) leaves (a) are shaped much like the distinctive leaves of (b) *Begonia ricinifolia* (meaning "leaves like *Ricinus*"), but these plants are not closely related.

helps biologists infer relationships by comparing organisms' characteristics. Cladistics (from the Greek word *klados*, meaning "tree branch") is a method for analyzing the characteristics of groups of organisms in order to infer their evolutionary relationships. Cladistic analyses yield diagrams known as **phylogenetic trees** (Figure 17.15) that have stems, nodes, and branches (shoots), and often roots, much like natural trees. Phylogenetic trees reflect concepts of evolutionary relationship among the organisms in the group being studied.

Structural, biochemical, or molecular (commonly DNA sequences) characteristics used in cladistic analyses are known as **characters**. Generally, the greater the number of characters and species examined, the better the result. DNA sequences are widely used to infer relationships among modern organisms because sequences contain very large numbers of characters. In cladistics, it is important to distinguish between ancient characters that may be widely present in a group and thus not very informative about evolutionary divergence, and

FIGURE 17.13   Related plants can have very different leaves Although these *Begonia* species are very closely related, their leaves are very different. The casual observer might be misled into believing they were not closely related.

shared, **derived characters**—present in some, but not all, members of a group because of inheritance from a common ancestor. A derived character, also known as an **apomorphy**, is one that has evolved from a more primitive character (plesiomorphy) occurring in an ancestor. A group of related organisms that is defined by at least one shared, derived character is known as a **clade.** A shared, derived character that defines a clade is called an **synapomorphy.** A clade is a **monophyletic group** because its members have all descended from a single (mono) common ancestor (as revealed by shared, derived characters).

Phylogenetic trees are constructed by using computer programs. It is important to recognize that not all branches of a phylogenetic tree may be equally well supported by the available data. A computer procedure called *bootstrapping* (from the expression "pulling oneself up by one's own bootstrap") yields a percentage of the time that the same branch reappears when the character database is resampled 100 or more times. The greater this number, the more confidence one has in the clade defined by a given branch. A bootstrap value of 95% or greater means that a clade is well supported, but a lesser value may indicate the need to add more character data or more species to the analysis for better resolution. Ideally, all of the classification groupings—species, species complexes, genera, families, orders, classes, phyla, kingdoms, and domains (described next)—should be well-supported clades.

FIGURE 17.14   Reproductive characters are useful in inferring evolutionary relationships Flowers of different *Begonia* species are very similar in form, reflecting close relationship.

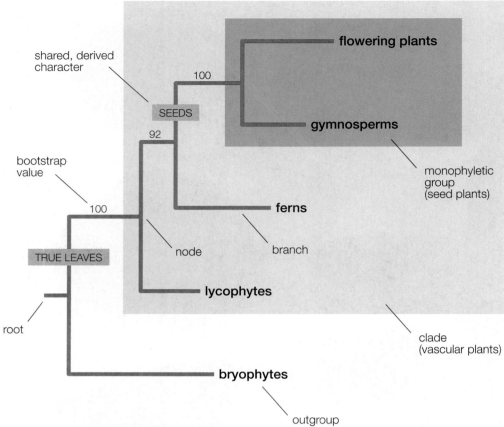

FIGURE 17.15 **An example of a phylogenetic tree illustrating the relationships of land plant groups** The major land plant groups are defined by shared, derived characteristics such as "true leaves" (characteristic of all groups except bryophytes) and "seeds" (characteristic of gymnosperms and flowering plants, but not bryophytes, lycophytes, and ferns). In this tree, the branches extend to the right, but phylogenetic trees can have other orientations.

Species can be grouped into species complexes, genera, families, orders, classes, phyla, and domains

Linnaeus' classification system, based primarily on one floral feature (the number of stamens), has been replaced by the use of many types of information, including secondary chemical compounds (Chapter 3) and fossils, in addition to DNA and other molecular evidence, and structural characteristics. These data are used to group related **species** into **species complexes** and **genera.** Species complexes are groups of species that are particularly closely related and thus may be important in breeding new types of plants. The wild grass *Zea diploperennis*, which is so closely related to corn (*Zea mays*) that the two species can interbreed, contains valuable genes that can be bred into corn for resistance to viruses. *Zea diploperennis* and *Zea mays* are members of the same species complex. Another example is the tomato, *Solanum lycopersicon (formerly Lycopersicon esculentum)*, with two species complexes—one named Esculentum (meaning "edible") and the other named Peruvianum (meaning "from

Peru"—where the plants are found). The cultivated tomato is classified in the Esculentum complex, but breeders incorporate genes from wild relatives of the Peruvianum complex into hybrid seeds, because they confer greater disease resistance and stress tolerance.

Related genera are gathered into larger aggregates known as **families** (Table 17.2). Related families are collected into orders, and related orders are assembled into larger groups known as **classes.** Classes are distributed among several **phyla.** (Some plant biologists use the term *division* instead of phylum, but the latter term is widely used today for plants as well as animals and microbes.)

Each species of plant (or other organism) belongs to a specific family, order, class, and phylum (Table 17.3), and each taxonomic level includes more organism types than the one below. For example, more species of plants will be contained in a phylum than any family within it, because a phylum is a much more inclusive group.

Several plant phyla (some consisting entirely of extinct species) together form the plant kingdom, formally known as the Kingdom Plantae. Other kingdoms include Animalia (the

DNA SCIENCE

TABLE 17.2   Examples of Economically Important Flowering Plant Families

Family	Characteristics	Example genus
Asteraceae	many flowers are grouped to form a head	*Helianthus*, sunflower
Brassicaceae	flower petals arranged in a cross shape	*Brassica*, broccoli
Cactaceae	most have fleshy green, spiny stems	*Opuntia*, prickly pear
Fabaceae	the fruit is known as a legume	*Pisum*, garden pea
Liliaceae	flower parts occur in threes	*Lilium*, Madonna lily
Nymphaeaceae	water lilies (except *Lotus*)	*Nymphaea*, water lily
Orchidaceae	ornate, usually colorful flowers	*Cattleya*, corsage orchid
Poaceae	have fruits known as grains	*Saccharum*, sugar cane
Rosaceae	flowers usually showy, with 5 petals	*Fragaria*, strawberry
Solanaceae	many produce toxic alkaloids	*Solanum*, potato

TABLE 17.3   Classification of the Plant Maize (Corn)

Domain	Eukarya	composed of cells having a nucleus
Kingdom	Plantae	land-adapted plants having an embryo
Phylum	Anthophyta	flowering plants
Class	Monocotyledones	flowering plants whose embryos have one seed leaf
Order	Poales	a fibrous-leaved group
Family	Poaceae	the grass family
Genus	*Zea*	grasses with separate "male" and "female" flowers
Species	*Z. mays*	maize or corn

domains are known as prokaryotes. Although virus genera are grouped into families and orders, higher taxonomic categories (classes, phyla, or domains) have not been defined for viruses.

Plant and microbe classifications change as new discoveries are made

In classification (systematics), concepts of relationships change when new discoveries are made. For example, water lilies were once thought to form a coherent group, the family Nymphaeaceae. But, surprisingly, DNA data has revealed that the water lily genus *Nelumbo* (*N. nucifera* is known commonly as "sacred lotus" or "padma") is not closely related to other water lilies. Another amazing discovery arising from the use of DNA characteristics is that tomatoes (formerly genus *Lycopersicon*) and potatoes (genus *Solanum*) are so closely related that they should not be classified as separate genera. Tomatoes have been renamed as species of *Solanum*. Conversely, sometimes species are discovered to be so different from presumed relatives that a new genus is constructed for them. Just as many people are intensely interested in tracing family trees to learn about their own ancestors, biologists are fascinated with discoveries about relationships among animals, microbes, and plants, because these lead to new ideas about how evolution has influenced all of Earth's living things. Biologists have assembled their concepts of the relationships among all living things into a Tree of Life (see Chapter 1). The Tree of Life demonstrates that all life on Earth is related. As new information becomes available about relationships of known organisms, and new organisms are discovered, the Tree of Life grows and its shape changes.

animals, or metazoa), Fungi, and Protista (algae and protozoa). Protista is a large, complex group that will probably be broken up into multiple phyla. All of these kingdoms together form the Domain Eukarya—the eukaryotes. Eukaryotes are composed of cells that are internally complex and have a nucleus. There are two other domains, the Bacteria and the Archaea, both of which contain organisms whose cells are simpler and lack a nucleus; members of these two

HIGHLIGHTS

17.1 Binomial (two-part) scientific names, originated by Linnaeus more than 200 years ago, are more precise than common names and are essential in agriculture, commerce, medicine, and ecology. Each kind of organism has a unique scientific name that provides useful information and is written in a particular style (e.g., *Zea mays*, for maize or corn).

17.2 Collections of expertly identified plant specimens that are often dried and pressed onto paper sheets are stored in herbaria and used for reference, research, teaching, and learning. Botanical gardens are collections of living plants. Plants are identified using taxonomic keys, among other resources. There is a formal process for naming newly discovered plants or new commercial varieties.

17.3 Plants are classified according to their relationships, as are animals and microbes. Reproductive characteristics such as DNA and flower structure are particularly useful in deducing evolutionary relationships. As new discoveries are made, plant classifications and names may change.

REVIEW QUESTIONS

1. Both common names and scientific names are used to identify plants and animals. Discuss the advantages and disadvantages of both types of names. Why are scientific names preferred in science, medicine, agriculture, and commerce?

2. What are the parts of a scientific name? Which part is a noun, which an adjective, and which may be used by itself? Are both parts always capitalized as well as underlined or italicized?

3. Distinguish between species, species complexes, and hybrids. What special convention distinguishes hybrid plant names from nonhybrid names? What subdivisions are used to further divide a species, and how do these subdivisions differ from each other?

4. What is an herbarium, where are they generally found, why are they important, and how do they differ from a botanical garden?

5. What are some of the characteristics upon which modern plant classification is based? Why are reproductive characters, such as the structure of the flowers, used more heavily than root, stem, and leaf anatomy?

6. Listed here, in no particular order, are the major categories that form the hierarchy used to classify life on Earth. Place these categories in order from the most inclusive (i.e., highest level, containing the most species) to the most exclusive (i.e., lowest level, containing only a part of a species): Order_ Family_Domain_Species_Class_ Phylum_Kingdom_Genus_Variety_

7. Distinguish between plant taxonomy and plant systematics.

8. Which scientist was the first to use binomials (two names) as scientific names, and when did he start to do this? How was it that he began using this format? Why did he choose to write his scientific names in Latin?

APPLYING CONCEPTS

1. Construct a dichotomous key to the following closed geometric figures: a circle, a trapezoid, a pentagon, an equilateral triangle, a scalene triangle, and a rectangle.

2. Imagine this hypothetical scenario. Dr. James is a professor of botany at a major university. One day he collects an orchid that he does not recognize. This plant appears similar to other orchids in the genus *Habenaria*, although it does not quite fit the description of any known species. This plant is new. Dr. James places a specimen in the university's herbarium. He even publishes a scientific paper where he casually mentions the new plant by the name he has tentatively assigned to it—*H. missouriensis*, in honor of the state where he first collected it. Six months later, Ms. Smith, an amateur botanist, also collects this orchid and recognizes it as new. She deposits a specimen in a well-known herbarium and publishes a scientific paper in which she describes the new plant as *H. ciliaris*. Her paper includes a Latin description of the new plant. Whose name is valid? Soon thereafter, Professor Chang points out that nearly 200 years ago Robert Brown had used the name *H. ciliaris* to describe another, separate species of *Habenaria*. He renames the plant described by Ms. Smith as *H. smithii*. Now whose name has validity, since *H. ciliaris* as a name has been occupied for almost 200 years? Which scientific rules govern nomenclatorial entanglements such as these? After all this, what would be the complete name of this *Habenaria*, including author(s)? Why would it be important to include the author's name when referring to *H. ciliaris*?

3. Describe several benefits, both scientific and nonscientific, of botanical gardens and also of herbaria.

4. Describe the purpose and several benefits of the International Code of Botanical Nomenclature.

18

Prokaryotes and the Origin of Life on Earth

Cyanobacterial bloom

- Experiments and scientific observations provide information about the origin of life on Earth.

- Prokaryotes are Earth's oldest forms of life.

- Prokaryotes have simpler cells, bodies, and reproduction than Earth's other life forms—eukaryotes.

- Prokaryotes are more numerous and reproduce faster than eukaryotes.

- Prokaryotes strongly influenced Earth's past atmosphere and climate and the evolution of algae and plants.

- Prokaryotes play important roles in the cycling of carbon and nitrogen in modern ecosystems.

- Agriculture, medicine, food and other industries, biotechnology, and environmental monitoring depend on a knowledge of prokaryotes.

nabaena—a beautiful blue-green, necklace-shaped bacterium—is related to the most ancient fossil life-forms known. This photosynthetic bacterium is also important in the modern world, performing an invaluable service to humans. *Anabaena* commonly lives within leaf cavities of the floating water fern *Azolla*, which covers surfaces of tropical waters, including rice paddies, around the world. The water fern provides *Anabaena* with resources needed to convert useless nitrogen gas in the air into ammonia, which acts as nitrogen fertilizer. Nitrogen fertilizer is essential to the survival, growth, and reproduction of all plants and algae. But in nature, only certain types of bacteria—*Anabaena* among them—can transform nitrogen gas into ammonia fertilizer. Its water-fern partner uses some of the fertilizer produced by *Anabaena*, but additional fertilizer is released into the water, where it is available to other algae and aquatic plants, including rice. *Anabaena* and *Azolla* are thus key to the high food productivity of rice paddies, used to grow food for millions of people. Without the fertilizer produced by *Anabaena*, paddies would produce much less rice, because tropical farmers often cannot afford to buy commercial fertilizer.

On the other hand, *Anabaena* can exhibit less beneficial behavior. *Anabaena* is one of several types of blue-green bacteria that form floating scums on the surfaces of fertile lakes around the world. People dread seeing these scums, known as water blooms, because they smell bad when they decay. Worse, the dying blue-green bacteria often release toxins into the water that kill fish and poison waterfowl, livestock, and other domestic animals that drink the contaminated water. Some human cases of illness and death have also been attributed to toxic blue-green bacteria. The scums also shade underwater forests of water plants, preventing them from obtaining enough light for photosynthesis. The water plants consequently die, eliminating essential habitat for fish and other aquatic life.

In this chapter, you will learn additional reasons why prokaryotes are important to people and other organisms. You will also learn why scientists regard prokaryotes as Earth's most ancient life-forms and how scientists study the origin of Earth's life. ■

Anabaena

18.1 | What is life and why does it occur on Earth?

Distinguishing living things from nonlife is not always easy. Take viruses, for example (see Essay 17.2). They have some of the characteristics associated with life—reproduction and evolution—but they lack other features considered essential to life, namely cell structure and metabolism. Recall that metabolism is a set of processes that use energy to produce and maintain organization. Traits of life also include adaptability—the capacity of an individual to cope with change—and the ability to defend against injury and respond to environmental stimuli. Earth's living organisms—prokaryotes, protists, fungi, animals, and plants—have all of these properties, including being composed of cells. Cells are the smallest structures that have all of the properties of life, so many studies of life's origins focus on the origin of cells.

Among Earth's organisms, prokaryotes have the simplest cell structure (Figure 18.1). For example, prokaryotes do not usually possess a membrane-enclosed nucleus (Figure 18.2a). This is the basis for the term **prokaryote**, which means "before" (pro) the "kernel" (karyote) (the word *kernel* refers to the nucleus). All other groups of organisms are **eukaryotes,** meaning "true (eu) kernel," because most of their cellular DNA is contained within a nucleus (Figure 18.2b). Protists, fungi, animals, and plants are eukaryotes, which are more sophisticated descendants of prokaryotic ancestors. Thus, the study of prokaryotes is not only important in understanding life's origin but also helps us to understand how eukaryotes evolved.

As yet, we have no evidence that life occurs anywhere other than Earth in our solar system. Even Mars, our solar system's most Earth-like planet, lacks life, so far as we know. Though NASA scientists found some very old prokaryote-like structures inside meteorites that originated on Mars, there is as yet no unequivocal evidence that these structures represent ancient life on that planet (Essay 18.1, "Did Life Evolve on Mars?"). Why then has life flourished on Earth? This question has a two-part answer. First, Earth has been a cradle for life

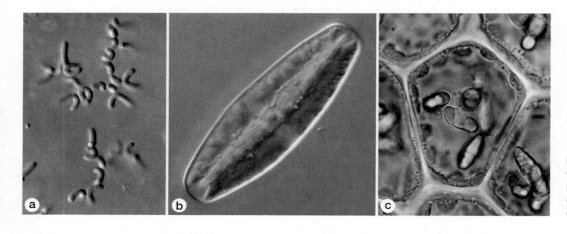

FIGURE 18.1 **Cells** Microscopic views of (a) bacterial cells from a water sample, (b) a unicellular algal cell (the green protist *Netrium*), and (c) a plant cell from a multicellular bryophyte.

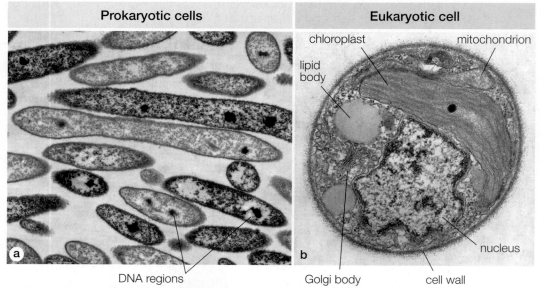

Prokaryotic cells | Eukaryotic cell

chloroplast — mitochondrion

lipid body

nucleus

DNA regions | Golgi body | cell wall

FIGURE 18.2 **Prokaryotic versus eukaryotic cell structure** The internal structure of prokaryotic cells (a) is simpler than that of even tiny eukaryotic cells (such as this green alga) in which (b) nuclei and other organelles become visible with the use of a transmission electron microscope.

ESSAY 18.1   DID LIFE EVOLVE ON MARS?

In 1999, several NASA scientists held a press conference to announce evidence that life might once have occurred on our neighbor planet Mars. This evidence came from a meteorite collected in 1984 from the Allen Hills ice field in Antarctica (Figure E18.1A). The announcement was widely covered by the news media, because people seem to be intrinsically interested in the origin of life. The meteorite had clearly come from Mars, no doubt about that. During its formation about 4.5 billion years ago, gas bubbles indicative of Martian origin were trapped in the rock. About 16 million years ago, an asteroid impact blasted the rock into space, and it landed in Antarctica about 13,000 years ago. The NASA scientists had opened the rock, being very careful to avoid contaminating it with Earthly materials. Inside they found globules of carbonate minerals, and within these they found three materials that suggested life.

The first of these three materials were complex hydrocarbons of the type that result when dead organisms are heated. Similar compounds are formed on the surfaces of burgers that have been left for too long on the grill. These hydrocarbons suggested the original presence of molecules characteristic of living organisms. Secondly, the scientists interpreted bacteria-shaped structures (Figure E18.1B) as possible fossil prokaryotes. Finally, and most compellingly, the meteorite con-

Could ancient Mars once have harbored similar prokaryotes whose metallic magnetite particles persisted in the meteorite?

tained magnetite crystals (Figure E18.1C). These crystals were amazingly similar to those known to occur in modern Earth bacteria that use particles of magnetite (Fe_3O_4) to detect gravity (Figure E18.1D). In cells of *Magnetospirillum magnetotacticum*, 5 to 40 magnetite particles are arranged in a row, which helps the cell align along the lines of the Earth's magnetic field. When an artificial magnetic field is applied to these bacteria, they rapidly move, by means of their flagellum, so that their long axis is aligned in a north-south orientation. Cells in the Northern Hemisphere move northward, whereas cells located in the Southern Hemisphere move southward. Bacteriologists are unsure how this behavior benefits these bacteria, but think it possible that magnetic orientation may help them remain in sites having low oxygen content, preferred by these bacteria. Magnetic bacteria have been found in such places as water treatment facilities. Much geological evidence indicates that Mars once had a much warmer and wetter environment than it now does. Could ancient Mars once have harbored similar prokaryotes whose metallic magnetite particles persisted in the meteorite?

More recently, skeptics have argued that the complex hydrocarbons obtained from the Martian meteorite could have

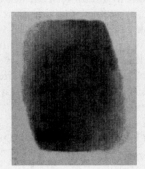

E18.1C   Magnetite crystal from the Martian meteorite

formed in the absence of life. Some claim that the bacterial "fossils" were too small to contain the components of metabolism characteristic of Earthly bacteria. Scientists have also demonstrated that magnetite crystals like those found in the meteorite can be produced in nonliving systems. These observations show that the complex hydrocarbons, bacterial "fossils," and magnetite crystals found in the Martian meteorite are not necessarily signs of life. But neither do they prove the absence of life on Mars, past or present. Geological surveys of Mars continue. If life did once exist there, scientists may discover more convincing traces of life.

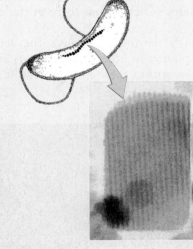

E18.1A   Martian meteorite ALH84001

E18.1B   Prokaryote-shaped structure from the Martian meteorite

E18.1D   Magnetite crystal from the bacterium *Magnetospirillum*, in which they can form a linear array

because of its position relative to the Sun. Second, once life began, simple early life-forms slowly but inexorably altered the environment in a manner that not only maintained life but also paved the way for later, complex life-forms. In other words, early prokaryotes changed the world's environment. These changes, which are discussed more completely later in this chapter, allowed eukaryotes to evolve and thrive. We eukaryotes owe our very existence and life-supporting environment to early prokaryotes. But what factors allowed Earth's first life-forms to exist?

The Earth's position in space is important to life

Earth's orbit lies between that of Venus and Mars (Figure 18.3). To understand what this means for the origin of life, think of the Sun as a campfire. The closer you are to the fire, the warmer you are. Even on a cold night, you can get too close for comfort, and it is easy to stand so far away that the fire does you little good. Most people will settle around the fire in a zone where they are most comfortable, which is slightly farther than the length of a good marshmallow-toasting stick. Planetary scientists use the term **habitable zone** to describe the comfortable zone around a star like our Sun. The habitable zone corresponds to the region around a star where water can be stable as a liquid. Recall that water has many properties that make it essential to life (Chapter 3). Inside the zone, water boils away; outside the zone, it turns to ice. Venus is outside the habitable zone; all its water has boiled off, while Mars is at the outer edge, and most of its water is locked up in ice. The Earth orbits in between and maintains a balance between water as a vapor, liquid, and solid. The importance of water to life on Earth is illustrated by the near absence of even the simplest known life-forms—prokaryotes—from surface soils of the extremely dry Atacama Desert in Chile (Chapter 26). This desert's surface seems to be at the dry limit of life as we know it.

Today's astronomers have the technology needed to observe planets in other solar systems and determine if these planets lie within their star's habitable zone. This is a very

active field of study because humans are naturally curious to know if life could have arisen elsewhere in the universe. As far as we know, the best place to look for life is where there is liquid water. But water is not the only critical factor; a planet's atmospheric composition is also important in fostering life.

Early life changed Earth's atmospheric composition, fostering modern life

Earth's earliest atmosphere contained several gases: molecular hydrogen, water vapor, ammonia, nitrogen gas, methane, and carbon dioxide, but no oxygen. Gas mixtures belched from present-day volcanoes resemble this early atmosphere, suggesting its origin from volcanic eruptions. In Earth's earliest atmosphere, methane and carbon dioxide occurred at much higher levels than at present—a circumstance that was favorable for early life. Recall that methane and carbon dioxide are greenhouse gases that warm atmospheres by retarding loss of heat to space (see Essay 1.2). These two gases kept Earth warm during the Sun's early history, when the Sun did not burn as brightly as it now does. (An early dim period, with later brightening, is normal behavior for stars of our Sun's type.)

Earth's modern atmosphere, which is 78% nitrogen gas, 21% oxygen, and about 1% argon, water vapor, ozone, and carbon dioxide, differs dramatically from the earliest atmosphere just described. The modern atmosphere supports many forms of complex life that would not have been able to exist in Earth's first atmosphere because the O_2 level was too low. Also, if atmospheric methane and carbon dioxide were now as abundant as they were in Earth's earliest atmosphere, the planet's temperature would likely be too hot for most life. It is a good thing for modern life that Earth's atmosphere gained oxygen and lost much of its carbon dioxide and methane. How and when did the atmosphere change?

The answer to this riddle lies in the metabolic activity of early prokaryotic life-forms that slowly but surely transformed the chemical composition of Earth's atmosphere. Some of these early organisms were photosynthetic relatives of

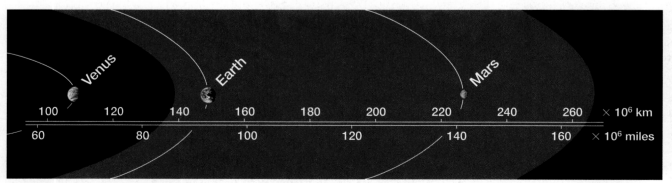

FIGURE 18.3 The habitable zone The green color denotes the habitable zone, and the number scales indicate the distances in kilometers (upper) and miles (lower) from the Sun. Note that Venus is not within this zone but that Mars is within its outer edge. This means that Venus is too hot and that all its water has boiled away. Mars is too cold; its water is frozen at the poles and as ice in the ground.

modern *Anabaena*, known as **cyanobacteria** (meaning blue-green bacteria). Recall the process of photosynthesis: carbon dioxide gas (CO_2) combined with water (H_2O) yields oxygen gas (O_2) plus sugar (CH_2O) (Chapter 5). In Earth's early days, all over the planet countless cyanobacteria performed photosynthesis. Together, these ancient cyanobacteria removed massive amounts of carbon dioxide gas from Earth's atmosphere by converting it to solid organic carbon. These ancient cyanobacteria also released huge quantities of oxygen into the atmosphere. Other ancient bacteria consumed methane, greatly reducing its amount in the atmosphere. When our Sun later became hotter, the continued removal of atmospheric carbon dioxide and methane by early prokaryotes kept Earth's climate from becoming too hot to sustain life. Modern prokaryotes still provide these valuable services today!

Cyanobacterial oxygen release improved conditions for life in two ways. First, because oxygen is useful in the metabolic process known as cell respiration, allowing cells to more efficiently harvest energy (as ATP) from organic food (see Chapter 5). Second, because oxygen in the upper atmosphere is used to form a protective shield of ozone. Earth is constantly bombarded by harmful **ultraviolet (UV) radiation** from the Sun; sunburn, skin cancer, DNA mutations, and even death are some consequences of excessive exposure to UV radiation. Today, Earth's upper-atmosphere **ozone** (O_3) shield absorbs enough UV to allow diverse forms of life to survive. But because the early Earth lacked oxygen in its atmosphere, it also lacked a protective ozone barrier. As a result, early life on Earth was confined to the oceans, where the water absorbed UV radiation. Only after oxygen released by ancient cyanobacteria drifted up into the upper atmosphere and reacted with other oxygen molecules to form a protective layer of ozone could life flourish at the surface and on land. The absence of an oxygen atmosphere on Mars and other planets in our solar system means that these planets also lack an ozone shield that would protect surface-dwelling life from UV radiation. The surface of Mars is bombarded with deadly radiation; if any modern life exists on Mars, it would almost certainly be subterranean.

Our reliance on the protection provided by ozone is why there is so much concern today about the destruction of the Earth's ozone layer by industrial gases, such as chlorofluorocarbons used in refrigeration. People living in areas affected by polar ozone "holes" (places where upper-atmosphere ozone levels are reduced) experience greater UV radiation levels and there is an increased incidence of severe sunburn, skin cancer, and DNA mutation.

18.2 | When did the Earth form, and when did life first appear?

With modern telescopes we can observe vast clouds of gas and dust around other stars, including some known to have planets. These observations support the hypothesis that our entire solar system formed by condensing out of a similar cloud of gas and dust some 4.6 billion years ago. (The age of

TABLE 18.1	Timetable for the History of the Earth	
Event	**Years Before Present**
Origin of the Earth and solar system	4,600,000,000
End of the period of heavy bombardment by asteroids and comets	3,800,000,000
First fossils of prokaryotes (found in Australia)	3,500,000,000
Major rise in atmospheric oxygen to 10% of present levels	2,200,000,000
Second major rise in atmospheric oxygen to 15% of present levels Oceans teem with multicellular life	600,000,000
Oxygen reaches present levels (21%) Early plants widespread on land	450,000,000
Origin of human genus *Homo*	2,000,000
Origin of agriculture and beginnings of civilization	About 10,000

our solar system—including Earth—is determined by measuring how much of a radioactive isotope of uranium (U) has decayed to form lead (Pb) in meteorites. Meteorites represent debris that is left over from the solar system's formation, and so they are the same age as the oldest materials in the solar system.)

Even after Earth was formed, there was a delay of nearly a billion years before life appeared. The first indications of life on Earth do not appear until 3.7 to 3.85 billion years ago (Table 18.1). What explains this delay?

A period of bombardment preceded the establishment of life

Anyone who has ever looked through a telescope at the Moon knows that it is heavily cratered, as are Mars and Mercury. By contrast, the surface of the Earth is not heavily cratered, because its surface is constantly being eroded and reshaped by geological forces. Craters on the Moon, Mars, and Mercury represent collisions between each planet and bits of debris left over from the formation of the solar system. Earth's Moon was formed by such an impact event. An object larger than Mars slammed into Earth (an event known to astronomers as "the big splat"), vaporizing surface material that aggregated into the Moon. Having a Moon is important to the persistence of life on Earth because it stabilizes our planet's orbital movements, thereby preventing severe environmental fluctuations. The number of such celestial collisions was quite intense for a long time after our solar system

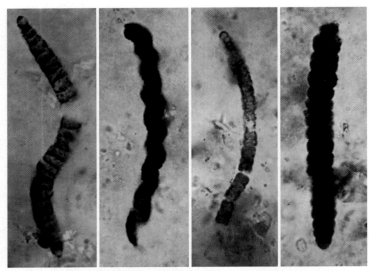

FIGURE 18.4 Structures that may be Earth's oldest fossils These potential fossils resemble modern cyanobacteria and are therefore potential signs that photosynthesis was present on Earth 3.5 billion years ago.

formed. The violence of the bombardment is thought to have prevented the establishment of life on Earth during this period. Fortunately for the future of Earth's life-forms, the period of heavy bombardment ended around 3.8 to 3.9 billion years ago (see Table 18.1).

Following the bombardment period, newly evolving life-forms left behind chemical evidence in the form of 3.7-billion-year-old carbon masses thought to have been produced by living organisms. Mineral tubes that formed about 3.5 billion years ago, which are similar to glassy tubes produced by modern deep-ocean bacteria, also provide evidence that prokaryotes lived by that time. Some experts have interpreted structures present in 3.5-billion-year-old Australian rocks (Figure 18.4) as fossil cyanobacteria (though some others have suggested that these structures might be minerals of nonliving origin). Large, rocky mounds called **stromatolites** (Figure 18.5)—produced

FIGURE 18.5 Stromatolites Stromatolites are layered deposits of chalk (a mineral of calcium carbonate) formed by colonies of cyanobacteria or photosynthetic protobacteria. They have a geologic history dating back more than 3 billion years.

even today in warm, shallow oceans off the shores of Australia, the Bahamas, and other places—indicate that cyanobacteria grew luxuriantly for billions of years. Chemical evidence from stromatolites in Australia reveals that photosynthesis was occurring by at least 2.7 billion years ago. However, prior to 2.2 billion years ago, oxygen produced by the cyanobacteria was bound with iron in the oceans and in the Earth's crust before it could escape into the atmosphere. We can observe evidence of this process in ancient, red, banded-iron formations, which occur in the Grand Canyon and other places. Once all this iron had become saturated with oxygen (forming iron oxide, otherwise known as rust), oxygen was released into the atmosphere and began to accumulate there.

Subsequent events in the development of life on Earth are linked to the rise of the level of oxygen in the Earth's atmosphere. Around 2.2 billion years ago, oxygen levels appear to have jumped to about 10% of present levels. These levels produced sufficient ozone to shield the ocean surfaces from UV radiation. This rise in atmospheric oxygen corresponds with the appearance of fossils thought to represent the first eukaryotes. The oxygen level of the atmosphere rose again around 600 million years ago, when multicellular life arose in the oceans. Atmospheric oxygen may have reached modern levels by the time early plants were widespread on land, about 450 million years ago.

18.3 | How did life originate?

Scientists approach the question of life's origin by using the scientific method (Chapter 1), which involves testing reasonable explanations, known as hypotheses. One hypothesis is that simple life-forms originated elsewhere and were subsequently transported to Earth via comets or meteorites—an idea known as panspermia ("seeding across"). At present, it is unclear how the panspermia concept could be tested. A more readily testable explanation is that life originated on Earth itself, from inorganic molecules present on the early Earth. The components of this concept are together known as the **chemical-biological theory** of life's origin on Earth.

The chemical-biological theory explains the origin of life as a series of stages

The chemical-biological theory of the origin of life features a series of stages of increasing levels of organization. The first stage is the formation of simple organic molecules such as sugars, fatty acids, amino acids, and nucleotides from inorganic molecules. The second stage is the construction of the macromolecules of life—carbohydrates, lipids, proteins, and nucleic acids (Chapter 3). The third stage is the formation of the first cell-like structures—**protocells**. Each stage can be experimentally reproduced in the laboratory.

Organic compounds can be formed from inorganic molecules

In the 1930s, Alexander Oparin in Russia and J. B. S. Haldane in England proposed that the atmosphere of the early Earth must have lacked oxygen if organic compounds were to persist. If oxygen had been present, any organic compounds formed would have broken down to carbon dioxide and water. They suggested that the early atmosphere was rich in hydrogen, methane, and ammonia.

In the 1950s, Stanley Miller and Harold Urey at the University of Chicago decided to test this hypothesis by performing an experiment. They constructed an apparatus in which they could mimic atmospheric and oceanic conditions thought at that time to have been present on the Earth before life evolved (Figure 18.6). A lower flask contained water, which was heated to produce water vapor. This water vapor then mixed with an atmosphere of hydrogen, methane, and ammonia in an upper flask, in which electric discharges simulated lightning, causing the gases to interact. The products were cooled in a condenser and reentered the water in the lower flask. By the following morning the water had turned yellow. The scientists had made a broth of short-chain fatty acids and amino acids! In later experiments of the same type, using updated information about the composition of Earth's early atmosphere, scientists have been able to generate all of the amino acids that occur in living things.

Scientists have also reproduced the formation of bases found in nucleotides and simple carbohydrates from inorganic

compounds in the presence of energy. In 1961, Juan Oro at the University of Houston discovered that if he mixed hydrogen cyanide (HCN) and ammonia with water, he could obtain amino acids and the base adenine. Recall that adenine is a component of DNA, RNA, and ATP—essential materials for life. Later studies revealed that the other bases present in DNA and RNA could be formed by reactions between HCN and other nitrogen gases likely to have been present in the atmosphere of the early Earth. Other research showed that a mixture of simple carbohydrates (sugars) could be obtained from formaldehyde if a catalyst such as clay were present. Thus, there is evidence that all of the types of simple organic molecules found in living things can form from simple inorganic molecules.

Macromolecules can form from simple organic compounds

In the first stages of the chemical-biological theory, simple organic molecules—such as sugars, amino acids, fatty acids, glycerol, and the bases found in nucleotides—are formed out of inorganic molecules. These simple organic molecules are the building blocks from which are assembled the larger macromolecules such as complex carbohydrates, lipids and phospholipids, proteins, and nucleotides. The next step proposed in the chemical-biological theory is the assembly of these macromolecules from simple organic molecules. In laboratory experiments, when amino acids are heated, they link up to form polypeptides, the macromolecules from which proteins are formed. Fatty acids, phosphate, and glycerol will condense into phospholipids under dry-heat conditions in the presence of cyanamide (H_2NCN). Similarly, if bases, sugars, and phosphates are present together under dry-heating conditions, they will condense into the molecules called nucleotides. These experimental conditions mimic a drying pond—an environment that could have been common on the early Earth.

Macromolecules in water can form cell-like structures

Several researchers have considered subsequent steps in the chemical-biological theory. Could macromolecules mixed in water form cell-like entities? In some early experiments, Oparin mixed concentrated solutions of proteins, lipids, and carbohydrates in water. The mixture rapidly formed colloidal particles (a suspension of gelatinous particles too small to settle out of water). These particles exhibited some of the features of true cells; they were surrounded by a membrane-like layer of phospholipid molecules, and they could bud off smaller particles, a simple type of reproduction. In the 1950s at the University of Miami (Florida), Sidney Fox heated amino acids, and aggregates called *proteinoids* formed. When heated a second time with added water, the proteinoids formed spheres in the size range of actual cells (Figure 18.7). Fox termed these spheres *protocells* and showed that they exhibit responsiveness to external stimuli, as do living cells. More recently, protocells consisting of lipid vesicles that

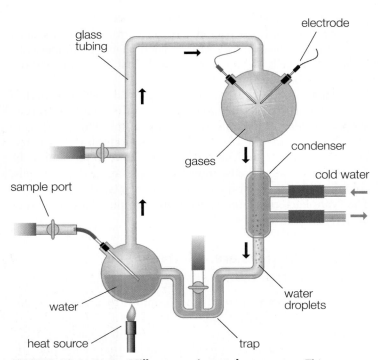

FIGURE 18.6 Urey–Miller experimental apparatus This apparatus was designed to test the hypothesis that simple organic compounds could have arisen from inorganic compounds present in the atmosphere of the early Earth. The upper chamber contained an atmosphere of hydrogen, methane, and ammonia gases to mimic what was thought to be the composition of Earth's first atmosphere. The lower chamber contained water that was heated to send vapor into the upper chamber. Water vapor interacted with the other gases in the upper chamber in the presence of electric discharges that simulated lightning. Within 24 hours, amino acids and fatty acids appeared in the lower chamber.

FIGURE 18.7 **Protocells** These cell-like spheres are made of proteins and range in size from 0.5 to 7 micrometers (μm) in diameter, the same size range as actual cells.

contain nucleic acids and conduct simple forms of metabolism have been produced in laboratories. Some scientists are working to produce artificial cells designed to perform technologically useful metabolic tasks.

18.4 | Bacteria and Archaea are Earth's smallest, simplest life-forms

So small that you cannot see them without using a microscope, billions of prokaryotes of two main types—Bacteria and Archaea—are everywhere around you. Prokaryotes occur in the air you breathe; on all types of surfaces, including human teeth, skin, and intestines; and in food and soil. These microbes are extremely numerous—a pinch of garden soil contains about 2 billion individuals. Globally, their numbers are estimated at more than 10^{30} and there are a million or so species! Fifty percent of the carbon that occurs in living cells on Earth occurs in prokaryotes. Multitudes of prokaryotes are associated with plants, some having highly beneficial effects and others causing disease. Bacteria and Archaea are essential to the maintenance of Earth's atmosphere, climate, soil, and water purity, providing ecological benefits upon which all life depends. Medicine, agriculture, industry, biotechnology, and environmental restoration are among the many areas of human endeavor that rely upon an understanding of these microbes. The study of microscopic organisms, including prokaryotes, is microbiology. Biologists who specialize in the study of prokaryotes are known as microbiologists or bacteriologists.

Prokaryote bodies are smaller and simpler than those of eukaryotes

The cells of most Bacteria and Archaea are very small—usually between 1 and 5 micrometers (μm; a micrometer is one-millionth of a meter), and their simple bodies are correspondingly small (Figure 18.8). Microscopes of various types, especially fluorescence microscopes and transmission electron microscopes (see Chapter 4) must be used to see them. However, cells and bodies of a few prokaryotes are unusually large, sometimes visible with the unaided eye. Among the largest-known prokaryotes is

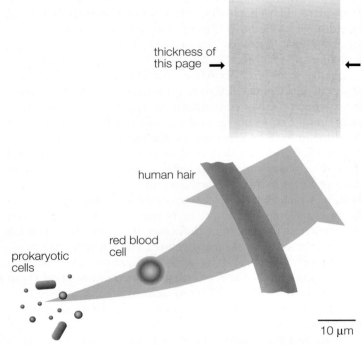

FIGURE 18.8 **Prokaryotic cells are typically small in size**

Thiomargarita namibiensis (whose name means "sulfur pearl of Namibia"), recently discovered in ocean-bottom sediments along the southwestern shore of Africa. These giant, gleaming-white bacteria have cells half a millimeter wide that are visible without using a microscope.

Prokaryotic cells are most commonly shaped as spherical **cocci**, cylindrical **rods**, comma-shaped **vibrios**, rigid corkscrew-shaped **spirilli**, or flexible spiral-shaped **spirochaetes** (Figure 18.9). *Anabaena* and *Thiomargarita* are examples of prokaryotes that have spherical cells. Prokaryotic bodies may be composed of single cells, **unicells;** simple aggregates of cells, **colonies;** or linear arrays of cells, **filaments.** (Figure 18.10). The cyanobacterium *Anabaena* is an example of a filamentous prokaryote.

Prokaryotic cells are relatively simple in structure

Prokaryotic cells are internally much simpler than are eukaryotic cells (see Figures 18.1, 18.2). Nucleic acids (DNA and RNA), enzymes and other proteins, ribosomes (small particles of RNA and protein that generate proteins), and internal membranes are the major constituents.

Prokaryotic DNA The DNA of prokaryotes typically occurs as one or a few circular molecules known as **genophores** or **chromosomes** (though some biologists prefer to use the term **chromosome** only when discussing eukaryotes). The DNA of prokaryotes is not abundantly coated with protein. This is in contrast to chromosomes in nuclei of eukaryotes, which are composed of DNA associated with abundant protein.

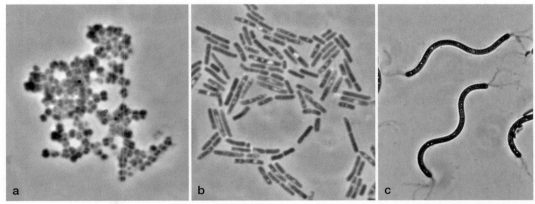

FIGURE 18.9   Prokaryotic cell shapes (a) Cocci, (b) rods (also referred to as *bacilli* in reference to the rod-shaped genus *Bacillus*), and (c) spirilli.

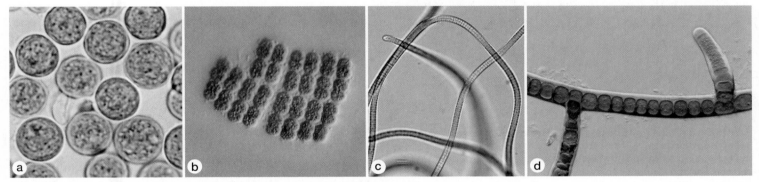

FIGURE 18.10   Cyanobacteria illustrate major prokaryote body types *Prochloron* (a) is a unicell, *Merismopedia* (b) is a flat colony of cells held together with mucilage, *Oscillatoria* (c) is an unbranched filament, and *Stigonema* (d) is a branched filament.

Prokaryotic cells often contain additional smaller circles of DNA known as **plasmids** (see Figure 15.6). Plasmids often carry genes that produce toxins or confer resistance to heavy metals and antibiotics. Bacteria can transfer plasmids from one cell to another, leading to the rapid spread of genes through populations. One of the many examples of the importance of plasmid gene transfer is the recent spread of antibiotic resistance among *Erwinia* bacteria that cause fire blight—a leading killer of fruit trees. Every spring, farmers apply thousands of kilograms of the antibiotic streptomycin to apple and pear orchards in attempts to control fire blight. With the spread of resistance, this effort will become less effective and crop productivity may drop. Overuse of antibiotics to prevent disease in farm animals is common in the United States and is associated with the spread of antibiotic resistance. Antibiotic overuse creates conditions in which resistant bacteria can grow more readily than nonresistant forms. Some bacteria that cause human disease have become resistant to nearly all known antibiotics, a source of major concern in medicine.

Prokaryotic enzymes   The particular enzymes that are present in a bacterial cell determine which type(s) of metabolism the cell possesses. *Anabaena* cells, for example, contain all of the enzymes needed for photosynthesis, but *Erwinia* does not—hence its dependence on organic food obtained by attacking crops and other plants. Though most of the enzymes in bacterial cells are difficult to visualize, large aggregations of the photosynthetic enzyme rubisco (Chapter 5) can be seen in cyanobacterial cells viewed with an electron microscope (Figure 18.11).

Other cell components   Cells of photosynthetic bacteria, such as those of *Anabaena* and other cyanobacteria, usually contain internal arrays of membranes known as **thylakoids** (see Figure 18.11). Thylakoid membranes anchor chlorophyll and other molecules needed to harvest light energy and are also present in the chloroplasts of algal and plant cells (Chapter 4). Internal membranes also occur in various nonphotosynthetic prokaryotes, where they serve as surfaces for attachment of enzymes important in metabolism.

Cells of *Anabaena* and some other bacteria that occupy aquatic habitats often contain stacks of hollow cylindrical structures made of protein that keep them buoyant. Known as gas vesicles, these lightweight structures don't really contain amounts of gas different from the rest of the cell, but occupy a lot of cell space. When many gas vesicles are present, the cells containing them float easily to the water surface,

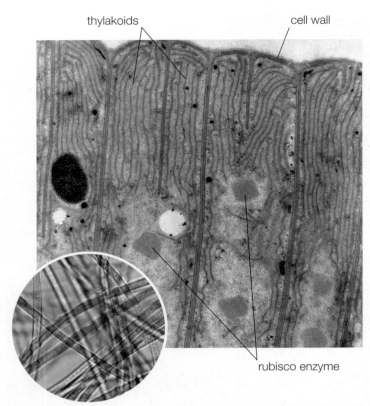

FIGURE 18.11   **Cyanobacterial cells viewed by transmission electron microscopy**   The cytoplasm of the cells in the filamentous cyanobacterium contains thylakoid membranes and aggregates of rubisco enzyme. The inset shows a similar cyanobacterium with light microscopy.

an advantage in obtaining light for photosynthesis. Gas vesicles are largely responsible for the fact that cyanobacterial water blooms often form surface scums, as described in the chapter opening.

Some types of the bacterium *Bacillus* contain protein crystals that are toxic to larvae (young stages) of insects such as mosquitoes, black flies, and potato beetles, a crop pest. The crystals do not become toxic until they enter the larval gut, whereupon the toxin becomes activated and destroys the insect's intestine. These protein crystals are known as Bt toxin, after *Bacillus thuringiensis*. Because the toxin is not harmful to humans, the genes for Bt toxin have been incorporated by genetic engineering techniques into some crop plants to increase their resistance to insect attack (Chapter 15).

Some prokaryotes can swim or glide

Prokaryotes may have one flagellum or several flagella, threadlike cell extensions enabling them to swim through liquids (Figure 18.12). The flagella of *Pseudomonas* and *Erwinia*, two examples of bacteria that cause plant diseases, allow them to spread within plant tissues. Prokaryotic flagella are simpler in structure and operate differently from eukaryotic flagella. Prokaryotic flagella twirl by rotating at their bases

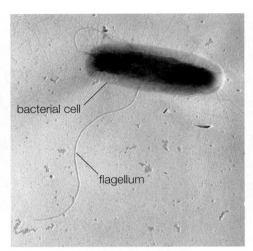

FIGURE 18.12   **A prokaryote with a flagellum**

like a propeller, whereas eukaryotic flagella repeatedly bend and straighten. Some prokaryotes that lack flagella, including short filaments of *Anabaena*, are able to glide across surfaces by mechanisms that are not completely understood. Motility allows prokaryotes to move in response to stimuli such as light, nutrients, or magnetic fields.

Slimy polysaccharides often coat prokaryote surfaces

Many prokaryotes produce an external coating of polysaccharides known as a capsule, mucilage envelope, extracellular matrix, or slime sheath. The capsules of disease-causing bacteria allow them to evade the defense systems of their hosts. *Anabaena* and many other cyanobacteria have a slimy envelope that aids in flotation, obtaining nutrients from lake water, and repelling microscopic animals that might eat them (Figure 18.13).

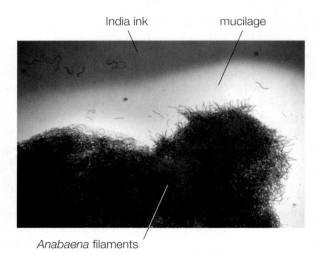

FIGURE 18.13   **Prokaryotic mucilage envelope**   The transparent envelope of *Anabaena* can be detected by using a solution of ink particles to outline the mucilage boundaries (the ink particles are too large to penetrate the mucilage). A firmer sheath of brownish mucilage surrounds cells pictured in Figure 18.10d.

Rhizobium and *Agrobacterium* are bacteria that often live in association with plants. These bacteria secrete the polysaccharide cellulose (also a component of plant cell walls) as an adhesive for attachment to plant surfaces. *Acetobacter*, an oxygen-loving bacterium, is able to grow in acid solutions such as vinegar. This microbe secretes a visible scum of cellulose that acts as a raft, floating the bacteria at the liquid surface where oxygen is more plentiful.

A mucilage coat also helps glue many types of prokaryotes onto surfaces of protists, animal tissues (where they can be of medical or ecological importance), sand, soil, and even pipelines (where the attached microbes can contribute to corrosion). Attached bacteria, together with their adhesive mucilage (sometimes also including small eukaryotic cells) are known as **biofilms.** Dental plaque is an example of a destructive biofilm. Medical scientists and bioengineers are seeking ways to prevent destructive effects of microbial biofilms or put them to use. For example, bacteria that consume methane can be grown within plastic tubing, where they use mucilage to attach themselves to the tubing walls. Waste methane can be passed through the tubing, where the bacteria remove it. This prevents waste methane from entering the atmosphere, where it has undesirable effects as a greenhouse gas (Chapter 1).

Prokaryote cell walls differ in structure and chemistry

Although a few types of prokaryotes lack walls, most have a rigid **cell wall** (see Figure 18.11) that encloses the cell membrane and cytoplasm within. Prokaryotic cell walls lie inside the mucilage capsule or sheath, if one is present. Cell walls help prokaryotes avoid exploding by taking in too much water, a problem that can occur when water is more abundant outside the cell than it is inside (Chapter 4). Prokaryotic cell walls may also help protect against viral infection, though some viruses have the ability to inject infectious genetic material through prokaryotic cell walls.

Peptidoglycan forms part of bacterial cell walls

Most bacteria have a cell wall containing at least some **peptidoglycan**—a substance composed of protein and carbohydrate. Though cell walls of archaeans lack peptidoglycan, some do possess a similar material. Peptidoglycan is very important to the biology of bacteria and human medicine, because the amount of this material present in bacterial walls affects their susceptibility to certain antibiotics. Penicillin and some other antibiotics work by interfering with bacterial peptidoglycan synthesis; these antibiotics are more effective against bacteria with peptidoglycan-rich cell walls. Penicillin and other antibiotics that work by inhibiting bacterial cell-wall production are not harmful to eukaryotes (except for those humans that are allergic to them), because eukaryotic cells lack peptidoglycan walls.

Some bacteria have a thin, outer phospholipid envelope that surrounds the cell wall in addition to the inner phospho-

lipid cell membrane that encloses the cytoplasm of all cells. The outer envelope helps bacteria that have it to resist the entry of antibiotics.

The Gram stain is useful in describing bacteria and predicting responses to antibiotics The Gram stain is a laboratory procedure that reveals differences in the cell coverings of bacteria. Gram staining is a common first step in the process of identifying bacteria, which is particularly important in medical diagnosis of infections. Differences in bacterial Gram staining are useful in predicting which antibiotics are likely to be most effective in treating infections of animals or plants. Bacteria that have an outer phospholipid envelope and lower levels of peptidoglycan in their walls stain pink when processed by the Gram staining procedure. Such bacteria are described as Gram-negative (Figure 18.14a). Because these bacteria have less peptidoglycan, they are generally less sensitive to antibiotics such as penicillin. Almost all of the bacteria that cause plant diseases are rod-shaped, Gram-negative bacteria; *Erwinia* and *Pseudomonas* are examples. A Gram-negative bacterium that is important in human health is *Escherichia coli* (also known as *E. coli*). This microbe normally inhabits the human intestine but can cause disease if it infects other parts of the body. Most *E. coli* do not cause harm when consumed, but a genetic variant (strain) known as *E. coli* O157:H7 has been prominent in the news because it causes a very serious, and sometimes deadly, illness in people who have eaten contaminated food.

Bacteria that lack an outer lipid envelope and have relatively more peptidoglycan in their cell walls stain purple when exposed to the Gram stain process. Such bacteria, described as Gram-positive (Figure 18.14b), are consequently more vulnerable to antibiotics that interfere with bacterial wall production than are Gram-negative bacteria. An example of a medically important Gram-positive bacterium is *Streptococcus*. Strains of this organism cause strep throat, pneumonia, bacterial meningitis (an infection of the spinal

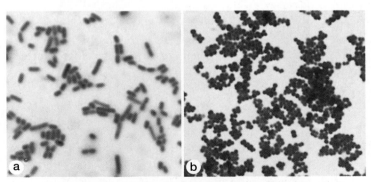

FIGURE 18.14 Gram-negative and Gram-positive bacteria
Escherichia coli (a) is an example of a common and important Gram-negative bacterium; *Streptococcus* (b) is an example of a common, important Gram-positive bacterium.

fluid), and eye infections. Streptococci include the infamous "flesh-eating" bacteria whose disease progression is notoriously difficult to control. The entire DNA sequence of *Streptococcus pneumoniae* has been determined, revealing the presence of a host of genes conferring the ability to break down many types of molecules in human tissues for use as food. The presence of these genes explains why *Streptococcus* can cause so many types of diseases.

Prokaryotes reproduce primarily by binary fission

One reason that most prokaryotes reproduce rapidly is that they have comparatively simple cell division processes. Prokaryote populations grow by repeated division of a single parent cell into two progeny cells; this is known as **binary fission.** A more complete description of binary fission and comparison to the more complex cell division processes of eukaryotes can be found in Chapter 7.

People often need to count the number of prokaryotes that are present in a sample of water, food, or body fluids. These counts are used to assess the degree of an infection, assure food safety, or determine if water is safe for drinking or swimming. Binary fission is the basis for one widely used counting method. Samples are placed into plastic dishes filled with nutrients in a natural material known as agar. Individual prokaryotes in the sample may grow by repeated binary fission into a population of cells known as a colony, visible with the unaided eye (Figure 18.15). Since colonies are large, they are easy to count; each colony represents a single cell in the original sample.

One problem with the counting method just described is that the nutrients provided may not allow the growth of all of the different types of prokaryotes that are present in the sample. If particular prokaryotes do not have the nutrients they need to grow and divide, they will not produce a countable colony. An alternative counting method overcomes this problem. Samples are treated with a stain that binds prokaryotic

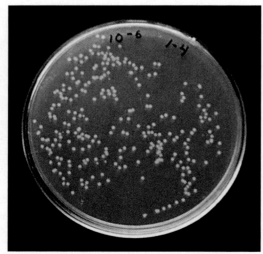

FIGURE 18.15 Colonies of prokaryotes develop from single cells on nutrient plates

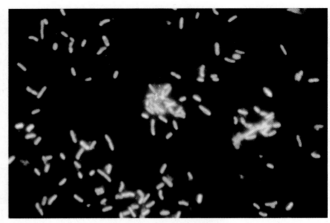

FIGURE 18.16 Fluorescently stained bacteria Prokaryotes can be counted microscopically after staining with a fluorescent dye that binds to cell DNA.

DNA and glows brightly when illuminated with ultraviolet light. The cells then become easily visible when viewed with a fluorescence microscope (Figure 18.16).

Many prokaryotes survive harsh conditions as tough spores

Thick-walled cells, known as **spores,** are produced by many kinds of prokaryotes when they are under stress, such as lack of nutrients or low temperatures. Spores are capable of remaining alive, though in a nongrowing, dormant state for long periods, then germinating into typical cells when conditions improve. Amazingly, bacterial spores have been revived from the guts of bees that were trapped in amber (hardened tree resin) more than 25 million years ago! *Anabaena* and related cyanobacteria produce spores known as **akinetes** when winter approaches. The akinetes are larger than photosynthetic cells and are filled with storage materials (Figure 18.17a). Akinetes survive winter at the bottoms of lakes and are able to produce new photosynthetic filaments in spring when resuspended by water currents to the brightly lit surface.

Endospores are another type of stress-tolerant cell; they are produced inside bacterial cells and are released when the parental cell dies and breaks open (Figure 18.17b). Endospores are one reason that bacterial infections and food contamination are such widespread problems. The endospore-producing bacterium *Bacillus anthracis* causes the disease anthrax, which is often in the news in relation to bioterrorism and potential use in germ warfare. Most cases of human anthrax result when spores of *B. anthracis* enter wounds, causing skin infections that are relatively easily cured by antibiotic treatment. But sometimes the spores are inhaled or consumed in undercooked, contaminated meat. In this case, the spores germinate in the lungs or gastrointestinal systems, producing actively growing cells that cause infections that are more difficult to control. This is partly because these cells produce toxins that continue to damage internal body tissues even after the bacteria themselves have been killed by antibiotics.

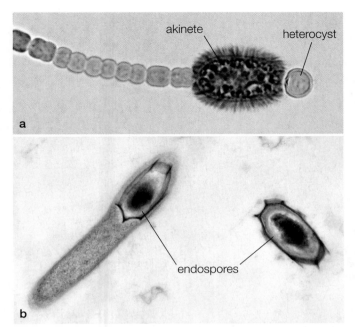

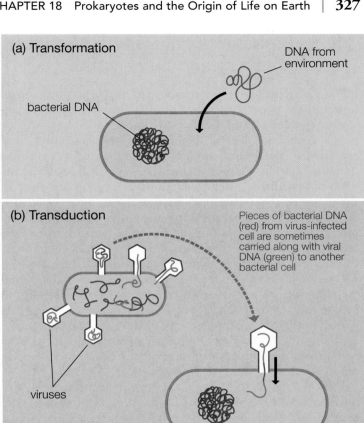

FIGURE 18.17 Bacterial spores include akinetes and endospores (a) Akinetes of *Cylindrospermum* (a relative of *Anabaena*) are larger than other body cells, have thick walls and food storage, are produced at the end of a favorable growth period, and generate rapidly growing filaments when conditions improve. (b) Two thick-walled endospores are visible in this picture taken with a transmission electron microscope. The endospore on the left is still contained within a bacterial cell; the one on the right has been released into the environment upon the death of a parent cell.

The food-contaminating bacterium *Clostridium botulinum* produces spores that can withstand the temperatures used in the pasteurization process (heating to 80°C for 10 minutes). These spores germinate into cells that multiply in the food and produce nerve toxins. When consumed, these toxins cause the paralysis known as botulism. Consumers and home canners are well advised to reject cans with bulging lids, the telltale sign of the accumulation of gas produced by *C. botulinum* growth. *Clostridium tetani* produces a nerve toxin that causes lockjaw (tetanus) when bacterial cells or endospores enter wounds from soil. The ability of *Bacillus* and *Clostridium* to produce resistant spores explains their widespread presence in nature and their danger to humans.

Prokaryotes lack sex, but can exchange DNA

Prokaryotes are distinguished by the absence of sexual reproduction involving gametes, zygotes, and the division process meiosis (Chapter 13), which is common among eukaryotes. However, prokaryotes can exchange genetic material by other means. Prokaryotes can take up pieces of DNA from their environments—a process known as *transformation* (Figure 18.18a), which is also useful in genetic engineering (Chapter 15). Prokaryotes may also obtain DNA that has been brought into cells by infecting viruses—a process known as *transduction* (Figure 18.18b). Some prokaryotes have the ability to receive DNA from other prokaryotes by a mating process known as *conjugation* (Figure 18.18c). Thin

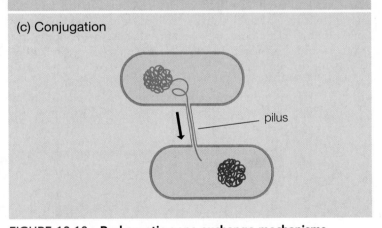

FIGURE 18.18 Prokaryotic gene-exchange mechanisms (a) Transformation by environmental DNA, (b) transduction by viruses, (c) conjugation (mating via pili).

tubes known as **pili** (singular: **pilus**) link conjugating cells and allow DNA to move from one cell to another, compatible cell. All of these gene-exchange processes are referred to as *horizontal gene transfer*. Genes acquired by horizontal transfer are thus distinguished from genes that were inherited from a parental cell (during binary fission). DNA exchange in prokaryotes provides the same advantage that sex gives to eukaryotes—more rapid adaptation to environmental change. Gene exchange allows prokaryotes to acquire useful new traits 10,000 times faster than if each species had to adapt on its own.

18.5 | Prokaryotic diversity is important in nature and human affairs

DNA differences are commonly used to classify organisms into species, genera, families, orders, classes, phyla or divisions, kingdoms, and domains (Chapter 17). But because prokaryotes have several methods for DNA exchange, they readily transfer genes among different taxa, even across domain boundaries. For example, 25% of *E. coli*'s genes were obtained from other bacterial species, and one-third of the genes of the archaean *Methanosarcina mazei* originally came from bacteria! This rampant, promiscuous DNA swapping can be a problem in prokaryote classification. Experts are working to identify sets of genes that are less commonly transferred, to increase the odds that classification schemes will more accurately reflect evolutionary descent. DNA samples taken from nature are used to explore the diversity of prokaryotes that have not been cultivated and thus were previously unknown. A recent study of the Sargasso Sea uncovered about 150 groups of unknown bacteria and 1.2 million new genes, and similar new diversity is being discovered in other habitats.

According to DNA data, Domain Archaea occupies a distinct branch of the prokaryotic family tree, with most known prokaryotic diversity falling within Domain Bacteria (Figure 18.19). Members of these domains are known informally as bacteria and archaea. Archaea are famous for the fact that many live in extremely harsh environments, such as waters of very high salt content, methane levels, or temperatures that would kill most bacteria and eukaryotes. *Pyrolobus fumarii*, for example, is an archaean that can grow in scalding-hot water of 113°C. However, archaea can also be found in more moderate environments, such as soils. Some

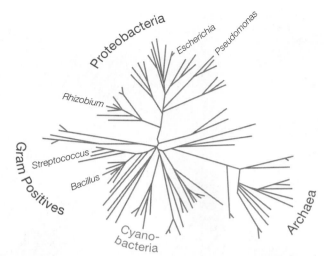

FIGURE 18.19 Phylogenetic tree showing major groups of prokaryotes

traits of archaea are more similar to eukaryotes (than bacteria), while other archaean features are held in common with bacteria (Table 18.2).

In the rest of this chapter, we have focused more closely on bacteria because these prokaryotes are more commonly involved in beneficial or harmful associations with plants and other organisms than are archaea. Three examples of bacterial groups that are particularly important in plant or human biology are proteobacteria, cyanobacteria, and Gram-positive bacteria. A few example genera from each of these bacterial groups, and their relevance to plant biology or human disease, are described in the following sections.

TABLE 18.2 Some of the Features Differentiating Bacteria, Archaea, and Eukarya

Feature	Bacteria	Archaea	Eukarya
Prokaryotic cell structure	yes	yes	no
Circular chromosome(s)	yes	yes	no
DNA usually bound to proteins	no	yes	yes
Peptidoglycan cell wall	yes	no	no
Plasmids	yes	yes	rare
Some are methanogenic	no	yes	no
Some are nitrogen fixers	yes	yes	no
Some are photosynthetic	yes	no	yes (those with chloroplasts)
Some produce gas vesicles	yes	yes	no
Some grow at 80°C or above	yes	yes	no

Proteobacteria are related to eukaryotic mitochondria

During the origin of eukaryotic cells, an ancient proteobacterium gave rise to the mitochondrion (Chapter 19). Recall from Chapters 4 and 5 that mitochondria are organelles that provide eukaryotic cells with high-efficiency systems for ATP production. In other words, mitochondria allow eukaryotic cells to obtain more energy from organic food than they would if they lacked these organelles. The acquisition of mitochondria was thus a pivotal event in the evolutionary history of eukaryotic cells. Proteobacteria remain important to eukaryotes today. For example, *Pseudomonas* is a proteobacterium that causes disease in both plants and animals. Legume plants can be attacked by aphid insects, which use their mouthparts to suck plant juices; certain proteobacteria that live within these aphids actually determine which legume plants will be attacked! *Escherichia*, a common inhabitant of the human gut, also much used in biotechnology (see Chapter 15), is likewise a member of this group. Methane-consuming bacteria are also found among the proteobacteria; these help reduce the global warming impact of methane gas in Earth's atmosphere. Some proteobacteria contain a bacterial type of chlorophyll and are photosynthetic, though they do not emit oxygen gas as do cyanobacteria, algae, and plants.

Cyanobacteria are related to eukaryotic plastids

Cyanobacteria are Gram-negative bacteria that are abundant in oceans, freshwaters, wetlands, and in soils, including desert soils. They are the only bacteria that have photosynthesis involving oxygen emission. The oxygen is released when cyanobacteria break apart water during photosynthesis (see Chapter 5). Though the hydrogen and electrons released from water are used during photosynthesis, oxygen gas is a waste product that easily diffuses from cyanobacterial cells. (In contrast, solid waste products build up in the cells of photosynthetic proteobacteria, eventually slowing their growth.) An ancient cyanobacterium was the ancestor of the chloroplast, endowing algae and plants with both photosynthetic ability and the ability to produce oxygen. The evolution in early cyanobacteria of oxygen-producing photosynthesis was a key breakthrough in the continued evolution of life on Earth. Without it, an oxygen-rich atmosphere and higher forms of life that require oxygen could not have arisen. Cyanobacteria also possess chlorophyll *a* attached to internal membranes—thylakoids. It is no coincidence that chlorophyll *a* is also the major photosynthetic pigment in algae and plants. Chloroplasts of algae and plants inherited their oxygen-producing abilities, chlorophyll *a*, thylakoids, and other photosynthetic attributes from cyanobacteria.

Many modern cyanobacteria are valued for contributing fixed nitrogen to their environments, as in the case of *Anabaena* enclosed within leaves of the aquatic fern *Azolla*, described in the chapter opening. However, some cyanobacteria are of concern because they produce toxins that affect animals, including humans. In addition to *Anabaena*, common toxin-producing genera of freshwaters include *Aphanizomenon*, *Microcystis*, and *Cylindrospermopsis*.

Gram-positive bacteria include important disease agents and antibiotic producers

Gram-positive bacteria include *Clostridium*, *Staphylococcus*, *Streptococcus*, and *Bacillus*, some species of which were earlier noted as important disease agents in humans. But not all species of *Bacillus* are harmful. For example, the soil bacterium *Bacillus cereus* helps plant roots form beneficial mycorrhizal associations with fungi (Chapters 10, 20). Actinobacteria are Gram-positive bacteria that often form branching filaments which are superficially similar to some fungi. Examples of actinobacteria include *Mycobacterium tuberculosis*, the agent of tuberculosis in humans, and *Deinococcus*, famed for its ability to resist the debilitating effects of nuclear radiation. Microbiologists study *Deinococcus* with the hope of discovering ways to better treat radiation sickness in humans. Many actinobacteria are notable producers of antibiotics; over 500 different antibiotics are known from this group. For example, the actinobacterium *Streptomyces* living within the tissues of the snakevine plant, *Kennedia nigriscans*, produces antibiotics known as munumbicins. These antibiotics are active against some cancer cells, fungi that cause plant diseases, and the tuberculosis agent *Mycobacterium*.

Prokaryotic nutrition is ecologically important

Prokaryotes have many types of metabolism, each playing an important role in Earth's ecology. Some prokaryotes are **photoautotrophs**, meaning that they use light as a source of energy for producing organic compounds. Other prokaryotes produce organic compounds by using energy obtained by means of chemical modifications; these **chemoautotrophs** are the agents of important mineral deposits and transformations. Together, these **autotrophs** ("self-feeders") produce organic food used by other organisms.

Heterotrophic prokaryotes ("other feeders") must absorb at least one (and often more) organic nutrient(s) from their environment. Some heterotrophic prokaryotes are photoheterotrophs, meaning that they are able to use light energy to generate ATP (but not organic food). Other heterotrophs are **chemoheterotrophs**; these prokaryotes must obtain organic carbon for both energy and carbon compounds. Chemoheterotrophs that break down dead organisms and organic compounds are known as **saprobes**. These prokaryotes (together with fungi) play an important role in decomposition in natural environments; otherwise, dead organisms would not decay. Heterotrophic prokaryotes and fungi also release mineral nutrients that would otherwise remain bound up in dead matter. Parasitic prokaryotes are heterotrophs that use organic compounds in living tissues of plants, animals, protists, or fungi. Parasitic prokaryotes that cause disease symptoms are known as **pathogens**. Some prokaryotes live within the bodies of eukaryotes as mutualistic symbionts, providing benefits as well as receiving them. For example, legume plants receive nitrogen compounds from symbiotic bacteria that live in root nodules (see Chapter 10). Humans obtain vitamins B_{12} and K from intestinal bacteria.

The balance of the greenhouse gas methane in Earth's atmosphere has been maintained by the activities of two broad groups of prokaryotes. Methane producers are archaeans known as **methanogens** ("methane generators"). Methane gas easily diffuses from the cells into the atmosphere. Methane produced by early archaeans helped warm Earth during the Sun's early dim period, but increases in modern methane emissions contribute to undesirable global climate change. Methanogens typically live in low-oxygen (anaerobic) environments. Marsh gas is mostly methane, and there are very large quantities of methane in deep-sea and Arctic subsurface deposits. Cattle and other ungulate animals emit methane, because methanogens live within their rumens. Bacteria known as **methanotrophs** and methylotrophs are able to consume methane. These important bacteria require oxygen, and thus live in oxygen-rich (aerobic) environments. Methanotrophic bacteria are symbiotic with marine invertebrates living in deep-sea communities. These bacteria also commonly occur with roots of wetland vascular plants and cells of wetland mosses. The plant roots and moss cells release oxygen needed by their methanotrophic bacterial associates. Ecologists have emphasized that preserving wetland plant communities will help prevent methane produced in deeper waters from escaping into the atmosphere.

Prokaryotes have useful applications in human affairs

Agricultural applications Each spring, the delivery of ammonia nitrogen fertilizer to farms is a common sight because fertilizer additions to soil greatly increase crop production. Ammonia fertilizer is produced by combining nitrogen gas—which is abundant in Earth's atmosphere—with hydrogen. During this process, the nitrogen gas is "fixed" or "combined," as it is converted to liquid ammonia or solid ammonium salt. **Industrial nitrogen fixation** requires very high temperatures and pressures and is costly in terms of energy requirements. Nitrogen fertilizer is essential to maintaining high crop productivity in the United States and other developed countries. However, many farmers throughout the world cannot afford industrially produced nitrogen fertilizer. This limits crop production and our capacity to fully nourish the world's current human population. For this reason, crop scientists are very interested in harnessing **biological nitrogen fixation**, the process carried out naturally by some prokaryotes.

Some bacteria and archaeans are able to fix nitrogen at normal temperatures and pressures because they possess the enzyme **nitrogenase**—a biological catalyst that vastly increases the rate at which ammonia can be generated from nitrogen gas. Ammonia and nitrogen-containing compounds derived from it, together known as **fixed nitrogen**, can be used by plants and microbes to produce the amino acids, proteins, and nucleic acids needed by all living things. All other organisms rely on nitrogen-fixing prokaryotes to produce fixed nitrogen because they are not able

to do it themselves (human industrial processes being the only exception).

Microorganisms living in soil and water can convert ammonia into other nitrogen compounds such as nitrate and urea, which can be used by crops and other plants. If you examine the label on houseplant or garden fertilizers, you will discover that a large percentage of their content is "ammoniacal nitrogen" or urea. Because many agricultural and garden soils are low in fixed nitrogen, and houseplants confined to small pots rapidly exhaust soil nitrogen supplies, adding nitrogen fertilizer substantially increases plant growth.

A worsening global environmental problem is that much of the nitrogen fertilizer applied to croplands washes from soil into natural waters, where it can cause water pollution and foster toxic blooms (large growths) of cyanobacteria (such as *Anabaena*) or eukaryotic algae. Combined nitrogen that has been naturally produced is less potentially harmful to the environment, in part because prokaryotes fix nitrogen only when it is not available in their environment. In addition, nitrogen-fixing symbionts associated with plants deliver fixed nitrogen directly to roots, reducing the chance that it will be washed from soil. A major reason for the importance of legumes to human nutrition and agriculture is that they are particularly adept at attracting nitrogen-fixing bacterial partners and harboring them in plant roots (Chapter 10). A variety of nonlegume plants also partner with nitrogen-fixing bacteria: for example, liverworts, hornworts, *Azolla* ferns of rice paddies, and alder trees.

Biotechnologists have been working toward incorporating biological nitrogen-fixation systems into the many crops that lack symbiotic bacteria as a way to increase crop productivity without applying industrial fertilizers. One problem is that most known forms of the nitrogen-fixation enzyme nitrogenase are poisoned by oxygen. (The O_2 molecule is about the same size and shape as the N_2 molecule and thus binds tightly within nitrogenase's active site, preventing entry of N_2. Many poisons work by binding cellular enzymes so that enzyme function is impaired or lost.) Nitrogenase first arose when Earth's atmosphere was nearly devoid of oxygen, but today's atmosphere contains about 21% O_2. For more than 2 billion years since an oxygen-rich atmosphere has existed, nitrogen-fixing microbes have been challenged to find ways of protecting their nitrogenase from oxygen. Some cyanobacteria have met this challenge by producing special cells—**heterocysts**—in which nitrogen fixation occurs but the oxygen-generating reactions of photosynthesis have been turned off (Figure 18.20). *Anabaena* cells communicate with each other by chemical signals that result in heterocyst development every 10 cells or so along the filaments. By spacing its heterocysts, *Anabaena* maximizes nitrogen fixation without losing too much photosynthetic capacity. Legumes have taken a different approach to protecting their nitrogen-fixing partners—*Rhizobium* and related bacteria—from oxygen. These plants harbor nitrogen-fixing bacteria within root nodules that also contain leghemoglobin (legume hemoglobin), a molecule very similar to the oxygen-carrying hemoglobin in animal blood. Leghemoglobin regulates the oxygen level within nodules so that it remains high enough

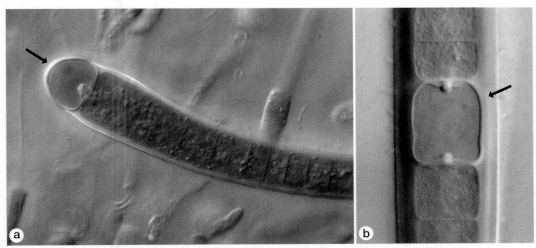

FIGURE 18.20   Cyanobacterial heterocysts are specialized sites of nitrogen fixation
Heterocysts (arrows) from two different species of cyanobacteria are shown, demonstrating that they can be formed at the end position of a filament (a) or between normal photosynthetic cells (b).

for respiration, but low enough to prevent nitrogenase poisoning.

Industrial applications   Some bacteria produce chemical changes in food that improve consistency or flavor and are widely used in the food industry—to make yogurt, for example. Other bacteria are grown in giant industrial vats for commercial production of useful chemicals such as vinegar, amino acids, enzymes, vitamin B_{12}, insulin, vaccines, and antibiotics. Some common antibiotics derived from bacteria include streptomycin, tetracycline, kanamycin, gentamycin, bacitracin, polymyxin-B, and neomycin. You probably have antibiotic creams containing mixtures of some of these compounds in your first-aid supplies.

Environmental remediation and monitoring   Bacteria have also proven useful in processing wastewater in sewage treatment plants and cleaning up toxic waste, oil spills, explosives, and other materials that endanger human environments. These are examples of environmental remediation. The proteobacterium *Geobacter*, for example, is useful in precipitating metals such as uranium from contaminated water. Bacteria are also used to monitor environments for hazardous chemicals that cause mutations and cancer.

Though prokaryotes come in small packages, they have had a powerful impact on Earth's environment and life for billions of years and continue to wield their power today. The future of people and plants alike depends on these tiny microbes.

HIGHLIGHTS

18.1   Living things are characterized by cell structure, metabolism (the ability to use energy to produce organization), growth, reproduction, (ability to defend themselves), response to stimuli, adaptation to change, and evolution. Life exists on Earth because this planet is positioned within the Sun's habitable zone, where water can exist as a liquid as well as vapor and ice and because early prokaryotes modified the environment so that more complex organisms could exist.

18.2   Formation of the Earth and its solar system occurred 4.6 billion years ago. Chemical and fossil evidence indicates that life first occurred 3.85–3.7 billion years ago. Photosynthetic cyanobacteria generated an oxygen-rich atmosphere from which a UV-protective ozone shield was formed.

18.3   Experiments supporting the chemical-biological theory suggest that life could have arisen on Earth from nonliving inorganic compounds via a series of stages exhibiting increasingly greater complexity from inorganic molecules to organic molecules to macromolecules and protocells.

18.4   Prokaryotes are organisms that lack membrane-bound nuclei and other organelles, and are classified in the domain Bacteria or domain Archaea. Prokaryotes lack sexual reproduction typical of eukaryotes, but can exchange genetic material in several ways. Bacteria and Archaea are distinguished primarily by DNA differences, but also by cell-wall chemistry. Prokaryotic cells are typically small and mostly occur as one of a few types of shapes—spheres, rods, or spirals. Prokaryote bodies are microscopic and occur as unicells, colonies, or filaments; some are motile, but others are not. Polysaccharide mucilage coats are common. Prokaryote species may differ in relative peptidoglycan content of their cell walls and presence of an external lipid envelope, properties reflected by Gram staining that also have medical importance. Prokaryotes reproduce by binary fission (dividing in half), and many can survive adversity as tough spores.

18.5 Diverse prokaryotes play many important roles in Earth's ecosystem and human affairs. Archaea are notable for living in extremely hot or acidic habitats. Bacteria are more commonly associated with plants and other organisms. Examples of major prokaryote groups are proteobacteria, cyanobacteria, and Gram-positive bacteria. An ancient proteobacterium was the ancestor of modern mitochondria. An ancient cyanobacterium was the ancestor of plastids of modern algae and plants. Some Gram-positive bacteria are disease agents, whereas others are notable for production of antibiotics. Some prokaryotes generate or consume the potent greenhouse gas methane, some are photosynthetic oxygen producers, and some are able to convert atmospheric nitrogen gas into fixed nitrogen, a form useable by plants to produce protein. Legumes are protein rich because they have nitrogen-fixing symbiotic bacteria in root nodules. Photosynthesis originated in prokaryotes and was inherited from them by plants.

REVIEW QUESTIONS

1. Explain why it is unlikely that life presently exists on Mars or Venus.

2. Why is ozone so important, and harsh ultraviolet light so harmful, to life on Earth? Why was early life confined to the oceans?

3. List at least five characteristics by which you can recognize living things.

4. Summarize the early history of the Earth, from the birth of the planet to the appearance of land plants and present-day levels of atmospheric oxygen.

5. Describe the first life-forms on Earth. What kind of organisms were they, when did they first appear, and how do we know of their existence?

6. Summarize the three stages by which the Chemical-Biological Theory proposes that the first cells arose from nonliving inorganic molecules.

7. What are the three common shapes of prokaryotic cells, and what type of higher-order aggregates might these simple cells form?

8. What are thylakoids, and which prokaryotes have them? Do plants?

9. Distinguish between prokaryotic endospores and akinetes. What service do these cells provide those species that can produce them?

10. How does biological nitrogen fixation differ from industrial nitrogen fixation? What is the importance of nitrogen fixation to plant life?

11. Why is it important that most nitrogenase enzymes be maintained in a low-oxygen environment? Discuss how this is accomplished in cyanobacteria and legumes.

12. What is a Gram stain? When examining stained cells, what is the difference in appearance between Gram-positive and Gram-negative cells? What are the cellular differences between Gram-positive and Gram-negative cells? What are the medical implications of Gram-negative cells with respect to antibiotics like penicillin?

APPLYING CONCEPTS

1. At the present time on Earth, why is it unlikely that we could observe stages in the early evolution of life occurring in nature, such as the production of organic compounds from inorganic ones? (*Hint*: Consider the conditions used in the Urey–Miller experiment.)

2. Today there is considerable concern regarding the destruction of ozone in the upper atmosphere by artificial chemicals such as chlorofluorocarbons. What is the reason for this concern, and which humans are most affected?

3. When some prokaryotes are grown under optimal conditions, they can double in number every 30 minutes using binary fission. If one such cell were plated on a nutrient plate (and thus provided optimal growing conditions) and that growth rate were strictly maintained, how many cells would be present after (a) 1 hour? (b) 3 hours? (c) 6 hours? (d) 12 hours? (e) 1 day? Obviously, this growth rate cannot continue indefinitely; what would normally limit it?

4. Discuss the importance of the nitrogen fixers (such as the bacteria described in this chapter) to the other inhabitants of the biosphere. Do they benefit only a small number of species, a moderate number of species, or virtually all species on this planet?

5. Genetic engineers use bacteria virtually every day in their research. Can you suggest reasons why bacteria have so many uses in the laboratory, and also suggest some ways in which modern scientists use bacteria?

6. Penicillin, you may recall, interferes with a Gram-positive bacterium's ability to make a functional cell wall. When a pathogenic bacterium is Gram negative or becomes resistant to penicillin, a different antibiotic must be used to overcome an infection. What other cell processes might other antibiotics inhibit or block besides cell-wall formation?

7. Most commercial antibacterial antibiotics are produced by either fungi or bacteria. Because resources and energy are required to produce antibiotics, what advantage would the ability to produce antibiotics confer upon natural populations of bacteria or fungi?

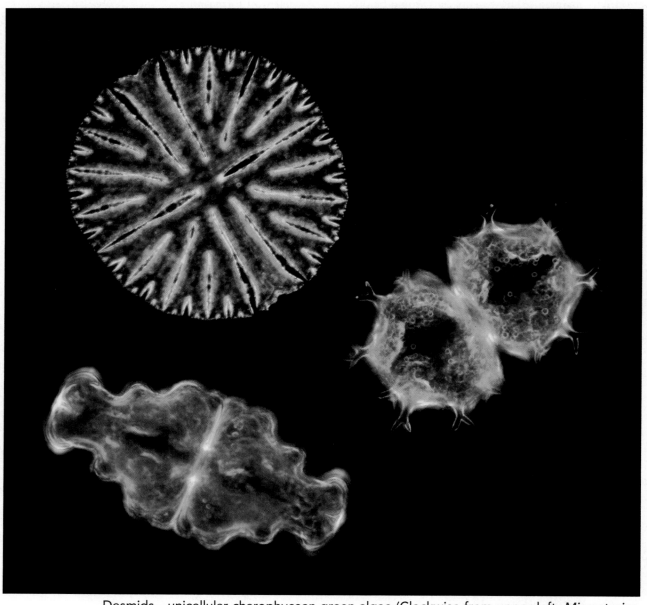

Desmids—unicellular charophycean green algae (Clockwise from upper left: *Micrasterias,*
Xanthidium, Euastrum)

KEY CONCEPTS

- Protists are eukaryotes that are not included in the plant, fungal, or animal kingdoms.

- Protists are mostly aquatic and often microscopic in size.

- Photosynthetic protists are commonly known as algae.

- Land plants arose from ancestral green algae; fungi and animals arose from nonphotosynthetic protists.

- Algae illustrate key evolutionary events in the history of plants, such as plastid acquisition.

- Algae illustrate the importance of photosynthesis in nature, as do plants.

- Structural and reproductive adaptations favor survival of algae and other protists in nature.

Have you ever wondered where that expensive gasoline in your car's tank originated? Would you believe dead algae? Gasoline is refined from oil derived from tiny algae that lived and photosynthesized in ocean waters long ago, then died, sinking to the ocean floor. Accumulating in ocean sediments over the passage of hundreds of millions of years, these algae were eventually pressure-cooked into oil by geological forces. One scientist has estimated that 90 metric tons of dead algae has gone into the formation of each gallon of gasoline and that the average car's gas tank contains the remains of more than a thousand tons of algae. Modern algae could potentially be grown as a renewable source of oil and gasoline, but it would take at least 400 years to grow enough algae to produce the same amount of fuel that people burn in just 1 year.

In addition to their importance in the origin of fossil fuels, algae are indispensable sources of food for aquatic animals and influence Earth's climate. Algae generate 50–70% of Earth's atmospheric oxygen and sometimes cause harmful water blooms. Humans use algae as food—think of sushi wrappers—and in many types of industrial products. For example, algae are the best sources of some fatty acids that are added to baby formula to foster infant brain growth and development. In science, the study of algae has been essential to understanding the molecular basis of photosynthesis and the evolutionary origin of eukaryotic cells and the plant kingdom.

Algae are classified among the protists, together with many kinds of nonphotosynthetic microorganisms that are not animals, plants, or fungi. Nonphotosynthetic protists play important ecological roles and illuminate the evolutionary origins of the fungal and animal kingdoms. This chapter reviews protist biology, with a special focus on algae and their ecological, evolutionary, and technological significance. ■

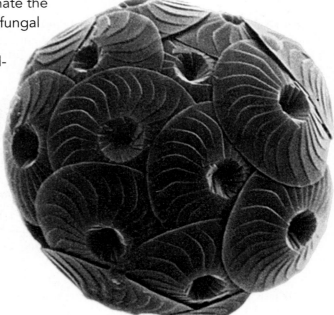

Calcidiscus, a haptophyte

19.1 | What are protists and where do they occur?

Protists are organisms that are (1) not included in the land plant, fungal, or animal kingdoms; (2) often microscopic in size; and (3) occur most abundantly in moist habitats. Groups of protists whose members generally contain one or more plastids and are usually photosynthetic are informally known as **algae** (Figure 19.1). Photosynthetic organisms are **autotrophic** ("self-feeding"), meaning that they can produce their own organic food. **Protozoa** is the informal name for various groups of protists that lack plastids and are not photosynthetic. Protozoa are single-celled, often mobile organisms first thought to be simple animals, but it is now known that most protozoa are not very closely related to animals (see Figure 19.1). Protozoa are fundamentally **heterotrophic** ("other feeding"), meaning that they must obtain organic food from outside their bodies. However, some protozoa contain algal cells that function like plastids, providing organic food. **Slime molds** and **oomycetes** are additional types of protists that lack plastids and photosynthesis. Slime molds and oomycetes are often described as funguslike protists because they often look or behave somewhat like fungi, but they are not closely related to fungi. Although the terms *algae* and *protozoa* are useful in ecological discussions, neither the algae nor the protozoa are a single group of closely related protists. There are, in fact, many distinct groups of algae, as well as many distinct groups of protozoa and several groups of funguslike protists.

How can protists be distinguished from plants, fungi, and animals?

Protists are eukaryotic organisms that are not included in the land plant, fungal, and animal kingdoms, because they lack characteristics that typify these kingdoms (Table 19.1).

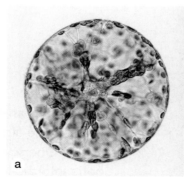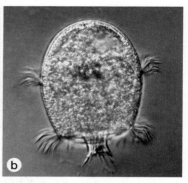

FIGURE 19.1 Algae and protozoa (a) The single-celled photosynthetic alga *Eremosphaera* containing numerous plastids. (b) The plastidless ciliate protozoan, *Didinium*.

Structural and reproductive features, as well as DNA sequence information, are used to group algae with protists, as we have done in this book. However, some experts include all photosynthetic eukaryotes, or at least some of the algal groups, in their concept of plants.

Some modern protist groups share specialized characters with plants, fungi, or animals, reflecting close relationship to ancestors of the latter groups. For example, a group of mostly aquatic green algae known as charophyceans (Figure 19.2a) is very closely related to the ancestors of land plants. However, charophyceans lack a set of adaptations to terrestrial life that characterize all of the land plants (Chapter 21). Similarly, the plastid-less protists known as choanozoa (Figure 19.2b) are very closely related to the ancestors of the animal kingdom but are not included in the animal kingdom based on absence of characters (such as multicellularity) that define animals. Charophyceans and choanozoa are of particular importance to evolutionary biologists interested in deducing the evolutionary origin of animals and land plants.

EVOLUTION

TABLE 19.1 Features that Distinguish Protists, Fungi, Animals, and Plants

Group	Habitat	Nutrition	Rigid Cell Walls	Body	Reproduction
Protists	Mostly aquatic; some terrestrial	Autotrophy heterotrophy, or mixotrophy	Variable in presence and composition	Unicellular or multicellular; colonies, filaments, or tissue	All known sexual cycle types represented; asexual common
Fungi	Mostly terrestrial; some aquatic	Heterotrophy	Chitin	Unicells or filaments	Zygotic sexual life cycle only; asexual common
Animals	Terrestrial and aquatic	Heterotrophy	none	Multicellular, tissues	Gametic life cycle only; asexual uncommon
Plants	Mostly terrestrial; some aquatic	Mostly autotrophy; some heterotrophy	Cellulose and pectin	Multicellular, tissues	Sporic life cycle only; asexual common

FIGURE 19.2 Some protists reflect the ancestry of the plant and animal kingdoms (a) The green alga *Micrasterias* is a member of the charophyceans, closely related to land plants. (b) *Salpingoeca* (arrow) is a choanozoan protist closely related to the animal kingdom. It is attached to a diatom cell here.

Microscopes must be used to observe most protists

Protists are commonly of microscopic size. Protist bodies may be (1) unicellular (Figure 19.3a), (2) aggregates of cells—known as colonies—that are held together by adhesive substances (Figure 19.3b), (3) cells joined end-to-end to form filaments (Figure 19.3c), or (4) composed of tissues whose cells communicate chemically amongst themselves and may become specialized (Figure 19.3d). The evolution of bodies having multiple coordinated cells, known as **multicellularity**, is key to the origin of the land plant, fungal, and animal kingdoms. Multicellular bodies evolved independently in several different protist groups, another reason for protists' evolutionary significance.

Some algae have bodies composed of very large single cells or many cells that are often macroscopic—visible with the unaided eye. These organisms represent exceptions to our characterization of protists as typically microscopic. Seaweeds, common on marine shores around the world, are examples of macroscopic protists (Figure 19.4). Although seaweed bodies sometimes resemble those of land plants, and they are sometimes referred to as "marine plants," seaweeds are more closely related to microscopic algae than they are to the land plants. There are approximately 7,000 species of brown-, red-, and green-pigmented seaweeds; some experts

think that species of microscopic protists may number in the millions.

Protists are common and numerous in aquatic and moist habitats

When the first microscopes were invented, biologists observed an incredible array of protists swarming through pond-water samples. Natural water samples remain the best place to observe a wide variety of protist types. Small protists that swim or float in water are members of the **plankton**. Photosynthetic plankton are known as **phytoplankton** (meaning "plantlike" plankton). Plankton also includes protozoa and small animals—the zooplankton.

Why are protists mostly found in watery habitats? Protists' generally small bodies have a high surface-to-volume ratio. This is advantageous when it comes to obtaining dissolved materials from water, but it also makes protists dry out much faster than larger organisms. Consequently, most protists reproduce and grow best in moist habitats. These include freshwater ponds, lakes, and streams; ocean waters; soil water; and moist tissues and cells of animal or plant bodies. Although various protists inhabiting animal or plant bodies are parasites that cause disease by destroying tissues (see Essay 19.1, "It's Not Easy Being Nongreen: Algae as Parasites"), others live cooperatively within hosts, often providing some benefit. One example is the presence of dinoflagellate algae within the tissues of corals and other marine animals (Figure 19.5). This is an example of **symbiosis**, an intimate ecological association between two (or sometimes more) organisms. Organisms that live within the bodies of other organisms are known as **endosymbionts** (the prefix *endo* means "inside"). The endosymbiotic dinoflagellates provide organic food and oxygen, both needed by the coral animals, which are heterotrophic. This partnership favors extensive growth of biodiversity-rich coral reefs in tropical regions around the world (Chapter 28).

Microscopic protists move in several ways

Many protists bear one or more **flagella**—long, propelling extensions from cells (Chapter 4). Flagella rapidly bend and straighten, producing a swimming motion. Some protists have just one flagellum, but others have two or more. Protists that rely on flagella to move in fluid are known as **flagellates** (Figure 19.6). Flagellate protists are typically small, 2–20 micrometers (μm) long. If they were any larger, their flagella would not be able to keep them suspended in the water due to the increased impact of gravity. Some flagellate protists live sedentary lives, attached to underwater surfaces; in this case, the flagella serve to collect bacteria and other small particles for consumption as food. Many protists that have immobile or multicellular bodies for much of their lifetime produce flagellate reproductive cells.

Cilia are structures that are internally similar to flagella but are shorter and more abundant on cells (Figure 19.7). Cilia extend from part or all of the surfaces of the nonalgal protists known as **ciliates**. Having more mobility-generating structures gives ciliates more swimming power, so they are often larger

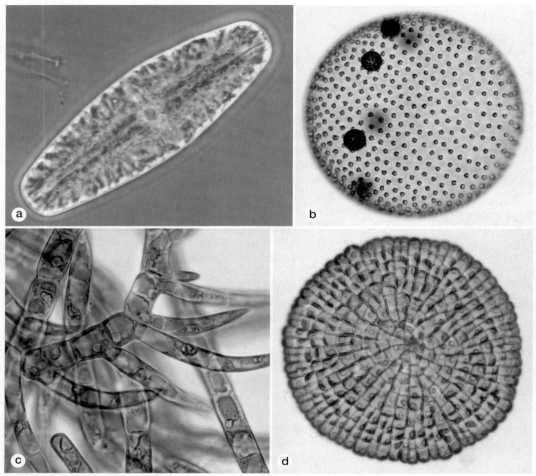

FIGURE 19.3 **Algal bodies occur in diverse types** (a) The green alga *Netrium* has a unicellular body. (b) *Volvox* is a colonial green alga whose individual flagellate cells are bound together to form a sphere. (c) *Chaetophora* is a green alga with a branched filamentous body. (c) *Coleochaete orbicularis* has cells interconnected to form a coherent tissuelike body.

than flagellates. The coordinated movement of cilia allows some ciliates to scamper across underwater surfaces rich in food materials and others to generate hurricane-like water funnels that concentrate food particles.

Some algae and other protists lacking mobility structures attach themselves to underwater surfaces; they depend on water motions for food supply or to move their reproductive cells from one place to another. Seaweeds, for example, grow attached to rocks or other surfaces that are underwater at least part of the time. Branches, holes, or surface irregularities in their bodies (see Figure 19.4) increase water flow and help seaweeds obtain minerals from the surrounding water.

19.2 | Protists include diverse groups whose relationships are not completely known

Biologists examine protist diversity and relationships by comparing their cell structure and DNA. The data from these studies indicate that all eukaryotes (protists, fungi, animals, and plants) may have a single common ancestor. But protists by themselves do not form a single clade (a group of organisms descended from a common ancestor—see Chapter 17). Rather, there are several distinct lineages of protists whose relationships are currently unclear (Figure 19.8). For this reason, in this book we have avoided classifying algae and other protists into kingdoms or other formal groups. We have also emphasized photosynthetic protists—major groups of algae and their distinguishing features (Table 19.2).

Euglenoids, together with some other groups of flagellates, including kinetoplastids (see Essay 19.1), are known collectively as the discicristates. This is because their mitochondrial cristae (internal membranes) are disk-shaped. Euglenoids feature a unique cell covering formed of interlocking ribbonlike protein strips (see Figure 19.6a). Most euglenoids also possess conspicuous granules of the carbohydrate paramylon. *Euglena* and some relatives have green plastids and are thus autotrophic, but other euglenoids are heterotrophic because they lack plastids altogether, or have colorless plastids. Some euglenoids maintain endosymbiotic cyanobacteria within their cells. These cyanobacteria function like plastids, providing photosynthetic products to the host cell. Some euglenoids have a distinctive

FIGURE 19.4 **Seaweeds** (a) *Postelsia*, also known as sea palm, is found on the eastern Pacific coast, from British Columbia to California. (b) *Sargassum* can be attached or float in loose masses. The small spherical structures provide flotation, and the high degree of branching increases contact with water for absorption of minerals. The leaf-shaped blades are the main photosynthetic organs.

kind of oozing motion—**metaboly**—that allows them to wriggle through moist soil or sand (see Figure 4.2). Ecologically, euglenoids are very important members of wetland food webs (Chapter 28). Autotrophic euglenoids provide organic carbon to ecosystems, and some of the heterotrophic euglenoids obtain their food by consuming bacteria and algae.

Cryptomonads are mostly single-celled flagellates (Figure 19.6b) with red, blue-green, olive, or colorless plastids, but the genus *Goniomonas* is plastidless. Cryptomonads, whose name means "hidden single cells," commonly occur in fresh and marine waters but are sometimes hard to observe because they are readily eaten by other protists and microscopic animals. Cryptomonads are important members of planktonic food webs for two major reasons. Their small size makes them easy for zooplankton to consume, and they are very nutritious. They contain fatty acids that are required by small aquatic animals.

Alveolates include the **dinoflagellates** (Figure 19.6c), together with ciliate protozoa and a group of parasites known as apicomplexans (see Essay 19.1). These organisms are grouped

together because they all have membranous sacs known as alveoli at the edges of their cells. Some dinoflagellates have plates of cellulose deposited in the alveoli, which together form an armor-like cell enclosure. These are known as the armored dinoflagellates. Dinoflagellates that have little or no cellulose in their alveoli are the unarmored, or "naked," dinoflagellates. Many species in about 70 genera of dinoflagellates lack chloroplasts and are heterotrophic. Most heterotrophic dinoflagellates feed by ingesting other cells as food. Many dinoflagellates have plastids and are therefore photosynthetic (see Figure 19.17a).

Dinoflagellate plastids originated from a variety of eukaryotic algae that were originally ingested as food by the dinoflagellates. Most of the cell components of the ingested algae were eventually digested, but the plastids (and, in some cases, additional algal cell components) were retained as endosymbionts. For example, some dinoflagellates have plastids that were obtained from cryptomonads and thus have photosynthetic pigments that are characteristic of cryptomonads. Other dinoflagellate plastids were obtained by consuming red algae, green algae, or diatoms. In addition to dinoflagellates' importance as endosymbionts in marine animals (see Figure 19.5), some photosynthetic marine dinoflagellates generate harmful ocean blooms (red tides) that are responsible for cases of human poisoning and marine wildlife deaths (see Essay 19.2, "Killer Algae"). Dinoflagellate poisoning occurs by production of various toxins that cause amnesia, nervous system damage, paralysis, or diarrhea in humans. Harmful blooms result when there are excess nutrients in the water. These human-generated pollutants arise from agricultural field runoff and disposal of untreated sewage (Chapter 28).

Haptophytes are flagellates or colonies of floating cells having golden plastids that occur mostly in ocean waters. Some

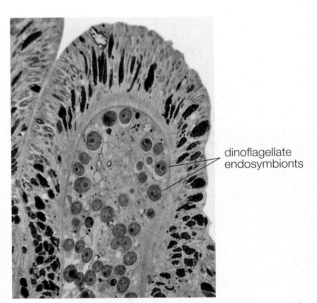

dinoflagellate endosymbionts

FIGURE 19.5 **Dinoflagellate cells living within tissues of corals** The algae provide organic food and oxygen to animal cells and receive carbon dioxide and minerals in return. Many corals can live only in shallow ocean waters because their algal partners require sunlight for photosynthesis.

ESSAY 19.1   IT'S NOT EASY BEING NONGREEN: PARASITIC ALGAE

Prototheca is a unicellular member of the "green" algae whose cells have a plastid that produces starch but has no photosynthetic pigments (Figure E19.1A). DNA evidence indicates that its ancestors were green-pigmented. Why would this protist give up the ability to photosynthesize? *Prototheca* usually grows in moist soil, where it is able to obtain sufficient organic compounds for growth and reproduction. But cattle—and humans with inadequate immune protection—can become infected with this alga. Cattle become infected via wounds in contact with the soil, and humans may breathe in *Prototheca* cells dispersed in dust. Within the animal body, the alga lives on organic molecules absorbed from the bloodstream. Infected humans can be treated with antibiotics to cure this infection, but cattle that become infected with *Prototheca* are typically destroyed.

Plasmodium infects about 40% of the world's population, . . .

Cysts (resistant stages) of *Helicosporidium*, an insect parasite, are able to survive in air until ingested. Once inside the insect's body, the cysts (Figure E19.1B) germinate into few-celled parasites that can eventually kill the host. *Helicosporidium* is heterotrophic, but DNA analysis reveals that *Helicosporidium* is actually a colorless member of the green algae, a *Prototheca*

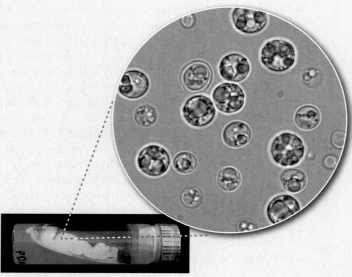

E19.1A *Prototheca*

E19.1B *Helicosporidium* cysts

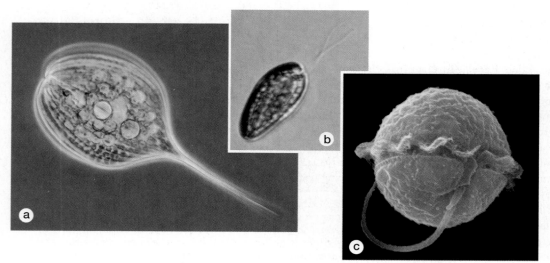

FIGURE 19.6 Flagellate algae
(a) *Phacus* is a euglenoid alga, (b) *Cryptomonas* is a cryptomonad, (c) *Peridiniopsis* is a dinoflagellate. Flagella are not visible on *Phacus*. The number and arrangement of flagella are often diagnostic for a particular group. Note that the dinoflagellate (c) has a coiled flagellum in addition to a more typical one, which is particularly obvious with scanning electron microscopy.

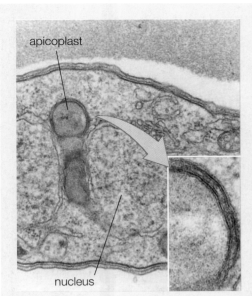

apicoplast

nucleus

E19.1C *Apicoplast of Garnia*

about 40% of the world's population, with up to 500 million cases and 2.7 million deaths occurring each year. Though insecticides can be used to control mosquito populations, and antimalarial drugs exist, some strains of *Plasmodium* have become resistant to drugs, and experts are concerned that cases may double in 20 years. Cells of *Plasmodium* and the close relatives *Toxoplasma* and *Garnia* contain a structure known as an apicoplast (Figure E19.1C), which is actually a modified plastid obtained from red or green algae! DNA sequencing studies have revealed that about 550 (some 10%) of *Plasmodium's* proteins are likely imported into the apicoplast, where they are needed for fatty-acid metabolism, among other things. Because mammals lack plastids, proteins that function in the apicoplast are possible targets for new types of drug therapy. Drugs that impair the apicoplast would kill *Plasmodium* but would not affect humans.

Trypanosomes are a group of non-photosynthetic parasites that infect mammals, fish, and plants. These parasites are closely related to euglenoid algae (both are discicristates). *Trypanosoma* (Figure E19.1D) causes sleeping sickness (transmitted by tsetse flies) and Chagas' disease, and a close relative (*Leishmania*) is the agent of leishmaniasis. These dire diseases primarily affect poor people in tropical or subtropical countries, and few treatments are available. *Phytomonas* is an insect-

transmitted trypanosome that infects plants, sometimes killing them. Recently, scientists made the unexpected discovery that trypanosomes contain photosynthesis genes that otherwise occur only in algae and plants! These scientists suggest that trypanosomes once had plastids, in which case they would have been considered algae. Alternatively, the genes might have been acquired in some other way. In either case, this new knowledge is useful because the photosynthesis genes are targets for development of treatment drugs.

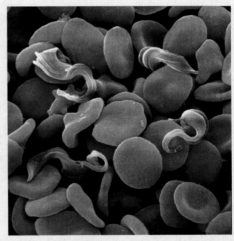

E19.1D *Trypanosoma* amidst red blood cells

relative. Some experts have proposed that *Helicosporidium* might be useful in controlling nuisance insect populations.

Though *Plasmodium* is a parasite, it is closely related to the dinoflagellate algae (both are members of the alveolates). *Plasmodium* belongs to a group known as the apicomplexans, which have a distinctive structure at their cell apices (front ends). *Plasmodium* is the agent of malaria, a serious disease of humans. Transmitted by mosquito bites, *Plasmodium* infects

members have a modified flagellum, known as a haptonema, for which this group is named. The haptonema may aid in feeding on bacteria, serve as an obstacle sensor, or attach cells to surfaces. Some members have a covering of exquisitely detailed calcium carbonate scales, known as **coccoliths** ("round stones") (Figure 19.9). Coccoliths give masses of haptophyte cells a milky-white appearance that can be seen in satellite images of oceans (Chapter 28). Haptophytes are important producers of sulfur-containing aerosols that influence Earth's climate. Together with carbonate-shelled protozoa, haptophytes produced huge chalk deposits about 100 million years ago; these formed the white cliffs of Dover in the United Kingdom and other notable geological formations.

Stramenopiles are a highly diverse lineage that includes tiny flagellates as well as giant seaweeds. Stramenopiles are named for the distinctive strawlike hairs on the flagella (*stramen* means "straw"; *pila* refers to hairs). The flagellar hairs occur only on the longer of the two flagella typically present on the flagellate body or reproductive cells. Because their two flagella are differently structured, members of this

FIGURE 19.7 *Climacostomum,* **a ciliate with endosymbiotic green algae**

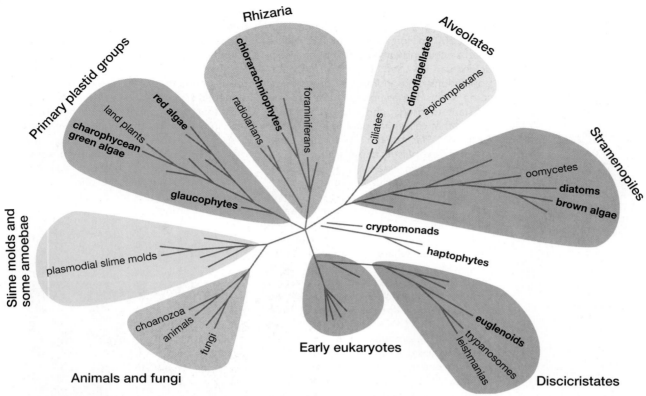

FIGURE 19.8 Evolutionary relationships of algae and other protists This diagram, based on comparative DNA sequences, shows current views that (1) the algae (in bold) do not form a coherent kingdom; (2) green algae, particularly charophyceans, are closely related to the land plants; and (3) some algal groups are more closely related to plastidless protists such as oomycetes than they are to other algae.

TABLE 19.2 Features that Distinguish Major Algal Groups

Group	Body Type	Rigid Cell Wall	Storage	Major Accessory Pigments
Euglenoids (in Discicristates)	Single cells, mostly flagellates	None	Paramylon (carbohydrate)	Similar to green algae
Cryptomonads	Single cells, flagellates	None	Starch	Similar to red algae
Dinoflagellates (in Alveolates)	Mostly single cells, mostly flagellates	None or cellulose	Starch	Mostly peridinin
Haptophytes	Mostly single cells	Many have $CaCO_3$ scales	Soluble carbohydrate	Fucoxanthins
Golden and brown algae (in Stramenopiles)				
Golden	Single cells or colonies	Silica	Lipid droplets or soluble carbohydrates	Fucoxanthin
Brown	Multicellular; filaments or tissue	Cellulose, Algin	Same	Same
Red algae	Single cells or multicellular	Carrageenan, agar	Nonstarch polysaccharide	Phycobilins
Green algae	Single cells or multicellular	Cellulose	Starch	Chlorophyll *b*, beta-carotene, lutein

ESSAY 19.2 KILLER ALGAE

Though most algae play beneficial roles in Earth's ecology, some have undesirable effects. The dinoflagellate known as *Pfiesteria* is one of several kinds of marine dinoflagellates that produce toxins at levels that harm fish, wildlife, and humans (Figure E19.2A). *Pfiesteria* has been labeled as a "killer alga" by the media because it consumes the decaying flesh of fish killed by its toxin. In recent years, *Pfiesteria* has become so abundant in Maryland's Chesapeake Bay that the airborne toxin it releases has caused amnesia in fishermen. Scientists who studied *Pfiesteria* before the dangers of working with this organism were fully realized have also suffered nervous system damage.

Even algae that do not produce toxins can be harmful. Several seaweeds have become notorious for rampant weedy

In most cases, harmful algal growths can be traced to human actions—

growth that crowds out natural marine communities. The most famous example is an invasive, weedy form of the green seaweed *Caulerpa taxifolia* (Figure E19.2B) that smothers corals and other marine life in the Mediterranean. Growths of this seaweed, also described in the news media as "killer algae," have recently invaded southern California waters. It may have come from water dumped by aquarium owners, who sometimes use *Caulerpa* for tank decoration.

Excessive growths of marine phytoplankton algae may also result in depletion of oxygen from ocean water as they decay. Bacteria and fungi consume oxygen when they use respiration to degrade organic compounds in dead algae. If decaying algae are extremely abundant, the oxygen depletion results in the deaths of many kinds of ocean

animals, including those important to coastal fisheries. Coastal waters that suffer these effects are known as "dead zones," and their numbers and sizes have been increasing around the world in recent years.

In most cases, harmful algal growths can be traced to human actions—pollution of coastal ocean waters with sufficient nitrogen and phosphorus minerals to support nuisance-level populations of algae. Excess amounts of these minerals originate from agricultural fertilizers that dissolve in rainwater and then run off into streams and rivers, eventually reaching the ocean. Sewage pollution and erosion are additional sources of nitrogen and phosphorus that support harmful algal growths. Coastal nations around the world now must closely monitor their offshore waters for the presence of high numbers of harmful algae and the excess nutrients that are responsible for their growth.

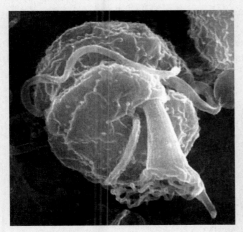

E19.2A *Pfiesteria*

E19.2B *Caulerpa taxifolia*

group are also known as the **heterokonts** (*hetero* means "different"; *kont* refers to flagella). Stramenopiles include plastidless protists, such as the single-celled flagellate *Cafeteria*, and funguslike **oomycetes**. Some oomycetes are parasites that cause widespread and serious diseases of seaweeds, molluscs, fish, and terrestrial crop plants. The oomycete *Phytophthora* (Figure 19.10a) caused the historic Irish potato crop failure that dramatically influenced immigration to the United States. *Phytophthora* continues to cause serious crop diseases today.

Stramenopiles also include many types of algae having golden-brown-colored plastids. Chlorophyll is present in these plastids but is often hidden by large amounts of gold or brown pigments that help chlorophyll absorb sunlight.

Photosynthetic stramenopile algae include the glass-walled **diatoms** (Figure 19.10b) and the giant **brown algae** (Figure 19.10c). Very large brown algae—known as **kelps**—form extensive forests in cold and temperate coastal oceans. The important global roles of diatoms and kelps are discussed more fully in Chapter 28.

Rhizaria are several types of protists whose cells have thin cytoplasmic extensions (the term *rhiza* refers to roots, which the thin extensions vaguely resemble). A photosynthetic group of rhizaria, the **chlorarachniophytes**, is aptly named for their spider (arachnid)-shaped cells with green (*chloro*) plastids (Figure 19.11a). Other rhizaria are equally fascinating. Exquisite mineral shells encase the marine radiolarians (Figure 19.11b), and ocean foraminiferans (Figure 19.11c)

FIGURE 19.9 Coccolithophorid The organism shown here with scanning electron microscopy is *Emiliania huxleyi*, a very common marine alga.

deposit beautiful calcium carbonate shells around their cells. Although neither radiolarians nor foraminiferans are themselves photosynthetic, both often have symbiotic algal partners.

Red algae (rhodophytes), **green algae (chlorophytes)**, and **glaucophytes** are other important protists. The simplest red algae are microscopic single cells (Figure 19.12a), some of which occur in hot, acidic freshwaters. However, most red algae are large enough to see easily and live in ocean waters (Figure 19.12b). Red algae are notable for their inability to produce flagella in any life stage. During sexual reproduction, the sperm cells of red algae rely upon water currents to carry them to female gametes. Red algae are rosy red, deep purple, or nearly black in color because they contain large amounts of a red pigment that aids in harvesting light energy for photosynthesis. The red pigment is very similar to the blue-green photosynthetic pigment that colors cyanobacteria (see Chapter 18) and glaucophyte algae (see later). Though green chlorophyll is present in red algae—as it is in all photosynthetic plants and algae—the red pigment hides chlorophyll's green.

Diverse green algae (see Figure 19.3)—many of which are flagellate or have flagellate reproductive stages—occur in freshwaters, in the ocean, and on land. Green algae are primarily colored by chlorophyll. Though additional photosynthetic pigments are present, they are usually not abundant enough to hide the green color of chlorophyll. As earlier noted, one subgroup of the green algae—the charophyceans—was ancestral to land plants. This relationship is illustrated by many similarities in cell structure, cell division, biochemistry, and DNA of charophycean green algae and land plants. Comparative studies of charophycean green algae and modern relatives of the earliest land plants help us to understand how the first plants colonized land (Chapter 21). The origin of land plants from green algae explains why the photosynthetic portions of plants are green. Imagine land plants evolved from brown or red algae. In

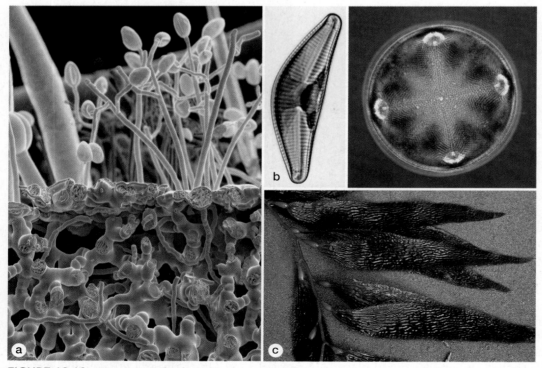

FIGURE 19.10 Stramenopile diversity is extraordinary Representatives include (a) the plastidless parasite *Phytophthora* growing within and extending from the surface of a leaf, (b) the diatoms *Cymbella* and *Aulacodiscus kittonii*, which have silica cell walls, and (c) the giant kelp *Macrocystis*.

FIGURE 19.11   Rhizaria (a) *Lotharella* with green plastids, (b) mineral shells of radiolarians, (c) mineral shells of foraminiferans.

FIGURE 19.12   Red algae (a) The unicellular *Porphyridium*. (b) A multicellular red alga, *Delisea*.

such a case, the terrestrial world would be dominated by brown or red plants, rather than green ones.

Glaucophytes comprise a small group of freshwater algae having blue-green plastids that resemble cyanobacteria as well as the plastids of red and green algae (see Figure 19.16). Some experts propose that glaucophytes, red algae, green algae, and land plants descended from a single plastid-containing ancestor. If this did happen, the glaucophytes, red algae, green algae, and land plants together form a monophyletic kingdom. But other experts argue that glaucophyte, red, and green algae could have independently acquired plastids when unrelated colorless protists ingested and retained similar cyanobacteria as endosymbionts. Protist experts are addressing this controversy by comparing the nuclear DNA of these three eukaryotic algal groups to determine how closely related they are to each other.

19.3 | Algal diversity reflects the occurrence of key evolutionary events

Algae and other protists have been useful in the discovery of how and in what order eukaryotic cells acquired structural features that distinguish them from prokaryotic cells. Recall that eukaryotic cells have a nucleus and an endomembrane system. The nucleus and cytoplasm of eukaryote cells are thought to have originated from a member of the Archaea. Several similarities linking Archaea with eukaryotes provide evidence for this hypothesis (see Chapter 18). The eukaryotic nuclear envelope and a simple endomembrane system may have originated by infolding of the archaean cell membrane (Figure 19.13).

Additional features of eukaryotic cells include peroxisomes, eukaryotic-sized cytoplasmic ribosomes, flagella having a 9 + 2 microtubular organization, the cytoskeleton and motor proteins, and the processes of mitosis and endocytosis (Chapters 4, 7). Recall that endocytosis is a process by which cells can move large molecules or particles into the cytoplasm from the outside. During endocytosis, the cell membrane forms pockets containing outside materials. Poking your

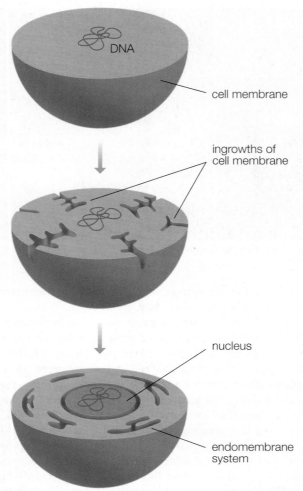

cell membrane

ingrowths of
cell membrane

nucleus

endomembrane
system

FIGURE 19.13 **The eukaryotic nucleus and endomembrane system are thought to have originated by infolding of the cell membrane in an ancestor of modern eukaryotic cells**

fingers into the surface of a balloon produces similar pockets. The cytoskeleton and motor proteins of eukaryotic cells, especially the actin-myosin (minimuscle) system, are necessary to form the membrane pockets. These pockets close up, producing spherical membrane vesicles that enclose parti-

cles. These vesicles then pinch off into the cytoplasm, thereby ferrying their cargo into the cell. Endocytosis does not occur in prokaryotes because they have such rigid cell walls that the cell membrane does not contact outside particles, and prokaryotes lack actin-myosin minimuscles. In contrast, many protists lack rigid cell walls, so their cell membranes can make contact with particles on the outsides of their cells. Many protists use a form of endocytosis—known as phagotrophy—to feed on particles, including other cells (Figure 19.14). Within the cell, digestive enzymes enter food vesicles and break the food particles into smaller components that can enter cell respiration pathways. *Phagotrophy* means "particle feeding" and is thought to have been a necessary first step in the process by which organelles such as mitochondria and plastids (and possibly also peroxisomes) originated by endosymbiosis.

Mitochondria are derived from endosymbiotic proteobacteria

Because mitochondria (or structures likely derived from them) occur in most eukaryotes, these energy organelles may have been present in very early protists. Modern protist lineages, fungi, animals, and plants inherited mitochondria from early protists. Much molecular evidence supports the concept that mitochondria originated from endosymbiotic proteobacterial cells (Chapter 18). A protist host cell engulfed a proteobacterial cell via phagotrophy but did not digest it (Figure 19.15). Rather, such bacterial cells became endosymbionts, adapted to life inside their hosts. During the process of evolving into mitochondria, the bacteria lost their peptidoglycan wall and transferred most of their genetic information to the host's nucleus. This explains why mitochondria cannot live or reproduce outside their host cells. However, some essential genes and protein synthesis capability are retained by mitochondria. (This explains why certain human genetic diseases result from mutations in mitochondrial DNA.) With the aid of host-translated proteins that are imported into mitochondria, mitochondria

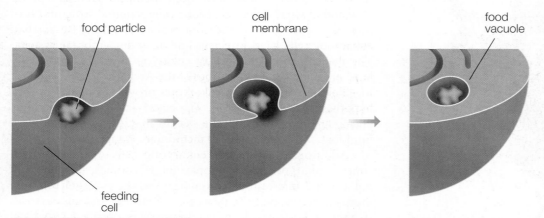

food particle

cell
membrane

food
vacuole

feeding
cell

FIGURE 19.14 **Phagocytosis** Phagocytosis is a type of protist feeding that involves enclosure of particulate food within a vacuole that pinches off from the cell membrane. Digestive enzymes supplied to the food vacuole break down the food, much as in the human digestive system. Endosymbiotic incorporation of cells resulted when prey cells were enclosed within vacuoles but digestion did not occur.

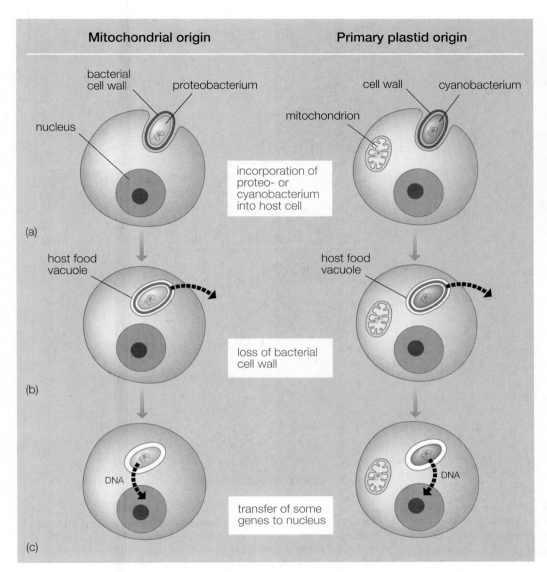

Mitochondrial origin   Primary plastid origin

bacterial cell wall   proteobacterium   cell wall   cyanobacterium

nucleus   mitochondrion

(a)   incorporation of proteo- or cyanobacterium into host cell

host food vacuole   host food vacuole

(b)   loss of bacterial cell wall

DNA   DNA

(c)   transfer of some genes to nucleus

FIGURE 19.15   Origin of mitochondria and primary plastids by endosymbiosis (a) Mitochondria arose from proteobacteria that were taken into an early eukaryotic cell, and primary plastids from cyanobacteria taken into eukaryotic cells that likely already possessed mitochondria. Following loss of the endosymbiont cell wall (b), most genes were transferred to the host's nucleus (c). This gene transfer explains why mitochondria and primary plastids can no longer exist on their own.

carry out essential cell metabolic activities. Mitochondria also divide by binary fission under control of the host cell's division cycle.

Proteobacterial endosymbionts were likely retained rather than being digested as food because the bacteria provided their host with a particularly efficient system for harvesting ATP from organic food. Recall that components of the cell respiration process that occur in mitochondria include the Krebs cycle, electron transport, and oxidative phosphorylation (Chapter 5). Endosymbiotic proteobacteria endowed their host cell with these valuable energy processes.

Primary plastids originated from endosymbiotic cyanobacteria

Because all modern eukaryotic organisms known to contain plastids also possess mitochondria, it is thought that algae acquired plastids after mitochondria, by the same method—endosymbiosis. Plastids that arose from endosymbiotic cyanobacteria are known as **primary plastids**. During the

process of evolving into plastids, cyanobacterial endosymbionts lost their cell wall and transferred most of their genes to the host's nucleus (see Figure 19.15), similar to the process that occurred during the origin of mitochondria. The process by which an endosymbiotic cyanobacterium is engulfed and transformed into a plastid is known as **primary endosymbiosis**. Primary plastids have an envelope consisting of two membranes. At least one of these is derived from cyanobacterial cell membrane, and the other possibly originated from the host's engulfing cell membrane. Modern glaucophytes, red algae, green algae, and green plants are derived from ancestors having primary plastids. Glaucophytes illustrate an early stage in plastid evolution; their plastids are blue-green in color (Figure 19.16) as are many cyanobacteria, and glaucophyte plastids still have a bacterial cell wall. When host cells obtained plastids, they also acquired autotrophy—the ability to produce their own organic food from inorganic compounds. Photosynthesis is such a valuable process that some protists have acquired plastids from *eukaryotic* algal cells, a process known as **secondary** or **tertiary endosymbiosis**.

Secondary plastids originated when eukaryotic protists engulfed, but did not digest, red or green algal cells. In many

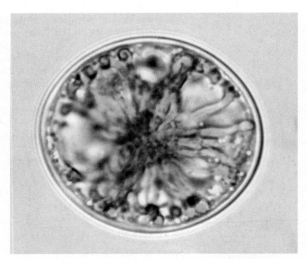

FIGURE 19.16 *Glaucocystis,* **a glaucophyte alga** The blue-green-colored plastids in this organism illustrate an early stage in the transformation of a cyanobacterial endosymbiont to a plastid.

modern protists having secondary plastids, only the plastid of the ancient algal endosymbiont has been retained (Figure 19.17a). But some other protists have also retained part of the endosymbiont's cytoplasm and nucleus. In these cases, the endosymbiont's nucleus is small, bears few genes, and is known as a nucleomorph (Figure 19.17b). Secondary plastids have more than two plastid envelope membranes, reflecting the additional endosymbiotic processes in their origins.

The photosynthetic euglenoids originated when an ancestor acquired secondary plastids from an ingested green alga (Table 19.3). Evidence for this is the fact that green euglenoid plastids have the same pigments and molecular features as plastids of green algae and have three envelope membranes. Chlorarachniophytes likewise obtained their plastids from a green algal endosymbiont but also have a nucleomorph rem-

TABLE 19.3 Plastid Origins

Group	Type of Plastid	Plastid Origin
Red algae	Primary	Cyanobacterium
Green algae	Primary	Cyanobacterium
Glaucophytes	Primary	Cyanobacterium
Cryptomonads	Secondary	Red alga
Haptophytes	Secondary	Red alga
Golden & Brown	Secondary	Red alga
Euglenoids	Secondary	Green alga
Chlorarachniophytes	Secondary	Green alga
Apicomplexans	Secondary	Red or Green alga
Dinoflagellates	Secondary or Tertiary	Red alga, Green alga Diatom, Cryptomonad

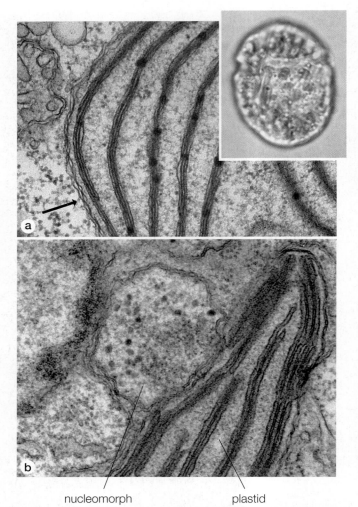

nucleomorph   plastid

FIGURE 19.17 **Secondary plastids arose from endosymbiosis of eukaryote cells** In some cases, such as in most dinoflagellates (a), all that is left of the eukaryotic endosymbiont is its plastid(s). These dinoflagellate secondary plastids have three bounding membranes (arrow) as is often the case with other such plastids. (b) In cryptomonads and chlorarachniophytes, some of the endosymbiont's cytoplasm, including a reduced form of the nucleus—the nucleomorph—remains in addition to the plastid. Shown here is a cryptomonad nucleomorph, lying adjacent to the plastid. Note that it has a double membrane envelope, as do nuclei.

nant of the endosymbiont's nucleus. Plastid-bearing haptophytes, dinoflagellates, and stramenopiles originated by endosymbiosis of a eukaryotic red alga, of which all cell components except plastids were lost. Cryptomonads retain not only the plastids of an ancient endosymbiotic red alga but also a nucleomorph. Some dinoflagellates have **tertiary plastids,** derived by endosymbiosis of a cryptomonad or diatom.

19.4 Protist nutritional variation explains diverse ecological roles

Protists obtain their food in one of four ways: (1) **phagotrophy**—particle feeding, (2) **osmotrophy**—uptake of small dissolved organic molecules such as sugars or amino acids, (3) **autotrophy**—photosynthesis, or (4) **mixotrophy**—

FIGURE 19.18 The underwater light environment experienced by many algae is blue-green

photosynthesis combined with osmotrophy and/or phagotrophy. This nutritional variation explains the diverse ecological roles played by protists, including algae.

Most algae are autotrophic (meaning "self-feeding"). Autotrophic algae are photosynthetic; they use sunlight energy to synthesize organic compounds from water and carbon dioxide (Chapter 5). Algae also need a variety of minerals to grow, especially combined nitrogen and phosphate. Water absorbs yellow, orange, and red light, so algae growing more than a few meters deep inhabit a blue-green world (Figure 19.18). Recall that chlorophyll *a*, present in all photosynthesizing organisms (Chapter 5), absorbs red and blue light. Since red light is not available to deep algae, different groups of algae also have characteristic **accessory pigments** (see Table 19.2), which help absorb the available green or blue light present at depth and transfer that energy to chlorophyll reaction centers. As was earlier noted, these accessory pigments are often so abundant that they hide the green chlorophyll and confer brown, red, or other colors to the algae.

Peridinin is an accessory carotenoid that absorbs blue light and confers a rich brown color to many dinoflagellates. Recall that carotenoids are orange- or yellow-pigmented lipids (Chapter 5). Brown algae and diatoms (see Figure 19.11) are rich in the blue-light-capturing carotenoid known as fucoxanthin, which gives plastids a golden to brown color. Red algal plastids (see Figure 19.12) possess phycobilin pigments, which harvest green light and give these algae their pink, red, or deep purple coloration. Green algae (see Figure 19.3) possess a green accessory plastid pigment (chlorophyll *b*) and the red-orange carotenoids lutein and **beta-carotene**. Because plant plastids are derived from those of green algae, plant plastids also possess chlorophyll *b*, lutein, and beta-carotene. Beta-carotene provided by algae and plants is particularly important to animal nutrition, including aquatic animals that consume algal food. In the animal body, beta-carotene is converted to vitamin A (retinol). In humans, vitamin A is essential to good skin health as well as to the formation of light-perception pigments in the eye. Lack of beta-carotene in the human diet causes the incurable form of blindness known as xerophthalmia, which is common in some parts of the world but completely preventable with carotene-rich foods or supplements. Vitamin A also plays a role in animal limb development. For this reason, pregnant women must avoid a chemically related skin medication commonly used to treat acne, because it can cause birth defects.

Protists that are exclusively phagotrophic or osmotrophic (or both) are heterotrophic. Heterotrophic protists that feed on living cells are **parasites;** those feeding on nonliving organic material are **saprobes.** Some autotrophic protists must take up certain organic compounds, such as vitamins, from their environments—this is known as **auxotrophy.** Some phagotrophic protists and mixotrophic protists are consumers of bacteria and are consumed by aquatic animals. In this way, protists link the considerable productivity of bacteria to higher food-web components. Other phagotrophic protists consume larger, eukaryotic prey. In summary, protists play a variety of ecological roles because as a group they exhibit so many ways of obtaining food.

19.5 | Structural and reproductive adaptations aid in protist survival

Common protist structural adaptations include distinctive cell coverings and food storage products; here we focus primarily on those of algae (see Table 19.2). Reproductive adaptations include asexual reproduction and tough-walled cysts by which cells survive stressful conditions, and sexual reproduction involving gametes, zygotes, and life cycles. Variations in protists' structure and reproduction illustrate many ways in which these organisms have adapted to their environments.

Cell coverings

Many algae have distinctive cell coverings that often provide some degree of protection from attack by bacteria, viruses, fungi, or predators. Cell coverings may also help to prevent osmotic damage in freshwaters or enhance flotation in water. Cellulose walls are common in brown and green algae; green algae bequeathed this feature to land plant descendents. Distinct polysaccharide polymers form a gooey coating around red algal cells; some of these materials are of significant economic value. For example, agar—used to solidify the nutrients used to grow cultures of algae, bacteria, fungi, and plant tissues—is extracted from red algal coverings. Agarose, used in the study of DNA, also comes from red algal surface materials. Agar and agarose are too complex for industrial synthesis; their only source is natural populations of red algae. Calcium carbonate armors many single-celled haptophytes, as well as some red, brown, and green algae. Such algae are described as calcified. Much of the beautiful white carbonate sand on many beaches worldwide was originally produced by calcified green algae. Silica (glass) forms the covering of diatoms, and some euglenoids are enclosed by iron and manganese crystallized within a mucilage matrix.

Organic food storage

Algae store organic food in various chemical forms (see Table 19.2). Food stores tide cells over when resources such as light limit photosynthesis but also make algae desirable food sources for heterotrophs. Oil droplets are major food stores in diatoms and some other algae. As illustrated in the chapter opening, oil and some coal deposits are derived primarily from the organic components (including food storages) of ancient algae.

Starch, the major food storage in plants, was inherited from ancestral green algae. All green algae and land plants store starch within their plastids. Starch is normally transparent when viewed with the light microscope but can be stained blue-black with a solution of iodine. Iodine also stains the starch of colorless members of the "green" algae, revealing their true identities (see Essay 19.1). Dinoflagellate starch, which occurs in the cytoplasm (not in plastids), can also be stained with the use of iodine. Other algae produce carbohydrates that are distinct from starch and do not stain blue-black with an iodine solution. Examples include solid grains of paramylon in the cytoplasm of euglenoids, polysaccharide lumps in the cytoplasm of red algae, and dissolved carbohydrates in vacuoles of haptophytes and stramenopiles.

Asexual reproduction

All protists are able to reproduce themselves by asexual means that involve mitotic cell divisions. Unicellular protists use mitosis to generate two or more offspring cells that resemble their parent. Asexual reproduction is responsible for the growth of protist populations in nature and in laboratory cultures (see Chapter 13). Asexual structures aid the survival of protists by generating large numbers of genetically identical offspring. Even if some of these are consumed or die from lack of resources, others are likely to survive long enough to reproduce. **Spores** are single-celled reproductive structures often produced asexually in protists.

Multicellular protists often produce spores that are released from the parent body in order to disperse the organisms in their environment. This prevents excessive competition between parents and offspring. Red seaweeds often disperse **monospores**, single nonflagellate cells that drift with the currents until they encounter a suitable surface where they can attach and grow into new seaweeds. As previously mentioned, many other protists produce flagellate cells that swim through water films, dispersing in the environment (Figure 19.19a). Flagellate asexual reproductive cells are known as **zoospores** (*zoo* refers to animals, which are typically mobile, like these cells). Zoospores are an important way in which parasitic protists, such as oomycetes, locate new hosts. Many protists produce unicellular, tough-walled cells known as **cysts** that can survive harsh conditions until environmental conditions improve (Figure 19.19b).

Sexual reproduction

Many, but not all, protists are known to exhibit sexual reproduction, which has the evolutionary advantage of producing diverse genotypes. Sex confers the potential for faster evolutionary response to environmental changes. Protists vary in the

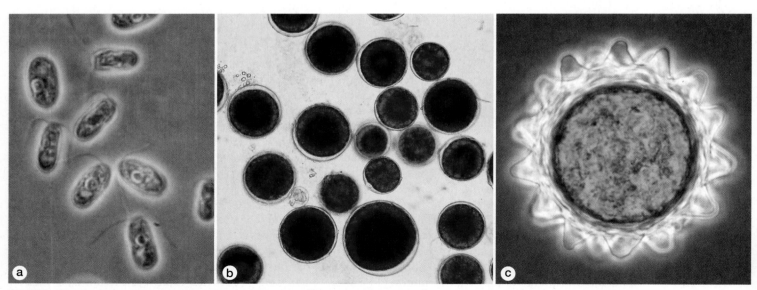

FIGURE 19.19 Reproductive cells play important roles in protist survival (a) Zoospores of the green alga *Chlorococcum* are flagellate cells released from parent cells that allow asexual reproduction. (b) These cysts of the green algae *Haematococcus* are thick-walled cells that can survive adverse conditions, then germinate when conditions improve. The orange pigment protects cells against damaging light. (c) The zygote of *Volvox* has a thick wall that aids in cell survival under stressful conditions. The role of the bumpy wall ornamentations is unknown.

type of sexual reproductive processes they display. This variation illustrates multiple ways of adapting to stressful conditions. For example, in some protists, sexual reproduction results in the formation of a thick-walled zygote (Figure 19.19c). Such zygotes, like the resistant cysts just described, are able to survive stressful conditions, then germinate when conditions improve. Other protists have sexual reproductive cycles that involve two or three kinds of multicellular bodies, each adapted to particular environmental conditions.

Examples of the three main types of sexual life cycles (zygotic, sporic, and gametic—Chapter 13) are found among algal protists and are illustrated here. The filamentous green alga *Spirogyra* provides an example of the zygotic sexual life cycle (Figure 19.20). In this life cycle, most cells have only one set of chromosomes; only the single-celled zygotes are diploid (have two chromosome sets). *Spirogyra's* zygotes are thick-walled and resistant to environmental stress. These zygotes can remain dormant in mud until conditions are suitable for photosynthesis and growth of the filamentous body. This explains how *Spirogyra* can rapidly form spring blooms on pond surfaces—many zygotes simultaneously germinate to form the lush algal growths.

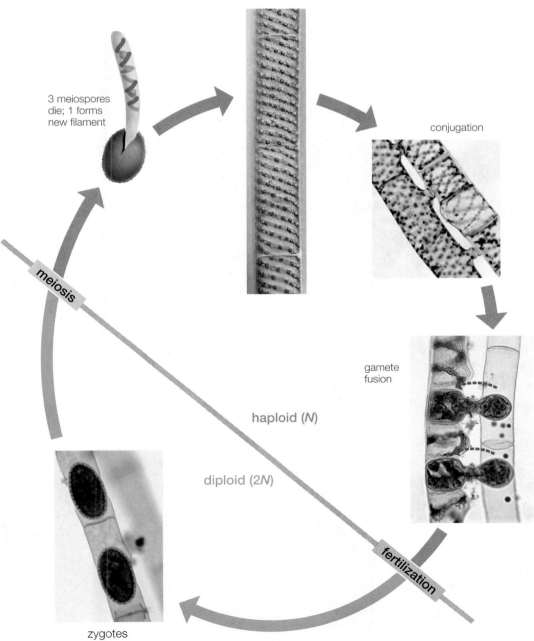

3 meiospores die; 1 forms new filament

conjugation

gamete fusion

meiosis

haploid (N)

diploid (2N)

fertilization

zygotes

FIGURE 19.20 **The green alga *Spirogyra* has a zygotic life cycle** The only diploid cells in the life cycle are the zygotes.

The brown kelp *Laminaria* is a common seaweed having a sporic life cycle (also known as alternation of generations) (Figure 19.21). The large seaweed is a spore-producing body—a **sporophyte**. Spores dispersed from the sporophyte grow into a microscopic body attached to submerged surfaces, often in deep, dark waters. The microscopic body is a **gametophyte** because it produces gametes; if these mate, the resulting zygote grows into a large seaweed that is able to obtain surface light. The sporophyte and gametophyte of *Laminaria* are differently adapted for survival in distinct habitats. Land plants have a similar life cycle that evolved independently from that of *Laminaria* and its brown algal relatives.

The seaweed *Fucus* has a gametic life cycle (Figure 19.22) similar to the life cycle of animals but independently evolved. The only cells in the *Fucus* life cycle that are haploid (having only one set of chromosomes) are the gametes. This type of reproduction benefits *Fucus* because most body cells, being diploid, have redundant genes. Even if deleterious mutations occur, normal alleles produce normal structure and function. Animals benefit similarly from having a gametic life cycle.

Most red algae have a life cycle involving three distinct multicellular bodies (Figure 19.23). Their unusual life cycle arose as a response to loss of flagella early in red algal evolution. Red algae often produce distinct male and female game-

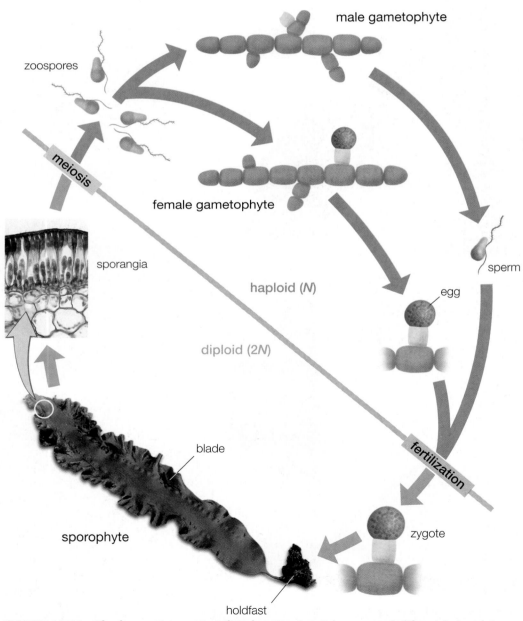

FIGURE 19.21 The brown (stramenopile) alga *Laminaria* has a sporic life cycle involving alternation of distinct multicellular sporophyte and gametophyte generations

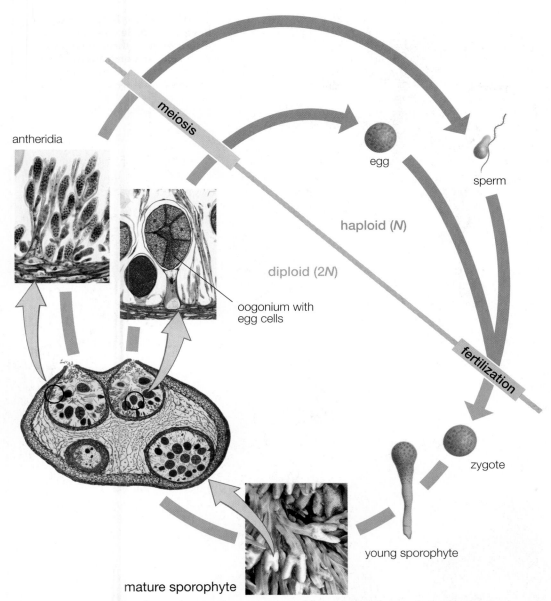

antheridia

meiosis

egg

sperm

haploid (*N*)

diploid (2*N*)

oogonium with
egg cells

fertilization

zygote

young sporophyte

mature sporophyte

FIGURE 19.22   The rockweed *Fucus* has a gametic life cycle similar to that of animals The only
haploid cells in the cycle are gametes.

tophytes—haploid bodies that each produce only one type of gamete, male or female. The female gametes remain attached to the parental female gametophyte, but male gametes are dispersed into the water. Since male gametes lack flagella, mating depends largely on water currents to transport male gametes to female gametes. In this situation, zygotes may be produced relatively rarely, and they remain attached to the parental gametophyte. Most red algae also produce a diploid body—a carposporophyte—that develops from the zygote by repeated mitosis. A carposporophyte remains attached to the body of the parental female gametophyte and receives food from it. This nutritional situation is very similar to the growth of plant embryos within the bodies of parental female gametophytes (Chapter 21). The red algal car-

posporophyte produces many copies of the zygote's diploid nucleus and disperses them within spores known as carpospores. This process amplifies the rare chromosome combination that originated with the zygote. Carpospores grow into diploid bodies called tetrasporophytes; these often live independently (as shown in Figure 19.23) but sometimes grow on the carposporophyte/gametophyte that produced them. Carposporophytes produce many carpospores by means of meiosis. These spores may each grow into a male or female gametophyte, completing the cycle. The carposporophyte further amplifies the genetic combination that began with a rare zygote. This amazingly complex life cycle effectively solves the problem encountered by the red algae when they lost the ability to produce flagella!

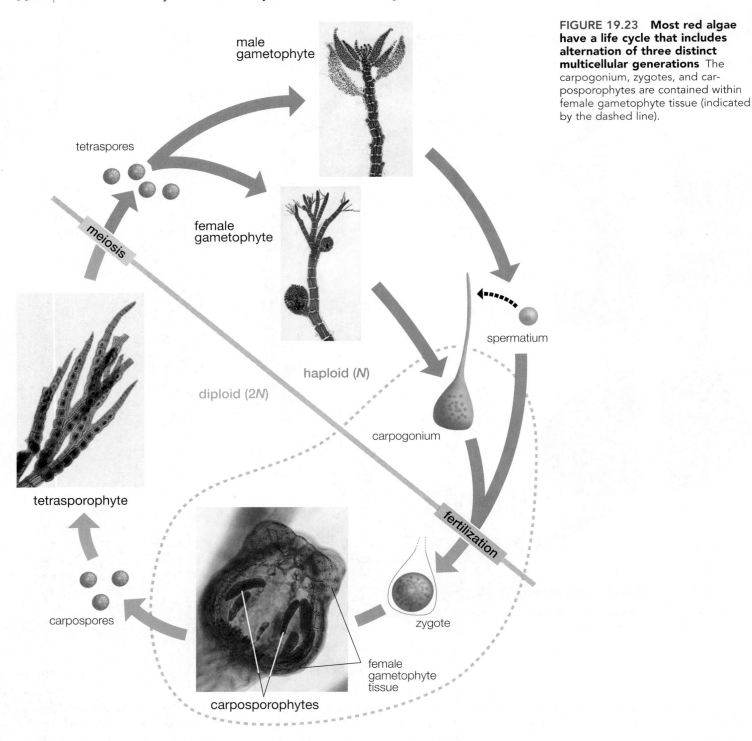

FIGURE 19.23 **Most red algae have a life cycle that includes alternation of three distinct multicellular generations** The carpogonium, zygotes, and carposporophytes are contained within female gametophyte tissue (indicated by the dashed line).

19.6 | Algae have useful biotechnological applications

Industrial and food products

Applications of algae include the use of diatoms in filters, reflective paint, and insecticides. **Alginates** are polysaccharide cell-wall polymers extracted from brown algae that are useful in making paint, linoleum, and dental supplies. As earlier mentioned, agar derived from red algae is used for laboratory growth of microorganisms; agar is also used in production of some foods. **Carrageenan**, an emulsifier ingredient in many dairy products, is also derived from red seaweeds. Particular red, brown, and green seaweeds are consumed as foods in many cultures. For example, nori, the seaweed wrapper used in sushi, is the cultivated red alga *Porphyra* (Figure 19.24).

(a)   (b)

FIGURE 19.24 **The economically valuable red seaweed *Porphyra* (nori)** (a) *Porphyra* growing on rocks viewed at low tide; (b) commercially packaged nori sheets.

Water quality improvement systems

Natural populations of protists are important components of sewage treatment systems to clarify water. "Algal scrubber" systems composed of algae and other protists have been engineered into systems for removing nitrates and phosphates from effluents, thereby increasing water quality. A system of this type has been used to clean aquaria on display at the Smithsonian Institution in Washington, D.C.

Laboratory model systems

Cellular slime molds (which lack plastids), such as *Dictyostelium discoideum*, produce amoebae that first feed on soil bacteria, then aggregate into a multicellular slug. The slug transforms into a mass of dispersible spores borne on a stalk (Figure 19.25). This amazing behavior in a protist illustrates a simple form of intercellular chemical communication that influences development. *Dictyostelium* thus has been a useful lab model system for analysis of eukaryotic gene expression in development.

The green algae *Chlamydomonas* and *Volvox* (see Figure 19.3b) have also been used as model systems for the study of gene expression in photosynthesis and flagellar development, in particular. For genetic analyses, these protists have the advantage of being haploid (zygotic life cycle type), so their phenotype directly reflects their genotype. Mutations are easily visible as changes in cell structure or behavior. The Nobel Prize winner Melvin Calvin used the green alga *Chlorella* to elucidate the light-independent reactions of photosynthesis.

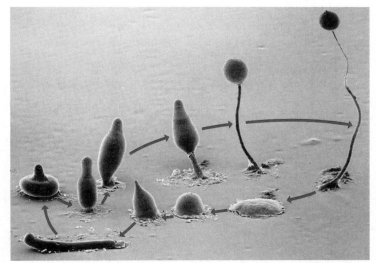

FIGURE 19.25 **The asexual developmental cycle of the slime mold *Dictyostelium***

HIGHLIGHTS

19.1 Protists are eukaryotic organisms that (1) are not included in the land plant, fungal, or animal kingdoms; (2) are often microscopic; and (3) occur primarily in moist habitats. Plastid-containing protists are commonly known as algae.

19.2 Protists include diverse groups whose relationships are not completely known. Major groups that include algae are the euglenoids among the discicristates, cryptomonads, alveolates, haptophytes, stramenopiles, rhizaria, rhodophytes, chlorophytes, and glaucophytes. Land plants evolved from a specific group of ancestral green algae known as charophyceans.

19.3 Algal diversity reflects the evolution of key structures and processes. These include endosymbiosis and the origin of primary, secondary, and tertiary plastids.

19.4 Algal nutritional variation explains their diverse ecological roles as autotrophs, heterotrophs, or mixotrophs. Heterotrophs include saprobes that consume nonliving organic materials, and parasites that feed on living tissues. Accessory pigments that help autotrophic algae absorb light filtered by water explain the various color groupings of algae.

19.5 Algal structural and reproductive adaptations include various types of cell coverings, organic food storages, asexual reproduction, sexual reproduction, and life cycles.

19.6 Algal biotechnological applications include industrial and food products, water quality improvement technologies, and laboratory model systems.

REVIEW QUESTIONS

1. List several similarities and differences between algae and terrestrial (land) plants.

2. Algae as a group may be described as "predominantly photosynthetic eukaryotes living primarily in aquatic environments and possessing relatively simple bodies." Although this general definition encompasses most algae, some algal species do not quite conform to it. Describe some of the more notable exceptions.

3. Discuss the difference between autotrophs and heterotrophs, and provide some examples of each.

4. Algae as a group exhibit a diverse array of body forms. Name and describe some of the more common algal body forms.

5. Distinguish between sporophytes and gametophytes. Do all algae produce a sporophyte and a gametophyte?

6. Name some of the major algal photosynthetic pigments and describe their distribution within the algae. Which pigment is found in all photosynthetic algae?

7. Discuss some of economic uses of algae, as well as some of their ecologically important roles, both beneficial and detrimental.

APPLYING CONCEPTS

1. Several parasitic algae cause diseases of people and animals. How could antibiotics be chosen that would retard the growth of these protists without harming people or algae? (*Hint:* What structures or metabolic processes do algae typically have that animals lack?)

2. New species of algae are being discovered constantly, but algae are often so small that it can be difficult to determine which algal groups are their closest relatives. What characteristics and methods might be used to determine a new alga's relationships?

3. What are the advantages of being a single-celled flagellate? What are the disadvantages? What are the advantages of being a multicellular alga (for example, being a seaweed)? What might be some disadvantages of large seaweed size?

4. Suppose that a large "dead zone" is reported each year in coastal waters near your home. What steps would you recommend to reduce the size and frequency of occurrence of such "dead zones"?

CHAPTER 20 | Fungi and Lichens

Lichens

W hat is the largest organism in the world? Although you might guess elephants, whales, or perhaps giant redwood trees, the modern record holder is a fungus. An individual of *Armillaria ostoyae*, whose body lies hidden within 2,200 acres of Oregon forest soil, weighs hundreds of tons, and is thought to be more than 2,000 years old. Scientists discovered the size of this fungus when they found identical fungal DNA sequences—indicating that they came from a single individual—in samples taken throughout the forested area. News reporters labeled this organism the "fungus humongous." Other examples of very large fungal bodies have been found, though they are likewise underground and thus not visible to the casual observer.

Another giant fungus in the news is a fossil with a "trunk" so long and thick that it was once thought to be a tree. This fossil had long been a mystery because it was formed at a time when early plants were still quite small. Though it was much older than any other fossil trees, the fossil looked so much like a tree trunk that it had been named *Prototaxites*. "Proto" means "first," and "taxites" refers to the modern conifer tree *Taxus*, commonly known as yew. This huge fossil was revealed as a fungus when thin slices of the "trunk" were examined under a microscope, showing cells of a type unique to fungi. The fossil's chemical composition also indicated that it was likely a fungus. How this ancient fungus was able to grow so large and why it has such an unusual treelike trunk is still a mystery.

Armillaria and *Prototaxites* are newsworthy because of their dramatic size, but there are many other remarkable fungi. Forest growth, cancer-causing food toxins, sick-building syndrome, and mushroom pizza are just some of the examples of ways in which fungi impact Earth's environment and human lives. In this chapter, we will consider how fungi grow, reproduce, and form essential partnerships with plants and photosynthetic microbes (to form lichens). ■

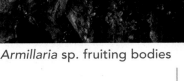

Armillaria sp. fruiting bodies

20.1 | Fungi are eukaryotes with distinctive cell walls and bodies

Although many fungi are microscopic in size and certain bacteria resemble them, fungi can be distinguished from the prokaryotic microbes by their cell structure. Fungal cells possess nuclei, mitochondria, endomembrane and cytoskeletal systems, and other typical features of eukaryotic cells. Unlike plants and algae, fungi lack plastids and thus cannot conduct photosynthesis. Fungal cells possess protective cell walls, also typical of plants and many algae. However, fungal cell walls are rich in a nitrogen-containing polysaccharide known as chitin, which is not present in walls of algae and plants. Fungi have threadlike, filamentous bodies quite different from those of plants, though similar to bodies of some protists (see Chapter 19). Fungi, like animals, store food within their cells as a polysaccharide—glycogen—that is distinct from the carbohydrate storages of algae and plants. Most fungi live on land, though some occupy aquatic environments (Table 20.1).

Two kinds of fungi—chytrids and yeasts—often occur as single cells (Figure 20.1a,b). But most fungi have bodies composed of microscopic threadlike, branched filaments known as **hyphae** (Figure 20.1c). Hyphae are so long and thin that they have a very large surface area, which is advantageous for obtaining food. Hyphae of many fungi are continuous tubes of cytoplasm containing many nuclei, but most fungi have hyphae with cross-walls (see Table 20.1). Cross-walls help prevent loss of cytoplasm if the hypha is wounded.

Hyphae grow at their tips. Cytoplasmic streaming and osmosis are important cellular processes in hyphal growth. Recall from Chapter 4 that osmosis is the entry of water into a solute-rich compartment through a semipermeable membrane. Water enters fungal hyphae because their cytoplasm is rich in sugars, ions, and other solutes. Water entry swells the hyphal tip, producing the force for tip extension. Masses of tiny vesicles carrying enzymes and wall materials made in the endomembrane system collect in the hyphal tip. The vesicles then fuse with the cell membrane, releasing wall materials at the hyphal tip, allowing it to grow. Fungal hyphae can thus extend rapidly through a food source—known as **substrate**—from areas where the food has become depleted to adjacent, food-rich regions.

A mass of interconnected fungal hyphae is a **mycelium.** A fungal mycelium can be loose and cottony, or it can form tissuelike, fleshy masses. Fungal mycelia occur nearly everywhere, but often are not conspicuous because individual hyphae are so delicate and may be hidden within substrates. Individual fungal mycelia may occupy very large volumes of soil or other materials.

Mushrooms and other fungal reproductive structures visible with the unaided eye are composed of tightly packed hyphae and are known as **fruiting bodies** (Figure 20.2). Lengths of hyphae may also lie in parallel to form visible white, yellow, or brown shoestringlike **rhizomorphs** ("structures that resemble roots") (Figure 20.3). Rhizomorphs transport water to portions of fungal bodies that occupy low-moisture habitats. To see rhizomorphs, go for a walk in the woods and look under the bark of rotten logs or on wet, dead leaves.

20.2 | Fungal nutrition is absorptive

How do fungi feed? The answer explains destructive fungal decomposition of materials as tough as shoe polish, newspaper, and peanut shells. Fungal feeding also explains how fungi cause diseases.

As fungi grow through food substrates, hyphae secrete enzymes that break down large organic molecules in the surrounding environment into sugars and other small organic molecules and mineral ions. Fungi absorb these molecules via membrane carrier proteins (Chapter 4). In the fungal cytoplasm and mitochondria, organic food is broken down in the process of aerobic respiration. Recall from Chapter 5 that oxygen is used in respiration to transform energy contained in the chemical bonds of organic food into chemical bonds of ATP, which fuels food absorption, cell

TABLE 20.1	Distinguishing Features of Fungal Divisions		
Division	**Habitat**	**Reproductive Spores**	**Hyphae**
Chytrids (Chytridiomycota)	Water and soil; some plant pathogens	Flagella present	No cross-walls
Zygomycetes (Zygomycota)	Terrestrial	No flagella; produced in sporangia	No cross-walls
AM fungi (Glomeromycota)	Terrestrial; mycorrhizal	Large, asexual, multinucleate, no flagella	No cross-walls
Sac fungi (Ascomycota)	Terrestrial; many in lichens; some mycorrhizal	No flagella; produced in sacs on fruiting bodies	Cross-walls present
Club fungi (Basidiomycota)	Terrestrial; some in lichens; many mycorrhizal	No flagella; produced on club-shaped structures on fruiting bodies	Cross-walls with clamp connections present

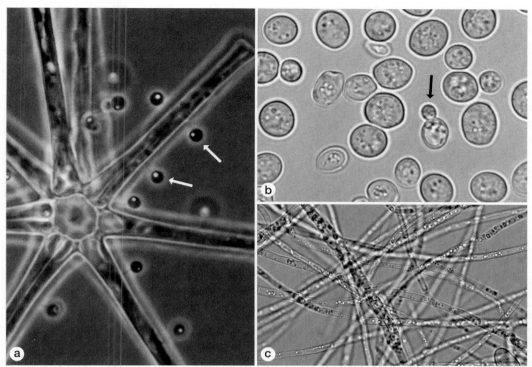

FIGURE 20.1 Fungal bodies (a) Single-celled chytrids (arrows) are feeding on a lake diatom. Chytrids can produce flagellate spores that are able to swim to new locations. (b) Yeasts occur as single cells that can produce progeny by budding (arrow) or grow as filaments. The yeasts illustrated are common baking yeasts. (c) Most fungi occur as branched filaments known as hyphae. Mushrooms and other solid fungal structures are composed of tightly appressed hyphae.

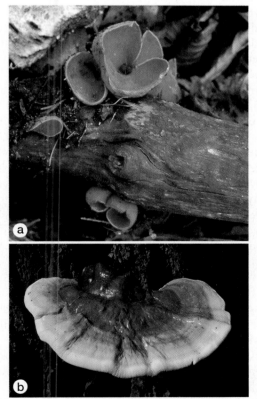

FIGURE 20.2 Fungal fruiting bodies (a) *Sarcoscypha* (scarlet cup) and (b) *Ganoderma*, a shelf or bracket fungus. Most of the mycelium is located within the rotting logs. Emergence of fungal fruiting bodies from the substrate allows spores to disperse in air currents.

FIGURE 20.3 Rhizomorphs on a rotting log The bark was removed to reveal the large number of rhizomorphs.

metabolism, growth, and reproduction. Also recall that yeasts can grow in low-oxygen environments by fermenting sugar to alcohol and carbon dioxide, a property useful in the brewing and baking industries. In addition to an organic carbon source, fungi also require water, several minerals, and vitamins.

Fungi differ in the kinds of enzymes that they secrete, so they also differ in the types of organic substrates that they can break down for food. Fungi that decompose nonliving organic materials are called **saprobes;** those that attack living tissues and cause disease are **pathogens.** Five thousand fungal species cause crop diseases, among them corn smut (Figure 20.4a) and black canker of cherry (Figure 20.4b). Athlete's foot and ringworm are common skin diseases caused by fungi, and yeast infections and some lung diseases are also of fungal origin. *Pneumocystis carinii* is a fungal pathogen that infects AIDS patients. The hyphae of pathogenic fungi grow through the tissues of plants and animals, absorbing food and minerals from them (Figure 20.4c). By this action, pathogenic fungi reduce host vigor, causing disease symptoms. Some soil fungi are **predators**—they trap tiny soil animals (nematodes) and digest their bodies (Figure 20.5).

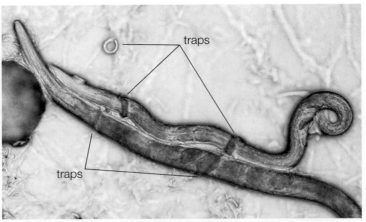

FIGURE 20.5 Predatory fungus that traps and consumes nematodes This fungus lives in soil, where nematode worms are abundant. Here, two small worms have been lassoed by hyphae that form specialized loops.

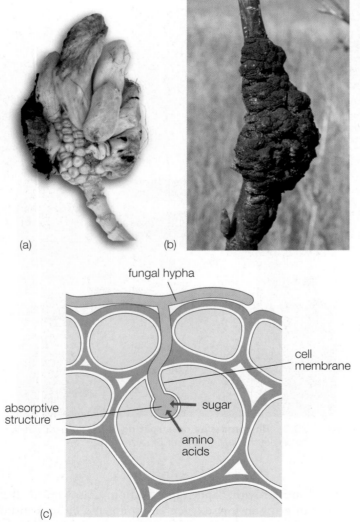

(a) (b)

fungal hypha

absorptive structure

sugar

amino acids

cell membrane

(c)

FIGURE 20.4 Fungal pathogens of plants (a) Corn smut, *Ustilago maydis*, and (b) *Dibotryon morbosum* (a member of the Ascomycota), which parasitizes black cherry trees. The fungal mycelium grows within the plant's tissue, taking nutrients from the plant and thereby weakening it. The fungus does not become obvious until the rather unappealing black fruiting bodies (cankers) appear on the surfaces of branches. (c) Diagram showing the movement of organic molecules from plant cells into parasitic fungal hyphae.

20.3 | Major fungal groups differ in reproduction

DNA and other data strongly indicate that all modern fungi descended from a single common ancestor and justify their placement into the kingdom Fungi (Figure 20.6). Fungi are closely related to the animal kingdom but differ from animals in lacking phagotrophy (ability of cells to ingest particles). DNA and cell characteristics have also revealed that several types of slime molds and **oomycetes**—which are often studied with fungi—are protists rather than true fungi (see Chapter 19). True fungi are classified into five phyla: chytrids, zygomycetes, AM fungi, ascomycetes, and basidiomycetes (see Table 20.1).

EVOLUTION

Chytrids, some of which cause plant diseases, are classified formally as Chytridiomycota (see Figure 20.1a). Chytrids are the only fungi that produce flagellate cells; these are used for dispersal in their aquatic or moist soil habitats. During sexual reproduction, gamete nuclei of chytrids and zygomycetes fuse shortly after the gamete cells fuse to form a zygote (the situation differs in most ascomycetes and basidiomycetes).

Black bread molds are included in the **zygomycetes** (Zygomycota). Zygomycetes have unique thick-walled zygotes that function as resistant spores (hence the term "zygospore") (Figure 20.7). **Spores** are one or few-celled reproductive structures, produced by parent fungi. Spores are adapted for dispersal in the environment. If water, food, and other resources needed for growth and mitosis are present, the spore will germinate, producing a new mycelium. The hyphae of zygomycetes lack cross walls.

The Glomeromycota is a group of fungi not known to grow separately from plant roots (or in one case, cyanobacteria). Within plant roots, the hyphae form highly branched structures (known as arbuscles) that function in exchange of materials with root cells. Such fungi are thus known as arbuscular mycorrhizal ("fungus roots") fungi, or **AM fungi** (see section 20.4). Most AM fungi reproduce by means of large, multinucleate, asexual spores.

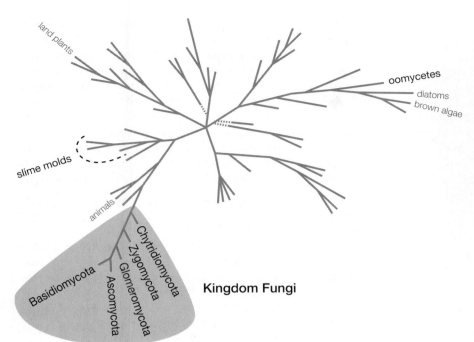

FIGURE 20.6   Relationships of the sexually reproducing groups of fungi
DNA and other evidence indicate that modern fungi arose from a single common ancestor. Thus, the fungi are grouped into a Kingdom. Fungi are more closely related to animals than to plants. Oomycetes and some other funguslike organisms were once thought to be fungi but are now grouped with diatoms and brown algae. Likewise, several groups of slime molds, previously thought to be fungi, are now known to be protists. The branches of the fungal groups are lengthened relative to others on the tree for clarity.

Mildews, chestnut blight, apple scab (Figure 20.8), and many other plant disease fungi; edible truffles and morels; and many yeasts are classified in the Ascomycota, commonly known as **ascomycetes** or sac fungi. Their name derives from the fact that the reproductive spores are produced in chambers (asci = sacs) (Figure 20.9). Sac fungal hyphae are divided into cells by cross walls (formally known as septa). Each cell typically contains two nuclei derived from mated gametes. In ascomycetes, nuclear fusion of the gamete nuclei to form a zygote nucleus may not occur until just before production of spores by meiosis (Figure 20.10).

Many types of mushrooms, as well as the widespread crop diseases wheat rust and corn smut, are placed into the Basidiomycota, also known as the **basidiomycetes** or club fungi. Their name derives from the fact that spores are produced at the ends of clublike structures, also known as basidia (Figure 20.11). Club fungal hyphae have cross walls with associated features known as clamp connections. There are usually two nuclei per cell as the result of previous mating of parental hyphae. In club fungi, pairs of compatible nuclei exist together in the same cells for long periods (sometimes hundreds of years) before finally combining to form a zygote

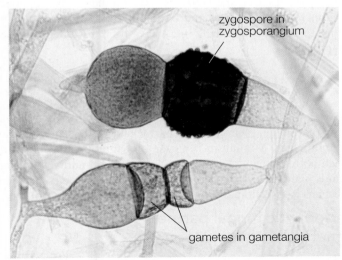

FIGURE 20.7   Zygomycetes Zygomycota are represented by *Rhizopus*, a bread mold. The zygospore is formed within a thick-walled structure (the zygosporangium) by the fusion of nonflagellate gametes, which are produced in structures called gametangia.

FIGURE 20.8   Apple scab disease on crabapple leaves

ascospores

FIGURE 20.9 Ascomycete spores occur in sacs The sac fungi (Ascomycota) produce sexual spores—ascospores—(usually eight) in asci, which are shaped like tubular sacks. When mature, the spores are released into the air for dispersal.

nucleus. The zygote nucleus then produces sexual spores of diverse genetic types for dispersal (Figure 20.12).

Sexual spores allow fungi to colonize new habitats

Sac and club fungi produce sexual spores on diverse types of fleshy **fruiting bodies** that usually emerge from the food substrate into the air. Sexual reproduction has the advantage of generating new genetic types that can colonize habitats differing from those of the parents (see Chapter 13). Emergent fruiting bodies promote dispersal of the spores by wind, water, or insects. The fruiting bodies of club fungi often have a distinct stalk and cap (Figure 20.13). Because of this structure, people sometimes refer to club fungi as "toadstools," but this term has no scientific meaning. Examples of club fungi whose fruiting bodies may be familiar to you are puffballs, earthstars, stinkhorns, bird's nest fungi, shelf fungi, chanterelles, coral fungi, jelly fungi, and disease-causing rusts and smuts (Figure 20.14). Variations in the shapes and colors of fruiting bodies represent adaptations for spore dispersal by various means. Some fungi produce their fruiting bodies underground. Such underground fruiting bodies, whose spores are dispersed by digging animals, are known as truffles (Essay 20.1, "Fungal Gold: Mining Truffles").

Dispersal of abundant spores is necessary in order for fungi to reach a new supply of needed food substrate, which may be distantly located. In some cases, fruiting bodies are adapted to attract insects, which carry spores directly to an appropriate food material (such as dung). However, most fungal spores are dispersed by wind, so it is a matter of chance whether or not a fungal spore lands at a site suitable for germination and growth. Fungi are thus adapted to produce many spores. A single fungus can produce as many as 40 million spores per hour over a period of two days!

Devastating outbreaks of destructive crop diseases can arise when fungal spores are dispersed over great distances, even between continents. Coffee leaf rust, sugarcane rust, and black leaf streak of banana and coffee are examples of such disease outbreaks, which are difficult to predict and control. Sugarcane rust was probably transported to the Dominican Republic from Cameroon in West Africa by cyclonic winds. Similarly, coffee leaf rust was likely carried by transatlantic winds from Angola to Brazil. Spores can also be transported by human travel; wheat yellow rust was probably carried on clothing from Europe to Australia. The accidental introduction of disease fungi into a country can cause devastating crop losses. For this reason, immigration officials often monitor the entry of plants, soil, foods, and other materials that might harbor disease fungi.

Asexual spores are used to disperse well-adapted genetic types

Fungal spores may also result from asexual reproduction, which does not involve mating. Asexual reproduction has some advantages, including preservation of gene collections that are particularly advantageous (see Chapter 13). Some fungi produce asexual spores within larger cells called **sporangia** (Figure 20.15a). The word *sporangium* means "spore container." The presence of billions of dark-pigmented spores in sporangia is what gives black bread mold its ominous color. Other fungi generate asexual spores known as **conidia** from the tips of hyphae (Figure 20.15b). Asexual spores and conidia grow into fungi that are genetically identical to their parents, which spread widely in the environment. It is no wonder that fungal spoilage of food and other materials is such a widespread and pervasive problem. Fungal allergies are also explained by high spore abundances in air (Figure 20.16). Prolific fungal spore production also explains the rapid spread of harmful molds within moist areas of buildings, a problem of increasing concern. Some 17,000 species of fungi appear to reproduce primarily by conidia. They include the athlete's foot fungus (*Epidermophyton floccosum*), infectious yeast (*Candida albicans*), some fungi that cause serious plant diseases, and the fungus from which the antibiotic penicillin is produced (*Penicillium chrysogenum*). Fungi for which sexual reproduction has not been observed are classified into the **imperfect fungi**, known as deuteromycetes. Cell and DNA characters indicate that imperfect fungi include ascomycetes and basidiomycetes for which loss of ability to reproduce sexually may have been advantageous (see Chapter 13).

20.4 | Fungi live in beneficial associations with most plants

Fungi that obtain their food in a partnership with algae or plants are **mutualists**. Mutualism is a type of symbiotic association in which both partners reap benefits. Examples of mutualistic fungi include plant endophytes, mycorrhizal fungi, and fungal partners in lichens (see section 20.6).

Endophytic fungal partners provide benefits to plants

Endophytes are fungi that live compatibly in the tissues of plants, particularly grasses. Endophytic fungi obtain organic food from plants and, in turn, contribute toxins

ECOLOGY

ECOLOGY

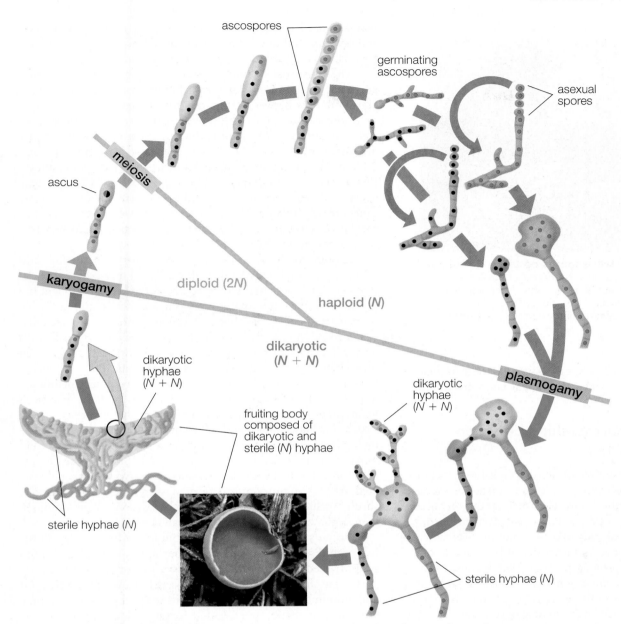

FIGURE 20.10 **Ascomycete life cycle** Sexual reproduction involves the cytoplasmic fusion (plasmogamy) of specialized cells of two genetically distinct parents. From the fused cells there emerge hyphae whose cells are described as dikaryotic ($N + N$) because they contain two nuclei, one nucleus contributed by each parent. Dikaryotic hyphae form the fertile (reproductive) portion of a fruiting body; hyphae with cells having single nuclei make up the sterile (nonreproductive) portion of the fruiting body. In surface cells of the fruiting body the two nuclei finally fuse (karyogamy) to form a diploid ($2N$) zygote nucleus. The zygote undergoes meiosis, and, often, a mitotic division follows to form a total of eight haploid (N) spores termed ascospores. Spores are dispersed and germinate to produce new hyphae (whose cells each contain a single haploid nucleus), completing the cycle. These hyphae may also reproduce asexually via spores.

or antibiotics that deter foraging animals, insect pests, and microbial pathogens. As a result, plants with endophytes often grow better than plants of the same species without endophytic fungi. Chemists have learned that endophytic fungi can be sources of useful drugs. One example is taxol, an anticancer drug having a market value of over a billion dollars (see Chapter 7). Taxol was first discovered as a product of the yew tree, *Taxus*, but is also produced by an endophytic fungus, *Taxomyces andreanae*. *Pestolotiopsis* is a fungus that often occurs in tropical plants. As

an endophyte in the Australian Wollemi pine, this fungus produces taxol, as do several other tropical endophytic fungi. It is possible that taxol was originally a product of endophytic fungi whose taxol biosynthesis genes were transmitted to yew. Taxol kills disease-causing protists (oomycetes) in the same way that human cancer cells are killed by this compound. Thus, taxol may have first evolved in fungi as an antibiotic. Fungal endophytes can provide plants with other benefits. For example, the plant *Dichanthelium lanuginosum*, which grows in hot soils at

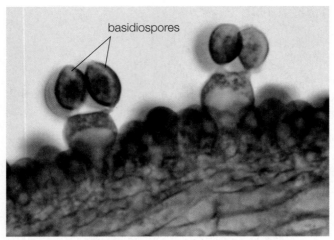

FIGURE 20.11 Basidiomycete spores occur on club-shaped structures The club fungi produce spores called basidiospores (usually four—some are just out of view in this photograph) from the top surface of club-shaped cells (basidia). When they are mature, these sexual spores will be released into the air for dispersal.

geothermal sites such as Yellowstone National Park, can tolerate much higher temperatures when the fungal endophyte *Curvularia* is present.

Mycorrhizae and partnerships between fungi and plant roots

An extremely important example of mutualism is the association of particular soil fungi with roots of plants to form **mycorrhizae** ("fungus roots"). Mycorrhizae are extremely common, occurring in about 80% of plants. Plants and their mycorrhizal fungi have influenced each other's evolution over time, so that the partners have come to depend on one another. Mycorrhizal fungi provide plants with minerals, particularly phosphorus, and sometimes with fixed nitrogen and water, obtained from soil. The extensive fungal mycelium is able to absorb minerals from a much larger volume of soil than can plant roots. The total length of mycelium in 1 cubic meter (m³) of soil can reach 20,000 kilometers (km)! Because mycorrhizal hyphae are very narrow and long, the fungi can reach into tiny soil spaces to obtain nutrients. Thus, the fungi increase the efficiency of soil mineral harvesting. By binding soils, the fungi also reduce water loss and erosion, and they help protect plants against pathogens and toxic wastes. In return, plants provide the heterotrophic fungi with organic food, sometimes contributing as much as 20% of their photosynthetic products. For plants, this is a very good investment. Experiments show that mycorrhizae greatly enhance plant growth by comparison to plants lacking fungal partners. Fungi play an important role in plant succession (Chapter 25), helping to determine which plants are able to thrive in new sites. For this reason, mycorrhizal fungi are becoming used in plant community restoration projects.

Fossils of mycorrhizal fungi have been found in close association with fossils of early plants. The earliest-known fossil fungi belong to the AM fungi (Glomeromycota), which today exist only in close partnership with land plants. Since early soils were thin and poor in nutrient content, mycorrhizae and other close plant-fungal associations may have been key to the success of early plants on land (Chapter 21).

The two most common types of mycorrhizae are ectomycorrhizae and endomycorrhizae. **Ectomycorrhizae** are beneficial symbioses between temperate forest trees and ascomycetes or (more commonly) basidiomycete fungi whose hyphae coat tree-root surfaces. The fruiting bodies of ectomycorrhizal fungi are often visible on the ground beneath host trees (Figure 20.17a). The intimate association of hyphae with roots allows the partners to exchange materials (Figure 20.17b,c). Some species of oak, beech, pine, and spruce trees will not grow unless their ectomycorrhizal partners are also present. Such information is essential to the success of commercial nursery tree production and reforestation projects.

Endomycorrhizae are partnerships between plants and AM fungi (Glomeromycota) (Figure 20.18a) that grow more extensively into root tissues than ectomycorrhizae. The fungal hyphae grow into the spaces between root cells and are able to penetrate cell walls (but not cell membranes) by secreting cellulose-degrading enzymes. In the space between the cell walls and cell membranes of the root, endomycorrhizal fungi often form highly branched, bushy **arbuscules** (meaning "treelike" structures). As the arbuscules develop, the root-cell membrane also expands. Consequently, the arbuscules and the plant-cell membranes surrounding them have very high surface areas, which facilitate rapid and efficient exchange of materials—minerals flow from fungal hyphae to root cells, and organic food moves from root cells to hyphae (Figure 20.18b,c). These fungus-root associations are known as **arbuscular mycorrhizae**, abbreviated **AM**. AM fungi are associated with trees such as apple, shrubs including coffee, and many herbaceous plants—among them legumes, grasses, tomatoes, strawberries, and peaches.

Orchids also depend on fungi, but in a different way. Dust-like orchid seeds are so small that they have little stored food to support seedling growth, in contrast to most seeds. In nature, particular club fungi supply orchid seedlings with organic food and vitamins. Recent studies have shown that even though they are photosynthetic, green-leafed forest orchids can receive as much as 90% of their carbon from fungi.

Some heterotrophic plants obtain organic food from fungi

Some orchids and other plants have lost photosynthetic ability (Figure 20.19); they have come to depend upon organic compounds produced by neighboring plants, transmitted by mycorrhizal fungi. This mode of plant nutrition is known as **mycoheterotrophy**. The fungi function like a nutrient pipeline, linking the photosynthetic plants with non-photosynthetic ones. Extensive underground subway-like networks may link a wide variety of plants growing in the same area. A large oak tree may be feeding not only its fungal partner but also other plants nearby. Seedlings can also tap into this pipeline on a temporary basis. In forests, light levels may be too low for short seedlings to photosynthesize. Organic compounds in the mycorrhizal pipeline can power seedling growth to a size large enough to intercept sunlight for photosynthesis.

ECOLOGY

EVOLUTION

ECOLOGY

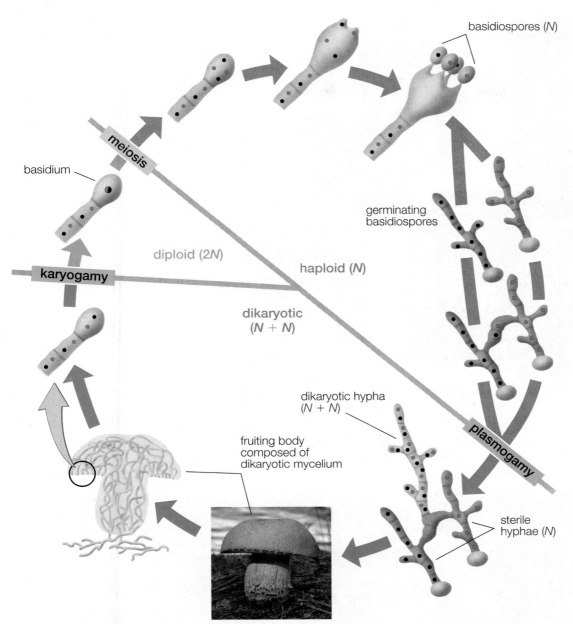

FIGURE 20.12 **Basidiomycete life cycle** Mycelia that develop from germinating basidiospores have cells with only a single nucleus. Typically, hyphae of compatible mating types fuse, forming a dikaryotic ($N + N$) mycelium (meaning that each cell contains two nuclei, one from each parent). The dikaryotic mycelium forms a fruiting body. Some cells at the surface of the fruiting body undergo nuclear fusion, then meiosis, after which spores called basidiospores are produced on a structure called a basidium. The spores are dispersed, completing the life cycle.

20.5 | Fungi are relevant to humans in many ways

From the human perspective, fungi have many beneficial activities. For example, yeasts and some other fungi are useful laboratory model systems for understanding the function of all life. Much of what we know about the genetic control of cell division in eukaryotes comes from the study of yeast DNA. Humans also value the ecological role of fungi as decomposers and their use in the food and chemical industries.

Nonetheless, humans regard fungi as having harmful effects as well.

Fungi function as decomposers

Together with bacteria, fungi are Earth's **decomposers**. Decomposers break down the bodies of dead organisms, wastes, and other organic matter that would otherwise accumulate in nature. By their actions, decomposers also release nutrients such as CO_2, combined nitrogen, and phosphate minerals that are continuously needed by plants and other living

ECOLOGY

FIGURE 20.13 **Basidiomycete fruiting body** Fruiting bodies of basidiomycetes are commonly known as mushrooms. (a) Mushrooms often have a cap (pileus) and stalk (stipe). (b) The surfaces of mushroom caps have toothlike structures, pores, or gills (as in this example) upon whose surfaces the spores are produced. (c) Teeth, pores, or gills (this example) greatly increase the surface area for spore production. (d) By placing caps of an unidentified mushroom onto both white and black sheets of paper, one can obtain a spore print, which indicates the spore color—a characteristic used in mushroom identification keys.

FIGURE 20.14 **Basidiomycete fruiting bodies are adapted in various ways for effective spore dispersal** (a) When mature, puffballs disperse spores into the air when they are touched (this is an immature specimen). (b) Raindrops disperse egglike containers of spores from the nestlike fruiting bodies of bird's nest fungi. (c) Shaggy-mane mushrooms (Coprinus) disintegrate into an inky liquid that contains spores; rain helps disperse spores. (d) Stinkhorn fungi such as Mutinus, shown here, use insects to unwittingly disperse fungal spores (the greenish portion). These fruiting bodies attract flies by releasing foul odors. The flies, expecting a source of decaying meat, land on the sticky surface. When they take off again, they carry a load of fungal spores on their bodies.

things. If decomposers were not present, these essential nutrients would remain unavailable to plants. White rot fungi are the only decomposers known to be able to completely break down lignin, an extremely tough plant material. Materials that are decomposable by bacteria and fungi are **biodegradable** materials.

Unfortunately, fungi also decompose substances such as food products, wood building materials, coatings on camera lenses, and other materials valued by humans. Additives, such as proprionic acid often listed on bread labels, may be used to retard molding.

ESSAY 20.1   FUNGAL GOLD: MINING TRUFFLES

Fruiting bodies described as truffles, resembling small dirty rocks or potatoes, are produced by several types of fungi. But *Tuber* (Figure E20.1) is the truffle fungus most prized by cooks for its exquisite flavor. Truffles are cultivated in southern Europe, New Zealand, Australia, and the United States in large tree plantations. This is because *Tuber* is an ectomycorrhizal associate of forest trees such as oak, birch, and pecan and thus receives essential nutrients from these plants. Despite extensive efforts to cultivate *Tuber* in the laboratory, people have not yet been able to grow marketable quantities of truffles separately from their host plants.

To locate these buried treasures, people have long used pigs or dogs that have been trained to detect truffles.

Roots of young oaks can be inoculated with a mycelium of the appropriate fungal species, then transplanted to plantations; or a truffle mycelium can be added to plantation soil after trees have become established. One problem with truffle cultivation is that competing ectomycorrhizal fungi can also become established in plantations, reducing the supply of food supplied by trees to truffles. Truffle development to harvestable size—a few millimeters to more than 10 centimeters—then takes about 10 years. Truffle locations must be determined before they can be dug up. In natural forests, digging animals disperse *Tuber*. The animals are attracted to the smell of chemical compounds emitted from the fruiting bodies. As they dig, the animals break the fruiting bodies up, thereby spreading the spores. To locate these buried treasures, people have long used pigs or dogs that have been trained to detect truffles.

Since truffle cultivation and harvest are so intensive, the demand for truffles exceeds the supply—only about 20 tons each year—and so truffles are expensive;

E20.1   Truffles

white truffles cost as much as $3,000 per kilogram. Consequently, truffle growers are interested in ensuring that the fungal inoculum they purchase is the correct species and free from fungal species that would compete with truffles. These problems are now being solved with DNA technology. Molecular tools are available for distinguishing truffle species and detecting the presence of other fungi even when only small bits of mycelium are available.

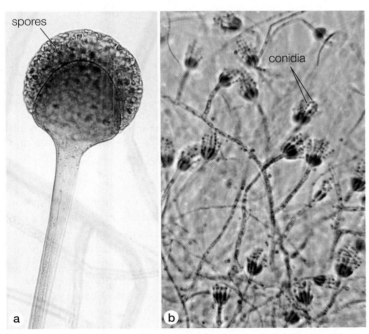

FIGURE 20.15   Asexual spores of fungi (a) Black bread mold reproduces asexually by means of sporangia—structures that produce, then release, many spores. This sporangium has not yet released its brownish spores. (b) Other fungi reproduce asexually by means of conidia, as illustrated here by *Penicillium*, which is used to produce the antibiotic penicillin.

FIGURE 20.16   Fungal spores These spores are from *Alternaria*, a common allergy-causing fungus. Allergenic fungal spores can be so abundant in air that news media may report their numbers in air-quality listings. The red color results from a stain used to make these spores more visible.

Fungi are useful as foods and in industrial production

Several nutritious and flavorful fungal species are cultivated for use as human food: the common pizza and portabella mushroom *Agaricus*, the shiitake mushroom *Lentinula*, and the morel *Morchella* (Figure 20.20), among others. Athletes may increase their protein consumption by drinking shakes

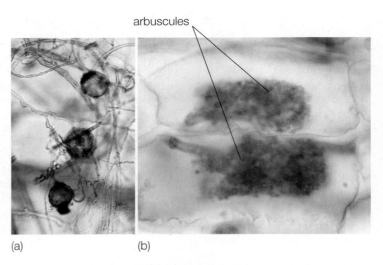

arbuscules

ectomycorrhizae

root

(a)

(b)

(a)

(b)

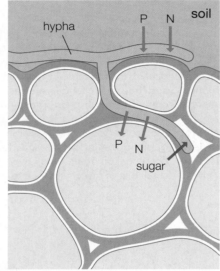

FIGURE 20.17 **Ectomycorrhizae** (a) The chanterelle
(Cantharellus) is an ectomycorrhizal fungus associated with
forest tree roots. (b) Microscopic view of a sheath of fungal hyphae
around plant roots. (c) Diagram showing movement of minerals
from soil into hyphae (green arrows), then into root cells (blue
arrows), and movement of sugar from root cells into fungal
hyphae (red arrow).

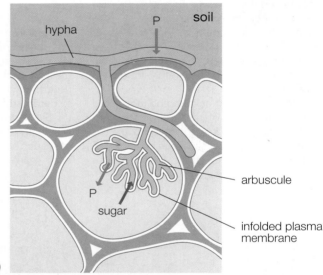

FIGURE 20.18 **Endomycorrhizae** (a) *Glomus* is a common
endomycorrhizal fungus that produces large spores. (b) Microscopic
view of blue-green stained endomycorrhizal arbuscules in spaces
between root-cell walls and membranes. (c) Diagram showing
movement of phosphorus from soil into hyphae (green arrow), then
into root cells (blue arrow), and movement of sugar from root cells
into fungal hyphae (red arrow). Note that the plasma membrane of
the root cell is greatly increased in surface area as it surrounds
the arbuscule.

made with protein extracted from yeast. The brewing industry depends on yeast for alcohol generation and carbonation,
which adds fizziness. The baking industry depends on yeast
CO_2 production, which leavens bread and other products
(see Chapter 5). Several types of cheeses and protein-rich
soybean tempe are also produced industrially with the use of
fungi. Some fungi are grown in large industrial vats for
extraction of antibiotics and other medicines for human use.
Cyclosporin, a drug that is widely used to prevent rejection
in transplant patients, and the antibiotic penicillin are important examples.

Some fungi are poisonous or hallucinogenic

Many wild fungi, like many plants, contain protective poisonous compounds that deter herbivory—consumption by hungry animals. There is no reliable test for poisonous fungi, and

people often confuse poisonous fungi for similar-looking edible forms. For example, the false morels *Helvella* and *Gyromitra* (Figure 20.21) resemble the edible morel mushroom
Morchella and are mistakenly picked and consumed. Some
species of the false morels contain monomethylhydrazine (a
component of used rocket fuel) that is poisonous to humans.
In 1998, there were nearly 10,000 reported cases of mushroom poisoning in North America alone, some causing liver
destruction so complete that victims needed transplants. One
such mushroom is *Amanita virosa*, the "Destroying Angel"
(see Figure 20.13a). Experts recommend that foraging for
edible mushrooms should be done in the grocery store, rather
than the woods.

FIGURE 20.21 **A false morel, *Gyromitra***

FIGURE 20.19 ***Corallorhiza*, a heterotrophic orchid** This orchid lacks chlorophyll and is thus nonphotosynthetic. It is an example of many types of flowering plants that tap mycorrhizal fungi as a source of nutrients. Such plants are connected by mycorrhizal fungi to neighboring green plants, which provide organic food for both the fungal partner and the "free-loader."

Some fungi attack stored grains, fruits, and spices, producing liver cancer-causing toxins, known as **aflatoxins**, which have become a major public health concern around the world. In 1998, U.S. corn growers lost $100 million to aflatoxin contamination. Dairy products can become adulterated with these toxins when cattle are fed contaminated grain.

Some fungal compounds have hallucinogenic effects in humans and have thus been used in spiritual ceremonies by cultures around the world. One example is the "magic mushroom" *Psilocybe*, which produces a compound related to LSD (lysergic acid diethylamide), and is a legally controlled substance. *Claviceps*, which causes a disease of rye crops known as ergot, produces a similar mind-altering compound. Records from the Middle Ages describe cases of human infertility, convulsions, hysteria, and a burning sensation of the skin that are now attributed to human consumption of *Claviceps*-contaminated rye. Events associated with the Salem witch trials may also have been the result of ergot poisoning. In general, it is unwise to consume hallucinogenic fungi, because the amount required for psychogenic effect is perilously close to the toxic dose.

FIGURE 20.20 **Edible, cultivated mushrooms** Cultivated mushrooms that are safely used in cooking include (a) the common mushroom (*Agaricus*), (b) Portabella mushroom (*Agaricus*), (c) shiitake (*Lentinula*), and (d) morels (*Morchella*). Morels, which grow in woodlands in association with trees such as oaks, have only recently been brought into cultivation. Although they are delicious, they should be cooked, never eaten raw.

20.6 | Lichens are partnerships between fungi and photosynthetic microbes

Lichens illustrate the way in which Earth's autotrophs and heterotrophs depend upon each other, each providing resources needed by the other. Lichen bodies are masses of fungal hyphae with layers of autotrophic (photosynthetic) green algae or cyanobacteria, or both (Figure 20.22). The algae and cyanobacteria provide lichen fungi—**mycobionts**—with organic food and oxygen. The fungi provide their autotrophs—**photobionts**—with carbon dioxide from respiration, water, minerals, and protection. Because lichens often occupy exposed habitats of high light intensity, their fungi often produce bright yellow, orange, or red-colored compounds that help prevent damage to the photobiont's photosynthetic apparatus (Figure 20.23). Lichen fungi also produce distinctive organic acids and other compounds that deter hungry animals and microbial attacks.

Lichen evolution and diversity

There are at least 25,000 lichen species, but these did not all descend from a common ancestor. In fact, DNA studies have shown that lichens evolved at least five separate times. Lichen fungi form lichens only with particular algal or cyanobacterial species. Several large groups of ascomycete fungi that no longer form lichens evolved from lichen-forming ancestors. Lichen bodies take one of three major forms: crustose—flat bodies that are tightly adherent to an underlying surface; foliose—flat, leaflike bodies; or fruiticose—bodies that grow upright or hang down from tree branches (Figure 20.24).

Lichen reproduction and development

Fungi of about one-third of lichen species can only reproduce asexually. Asexual reproductive structures include **soredia**, small clumps of hyphae surrounding a few photobiont cells that can disperse in wind currents (Figure 20.25). Soredia are lichen clones. By forming soredia, lichen fungi can disperse along with appropriate photosynthetic partners.

Although most photobiont algae have not been observed to reproduce sexually, DNA data suggest that the green algae most commonly found in lichens do exchange genes, probably by sexual means. The fungal partners of many lichens can

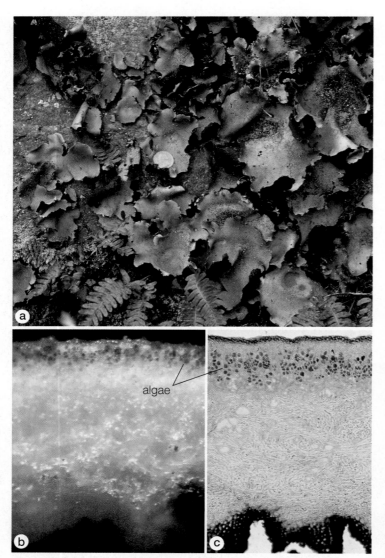

FIGURE 20.22 Lichens are composed of fungal hyphae and a layer of autotrophic green algae or cyanobacteria The lichen *Umbilicaria* is seen growing on a rock face in (a). A quarter was placed in the photo for scale. (b) A handmade slice of *Umbilicaria* showing the green layer of algal cells lying just below a protective surface of tightly appressed fungal hyphae. (c) A stained section of *Umbilicaria* showing the position of the algal cells.

FIGURE 20.23 The bright colors of many lichens are due to compounds that protect photobionts from damage by excess light

FIGURE 20.24 Lichen body types (a) Crustose lichens are thin and tightly appressed to surfaces. (b) Foliose (leaflike) lichens are flat, but not appressed to surfaces. This example is from the genus *Lobaria*. (c) *Cladonia*, or reindeer lichen, is an example of a fruticose lichen.

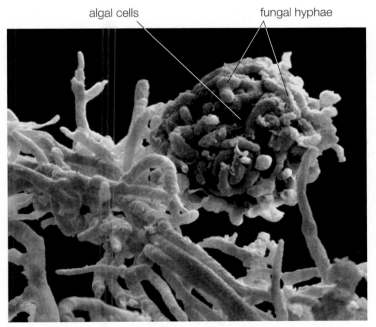

algal cells   fungal hyphae

FIGURE 20.25 An asexual lichen reproductive structure Lichen soredia, one of which is seen here with a scanning electron microscope, are small clumps of fungal hyphae wrapped around a few algal cells that break off the surface of lichens and serve in dispersal.

undergo sexual reproduction, producing fruiting bodies and sexual spores much like those of related fungi that do not form lichens. DNA studies have shown that some lichen fungi can self-fertilize. This is advantageous in harsh environments where potential mates may be few and far-between. To produce new lichens, hyphae that grow from sexual spores must acquire new photobionts—not always an easy task. Only particular green algae or cyanobacteria are able to serve as photobionts. Some lichen fungi steal desirable photobionts from other lichens growing nearby, and DNA data have shown that photobiont partner switching is common.

Even if compatible photobionts and mycobionts are present together, they may not form lichens. Laboratory experiments have provided insight into the reason. Lichen fungi, cyanobacteria, and algae can grow separately in nature and in the laboratory, but it is difficult to synthesize a lichen from its components in the lab. This synthesis has been accomplished only when the autotroph and fungus are exposed to stressful conditions such as alternate wetting and drying. These results suggest that lichens form in nature under similarly challenging conditions.

Lichen ecology

Lichens often grow on rocks, buildings, tombstones, tree bark, soil, or other surfaces that can dry out. When water is not available, the lichens become dormant until moisture returns. Thus lichens may spend much of their time in an inactive state, and for this reason often grow very slowly. But because they can persist for long periods, lichens can be very old; some are estimated to be more than 4,500 years of age.

Lichens are recognized for their soil-building activities, which occur over very long time scales. Lichens produce acids that help to break up the surfaces of rocks, beginning the process of soil development. Lichens having cyanobacterial photobionts can also increase soil fertility by adding fixed nitrogen. One study showed that such lichens released 20% of the nitrogen they fixed into the environment, where it then became available for uptake by plants.

Though lichens occur in diverse types of habitats, some lichens grow in some of the most extreme, forbidding terrestrial sites on Earth—deserts, mountaintops, and the Arctic and Antarctic—places where most plants cannot survive (Chapter 26). In these locations lichens serve as a food source for other hardy organisms such as reindeer.

Human uses of lichens

People use lichens as air-quality monitors because lichens are particularly sensitive to air pollutants such as sulfur dioxide. Air pollutants severely injure photobionts, without which lichens cannot survive. The disappearance of lichens serves as an early warning system of air pollution levels likely to eventually affect humans. Lichens are also good monitors of atmospheric radiation levels because they accumulate radioactive substances from the air. After the Chernobyl nuclear power plant accident, lichens in nearby countries became so radioactive that reindeer, which consume lichens as a major food source, became unfit for human use as food or in milk production. Although lichens don't taste good to people or most animals due to the presence of protective compounds, lichens are not known to cause poisoning or other harm. They have served as survival foods for aboriginal peoples in times of crisis. Lichens are also used as clothing dyes and can be found in some natural personal-care items.

HIGHLIGHTS

20.1 Fungal bodies are mycelia composed of threadlike hyphae that are abundant in nature. Mycelia are often invisible to the unaided eye because they are embedded within food substrates. Mycelia may produce fruiting bodies—large, visible masses of compressed hyphae that emerge from substrates and produce and disperse sexual spores. Fungi may also reproduce by asexual spores.

20.2 Fungi are heterotrophs that obtain food from organic substrates by excreting digestive enzymes and absorbing the resultant small organic molecules. Saprobic fungi feed on dead organic material, whereas parasites obtain food from living plant or animal tissues, thereby causing disease.

20.3 The five major groups of true fungi have distinctive reproduction. Chytrids, which are often aquatic, are the only fungal group that produces flagellate cells. The other four groups primarily live in terrestrial habitats. Zygomycetes produce resistant zygospores formed as the result of sexual reproduction. AM fungi (Glomeromycota) produce very large, multinucleate, asexual spores. Ascomycetes produce sexual spores in sacs, and basidiospores produce sexual spores on club-shaped structures. Some fungi, classified as imperfect fungi, have apparently lost the ability to reproduce sexually but can reproduce by means of asexual spores or conidia. Long-range dispersal of spores of plant pathogenic fungi, which can cause unpredictable and devastating crop losses, is an emerging societal concern.

20.4 Most land plants have beneficial fungal mutualists. Endophytic fungi live within the tissues of some plants, often contributing antibiotics, herbivore resistance, or other benefits. The plant root/fungal partnerships known as mycorrhizae are very common, increasing plants' ability to obtain soil nutrients. Some nonphotosynthetic plants get organic compounds from photosynthetic plants via mycorrhizal networks within the soil. Plant-fungal associations probably fostered the ability of earliest plants to live on land.

20.5 People use fungi as food and in industry. Many wild fungal fruiting bodies are poisonous, and some are deadly.

20.6 Lichen bodies are composed of fungi and photosynthetic green algae or cyanobacteria. Lichens are adapted to live in harsh environments and they typically grow only very slowly. Lichens are useful air-quality monitors.

REVIEW QUESTIONS

1. Describe the four divisions of sexually reproducing fungi along with some of their identifying characteristics or representative species.

2. Discuss some of the economic uses of fungi.

3. Describe some of the ecologically important roles of fungi.

4. Describe ways in which fungi can be harmful to people or property.

5. Some fungi are especially adept at forming close, mutualistic associations with plants. Discuss these symbioses and the benefits to each of the partners.

6. Name the three growth forms of lichens, and describe their growth habits.

APPLYING CONCEPTS

1. While examining a soil sample with a light microscope, you see what appears to be a fungal hypha without any reproductive structures or spores. How could you determine to which fungal division this organism belonged? (*Hint*: Think about cross walls.)

2. Some species of the fungal genus *Penicillium* produce the well-known antibiotic penicillin, which has had an enormous impact on human health care. What benefit does the fungus derive from making penicillin?

3. Imagine that somehow a new virus arose which eliminated all fungi from the planet. Besides the devastation in scientific research, in the food trades, and in the biomedical production of pharmaceuticals, what would be the effect of the loss of fungi on the environment? In answering this question, consider all the ecological roles that fungi play, as discussed in this chapter.

4. Imagine that you are part of a team of explorers in a dry, cold region of the world lacking green plants of any type. What photosynthetic, eukaryotic life-forms might be present in such a forbidding place, and how would you recognize them?

Sphagnum moss

Invasive plant species are often in the news because they tend to form large populations that replace or damage natural vegetation. Such disturbance affects associated animals and microbes because invasive species do not provide equivalent food and habitat. Most invasive species are exotic; that is, they are not native to the regions they invade. Colonization of new habitats allows invasive plant species to escape from pests and disease organisms that control such plants in their native habitats. Released from inhibition by pests and diseases, invasive plants grow rapidly, challenging human attempts to control them. Invasive plant species also commonly reproduce rapidly, contributing to their invasive abilities. Some seedless plants, including bryophytes and ferns, can be serious invaders. Seedless plants have the advantage of very lightweight reproductive structures that can travel very long distances in the wind.

Many species of *Sphagnum* or peat moss occupy moist habitats around the world, providing ecosystem benefits. But a few species are invasive—sometimes able to grow up into the treetops—and have displaced natural vegetation in some areas of New Zealand and Hawaii. In Europe, several other mosses are regarded as weedy, invasive species that threaten other plant species. Liverworts, another group of bryophytes, also include some invasive species (*Lunularia cruciata*, *Marchantia polymorpha*, and *Lophocolea semiteres*). Kariba weed—the fern *Salvinia*—can completely blanket water bodies in tropical regions. Such growth cuts off the sunlight supply to the underlying water, preventing growth of algae that feed many animal species. In Florida, two species of the viny fern *Lygodium (L. microphyllum* and *L. japonicum)* form kudzulike mats up to a meter thick that smother and displace native vegetation. By colonizing the southeast United States, *Lygodium* has escaped the enemies that control its growth in its native habitat tropical regions of Asia and Africa.

Seedless plants are important to people in many other ways. In this chapter, we will focus on the structure, reproduction, and diversity of seedless plants. This will explain why these plants are useful in gardens and landscaping, help moderate Earth's climate, and aid in understanding how seed plants arose. We begin with an overview of plant diversity and evolution, in order to understand how seedless plants are related to other plant life on Earth. ■

Lygodium japonicum

21.1 | What are plants?

Modern terrestrial plants include an astonishing array of more than 250,000 species. These range from relatively small and simple **bryophytes**, such as mosses, to larger and more complex **vascular plants**. Vascular plants—those having lignified conducting tissues (Chapter 8)—include ferns and close relatives, which do not produce seeds, plus the seed plants. The modern groups of seedless vascular plants are **lycophytes** and **pteridophytes**. Thus, the seedless plants include bryophytes, lycophytes, and pteridophytes. Seed plants include the **gymnosperms** (such as conifer trees) and flowering plants—the **angiosperms** (Figure 21.1). Several features unite these diverse types of plants.

Plants are multicellular autotrophs that are adapted to life on land

One characteristic of plants is that they are multicellular—their adult bodies are composed of more than one cell. Plants are large enough to be seen without the use of a microscope. Photosynthesis is another general plant characteristic. Unlike fungi and animals, which cannot make their own food, the vast majority of plants are able to produce their own organic food. There are, however, examples of heterotrophic bryophytes and flowering plants that have lost photosynthetic capacity.

It is not as easy to distinguish plants from green algae, which are also mostly photosynthetic, often multicellular, and sometimes as large as land plants. Green algae and plants have similar chlorophyll pigments and produce starch in their chloroplasts, indicating a close relationship. Thus, some experts include green algae in their definition of plants. Other biologists extend the definition of plants to include brown and red seaweeds, even though their photosynthetic pigments, food storage molecules, reproduction, and other features differ from those of plants. In this and some other textbooks, groups of plants that mostly live on land, and are adapted in many ways to life on land, are classified in the kingdom Plantae.

All groups of land-adapted plants—bryophytes, lycophytes, pteridophytes, gymnosperms, and angiosperms—have in common a set of characteristics that is absent from even closely related green algae. For example, plant bodies are composed of tissues produced by an **apical meristem** (Figure 21.2a). Land plants produce **spores** with incredibly tough walls that help them survive during dispersal in dry air (Figure 21.2b). Land plants also have a life history known as **alternation of generations**. This means that in addition to a haploid gametophyte generation, plants also possess a multicellular, diploid, spore-producing generation known as the sporophyte. In plants, the early-stage sporophyte—the **embryo** (Figure 21.2c)—is always dependent on maternal (female gametophyte) resources and, for this reason, plants are known as **embryophytes**. All plants also produce their spores in multicellular enclosures known as **sporangia** (Figure 21.2d).

Apical tissue-producing meristems, embryos, and the sporophytes that develop from them, sporangia, and walled spores are defining features of land plants. These features—present in all plant groups from bryophytes to angiosperms—are regarded as adaptations to life on land. Other traits characterize each of the five modern plant groups (Table 21.1), reflecting ways in which plants have become increasingly adept at solving the problems associated with life on land. Although seagrasses and some other plants are aquatic, these are descended from land-adapted ancestors. This is analogous to the evolution of dolphins and whales from land-adapted animal ancestors.

Plant diversity is important in global ecology and human affairs

Why is it important to know something about plant diversity? The answer is that all plant groups are important to humans and other animals. Humans are most dependent upon the seed plants for food and other materials, but seedless plants—bryophytes,

FIGURE 21.1 Representative examples from the major modern plant groups (a) A bryophyte (the moss *Mnium*), showing both the "leafy" gametophyte and dependent sporophyte generations; (b) the lycophyte *Lycopodium*, also known as "club moss," showing cones ("clubs") bearing sporangia and spores; (c) the fern *Osmunda* (a pteridophyte), seen in early spring; (d) Ponderosa pine (*Pinus ponderosa*), a gymnosperm; (e) Moccasin flower orchid (*Cypripedium acaule*), an angiosperm (flowering plant).

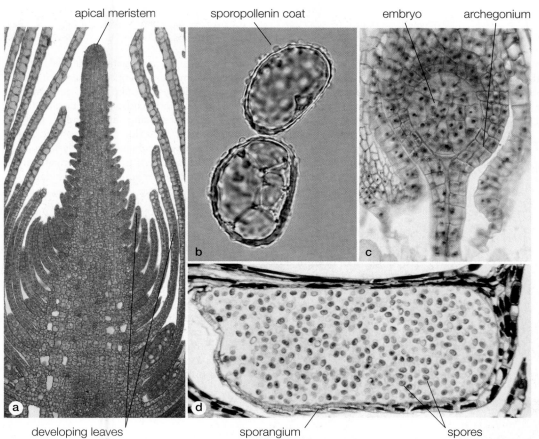

FIGURE 21.2   **Traits common to all plants** (a) Apical meristem (of the flowering plant *Elodea*).
(b) Sporopollenin-coated spores (from a fern). (c) Embryo (of the liverwort *Marchantia*). (d) Sporangium
(a microsporangium of the gymnosperm white pine, *Pinus strobus*).

TABLE 21.1	Distinguishing Characteristics of Modern Streptophytes (the land plants and their closest algal relatives)
Charophycean Algae	Apical meristems, when present, do not produce tissues; zygotes the only diploid cells; sporangia absent; spores lack sporopollenin walls
Embryophytes	Apical-tissue-producing meristems; multicellular sporophytes; sporangia; sporopollenin-walled spores
Nonvascular Plants	
Bryophytes (liverworts, mosses, hornworts)	No lignified vascular tissue; no true roots, stems, or leaves; sporophytes unbranched, and cannot grow independently of gametophytes
Vascular Plants	Lignified vascular tissue; sporophytes branched and become independent of gametophytes
Seedless Vascular Plants	Seeds absent; adventitious roots present; embryonic roots absent
Lycophytes	Leaves lycophylls
Pteridophytes (ferns)	Leaves euphylls (some lack leaves and/or roots)
Seed Plants	Seeds present; all leaves are euphylls; embryonic roots present
Gymnosperms	Flowers and fruits absent; seeds lack endosperm
Angiosperms	Flowers and fruits present; seeds possess endosperm at least early in development

ESSAY 21.1 THE PLANTS THAT CHANGED THE WORLD

Seedless plants have literally changed our world. The first plants, though small and similar to today's simple bryophytes, were nevertheless abundant enough by 450 million years ago to help build Earth's first soils. These early plants thus set the stage for the larger and more complex rooted vascular plants that came later. Early bryophyte-like land plants also helped to change the world in another way. Rotting experiments show that modern bryophytes produce organic materials that are quite resistant to decay. These materials help protect plants from attack by soil bacteria and fungi. The fact that very ancient fossils closely resemble rotted remains of modern bryophytes (Figure E21.1A) suggests that early plants also had decay-resistant tissues. When organic material is not completely degraded to CO_2, it can be buried for millions of years. As more organic carbon is buried, atmospheric CO_2 content is reduced. Measurements of stable carbon isotopes in ancient rocks show that atmospheric carbon dioxide was much higher (more than 16 times the present level) when early plant fossils first began to become abundant. If atmospheric CO_2 were this high now, Earth's climate would be much too hot for most modern life. Plants helped reduce this CO_2, moderating Earth's climate.

Plant photosynthesis continued to transform carbon dioxide into organic carbon, but the reverse process was so limited that atmospheric carbon dioxide levels plummeted.

Amounts of decay-resistant materials produced by modern bryophytes suggest that ancient relatives could have generated enough buried carbon to begin the process of reducing atmospheric CO_2. Even greater reduction of atmospheric carbon dioxide was achieved by early vascular plants. These plants trapped so much carbon in the form of lignin and other decay-resistant organic compounds that atmospheric CO_2 reached an historic low (Figure E21.1B). A resulting dramatic climate change changed the world forever. If plants had not accomplished this change in atmospheric chemistry, humans might not have appeared.

The coal age (Carboniferous) was named for fossil organic carbon originally produced by luxurious growth of many species of tree-sized seedless plants that lived several hundred million years ago (Figure E21.1C). Related to modern lycophytes and pteridophytes,

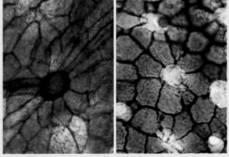

E21.1A Lower surface of a modern liverwort (left) and fossil (right) with similar arrangement of cells A rhizoid emerges from a rosette of cells in the liverwort; the clear circles at the center of similar rosettes in the fossil may represent sites where rhizoids were once attached.

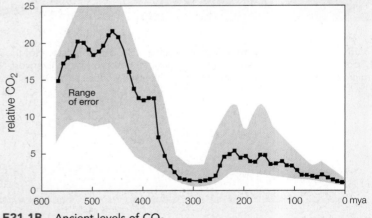

E21.1B Ancient levels of CO_2

lycophytes, and pteridophytes—are also useful to people and essential to the healthy functioning of ecosystems. For example, the bryophytes known as peat mosses are widely sold as soil conditioners for gardening, and the long-standing Asian tradition of moss gardening is becoming popular in the United States. Vast regions of the Northern Hemisphere that are dominated by peat mosses act as a global thermostat, helping reduce fluctuations in the Earth's temperature. Ancient forests of ferns and lycophytes were the main sources of modern coal deposits and also caused dramatic changes in Earth's climate (see Essay 21.1, "Plants That Changed the World"). Ferns are still very common and diverse in modern environments—particularly in tropical forests. Many people who keep houseplants have at least one fern, often the Boston fern, and ferns are valued components of landscape design. Like all plants, seedless plants are components of food webs that support myriad other organisms.

Another way in which seedless and early seed plants are important is that they were critical to the evolution of modern ecosystems. Each of these plant groups, as well as several extinct plant groups known only from fossils, appeared at distinct times during the Earth's history, each setting the environmental stage for the appearance of the next and more complex group.

E21.1C Coal-age forest reconstruction

these plants left many gigantic fossils. When the coal-age plants died and became submerged in the swampy water, their abundant woody tissues did not readily decompose into carbon dioxide. Since decay processes use oxygen (in aerobic cell respiration), and plant decay was inhibited, atmospheric oxygen levels rose to record levels. This allowed the existence of giant dragonflies and other incredible insects that could not exist in today's lower oxygen conditions. Huge amounts of dead-plant organic carbon were buried, compressed, and chemically altered into coal. Plant photosynthesis continued to transform carbon dioxide into organic carbon, but the reverse process was so limited that atmospheric carbon dioxide levels plummeted. Because carbon dioxide is a greenhouse gas that helps warm Earth's climate, when CO_2 declined the climate became substantially cooler. In the face of global climate change, many seedless plants (and other organisms) that had previously been abundant became extinct. Seedless plant fertilization requires liquid water, which was abundant during the warm, wet coal age. But as the climate cooled, it also became much drier (because more water occurs in the frozen form). The size of arid regions—in which water was less available for fertilization—increased.

Though the climate change worked against seedless plants, it favored the rise of early seed plants. These included several extinct groups of plants that had fernlike leaves but also seeds ("seed ferns") and early gymnosperms. Seed plants were successful in the new conditions because they can achieve fertilization in the absence of liquid water. Gymnosperms dominated Earth's flora for the next 200 million years, supporting giant plant-eating dinosaurs. During this period, early flowering plants and mammals existed but were only minor components of Earth's ecology. This situation might have continued indefinitely, but 65 million years ago a massive object from space struck Earth. This impact produced fires and resulting atmospheric soot that cooled the Earth for a long time. As a result, many seed plant groups became extinct, along with the dinosaurs. As the world recovered, flowering plants became dominant, triggering another major alteration in Earth's ecology and leading to modern life. Flowering plant diversification fostered the evolutionary radiation of birds and mammals, among them the primate ancestors of humans.

21.2 | DNA and fossils help trace the history of plants

Plant scientists have used many tools to deduce the evolutionary history of plants. Among these, DNA sequences and fossils have been particularly useful. DNA, fossils, and other characteristics are used to construct plant **phylogenies**—branching, tree-shaped diagrams showing relationships and patterns of ancestry (Figure 21.3). Phylogenetic information can be used to determine when and in which order adaptations to life on land evolved.

DNA data reveal relationships of modern plants

Plant scientists can extract DNA from nuclei, mitochondria, and chloroplasts of modern plants and even from some fossils. Particular DNA sequences (genes) that code for proteins or nucleic acids can then be amplified—copied many times (Chapter 15). DNA sequences can be compared to those from other organisms. The genes commonly used to reconstruct plant evolutionary history include those that encode tubulin—the protein that makes up microtubules; rubisco—the enzyme that plants and algae use to incorporate CO_2 into organic carbon; phytochrome—the pigment used by plants to detect light and day length; and the RNA in the small subunit of ribosomes—involved in protein synthesis. Plant scientists then use DNA sequence data (often along with structural or biochemical information) to reconstruct evolutionary relationships, a process carried out with specialized computer applications.

Land plants evolved from charophycean green algae

DNA and other cellular characteristics indicate that land-adapted plants—including all of the modern bryophytes, pteridophytes, gymnosperms, and angiosperms—form a **monophyletic** group. In other words, plants originated from a single common ancestor (see Figure 21.3). These data also show that particular green algae, known as charophyceans (Figure 21.4), are the modern protists most closely related to plants. The term **streptophytes** refers to a combination of embryophytes and charophyceans. It is important to realize that modern charophyceans are not actually the ancestors of plants. Instead, plants arose from an ancient ancestor that probably resembled modern charophycean green algae in many ways. Many generations of algae have come and gone since the common ancestor of the streptophytes lived.

DNA SCIENCE

EVOLUTION

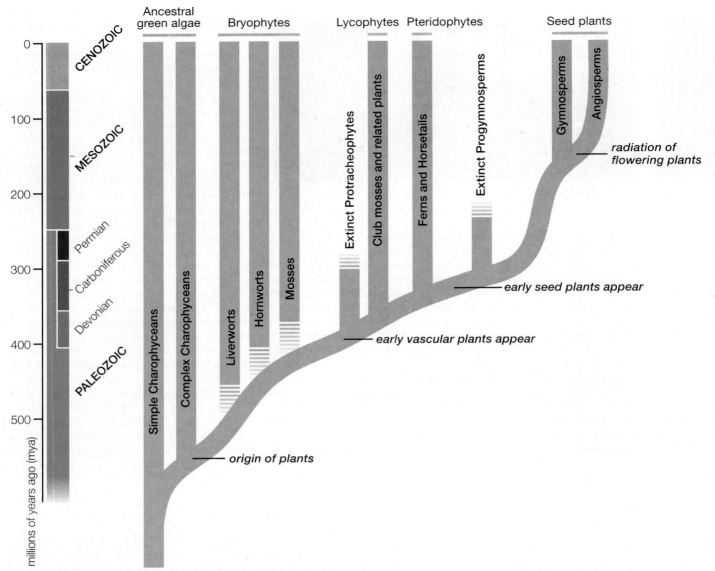

FIGURE 21.3 **Diagram of plant phylogeny on a geological timescale** Land plants evolved from ancestors that resembled modern, complex, charophycean green algae more than 500 million years ago (mya). Early vascular plants made an appearance around 400 mya, and early seed plants, about 365 mya. Because the exact times and order of branches indicated by the broken lines are uncertain, some changes in branching order may occur as additional data are obtained.

By comparing charophycean algae to plants, researchers have learned how some basic plant structures and processes evolved. For example, *Coleochaete* and *Chara* (Figure 21.4c,d) divide their cells very much as do land plants, indicating that the plant division system was inherited from algal ancestors. However, these algae lack particular carbohydrates (xyloglucans) that are the major component of gluey cell wall hemicelluloses (Chapter 4) in all groups of land plants. Thus, the evolutionary origin of plant cell division occurred in the water, whereas later changes in plant cell-wall biochemistry are correlated with the terrestrial lifestyle.

DNA evidence reveals the order in which the modern plant groups appeared DNA and other evidence suggests that bryophytes evolved later than charophycean

green algae but earlier than vascular plants and that seedless vascular plants arose prior to the origin of seed plants (see Figure 21.3). By knowing the order in which the groups evolved, we can compare particular features of plants within and between groups and thereby gain some insights into how they changed as plants adapted to life on land.

Seedless vascular plants include the familiar ferns and the less-familiar lycophytes, such as club mosses (see Figure 21.1b). DNA evidence indicates that lycophytes are the earliest-appearing modern vascular plants and that among bryophytes, mosses are probably most closely related to lycophytes. Bryophytes have unbranched sporophytes (see Figure 21.1a) and lack lignin-reinforced, water-conducting tissues. In contrast, lycophytes, like all other vascular plants,

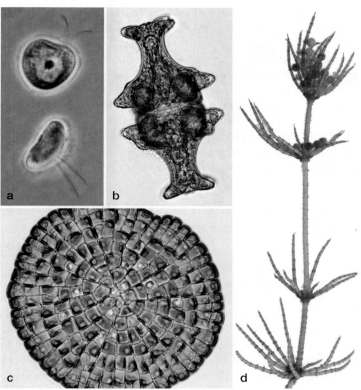

FIGURE 21.4 Representative charophyceans Simple charophycean green algae include (a) *Mesostigma* and (b) *Euastrum*, both of which are single-celled organisms. *Mesostigma* can move by use of its two flagella, but *Euastrum* lacks flagella. More complex, multicellular charophyceans include (c) *Coleochaete* and (d) *Chara*, for which this group of green algae is named. Sperm cells are produced within the orange structures.

possess both branched sporophytes and true xylem, defined by lignin-reinforced cells. Comparing the DNA of mosses and lycophytes may reveal the genetic basis for evolution of branched sporophytes and xylem, two extremely important plant traits.

Fossils reveal important events in the early evolutionary history of plants

Several major groups of plants that were important in global ecology and plant evolutionary history left no representatives in the modern plant world. The only way to know about these plants is from fossils, which are organic or mineralized plant remains or impressions of ancient plants left in soft materials that later hardened into rocks. **Paleobotany** is the study of plant fossils, which are very useful for estimating the ages of plant groups. Discoveries of new fossils that change our views about the ages of plants (and other organisms) are exciting events that are widely reported in the news media.

Although plants lack bony skeletons, which are a major source of animal fossils, plants produce tough polymers, such as lignin, sporopollenin, and cutin. These materials resist

microbial decay and thus favor the formation of fossils. The single most resistant biological material is **sporopollenin**, which coats the walls of plant spores. Sporopollenin helps spores retain precious water and resists attack by microbes. The toughness of sporopollenin explains the occurrence of plant spore fossils in rocks. **Lignin**, a tough, plasticlike material (Chapter 3) strengthens and waterproofs the walls of tracheids, vessels, and some other types of plant cells. **Cutin** is a polyester material that coats plant surfaces, providing resistance to microbe attack. Lignin and cutin are largely responsible for the preservation of plant parts and entire plants as fossils (Figure 21.5).

Plant evolution involved four key events. The origin of land-adapted plants and the rise of vascular plants are covered in this chapter. The evolution of seed plants and the diversification of flowering plants are covered in Chapters 22 and 23. Fossils help scientists not only to determine when these events occurred but also to deduce the forces that influenced them.

The origin of land-adapted plants In comparison to the watery habitat of their green algal ancestors, the land environment offered early plants plentiful carbon dioxide and sunlight for photosynthesis, and there were few competitors or herbivores. It is not clear when the first land plants appeared. However, fossil spores indicate that by 460 million years ago, early plants had become abundant and widespread around the world. Fossil sporangia enclosing bryophyte-like spores suggest that early plants also possessed tissue-producing meristems and sporophytes by 440 million years ago. Major challenges to plants at this time were heat, desiccation, and damage by the Sun's ultraviolet radiation.

The rise of vascular plants A group of fossil plants that first appeared about 400 million years ago helps explain the origin and diversification of vascular plants. These plants, the protracheophytes (meaning "before vascular plants"), were like bryophytes in being relatively small and lacking

FIGURE 21.5 Plant fossil Compressed remains of a fern.

lignified vascular tissue. But unlike bryophytes, protracheophytes had branched sporophytes that were able to grow separately from parental gametophytes. The sporophyte's ability to branch allowed protracheophytes (and later vascular plants) to harvest more light energy through photosynthesis, reproduce more prolifically, and better survive attack by herbivores. If an animal bites off one or even several shoot tips of a vascular plant, other branches are able to continue growth and reproduction. But moss sporophytes, which have only a single growth tip, are more vulnerable. Protracheophytes are viewed as evolutionary intermediates between the body forms represented by modern bryophytes and vascular plants. But they were not better adapted than bryophytes, as judged by the fact that protracheophytes are all extinct, whereas bryophytes have been very successful up to and including modern times.

By 408–360 million years ago, there were many types of vascular plants that left an abundant fossil record but have since become extinct (see Figure 9.2). These earliest vascular plants had no leaves or roots, but they did have stems with a small central core of lignified tracheids for water and mineral conduction, a tough outer cuticle that provided protection from drying and UV damage, and epidermal stomata that allowed for gas exchange. These features suggest that early vascular plants had achieved the ability to maintain a stable internal water level—a general characteristic of modern vascular plants that allows them to continue photosynthesis, growth, and spore production even when water availability varies. In contrast, the water content of bryophyte bodies varies directly with that of the environment. When exposed to dry conditions, bryophytes become dormant. The metabolism of dry bryophytes remains suspended until environmental moisture levels become high enough to allow photosynthesis, growth, and reproduction to resume.

During the coal age, also known as the Carboniferous era (360–256 million years ago), there were vast forests of plants. Coal-age forests were dominated by giant lycophytes, ferns, and other relatives of modern seedless vascular plants. The first seed plants also appeared during this time. Coal-age plants were able to grow abundantly because the climate was very warm and moist during this period. Liquid water was abundant, facilitating plant reproduction by spores and flagellate sperm. Coal-age plants mostly grew in water-filled swamps, which were globally widespread. Dead plants that fell into low-oxygen swamp bottoms did not decay completely, since many decay organisms require oxygen for respiration. Under these conditions, extensive coal deposits and a wide array of fossils were formed (see Essay 21.1).

ECOLOGY

21.3 | Early plant evolution illustrates the concept of descent with modification

Studies of many groups of organisms have revealed that evolution has built complex structures and processes onto the genetic foundations provided by simpler ancestors. For example, hemoglobin—the essential red molecule that ferries oxygen in our bloodstream—originated long ago in bacteria and has since been inherited and modified by protists, fungi, plants, and animals. This principle was first described by the father of evolutionary biology, Charles Darwin, as "descent with modification." It explains the diverse properties of modern living things, including the origin of features that define plants and the major plant groups. Examples of seedless plant features that arose by descent with modification are (1) the sporophyte generation and (2) large leaves with branched vascular systems.

EVOLUTION

The plant sporophyte probably originated by delaying zygote meiosis

How did plants acquire their distinctive sporophytes? Most likely by modifying traits inherited from their green algal ancestors. Although modern charophycean green algae lack sporophytes, some species retain and appear to nourish zygotes on the maternal bodies, as do plants. This behavior is advantageous for algae growing in nearshore pond or lake waters where these algae typically live, because it tends to keep progeny in well-illuminated shallow water where they are better able to survive and grow. Provision of nutrients to zygotes allows them to store materials that supply the needs of development in spring, much as the storage lipids, proteins, and carbohydrates present in seeds aid in seedling germination and growth.

A reasonable explanation for the origin of plant embryos and sporophytes is that an ancestral charophycean green alga acquired a mutation that resulted in delay of zygote meiosis until one or more mitotic divisions had already occurred. This mutation would result in a small, but multicellular, diploid sporophyte and would increase the number of cells that could undergo meiosis (Figure 21.6). Food provided to the young sporophyte by its parent gametophyte would help fuel increase in sporophyte size. The greater the number of meiotic cells in a sporophyte, the greater the number of spores that can be produced per zygote. This increase in the number of sporophyte cells and spores would maximize the output of sexual reproduction in terrestrial environments, where water shortage decreases the chance that swimming sperm can reach and fertilize eggs. Evidence for the adaptive value of sporophytes and their increased size is the fact that sporophytes are the larger, dominant generations of all modern vascular plants.

Leaves of ferns arose from branched stem systems

Fossils reveal that the earliest vascular plants lacked leaves. Although small, leaflike structures are produced by some bryophytes, these lack lignin-reinforced vascular tissue and thus are not true leaves. Lycophytes have true leaves, but these contain only a single, unbranched vein and are known as **lycophylls** (Figure 21.7a). Because the leaves of modern lycophytes are small, their leaves are also known as

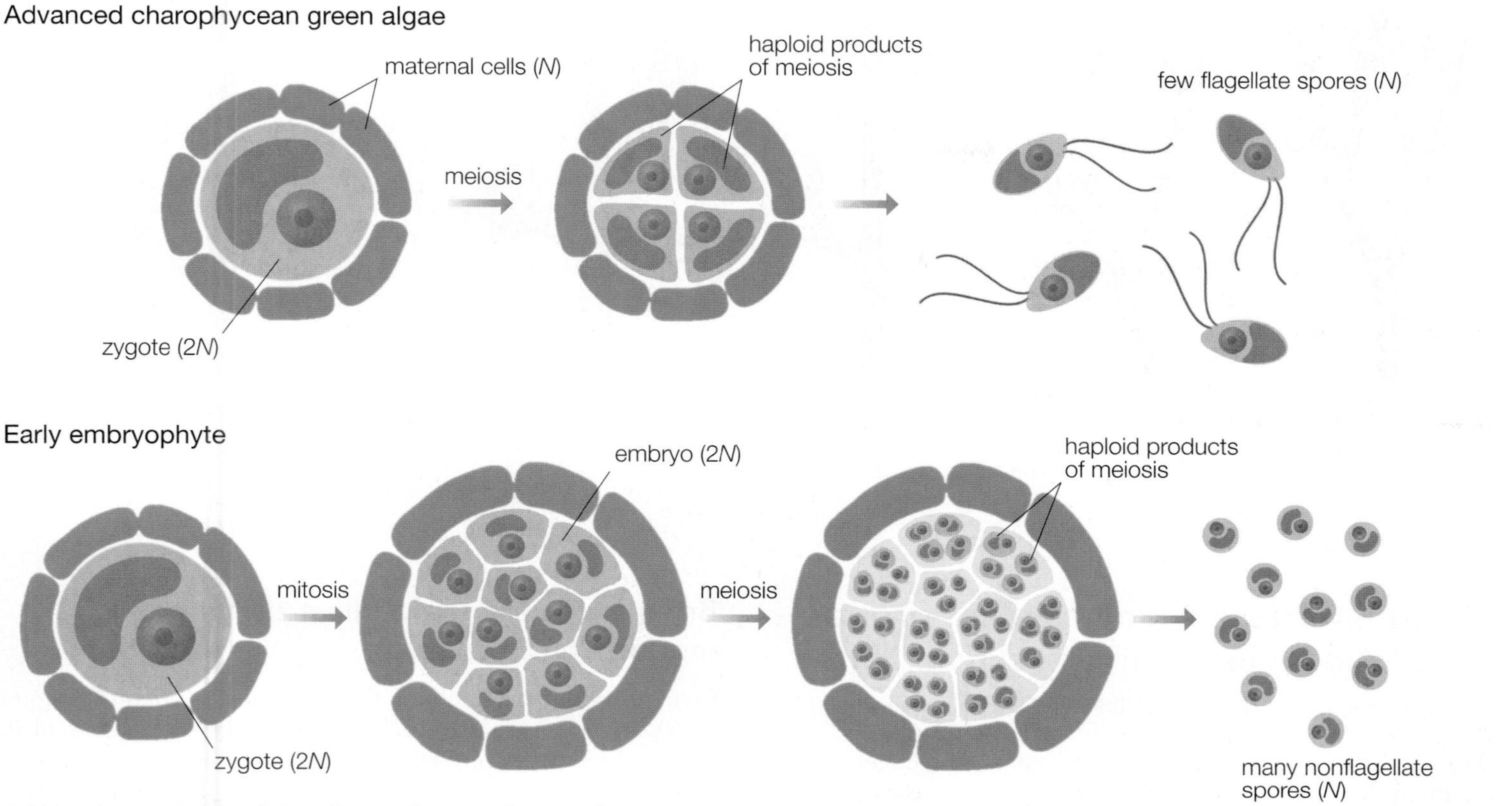

Advanced charophycean green algae

maternal cells (N)

haploid products
of meiosis

few flagellate spores (N)

meiosis

zygote (2N)

Early embryophyte

embryo (2N)

haploid products
of meiosis

zygote (2N)

mitosis

meiosis

many nonflagellate
spores (N)

FIGURE 21.6   Origin of the plant embryo and sporophyte In advanced charophycean green algae, the zygote is retained on the parental haploid thallus, from which it may receive nourishment. The zygote is the only diploid cell in the life cycle and undergoes meiosis, forming haploid flagellate spores (in *Coleochaete*). A simple hypothetical scenario of plant embryo/sporophyte evolution is that meiosis was delayed until after some mitotic divisions had first taken place. Land plant spores lost flagella and acquired a sporopollenin coat that provided protection during dispersal in air.

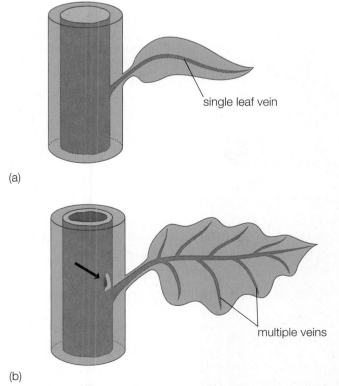

single leaf vein

(a)

multiple veins

(b)

FIGURE 21.7   Lycophylls versus euphylls (a) Lycophylls have only a single vascular strand, whereas euphylls (b) contain a more extensive network of veins. Stems with euphylls are characterized by the presence of a leaf gap (arrow).

microphylls (meaning "small leaves"). In contrast, leaves of most ferns have extensively branched vascular systems, as do those of seed plants, allowing such leaves to become larger (Figure 21.7b). Since leaves are the "solar collection panels" of the plant world, large size confers the ability to collect more sunlight. The leaves of ferns and seed plants are known as **euphylls** (meaning "true leaves") or megaphylls (meaning "large leaves"). Euphylls are defined by the occurrence of a leaf gap, a break in the vascular tissue in the stem, above the point where a strand of vascular tissue or "leaf trace" emerges into the leaf. Plants with microphylls do not have leaf gaps.

Study of fossils suggests that fern euphylls arose from leafless, branched stem systems by a series of steps. First, branch systems became flattened, then one branch assumed the role of the main axis, and finally the spaces between the branches became filled with green tissue (Figure 21.8). But plant evolutionary biologists also suspect that euphylls arose several times, and it is unclear whether the leaves of seed plants arose in the same way as those of ferns.

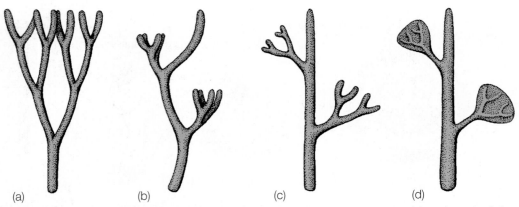

FIGURE 21.8 Leaf evolution Fossils suggest that fern euphylls likely evolved from branched shoots (a). Stages in leaf evolution may have included (b) dominance by one branch (overtopping), (c) flattening (planation), and (d) development of leaf tissue between branches (webbing).

21.4 | Modern seedless plant groups include bryophytes, lycophytes, and pteridophytes

How can the modern seedless plant groups be recognized and distinguished? A survey of the characteristics of each group is useful in identifying these different plants in nature and understanding their ecological roles.

Bryophytes are the earliest-divergent modern land plants

Living bryophytes include three groups of plants (Figure 21.9). **Liverworts** (also known as hepatics) are classified in the phylum (division) Hepatophyta. **Hornworts** are classified as Anthocerophyta, and **mosses** as Bryophyta. Most bryophytes grow on soil, rocks, or the surfaces of plants, although some mosses and liverworts live in freshwater lakes,

FIGURE 21.9 Bryophytes (a) *Pallavicinia*, a flat-bodied "thalloid" or "thallose" liverwort, (b) a "leafy" liverwort, (c) the hornwort *Anthoceros*, showing horn-shaped sporophytes growing from a flat gametophyte body, (d) the "leafy" gametophyte of *Polytrichum*, a moss.

streams, or ponds. Liverworts and hornworts are small herbaceous (nonwoody) plants named for their liverlike or hornlike shapes. The suffix -*wort* is an old Anglo-Saxon word meaning "herb." The short, leafy bodies of mosses often form extensive short turfs or mats that are familiar sights in forests or shady lawns. Several types of organisms commonly described as "mosses" are not actually bryophytes; these include Irish moss (a red seaweed), reindeer moss (a lichen), club mosses (lycophytes), and Spanish moss (a flowering plant).

Bryophyte bodies are simpler than those of vascular plants

Bryophyte gametophytes are only one or a few cells thick (Figure 21.10), allowing them to take up water and minerals directly from the soil, air, or water. Although some bryophytes produce specialized tissues that function in water and solute conduction, their walls lack the lignin coating that is characteristic of vascular plant tracheids and vessels. The absence of lignified vascular tissues also limits the height of bryophytes, most of which are only a few centimeters tall. Thus, most bryophytes grow close to the ground, anchored by delicate, colorless **rhizoids** (see Figure 21.10). Rhizoids are long, tubular, single cells (liverworts and hornworts) or multicellular filaments (mosses). Unlike roots, rhizoids are not composed of tissues; they lack specialized conducting cells, and they probably do not play a primary role in water and mineral absorption.

The gametophytes of hornworts and some liverworts are flattened (see Figure 21.9a,c) and grow close to the ground—an adaptation useful in absorbing water directly into their photosynthetic tissues. Moss gametophytes and those of some liverworts are described as "leafy," because they have stemlike structures that bear many thin, leaflike appendages (see Figure 21.9b,d). These are not true stems and leaves, however, because they lack lignin-coated conducting cells. Moss "leaves" are usually only one cell thick and (with some exceptions) lack a cuticle—features that permit water and minerals to be easily absorbed from the environment. Recall

that the plant cuticle consists of a surface layer of protective polyester cutin, coated by waxes that help reduce water loss from plants (see Chapter 8).

Bryophyte sporophytes are much simpler than those of vascular plants, and they are much less noticeable than the bryophyte gametophyte. Bryophyte sporophytes consist of (1) a basal foot that is always embedded in the tissues of the maternal gametophyte, from which materials are obtained; (2) a seta or stalk (not present in all bryophytes), which may raise the sporophyte into the air for greater spore dispersal efficiency; and (3) a capsule, or sporangium, which produces and disperses the spores (Figure 21.11). Bryophyte sporophytes are not branched and never grow independently of the gametophyte.

Bryophytes reproduce by wind-dispersed spores, breakage, or asexual structures

Bryophytes can reproduce and disperse themselves in several ways. If bryophyte spores are dispersed to favorable habitats, they may germinate and grow into gametophytes by repeated mitosis

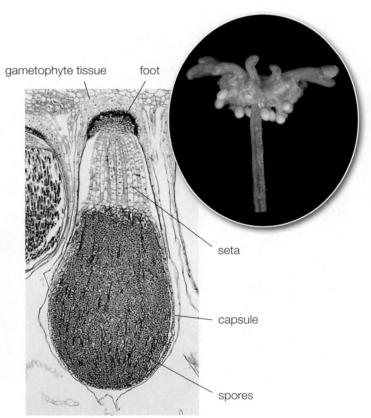

FIGURE 21.11   Liverwort sporophyte (of *Marchantia polymorpha*) The three main parts of the sporophyte body are the foot, seta, and capsule. The foot is an absorptive organ through which food, water, and minerals are supplied by the parental gametophyte. The seta helps extend the capsule (a sporangium) into the air, for more effective dispersal of the many spores (dark dots). The foot is shown at the top of the photograph because in *Marchantia* the sporophytes, with their bright yellow spores, hang downward from the umbrella-shaped gametophytic structure on which they are borne (inset).

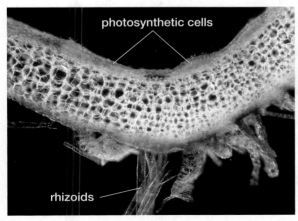

FIGURE 21.10   Liverwort body showing rhizoids Photosynthesis in this liverwort is confined to the upper few cell layers. Rhizoids—thin, colorless extensions from epidermal cells on the bottom surface of the liverwort—help anchor the plant to the substrate.

(Figure 21.12). Germinating moss spores typically produce a mass of green, branched, one-cell-thick filaments, each known as a **protonema** (meaning "first threads"), which can sometimes be mistaken for algae. Protonemata (plural of protonema) have a high surface area, which enhances the absorption of water and minerals. When sufficient resources are available, a protonema produces buds with a tissue-producing meristem. This meristem generates the mature, gamete-producing structure, which is known as a **gametophore** (gamete bearer). Together, the protonema and gametophore make up the gametophyte body of a moss.

Mature bryophytes produce gametes in specialized structures known as **gametangia**. Eggs are produced singly in vase-shaped **archegonia**, and multitudinous sperm are produced in elongate **antheridia**, both of which have an enclosing layer of tissue known as the *jacket* or *sterile jacket layer* (Figure 21.13). Bryophyte gametophytes can produce many archegonia and antheridia; in mosses, these are typically borne on separate male and female plants. Sperm are released into water films, in which they swim toward eggs, passing down the openings of archegonia in response to chemical attractants. Eggs are not released but, rather, remain within archegonia. Gamete fusion

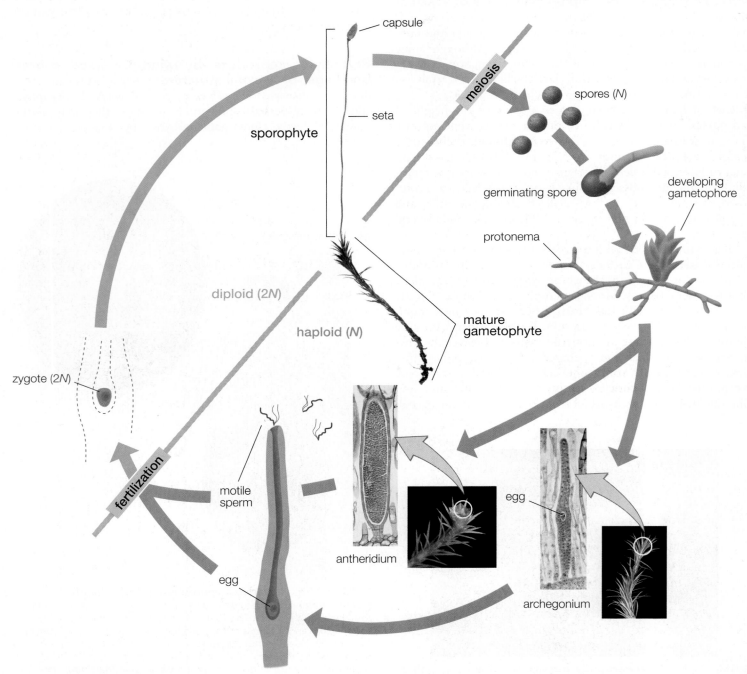

FIGURE 21.12 Bryophyte life cycle The life cycle of a moss such as *Polytrichum* is shown. Note that the sporophyte is small and attached to the larger gametophyte, which represents the dominant stage in the life cycle.

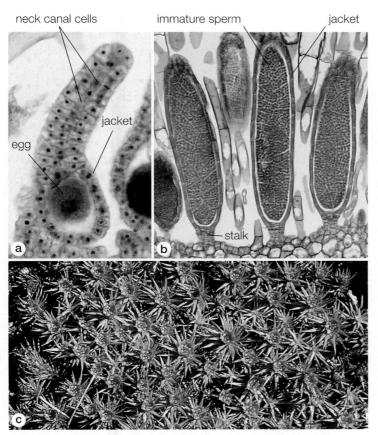

FIGURE 21.13 **Bryophyte archegonia and antheridia**
(a) Archegonium of the liverwort *Marchantia*, as viewed with a light microscope. The egg lies in the swollen basal portion of the archegonium, surrounded by the protective jacket tissue. The narrow portion is the neck. The central portion of the neck will become the neck canal (after the "neck canal cells" break down); the sperm swim through the neck canal to reach the egg. (b) Antheridia of the moss *Mnium*. Note the surrounding jacket of cells and the short supporting stalks. The antheridia are filled with numerous sperm that will be discharged through a hole in the jacket. Sperm require liquid water as a medium to swim to eggs. (c) Antheridial "heads" of the moss *Polytrichum* are visible with the unaided eye. Numerous antheridia are clustered to form each head.

results in the formation of zygotes, which also remain within the archegonia during their development into young sporophytes (embryos). As in all plants, organic food, water, and minerals are transported from the parental gametophyte to embryos.

Among bryophytes, liverworts have the least conspicuous sporophyte; hornwort and moss sporophytes are larger and more complex. Hornwort sporophytes superficially resemble grass blades and likewise grow from their bases and possess a cuticle (see Figure 21.9c). Sporophytes of hornworts and mosses possess epidermal stomata similar to those of vascular plants. The moss capsule (sporangium) is the site of meiosis and spore production. A single capsule can generate as many as 50 million spores. A hatlike covering composed of gametophyte tissue, the calyptra, protects immature moss capsules (Figure 21.14a). The calyptra and the upper lidlike portion of the capsule (the "operculum," Figure 21.14b) are lost when

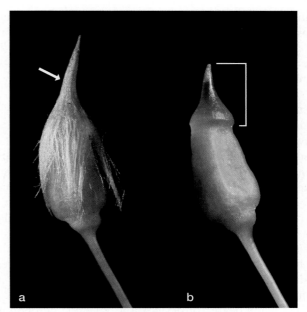

FIGURE 21.14 **Calyptra of the moss *Polytrichum*** (a) Calyptra (arrow) covering an immature capsule. The capsule on the right has had the calyptra removed, revealing the lidlike operculum (indicated by the bracket). Both the calyptra and operculum fall off when the capsule and spores are mature.

the capsule is ready to release spores. This exposes a series of teeth-shaped structures called a **peristome**, which is specialized to enhance spore discharge (Figure 21.15). Thanks to the peristome, spores are gradually sprinkled from the capsule, rather than dumped out all at once. This allows mosses to take advantage of periodic wind gusts, which can carry spores long

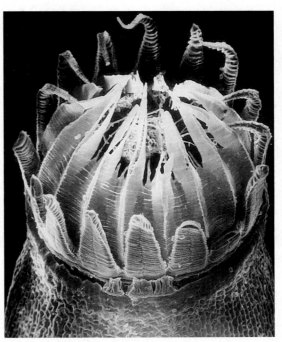

FIGURE 21.15 **Peristome of the moss *Brachythecium rivulare* as seen with the scanning electron microscope** The complex sporangial ornamentation helps disperse spores gradually into the wind.

distances, even between continents. Many bryophytes also, or exclusively, reproduce asexually by production of small tissue pieces including those known as **gemmae** (Figure 21.16). These pieces break off and are dispersed by wind or water, splashing to new habitats where they may grow into adult plants.

Mosses are diverse, ecologically significant, and economically useful

There are only about 100 species of hornworts, and perhaps 6,500 liverwort species, but mosses include some 12,000 species. Mosses are widely distributed around the world and are particularly common and diverse in alpine, boreal, temperate, and tropical forests. Some mosses inhabit stressful high-altitude mountaintops, Arctic and Antarctic tundra, and deserts—extreme habitats similar to those likely experienced by the earliest plants. Mosses are able to exist in very cold or dry habitats because many are able to lose most of their body water without dying, then readily rehydrate and reactivate their cells when moisture again becomes available. In contrast, few other plants can recover from the same degree of desiccation. Phenolic compounds in moss cell walls absorb damaging levels of ultraviolet light present in deserts or at high altitudes and latitudes. These compounds also likely protect mosses against attack by disease microbes.

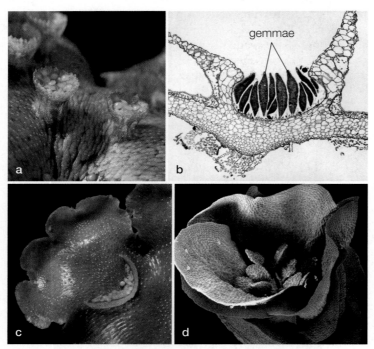

FIGURE 21.16 Bryophyte asexual structures—gemmae
(a) Numerous small, green, disk-shaped gemmae are produced in cups on the surface of the liverwort *Marchantia*. When these are dispersed, each can grow into a new liverwort gametophyte that is genetically identical to its parent plant. The dots on the surface of the plant are pores, through which gas exchange can take place. (b) A section through a *Marchantia* gemma cup with darker blue-stained gemmae, viewed with the light microscope. (c) Gemmae in the lunate (crescent-moon-shaped) cups of the liverwort *Lunularia*. (d) Scanning electron microscope view of disk-shaped gemmae of the moss *Tetraphis*.

Sphagnum is a wetland moss (Figure 21.17) that forms extensive deposits of incompletely decayed organic material known as peat. *Sphagnum* is known as peat moss, and boggy regions dominated by this moss are known as peat bogs or peatlands. *Sphagnum* and the peat formed from it do not readily decay because the moss's cell walls are impregnated with resistant phenolic compounds. Living *Sphagnum* moss excretes acidic and phenolic compounds that may help prevent microbial attack. The low temperatures and nutrient levels typical of peat bogs also inhibit microbial decay activity. Most of the world's peatlands are located in the northern hemisphere and occupy about 3% of the Earth's land surface. These northern peatlands were formed about 10,000 years ago. An estimated 400 gigatons (million tons) of organic carbon are stored in the world's peatlands, and they are also major sources of methane. For these reasons, peatlands play very important roles in Earth's global carbon cycle. As carbon reservoirs, peatlands play an important role in stabilizing the Earth's atmospheric carbon dioxide levels and climate. When carbon dioxide levels drop, moss growth is slowed, with the result that less carbon dioxide is converted to organic peat. The climate warms a bit because carbon dioxide is a "greenhouse gas"—it absorbs heat, causing atmospheric warming as in a greenhouse. When carbon dioxide levels rise, the mosses grow more abundantly, and more peat is stored, causing a drop in carbon dioxide levels and climate cooling. Variations in moss growth rate thus help keep climate from swinging from one extreme to the other.

Sphagnum mosses, which emit oxygen as a product of photosynthesis, harbor oxygen-loving bacteria that consume methane. Produced by different, submerged bacteria, methane bubbles through the moss layer of peatlands. The methane-destroying bacteria thus reduce the amount of methane that reaches the atmosphere. Since methane is another "greenhouse gas," the moss-associated bacteria help to prevent excess climate warming. Climate scientists are concerned that global warming caused by burning fossil fuels might speed the drying and oxidation of northern peatlands. This would release huge amounts of carbon dioxide into the atmosphere. Even if reduction in the area occupied by peatlands caused methane emissions to also decline, climate warming could accelerate.

In the past, *Sphagnum* moss was widely used by aboriginal people for diapers and during wartime as a naturally antiseptic packing material for wounds. Today, it is widely harvested for use as fuel, soil conditioner, and packing plant roots during shipment. The usefulness of this moss is based on the presence of large, dead, empty cells (Figure 21.17d), which allow dry moss to absorb 20 times its dry weight in water. Ecologists have expressed concern that overharvesting *Sphagnum* moss may bring about environmental damage, due to the loss or reduction of its beneficial ecological effects.

Many types of mosses grow best on mineral-poor soil and can tolerate shade. Thus, mosses can be conspicuous in poor lawns or in shady areas that are bare of other vegetation. Some people find lawn mosses objectionable and try to eliminate them by using chemical treatments. However, bryophytes are not harmful to people or pets, and their delicate rhizoids do

ECOLOGY

FIGURE 21.17   *Sphagnum* moss (a) Peat bog, showing the rim of peat moss surrounding the open water. (b) The top of a *Sphagnum* moss plant, showing "leafy" branches. (c) New green growth of the moss extends from older, brown tissue, thus forming large hummocks that may rise a meter or so above the water table. (d) Microscopic views of *Sphagnum* "leaves," showing the thin, green, photosynthetic cells (arrowheads point out the chloroplasts in the higher magnification inset) and larger hollow cells with pores that allow entry and storage of water. This structure accounts for the useful water-holding ability of peat moss, which is often harvested and sold in garden stores for soil conditioning. (e) Sporophytes of *Sphagnum* are small brown-black structures when mature. In the case of *Sphagnum*, the greenish stalk that supports the sporophyte belongs to the gametophyte generation. The operculum is ejected with an audible pop, releasing the spores. The tiny spores are transported by wind and may grow into new moss plants.

not compete for nutrients with the deeper-growing roots of vascular plants. Unless the soil is rich and sunny enough to support the growth of vascular plants, mosses will eventually return to the treated areas. It is best to leave mosses alone and learn to appreciate these hardy pioneers or encourage their growth in moss gardens. Mosses are widely used in Asian-style gardens to add interesting texture and colors to the landscape composition.

Lycophytes and pteridophytes are modern phyla of seedless vascular plants

Seedless vascular plants include the lycophytes such as club mosses (see Figure 21.1b) (formally the Lycophytina), and the pteridophytes (formally Moniliformes). Pteridophytes include

ferns, horsetails, and whisk ferns. Like gymnosperms and flowering plants, lycophytes and pteridophytes have a dominant, branched sporophyte generation with food transport tissue (phloem) as well as water-conducting xylem tissues having lignified cells. Unlike gymnosperms and angiosperms, lycophytes and pteridophytes do not produce seeds.

Lycophytes have simple leaves   Modern lycophytes are small relics of a far more magnificent past. During the Carboniferous period, 40-meter-tall tree lycophytes coexisted with small herbaceous lycophytes. The tree lycophytes thrived for millions of years, becoming extinct when the climate cooled at the beginning of the Permian period (see Essay 21.1). But some of the small lycophytes survived and gave rise to about 1,000 modern species.

Many lycophytes are tropical plants that grow in the branches of trees; others grow on temperate forest floors. Lycophytes that grow in trees often have stems that hang downward (Figure 21.18a); those that grow on the ground have upright stems as well as horizontal stems that grow along the surface with roots extending downward from them (Figure 21.18b). Stems bear many small green leaves with a single unbranched vein. Recall that such leaves are lycophylls or microphylls. Sporangia occur on sporophylls, which are often clustered to form club-shaped **cones**, also known as stobili (Figure 21.19). When mature, the oil-rich, flammable spores are released for air dispersal. In the early days of photography, before flash units, photographers would ignite lycophyte spores to produce a flash of light. In some lycophyte species, all of the spores are of the same size; in others, spores of two sizes are produced. Lycophyte spores develop into inconspicuous haploid gametophytes, on which antheridia and archegonia are produced (Figure 21.20). Depending upon the species, these tiny gametophytes may be green and grow aboveground, or they may be colorless, nonphotosynthetic

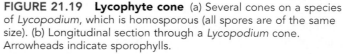

FIGURE 21.19 Lycophyte cone (a) Several cones on a species of *Lycopodium*, which is homosporous (all spores are of the same size). (b) Longitudinal section through a *Lycopodium* cone. Arrowheads indicate sporophylls.

FIGURE 21.18 Tropical and temperate lycophytes (a) A tropical species that hangs from trees. (b) A temperate species that grows on the forest floor.

plants that may live underground for several years, nurtured by symbiotic fungi.

Most pteridophytes have leaves with branched vascular systems There are more than 12,000 species of pteridophytes, which include ferns, horsetails, and whisk ferns (Figure 21.21). Most of these have large leaves with branched vascular systems, but the whisk ferns and horsetails are exceptions. The whisk fern, whose scientific name is *Psilotum*, is a

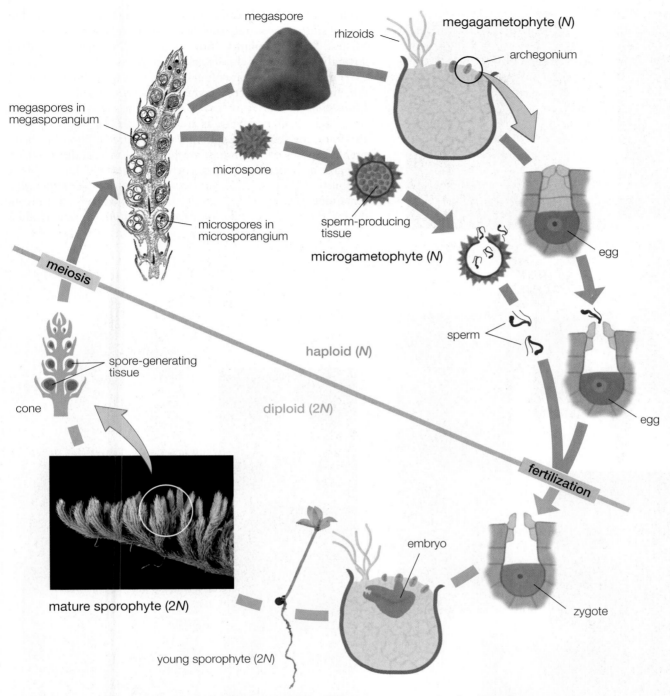

megaspore

megagametophyte (*N*)

rhizoids

archegonium

megaspores in megasporangium

microspore

microspores in microsporangium

sperm-producing tissue

microgametophyte (*N*)

egg

meiosis

haploid (*N*)

sperm

spore-generating tissue

cone

diploid (2*N*)

egg

fertilization

mature sporophyte (2*N*)

embryo

zygote

young sporophyte (2*N*)

FIGURE 21.20   Life history of a lycophyte *Selaginella* is heterosporous, meaning that it produces two types of spores within its cones, each able to produce a gametophyte. The small microspores produce male gametophytes (microgametophytes), which form an antheridium within the microspore's wall. When mature, the sperm are released from the antheridium. The large megaspores produce female gametophytes (megagametophytes), which develop within the megaspore wall. The megaspore wall opens up at one end to expose the archegonia so that fertilization can occur. The zygote develops into an embryo, which is surrounded and nourished by gametophytic tissue in the megaspore.

tropical plant that lacks both leaves and roots (Figure 21.21e). Upward-growing *Psilotum* stems are the green, photosynthetic organs; these emerge from a nongreen, horizontal stem (rhizome). Like several other ferns, the whisk fern lives on the surfaces of tropical plants, such as palms. In such a habitat,

roots are not essential. Yellow sporangia may be visible on the bifurcating stems. *Psilotum*'s tan, heterotrophic gametophytes grow underground, in association with fungi. They bear antheridia that produce multiflagellate sperm, as well as archegonia that produce an egg and foster the embryo resulting

from fertilization. Because it lacks leaves and roots, in the past, *Psilotum* was regarded as a modern descendent of the earliest vascular plants. But modern DNA sequence data strongly suggest it is closely related to ferns.

Horsetails, which are often brushy, thus resembling a horse's tail, are classified as the genus *Equisetum* (Figures 21.21d). *Equisetum* has very tiny leaves, though these are euphylls with branched vascular systems, as are fern leaves. As is the case with *Psilotum*, the stems of *Equisetum* are the green, photosynthetic organs. Spores of *Equisetum* are produced in sporangia, which are grouped together to form a cone (formally known as a strobilis) (Figure 21.22). *Equisetum* is the only representative remaining of a once large group that included tree-sized plants during the coal age. Like modern *Equisetum*, ancient treelike

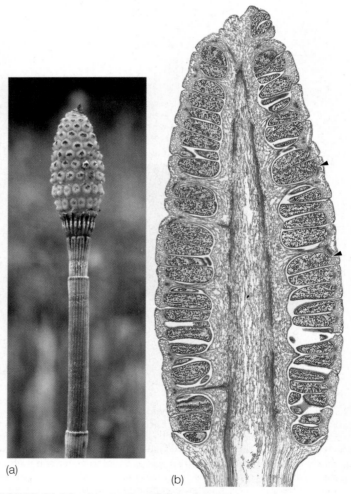

FIGURE 21.21 **Examples of pteridophytes** (a) Bird's-nest fern *(Asplenium)*. (b) Staghorn fern *(Platycerium)*. (c) Maidenhair fern *(Adiantum)*. (d) A branched species of *Equisetum*. (e) The whisk fern *(Psilotum)* is unusual in lacking both leaves and roots. The inset shows a close-up view of an immature (yellow) and mature (tan) sporangium.

FIGURE 21.22 **A cone (strobilis) of *Equisetum*** (a) Cones are collections of sporophylls (leaves bearing sporangia) that occur on green shoots in some species (such as the unbranched species shown here). Also known as scouring rushes, *Equisetum* plants were in earlier times used to clean pots and pans because of the presence of abrasive silica particles in their ribbed stems. (b) Longitudinal section through a cone. Arrowheads indicate sporangia.

relatives had leaves and branches emerging in circles (whorls) at stem nodes. The tree forms became extinct when the climate changed at the end of the coal age (see Essay 21.1). Today's *Equisetum* is herbaceous, although a giant species found along streamsides in Chile grows taller than an adult human. Gametophytes of *Equisetum* are green and grow aboveground. The sex of *Equisetum* gametophytes depends on the amount of sugar present in the environment. If sugar is present, gametophytes bear antheridia (with sperm), or both antheridia and archegonia. But when sugar is not available, only archegonia (with eggs) are produced. In the past, whisk ferns, horsetails, and ferns were thought to be separate pteridophyte divisions (phyla). However, recent DNA evidence has revealed that these groups are actually very closely related.

Ferns are most diverse in the tropics, but there are many representatives in temperate forests, and some grow in arid habitats. Experts used to assume that fern diversity decreased after the seed plants evolved, but molecular and fossil evidence instead indicate that most modern ferns originated well after major groups of angiosperms appeared. There are about 12,000 species of modern ferns, so they are very successful today. Ferns often have horizontal rhizomes (underground stems) from which grow leaves with extensively branched vascular systems (euphylls). Fern leaves are commonly known as fronds and are often compound—divided into many leaflets. The fern frond grows by unfurling its coiled tip, the fiddlehead (Figure 21.23). Ferns produce clusters of sporangia known as **sori** (singular, sorus) on the backs of the green leaves or on specialized, nongreen leaves (sporophylls). Sori can be arranged in various patterns, such as parallel lines or dots—characteristics useful in fern identification (Figure 21.24). Fern sporangia can catapult spores several meters. Once airborne, spores can be blown by the wind far from their origin, and if they encounter suitable habitat, will grow into gametophytes. Most ferns have aboveground, green gametophytes (Figure 21.25); but some produce white or tan, half-inch-long, potato- or carrot-shaped subterranean gametophytes that rely on fungal partners for their nutrition.

Lycophyte and pteridophyte reproduction illustrates early steps toward seed evolution

Most lycophytes and pteridophytes produce a single type of spore that grows into a free-living, bisexual gametophyte. This situation is known as **homospory**, meaning that all of the spores are of approximately the same size and germinate into similar gametophytes. "Bisexual" means that both the egg-bearing archegonia and sperm-containing antheridia are present on the same gametophyte (see Figure 21.25). Homosporous plants have mechanisms for preventing self-fertilization, which would lead to inbreeding and a resultant reduction in genetic variation and loss of evolutionary potential. Many ferns are known to produce antheridia and archegonia at slightly different times, in order to increase the odds that sperm will fertilize eggs of another gametophyte whose genotype may differ.

As an alternative strategy to reduce inbreeding, some lycophytes, some ferns, and some plants known as progymnosperms

FIGURE 21.23   Fern fiddleheads Fiddleheads are young fern leaves that have yet to unroll to their final size. Fiddleheads of some fern species are considered delicacies.

(probably including the ancestors of seed plants) independently evolved **heterospory**. Heterosporic plants produce two types of spores that differ in size and in the gender of gametophytes (Figures 21.21 and 21.26). Among modern vascular plants, heterospory always involves **endosporic** ("inside the spore") gametophytes. This means that gametophytes of heterosporous vascular plants are produced within the confines of the spore wall and do not grow outside it. **Microspores** are tiny spores whose tough sporopollenin walls protect enclosed **male gametophytes**, which produce sperm. Pollen grains are microspores. **Megaspores** are larger spores that contain **female gametophytes**, in which are contained archegonia and eggs. Since male and female gametes are not produced on the same gametophyte, heterospory increases the chances for outbreeding. In addition, this strategy allows male gametophytes to be transported in air, which enables them to spread much greater distances than the water-dispersed sperm of homosporous plants. In order to fit within the confines of spore walls, gametophytes of heterosporous plants are necessarily very small, reflecting the previously mentioned trend toward the reduction of plant gametophytes.

Megaspores are produced in **megasporangia,** and microspores in **microsporangia**. These sporangia are produced on the surfaces of leaflike structures known as sporophylls—**megasporophylls** and **microsporophylls** (see Figure 21.17). Because the heterosporous ancestors of modern seed plants are all extinct, the study of living heterosporous ferns and lycophytes may be helpful in understanding the kinds of molecular changes that were required to initiate the process of seed evolution. Seed evolution was a particularly important step in plant evolution that is discussed more fully in the next chapter.

(a)

(b)

(c)

(d)

(e)

(f)

FIGURE 21.24 Fern sori Ferns produce clusters of sporangia, known as sori, on the lower surfaces of leaves (shown in black). People sometimes mistake sori for insect infestations or diseases, but they are a natural part of the fern reproductive process. Ferns can be distinguished in part by the patterns of their sori, which may occur in dots (a), lines (b, d), or solid patches (c). Some ferns produce sporangia on specific portions of leaves (e) or on special, nongreen leaves (f).

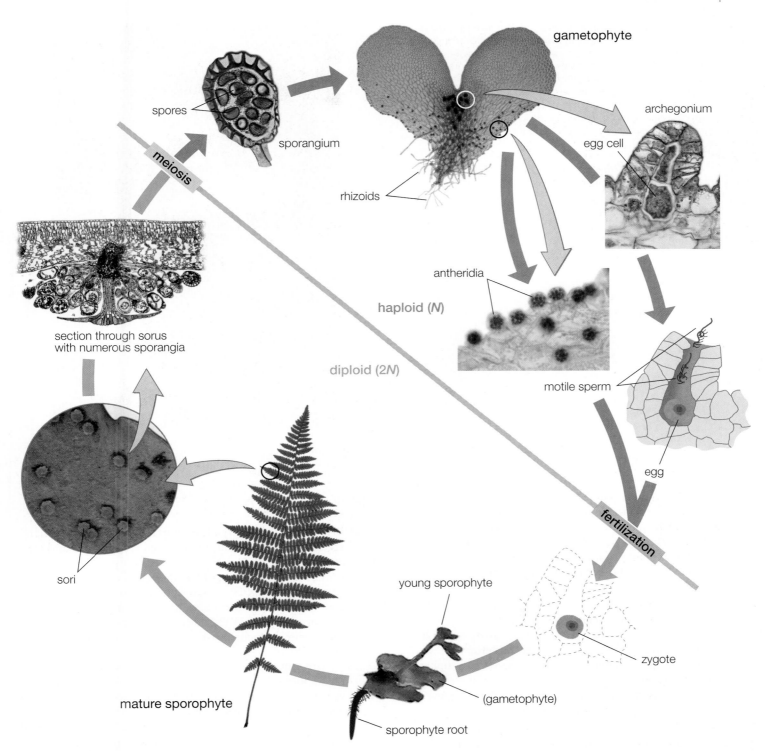

FIGURE 21.25   Life cycle of a homosporous fern As is true for all embryophytes, ferns have
an alternation of sporophyte and gametophyte generations. Thumbnail-size gametophytes produce eggs
in archegonia and sperm in antheridia. Fertilization produces zygotes that may develop into embryos,
then into young sporophytes. When mature, sporophytes produce haploid spores by meiosis. Spores
grow into gametophytes by mitosis.

FIGURE 21.26   A heterosporous fern (a) The fern *Regnellidium* has leaves with two leaflets, grows in moist places, and is one of the "water ferns." Special leaves produce and disperse into the water two types of spores, large megaspores and tiny microspores. (b). Megaspores contain female gametophytes, archegonia, and eggs, which are eventually exposed by a break in the spore wall, as in the case of *Selaginella*. Fertilization can then take place by flagellate sperm released from male gametophytes produced inside the microspores. The spore walls protect delicate gametophytes during their development.

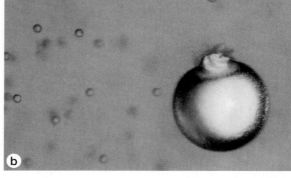

HIGHLIGHTS

21.1 Plants, also known as embryophytes, can be defined as multicellular photosynthetic organisms that are adapted to life on land. Plant characteristics include tissue-producing meristems, sporophytes, sporangia, and walled spores. Major living plant groups include bryophytes, lycophytes, pteridophytes, gymnosperms, and angiosperms (flowering plants).

21.2 Plant evolutionary history is deduced by studying fossils as well as the DNA and other characteristics of living plants. Plants evolved from ancestors that, if still alive, would be classified as charophycean green algae. Streptophytes are charophyceans plus embryophytes. Bryophytes, lycophytes, and pteridophytes are the nonseed plants. Extinct plant groups such as protracheophytes and progymnosperms also provide information about plant evolution.

21.3 Plant evolution illustrates Darwin's concept of "descent with modification." Sporophytes likely originated by delay of zygotic meiosis. Leaves of ferns likely evolved from branched vascular systems.

21.4 Bryophytes (liverworts, hornworts, and mosses) are characterized by dependent, unbranched sporophytes and the absence of lignified vascular tissue. Bryophytes lack true roots, stems, or leaves. *Sphagnum* moss is an important climate regulator. Lycophytes possess dominant branched sporophytes and have leaves with unbranched vascular systems (lycophylls). Pteridophytes (ferns) usually have leaves with branched vascular systems (euphylls), but lack seeds.

Ancient relatives of modern lycophytes and pteridophytes formed large forests and coal deposits during the coal age. These plants also reduced atmospheric carbon dioxide to historic low levels, precipitating climate change and the rise of seed plants. Some ferns and lycophytes are heterosporous and thus illustrate early stages in the evolution of seeds.

REVIEW QUESTIONS

1. What common characteristics set plants apart from other organisms, including the closely related green algae?

2. Describe several ways in which each of the following groups of land plants affect human life today: (a) bryophytes, (b) lycophytes, (c) pteridophytes.

3. How are DNA data used to construct phylogenies of modern plant groups? Which genes are commonly compared?

4. What is paleobotany? Which compounds are largely responsible for the formation of most plant fossils, and where do these compounds occur in plants?

5. What were the four key events in the evolution of land plants? Approximately when did these events occur in geologic time?

6. What is the difference between lycophylls and euphylls? Among the groups of modern land plants, which possess lycophylls and which possess euphylls? Why are bryophytes considered to lack true leaves?

7. Briefly summarize the major characteristics of bryophytes.

8. Briefly summarize the major characteristics of lycophytes.

9. Briefly summarize the major characteristics of pteridophytes.

APPLYING CONCEPTS

1. Both paleobotanists and plant molecular biologists often study relationships between plants. Paleobotanists glean this information from fossils, whereas plant molecular biologists obtain it from DNA sequences. What are some strengths and weaknesses of each approach?

2. In the latter part of the Paleozoic era, a group of plants called progymnosperms arose. These plants have been described as having "gymnospermous anatomy with pteridophytic reproduction." What does this mean exactly? Did these plants produce flowers? Seeds? Pollen? Did they exhibit secondary growth?

3. Mosses and ferns both require liquid water in order to reproduce. At what stage of the life cycle is water required?

4. This chapter begins with an overview of the process by which land plants evolved from aquatic, green algal (charophycean) predecessors. Name several features that land plants share with their charophycean relatives. What problems did (and do) land plants face that their aquatic, algal relatives did not? How have land plants solved these problems?

Gymnosperms, the First Seed Plants

Cycas revoluta

The native Chamorro people of Guam and Rota (Mariana Islands) suffer higher-than-usual rates of Alzheimer's and other dementias, which are brain deterioration diseases. These diseases are linked to their consumption of tortillas made with flour produced from the seeds of a native plant, the cycad *Cycas micronesica*. The leaves and seeds of other cycads are also toxic, causing livestock poisoning in tropical and subtropical regions. Cycads are members of the gymnosperms—plants that produce seeds, but not fruits. Birds, squirrels and even people eat harmless and nutritious seeds produced by other gymnosperms, such as pine and other conifer trees. Why are cycad seeds and leaves so unusually toxic?

In 2003, experts revealed that the brain damage observed among the Chamorro was caused by an amino acid that is not normally found in cellular proteins. This amino acid, known as BMAA (short for beta-methylamino-L-alanine), is actually produced by cyanobacteria that typically live within the roots of cycads, but not other gymnosperms. BMAA is transported from roots to other parts of the cycad plant, including seeds. Even when cycad seed flour is washed, in a traditional procedure designed to remove toxic materials, BMAA can remain in the flour. BMAA toxin has several harmful effects on brain cells. For example, during brain cell protein synthesis, erroneous incorporation of BMAA in place of normal amino acids causes proteins to stop elongating before they reach normal length. Such proteins are too short to function, causing brain cell malfunction. The experts who discovered BMAA's effects in the Chamorro have also detected BMAA in the brain tissues of Alzheimer's patients who do not live in Guam. Because cyanobacteria closely related to those in cycad roots occur widely in freshwaters, these experts suggest that BMAA-contaminated drinking water might be linked to dementia more generally. The discovery of BMAA in cycad seeds is exceptionally important because it has the potential to influence public health on a worldwide basis.

This chapter provides an overview of the gymnosperm plants, revealing many other ways in which this group of seed plants is relevant to modern human concerns and has influenced Earth's past.

Cycad ovules

22.1 | Gymnosperms include four modern groups and diverse extinct forms

Gymnosperms are plants that produce seeds, but not flowers or fruits. Most people are familiar with cone-bearing conifers, an economically and ecologically important group of modern gymnosperms. But some modern gymnosperms are less well known, and the importance of these and many groups of extinct gymnosperms is not widely appreciated.

Modern gymnosperms are classified into four major groups

Cycads, such as *Cycas micronesica* featured in the chapter opening, somewhat resemble palms, which are flowering plants. The common street-side tree *Ginkgo biloba* is the single modern representative of a once large group known primarily as fossils. Conifers, such as pine *(Pinus)*, are trees or shrubs whose seed cones are distinctively complex in structure. Cycads, *Ginkgo* relatives, and conifers were dominant vegetation in past ecosystems, including the age of dinosaurs. Gnetophytes, named for the tropical vine or tree *Gnetum*, includes two other genera of unusual gymnosperms, the desert shrub *Ephedra* and African *Welwitschia* (Figure 22.1).

Conifers are the dominant plants of mountain and high-latitude (boreal) forests (Chapter 27) and are important to people as sources of wood and paper pulp. Various gymnosperms are valued houseplants or components of landscape design. Though cycads can be quite toxic (as noted in the chapter opening), some gymnosperms are the sources of useful medicinal compounds. The cancer drug taxol, derived from the conifer *Taxus*, is one example (Chapter 7). Early settlers of the western United States used *Ephedra*, also known as Mormon tea, for medicinal purposes. The modern decongestant drug pseudephedrine, is based on the chemical structure of the alkaloid ephedrine obtained from *Ephedra*. Ephedrine has been featured in recent news media because of risks associated with its use in weight reduction and sports performance enhancement. In 2003, the U.S. Food and Drug Administration ruled that products containing ephedrine cannot be sold as food supplements, because of the health dangers.

Diverse groups of extinct gymnosperms were ecologically significant in the past

Extinct gymnosperms include (1) diverse groups of seed ferns, whose lacy leaves resembled those of ferns, but produced seeds (true ferns are seedless); (2) cordaites, trees with large, strap-shaped leaves; and (3) cycadeoids (also known as Bennettitales), whose bodies resembled those of cycads but whose reproductive structures somewhat resembled flowers. These gymnosperm groups are known only from their fossils (Figure 22.2), because no representatives are known to have survived to the present day. But they were important components of ancient forests, and some may be closely related to the ancestors of modern seed plants. They help us to understand how modern ecosystems and plant groups evolved.

FIGURE 22.1 Gymnosperms (a) Leaves and cones of a cycad (*Encephalartos*); (b) attractive, fan-shaped leaves and vile-smelling seeds (not fruits) of the tree *Ginkgo biloba*; (c) *Pinus* (pine); (d) *Gnetum* is a tropical vine; (e) *Ephedra* is a shrub native to North American deserts; (f) *Welwitschia*, a bizarre plant that grows only in coastal areas of the South African Namib desert.

22.2 | Gymnosperms produce ovules and seeds in cones, rather than within fruits

The word *gymnosperm* is derived from two Greek terms, *gymnos*—meaning "naked" (referring to the unclothed state of ancient Greek athletes)—and *sperm*—referring to seed. Gymnosperm plants produce seeds that are not enclosed within a fruit. Instead, gymnosperm seeds are produced on the surfaces of leaflike structures (sporophylls) that are aggregated to form **cones**. Mature gymnosperm seeds lie exposed to the environment into which they are dispersed. Thus, gymnosperm seeds are regarded

FIGURE 22.2 **An extinct gymnosperm, the seed fern *Scutum*** (a) Reconstruction of the tree; (b) reconstruction of leaf and ovulate cone with associated modified leaf (bract); (c) fossil leaves and cone.

as naked, in contrast to seeds of angiosperms (flowering plants), which are produced within fruits (Figure 22.3). Gymnosperms lack fruits because they do not produce flowers. More specifically, gymnosperms lack the flower structure—the ovary—from which angiosperm fruits develop. Several other distinctive reproductive features of flowering plants, described in Chapter 23, are also lacking from gymnosperms. Even so, gymnosperms have been very successful in the past, and continue to be so today, in large part because they produce seeds. Seeds develop from structures known as ovules, after egg cells contained within ovules have been fertilized. In order to appreciate the importance of seeds, we will first examine gymnosperm ovule and seed structure and development. We will also consider how gymnosperm ovules/seeds might have originated.

An ovule is an integument-covered megasporangium

The botanical definition of an ovule—an integumented megasporangium—is quite a mouthful! In order to comprehend this statement, we first review the meaning of the terms

heterospory and *megasporangium* and then define the new term *integuments*.

Heterosporous plants produce two types of spores. Recall that all seed plants and certain seedless plants are heterosporous (Chapter 21). Also recall that **heterospory** means the production of two different types of spores—very tiny microspores that will produce male gametophytes with sperm, and larger megaspores that produce female gametophytes with eggs. The lycophyte *Selaginella* and the water fern *Regnillidium* were illustrated in Chapter 21 as examples of heterosporous seedless plants. You may remember that all known heterosporous plants produce endosporic gametophytes, that is, gametophytes that mature within the confines of the microspore or megaspore wall. This means that heterosporous plants, including seed plants, necessarily have microscopic gametophytes. Tiny female gametophytes are less able than larger ones to feed the young sporophytes (embryos) that develop from fertilized eggs. Recall that land plant embryos are nutritionally dependent on their mother gametophytes, at least for some period of time during early development. So how do tiny seed plant female gametophytes manage to feed their

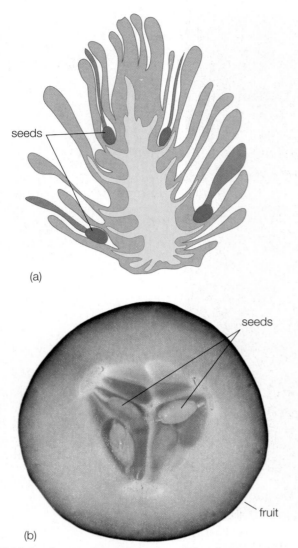

(a)

(b)

FIGURE 22.3 Comparison of a cone with a fruit (a) Diagram of a longitudinally sectioned pine cone with exposed seeds; (b) cross sectioned cucumber fruit with enclosed seeds.

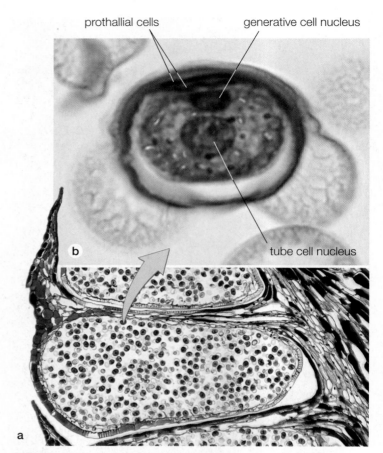

FIGURE 22.4 Gymnosperm microsporangium and pollen (a) Microsporangium of pine, containing many microspores; (b) mature pollen grain of pine.

embryos? The answer is that mother seed plant gametophytes get help from the previous generation.

Plant spores are produced within sporangia. A megasporangium is a structure in which megaspores arise by meiosis, while microspores originate by meiosis within the microsporangium. Megasporangia of seedless heterosporous plants (for example, *Selaginella* and *Regnillidium*) release their spores into the environment, where spore walls crack open. This releases sperm from microspores and allows fertilization of eggs within the megaspores. Seed plant microsporangia likewise release microspores for dispersal by wind or other means (Figure 22.4a). Seed plant microspores, with developing male gametophytes inside, are known as pollen grains (Figure 22.4b). But seed plant megaspores differ.

Seed plant megaspores are not dispersed. Seed plant megaspores are retained within the megasporangia that produced them. Furthermore, megasporangia remain attached to

their parent sporophyte plant. If you think of seed plant megaspores as behaving like young adult humans reluctant to leave the resources available at home, you can probably figure out the advantage of this behavior. Seed plant sporophytes are large, usually photosynthetic, and able to provide a continuous supply of organic nutrients, minerals, and water to the attached megaspores—and the tiny gametophytes growing within them. This fosters successful development of the next generation, the embryo sporophyte. Think of this process as working something like generous grandparents who supply their less wealthy children with financial assistance for the cherished grandchildren's expenses. Close association of the generations facilitates resource flow between them. That's the advantage of seed plant retention of megaspores and megasporangia on parental sporophytes. But what physically inhibits detachment of megasporangia and megaspores? The answer is an integument, which means a covering, produced by the parent sporophyte. Think of integuments as functioning like metaphoric "apron strings" that keep the young close to home.

In summary, seed plant megasporangia (also known as nucelli) are attached to the parental sporophyte by means of enclosing, leaflike integuments. The definition of seed plant ovules as integumented megasporangia should now seem comprehensible. Seed plant ovules may be covered by either one (Figure 22.5a) or two integuments. A gap in integument coverage, known as a micropyle, allows access for pollen, so that pollination can occur. In gymnosperms, pollination is the transfer (by wind or other means) of dispersed pollen to ovule-bearing cones. Often, a droplet of liquid will be secreted by the ovule. As this liquid dries, it pulls pollen grains into the micropyle, toward the egg. Under these circumstances, pollen may germinate, extending a pollen tube through part of the pollen wall. The pollen tube grows by adding wall material at its tip, a process known as tip growth. The pollen tube eventually releases sperm in the vicinity of egg cells (Figure 22.5b). If fertilization occurs, the ovule will start developing into a seed.

Seeds develop from ovules whose egg cells have been fertilized

Fertilization triggers seed development. During seed development, various parts of the ovule develop into seed structures. The integument, or part of it, develops into the protective, often hard and tough seed coat (Figure 22.5c). The gymnosperm female gametophyte stores nutrients obtained from the sporophyte, prior to fertilization. These nutrients fill the energy needs of the developing embryo, much as a developing mammalian fetus is fed by its mother's placenta. A gymnosperm seed's structure reflects the plant life cycle—alternation of gametophyte and sporophyte generations (Chapter 13). The seed coat, derived from parental

sporophyte tissues, is diploid. The female gametophyte within is haploid, as is typical of plant gametophytes. The embryo is the first stage of the next sporophyte generation, and thus diploid. Seed maturation typically involves dehydration, which helps seeds to survive a period of **dormancy**, when growth does not actively occur.

Ovule evolution illustrates Darwin's concept of descent with modification

Ovules (and the seeds that develop from them) originated by a sequence of changes to preexisting structures and processes. Thus, ovule and seed evolution provide examples of Darwin's concept that evolutionary change occurs by modification of inherited structures and processes. Recall that this principle was introduced in a previous discussion of seedless plant evolution (Chapter 21). The ancestors of modern seed plants likely belonged to an extinct group of plants known as the **progymnosperms** (Figure 22.6), which originated toward the end of the coal age (about 300 million years ago). *Archaeopteris* is one example; it was a tree-sized plant that formed ancient forests (Figure 22.7). Progymnosperms did not produce seeds, but they had secondary growth by means of a vascular cambium that produced wood (see Chapter 9) much like that of modern conifers. Progymnosperms such as *Archeopteris* also reproduced by means of two different types of spores (heterospory), which was a necessary precursor to the origin of seeds (see Chapter 21). The advantages of heterospory and other changes involved in seed evolution are described next.

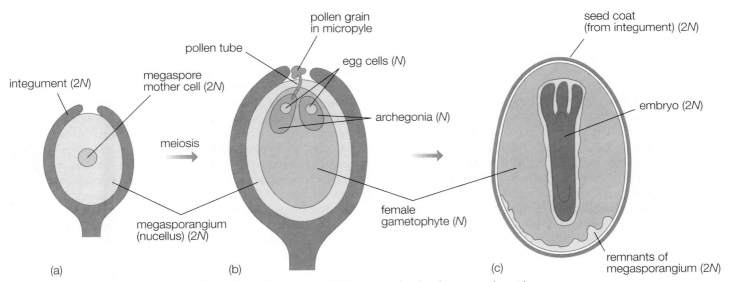

FIGURE 22.5   Comparison of ovule and seed structure (a) Diagram of a developing ovule, with integuments surrounding an enclosed megasporangium (nucellus) and megaspore mother cell, which will divide meiotically. All of these structures belong to the parental (2N) generation. (b) Mature ovule in which a haploid (N) female gametophyte with archegonia and eggs has developed from the single surviving megaspore, at the time of pollination. (c) Mature seed, which is encased by a diploid (2N) coat arising from the integuments. The haploid (N) female gametophyte serves as a food reserve for the diploid (2N) embryo, which develops from the fertilized egg.

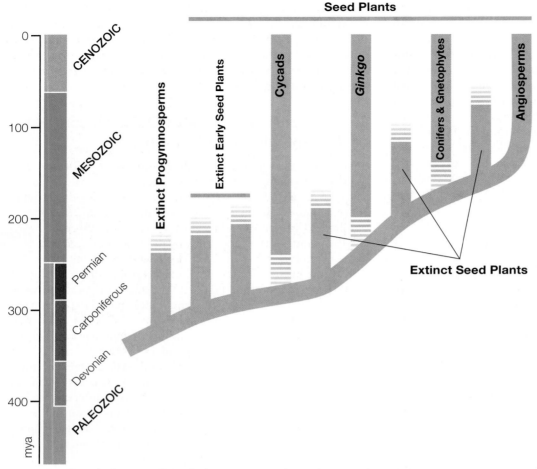

FIGURE 22.6 Phylogeny of seed plants on a geological timescale The relative times of origin and branching order of living and extinct gymnosperm groups and early-evolved angiosperms are still unclear.

An early step in the origin of seeds was a reproductive change from homospory (one kind of spore) to heterospory (production of two different kinds of spores with endosporic gametophytes). Since gametophytes of heterosporous plants produce only eggs or sperm (but not both), the chances of outcrossing are increased. Recall that outcrossing involves mating of parents having different genotypes (see Chapter 14) and contrasts with self-crossing (male and female gametes have the same genotype). Combining different parental genotypes provides progeny with greater genetic diversity, which contributes evolutionary flexibility to meet environmental challenges (see Chapter 13). The advantages of heterospory also include protection of vulnerable haploid gametophytes.

Another early change in ovule/seed evolution would have been retention of megaspores within megasporangia, perhaps by inhibition of the process by which seedless plants disperse spores from sporangia. A further change occurred in the number of spores produced in megasporangia. In modern seed plants, only one megaspore mother cell occurs per sporangium. A **megaspore mother cell** is a cell that undergoes meiosis to produce four spore products. The next step toward ovule/seed evolution involved the death of three of the four mother cell products, so that only

one megaspore survives to produce a female gametophyte. The advantage is that available resources are not split four ways. This process involves controlled cell death, an important feature in the development of animals and plants (see Chapter 8).

Integument evolution likely involved the appearance of several new regulatory genes that control DNA transcription. In flowering plants, activity of the genes named *ANT*, *WUS*, and *INO* is required for normal formation of both integuments, and it is likely that similar genes function in gymnosperms. Change in the regulation of integument growth to allow for an opening (micropyle) is another step that occurred during ovule evolution. Gene activity associated with seed maturation, which typically involves controlled dehydration, must also have been added to the sequence of events that occurred during seed evolution. Recall that seedless plants such as giant lycophytes, horsetails, and ferns dominated Earth's vegetation during the coal age (Carboniferous, 360–300 million years ago) (Chapter 21). The earliest known ovules or seeds appeared about 365 million years ago, just prior to the coal age. Evidence for this is the occurrence of two fossils, *Elkinsia polymorpha* and *Archeosperma arnoldii*. These fossils illustrate stages in the formation of complete integument coverings.

DNA SCIENCE

dispersal agents (Figure 22.9). In addition, seeds—composed of many cells and several types of tissues—are much larger than unicellular spores. Seeds can thus store much more food than spores. Seed food storage helps plant seedlings grow large enough to compete for light, soil water, and minerals. Seed food storage is also useful to humans and other animals.

FIGURE 22.7 **Reconstruction of the progymnosperm** *Archeopteris*

22.3 | Seeds provide ecological advantages to present and past plants

Nonseed plants reproduce solely by spores (or asexual processes). In contrast, seed plants reproduce using both spores and seeds (and/or use asexual methods). It is important to recognize that seed plants have not replaced spores with seeds. Rather, during evolution, seeds were added to an ancestral life history that involved spores. Reproduction of all plants involves spores.

The addition of seeds to a plant's reproductive repertoire provides numerous ecological advantages. First, seeds can remain dormant in the soil for many years, until conditions are favorable for germination and seedling growth. In contrast, spores have much shorter lifetimes. Though spore walls are tough, they are thinner than the multicell thick seed coat and thus more vulnerable to mechanical damage and pathogen attack. Seed coats can be modified in many ways to facilitate dispersal. They can be shaped like wings that disperse efficiently in wind (Figure 22.8) or have fleshy coatings that appeal to animal

ECOLOGY

FIGURE 22.8 **Winged seeds of gymnosperms**

FIGURE 22.9 **Some gymnosperms have fleshy seed coats** (a) Juniper seeds and (b) brightly colored yew "berries" look like fruits, but they are not.

By gathering seeds for food, animals often help disperse plants. A further seed advantage is that most seed tissues are diploid, in contrast to haploid spores. Diploidy, the presence of two copies of all essential chromosomes, confers increased genetic variability (see Chapter 13). In most seed plants, liquid water is not required as a transport medium for swimming sperm, so fertilization is not limited by lack of abundant water. Seeds thus give plants greater flexibility in coping with stressful environmental conditions, allowing plants to colonize seasonally cold or dry habitats.

Because they had seeds, early gymnosperms were able to thrive even when the warm, moist conditions characteristic of most of the coal age radically changed. The late coal age and the subsequent early Permian period (which began about 300 million years ago) were ice ages. The change to a cold climate resulted from a dramatic decline in the amount of the greenhouse gas carbon dioxide in Earth's atmosphere. Recall that atmospheric CO_2 dropped because the organic remains of dead coal age plants did not readily decay (see Essay 21.1). The Earth's climate continued to remain cooler and drier for an additional 50 million years. This long-term climate change spelled doom for many of the seedless plants that had been abundant during the coal age. The advantages provided by seeds explain why seed plants have dominated Earth's plant communities for the past 300 million years. During the Triassic period (250–200 million years ago) climate conditions

warmed, eventually becoming hot and dry. Forests of that time were composed of relatives of modern *Ginkgo* and conifers, as well as tree-sized seed ferns such as *Dicroidium* (Figure 22.10). Later Jurassic forests (remember the movie *Jurassic Park*?) of 200–145 million years ago were warm and moist, dominated by conifers, cycads, seed ferns, and *Ginkgo* relatives, with low-growing lycophytes and horsetails beneath. Dinosaurs of many types were present then and in the next period. The cooler Cretaceous period (145–65 million years ago) was a time of change. Though conifers and cycads continued their dominance, early angiosperms appeared (Figure 22.11) and spread rapidly. Then disaster struck from the sky.

About 65 million years ago at least one large meteor or comet crashed into the Earth near the present-day Yucatan Peninsula in Mexico. This 100-million-megaton impact caused global fires, acid haze, and smoke. Many groups of marine and terrestrial animals—including the dinosaurs—quickly became extinct, and so did many previously dominant gymnosperms and early angiosperms. But some seedless and seed plants survived to serve as the ancestors of the modern plants, which are now so important to global ecology and human life. After the extinction event, angiosperms diversified in many directions, becoming the dominant plant life in most of Earth's terrestrial ecosystems. The fossil record also indicates that a few of the early mammals that

(a)

(b)

FIGURE 22.10   A Triassic (225-million-year-old) seed fern, *Dicroidium* (a) Reconstruction of *D. zuberi* showing the fernlike leaves, microsporangiate (male) cones (orange structures), and ovulate (female) cones (yellow structures). The artist has guessed at cone colors, as these would not be apparent from the fossils. (b) Fossil *D. dubium* leaf.

primate ancestors of humans. It is sobering to reflect that if this catastrophe had not occurred, the evolution of mammals would have been greatly different, and humans might not have appeared at all. Realizing the destructive magnitude of asteroid or comet impacts and the potential for another such collision to occur and change the evolutionary trajectory of life on Earth, astronomers are now carefully monitoring the orbits of hundreds of objects in space that have the potential to strike Earth. Humans are not anxious to become the next victims of a similar global extinction event.

22.4 | Diversity and utility of modern gymnosperms

In this section, we survey the characteristics of modern gymnosperms (Table 22.1). This is useful in understanding their evolutionary relationships and how gymnosperms are otherwise of concern to people.

Cycads are widely planted, but endangered in the wild

Nearly 300 species and 11 genera of cycads have been described. Cycads primarily grow in tropical and subtropical regions where they are also useful in outdoor landscape plantings. Cycads don't survive outdoors in regions with harsh winters, but they can be grown as houseplants. People cultivate cycads because they have interesting shapes, often-colorful cones, and spreading, palmlike leaves (Figure 22.12). The word *cycad* comes from a Greek word meaning "palm." People sometimes have trouble distinguishing cycads from palms. *Cycas revoluta*, for example (see the chapter-opening photograph), is a cycad that is commonly known as the Sago palm. However, true palms are flowering plants. Cycad stem surfaces are rough, covered with an armor of persistent leaf bases and scaly leaves. In contrast, palm stems are often (though not always) smooth.

The stems of cycads are either emergent and treelike (some reaching 50 feet in height) or subterranean. Both types bear a crown of large pinnate leaves at the top. (If you don't

FIGURE 22.11 **Plant fossil** Compressed remains of an early flowering plant, *Archaefructus liaoningensis*, from Chinese deposits about 150 million years of age. (Inset shows a reconstruction of this plant.)

were present before the disaster 65 million years ago managed to survive. In the absence of dinosaurs, which had previously dominated the Earth's fauna, these mammals diversified into the many types that exist today, including the

TABLE 22.1	Characteristics of Living Seed Plant Groups					
Seed Plant Group	Flowers/Fruits	Flagellate Sperm	Double Fertilization	Endosperm	Complex Seed Cones	Xylem Vessels
Cycads	no	yes	no	no	no	no
Ginkgo	no	yes	no	no	no	no
Conifers	no	no	no	no	yes	no
Gnetophytes	no	no	yes	no	yes	yes
Angiosperms	yes	no	yes	yes	no	yes

FIGURE 22.12 **The cycad *Zamia***

FIGURE 22.13 **Cycad coralloid roots** Cycads often produce aboveground roots that resemble miniature branching corals; these roots are thus known as coralloid roots. The pale green roots (seen here approximately life-size) contain cyanobacteria, which provide fixed nitrogen to the cycad plant. Inset: a microscopic view of a cross-sectioned root showing the blue-green ring of cyanobacteria.

remember what pinnate leaves look like, check Chapter 11.) Young cycad leaves may be fuzzy, covered with hairy trichomes, but older leaves are smooth and glossy. The mature leaves of the African cycad *Encephalartos laurentianus* can reach 29 feet in length! Though most cycads are rooted in the ground, *Zamia pseudoparasitica* grows on trees in the moist tropical forests of Panama. In addition to underground roots that provide anchorage, soil water, and minerals, many cycads produce **coralloid roots**. Such roots are so called because they emerge from the soil surface and their branching shapes resemble corals (Figure 22.13a). Coralloid roots grow into the light because they harbor light-dependent, photosynthetic cyanobacterial partners within their tissues. The cyanobacteria can be seen as a bright blue-green ring beneath the root surface when coralloid roots are sliced across (Figure 22.13b). The cyanobacteria are also nitrogen fixers, which provide their plant hosts with nitrogen fertilizer. As noted in the chapter opening, cyanobacterial partners also produce a toxic amino acid that is distributed to the host plant's leaves and seeds, rendering them toxic to herbivores.

Cycad seed cones are often very large, as much as 30 inches (75 cm) long and weighing 88 pounds (40 kg). Individual cycad plants produce either ovulate (female) cones or pollen (male) cones, but not both. Sometimes cycads change sex if they've been damaged by cold, heat, or drying. Ovulate cones usually produce two ovules per leaflike scale, and pollen cones produce numerous microsporangia. Cones can be various colors, including white, brown, red, orange, and yellow. In some cycads, the ovules/seeds are exposed on scales that are not grouped into dense cones (see photograph accompanying the opening story). When mature, both types of cones emit an odor that attracts beetles. These beetles carry pollen from the male cones to the female cones. Thus, cycads are an exception to the general rule that gymnosperm pollen is transferred from one plant to another by wind. In using

animal pollen carriers, cycads are like many flowering plants (Chapter 24).

Archeological remains suggest that Australian aborigines have been using cycad seeds for food for the past 6,000–7,000 years. Amerindians and European settlers in Florida likewise used local *Zamia* as food. Commercial mills that extracted arrowroot flour from *Zamia* seeds closed only when too few plants were left in Florida to harvest. Because all cycads produce toxins harmful to human and animal health, and recent studies show that toxins cannot be completely removed from cycad flour by traditional methods, these plants should not be used as food. All cycads are regarded as endangered species, and cycad commerce is regulated by CITES (Convention on the International Trade of Endangered Species).

Ginkgo biloba is the maidenhair tree

Ginkgo biloba is commonly known as the maidenhair tree because its notched, fan-shaped leaves resemble leaflets of the maidenhair fern. It is the single remaining species of a once larger group that originated in Permian times (286–251 million years ago). Many of *Ginkgo biloba*'s ancient relatives became extinct as the result of an ancient global warming event that occurred about 206 million years ago (see Essay 22.1, "Plant Survivors"). Valued for its ornamental appearance and resistance to air pollution, *Ginkgo* trees can grow to 30 meters in

ESSAY 22.1   PLANT SURVIVORS

Several gymnosperm species should win prizes for long-term survival against all odds. These plant survivors, known as "living fossils," differ little from ancestors that lived 10s to 100s of millions of years ago. This conclusion is based on comparison of modern plants with ancient fossils. Take *Ginkgo biloba* for example.

In many places in the world, plant species are being lost as the result of human activities.

During Triassic times (250–200 million years ago) diverse gymnosperm trees related to modern *Ginkgo* were widespread and abundant. Fossil leaves of these plants are easily recognized because, like modern *Ginkgo* (Figure E22.1A), they are fan-shaped and their veins forked into two (a pattern known as dichotomous venation). Even with these common features, Triassic ginkgo leaves varied, revealing the occurrence of many different species. Some of the fossil leaves were only slightly fan-shaped, but entire. Others were more broadly fan-shaped but were divided into few to many segments (Figure E22.1B). Some were notched at the broad edge, like modern *Ginkgo*. Then a major environmental crisis occurred at the end of the Triassic. In a mass extinction event, 30% of marine genera, 50% of four-footed animals, and more than 95% of plant species were lost. The crisis has been linked to the breakup of the continental landmass known as Pangea, associated with extensive volcanic activity. Likely expelled by the volcanoes, CO_2 levels rose to an estimated 4 times that of previous levels. From numbers of stomata on fossil ginkgo and cycad leaves, experts deduced that this increase in CO_2 caused global temperatures to increase by 3–4°C. Fossil evidence shows that ginkgos responded dramatically to this climate change. *Ginkgo* species with large, undivided leaves were replaced by species having leaves that were divided into strips. The best explanation for the replacement is that large, entire leaves were less effective than divided ones in shedding heat well enough to continue photosynthesis. Even though some ginkgo species survived the end-Triassic global warming crisis, only one—*Ginkgo biloba*—has survived to the present time. It has probably become extinct in the wild but was maintained in Asian temple gardens until its discovery by European explorers in 1690. Now, *Ginkgo biloba* has been planted as an ornamental in many places. Humans saved this plant survivor from extinction.

The dawn redwood conifer, *Metasequoia glyptostroboides* (Figure E22.1C) is another survivor. Fossils reveal that this tree was very common and abundant in the Northern Hemisphere for a long period, until a few million years ago. Botanists believed that this plant had become extinct until a Chinese forester found a living dawn redwood tree at a remote temple in 1944. A few years later, an American-led expedition located remnant wild forests of dawn redwood in deep valleys of the Hupeh province of China. Seeds from these survivor trees have now been planted all over the world. A small population of the wollemi pine (*Wollemia nobilis*), discovered in 1994 in an Australian national park, is another such survivor. Might more plant survivors exist in remote regions, just waiting to be found?

In many places in the world, plant species are being lost as the result of human activities. For example, Chilean forests of monkey puzzle trees (*Araucaria araucana*), a species of conifer named for its intricate crown of branches, are threatened with extinction, in this case by fires. Humans may have rescued some plant survivors but are also causing the loss of even more species.

E22.1A Notched leaves of modern *Ginkgo biloba* in fall color.

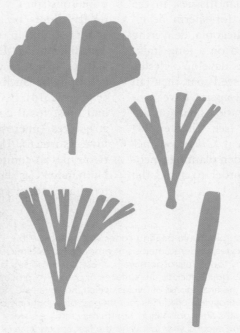

E22.1B Variation of extinct ginkgo leaves.

E22.1C Dawn redwood (*Metasequoia glyptostroboides*).

TABLE 22.2 Families of Modern Conifers and Representative Genera

Family Name	Geographical Distribution	Major Characteristics	Example Genera
Pinaceae	Mostly Northern Hemisphere	Cone scales woody or leathery, leaves needle-shaped, 2 ovules/seeds per scale, seeds generally winged	*Pinus* (pine), *Larix* (tamarack), *Pseudotsuga* (Douglas fir)
Cupressaceae	Mostly Northern Hemisphere	Cone scales woody or leathery, scalelike or needle leaves, 1–20 ovules/seeds per scale	*Cupressus* (cypress) *Sequoia sempervirens* (coast redwood) *Taxodium* (bald cypress)
Taxaceae	Mostly Northern Hemisphere	Cones small, with just one ovule/seed, seeds surrounded by fleshy, red covering (aril)	*Taxus* (yew)
Podocarpaceae	Southern Hemisphere	Seeds mostly covered by a modified cone scale, commonly with juicy, colored bracts	*Podocarpus*
Araucariaceae	Mostly Southern Hemisphere	Woody cones, one seed per cone scale	*Araucaria* (Norfolk Island "pine," monkey puzzle tree), *Wollemia* (wollemi "pine")

height, and individuals can live for more than a thousand years. Individual trees produce either ovules (and seeds) or pollen, not both. People often refer to ovule-producing trees as female trees and to pollen-producers as male trees. The basis for this difference is the presence of sex chromosome combinations much like those of humans. Ovulate (female) trees have the genotype *XX*, whereas pollen (male) trees are of the genotype *XY*. *Ginkgo* and cycads are a bit unusual for gymnosperms in that their large sperm have many flagella (as do pteridophyte sperm). *Ginkgo* sperm are delivered by unusual, branched pollen tubes. For several months, these tubes grow through the ovule's megasporangial (nucellar) tissues, in the process absorbing nutrients that are used for sperm development. Eventually, the tube bursts near archegonia, delivering mature sperm. *Ginkgo's* ovules occur paired on a long stalk. Following wind pollination, fertilized ovules develop a fleshy, bad-smelling outer seed coat and hard inner seed coat, then fall to the ground before fertilization occurs. Experts have proposed that the unusual seed coat and fertilization behavior are adaptations to an animal dispersal agent (see Chapter 24) that has become extinct. In Asia, people eat *Ginkgo* gametophytes and embryos. For street-side or garden plantings, people usually choose pollen-bearing trees in order to avoid the stinky seeds.

Conifers are the most diverse living gymnosperms

Conifers (named for their distinctive seed cones) have been important components of terrestrial ecosystems since the end of the coal age, 300 million years ago. Modern conifers (the order Coniferales) include several families (Table 22.2), with more than 500 species in 50 genera. The modern families have existed for about 200 million years.

The ovule-bearing cones of modern conifers have a distinctively complex structure. These cones have a stem axis bearing numerous spirally arranged branch systems known as short or dwarf shoots because the stem nodes are very short. Each shoot occurs in the junction between the stem axis and a leaflike bract. Each shoot consists of a scale that bears a leaflike structure with ovules on the upper surface (a megasporophyll). Such shoots are known as **ovuliferous scales** (Figure 22.14). In contrast, the pollen-bearing cones of modern and most fossil conifers are simpler in construction. Leaflike structures (microsporophylls) bear microsporangia on their lower surfaces. Until recently, no one understood why the two types of conifer cones were so different. In 2001, Genaro Hernandez-Castillo and associates described an early (coal-age) fossil conifer pollen cone that amazingly had the

FIGURE 22.14 Pine life cycle In spring, small, purple, ovule-bearing cones and pink or yellow, pollen-bearing cones are produced. On the pollen cones, diploid microspore mother cells (microsporocytes) undergo meiosis to produce haploid microspores within sac-shaped microsporangia. Pollen grains are immature male gametophytes enclosed by tough microspore walls; these are dispersed by wind in copious numbers. Bladderlike wings on pollen walls facilitate wind transport. Some pollen grains stick to the surfaces of ovules—a process known as pollination. Pollen germinates to form pollen tubes that eventually carry sperm to eggs in the archegonia of ovules, which, however, are slow to develop. About a month after pollination, meiotic division of megaspore mother cells (megasporocytes) produces four megaspores, but only one grows into a mature female gametophyte; this process can take well over a year. About 15 months after pollination, sperm may fertilize several eggs; but often only one embryo fully develops. In pines, seeds are typically dispersed in the fall of the second year following the first appearance of ovulate and pollen cones. Seedlings typically emerge from the soil carrying the remains of the seed coat on the several seed leaves (cotyledons). Cotyledons eventually wilt as the shoot grows upward.

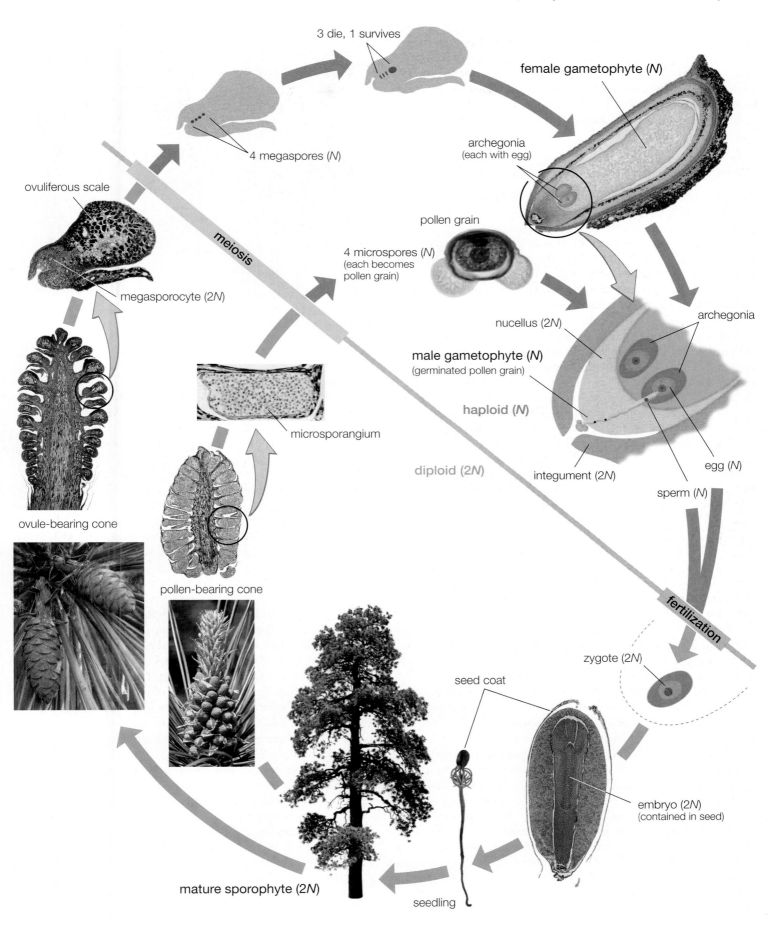

3 die, 1 survives

female gametophyte (*N*)

4 megaspores (*N*)

archegonia
(each with egg)

ovuliferous scale

meiosis

pollen grain

4 microspores (*N*)
(each becomes
pollen grain)

megasporocyte (2*N*)

archegonia

nucellus (2*N*)

male gametophyte (*N*)
(germinated pollen grain)

haploid (*N*)

microsporangium

diploid (2*N*)

integument (2*N*)

egg (*N*)

ovule-bearing cone

sperm (*N*)

pollen-bearing cone

fertilization

seed coat

zygote (2*N*)

embryo (2*N*)
(contained in seed)

mature sporophyte (2*N*)

seedling

same type of complex structure as ovulate cones. This discovery suggests that the earliest conifer pollen and ovulate cones were similar and that pollen cones acquired their simpler structure very early in the history of conifers. Hernandez-Castillo's discovery also links the evolutionary history of conifers to that of extinct cordaites and fossil and modern gnetophytes. This is because cordaites and gnetophytes have similarly complex ovulate and pollen cones.

Pines illustrate the major features of conifers

The pine *(Pinus)* life cycle (see Figure 22.14) is similar to that of conifers in general. There are separate ovulate (female) and pollen (male) cones, typically on the same tree, and sometimes on the same branch. In pine, the yellowish pollen cones appear in early spring on lower branches. The pink or purplish ovulate cones, with open scales, also appear in spring. Ovulate cones often occur on higher branches, which favors pollination by pollen blown from the microsporangiate cones of a different tree (outcrossing). The walls of mature pine pollen grains include two balloon-shaped wings that aid in wind transport. When shed, the male gametophyte has not fully developed; it consists of two tiny prothallial (prebody) cells, a generative cell, and a larger tube cell (see Figure 22.4b). After pollination, the ovulate cone scales grow together, trapping pollen within. Megaspores are not produced until a month after pollination, and mature female gametophytes may not be present until a year later. At this time, the pollen grain's male gametophyte matures. The generative cell divides to produce a stalk cell plus a cell that further divides to produce two nonflagellate sperm. But these sperm have to wait, because the eggs within the archegonia are not ready for fertilization until 3 months later. Two to five archegonia may be produced by each female gametophyte. Though more than one egg may be fertilized to produce several embryos, usually only one survives in the mature seed. Several more months are required for full seed maturation. All together, more than 2 years elapse between the appearance of pollen and ovule cones and seed dispersal in pine (see Figure 22.14).

Seed coats of most pines develop wings that serve in wind dispersal. Birds also aid in dispersal when they transport and store the seeds. Gametophyte tissue within the pine seed is rich in protein and lipid stores that fuel seedling development. The pine "nuts" that people harvest from *Pinus edulis* are actually not nuts. A nut is a type of fruit, and only angiosperms produce fruits (Chapter 23). When eating a delicious pine "nut" you are consuming mostly female gametophyte tissue.

All modern gymnosperms are woody plants: trees, shrubs, or vines. Pinewood, like that of most gymnosperms, is composed primarily of tracheids and lacks fibers (Figure 22.15). Canals that contain sticky resin are present in pinewood (and pine leaves); resin is characteristic of most conifers. Resin exuded from conifers may trap insects and other materials, then harden into amber. Conifer wood is often called "softwood" because cutting and nailing this type of wood is relatively easy. This contrasts with "hardwoods," produced by angiosperm trees such as oak. However, some angiosperm woods are quite soft (think of balsa), and some conifer woods can be relatively hard.

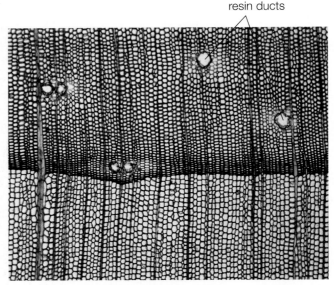

resin ducts

FIGURE 22.15 Pinewood As is typical for gymnosperm wood, the tissue is composed primarily of narrow tracheids. Resin ducts are characteristic of pine.

Pines also illustrate ways in which various conifers are adapted to the cool or cold, mountainous or high latitude habitats in which they are most numerous. Many cold-climate conifers have conical shapes and flexible branches. These adaptations allow the trees to shed snow, helping prevent limb damage. In cold climates, water occurs as snow and ice much of the time, and thus is not available for leaf photosynthesis and other plant needs. Thus, cold creates drought conditions. Pine leaves (Figure 22.16) are well adapted to resist drought damage. Their needle shape reduces the area of surface from which water could evaporate. The surface of pine needles is coated with a thick, waxy cuticle, which retards water loss and helps prevent the entry of disease microbes. Stomata are sunken below the leaf surface, helping reduce exposure to drying winds. A layer of thick-walled cells just under the cuticle, the hypodermis, is also protective of underlying photosynthetic tissue, the mesophyll. Walls of mesophyll cells have inward projections that increase cell surface area. Resin ducts extend through the mesophyll. The resin they contain helps to protect the leaves against herbivores and pathogens. One or two vascular bundles at the leaf center are surrounded by a mass of parenchyma cells and tracheids that may aid in conducting materials to the mesophyll. An endodermis ("inside skin") separates the leaf center from the mesophyll.

Most pine needles live for 2–4 years before being shed. Pines and most other conifers are evergreen, meaning that their leaves are not deciduous, which is to say they are not shed all at once at a particular time of the year. Keeping leaves through winter helps conifers start up photosynthesis quickly in spring. This can be advantageous because the growing season is short in alpine or high-latitude habitats. However, not all conifers are evergreen. Bald cypress *(Taxodium distichum)* of the southern U.S floodplains, tamarack *(Larix laricina)* of northern bogs, and dawn redwood *(Metasequoia glyptostroboides)* are exam-

ECOLOGY

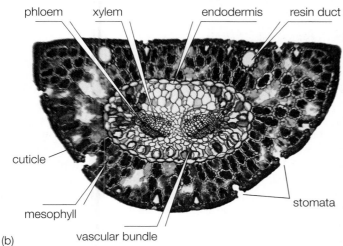

FIGURE 22.16 Pine leaves (a) Pines typically produce their needle-shaped leaves in clusters, which are short, compressed shoots. Brown scale leaves enclose the bases of these needle clusters. (b) Cross section of a pine needle.

photosynthetic proteins. Conifers also increase leaf xanthophyll pigments in winter; these yellow pigments help absorb damaging light energy. Though Ponderosa pine needles can photosynthesize in winter at more than half the summer rate, winter photosynthetic rates in Douglas fir *(Pseudotsuga menziesii)* are close to zero. Photosynthetic shutdown helps to keep most of the chlorophyll and other photosynthetic components intact so that photosynthesis can start up quickly in spring. Only a few proteins must be synthesized for the system to work again. Douglas fir exhibits its highest photosynthetic rates in spring and can do so because it does not have to repair winter-damaged photosynthetic systems.

Gnetophytes are of evolutionary significance

Gnetophytes comprise three genera of unusual plants that share certain features with angiosperms. Gnetophytes thus have attracted botanists interested in understanding the origin of flowering plants, which Darwin called "an abominable mystery." Gnetophyte vascular tissues include vessels in addition to tracheids, as do those of most angiosperms (see Chapter 9). In contrast, the vascular tissues of other gymnosperms contain only tracheids. Vessels likely evolved several times, because they also occur in some lycophytes and pteridophytes.

EVOLUTION

At least some gnetophytes have a kind of **double fertilization**, a process that is also characteristic of angiosperms. Double fertilization is the fusion of a single pollen tube's two sperm with different cells of the female gametophyte. In gnetophytes, the nonflagellate sperm both fuse with egg cells to produce embryos (in other gymnosperms only one sperm does, while the other disintegrates). Double fertilization of flowering plants differs. In many angiosperms, of the two sperm provided to an ovule by each pollen tube, one sperm fertilizes the single egg cell, while the other sperm fuses with two other gametophyte nuclei to form a triploid cell (Chapter 23). In angiosperms, this triploid cell develops into a tissue rich in stored organic food, the **endosperm**. This endosperm fuels development of the embryo. Endosperm is a defining characteristic of angiosperms. Endosperm is absent from all gymnosperms, including gnetophytes. The advantage of endosperm production by double fertilization is that it links formation of food storage directly with embryo production. In gymnosperms, food is stored in the female gametophyte prior to fertilization. If gymnosperm ovules do not develop into seeds, the food stored in them is wasted. The endosperm production system of angiosperms prevents such storage losses. If an angiosperm ovule's egg is not fertilized, no food-rich endosperm forms either.

Another feature that some gnetophytes hold in common with angiosperms is absence of archegonia (female gametangia). Archegonia were likely lost as female gametophytes decreased in size during evolution. The presence of angiosperm-like features—vessels, a kind of double fertilization, and loss of archegonia—in gnetophytes has led some experts to propose that flowering plants might have arisen from a gnetophyte ancestor. Currently, DNA sequences and other data suggest that gnetophytes had conifer ancestors. Fossils suggest that angiosperms evolved from a different

ples of deciduous conifers. In general, needle loss in these gymnosperms is regarded as an adaptation to drought stress. Dawn redwood is thought to have evolved near the Arctic during a warm period 95 million years ago. At this time, temperatures in high-latitude lowlands rarely went below freezing. Winters at high latitude would have been (and still are) too dark for photosynthesis, yet needles could still respire. Needle loss may have helped dawn redwoods avoid losing too much carbon by needle respiration.

Even though needles of most conifers persist throughout the year, photosynthesis occurs at lower levels during winter in cold regions. Recall that cold conditions are equivalent to drought. Keeping stomata closed helps to preserve internal water, but it also prevents entry of needed carbon dioxide. Also, during winter air bubbles form in tracheids during freeze-thaw cycles, interrupting water transport (see Chapter 9). Lacking sufficient water and carbon dioxide, leaves cannot utilize much of the light energy they receive. Under these conditions, light energy causes oxygen molecules to split into reactive forms that damage chlorophyll and other cell components. Conifers and some other plants that retain leaves in winter adapt to this situation by shutting down their photosynthetic apparatus in a controlled fashion. This is accomplished by destroying some key

FIGURE 22.17 **The gnetophyte *Welwitschia*** (a) The two leaves are evident as are pollen-producing (microsporangiate) cones; (b) ovulate cone.

group of seed plant ancestors, now extinct. The gain of vessels and double fertilization and the loss of archegonia may have occurred independently in the ancestors of gnetophytes and angiosperms. Whether gnetophytes were ancestral to angiosperms or not, they remain very important in understanding how plant traits evolved.

More than 30 species of *Gnetum* occur as vines (see Figure 22.1d), shrubs, or trees in tropical Africa or Asia. *Gnetum* has broad leaves that are very unusual among modern gymnosperms. Though *Gnetum*'s leaves look much like those of many tropical flowering plants, broad leaves were also characteristic of some extinct gymnosperms (see Figure 22.2).

Ephedra is another gnetophyte, native to arid regions of the southwest United States. The leaves are tiny, brown scales, and the green stems are the plant's photosynthetic organs (see Figure 22.1e). *Welwitschia* is the third gnetophyte genus, a strange-looking plant that grows in coastal deserts of southwestern Africa. These plants may obtain much of their water from coastal fog. A long taproot anchors a stubby stem that barely emerges from the ground, from which emerge two very long leaves (Figure 22.17a). The leaves become shredded by wind into many strips. Pollen (microsporangiate) cones (Figure 22.17a) produce pollen that is wind-dispersed to ovulate cones (Figure 22.17b), which produce seed.

HIGHLIGHTS

22.1 Gymnosperms are seed plants that lack flowers and fruits. There are four groups of living gymnosperms: cycads, *Ginkgo*, conifers, and gnetophytes. Several extinct groups of seed plants were abundant in the past.

22.2 An ovule is an integument-covered megasporangium. A seed develops from an ovule in which one or more eggs have been fertilized. Ovule/seed evolution is an example of Darwin's concept of descent with modification.

22.3 Ecological advantages of seeds include the potential for dormancy and greater protection of delicate gametophytes enclosed within.

22.4 Cycads are diverse, though endangered in tropical and subtropical habitats. *Ginkgo biloba*, dawn redwood, and several other gymnosperms are regarded as living fossils, because these modern gymnosperms are very similar to ancient fossils. Conifers are the most diverse modern gymnosperm group. Pines illustrate many ways in which the structure and reproduction of conifers is adaptive. Gnetophytes are central to studies of angiosperms' origin.

REVIEW QUESTIONS

1. Briefly summarize the major characteristics of gymnosperms.

2. How do gymnosperms differ from pteridophytes (ferns) and bryophytes (mosses)?

3. Briefly describe the four modern groups of gymnosperms.

4. Discuss several reproductive and survival advantages of seeds over spores.

5. Describe the two major stages in the evolution of seeds.

6. In what ways are gnetophytes similar to angiosperms? Are gnetophytes considered to be ancestors of flowering plants?

7. Briefly describe some positive and negative feature or uses of gymnosperms as far as humans are concerned.

APPLYING CONCEPTS

1. Flowering plants are well known for their interactions with animals during their reproductive processes. Are animals involved in pollination or seed dispersal in any gymnosperms? If so, give some examples. If not, how do gymnosperms accomplish these processes?

2. Usually we think that evolution selects for positive characteristics such as larger or more numerous seeds. But, in this chapter, we learned that cell death and dehydration occurred during seed evolution. In some gymnosperms, roots grow right up out of the ground and in some conifers all the needles fall off the tree at some times of the year. Explain why these apparently negative characteristics might have been favored by natural selection.

3. Being seed plants both gymnosperms and angiosperms produce pollen, ovules, and seeds. How then do they differ in their reproductive life cycles?

23

Angiosperm Diversity and Reproduction

Wild rose (*Rosa*) flower

- Humans have used flowers as symbolic gifts and flowering plants for food and medicine for longer than modern humans have existed.

- Flowers are the reproductive organs of flowering plants, and they consist of whorls of four different parts: sepals, petals, stamens, and carpels.

- Flowering plants vary greatly in the number and arrangement of flower parts; this variation is related to the environmental conditions in which the plant grows and is under genetic control.

- Fertilization leads to the production of seeds and fruits, upon which humans and many other organisms depend.

- Fruits vary greatly in color, size, and the means by which their seeds are dispersed.

- Flowering plants have a complex life cycle with alternation of generations.

In a remote region of northern Iraq lies the cave site called Shanidar. This site was excavated in the 1950s by a group of archeologists who found the remains of nine Neanderthals, including some who had died accidentally in rock falls and others who had been intentionally buried. About 50 feet inside the cave mouth, the buried skeleton of an adult male was found. It was subsequently dated to around 60,000 years ago. The discovery, by itself, is not remarkable. But when soil samples from this grave were analyzed in 1968, they revealed the presence of clusters of pollen grains from flowering plants; moreover, these clusters preserved the form of the original "male" flower parts or anthers that had once held them. These clusters revealed that actual bunches of flowers had been intentionally placed in this Neanderthal grave. Was this the first discovered example of the use of flowers as a symbolic gift at a burial ceremony?

Another explanation developed when the identity of the plants in the grave was determined. There were six species of flowering plants and one gymnosperm, *Ephedra*. All of the plants found have known medicinal properties and are still used in folk medicine today. Hollyhock (*Althea*) is a pain reliever, groundsel (*Senecio*) can be used to prevent bleeding, yarrow or *Achillea* is effective against dysentery, and *Ephedra* is useful in the treatment of coughs and asthma. Perhaps these plants were placed in the grave so that their healing properties could help the deceased in the next world. We can never know for sure what was in the minds of these Neanderthal humans. We can be certain only that the flowers were placed there deliberately. If the placement of flowers was an aesthetic expression, it is a custom that has endured for over 60,000 years. If the plants selected also indicate medicinal usage, then early humans may have been better botanists than most modern humans.

Today, we still give flowers to mark births, weddings, birthdays, holidays such as Easter and Valentine's Day, funerals, and other occasions. A significant fraction of the medicines we depend on for our health is derived from flowering plants, and, of course, much of our food comes from the fruits and seeds that develop from fertilized flowers.

In this chapter we will focus on the structure and diversity of flowers, the life cycle and reproduction of flowering plants, the variety and formation of fruits and seeds, and the development of flowering plants from seed to mature plant. ■

Achillea (yarrow)

23.1 | Flowering plants comprise an enormous number and diversity of species

The flowering plants make up the phylum Anthophyta, which includes some 235,000 species classified into about 350 families. The phylum name Anthophyta derives from the Greek words *anthos*, for "flower," and *phyton*, for "plant." Flowering plants represent an enormous range of sizes—from towering eucalyptus trees to tiny duckweeds, which may be only a millimeter across. In their growth form, flowering plants may be trees, shrubs, vines, or herbs. They may grow as spiny desert plants or as soft, submerged, aquatic plants like coontail (Figure 23.1). Some have even ceased to be free living and have become parasites, such as yellow dodder *(Cuscuta salina)* or saprophytes (which live off decaying organic matter) such as Indian pipe (*Monotropa uniflora*; Figure 23.2).

Flowering plants are all vascular plants that reproduce sexually by forming flowers, though some flowering plants such as dandelions and hawkweeds reproduce by an asexual process called *apomixis*. Flowering plants carry out a unique process known as **double fertilization**, in which one fertilization produces a zygote and the second a nutritive tissue called **endosperm**. Among the gymnosperms (Chapter 22), only *Ephedra* and *Gnetum* exhibit double fertilization; but in their case, the double fertilization produces two zygotes rather than a zygote and endosperm. Another defining characteristic of flowering plants is that the **ovules**, which develop into seeds after fertilization, are contained within a **carpel**. The formal term for

FIGURE 23.2 **The saprophytic flowering plant, Indian pipe (*Monotropa uniflora*)**

flowering plant is **angiosperm**, a word that derives from the Greek *angeion*, meaning "a vessel," and *sperma*, or "seed." The vessel is the carpel, which after fertilization ripens into a fruit, enclosing seeds. Fruits are another defining characteristic of flowering plants. Gymnosperms produce no flowers, carpels, or fruits.

The phylum Anthophyta was traditionally divided into the classes Dicotyledones (dicots) and Monocotyledones (monocots). A variety of features distinguish the two groups (Figure 23.3). The embryos of dicots have two **cotyledons**, which are embryonic seed leaves containing the nutritive tissue derived from the endosperm; monocot embryos have only one cotyledon. Dicots include plants such as oaks, maples, roses, daisies, peas, and cacti; monocots include plants such as grasses, orchids, tulips, lilies, and palms. Dicots typically have broad leaves with netted venation (Chapter 11). Their flower parts are usually in fours or fives or multiples thereof. The pollen grains of most dicots have three pores or furrows on their surface. In contrast, monocots typically have long, narrow leaves with parallel venation, and their flower parts are usually in threes or multiples of three. Monocot pollen grains have only one pore or furrow.

Recent molecular studies—in which five genes were sequenced in 105 species of flowering plants—do not support the simple division of the angiosperms into dicots and monocots (Figure 23.4). There are, in fact, several distinct lineages of flowering plants—the monocots and several groups of dicots. Two early-divergent lineages include one represented by *Amborella* (Figure 23.5a), a shrub native to New Caledonia, and a second containing water lilies in the order Nymphaeales (Figure 23.5b). Another lineage of flowering plants includes the monocots (and, according to some studies, the aquatic plant *Ceratophyllum*—see Figure 23.1b). Subsequent to the divergence of the monocots, several other dicot groups

DNA SCIENCE

EVOLUTION

FIGURE 23.1 **Flowering plants of varied habitats** (a) *Opuntia*, a desert cactus; (b) *Ceratophyllum*, an early-divergent flowering plant that grows in aquatic habitats. The inset shows the small protrusions on the leaves that give the plant its name, which literally means "leaf with horns."

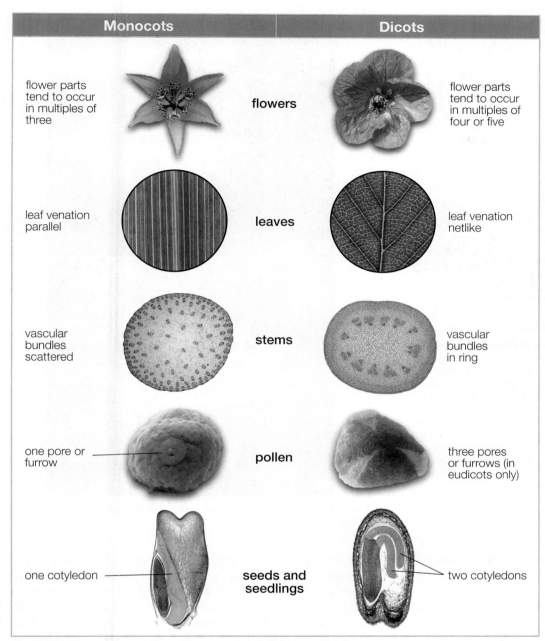

Monocots		Dicots
flower parts tend to occur in multiples of three	**flowers**	flower parts tend to occur in multiples of four or five
leaf venation parallel	**leaves**	leaf venation netlike
vascular bundles scattered	**stems**	vascular bundles in ring
one pore or furrow	**pollen**	three pores or furrows (in eudicots only)
one cotyledon	**seeds and seedlings**	two cotyledons

FIGURE 23.3 Characteristics of monocots and dicots

appeared, including the **eudicots**, where *eu* means "good" or "true." The eudicots are the largest and most diverse group of flowering plants. Most modern species of flowering plants are either monocots or eudicots.

23.2 | The parts of flowers are arranged in whorls

A flower is a reproductive shoot whose parts are evolved from leaves and arranged in whorls or circles on the end of a stalk called a **peduncle** (Figure 23.6). A peduncle may bear a single flower or a cluster of flowers called an **inflorescence**. The end of the peduncle forms an enlarged area known as a **receptacle**, to which the flower parts are attached. Most flowers have an outermost whorl of flower parts known as **sepals**, which are leaflike and often thick and green. They protect the flower when it is still in the bud stage. The whorl of sepals is called the **calyx**. Arising just above the sepals is the next whorl of parts, the **petals**. Like sepals, the petals are leaflike but thinner and often brightly colored. They serve to attract pollinators. Petals may be separate or fused together into a tube. Collectively, the whorl of petals is called the **corolla**. The calyx and corolla together form the **perianth**, the two

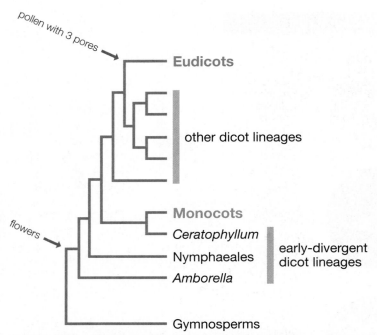

FIGURE 23.4 **Phylogenetic relationships among the major groups of flowering plants** Modern phylogenetic studies using DNA sequencing techniques have shown that the old schemes in which flowering plants were divided into monocots and dicots were incorrect. There are actually several lineages of dicots. One of the earliest-divergent of these is represented by *Amborella*. Another group that diverged early from the flowering plant lineage includes the water lilies belonging to the order Nymphaeales. The pondweed *Ceratophyllum* groups with the monocots. There are several other distinct lineages of dicots, including the eudicots. The eudicots are distinct in having pollen with three pores. The cotyledons in the other dicot lineages are often very small and indistinct.

FIGURE 23.5 **Two early-divergent plants** (a) *Amborella*, (b) the water lily *Nymphaea*

whorls of sterile (nonreproductive) parts of the flower. Early-divergent angiosperms commonly have flowers whose petals and sepals are similar in shape and color; these are known as **tepals**. For this reason, it is thought that the earliest flowers probably also had tepals.

The remaining two whorls of flower parts are fertile and reproductive. Inside and above the whorl of petals lie the pollen-producing organs, the **stamens** (Figure 23.6). A stamen consists of a **filament** terminated by an **anther**, which contains four pollen sacs arranged in two pairs (Figure 23.7). The pollen sacs produce the pollen grains. Mature pollen grains may contain starch or oils and are an important source of food for many animals (Chapter 24).

Pollen grains vary in size from less than 20 μm (micrometer, 1 micrometer equals one-thousandth of a millimeter) to over 250 μm in diameter. They also differ in the number and arrangement of the pores and furrows through which the pollen tube will grow (see Figure 23.3). Pollen grains are covered in a tough outer coat composed of the resistant substance sporopollenin. Almost all families, genera, and many species of flowering plants can be identified by the

sculpturing of their pollen grains and the number and arrangement of pores and furrows. Because sporopollenin is so resistant to degradation, pollen is abundant in the fossil record and can therefore be used to determine the composition of plant communities and climates of the past. Pollen grains also are associated with hay fever and allergies. The familiar symptoms of watery eyes and itchy nose are most often caused by allergic reactions to pollen from wind-pollinated plants such as oak trees, grasses, and ragweed, which produce much larger amounts of pollen than do insect-pollinated plants.

The **carpels** lie inside the whorl of stamens (Figure 23.6). The carpels evolved from leaves that folded lengthwise to enclose the ovules. After fertilization, the ovules develop into seeds. A flower may have one or more carpels. If several carpels are present, they may be separate or fused together in whole or in part. In most flowers, the carpels are differentiated into a broad, round lower **ovary**, where the ovules lie (Figure 23.7), a column of tissue called the **style**, through which the pollen tubes grow, and a terminal **stigma**, to which pollen grains adhere (see Figure 23.6). The entire carpel

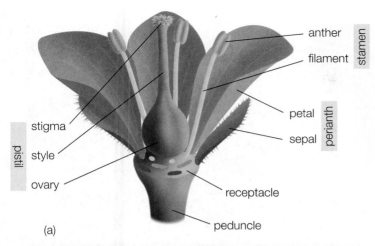

(b)

FIGURE 23.6 Parts of a typical flower (a) Typical flower parts are indicated, including the peduncle, receptacle, sepals (collectively termed the calyx), petals (collectively termed the corolla), stamens (filament and anther), and pistil (ovary, style, and stigma). (b) A flower from the cactus *Pereskia*. All four whorls are present (this is a perfect flower).

including ovary, style, and stigma is sometimes referred to as the **pistil**, a word derived from *pestle*—the device used by pharmacists to grind substances into powders. Inside the ovary, the ovules are attached at a region known as the *placenta*. The distribution of placentae and ovules varies among flowering plants and is used in identification and classification.

Flowers vary greatly in the numbers, positions, and arrangements of parts

Flowers that possess all four floral whorls—sepals, petals, stamens, and carpels—are termed **complete** flowers. Any flower lacking one or more of the four basic whorls is therefore **incomplete.** Most flowers have both stamens and

carpels; such flowers are termed **perfect** and are bisexual. If either stamens or carpels are lacking, a flower is **imperfect** and unisexual. Thus, a flower can be incomplete but still perfect if it lacks only sepals or petals. An imperfect flower, therefore, would be incomplete. For a discussion of the genes involved in flowering and the four basic floral whorls, refer to Essay 23.1, "The ABCs of Floral Organ Development."

An imperfect flower is **staminate** if it lacks carpels, and **carpellate** if it lacks stamens. If both staminate and carpellate flowers are present on the same plant, as in oaks *(Quercus)*, cattails *(Typha)*, and corn *(Zea mays)*, the plant is **monoecious**, from the Greek *monos*, "one," and *oikos*, "house." In corn, the staminate flowers form an inflorescence called a *tassel* at the top of each plant (Figure 23.8a). The carpellate flowers make up the ears, surrounded by leafy bracts or husks, and the corn silks are actually long stigmas (Figure 23.8b).

If staminate and carpellate flowers are found on separate plants—as in holly trees, date palms, willows, and cottonwoods (Figure 23.9)—the plants are **dioecious**, meaning "two houses." In dioecious species, pollen must be transferred from staminate flowers on one plant to another plant bearing carpellate flowers for reproduction to occur. Dioecious species therefore reproduce by **cross-pollination**, the transfer of pollen between different plants. Corn, in contrast, may reproduce by **self-pollination**, as pollen from the tassels falls onto the silks of the ears. Self-pollination fosters inbreeding, whereas cross-pollination promotes outcrossing and new gene combinations (see Chapter 24).

The whorls of sepals, petals, and stamens may be attached at different levels relative to the ovary or ovaries. If sepals, petals, and stamens are attached to the receptacle below the ovary, as in lilies, the ovary is termed **superior** (Figure 23.10a). An ovary or ovaries are called **inferior** if the sepals, petals, and stamens are attached above them (Figure 23.10b).

Flowers also vary in their symmetry. If all the petals are of similar size and shape and evenly spaced out around the ovary, the flower is radially symmetrical (Figure 23.11a). Poppies and tulips exhibit radial symmetry and are called **regular** flowers. Other flowers show bilateral symmetry and have one or more parts of a whorl different from others in the whorl. Orchids and snapdragons show bilateral symmetry and are termed **irregular** flowers (Figure 23.11b). The symmetry of flowers and the number, size, location, and shape of flower parts vary in relationship to the pollinators that visit them (Chapter 24).

Flowers have evolved many different types of inflorescence for pollination

Flowers may be solitary, as in the garden tulip, but often they are grouped into clusters called *inflorescences*. Variation in the form of inflorescences reflects adaptation to different means of pollination. An inflorescence has a main flower stalk, the

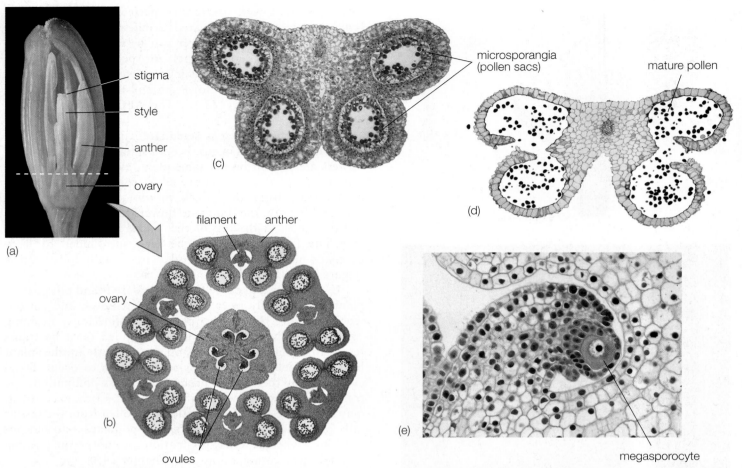

FIGURE 23.7 The reproductive organs of flowers (a) Free-hand longitudinal section through a young day-lily flower bud. The flower parts are very short compared to the mature flower. (b) A cross section through a stained lily flower bud at approximately the level indicated by the white dashed line in (a). Note the six stamens surrounding the ovary with ovules. (c) Close-up view of a cross-sectioned anther, in which there are four microsporangia or pollen sacs. Meiosis was taking place at the time this material was prepared. The filament, which supports the anther, is also seen in cross section. (d) An anther that has opened to release the pollen grains. When the anther is mature, the partitions between adjacent pollen sacs break down and the pollen grains are released. (e) An ovule seen at higher magnification, which is at about the same stage of development as those shown in part (b). Note the large cell with prominent nucleus; this is the megasporocyte, which will divide by meiosis to form four megaspores (discussed in section 23.3). The layers of cells (integuments) that surround the ovule are beginning to form.

ESSAY 23.1 THE ABCs OF FLORAL ORGAN DEVELOPMENT

The switch from a shoot meristem producing leaves to a floral meristem producing flowers is one of the most dramatic changes in plant development. The switch is brought about by environmental cues such as day length and internal chemical signals. Most of the genetic work on flowering has been done on the tiny wild mustard *Arabidopsis thaliana*, snapdragon (*Antirrhinum majus*), and corn (*Zea mays*). The number of genes involved in the

Flowers are produced by floral meristems at the ends of a single shoot or the multiple shoots that make up an inflorescence.

process indicates the complexity in the switch to flowering; more than 40 flowering regulatory genes are now known in *Arabidopsis*. This discussion focuses on two groups of flower development genes—the meristem identity gene and the floral organ identity genes.

LEAFY (*LFY*) is a major meristem identity gene. Flowers are produced by floral meristems at the ends of a single shoot or the multiple shoots that make up an inflorescence. *LFY* plays an

important role in mediating the switch between an inflorescence shoot meristem and a floral meristem (Figure E23.1A). *LFY* also activates the floral organ identity genes and thus appears to play a pivotal role in the control of flowering.

There are three classes of floral organ identity genes—A, B, and C. The interactions of these genes is referred to as the **ABC model** of flower developmental control (Figure E23.1B). Class A genes include APETALA 1 (*AP1*) and APETALA 2 (*AP2*), which control the formation of sepals and petals. Class B genes—APETALA 3 (*AP3*) and PISTILLATA (*PI*)—control petals and

DNA SCIENCE

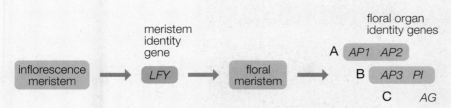

E23.1A Genetic control of flowering *LFY* mediates the transition of an inflorescence meristem to floral meristem and activates the three classes of floral organ identity genes (A, B, and C), which control the development of sepals, petals, stamens, and carpels in the correct positions within the flower.

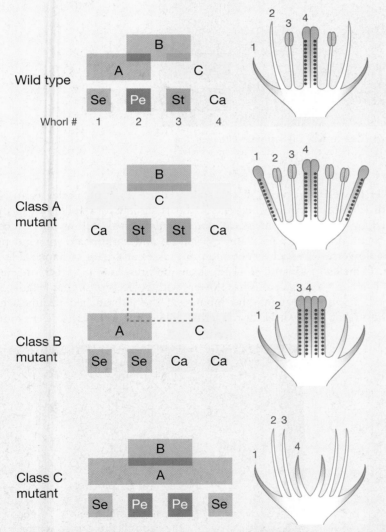

E23.1B Wild type (normal) plant and mutants in the floral organ identity genes A mutation in class A genes results in flowers without sepals or petals. Mutants in the class B genes lack petals and stamens. Class C mutants in the gene AG lack both stamens and carpels. Se: sepals, Pe: petals, St: stamens, Ca: carpels.

stamens. The class C gene AGAMOUS *(AG)* controls the formation of stamens and carpels. The function of each class of gene overlaps with the function of the next class. Sepals are controlled by class A genes alone, but both class A and B genes are necessary for petal development, and both class B and C are needed for expression of stamens. Class C genes alone can specify carpels. All three classes determine the normal arrangement of whorls—sepals, petals, stamens, and carpels.

Mutants in class A genes result in flowers with no sepals or petals (Figure E23.1B). Mutant class A flowers have the whorls carpels, stamens, stamens, carpels. Class B mutants *pi* and *ap3* have the whorls sepals, sepals, carpels, carpels—because class A and C genes do not overlap in function, and class B is necessary for formation of petals and stamens. A mutant in the class C gene (*ag*) results in flowers without stamens or carpels and the whorls sepals, petals, petals, sepals. Similar sets of regulatory genes and mutants have been found in snapdragons (*Antirrhinum*). The same three basic classes of floral regulatory genes are probably widespread throughout flowering plants.

Recent research has revealed that a fourth class of floral organ identity genes exists, called class E. Class E includes three SEPALLATA genes, abbreviated as *SEP1*, *SEP2*, and *SEP3*. In mutants where these three genes lose their function, all floral organs resemble sepals. Thus, expression of petals requires gene classes A, B, and E. Stamens need classes B, C, and E, and carpels require classes C and E.

Several species of gymnosperms possess genes similar to the class B and C genes of flowering plants, suggesting that they were present in a common ancestor of angiosperms more than 300 million years ago. Origin of the class A genes may explain how flowering plants evolved separate sepals and petals from ancestral tepals.

EVOLUTION

FIGURE 23.8 Staminate and carpellate flowers in corn (Zea mays) (a) Staminate flowers (the "tassel"). (b) Carpellate flowers in a young "ear" (10–12 cm long). Note the long stigmas ("corn silk"), one attached to each carpellate flower.

peduncle, from which emerge many secondary smaller flower stalks, termed *pedicels*, each with a flower at its tip (Figure 23.12). The shape of an inflorescence is determined by the arrangement of pedicels on the main peduncle. If the main peduncle is branched and the branches bear clusters of flowers, the inflorescence is called a *panicle* (Figure 23.12a). Panicles are found in cereal grasses such as oats and rice.

In most cases, the main peduncle is unbranched. If an unbranched main axis bears flowers with either very short or no pedicels, it is a spike (Figure 23.12b). Barley is an example. A catkin (Figure 23.12c) is similar to a spike but has flowers of only one sex. Woody plants such as willows and birches produce catkins. Panicles, spikes, and catkins aid in wind pollination, which is common in grasses and trees (Chapter 24). A raceme (Figure 23.12d) is an inflorescence in which the unbranched main axis has flowers with pedicels of similar lengths, as in snapdragons and lupines. If the unbranched peduncle has all the pedicels at the terminal end, the inflorescence is a simple umbel

FIGURE 23.9 Staminate and carpellate flowers in willow (Salix), a dioecious plant (a) Staminate flowers. (b) Carpellate flowers. Unlike in corn, where staminate and carpellate flowers occur on the same plant, in willow they are found on separate individual plants. Both types of flowers are arranged in clusters (inflorescences) called catkins.

FIGURE 23.10 Superior and inferior ovaries A citrus flower (a) has a superior ovary; *Fuchsia* (b) has an inferior ovary (arrowheads point to ovaries).

(a)

(b)

FIGURE 23.11 **Regular versus irregular flowers** (a) The regular (radially symmetrical) flower of a cactus. (b) The irregular (bilaterally symmetrical) flower of an orchid. You could cut the regular flower at any angle and obtain more or less equal halves, whereas the irregular flower can be cut only along one plane to produce equal halves.

(Figure 23.12e). Onion, geranium, and milkweed are examples. Umbels can be grouped into compound umbels (Figure 23.12f), which are found in carrots, dill, and parsley.

In the inflorescence called a head, the end of the peduncle is expanded to bear numerous tiny flowers without pedicels (Figure 23.13). Members of the sunflower family (Asteraceae), including such familiar garden plants as zinnias, daisies, sunflowers, and marigolds, as well as lawn weeds such as thistles and dandelions have heads. The "flower" of all these plants is really a head composed of hundreds of tiny flowers. Many heads contain two types of flowers—disk flowers, which are perfect and located in the center of the inflorescence, and ray flowers, which are carpellate or sterile and found at the margins. In the daisy and black-eyed Susan, the ray flowers often have their petals fused into a single, long, straplike "petal." Umbels and heads form flower platforms that enable insects to move easily from flower to flower, transferring pollen as they go (Chapter 24).

23.3 | The angiosperm life cycle involves an alternation of generations

Like mosses, ferns, and gymnosperms, the angiosperms have a life cycle consisting of an alternation of generations (Figure 23.14). The often large, flowering plant we are familiar with

represents the diploid sporophyte stage in the life cycle. The haploid gametophyte stages are microscopic and nutritionally dependent on the sporophyte or stored food, in the case of pollen. The diploid sporophyte generation produces spores by meiosis; the spores give rise to the haploid gametophytes, which then produce the haploid gametes by mitosis. Like gymnosperms and some pteridophytes, flowering plants are heterosporous, meaning they produce two types of spores—**megaspores** and **microspores**.

Double fertilization produces a zygote and an endosperm

In Figure 23.14, the life cycle begins when a seed germinates (bottom center). The seed grows into a sporophyte that produces flowers. Each anther contains four pollen sacs or microsporangia containing numerous **microsporocytes**. Microsporocytes undergo meiosis to produce four haploid microspores. Each microspore then divides by mitosis to form two cells within the microspore wall—a large tube cell and a small generative cell (Figure 23.15). This two-celled stage is a pollen grain or microgametophyte (Figure 23.14, top left). When the anthers open, the pollen grains are shed and transported to the stigma of the same flower or another flower of the same species. Pollen transport is known as pollination (mechanisms of pollination are discussed in Chapter 24). If the pollen grains reach the stigma of a suitable flower, they germinate, and the tube cell sends out a pollen tube through the pore or furrow in the wall of the grain. Before or during pollen grain germination, the generative cell divides by mitosis to form two haploid sperm cells that then move down the pollen tube. The pollen tube grows down the style into the ovary, where it grows toward an ovule (upper right).

The ovary contains the ovules (Figure 23.14, center). The ovule consists of a tissue called the **nucellus** (the megasporangium) wrapped in two layers—the integuments—with a small opening, the micropyle. A single diploid **megasporocyte** arises from the nucellus in each ovule. The megasporocyte then divides by meiosis to form four haploid megaspores. Commonly, three of the four megaspores disintegrate, and the surviving megaspore develops into a megagametophyte. The megaspore enlarges and divides by mitosis, to produce a total of eight nuclei in most flowering plants. In general, the eight nuclei initially lie in two groups of four—one group at the micropyle end and the other at the opposite end. One nucleus from each group migrates to the center of the megagametophyte. These two nuclei are then called the **polar nuclei** (Figure 23.14, right center). Cell walls form around the three nuclei at the micropyle end; the resulting three cells are called the *egg apparatus*, which consists of the middle egg cell and lateral synergids. Cell walls also form around the three nuclei at the opposite end, and they become the antipodal cells. The two polar nuclei lie in the center of the large central cell. The mature female gametophyte or megagametophyte (also called the **embryo sac**) of the lily and some other flowering plants contains seven cells with eight nuclei, the central cell containing the two polar

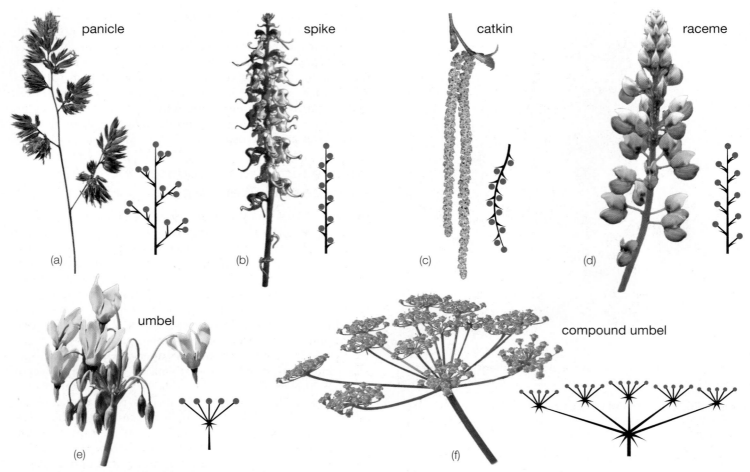

FIGURE 23.12 Types of inflorescences (a) Panicle from a grass plant. (b) Spike (of elephant's head, *Pedicularis groenlandica*). (c) Catkin (of birch tree, *Betula*). (d) Raceme (of lupine, *Lupinus perennis*). (e) Simple umbel (of shooting star, *Dodecatheon meadia*). (f) Compound umbel (of wild parsnip, *Pastinaca sativa*). The red dots in the stylized diagrams represent individual flowers.

nuclei. Other angiosperms, however, may have embryo sacs with different numbers and arrangements of cells.

The pollen tube enters the ovule through the micropyle (Figure 23.14, center right). One sperm nucleus enters the egg cell and the other enters the central cell to fuse with the polar nuclei. This process is called **double fertilization** because both sperm nuclei fuse with separate nuclei (lower right). The fusion of the sperm nucleus with the two polar nuclei is a triple fusion producing a (usually) triploid (3N) endosperm nucleus. The triploid endosperm nucleus then develops into endosperm, a nutritive tissue for the plant embryo. The union of sperm and egg nuclei produces a diploid zygote (2N), which develops into the plant embryo. The integuments of the ovule become a seed coat around the embryo and endosperm. The ovary and, in some cases, other related structures such as the receptacle, develops into a fruit. The fruit or seeds are dispersed, and the cycle begins again with the germination of a seed.

Apomixis produces seeds without fertilization

Some flowering plants have the capacity to produce seeds without fertilization. This process of **apomixis** has evolved independently a number of times and is found within 40 flowering plant families. It is common among members of the sunflower family (Asteraceae), rose family (Rosaceae), and the grass family (Poaceae). Apomixis is a form of asexual reproduction in which the embryo is genetically identical to the parent.

There are several different mechanisms of apomixis, but we will consider only one. As described earlier, in sexual reproduction, the megasporocyte undergoes meiosis to form haploid megaspores. In the form of apomixis called *diplospory*, meaning diploid spores, the megasporocyte fails to undergo meiosis but develops instead into a diploid spore (Figure 23.16). The diploid spore divides by mitosis to form a diploid megagametophyte with eight diploid nuclei. The egg cell (2N) and polar nuclei (2N) develop directly into an embryo and endosperm, forming apomictic seeds.

Apomixis appears to be advantageous in environments such as the polar regions, where animal pollinators are scarce, unreliable, or absent. The control of apomixis would be highly valuable in plant breeding. As discussed in Chapters 14 and 16, hybrid vigor is a common phenomenon, and hybrid seeds are widely used in agriculture to boost crop production. But hybrids lose their vigor when they reproduce sexually. New hybrid seed must be produced each year. If hybrid plants could

FIGURE 23.13 The head, another common type of inflorescence In members of the sunflower family (Asteraceae), what is often called the "flower" is actually an inflorescence of many small flowers attached directly to a large peduncle. (a) In black-eyed Susans (Rudbeckia hirta), the head consists of central disk flowers, which are perfect and have a very reduced corolla, and peripheral ray flowers, which are carpellate and have their corolla modified into a long, petallike strap. The inset shows an unsectioned head. (b) In dandelion (Taraxacum), there are no disk flowers. All flowers are perfect, and each has a straplike corolla. The inset shows a mature head and the small brown fruits, each with a tuft of bristles.

be propagated by apomixis, however, their vigor could be maintained because the progeny would be the same genetically as the parents.

23.4 | The development of the embryo and seed follows double fertilization

Following double fertilization, the early stages in the development of the embryo are much the same throughout flowering plants. The triploid nucleus of the central cell divides and begins to form the endosperm tissue, which will provide nutrition to the embryo. The diploid zygote divides first by mitosis in an unequal crosswise or transverse direction to form a small upper apical cell and large lower basal cell (Figure 23.17a). This first division establishes

the polarity of the embryo. The upper apical cell will give rise to most of the embryo (Figure 23.17b,c). The lower basal cell forms the suspensor, a stalk that attaches the embryo at the micropylar end of the ovule (Figure 23.17c). Nutrients pass through the suspensor from parent plant to embryo.

Plant embryos pass through a number of developmental stages

The apical cell divides until it forms a nearly spherical ball of cells—the "globular" stage (Figure 23.18a)—and the formation of the basic plant tissues begins. The outermost cells of the embryo divide to become the protoderm, the future epidermis. The inner cells divide to form the procambium, the future vascular tissues (xylem and phloem), and the ground meristem, the future ground tissues such as

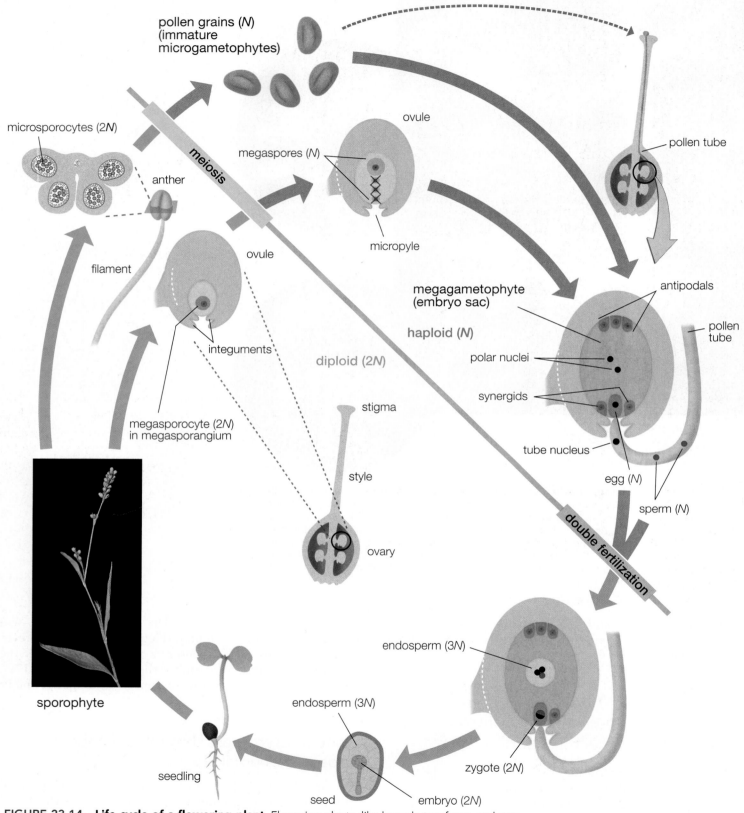

FIGURE 23.14 **Life cycle of a flowering plant** Flowering plants, like bryophytes, ferns, and gymnosperms, have a life cycle that consists of an alternation of generations. The large, familiar flowering plant is the diploid sporophyte; the haploid gametophyte stages are microscopic. The unique feature about the life cycle of flowering plants is a double fertilization that produces a diploid zygote and a triploid endosperm or nutritive tissue. Steps in the life cycle are described in detail in the text.

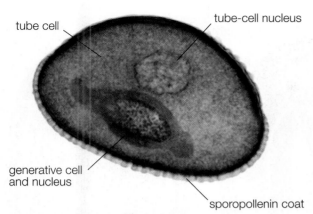

FIGURE 23.15 Pollen grain of lily The tube cell (and tube-cell nucleus) and generative cell (which divides to form the two sperm) are both evident. The larger tube cell surrounds the generative cell.

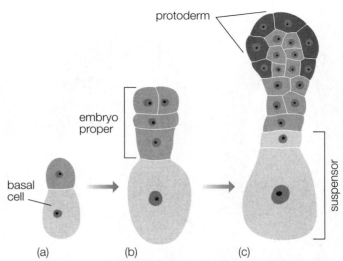

FIGURE 23.17 Early stages in the development of the angiosperm embryo (a) Two-celled stage showing upper apical cell and lower basal cell. (b) Four-celled stage of embryo and lower basal cell. (c) Formation of protoderm at the apical end of the embryo and the two-celled suspensor.

parenchyma. These three embryonic tissues eventually give rise to the primary meristems for both root and shoot (Chapter 8).

As the embryonic tissues—the protoderm, procambium, and ground meristem—develop, the plant embryo follows different patterns in eudicots and monocots, largely because eudicots have two cotyledons and monocots have only one. Eudicots assume a bilobed or heart-shaped form, the "heart" stage (Figure 23.18b). The lobes of the heart are the developing

cotyledons or seed leaves. Since monocots form only one cotyledon, their embryos assume a cylindrical shape. At this stage, the embryo is elongating and is divisible into shoot apical meristem, cotyledon or cotyledons, embryonic root, and root apical meristem. In eudicots, the shoot apical meristem is located between the two lobes of the cotyledons. In monocots,

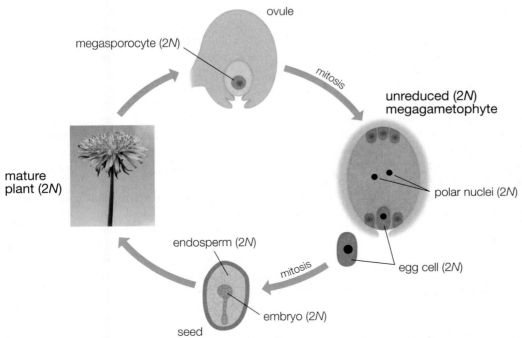

FIGURE 23.16 Apomixis by diplospory In this form of apomixis (a type of asexual reproduction), the diploid (2N) megasporocyte does not undergo meiosis and essentially functions as a diploid spore (hence the term *diplospory*). It divides by mitosis to form a megagametophyte that is also diploid. The diploid egg develops into an embryo (much as would a diploid zygote), which in turn gives rise to a mature sporophytic plant.

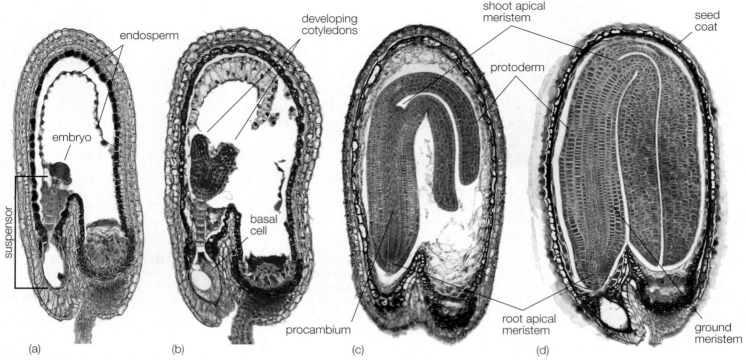

FIGURE 23.18 **Later stages in embryo development** These micrographs show embryos of *Capsella bursa-pastoris* (shephard's purse—a eudicot). (a) Globular stage; the protoderm, which will develop into the epidermis, is developing. (b) Heart stage of development. The lobes of the heart are the developing cotyledons. (c) As the embryos continue to develop, the cotyledons of many eudicots curve. (d) A mature embryo. The shoot apical meristem lies between the two cotyledons, and the procambium has formed between the shoot and root apical meristems. The procambium is surrounded by ground meristem.

the shoot apical meristem is located in a notch in the single cotyledon.

The cotyledons typically become curved within the seed as the embryo matures (Figure 23.18c,d). The suspensor degenerates and is absent from the mature seed. In seeds of monocots such as corn, the endosperm with its reserves of stored food is present in the mature seed and supplies the energy for germination and establishment of the seedling. Coconut milk is liquid endosperm. The monocot cotyledon digests, absorbs, and transfers the food reserves in the endosperm to the embryo during germination. In many eudicots, such as sunflowers and peas, the food reserves of the endosperm are transferred into the cotyledons. But in other eudicots, such as castor bean, endosperm is found in the mature seed. The cotyledons then function like those of monocots and transfer the food in the endosperm to the embryo upon germination.

The mature seed is nutritionally independent of the parent plant

As the seed reaches full maturity, the stalk connecting the ovule to the ovary wall separates, and the seed becomes a nutritionally closed system. The seed loses water until water constitutes only 5 to 15% of its weight. The integuments of the ovule form a hardened seed coat enclosing the embryo and stored food. The embryo ceases to grow.

The mature embryo of eudicots consists of an embryonic axis attached to two cotyledons. In some eudicots like castor bean, the shoot apical meristem is the only part of the embryonic axis extending above the attachment point of the cotyledons (Figure 23.19a). In others, such as the garden bean and sunflower, an embryonic stem called the **epicotyl**, with one or more tiny leaves and a shoot apical meristem, extends above (epi-) the cotyledons (Figure 23.19b). This embryonic stem is called the **plumule**. The region of the embryonic axis below (hypo-) the cotyledons is the **hypocotyl**, which may possess an embryonic root, the **radicle**, or may terminate in only a root apical meristem and root cap.

In monocots such as onion, the embryonic axis consists of a single cotyledon, shoot apical meristem, and a hypocotyl with a terminal root apical meristem and cap (Figure 23.19c). In grasses such as corn and wheat, however, the embryo is highly specialized. The single cotyledon, termed a *scutellum* from the Latin *scutella*, meaning "a shield," is thin and has a large surface area (Figure 23.19d). The scutellum absorbs nutrients from the endosperm during seed germination. Both epicotyl and hypocotyl of the monocot embryo are enclosed in sheaths called the *coleoptile* and *coleorhiza*, respectively, which protect them during germination.

All seeds are enclosed in a seed coat that is formed from the integuments that surrounded the ovule. The seed coat is usually very hard and provides protection for the embryo from damage.

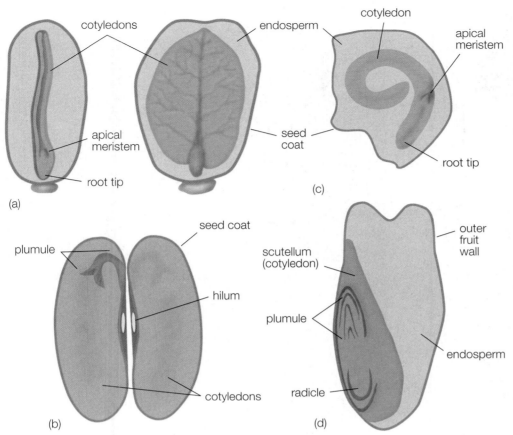

FIGURE 23.19 **Seeds of common eudicots and monocots** (a) In the seeds of castor bean (*Ricinus communis*), a eudicot, the embryo has only an apical meristem above the cotyledons. Food is stored in the endosperm. (b) In garden bean (*Phaseolus vulgaris*), another eudicot, the embryo has a plumule above the two cotyledons. The plumule consists of a short stem (the epicotyl), two small leaves, and an apical meristem. Food is stored in the cotyledons. (c) In seeds of onion (*Allium cepa*), a monocot, the embryo has a single large cotyledon with the apical meristem located in a notch at its base. (d) Maize (*Zea mays*), a monocot, has embryos with a well-developed cotyledon called the *scutellum* and a well-developed plumule and radicle. In both types of monocot seeds food is stored in the endosperm.

On many seeds the remains of the micropyle of the ovule can be seen on the seed coat as a small pore, often associated with a scar (the hilum) left by the stalk that once attached the seed to the ovary wall.

23.5 | A fruit is a mature ovary containing seeds

As the ovules develop into seeds, the ovary or ovaries develop into a fruit. The fruit evolved as a mechanism for the dispersal of seeds. Fruit and seed dispersal are discussed in the next chapter; in this chapter we consider the diversity of fruit structure. A **fruit** is a mature ovary. If additional floral parts are included, the fruit is called an **accessory fruit**. Apples and pears are examples of accessory fruits, in which the "core" is the ovary and the fleshy tissue is derived from floral parts and the receptacle. Fruits can be classified as simple, aggregate, or multiple.

Simple fruits are the most common type

Simple fruits develop from a single carpel or several fused carpels. Simple fruits may be fleshy or dry. Fleshy simple fruits can be classified as berries, drupes, hesperidiums, pepos, or pomes. **Berries** have one to several carpels—each with one to many seeds—and the flesh is soft throughout the fruit. Tomatoes, peppers, and grapes are examples of berries (Figure 23.20a). Dates and avocados are also berries, although they have only one seed. In **drupes**, there may also be one to several carpels; but each carpel has only one seed, and the inner layer of the fruit forms a hard, stony pit that is bound tightly to the seed. Peaches, cherries, plums, almonds, and olives are drupes (Figure 23.20b). Citrus fruits are special types of berries called **hesperidiums**, which have leathery rinds containing oil packets (Figure 23.20c). A **pepo** is similar to a berry but with a thick outer rind. Watermelons, pumpkins, and cantaloupes are pepos (Figure 23.20d). **Pomes** are fleshy fruits whose flesh comes from an enlarged floral tube and receptacle. Apples and pears are pomes as well as accessory fruits (Figure 23.20e).

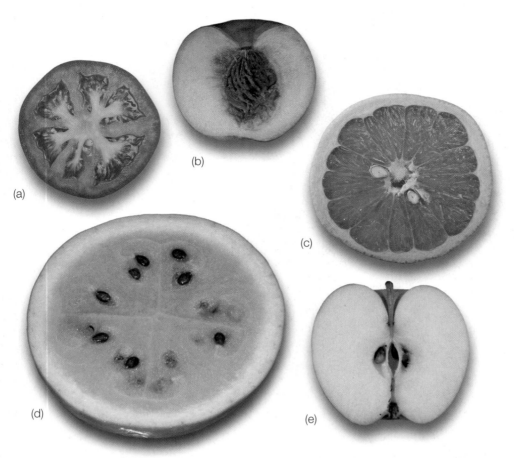

FIGURE 23.20 Simple fleshy fruits (a) Berry (of tomato). (b) Drupe (peach). (c) Hesperidium (grapefruit). (d) Pepo (watermelon). (e) Pome (apple).

Simple dry fruits are either dehiscent, meaning the walls of the mature ovary break open to free the seeds, or indehiscent, in which case the seeds remain inside the fruit after it is shed from the plant. There are four basic types of dehiscent, simple dry fruits. The **follicle** is a dry fruit formed from a single carpel that splits down one side along a seam to release seeds. Follicles are found in milkweeds, columbines, magnolias, and peonies (Figure 23.21a). A **legume**, which is the characteristic fruit of the pea family (Fabaceae), is similar to a follicle but splits along two seams. Thousands of species—including peas, beans, lentils, and mesquite—have legumes as fruits (Figure 23.21b), and members of the pea family are therefore commonly known as "legumes." Plants in the mustard family (Brassicaceae), such as *Arabidopsis*, broccoli, and cabbage, bear a fruit called a **silique**, which is formed from two fused carpels (Figure 23.21c). At dehiscence, the silique splits open on two sides, and the seeds are left attached to a central region. The most common type of simple, dry, dehiscent fruit is the **capsule** (Figure 23.21d). Capsules consist of two or more fused carpels and can split open in a variety of ways. Capsules may split along the length of the carpels, or they may form a cap on one end that pops off.

Other types of capsules form rows of pores, through which the seeds shake out. Irises, orchids, poppies, snapdragons, and cotton all form capsules.

Indehiscent simple dry fruits include the achene, samara, schizocarp, nut, and grain or caryopsis. The most common type is the **achene**—a small simple dry fruit containing a single seed (Figure 23.22a). The seed is attached to the fruit wall at a single point, making it easy to separate the seed from the fruit wall. Sunflower fruits are achenes, and the edible seed is readily separated from the fruit wall or husk. In elm and ash trees, the fruit is called a **samara** and is like an achene but with a wing around each seed (Figure 23.22b). In the **schizocarp**, the fruit opens into two or more parts, each with a seed and a wing. Schizocarps are the fruits in carrots, celery, and maples (Figure 23.22c). Grasses in the family Poaceae, which includes cereal crops such as wheat, barley, and corn, have fruits called **grains** or **caryopses**, which are similar to achenes but have the outer seed coat fused to the fruit wall (Figure 23.22d). In **nuts** such as acorns, chestnuts, and hazelnuts (filberts), the fruit wall is stony and the nut is derived from multiple carpels in a compound ovary (Figure 23.22e). Most so-called nuts are not nuts in a botanical sense. The peanut is a legume. Many "nuts" are actually drupes,

FIGURE 23.21   Simple dehiscent dry fruits (a) Follicle (of milkweed). (b) Legume (lupine). (c) Silique (garlic mustard). (d) Capsule (white campion).

FIGURE 23.22   Simple indehiscent dry fruits (a) Achenes (sunflower). The narrow end of the sunflower seed represents the single point of attachment to the fruit. (b) Samara (elm). (c) Schizocarp (maple). (d) Grain (corn). (e) Nut (hazelnut).

including almonds, coconuts, walnuts, pecans, and cashews. Brazil "nuts" are seeds from a capsule.

Complex fruits develop from multiple pistils or multiple flowers

Thus far we have described only simple fruits. The remaining types of fruits can be described as complex, in that they develop from multiple pistils or multiple flowers. **Aggregate fruits** are formed from a single flower with multiple pistils. Each pistil develops into a fruitlet, and all the fruitlets mature on a single receptacle. Raspberries, blackberries, strawberries, and magnolia fruits are aggregate fruits (Figure 23.23a). In strawberries, the receptacle becomes the large, fleshy red "fruit," and the "seeds" on the surface are achenes formed by the individual pistils (Figure 23.23b). Most "berries" like raspberries and strawberries are not berries in the botanical sense. The strawberry is both an aggregate fruit and an accessory fruit because of the development of the receptacle.

Multiple fruits are formed when the fruitlets of individual flowers in a single inflorescence fuse together to make a single large fruit (Figure 23.23c). Examples of multiple fruits are mulberries, Osage orange, figs, and pineapples.

23.5 | Seed germination and the formation of the adult plant

After a seed has reached maturity and been dispersed, it frequently undergoes a period called **dormancy**, during which the embryo does not grow. The germination of the seed marks the resumption of growth of the embryo. After germination, plants follow different patterns of development toward the adult form.

Germination requirements are closely linked to the environment

Biochemical changes occur within the seed during dormancy. In temperate regions, it would not be adaptive to disperse seeds in the fall if the seeds germinated immediately and then froze in the coming winter. Dormancy, and the accompanying biochemical changes, ensures that environmental conditions will be suitable for growth of new seedlings when germination occurs.

Germination is triggered by a number of factors, both external and internal. Among the external or environmental factors, the most important are temperature, water availability, and oxygen levels. The seeds of a particular species of plant will have a temperature range above or below which they will usually not germinate. The optimum temperature for most plants is around 25° to 30°C and the minimum is about 5°C. Most seeds are dry and have a water content that is 5 to 15% by weight. Germination requires that seeds

FIGURE 23.23 Complex fruits (a) Raspberries are aggregate fruits. (b) Strawberries are both aggregate and accessory fruits. The fleshy portion is composed largely of receptacle tissue and the "seeds" are actually achenes. (c) Multiple fruit of pineapple.

absorb water to resume metabolic activity. The water, in turn, activates enzymes that mobilize the stored food. Cells enlarge and resume growth. Initially, energy may be obtained by fermentation, but once the seed coat is broken, metabolism switches to respiration and oxygen is necessary. If the soil is deficient in oxygen, as in waterlogged soils, the seeds of many plants will fail to grow.

The germination requirements of plants are linked to the conditions of the environments in which they grow. Many temperate plant seeds require a period of winter cold before they will germinate. In deserts, seeds often will germinate only if there is sufficient rainfall to leach out chemical inhibitors. The level of rainfall needed to remove the chemical inhibitors is sufficient to ensure that the plant can grow to maturity. Waterlogged soils kill most seeds, but those of the cattail *(Typha)* will germinate only if the waterlogged soil lacks oxygen. In Mediterranean-type environments, such as California chaparral, some seeds require fire to release them for germination. Once fire has burned through the area,

another fire is unlikely for years, and seeds can germinate and plants grow to maturity.

After germination, plants follow various patterns of development

After germination, the first structure to emerge from the seed is the embryonic root or radicle. This primary root anchors the seed in the soil and allows it to absorb water. In eudicots, the primary root becomes the **taproot**, which grows down into the soil and develops lateral roots. The lateral roots form more lateral roots, and a complex branching system of roots results. By contrast, the primary root of monocots is temporary and is replaced by other roots at a later stage of growth.

The primary shoot also emerges from the seed in ways that differ among plant species. In the garden bean *(Phaseolus vulgaris)*, the radicle emerges first, followed by the hypocotyl, which is the portion of the embryo lying between the radicle and the cotyledons. The hypocotyl elongates and bends toward the soil surface (Figure 23.24a). When the hypocotyl reaches the surface, it straightens out and draws the cotyledons and plumule out of the soil. The shoot tip is protected from damage by being pulled out of the soil rather than being pushed through the soil. This growth pattern, in which the cotyledons are pulled out of the ground, is called *epigeous*, where *epi* is Greek for "above" and *geo* means "Earth." The food stored in the cotyledons is digested and used to support the growth of the seedling. By the time the cotyledons wither and fall off, the seedling will be established as an independent, photosynthesizing plant.

The growth pattern of the pea *(Pisum sativum)* differs from that of the garden bean. In peas, the embryonic stem (the epicotyl) elongates and forms the bend that draws the plumule out of the soil. The cotyledons remain behind in the soil where they slowly decompose after the stored food has been used (Figure 23.24b). Seed germination in which the cotyledons remain behind in the soil is called *hypogeous*, where *hypo* is Greek for "under." In most eudicots the cotyledons act as a source of stored food and do not carry out any photosynthesis.

In the majority of monocot seeds, the stored food is located in the endosperm, and the cotyledon digests and absorbs the stored food and passes it on to the embryo. In some monocots, like the onion *(Allium cepa)*, the single tubular green cotyledon follows the primary root out of the seed and forms a bend (Figure 23.24c). As the cotyledon straightens, it pulls the seed coat and endosperm out of the ground. The onion therefore has an epigeous pattern of seed germination. The developing onion seedling receives food from the endosperm by way of the digestive activities of the cotyledon and also directly from the cotyledon, which is photosynthetic.

In other monocots such as corn *(Zea mays)*, the large cotyledon, the scutellum, remains beneath the soil surface within the seed, where it obtains nutrients from the endosperm and passes those nutrients on to the embryo. Corn has a hypogeous form of seed germination. The coleorhiza, which is the sheath

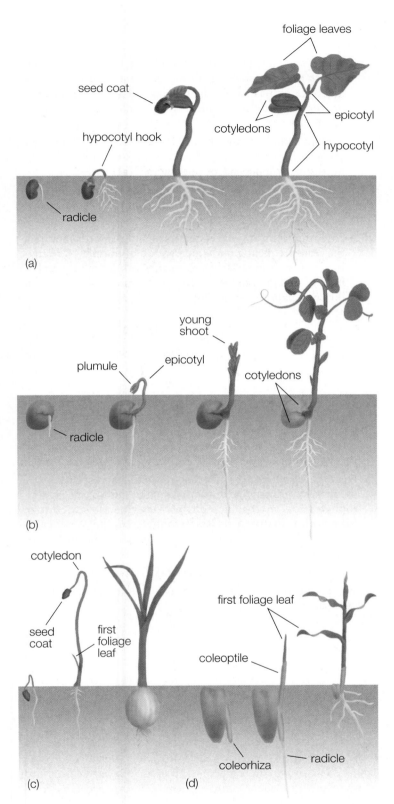

that encloses the radicle, is the first structure to emerge from the seed (Figure 23.24d). The radicle then grows out rapidly through the coleorhiza and additional roots emerge from the seed. The coleoptile, the sheath surrounding the plumule, then elongates upward toward the soil surface. When the coleoptile reaches the surface, it spreads open, and the first leaves of the plumule appear.

After the seedlings have emerged from the ground, the shoot apical meristem proceeds to form an orderly pattern of leaves, nodes, and internodes. Apical meristems form in the axils of leaves, in turn giving rise to more shoots and leaves until the adult plant is formed (Chapter 8).

FIGURE 23.24   Epigeous and hypogeous seed germination
(a) In garden bean (*Phaseolus vulgaris*), the cotyledons are pulled aboveground as the seed germinates (epigeous germination). (b) In pea (*Pisum sativum*), the cotyledons remain beneath the soil surface (hypogeous germination). (c) In onion (*Allium cepa*), the single green cotyledon draws the seed coat and endosperm out of the soil (epigeous germination). (d) In maize (*Zea mays*), the endosperm and cotyledon remain below the soil surface and the coleoptile extends upward to carry the plumule above the ground.

HIGHLIGHTS

23.1 Flowering plants occur in an enormous range of sizes and growth forms. Most flowering plants, or angiosperms, reproduce sexually by forming flowers and carry out a unique process of double fertilization in which one fertilization produces a zygote and the other a nutritive tissue. Most modern angiosperm species are either eudicots or monocots.

23.2 A flower is a reproductive shoot whose parts evolved from leaves. These parts are grouped into four whorls: the outermost sepals; the petals—brightly colored to attract pollinators; the stamens—the reproductive organs that produce pollen in their anthers; and the carpels—the reproductive organs that contain the ovules. Flowers may be grouped into different types of inflorescence depending on their mode of pollination.

23.3 In most flowering plants a double fertilization results in the formation of a diploid zygote that develops into a plant embryo and a triploid endosperm tissue that provides nutrition to the embryo. Some flowering plants reproduce asexually by apomixis, in which both egg cell and polar nuclei are diploid and develop directly into embryo and endosperm.

23.4 A plant embryo passes through a number of developmental stages before a mature seed is formed. A mature seed is nutritionally independent of the parent plant.

23.5 A fruit is a mature ovary containing seeds. Fruits come in many forms but may be classified as simple, aggregate, or multiple.

23.6 After the fruits have been dispersed and the seeds matured, the seeds frequently undergo a period of dormancy. Many kinds of environmental factors can break dormancy and initiate seed germination. After germination, plants follow different patterns of development to maturity.

REVIEW QUESTIONS

1. How do simple fruits differ from complex fruits, and aggregate fruits from multiple fruits? Give an example of each.

2. Describe some of the external (environmental) requirements that are linked to seed germination.

3. What does *angiosperm* mean? How do angiosperms differ from (a) gymnosperms? (b) pteridophytes (ferns)? (c) bryophytes (mosses and liverworts)?

4. Distinguish monocots from dicots by describing some typical features of each.

5. Distinguish between the following words, all of which are concerned with flower structure: (a) *peduncles* and *pedicels*; (b) *petals, sepals,* and *tepals*; (c) *calyx* and *corolla*; (d) *stamens* and *carpels*; (e) *filaments* and *anthers*; and (f) *ovary, style,* and *stigma*.

6. Distinguish between the following types of flowers, based on the arrangements of floral parts: (a) complete versus incomplete; (b) perfect versus imperfect; (c) staminate versus carpellate; (d) monoecious versus dioecious; (e) superior versus inferior ovaries; and (f) regular versus irregular flowers.

7. Distinguish between (a) megaspores and microspores; (b) megasporangia and microsporangia; and (c) megagametophytes and microgametophytes. Describe the structure of (d) a typical mature microgametophyte, and (e) a typical mature megagametophyte. Lastly, (f) what is meant by double fertilization?

8. What is a fruit? What is an accessory fruit?

9. What is the difference between a dehiscent and an indehiscent fruit?

APPLYING CONCEPTS

1. Imagine that you are in the produce section of your local grocery or health food store. Identify some of the fruit types you might find there.

2. In many flowers, pollen is produced either well before or considerably after the carpels mature and develop an embryo sac. What is the advantage to the plant of this kind of shift in maturation times?

3. Are all imperfect flowers also incomplete? Are all incomplete flowers also imperfect? Explain your reasoning.

4. Tomatoes are often referred to as vegetables. Does this make sense botanically?

5. Are all flowering plants either monoecious or dioecious? Explain your reasoning.

6. Construct a simple dichotomous key to the following fruit types: berries, drupes, pomes, follicles, legumes, capsules, achenes, samaras, grains, and nuts.

CHAPTER 24

Flowering Plant and Animal Coevolution

Animal seed dispersal

- Flowering plants and many types of animals have changed over time in response to each other, providing many examples of the process known as coevolution.

- Pollinators help plants outbreed, which confers genetic and evolutionary advantages. Flower characteristics are correlated with those of their pollinators.

- Animals that disperse plant seeds prevent competition between plants and their progeny and help plants colonize new habitats. Plant fruit and seed characteristics are often correlated with the behavior of animal dispersal agents.

Hosts of attractive butterflies flitting among beautiful flowering plants—a common and appealing sight, but not a coincidence. Butterflies and flowering plants are natural partners. They form alliances that aid each other's survival and reproduction. If you watch butterflies over a period of time, you may notice that different butterfly species tend to visit particular flowers. Monarch, black swallowtail, and great spangled frittillary, for example, tend to favor flowers of the butterfly weed (*Asclepias tuberosa*) and butterfly bush *(Buddleia davidii)*. Flowers of these plants have colors, shapes, and fragrances that specifically attract these and other butterflies and provide nectar—butterflies' major food. When butterflies visit flowers to harvest nectar, they also transport pollen from one flower to another of the same type, which helps the plants accomplish mating. Consequently, butterflies and flowering plants have evolved together—coevolved—in ways that foster the continuation of this beneficial alliance.

Animals and plants are also coevolved in ways that help plants disperse their seeds and protect them from damage. Coevolution is important because it helps explain why there are so many species of flowering plants, butterflies and other insects, hummingbirds and other birds, and mammals on Earth today. ■

Butterfly pollination

24.1 | Coevolutionary interactions between flowering plants and animals are common and important in nature and human affairs

Think of bees burrowing into one flower after another in a flowerbed or homing toward the lush blossoms of an apple orchard. Or recall the common sight of birds searching out the most lushly ripe berries, to the annoyance of fruit growers. In both cases, the plants receive benefits; the bees aid flowers in pollination, while birds help disperse seeds contained within delicious fruits. In tropical regions many people know that if they disturb the bullhorn acacia, masses of angry ants that live in the acacia's large, hollow thorns and receive special food supplies from the plant (Figure 24.1) will rush to defend their home. Not only do the ants chase off animals that might consume the acacia but the ants will also bite off the growing tips of nearby plants that invade the acacia's space. These are all examples of **coevolutionary plant-animal associations**—alliances that have influenced the evolution of

both partners. In these and other examples provided in this chapter, the plants have acquired traits that are attractive to particular animals, and the animals' behavior has evolved in ways that benefit plants.

Because they cannot move in order to mate, disperse to new locations, or avoid animals that would eat them, plants must accomplish these functions in other ways. Some plants use wind or water to carry out mating and seed dispersal, and many plants utilize chemical deterrents to discourage herbivory (Chapter 3). But alternative evolutionary strategies involve the use of mobile animals to transport pollen and seeds or perform as security guards. Plants usually offer food rewards for such services. For example, the bullhorn acacia produces special ant-sized protein and lipid-rich food nuggets (known as Beltian bodies) at the tips of leaflets (see Figure 24.1). These food bodies, together with nectar secreted from leaf petioles, provide a nutritious diet for their ant protectors. Flowering plants typically attract animal pollinators with colorful, fragrant flowers, and they attract seed dispersers with colorful, odorous, nutritious fruits. Animal sensory systems are coevolved to detect these color or fragrance cues. Humans have similar sensory systems, explaining why many flowers and fruits are also attractive to people. Humans use flower and fruit essences in perfumes and they also use flowers and fruits as motifs in art, architecture, clothing design, and interior decoration.

Coevolutionary relationships are not always mutually beneficial. There are many plants that "trick" animals into providing services without reciprocating with nutritional rewards. For example, some flowers look or smell so much like female insects that males of the same insect species will attempt to mate with the flowers, thereby accomplishing pollen transfer. Plants and their bacterial, fungal, and viral pathogens engage in coevolutionary warfare. As pathogens acquire new attack capabilities, plants respond by evolving defenses (see Chapter 12). In this chapter we focus on pollination and dispersal coevolutionary interactions involving flowering plants and animals. Understanding these relationships is particularly important in agriculture as well as conservation of biological diversity.

FIGURE 24.1 Bullhorn acacia This acacia species produces large hollow thorns, in which colonies of ant defenders live. The plant also produces protein and lipid-rich food nuggets—the numerous light-colored bodies at the tips of leaflets (known as Beltian bodies). These, together with sugary nectar generated by petiole nectaries (arrow), help fulfill the ants' food requirements.

Flowering plant-animal coevolutionary interactions are important in agriculture

Fleshy, sugar-rich fruits, many of which humans have developed into valued crops, are the evolutionary product of plant-animal dispersal relationships. Delicious fruits such as blueberries, strawberries, and raspberries evolved as a food reward for the birds and mammals that disperse their seeds. Thus, fruit crops are adapted by nature in ways that also make them useful as human food. Production of apple, cherry, and many other fruit crops depends on a reliable source of bee pollinators. Many growers use the services of beekeepers, who move their hives from orchard to orchard to perform essential pollination services, with honey as an additional valued product (Figure 24.2). Gardeners also rely on insect pollinators. The unisexual flowers of squashes, for example, cannot produce fruit without the pollination services of bees.

FIGURE 24.2   Beehives are often brought to orchards to increase pollination

FIGURE 24.3   *Brighamia* The specialized pollinator of this endangered flowering plant has apparently become extinct.

The widespread use of insecticides to control insects that damage crops or carry diseases poses a danger to insect pollinators important in agriculture, because chemical sprays typically kill a wide spectrum of insects. For example, chemical sprays applied by airplane to the forests of eastern Canada to control destructive spruce budworm also killed native bees needed to pollinate cultivated blueberries, causing high losses to fruit growers. When farmers and gardeners reduce or eliminate the use of chemical insecticides, beautiful pollinators such as butterflies are more abundant and plants can be more fruitful.

The development of genetically modified (GM) crops that produce insect-killing proteins in their tissues is a strategy designed to provide crops with protection from harmful insects, while at the same time reducing the need to use chemical sprays. However, some ecologists are concerned that GM plants or pollen harm butterflies and other pollinators. Further, there is a danger that foreign genes in GM crops might be transmitted to wild plants by cross-pollination (Chapter 15). Scientists who aim to assess the ecological risks posed by GM crops must be aware of pollination processes.

DNA SCIENCE

Coevolution is important in global ecology

Continued existence of most flowering plants in nature depends on the presence of appropriate pollinators or dispersal agents. Some plants rely on a single species of pollinator or disperser. If the animal upon which they depend becomes locally or globally extinct, such dependent plants are more vulnerable to extinction and so are any animals that rely upon them as a food source. For example, a beautiful Hawaiian flowering plant (*Brighamia*) (Figure 24.3) has stopped

ECOLOGY

reproducing naturally because it can be pollinated in nature only by one type of insect, a moth that has become extinct. Only a few individual *Brighamia* plants are left, and this plant continues to reproduce in the wild only because volunteer humans rappel down steep cliff faces to hand-transfer pollen between plants. Understanding pollination, dispersal, and defense coevolutionary interactions is essential to developing strategies for preserving endangered plant and animal species, restoring damaged habitats, and maintaining high global biodiversity.

24.2 | Cross-pollination is the transfer of pollen from one flower to another of the same species

Flowering plants must have mechanisms for bringing pollen to the stigmas of flowers in order for them to reproduce sexually. Such pollen transfer is known as **pollination** (Chapter 23). If pollination is accomplished and pollen and stigma are compatible, a pollen grain will germinate to form a pollen tube, which transports sperm cells to eggs contained within ovules in the ovary. For most seed plants, seeds are essential for survival of the species, and pollination is necessary for seed production. The flower is an adaptation that increases the efficiency of pollination and is one of the traits that explain the dominance of flowering plants on Earth today.

Most angiosperms produce **bisexual flowers** that possess both pollen-producing stamens and ovule-containing carpels. Bisexual flowers are both common and ancient. The oldest living groups of flowering plants (*Amborella* and some water lilies) have bisexual flowers, suggesting that earlier angiosperms also had them. Some plants are self-fertile, that is, capable of self-pollination and self-fertilization. This being the case, why do so many plants cross-pollinate—exchange pollen among individuals of the same species? The answer is that pollen exchange facilitates outbreeding, also known as outcrossing. How is outbreeding advantageous to plants?

Outbreeding provides greater genetic variability than inbreeding

Human societies often prohibit marriage between very close relatives as a way to reduce the deleterious effects of **inbreeding**—the combination of two genomes that are genetically very similar. Similarly, many plants have mechanisms to avoid self-pollination, because the progeny of a self-mating may have too many deleterious recessive genes (Chapter 14), which could result in poor condition or death. **Outbreeding**—mating of sperm and egg from genetically different organisms of the same species—is typically advantageous. One reason is that the number of deleterious recessive genes can be lower. Outbreeding also generates progeny that are more genetically diverse, providing the evolutionary flexibility needed to survive environmental change or colonize new habitats (Chapters 13 and 16).

Flowering plants have several ways to reduce the incidence of inbreeding. One of these is genetically controlled self-incompatibility, first described in plants by Charles Darwin, the father of evolutionary biology. In some flowering plants, genes expressed in cells at the stigma surface allow genetically distinct pollen from flowers of the same species to germinate, but not genetically identical pollen. In some other angiosperms, a flower's own pollen can germinate on the stigma, and the pollen tube can begin to grow through the style toward the ovary; but enzymes produced within the pollen tube destroy pollen-tube RNA, which causes the tube to stop growing (Figure 24.4). These early roadblocks prevent the formation of unhealthy, inbred embryos, allowing eggs the chance to successfully outbreed.

DNA SCIENCE

In other flowering plants, stigmas become receptive to pollen before stamens of the same flower become mature, or anthers mature before the carpels. In *Lobelia* flowers, for example, pollen is produced before the ovules are formed; but in oats, the stigmas mature before pollen is produced. Because flowers of different ages are produced by such plants, pollen from any given flower is more likely to reach a receptive stigma on a different plant, thereby promoting outbreeding.

Other plants promote outbreeding by forming **unisexual flowers**—"male" flowers that possess stamens but not carpels, and "female" flowers that have carpels but not stamens. The terms **male** and **female** are in quotes because flower organs are structures belonging to the sporophyte body. Recall that plants have two multicellular bodies—sporophytes, which produce spores, and gametophytes, which produce gametes. Hence, gametophytes can be male or female, but sporophytes are not of one gender or the other. However, flower organs can contain male or female gametophytes within their tissues. DNA-sequence-based studies of plant relationships reveal that unisexual flowering plants evolved many times from ancestors having bisexual flowers. Examples of crop plants that produce unisexual flowers are spinach, hemp, asparagus, squash, and maize (corn).

(a)

(b)

FIGURE 24.5 Some plants may produce seed by self-pollination
Violets are plants that can produce seed by both cross-pollination or self-pollination. Self-pollination is accomplished by flowers that remain closed and are found on or just beneath the soil surface (a). (b) A typical, cross-pollinating open violet flower (left) is compared to a closed flower (right).

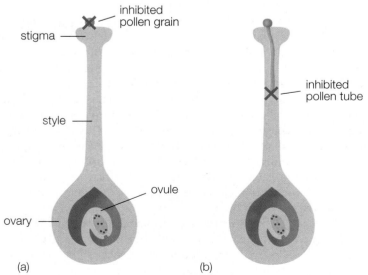

(a)   (b)

FIGURE 24.4 Flower mechanisms that promote outcrossing
(a) Pollen that is incompatible, or too genetically similar to ovules, does not germinate on the stigma. (b) Pollen tube growth may be inhibited as it grows through the style.

Even though outbreeding would seem preferable, inbreeding does occur in some plants. DNA-based studies of plant relationships indicate that some flowering plants that had self-incompatible ancestors have subsequently lost this trait and are now self-fertile. Some other plants, such as *Impatiens* (jewelweed) and violets *(Viola)*, produce open flowers that can outbreed and/or closed flowers that are forced to inbreed (Figure 24.5). Such plants have the "fail-safe" advantage of being able to produce seeds if cross-pollination is not possible.

Animal pollinators offer precision, high-fidelity pollination services

Many seed plants achieve pollination by the use of wind or water (Essay 24.1, "Pollination by Wind and Water"). However, most flowering plant species are pollinated by animals. Nearly 300,000 animal species serve as pollinators, including insects such as bees, butterflies, moths, wasps, thrips, and flies, as well as many bats, birds, and some mammals (certain rodents, marsupials, and primates). Most plants are attractive to more than one

---

ESSAY 24.1 POLLINATION BY WIND AND WATER

Many plants have not relied on pollination coevolution. Rather, they are adapted to use the physical properties of wind or water to disperse their pollen. About 10% of flowering plants are wind pollinated. An advantage of wind pollination is that plants do not have to expend precious resources to produce brightly colored petals or fragrances that are attractive to animal pollinators. However, direction, distance, and timing of pollen transport by wind vary greatly—a disadvantage of this pollination mechanism. Tests using sticky microscope slides to trap pollen reveal that wind-borne pollen often travels only short distances—

About 10% of flowering plants are wind pollinated.

only 6–9 meters in the case of maize (corn). In some cases, however, pollen may travel up to 300 km per day. In addition, the stigma is a very small target in comparison to the large volume of air into which pollen can be dispersed. For this reason, flowers that disperse pollen via wind must produce very large amounts of it—an estimated 1 million pollen grains per square meter of habitat—in order to achieve necessary pollination rates. A single birch tree catkin ("male" inflorescence; Figure E24.1A) produces more than 5 million pollen grains. This abundance of pollen explains the tendency of many wind-pollinated plants, including trees and grasses, to cause pollen-allergy problems in humans. The flowers of wind-pollinated plants typically lack showy petals and are often unisexual and grouped into inflorescences that dangle in the wind.

Several flowering plant groups have members that live in water—freshwater ponds, streams, and lakes, or shallow marine waters along seacoasts. These plants descended from land-adapted plants but now are adapted to life in the water. Most aquatic flowering plants produce flowers above the water surface that are wind pollinated or attract flying pollinators, much like their land-dwelling relatives. But more than 30 genera of aquatic flowering plants use water to transport pollen, even though water can greatly dilute pollen, decreasing the chances of successful pollination. To cope with this problem, most water-pollinated flowering plants have flowers that float at the water surface, where pollen transfer occurs. For example, tiny "male" flowers of the freshwater ribbon-weed *Vallisneria* break off from

the plant and float to the water surface (Figure E24.1B). These flowers produce globules of pollen on stiff stamens that function like sails, and water movements bring them to water-surface depressions in which lie "female" flowers whose stigmas project above the surface. Once pollinated, female flowers are pulled beneath the water surface by coiling of the peduncle that attaches them to the rest of the submerged plant. The sea grasses *Zostera* and *Halodule* produce unusually long, thin, spaghetti-like pollen grains that drift in spider-web-like rafts along the water surface until they reach floating "female" flowers. The female flowers have elongate stigmas that snag the stringy pollen from the waves.

Only a few aquatic flowering plants are pollinated underwater. *Ceratophyllum*, a freshwater descendant of an ancient lineage of flowering plants, has "break-away" stamens that float to the water surface where pollen is released. The pollen then sinks, eventually contacting stigmas of submerged flowers.

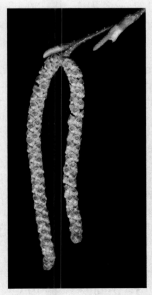

E24.1A Birch *(Betula)* catkin (male inflorescence)

E24.1B *Vallisneria* male and female flowers

species of pollinator (Figure 24.6). For example, 334 species of insects belonging to 37 different families were observed to visit carrot flowers in a Utah field. In addition, most pollinators are attracted to several or many plant species. Honeybees are a good example—they pollinate many species of wildflowers as well as numerous types of fruit crops. This redundancy increases the chances of survival should a particular pollinator or plant species disappear. Desert and tropical forest plant populations are often sparse or widely scattered (Chapters 26 and 27), as are nongrass herbs of prairies and savannas and herbaceous plants of temperate forest floors (Chapters 26 and 27). In these habitats, animal pollinators are typically more effective than wind, because animals offer greater precision in transporting tiny pollen grains to small stigma targets. Flying insects—particularly beetles, bees, flies, butterflies, and moths—match the sizes of inflorescences, flowers, and flower parts and are large enough to effectively transport masses of pollen from flower to flower. Flying vertebrates—birds and bats—are important pollinators in deserts and tropical forests. These long-distance travelers, together with large bees, are capable of transporting pollen between widely separated individuals of the same plant species. Large vertebrates require a greater amount of food reward than insects and, thus, vertebrate-pollinated flowers are often larger and capable of producing more food than those pollinated by insects.

Plant food rewards to animal pollinators include nectar, pollen, and oil

Nectar is a solution of sugar, amino acids, and other substances; it is produced in **nectaries**, glands that may occur in various locations within flowers (Figure 24.7). Flowers differ in the types of sugar (typically glucose, fructose, or sucrose) or amino acids present in nectar, as well as in nectar concentration and amount. These variations often correlate with the dietary requirements of pollinators. Pollen is rich in protein (some pollen is 16–60% protein) and lipid (3–10%), and oils have twice the caloric value of carbohydrates. Flowers such as poppies *(Papaver)*, peonies *(Paeonia)*, and kiwifruit *(Actinidia deliciosa)* produce only pollen as a food reward. Other flowers offer primarily nectar or oil, and some provide a mixture of

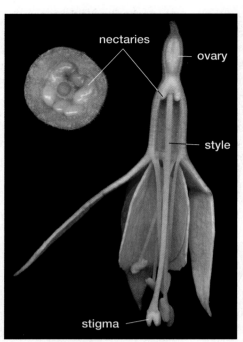

FIGURE 24.7 *Fuchsia* **nectaries** The nectar-secreting glands (nectaries) of *Fuchsia* are the yellow-green objects at the base of the floral tube. The tube accumulates nectar, which can be harvested by hummingbirds.

foods. This variety makes sense because some pollinators require only sugary nectar, while others also need protein and/or oil in their diets.

Flowering plant species that occur in the same habitat may benefit from each other's presence. For example, Queen-of-the-night cacti *(Peniocereus* sp.*)* are rare plants that bloom for only a few nights per year. These cacti benefit from the fact that abundant neighboring *Datura* (sacred Datura or jimson weed) plants have similar trumpet-shaped flowers that keep the hawkmoth and bee pollinators of both plants present from spring to fall (Figure 24.8).

Plants attract pollinators by flower scent, color, shape, and arrangement

Flowers can be very effective advertisers. Because pollinators vary in their ability to detect odors, colors, and shapes as well as in their pollination effectiveness, flowers are adapted in ways that attract the animals best adapted to pollinate them.

Flowering plants use more than 700 compounds as floral odors. Flowers that are similar in appearance may have very different odors. The chemical composition of flower odors can be analyzed and fragrance compounds placed into three major categories: (1) flowery scents, (2) compounds that mimic the sex attractants (pheromones) of insects, and (3) dung or rotten-meat smells. Flowery fragrances result primarily from terpenes such as geraniol (produced by roses) and limonene (of *Citrus*), or phenolics such as vanillin (of *Vanilla* orchids) and eugenol (cloves). Sex-attractant odors are conferred by hydrocarbons, alcohols, terpenoids, and other compounds. They are used to trick insects into trying to mate with flowers rather than offer-

FIGURE 24.6 Plants may be visited by multiple pollinators

FIGURE 24.8 **The Queen-of-the-night cactus (*Peniocereus*) and *Datura*** The rare cactus (a) depends upon pollinators attracted by more abundant *Datura* flowers (b).

ing them food rewards! For example, the Australian orchid *Chiloglottis trapeziformis* attracts its male wasp pollinators, *Neozeleboria cryptoides*, by emitting the same perfume that is used by female wasps of this species to attract mates. Flowers that smell like dung or carrion fool pollinators such as flies into mistaking them for food sources with compounds (such as amines, ammonia, and indoles) that are also produced during the decay of protein in dung and dead meat. The odor of such flowers is really offensive to people (you are well advised not to take a deep whiff of dung or carrion flowers or to offer them as a token of love or respect!).

Flower color compounds include anthocyanins and anthoxanthins (which are flavonoids), betalains (related to alkaloids), and carotenoids (related to terpenes) (Chapter 3). Anthocyanins generate blue, purple, and pink flower colors, including the widely admired blue of cornflower (*Centaurea cyanus*), the common roadside plant chickory (*Chichorium intybus*), and *Hydrangea macrophylla* (Figure 24.9). Anthoxanthin colors range from pale ivory to yellow. Some white flowers lack pigments altogether; the white effect is caused by reflection and refraction of light at many cell surfaces, as is also the case with snow. Examples of the yellow, orange, or red pigments conferred by betalains are yellow *Portulaca* and the red-

FIGURE 24.9 **Blue flowers** (a) Chickory (*Cichorium*) and (b) *Hydrangea*

to-purple flowers of cacti and the tropical or subtropical ornamental vine *Bougainvillea* (Figure 24.10). Daffodil and marigold petals are colored yellow, orange, or red by

FIGURE 24.10 **Yellow and red-purple pigments** (a) *Portulaca* flower. (b) *Bougainvillea* has colorful leaflike bracts that attract pollinators to inconspicuous flowers hidden within them (inset).

carotenoids. The daffodil gene for beta-carotene was used to genetically engineer rice so that this widely planted crop has a greater nutritional value for people (Chapter 15). Recall that beta-carotene is required in animal and human diets for normal eye function and skin health (Chapter 19).

Flower pigments are located either in cell vacuoles or in chromoplasts, which are modified chloroplasts. Some flowers have combinations of pigments in their cells or produce different pigments in separate tissue regions, which accounts for the color patterns on flower petals. Patterns of colored spots or lines often function as "highway signs" or "runway lights," guiding pollinators to food rewards and the reproductive parts of the flower, and are known as nectar guides. Some of these patterns are not visible to humans, but they are visible to pollinators having different visual capabilities. For example, some flower petals appear pure yellow to human eyes. But the centermost part of the petals is pigmented with a flavonoid that strongly absorbs ultraviolet light. This produces a color contrast that guides bees to the location of nectar and pollen (Figure 24.11).

Flowers control pollinator access by flower shapes and positions

Some pollinators are able to hover, whereas others need a place to land—a landing platform. Only animals that are able to hover can obtain rewards from and pollinate flowers that hang upside down (such as the common ornamental *Fuchsia*). Flowers that attract landing animals provide them with a strong foothold in the form of rigid flower parts that they can grasp or a broad flat surface upon which to stand—either large single flowers that face upward such as roses, or inflorescences such as marigolds, dandelions, and asters. Small insects can wander over the surfaces of inflorescences, taking rewards from each of many small flowers in turn and pollinating them in the process (Figure 24.12).

Many flowers have complex structures that prevent entry by pollinators unless they are strong enough to push flower parts out of the way. Flowers of snapdragon, as well as certain *Penstemon* and orchid species, are examples (Figure 24.13); only bees are strong enough to enter. Other flowers tailor the dimensions of tubes formed by petal fusion to allow entry only by favored pollinators or the mouthparts they use to obtain food. Some flowers,

FIGURE 24.12 Landing platforms (a) Rose flower. Inflorescences of aster (b) and dandelion (c).

FIGURE 24.11 Nectar guides (a) Inflorescence photographed with regular film. (b) Same photograph, but with UV-sensitive film. The pollinator is able to see the UV-reflecting inner petals, which helps it locate pollen and nectar in the inflorescence center.

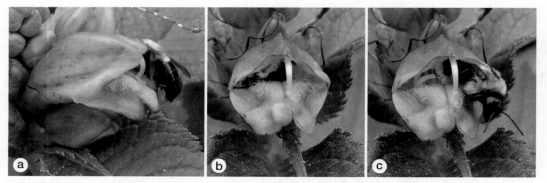

FIGURE 24.13 *Penstemon* has intricate flowers that require forced entry After a bee lands on this flower (a), it forces its way deep into the flower (b), brushing up against the stamens and stigma of the flower, effecting pollination. When the bee leaves (c), the flower's "mouth" again closes, preventing smaller insects (which would be less likely to successfully pollinate the plant) from entering and "stealing" the nectar reward.

such as jack-in-the-pulpit (*Arisaema triphyllum*; Figure 24.14), occur within traps formed by leaflike structures that enclose pollinators for a time, increasing the chances that pollen transfer will occur.

Animal pollinators can learn to recognize the features of favored flowers

Flower colors, shapes, and odors allow pollinators to recognize and remember the sources of valued food rewards. Animal pollinators differ in sensory perception, behavior, size, and ability to transport pollen. Hence, flowers differ in petal or sepal color, fragrance, shape, and other features.

The ability of pollinators to recognize and repeatedly visit favored flower species is a characteristic known as **constancy,** or fidelity. Flowers that are highly specialized in some easily recognizable way are likely to attract constant pollinators. Pollinator constancy is favorable to the plant because the pollinator spends less time visiting other flower species, and so pollen brought to the stigma is more likely to be of the appropriate species. Constancy is also favorable to the animal, which requires less time to find the food reward in a familiar flower than in an unfamiliar flower. The less time an animal is exposed during feeding, the smaller the chance that its enemies will find and eat it.

Flowers and pollinators have coordinated traits known as pollination syndromes

It is often relatively easy to deduce pollinator type from the appearance of flowers (and vice versa) with a knowledge of **pollination syndromes**—the coordinated traits of flowers and pollinators (Table 24.1). Botanists use pollination syndromes to develop testable predictions about plant and animal pollination relationships. For example, South African botanists noticed that the African lily *Massonia depressa* had flowers similar to those of *Protea* plants—which are known to be pollinated by rodents. So these scientists hypothesized that *Massonia* was also pollinated by rodents. By using cages that excluded rodents from *Massonia* flowers, and other experiments, they demonstrated that this plant is indeed pollinated by rodents, including gerbils, which were not previously known to serve as pollinators (Figure 24.15). Knowledge of pollination syndromes is useful and enjoyable in everyday life, as well. For example, if you want to see more butterflies, hummingbirds, or beautiful moths in your garden, you can plant the particular flowers that are most likely to attract them.

Beetles and beetle-pollinated flowers

Beetles represent a very ancient group of insects that are still very diverse today, with thousands of modern species. Their high diversity has been attributed to extensive coevolution with flowering plants. Beetles are particularly important pollinators in dry habitats and in tropical forests. Some beetles pollinate certain gymnosperms (which originated prior to the first flowering plants) as well as members of some early lineages of flowering plants. Together with fossil evidence, this pattern suggests that beetle pollination arose before the first flowering plants. The first flowering plants were probably pollinated by a variety of insects, including flies, wasps, and moths, in addition to beetles.

Beetles consume ovules when they are accessible, as well as pollen and nectar. Some experts regard ovule consumption as evidence that beetle herbivory might have driven evolution of the first carpels—structures that protect ovules and are unique to flowering plants. Beetles have a good sense of smell, and at least some pollinating beetles appear to have color vision. But beetles' bodies are so hard and smooth that they are not the most efficient pollen carriers.

Flowers that are attractive to beetles tend to be large and sometimes flat, with easy access (Figure 24.16). Beetle-pollinated flowers offer excess pollen as a nutritional reward. Colors include white and purple, and the flowers often have a strong fruity odor. A number of beetle-pollinated plants produce flowers that generate heat (thermogenesis), which aids in diffusing odors, much as human body heat volatilizes perfumes. Thermogenesis also provides warmth to beetles and other insect pollinators (which are cold-blooded). *Magnolia tamaulipana*, whose flowers open in the evening, is a threatened species of the tropical cloud forests in Mexico. This *Magnolia*'s flowers heat by as much as 9°C, thereby attracting pollinating beetles by odor. Common, cultivated *Magnolia* species (see Figure 24.16) and skunk cabbage are also thermogenic and pollinated by beetles.

Bees and bee-pollinated flowers

Bees are excellent pollinators. Bees are strong enough to enter intricate flowers and often use landing platforms. Bees have "high blossom intelligence"; that is, they are able to recognize flowers that produce copious nectar and pollen (required for development of their young). Bees thus exhibit constancy, which benefits both plant and pollinator. Social bees are able to communicate the distance, direction, type, and

FIGURE 24.14 Jack-in-the-pulpit (*Arisaema triphyllum*) (a) The inflorescence of this plant consists of a "spadix" (jack) surrounded by a leaflike "spathe" (the pulpit). A cutaway view of the tiny flowers (arrowheads) that are borne on the basal portion of the spadix is shown in (b). Pollinators may be momentarily trapped in the confined space surrounding these flowers, increasing the odds for pollination success.

TABLE 24.1 Common Pollination Syndromes

Beetles	Beetle-pollinated Flowers
Ancient insect group	Ancient plant groups
Good sense of smell	Strong, fruity odors
Some have color vision	White and purple common
Hard, smooth bodies	Large, often flat, easy access
	Some thermogenic (heat-producing)
Bees	**Bee-pollinated Flowers**
Strong, with medium-length tongues	May be complexly structured
Color vision into UV, but not red	Colors typically blue, purple, lavender, yellow, or white (not red)
Good sense of smell	Nectar guides visible in UV may be present
Require nectar and pollen for feeding brood	Fragrant
Pollen carried in crop or baskets on legs	Nectar and pollen offered as reward
High blossom intelligence, communication within social group possible	
Nectar-feeding Flies	**Nectar-feeding Fly-pollinated Flowers**
Sense nectar with feet	Simple with easy access
Suck nectar with tubular mouthpart	Red or light color, little odor
Carrion Flies	**Carrion Fly-pollinated Flowers**
Attracted by heat or odors and colors of carrion or dung	Colored to resemble dung or carrion
Food in the form of nectar or pollen is not required	Produce heat or foul odors
	No nectar or pollen reward offered
Butterflies	**Butterfly-pollinated flowers**
Good color vision	Blue, purple, deep pink, orange, red
Sense of smell with feet	Fragrant; with light, floral scent
Feed with tubular proboscis	Nectar often in narrow deep tubes
Require landing platform	Landing platform provided
High blossom intelligence	

relative abundance of such flowers to other members of the hive by means of dances performed in the dark and felt by other bees, as well as by flower scent brought into the hive.

Bees have medium-length tongues that are able to collect nectar from many types of flowers. Nectar is carried in the bee's crop (a pouchlike enlargement of the gullet), where enzymes convert it to honey. Some bees also carry pollen in the crop; others use their legs to transfer pollen from flowers to a brushy region of their abdomen or carry it in basket-shaped structures on their legs (Figure 24.17). Bees have good color vision, which extends into the ultraviolet, but they are red-blind. Bees also have a good sense of smell, detecting odors (via antennae) in concentrations 10 to 100 times weaker than can be smelled by humans.

Bee-pollinated flowers are typically blue (Figure 24.18), purple, lavender, or white—but not red—and they may have nectar guides that are visible as spots (Figure 24.19). If you are allergic to bee stings, it is not a good idea to dress in floral patterns or in bee-flower colors! Archeologists and others who

TABLE 24.1 Common Pollination Syndromes	
Moths	**Moth-pollinated Flowers**
Many active at night	May be open at night
Good sense of smell	White (reflects moonlight) or bright colors
Feed with long, narrow tongues	Fragrant; with heavy, musky scent
Some require landing platforms	Nectar in narrow, deep tubes
Some can hover while feeding	Landing platforms often provided
Birds	**Bird-pollinated Flowers**
Most require rigid perching site	Large and damage-resistant
Good color vision, including red	Often red or other bright colors
Poor sense of smell	Not particularly fragrant
Feed in daytime	Open during day
High energy (nectar) requirements	Copious nectar produced within tubes
Bats	**Bat-pollinated Flowers**
Active at night	Open at night
High food requirements	Plentiful nectar and pollen offered
Color blind	Light or dingy colors
Good sense of smell	Strong, often batlike odors
Use echolocation to navigate (cannot fly in foliage)	Open shape, easy access, often pendulous or occur on trunks of trees
High blossom intelligence	
Nonflying Mammals (Primates, marsupials, or rodents)	**Nonbat Mammal-pollinated Flowers**
Relative large size	Robust, resistant to damage by pollinator
High energy requirements	Copious, sugar-rich nectar
Color vision may be lacking	Dull-colored
Good sense of smell	Odorous (but not necessarily fragrant)
	Located in canopy (primates) or on ground (rodents)

work in remote areas where bees can be a hazard—and medical treatment hard to find—purposely dress in khaki, olive, or other dull colors that do not attract bees.

Many bee-pollinated flowers are **bilaterally symmetrical**—the flowers can be divided into two equal parts by only a single plane (see Figures 24.13, 24.19). The terms **zygomorphic** and **monosymmetric** are also applied to bilaterally symmetrical flowers. In many cases, bilaterally symmetrical shapes appear to have evolved from **regular** flowers (Figure 24.20) as a result of coevolution with bees. (Regular flowers are also described as **actinomorphic** or **polysymmetric**.) Bee flowers sometimes occur in inflorescences where bees first visit the lowermost, mature flowers and then climb upward to younger flowers, eventually finding the best sources of nectar and pollen. As you would expect, bee-adapted flowers offer both nectar and pollen as rewards.

Nectar-feeding flies, carrion flies, and fly-pollinated flowers

Nectar-feeding flies are attracted to simple flowers that are often red or a light color, produce little odor, and have easy access. Nectar is the primary reward in such flowers (Figure 24.21). Flies sense sugar in nectar with their feet and suck it up with a tubular mouthpart. They walk around the surfaces of flowers probing for nectar, and in the process they transfer pollen.

FIGURE 24.15 **The African lily *Massonia* is pollinated by rodents**

FIGURE 24.17 **Bee bodies are adapted for carrying pollen**
Note pollen on the fuzzy body and yellow pollen baskets on legs.

carpels   stamens

FIGURE 24.16 ***Magnolia* flower** Note the broad, open petals, which allow easy access by beetles. There are numerous stamens with large amounts of pollen, some of which is consumed by the beetles.

FIGURE 24.18 **Bee flowers** Examples of bee-pollinated flowers include bluebells (*Mertensia virginica*), upon which a bumblebee is feeding in this photograph.

Carrion flies pollinate flowers that often look and smell like rotten meat or dung (Figure 24.22). Many carrion- or dung-fly–pollinated flowers attract pollinators by heat or by odor molecules diffused by heat, but they often provide no food rewards. Tropical botanists trying to locate the weird, giant flowers of endangered *Rafflesia* (Figure 24.23) in Malaysia only have to follow their noses, because the flowers' strong carrion odor travels long distances through the jungle. The powerful odor motivates scarce pollinators to move among the widely scattered *Rafflesia* flowers. The titan arum (*Amorphophallus titanum*), native to the island of Sumatra, produces a huge inflorescence that attracts pollinators through an extremely foul odor and a color reminiscent of rotting flesh. This spectacular structure is pictured in Chapter 1.

FIGURE 24.19 Bee-pollinated flowers with bilateral symmetry
Flowers of foxglove (*Digitalis purpurea*) have bilateral symmetry; they can be divided into two equal parts by a single plane. Note the white spots on the petals, which serve as a nectar guide.

FIGURE 24.20 Bee-pollinated flowers with radial symmetry
Shooting star (*Dodecatheon*) flowers are coevolved with bumblebees that use vibration to obtain pollen. Their downward-directed orientation aids the bees in shaking pollen from the anthers.

FIGURE 24.21 Fly-pollinated flowers Note the numerous flies on this inflorescence of small, nectar-producing flowers.

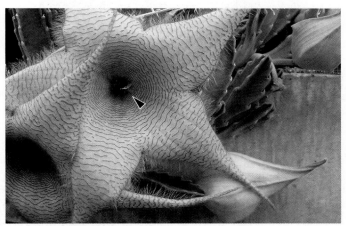

FIGURE 24.22 *Stapelia*, a carrion fly flower This member of the milkweed family produces flowers that have a foul odor mimicking rotten flesh, which attracts its pollinators. A fly (arrowhead) is visible in the center of the pale-colored flower. The cactuslike stems of this plant are visible in the background, as are two flower buds.

Butterflies, moths, and coevolved flowers Butterflies have good color vision, extending from ultraviolet to red. They find flowers by vision and then smell them with their feet and antennae after landing to determine if the flowers are a likely nectar source (Figure 24.24). They have nearly as much "blossom intelligence" as bees. Like bees, butterflies are able to remember characteristics of desirable flowers and find others of the same type, exhibiting constancy. Butterfly flowers are fragrant and are mainly blue, purple, deep pink, orange, or red. Lilacs, lavender, carnations, and spicebush (*Viburnum*; see Figure 24.24) are among the many butterfly-pollinated plants that people plant in gardens and appreciate in floral arrangements for their wonderful fragrances.

Night-flying moths are particularly attracted to night-blooming, white flowers such as moonflower (*Ipomoea alba*; Figure 24.25) that are more conspicuous at night. Moths can find flowers in very low light by tracking flower odor until they

FIGURE 24.23 *Rafflesia* **flower** This tropical plant produces huge flowers that attract pollinators with foul odors and petals the color of dead meat.

FIGURE 24.24 **Butterfly and butterfly-pollinated flowers** (a) Red admiral butterfly on the tubular flowers (arrowhead) of the Korean spicebush (*Viburnum*). (b) The proboscis, through which the butterfly draws the nectar, is visible (arrow).

find the strongest source. The obligate association between *Yucca* plants and the yucca moth *(Tegeticula yuccasella)* is one of the most famous and well-studied pollination partnerships of North America. *Yucca* produces large inflorescences of white flowers that are most fragrant at night (Figure 24.26). Yucca moths do not feed from the flowers but, rather, enter them in order to determine if the flower is suitable for laying eggs within the ovaries. The moths examine the ovaries for eggs deposited by previous visitors. If the moths find eggs, they move to another plant's inflorescence, carrying pollen along and thereby accomplishing outbreeding. But if a flower is judged suitable, the moth places an egg into carpels of the flower's ovary and pollen on the stigma. When the moth's larvae emerge, they will feed on some of the ovules, leaving others to mature into seed. Moth larvae exit flowers at the time when *Yucca* seeds are maturing.

Many moths can pollinate and gather nectar from brightly colored flowers such as *Petunia*, flowering tobacco *(Nicotiana)*, and *Cleome* that also attract butterflies (and hummingbirds). Moths and butterflies use their long, thin tongues to suck nectar from flowers whose fused petals form tubes too narrow for bees to penetrate. As they feed, they transfer pollen that has been dusted on their heads. Butterflies and most moths require a landing platform. However, some moths obtain nectar while hovering, for example the hawkmoths—abundant in the tropics (Figure 24.27)—and day-flying temperate moths such as the hummingbird clearwing *(Hemarius thysbe)*, which looks and behaves remarkably like a hummingbird.

Birds and bird-pollinated flowers Several types of birds are particularly important pollinators in tropical regions. Examples include hundreds of species of hummingbirds, honeycreepers, sunbirds, honeyeaters, flower-peckers, sugar birds, bananaquits, lorikeets, hanging parrots, Hawaiian finches, and American orioles. Flower-visiting birds feed in the daytime, and many require

FIGURE 24.25 **Moonflower, a moth-pollinated plant**

FIGURE 24.26 **Yucca plant—inflorescences and moth-pollinated flowers**

FIGURE 24.27 **A hawkmoth visiting the flower of Adansonia (baobab)**

FIGURE 24.28 **Hummingbird flower** Hummingbird behavior and anatomy are adapted for gathering nectar from hanging flowers with deep tubes.

a place to perch. Birds have good color vision, including the ability to see red objects against a green foliage background, but flower-visiting birds do not have a good sense of smell.

Hummingbirds are especially well known for their associations with flowers (Figure 24.28). Over 300 species of hummingbirds occur in North and South America. Hummingbirds' needlelike bills (and long tongues, which can be rolled into an extended tube) are thus adapted for taking nectar from flowers having deep tubes. Hummingbirds do not actually suck nectar; rather, they hold nectar in their tongue and swallow it only after returning the tongue to their mouth. In addition to nectar, the birds consume small insects in the vicinity of flowers. Hummingbirds' unique wing structure, with a rotating shoulder joint, allows them to hover and fly both backward and forward, abilities useful in feeding quickly from dangling flowers. As they feed, their heads are dusted with pollen, which is transferred to the stigmas of subsequently visited flowers.

Not surprisingly, bird-pollinated flowers are open in the daytime, are often bright red or deep pink in color or have gaudy "parrot" color combinations, and lack scent. Color and lack of fragrance prevent attraction of bees (which cannot see red), preserving the nectar reward for birds. In addition, the nectar offered by bird-adapted flowers is more dilute than the richer nectar preferred by bees. Corollas of bird-pollinated flowers are often fused to form tubes capable of holding the copious amounts of nectar required by birds. Fusion of tissues belonging to separate organs during development is a trait nearly unique to flowering plants, and its evolutionary origin is linked to pollination associations. Striking correlations of flower color, nectar viscosity, and nectar tubes with the vision and tubular feeding structures of birds provide compelling examples of the powerful reciprocal effects of coevolution.

Popular garden plants pollinated by hummingbirds include *Salvia*, red columbines (*Aquilegia*), bee balm (*Monarda*), and

Fuchsia (Figure 24.29). Bird-of-paradise plants *(Strelitzia)*, which are related to banana, have striking, parrot-colored orange and blue flowers (Figure 24.30). The *Strelitzia* flower is adapted for pollination by African sunbirds, which must perch on flowers to reach the nectar. The bird's weight pushes apart the blue petals, exposing stamens, which dust the bird's underside with pollen.

Bats and bat-pollinated flowers Bats are nocturnal, able to fly long distances (and thus have high food requirements), and have excellent memory for the location of favored plants. They are color-blind but have a good sense of smell, and they use echolocation (sonar) to navigate. Bats cannot fly as well in foliage because their emitted sounds bounce off multitudes of leaves, resulting in confusing signal reception. Flowers that are coevolved with bat pollinators are open at night and have light or dingy colors that do not

FIGURE 24.29 Several commonly cultivated flowers are hummingbird-pollinated Examples include (a) bee balm or bergamot *(Monarda)* and (b) *Fuchsia*.

FIGURE 24.30 *Strelitzia* **produces parrot-colored, bird-pollinated flowers**

FIGURE 24.31 Bats are common pollinators in the tropics

attract other pollinators. Bat-pollinated flowers also produce strong fragrances that imitate the smell of bats, and they have a wide bell shape for easy access. The flowers often hang beneath the foliage or are located on the trunks of trees where bats can easily locate them. The flowers are strong enough to survive clutching by bats, some of which need landing platforms. Plentiful nectar and pollen are offered as food rewards.

Bats are very important pollinators in the tropics. *Hymenaea stigonocarpa*, a tropical and subtropical legume tree valued for resin production, has large white flowers that are open for only one night and produce large amounts of nectar through the night hours. Its most common visitor is the bat *Glossophaga soricina*, a specialized nectar feeder. The bat hovers in front of the flowers and inserts its head to lap nectar and lick the anthers for pollen. Other tropical plants pollinated by bats are the kapok tree (from which a useful fiber is obtained) and wild banana *(Musa)* (Figure 24.31).

24.3 | Plants have also coevolved with animals to accomplish seed dispersal

Many flowering plants (and some gymnosperms) depend on animals to disperse their seeds, often offering fruit nutritional rewards for this service (Figure 24.32). Seed dispersal is advantageous to plants because it reduces competition between parent and progeny plants for the same resources, lowers the probability that seed predators will find and destroy all of

FIGURE 24.32   Animal seed disperal agent Note that the macaw is attracted to the ripe fruit rather than the green, unripe fruits.

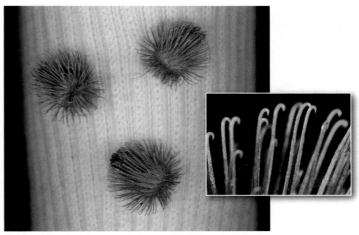

FIGURE 24.34   Animal-dispersed fruits and seeds The fruit of burdock (*Arctium*), a member of the sunflower family. The hooked brackets (inset) are highly effective at grabbing onto the fur (or socks) of passing animals. Fruits developed from the entire inflorescence break off the plant, carrying within them a sizable number of seeds.

a plant's yearly seed production, and allows plants to colonize new areas.

Although many plants utilize wind (Figure 24.33), and some (such as coconuts) use water to disperse their fruits and seeds, using animals to do this job has many advantages. Animal movements through the day or during migrations are predictable, and the fruit and seed production of coevolved plants is timed to coincide with the presence of animals. Furry animals can transport fruits having surface burrs that stick (the burrs also annoyingly attach to hikers' socks and pants; Figure 24.34). Rodents and ants often transport seeds underground to food hoards, where some of the seeds are able to survive and germi-

nate. Larger animals can ingest large, heavy fruits produced by many tropical tree species and thus transport the often correspondingly large seeds long distances. Large seeds are needed by such trees to provide embryos with the significant amounts of food needed to support germination and seedling growth on the deeply shaded forest floor.

As seeds pass through an animal's digestive system, they are eliminated, together with a useful pile of fertilizer—further favoring the growth of seedlings. For example, forest elephants disperse 50 or so species of African trees by consuming their large, fallen fruits and depositing them as much as 50 kilometers (km) from the parent tree. One example is the fruit of *Omphalocarpum*, which is the size and hardness of a human skull (Figure 24.35). In nature, only an elephant is able to break this fruit open, thereby releasing its seeds. A single elephant dung pile may contain hundreds of intact or germinating

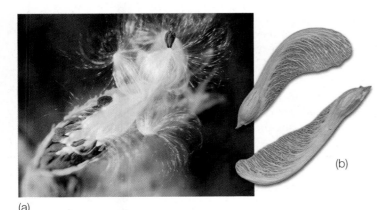

FIGURE 24.33   Wind-dispersed seeds and fruits (a) Milkweed (*Asclepias*)—the tufted seeds are being released from the fruit (a follicle). (b) Winged fruits (schizocarps) of silver maple (*Acer saccharinum*); each contains a single seed.

FIGURE 24.35   The large fruits of *Omphalocarpum* are dispersed by elephants

seeds. Finally, as in the case of pollination, animals have food preferences that endow them with constancy—repeated visits to the same plant species.

Color and odor are major cues that inform animals about the presence of ripe fruits whose seeds are mature and ready for dispersal. For example, African elephants find *Omphalocarpum* and other desirable fruits on the jungle floor by their strong odor (and by hearing the tremendous thud the sizable fruits make as they crash to the forest floor). Plants often signal that ripe fruit is present by a color change to red, orange, yellow, blue, or black (Figure 24.36). Birds and other fruit-eating animals having color vision do not notice green, immature fruits against a green foliage background and thus leave fruits alone until they have fully ripened. This condition favors dispersal of mature seeds that are more likely to germinate than are immature seeds. Color change also favors speedy dispersal before fungi and bacteria are able to rot vulnerable ripe fruits.

In temperate forests, about 300 plant species are adapted for bird dispersal of their seeds. These plants typically produce fleshy fruits that provide a nutritional reward, with hard seeds that are small enough to easily pass undamaged through a bird's intestinal tract. Summer-ripening fruits attract resident birds by providing a sweet, juicy, carbohydrate-rich reward. Examples include blueberries, cherries, strawberries, and blackberries. Some of these fruits go through a double color change—from green to pink or red, and then to blue or black. Resident birds repeatedly visit and monitor the condition of fruits and thus are able to predict where and when ripe fruits are likely to be present. Humans who admire the rich diversity of Earth's flowers and appreciate fruits owe a special vote of thanks to the natural process of coevolution.

FIGURE 24.36 Fruit color changes are a signal to animal dispersers that fruits are ripe In pokeweed *(Phytolacca)*, flowers (a) give rise to fruits that are initially green (b), after which they turn purple-black (c).

HIGHLIGHTS

24.1 Organisms that have influenced each other's evolution over a period of time are said to have coevolved. Not all coevolutionary interactions are beneficial, but many associations between flowering plants and animals are mutually beneficial. These include pollination, dispersal, and defensive coevolutionary interactions. Coevolution is an important aspect of agriculture and ecology.

24.2 Cross-pollination, the transfer of pollen from one flower to another of the same species, benefits plants by favoring outbreeding. Outbreeding generates increased genetic variation, useful in adapting to environmental change. Plants have a variety of mechanisms for promoting outbreeding, but some plants are able to self-pollinate. Animal pollinators offer precision, high-fidelity cross-pollination services. Plants usually provide rewards such as sugary nectar and/or lipid- and protein-rich pollen. Plants attract pollinators with flower colors and scents.

Flowers control pollinator access by shape and arrangement of flowers in inflorescences. Animal pollinators can learn the characteristics of favored flowers and repeatedly visit flowers of the favored type. Repeated visits are advantageous to both plant and pollinators. The traits of flowers are coordinated with those of pollinators. Important pollinators are beetles, bees, flies, moths, butterflies, birds, and bats. Each visits flowers having a particular color, odor, shape, and other traits.

24.3 Plants and animals are also involved in seed-dispersal coevolutionary relationships. Seed dispersal benefits plants by spreading the progeny of parents to new environments. Fruits or excess seeds are provided as the nutritional reward for animal dispersal services. Color and odor are major cues to animals that ripe fruits (with mature seeds) are available. Fruit crops are adapted in ways that attract animals, including humans.

REVIEW QUESTIONS

1. How does inbreeding differ from outbreeding? What is a disadvantage of inbreeding? Describe several strategies used by flowering plants to decrease the likelihood of inbreeding.

2. Does inbreeding serve any positive functions?

3. Describe some food rewards that plants provide to animal pollinators.

4. Discuss some strategies used by plants to attract pollinators.

5. Describe flower constancy, or fidelity, with regard to flower pollination. What is the advantage of pollinator constancy to the plant? To the pollinator?

6. Discuss the interactions between beetles and beetle-pollinated flowers. How is each organism modified for this interaction?

7. Discuss the interactions between bees and bee-pollinated flowers. How is each organism modified for this interaction?

8. Discuss the interactions between flies and fly-pollinated flowers. How is each organism modified for this interaction?

9. Discuss the interactions between butterflies and moths and the flowers they pollinate. How is each organism modified for this interaction?

10. Discuss the interactions between birds and bird-pollinated flowers. How is each organism modified for this interaction?

11. Discuss the interactions between bats and bat-pollinated flowers.

APPLYING CONCEPTS

1. Figs produce their flowers in closed inflorescences called syconia. A fig syconium will be visited by a single female fig wasp carrying pollen on her thorax. She bores into the syconium and pollinates the fig flowers as she lays her eggs in some of the fig ovaries. The resulting wasp larvae feed off the developing seeds, one larva per seed; the ovules without larvae eventually mature into normal fig seeds. After hatching, the males mate with the females, who then crawl around the syconium collecting pollen from the male flowers. Eventually, the females fly off to a new syconium to start the cycle over. If a female is not carrying any pollen when she enters the next syconium, none of her offspring will survive, because the required seeds will never develop. Conversely, if no female wasp visits a syconium, there will be no fig seeds set, because no flowers will be pollinated. The fig and the wasp are intimately dependent on each other to complete their life cycles. Describe some of the advantages and disadvantages to each organism of such a close association.

2. In 1877, Charles Darwin forwarded a letter from Fritz Müller, a naturalist who had been observing the multicolored *Lantana* flowers growing in the Brazilian forest. He noted, "We have here a *Lantana* the flowers of which last three days, being yellow on the first, orange on the second, purple on the third day. The plant is visited by several butterflies. As far as I have seen the purple flowers are never touched. Some species inserted their proboscis both into yellow and into orange flowers; others ... exclusively into the yellow flowers of the first day. If the flowers fell off at the end of the first day the inflorescences would be much less conspicuous" (*Nature* 17[78]: 1877). What do you think is happening here? Are the plants communicating with the butterflies?

3. Describe some of the ways that bat-pollinated flowers typically differ from bee-pollinated ones.

4. Smyrna figs are delicious fruits that are quite suitable for drying and eating. These plants propagate easily by cuttings and were introduced into California in the 1700s. Large orchards were planted in the early 1800s, and although the trees grew very well, they did not set fruit. In the 1860s, a wild fig was planted in a fig orchard on the Gates farm near Modesto, California, and the Smyrna figs there soon began producing delicious fruits. At that time, other orchards around the state still were not setting fruits. Considering Question 1, what do you think happened here? Why did the mature fig trees suddenly begin bearing fruit?

25 Principles of Ecology and the Biosphere

Earth

The main feature that strikes a visitor to a desert is the heat. Although not all deserts are hot, a summer day in a hot desert can be like an oven with temperatures above 50°C (122°F). The second feature is light. A desert on a summer day can be blindingly bright. But neither heat nor light characterizes a desert. Deserts are defined by their scarcity of water. Deserts are fascinating environments because all life in them—microbial, plant, and animal—is shaped by this limitation of water.

Desert plants have evolved a number of adaptations to water scarcity. Many desert plants avoid harsh conditions by waiting in a dormant state as dry seeds in the soil, underground bulbs or storage roots, or seemingly dead clusters of aboveground stems. When rains come these plants can burst out in leaves and flowers to cover the desert in a living carpet. Other plants grow in low-lying areas such as dry streambeds, where their roots can reach underground water. The most familiar desert plants, however, are the thick-stemmed cacti and thick-leaved agaves that endure the harsh desert climate all year. These plants tolerate high temperatures and can take up and store large volumes of water in their thick stems or leaves.

Animals are also present in deserts, and plants interact with them in many ways. Cacti and other desert plants protect their stems from herbivores with sharp spines. Bees pollinate desert flowers, and ants and rodents forage on seeds. Birds nest among the spiny stems of cacti and hunt for unwary rodents. At night many more animals emerge from burrows to forage or hunt for food.

A desert is one example of an ecosystem. An ecosystem consists of the community of organisms in a defined area plus the physical environment of that area. Like most ecosystems today, deserts are threatened by human exploitation. Knowledge of ecology, the field of biology covered in this chapter, is essential to understanding, using, and sustaining the natural world upon which humans depend. ■

Desert plants

25.1 | Ecology focuses on populations, communities, ecosystems, biomes, and the biosphere

The word **ecology** is derived from two Greek words—*oikos*, meaning "house," and *logos*, meaning "knowledge of" or "study." Thus ecology literally means knowledge of the house, where "house" is used in a poetic sense to refer to nature. In practice, ecology is the study of organisms and their interactions with each other and with the physical and chemical components of their environment. Ecology deals with biological organization above that of the individual organism. A **population** consists of all the individual organisms of a single species in a defined area, such as all the dandelions in a meadow. Most ecologists work on populations.

A **community** consists of all the organisms of different species in a particular area. An area of temperate forest would contain a community consisting of all microorganisms, plants, and animals within that area. **Ecosystem** is a more general ecological term, which is defined as the community plus its physical environment. An ecologist studying a grassland ecosystem might look at how the seasonal pattern of rainfall, temperature, and nutrients affects biomass accumulation and interactions between species.

Ecosystems with similar climates are grouped into **biomes.** The tropical rain forest is an example of a biome whose climate is characterized by warm temperatures and high levels of rainfall. Tropical rain forests occur in Central and South America, Africa, and southeastern Asia. Although these rain forests share a similar climate, the communities are composed of different species in each area. Physical matter, such as water and carbon, flows in global cycles between organisms and the physical environment of the Earth. Together, all of the biomes make up the **biosphere**, which includes all organisms on Earth. Ecologists study each of these biological systems—population, community, ecosystem, biome, and biosphere—to look for general principles of biological organization.

25.2 | Populations show patterns of distribution and age structure, grow and decline, occupy specific niches, and interact with other populations

A number of general ecological principles apply at the population level. Individuals within a population may be distributed in space in distinct patterns that are described as random, uniform, or clumped. Individual organisms are born, grow, age, and die. Populations also exhibit growth and decline and have an age structure. Populations of organisms live in specific types of habitats, such as marshes or rain forests. The habitat in which a population lives can be compared to an address on a street. The ecological niche is similar to an occupation; the niche describes what a population does and how it fits into its community. A niche includes both abiotic (nonliving) factors, such as temperature and moisture, that determine where a population occurs and biotic (living) factors, such as other organisms with which that population interacts. Populations of different species interact in many different ways. Some populations live together in a mutually beneficial relationship, while other populations may compete with each other for space or other resources.

Plants in a population may be distributed in a random, uniform, or clumped pattern

Within a population individuals can space themselves at various distances from their nearest neighbors. Ecologists consider three basic kinds of spacing: random, uniform, and clumped (Figure 25.1). A random pattern of individual organisms in a population may occur when environmental conditions are nearly uniform, resources are steady, and individual organisms do not attract or repel each other. Dandelions (*Taraxacum officinale*) growing in a park or lawn are an example. The lawn environment is uniform and nutrients steady. The dandelion seeds are dispersed randomly across the habitat by wind, producing a random pattern of flowering plants.

A uniform distribution of plants in a population is more common in nature. A planted crop such as corn (*Zea mays*) and an apple orchard are examples of uniform distribution patterns. A forest of pine trees is a natural example. Adult trees shade out seedlings underneath them and compete for space with other mature trees to generate a uniform distribution of trees in the forest. In the Mojave Desert of California, creosote bushes (*Larrea tridentata*) space themselves out in a uniform pattern partly due to root competition for moisture (Chapter 26). Ants and rodents shelter under the mature bushes and consume any seeds dispersed outside of mature shrubs. The only seeds that survive are those that germinate inside old declining or dead creosote bushes, where they are hidden from herbivores.

Clumped patterns of distribution are the most common because resources are often clumped in nature. Certain plants such as cattails (*Typha latifolia*) grow best in moist locations. Low-lying areas contain more moisture. Seeds of cattails will therefore grow best in low-lying areas where they will clump together. The mature plants may also drop their seeds near themselves or reproduce by runners within that favorable habitat. Distribution patterns indicate important aspects of the biology and interactions of plants in nature.

Age distribution and survivorship curves describe the age structure of populations

Many plants, called annuals, live only a single year. Other plants may live from 2 years (biennials) up to hundreds and even thousands of years, as do the bristlecone pines (*Pinus longaeva*) in the southwestern Rocky Mountains. The age structure of a population can be characterized by an age distribution graph (Figure 25.2). Age distribution graphs show the numbers or proportions of individuals of different ages in the population. The graphs indicate the most abundant age classes and the age at which the highest death rate occurs.

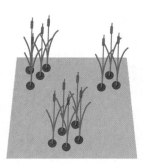

(a) random   (b) uniform   (c) clumped

FIGURE 25.1   Distribution patterns of individual organisms in a population (a) Random, (b) uniform, and (c) clumped.

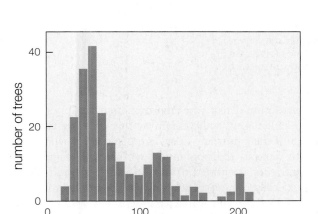

FIGURE 25.2   Age structure of a population of pines (Pinus sylvestris) in Sweden

Survivorship curves show how the death rate in a population changes with the age of the individuals in the population. There are three basic types of survivorship curves (Figure 25.3). In Type I curves, the death rate in the population is very low for young and middle-aged individuals but increases sharply for older individuals. A Type I survivorship curve is typical for humans and other vertebrates (animals with backbones) that perform parental care of their young and for the phlox plant *(Phlox drummondi)*. Type II curves show a constant death rate with age. A few types of birds and invertebrates (animals without backbones) and the common buttercup *(Ranunculus acris)*

have Type II survivorship curves. In Type III curves, the death rate is very high for the youngest age groups. Sweet clover *(Melilotus alba)* and the desert giant saguaro *(Carnegiea gigantea)* have Type III curves. Seeds and seedlings are vulnerable to harsh growing conditions such as water shortages and are heavily eaten by herbivores. A single saguaro cactus over its life span can produce about 40 million viable seeds. Less than one-fourth of those viable seeds land in an environment favorable for germination. Of that one-fourth, only about 0.4% escapes herbivores such as ants, birds, and rodents and becomes seedlings. Of those seedlings, animals eat 99% in the first year and leave only a few survivors concealed under other plants or among rocks. Once a saguaro seedling reaches a certain size and age, it becomes more resistant to herbivores as plant secondary compounds build up and spines develop, and the death rate declines.

Populations show distinct patterns of growth

One of the most basic aspects of the biology of populations is that of growth. An individual organism is born (or germinates, in the case of spores and seeds), grows, and finally dies. Populations have birth rates and death rates. The birth rate of a population is the number of births or germinations in that population in an interval of time. Similarly, the death rate is the number of deaths in the population in the same time interval. The population growth rate or, more accurately, the net growth rate, is the difference between the population birth rate and the population death rate. In algebraic form, this relationship is written as $r = b - d$, where b is the population birth rate, d is the population death rate, and r is the net growth rate. The net growth rate will be negative if the death rate exceeds the birth rate. A population with a negative value of r will decline over time (Figure 25.4a). The

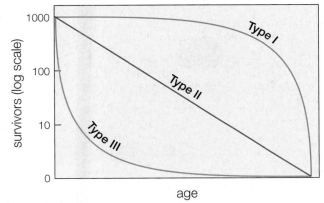

FIGURE 25.3   Survivorship curves Type I includes humans and other mammals. Type II has been found among some birds and invertebrates. Type III includes many plants and invertebrates.

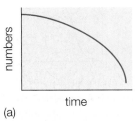
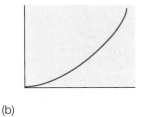

(a)   (b)

FIGURE 25.4   Growth rate (a) A population showing a net negative growth rate. (b) A population with a net positive growth rate.

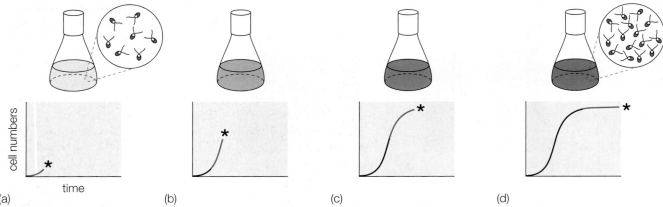

FIGURE 25.5 **A typical growth curve** A population of algae in a flask at different stages (in red) in its growth: (a) At first, there are few cells present and numbers increase slowly. (b) Numbers increase at the maximum rate. (c) Growth begins to slow. (d) As nutrients become limiting, the population reaches a maximum.

net growth rate is positive if the birth rate exceeds the death rate. If r is positive, the population size will increase over time (Figure 25.4b). Obviously, no population can continue to increase indefinitely over time, or it would soon overwhelm its environment.

The easiest way to follow the process of population growth is by an example. Imagine a 1-liter (L) flask filled with a nutrient-rich culture medium suitable for growing algae. At time zero we introduce a single motile algal cell that proceeds to reproduce asexually by mitosis. Under these conditions the death rate is negligible, and so the net growth rate equals the birth rate. At first, the population increases slowly because there are so few algae in the culture (Figure 25.5a). As numbers increase, the population growth rate also increases until it reaches the maximum the population can sustain (Figure 25.5b). At this point, the population has entered the phase of maximum net growth. If the population could sustain this rate, the flask would become packed with algae. But this does not happen. As a nutrient such as phosphorus is depleted, the net growth rate of the population begins to decrease (Figure 25.5c). The growth rate approaches zero as phosphorus is removed, and the population reaches a maximum size called the **carrying capacity** (Figure 25.5d), which

is the maximum density the environment can support. The population of algal cells will remain at this carrying capacity for a while until their nutrient stores are depleted. Then the cells will begin to die and the population will decline (Figure 25.6a). If fresh nutrients are added to the culture vessel and old medium and algal cells removed each day, the population will maintain some level of carrying capacity indefinitely (Figure 25.6b). If a higher level of nutrients is added each day, the population will reach a new carrying capacity (Figure 25.6c). The magnitude of the carrying capacity is determined by limiting factors. In the algal culture, the level of phosphorus could be one limiting factor. If excess phosphorus is added, the algal population will increase until a new limiting factor takes control, such as the amount of nitrogen or carbon.

The same principles of population growth apply to flowering plants as well as to algae. Imagine an empty plowed field surrounded by woods. Scatter a few sunflower seeds around the field. The first year, those few sunflowers will grow and produce more seeds, which will be scattered about the field by wind, rain, and animals. Many more sunflowers will grow in the next year to produce and disperse seeds. Like the algae in the flask, the sunflowers will reach a carrying capacity determined by some

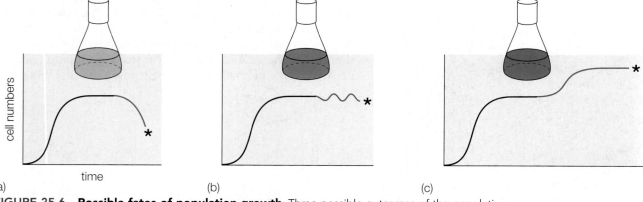

FIGURE 25.6 **Possible fates of population growth** Three possible outcomes of the population of algae shown in Figure 25.5: (a) As the nutrients in the culture are used up, the cells start to die and the numbers decrease. (b) If nutrients are periodically added to compensate for those lost, the carrying capacity of the population fluctuates. (c) If a richer medium is added, the population reaches a new, higher carrying capacity.

limiting factor in the field. Add fertilizer and a new carrying capacity will be reached. All of those sunflower seeds, however, could cause an explosion in the population of field mice and a decline in sunflower population in a subsequent year!

Population growth is of more than ecological interest. Ecologists use mathematical models to predict growth of human populations as well as that of many species of wildlife and marine fisheries for purposes of long-range planning.

The ecological niche includes the abiotic factors that determine the area the population occupies

The niche includes all the abiotic factors that define or limit the area the population occupies. Abiotic factors include physical and chemical features of the environment such as light, temperature, moisture, pH, salinity, winds, and water currents. All photosynthetic organisms need light, but the amount of light required varies greatly. Trees in temperate deciduous forests differ considerably in their need for light. Trees at the top or canopy of the forest have the highest requirements for light, but they create shade beneath them that prevents their own seedlings from growing. Different species of trees that grow at intermediate light levels thrive below the tallest species, and shade-loving species grow on the forest floor. The temperate deciduous forest is therefore arranged in layers or strata according to the decline in light levels. Light is absorbed as it passes through bodies of water. Photosynthetic organisms are therefore restricted to the euphotic (good light) zone near to the surface where light is adequate for photosynthesis.

Among terrestrial plants, the most common limiting abiotic factors are moisture and temperature. In the eastern United States, the temperate deciduous forest gives way to grasslands when annual precipitation falls below about 85 cm (33 in.) (also see Chapter 27). Grasslands yield to desert when annual precipitation falls below 20 cm (8 in.). Desert plants are clearly adapted to live with much more limited water than temperate forest plants or wetland plants. The temperature range for most plants lies between 0°C (32°F) and 40°C (104°F). During winter dormancy spruce and fir trees of the boreal forest can survive temperatures below −40°C (−40°F). Many desert plants routinely survive temperatures far above 40°C. The giant saguaro cactus (*Carnegiea gigantea*) of the southwestern United States (Chapter 26) can tolerate temperatures of 64°C (147°F), but it cannot survive if it is subjected to more than 36 hours of below-freezing temperatures. Many other abiotic factors may be significant in determining a niche, such as the occurrence of fire, the acidity or alkalinity of the environment, or the presence of important nutrients in water or soil.

The ecological niche includes interactions between populations of different species

Symbiosis is any close association between two or more species. The association may be harmful or beneficial to one or both species. Symbiotic interactions can take many different forms, including mutualism, parasitism, predation, and competition.

In mutualism, two populations exchange benefits

There are numerous examples of mutualism. Many protozoa and aquatic invertebrate animals contain symbiotic algae. The algae provide products of photosynthesis to the host, which in turn supplies nutrients such as nitrogen and phosphorus to the algae (Figure 25.7). Flowering plants and their insect pollinators provide many examples of mutualism in which the species show a high degree of coevolution (Chapter 24).

Commensalism is a form of symbiosis in which one species benefits but does no harm to the other. Many orchids in tropical rain forests grow as *epiphytes* (meaning "upon plants") on the upper branches of trees. They obtain access to light and rainwater in the forest canopy and apparently do no harm to the trees. Plants that disperse their fruits or seeds by attaching them to the fur or feathers of animals may be acting as commensals (Chapter 24).

In parasitism, herbivory, and predation, one population benefits and the other is harmed

In **parasitism**, individuals of one population (the **parasite**) live in or on members of the other (the **host**) and feed upon it without killing it, at least in the short term. Animals feed upon plants (herbivory) and, although many plant parts may be damaged, the damage is usually not lethal. Predation is commonly thought of as a property of animals, where one individual kills another. However, a number of plant species are effective predators on animals, particularly insects.

Among flowering plants there are more than 3,000 parasitic species, and some cause serious losses among agricultural and commercial crops. Mistletoes are the best-known parasitic plants because they are used as Christmas decorations. Most

FIGURE 25.7 Mutualism Shown here is a freshwater sponge; its green coloration is the result of the green alga *Chlorella* living within its tissues. *Chlorella* is a common partner in freshwater mutualistic relationships.

species grow as epiphytic parasites on the branches of trees and shrubs. Their roots grow into and connect with their host's vascular tissue, from which they derive water and mineral nutrients but, in most cases, no organic substances. In the western United States, dwarf mistletoes *(Arceuthobium)* are a serious parasite on conifers (Figure 25.8) and hardwoods grown for timber. Unlike most mistletoe species, *Arceuthobium* derives more than 60% of its carbon from its host trees, which become malformed and of no commercial use.

In **herbivory**, animals eat leaves or other plant parts without fatal consequences to the plant. Since plants cannot flee from herbivores, ecologists have long wondered why animals do not consume all the plants. First, plants are not just a "wimpy" bowl of salad greens; they have a variety of defenses—both chemical and physical—that prevent herbivores from taking full advantage of them. Second, herbivores are subject to predation, which restricts their populations, usually to levels below those at which they could deplete plant populations.

Plant chemical defenses are of two general types—quantitative and qualitative. A quantitative chemical defense is not toxic in small doses, but as the animal consumes the plant, the dosage builds up in the gut to prevent digestion. The herbivore is forced to stop eating. Tannins and resins are quantitative chemical defenses. Qualitative chemical defenses, such as alkaloids, terpenes, and quinones, are toxic at small doses. Quantitative and qualitative chemicals are all plant secondary compounds (Chapter 3).

Plants also employ various physical defenses against herbivores. Thorns and spines deter many herbivores. Many plants also have special hairs (trichomes) that can poison herbivores. The stinging nettle *(Urtica dioica)* has hairs that break on contact and inject poison into animals (Figure 25.9). Thick layers of wax on leaves may also deter herbivores.

FIGURE 25.9 The stinging nettle, *Urtica dioica* (a) A small group of plants in flower. (b) A leaf petiole with stinging hairs, one of which is shown at higher magnification (inset).

In **predation**, one population benefits and the other suffers mortality. Carnivorous plants occur in ecosystems where the supply of nitrogen is limited. Predation on animals provides these specialized plants with a nitrogen source. Carnivorous plants capture their prey in traps made from modified leaves, where they also digest the prey (Chapter 11).

In competition, individual organisms have a negative impact on each other

Competition occurs when individuals or populations attempt to use the same limiting resource. The principal limiting resources for plants are light, water, mineral nutrients, and space. Competition can be intraspecific, that is, between members of the same species, or interspecific—between members of different species. Based on observations of population interactions in controlled environments, ecologists determined that if two species compete for the same resource in the same area, one species would eventually drive out the other. This observation is called the *competitive exclusion principle.* How the principle applies to natural communities varies greatly. In nature, plants generally compete with each other by preempting space, overgrowing one another, or attacking each other chemically.

In temperate forests, herbaceous flowering plants compete by preempting space or overgrowing each other. In dry, infertile oak forests, low resources severely limit the density of herbaceous plants. Because the herbs are widely spaced, they do not shade each other, and there is little advantage to growing taller. Herbs such as rattlesnake plantain (Figure 25.10) grow close to the ground and spread laterally by vegetative growth. Herbs growing on infertile sites compete by preempting space. On moist, fertile sites within temperate forests, however, high resource levels favor dense stands of herbs that shade each other. The herbs gain an advantage by growing taller because they intercept more light for photosynthesis. Herbs such as nettle *(Urtica)* and goldenrod *(Solidago)* (Figure 25.11) reach 1 meter (m) or more in height. They will grow taller until the energy they gain from photosynthesis is balanced by the

FIGURE 25.8 The parasite *Arceuthobium* (dwarf mistletoe) This golden-colored flowering plant is a common parasite of conifers in the western United States.

FIGURE 25.10   A rosette plant Rattlesnake plantain (*Goodyera pubescens*), a member of the orchid family, has a whorl of distinctive leaves that lie close to the ground.

FIGURE 25.11   A tall forest herb This group of goldenrods (*Solidago*) was growing in a site having flecks of sunlight.

energy they must expend in making unproductive stems. Thus herbs in dense stands compete by overgrowing each other. Taller herbs shade out shorter species.

Plants also compete by releasing chemicals into their environment that inhibit the growth of other individuals of the same or different species. This type of interaction is called **allelopathy.** The best-known example of plant allelopathy is that of black walnut *(Juglans nigra)*. Many species of plants wither and die if planted too close to black walnut because it releases toxic chemicals from its roots and seeds. The desert creosote bush is another example of a plant that uses allelopathic chemicals to inhibit growth of nearby species (Chapter 26).

25.3 | Communities are composed of individuals of many different species

In community ecology the focus is on individuals of many different species interacting with each other in a defined area. The defined area may be relatively small, such as a small pond or a patch of forest surrounded by grassland, or it may be quite large, such as the entire Sonoran Desert. In either instance the area is defined by some physical property of the environment such as the dominant plants or the edges of the pond. Communities may be studied in terms of their species richness and diversity. The species composition of communities also changes over time in a dynamic process called *succession*.

Communities can be characterized by their species diversity

A population consists of a number of individual organisms. Similarly, communities consist of a number of individual species. The higher the number of species, the greater the species richness. **Species richness** is simply the number of species present in the community. In most communities, a few species are abundant and many are relatively rare. **Evenness** describes the distribution of individual organisms among the species in a community. Thus a community with 10 species, each having 10 individuals, would have the maximum evenness possible. The concept of **species diversity** includes both species richness and evenness.

Consider two forest communities in West Virginia (Table 25.1). The first community (plot one) has 24 species of trees and 256 individuals. Two species represent about 44% of the total number of trees. Nine more species contain an additional 44% of the total number of trees. In the second community (plot two), there are 10 species of trees and 274 individuals. Two species represent almost 84% of the total number of trees. The first forest community has higher species diversity than the second because of more species and higher evenness.

Communities differ in their species diversity. Species diversity is lowest among polar communities, increases through the mid-latitudes, and is highest in the tropics. Species diversity is lower at high altitudes than at lower elevations. The diversity of tree species across the United States and Canada is highest in the southeastern United States and decreases westward with decreasing rainfall and northward with declining average temperatures (Figure 25.12).

Ecologists have long been intrigued by such patterns of diversity and have offered many hypotheses to explain them. Some communities have high species diversity because they contain a large number of habitats. Rocky intertidal zones along ocean shores may have high species diversity in part because they experience heavy wave and storm action at fairly frequent intervals. Storms dislodge organisms and open up new space for colonizers. These pioneer species will in time be displaced by other species that are superior competitors, but they persist in the community because disturbances continue to open new areas for colonization.

Herbivory appears to play an important role in diversity. Herbivores can prevent potentially competing plant species from

TABLE 25.1   Species of Trees Present in Two Deciduous Forest Plots in West Virginia

Species	Plot One		Plot Two	
	Number (n_i)	Proportion (n_i/N)	Number (n_i)	Proportion (n_i/N)
Yellow poplar (Liriodendron tulipifera)	76	0.297	122	0.445
White oak (Quercus alba)	36	0.141	0	0.000
Black oak (Quercus velutina)	17	0.066	0	0.000
Sugar maple (Acer saccharum)	14	0.054	1	0.004
Red maple (Acer rubrum)	14	0.054	10	0.036
American beech (Fagus grandifolia)	13	0.051	1	0.004
Sassafras (Sassafras albidum)	12	0.047	107	0.390
Red oak (Quercus rubra)	12	0.047	8	0.029
Mockernut hickory (Carya tomentosa)	11	0.043	0	0.000
Black cherry (Prunus serotina)	11	0.043	12	0.044
Slippery elm (Ulmus rubra)	10	0.039	0	0.000
Shagbark hickory (Carya ovata)	7	0.027	1	0.004
Bitternut hickory (Carya cordiformis)	5	0.020	0	0.000
Pignut hickory (Carya glabra)	3	0.012	0	0.000
Flowering dogwood (Cornus florida)	3	0.012	0	0.000
White ash (Fraxinus americana)	2	0.008	0	0.000
Hornbeam (Carpinus caroliniana)	2	0.008	0	0.000
Cucumber magnolia (Magnolia grandiflora)	2	0.008	11	0.040
American elm (Ulmus americana)	1	0.004	0	0.000
Black walnut (Juglans nigra)	1	0.004	0	0.000
Black locust (Robinia pseudoacacia)	1	0.004	0	0.000
Sourwood (Oxydendrum arboreum)	1	0.004	0	0.000
Tree of heaven (Ailanthus altissima)	1	0.004	0	0.000
Black maple (Acer nigrum)	1	0.004	0	0.000
Butternut (Juglans cinerea)	0	0.000	1	0.004
Totals	256	1.000	274	1.000

Data adapted from Robert L. Smith, *Ecology and Field Biology*, 1996.

eliminating each other. In the chalk grasslands of southern England, rabbits were the main herbivores. Their activities cropped down the dominant grasses and tall herbs and permitted other species of grasses and herbs to coexist. When a severe epidemic swept through the rabbit population, the dominant grasses and tall herbs took over, and many other plants became rare or absent.

Ecological succession is the change in community composition over time

Communities are not static, unchanging entities. The composition of species in a community changes over time—a process called *ecological succession*. Ecological succession is most often

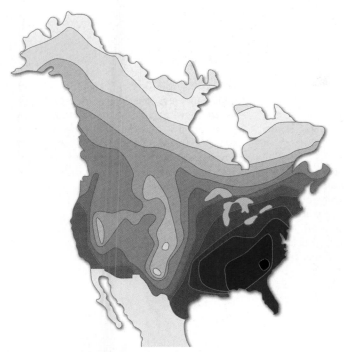

FIGURE 25.12   Map of tree species richness across North America The highest concentration of tree species occurs in the U.S. Southeast. The extreme north lacks trees altogether.

passes through a series of stages until the original community is restored or a new, mature community is established. Stages in succession are usually recognized based on the plant species composition, although animal species change as well. Ecologists recognize two basic types of succession—primary succession and secondary succession.

Primary succession begins on areas not previously occupied by organisms In Alaska, glaciers have retreated by as much as 100 kilometers (km; 62 miles) over the past 200 years. Stages in primary succession can be observed by simply hiking from the base of the glacier down its valley (Figure 25.13). Close to the glacier, recently exposed land, which consists of pulverized stones called *glacial till*, is poor in nitrogen and organic matter. The **pioneer species** are cyanobacteria, lichens, and liverworts. The cyanobacteria fix nitrogen, the lichens secrete organic acids that break down rock, and all organisms add organic material to the glacial till. In the next stage, located farther from the glacier, herbaceous flowering plants such as fireweed *(Epilobium)* and mountain avens *(Dryas drummondi)* appear. Alder trees *(Alnus sinuata)*, which have symbiotic nitrogen-fixing bacteria, enter the community in the next stage and establish dense stands. Soil nitrogen increases. On sites that have been exposed for about 50 years, spruce trees shade out the alders and form a dense spruce forest after about 120 years.

A similar process of succession took place on a much larger scale in North America at the end of the last ice age. As the glaciers retreated, plant communities migrated northward; their movements can be reconstructed by examining the pollen grains left in lake sediments (refer to Essay 25.1, "Determining Past Climate and Vegetation from Pollen Data").

set into motion by some natural disturbance to a community—such as floods, storms, or fires—and, in geologically active regions, earthquakes, landslides, and volcanic eruptions. Any of these disturbances may remove the original species and allow new species to move into the area. The species composition

FIGURE 25.13   Primary succession in a glacial valley In this series of photographs taken at Glacier Bay, Alaska, fireweed (a) is the first plant to appear. Mountain avens (b) is seen next, followed by alder trees (c), whose symbiotic nitrogen-fixing bacteria improve the soil, allowing for the later occurrence of spruce trees (d).

ESSAY 25.1 DETERMINING PAST CLIMATE AND VEGETATION FROM POLLEN DATA

Year after year throughout the history of a lake or bog, sediments and pollen grains accumulate at the bottom. The sediments preserve the pollen grains, which record the vegetation present in the surrounding landscape and, by inference, the past climate as well. From knowledge of the temperature ranges tolerated by these species, the mean annual temperatures can be estimated. A sample of the sediments in a lake can be removed by coring (Figures E25.1A and B). The ages of the various levels in the core can be determined by radiocarbon techniques, and the pollen grains in those same levels can be recovered and analyzed.

Presence of spruce pollen implies a cold climate, grass suggests a prairie environment, and oaks a temperate forest.

As climate and pollen data built up in the 1970s, a group called COHMAP emerged to examine changes over the past 18,000 years, since the end of the last great glacial episode. COHMAP—Cooperative Holocene Mapping Project (*holocene* refers to the geologically recent period since the last glacial)—analyzed data from more than 100 palynologists (scientists who study fossil pollen) working at 328 sites across eastern North America. Each site presented a local history of changes in climate and vegetation. Together, all the sites gave a series of portraits of eastern North America at 3,000-year intervals over the past 18,000 years. Figure E25.1C shows changes in oak (*Quercus*) distribution since the last glacial period. Oak was confined to southern regions 18,000 years ago, when temperatures were much lower than today. As the glaciers withdrew, temperatures warmed and oaks migrated north, reaching the southern Great Lakes

by 9,000 years ago. Beginning around 6,000 years ago, oaks were displaced from the southern portion of their range by the development of southern pine forests, whose present distribution is shown in Figure E25.1D. Pollen grains of oak and pine are shown in Figures E25.1E and E25.1F, respectively.

Pine (*Pinus*)

present day

E25.1D Current distribution of pines

E25.1A The coring process

E25.1B Removing the core from the coring tube

Oak (*Quercus*)

ice sheet

18,000 years ago   12,000 years ago

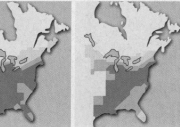

6000 years ago   present day

E25.1C Distribution of oaks since the last glacial period

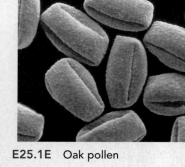

E25.1E Oak pollen

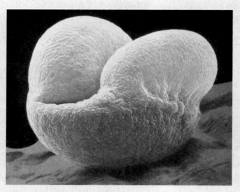

E25.1F Pine pollen

Secondary succession occurs on areas where a community has been removed The most common sites for secondary succession are abandoned farmland and clear-cuts in forests; other examples are highway roadsides and areas cleared by fires. In each of these cases, a true soil is already present. One of the best-documented studies of secondary succession involved abandoned farmland. Over the first 3 years, the abandoned field was invaded by a sequence of weeds and grasses, such as crabgrass and ragweed. Within 5 to 10 years, pine trees were shading out the grasses and weeds. In the final stages, oak and ash grew up among the pines and eventually replaced them.

Community ecologists originally believed that the final stage of ecological succession was a stable, mature community called a **climax community**. A climax community was thought to be self-perpetuating, in the sense that its species composition would replace itself and thus remain stable. Tests of the stability of mature ecosystems are rare, because few such systems exist. One such mature area of old eastern temperate forest lies in the Dick Cove Natural Area in Tennessee. Over a 9-year period, the apparent climax community changed from one dominated by oak and hickory to one becoming dominated by sugar maple, yellow poplar, and hickory. Ecologists have learned that climax communities change, but the changes occur so slowly they may not be noticed.

Ecological succession is more than a description of changing communities. Increasingly, human populations are affecting natural systems. **Restoration ecology** aims to understand the process of ecological succession and to use that understanding to restore natural communities that have been severely disturbed by human activities.

25.4 | Ecosystem studies focus on trophic structure and energy flow

When ecologists study ecosystems, they group individual organisms into functional categories based on how they obtain their energy. There are three basic functional categories—producers, consumers, and decomposers. These functional categories are arranged on different levels in a hierarchy called a *trophic structure*. Energy flows through this trophic structure in a linear manner from producers to consumers and finally to decomposers. Trophic structure and linear energy flows have a number of consequences for ecosystem dynamics.

Organisms may be grouped into functional categories

Plants and photosynthetic microorganisms are primary producers or autotrophs (meaning self-feeders), which in most ecosystems derive their energy from sunlight by photosynthesis. In deep-sea vent ecosystems, however, the primary producers derive their energy from geothermal sources. All other organisms derive their energy—directly or indirectly—from primary producers. Consumers are all heterotrophs (meaning "other-feeders") and obtain their energy by consuming other organisms. Primary consumers or herbivores eat primary producers. A sheep eating grass or protozoa eating algae are examples of primary consumers. Secondary consumers eat primary consumers. Lions and wolves are secondary consumers, or carnivores. Some organisms do not fit into a single category. Bears and humans are omnivores and act both as primary and secondary consumers. Other organisms are detritivores and consume detritus, which is dead organic matter. Worms and snails are detritivores. Organisms that break down organic matter for a source of energy are decomposers. Bacteria and fungi are decomposers, which may in turn be ingested by consumers. Ecosystems must contain a balance of organisms in these different categories. Without primary producers, there would be no input of energy into ecosystems. Without decomposers, organic matter would pile up in the environment and essential mineral nutrients would not be recycled back into the ecosystem.

The flow of energy through a food chain is linear

In any ecosystem, the organisms that occupy these different functional categories can be organized into a trophic structure called a **food chain** (Figure 25.14). Energy from the Sun enters the food

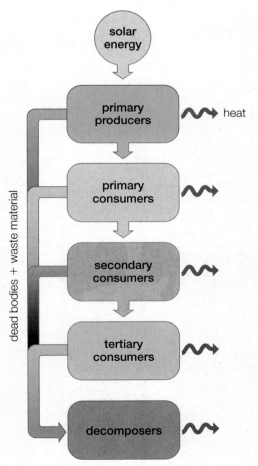

FIGURE 25.14 Schematic food chain Energy from the Sun is converted into chemical energy by primary producers (mainly plants and algae). Some of this energy is passed along to primary consumers when the primary producers are eaten. Energy is transmitted from primary to secondary to tertiary consumers. At each stage, most energy is lost as heat through the process of respiration. Also, some of the original solar energy is passed along to decomposers when organisms produce waste or die.

chain at the level of the primary producers. The primary producers convert solar energy into the chemical bonds of organic molecules. When they use this energy to perform work, grow, or reproduce, the process of respiration releases that energy, and most of it is dissipated as heat into the environment. Primary consumers feed upon primary producers for their energy and are linked to them in the food chain. Secondary consumers are linked to primary consumers and feed upon them. Tertiary consumers feed on secondary consumers. Thus the energy bound in organic molecules by primary producers passes up the food chain to the highest-level consumers. Each of the different links in the food chain is called a *trophic level*. The energy in the organic molecules, which remains in wastes and dead bodies, passes to decomposers, after which it is finally dissipated as heat.

In nature, the community is seldom as simple as a single linear food chain. Normally there are many different species of primary producers, and many species of primary consumers

that feed on them. All the different populations at different trophic levels are linked together in a **food web** (Figure 25.15).

Only a small fraction of energy passes between trophic levels

Every year, the activities of primary producers generate a great deal of organic matter or biomass on Earth. Estimates place this annual primary production at around 200 billion metric tons. Although this amount of production is enormous, the process of photosynthesis is not very efficient. Plants and other photosynthetic organisms capture only about 1–3% of the light energy that falls on them and convert it into chemical energy. When a primary consumer eats plants, much of the energy in the plant biomass is lost as heat or waste, and only a fraction is converted into new primary consumer biomass. In general, only about 10% of primary producer biomass is converted into new

GREAT PLAINS PRAIRIE

THIRD IN A SERIES

NATURE OF AMERICA

FIGURE 25.15 Terrestrial food web This sheet of United States Postal Service stamps depicts a grassland ecosystem in the Great Plains. Yellow arrows have been added to indicate some of the trophic interactions. Bison and pronghorn antelope are primary consumers of grasses and herbs, which are the primary producers in the prairie. The western meadowlark is a secondary consumer, feeding on the grasshopper in its beak. The rattlesnake is also a secondary consumer of gophers, kangaroo rats, and spadefoot toads.

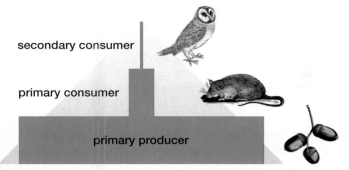

FIGURE 25.16   Pyramid of biomass In a simple example of a food chain, a population of plants that produce fruits and/or seeds eaten by rodents can support a rodent biomass considerably smaller than that of the seeds. A relatively large biomass of rodents, in turn, would be necessary to provide sufficient food for an owl (whose biomass would be a small fraction of that of the rodents).

primary consumers. Similarly, only about 10% of primary consumer biomass is incorporated into new secondary consumers. For example, 1,000 kilograms (kg) of plant biomass as seeds would yield about 100 kg of seedeaters such as rodents. That 100 kg of rodents could support, at most, about 10 kg of owls. This example illustrates one of two ways in which food chain dynamics can be represented graphically as an ecological pyramid (Figure 25.16). The owls, rodents, and seeds make up a pyramid of biomass. Each block represents the total biomass at each trophic level. Another way to represent food chain dynamics is the pyramid of energy (Figure 25.17). In this example, the energy in kilocalories (kcal) is shown for each trophic level in Lake Cayuga in New York.

In general, you expect more biomass or energy at the level of primary producers than at that of primary consumers and more

at the level of primary consumers than at that of secondary consumers. This is because the energy or biomass available to produce new organisms is reduced sharply between trophic levels. Most pyramids are upright with a wide base and narrow top.

The trophic structure of ecosystems and the inefficiency of energy transfer between trophic levels have important consequences for ecosystems. Food chains are normally limited to three or four links because the amount of energy left at the end is too little to support a viable population of higher consumers. The top consumer also has to be large enough to attack and kill the consumer below it. This size factor places a further constraint on the number of top consumers.

Because species are linked together in food chains and webs, some species can have far-reaching effects on entire systems. Such species are called **keystone species**, and their effects are called a **trophic cascade**. Rabbits were a keystone species in the chalk grasslands of southern England. In northern freshwater lakes, the keystone species may be a pike. These carnivorous fish prey on sunfish and perch, which eat zooplankton crustaceans—tiny animals related to crabs and shrimp. The zooplankton crustaceans consume the algae, which are the primary producers in the lake. An increase in numbers of pike would mean more predation on sunfish and perch and a reduction in their population sizes. Zooplankton crustaceans would then increase due to reduced predation on them. Finally, increased grazing from zooplankton crustaceans would reduce the algal populations. The effects of the tertiary consumer thus cascade down the trophic levels of the food chain.

A third consequence of food chains is that chemical pollutants can become concentrated as they pass up through trophic levels. The insecticide DDT, which is outlawed in the United States but not in all other countries, may be applied to crops to reduce insect damage or control insects that transmit human diseases. Rodents consume contaminated vegetation and concentrate DDT in their fatty tissues. Predatory birds consume the rodents, and the insecticide reaches toxic levels in their tissues and eggs. In countries where DDT has been banned, there has been a marked recovery of predatory birds such as falcons and eagles. Many other substances may be concentrated up a food chain.

25.5 | Global climatic patterns determine the distribution of biomes

Biomes are ecosystems that share similar climatic conditions. In later chapters we will examine specific biomes in detail.

The distribution of biomes is determined primarily by global patterns of atmospheric circulation

The Earth's surface receives energy input from the Sun. That radiant energy warms the surface but, because the Earth is not flat, the amount of solar energy received at the surface varies

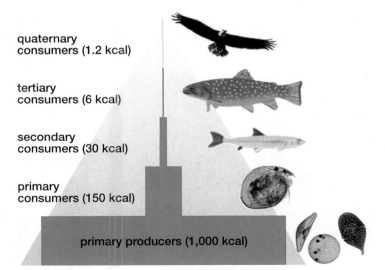

FIGURE 25.17   Pyramid of energy for Lake Cayuga In a real example from a freshwater lake, for every 1,000 kilocalories (kcal) of sunlight harvested by algae, some 150 kcal find their way to small aquatic animals. Only 30 kcal are transferred to smelt (a small fish) when they eat the small animals. If a trout were to eat a smelt, about 6 kcal would be transferred into this larger fish. If an eagle or human were to eat the trout, only about 1.2 kcal would be gained (in other words, some 999 of the original 1,000 kcal of energy harvested by the algae would have been lost).

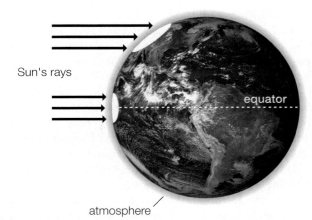

Sun's rays

equator

atmosphere

FIGURE 25.18 Solar energy at the equator compared to higher latitudes The energy at high latitudes is reduced for two reasons: (1) the Sun's rays strike the Earth's surface at a more oblique angle, resulting in the same amount of solar energy being spread out over a greater surface area; and (2) the Sun's rays travel a greater distance through the atmosphere, which results in more energy being either scattered or reflected back to space or absorbed before hitting the surface.

with latitude north or south of the equator (Figure 25.18). The same amount of solar energy is spread over a much larger land area in a polar region than in the tropics. Consequently, the total annual solar energy being received at the poles is less than half of that reaching the equator.

Furthermore, the axis about which the Earth rotates is not straight up and down but is tilted at an angle of 23.5° to a line perpendicular to the plane of the Earth's orbit (Figure 25.19). Because of this tilt, the Earth passes through a cycle of seasons as it moves in its orbit about the Sun. At the solstice on June 20 or 21, the axis of rotation in the Northern Hemisphere is pointed toward the Sun as far as possible, and the Northern Hemisphere enters summer with the longest day of the year. In the Southern Hemisphere, however, the axis of rotation points away from the Sun, and the Southern Hemisphere enters winter with the shortest day of the year. On September 22, the Earth completes one-quarter of its orbit. The Sun is now directly overhead at the equator, and the lengths of day and night are equal at 12 hours each. In the Northern Hemisphere this date marks the autumnal equinox (literally "equal night"), but in the Southern Hemisphere it is the vernal (spring) equinox. Three months later—at the solstice on December 21 or 22—the Earth has completed one-half its orbit. The axis of rotation in the Southern Hemisphere now points as far as possible toward the Sun, and the Southern Hemisphere enters summer. In the Northern Hemisphere the axis of rotation is pointing away from the Sun, and the Northern Hemisphere enters winter. Three months later on March 21, the Earth reaches an equinox as the Sun again passes directly over the equator. In the Northern Hemisphere this is the vernal equinox, and in the Southern Hemisphere it is the autumnal

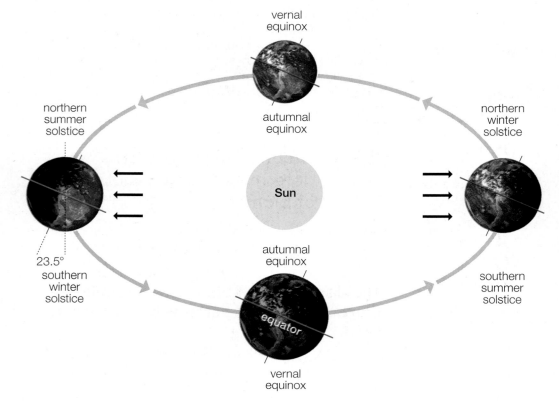

vernal equinox

northern summer solstice

autumnal equinox

Sun

northern winter solstice

23.5°
southern winter solstice

autumnal equinox

southern summer solstice

vernal equinox

FIGURE 25.19 Tilt of the Earth's axis and the seasons Because the axis about which the Earth rotates is tilted 23.5° with respect to a line perpendicular to the plane defined by Earth's orbit around the Sun, the Earth has distinct seasons. The Sun's solar radiation strikes the Earth at different angles depending on latitude and the season. For example, when the Northern Hemisphere is tilted toward the Sun, it receives more radiation than the Southern Hemisphere and experiences summer. If the axis were exactly perpendicular to the Earth's orbital plane, the Earth would have no seasons, but the climate would be cooler as one moved away from the equator.

equinox. The Earth completes its orbit in 3 more months to return to the June 21 solstice. If the axis of rotation were not tilted, there would be no seasons and no cyclical change in climate at higher latitudes.

Because the Earth is a tilted, rotating sphere, its surface and atmosphere are not heated uniformly. Differences in atmospheric heating create winds and drive global patterns of atmospheric circulation (Figure 25.20). The region of the Earth lying between 23.5° N and 23.5° S latitude receives the most direct solar radiation and consequently has the highest average annual temperature. At the equator, moisture-laden air is heated and rises in the atmosphere. As it rises, it cools and the moisture condenses into rain that nurtures tropical rain forests (Figure 25.21). Pushed by rising air from below, the cool air flows at high altitude toward the poles and radiates more heat into space. It descends at around 30° N and 30° S of the equator. As it descends, the air warms and picks up moisture, drying out the land below and forming two belts of desert lands centered at 30° N and S. The descending air divides at the surface; part flows back to the equator, and the rest flows toward the poles. The part flowing back to the equator completes a circular pattern of airflow known as a *Hadley cell*, which acts as a giant conveyor that moves moisture from latitudes near 30° toward the equator.

The portion of the descending air at 30° latitude that flows toward the poles is deflected eastward by the Earth's rotation. Thus, in the Northern Hemisphere, weather patterns flow mainly from west to east. Around 60° N and S latitude, this surface flowing air rises again, cools, and drops its moisture, nurturing grasslands and temperate forests. At high altitudes,

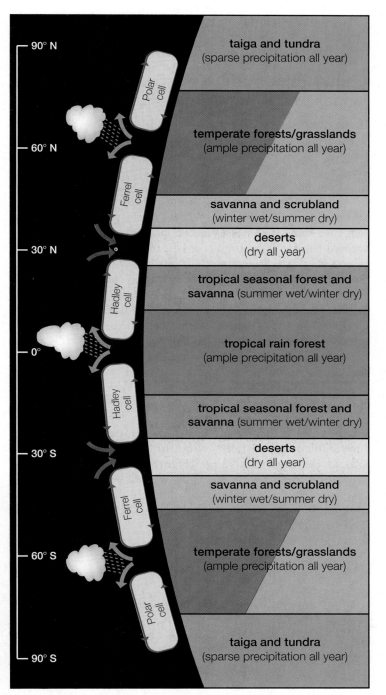

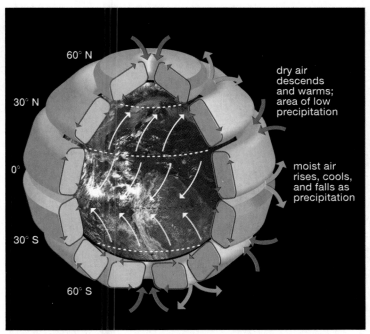

FIGURE 25.20 Air cells Six rotating belts of air currents (in gray), called *air cells*, cover the surface of the Earth and determine global patterns of wind and rain. Magenta arrows indicate the direction of airflow in each cell. Where air currents in adjacent air cells are rising, moisture condenses as rain (blue arrows). Where the air currents in adjacent air cells are descending (red arrows), the air is dry and the land is arid. The Earth's rotation deflects the surface air currents in the air cells, as shown by the white arrows on the globe of the Earth.

FIGURE 25.21 Relationship of air cells and biomes The left portion of this diagram shows the air cells that were described in Figure 25.20. The right portion shows the correspondence of biome types to the atmospheric circulation patterns (e.g., deserts occur where there are descending dry air masses, and moist biomes are found where ascending air forms clouds and rain falls). Biome colors correspond to those used in the map in Figure 25.22.

this cool air again spreads out and divides, part returning toward the equator and part moving toward the poles. The portion flowing back toward the equator completes a second set of air cells known as *Ferrel cells* (see Figures 25.20 and 25.21). The Ferrel cells carry moisture to supply the temperate biomes of both hemispheres. The air that flows toward the poles forms part of the last set of air cells, called the *polar cells*.

This cold air descends at the north and south poles, where it flows along the surface back to the temperate zones. Moisture levels tend to increase from the poles toward lower latitudes.

Continentality, ocean currents, and mountain ranges also affect the distribution of biomes

If the landmasses of Earth were uniformly distributed and without mountains, the distribution of the world's biomes would be determined by the pattern of air cells as described in Figure 25.21. The biomes do not strictly follow this pattern, however, because three additional factors play major roles in the distribution of biomes—rain shadows, ocean currents, and continentality (Figure 25.22). Continentality refers to the fact that some landmasses are so large that interior areas are located too far from the oceans to receive moist winds. Such inland areas are drier than their latitude would predict.

Wind patterns and the Earth's rotation create the ocean currents. The major ocean currents rotate in vast circles, turning in a clockwise direction in the Northern Hemisphere and counterclockwise in the Southern Hemisphere (Figure 25.23). These currents affect climate and therefore modify the distribution of the world's biomes. The Gulf Stream carries warm water from the Caribbean to Western Europe, where it warms the climate. Conversely, the cold Benguela Current flows north along the southwest coast of Africa, where it cools the coastal climate.

As moist air flows up the slopes of mountain ranges, the air cools by 1°C for each 100-m increase in altitude. Moisture condenses and falls as rain or snow on the windward side of the mountain range (Figure 25.24). Once over the range, the cold, dry air descends, warms, and picks up moisture from the land. This land lies in the rain shadow of the mountain range and is drier than its latitude would otherwise predict.

25.6 | Matter moves between biomes and the physical environment in large-scale biogeochemical cycles

All the biomes of Earth together make up the biosphere. Any element or molecule used by living organisms is involved in a biogeochemical cycle at the local and regional scale and at the global scale. The Earth's biomes are all linked through global biogeochemical cycles. The most important biogeochemical cycles are those involving water, nitrogen, and carbon dioxide.

Water cycles through the oceans, atmosphere, lands, and organisms

All water on Earth, whether in glacial ice, oceans, or freshwater lakes and rivers, makes up the hydrosphere. The hydrosphere is therefore the main reservoir of water on Earth. Water passes from the hydrosphere into the atmosphere, the gaseous layer surrounding the Earth, by evaporation—the change from liquid water to its gaseous phase (Figure 25.25). Global wind patterns carry water vapor around the world. Whenever moist air rises and cools, gaseous water condenses and falls to the surface of the land as precipitation—either as rain, snow, sleet, or hail. Once on land, water may evaporate directly back into the atmosphere or it may travel across land and enter rivers that empty into the oceans. Water that moves across the land and back to the oceans is called *runoff*. Water on land may also seep down through the soil to become groundwater. Eventually, all water on land returns to the oceans.

Living organisms consist mostly of water, which is the medium for all biochemical reactions. Plants acquire water through their roots and expel it as vapor through their stomata

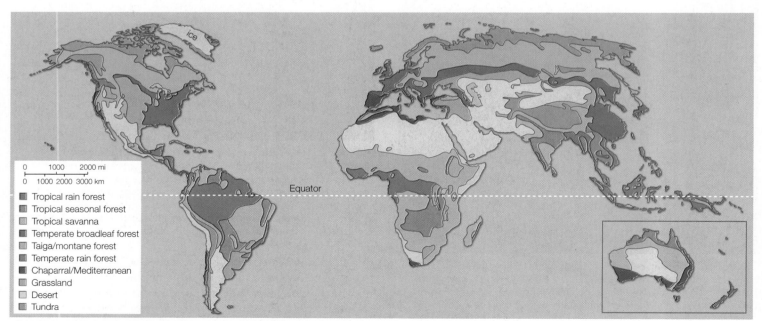

FIGURE 25.22 Global distribution of the Earth's biomes Because of the map's scale, biome contours are only approximate. Also, boundaries between biomes may not be sharp, as indicated here, because there are commonly broad zones where one biome changes into the next.

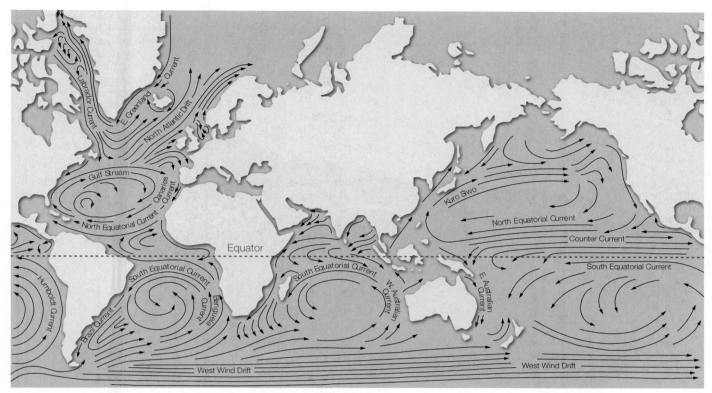

FIGURE 25.23 Major ocean currents The ocean currents affect regional climate and modify the distribution of biomes. The cold Humboldt (Peru) Current, for example, cools the climate of the Atacama Desert in Chile and delivers moisture in the form of fog.

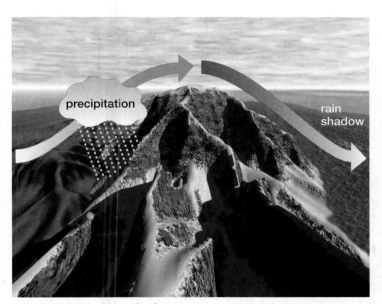

FIGURE 25.24 Rain shadow As moist air rises over a mountain range, the air mass cools and moisture condenses as rain on the windward side. Once over the mountain range, the dry air mass descends and warms up, drawing moisture out of the land and creating a rain shadow.

during transpiration. Water lost by transpiration returns to the atmosphere. Plants play a crucial role in retention of water on land. If the plant cover of an area is removed by deforestation (Chapter 27) or desertification (Chapter 26), any water falling on that area will evaporate or run off much faster than if a plant cover were present. Wetlands, with their characteristic vegetation, act to retain water that might otherwise contribute to floods (Chapter 28).

Microorganisms largely control the nitrogen cycle

The Earth's atmosphere is composed of 78% nitrogen gas (N_2) and is the main reservoir of nitrogen on Earth. All organisms require nitrogen for proteins and nucleic acids, but the vast majority of organisms need nitrogen in the oxidized form— nitrate (NO_3^-) or nitrite (NO_2^-)—or in the reduced form, ammonia (NH_3). Atmospheric nitrogen gas is highly stable, and only specific kinds of bacteria and cyanobacteria are capable of fixing it into the oxidized or reduced forms that all other organisms require. These microorganisms therefore largely control the cycling of nitrogen.

There are five main processes in the nitrogen cycle: (1) nitrogen fixation, (2) nitrification, (3) assimilation, (4) ammonification, and (5) denitrification (Figure 25.26). Bacteria and cyanobacteria direct all these processes except for assimilation, which all organisms carry out.

Nitrogen fixation is the conversion of gaseous nitrogen into ammonia. Lightning discharges fix some gaseous nitrogen. Human industrial processes fix atmospheric nitrogen for fertilizers, and these processes have about doubled the input of nitrogen to soils. Human activity therefore has a major impact on the fixation portion of the nitrogen cycle. Bacteria and cyanobacteria in soil and water carry out all biological nitrogen fixation.

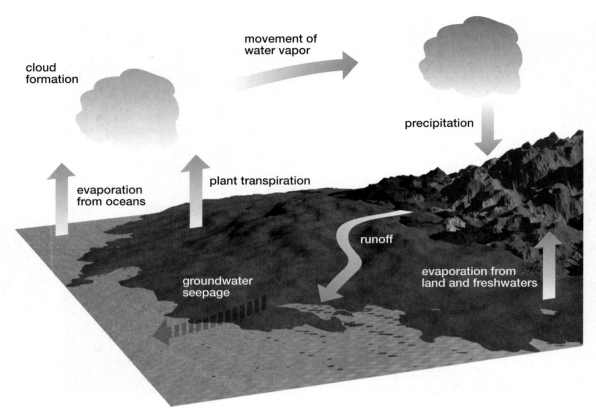

FIGURE 25.25 The water cycle Water enters the atmosphere as vapor by evaporation from land and water surfaces and transpiration from plants. Global air cells carry water vapor around the Earth. Vapor condenses as it rises, and falls back as precipitation on water and land. Water on land flows across the surface as runoff and enters streams, rivers, and the oceans; or it may enter the groundwater table, from which it eventually enters streams, rivers, and oceans by seepage.

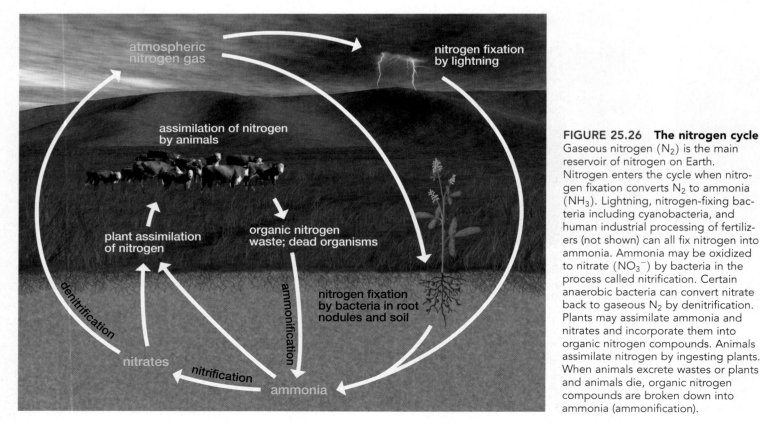

FIGURE 25.26 The nitrogen cycle Gaseous nitrogen (N_2) is the main reservoir of nitrogen on Earth. Nitrogen enters the cycle when nitrogen fixation converts N_2 to ammonia (NH_3). Lightning, nitrogen-fixing bacteria including cyanobacteria, and human industrial processing of fertilizers (not shown) can all fix nitrogen into ammonia. Ammonia may be oxidized to nitrate (NO_3^-) by bacteria in the process called nitrification. Certain anaerobic bacteria can convert nitrate back to gaseous N_2 by denitrification. Plants may assimilate ammonia and nitrates and incorporate them into organic nitrogen compounds. Animals assimilate nitrogen by ingesting plants. When animals excrete wastes or plants and animals die, organic nitrogen compounds are broken down into ammonia (ammonification).

Cyanobacteria and other bacteria that fix gaseous nitrogen use the enzyme **nitrogenase** to split molecular nitrogen and combine it with hydrogen. Nitrogenase, however, can function only if oxygen is absent. Some nitrogen-fixing microorganisms grow in anoxic (oxygen-free) environments. Most cyanobacteria, which perform both oxygen-evolving photosynthesis and nitrogen fixation, confine their nitrogen fixation to cells called **heterocysts**, whose thick walls exclude oxygen (Chapter 18). Nitrogen-fixing cyanobacteria are especially common in aquatic systems, where some may form symbioses with aquatic plants (Chapter 28). Others live within special roots of particular terrestrial plants (Chapter 27). Certain nitrogen-fixing bacteria, such as *Rhizobium*, form symbioses with legumes. The legumes house the bacteria in special nodules on their roots, where a low-oxygen environment is created (Chapter 10).

Nitrification is the process by which aerobic bacteria convert ammonia (NH_3) into nitrate (NO_3^-). The reaction is a two-step oxidation. The first step is the conversion of ammonia into nitrite (NO_2^-) by bacteria in the genus *Nitrosomonas*, which occur in soil, freshwaters, and the oceans. In the second step, aerobic bacteria in the genus *Nitrobacter* oxidize nitrite to nitrate. Under aerobic conditions, nitrification provides these bacteria with energy to fix CO_2 into organic molecules.

All organisms carry out the process of **assimilation** in some form. Plants may take up ammonia or nitrate from the soil, using special enzyme systems. Animals acquire their nitrogen by ingesting plants and incorporating plant organic nitrogen.

Plants and animals die, and their organic compounds decompose in the environment. Animals expel wastes that contain organic nitrogen compounds, such as urea or uric acid. Ammonification is the process by which bacteria in soil and water reduce these organic nitrogen compounds to ammonia. The ammonia released is then available for direct assimilation by plants or for nitrification to nitrate.

The final process in the nitrogen cycle—denitrification—closes the cycle. Here, nitrate is reduced to gaseous nitrogen, which returns to the atmosphere. Anaerobic bacteria perform denitrification in low-oxygen environments such as waterlogged soils, sewage, heavily fertilized soils, or deep underground near the water table.

Carbon dioxide cycles between the atmosphere and the biosphere

Carbon is present in the atmosphere in the form of the gas carbon dioxide (CO_2). It occurs in relatively low amounts (0.037%) compared to oxygen (21%) and nitrogen (78%).

Additional carbon is dissolved in the oceans as carbonate (CO_3^{2-}) and bicarbonate (HCO_3^-) ions, but the bulk of the Earth's carbon is locked up in limestone rock, where it is unavailable to organisms. All organisms must have a source of carbon because all of the organic molecules essential for life are built of carbon.

In the carbon cycle, carbon, as CO_2, is removed from the atmosphere when plants and photosynthetic microorganisms fix it into organic molecules (Figure 25.27). Each year, primary producers remove about one-seventh of the CO_2 in the atmosphere. When those same organisms carry out respiration to obtain energy for their metabolism, organic molecules are broken down into CO_2 and water, and the CO_2 returns to the atmosphere. When consumers ingest primary producers, they use the organic molecules to fuel their own metabolism, and their own respiration returns CO_2 to the atmosphere. Thus, cellular respiration returns CO_2 removed by photosynthesis back to the atmosphere.

Not all carbon fixed by photosynthesis returns to the atmosphere by respiration in the same year. A great deal of carbon is stored in the wood of trees, in peatlands, and in soils, where it may remain for hundreds of years. Fossil fuels such as coal, oil, and natural gas are all derived from organic molecules fixed by photosynthesis millions of years ago. When these fossil fuels are burned, their carbon is released back into the atmosphere. Over the past 200 years, human burning of fossil fuels has raised the level of CO_2 in the atmosphere from 0.028% to the present 0.037%, with global warming as a consequence (Chapter 1 and Chapter 29).

The carbon locked up in limestone rock slowly enters the carbon cycle as it is exposed to weathering, which releases mineral carbon into the oceans and atmosphere. Marine organisms may remove carbon from the cycle by incorporating it into their shells and bones. When these organisms die, their remains sink into the sediments that build up over eons of time and are converted into limestone rock. After more eons of time, that limestone rock may be uplifted by geological forces and become subject to weathering again. Thus, the carbon cycle operates on two very different time scales. Carbon dioxide may be rapidly exchanged between the atmosphere and biosphere. The cycling of carbon between limestone by weathering and back to rock by sedimentation, however, is a very long-term process.

Biogeochemical cycles such as those of water, nitrogen, and carbon dioxide involve interactions among all biomes of the Earth. Ecology at the level of the biosphere brings together the sciences of biology, geology, and meteorology.

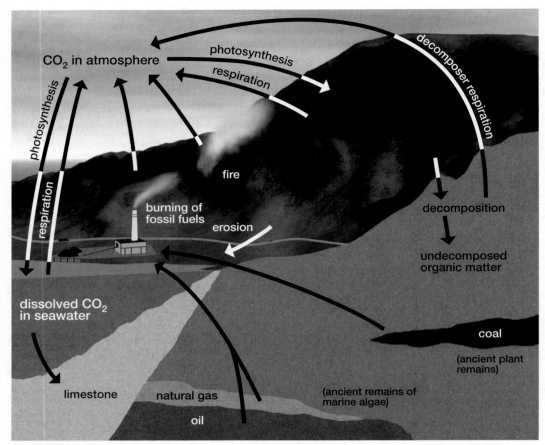

FIGURE 25.27   The carbon cycle   Carbon cycles rapidly between the atmosphere and biosphere. Photosynthetic organisms on land and in water remove CO_2 from the atmosphere. Respiration returns most of this carbon to the atmosphere, but some remains in organic matter until fires or decomposers break it down. Organic matter may also enter sediments, where over eons of time it may be converted into rock. Coal, oil, and natural gas are the products of photosynthesis in the ancient past. When humans burn these fossil fuels, their carbon returns as CO_2 to the atmosphere. Carbon in limestone enters the carbon cycle when erosion slowly dissolves the rock and releases the carbon as CO_2 in water or air. Arrows show movement of carbon atoms.

HIGHLIGHTS

25.1 Populations, communities, ecosystems, biomes, and the biosphere provide different ways to examine biological systems from an ecological perspective.

25.2 Ecological studies of populations include distribution patterns, age structure, population growth, and niches. Population growth includes the concepts of net growth rate, carrying capacity, and limiting factors. Niches are defined by abiotic factors such as moisture and biotic interactions such as mutualism, parasitism, herbivory, predation, and competition.

25.3 Community ecology focuses on individual organisms within a defined area. Communities contain different degrees of richness and diversity of species. Community composition changes over time in the process of succession.

25.4 Ecosystem studies group organisms into functional categories arranged in levels in a hierarchy called a trophic structure. Food chains and webs are trophic structures through which energy flows in a linear manner.

25.5 Ecosystems sharing similar climates are grouped into biomes. Global weather patterns, ocean currents, mountain ranges, and continents determine the distribution of ecosystems with similar climates.

25.6 All biomes together constitute the biosphere. There is a constant flow of materials in cycles at local, regional, and global scales. The most important of these biogeochemical cycles are those of water, nitrogen, and carbon dioxide.

REVIEW QUESTIONS

1. Define each of the following words: (a) ecology, (b) population, (c) community, (d) ecosystem, (e) biome, and (f) biosphere.

2. What is the growth rate of a population? How is it defined? What are the phases of the typical growth curve?

3. What is an ecological niche? Discuss some of the important abiotic factors that help determine the area occupied by a particular population in a terrestrial system such as a grassland, desert, or deciduous forest.

4. What is a symbiosis? Define each of the following symbioses, and provide a specific example. For each example, discuss the benefit or harm that each member derives from the interaction: (a) mutualism, (b) commensalism, (c) parasitism, (d) herbivory, and (e) predation.

5. Discuss some of the chemical and physical adaptations plants have made to deter herbivory.

6. What is ecological succession? How does primary succession differ from secondary succession? What are pioneer species?

7. What is a climax community, and is it considered a valid concept today?

8. How do autotrophs differ from heterotrophs? Define each of the following functional groups based upon their position in the trophic hierarchy: (a) primary producers, (b) primary consumers, (c) secondary consumers, (d) detritivores, and (e) decomposers.

9. How do global patterns of atmospheric circulation affect the distribution of biomes?

10. How do continentality, ocean currents, and mountain ranges affect the distribution of biomes?

11. Describe the nitrogen cycle. What is the main nitrogen reservoir on Earth, and what are the five main processes in the nitrogen cycle?

12. Briefly summarize the main components of the water cycle as well as its primary reservoir on Earth.

APPLYING CONCEPTS

1. Classify each of the following as a population, community, ecosystem, or biome, based on the definitions given in this chapter: (a) The Joshua trees *(Yucca brevifolia)* in Joshua Tree National Park, California. (b) The desert of Death Valley National Park, California. (c) The plants and animals of the Okefenokee Swamp in southern Georgia and northern Florida. (d) The cattails of the Okefenokee Swamp. (e) The alpine tundra in Rocky Mountain National Park, Colorado. (f) The tundras of the world. (g) The life in Lake Ponchartrain, Louisiana.

2. In "Dinosaur Renaissance" (*Scientific American*, 1975, 232:58–78), paleontologist Robert Bakker argues that dinosaurs were not cold-blooded ectotherms but rather warm-blooded endotherms. One line of evidence he presents is that in Mesozoic dinosaur communities, predator-prey bio-mass ratios (as gleaned from fossil data) were typically from 1 to 3% (i.e., 30–100 times as much prey biomass as predator biomass). Why are predator-prey biomass ratios correlated with predator ectothermy/endothermy status? How would you expect these numbers to change if dinosaurs were ectotherms?

3. Imagine a freshwater lake and its quiet inshore waters. Following is a list of organisms that you might encounter in this ecosystem; the text in parentheses indicates either the type of organism or how/what it eats. Diagram the food web of this ecosystem, placing each organism on one of four trophic levels: (1) secondary or tertiary consumer (STC),

(2) primary consumer (PC), (3) primary producer (PP), and (4) detritivore (Dt) and decomposer (Dc) at the bottom of your diagram.

crayfish	(an animal that eats dead organic matter)
Pediastrum	(a photosynthetic green alga)
Potomogeton	(a rooted plant)
northern pike	(a fish that eats medium-sized animals)
mallard duck	(a bird that eats plants)
clam	(a filter feeder)
bass	(a fish that eats insects)
giant water bug	(an insect that eats insects and small fish)
water boatman	(an insect that eats algae)
Saprolegnia	(an aquatic, colorless protist that feeds off organic matter)

4. If the birth rate and the death rate of a population are equal, what happens to the growth rate? What if the birth rate is slightly lower than the death rate?

5. Are food chains the same as food webs? Do humans participate in a food web?

6. It has been said that it is more eco-friendly to consume lower on the food chain—to consume more fruits, vegetables, and grains and fewer meats. What is the logic behind this statement, and is it justified considering what you have read in this chapter?

Arid Terrestrial Ecosystems

Prairie scene with bison

KEY CONCEPTS

- Polar deserts include the most severe climates on Earth.

- Water is the dominant limiting factor shaping the biology of deserts.

- Plants in temperate and tropical deserts show remarkable adaptations to high temperatures and water stress, including tolerance and avoidance, succulence, and crassulacean acid metabolism.

- Grasslands are important because they occupy large areas of the Earth's surface, store vast amounts of organic carbon in their soils, contain the world's most productive agricultural land, and have shaped the evolution of humans and grassland animals.

- Grasses, thousands of species of nongrass herbaceous plants, and a few types of trees and shrubs in grasslands are adapted to relatively low precipitation, seasonal extremes of temperature, frequent fires, and large herds of grazing animals.

- The chaparral of western North America has a climate of hot, dry summers and cool, wet winters and is dominated by evergreen shrubs. Fire is a major factor in chaparral ecology.

"There is no timber in this part of the country; but continued prairie on both sides of the river. A person going on one of the hills may have a view as far as the eye can reach without any obstruction; and enjoy the most delightful prospects." This description of a first encounter with the vast grasslands of the U.S. Great Plains was recorded in a journal kept by Patrick Gass, a member of the famous Lewis and Clark expedition sent in 1804 by President Thomas Jefferson to explore the new lands acquired in the Louisiana Purchase. Leader Meriwether Lewis wrote, "this scenery, already rich, pleasing and beatiful, was still farther hightened by immence herds of Buffaloe, deer, Elk ... which we saw in every direction feeding on the hills and plains. I do not think I exagerate when I estimate the number of Buffaloe which could be compre[hend]ed at one view to amount to 3,000." The expedition found and described many animals and plants new to science, including the prairie dog and the bitterroot plant, *Lewisia rediviva,* the state flower of Montana, whose scientific name honors Lewis and from which a large western U.S. mountain chain and major river take their names. (The accompanying photograph shows a portion of the type specimen of *Lewisia*—the herbarium specimen used in first describing it as a new species.)

If Gass, Lewis, or Clark could visit the same grasslands today, they would see a vastly changed landscape. The same wide-open spaces they found so amazing still remain but are now covered by cornfields, wheat, or other crops and sometimes large expanses of noxious, inedible, nonnative plants. The vast buffalo (bison) herds, estimated to have numbered 60 million animals in the 1880s, are gone—replaced by fenced herds of cattle. The many Amerindian tribes the explorers met no longer occupy the same lands, and some hover near the brink of extinction. Grasslands around the world are or have been equally as wonderful and important but have also undergone dramatic changes under the influence of ever-larger human populations.

Grasslands are one example of an arid terrestrial ecosystem. This chapter focuses on four types of arid terrestrial ecosystems—polar deserts, temperate and subtropical deserts, grasslands, and chaparral—and describes their physical characteristics, the adaptations of the organisms that live in them, and their importance to humans and the impact of humans on them.

Type specimen of *Lewisia*

26.1 | Arid terrestrial ecosystems are diverse

Arid terrestrial ecosystems include a diverse group of ecological systems that have in common a relatively low average annual amount of precipitation. The upper limit for arid ecosystems is about 100 cm (39 in.) per year. The lower limit can be effectively zero; no measurable precipitation may occur in a period of one or more years. Arid terrestrial ecosystems include polar deserts, temperate and subtropical deserts, grasslands, and chaparral. Polar deserts occur in the extreme northern parts of North America, Europe, and Asia above 72–73° N latitude and in the continent of Antarctica. The climate in polar deserts is bitterly cold, and annual precipitation as snow is less than 1 to 2 cm (0.4 to 0.8 in.) per year. Temperate and subtropical deserts range in precipitation from 2.5 cm (1 in.) per year up to 60 cm (24 in.) per year. Most precipitation occurs during one or two rainy seasons of the year. At the upper end of this range of annual precipitation, deserts overlap with grasslands and chaparral. Grasslands receive between 25 cm to 100 cm (10 to 39 in.) of annual precipitation, but, compared to deserts, the precipitation in grasslands is more evenly distributed through the year. Chaparral, or Mediterranean scrub, receives 20 to 100 cm (8 to 39 in.) of precipitation per year, but that precipitation occurs almost entirely during the winter. Chaparral has a climate of hot, dry summers and cool, wet winters.

Arid terrestrial ecosystems also share a general lack of trees. Even the maximum levels of annual precipitation are too low to support trees. In temperate and subtropical deserts, trees may appear in local areas where moisture is greater, such as along seasonal streams where water may flow once or twice per year. In grasslands and chaparral, trees may also occur along streams and in low-lying areas where greater moisture collects. The importance of these arid ecosystems to humans is directly related to their level of annual precipitation; grassland and chaparral ecosystems are more important to humans because they receive enough precipitation to be used for agriculture.

26.2 | Polar deserts have the most severe climates on Earth

In the Northern Hemisphere, polar deserts have a climate classified as cool-polar, where the highest average monthly temperature in the warmest month of summer (July) lies between 3° and 7°C (37°–45°F). These polar deserts, known as herb barrens, occupy thousands of square kilometers across the High Arctic of Canada and Greenland, where they occur from near sea level to elevations of 200–300 meters. In the Southern Hemisphere, the area of polar deserts is more restricted. More than 98% of Antarctica is buried under glacial ice, which is as much as 4 km (2.5 mi) thick. The continent is bordered by coastal mountain ranges that shelter ice-free valleys (Figure 26.1). Polar deserts are restricted to these ice-free valleys, offshore islands, the peninsula of Antarctica, and some scattered islands in the Southern Ocean.

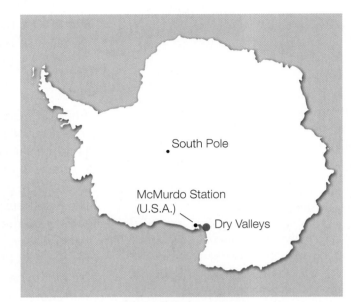

FIGURE 26.1　**Location of Dry Valleys in Antarctica**

Antarctica contains two types of polar ecosystems—the maritime Antarctic and the continental Antarctic. Both may be considered as polar deserts although their climates are more severe than those in the Northern Hemisphere. The maritime Antarctic is a cold polar environment where the mean monthly air temperatures in the southern summer (January) are only about 0°–2°C (32°–36°F). The maritime Antarctic consists of the west coast of the Antarctic Peninsula and some islands in the Southern Ocean. Continental Antarctica consists of the ice-free valleys, called the Dry Valleys, and some offshore islands. It has a harsh climate classified as frigid polar. Mean monthly summer temperatures reach only 0°–1°C (32°–34°F) in some areas, and low moisture is a major limiting factor. Very few plants have been able to colonize Antarctica because of its isolation from any other landmass. South America is more than 850 km (525 mi) across the stormy Southern Ocean.

Arctic herb barrens contain few species of plants

In the High Arctic, the most widespread plant community in the polar desert is called an herb barren. An herb is a nonwoody plant with a short-lived stem or stems. Grasses are herbs with narrow leaves, and forbs are herbs with broad leaves. Herb barrens are so named because they contain vast areas of bare rocky ground, and herbs usually occupy no more than 5% of the surface area (Figure 26.2). Soils are very dry and very low in nutrients, and the growing season is short (1–1.5 months). Only about 16–17 species of flowering plants occur on these barrens. The number of species of mosses, lichens, fungi, and cyanobacteria is greater than of flowering plants, but these species are restricted to sites with higher levels of moisture. Such sites occur below large snowfields, where summer meltwaters allow the development of moss mats and crusts of algae, lichens, and mosses. Flowering plants become

FIGURE 26.2 **Herb barren in the High Arctic**

established in these sites, which are conspicuous against the otherwise barren landscape.

Continental Antarctica contains only sparse populations of mosses, lichens, and algae

In the continental Antarctic, populations of mosses, lichens, and algae are widely dispersed, and extensive areas of land are devoid of any visible vegetation. Mosses and lichens are abundant only in scattered coastal locations. The most severe conditions occur in the ice-free Dry Valleys, where the maximum air temperatures rarely rise above 0°C (32°F) and the soil moisture levels are comparable to that in Arizona deserts. Dry Valley soils contain high levels of salts and may be acidic. Most of the soil is permanently frozen as **permafrost,** while the surface may undergo rapid freeze-thaw cycles. Once thought to be barren, Dry Valley soils are now known to contain bacteria, algae, fungi, and protozoa.

Two special ecosystems in the Dry Valleys deserve special mention—lakes and lithic (rock) ecosystems. Lakes provide some of the most favorable ecosystems for microbial life in continental Antarctica. Most lakes in the Dry Valleys are permanently frozen over by ice sheets that may be 3 to 6 m (10 to 20 ft) thick (Figure 26.3a). Beneath these permanent ice sheets, the deeper lakes contain liquid water and a microbial community of cyanobacteria and small flagellates. The flagellates belong to genera commonly found in temperate lakes—such as *Cryptomonas* (Chapters 19 and 28). Dense mats of cyanobacteria often cover the sediment surface at the bottoms of these lakes (Figure 26.3b and c). Many inland lakes are derived from the ocean and are hypersaline. Streams bring water from glaciers to many lakes, but the streams are frozen over for all but a few weeks during the southern summer.

Lithic ecosystems are the most unusual microbial habitat in continental Antarctica. Rocks readily absorb liquid water and can store it for days. They can also absorb heat on sunny days and maintain a temperature above that of the surrounding air for extended periods. These features of rocks make them a favorable environment for microbes in the Dry Valleys.

FIGURE 26.3 **Dry Valleys** (a) Dry Valley with frozen lake. (b) An underwater view showing a diver collecting samples from a benthic cyanobacterial mat in a Dry Valley lake. (c) The mat coating the lake bottom.

A variety of microscopic algae, cyanobacteria, fungi, and heterotrophic bacteria are found in association with light-colored, translucent rocks such as marble, granite, and sandstone (Figure 26.4a). The fungi and algae are often associated together as lichens. These microorganisms may occur on the surfaces of rocks, in fissures and cracks in rocks, inside rocks among the mineral grains, and underneath translucent rocks. In the Dry Valleys, the microbial communities inside rocks occur in distinct layers (Figure 26.4b). The topmost few millimeters are free of microbes, the first layer consists of black lichens, and the second is a white layer of the fungal mycelium of the lichen without the symbiotic algae. Beneath these layers is a green layer of algae.

The Dry Valleys of Antarctica are among the harshest environments on Earth in which life can still be found. Above 1,000 m in altitude (3,280 ft) in these valleys, microbes disappear from rock surfaces; above 2,000 m (6,560 ft), even microbes within rocks are absent. For this reason, the Dry Valley microbial ecosystems are thought to be the nearest terrestrial equivalent to the environment of Mars. Researchers use them as an outdoor laboratory for the development of approaches to the biological exploration of Mars and other worlds. For more on polar biomes and Mars, refer to Essay 26.1, "Building Biomes on Mars."

ESSAY 26.1 BUILDING BIOMES ON MARS

The *Mariner* and *Viking* spacecraft that visited Mars in the 1970s returned images clearly indicating that Mars once had a much warmer and wetter climate. Mars today is extremely cold. Its average surface temperature is −60°C (−76°F) compared to that of Earth at +15°C (59°F). The atmosphere of Mars is so thin (less than 1/100th that of Earth) that liquid water cannot exist. Water is present either as ice or vapor. Yet the early

Thus, the plant communities of terrestrial polar and alpine biomes may one day make other worlds habitable.

spacecraft gathered evidence that water had once flowed through channels and drainage systems. Recent observations by *Mars Global Surveyor* have strengthened this evidence, indicating that a vast ocean may once have existed in the Northern Hemisphere and that, on a local scale, water in some form may still occasionally flow on Mars.

With the realization that Mars had once been a more hospitable planet for life, scientists and authors began to

speculate on how Mars might be returned to its prior warmer climate by some global engineering techniques. This science of terraforming, by which another world might be made more Earthlike, has developed slowly since the 1970s. It is now part of the new science of astrobiology, which is directed toward the study of organisms in extreme environments on Earth, the search for life on other worlds, and the potential extension of terrestrial life to other worlds.

The terraforming of Mars would be a long-term program. Assuming there is no indigenous life on Mars, terraforming might begin some 25 to 50 years after the

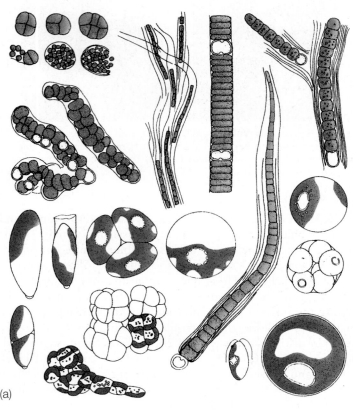

(a)

black layer
white layer
algal layer

(b)

FIGURE 26.4 Some microbes found in association with rocks in Antarctica (a) A variety of cyanobacteria (blue-green algae) and green algae are present in these habitats. (b) Layers of microbes in a Dry Valley rock.

In the maritime Antarctic, bryophytes and lichens are dominant

Vegetation in the maritime Antarctic is more widespread and diverse in form and species than in the continental Antarctic. Lichens are common on rock surfaces and may form closed stands on drier sites (Figure 26.5). On moister sites, mosses form a number of different communities, which can be described as mats, carpets, hummocks, or turfs. The turfs may lie over as much as 2 m (6.5 ft) of peat, all but the top 20 cm (8 in.) of which is frozen permanently.

The maritime Antarctic contains the only two flowering plants on the continent. The Antarctic hair grass *(Deschampsia antarctica)* forms low mats and is the more widespread of the two (Figure 26.6a). The other, the Antarctic pearlwort, *Colobanthus quitensis,* is a member of the carnation family and grows as small cushions (Figure 26.6b). Both occur at low altitude on level or north-facing slopes where solar radiation is greatest. Both species are showing effects of global warming. Since 1964, summer temperatures in the Antarctic Peninsula have risen by 2°C, and the growing season has increased by about 2 weeks. As a result, both flowering plants have dramatically increased in number and in the number of sites they occupy.

26.3 | Temperate and subtropical deserts are characterized by low annual precipitation

Deserts occur on the Earth's surface wherever there is either low rainfall or high evaporation from land surfaces together with high **transpiration** from plants (Chapter 11), or a combination of both under present climatic conditions. The most widely used definition of deserts was developed for the United Nations in 1953, and it classifies them into semiarid, arid, and hyperarid lands. Together, these three categories represent one-third of the Earth's land area. By their definition, semiarid lands receive between 200 and

first human settlement on Mars. Ozone-safe greenhouse gases, such as compounds of carbon and fluorine, would be manufactured to raise the average surface temperature of Mars. As Mars became warmer, frozen carbon dioxide would enter the atmosphere and thicken it. Liquid water would become stable on the surface. Hardy microorganisms derived from those in Antarctica would be grown and spread across the Martian surface. Photosynthetic microbes would add organic compounds to the Martian surface material, converting it into a true soil, and begin converting carbon dioxide in the atmosphere into oxygen. Other microbes would initiate a nitrogen cycle. When sufficient oxygen had built up in the atmosphere, it would be possible to introduce hardy polar and alpine mosses and lichens from Earth to the surface of Mars (Figure E26.1). These bryophytes would be capable of effecting a major change in the atmosphere over several hundred years. By removing large amounts of carbon dioxide and fixing it

into decay-resistant organic compounds in the form of peat, bryophytes would significantly increase the oxygen levels of the Martian atmosphere. Flowering plants and simple animal life could then follow. Thus, the plant communities of terrestrial polar and alpine biomes may one day make other worlds habitable.

E26.1   Artist's representation of an early stage in terraforming of Mars

FIGURE 26.5   Antarctic lichens

600 millimeters (mm) (8 to 24 in.) of rainfall per year. Arid lands receive between 25 and 200 mm (1 to 8 in.), and hyperarid less than 25 mm (1 in.) per year. If a semiarid land receives about 300 mm (12 in.) of rainfall per year, it could support grassland if that rainfall were evenly distributed throughout the year, with some rain falling every month. But if that rainfall is concentrated into 1 or 2 months, that land will be a desert. Thus, temporal unevenness of rainfall is a factor in determining the presence of deserts.

Four physical factors determine the locations of temperate and subtropical deserts

The distribution of the world's deserts is shown in Figure 26.7. One or more of four physical factors are responsible for their locations—subtropical high-pressure zones, continentality, cold ocean currents, and rain shadows. Present locations are a direct result of present-day climatic conditions. **Subtropical high-pressure zones** result from airflow patterns called Hadley cells (Chapter 25). Air masses circulating in the Hadley cells sink back to the surface at 30° N and 30° S latitudes and create the high-pressure zones. As they sink the air masses warm and take up moisture while drying out the land beneath them. If the landmasses of Earth were uniformly distributed and without mountains, the Hadley cells would generate two complete belts of desert centered at 30° N and 30° S encircling the planet. African deserts fit this pattern, but other physical factors alter it. Continentality means that some land areas can be located so deep within the interiors of continents that moist winds from oceans cannot reach them. The deserts of central Asia are continental deserts. Cold ocean currents moving from the poles to the tropics intensify the aridity of coastal deserts (see Figure 26.7). Less moisture evaporates from cold water than warm. Thus, the Humboldt Current dries out the Atacama Desert of Chile and Peru. As air masses move up over mountain ranges, they cool and their moisture condenses into rain on the windward side of

FIGURE 26.6 The two flowering plants found in Antarctica
(a) *Deschampsia antarctica*, growing among rocks. (b) A cushion of *Colobanthus quitensis*.

the mountains. Passing over the range, the now cold, dry air descends, warms, and draws moisture from the land, forming an arid land in the **rain shadow** of the mountain range.

Rain shadow effects are particularly important in the formation of the four major North American deserts (Figure 26.8). The **Great Basin Desert** of Nevada and Utah is a high-altitude cold desert that lies between the Sierra Nevada and Cascade Ranges of the West Coast and the Rocky Mountains of Colorado. The high-elevation **Mojave Desert** (which includes Death Valley National Monument) lies in southern California and Nevada. The **Sonoran Desert**, which contains Saguaro National Monument, extends across southern California and Arizona into Mexico. The fourth North American desert is the **Chihuahuan Desert**, which covers much of north-central Mexico and extends into New Mexico and Texas. Big Bend National Park lies in the Chihuahuan Desert.

Desert plants have adapted to acquire water

In deserts the main limiting factor on plants is the lack of water. Most plants have adapted to acquire water in the brief periods when it is available, and many are adapted to conserve it when it is scarce. There are two sources of water in desert regions—surface water and the deep-water table.

Plants using the deep-water table must put down long roots

Even in deserts where the surface soil may be extremely dry, there will likely be a significant amount of water located 9 to 60 m (30 to 200 ft) deep in the ground, where it forms a water table. Plants that utilize this deep-water source must have roots long enough

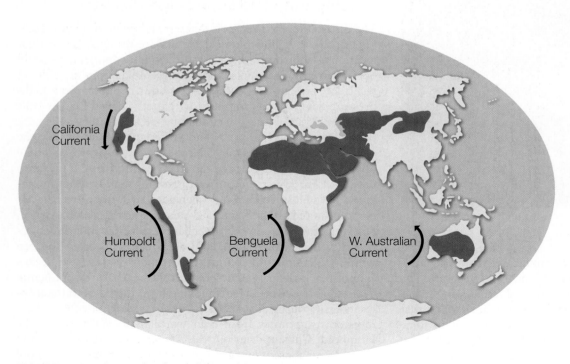

California
Current

Humboldt
Current

Benguela
Current

W. Australian
Current

FIGURE 26.7 Desert lands of the world

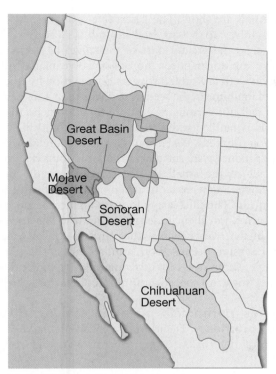

FIGURE 26.8   **The four American deserts**

FIGURE 26.9   **Two desert trees with long root systems that draw water from the deep-water table** (a) Paloverde (*Cercidium*), (b) mesquite (*Prosopis*).

to reach it. They essentially escape harsh surface conditions by sending their roots into a more favorable environment.

Trees adapted to use the deep-water table are generally found within **arroyos**—watercourses that are dry except for brief periods of the year. In arroyos the water table is closer to the surface than in upland sites. Therefore trees growing there do not have to send their roots down as far to reach the water table. One of the most common trees in the Sonoran Desert is the paloverde, whose common name is Spanish for "green stick (Figure 26.9a). Paloverdes (*Cercidium*) are members of the pea family (Fabaceae) and have greenish photosynthetic bark. They bloom in April or May, after the winter rainy season, and are covered with lemon-yellow flowers. Paloverdes produce leaves only in the rainy seasons. The leaves are pinnately compound (Chapter 11), and the leaflets are small in order to easily shed heat. Leaves are dropped when dry weather begins, and the trees rely on photosynthesis in their bark. Paloverdes make another use of the arroyos where they grow. In desert regions when rainstorms do occur, they are often intense, and lack of vegetation causes much of the water to run off rather than soak into the land. Dry washes can fill with flash floods in minutes. The paloverdes use these flash floods to disperse their seeds.

Another desert tree that may access the subterranean water table is the mesquite *(Prosopis)*. Like the paloverde, mesquite is a member of the Fabaceae (Figure 26.9b). In dry habitats mesquite is a shrub. In arroyos, however, it may reach the height of a tree. Nearly every desert animal, including humans, can eat the beans of mesquite, which protects itself by covering its branches with long thorns. Mesquite wood makes a delightfully aromatic fire, a fact known to desert dwellers long before it became a feature of steakhouses.

EVOLUTION

A great variety of desert plants use surface water

Desert algae, mosses, and lichens are inconspicuous but widely prevalent members of the desert community that use surface water. Desert annuals and perennials complete their life cycles when surface water is available. Deciduous perennials drop their leaves during water-stress periods and survive in a state of dormancy. Succulents—the thick plants we most often associate with deserts—store large quantities of surface water to endure periods of water stress.

Desert algae, mosses, and lichens are tolerant of high temperatures   In deserts algae occur in two main forms—as cryptogamic crusts and as cryptoendolithic communities. Cryptogamic crusts (or soil-algal crusts) cover large areas of arid and semiarid lands throughout the world. They are aggregates of algae, fungi, lichens, and mosses that form most often in areas where temporary ponds occur. Crusts are common in the cool Great Basin desert. Cryptoendolithic algae live inside the pore spaces of granular light-colored rocks. Both cyanobacteria and green algae occur in this habitat, but the most widespread organism is a cyanobacterium called *Chroococcidiopsis.* Cryptoendolithic algae are very tolerant of high temperatures and drying. They rapidly become active when moisture is available.

Desert mosses, such as the genus *Tortula,* can also survive high temperatures and prolonged water stress. They occur in crevices and on slopes where they are partially sheltered from direct sunlight. Museum specimens have been reported to recover after 250 years without water. Lichens—a symbiosis between an alga and a fungus—are even more widespread in deserts than are mosses. Crustose lichens are common on stable rock surfaces (Figure 26.10). Other lichens live inside rocks in the pore spaces. Lichens are also tolerant of high temperatures. One, *Ramalina maciformis,* can survive temperatures up to 85°C (185°F).

Desert annuals and herbaceous perennials grow when water is available

Desert annuals can be divided into two groups based on their season of growth. Summer annuals (or ephemerals) germinate following rains when temperatures exceed 25°C; winter annuals germinate when cooler temperatures prevail. In desert locales with only one rainy season, only the appropriate annuals will occur, but in areas that receive two rainy seasons—such as southern Arizona—both types of annuals may be abundant.

Desert annuals may have very short life cycles—the time from seed germination to seed set may be as short as 3 weeks. Consequently, there is little time for vegetative growth, and these desert annuals are short. The desert five-spot *(Eremalche rotundifolia)* is only 7 to 10 cm (3–4 in.) tall (Figure 26.11). Desert annuals spend the greater part of their lives as dormant seeds. Seeds have lain dormant for more than 10 years and still germinated when wetted. Rainfall triggers germination by leaching out a chemical inhibitor in the seed coat. If less than 1 in. of rain falls, the seeds will not germinate. If more than an inch falls, enough moisture is available for the annual to complete its life cycle, and germination proceeds. When enough rain accumulates, desert annuals can produce a spectacular carpet of blooms (Figure 26.12).

Not all of the small flowering plants in the desert are annuals. Some are herbaceous perennials, like the mariposa lily and desert lily. The lilies grow long enough to accumulate new supplies of food in their underground bulbs. The aboveground portions of the plants then wither, and the bulbs wait for the next rainy season. Because both annuals and herbaceous perennials escape water stress by growing only when conditions are moist, they show no specialization for conserving water.

Deciduous perennials maintain significant aboveground biomass

Deciduous perennials have extensive surface root networks to capture rainwater. During water-

FIGURE 26.11 **Desert five-spot *(Eremalche rotundifolia)*** This annual plant is a member of the mallow family and blooms from March through May. It is also known as the lantern flower because light passing through the thin petals causes the flower to glow.

FIGURE 26.10 **Desert crustose lichens** Those pictured are in the Namib Desert of southern Africa.

FIGURE 26.12 **Carpet bloom of desert annuals**

limiting periods, they drop all their leaves but, unlike desert annuals and herbaceous perennials, they retain significant aboveground biomass year-round. Dormant buds on the shoots allow them to leaf out quickly in response to any significant rainfall.

The ocotillo *(Fouquieria splendens)* looks like a spray of up to 20 long, spindly stems emerging from a common base (Figure 26.13a). The individual stems may be 3 to 6 m (10 to 20 ft) long and are lined with thorns. Following rains, leaves appear in little bunches along the stems, and in spring the stems are topped with clusters of scarlet flowers. When dry weather sets in, leaves are dropped, but their petioles remain and function as thorns. The stems are green but do not carry out photosynthesis. Ocotillo may set and drop leaves as many as six times in a year. In the past, stems of ocotillo were cut off and stuck in the ground to form fences. Frequently the stems rooted and formed a living green fence.

Perhaps the strangest plant in the desert is a relative of the ocotillo called the boojum tree *(Fouquieria columnaris)*. The boojum looks like a giant white carrot turned upside down and stuck in the desert soil (Figure 26.13b). The boojum has short, thin branches covered in small **xeromorphic** (dry-form) leaves (Chapter 11). Like ocotillo, the boojum drops its leaves during periods of limited water and may do so many times in a year. It has a vast, shallow root system to capture rainwater; a thick, light-colored bark to reflect solar radiation; and a semisucculent trunk interior, to store water. Originally the boojum was found only in the Baja Peninsula of Mexico. Recently, however, it has become a popular ornamental in southwestern gardens.

Another bizarre desert plant is the Joshua tree *(Yucca brevifolia)*, which is found only in the Mojave Desert (Figure 26.13c). Joshua trees often grow in groves, where their shallow root systems capture rainwater. Individual trees may reach 4.5 to 12 m

FIGURE 26.13 **Shrubby deciduous desert perennials** Both the ocotillo (a) and the boojum tree (b) are members of the genus *Fouquieria*. A small individual cardon cactus is seen in the background of (b). (c) The Joshua tree *(Yucca brevifolia)*. (d) The creosote bush *(Larrea tridentata)*.

(15 to 40 feet) tall and live up to 200 years. A member of the lily family (Liliaceae), the Joshua tree may bear clusters of creamy yellow and green, bell-shaped flowers that emit a foul odor. Flowers are pollinated by one species of yucca moth.

The creosote bush *(Larrea tridentata)* is widespread in the Mojave, Chihuahuan, and Sonoran Deserts (Figure 26.13d). The leaves are covered in resin that slows water loss but also emits a pungent odor from which the shrub derives its name. Probably because it tastes terrible, desert animals will not eat it, and it is one of the few desert plants without thorns. The plant drops its leaves during periods of water stress. The roots of the creosote bush secrete a substance that inhibits root elongation of other creosote bushes and other species of shrub near it. This root substance is an example of **allelopathy**, the release of a chemical into the environment that inhibits growth of other individuals of the same or different species. The allelopathic substance in creosote bushes spaces out the shrubs, so that competition for surface water is reduced.

Desert succulents have a number of adaptive features to survive aridity

Succulents may consist of thick, heavy leaves (leaf succulents) or thick, heavy stems (stem succulents). Under desert conditions they are slow-growing perennials that use surface water. A number of plant families have developed a succulent growth form as an adaptation to low moisture—an example of convergent evolution (Chapter 16). Succulents have four basic adaptive features to survive aridity—efficient water conservation, high capacity for water storage, high temperature endurance, and high capacity for rapid uptake of surface water when it is available.

Succulents conserve water by a low surface-to-volume ratio and CAM metabolism In temperate plants, leaves have their stomata—the minute pores through which gas exchange takes place (Chapter 11)—open during the day to take in CO_2 for photosynthesis. As CO_2 enters, water vapor moves out of the leaves. This loss of water by transpiration is not a problem in a temperate climate with frequent rainfall, but a high rate of transpiration would quickly be lethal in a desert. How do desert plants reduce water loss?

Because succulents have thick, fleshy stems or leaves, they have a very low surface-to-volume ratio. This means there is relatively little surface from which transpiration can take place. A barrel cactus, for example, has about 9 cm^3 of tissue for each 1 cm^2 of surface. In addition, a waxy cuticle covers the outer cell layer (epidermis) of cacti and prevents water vapor from leaving the underlying tissues. Desert plants in general have fewer stomata than do temperate plants. But the most important adaptation to conserve water is crassulacean acid metabolism (CAM), which has evolved in cacti and some other plant families.

As discussed in Chapter 5, CAM plants are able to open their stomata at night to allow entry of CO_2, which they then fix and store as malate. As dawn approaches, the stomata close and CO_2 is split from malate to supply the Calvin cycle in photosynthesis. Because the stomata are open only at night when

the desert is cool, water losses are greatly reduced. If drought is severe, the stomata may remain closed even at night. Cacti then take the CO_2 released from respiration and fix and store it as malate. This closed state, in which CO_2 is recycled, is termed "CAM idling." The capacity to go into idling is the main reason cacti can survive for several years without water.

Much of the volume of succulents is available for water storage Figure 26.14 shows the anatomy of a typical cactus. The main water-storage tissues are the cortex and pith, which are separated by the woody vascular bundles of the plant. The large spheroidal parenchyma cells of the pith and cortex store water in their large central vacuoles. Many cacti contain foul-tasting chemicals such as alkaloids and mucilage, which are used for wound repair and insect defense. Any tissue-water they contain would be undrinkable. Some species of barrel cacti, however, lack mucilage, and liquid can be squeezed out of their cortex in an emergency.

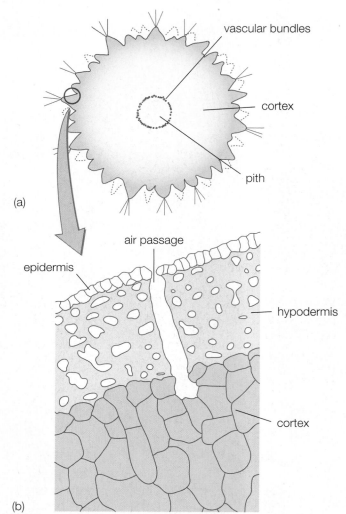

FIGURE 26.14 Cactus stem anatomy (a) Cross section of a cactus stem showing the thin layer of photosynthetic tissue (green shading), the extensive cortex, vascular tissue, and inner pith. Cortex and pith store water. (b) Enlargement of a section of the outer portion of the stem showing the epidermis, thick hypodermis, and the outermost layer of cortex, which is photosynthetic. The air passage allows gas exchange at night.

Cacti are extremely tolerant of high temperatures A temperature of 45°C (113°F) kills most nonsucculent plants. Temperate-plant leaves can dissipate excess summer heat by transpiration of water vapor. Desert plants cannot afford this luxury. Under field conditions, the surface of cacti can rise above 50°C (122°F) without harm. Some barrel cacti can tolerate up to 69°C (156°F)! This remarkable tolerance is achieved through three mechanisms. First, light-colored outer surfaces or dense coatings of white spines on many cacti reflect solar radiation rather than absorb it. Second, the outer layers of cacti may act as very efficient insulation. During the day, the interior of a cactus may be as much as 20°C cooler than the surface. Third, succulents open their stomata at night and exchange gases when the desert air is cool. This exchange then cools the interiors of the succulents.

Water uptake in desert succulents may be very rapid Most cacti have shallow, horizontal roots lying about 5 to 15 cm (2 to 6 in.) below the soil surface and spreading out for many feet around the plant. These shallow roots are ideally situated to pick up any surface water that may fall. When rain falls, the main roots absorb water and thin **rain roots** quickly form. These rain roots are visible within a few hours, and after a few days they may be quite extensive. In those few days, a barrel cactus can take up about 20 liters (5 gal) of water. When the soil dries out again, the rain roots wither, and the main roots reseal themselves until the next shower.

To accommodate rapid water uptake, many columnar and barrel cacti have a pleated skin. Underlying their epidermis (see Figure 26.14) is a hypodermis of cells containing **pectin**, the same substance we add to fruit to make jam. When the pectin takes up water, it becomes firm but flexible. The flexible hypodermis allows the pleats to expand when water is taken up and contract when used up, without damage to the skin (Figure 26.15).

Stem succulents have cylindrical, globose, or paddlelike stems

Barrel cacti *(Ferocactus)* are globose to cylindrical (Figure 26.16a). The prefix *Fero* in their botanical name means "ferocious" and refers to their red spines. Spines on cacti are actually modified leaves thought to serve in defense. Few modern mammals eat cacti, however, other than rodents, rabbits, and bighorn sheep.

Most cacti have erect stems, but the stem of the caterpillar cactus, or creeping devil *(Stenocereus eruca)*, lies on the ground (Figure 26.16b). As it grows along the ground, it puts out roots and dies back at the rear. If it encounters an obstacle, it simply grows over it.

Cholla (pronounced "chaw-yuh") cacti consist of jointed, cylindrical stem segments, varying in diameter from pencil-size to salami-size (Figure 26.16c). The chollas *(Cylindropuntia)* are extremely spiny, but they bear beautiful red or yellow flowers. Commonly called jumping cactus, the chollas do not actually jump, but the segments are so loosely attached that the slightest contact will cause them to stick to any unfortunate passerby.

The prickly pear cacti *(Opuntia;* Figure 26.16d) grow wild in every state in the United States except Maine, Vermont, and New Hampshire; they are even found in Canada. Their flat, paddlelike

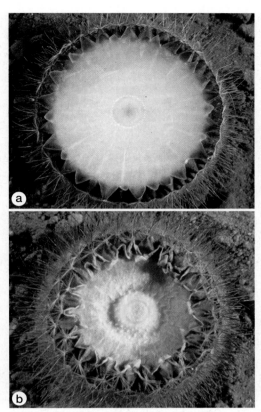

FIGURE 26.15 Barrel cactus Plants have been cut open under (a) hydrated and (b) dehydrated conditions. The pleated surface allows for expansion and contraction of the stem.

stem segments have earned them the nickname "beaver-tail cactus." Most *Opuntia* produce yellow, bee-pollinated flowers and bear abundant edible fruit. Some species, such as *Opuntia ficus-indica*, are grown commercially for their red fruits.

No cactus typifies the Sonoran Desert more than the giant saguaro *(Carnegiea gigantea;* Figure 26.17a). Adult saguaros (pronounced "sah-wah-ro") stand about 15 m (50 ft) tall, weigh about 10 to 15 tons, and live over 250 years. A 50-ft saguaro will have a shallow root system spreading out about 20 m (65 ft) in every direction. Saguaros are slow-growing; they reach 0.6 m (2 ft) at about 25 years old, 2 m (6 ft) by 50 years, and develop their first arm by 75 years (Figure 26.17b). Although most cacti are bee pollinated, the giant saguaro produces large, waxy, white flowers that are visited by bats at night (Figure 26.17c). If human settlement in the desert leads to the elimination of bats, the giant saguaros will lose their primary pollinators and may die out. Of the thousands of seeds produced by a giant saguaro, only a few survive to reach maturity. Rodents eat most of the seeds (see Chapter 25).

Leaf succulents include the agaves, aloes, and stone plants

In the deserts of North and South America, the prominent leaf succulents are the agaves (pronounced "ah-gah-vay"; Figure 26.18a). Like cacti, the agaves are CAM plants with reduced numbers of stomata and extensive shallow root networks. Agaves are commonly called century plants, although they

FIGURE 26.16 **Common cacti of the Sonoran Desert** (a) Barrel cactus (*Ferocactus covillei*), (b) caterpillar cactus (*Stenocereus eruca*), (c) cholla cactus (*Cylindropuntia* sp.), and (d) prickly pear cactus (*Opuntia* sp.).

FIGURE 26.17 **Saguaro** (a) Adult giant saguaro cactus, and (b) a series of young individuals. (c) A long-nosed bat about to pollinate a saguaro flower.

(a)

(b)

FIGURE 26.18 Leaf succulents (a) Vegetative *Agave* plants, (inset) flowering stalk. (b) Flowering *Aloe sinkatana* from Sudan.

Human impacts on deserts include mining, depletion of aquifers, and urban sprawl

Some 13% of the world's population lives on semiarid lands. The mining of deserts for oil, gas, and minerals has, as an immediate consequence, the increase of human population in desert lands and, subsequently, an increase in pollution.

In the United States, there has been a massive long-term shift in population toward the "Sun Belt" states of California, Arizona, Nevada, and New Mexico. Originally much of the shift represented so-called snowbirds, wintertime residents fleeing the cold conditions of the Upper Midwest and East. The widespread use of home and commercial air-conditioning has made permanent residence more attractive. The result has been a rapid spread of suburban homes across desert lands and an increase in demand for water. That demand has often been met by the pumping out of the deep-water table, which has consequently dropped by as much as 60 m (200 ft) in some areas and caused the land surface to collapse. Reduction in the water table can also have serious effects on desert trees and shrubs in arroyos if their roots can no longer reach it.

The spread of suburbia in the desert has a more direct adverse effect on desert plants. All those suburban homes require landscaping. Desert perennials are slow-growing, and it takes time for plant nurseries to raise succulents for home and commercial landscaping (Figure 26.19). A big saguaro or organ pipe cactus can be worth thousands of dollars. Such sums of money have led to cactus rustling. Rustlers go out into private lands and national parks and dig up cacti for resale. States slow to enact laws and hire "cactus cops" have seen lands around desert cities—such as Borrego Springs and the Palm Desert in California—dug up and the cacti hauled away. New laws and enforcement are helping, and nurseries are now growing desert plants in greater numbers.

Aside from the demand for desert plants, the increased human population also demands recreational space. In many areas, this means the use of off-road vehicles. Deserts are fragile and take a long time to recover after they are eroded. Off-road vehicles also take people into wilderness areas where some have destroyed ruins of Native American dwellings by digging for

actually live about 10 to 75 years, and bloom just once in their lives. The flowering stalk, which resembles a giant asparagus tip, can grow as much as 0.3 m (1 ft) per day and reach a height of up to 4.5 to 9 m (15 to 30 ft, Figure 26.18a). The pale yellow flowers are bat pollinated. In Mexico, agaves are grown commercially for the sap, which is made into tequila, and for their fiber.

In Africa, Madagascar, and Arabia, the common leaf succulents are members of the lily family—the aloes. Hummingbirds and insects pollinate the bright flowers (Figure 26.18b). The jellylike juice of *Aloe vera* eases the pain of burns and insect bites. In southern Africa, stone plants belonging to the genera *Lithops* and *Pseudolithops* (*lithos* means "stone") are small leaf succulents colored so as to blend in with a background of rocks (see Figure 16.8). These cryptic plants do not need thorns to deter herbivores (or plant collectors).

If you wish to see some desert plants firsthand, most universities and colleges have greenhouses with some desert plant specimens. Large cities may have a botanical garden with a desert collection. In the southwestern states, there are many fine desert botanical gardens, such as the Desert Botanical Garden of Phoenix, Arizona. Alternatively, desert ecosystems can be visited in such national parks and monuments as Saguaro National Monument near Tucson, Arizona.

FIGURE 26.19 An example of desert landscaping (xeriscaping)

artifacts. In the long run, conservation of arid lands with their resources and diversity of plants will require an integrated approach of management practices, institutional development, public awareness and involvement, and legal enforcement.

26.4 | Grasslands are temperate areas dominated by grasses

Grasslands are ecosystems with a temperate climate that are dominated by grasses (Poaceae). Grasslands receive more annual precipitation than deserts, and that precipitation is generally distributed in a more uniform manner through the year than in deserts. Trees occur sporadically in grasslands, particularly along streams and rivers and in low-lying areas where moisture collects. **Prairies** are grasslands that have few or no trees (Figure 26.20). **Savannas** are grasslands having more trees, but it is sometimes difficult to distinguish prairies from savannas in which trees are sparse. Small, grassy areas that occur within forest gaps are known as meadows or glades.

Grasslands are ecologically, evolutionarily, and economically important

Grasslands are important to people for four main reasons: (1) Grasslands occupy vast areas of land and support immense populations of animals. (2) Grassland soils store huge amounts of organic carbon. (3) Areas that were formerly grasslands now support the world's most productive agricultural lands. (4) Grasslands have greatly influenced animal and human evolution as well as the development of many cultures. Understanding the ecology of grasslands is critical to preserving biodiversity and feeding the burgeoning human populations of the world.

Grasslands occupy vast areas of land and support immense populations of animals Together, the world's grasslands occupy an area of 4.6 billion hectares (1 hectare equals about 2.5 acres). Some of the world's important grasslands include the Eurasian steppes, the pampas in Argentina, the thorn forest (cerrado) of Brazil, the East and South African veldts, and the Great Plains and Corn Belts of the United States (see Figure 25.22). Before the arrival of European settlers, grasslands formed the largest of the North American biomes, covering more than 300 million hectares in the United States alone (Figure 26.21). Even today, natural grasslands, primarily occurring in western rangelands, form the largest biome in the United States and cover about 125 million hectares.

More than 3,000 species of mammals, birds, reptiles, fish, and amphibians in the United States alone depend on grasslands. Before modern times, all of the world's major grasslands supported large herds of grazing mammals and their large predators. Herds of antelope, giraffe, and other grazers—as well as lions and other large predators—still live on the Serengeti Plain and other African grasslands. In addition to the famous buffalo (bison), past U.S. grasslands supported enormous herds of pronghorn antelope (which are not closely

FIGURE 26.20 **A restored prairie in southern Wisconsin**

related to African antelope), an estimated 60–75 million elk, and many types of burrowing animals, such as prairie dog, badger, ground squirrel, and western harvest mouse. Today, much-reduced populations of these animals survive primarily in public and private reserves, and most people have observed them only in zoos.

Grasslands store vast amounts of organic carbon in their soils Grassland soils are naturally high in organic material, which supplies nutrients to plants, increases soil aggregation, limits erosion, and increases water-holding capacity. Grasslands contain 12% of the world's soil organic carbon but much has been lost due to overgrazing and conversion of natural grasslands to crop fields. Improved management, restoration, and preservation of grasslands can help prevent soil organic carbon loss and the resulting effects on the climate.

Grasslands support the world's most productive agriculture Throughout the world, the most productive rangelands and most corn, wheat, barley, and other cereal crops grow on lands once occupied by grasslands. This is no accident. All cereals are grasses that are well adapted to the climatic and biotic conditions found in grasslands, yielding abundant crops and forage that are well suited to human needs. The primary productivity of grasses and associated plants is very high. This high productivity explains both the ability of natural grasslands to support large wild animal populations and the capacity of modern Midwestern farms and Western ranches to provide crops and meat to a growing U.S. human population and much of the rest of the world.

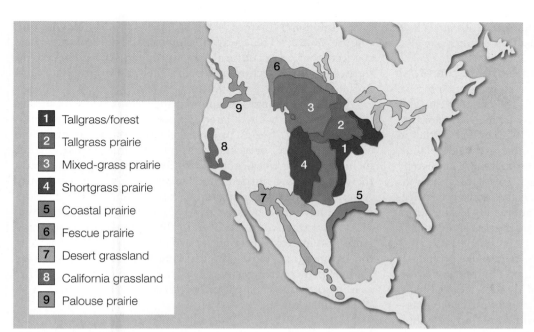

FIGURE 26.21 **Pre-European settlement grasslands of North America**

Legend:
1. Tallgrass/forest
2. Tallgrass prairie
3. Mixed-grass prairie
4. Shortgrass prairie
5. Coastal prairie
6. Fescue prairie
7. Desert grassland
8. California grassland
9. Palouse prairie

The deep, rich soils of grasslands are a major reason for their utility as croplands. For reasons that are detailed later in this chapter, grasses and other native grassland plants (and the animals and microbes with which they are associated) are responsible for soil fertility. But because grasslands are so fertile, many of the world's natural grasslands have been transformed into farmland (see Figure 2.8) or pasture. For example, the highly productive type of U.S. prairie that is dominated by tall grasses (and thus known as tallgrass prairie) has been reduced from a pre-European settlement area of 68 million hectares to less than 5% of that value. Transformation of natural grassland into agricultural land brings the risk of land degradation by overgrazing, invasion by noxious nonnative weeds, and loss of topsoil to wind or rain erosion. For example, prior to settlement, extensive grasslands covered the Central Valley of California and nearby areas, but ranching and fire suppression resulted in invasion by alien, mostly inedible plants. Now the Central Valley is dominated by croplands, urban development, and industry. Excessive grazing on the Palouse prairie of eastern Washington and Oregon has favored the spread of a nonnative grass that is much less suitable for forage than the native grasses it displaced.

In the 1930s, extensive wind erosion and other factors created the Dust Bowl—portions of Colorado, Kansas, Texas, and Oklahoma in which farming of dry grasslands failed. Dust Bowl conditions led to the abandonment of farms and extensive human migration, described by author John Steinbeck in his famous book *The Grapes of Wrath*. (Since then, soil conservation and other measures have decreased the rate of topsoil loss.) Dry grasslands are vulnerable to desertification—the transformation of grassland into desert, which in recent years has led to starvation in the Sahel region of Africa. Grassland scientists work to obtain information that will help preserve the integrity and productivity of grasslands, in order to prevent more such disasters and contribute to the sustainability of the human species.

Grasslands have had a major impact on the evolution of grazer animals and humans American and other modern grasslands first appeared about 20 million years ago, replacing forest vegetation that retreated in the face of increasing drought—a natural climate-change event. Fossils reveal that mastodons, giant rhinoceroses, and peculiar four-horned deer were among the herds of animals that grazed U.S. grasslands at this time. Approximately 7 to 8 million years ago, a dramatic increase in area of the world's grasslands favored the success of many types of modern grazing and burrowing animals.

EVOLUTION

Then, 3 to 4 million years ago, alteration in ocean circulation patterns arising from the closure of the isthmus of Panama and the Indonesian Seaway contributed to the onset of the Pleistocene ice ages and led to the increased aridity of the African continent. In East Africa—home of the forest-dwelling primate ancestors of humans—the area of tropical forest shrank and savanna area enlarged. This event has been correlated with a dramatic change in the habitat of human ancestors—from life in the treetops to life on open grasslands. The presence of large, dangerous grassland predators such as the lion is thought to have stimulated early humans to use fire as a defense (and as a way to improve the edibility of food by cooking), as well as to form protective social groups, which in turn fostered the development of language. The savanna environment of East Africa appears to have shaped humans in many ways, explaining some of our physical features and behaviors.

Climate, fire, and herbivores shape grassland environments

Grasslands, like other ecosystems, have characteristic physical and biological features and can be recognized by typical climate features, plants, and animals. As is the case for all ecosystems, the physical environment determines the types of organisms that can survive there, but plants and associated microbes are particularly important in shaping soil characteristics as well as other physical aspects of their environments.

Grassland climate is drier and more extreme than that of most forests

In comparison to forests, grasslands receive substantially less precipitation, from 25 to 100 cm per year. This is one reason that trees, which typically require high soil moisture levels, occur only sparsely in grasslands. Total annual rainfall in grasslands and some deserts is about the same; but in grasslands, precipitation is more evenly distributed through the year, thereby allowing for a more extensive plant cover. However, grasslands do have distinct wet and dry seasons. One advantage of relatively low precipitation is that nutrients are not readily leached from the soil. Soil fertility is maintained better in grasslands than in moist forests, for example.

Because most grasslands lie in the center of continental landmasses and coastal mountain chains can prevent west-to-east air movements, as in the case of the U.S. Great Plains, air masses from the north and south strongly influence grasslands. The U.S. Great Plains receive cold Arctic air in winter and hot tropical air in summer. Grassland climates typically include hot summers with extended periods of drought and cold winters with unpredictable amounts of snow cover. As a consequence, grassland plants experience low soil moisture and, in addition, are exposed to full sun, strong drying winds, and extreme summer heat and winter cold. Grassland plants must adapt to these conditions.

In the 1970s, the National Science Foundation began a Long-Term Ecological Research (LTER) program to support research on many types of ecosystems. Under the LTER program, scientists working at distant sites can share data and collaborate to identify and understand broad ecological patterns. One of these LTER sites is the Konza Prairie Research Natural Area located in the Flint Hills of northeastern Kansas. At this grassland site scientists made an important discovery about the influence of precipitation on plant productivity, the fuel on which ecosystems run. The researchers found that grasslands respond more strongly than any other ecosystem type to pulses in rainfall. Episodic rainfall in grasslands results in dramatic bursts of plant growth and substantial increases in primary productivity. This pattern indicates that grasslands have a high underlying growth potential that is realized when enough water suddenly becomes available. As a result of these findings, LTER scientists have proposed that grassland annual primary productivity will be a useful indicator of global climate change. Like the canaries miners used to carry into mines to warn of poisonous gases, grasslands may provide a warning of climate change and how it is affecting plants and humans.

Fire plays an important role in maintaining grasslands

During summers, dry grasses and their dead remains that have accumulated at the soil surface over previous years (known as mulch) make a highly combustible fuel for fires started by lightning or by people. Dry lightning, which occurs in the absence of rain, is a common occurrence in grassland areas, as is the accidental escape of campfires or other fires started by humans. Grassland fires can burn for many kilometers before being stopped by wet areas or rain (Figure 26.22). The study of charcoal buried in soil layers suggests that under natural (pre-European settlement) conditions, any given hectare of North American prairie may have burned once every 5 to 30 years.

Grassland plants are adapted in various ways to survive fire (a topic that we will later explore in more detail), but, just as importantly, fire is an integral component of grassland ecosystems. Fire prevents invasion by fire-sensitive plants, including most trees, and clears away dead plant material, thereby releasing nutrients that facilitate new growth. Native Americans often set grassland fires to stimulate the development of new grass because it was favored by their horses and was also attractive to the buffalo they hunted. Burning grasslands also increased productivity of the Native Americans' wild food plants, improved visibility (and thus security), and may have helped control pests such as ticks. Today, people use fire as a major tool to restore and preserve grasslands (Essay 26.2, "Restoring Prairies").

Large animal grazers also influence grassland environments

Grazing by buffalo and other natural herbivores, as well as by livestock on rangelands (Figure 26.23), is a disturbance to which grassland plants are also adapted in ways that will later be described. Grazing is more important in grasslands than in other biomes. Up to 60% of grassland materials and energy flows through primary and other consumers, in contrast to other terrestrial ecosystems, in which less than 5% typically flows through consumer food webs. This explains the presence of high animal populations in grasslands.

Grazing benefits lower-growing plants by preventing shading by tall species, and grazer excrement is a rich source of plant nutrients and organic matter, which helps retain soil nutrients. Both grazing and fire help maintain the high natural plant diversity found in grasslands by preventing just a few plant species from dominating. The results of experiments conducted by David Tilman and associates at the University of Minnesota indicate that high plant diversity has been integral to establishing and maintaining the high fertility typical of natural grassland soils.

FIGURE 26.22 Fire is a natural and essential component of grassland environments

ESSAY 26.2   RESTORING PRAIRIES

If you live in an area where grasslands were formerly the dominant vegetation, there are many benefits to transforming an old field, roadside, or lawn into a prairie. Native prairie plants (those that grew in grasslands before European settlement) have evolved over thousands of years to natural conditions, so they do not succumb to drought, poor soils, and insect attack as readily as nonnative plants; yet many are just as attractive. Less chemical fertilizer and pesticide is required to maintain a natural prairie lawn or garden, fostering increased numbers of attractive native butterflies, many of which lay their eggs on or feed on nectar from grassland plants. Native plants provide more appropriate food rewards for wildlife than do commercial hybrid flowering plants, whose flowers may be so altered as to be unattractive or unrewarding to pollinators. Restoring a prairie helps maintain natural biodiversity of native plants and animals. A prairie garden can serve as a "rain garden," which absorbs excess rainfall, preventing overflow into sewers. Rain gardens help prevent flooding and keep nutrient-polluted water out of lakes and streams.

Restoring a prairie helps maintain natural biodiversity of native plants and animals.

Suppliers of prairie-plant seed can often provide expert advice about the steps involved in starting or restoring a prairie. These involve preparing the soil by removing weeds and other vegetation by one of several methods: (1) using a short-duration herbicide, (2) covering the soil with black plastic or other material that shades the soil, or (3) plowing the soil with a mechanical cultivator. Another technique is to plant a cover crop of oats or bluegrass, into which prairie seeds can later be added. Prairie-plant seeds can be distributed onto small areas of soil by hand or by using a mechanical device, in the case of larger fields. Adding nitrogen-fixing bacteria, widely available in seed stores, aids in the establishment of legumes. As the prairie first develops, it is often desirable to mow it at intervals, which prevents tall plants from shading shorter ones. Finally, after two growth seasons, planted prairies need to be burned annually for several years. Only someone trained in performing controlled burns should conduct the burning; a permit is required in most areas. Once the prairie has become established, burning every 2 to 4 years is sufficient.

Some attractive plants to consider for prairie gardens include the black-eyed Susan (*Rudbeckia hirta*; Figure E26.2A), which blooms in summer and fall and tolerates drought, full or part sun, and poor soil. Its nectar and bright yellow petals attract butterflies and bees. Butterfly weed (*Asclepias tuberosa*; Figure E26.2B) has orange flowers, blooms in midsummer, and has interesting fruits. It is a host plant for monarch butterflies and other butterflies, and bees feed on its flower nectar.

E26.2A   Black-eyed Susan (*Rudbeckia hirta*)

E26.2B   Butterfly weed (*Asclepias tuberosa*)

FIGURE 26.23   Grazing animals have long been integral to grassland habitats, and they remain so today

Grassland plants are adapted to cope with environmental stresses

There are about 7,500 plant species in North American grasslands alone. These include 600 types of grasses, thousands of species of nongrasses known as forbs, and a few tree or shrub species. There are so many different kinds of plants in grasslands that many can be observed to bloom or fruit at any given time during the growing season, contributing to the great beauty and bounty of these ecosystems. Why are there so many grassland plant species, and how are they adapted to cope with drought, temperature, fires, and grazing stresses?

Dominant grass species vary through the year and by region   Although there are many species of grasses worldwide, at any particular time and place grasslands are typically dominated by only three or four grass species that produce most of the biomass. In a particular region, species dominance shifts between the cold and warm seasons. Several grasses having C_3 photosynthesis (Chapter 5) are most abundant during the cool season and are thus known as cool-season grasses. Barley *(Hordeum)*, rye *(Secale)*, and wheat *(Triticum)* are cool-season grasses native to Eurasia that are important crops. During summer, the warm-season grasses become dominant. Warm-season grasses typically have C_4 photosynthesis, an adaptation that improves productivity during hot, dry conditions. Corn *(Zea mays)* is a C_4 grass. It is thought that C_4 photosynthesis first arose during the period of global aridity that occurred 7–8 million years ago, which, as previously noted, led to the expansion of grasslands on a global basis.

Major mid-continent grasslands of the United States are an easternmost tallgrass prairie, a central mixed-grass prairie, and a westernmost shortgrass prairie (see Figure 26.21), reflecting decreasing moisture levels moving east to west. Tall and shortgrass prairies each have characteristic grass species. For example, big bluestem *(Andropogon gerardi;* Figure 26.24a), a warm-season species of the tallgrass prairie, grows to 2.5 m in height. European settlers recorded that they sometimes lost their horses in these grasses. Several species in the genus *Bouteloua* (Figure 26.24b) are common in shortgrass prairies. Mixed-grass prairies combine species of tall and shortgrass prairies.

Grass plants are adapted for fast growth, high productivity, and resistance to fire and grazing Big bluestem and other prairie grasses are perennial monocots, having lifetimes of more than 2 years. The leafy aboveground portions sprout each spring from an underground horizontal stem (rhizome) that is protected from winter cold by an insulating soil layer. The growing points (apical meristems) of grasses are thus underground, sheltered from damage by fire, grazers, or drought. (Recall from Chapter 9 that this is why cutting your lawn with a mower does not kill the grass.) The rhizome can branch prolifically, and each branch may produce many leaves. In this way, grasses can rapidly colonize large areas. The rhizome also produces an extensive root system that can expand rapidly when moisture is available (Figure 26.25). Grass roots are known as adventitious roots because they emerge from the stem, which, as you may remember, is common for monocots. Grass roots can extend as far as 4 m into the soil and can thus act as conveyors, bringing deep nutrients to the surface and enriching the topsoil. These deep roots also provide access to subsurface water during droughts. Grass roots are also very tough, which enables them to avoid breaking should the soil crack upon drying. You may have encountered tough grass roots and rhizomes firsthand while weeding a garden or attempting to dig weedy grasses such as crabgrass from your lawn.

FIGURE 26.24 **Representative grasses** (a) Big bluestem *(Andropogon gerardii)*, common to tallgrass prairies, is easy to recognize by its height (up to 2.5 m tall) and its inflorescence, which resembles a turkey foot. Its common name is based in part on the blue cast to its stems. This grass tolerates full sun, drought, and many soil types. Birds consume its seeds, and some butterflies prefer to lay their eggs on its surfaces. (b) *Bouteloua* is a representative grass of shortgrass prairies.

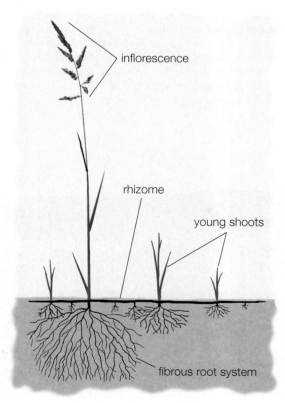

FIGURE 26.25 **Grass bodies include a horizontal rhizome that bears roots, leaves, and flowering shoots**

Have you ever wondered why grasses have such distinctive narrow leaves? Narrow leaf blades have a reduced surface-area-to-volume ratio, an adaptation that helps reduce transpiration rates and thus conserve water. Grass blades are vertically arranged and fairly tightly packed, vastly increasing the area available to capture light energy. Because of this arrangement, 1 hectare of grass has 5–10 hectares of photosynthetic surface area. Grass leaves contain particles of silica (glass), a development that is thought to have first evolved as a way of deterring herbivores. Although it is possible that silica inhibits feeding by some animals or may have done so in the remote past, large modern grazers such as buffalo and cattle have evolved large, tough teeth for which cropping and chewing grass leaves pose no problems.

Grass flowers and fruits are adapted for efficient reproduction Grass flowers and fruits are very distinctive in ways that adapt them to grassland environments. Adaptive fruit features have made several grasses particularly useful as human foods—corn, wheat, rye, barley, and rice are major examples.

Grass flowers are small, often greenish, and nonshowy (lacking conspicuous petals; Figure 26.26), from which it can be deduced that they are wind pollinated (Chapter 24). Wind pollination is effective when many individual plants of the same species occur close together, as is the case for grasslands. Grasses do not have to invest resources in producing large, showy, fragrant flowers because animal pollinators are not needed.

Adaptation to wind pollination has resulted in changes to grass flowers that make them appear quite different from most other flowers. Grass flowers are composed of an ovary with two feathery stigmas adapted to catch wind-borne pollen, three stamens, and a pair of small, swollen scales called lodicules (Figure 26.26c). It is thought that lodicules might have evolved from petals or sepals, but their function is quite different from that of most flowers' petals and sepals. Lodicules swell when the anthers and stigmas are mature, opening grass flowers so that release and capture of pollen can more easily occur. Each grass flower is enclosed by leaflike structures known as the lemma and palea, features not typical of most flowers. Individual grass flowers, known as florets, are borne either singly or in inflorescence

EVOLUTION

EVOLUTION

DNA SCIENCE

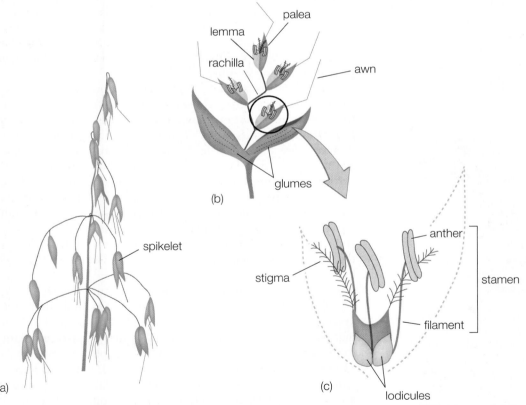

(b)

(a)

(c)

FIGURE 26.26 Grass flower parts Grasses have nonshowy flowers that are modified for wind pollination. An example is oats. (a) An inflorescence, with numerous spikelets, each containing several flowers (florets). (b) A single spikelet is shown with four florets attached to the stemlike rachilla (the florets are spread out somewhat for clarity). Each spikelet is originally covered by a pair of glumes, and each floret is enclosed by a pair of leaflike structures—the palea and the lemma. Awns, bristlelike extensions, are present on the paleas in oats. (c) The reproductive parts of an individual floret can be seen with the palea and lemma removed. These parts consist of a pair of feathery stigmas, which extend out of the ensheathing palea and lemma to catch pollen in the wind, and three stamens. A pair of bulbous lodicules is found at the base of the floret and helps open it.

clusters (known as spikelets) on a stemlike axis (rachilla). Each spikelet is enclosed by two papery glumes, and a bristle-like extension (awn) is present on lemmas or glumes of some grasses, such as wheat. Despite the many unusual features of grass flowers, the genes that control flower structure in grasses are similar to the *ABC* genes of other flowering plants (Chapter 23).

Following successful pollination and fertilization, the ovaries of grasses develop into one-seeded **grass fruits** known as **grains** (Figure 26.27). Grass fruits and seeds are dispersed by many types of grassland rodents, such as mice and ground squirrels, which hoard seeds for food.

During development, a grass grain becomes almost dry, a process that is controlled by the expression of specific genes. This adaptation fosters survival through extended periods of drought, which often occur in grasslands. Because dry fruits are easily stored, this adaptation contributed to human development of various grasses as food crops. The high nutrient content of grass grains—which include stored carbohydrates, proteins, and lipids—aids in seedling survival and makes the grains attractive as food for humans and other animals.

Grass seed germination has been intensively studied, in large part because barley grain germination is an important step in the production of beer. The result has been a better general understanding of seed germination. Grass seeds include an embryo and a food-storage tissue—the starchy endosperm. Grass seeds also contain aleurone—a tissue composed of living cells enclosing the endosperm—and scutellum—a leaflike organ (=the cotyledon) that lies at the side of the embryo, near the endosperm (see Figure 26.27). When grass seeds are moistened, the embryo becomes active, releasing the hormone gibberellin (Chapter 12). The scutellum's role is to transport gibberellin to the aleurone. Gibberellin is a signal for the aleurone to secrete enzymes that break down endosperm starch into sugar. (Human saliva contains similar enzymes that likewise participate in starch breakdown.) The embryo needs sugar as food, because it cannot yet photosynthesize. Sugar is respired to produce ATP, the carrier of chemical energy needed for embryo development and seedling growth. This highly coordinated system allows grass seeds to germinate rapidly when rains occur, a very useful adaptation in relatively dry grassland habitats.

Forbs are diverse grassland plants that are not grasses, trees, or shrubs

About 80% of grassland species are nongrass, herbaceous plants classified as forbs. They are much more diverse than the grasses that occur with them. Although some ferns and mosses occur in grasslands, most forbs are flowering plants. As many as 100 forb species may occur in the same 1-acre (0.4 hectare) area that harbors only a few grass species, with the result that individual forbs of the same species often are widely separated. Under these conditions, wind pollination is less effective than insect pollination (Chapter 24), so it is no surprise that grassland forbs display a stunning variety of beautiful flowers and inflorescences (Figures 26.28). Gayfeather or blazing star (Figure 26.28c) has such beautiful purple inflorescences that it has become a staple flower in the florist industry.

Grassland forbs, like the grasses, produce extensive, deep roots that collect nutrients from deep soil layers, ferrying them to the surface. For example, the compass plant (*Silphium laciniatum*; Figure 26.28b) has roots that may extend 3 meters deep or more. Some grassland forbs have nitrogen-fixing bacterial associates that help enrich the soil. The compass plant's vertically positioned leaves are oriented in a north-south direction, an adaptation that reduces water loss from stomata, which are more abundant on the leaves' lower sides. Amerindians and settlers used this plant as a navigational aid, hence its common name. Grassland forbs often have very hairy leaves; the hairy blanket helps prevent evaporation of water from the leaf surface or provides protection from the cold. The pasqueflower (*Anemone patens*) illustrates this feature (Figure 26.29).

Grassland forbs contain secondary compounds that deter grazers and disease microbes. Native Americans took advantage of this when they used extracts of many grassland plants to prepare medicinal treatments. For example, the rattlesnake master (*Eryngium yuccifolium*; Figure 26.30) was used to treat kidney disorders and snakebite. Some grassland plant products are commercially sold as herbal medicines today. One example is purple coneflower (*Echinacea*), widely sold as an herbal treatment for colds.

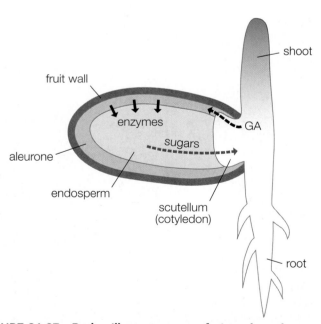

FIGURE 26.27 Barley illustrates grass fruit and seed structure and function Grass seeds germinate by means of a highly coordinated process involving specialized tissues (aleurone, endosperm, and scutellum). The plant hormone gibberellin (also known as gibberellic acid or GA) is instrumental in communication among tissues, with the result that stored food is rapidly converted by enzymes to sugars that are used by the developing seedling for growth.

Grassland trees and shrubs are adapted to survive fire and provide important resources for some grazers

Most trees are not able to survive the fires that are common to grasslands; an exception is the bur oak

EVOLUTION

DNA SCIENCE

FIGURE 26.28   Representative grassland forbs (a) Prairie dock (*Silphium terebinthinaceum*). This member of the sunflower family has broad, flat leaves (inset) and tall (up to 3 m), brightly colored inflorescences. (b) Another species of *Silphium*, the compass plant (*S. laciniatum*). The dissected leaves of the compass plant tend to orient in a north-south direction, thereby keeping themselves cooler by minimizing exposure of stomata-rich leaf undersides to sunlight. (c) Blazing star (*Liatris*) is a member of the sunflower family that is somewhat unusual in that the upper "flowers" (each is actually an inflorescence) mature before the lower ones. (d) Wild indigo (*Baptisia leucantha*), a legume, can grow as tall as 2 m. Fruits are shown in the inset.

(Figure 26.31), native to savannas of the United States. Its thick bark enables it to survive fire episodes. Baobab and acacia are examples of trees or shrubs that inhabit African savannas. Acacia is a mainstay for giraffes, which, unlike other grazers, are able to feed on its leafy branches.

Grassland improvement, restoration, and preservation yield benefits for people   Grassland improvement, restoration, and preservation are beneficial in maintaining a resource used by humans for food and forage crops. Improvement of grasslands also increases their ability to store organic carbon,

FIGURE 26.29 **Hairy surfaces help insulate leaves from cold, extreme heat, or drying winds** An example includes pasqueflower (*Anemone patens*).

preventing the release of carbon dioxide into the atmosphere, where it can contribute to global warming. Grassland improvements include irrigation, fertilization, more effective grazing management, addition of native earthworms, and planting of native grasses and legumes. Many people are interested in restoring prairies on old fields or other lands no longer used for agriculture or have transformed their lawns into natural prairie, thus reducing the need to mow as often (see Essay 26.2, "Restoring Prairies"). Grasslands can be classified as dry, mesic (of interme-

diate moisture level), or wet, depending on the average amount of soil moisture. Preservation strategies are specific to these different types of grasslands.

Dry grasslands can degrade into deserts As earlier noted, dry grasslands, which often border deserts, are easily converted to deserts if the plant cover is removed. Overgrazing is a common cause of desertification. Loss of grassland transpirational leaf surface area alters local rainfall patterns because less water vapor is able to enter the atmosphere. Preserving dry prairies may require building fences to exclude herds of domesticated animals.

Nonnative plant species such as spotted knapweed (*Centaurea maculosa* or *C. biebersteinii*) can outcompete native plants and are a particular problem in dry grasslands (Figure 26.32a). This invader is among the 94 plants that have been designated by the U.S. federal government as noxious weeds. Native to Eurasia, spotted knapweed apparently made its way to North America as a contaminant of alfalfa seed in the late 1800s and has since been very difficult to control. Brome grass (*Bromis inermis*) and leafy spurge (*Euphorbia esula;* Figure 26.32b) are also European invaders that threaten the biodiversity of native grasslands and cause billions of dollars per year in agricultural losses.

Mesic grasslands have mostly been transformed into farmlands Throughout the world, mesic grasslands have been widely converted to agricultural uses because of their rich soil and high productivity. Soil richness derives from the lack of extensive rain leaching, the action of deep-growing conveyor root systems, the presence of nitrogen-fixing forbs, the activities

FIGURE 26.30 **Rattlesnake master (*Eryngium yuccifolium*)** This member of the parsley (carrot) family has yuccalike leaves (a). The tiny white flowers occur in dense, spherical heads (b).

FIGURE 26.31 **Bur oak (*Quercus macrocarpa*)** This stately tree has a broad crown when it grows in the open. The leaves have round lobes, with the tip portion often being less indented. The thick, sculptured bark resists fires. The name comes from the acorns, which have a cap with numerous bristles ("burs").

FIGURE 26.32 **Grassland invaders** (a) Spotted knapweed (*Centaurea maculosa*). Like many invasive plant species, this member of the sunflower family threatens to displace native North American species. (b) Of the same genus as poinsettia, leafy spurge (*Euphorbia esula*) has yellow-green bracts beneath the small flowers.

of burrowing animals (which help mix soil layers), and mineral-supplying mycorrhizal associations. The mycorrhizal fungi associated with grassland plants also produce proteins having soil-stabilizing properties, which help prevent erosion. Despite a century of cultivation, farmland soils in the U.S. Midwest retain about half of their original depth and continue to be extremely productive. They represent a national treasure that deserves preservation as a way to ensure future food security.

Remnants of natural mesic prairies, small patches of land that escaped cultivation, provide us with a sample of the

FIGURE 26.33 **Wet prairie** The purple-pink flowering plants are species of *Liatris*.

magnificent tallgrass prairies experienced by Amerindians and early settlers. Inasmuch as they cannot be replaced, preservation of these prairie remnants is as important as preserving unique art and architecture. Studies conducted by Mark Leach and Tom Givnish at the University of Wisconsin revealed that mesic prairie remnants lose about 1% of their species per year when fire is suppressed. Particularly vulnerable are legumes, short plants, and those with small seeds. To preserve natural diversity, experts burn remnant mesic prairies. Like natural fires, controlled burning removes tree seedlings and other invaders as well as dead plant materials and releases mineral nutrients.

Wet grasslands provide valuable ecological services

Lush growths of grasses and forbs are found in wet grasslands, which often border wetlands. Like wetlands (Chapter 28), wet grasslands (Figure 26.33) serve the valuable functions of capturing phosphorus before it can pollute lakes and rivers and helping remoisten nearby land areas during droughts. Wet grasslands are uncommon today because they are often invaded by trees or have been drained for truck farming (cultivation of lettuce, mint, onions, or other crops). Unfortunately for the truck farmer, the soils of wet grasslands—like most wet soils—are typically low in oxygen, so decomposition and mineral nutrient recycling rates are also low. For this reason, without extensive fertilization efforts, wet grassland soils are able to support farming efforts for only a few years. It is probably best to leave wet grasslands uncultivated to serve as biodiversity storehouses and wildlife habitat.

26.5 | The chaparral ecosystem has hot, dry summers and cool, wet winters

Mediterranean scrub is a biome that is distinct and restricted in area throughout the world. Areas of Mediterranean scrub occur around the Mediterranean Sea, along the western coast of California, and in parts of the west coast of Chile, western South Africa, and western Australia (see Figure 25.22). These regions all lie within 30° to 40° N or S latitude and on the western coasts of continents, the Mediterranean Sea acting as one extended coastline. In each of these areas, the Mediterranean scrub has a regional name: chaparral in California, maquis around the Mediterranean Sea, matorral in Chile, fynbos in South Africa, and mallee scrub in Australia. The Mediterranean scrub in California received its name from Spanish explorers and colonists, who called it *chaparro* after the Basque word *chabarra*, for a type of scrub oak found in the Mediterranean scrub of the Pyrenees Mountains of Spain. Vegetation similar to the chaparral of California also occurs inland in mountainous parts of Arizona, but the climate there is quite different from that in California. Arizona has a pattern of summer rainstorms.

The chaparral extends from Oregon to Baja California (Figure 26.34) between sea level and an elevation of 2,000 m (6,500 ft). It receives between 20 to 100 cm (8 to 39 in.) of rainfall per year, most of which falls between November and April. Prolonged dry spells and extreme drought are common. Summer temperatures can exceed 40°C (104°F) but are milder along the coasts and at higher elevations. Average winter temperatures range from 0°C (32°F) in the mountains to 10°C (50°F) at lower elevations. In southern California, an additional climatic factor is the Santa Ana wind. In spring and fall the Santa Ana winds arise in the interior deserts and bring high temperatures and low humidity to the coasts at wind speeds often exceeding 100 km hr^{-1} (60 mph). Wildfires (a major factor shaping chaparral communities) often occur at these times.

FIGURE 26.34 **Distribution of chaparral and similar vegetation in western North America**

Evergreen shrubs with sclerophyllous leaves dominate the chaparral

The vegetation of the chaparral evolved from desert plants and mixed deciduous evergreen forest plants. Areas of chaparral often lie adjacent to oak woodlands, coniferous forest, grasslands, or even desert, forming a mosaic of plant communities. An herb layer is generally lacking in chaparral, except after a fire, and dense groves of trees such as pines *(Pinus)*, cypress *(Cupressus)*, oaks *(Quercus)*, and small trees such as ash *(Fraxinus)* may dominate on moist sites within chaparral.

In California, chaparral forms a nearly continuous cover of shrubs 1.5 to 4 m (5 to 13 ft) high. The shrubs are often so closely spaced that they are virtually impenetrable (Figure 26.35). Shrubs are the dominant form of vegetation, and convergent evolution has operated on them to produce similar growth forms and morphologies both among species in the chaparral and among species in different Mediterranean scrub communities. Many chaparral species show adaptations similar to desert plants. Most species have **sclerophyllous** (Greek *sclero,* "hard," and *phyllos,* "leaves") leaves that are small, tough, and covered with a thick cuticle. Sunken stomata on leaves are also frequent. These features all act to reduce water loss. Root systems are also similar to those of desert plants. Some shrubs spread their roots out under the soil surface, and others produce deep roots to reach the water table.

There are more than 100 species of evergreen shrubs in the California chaparral. Individual sites may be dominated by a

EVOLUTION

FIGURE 26.35   A dense cover of chaparral shrubs

single species or have as many as 20 species, depending on a number of factors, including soil moisture, the direction in which the slope is facing, the steepness of the slope, its distance from the coast, elevation above sea level, latitude, and fire history. Many species are very restricted in area and therefore endangered by human disturbance.

The most common species of shrub in California is chamise *(Adenostoma fasciculatum),* a member of the rose family with narrow leaves resembling those of a spruce tree (Figure 26.36a). The chamise is especially common in southern California at low elevations on dry or xeric sites, where it

FIGURE 26.36   Common chaparral plants (a) Chamise chaparral *(Adenostoma fasciculatum).* (b) Manzanita *(Arctostaphylos).* (c) Close-up of a manzanita shrub.

forms almost pure stands of chamise chaparral. The next most common chaparral shrubs are the manzanitas (*Arctostaphylos;* Figure 26.36b and c), which are members of the blueberry and cranberry family, and the ceanothus species (*Ceanothus;* Figure 26.37). Both genera contain many species that can form pure stands of manzanita chaparral and ceanothus chaparral. In northern areas and at higher elevations, they are more common than chamise chaparral. Ceanothus shrubs have root nodules with symbiotic nitrogen-fixing bacteria. Shrubs are generally taller in moist or mesic sites.

Chaparral shrubs follow a seasonal pattern of growth determined by temperature and moisture. Growth starts in the fall after rains begin but ceases as temperatures decline in the winter. Growth resumes in the spring when temperatures rise and continues until soil moisture is depleted in the summer. This pattern of growth promotes wildfires. Mild winters allow the buildup of dense growth, which summer drought turns into highly flammable fuel.

Fire is a major ecological force in the chaparral

Fires occurred in the chaparral long before humans were present. Lightning is still a major cause of fire, but now humans are more often the cause. On any given site, fires will occur at intervals of 20 to 30 years. Normally, they will kill all of the aboveground vegetation. In the first year following a wildfire, many species of herbs germinate, and "fire annuals" may produce spectacular flower shows. The seedlings of shrubs also germinate and grow; many shrubs resprout from underground stems (rhizomes) or special tubers called burls, which are rare outside the Mediterranean scrub biome. Within 3 to 4 years, the shrubs take over the community, and herbs and fire annuals decline. After 15 years, the community has resumed its prefire appearance. Plant community composition often does not change, because the plants are adapted to survive these wildfires. Chaparral shrubs follow one of two basic adaptive

patterns to survive and reproduce where fire is a frequent selective agent.

Some shrubs require fire to establish new seedlings
Shrubs that require fire to reproduce establish their seedlings in the first year following a wildfire but produce almost no seedlings in subsequent years. Seeds build up in the soil over a number of years and are induced to germinate by the heat or smoke from a wildfire. Once established, the seedlings will need 5 to 15 years to mature and begin producing new seeds. Because seed predators such as rodents take most of the seeds produced, many years must pass to build up the number of seeds in the soil, and those seeds must be long-lived. Seeds are not dispersed widely in space but are instead dispersed in time; they wait in the soil for a fire to signal germination.

Most species of the common chaparral shrubs *Arctostaphylos* and *Ceanothus* reproduce in this manner. If wildfires occur more frequently than at 5- to 15-year intervals, these shrubs may be eliminated from the plant community because the seedlings will not reach maturity and, therefore, will not form seeds. These plants are termed *obligate-seeding* shrubs because they must produce seeds and seedlings to persist in the chaparral after a fire. The chamise (*Adenostoma fasciculatum*) and a few species of *Arctostaphylos* and *Ceanothus* establish seedlings after fire, but they can also resprout from their roots or a basal burl. Such shrubs are called *facultative-seeding* shrubs. Two fires spaced just 4 years apart could eliminate obligate seeders from a chaparral community, but facultative seeders would survive.

Other shrubs require an absence of fire to establish seedlings   Some species of chaparral shrubs, such as species of oak (*Quercus*) and cherry (*Prunus*), do not establish seedlings after a wildfire. They survive fires and remain in the chaparral plant community by resprouting from underground roots, rhizomes, or basal burls. They are termed *obligate resprouters*.

In mature chaparral, these shrubs produce large seed crops that are widely dispersed. Their seeds are short-lived (less than 1 year) and germinate when moisture is adequate. Wildfires easily kill them. Seedlings develop underneath the mature shrub canopy, but the saplings remain stunted in growth by the mature shrubs until a fire sweeps through the area. When the aboveground vegetation is burned off, the saplings can resprout and reach maturity.

Trees in the chaparral are adapted to survive wild-fires   Several species of trees may form dense groves within chaparral shrub vegetation. Conifers such as pines (*Pinus*) and cypress (*Cupressus*) occur in such groves but cannot resprout after a fire. They survive by forming fire-resistant cones that are sealed by resins (Figure 26.38). The cones can be sealed for decades and the seeds remain capable of growth. The high temperature of wildfires melts the resin and opens the cones to allow the seeds to germinate. Trees such as oaks also form groves on more mesic sites, and after a fire, these trees can resprout from stems or roots.

FIGURE 26.37   *Ceanothus* plant

FIGURE 26.38 Resinous cypress cones from chaparral

Human impact has been severe on Mediterranean scrub ecosystems

Because Mediterranean scrub ecosystems are suitable for agriculture and human settlement, they have been heavily affected wherever humans have occupied them. In most areas of the maquis around the Mediterranean Sea, inhabitants have completely ruined the natural vegetation over thousands of years and converted many fertile lands into barren waste-lands. In contrast, the California chaparral is much more diverse in species, but many local or endemic species grow only in restricted areas and are threatened. The chaparral has been subjected to logging, mining, burning, and grazing. Areas of chaparral may be burned and sown with grasses for conversion to grassland. Adjacent woodlands are logged and replaced by chaparral.

Suburban sprawl is a major threat to chaparral. Areas are cleared for suburban homes, businesses, and malls; but the danger of fire remains. If low-intensity fires in the chaparral are suppressed, the danger of large-scale wildfires grows as flammable vegetation accumulates. Wildfires can then race through neighborhoods, causing millions of dollars in damage (Figure 26.39).

FIGURE 26.39 Wildfire sweeping through a community in California

HIGHLIGHTS

26.1 Arid terrestrial ecosystems include polar deserts, temperate and subtropical deserts, grasslands, and chaparral. Arid terrestrial ecosystems receive less than 100 cm of precipitation per year and generally lack trees except in more moist local sites.

26.2 Polar deserts have the most severe climates on Earth. In the High Arctic, polar deserts called herb barrens cover thousands of square kilometers. In the continental Antarctic, the most favorable sites for microorganisms include lakes that are permanently frozen over and the subsurface layers of porous rocks. In the less harsh maritime Antarctic, mosses and lichens are more extensive.

26.3 Four physical phenomena determine the locations of deserts: subtropical high-pressure zones, continentality,

cold ocean currents, and rain shadows cast by high mountain ranges. Desert trees send long roots into the deep-water table. Many algae, mosses, lichens, annuals, and perennials use surface waters when available. Desert succulents conserve water by having a low surface-to-volume ratio, reduced numbers of stomata, and CAM. Cacti are very tolerant of high temperatures, and they can rapidly take up and store large volumes of water after rains. Human effects on deserts include mining of minerals, gas, and oil. In the United States, expansion of cities in desert areas depletes water supplies, removes native vegetation, and degrades arid lands through recreational uses.

26.4 Grasslands occupy large areas of the world's land surface, support large animal populations, store vast quantities of soil organic carbon, support the world's most productive agriculture, and have influenced the evolution of humans and grassland animals. Fire and grazing animals play large roles in shaping grassland environments. Grassland soils are deep and rich due to the activities of grassland plants, microbes, and animals.

26.5 Grassland plants are adapted to survive drought, fire, and the effects of animal herbivory. Deep grass roots convey minerals to the surface, helping enrich topsoil. The leaf-blade shape and vertical orientation of grass help conserve water and increase photosynthetic productivity. Grass flowers have become highly modified for effective wind pollination, and grass seed germination involves

a high degree of coordination among seed tissues and organs, which is advantageous in relatively dry habitats. Grassland trees and shrubs are resistant to fire. Grassland preservation, improvement, and restoration are important in agriculture, soil conservation, and protection of biodiversity.

26.6 The North American chaparral is part of the Mediterranean scrub ecosystem. Chaparral has a climate of hot, dry summers and cool, wet winters. Evergreen shrubs dominate the chaparral vegetation and have many adaptations similar to those of desert plants. Fire is a major ecological force. Logging, mining, grazing, and suburban sprawl all threaten chaparral ecosystems.

REVIEW QUESTIONS

1. Which two polar desert ecosystems exist on the Antarctic continent, what are their physical characteristics, and what types of plants grow in each?

2. Describe the four physical phenomena that determine the occurrence of deserts.

3. What are the two main sources of water in deserts? Provide an example of how plants make use of each.

4. How do annuals, herbaceous perennials, and deciduous perennials differ in their adaptations to desert life?

5. Discuss some adaptations that desert succulents have evolved to cope with life in such an arid environment. Provide some examples of leaf succulents and stem succulents.

6. What are grassland forbs, and what are some of their adaptations to life in the grasslands?

7. List the four major reasons that grasslands are important to humans.

8. What roles do fire and grazing play in the maintenance of grasslands?

9. What are cool-season grasses, and how do they differ from warm-season grasses? Give species examples of each.

10. How are the vegetative parts of grass plants adapted to life on the prairie?

11. How is grass plant reproduction adapted to life on the prairie?

12. What is the current status of (a) dry, (b) mesic, and (c) wet grasslands—that is, how are these areas being put to human use, and what is the best preservation strategy for each?

13. Describe the physical features of chaparral, including climate (seasonality and rainfall patterns), winds, and geographic locations.

14. What types of plants dominate the chaparral, and what are their adaptations for life there?

15. What is the role of fire in maintaining the chaparral? Describe the two basic adaptive patterns that enable chaparral shrubs to survive and reproduce in this fire-prone biome.

APPLYING CONCEPTS

1. With the discovery by *Mariner, Viking,* and *Mars Global Surveyor* spacecraft that Mars was once much warmer and wetter, there has been a resurgence of interest in terraforming Mars. Describe some similarities and differences between the Martian climate of today and that of the continental Antarctic.

2. The atmospheric circulation patterns of Hadley cells cause dry air to descend toward the Earth's surface at 30° N and S. This region includes the southern United States, from the deserts of Arizona and New Mexico to the bayous of the southeastern seaboard. Why is the climate of Louisiana and Florida so different from that in Arizona, when both lie at approximately 30° N? In particular, why is the southern seaboard of the United States not a desert?

3. Yosemite Valley, along the western slopes of the Sierra Nevada in California, often receives extremely heavy snow-falls during the wet winter months. Just over the mountains to the east, a few miles away in Nevada, lie the sagebrush flats of the Great Basin Desert, an area that receives little precipitation during the year. Why are these areas, so close geographically, so different in climate?

4. The owner of a large, vacant piece of land decided to restore the property to the tallgrass prairie that once stood there. After planting appropriate prairie-plant seeds, she watched as the prairie plants established themselves and produced a magnificent prairie for a few years. Gradually, however, woody species began showing up as well—box elder and chokecherry trees, poison ivy and staghorn sumac shrubs, and so on. Eventually, these unwelcome invaders became so numerous that the landowner hardly recognized her original prairie. What went wrong?

5. Perhaps more than any other biome, the grassland biome has suffered substantial—and in some cases, complete—loss due to human activity. Why was this so? What characteristics of the grasslands made them so attractive to humans and so vulnerable to human activities?

6. Small woodland tracts used to be well distributed throughout the prairie biome. In 1965, P. Wells stated, "There is no range of climate in the vast grassland province of the central plains of North America which can be described as too arid for all species of trees native to the region" (*Science* 148:246). If trees were able to grow in the grasslands, what prevented them from turning this region into savanna or even forest?

Moist Terrestrial Ecosystems

Redwood forest

Tropical rain forests are home to some of the strangest, most unbelievable plants in the world. Rain forest plants hold many records. For example, the rotang palm *Calamus*, at 180 meters (1 m is 3.28 ft), is among the world's longest plants (590 ft). The tallest broadleaf tree ever recorded, at 83 m (272 ft), is *Koompasia excelsa*, which grows in the rain forest of Sarawak. The longest leaves in the world, more than 25 m (82 ft), are produced by the Raffia palm (*Raphia regalis*), which grows in the rain forests of Central Africa. The longest inflorescence record is held by the *Corpha* palm, which produces 10 million flowers per inflorescence! The plants having the world's largest seeds also grow in tropical rain forests. *Mora megistosperma*, which grows in Costa Rica, has the largest dicot seed, some 18 centimeters (7 in.) in diameter and weighing as much as 850 grams (1.8 lb). The "double coconut," *Lodoicea maldivica*, has the largest monocot seed, 40 cm (15.7 in.) across and weighing up to 20 kilograms (44 lb). The world's largest flower is produced by *Rafflesia arnoldii*, a parasite that grows in Borneo and Sumatra. This plant produces no leaves, stems, or roots—only a single giant flower, 1 m (3.28 ft) in diameter and weighing 7 kg (15 lb).

Charles Darwin, on seeing the shores of Brazil during the voyage of the HMS *Beagle*, wrote in his journal, "Here I first saw a tropical forest in all its sublime grandeur," and "Nothing but the reality can give any idea how wonderful, how magnificent the scene is ..." In recent years, we have learned how important tropical rain forests are to Earth's ecology. The entire world benefits from the products and ecological services provided by tropical forests. Though people have been influencing tropical forests from prehistoric times, in the past several decades, vast tropical rain forests around the world have been destroyed and degraded by logging, mining, and by clearing the land for farms, plantations, and cattle ranches. Many people around the world are concerned about protecting the remaining forest and attempting to restore degraded lands.

Tropical rain forests are one example of a moist terrestrial ecosystem. This chapter explores the moist terrestrial ecosystems of Earth, which include the vast treeless tundra of the Northern Hemisphere, the taiga or boreal coniferous forests that lie to the south of the tundra, the temperate deciduous forests that lie farther south, and the tropical rain forests of the equatorial regions of Earth. ■

Tropical rain forest

27.1 | Moist terrestrial ecosystems cover a wide range of climates

Moist terrestrial ecosystems extend over the full range of climates from polar to tropical. Tundra ecosystems generally receive only about 10 to 40 cm (4 to 16 in.) of annual precipitation, but because temperatures are low (averaging −7°C, or 19°F, annually) and evaporation is minimal, wet meadows and mires are extensive. South of the tundra, the boreal coniferous forest or taiga ecosystem extends around the Northern Hemisphere. Temperatures are warmer than in the tundra and annual precipitation higher at around 50 cm (20 in.). Because average annual temperatures are still low, evaporation is also low and lakes and bogs are common. The boreal coniferous forest gives way to the temperate deciduous forest around 48° N latitude in North America. The temperate deciduous forest has a climate of warm to hot summers, marked seasonality, winter cold and snow, and absence of drought. Annual precipitation can be as high as 125 to 150 cm (49 to 59 in.). The temperate deciduous forest begins to yield to tropical vegetation when winter temperatures generally remain above freezing, as in southern Florida. In tropical rain forests the average annual temperatures range from 22° to 27°C (72°–81°F), with monthly averages varying by only 2°–4°C from the annual mean. Annual precipitation usually exceeds 200 cm (79 in.) per year.

27.2 | Polar and alpine ecosystems have arisen since the retreat of the last glaciation

Polar ecosystems include the polar deserts of Antarctica and the herb barrens of the High Arctic (Chapter 26); the tundra of arctic Canada, Alaska, Europe, and Russia; and the **taiga** or **boreal coniferous forests**, which lie south of the tundra around the Northern Hemisphere (see Figure 25.22). Alpine environments and subalpine coniferous forests are high-altitude extensions of these high-latitude ecosystems. Polar and alpine ecosystems represent about 27% of the land surface area of the Earth. Of this total, about 9% is permanent ice. These ecosystems all share a climate consisting of a short growing season with cool or cold summers, low temperatures, and strong winds. Above latitudes of 66°33′ N and S, the Sun does not rise in midwinter, and during a comparable period in midsummer, there is continuous daylight.

Beginning around 1.8 million years ago, the Earth entered a series of four major glaciations, or ice ages. The most recent glacial interval, the Wisconsin or Würm glaciation, reached its maximum around 115,000 years ago. At that time, much of North America and Europe was buried under up to 4 km (2.5 miles) of glacial ice (Figure 27.1a). About 14,000 years ago, the climate began to warm and the vast ice sheets began to melt. The retreat of the glaciers left behind the Great Lakes and opened up a vast area of barren land, into which vegetation expanded from southern refuges in North America and Europe and northern refuges in Alaska (Figure 27.1b). Glacial ice melted off the mountain ranges, and vegetation spread upward to establish alpine ecosystems. Large areas of the Arctic in Cana-

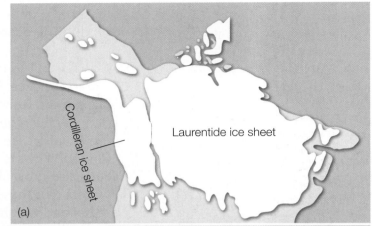
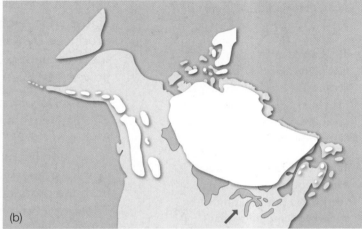

FIGURE 27.1 Ice sheets once covered North America (a) At 18,000 years ago, two major ice sheets covered a large portion of what would become Canada and the northern United States. (b) By 10,000 years ago, the Great Lakes (arrow) had formed.

da have become deglaciated only in the last 5,000 to 10,000 years. Present-day locations of polar and alpine ecosystems are therefore of recent origin.

Polar ecosystems now cover large areas of North America, Europe, and Asia. In the Northern Hemisphere, glaciation is restricted to Greenland, some far-northern islands, and mountains in Alaska. The land surface varies in relief from flat plains and plateaus to mountain ranges such as the Brooks Range in Alaska. In the Southern Hemisphere, in contrast, the area of polar ecosystems is much more restricted. The polar deserts of Antarctica were described in Chapter 26. Tundra and taiga are restricted to mountainous parts of the southern tip of South America and the South Island of New Zealand (see Figure 25.22).

27.3 | Tundra can be characterized by the absence of trees

The **tundra** ecosystem is widespread in the Northern Hemisphere but of limited extent in the Southern Hemisphere. Arctic tundra is divided into two regions, the High Arctic in the north and the Low Arctic to the south, with the dividing line drawn at 72–73° N latitude. Nutrient levels are low in tundra soils. The High Arctic has a cool polar environment, with the average temperature in the warmest month between 3° and 7°C (37°–45°F). Moisture is very limited, with annual precipitation averaging around 10 to 15 cm (4 to 6 in.), but in some areas the High Arctic may seem wet because evaporation is minimal at low temperatures, little runoff occurs across flat terrain, and drainage is poor due to the presence of permafrost. The Low Arctic is a mild polar environment, and the temperature in the warmest month averages 7° to 12° (45°–54°F). Moisture is more abundant, falling in the range of 15 to 40 cm (6–16 in.) per year. At its southern border, the tundra of the Low Arctic intermingles with the taiga or boreal coniferous forest, particularly along sheltered river valleys.

The most striking feature of the tundra is the lack of trees. Shrubs and herbs dominate the vegetation. A **shrub** is a woody plant with several main stems, and an **herb** is a nonwoody plant with a short-lived stem or stems. Common shrubs in the Arctic include birch *(Betula)*, bearberry *(Arctostaphylos*, Figure 27.2a), and willow *(Salix*, Figure 27.2b), which are often less than 1 m tall, and dwarf shrubs of blueberry *(Vaccinium)*. Grasses and sedges are common herbs. Mosses, lichens, and broadleaved herbs or forbs make up the balance of the typical vegetation.

The most widespread plant community in the High Arctic is the polar desert ecosystem called an herb barren, discussed in Chapter 26. Other types of plant communities, such as those dominated by grasses, forbs, or shrubs, occur only in sheltered sites where more moisture is available. In contrast, the diversity of plant communities is much greater in the mild polar environment of the Low Arctic. Large areas are covered in wet meadows and mires (Figure 27.3). Wet meadows are dominated by

FIGURE 27.3 **Wet meadow in Alaska** This scene is from Denali National Park.

FIGURE 27.2 **Two tundra shrubs** (a) Bearberry *(Arctostaphylos)*, a member of the blueberry and cranberry family, can form extensive mats. It is seen in fall color in this photograph, which shows Mount Denali (Alaska) in the background. (b) A small willow plant in the arctic tundra. Note the hairy leaves, which are helpful in retaining heat.

grasses and sedges rooted in a layer of mosses and lichens. Mires are extensive bogs that occur where the water table is high and mosses, such as *Sphagnum* (see Chapter 21), build up thick deposits of peat. On moderately moist sites, the vegetation may consist of dwarf shrubs less than 1 m tall. Drier sites are covered in grasses and sedges. Well-drained locations in sheltered river valleys may contain woodlands in the form of open stands of birch, larch *(Larix)*, cottonwood *(Populus balsamifera),* or spruce *(Picea).* Such woodlands merge with the boreal coniferous forests farther south in the tundra. Herb barrens do not occur in the Low Arctic.

Tundra plants show a number of adaptations in form and function

Plants in polar biomes must adapt to high winds, low temperatures, and a lack of available moisture. Winter winds damage plants in two ways. High winds drive snow and ice crystals into exposed plant parts and cause abrasion damage. They also increase the rate of water loss by transpiration at a time when the plant cannot obtain water from the frozen ground.

Many tundra plants grow only in sites where early winter snows will cover and protect them. Species that do not grow under snowbanks avoid wind by being short—wind speed is reduced close to the ground. Such exposed species may also retain dead leaves and stems, which take the brunt of the wind abrasion and collect snow and ice particles for further protection. Some species have such dense clusters of stems and leaves that they resemble a cushion (Figure 27.4). The wind speed inside such cushions can be reduced by up to 99%. Although strong winds are a survival problem for tundra plants, they use high winter winds for seed dispersal, since animals are suitable for seed dispersal may be scarce.

Because the growing season of 60 to 100 days is so short in tundra ecosystems, 98% of tundra plants are perennials. Perennials typically keep more than 50% of their biomass in below-ground storage organs such as roots and bulbs. Carbohydrates stored as food reserves supply the energy needed for a burst of growth in the spring, and food reserves are then replaced during the brief summer. Tundra plants may enhance their growth rate in spring by producing dark pigments. The red pigment anthocyanin can be used with chlorophyll to produce a very dark red. The dark pigment absorbs sunlight and raises the temperature of the plant—in one cushion-forming species, by as much as 15°C (27°F) above that of the surrounding air. Many species of polar and alpine lichens, and some mosses, are dark brown or black for the same reason—to absorb sunlight and heat. Other Arctic species raise their temperatures and reduce heat loss at night with a dense covering of white hairs, as in the woolly lousewort (*Pedicularis lanata*, Figure 27.5).

Dry conditions present another problem for survival in polar ecosystems. Snowbank plants grow only where a spring supply of water is assured. Some tundra plants can absorb moisture through their leaves. Lichens and mosses can dry out but resume metabolism shortly after becoming wet again.

Tundra plants show a number of reproductive adaptations

In the harsh climates of polar and alpine biomes, cold and aridity reduce the numbers of animal pollinators. Because animal pollinators are unreliable for pollination, plants may use them when available but must rely on other mechanisms to ensure reproduction.

Tundra plants may use one or more of four mechanisms for reproduction that do not make use of animals as pollinators. The simplest method is vegetative reproduction. Plants may send out aboveground stems called *runners*, which spread out from the parent plant and establish new plants at the nodes located along the runners. Many polar and alpine plants are apomictic, producing seeds without any sexual process (Chapter 23), for example the Antarctic grass *Deschampsia* (Chapter 26). Other tundra plants are capable of self-fertilization,

EVOLUTION

EVOLUTION

FIGURE 27.4 Cushion plants in alpine tundra of Colorado
Although they look like mosses, these small green cushions are actually flowering plants.

FIGURE 27.5 Woolly lousewort *(Pedicularis lanata)*

or autogamy. Antarctic *Colobanthus* is autogamous. Wind pollination, rather than pollination by animals, is often the agent bearing pollen to the ovules in tundra plants. Arctic willows *(Salix)* may be insect pollinated or wind pollinated. Some tundra plants combine more than one of these four mechanisms for reproduction. The whip saxifrage, *Saxifraga flagellaris*, bears yellow apomictic flowers and forms long, whiplike runners. Vegetative reproduction allows a plant to occupy the entire area around a favorable site while seeds formed by apomixis function in long-distance dispersal.

27.4 | Coniferous trees dominate the taiga

The boreal coniferous forest or taiga ecosystem in the Northern Hemisphere encircles the entire Earth across northern Eurasia, Alaska, and Canada (see Figure 25.22). The taiga parallels the tundra, with no sharp boundary between the two environments (Figure 27.6). The boundary area where trees disappear is called the *tree line*, also known as **timberline**. Taiga may extend up to 350 km (215 mi) into the tundra along sheltered valleys. Similarly, the tundra may extend into the taiga along windy, exposed ridges. The southern boundary of the taiga is even more obscure. In eastern North America, it blends into the temperate deciduous forest ecosystem. In the central plains of North America, the taiga gradually changes over into grasslands, and in the west, the boreal coniferous forest moves up into mountains, where it forms montane coniferous forest.

The climate of the taiga is less severe than that of the tundra, but it is still one of low temperatures, low precipitation, and a short growing season of from 90 to 120 days. Annual precipitation is around 40 to 50 cm (15 to 20 in.), but the landscape is dotted with lakes and bogs because evaporation is low and soil drainage poor. Glaciers across arctic Canada leveled the landscape, so surface relief is low. Soils are acidic and poor in nutrients.

Coniferous trees, including white and black spruce *(Picea glauca* and *Picea mariana)* and balsam fir *(Abies balsamea)* dominate the vegetation of the taiga. Plant species diversity is low. Trees at the tree line are short and shrubby but grow taller toward

the south. Coniferous forests have four layers of vegetation: (1) the canopy of coniferous trees, (2) a layer of shrubs such as willows and blueberries, (3) an herb layer, and (4) a surface layer of mosses on wet sites or lichens on dry sites. Black spruce *(Picea mariana)*, tamarack *(Larix laricina)*, and *Sphagnum* moss are abundant in bogs. Jack pine *(Pinus banksiana)* and two deciduous angiosperms—paper birch *(Betula papyrifera)* and quaking aspen *(Populus tremuloides)*—are common in areas cleared by past fires (Figure 27.7). Aspens also form a part of the transition between taiga and grasslands in central North America.

Boreal forest species must adapt to short growing seasons, snowfall, low temperatures, and limited moisture. Most conifers are evergreen, retaining their leaves or needles throughout the year. Leaf retention allows them to begin photosynthesis as soon as temperatures rise in spring, without waiting to generate new leaves. Conifers are typically cone-shaped and have flexible branches—features that allow snow to slide off the branches without breaking them. Conifer leaves (Chapter 22) are adapted to survive dry conditions in winter by having thick cuticles, a tough epidermis, and sunken stomata—features that reduce water loss. Conifers also enter a period of dormancy in winter, reducing their respiration to very low levels.

EVOLUTION

Conifers have shallow roots that can quickly absorb surface moisture as it becomes available in spring. Many conifers are associated with mycorrhizal fungi, which enlarge the root system to pick up nutrients and break down organic matter to make more nutrients available to the roots. The presence of these fungi is especially important in a soil where low temperatures retard decomposition. Black spruce can reproduce by a vegetative process; their lower branches can form roots where they touch the ground to produce new individuals.

FIGURE 27.7 Quaking aspen (*Populus tremuloides*) in fall color

FIGURE 27.6 Taiga-tundra boundary in Alaska

27.5 | Alpine tundra and montane coniferous forest are southern extensions of arctic tundra and taiga

Tundra and taiga are able to extend into southern latitudes by rising up into mountain ranges because the average annual temperature in mountains declines by about 1°C for each 100-m (328 ft) rise in elevation above sea level. Thus alpine tundra occurs down to central California in the Cascade and Sierra Nevada Mountains above elevations of 3,500 m (11,480 ft). The tundra also extends from the Rocky Mountains in the United States and Canada down to Guatemala, where it occurs above 4,500 m (14,760 ft). Alpine tundra occurs above 1,200 m (3940 ft) in the Appalachian Mountains of New Hampshire and Vermont.

Just below the alpine tundra lies the montane coniferous forest. The upper altitude limit of the montane coniferous forest is called *timberline*, where the trees grow short and shrubby before they disappear entirely. The altitude of timberline varies with latitude, slope, and moisture. In the Canadian Rockies, timberline is around 2,900 m (9,500 ft) (Figure 27.8). Montane coniferous forest occurs in the Appalachian Mountains as far south as the Great Smoky Mountains of Tennessee and North Carolina. It is widespread in the Rocky Mountains, where it extends from Alberta, Canada, south to New Mexico, and west across Utah to the Cascade Mountains. Montane coniferous forest has a narrower range in the Cascade and Sierra Nevada Mountains, where it extends from Canada to southern California.

Alpine tundra and montane coniferous forest have climates characterized by a short growing season, low temperatures, and strong winds, as do the Arctic tundra and taiga. But beyond these general features, the climates of alpine and arctic systems differ. In winter, alpine tundra has no period of continuous darkness. Day lengths in alpine and montane systems are the same as at lower elevations. Light intensities are also greater in alpine tundra and montane forests than in the Arctic because the Sun is more directly overhead and the atmosphere is thinner at high elevation. The thinner atmosphere is also

responsible for a higher level of ultraviolet (UV) radiation. In general, UV radiation increases by about 19% per 1,000 m of elevation above sea level. At 4,000 m (13,120 ft), therefore, the level of UV radiation is about twice that at sea level. The difference between daytime and nighttime temperatures at high altitude is extreme. Soil surface temperatures in summer may be over 40°C (104°F) in the day but plunge below 0°C (32°F) at night due to heat loss. In the Arctic, annual precipitation is low but drainage poor. In alpine tundra, however, annual precipitation is often very high but drainage good. Drought is more likely. Lakes and bogs are not as numerous in mountains as in the boreal coniferous forest.

Alpine tundra and arctic tundra have many species in common

Despite differences in climate, about 45% of the tundra plants in the Rocky Mountains of Montana also grow in the Arctic, and 75% of tundra plants in the Appalachian Mountains of New Hampshire are also found in the Arctic. The plant communities of alpine tundra vary depending on soil surface, moisture, and drainage. Wet meadows dominated by mats of sedges and grasses occur where moisture is available all through the growing season. Fell-fields, rocky areas where grasses and forbs grow between the rocks, support mosses, lichens, and cyanobacteria that grow on the rocks. Mosses and lichens dominate the most severe alpine environments.

Alpine tundra plants show many of the same adaptations as those in the Arctic, but high light intensities require some different responses. Many more alpine tundra plants have dense coverings of epidermal hairs than do plants in the Arctic. These dense coverings of hairs reflect light away from the plant and reduce heat loss.

EVOLUTION

Many lichens and cyanobacteria may be bright orange or yellow from high levels of pigments called *carotenoids*. Carotenoids shield these organisms from high light levels. Some mosses and lichens are brown or black. These dark colors may absorb heat to increase photosynthesis and screen out UV radiation. One dark brown moss called *Andreaea* grows on rocky substrates at high altitudes in the Rocky Mountains and in Antarctica. *Andreaea* may be the toughest organism on Earth. More than 80% of its biomass consists of phenolic compounds that can even stand up to boiling in acid. These phenolic compounds are located in thick cell walls, where they screen out UV radiation and protect the moss from freeze-thaw cycles, desiccation, and mechanical abrasion.

Dominant conifer species differ among mountain ranges

Spruces and firs dominate montane coniferous forests, as in the taiga, but the species are different in each of the mountain ranges. Species differences reflect past history—particularly glacial history—and subsequent speciation in geographically isolated mountain ranges. Appalachian montane forests are most similar to the taiga, although the climate is wetter and warmer. Balsam fir and red spruce (*Picea rubens*) are dominant,

FIGURE 27.8 Timberline in the Canadian Rockies

FIGURE 27.9 Rhododendrons in the Appalachian Mountains of western North Carolina

along with paper birch and yellow birch *(Betula lutea).* Azaleas and rhododendrons (Figure 27.9) are common flowering shrubs, and mosses and lichens form a thick carpet.

The montane forests of the Rocky Mountains are not continuous across their range. Dominant species vary depending on location, slope, moisture, and past history. Dominant species in Colorado include Engelmann spruce *(Picea engelmanii)* and subalpine fir *(Abies lasiocarpa)* (Figure 27.10). If an area has been disturbed by fire or logging in the past, lodgepole pine *(Pinus contorta)* and groves of yellow-green quaking aspen *(Populus tremuloides)* will occur. In the Southwest, bristlecone pines *(Pinus aristata* and *P. longaeva)* are dominant at timberline (Figure 27.11a). Bristlecone pines are well known for their longevity, with some individuals of *P. longaeva* over 5,000 years old. At lower elevations in the Southwest, ponderosa pine *(Pinus ponderosa)* forms open stands above an herb layer of grasses (Figure 27.11b).

In the montane coniferous forests of the Cascade and Sierra Nevada Mountains, the climate is Mediterranean, which means that the montane forests have cool, wet winters and warm, dry

FIGURE 27.11 Pines of montane forests (a) Bristlecone pines on Mount Goliath, Colorado. The harsh, windy conditions produce trees with a straggly (flagged) appearance. (b) Ponderosa pine. Note the grass understory in this scene from the Deschutes National Forest, Oregon.

summers. Precipitation varies with altitude, ranging from 25 cm (10 in.) per year at the base of the range to over 125 cm (49 in.) per year in the alpine zone. Most of the precipitation falls as snow in the winter. Common species include lodgepole pine and mountain hemlock *(Tsuga mertensiana),* but the most famous tree species is the giant sequoia *(Sequoiadendron giganteum),* which occurs in scattered groves in the central and southern Sierra Nevada at low elevations (Figure 27.12). Giant sequoias are among the largest living organisms on Earth. Individual trees may reach a weight of 6,000 tons and a height of 100 m (328 ft). Because of the warm, dry summers, fire is an important part of the ecology of these montane forests. Many species have fire-resistant bark and cones that open only after exposure to heat. Seeds of such cones germinate best after a burn, when minerals are available.

Mining, logging, grazing, and recreation affect polar and alpine ecosystems at local and regional scales

Historically, human populations have been very sparse in polar and alpine ecosystems because of the harsh climate. But because of declining resources and increasing pollution in other ecosystems,

FIGURE 27.10 Montane forest The coniferous montane forest is seen here surrounding Bear Lake in Rocky Mountain National Park, a popular destination of park visitors.

FIGURE 27.12 Giant sequoia These giant sequoias in Sequoia National Park, California, are among the largest organisms on Earth.

environmental pressures from human activities are increasing on polar and alpine ecosystems.

The main human impact on arctic tundra is the removal of mineral resources, particularly oil. The extraction of oil from the North Slope of Alaska is one familiar example. Tundra ecosystems are fragile and may take decades to recover from damage. The pipeline that was built to carry oil from northern Alaska to a southern port had to be constructed to minimize interference with wildlife migration routes and to avoid melting the permafrost, which could have caused the pipeline to sink. The full results of oil extraction may not be known for some time, but pressure to extract oil from additional sites is sure to continue as other resources dwindle. In the last decade, diamond mining has begun in the Northwest Territories of Canada. Because diamond mining requires the removal and washing of tons of earth, it is likely to have a serious effect locally.

Boreal coniferous forest has long been logged to produce pulp for paper products. Hydroelectric power plants are being built on rivers in these areas to provide power to southern cities. As human population continues to grow and as resources in other regions are consumed, the demand for resources in the taiga will also grow.

Because alpine and montane coniferous forests are more accessible than polar regions, they have been more affected by human activity. The montane forests of the Appalachian Mountains have been particularly hard hit. Most of the forests have been logged in the past. Over the past few decades, the construction of vacation homes and resorts has had a major impact. Sheep and cattle grazing has increased erosion in some areas of the Rocky Mountain montane forests. Logging is

also widely practiced and can increase the danger of fires and erosion. Resorts, vacation homes, and suburban sprawl all encroach on the montane forests. Recreational uses have had especially severe consequences in the Cascade and Sierra Nevada Mountains. Campgrounds may have population densities greater than cities and have comparable levels of crime, sewage, and air pollution. In these mountains forests of ponderosa pine, which are especially prone to air pollution, may be replaced by shrub lands.

27.6 | Temperate deciduous forests are ecosystems with seasonality and abundant precipitation

Temperate deciduous forest ecosystems occur as three distinct areas in the Northern Hemisphere: eastern North America, western and central Europe, and eastern Asia, including eastern China and Japan (see Figure 25.22). Although now widely separated, these areas were joined together about 55 million years ago into one vast forest that extended around the Northern Hemisphere. Continental drift, mountain building, and climate changes broke up this forest. Over the last 2 million years, the temperate deciduous forests of eastern North America have been repeatedly broken up and reassembled by a series of ice ages. Each ice-age advance drove the forest species into refuges in southern North America, from which they migrated northward during subsequent interglacials—warm periods between the ice ages. At the time Europeans first arrived in North America, the temperate deciduous forest formed a nearly unbroken canopy across all of North America east of the Mississippi and in parts westward (Figure 27.13).

The eastern temperate deciduous forest has four consistent climatic features over the entire area—a marked pattern of seasonal changes, frosts in winter, absence of droughts, and moderate to long growing seasons. Within this general framework, climate changes across the area are in east–west and north–south gradients. Precipitation decreases from east to west, from 125 cm (49 in.) along the Atlantic seaboard to 85 cm (33 in.) at the boundary between forest and grassland in the west. Precipitation varies from 75 cm (30 in.) around the Great Lakes to 150 cm (59 in.) along the Gulf Coast, where almost all precipitation falls as rain. The length of the growing season varies from 120 days in the north to 250 days in the south.

Temperature and precipitation define the limits of the eastern temperate deciduous forest. The northern limit, where it changes into boreal coniferous forest, occurs where the average minimum winter temperature is −40°C (−40°F), which occurs at about 48° N latitude. Precipitation determines the western boundary, where forest changes over to grassland. The southern limit occurs where minimum temperatures are warm enough to permit subtropical vegetation.

The structure of the temperate deciduous forest consists of five layers of vegetation. Trees make up the canopy and a lower, more open subcanopy consisting of immature and short trees such as dogwoods (Cornus) and redbud (Cercis canadensis). The three lower layers of understory plants are the shrub layer,

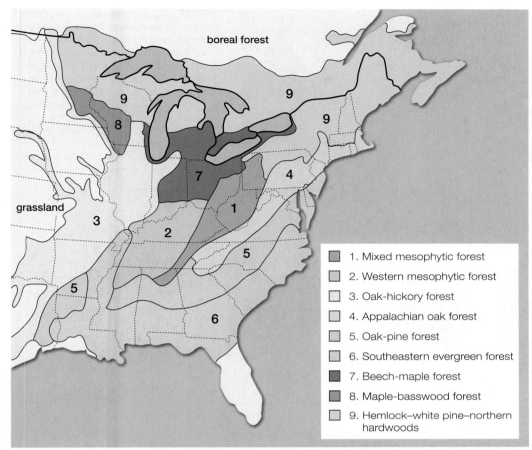

FIGURE 27.13 **The nine regions of eastern temperate deciduous forest**

Legend:
1. Mixed mesophytic forest
2. Western mesophytic forest
3. Oak-hickory forest
4. Appalachian oak forest
5. Oak-pine forest
6. Southeastern evergreen forest
7. Beech-maple forest
8. Maple-basswood forest
9. Hemlock–white pine–northern hardwoods

herb layer, and surface layer. Broadleaved perennials dominate the herb layer. The mosses and lichens of the surface layer occur mainly on rocks and tree trunks.

Eight genera of trees define the eastern temperate deciduous forest

Six genera of deciduous angiosperm trees dominate the canopy across the North American eastern deciduous forest ecosystem. These genera (Figure 27.14a–f) include the oaks *(Quercus)*, hickories *(Carya)*, the yellow poplar or tulip tree *(Liriodendron tulipifera)*, American beech *(Fagus grandifolia)*, sugar maple *(Acer saccharum)*, and the basswoods *(Tilia)*. Two species of conifers (gymnosperms) are also important (Figure 27.14g, h): the white pine *(Pinus strobus)* and the eastern hemlock *(Tsuga canadensis)*. Until the 20th century, the American chestnut *(Castanea dentata)* was also a dominant member of the eastern temperate deciduous forest, but a fungal blight introduced from Asia wiped it out between 1906 and 1940. Different members of these eight genera dominate different regions within the eastern temperate deciduous forest, where they form nine different plant associations (see Figure 27.13).

The most diverse plant association occurs in the mixed mesophytic forest region (region 1 in Figure 27.13). Almost all the taxa that grow as canopy dominants in the eastern temperate deciduous forest occur here as codominants. (For a list of

tree species occurring in two tracts of forest in this region, refer to Chapter 25, Table 25.1.) Oaks and hickories dominate the western mesophytic forest (region 2 in Figure 27.13), with red cedar occurring in forest openings called *glades*.

Oaks and hickories also dominate the oak-hickory forest, which is the most westerly region in the eastern temperate deciduous forest (region 3 in Figure 27.13). The Appalachian oak forest region extends along the eastern valleys and ridges of the Appalachian Mountains (region 4 in Figure 27.13). Rhododendrons are common shrubs. The American chestnut was once a dominant member of this plant association, but now oaks, pines, and hickories are codominants.

European settlers largely destroyed the original forests of the oak-pine forest region (region 5 in Figure 27.13), and pines make up the forests that have returned. The southeastern evergreen forest region covers the coastal plains of the southeastern United States except for the tropical vegetation of Florida (region 6 in Figure 27.13). Before European settlement, beech and southern magnolia *(Magnolia grandifolia)* dominated this region, but now pines are widely grown as a commercial crop. Fire is an important ecological factor here. The pines are fire resistant, and periodic burning kills hardwood trees that might compete with them.

The beech-maple association occurs only on land that was formerly glaciated during the last ice age (region 7 in Figure 27.13). When the glaciers retreated from this area,

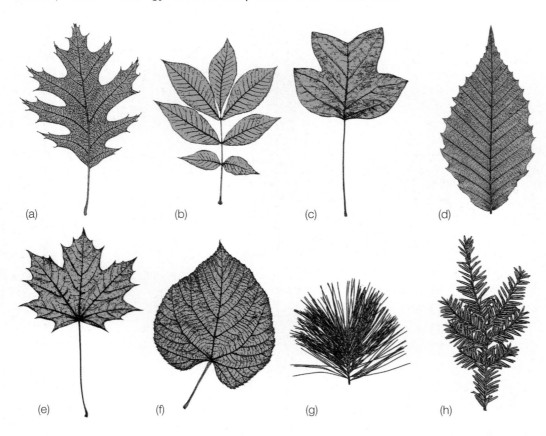

FIGURE 27.14 Leaves of selected genera of dominant trees found in temperate deciduous forests (a) Red oak *(Quercus rubra)*, (b) shagbark hickory *(Carya ovata)*, (c) tulip tree *(Liriodendron tulipifera)*, (d) American beech *(Fagus grandifolia)*, (e) sugar maple *(Acer saccharum)*, (f) basswood *(Tilia americana)*, (g) white pine *(Pinus strobus)*, and (h) eastern hemlock *(Tsuga canadensis)*. White pine and eastern hemlocks are gymnosperms but occur with the other trees, which are angiosperms.

coniferous forest trees migrated through it before beech and sugar maple arrived. Remnants of this boreal forest occur around bogs. The maple-basswood forest association is the smallest region within the eastern deciduous forest biome (region 8 in Figure 27.13). This area, which is called the "Driftless Area," was never covered in glacial ice and acted as a refuge for species during the last ice age. Before European settlement, much of the hemlock–white pine–northern hardwoods region (region 9 in Figure 27.13) was covered by white pines *(Pinus strobus)* 60 m (210 ft) tall, but these forests were cut down by the mid-19th century. Today, white and red pine *(Pinus resinosa)* and yellow birch *(Betula lutea)* are common in the region, along with species from the boreal forest such as quaking aspen *(Populus tremuloides)* and temperate deciduous species such as sugar and red maple *(Acer rubrum)*. Severe windstorms, called *downburst storms*, occur in this region at intervals of 60 to 100 years (Figure 27.15). These storms can destroy all the trees in an area of forest and open up space for new species to enter and grow.

FIGURE 27.15 Results of a downburst storm in a hemlock–white pine–northern hardwoods region forest

Plants in the temperate deciduous forest are adapted to cold winters and competition for light

Except for the evergreen conifers, the trees of the temperate deciduous forest shed their leaves in the fall. This deciduous habit is highly adaptive in a climate with marked seasonality.

Water is restricted in winter and if the deciduous trees retained their leaves, they would lose water at a higher rate than they could replace by root absorption from frozen ground. Deciduous angiosperms still have higher rates of water loss in winter than evergreen conifers, a factor that must affect their relative abundance at the northern limits of the temperate deciduous forest where winters are more severe.

Because deciduous angiosperm trees must produce new leaves each spring, there is a period when full sun reaches the forest floor before the canopy species leaf out. The mosses and lichens of the surface layer use this period of full sun to grow and reproduce. Herbaceous perennials then follow rapidly (see figures in Essay 27.1). Many of these species have underground storage organs such as bulbs or rhizomes, and spring photosynthesis produces a rapid buildup of starch in these structures. Once the canopy closes, many of these herbaceous perennials become dormant, but in the fall they will preform shoots to speed the process of leaf-out next spring. Competition for light and space among temperate deciduous forest herbs was discussed earlier (Chapter 25).

Many understory plants—including subcanopy trees, shrubs, vines, and herbs—produce seeds that are dispersed by animals such as ants, birds, and mammals. Ant-dispersed plants, such as wild ginger *(Asarum)* and violets *(Viola)*, produce seeds with attached lipid-rich structures as food rewards for their ant dispersers (Figure 27.16). Plants that produce fruits for consumption by birds and mammals fall into two seasonal categories. Summer-ripening fruits, such as blueberries, cherries, and blackberries, are sweet and rich in carbohydrates. These fruits attract resident bird and mammal species (Chapter 24). Fall-ripening fruits such as those of *Magnolia* (Figure 27.17a) and dogwood *(Cornus)*, however, are rich in lipids. These fruits are consumed and their seeds dispersed by migratory birds that need energy from lipids for flight. Such fruits are often bright red to attract birds, or they may be marked by foliar fruit flags, which are leaves that change color to guide birds to the fruits (Figure 27.17b).

Many species of deciduous angiosperm trees are wind pollinated. They produce flowers and complete pollination before they leaf out (Figure 27.18). If they produced leaves at the same time or prior to forming flowers, the leaves would

FIGURE 27.17 **Bright colors attract seed dispersers** (a) Bright red, lipid-rich fruits of *Magnolia*. (b) Foliar fruit flag of Virginia creeper *(Parthenocissus)*.

intercept the wind-borne pollen and reduce reproductive success. Once the leaves are formed, they represent an enormous surface area and the canopy is effectively closed. Foliage leaves are of two types, based on canopy position—outer sun leaves and interior shade leaves (Figure 27.19). Sun leaves are smaller, thicker, sometimes hairier, and more deeply lobed than shade leaves. These features of sun leaves act to shed heat and reduce water loss. Sun leaves may also be oriented on their branches at an inclined angle with respect to the Sun's rays rather than being perpendicular to them. This orientation

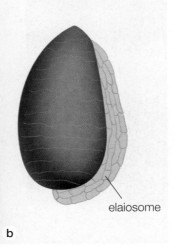

elaiosome

FIGURE 27.16 **Wild ginger *(Asarum canadense)*, an understory plant that employ ants for seed dispersal** (a) The flower of wild ginger (lower center) lies on the ground and has a fetid scent and reddish color, which attract flies that serve as pollinators. (b) Seeds have lipid-rich structures (elaiosomes) that provide nutrition to ants that collect them.

FIGURE 27.18 **Temperate forest trees are usually wind pollinated** (a) Pollen-producing flowers of red oak occur in catkins from which pollen is easily dispersed by wind. (b) A cluster of flowers of silver maple. (c) Close-up of a single flower. Note the two long stigmas, upon which windborne pollen is deposited. Maple trees may also be insect pollinated.

FIGURE 27.19 **Sun and shade leaves of basswood** (a) Sun leaves tend to be smaller but thicker than shade leaves. (b) The palisade and spongy parenchyma are much more developed in the sun leaf.

exposes less surface area to solar radiation, reduces heating from sunlight, and allows light to penetrate the canopy and reach the shade leaves. Shade leaves are larger than sun leaves and can reach maximum photosynthesis at lower light levels than sun leaves.

Humans have had a major impact on many features of the eastern temperate deciduous forest

Human impact on the temperate deciduous forest began with the arrival of Paleo-Indian populations in eastern North America around 12,000 years ago. These peoples initially contributed to the extinction of many species of large mammals in North America, extinctions that disrupted coevolutionary relationships between mammals and temperate deciduous trees and led to isolation of some species. Native Americans affected the temperate deciduous forest by their use of fire to clear land for agriculture, particularly in river valleys. (For a look at uses of temperate forest plants by Native Americans, refer to Essay 27.1, "Native American Uses of Temperate Forest Plants.") The major impact of humans on the temperate deciduous forest began with the arrival of Europeans in the 1600s. As they settled, first along the East Coast and then progressively further inland, they cleared most of the temperate deciduous forest, leaving only isolated patches of original forest. Nearly all of present-day temperate deciduous forest is the result of second, third, or fourth regrowth after successive cuttings.

Many exotic species have invaded the temperate deciduous forest As European settlers came to North America, they brought familiar plants from their homelands for use in fields and gardens, a process later accelerated by seed companies and nurseries. Unfortunately, exotic species do not restrict their growth to gardens. For example, in the late 1900s several honeysuckle *(Lonicera)* species were imported into the United States for use as garden shrubs. Some of these species (or hybrids between them) have escaped cultivation (Figure 27.20a). Birds eat the bright red berries and spread the seeds throughout

ESSAY 27.1   NATIVE AMERICAN USES OF TEMPERATE FOREST PLANTS

Native Americans living in forested regions used a wide variety of local plants and fungi as food, beverages, and medicines and to construct art, toys, tools, and lodgings. They believed that every bush, tree, and herb had a use. Woodland inhabitants discovered a modern cancer medicine. As a purgative and to treat skin disorders and tumors, they used the belowground stem (rhizome) of the mayapple plant *(Podophyllum peltatum),* which grows throughout deciduous forests of Canada and the eastern United States (Figure E27.1A). Modern extraction, chemical analysis, and testing procedures have validated these applications. Modern medicine uses the alkaloids podophyllin and alpha- and beta-peltatin to treat leukemia.

The Chippewa, who inhabited north-central forests of the United States and Canada, developed technology for maple sugar production. Each family had the use of a particular population of sugar

> *Woodland inhabitants discovered a modern cancer medicine.*

maple trees, and special huts were constructed for boiling the sap and storing utensils from year to year. Collection vessels were made of bark or carved from wood. Trees were tapped in mid-March by inserting 2–3 wooden spikes shaped like troughs into diagonal cuts made in the bark about 3 feet from the ground. The diagonal cuts collected the maple sap at the bottom, and the spike directed it into the collection vessel. The average family maintained about 900 taps. Before metal kettles were available for boiling syrup, the Chippewa concentrated the sugar by dropping red-hot stones into the vessels containing syrup, or they repeatedly froze the syrup, removing ice formed on top each day. The sugar was stored as cakes in fancy wooden molds.

Other useful plants included bloodroot *(Sanguinaria;* Figure E27.1B), from whose roots a red dye was produced, and horsetail *(Equisetum),* used to scour utensils. Plants from nearby wetlands were also used—*Sphagnum* moss was used for baby diapers, and young rhizomes and stems of cattails *(Typha)* were eaten fresh in the spring and mats were woven from the leaves.

Chippewa medicine was also sophisticated. There were guilds of professional healers who were forbidden to impart their medical secrets and who were not allowed to work without compensation. However, most people knew how to use various medicinal plants, such as the ladyslipper orchid *(Cypripedium,* Figure E27.1C). They

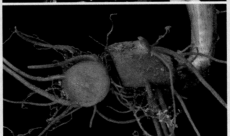

E27.1B *Sanguinaria* The lower photo shows the blood-red color in the rhizome.

E27.1A *Podophyllum*

E27.1C *Cypripedium*

used chopped rhizome and roots of this orchid to make a poultice for toothache. From rhizomes of *Trillium* (Figure E27.1D), they made eardrops and a medicine for rheumatism that was pricked into the skin using special needles. An extract of the young twigs of arborvitae (white cedar—*Thuja*, Figure E27.1E) was effective as a cough medicine. Today, all these plants are known to contain secondary compounds (Chapter 3), which explains their medicinal properties.

E27.1D *Trillium*

E27.1E *Thuja*

FIGURE 27.20 Invasive plants of temperate forests
(a) Honeysuckle is an exotic shrub in the temperate deciduous forest. The inset shows the fruits. (b) Garlic mustard (*Alliaria petiolata*) covering a forest floor.

the temperate deciduous forest, where these species now crowd out native shrubs.

Early settlers also brought in garlic mustard *(Alliaria petiolata)*, a rapid-growing herb used in cooking and herbal medicines. Molecular studies of populations from North America and Europe indicate that garlic mustard was brought from the British Isles about 125 years ago. The name derives from the smell given off from crushed leaves. The seeds are easily spread on the fur of mammals and soles of shoes. Garlic mustard becomes so dense over the forest floor that it shades out native wildflowers (Figure 27.20b).

DNA SCIENCE

Recently, researchers at the University of Wisconsin conducted a plant survey in stands of forest in northern Wisconsin and Michigan. Wisconsin botanist John Curtis surveyed these same stands in the 1950s. The current survey found that of the species present in the 1950s, 70% had declined by 2000 and that one in four species had become extinct. Invasive species and extensive browsing by increased whitetail deer populations had eliminated many rare native species, and the northern woods were becoming less diverse.

Plant rustling from national forests is becoming a serious problem The current popularity of herbal ingredients in health foods has led to widespread rustling of wild plants in our national forests and parks and is creating a serious problem for the U.S. Forest Service and Park Service. Goldenseal has a market value of $30 per pound, and ginseng root is priced at $270 to $600 per pound. Even the mosses of the surface layer are being looted. Logmosses (*Hypnum* species) are used in the floral business for decorative purposes, such as wire bunnies stuffed with moss. Looters in the Appalachian forests roll the moss off logs like pieces of carpet and load it onto trailer trucks. Its value is about $16 per pound. The value of the illegal moss harvested from the

Monongahela National Forest in West Virginia is estimated at about $1 million per year.

Research at Hubbard Brook was undertaken to determine how temperate forest ecosystems function

Today, the temperate deciduous forest is a managed ecosystem. Most management of land resources such as forests emphasizes maximizing production of a product such as lumber, not the impact on recreation, wildlife, or even long-term sustainability of the ecosystem. Lack of information about how any ecosystem functions limits our ability to devise wiser management plans. From an ecological perspective, every ecosystem may be studied as if it had a balance sheet or budget of important substances such as water and mineral nutrients. Water enters an ecosystem as precipitation and leaves as runoff in streams, evaporation from surfaces, or transpiration from plants. Minerals may enter the ecosystem by weathering of rock, in dust from the air, or as dissolved ions in rain or snow. Those minerals may be taken up and retained by plants for a time, and finally they may leave in runoff into streams, as plant litter, or in gases from fires. The various ways water and minerals enter an ecosystem are called *inputs*. The ways these substances leave an ecosystem are called *outputs*. In the absence of information about these inputs and outputs, it is difficult to predict the impact of management plans on ecosystems.

Early in the 1960s, a research program was begun to determine how the eastern temperate deciduous forest operates as an ecosystem. This multiyear program had three goals: (1) Determine a detailed budget for an undisturbed forest in terms of inputs and outputs of water and mineral nutrients. (2) Disturb a large area of forest by clear-cutting an entire watershed and measure the effects. (A **watershed** is a region where all water drains toward a common exit point such as a stream. Watersheds may be separated by ridges and can be thought of as separate units within the forest ecosystem; Figure 27.21a.) (3) Apply the results to improve management procedures. The research accomplished these goals but documented something even more important—the impact of acid rain produced by human industrial activities on ecosystems.

The area chosen for the research was called the Hubbard Brook Experimental Forest, located in the White Mountain National Forest of New Hampshire, which is within the hemlock–white pine–northern hardwoods region. The forest is dominated by sugar maple, American beech, and yellow birch, and it had been undisturbed by cutting since 1919. The Hubbard Brook Experimental Forest is an ideal ecosystem for measuring inputs and outputs; all of the watersheds are underlain by relatively impermeable bedrock. Therefore, all mineral nutrients that enter a particular watershed through precipitation can leave only through the stream that drains that watershed. To measure inputs, the researchers set up weather stations to collect precipitation. To measure the output from the ecosystem, they built a weir across the stream at the base of the watershed (Figure 27.21b). A weir is a small dam that allows

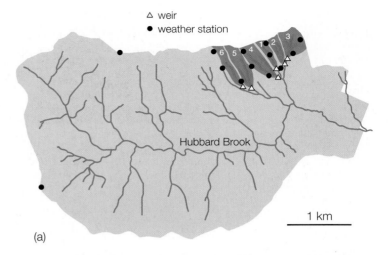
(a)

(b)

FIGURE 27.21 Hubbard Brook Experimental Forest (a) Map of the watersheds studied. (b) A weir measures stream outflow. Mineral nutrient concentrations are measured in the water behind the weir.

water flow to be measured. The researchers then measured the concentrations of various mineral nutrients in the precipitation and in the water leaving the watershed in the stream.

The water budget for the ecosystem showed an average annual precipitation of 137 cm (54 in.) and runoff of 89 cm (35 in.). The difference—48 cm (19 in.)—represents the return of water to the atmosphere by evaporation and transpiration through the plant cover. Knowing the amount of water and the concentrations of dissolved minerals in the water entering and leaving the ecosystems, the researchers were able to calculate a budget for each watershed in terms of kilograms of each mineral per hectare (1 hectare equals 10,000 m^2, or 2.47 acres) entering and leaving each year (Table 27.1).

After 2 years of data collection, the research team deforested one entire watershed and kept it barren of vegetation for 3 years while they measured the effects on the water and mineral nutrient budgets. Runoff jumped by over 40% because there was no vegetation to conduct transpiration. Since there was no vegetation to absorb mineral nutrients, losses rose 22 times for potassium, 9

TABLE 27.1 Mineral Budget for the Hubbard Brook Experimental Forest Watersheds

Values for calcium, potassium, sodium, and magnesium are in kilograms per hectare per year ($kg\ ha^{-1}\ yr^{-1}$) and are the averages for the period 1963 to 1974. These watersheds were undisturbed by the researchers. Output values include dissolved ions and particulate matter.

Mineral Element	Input from Precipitation	Output from Stream Flow
Calcium (Ca)	2.2	13.9
Sodium (Na)	1.6	7.5
Magnesium (Mg)	0.6	3.3
Potassium (K)	0.9	2.4

times for calcium, 6 times for magnesium, and 3 times for sodium (Table 27.2). The critical plant nutrient nitrogen leached out of the soil and contributed to algal blooms in the stream. Clear-cutting showed that an ecosystem could retain mineral nutrients only if an orderly cycling between living and nonliving components occurred. Destruction of the vegetation broke the cycle. Clear-cutting also revealed that land and stream systems were tightly coupled. A disturbance to the land caused a disturbance in the stream. The researchers concluded that the rapid restoration of vegetation after cutting was essential to minimize disturbance.

Acid rain is damaging the eastern temperate deciduous forest

The numbers in Tables 27.1 and 27.2 reveal that even in the control forest ecosystems, there was a net loss of mineral ions. The fertility of the soil in the so-called undisturbed ecosystems was declining. The cause was acid present in the precipitation falling on the eastern United States. Power plants emit large quantities of sulfur dioxide, which combines with water in the atmosphere to form sulfuric acid. Combustion of fuels oxidizes nitrogen in the atmosphere to nitrate, which combines with water in the air to form nitric acid. Normally, the positively charged mineral ions (cations) in soil are bound to negatively charged soil particles (Figure 27.22). When acid rain or snow falls, hydrogen ions (H^+) exchange with mineral cations (Na^+, K^+, Ca^+) on the soil particles, and these cations wash out of the system along with anions such as SO_4^{2-} and NO_3^-. These mineral nutrients then enter streams and lakes, which become more acidic.

The Clean Air Act of 1970, the passage of which was influenced by the Hubbard Brook Ecosystem study, together with 1990 amendments to the act, mandated sulfur reductions in power plant emissions. Over the three decades since 1970, sulfur emissions have dropped from the 1973 peak of 31 million tons per year to 21 million tons in 1994. Despite these cuts in sulfur dioxide emissions, the eastern deciduous forests and their associated aquatic systems are not showing improvement. Acid rain has leached the mineral elements essential for plant growth from the soil. The Hubbard Brook forest had nearly stopped accumulating biomass by 1982, possibly because the supply of calcium was limiting. The acidity of streams is still greater than normal, and many lakes in the Adirondack Mountains of New York are chronically acidic and have few or no fish. Effects on forests are increasing in the Northeast and becoming more apparent in the Southeast, where thicker soils and different soil chemistry had delayed their appearance until the past decade.

According to leading acid rain researchers, sulfur emissions must be cut by an additional 80% if some recovery is to take place by 2050. An additional problem is that nitrogen oxides such as nitrates are less regulated, and their concentrations are either level or increasing in the air. Ozone levels in the atmosphere are also damaging eastern temperate forests, particularly

TABLE 27.2 Comparison Between the Nutrient Budgets for the Undisturbed Watershed #6 and the Deforested Watershed #2

Values shown are in units of kilograms per hectare per year ($kg\ ha^{-1}\ yr^{-1}$) and are the averages for two field seasons, 1966–67 and 1967–68. Values for water inputs and outputs are in centimeters.

Element	Watershed #6 (Control)			Watershed #2 (Deforested)		
	Input	Output	Net +/−	Input	Output	Net +/−
Ca^{2+}	2.7	11.4	−8.7	2.5	85.2	−82.7
Mg^{2+}	0.6	3.2	−2.6	0.6	17.4	−16.8
K^+	0.7	2.0	−1.3	0.6	29.8	−29.2
Na^+	1.6	7.8	−6.2	1.3	18.6	−17.3
NO_3^-	4.9	2.0	+2.9	5.9	125.5	−119.6
Water	139	85		135	122	

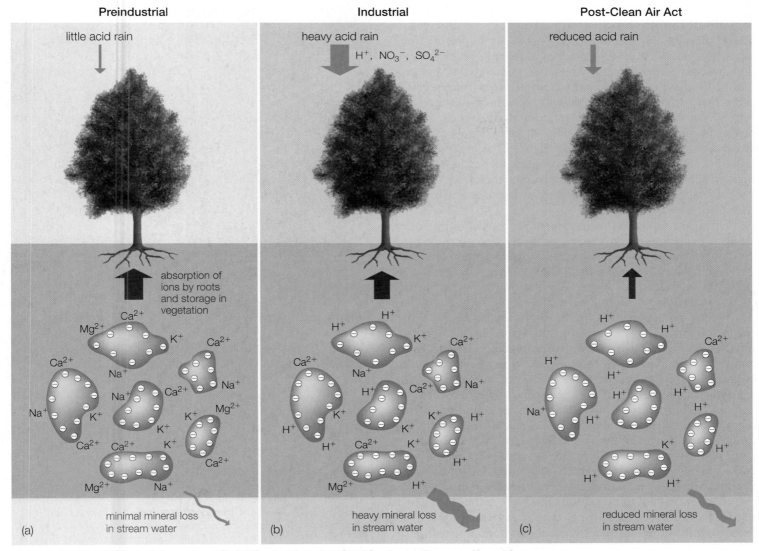

FIGURE 27.22 Soil particles, mineral nutrient cations, and acid rain (a) The preindustrial forest ecosystem received minimal inputs of acids from precipitation. Stream losses of nutrients were also minimal, because nutrients were tightly recycled. (b) The industrial-era forest received large inputs of acids such as sulfuric and nitric acids in precipitation. Mineral ions such as calcium and magnesium were displaced from soil particles and lost in streams. (c) Following the mandated controls of the Clean Air Act (1970), inputs of acids have declined but the soil is depleted of mineral nutrients after more than three decades of acid rain. Forests grow slowly, if at all, due to the lack of soil nutrients.

at higher elevations. Serious air pollution problems must still be solved before eastern ecosystems can begin recovery.

27.7 | Tropical rain forests have a nonseasonal climate and abundant precipitation

The climate of the tropical rain forest ecosystem is essentially nonseasonal. The average annual temperature of tropical rain forests ranges from 22° to 27°C (72°–81°F). Average monthly temperatures vary by only about 2°–4°C from these values. In fact, the temperature variation between night and

day in a tropical rain forest is greater than the average variation over an entire year. Annual precipitation is high, at least 170 cm (70 in.), but often much higher. One town in Nicaragua, for example, has an annual precipitation of more than 659 cm (259 in.).

Tropical rain forests have high biotic diversity and important global climate effects

Tropical rain forests are famous around the world for their high species diversity. There are several hypotheses why these forests have so much more biological diversity—**biodiversity**—than most other ecosystems. Tropical rain forests also play important roles in global climate dynamics.

Great age, rapid evolution, and complex structure may foster high biodiversity Tropical rain forests are among Earth's oldest ecosystems. Fossil leaves, pollen, fruit, seeds, and wood bits reveal that tropical forests similar to those present today have existed for at least 60 million years, and perhaps 200 million years. Ancient tropical forests were rich in some plants that are still present today—warmth-loving tree ferns (Figure 27.23a) and other pteridophytes, as well as various gymnosperms like today's giant *Agathis* (Figure 27.23b), *Gnetum* (a vine with broad leaves; Figure 27.23c), and cycads (Figure 27.23d). The great age of tropical forests has allowed them to accumulate many species, including these very old lineages as well as orchids and other plant groups of comparatively recent origin.

The year-round even temperatures of tropical forests offer another explanation for species richness. Reproduction can occur year-round, which means that natural selection and evolution also occur rapidly, yielding more species than in seasonal climates. When more species are present, the formation of coevolutionary partnerships, such as pollination and dispersal associations (Chapter 24), is more frequent. Flower and fruit diversification provides the opportunity for more pollinator and disperser animal species to exist, and vice versa. Storms, which can topple trees, leaving small to large gaps, provide an intermediate level of disturbance—just sufficient to open up space for new species to gain a foothold without being so extreme as to destroy basic conditions of the ecosystem. Layering of the forest (Figure 27.24), with tall canopy trees, several layers of shorter trees or herbs, and many types of plants that grow on them, is yet another reason why tropical rain forests are so diverse. The layering, known as **stratification**, provides a wide range of temperature, humidity, light, and nutrient conditions and thus many more niches (ways that organisms can make their living) than in less complex ecosystems. Stratification is an adaptive response of the tropical forest ecosystem to competition among plants for light, which arises when plant height is not constrained by water availability.

EVOLUTION

FIGURE 27.23 Ancient plants of tropical rain forests (a) Tree ferns. (b) *Agathis*. This coniferous tree is still abundant in New Guinea, New Zealand, and the forests of Queensland, in Australia. (c) *Gnetum*, a gymnosperm with unusual broad leaves. (d) Cycads, a group of gymnosperms that occur naturally only in the tropics and subtropics.

FIGURE 27.24 Tropical rain forest vegetation occurs as several layers Some trees emerge from the canopy, which is composed of typically fan-shaped trees. Shorter, understory trees, often with oval-shaped crowns, and giant herbs grow beneath the canopy, as does a relatively sparse ground cover.

Tropical rain forests store much of the Earth's carbon

Because woody plants are so abundant in tropical rain forests—as well as long-lived—their wood represents one of the Earth's largest, long-term carbon storage sites, which are termed **carbon sinks.** A tree is about 50% carbon. An area of 7,000 ha (hectares; 1 ha = about 2.5 acres) of tropical forest stores the organic equivalent of more than 3.8 million tons of carbon dioxide, approximately equal to yearly emissions from 750,000 automobiles. It has been estimated that if the Amazonian rain forest alone were cleared during the next 50 years, 50 billion tons (50 Gt)—equivalent to about 7% of the current amount of atmospheric CO_2—would be released. The addition of this much carbon dioxide to Earth's atmosphere could have a serious effect on climate. The tropics are also important in global carbon cycles because they are major sources of methane and carbon dioxide (produced both by bacteria and animals), which are greenhouse gases.

Tropical rain forests play an important role in global water cycling

Charles Darwin described an "extraordinary evaporation," in the form of smokelike vapor rising from tropical forest trees (Figure 27.25). This vapor reveals the important role of tropical plant transpiration (Chapter 11) in transporting water into the atmosphere, where it can form rain. Currently, one-half to three-quarters of Amazonian precipitation consists of water transpired by its rain forest plants; the rest arises from evaporation of ocean and other surface waters. Other world rain forests have similarly important effects on water cycling, with wider effects on climate worldwide. About half of the solar radiation received by tropical forests is used for transpiration. When tropical rain forest is cut down and the leaf surfaces for transpiration are removed, temperatures increase while humidity, rainfall, and cloudiness decrease dramatically. The rain forest creates the moist conditions needed for its own survival and helps moderate Earth's climate.

FIGURE 27.25 **Visible water vapor over a tropical rain forest**

Tropical forest vegetation is distinctive

The distinctive vegetation is one reason why Charles Darwin was so awed by his first view of a tropical forest, as are most visitors. In what ways do tropical forests differ from temperate and boreal forests?

Tropical forests are tall, evergreen, and layered Tropical rain forests are characterized by very tall, slender trees. The trees are at least 30 m tall (98 ft), average 45–55 m (180 ft) in height, and can be as much as 80 m tall (262 ft). Only California redwoods (*Sequoia*) and Australian gum trees (*Eucalyptus*) are taller. Tall rain forest trees are almost always angiosperms, and they are usually evergreen, meaning that they do not drop all their leaves at the same time. This pattern contrasts with boreal forests, which are dominated by gymnosperms, and with temperate forests, whose flowering trees lose their leaves in winter (i.e., are deciduous). Tropical rain forest trees occur in several layers: very tall, moderately tall, and shorter trees, whereas fewer canopy layers occur in temperate and boreal forests, where trees are usually about the same height.

Tropical forests are richer in tree species than other forests Another major way in which tropical rain forest trees differ from those of temperate and boreal forests is that rarely do one or a few species dominate. This is because tropical rain forests are so rich in tree and other species. It is not unusual for botanists to find 100–300 tree species per hectare in a tropical rain forest. In contrast, there are only about 30 species of trees in the richest forests in North America—those of the Great Smoky Mountains National Park in the southern Appalachians of the United States.

Tropical forests contain plant forms that are rare elsewhere In addition to trees, tropical (but not temperate) forests contain **giant herbs**—plants that are as tall as some trees (Figure 27.26) but lack wood produced by a vascular cambium (Chapter 9). Giant herbs include 15-m-tall tree ferns, as well as palms and bamboos that can be more than 18 m (60 ft) tall. Tree branches and trunks often bear masses of **epiphytes** (see Figure 27.26), which are plants that grow on the surfaces of other plants and possess a number of unusual features that aid in this type of existence. Epiphytes include thousands of species of bryophytes and ferns, as well as orchids, vase-shaped bromeliads, and other flowering plants (Figure 27.27a). The roots of epiphytic orchids (Figure 27.27b) are often photosynthetic and have a velamen—a thick outer layer of nonliving cells with large holes in the walls that allow water to enter. These water-storing cells help protect internal tissues from overheating and drying. Orchids are extremely diverse in tropical forests, much more so than in other ecosystems (Essay 27.2, "Fascinating Orchids"). Like the lightweight spores of ferns and bryophytes that can be dispersed by wind to new treetop habitats, the very tiny seeds of orchids are likewise carried by wind. Bromeliads, whose leaf bases overlap to form water tanks in which a high diversity of animals live, produce bright red, sticky fruits that are distributed to tree branches by perching birds. Dramatic woody vines or lianas (Figure 27.28), which get their start on tree limbs or

FIGURE 27.26 Royal palm, a giant herb Epiphytic orchids are growing on the lower portion of the trunk.

trunks and later root themselves in the soil, are found festooning tropical rain forests far more often than in temperate forests.

Because of the lush, tall vegetation, which intercepts most of the sunlight, a tropical forest floor is a very shady place, except for a few small spots of sunlight (sun flecks) and the occasional bright gap left where a giant tree has fallen. Consequently, there are not as many small forest-floor herbaceous plants as in temperate forests. Mosses and ferns (which are common on the floor of temperate forests) are rare on the ground of lowland tropical forests, except along the well-lit edges of streams or rivers.

Tropical rain forests are among the most productive ecosystems on Earth The productivity of a tropical rain forest ranges from 1,000 to 3,500 grams per square meter per year, whereas that of temperate forests is 600–2,500 grams per square meter per year. The higher production is mostly due to a higher rate of leaf photosynthesis. Wood production rates of temperate and tropical rain forests are about the same. This high productivity generates enormous amounts of plant biomass—60% of total global biomass occurs in tropical rain forests. What climate conditions favor such high primary productivity?

Warm, moist tropical climates favor lush plant growth

The equatorial position of tropical forests—roughly between 5° N and 5° S latitudes (see Figure 25.22)—provides them with a high level of solar energy throughout the year. The mean

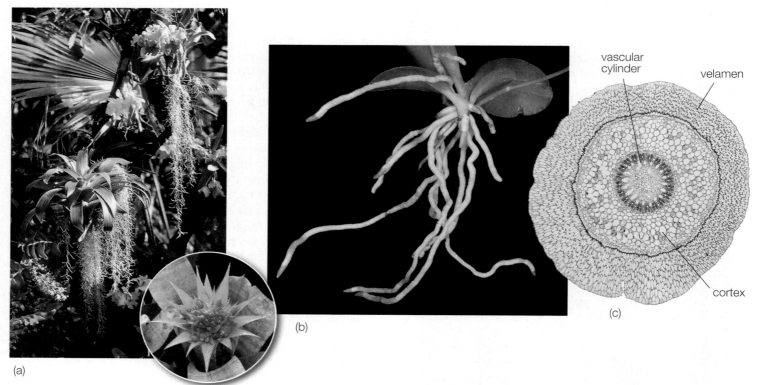

FIGURE 27.27 Tropical rain forest epiphytes (a) Epiphytic orchids and bromeliads clustered on a branch. The inset shows a bromeliad with flowers that will produce juicy red fruits attractive to birds, which disperse the sticky seeds to new branches. (b) The large, smooth, white roots often encountered on orchids are exposed to the air. Note the green tips of these photosynthetic roots. (c) Such a root, seen in cross section. An outer layer of cells (the velamen) protects the inner tissue from sun damage and helps it retain moisture.

Orchids have some of the most beautiful and desirable flowers on Earth. But orchid flowers are strangely different from those of related monocots, such as lilies. These differences reflect evolutionary adaptation of orchid flower parts for more effective pollination relationships with animals, commonly insects.

Orchid flower parts occur in threes, as in other monocots, but some of the parts of orchid flowers are difficult to recognize (Figure E27.2A). Though

The orchid Anagraecum sesquipedale, *native to Madagascar, is famous for its white, waxy flowers that are about half a meter across.*

orchid flowers have three sepals and three petals, one petal is usually larger than the others and is known as the lip (or labellum). The lip is really the uppermost petal, but most orchid flowers grow "upside down." In this position, the lip is at the bottom, where it can serve as a landing platform for pollinating insects. Most of the six stamens likely present in the ancestors of orchids have disappeared or have been highly modified during evolution. The remaining stamens have become fused with the three stigmas

to form a central column that projects above the lip. Usually, only one stamen produces pollen and only two stigmas receive pollen. The pollen is produced in masses known as pollinia, which enable pollinators to transport a great deal of pollen at a time. Orchid flowers often occur in groups (inflorescences), which is also an adaptation that increases pollination effectiveness. Orchid flowers—truly products of nature's laboratory of evolutionary experimentation—continue to hold the fascination of plant scientists and the general public alike.

The orchid *Anagraecum sesquipedale*, native to Madagascar, is famous for its white, waxy flowers that are about half a meter across. These orchids have an amazing 30-cm-long (nearly a foot) green tube (spur) that holds nectar. When Charles Darwin saw this orchid, he proposed that there must be a moth pollinator with a matching long tongue; but because such a moth was then unknown, people ridiculed his idea. But Darwin was right. At least two Madagascar hawkmoths are now known whose tongues are long enough to obtain nectar from the orchid. Darwin was also interested in *Coryanthes*, known as the bucket orchid because part of its lip forms a bucket containing liquid secreted by the flower (Figure E27.2B). Bees, attracted by the flower's scent, slip into the bucket and get their wings wet, so they are unable to fly away immediately. Eventually bees discover that the only way out is through a narrow opening that forces them to first pass the stigma and then the anther. The bees leave pollinia picked up from previously visited bucket orchids on the stigma and, as they leave, more pollinia become stuck to the bee's abdomen. You would think that the bees would learn to avoid the bucket orchid. But, bewitched by the fragrance of another *Coryanthes*, bees soon forget their traumatic dunking and repeat the experience.

column

labellum   pollinia   stigma

E27.2A Orchid flower stucture An undissected flower is shown on the left, where the elaborate labellum (lip) and column are labeled. The labellum was removed on the right-hand image, and the column is viewed from below. A cap covers the pollinia, which are exposed in the inset after removing the cap. The stigma is cup-shaped in this species.

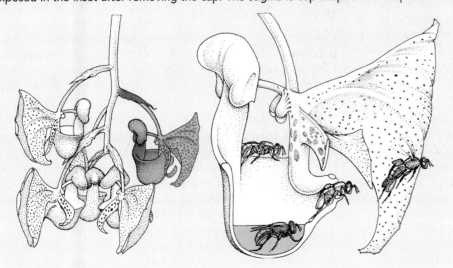

E27.2B The bucket orchid, *Coryanthes*

FIGURE 27.28 Lianas are woody vines that use trees as supports

temperature is at least 18°C (about 64°F) throughout the year (often much higher), so plants are not subjected to extended cold periods and are thus physiologically active year-round. However, there are some downsides for plants continually exposed to high temperatures. Nighttime respiration rates are high, which consumes carbon fixed during the day and thus lowers net productivity. In contrast, cooler nighttime temperatures of boreal and temperate forests help plants conserve fixed carbon. High tropical temperatures also favor reproduction of insect pests and disease microbes, which attack tropical plants throughout the year—not just seasonally, as in boreal and temperate forests.

Tropical plants compete for light Tropical day length changes little throughout the year, so tropical plants do not have to contend with light limitations posed by short days, as do those of boreal and temperate forests. However, tropical forests also do not experience the long days of higher-latitude summers. In addition, the many layers of canopy tree leaves filter out so much sunlight energy that forest-floor plants are quite shaded. Competition for sunlight has selected for evolution of tall trees, giant herbs, epiphytes, and lianas—each of these groups exhibit different ways plants compete for light.

Some tropical plants are adapted to drought High tropical rainfall results from the behavior of atmospheric air circulation belts, which are influenced by ocean currents and land-sea distribution patterns (Chapter 25). The annual average rainfall in the tropics is at least 1,700 mm (67 in.), usually above 2,000 mm (79 in.) per year, and often much more. The humidity is high through the year in part because warm air can hold more water than the cooler air of temperate and boreal forests. Anyone who has visited the tropics has experienced the fact that daytime humidity is often 60–80% and nighttime humidity reaches 95–100%. Tropical weather is often cloudy, which diffuses direct sunlight, helping prevent drying. In such conditions, water availability often is not limiting to photosynthesis of tropical forest plants. Low water stress means that the plants' stomata can remain open for absorption of carbon dioxide for longer periods than is possible for temperate and boreal forest plants.

However, water stress can occur in tropical forests. Some tropical forests are known as tropical seasonal forests (see Figure 25.22) because they experience a drier season, when plant photosynthesis can be constrained by water availability. Even in wet periods, the tallest canopy trees can suffer water stress on sunny days simply because they are more exposed and thus have high water needs. Because soil water in the forest floor is not available to treetop epiphytes, these plants are also likely to experience periods of drought.

Heavy rainfall and high winds can damage tropical plants Tropical storms (including hurricanes, monsoons, and cyclones) can topple tall rain forest trees and damage the epiphytes and lianas that grow on them. Lashing rainstorms can also shred large leaves and leave pools of rainwater standing on leaves, thus interfering with absorption of carbon dioxide and fostering growth of disease microbes. These selective pressures have influenced the structural evolution of rain forest plants, a topic addressed in more detail later.

Paradoxically, lush rain forests grow on poor soils

On viewing lush plant growth, most people would assume that the soil beneath it is rich in nutrients and thus suitable for temperate-style farming. In the case of grasslands, this assumption would be valid, but it is not valid in the case of tropical rain forests.

Tropical forest soils are low in nutrients and organic materials Although rain forest soils are highly variable, they have in common the fact that abundant rainfall washes mineral nutrients out of them and into streams. This process is known as **leaching.** Because of rain leaching, most tropical rain forest soils have low to very low mineral nutrient content, in dramatic contrast to mineral-rich grassland soils. Tropical forest soils also often contain particular types of clays that, unlike the mineral-binding clays of temperate forest soils, have low electrostatic charge and thus do not bind mineral ions well.

Aluminum is the dominant cation (positively charged ion) present in tropical soils; but plants do not require this element, and it is moderately toxic to a wide range of plants. Aluminum also reduces the availability of phosphorus, an element in high demand by plants.

High moisture and temperatures speed the growth of soil microbes that decompose organic compounds, so tropical soils typically contain far lower amounts of organic materials (humus) than do other forest or grassland soils. Because organic compounds help loosen compact clay soils, hold water, and bind mineral nutrients, the relative lack of organic materials in tropical soils is deleterious to plants. Plant roots cannot penetrate far into hard clay soils and, during dry periods, the soil cannot hold enough water to supply plant needs. Because the concentration of dark-colored organic materials is low in tropical soils, they are often colored red or yellow by the presence of iron, aluminum, and manganese oxides; when dry, these soils become rock-hard. The famous Cambodian temples of Angkor Wat, which have survived for many centuries, were constructed from blocks of such hard rain forest soils.

Tropical forest mineral nutrients are held within tissues of living organisms

Given such poor soils, how can lush tropical forests exist? The answer is that the forest's minerals are held in its living biomass—the trees and other plants and the animals. In contrast to grasslands, where a large proportion of plant biomass is produced underground, that of tropical forests is nearly all aboveground. Dead leaves, branches, and other plant parts, as well as the wastes and bodies of rain forest animals, barely reach the forest floor before they are rapidly decayed by abundant decomposers—bacterial and fungal. Minerals released by decay are quickly absorbed by multitudinous, shallow, fine tree feeder roots and stored in plant tissues. Many tropical rain forest plants (like those in boreal and temperate forests) have mycorrhizal (fungus-root) partners whose delicate hyphae spread through great volumes of soil, from which they release and absorb minerals and ferry them back to the host plant in exchange for needed organic compounds. The fungal hyphae are able to absorb phosphorus that plant roots could not themselves obtain from the very dilute soil solutions, and fungal hyphae can transfer mineral nutrients from one forest plant to another. Consequently, tropical rain forests typically have what are known as **closed nutrient systems,** in which minerals are handed off from one organism to another with little leaking through to the soil (Figure 27.29). When mineral nutrients do not spend much time in the soil, they cannot be leached into streams. Closed nutrient systems have evolved in response to the leaching effects of heavy tropical rainfall. Evidence for this conclusion is that nutrient systems are more open in the richest tropical soils and tightest in the poorest soils.

Nitrogen-fixing bacteria and mycorrhizal fungi help tropical rain forest plants cope with poor soils

Many species of tropical rain forest trees belong to the legume

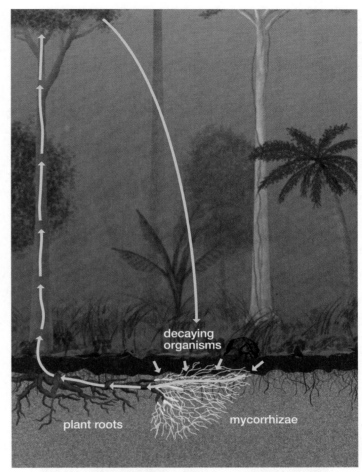

FIGURE 27.29 **Closed mineral cycling in tropical forests** A tropical rain forest's mineral nutrients mostly occur in the living organisms. Minerals (yellow arrows) are harvested from dead organisms by soil microbes, such as mycorrhizal fungi, that rapidly transfer them to plants, bypassing the soil. Closed mineral cycling reflects adaptation to the nutrient-depleted tropical soils, which result from mineral ions being washed away by heavy rainfall.

family, which is known for associations of nitrogen-fixing bacteria within root nodules. Also, cycads produce special aboveground roots that harbor nitrogen-fixing cyanobacteria (Chapter 22). By growing above the ground, the roots are exposed to sunlight, which the cyanobacteria require for growth. *Gunnera*, a flowering plant that grows on nutrient-poor volcanic soils in Hawaii, has nitrogen-providing cyanobacteria embedded in the petioles of its giant leaves (Figure 27.30). Nitrogen fixation by free-living bacteria in tropical soils is also beneficial. Thanks to nitrogen fixation, mycorrhizal partnerships, and closed nutrient cycling, lush plant growths can occur on poor tropical soils. But producing crops on such soils can be difficult.

Early tropical forest farmers learned to cope with poor soils

Early farmers in the humid tropics around the world learned that they could not produce abundant crops on tropical forest soils unless they first cut down most of the

FIGURE 27.30 Association of *Gunnera* and cyanobacteria (a) Giant leaves, which have cyanobacteria in the petioles. (b) Cyanobacterial colonies in *Gunnera* tissue.

FIGURE 27.31 Slash-and-burn agriculture In this traditional agricultural method, a patch of tropical forest in Panama is prepared for planting by cutting and then burning the vegetation. Minerals that were contained in plant tissues are thereby released to support crop growth. But fertility lasts only a few years, after which farmers must repeat the process in a new patch of forest.

trees in a small area (1–15 ha; 2.5–37.5 acres), left the fallen timber in place, and then burned it (Figure 27.31). Most of the nitrogen and some of the phosphorus in plant tissues is lost during burning, but most of the phosphorus and mineral cations needed by plants (such as calcium, magnesium, and iron) remain in the ashes. Some tropical crops grown by indigenous people included maize *(Zea),* cassava *(Manihot),* yams *(Dioscorea),* taro *(Colocasia),* bananas *(Musa),* and sweet potato *(Ipomoea).* Multiple crops were often grown together, a technique known as **polyculture.** To some degree, polyculture mimics the diversity of the natural forest, making it difficult for pests and disease microbes to find and destroy all of the crops. After a period of 2 or 3 years, the crops would have used up all the mineral nutrients in the ashes or weeds would have become too numerous to manage. Because decades are required for a tropical forest to regenerate, the people would have to move to an uncultivated part of the forest and repeat the process. Practiced in tropical areas around the world, this type of farming is known as **slash and burn,** swidden, shifting, or ladang agriculture. It represents cultural adaptation to the poor mineral content of tropical soils. In Belize, Amerindian farmers use the slash-and-burn technique to grow crops for 2 to 3 years and then allow the land to recover for years before recultivating it. The slash-and-burn technique was a masterful solution to the problem of crop production in the tropics, and it can support a population of about seven people per square kilometer, but no more than that. Slash-and-burn techniques generally cannot support the needs of higher modern populations, and wildfires that burn more forest than intended are often started by tropical farmers.

Temperate-style agriculture is often difficult to practice in the tropics

Because human populations have grown precipitously in tropical regions in recent decades, and because there is a worldwide demand for tropical crop products, attempts have been made to import permanent (nonshifting) temperate-style agriculture. Sedentary irrigated rice production in tropical lowlands can support hundreds of people per square kilometer, explaining the widespread importance of rice crops in tropical regions (Figure 27.32). Rice cultivation is an example of **monoculture**—the production of single crops in large areas—as is characteristic of temperate-style agriculture. Think of the vast corn and wheat fields of the U.S. Midwest growing on rich grassland soils, in areas where winter controls pest populations. But monoculture often does not work as well in the tropics. One reason is that tropical soils are typically so poor that farmers must add large amounts of expensive fertilizer. A second reason is that the continuously

FIGURE 27.32   Rice field in tropical region

high populations of competing weeds, herbivorous insects, and disease pathogens can readily wipe out large areas of crops unless large amounts of expensive herbicides, insecticides, and fungicides are applied frequently. In addition, nighttime temperatures are high enough to reduce crop productivity, because the resulting high respiration rates consume large amounts of the organic carbon produced by plant photosynthesis.

Tropical forest that has been cleared for cattle pasture often becomes infested with weedy plants that have low palatability, digestibility, and nutritional value, leading to abandonment. Cash crop plantations such as rubber, oil palm, tea, coffee, cacao, cashew, coconut, timber pine or eucalyptus, sugarcane, bananas, or pineapples often fail because they are so vulnerable to soil erosion, pests, weeds, and diseases or because pollinators are absent. Polyculture often works better as an agricultural strategy for tropical regions because it more closely mimics the naturally diverse, complex structure of tropical forest, which is more resilient to stresses such as nutrient deficiency, pests, and diseases. Examples of polyculture include growing coconut palms together with chocolate *(Theobroma cacao)*, or growing pepper and other spice plants within mahogany plantations. Polyculture allows people who live in tropical regions to use tropical ecosystems in a sustainable way.

Tropical forest plants are adapted to numerous environmental stresses

Tropical rain forest plants provide many fascinating examples of evolutionary adaptation to the selective pressures of their habitats—shady conditions, high or low moisture, storm damage, and high pest populations.

Canopy trees are well adapted to intercept light, but are vulnerable to fire and forest fragmentation

One way for plants to compete for light is to invest in wood production and become a tall tree. Great age (600 or more years) allows tropical trees to achieve the considerable heights necessary to intercept unfiltered sunlight. The most common causes of tree death are storms, lightning, landslides, and fire. Moisture levels in the forest are usually so high that fires are quickly extinguished. Under drought conditions, however, larger areas of trees may burn because the bark of tropical trees is not usually thick enough to be fire retardant.

The tall trees are also particularly vulnerable to fragmentation of the forest into small areas or "islands," as the result of human activities (Essay 27.3, "Restoring a Lost Forest"). One study of a 1,000-km^2 forest in Brazil showed that three times as many trees died at the edges of forest fragments, compared to the interior of forest "islands," and a higher proportion of the dead trees were of the largest size class. There are several reasons why the largest trees are so vulnerable. First, their tall, inflexible trunks make them prone to uprooting by wind, which is stronger at the edges of forest fragments. In addition, lianas (woody vines) occur more commonly near fragment gaps and can weaken the large trees on which they grow. Finally, leaves at the crowns of tall trees are more susceptible to drought when trees are more exposed at "island" edges, because there are more air currents to dry them out. Because centuries are required to regrow such forest giants, forest fragmentation is a stress from which some tropical trees may not be able to recover.

Tropical tree stem architecture is amazingly diverse

Botanists have defined 21 different tropical tree branching types, each implying a different strategy for coping with competition for light, heat load, and water stress, which differ among the canopy levels. Tropical trees often do not branch until very near the top. A few very tall trees emerge from the canopy. The next-tallest trees, which form the top canopy layer, often have crowns shaped like umbrellas, while shorter understory trees have narrower, tapering crowns (see Figure 27.24). The bark of tropical trees is commonly lighter in color than the bark of temperate trees, and it is usually very smooth (Figure 27.33), which may be an adaptation that helps reduce attachment of heavy climbing vines and epiphytes.

Buttress roots provide structural support Tropical trees typically have an extensive system of shallow feeder roots but do not possess taproots, which characterize trees of temperate forests. This is largely because tropical soils are thin or very hard to penetrate. Trees that lack taproots often have dramatic buttress roots (Figure 27.34)—flat, triangular plates of wood that begin growth at the base of trunks and develop by secondary growth of a lateral root, together with activity of the stem cambium. Buttress roots, produced more commonly by the tallest trees, are typical of trees belonging to the legume family but occur in many taxonomic groups as a parallel

ESSAY 27.3 RESTORING A LOST FOREST

Brazil's Atlantic Forest was once a wonder of the world, but only 7% of it remains, having been extensively cut by people since its discovery by the Portuguese in 1500. Even so, the forest has an exceptional concentration of rare species, such as the black-faced lion tamarin (*Leontopithecus caissara*), a primate that was first discovered in 1990 and of which only some 260 individuals remain. Large U.S. corporations are partnering with tropical ecologists and the Nature Conservancy to try to restore the Atlantic Forest for both biodiversity conservation and carbon storage. But restoration of this or other lands formerly occupied by tropical forests will not be easy.

When large tropical crop fields, pastures, or plantations are abandoned, they have sometimes become so degraded that reestablishment of the forest does not begin; rather, grassland, scrubby savanna, or fern-dominated bush develops. This vegetation is vulnerable to fire arising from lightning or human activities. Fire causes further soil nutrient loss

Large U.S. corporations are partnering with tropical ecologists and the Nature Conservancy to try to restore the Atlantic Forest for both biodiversity conservation and carbon storage.

and prevents trees from colonizing the area. Tropical forest areas that have been cleared must be protected from fire before regeneration can occur.

If a tropical forest does begin to regenerate on disturbed land, it does not resemble the original, primary forest. Instead, it follows the succession pattern that occurs in natural forest gaps or larger areas destroyed by storms. There are very few large trees in secondary tropical forest. The pioneer trees are no taller than about 30 m, and they cannot store as much carbon as primary forest trees. There are fewer layers and thus fewer ecological niches than in primary forest. The shorter vegetation is often very dense, tangled, and hard to penetrate, in contrast to the relatively open floor of a primary forest. Invasive trees flourish, such as *Cecropia* (Figure E27.3), which has soft, hollow wood that is not as useful as the hardwoods of the primary forest. Shrubs such as *Psidium* (guava) are also invasive species that can reproduce explosively in disturbed tropical forests, slowing forest regeneration. Finding

ways to control invasive plant species is important to forest restoration. Restoration ecologists know that it will take decades to restore the Brazilian Atlantic forest and that they must perform experiments to determine the best methods for planting and the best plants to start with. They also know their goal is a supremely worthy one.

E27.3 *Cecropia*, an invasive species of disturbed tropical forests

adaptation to the need for structural support. They help keep tall forest giants from tipping over in storms.

Leaves of tropical trees are surprisingly uniform in shape The leaves of thousands of species of tropical trees are amazingly uniform in size and shape, likely reflecting similar adaptation to the same ecological problems. Most tropical tree leaves are oval, simple (not divided into leaflets or compound), leathery, and with entire (not toothed) edges (Figure 27.35). In the Amazon River region, 90% of the trees and 87% of the shrubs have such leaves, compared to about 10% of trees and shrubs in broadleaf forests of the United States. The reason tropical tree leaves tend to be so uniform is unknown. Tropical plant leaves often live 18 months or longer—more than 3 years in some cases.

Another common feature of tropical tree leaves is a long, drooping tip (see Figure 27.35), commonly known as a "drip

tip" because one hypothesis for its function is that it drains rainwater. Drip tips are sometimes as long as 4 cm and are rare outside tropical forests. Another hypothesis explaining long leaf tips is that the tropical leaf apex grows rapidly before the rest of the leaf has fully expanded as a result of hot, humid growth conditions. This idea is supported by the results of an experiment on coffee plants. When grown in hot, humid conditions, the coffee plant produces leaves with extended tips; but tips are shorter when the plants are grown in cooler, drier conditions.

Tropical tree leaves are typically rich in tannins, alkaloids (Chapter 3), and other bitter-tasting substances that help reduce herbivory. Scientific expeditions into tropical rain forests must carry all needed food because, despite the lush plant growth, the only edible plant materials in the forest are likely to be fruits located so high up in the canopy that they are hard to obtain.

FIGURE 27.33 Tropical trees often have smooth bark Smooth bark such as found on this guava plant may be an adaptation that helps prevent the attachment of lianas.

FIGURE 27.35 Leaves of tropical trees Tropical tree leaves are typically oval with entire (nontoothed) edges and drip tips (arrows). Young leaves (inset) are often red or other colors due to the presence of protective pigments.

FIGURE 27.34 A tropical rain forest tree with buttress roots This example is a fig tree in Queensland, Australia.

Young leaves of tropical trees often appear limp and are colored red—or they may be purple (see Figure 27.35), blue, white, or pale green. The explanation for limpness is that lignin deposition has not been completed in young leaves. The explanation for unusual colors is that chlorophyll synthesis has just started by the time the leaves have just fully expanded; further, the cell vacuoles of young leaves are often filled with red or purple (anthocyanin) pigment. The pigment is thought to protect developing leaves from strong light and harmful ultraviolet radiation, which is particularly intense in the canopy at the equator. As the leaves mature, protective cuticle and internal air spaces develop, increasing the leaves' ability to reflect ultraviolet radiation, so red pigment production decreases.

Some experts think tropical forest primates that preferentially feed on tender young red canopy tree leaves have evolved trichromatic color vision as an adaptation that enables these animals to more readily locate their food. Trichromatic color vision is the ability to distinguish red from green (and blue), a trait that evolved by modification of a vision pigment gene in an ancestor of modern "Old World" primates. This trait also helps treetop primates find ripe fruits, which are often colored red or orange. Humans almost certainly inherited their trichromatic color vision from primate ancestors.

EVOLUTION

Tree reproduction is adapted to forest conditions

Most tropical rain forest trees reproduce primarily by means of seeds; asexual, vegetative reproduction is not very common. Wind pollination is also uncommon because there is little air movement within the forest, individuals of the same species are widely scattered, and insects, birds, bats, and other animals that serve as pollinators are available year-round.

The tallest canopy trees often have large, bright flowers that are visible to birds and large bees at long distances, whereas shorter trees generally have smaller, less conspicuous flowers that are not bird-pollinated. Shorter trees—chocolate (*Theobroma cacao*), for example—often produce their flowers and fruits on their trunks (Figure 27.36). This phenomenon, known as *cauliflory*, facilitates access by shade-loving pollinators and dispersers and those that cannot easily fly within a dense canopy.

FIGURE 27.36 Cacao (Theobroma) flowers and fruits (a) Flowers of the chocolate plant are cauliflorous (borne on the stem) and are pollinated by insects—thrips, aphids, midges, and ants. (b) Several fruits.

Flowering and seed production is usually seasonal—once to three times a year for the tallest canopy trees—but is almost continual lower in the canopy. Some of the tallest trees have wind-dispersed fruits or seeds, but the shorter trees rely on animal dispersers. Fleshy berries or drupes are readily eaten by birds, bats, rodents, monkeys, and gorillas, which serve as major seed dispersers. Parrots and hornbills are specialized for eating hard fruits such as nuts. Tree fruits tend to be good food sources for animals because they are large enough to contain hundreds of tiny seeds—in the case of figs, for example—or one to a few large, hard seeds, such as those of the Brazil nut *(Bertholletia excelsa)*. The tiny seeds of figs and hard seeds pass intact through the herbivore's gut. Tree seeds tend to be large because they must germinate on the dimly lit forest floor and thus are packed with ample storage carbohydrates, lipids, and proteins that fuel early seedling growth. Even so, few tree seeds actually grow into mature trees. Why is this the case?

Insects, small mammals, or fungi destroy most tropical forest tree seeds. Experts cite such seed destruction as a major cause of the widely scattered distribution of tropical tree species. Seed predators (and fungi) are able to find and destroy seeds that occur in high populations near parent trees. Only seeds that have been carried far from the parent are spared, because they are harder for predators and decomposers to find. In response, some trees produce huge numbers of seeds at long, irregular intervals—a process known as *mast fruiting*. This adaptation satiates the seed predators so that some seeds are left alone. Other trees load their seeds with protective, toxic secondary compounds.

Most tropical tree seeds cannot remain alive for long periods in the soil. Seeds of the chocolate plant *(Theobroma)*, for example, can live for only 3 months, and then only if kept moist. Seeds of certain pioneer tree species that colonize gaps and disturbed ground can survive as long as 2 years. Even if seeds are able to germinate, the seedlings are vulnerable to starvation by lack of light and competition with the more extensive roots of large trees for water and minerals. Their best opportunity lies where the fall of a large tree has opened up a sunny gap.

Lianas' growth and reproduction reflect their clinging lifestyle Woody vines that climb forest trees, known as lianas, occur in temperate forests—grape and ivy are familiar examples—but are larger and much more abundant in tropical forests (see Figure 27.28). In fact, 90% of all climbing plant species occur in tropical forests. There, liana stems can be as thick as your arm or thigh, and lengths of 70 m are common, with the record being 240 m (more than 700 feet). Liana seedlings germinate on the forest floor, where they tolerate shade, sometimes having leaves that look much different from those of the mature vine. Seedlings can survive for many years until a sunny gap opens, allowing them to grow tall enough to start climbing a nearby tree.

Some ferns are lianas, as is the gymnosperm *Gnetum* (see Figure 27.23c), but most lianas are flowering plants (Figure 27.37). There are several major groups of lianas, classified based on their climbing technique. Twiners revolve the tips of their young stems around a host stem, but this works only if the support is not very thick. Root climbers attach to a host tree via aerial roots that are adapted to cling. Tendril climbers use modified leaves to curl around and attach to a tree trunk's surface. Stems of some lianas produce many sharp spines that

FIGURE 27.37 Flowering liana Red jade vine *Mucuna bennettii* in Oahu.

help them climb trees, much as rock and ice climbers use spikes and crampons.

Vessels in liana stems are the widest known in nature, and liana xylem occurs in strands that are partially or completely separated and embedded in softer parenchyma tissue. This structure generates a flexible stem with great tensile strength. In fact, liana stems can support 4–6 times as much leaf biomass as trees of equivalent stem area. Lianas are able to produce a high leaf surface area that supports many flowers and fruits with much less wood than trees. But because lianas must use forest trees for support, they have been called "structural parasites."

Many lianas have conspicuous, bright flowers (see Figure 27.37) mostly pollinated by insects, birds, or bats. Some disperse winged seeds by air; others produce fleshy fruits attractive to birds, monkeys, and bats. Most lianas are also probably capable of asexual, vegetative reproduction by means of detached stem pieces that can stay alive for months until conditions favor their growth. This is possible because the parenchyma tissue of stems contains abundant food reserves.

Lianas are most abundant at the edges of tropical forests and in disturbed areas. In Amazonia—the basin of the Amazon River in South America—hundreds of square kilometers are covered with liana forest, consisting of a few large trees with many lianas growing on them. Some experts think that these areas may have resulted from ancient slash-and-burn agriculture. Lianas can benefit forest trees by binding crowns together, making them more resistant to damage by windstorms. But lianas are also destructive. A heavy load of vines can break tree limbs, and liana seedlings compete with those of trees for scarce light and mineral resources on the forest floor. Some trees have developed defenses against lianas, including smooth or shredding bark. The West African tree *Barteria fistulosa* partners with aggressive ants that bite off the growing tips of nearby liana and tree seedlings, leaving a cleared circular area around the trunk. Local people consider this evidence that the tree has magical powers. But it is another example of plant-animal coevolution, which is so common in the amazing tropical rain forest.

HIGHLIGHTS

27.1 Moist terrestrial ecosystems cover a wide range of climates from polar to tropical. In tundra and taiga ecosystems, average temperatures and annual precipitation are low, and because evaporation is low, wet meadows, mires, bogs, and lakes are widespread. Temperate deciduous forests have a climate marked by seasonality, including warm to hot summers and cold winters, and an abundance of precipitation. Tropical forests occur where winter temperatures do not drop below freezing, annual temperatures vary by only a few degrees, and precipitation generally exceeds 200 cm per year.

27.2 Present-day polar and alpine biomes have developed in their present locations since the end of the last glacial interval.

27.3 In the arctic tundra, trees are absent except along sheltered river valleys, and shrubs and herbs dominate the vegetation. Tundra plants are adapted to survive high winds, low temperatures, and a lack of available moisture. Because animal pollinators are scarce, the plants have developed other methods of reproduction.

27.4 South of the tundra, coniferous trees dominate the taiga ecosystem. Conifer trees are adapted to short growing seasons, heavy snows, low temperatures, and limited moisture.

27.5 Tundra and taiga extend southward along mountain ranges as alpine tundra and montane coniferous forest. Arctic and alpine tundra have many species in common, but the dominant conifer species differ in the montane forests of each mountain range. These differences reflect past history and evolution on isolated mountain ranges. Humans are affecting polar and alpine biomes at the local and regional scale through mining, logging, grazing, and recreational development.

27.6 The temperate deciduous forest ecosystem has a climate characterized by marked seasonal changes, frosts in winter, absence of droughts, and moderate to long growing seasons. Temperate deciduous forest trees compete for

space and sunlight in a closed canopy. Deciduous trees drop their leaves in fall as an adaptation to winter snows. Most are wind pollinated and complete pollination before they leaf out in spring. The major human impacts on the temperate deciduous forest include widespread clearance for farming, introduction of invasive species, and in the last four decades, acid rain produced by the burning of fossil fuels.

27.7 Tropical forests are known for their amazingly high biodiversity, which results from great age, rapid evolution, and complex structure. Tropical rain forests are also important for carbon storage and their role in water cycling, both of which help stabilize Earth's climate. Tropical rain forests contain distinctive types of plants that are rare in other forests, including tall, evergreen trees in layers, lianas that use trees as supports, giant herbs, and epiphytes that grow on tree trunks, branches, and leaves. Tropical rain forest plants compete for light and are adapted to cope with mineral-poor soils and high populations of herbivorous animals and pathogenic microbes. To exist on low-nutrient soils, the forest plants form partnerships with mycorrhizal fungi or with nitrogen-fixing bacteria. Most of the mineral nutrients in a tropical forest are bound into the tissues of living organisms. Traditional farmers use the "slash-and-burn" technique to release plant tissue minerals into the soil for crop production. Such techniques can support only a limited number of people. Temperate-style agriculture (monoculture) often does not succeed in the tropics. Polyculture, a traditional farming technique in which several types of crop plants are grown together, mimics the natural tropical forest and helps prevent crop loss to pests. Tropical forest trees are adapted in many ways to cope with the challenges of their environment. Canopy trees are shaped to maximize light capture, buttress roots help keep tall trees from toppling over during storms, young tree leaves contain pigments that protect against harmful amounts of light, and older leaves are protected from herbivory by toxic or bad-tasting secondary chemical compounds. Tree flowers and fruits are often co-evolved with animal pollinators and seed-dispersal agents. Seed predation and fungal decay destroy many plant seeds, explaining the wide spacing of plants in tropical rain forests.

REVIEW QUESTIONS

1. Why is it somewhat difficult to describe the exact extent of the boreal forest?

2. Discuss some of the nonreproductive adaptations of tundra plants to life in the tundra ecosystem.

3. Discuss some reproductive adaptations of tundra plants to life in the tundra ecosystem.

4. What is the geographic distribution of alpine tundra and montane coniferous forest in North America? What are the similarities and differences of these alpine systems to their polar equivalents?

5. Discuss some of the characteristic tree species growing in the montane coniferous forests of the Appalachian, Rocky, and Cascade/Sierra Nevada Mountains. Why do the dominant species differ so drastically among these mountain ranges?

6. Describe the physical features of the eastern temperate deciduous forest in North America, including climate (seasonality and rainfall patterns), geographic locations, and soil characteristics.

7. How does precipitation vary from east to west and from north to south within the eastern temperate deciduous forest? How does the length of the growing season vary?

8. What is acid rain, and what causes it? How has acid rain damaged the eastern temperate deciduous forest, and what is the outlook for remedying this damage within the next half-century?

9. What is an epiphyte? In the tropical rain forest, ferns, epiphytic orchids, bromeliads, and lianas are especially common. Discuss these groups, including a description of at least one adaptation each group uses for its epiphytic lifestyle.

10. What are some advantages and disadvantages to plants of growing in a tropical climate?

11. Discuss some of the physical factors—soil type, winds, sunlight, water availability, and so on—encountered by plants of the tropical rain forest.

12. Because tropical rain forest soils are typically very poor in nutrients, how do tropical plants satisfy their nitrogen and phosphorus requirements?

13. Describe the agricultural practices of indigenous tropical farmers, in particular, how they are able to grow crops on such nutrient-poor soils. What are some disadvantages of this type of agriculture?

14. Discuss some adaptations of trees to life in the tropical forest.

15. Provide several reasons why tropical rain forests are so species-rich.

APPLYING CONCEPTS

1. Why is tundra largely restricted to the Northern Hemisphere?

2. Even today, the tundra and taiga regions of North America are sparsely populated at best. Nonetheless, these ecosystems have considerable impact on the daily lives of North Americans. Discuss some effects of these ecosystems on your life.

3. Boreal coniferous forests and temperate deciduous forests are similar in that both support large trees. Yet significant differences exist. Compare these two biomes with regard to (a) seasonality, (b) precipitation amounts, (c) soil characteristics, and (d) annual temperatures.

4. The genus *Alligator* contains two species. One species, the American alligator, is a common inhabitant of the wetlands in region 6 (southeastern evergreen forest association) of the eastern temperate deciduous forest (and of the southern parts of Florida). Recalling the worldwide biogeography of the temperate deciduous forest, where might you suggest the other species of *Alligator* naturally lives?

5. Eastern deciduous forest is found in eastern North America, eastern Asia, and Europe. Today the forests in North America and Asia are quite rich in species, though Europe has suffered a disproportionate loss in species diversity because of glaciations over the past few million years. Looking at a geographical map of the regions, can you suggest a reason for this loss in species?

6. Imagine that you are walking into a tropical rain forest. From the descriptions given in this chapter, describe the ambiance of this tropical world—from the canopy to the forest floor. What kinds of plants predominate? How many layers of branches form the canopy? Is the forest floor crowded, and is it sunlit?

7. List several reasons why it is proving so difficult to conduct large-scale agriculture in cleared areas of tropical rain forest.

8. Discuss decomposition of dead organic materials on the tropical forest floor. Is the rate of decomposition fast or slow? What factors favor this rate? What are the consequences of this rate to mineral recycling and to overall soil quality?

Chilkat Bay, Alaska

- Freshwater plants, algae, and microbes are well adapted to aquatic life and provide essential food, oxygen, and habitat for aquatic animals.

- The quality of freshwater systems is endangered by pollution, especially by phosphorus, which fosters nuisance growths of aquatic weeds, algae, and cyanobacteria.

- Wetlands help protect lakes, rivers, and coastal ocean waters from pollution; provide fire and flood protection to surrounding lands; influence stream and river flow; and renew groundwater sources that humans use.

- Oceans moderate the global climate and provide the rain that supports life on land; the coastal zones support some 60% of the human population.

- Littoral and sublittoral ecosystems stabilize shorelines from erosion and shelter many species of marine organisms that are commercially valuable to humans.

- The littoral and sublittoral ecosystems of the oceans are threatened by human exploitation and pollution.

In the spring of 1993, floods devastated the U.S. Mississippi and Missouri River valleys, causing the loss of 46 human lives and an estimated $10 billion in damage. The news media daily reported the travails of people living along the rivers whose homes, schools, and businesses had been destroyed. Water levels were so high that the famed St. Louis Arch—symbolizing the gateway to the western United States—was partially underwater, a dramatic testimony to the enormity of the flood.

The 1993 flood also caused extensive soil erosion, and the Mississippi washed the lost soil into the Gulf of Mexico. Mineral nutrients in the soil and floodwaters fueled growth of extensive algal "blooms" in the Gulf. As the algae died and were decomposed by ocean microbes, dissolved oxygen levels plummeted. Fish, shrimp, and other animals suffocated in a huge "dead zone" that devastated commercial Gulf coast fisheries. Since 1993, the Gulf dead zone has persisted and even expanded, reaching an area of 20,000 square kilometers in 1999. Experts have concluded that the Gulf dead zone was created and is maintained by the persistent flow of mineral nutrients from Midwest soils and rivers. They also have evidence that the magnitude of recent floods and mineral flow is related to massive destruction of wetlands by development along river watersheds. The quality of Gulf coastal waters is linked to the condition of the freshwaters and wetlands that discharge into them. Like giant sponges, the lost wetlands would have absorbed much of the floodwaters. Wetland plants and algae could have trapped more of the mineral nutrients in their cells and tissues, helping prevent ocean algal blooms.

Newsworthy floods and dead zones have made it clear that agricultural and urban practices that result in erosion and mineral nutrient pollution are tightly linked to the condition of freshwater, wetland, and ocean ecosystems. The degradation of aquatic ecosystems in turn severely affects the health of humans and their commercial and recreational fisheries. The World Health Organization recently estimated that 1 in every 20 ocean bathers in "acceptable waters" will become ill after entering the ocean just once, a rate that produces 250 million cases of gastroenteritis and upper respiratory disease per year. Consumption of sewage-contaminated shellfish causes some 2.5 million cases of infectious hepatitis each year. Thus it is important for everyone to understand basic ecological features of freshwaters, wetlands, and the oceans with their plants and associated organisms that perform such useful services. ■

28.1 | Aquatic ecosystems are essential to humans

Throughout our history, humans have lived near lakes, rivers, wetlands, and the shores of oceans. The serenity and drama of these watery places has inspired great artists and writers. Examples of freshwater works include Tchaikovsky's *Swan Lake* ballet, Mark Twain's stories set along the Mississippi River, and Thoreau's book *On Walden Pond*. Ocean examples include Hemingway's *The Old Man and the Sea* and Coleridge's *The Rhyme of the Ancient Mariner*. Around the world, bodies of water have long been revered, excited people's imaginations, and shaped the cultures that developed around them—the Mediterranean Sea, the Nile River and the African Rift Lakes, Lake Titicaca of the Andes, the Indian Ganges River, and the North American Great Lakes are but a few examples. Before the advent of modern technology, rivers, lakes, and ocean coasts were the only practical highways for travel and trade in goods and ideas. For many people, the perfect vacation is a trip to a favorite lake, river, or seashore for fishing, boating, swimming, and other activities. Lake-, river-, and ocean-front properties are expensive because people often desire to live close to the water, even though flooding or storms can be a risk.

28.2 | People and wildlife depend on freshwaters and wetlands for many services

For many people throughout the world today, including about 80% of the U.S. population, surface freshwaters are the major source of water for drinking, irrigation, and industry. Though 1.5 billion people get their drinking water from groundwater, most of this water comes from the same sources that supply surface freshwaters. About 40% of the world's food is grown on irrigated land, and nearly 20% of the world's electricity is generated by hydropower. Cost analyses suggest that wetlands provide people with more than $150,000 per acre of fire and flood protection, water purification, and other services.

Wetlands, lakes, and waterways are also essential to fish, amphibians, reptiles, mammals such as beaver, and water birds. It has been estimated that 12% of all animal species live in freshwaters or wetlands, and many other animal species depend either directly or indirectly on freshwaters. In North America alone, more than 150 bird and 200 fish species depend on wetlands. The thousands of wetlands that dot the midcontinental United States and Canada provide critical resting and feeding sites for many species of migratory birds and are essential breeding habitat for an estimated 10 million ducks.

In small streams at the headwaters of major river systems, bryophytes, algae, and bacteria rapidly take up inorganic nitrogen that would otherwise cause downstream nuisance algal blooms and dead zones. These stream organisms thus act as water purifiers, allowing only half of the nitrogen they receive to travel downstream. Wetlands also absorb mineral nutrients, helping protect lakes, streams, rivers, and coastal oceans from pollution. Wildlife species, as well as people, depend on the purification services provided by freshwaters and wetlands.

28.3 | Lake ecosystems: seasonal changes, habitats, and primary producers

Natural lakes are bowl-shaped to steep-sided depressions that have been formed by a variety of natural processes including glacial (Figure 28.1) and volcanic action and river blockage.

Dam construction along river courses results in the formation of impoundments or reservoirs, which are artificial lakes. The number of large reservoirs on Earth—estimated at 10,000—has tripled in the past four decades. Scarcely a major river remains unmodified by dams. In the United States, Lake Powell and Lake of the Ozarks are well-known water bodies that were formed by river impoundment.

Lakes contain three major types of habitats and communities

If you like to go fishing or are interested in maintaining the beautiful appearance of clean lakes, a knowledge of lake habitats and organism community types will be useful. Lake habitats include surface and subsurface waters known as the **pelagic zone**. This shallow-water zone is illuminated well enough to support floating plants, algae, and photosynthetic bacteria. A second type of lake habitat is the **littoral zone**—shallow nearshore waters, where rooted aquatic plants and attached algae often occur. The third lake habitat type is the **benthos**—deeper, often dark, waters and bottom sediments formed by particles that have settled to the lake bottom and are occupied by fungi, bacteria, and other organisms (Figure 28.2). Benthic organisms include decomposers, which play an essential role in recycling the nutrients contained in fecal pellets produced by living fish and other aquatic animals. Minerals released by decomposer action become available for reuse by

FIGURE 28.1 Glacial lakes The upper Midwest region of the United States has many lakes that were formed by glacial action or melting of glacial ice. The lakes in this aerial photo are under study as part of the North Temperate Lakes Long Term Ecological Research program, supported by the U.S. National Science Foundation.

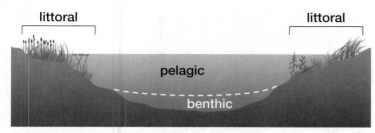

FIGURE 28.2   The different habitats in a lake Lake habitats include the open waters known as the pelagic zone, shallow near shore waters known as the littoral zone, and deep bottom waters and sediments that make up the benthic zone. Each zone is inhabited by distinctive organisms.

communities of lake primary producers. Recall that primary producers are photosynthetic organisms that produce organic food needed by primary consumers (Chapter 25).

There are several categories of primary producer communities in lakes. Photosynthetic bacteria include the cyanobacteria (also known as blue-green algae) (Chapter 18). Swimming or floating photosynthetic bacteria and heterotrophic bacteria are together known as the **bacterioplankton.** Lake primary producers also include floating or swimming algae known as **phytoplankton,** rooted or floating pteridophytes or flowering plants known as **macrophytes,** and algae and bacteria that are attached to the water plants—**periphyton.** Cyanobacteria are particularly important in plankton and periphyton communities because many of them fix nitrogen. Nitrogen is essential for the growth of aquatic plants and algae.

Pelagic primary consumers (organisms that consume primary producers) include heterotrophic protists (also known as planktonic protozoa), small crustacean animals (zooplankton), and various fish. Herbivorous insects that occur in the benthos or littoral zone and many waterfowl also consume primary producers. Secondary consumers such as fish eat the primary consumers; tertiary consumers include people and carnivorous fish such as northern pike. Freshwater plants, algae, and photosynthetic bacteria thus form the bases of aquatic food webs.

Mineral nutrient availability in temperate lakes varies with seasonal temperature change

Lake plants and algae grow most abundantly when mineral nutrients are plentiful, but the nutrient content of lake waters varies through the seasons. In temperate regions, many lakes un-

dergo an annual cycle of spring and fall circulation events, punctuated by summer and winter periods in which circulation does not occur (Figure 28.3). This cycle results from the physical properties of water—in particular, the fact that water is most dense at 4°C and less dense at temperatures above and below this value.

Spring   In spring, the temperature difference between surface and deeper waters is not great, so the density of water within the lake does not vary much. Thus, wind energy is sufficient to generate mixing, which distributes throughout the water the nutrients released from benthic decomposition. Using these minerals, lake plants and algae are able to begin rapid growth.

Summer   As the summer sun heats surface lake waters, the density difference between warmer surface and colder, deeper waters increases, so the energy required to mix water is much greater than can be accomplished by wind action. As a result, algae and plants growing in surface waters can deplete them of minerals (such as nitrate) that are necessary for further growth. Even if benthic decomposition is able to regenerate additional nitrate, it cannot be circulated to surface waters. Under these conditions, growth of plants and algae becomes limited. But cyanobacteria that prefer high temperatures and that are capable of nitrogen fixation can persist and produce blooms when phosphate is abundant (Figure 28.4).

Fall   Cooling of surface waters in fall reduces density differences to the point that wind energy can once again circulate minerals from the benthos to surface waters. This new supply of minerals enhances growth of phytoplankton species that are adapted to cool conditions.

Winter   Further cooling leads to the formation of an ice layer in winter, which prevents wind-induced circulation until the ice melts in spring. Many phytoplankton and periphyton continue to photosynthesize in the water underlying the ice, especially if excessive snow blankets do not shade them, producing oxygen needed by lake animals. If deep snow shades the ice, the algae in underlying waters may not receive enough light for photosynthesis. The consequent oxygen depletion may cause winter fish kills.

There are many variations of this cycle. Some lakes undergo only one circulation event per year, tropical lakes may circulate continuously, and permanently ice-covered lakes that occur in Antarctica do not circulate at all.

| Spring | Summer | Fall | Winter |

FIGURE 28.3   Temperate lakes often undergo a cycle of circulation events Wind-induced circulation in spring and fall helps mix mineral nutrients (represented by yellow triangles) regenerated by organisms in the benthos into the well-illuminated surface waters, thereby allowing summer plant and algal growth to occur. In winter, formation of surface ice prevents wind-induced circulation.

FIGURE 28.4 Cyanobacteria Buoyant scums of cyanobacteria are conspicuous in overfertilized lakes. Cyanobacteria such as *Microcystis* (inset) are not readily eaten by zooplankton, because the mucilaginous colonies are too large. Cells are dark because they contain numerous, tiny, light-refracting gas vesicles that help keep them afloat. *Microcystis* and some other freshwater cyanobacteria are notorious toxin producers and thus can be harmful to human health.

Freshwater algae and plants are adapted to aquatic habitats

Microbes and plants that grow in freshwaters do not have to contend with water stress, as do terrestrial plants. However, cyanobacteria, phytoplankton, and floating plants must have adaptations that prevent them from sinking because they depend on light, which is available in sufficient amounts only at or near the surface. They also must avoid being completely consumed by herbivorous animals. Rooted plants must contend with the fact that sediments are often very low in oxygen; yet oxygen is required for root respiration, which powers mineral nutrient uptake (Chapter 10).

Algae Phytoplankton algae are small, though most are large enough to distinguish with a microscope (Figure 28.5). Small size enables them to grow in well-illuminated surface waters by taking advantage of water motions that suspend and circulate them. This process exposes phytoplankton to a constant supply of dissolved mineral nutrients, which they absorb directly from the water. Many phytoplankton bear flagella that help them swim or spine-like projections that help them stay afloat. Spines, formation of large colonies of cells, and coatings of mucilage also help defend phytoplankton from some types of hungry aquatic animals.

Some algae grow attached to rocks, plants, or other surfaces in shallow waters as part of the periphyton community. In this way they obtain light for photosynthesis without the risk of sinking into deeper, darker waters where photosynthesis cannot occur. Periphyton algae may produce more biomass than the phytoplankton, especially in lakes with large shallow areas, and are important to the survival of many aquatic animals.

Floating plants In the summertime it is common for surface growths of floating ferns or flowering plants to occur in sheltered regions of lakes, wetlands, and canals. The duckweed, *Lemna*

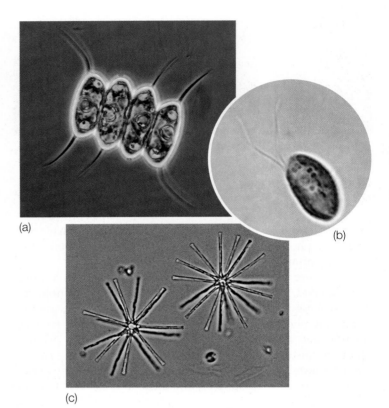

FIGURE 28.5 Phytoplankton species that are common in healthy lakes (a) *Scenedesmus*, a spiny, colonial green alga; (b) *Cryptomonas*, a cryptomonad with two flagella; and (c) *Asterionella*, a colonial diatom, are common members of the phytoplankton of healthy lakes. Each possesses structural features (spines, flagella, and a star-shaped colony, respectively) that help keep it suspended in the water.

(whose name derives from the Greek word *limnos*, meaning "lake") (Figure 28.6a), is a tiny, floating, flowering plant only a few millimeters across that is common in freshwaters worldwide. *Lemna* has only a few leaves and one root and primarily reproduces by vegetative methods. *Lemna*'s small size is an adaptation that allows it to remain afloat in an environment with plentiful sunlight, water, and dissolved mineral nutrients. Duckweed is of high nutritional value to ducks and muskrats, which may consume large quantities.

The water fern *Azolla* (also known as mosquito fern; Figure 28.6b) also forms floating masses with roots hanging down into the water. *Azolla* is valued for the presence of nitrogen-fixing cyanobacterial symbionts within cavities in its leaves and often grows in rice paddies. Some of the nitrogen fixed by the cyanobacterial symbionts is excreted into the water and, upon death of the water fern, even more fixed nitrogen is released. This nutrient fertilizes the rice crop, contributing significantly to rice production on a worldwide basis.

Rooted macrophytes Some mosses and certain large algae, such as *Chara*, form extensive "meadows" at the bottoms of lakes and streams, but they lack true roots and are attached to sediments by means of simple rhizoids. Rooted macrophytes include a few types of pteridophytes and a diverse array of flowering plants.

FIGURE 28.6 Floating plants (a) Duckweed (*Lemna*): Large populations may occur on the surfaces of quiet waters. This very small flowering plant has only a few leaves (inset). (b) *Azolla*: In warm regions, the water fern *Azolla* floats on the surfaces of water bodies, including rice paddies.

Rooted plants cannot grow deeper than about 10 m (33 ft) because at greater depths water pressure is so high that oxygen cannot be transported to roots, where it is required for mineral uptake from the lake sediments. Aquatic plants typically transport oxygen from their surface parts through air canals in tissues known as *aerenchyma* to submerged parts, including rhizomes and roots (Chapter 10).

Water lilies may be rooted at considerable depth, but their leaves and flowers extend to the surface by the growth of very long petioles (leaf stalks) and peduncles (flower stalks) from a rooted rhizome. Water lily leaves absorb not only carbon dioxide from the air but also oxygen, which is transported downward through petiole aerenchyma to the rhizome and roots at a rate of about 22 liters (L) per day. Water lily leaves (Figure 28.7a) appear shiny because they possess a thick, waxy cuticle that repels water, aiding in flotation and helping prevent microbial attack. Like a solar panel, the large leaf-surface area maximizes absorption of solar energy. The giant Amazonian water lily *(Victoria)* has leaves so large and buoyant (Figure 28.7b) that they can support the weight of a child. Some macrophytes are known as submergents because they grow almost completely under water, with flower stalks often emerging above the water surface. The arrowhead *(Sagittaria)* is referred to as an emergent macrophyte because much of the plant, including flowers and leaves, emerges above the water surface (Figure 28.7c).

FIGURE 28.7 Littoral macrophytes Macrophytes that are largely emergent from the water in which they grow include (a) the temperate water lily *Nymphaea*, (b) the giant Amazonian water lily *Victoria*, whose leaves can be more than a meter in diameter, and (c) arrowhead *(Sagittaria)*.

Human activities have degraded freshwaters

Because of the ever-increasing numbers and material needs of humans, they have modified the natural flow of streams and rivers. Worldwide, 41,000 dams modify the flow of almost 60% of rivers, leading to natural habitat fragmentation and loss of species. So much water is diverted for agricultural and other uses that the Colorado River in the western United States and the Yellow River in China no longer flow into the sea during the dry season.

Humans have also degraded freshwaters with sewage and industrial wastewater and agricultural pollutants. In developing countries, an estimated 90% of wastewater is released untreated to natural waters, and even in developed countries it often is only partially treated. Contaminated water supplies affect some 3.3 billion people. Freshwater fish are a valued source of food, accounting for 12% of all fish consumed by humans. But government agencies frequently post warnings that some lake and river fish should not be consumed, especially by children and pregnant women, because they have accumulated mercury or other dangerous substances to toxic levels.

Another way that humans have degraded freshwater quality is via acid rain—a result of air pollution (Chapter 27). Acid rain kills fish and other aquatic animals, causes population declines in fish-eating birds (such as the common loon), generates nuisance growths of acid-tolerant algae, and releases unhealthy levels of nitrate and toxic heavy metals into lakes and streams. Acid rain also causes chemical changes in lakes that allow more ultraviolet radiation to penetrate deeper into the water, affecting the organisms living there. Recall that ultraviolet radiation is a major cause of mutations in organisms' DNA. One of the most serious human pollution impacts has been the enrichment of freshwaters with phosphate, which fosters the growth of harmful populations of waterweeds, algae, and cyanobacteria. This process, which converts clean lakes into polluted ones, is known as **eutrophication.**

Oligotrophic freshwaters are low in nutrients and productivity but high in species diversity

When people think of clean, healthy lakes and streams, they imagine clear waters (Figure 28.8a), which scientists term **oligotrophic.** These waters are defined by low mineral nutrient content and low primary productivity (the amount of carbon fixed in photosynthesis per unit area per year), but high levels of oxygen and high species diversity. Although many different kinds of phytoplankton and other organisms may occur in oligotrophic waters, they are few and far between. Such waters produce desirable game fish and do not generate harmful algal blooms or excessive growths of weeds.

Eutrophic freshwaters are high in nutrients and productivity but low in species diversity

Eutrophic waters have a high mineral nutrient content and primary productivity, but low oxygen and low species diversity. Eutrophic waters are not known for good fishing unless they have been artificially stocked with fish, and people regard them as degraded waters having low quality. Eutrophic freshwaters may be crowded with fast-growing, weedy plants (Figure 28.8b) that may en-

tangle swimmers, interfere with boat traffic, and shade out phytoplankton in the underlying waters. Water hyacinth, water lettuce, kariba weed, and water milfoil are examples of weedy macrophytes that grow in abundance in eutrophic waters (Figure 28.9). Overgrowths of periphyton algae such as *Cladophora* or *Pithophora* can also occur in eutrophic waters (Figure 28.10). These growths wash up onto beaches, producing foul odors as they decay.

Eutrophic water (see Figure 28.8b) is often visibly cloudy because it contains large populations of a few types of suspended cyanobacteria. Lakes having persistent and enduring cyanobacterial blooms are described as **hypereutrophic.** When cyanobacteria become very abundant, about 70% of the time they produce toxins that are released into the water and can kill waterfowl, together with wild and domestic animals. Domestic animals should not be allowed to drink water having a visible blue-green bloom, and people should limit their exposure when swimming or water-skiing in such waters. It is important to understand that the culprits for this pollution are not the weedy macrophytes, overgrowths of periphyton, and blooming cyanobacteria. People have only themselves to blame when lakes become unfit for swimming, lose their good fishing qualities, or poison animals that drink from them. Much research has shown that harmful cyanobacterial growths result from phosphate pollution that is linked to human activities.

Phosphorus availability controls growth of freshwater plants, algae, and cyanobacteria

Why is phosphorus so important in causing freshwater algal blooms and nuisance

FIGURE 28.8 Lakes of different nutrient status (a) An oligotrophic lake in the Rocky Mountains of Colorado. (b) A eutrophic lake with growths of macrophytes and algae.

FIGURE 28.9 Some weedy aquatic macrophytes (a) *Eichhornia* (water hyacinth), (b) *Salvinia* (kariba weed), (c) *Pistia* (water lettuce), and (d) Eurasian water milfoil (*Myriophyllum*).

FIGURE 28.10   **Some weedy periphyton algae** Periphyton algae that may produce nuisance growths in overfertilized freshwaters include (a) *Cladophora*, which can form visible masses in a eutrophic lake (inset, microscopic view). (b) Large growth of *Pithophora* covering a lake (inset, microscopic view).

growths of aquatic weeds? Plants and algae require relatively large amounts of phosphate to produce ATP, phospholipids for cell membranes, the photosynthetic energy carrier NADP, the sugar-phosphate backbone of DNA and RNA, and many other cell constituents that contain phosphate. Although phosphate is in great demand, its supply in the unperturbed freshwater environment is relatively low. This is largely because phosphate forms insoluble complexes with abundant cations (positively charged ions) such as Al^{3+} (aluminum), Fe^{3+} (iron), and Ca^{2+} (calcium) in soil and lake sediments. When combined with these cations, phosphorus is not available for uptake by photosynthetic organisms. Thus, phosphorus limits cyanobacterial, algal, and macrophyte growth in most oligotrophic lakes and streams. (This is in contrast to open-ocean waters, which are often nitrogen-limited. Freshwaters are less likely to experience nitrogen limitation because higher concentrations of nitrogen-fixing cyanobacteria are typically present.) Addition of phosphate to freshwaters releases algae and plants from limitation by this substance, and their populations continue to grow until they become limited by some other requirement. This other requirement may be nitrogen or a micronutrient such as iron. In hypereutrophic lakes or those affected by acid rain, inorganic carbon may become the limiting factor.

The relationship between phosphate and algal or plant growth can be expressed by a hyperbolic function, in which there is a period of rapid, linear growth response at low nutrient concentrations, followed by leveling off of the response at higher concentrations (Figure 28.11). Small increases in phosphate can cause large increases in algal or plant biomass. That is why it is so easy to transform oligotrophic water bodies to eutrophic ones. Additionally, the function reveals that under eutrophic conditions, even very large decreases in phosphate may have little or no effect on algal or plant biomass. Thus it is difficult to restore eutrophic lakes to a more desirable oligotrophic condition. Because phosphate removal is expensive, preventing eutrophication may be economically preferable.

A remote area in Canada, occupying 30,000 square kilometers and containing hundreds of lakes, has been used for research into the causes of eutrophication. Scientists have experimentally manipulated dozens of these lakes, adding excess amounts of various nutrients and watching for algal

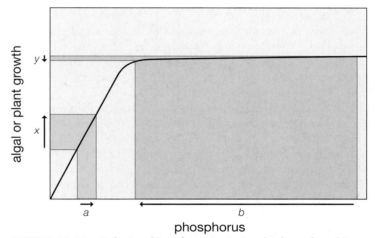

FIGURE 28.11   **Relationship of nutrients to algal or plant biomass** When phosphate concentrations are low, biomass is also low, and small increases in phosphate concentration (a) can result in large increases in biomass (x). When phosphate concentrations are high, biomasses are also high, but even very large decreases in phosphate concentration (b) may have little effect on biomass (y). This graph explains why it is so easy to decrease water quality and so hard to restore it.

blooms and other evidence of eutrophication, in comparison to control lakes that were not disturbed. They obtained convincing evidence that excess phosphate causes eutrophication. In the U.S. Great Lakes, the quality of Lake Erie and Lake Michigan has improved greatly since nearby states enacted laws prohibiting the sale of phosphate-containing detergents.

Lake and stream eutrophication can be prevented or reversed Nutrients enter lakes and streams at point sources, such as industrial or sewage effluent emerging at a localized position, or more diffuse (nonpoint) sources, such as soil erosion from construction sites or runoff from fertilized lawns and agricultural fields. "End-of-pipe" effluents can be treated with ferric chloride or aluminum salts to form precipitates of ferric or aluminum phosphate, which can be removed before the effluents are released to natural waters. Such a step can be added as a final stage of sewage treatment but is often considered too expensive by taxpayers. Erosion

FIGURE 28.12 **Weed cutters** One way to reduce mineral nutrient concentrations in a lake is to cut aquatic weeds with specialized weed-cutter watercraft. Removal of the cut weeds is essential to eutrophication prevention and control strategies.

from construction sites near water bodies can and should be controlled. Sedimentation (settling) ponds and vegetation buffer strips along streams are helpful in intercepting soil and nutrients arising from lawn or agricultural fertilizers before they are able to reach lakes and streams.

The extent to which weed growth and algal blooms will occur in a lake is a function of the nutrients stored in the sediments during the previous year (see Figure 28.3). Methods for reducing weed and algal growth include dredging up phosphate-rich bottom sediments, an option for smaller bodies of water. The removal of sediments reduces the amount of nutrients available to supply the next season's algal blooms and weed growths. Another strategy is removal of macrophytes; the phosphate stored in their tissues is thus unable to reach the sediments and contribute to the following year's plant and algal growth. Macrophytes can be removed with weed-cutting machinery (Figure 28.12). It is important to remove the cut weeds from the lake because leaving them will only result in the recycling of the nutrients they contained to next year's growths. An understanding of nutrient cycling also explains why the use of aquatic herbicides has only short-term benefits. Weeds, algae, and cyanobacteria killed by herbicides stay within the lake and are decomposed in the benthos, with the result that their nutrients are recycled and support more nuisance biomass in the following season. Finally, preserving or restoring natural wetlands adjacent to lakes and streams is a widely approved method for preventing eutrophication.

28.4 | Wetland ecosystems

Swamps, marshes, bogs, fens, mires, sedge meadows, prairie potholes, bayous, and sloughs are just some of the many types of wetlands—places where you are likely to get your feet wet. Wetlands are the zones of transition between typical terrestrial and aquatic habitats, occurring in places where rainfall or stream input exceeds the rate of water loss by runoff or evaporation. In wetlands, the water table is very high and the soil is typically saturated with water for at least part of the year or actually lies beneath a shallow layer of water. Consequently, wetland soils are often low in oxygen, which leads to low rates of mineral-releasing decomposition and causes low fertility. The various kinds of wetlands are each characterized by typical vegetation and microbial communities that are often highly adapted for waterlogged soil conditions.

Common freshwater wetlands include riparian wetlands, deep-water swamps, marshes, acid bogs, and sedge meadows

Riparian wetlands are those that border streams and rivers and receive their water and nutrients from upstream sources. They can occur even in desert regions, supporting plants as large as willow and sycamore trees. Warm-region, **deep-water swamps** also typically contain large trees, such as *Taxodium* (cypress) in the southern United States.

Freshwater marshes can be identified by the presence of cattails *(Typha)*, sedges, and certain grasses. These plants occur where the water is not acidic and where calcium levels are relatively high. In **acid bogs** (Figure 28.13a), dominated by *Sphagnum* moss, shrubs, and trees such as black spruce and tamarack, the water is low in mineral nutrients. **Sedge meadows** (Figure 28.13b), often located near lakes or other bodies of water, are dominated by hummocky clumps of sedges—grasslike plants that usually have distinctive triangular stems. The Florida Everglades is a very large sedge meadow.

Wetlands play important roles in global carbon cycling

Wetlands have very high primary productivity, higher than that of tropical rain forests, grasslands, pine forests, or cultivated land. Thus they generate large quantities of organic carbon, which is stored in their soils as incompletely decomposed plants, known as peat. Over many years, deep layers of peat may accumulate, representing a large store, or "sink," of organic carbon. Peat accumulation is attributed to reduced rates of microbial decomposition—due to cool climates, acidic waters (in some cases), anaerobic conditions and the presence of decay-resistant phenolic compounds such as lignin in plant tissues.

Because of their large geographical coverage—perhaps 5% of the Earth's surface—and high carbon-storage capacity, northern peat-producing wetlands are regarded as being particularly important in global carbon cycles. They are believed to function as a giant thermostat that is able to buffer global changes in climate by storing either more or less carbon dioxide as peat. When the temperature is a bit warmer, the plants photosynthesize at a faster rate (and for a longer growing season), in the process producing more decay-resistant organic carbon, which is not easily decomposed by fungi and bacteria. This process removes greater amounts of CO_2 from the atmosphere, storing it for thousands of years. Reduced atmospheric CO_2 content then has a slight cooling effect on the climate, which in turn slows the rate of wetland peat carbon storage. When less carbon dioxide is stored, more enters the atmosphere, causing CO_2-based climate warming. Wetlands are also important locations for the microbial production and

FIGURE 28.13 Common freshwater wetlands (a) Acid bog: The grassy area adjacent to the water is dominated by *Sphagnum* moss and floats on a layer of water. After taking a few steps here, it is clear why the term *quaking bog* is sometimes used to describe this floating mat of live and dead plant material. (b) Sedge meadows: Sedge meadows are often dominated by tussock-forming sedges. Though grasslike in appearance, sedges constitute a different group of plants.

destruction of methane, which is an even more effective climate-warming "greenhouse gas" than carbon dioxide.

Wetland plants are adapted in ways that help them overcome stresses of wetland habitats

Representatives of all the major groups of land plants—bryophytes, lycophytes, pteridophytes, gymnosperms, and angiosperms—can be found in wetlands and are adapted to these habitats.

Sphagnum moss (Chapter 21), which dominates acid bogs, and other bryophytes are common in wetlands. In fact, wetlands are among the best places to observe a wide diversity of bryophytes. Wetland nonvascular plants do not tolerate drying, but *Sphagnum* moss leaves have large water-storage cells

that help prevent desiccation of the plant. Moss cell walls are rich in phenolic compounds that help them resist decomposition, because bacteria and fungi are abundant in wet habitats.

Tamarack (larch, *Larix laricina*) is a gymnosperm tree (Figure 28.14a) commonly found in bogs that provides forage for birds, rabbits, and deer. Tamarack is unusual for a conifer in that it is deciduous, losing its leaves in the fall and regenerating them in the spring, like many angiosperm trees of cool climates. Losing leaves reduces water loss through transpiration. This process is an advantage when soil water is frozen in winter and thus unavailable for uptake by the plant. Bog trees are particularly vulnerable to cool-season water loss, because bogs—often occurring in low-lying depressions—cool faster in the fall and warm more slowly in the spring, compared to uplands.

Wetlands boast a wide variety of interesting and attractive flowering plants, including the showy grass pink orchid (Figure 28.14b) and several types of insectivorous plants (Chapter 11) that occur in bogs. Because bog soils are low in mineral nutrients, leaves of insectivorous plants are adapted to harvest minerals from the bodies of insects such as mosquitoes. Many flowering shrubs of bogs such as *Chamaedaphne calyculata*, the leatherleaf, respond to transpirational water loss during cold weather by having evergreen, drought-resistant leaves (Figure 28.14c). Such leaves are morphologically adapted to reduce water stress in that they are typically small and leathery, with a thick outer cuticle. The reduced surface-area-to-volume ratio of such leaves helps them retain water, as does the cuticle.

Cattails *(Typha)* are familiar and easily recognizable flowering plants that are adapted in several interesting ways for life in marshes and waterlogged ditches (Figure 28.15a). The broad-leaved cattail, *Typha latifolia,* grows to around 3 m in height, often forming large populations that develop from the same horizontally growing underground stem (rhizome). Aerenchyma tissues conduct oxygen from aboveground plant parts to the oxygen-starved, submerged rhizome and roots, allowing them to generate the energy needed to take up essential minerals. Leaves of *Typha* are adapted to withstand wind forces and to stand erect even though they are very thin and only about an inch wide. Their structure has been compared to that of an airplane wing, where high stiffness and low weight are also advantageous. Like airplane wings, *Typha* leaves are flat on their inner surfaces but curved on the outer surfaces (Figure 28.15b). Internally, *Typha* leaves are compartmentalized to form columnlike structures and are strengthened widthwise by alternating thick and thin diaphragms. Lengthwise, the blades are strengthened by rows of parallel, fibrous strands, which are visible as fine lines on the outer surfaces. Further, the leaves are slightly twisted in a way that minimizes lift and drag forces encountered under windy conditions.

The flowers of *Typha* are located in the terminal cattail (Figure 28.15c). The uppermost, light-tan flowers are the staminate, or "male" flowers; the lower, darker flowers are pistillate ("female") flowers. Pollen is wind dispersed, an advantage in situations where many individuals of the same species are closely spaced. Consequently, the flowers lack showy petals. After pollen dispersal, the upper flowers disintegrate, and the stem that bore them remains as a withered stalk. Ovaries of the pistillate flowers develop into small

FIGURE 28.14 Wetland plants (a) The tamarack *(Larix laricina)* is a common conifer of bogs and other wetlands. It is shown in its leafless winter condition and with its soft, needlelike leaves (inset). (b) The grass pink orchid *(Calopogon)* blooms during spring in acid bogs. Its showy flowers attract insect pollinators. (c) The leatherleaf, *Chamaedaphne*, is a common shrub of peatlands and has leathery leaves that are adapted for withstanding desiccation due to a cold climate.

FIGURE 28.15 The cattail, *Typha* This common and familiar plant occurs in large populations (a) that are generated from an underground rhizome. The leaves, one of which is seen in cross section in (b), are constructed much like airplane wings for lightness together with maximal strength. Pollen-bearing staminate flowers are situated above a group of ovule-bearing flowers (c). Fruits are small and covered with hairs that facilitate wind dispersal (d).

FIGURE 28.16 **Sedge (*Carex* species)** Sedges are common plants of marshes and sedge meadows.

Humans have destroyed many of the world's wetlands

Humans have greatly reduced the number and area of wetlands in the United States (Figure 28.17) and around the world. Wetlands are destroyed when the flow of rivers is altered, such as in dam construction or they are deliberately drained by digging drainage channels. Wetlands are frequently, but wrongly, regarded as wastelands that must be reclaimed by filling them for construction of homes and other buildings or using them to grow crops. For example, many acid bogs and marshes in Chile and the Northeast and Upper Midwest in the United States have been converted to commercial cranberry production (Figure 28.18). The native vegetation is highly disturbed by changes in the water regime, such as flooding to facilitate survival of cranberry vines in winter and in the fall to assist with the harvesting process. Cranberry production provides an example of the many situations in which humans must carefully balance their needs for commodities with the consequences of habitat destruction.

Wetland delineation, invasive species, and restoration are issues in wetland protection and restoration

Though many people appreciate the value of wetlands and accept the need for laws that restrict destruction of wetlands, controversy often surrounds the criteria used to define wetlands. Many wetland soils are saturated with water for only part of the year. Cattails and other wetland plants can transpire very large amounts of water during the summer, with the result that their habitats may become relatively dry for months at a time. Thus wetland scientists recommend that, for the purposes of regulation, wetlands should be defined by the types of plants present rather than by soil moisture levels.

Exotic, invasive species are nonnative plants that, in the absence of control by natural diseases and herbivores, have spread widely in new environments, often with the help of

seedlike fruits (Figure 28.15d) with hairs that form a fluffy mass dispersed by the wind during winter. *Typha* is a favorite source of food for muskrats, which graze on spring shoots, eat leaves and aboveground stems in summer, and consume rhizomes in the fall and winter. Muskrats also use *Typha* leaves and stems to construct dens.

The grasslike plants known as sedges are common inhabitants in a variety of wetlands but especially sedge meadows. Most, but not all, belong to the genus *Carex* (Figure 28.16). Sedges occur in dense clumps that are useful places to step when trying to walk through sedge meadows (see Figure 28.13b). The flowers, typical of wind-pollinated plants, are small and nonshowy.

FIGURE 28.17 **Wetland loss** The number and area of wetlands have declined alarmingly on a worldwide basis. These maps illustrate wetland losses that have occurred in the United States from the 1780s through the 1980s.

FIGURE 28.18 Cranberries
Cranberries occur in bogs as creeping vines (a). They have small, desiccation-resistant leaves and tiny pink flowers. The tasty fruits appear in early autumn. Cultivated cranberries are grown in specially constructed or natural wetlands that have been hydrologically modified to allow control of water levels. At harvesting time, the fields are flooded and the berries dislodged from the vines by special machines (b). Cranberries are then corralled with booms and loaded onto trucks by conveyors (c) or a vacuum system. Truckloads are dumped at receiving centers (d), where the cranberries are packaged or shipped on for further processing.

humans. An example of an exotic, invasive species that has caused much damage to U.S. wetlands, by replacing native plants upon which wildlife depend, is purple loosestrife (*Lythrum salicaria*, Figure 28.19). A native of Eurasia, purple loosestrife was imported as a garden plant because of its attractive purple inflorescences but is now illegal in many states. Experts in wetlands preservation and restoration spend much time studying the traits of invasive species in the attempt to devise strategies for controlling them.

Federal policy in place since 1980, with the aim of zero net loss of wetlands, allows developers to build over wetlands if they agree to create a replacement marsh elsewhere in the area. However, in 2001, a National Research Council expert panel,

led by famed wetland ecologist Joy Zedler of the University of Wisconsin, found that many man-made marshes do not contain vegetation natural to the region or do not provide the same ecological benefits as natural wetlands. Finding ways to preserve the world's precious wetlands will continue to be a challenge.

28.5 | Oceans are essential to humans and life on Earth

The coastal zones of the world's oceans consist of areas of mainland, islands, and shallow seas extending from 200 m above sea level to 200 m (656 ft) below sea level. Defined in this way, coastal zones represent about 18% of the surface of the Earth. About 60% of the human population of the world lives in this relatively small but productive and sensitive area. This 18% of the world's surface supplies 90% of the total fish caught per year and also produces about 25% of total global primary production. At the same time, this essential area is among the most endangered from human population growth and development, eutrophication, pollution, and overfishing.

The oceans moderate the climate of Earth and provide the landmasses with the precipitation that sustains life on land. The vast volume of ocean water absorbs heat and releases it slowly over decades to moderate climatic variability. Without the oceans, the landmasses would be barren. Moisture evaporates from the surface of the oceans and enters the atmosphere. From the atmosphere the moisture cools and condenses as rain that falls as weather fronts move across the landmasses. The resulting precipitation supports life on land and ultimately makes its way by streams and rivers back to the oceans in a vast global water cycle (Chapter 25). The oceans also act as the final waste disposal site for our entire planet. Whatever is dumped on the land and does not degrade there eventually finds its way down streams and rivers to the oceans. The coastal zones bear the brunt of this human pollution.

FIGURE 28.19 An alien invader of wetlands Purple loosestrife has invaded marshlands throughout the United States. The inset shows the same area, about 20 years earlier.

28.6 | Oceans cover most of the surface of the Earth

About 71% of the surface of the Earth is covered in oceans. The average depth of the oceans is 3.8 km (2.4 mi), and the total volume of seawater is 1370×10^6 km^3 (326.5×10^6 mi^3). Because living organisms exist throughout this entire volume, the oceans are the single largest biome on the Earth. Marine organisms are extremely varied and represent virtually every known phylum. All these varied organisms are subject to and adapted to the properties of the seawater that surrounds them. Before considering the various marine ecosystems, we need to examine the general chemical and physical properties of seawater, its motions, and geographical realms.

Seawater and freshwater differ in the amount of dissolved substances

Because water molecules are polar molecules, water dissolves many substances, such as salts, polar organic molecules, and gases, and has a high heat capacity (Chapter 3). Seawater differs from freshwater mainly in the amount of dissolved solids and gases. A 1,000-g sample of seawater contains 35 g of dissolved substances. The total amount of dissolved substances is called the **salinity**, and salinity is measured in **practical salinity units (psu)**. An open-ocean seawater sample with 35 g of dissolved substances in 1,000 g of seawater has a salinity of 35 psu. By comparison, the Great Salt Lake in Utah has a salinity of 220 psu, an average eutrophic lake 0.18 psu, and an oligotrophic lake 0.001–0.006 psu. The salinity of ocean waters away from coastal zones normally varies from 34 to 37 psu. Nearshore areas are more variable because of discharge of freshwaters from rivers.

The bulk of dissolved substances in seawater consists of inorganic salts in the form of ions such as chloride (Cl^-), sodium (Na^+), sulfate (SO_4^{2-}), magnesium (Mg^{2+}), calcium (Ca^{2+}), and bicarbonate (HCO_3^-). Seawater can be thought of as largely a solution of sodium chloride and magnesium sulfate. Among the remaining dissolved substances in seawater are three ions of critical importance to marine organisms: nitrates (NO_3^-), phosphates (PO_4^{3-}), and silicon dioxide (SiO_2). Silicon dioxide is essential for protists such as diatoms and radiolarians to construct their frustules or external skeletons (Chapter 19). Nitrates and phosphates are essential to all organisms as components of phospholipids, proteins, and nucleic acids (Chapter 3). One or both may be limiting in ocean surface waters depending on biological activity and may limit photosynthesis and primary productivity. Other elements such as iron, manganese, cobalt, and copper are also essential to organisms but may be present as ions in only trace amounts. In some parts of the ocean where nitrate and phosphate are abundant, primary production can be limited by a lack of trace amounts of iron.

The salt content of seawater affects its physical properties. As temperature declines in freshwaters, the density of water increases to a maximum at 4°C (39°F). Below 4°C, the density of water decreases until it freezes at 0°C (32°F). In seawater, however, the amount of dissolved salts depresses the freezing point. Seawater with a salinity of 35 psu has a freezing point of −1.9°C (28.6°F), and the density of seawater increases down to this freezing point. When seawater does freeze, the density of the ice is less than the surrounding water and it thus floats. The dissolved salts are excluded from the ice during freezing but may form pockets of brine within the ice mass. At high latitudes, the formation of very cold and very dense seawater leads to masses of surface water sinking into the deep-ocean basins. There this cold, dense water sets up deep currents and forms the environment of deep-sea organisms.

Carbon dioxide and oxygen are important dissolved gases in seawater. The lower the temperature of seawater, the more gas can dissolve in it, but seawater holds relatively little dissolved oxygen as compared to air. At 0°C (32°F) seawater with a salinity of 35 psu contains only 8 ml O_2 L^{-1} while air has 210 ml L^{-1}. Oxygen content also shows a distinct pattern with ocean depth (Figure 28.20). The maximum level of oxygen usually occurs in the top 10–20 m, where photosynthesis is active and atmospheric oxygen can diffuse across the surface. With increasing depth, the oxygen level declines and reaches a minimum near zero at around 500–1,000 m depth. This oxygen minimum results from biological activity that depletes the dissolved oxygen concentration at a depth where no photosynthesis or atmospheric diffusion can occur. At greater depths, dissolved O_2 levels rise due to the flow of cold, oxygen-rich waters from high latitudes.

Unlike oxygen, carbon dioxide reacts with seawater. Initially, dissolved CO_2 reacts with water to form carbonic acid.

$$CO_2 + H_2O \longrightarrow H_2CO_3$$

The carbonic acid quickly dissociates into a hydrogen ion and a bicarbonate ion.

$$H_2CO_3 \longrightarrow H^+ + HCO_3^-$$

Bicarbonate ion may further dissociate into another hydrogen ion (H^+) and a carbonate ion (CO_3^{2-}), but the bicarbonate ion

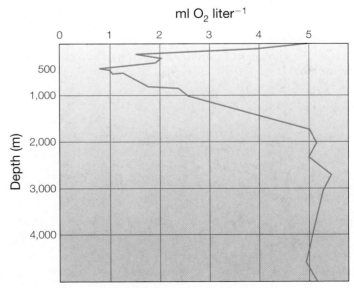

FIGURE 28.20 Oxygen levels as a function of depth in the ocean Shown here is a graph depicting the oxygen levels versus depth for the Atlantic Ocean.

is the most abundant form in seawater. Carbon dioxide is more abundant in seawater than in air, and therefore it is not limiting to photosynthesis in protists and plants. Because of this abundance of bicarbonate ions, the pH of seawater is slightly alkaline, with a range of 7.5 to 8.4 (Chapter 3). The carbonic acid–bicarbonate ion system acts as a **buffer** that maintains the pH of seawater within this narrow range. The chemical reactions shown previously shift to the right if there are too few H^+ ions, and to the left if there are too many.

Ocean basins contain a varied terrain

There are four major oceans—the Pacific, Atlantic, Indian, and Arctic. All are interconnected into one world ocean. Seas are smaller areas partially cut off from the major oceans, such as the Mediterranean, Caribbean, and Red Seas.

The oceans are very shallow along the margins of the major landmasses where they cover submerged extensions of the continents called *continental shelves* (Figure 28.21). The **continental shelf** slopes gently down from shoreline to a depth of 100–200 m. The width of the shelf varies from as much as 400 km (248 mi) off eastern Canada to as little as a few kilometers off the coast of California. At the outer edge of the shelf, the bottom drops off steeply to form the **continental slope**. The continental slope descends to a depth of 3,000–5,000 m (9,840–16,400 ft), where it levels out into the **abyssal plain**, a vast sediment-covered expanse that extends over most of the ocean floor. The abyssal plains are interrupted by three types of geological formations. **Submarine ridges** are midocean mountain ranges that extend throughout all the oceans. The mid-Atlantic ridge bisects the Atlantic Ocean into east and west basins and runs from Iceland in the north Atlantic down to the south Atlantic, where it joins a submarine ridge from the Indian Ocean. In places, the submarine ridges rise above the ocean surface to form islands such as Iceland. Submarine ridges are often sites of volcanic activity. Abyssal plains may also be broken up by deep oceanic **trenches**, most of which occur in the Pacific Ocean. Trenches reach depths of 7,000 (22,960 ft) to over 11,000 m (36,080 ft). Finally, isolated volcanoes may rise from the abyssal plain to form islands, such as the Hawaiian Islands or, if below sea level, submarine **sea mounts**.

Atmospheric circulation and the Coriolis force drive ocean currents

Currents are the horizontal movements of large water masses. Wind belts that are arranged by latitude around the globe produce the major ocean currents (see Figure 25.20). These wind belts are generated by differential heating of the atmosphere and are deflected by the Coriolis effect. The Coriolis effect results from the rotation of the Earth on its axis from west to east. Warm air masses rise at the equator and spread out at altitude to the north and south. The rising air masses that move north are deflected to the right by the Coriolis effect. When they sink back to the surface at 30° N latitude, they form the northeast trade winds that blow from northeast to southwest. The rising air masses that move south from the equator are deflected to the left by the Coriolis effect. When they sink back to the surface at 30° S latitude, they produce the southeast trade winds that blow from the southeast to the northwest. Together, the trade winds impart a net flow of surface seawater at the equator from east to west (see Figure 25.23). This net westward flow of equatorial water piles up water at the western side of the ocean basins, where it is deflected north and south toward the poles. Above 30° N and S latitude, these poleward flows of seawater encounter the westerlies. In the Northern Hemisphere the westerlies blow from southwest to northeast, and in the Southern Hemisphere the westerlies blow northwest to southeast (see Figure 25.20). The westerlies drive the seawater masses back east across the ocean basins. Continental landmasses on the eastern side of the ocean basins deflect these eastward currents back toward the equator. The resulting cycle of ocean currents consists of five huge circular patterns called **gyres** (see Figure 25.23). Gyres in the Northern Hemisphere turn clockwise and those in the Southern Hemisphere rotate counterclockwise. These surface ocean currents affect the distribution of marine organisms and the geographical locations of terrestrial ecosystems.

The deep-water masses of the oceans move in a different manner from the surface waters. Wind is not a factor. In the North Atlantic, warm tropical seawater flows north in the Gulf Stream. In the area of Iceland, these warm waters meet cold waters from the Arctic, where they cool, increase in

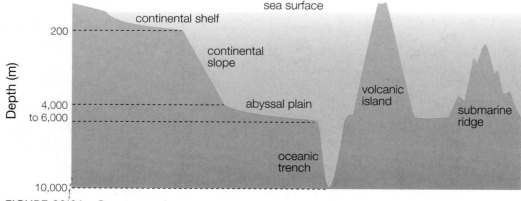

FIGURE 28.21 **Ocean terrain**

density, and sink into the deep ocean. The cold water mass then slowly flows south toward Antarctica over a period of hundreds of years. Similarly, warm surface waters moving south in the Southern Hemisphere reach Antarctic waters, where they cool and sink into the deep ocean. The mass of cold water around Antarctica becomes the cold bottom water in most oceans. These mass flows of ocean water are one major avenue for heat transport from the equator to the poles.

Ocean temperatures vary with depth, season, and latitude

Ocean temperatures show a distinct pattern with depth in the water column. The warmest temperatures occur at the surface, and tropical surface waters may average 20°–30°C (68°–86°F) for the year. Temperatures decline below the surface, and at a depth of 50–300 m (164–984 ft) in the water column, they drop rapidly. This zone of rapid temperature decline is called the **thermocline**. Temperature continues to decline below the thermocline but at a slow rate. Thermoclines are permanent features in tropical oceans, occur in summer in temperate oceans, but are absent from polar seas.

Polar seas are well mixed most of the year, and nutrients are not limiting. A thermocline is either weak or absent entirely (Figure 28.22a). Algal growth is mostly confined to the summer, usually only July and August in the Arctic when light is available. Lack of light limits algal growth.

In temperate oceans, which lie between 30° and 60° N and S latitude, the amount of solar energy striking the ocean surface varies seasonally. Seasonal variation in solar energy input means that surface temperatures also vary seasonally (Figure 28.22b). In winter, solar energy is at a minimum, and the water column and nutrients are well mixed. With the arrival of spring, solar energy input increases, surface waters warm, winds decline, and a thermocline may begin to form. A spring bloom of algal growth may occur. The input of solar energy is at a maximum in summer, and a strong thermocline is established. Because mixing cannot occur throughout the water column, algal growth can deplete the surface waters of nutrients. In fall, the surface layers cool as solar energy input declines, the thermocline weakens, and mixing can resume. Algal growth may produce a small secondary bloom in fall.

In tropical oceans, which lie between 30° N and S latitude, surface waters receive high levels of solar energy year-round. A permanent thermocline prevents mixing of the water column (Figure 28.22c). Both nutrient levels

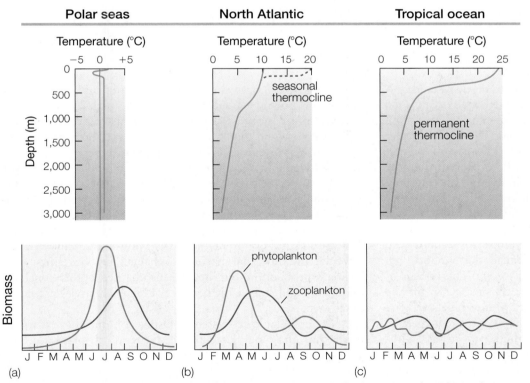

FIGURE 28.22 Seasonal changes in thermal structure and plankton populations in three oceans (a) Polar ocean. In polar oceans a thermocline is absent or weak, waters are mixed, and nutrients are plentiful. Phytoplankton grow mainly in the summer when surface waters are warmer. A late summer bloom of zooplankton reduces the phytoplankton. (b) Temperate ocean. In temperate oceans mixing begins in fall and continues through winter. A seasonal thermocline (dashed line) occurs in summer. Phytoplankton bloom in the spring and again in the fall when mixing makes nutrients available. Zooplankton track the blooms of phytoplankton. (c) Tropical ocean. In tropical waters a thermocline is always present and phytoplankton are nutrient-limited. Nutrient recycling is important in maintaining productivity.

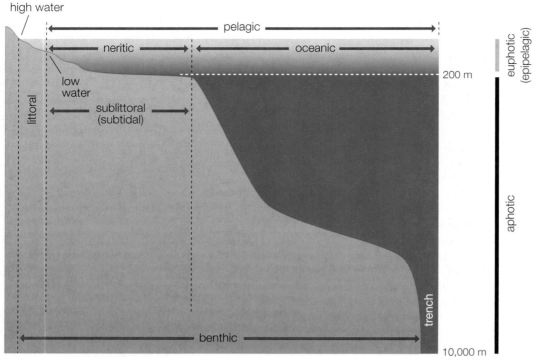

FIGURE 28.23 Ocean realms

and algal growth are low. Lack of nutrients limits primary production.

The oceans can be divided into realms

The vast volume of the oceans can be divided vertically and horizontally into realms. The entire volume of the open ocean is the **pelagic** realm (Figure 28.23). Pelagic organisms are those organisms that live away from the bottom. The **benthic** realm refers to the zones of the ocean bottom and the organisms that live there. The pelagic realm can be divided horizontally into the neritic zone, which includes the water that overlies the continental shelves, and the oceanic zone, which includes the waters that cover the continental slopes and abyssal plains. The pelagic realm may also be divided vertically into a number of zones. The **epipelagic** zone is the uppermost zone. Since this zone is well lighted for photosynthesis, it is also called the euphotic (*eu* means "good" and *photos* means "light") zone. This zone is of major importance because all the primary production in the oceans occurs there. The lower boundary of the epipelagic zone lies around 100 to 200 m (328 to 656 ft) deep. All the pelagic zones beneath the epipelagic lie in the aphotic (*a* means "without") zone, where photosynthesis cannot occur. These deep pelagic zones represent the majority of the volume of the oceans, but the organisms that live within them are dependant upon the productivity of the epipelagic for their survival. The benthic realm is similarly divided into different benthic zones. The benthic zone beneath the neritic zone is the **sublittoral** or subtidal zone. The sublittoral includes three important marine ecosystems: seagrass beds, kelp forests, and coral reefs. The **littoral** zone lies above the sublittoral zone and occupies the area between the extremes of high and low ocean tides.

28.7 | The epipelagic ecosystem contains plankton and nekton communities

The epipelagic ecosystem consists of all the surface waters of the oceans down to a depth of 200 m (656 ft). The epipelagic ecosystem is enormously important because all the photosynthetic activity and therefore primary production in the oceans occurs in this one zone. From the epipelagic there is a constant rain of living and dead organisms and detritus into the deeper ocean zones. All the deeper pelagic systems and benthic ecosystems depend on this rain of food to supply their organisms. As depth increases, the number of fish and other organisms decreases. Animals such as sponges, jellyfish, mollusks, and starfish dominate the benthic zones. Shallow-water subtidal and intertidal marine ecosystems may have their own primary production, but they are also bathed in the epipelagic ecosystem and receive the benefits of its productivity as well as their own.

The epipelagic ecosystem contains communities of organisms that may be described in general as **plankton** and **nekton**. Planktonic organisms are free-floating or weakly swimming. The word *plankton* comes from a Greek word that means "wanderer." Planktonic organisms come in a wide range of sizes, although most are small in greatest linear dimension. Nektonic organisms are large, strong-swimming animals of the open ocean. The nekton include fish, reptiles such as sea turtles, birds, mammals, and squids. Planktonic organisms occupy the lower trophic levels of the epipelagic food web, and the nektonic organisms occupy the higher levels.

Plankton may be divided into groups based on function and size. **Viroplankton** are free-floating viral particles. The **bacterioplankton** include the prokaryotic plankton—the bacteria (including cyanobacteria) and archaea. **Phytoplankton** are free-floating or weakly swimming algal protists capable of photosynthesis. **Protozooplankton** are heterotrophic protists, and **zooplankton** are planktonic animals belonging to many different groups. Planktonic organisms may also be grouped into size classes. The viroplankton are included in the femto-plankton ($0.02-0.2\ \mu m$); the bacterioplankton are in the **picoplankton** ($0.2-2.0\ \mu m$). Some cyanobacteria are large enough to be included in the **nanoplankton** ($2-20\ \mu m$). Phytoplankton and protozooplankton cover a wide range of sizes from nanoplankton through microplankton ($20-200\ \mu m$) to mesoplankton ($0.2-20$ mm). The two sets of terms may be combined; nanophytoplankton refers to phytoplankton in the size range of $2-20\ \mu m$. Standard plankton nets, used to collect marine organisms, usually retain only organisms larger than $20\ \mu m$, and these organisms are called **net plankton**. Smaller organisms are collected by filtration, centrifugation, or settling in chambers.

Bacterioplankton are the most important group in terms of productivity

Bacterioplankton exist in ocean waters as free-floating cells or attached to various particles. Recent marine research has established the importance of bacterioplankton in the productivity of the oceans. This importance is based on three major roles: primary producers, consumers of dissolved organic carbon, and recyclers of mineral nutrients.

Primary producers include the cyanobacteria and prochlorophytes. Cyanobacteria are photosynthetic prokaryotes containing chlorophyll *a*. The cyanobacterium *Synechococcus* is one of the most important primary producers in the oceans. Prochlorophytes are photosynthetic bacteria that contain both chlorophyll *a* and *b*. They are small ($0.6-0.8\ \mu m$ in diameter), numerically abundant (up to 10^6 cells ml^{-1}), and may represent up to one-third of all the chlorophyll *a* in the oceans. Together, the cyanobacteria and prochlorophytes are responsible for a large part of the primary productivity in the seas.

Bacterioplankton also take up dissolved organic compounds. The oceans represent one of the largest reservoirs of organic carbon on Earth. That organic carbon is present as dissolved organic carbon (DOC) and particulate organic carbon (POC). DOC and POC are washed into the oceans from the land and are also produced in the oceans. Photosynthetic organisms leak about 25% of the carbon they fix in photosynthesis into the surrounding seawater as DOC. Heterotrophic bacteria take up the bulk of this DOC and convert it to POC in the form of their cells. These bacteria are then consumed by nanoflagellates and ciliates, which are in turn eaten by larger zooplankton. Thus the DOC leaked from phytoplankton is recaptured as bacterial POC and returned to the planktonic food web. This microbial food chain is termed the **microbial loop**.

Bacterioplankton also break down organic matter and regenerate mineral nutrients in the euphotic zone. This regeneration is very important in tropical ocean waters where a thermocline is always present, because the thermocline prevents mixing of the surface waters with deeper, nutrient-rich waters. Nutrient regeneration allows the phytoplankton to continue to grow.

Phytoplankton are very diverse in form

Traditionally in ocean research, only net phytoplankton ($>20\ \mu m$) were considered to be phytoplankton. It is now understood that these larger algal cells are not dominant either numerically or in terms of primary productivity, and photosynthetic organisms in the picoplankton and nanoplankton are more significant in the oceans. Net phytoplankton come in a wide range of morphologies, but the species mostly belong to two groups of algae, the diatoms and dinoflagellates.

Diatoms Diatoms are free-floating algae with golden-brown plastids enclosed in a glass case that resembles a minute Petri dish (Chapter 19). The glass case consists of two valves, one inside the other, and is composed of silicon dioxide (SiO_2), the main material in windows and bottles. The valves are highly ornamented with grooves, ridges, and pits that are species specific (Figure 28.24). Diatoms have two basic shapes: centric (resembling a cylinder) and pennate (various shapes, all of which are bilaterally symmetrical). When they divide, each of the daughter cells retains one parental valve and secretes one new valve. Diatoms are able to glide slowly over surfaces by excreting mucilage through pores in their valves. Their speed range is slow, about 0.2 to 25 μm s^{-1}. Diatoms grow as single cells, colonies, or as chains and are grazed on by a variety of zooplankton. They generally prefer turbulent waters and may be abundant in surf zones of beaches.

Dinoflagellates Dinoflagellates have two flagella that propel them through the water. They are the fastest known swimming flagellate algae, with speeds of 200 to 500 μm s^{-1}. They are often armored with plates made from cellulose (Chapter 3). Most dinoflagellates exist as single cells. Many dinoflagellates possess chloroplasts and are autotrophic (self-feeding) through photosynthesis, but others lack chloroplasts and are phagotrophic, ingesting prey as food. The large dinoflagellate *Noctiluca* feeds by engulfing food particles such as diatoms, copepods, and even fish eggs that it captures with its tentacle

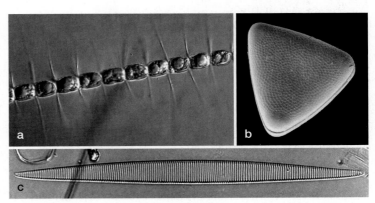

FIGURE 28.24 Common marine diatoms (a) *Chaetoceros*, a chain-forming diatom. (b) *Triceratium*. (c) *Pseudo-nitzschia*. The species shown is a toxin-producing diatom.

(Figure 28.25). *Noctiluca* is also bioluminescent and can light up waves on a beach or patches of ocean surface at night. Easily the most bizarre dinoflagellates are those that bear a complex "eye" or ocellus that has a remarkable resemblance to the eyes of metazoans (Figure 28.25c). In *Erythropsidinium* (the "red-eyed dinoflagellate"), the ocellus contains a lens surrounded by fibers that can change the shape of the lens. Behind this lens is a chamber lined by a cup-shaped *retinoid* that is backed by a reddish-black layer of droplets. Presumably, the ocellus can form images that allow this dinoflagellate to see its prey, which it then captures with a tentacle, but how the image is used to direct the behavior of the organism is unknown.

Dinoflagellates in the oceans do not respond well to turbulence and prefer calm seawaters. Some dinoflagellates can produce toxins, and, if these species form blooms, the toxin level can become high enough to cause massive localized fish kills (see Essay 19.2 "Killer Algae"). Some dinoflagellates live as endosymbionts in the tissues of corals, sea anemones, and giant clams; others are parasites.

Haptophytes Haptophytes are nanoplankton rather than net plankton, but they are abundant and widespread members of the marine phytoplankton. They are small flagellates with two flagella and a unique appendage called a *haptonema* (Chapter 19). The haptonema can bend and coil and is involved in collision avoidance, attachment to substrates, and capture

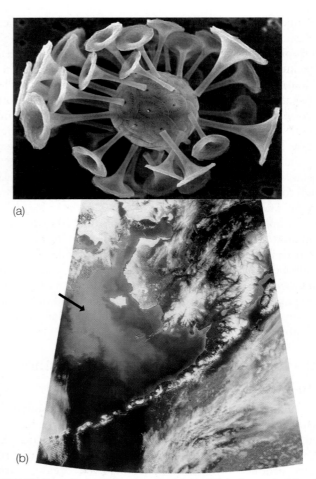

(a)

(b)

FIGURE 28.26 **Coccolithophores** (a) An SEM of a coccolithophore to show the platelike coccoliths (which have trumpet-shaped extensions in this species). (b) Aerial photo of a bloom of coccolithophores (arrow).

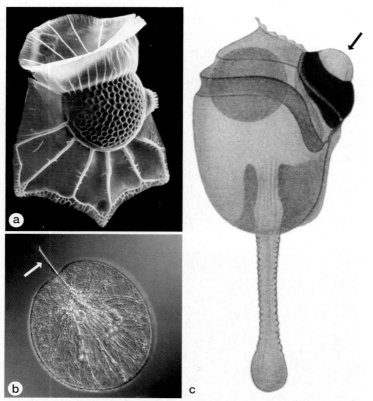

FIGURE 28.25 **Marine dinoflagellates** (a) *Ornithocercus* ("bird tail"), a spectacular dinoflagellate seen here with scanning electron microscopy. (b) *Noctiluca*. Note the tentacle (arrow) extending from the cell. (c) *Erythropsidinium*. The ocellus (eyelike structure) is indicated by the arrow. A tentacle, which is used in feeding, extends from the cell.

and ingestion of food particles. Most species of haptophytes produce external body scales composed of calcium carbonate called *coccoliths* (Greek *kokkos*, "berry," and *lithos*, "stone"); these species are known as **coccolithophores** or **coccolithophorids** (Figure 28.26). Coccolithophores are a major source of primary production in many areas of the ocean and may form extensive blooms that turn the ocean surface cloudy. When coccolithophores die, they sink into the ocean depths, carrying large quantities of carbon with them. Coccoliths contribute about 25% of the total annual transport of carbon to the deep ocean.

A few other groups of algae contribute to primary production, although they are less abundant, including cryptomonads, motile green algae, and in tropical waters, filamentous cyanobacteria.

The planktonic food web begins with picoplankton

At the base of the marine planktonic food web lie the picoplankton, the heterotrophic bacteria, and the photosynthetic cyanobacteria and prochlorophytes (Figure 28.27). Nanoplanktonic grazers—small flagellates and ciliates less than 20 μm long—consume these tiny cells. Larger protozoa in the microplankton size range (20–200 μm) feed on these grazers and various nanophytoplankton. This portion of the planktonic food

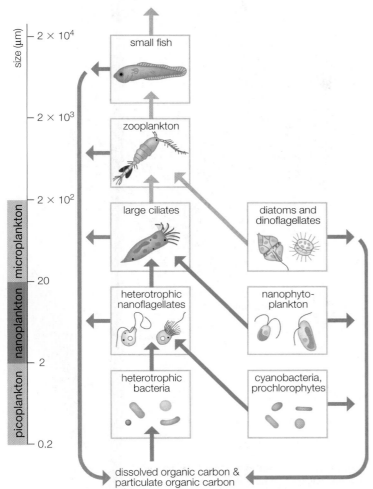

FIGURE 28.27 **Marine planktonic food web** The microbial loop (green arrows) contains all the organisms less than 20 μm in greatest linear dimension. The "classic" food chain (orange arrows) includes all organisms larger than 20 μm in length.

web is called the *microbial loop*. Zooplankton (0.2–20 mm) then consume the protozoa and algae in the microplankton size range. This step in the food web represents the "classical" planktonic food web before the importance of picoplankton and nanoplankton was understood. The most common and widespread zooplankton are the aquatic crustaceans called copepods. Copepods, which are related to shrimp and crabs, are small (1–4 mm) grazers on phytoplankton and protozooplankton. Marine zooplankton are extremely diverse and represent a large number of animal phyla as either adults or immature larval stages. Most are filter feeders. Fish and other members of the nekton then feed on the zooplankton, completing the marine epipelagic food web.

The productivity of the epipelagic ecosystem varies with latitude and proximity to coastal zones. Productivity is measured in grams of carbon fixed per square meter of ocean surface per year ($g\,C\,m^{-2}\,yr^{-1}$). The productivity of continental shelf waters ranges from $100{-}160\,g\,C\,m^{-2}\,yr^{-1}$, open waters of temperate oceans from 70 to $120\,g\,C\,m^{-2}\,yr^{-1}$, and tropical oceans only $18{-}50\,g\,C\,m^{-2}\,yr^{-1}$. Tropical oceans are less productive because the permanent thermocline prevents mixing of the water column. Productivity measurements are a useful way to compare ecosystems.

The food web of the marine epipelagic zone is very different from any terrestrial ecosystem. Terrestrial ecosystems are generally characterized by the dominant plant forms, such as temperate forests or grasslands. The primary producers are large and composed of a significant amount of rigid structural material such as lignin and cellulose. The primary consumers on land are large herbivores, which normally remove only a fraction of the plant biomass. In contrast, in the epipelagic ecosystem the primary producers are microscopic. Of necessity, the primary consumers must also be small. Most primary consumers are filter feeders, a method that is nonexistent in air. There are no large herbivores. The only exceptions are kelp forests, where there are large primary producers but even here there are no large herbivores. In the marine epipelagic ecosystem, large organisms are only reached at the tertiary and quaternary consumer level. On land, humans consume primary producers (corn, beans) and primary consumers (cattle, chickens), but from the ocean, humans eat carnivores at the tertiary or quaternary level, such as tuna and salmon. The next sections will consider several marine ecosystems that are bathed in the epipelagic system but are distinct from it.

28.8 | The sublittoral zone includes kelp forests, seagrass beds, and coral reefs

The sublittoral (or subtidal) zone includes the area from the level of the lowest tide along the shoreline down to the edge of the continental shelf (see Figure 28.23). Because the sublittoral zone lies within the euphotic zone, it is well lighted for photosynthesis. There are three major vegetated ecosystems in this sublittoral zone: kelp forests, seagrass beds, and coral reefs. There is one additional ecosystem that is not vegetated—that of soft sediments. Soft sedimentary systems represent most of the sublittoral area in the oceans and are dominated by animals such as worms, mollusks, and starfish that depend on the neritic plankton for primary productivity. This section focuses on the vegetated ecosystems.

Kelp forests are dominated by large photosynthetic protists

Kelps are large brown algae (Chapter 19) that grow on hard surfaces in the sublittoral zones of cold temperate oceans (Figure 28.28). Kelp forests are common along the west coasts of North and South America because of cold currents in these areas. A community of kelp is called a *kelp forest* if the brown algae form a closed canopy at the surface of the ocean (Figure 28.29) and a *kelp bed* if they do not. Kelp forests are highly productive systems with primary productivity in the range of $800\,g\,C\,m^{-2}\,yr^{-1}$.

Kelps grow attached to hard surfaces by a structure called a *holdfast* (Figure 28.30). The holdfast does not take up nutrients, as do the roots of terrestrial plants. The entire kelp takes up nutrients from the surrounding seawater. From the holdfast a long stemlike process called a *stipe* extends toward the ocean

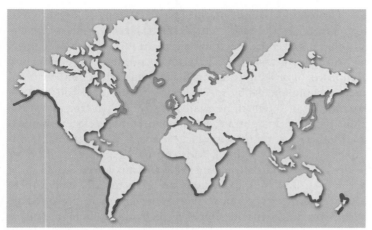

FIGURE 28.28 Map of distribution of kelp forests and kelp beds *Macrocystis* (in brown) dominates the kelp forests, and *Laminaria* (in yellow) dominates the kelp beds.

surface. Broad, flat leaflike blades emerge along the length of the stipe or at the end of the stipe. In some kelps there is a gas-filled float called a *pneumatocyst* at the base of each blade or group of blades that holds the blades up in the water column. *Macrocystis* and *Nereocystis* dominate the kelp forests of the Pacific coasts of North and South America and Africa, as well as Australia and New Zealand. *Laminaria* dominates kelp beds in the Atlantic and in the ocean waters around Japan (see Figure 28.28). *Macrocystis* and *Nereocystis* may reach lengths of 20–30 m (65–98 ft), which is comparable to terrestrial trees, and form a closed canopy. Beneath the canopy there is an understory of various red and brown algae. The kelp forest

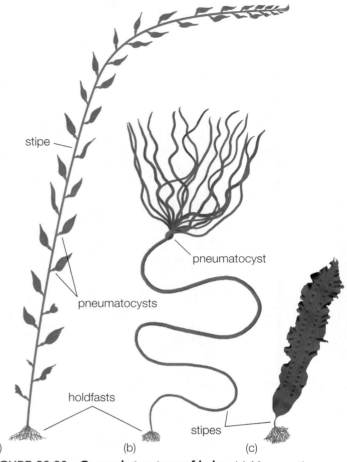

FIGURE 28.30 General structure of kelps (a) *Macrocystis*, (b) *Nereocystis*, (c) *Laminaria*.

FIGURE 28.29 *Macrocystis* kelp forest (a) Surface view, (b) interior view.

shelters a wide range of marine animals, including invertebrates, fishes, and the sea otter.

A complex relationship exists between kelp forests and sea otters *(Enhydra lutris)* along the Pacific coast of North America. The fur trade had decimated the sea otter populations in the coastal waters of Alaska, where the sea otter had been the main predator on sea urchins. In the 1970s, researchers noted that where otters were present, sea urchin populations were low and kelp forests were dense. Where otters were absent, sea urchins were abundant and kelps largely absent. Sea urchins consumed the holdfasts of the kelps, which were then washed out of the area. Following this discovery, sea otters were given protected status. By the 1990s, sea otters had recolonized many areas of Alaskan waters, the sea urchin populations had dropped 50 to 99%, and the kelp forests had returned. Now the sea otters and kelp forests are again threatened. Overfishing in the area has deprived killer whales of their normal food supply. The killer whales have turned to eating otters, with the potential for affecting sea urchins and kelp forests again.

Aside from the fish they shelter, kelp forests are of direct economic value. The growth rate of *Macrocystis* is very high, up to 50 cm day^{-1} along the California coast. *Macrocystis* is a source of alginates—complex algal polysaccharides used in textiles, printing, dental creams and adhesives, shoe polish, and as thickening agents in ice creams, puddings, jams, and sauces. Too complex to synthesize, the polysaccharides can be obtained from the natural source, which is commercially valuable. *Macrocystis* can be harvested by mowing to a depth of about 1 m without harm to the forest. They cannot be harvested, however, if sea urchins destroy them.

Seagrasses stabilize soft, sandy sediments and provide shelter for many marine animals

Seagrasses form submarine meadows populated by about 50 species of plants that resemble grasses but are not actually related to them. They originated from salt marsh relatives that had begun the process of adaptation to saltwater. Seagrasses are able to tolerate a wide range of salinity and, like salt marsh plants, often occur in estuaries where salt and freshwaters mix. Some, like *Phyllospadix* on the U.S. Pacific coast (Figure 28.31a), can grow on rocks by penetrating crevices with tough roots, but most seagrasses, like the eelgrass *Zostera* on the U.S. Atlantic coast (Figure 28.31b), grow in sandy or muddy sediments.

Like many wetland and salt marsh plants, seagrasses have horizontally growing rhizomes that branch extensively, putting up clusters of leaves and sending down masses of roots. Air canals (aerenchyma) penetrate seagrass rhizomes and allow transport of oxygen to roots. Fiber cells strengthen seagrass leaves along their margins and help the leaves resist tearing. Leaves also have tannin cells that are distasteful to herbivores. The flowers, which are pollinated underwater, often occur singly and usually produce only one single-seeded fruit per plant.

Seagrass meadows are among the world's most productive ecosystems. Temperate seagrass beds dominated by the eelgrass

FIGURE 28.31 Two species of seagrasses (a) *Phyllospadix* on the U.S. West Coast, and (b) *Zostera* on the Atlantic coast.

Zostera have primary productivities in the range of 500 to 1,000 g C m^{-2} yr^{-1}. Despite this high level of productivity, relatively few species of animals directly consume seagrasses. Endangered species such as the manatee and green sea turtle are among the few that do. Most of the seagrass productivity is consumed as detritus and dissolved organic carbon. Seagrass meadows are a major source of food in these forms to other shallow-water ecosystems.

Seagrass beds stabilize soft and sandy sediments by developing dense mats of rhizomes and roots. These mats are strong enough to withstand storms—even hurricanes. The leaves of seagrasses dampen wave action and slow currents within the beds, allowing sediments to accumulate and preventing shoreline erosion. Seagrass beds form a nursery and shelter for many commercially valuable animal species, including shrimp and scallops. Destruction of seagrass beds for reclamation and housing construction can increase shoreline erosion and have serious impacts on commercial fisheries.

Coral reefs are among the most beautiful and diverse ecosystems on Earth

Even a brief dive around a coral reef in the waters off Florida or the Bahamas reveals the most fascinating aspect of coral reefs—their enormous natural beauty and diversity of life (Figure 28.32). Coral reefs are unique because these massive structures are built entirely by biological activity. Reefs are deposits of calcium carbonate (limestone) produced by animals called *corals*, which are related to jellyfish and sea anemones. Reef-forming corals are colonial, and each individual coral animal or polyp lives in a cup within a massive colonial skeleton. The corals extract calcium from seawater and combine it with carbonate to form the limestone ($CaCO_3$) that makes up their external skeletons. The reefs are made up of the external skeletons of many coral colonies. **Coralline algae**, which are red algae that also precipitate calcium carbonate, grow and spread out as encrusting sheets that cement the individual coral colonies together. Most reef-forming corals have dinoflagellates living symbiotically within their tissues. These symbiotic algae provide the coral animals with food that supplements the zooplankton the coral animals capture with their feeding tentacles. Although corals are present in all oceans, reef-forming corals are found only in warm, shallow, clear waters of the tropics, where they cover millions of square miles.

The global distribution of coral reefs provides a good illustration of the importance of abiotic factors in determining the distribution of distinct populations (Figure 28.33). Reef-forming corals cannot tolerate temperatures cooler than about 18°C or warmer than 36°C (64° to 97°F), and the optimal temperature range is 23°–25°C (73°–77°F). Corals lose their algae (bleach) when exposed to extreme temperatures. They are also limited by water depth. Most reef-forming corals grow at depths of 25 m (80 ft) or less because corals require light for the algae that live within them. The maximum depth for coral reefs is 50–70 m (160–230 ft). Coral animals cannot stand exposure to air for more than 1 to 2 hours and must remain below the level of the lowest tides. Reef-forming corals need a nearly constant level of salinity in the seawater and therefore do not occur near river mouths. Corals also require strong wave action to remove sediments and bring in fresh seawater.

Tropical ocean waters have few nutrients and very low rates of primary productivity (18–50 g C m^{-2} yr^{-1}), but reefs teem with life. Every inch of space is covered in layers of organisms, and the rates of productivity ($1,500$–$5,000$ g C m^{-2} yr^{-1}) are among the highest in the world. How is this possible? The answer appears to be that coral reefs act as accumulators, or sinks, for any nutrients brought in by the ocean waves. The algae living within the coral tissues remove waste products, such as ammonia, phosphate, and carbon dioxide, and provide the corals with food from their photosynthesis. Since the algae are inside their coral hosts, they cannot be swept off the reef. The red coralline algae are encrusting and similarly cannot be swept off the reef. Coral reefs are productive because the corals are tightly coupled to algae that retain and recycle nutrients brought to the reef.

FIGURE 28.32 A coral reef (a) Underwater view of a Pacific coral reef. (b) Close-up of coral polyps showing individual animals with feeding tentacles.

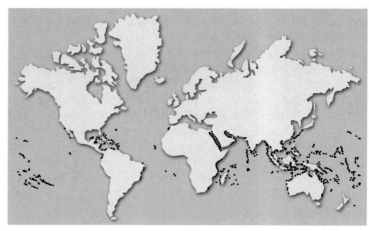

FIGURE 28.33 Global distribution of coral reefs

High productivity, however, does not explain high diversity. Seagrass beds are highly productive but are not very diverse. The diversity of coral reefs appears to arise from the large number of different habitats that occur in and around the reefs. Coral reefs can be divided into a number of different zones with different physical features and different species living in them. The ocean face of a coral reef consists of a deep fore reef, a fore reef, and a reef crest, each with different amounts of wave energy, sunlight, slope, and coral species. Behind the reef crest there may be an algal ridge dominated by coralline algae (Figure 28.34) and a lagoon. The lagoon may have many different habitats, including areas of coralline sand, beds of seagrasses, and beds of green algae such as *Halimeda* and *Cymopolia*, which generate much of the white coralline sand (these algae also encase themselves in calcium carbonate). The massive wave-resistant reefs provide living space and diverse habitats for large numbers of invertebrates and fishes, which make the reefs rich in species, a valuable source of food for local people, and an exciting place for tourists to visit.

Unfortunately, coral reefs are the most threatened ecosystem on Earth, and many researchers believe they will disappear entirely in a few decades unless massive intervention to prevent human exploitation occurs. Increasing human populations in the coastal areas of coral reefs has led to increasing amounts of sediments, sewage, nutrients, and pollution in nearshore waters. The increase in sediment loads suffocates coral animals. The added nutrients trigger rapid growth of phytoplankton and attached algae, which can also suffocate or grow over the corals, killing them. Increasing human populations and tourists place increased demands on local fisheries, resulting in overfishing. Overfishing removes the herbivorous fish that would otherwise keep the populations of algae in check. For the result of all these factors on the reef system in Jamaica, see Essay 28.1, "Jamaican Coral Reefs: Going, Going, . . . Gone." In addition, global warming (Chapter 29) has raised the ocean temperatures in the tropics to the point where waves of coral bleaching now occur frequently (Figure 28.35), threatening the continued existence of coral reefs and the hosts of species they shelter.

28.9 | The littoral zone includes estuaries, salt marshes, and mangrove forests

The littoral zone occupies the area between the high and low tide water levels (see Figure 28.23). Littoral ecosystems form a transition between terrestrial and marine environments. Estuaries and salt marshes are ecosystems at the interface between freshwaters and oceans; mangrove forests form a transition between tropical terrestrial systems and the oceans.

Estuaries have relatively few species of organisms

An estuary is a partially enclosed coastal area where freshwaters and seawater meet and mix (Figure 28.36a). Consequently estuaries show a gradient of salinity from freshwaters at the mouths of rivers to full seawater at the openings to the ocean, and the gradient shifts with the level of freshwater flow and tides. Estuaries therefore go through large variations in physical conditions. These variations cause stresses that limit the types of organisms that can live permanently in estuaries. Marine worms, oysters, clams, crabs, and shrimps are examples of estuarine animals. Vegetation is also limited. Estuaries are generally places of calm waters where bottom surfaces are soft and muddy. Suspended sediments reduce light levels in the water column and limit vegetation. Where water clarity permits, seagrass beds occur in the sublittoral areas, and salt marshes fringe the shores. Both ecosystems export productivity to the estuary in the form of detritus. Primary production in the estuary is due to phytoplankton in the water, diatoms and mats of cyanobacteria on the mud

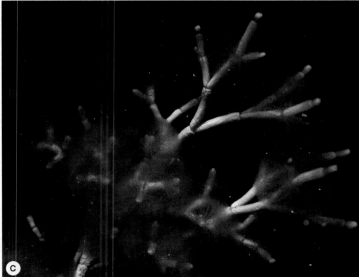

FIGURE 28.34 Green and coralline algae (a) The coralline red alga *Mesophyllum*. (b) The calcareous green alga *Halimeda*. (c) *Cymopolia*.

ESSAY 28.1 JAMAICAN CORAL REEFS: GOING, GOING, . . . GONE

Jamaica is the third-largest island in the Caribbean Sea. The human population of Jamaica has been increasing rapidly, and by the 1960s, fishing pressure was at least twice the level that could be maintained. The numbers of herbivorous fish that eat attached algae were greatly reduced, but the reefs did not show an immediate response to overfishing because the population of the herbivorous sea urchin *Diadema antillarum* initially kept the attached algae in check (the intense fishing also reduced populations of fishes that prey on the urchin). In the early 1980s, *Diadema* near the Panama Canal began to die from a disease. Within 1 year, the disease swept through the Caribbean and eliminated 95–99% of the sea urchins. The cause of this demise of *Diadema* is unknown, but one potential cause has been suggested. Since the 1960s, satellite images have detected dust clouds from northwest Africa moving across the Atlantic Ocean. These dust clouds represent hundreds of millions of tons of dust from the Sahara Desert and contain enormous numbers of microorganisms. Peak years for dust transport correspond to outbreaks of disease on the reefs. After the loss of the sea urchins, the Jamaican reefs underwent a major change in community composition, and that change has persisted to the present.

Species-poor algal beds replaced species-rich coral reefs.

Between 1977 and 1994, coral cover declined from 52% to 3% and attached algal cover increased from 4% to 92% (Figure E28.1). Species-poor algal beds replaced species-rich coral reefs. The nearly complete cover of algae prevented recruitment of new corals. Although the few remaining sea urchins had abundant algal food, their numbers were too low for successful breeding and population growth. Since the change occurred, there has been no indication of recovery. Recovery may take decades, if it can happen at all.

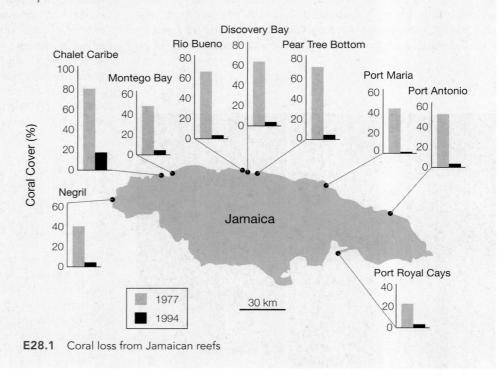

E28.1 Coral loss from Jamaican reefs

surfaces, and large green algae on the intertidal mudflats. Estuaries are important feeding grounds for migratory birds and fishes and nurseries for many species that live elsewhere as adults. Human populations heavily impact estuaries: Major world cities are located on them, they are used as dumps, and they are often filled for development.

FIGURE 28.35 Bleached corals

Salt marshes contain herbs, grasses, and shrubs rooted in soils washed by tides

Salt marshes are wetlands that occur at the land-sea interface in temperate regions, where wave energy is low and sediment is plentiful. They dominate U.S. shorelines from northern Alaska to Florida and Texas, forming a transition between freshwater and marine systems. Salt-tolerant herbs, grasses, and shrubs dominate salt marshes (Figure 28.36b). Once considered unimportant wastelands, salt marshes are valued for their extremely high primary productivity (up to 1,600 g C m^{-2} yr^{-1} for the emergent herbs), wildlife habitat, and association with productive fisheries.

FIGURE 28.36 Littoral marine ecosystems in the temperate zone (a) An estuary, (b) a salt marsh.

They help stabilize coastlines, preventing erosion due to storms. Salt marshes also filter mineral nutrients and sediments from freshwater runoff, helping protect the ocean from pollution. In tropical and subtropical regions, mangrove forests replace salt marshes and serve similar functions.

Salt marsh plants must be able to tolerate extremes of physical and chemical conditions arising from tidal changes and ice scour, high salt concentrations that would be toxic to most plants, and low sediment oxygen. The vegetation, low in species diversity, consists of herbaceous plants such as *Spartina* and some types of shrubs. *Spartina* has a tough, perennial rhizome lying beneath the surface sediments. The rhizome and the roots that emerge from it help anchor the plant. *Spartina* and other salt marsh plants can also reproduce asexually, by sprouting of fragments, to form huge clones. Their seeds are able to germinate even after extended periods in saltwater and are dispersed by tides and currents.

Humans damage salt marshes much as they do estuaries and seagrass beds. Salt marshes may be surrounded by dikes and drained, dredged for channels, used as dumps for waste and pollutants, or filled-in for construction.

In mangrove forests the trees grow in shallow seawater

Mangrove forests form a transition between terrestrial systems and the marine environment in tropical and subtropical coastal areas sheltered from wave action (Figure 28.37). From 60 to 75% of the coastline of tropical regions is covered in mangrove forests, which extend over a larger area than coral reefs. Mangroves are shrubs and trees belonging to 12 genera of flowering plants. The most conspicuous feature of mangroves is the peculiar prop roots that are located above the surface of the mud in which they grow (see Chapter 10). The prop roots are an adaptation to growth in mud, where high bacterial activity and low water movement produce anoxic (low-oxygen) conditions.

Most mangrove forests show zonation of species from the ocean edge inland toward true terrestrial plant communities. Thus some genera of mangrove are adapted to the ocean edge, others to the intertidal, and still others to the high-water level. Mangrove forests are unique in that the trees allow terrestrial organisms to live in the branches and marine organisms to live among the roots. Common marine animals in mangrove forests include mollusks, crabs, shrimp, and peculiar fish that can walk on the mudflats and even climb mangrove roots. These fish, called mud skippers, act very much like frogs and dig burrows for shelter. Like other transitional ecosystems, mangrove forests function as a nursery for commercially valuable fish and shellfish. Mangroves also are the preferred habitat of a number of endangered species such as the American crocodile and brown pelican. Humans destroy mangrove forests by filling them with dirt, cutting them for construction materials, dredging them for channels, and constructing ponds for culture of shrimp and fish. They are also used as dumping grounds for sewage and industrial waste.

FIGURE 28.37 Mangrove forest in the tropical littoral

HIGHLIGHTS

28.1 Aquatic ecosystems have inspired human imagination and shaped the cultures that developed along their shores through travel and trade in goods and ideas.

28.2 Freshwater lakes, streams, rivers, and wetlands are used in many ways by people and support a wide array of wildlife.

28.3 Lakes contain three major habitat types, each populated by distinctive plants and microbes. Floating and swimming algae are the phytoplankton of the pelagic zone and are important food for aquatic animals. Some plants also float at the water's surface, obtaining mineral nutrients directly from the water. Rooted vascular plants and certain algae known as macro-phytes occupy the nearshore littoral zone. The periphyton is a highly productive mixture of algae and other microbes that is attached to littoral rocks and plant surfaces. Decomposer microbes occupy the deep benthic zone, where they recycle nutrients from dead animals, plants, and algae. The properties of water cause the formation of a warm layer of lake water in summer; this layer can become depleted of mineral nutrients by algae and macrophyte growth. In fall and spring, wind circulates the deeper, mineral-rich benthic waters to the surface; these support algae and plant growth in warm seasons. Fresh-water algae and plants are adapted to this cycle of nutrient availability. Many freshwaters have become degraded as the result of human agricultural, industrial, and dam-construction activities. Eutrophication is the conversion of low-nutrient clear waters with high biodiversity into high-nutrient waters that are clouded or clogged by excessive growth of weedy algae and plants. It is easy to degrade freshwaters but difficult to reverse eutrophication.

28.4 Wetlands of many types are found near freshwaters and help protect them from eutrophication. Wetlands also protect nearby land areas from drought, floods, and fires. Wetlands store large amounts of carbon dioxide as organic peat, thereby helping stabilize Earth's atmospheric chemistry and climate. Wetland plants are adapted to low levels of soil oxygen and other features of wetland habitats. Humans have destroyed large areas of wetlands around the world, and invasive plant species also threaten wetland habitats.

28.5 Oceans are essential to humans and life on Earth. Coastal zones represent 18% of the Earth's surface but support 60% of the human population. Oceans moderate the climate of Earth and provide the rain that nurtures life on land.

28.6 Oceans cover 71% of the Earth's surface. Ocean surface currents are determined by the prevailing wind patterns, which are deflected by the rotation of the Earth. The oceans contain complex terrain and may be divided into realms.

28.7 The epipelagic ecosystem contains plankton and nekton communities, and the epipelagic zone produces all the primary production in the oceans. The bacterioplankton are the single most productive group in the oceans. Diatoms, dinoflagellates, and haptophytes dominate the phytoplank-ton. The planktonic food web includes a microbial loop connecting picoplankton to zooplankton and the nekton such as fish. The epipelagic food web is very different from terrestrial food webs.

28.8 The subtidal zone includes kelp forests, seagrass beds, and coral reefs. Large brown algae dominate the kelp forests that grow in shallow waters of cold temperate oceans. Sea-grasses stabilize soft, sandy sediments along shorelines and shelter many marine animals. Coral reefs are among the most diverse ecosystems on Earth. All three subtidal ecosys-tems have high productivity, and all three are threatened by human exploitation.

28.9 Estuaries, salt marshes, and mangrove forests occur in the littoral zones of oceans. Estuaries are subject to wide varia-tions in salinity and contain few permanent resident species. Salt marshes contain grasses, herbs, and shrubs rooted in soils washed by ocean tides. Mangrove forests grow along tropical and subtropical shores where waves are calm. All three systems serve as nurseries for many valuable commer-cial species of fish and shellfish. Salt marshes and man-groves also stabilize shorelines against erosion.

REVIEW QUESTIONS

1. Describe the three habitat zones found in a typical lake.

2. Define the following groups of organisms that are found in a typical lake or waterway: (a) bacterioplankton, (b) phyto-plankton, (c) macrophytes, (d) periphyton, (e) protozoo-plankton, and (f) zooplankton.

3. Define the following four trophic groups that form the food chain of a lake: (a) primary producers, (b) primary con-sumers, (c) secondary consumers, and (d) tertiary consumers. What types of organisms occur in each of these groups?

4. Describe the seasonal circulation of water and nutrients that occurs in a typical temperate lake over the course of a year. What is responsible for these circulation patterns?

5. Describe some of the adaptations made by bacterioplankton and phytoplankton to help them stay afloat in the water col-umn and avoid herbivores.

6. What are the characteristics of oligotrophic, eutrophic, and hypereutrophic waters?

7. Match the following wetland types with the statement that best describes them.

   A. sedge meadows   **D.** riparian wetlands

   B. freshwater marshes   **E.** acid bogs

   C. deep-water swamps   **F.** salt marshes

   ____**1.** Water low in nutrients; dominated by *Sphagnum* moss, shrubs, and trees such as black spruce *(Picea marina)* and tamarack *(Larix laricina)*.

   ____**2.** Often located near other wetlands; dominated by clump-forming sedge plants.

   ____**3.** Distributed in the southern United States; typically contain large trees, such as bald cypress *(Taxodium)*.

   ____**4.** Border streams and rivers, from which they receive their nutrients.

   ____**5.** Neutral to alkaline in pH; calcium abundant, and characterized by cattails *(Typha)*, sedges *(Carex* spp. and others), and certain grasses.

   ___**6.** Occur in coastal areas where wave action is low and sediment abundant; often contain *Spartina* grass.

8. Many wetland plants possess a variety of adaptations to their life in such a windy, watery habitat. Using cattail *(Typha)* as an example, discuss some of these adaptations.

9. List several ways in which human activity has degraded freshwater wetlands.

10. Oceans are critically important to life on land. Describe some ways that the oceans and their inhabitants support terrestrial life.

11. Define the following terms that are used to describe the terrain of ocean basins: a) continental shelf, (b) continental slope, (c) abyssal plain, (d) submarine ridge, (e) sea mount, and (f) trench.

12. The ocean waters in the North Atlantic move in a clockwise direction in a huge circular pattern called a gyre. Describe the pattern of winds and Coriolis force that generate this gyre.

13. In temperate oceans the amount of solar energy striking the surface varies seasonally. Describe the effect of this variation in solar energy on temperate ocean water temperatures and mixing. What effect does this pattern have on the abundance of phytoplankton and zooplankton? Do tropical oceans show a seasonal pattern?

14. Define the following terms used to describe ocean realms: (a) pelagic, (b) epipelagic, (c) littoral, (d) sublittoral, and (e) benthic. In which realm(s) does the primary productivity occur in the ocean?

15. Planktonic organisms are important in the oceanic food web. How are they categorized?

16. What is meant by the microbial loop in the planktonic food web? How does this differ from the classic food web?

17. Briefly describe three littoral ecosystems. What is their importance to humans and other animals?

APPLYING CONCEPTS

1. Imagine filling two sturdy glass containers with a dense suspension of a cyanobacterium such as *Microcystis*. The openings are stopped with large corks, and one bottle is set aside for reference. Using a rubber mallet, the cork on the second bottle is given several sharp taps and then set aside as well. After 15 minutes or so, many of the cyanobacteria in the second (tapped) bottle have settled to the bottom, while those in the first bottle are still in suspension. What happened?

2. In 1950, the body of a man with a noose around his neck was found in a Danish peat bog. Although this individual died 2,000 years ago, his body was remarkably well preserved. What conditions in the peat bog allowed this preservation?

3. Wetlands provide many valuable services to humans and to the environment. Can you name at least five of these services?

4. Lake Davis is a clear lake in the Sierra Nevada mountain range of northern California. Recently, a nonnative predator fish from eastern North America, the northern pike, was found in this lake. Hoping to eradicate the invader, in 1997 biologists at the California Department of Fish and Game poisoned the entire lake with rotenone, a plant-derived substance that is toxic to fish. Why were state biologists so concerned about this one species?

5. Suppose that you are a scientist interested in what life exists throughout the ocean. You descend in a submersible vehicle through a tropical ocean and make observations from the surface to the bottom at a depth of 5,000 m. What would you notice about the temperature, light, and oxygen levels as you descend? What kinds of life would you expect to find at different depths? Would there be any living organisms on the bottom?

6. All organisms in a community are interrelated on some level even though this may not always be apparent. In the cases of kelp forests and coral reefs, describe how human activities that primarily affected one type of organism ended up changing the whole community.

CHAPTER 29

Human Impacts and Sustainability

Kenyan environmentalist Wangari Maathai

In 2004 Wangari Maathai was awarded the Nobel Peace Prize "for her contribution to sustainable development, democracy and peace." For nearly 30 years she has organized poor women to plant trees in Africa.

At the dawn of the last millennium, two North American cultures reached their peaks of population and power. The Anasazi or "ancient ones," who occupied much of the Southwestern United States, are known for multi-storied cliff dwellings as well as baskets and pottery of great artistry. The Mississippian "mound builders" were centered at the southwestern Illinois site known as Cahokia—a planned city whose population in 1200 was greater than that of London. Both the Anasazi and the Cahokians were skilled agriculturalists who cultivated corn, squash, and other crops. But in the late 1200s, both cultures declined precipitously. By 1300, Cahokia was deserted, as was the last-known Anasazi-occupied site—Canyon de Chelly in northeastern Arizona.

For many years, the disappearance of these cultures was a mystery; but archeological studies have since revealed that population pressures, coupled with environmental change, led to hunger and malnutrition, internal strife, political repression or warfare, and dispersal of the people. Despite their carefully engineered irrigation systems, the Anasazi were unable to cope with drought arising from climate change. The Cahokians had deforested their land, causing the loss of habitat for animals that were important as sources of dietary protein. Silt from the denuded hillsides polluted their fishing streams, and smoke from cooking fires polluted their air, leading to respiratory disease.

The archeological remains of these two great cultures offer an invaluable lesson—human survival depends on the maintenance of healthy environments. Modern humans depend on clean air and water, rich soil, and a wide diversity of plants and other organisms that provide these resources, as well as food and other products. But on a global basis, burgeoning human populations are polluting Earth's air and water, changing its climate, degrading its lands, and causing the extinction of many organisms. Will modern humans be able to escape the fate of the ancient Anasazi and Cahokians? Will we be able to control our populations, grow enough food, improve air and water quality, reverse climate change, restore degraded lands, and protect habitats and organisms? Maintenance of a reasonable standard of living for humans within a healthy environment—sustainability—is the topic of this chapter. ■

Anasazi ruins

29.1 | Sustainability is the maintenance of humans together with healthy environments

In 1968, ecologist Garret Hardin published an essay in *Science* entitled "Tragedy of the Commons" in which he discussed the impact of human population growth on the exploitation of natural resources. By the term "commons" he meant resources that are shared by everyone and owned by no one, such as water, air, and the oceans. The "tragedy" was (and still is) that since no one owns these resources, everyone tries to exploit them beyond their ability to recover in order to maximize their personal or national gain. This conflict between human population growth and the exploitation of natural resources has grown more serious in the years since his essay was published. In 1968, the human population of the world stood at 3.5 billion; it has now reached 6.3 billion. That population growth, accompanied by increasing technology, economic activity, and rising global levels of affluence, has increased pressures on "common" resources such as air, freshwater, and the oceans. The result of these pressures has been species extinctions and loss of biodiversity, degradation of ecosystems and their vital functions, global climate change, pollution of freshwaters, loss of soil fertility, and emergence of new global diseases such as AIDS and SARS. Rising human populations and resource depletion threaten armed conflicts over scarce resources. Evidence is building that the human population as a whole is endangering its own long-term interests by living beyond the means of the Earth to support it.

The depletion of natural resources and the degradation of the environment over the past decades have not gone unnoticed and have led to the development of the ideas of **sustainability**, sustainable development, and a science of sustainability. Sustainability incorporates the concept that resources can be extracted from a high-quality environment, providing an acceptable standard of living to Earth's human population, on a long-term basis without a reduction of value to future generations. Value has more than one meaning. The value from agricultural land may be in the form of food. Value from a temperate forest ecosystem can be in lumber but may also be in recreation activities such as camping, hiking, and hunting. Sustainability requires that future generations of humans enjoy the same or higher level of all these values than the present generation.

Humans require fertile soils, clean water, and unpolluted air. For food and other materials, people also need a wide variety of plants, microbes, and animals—Earth's **biodiversity**. Diverse organisms also provide ecosystem services that benefit people as well as other organisms. **Ecosystem services** include soil formation and enrichment, water purification, oxygen production, carbon dioxide storage, temperature control, and crop pollination; human activities affect these processes in many ways. People must also have sources of energy that minimally harm soil, water, and air quality. The availability of these resources maximizes human health and minimizes social and political strife and armed conflict.

Sustainability science offers options for combining human economic well-being with environmental sustainability. Carbon dioxide from the burning of fossil fuels (coal, oil, and natural gas) can be captured and stored. Alternate energy sources such as solar, wind, and hydrogen fuel can be developed and employed. Agricultural productivity can be sustained and enhanced through sustainable agricultural techniques. Freshwaters do not have to be polluted. Marine ecosystems can be restored and coastal waters cleaned to revive ocean fisheries, which can be managed on a sustainable basis. Threatened species and ecosystems can be saved and the biodiversity of the Earth maintained through conservation biology.

These options are not mere dreams of a technological utopia. Realization will require time, funding for research, development and implementation, and new systems of global management and governance. Current funding for these technologies is only a small fraction of global military spending (which was estimated at between $900 billion and $1 trillion in 2004). The public generally does not understand that real scientific options exist for the future of society. The real question is not if science can help with sustainable development but if scientific evidence and technological options can succeed against economic and political resistance. This chapter will first present the global human impacts that are affecting the Earth and then discuss a number of scientific and technological options that could reduce these impacts.

29.2 | Humans impact the global environment in many ways

Humans affect the global environment through their increase in population size, which results in transformation of much of the Earth's land surface area and a degradation of the environment. With the increase in human population, there has been an increase in per capita energy consumption through burning of fossil fuels (coal, oil, and natural gas) and, in developing countries, through burning of wood along with removal of forests. Consumption of these carbon-based energy sources has lead to massive release of carbon dioxide into the atmosphere, resulting in global warming. Global warming has affected every ecosystem on the Earth and will do so with increasing force in the future. In addition, human industrial activity has released large amounts of sulfates and nitrates into the atmosphere, producing acid rain in every region on the globe where heavy industry is common. Intensive agricultural development has depleted soil fertility through loss of nutrients and increased erosion of topsoil. The resulting lost nutrients and soil enter rivers where they are joined by urban and industrial pollutants, all of which flow into the coastal zones of the oceans. The coastal zones consequently suffer from pollution and develop dead zones where fish cannot survive. Pollution and overfishing have depleted marine fish stocks, which are mostly located in coastal areas. Finally, human population growth and economic activities have transformed and degraded ecosystems, resulting in loss of species and biodiversity on a global scale.

How many people can Earth sustain?

A fundamental sustainability issue is the level of human population that Earth can support, and at what standard of living. The global human population currently exceeds 6 billion, and is forecast to be 7.5 billion by 2020, and about 9 billion by 2050 (Figure 29.1). Imagine 1.5 times the number of people occupying your classrooms, shopping malls, roadways, and airports—this is what life could be like in another half-century.

Of the current Earth population, 1 billion people are malnourished, and 2 billion live in what U.S. residents would describe as poverty. Some experts suggest that Earth might be able to support as many as 20 billion people, but only if they all lived at the lowest possible standard of living, a prospect that most people would not find attractive. Ecologist David Pimentel has assembled data indicating that if all the people in the world had a living standard equivalent to one-half the current U.S. standard, the Earth's carrying capacity (Chapter 25) would be 2 billion people, and that of the United States would be 200 million. The current U.S. population is approximately 285 million. In short, Dr. Pimentel concludes that humans have already exceeded the carrying capacity of both the United States and the Earth.

Dr. Pimentel's work also indicates that current large populations are precariously subsidized by the nonrenewable fossil fuel energy used to produce food. Energy required for modern food production is 10 times the amount contained in

the food and is a much greater energy investment than was made by farmers in preindustrial times. Some experts predict that fossil fuels will be exhausted or too expensive to mine by the year 2025. If the energy subsidy for food production were to end, four times as much farmland would be required for everyone to raise their own food. But the lands most suitable for farming are already under cultivation, and conversion of remaining forests and grasslands to farmlands will drastically reduce the environmental services that these ecosystems currently provide.

Human population growth is correlated with environmental degradation

Environmental degradation provides another indication that human populations have exceeded Earth's carrying capacity. The link between increasing human population and environmental degradation in ancient times has been well established by archeological studies. In addition to the demise of the Anasazi and Cahokians, ecological decline also led to collapse of the 600-year-old Mayan state in the Yucatán region of Mexico. In the highland valley of Mexico, three major civilizations—including the Aztecs—rose and fell within 1,500 years. Each underwent population expansion followed by environmental depletion and decline. Past cultures made technological improvements that provided short-term solutions to the need for increased food production, but these did not solve the problem of environmental degradation. For example, 3,000 years ago the Sumerians (who lived in what is now Iraq) were forced to replace highly productive wheat crops with barley, which was less desirable but more tolerant of high soil salt levels resulting from intensive irrigation.

Modern humans have transformed 40–50% of the Earth's land from natural to agricultural or other human-oriented landscapes. Population projections indicate that an additional one-third of natural lands will be modified within the next 100 years. Land transformation is expected to occur predominantly in Latin America and sub-Saharan central Africa, causing loss of about one-third of existing tropical and temperate forests, savannas, and grasslands. Within the next century, agriculture is projected to result in a nearly threefold increase in nitrogen and phosphorus water pollution, causing even more extensive eutrophication of natural waters than now occurs (Chapter 28). Humans currently intercept for their own use more than 50% of freshwater resources and will likely demand more than 70% by 2025, reducing water sources for other organisms. The need for wood and other forest products is projected to double in the next 50 years. As a result of this pressure, natural forests are rapidly disappearing.

The balance of evidence shows that humans have influenced climate by generating heat-absorbing gases while burning fossil fuels in furnaces and car engines. The United Nations-sponsored Intergovernmental Panel on Climate Change (IPCC) has concluded that our planet could be nearly 6°C warmer in 2100 than it is today. However, atmospheric scientists also recognize that microscopic particles of sulfate, soot, and

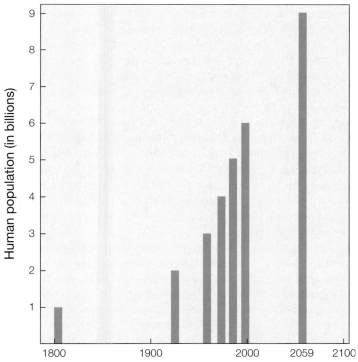

FIGURE 29.1 Human population growth It took from the dawn of humankind until around 1804 to reach a global population of 1 billion. Since then, however, the amount of time it has taken to reach additional billions has decreased dramatically. A population of 9 billion people may be reached around the year 2059.

organic material produced by burning fossil fuels and volcanoes influence Earth's temperature. These particles may have a cooling effect that partially offsets the warming influence of carbon dioxide and other "greenhouse gases." All these factors have to be taken into account in efforts to predict the future climate of the Earth.

Global warming is affecting every ecosystem on Earth

In 1958 Charles Keeling began measurements of atmospheric carbon dioxide from the top of Hawaii's Mauna Loa volcano (see Chapter 1). As measurements accumulated, he noticed regular seasonal changes in the atmospheric levels of carbon dioxide and also a steady rise in the levels over succeeding years. In 1958 he found 315 parts per million (ppm) of CO_2 in the atmosphere. Present levels are now at 375 ppm. Each year fossil fuel consumption vents 25 billion tons of CO_2 into the atmosphere. The United States produces about 5 billion tons of carbon dioxide (20%) of this total per year. Average global temperature is up by 0.6°C (1°F) over the past century. That increase may not seem like very much but it is having global effects, especially in high-latitude polar environments and alpine environments where the temperature rise is more severe. The city of Barrow, Alaska, has recorded an increase in temperature of 2.3°C (4.2°F) over the past three decades. Over the same time period, Arctic sea ice has declined by 10% in area, and woody shrubs have invaded the tundra as winters have warmed. Global warming has also been noted in Antarctica, where the average temperature in the Antarctic Peninsula has risen 5°C (9°F) over the last 50 years. Researchers in the Antarctic have documented changes in food webs such as declines in certain penguin populations and invasions of new species from subantarctic areas farther north. The only two flowering plants in Antarctica, *Deschampsia* and *Colobanthus*, have significantly increased their range and numbers over the past decades (Chapters 26, 27). Alpine areas are showing similar effects around the Earth. In 1910 when Glacier National park was established in Montana, the park contained 150 glaciers. Today, it holds fewer than 30, and those glaciers have decreased in area by over 60% (Figure 29.2). The snows of Kilimanjaro, made famous by Ernest Hemingway, have declined by over 80% since 1912. In the mountains of the western United States, the level of winter snowfall is critical to the freshwater supply for the spring and summer. The level of snowpack has decreased by as much as 60% in some areas, reducing summer stream flows and threatening an increase in fires.

In temperate regions, global warming effects have also been detected, but the observed effects are less extreme. In temperate freshwater lakes, the date of autumn freeze-up is 10 days later and the date of spring ice breakup 9 days earlier than 150 years ago. The number of days that Lake Mendota in Madison, Wisconsin, has been frozen over has been recorded since 1860 (Figure 29.3). Lake Mendota now averages 39 fewer days frozen over (80) than it did 144 years ago (119). Similar changes have been reported for lakes in Europe.

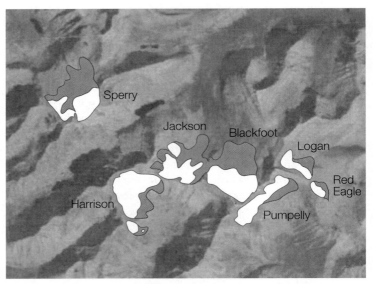

FIGURE 29.2 **Glacier retreat in Glacier National Park** The glaciers in the Mount Jackson area occupied the areas in white in the mid-1990s. In the mid-1800s, they also included the area shaded purple. In the early 20th century, the park contained about 150 glaciers, less than 30 of which currently survive. Some experts predict that within 30 years few, if any, glaciers will remain.

Changes in the behavior of plants and animals have been noted. In Europe many plants now flower about 1 week earlier and shed their leaves 5 days later than 50 years ago. The frequency of heat waves in temperate areas appears to be increasing under the influence of global warming.

Tropical regions also are showing effects of global warming. In tropical mountains a unique type of forest called cloud forest lies between 2000 and 3000 m (6,600 and 9,800 ft) altitude where the forests are nurtured by clouds of mist. Cloud forests harbor many unique species. Under global warming, the clouds of mist are forming farther up the mountains, causing some species to move to higher altitudes. The most pronounced effect of global warming in the tropics is on coral reefs. Coral reefs are sensitive to temperature, and elevated temperatures cause the corals to lose their symbiotic algae and bleach (Chapter 28). Global warming is increasing the number of these bleaching events and threatening the survival of the reefs.

The oceans as a whole are showing the effects of global warming. Ocean temperatures are increasing in most surface waters and at considerable depth. As ocean temperatures rise, seawater expands, and freshwater enters the oceans from the melting of polar ice. As a result of these changes, global sea levels have risen 10–20 cm (4–8 in.) over the past 100 years. Rising sea levels pose a threat to the lives of the 100 million people who live within 1 m of mean sea level. There is growing evidence that global warming may be causing an increase in the frequency and intensity of tropical storms. Tropical storms threaten the lives and property of people living in coastal areas. Global warming is also affecting ocean circulation patterns. The ocean phenomena known as El Niño occurs in the South Pacific at intervals of

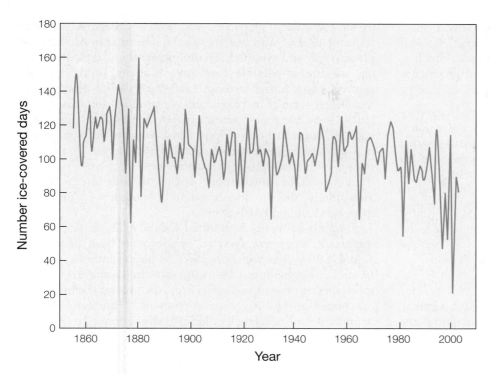

FIGURE 29.3 Ice cover data for Lake Mendota The number of days that ice covers this southern Wisconsin lake has declined markedly since the mid-1800s.

10 or more years. Normally, deep, cold, nutrient-rich waters rise to the surface along the west coast of South America, where they support a rich fishery. During an El Niño event, these nutrient-rich waters are shut off and the food web collapses. The extreme climate of the El Niño of 1997–1998 affected 117 million people, killed 21,000, drove 4.9 million from their homes, and cost $14 billion in damages. Global warming may be causing the El Niño phenomenon to occur more frequently and with greater severity than previously.

Humans impact the global environment in a number of other ways

In addition to global warming, human activities affect the world's environment through the production of acid rain, soil erosion and depletion of nutrients, clearing of forests, pollution of freshwaters and coastal ocean zones, overfishing, and the transformation and degradation of ecosystems, with a corresponding loss of species diversity.

Acid rain When fossil fuels are burned in cars and industries, more gases than just carbon dioxide are released into the atmosphere. Combustion also releases sulfates and nitrates, which combine with rainwater to form sulfuric and nitric acids. These acids enter lakes and streams, where they acidify freshwaters. When these acids fall on soils, the hydrogen ions (H^+) exchange with mineral ions such as calcium that are associated with soil particles (Chapter 27). The mineral ions are washed out of the soil, decreasing soil fertility and adding to the nutrients in freshwaters, where they can lead to eutrophication. Acidification of freshwaters can also eliminate fish from lakes. Acid rains threaten the fertility of agricultural soils, forest productivity, and the health of freshwater ecosystems.

Depletion of soil fertility and erosion Humans have converted vast areas of forest and grassland into agricultural land, resulting in a loss of biodiversity. Modern mechanized agriculture emphasizes growth of single crop species in large fields (monoculture) with intensive use of chemicals and machinery. Large areas of soil are left exposed to the elements for all or part of the year by farming in distinct rows and removing plant litter from soil surfaces. Such practices allow erosion of soil particles by exposure to rain and wind, loss of soil mineral content in water runoff, and loss of soil organic carbon due to exposure to air and runoff. Lost minerals must be replaced with chemical fertilizers. Growth of crops in monoculture favors large populations of insects that feed on that crop, requiring use of chemical pesticides. As a result, topsoil erodes from fields and flows into streams and rivers, accompanied by soil minerals, soil organic carbon, and residues of pesticides and herbicides. Freshwater resources are consequently degraded.

Pollution of coastal zones The soil particles, minerals, and agricultural chemicals washed into freshwater streams and rivers find their way to the coastal zones of the oceans. Along the way to the ocean, they are joined by many types of urban runoff such as sewage and a host of industrial waste products that may include heavy metals, toxics, acids and bases, organics, and even radioactive substances. More than 60% of U.S. coastal rivers and bays are moderately to severely degraded by nutrient runoff. In the coastal zones, the excess of nutrients from fertilizers results in blooms of algae. When these algae die, their decomposition depletes oxygen in the coastal waters. When these effects are severe, the depletion of oxygen creates a **dead zone** where marine life is killed by suffocation. Some 146 dead zones have been recorded (43 in coastal U.S. waters), and the number is increasing with each passing decade. Every summer a vast dead zone develops in the Gulf of Mexico where nutrients enter from the Mississippi River (Figure 29.4). This area now covers up to 21,000 square kilometers, which is larger than the state of New Jersey, and disrupts a productive shrimp harvesting industry.

Overfishing   Dead zones are a serious threat to the fishing industry, but the fishing industry also harvests fish in a way that is not sustainable. More than 30% of the fish populations that have been assessed are being harvested in an unsustainable manner. The biomass of large predatory fish is now only about 10% of pre-industrial levels. Many marine species are being driven to extinction as the tragedy of the commons continues. According to a report by the United Nations, "Fisheries are a shambles, grossly mismanaged and overexploited almost everywhere."

Transformation of ecosystems leads to loss of biodiversity

The rapid increase in human population has resulted in a global-scale transformation of natural ecosystems into agricultural and other human-modified environments. Half of all the land surface area of Earth has already been transformed and an additional third will likely be transformed in the next century. Humans also degrade ecosystems by using them as dumping grounds for their wastes. Freshwater lakes, streams, and wetlands as well as marine estuaries, salt marshes, and mangrove forests are often degraded by draining, dredging, dumping, or "reclamation" because they are mistakenly seen as having little value. Transformed and degraded ecosystems have less biodiversity than natural ecosystems, and they are less capable of carrying out vital ecosystem services and functions.

Ecosystem (or environmental) services are processes performed by Earth's organisms that help maintain the stability of Earth's environment. Soil enrichment, for example, results from the activities of nitrogen-fixing bacteria and mycorrhizal fungi, as well as soil animals such as nematodes and earthworms. Bacteria, algae, and wetland plants carry out water purification—the removal of toxic compounds and absorption

FIGURE 29.4   Dead zone in the Gulf of Mexico The Gulf of Mexico contains the second-largest dead zone in the oceans, covering an area larger than the state of New Jersey. The dead zone is marked by lighter blue and tan (sediment-laden) waters.

of excess mineral nutrients that would otherwise stimulate harmful algal or waterweed growths. The oxygen produced by plant, algal, and cyanobacterial photosynthesis replenishes our air, making it possible for fungi, animals, and other heterotrophic organisms to exist. Long-lived woody plant tissues, fossil fuels formed by plants and algae, moss-dominated peatlands, and limestone (calcium carbonate) deposits formed by algae all store carbon for long periods, maintaining a balance between atmospheric carbon dioxide and oxygen that is compatible with modern life. In the absence of such carbon storage, there would be much more carbon dioxide in the Earth's atmosphere, and the climate would be too hot for humans and most other life on Earth to exist.

High biodiversity is essential for the maintenance of these important ecosystem services. Species that have particularly valuable ecosystem functions, such as photosynthesis, nitrogen fixation, or pollination, are members of **functional groups.** The more species present in a given area, the greater the number of functional groups likely to be present. Transformations or degradation of ecosystems that affect functional diversity and composition are likely to have large impacts on ecosystem services. Some species have ecological effects much greater than expected, given their biomass, and are thus known as **keystone species.** They are analogous to the keystone at the top of an arch, which maintains the structural integrity of the arch. Loss of keystone species has resulted in some dramatic ecosystem changes, much as removal of an arch's keystone would result in collapse. Nitrogen-fixing bacteria, mycorrhizal fungi, extensive offshore forests of giant kelps, and the palms and figs that feed many tropical animals are among the keystone species described in this book.

As a result of ecosystem transformations and degradation, current species extinction rates range from 100 to 1,000 times greater (depending on the group of organisms and geographical region) than occurred before modern humans appeared about 40,000 years ago. Studies of fossils and archeological remains suggest that human hunting is the major factor explaining disappearance of one-quarter of all modern tropical bird species. Archeologists have also linked evidence for extensive human hunting with the demise of mastodons, giant sloths, and other amazing animals that lived in North America about 12,000 years ago—a time when climate was also changing rapidly. At present, species losses are occurring 1,000 times faster than new species are originating. In fact, modern extinction rates rival those following the famous asteroid collision 65 million years ago, when 70% of Earth's species disappeared forever (Chapters 21, 22). Today, more than 1 in 10 plants are on the brink of extinction, and 34,000 plant species are at risk, according to the Threatened Plants Report from the World Conservation Union. One-third of lily and palm species and 15 of 20 yew species are endangered. In the United States alone, 30% of plant species are in peril.

One of the ways human activities reduce biodiversity and lead to species extinctions is through habitat fragmentation. Human activities break up or fragment natural habitats, creating "islands" (Figure 29.5) that harbor small populations of organisms that were previously part of larger populations extending over a wider continuous area. When separated from

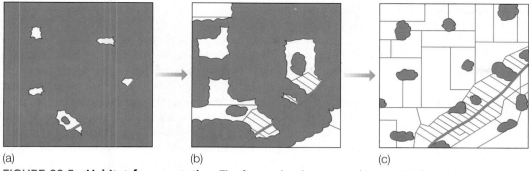

(a)   (b)   (c)

FIGURE 29.5 **Habitat fragmentation** The forest plot diagrammed in (a)–(c) is becoming more fragmented over time as the land is converted to agricultural plots.

each other, members of subpopulations cannot easily mate with individuals outside the local group. Individuals in small, isolated subpopulations have few potential mates, leading to inbreeding. Inbreeding depression occurs when small populations become inbred, and vitality decreases as deleterious genes become more common. Inbreeding depression increases the probability that a small population will go extinct. Genetic drift is another mechanism that can lead to extinction of small populations. Genetic drift occurs when particular alleles of genes within a small population are lost due to chance events alone. Genetic drift results in a decrease in genetic diversity, and low genetic diversity increases the chance of extinction. Populations may gain genetic diversity, and thus reduce the chance of extinction, if organisms can move among habitat "islands." If the colonization rate is greater than the extinction rate, a larger population can persist, so dispersal has an important effect on extinction. Organisms that cannot disperse very far are more vulnerable to extinction than those having effective longer-distance dispersal ability. Organisms that occur only in particular areas (i.e., have low dispersal ability) and are often rare are known as **endemic species** (Figure 29.6). Regions that contain many endemic species are at risk for loss of biodiversity. In other words, endemism is a predictor of extinction potential.

Habitat fragmentation is one type of disturbance of a natural environment. When natural ecosystems are disturbed, they become more susceptible to invasive species. Invasive species are usually nonnative organisms that exhibit aggressive growth when introduced into areas lacking their normal predators and diseases that might control them. Examples of invasive species introduced earlier include spotted knapweed and leafy spurge (Chapter 26), honeysuckle and garlic mustard (Chapter 27), and purple loosestrife (Chapter 28). Invasive species can cause the extinction of native species through aggressive competition for space and nutrients. Invasive species have certain common features. They often develop rapidly from seedling to reproductive adult stage, produce many progeny, have efficient dispersal mechanisms, tolerate a wide range of environmental conditions, and vary in their appearance and behavior. Many invasive plant species are able to reproduce asexually (Chapter 13), for example, pampas grass (Figure 29.7), a native of Argentina that has become a pest in California. Invasive species that can reproduce asexually seem to be more susceptible to biocontrol through planned introduction of natural enemies. This is because

FIGURE 29.6 **An endemic plant** This species of baobab tree is found only in Madagascar.

FIGURE 29.7 **Pampas grass is a common nonnative invader**

populations derived mainly from asexual reproduction have little potential for genetic variation and therefore do not readily become resistant.

29.3 | The concept of sustainability has many different dimensions

Many people believe that human ingenuity will find ways to produce more food and otherwise sustain humans without further degrading nature and to develop methods for restoring degraded ecosystems. Human needs for energy have caused air pollution, climate change, and acid rain. Scientists and engineers have developed improved and alternative energy technologies that offer hope of cleaner energy production without the release of greenhouse gases. Expansion of agricultural activities is recognized as one of the most significant ways by which humans alter global environments. Agricultural productivity increased nearly 500% in the years from 1700 to 1980, thanks to technological improvements such as high-yielding crop varieties, chemical fertilizers and pesticides, irrigation, and farm and food processing machinery. However, the environmental price for increased productivity has included more soil erosion, lower soil fertility, reduced biodiversity, pollution of groundwater, eutrophication of rivers and lakes, and impacts on atmosphere and climate. Achieving sustainability in food production will require technological advances in both crop biology and pollution control. Techniques for restoring degraded environments offer the prospect of repairing ecosystems so that they can more effectively perform environmental services.

Stable human populations are necessary for sustainability

The current world human population stands at 6.3 billion and is projected to reach 8.9 billion by 2050. Every year the human population adds 77 million new people to the world, posing major challenges for providing food, housing, education, health, employment, and public order. The human population passed through a significant stage in 1965–1970 when its growth rate reached a maximum of 2.1% per year. The global population growth rate has since declined to 1.2% in 2002. In the coming decade, more than half of all people will live in cities for the first time in history. These are projections based on assumptions about human population growth. The problem with demographics—the science that focuses on human population growth and economic activities—is that the models used to project human population growth do not include food, water, housing, education, health, religious values, family structure, and the physical and biological environment upon which humans depend for their support. Humans are treated as if they were free of the constraints that control all other species such as carrying capacity (Chapter 25). Only recently have demographic models included the impact of such diseases as AIDS.

Demographic projections should therefore be taken with a dose of skepticism.

From the standpoint of sustainability, the size of the global human population is important in terms of the per capita living standard. If all resources were managed in a sustainable manner, there would be some total amount of resources that would be available for the human population each year. The average standard of living of that human population would then depend on the number of people among whom that amount of resources had to apportioned. If the human population was stable at some level and resources were managed in a sustainable manner, the standard of living would be stable except for technological innovations. If resources were managed on a sustainable basis but human population continued to increase, the average standard of living would have to decline. On the other hand, if human population continues to grow and resources are exploited in an unsustainable manner, the resources would eventually fail and the human population would crash. Sustainability science looks to a future stable global human population. Harvard University biologist E. O. Wilson predicts that the global human population will pass through a "bottleneck" around 2100, after which the world might reach a stabilized and perhaps slowly declining global population that would be supported by technological systems consistent with long-term environmental sustainability.

Innovative technologies may improve energy sustainability

Renewable energy sources include wind, sun, flowing waters, the Earth's heat, and biomass derived from plants and animals. For example, on some farms animal manure is converted to methane and then burned to generate energy, in the process preventing water pollution.

Wind farms are a possible cost-competitive renewable energy source (Figure 29.8). Wind farms are possible over

FIGURE 29.8 Wind farm

much of the United States, and the same land can be used for farming, ranching, and forestry. Solar cells (Figure 29.9) and hydroelectricity produced by dammed rivers currently make the greatest contribution to renewable energy in the United States. A square photovoltaic (solar panel) system about 100 miles (160 km) on a side would produce the annual energy needs for the entire United States, and if wind and geothermal energy sources were added, even less solar collection area would be needed. Photovoltaic systems can be installed on the roofs of existing houses to provide power and heat. In the Netherlands, one community includes 500 houses built with photovoltaic panels that supply 60% of the community's power needs. Solar energy systems pay for themselves within 3–4 years, and wind-farm equipment pays for itself within 3–4 months. After 20 years of development, wind and solar energy systems have seen a tenfold decline in cost, but they still represent less than 1% of total U.S. electricity generation. Much more needs to be done with these sources of energy, which could power homes and businesses without producing greenhouse gases.

Most energy is still derived from fossil carbon compounds (coal, oil, and gas) that release carbon dioxide when consumed. The world mainly runs on a carbon economy. An alternative, which has been receiving a great deal of attention, is hydrogen. When hydrogen is burned, water is produced. Hydrogen is a versatile energy-storage system and strong energy carrier, but there are problems. The United States alone would need 150 million tons of hydrogen each year to power its cars and trucks. Hydrogen cannot be mined. It must be extracted from biomass or water, and either process consumes energy. If hydrogen is made from fossil fuels or biomass, carbon dioxide is still released. Solar energy or wind systems can produce hydrogen by splitting water (electrolysis) without producing carbon dioxide, but these processes are more expensive at present. Hydrogen can be transported by gas pipelines or can be generated on-site,

but a hydrogen economy would also require a nationwide network of hydrogen fueling stations and cars and trucks with engines designed to use hydrogen.

The vast majority of current vehicles use gasoline-fueled internal combustion engines (ICE), but hybrid cars are available now that combine ICE power with electric power to achieve higher fuel economy. There are two types of hydrogen-fueled vehicles proposed. One is an internal combustion engine powered by hydrogen (H2 ICE) and the other is powered by hydrogen fuel cells (FC). Fuel cells are expensive, and FC technology is not yet available commercially. The best current strategy for the United States might be to move toward hybrid vehicles now for greater fuel economy and reduced CO_2 emissions and hydrogen-powered vehicles in the future. One country, however, is in a position to move toward energy independence in the immediate future (see Essay 29.1 "Fossil Fuel-Free in Iceland").

Sustainable agriculture reduces erosion, nutrient and organic carbon loss, and chemical use

Modern conventional farming has produced a great increase in productivity but at the expense of using large amounts of energy in the form of fossil fuels and numerous agricultural chemicals. Topsoil losses are high from erosion, and the resulting loss of nutrients and soil organic carbon depletes soil fertility and requires use of fertilizers to replace lost minerals. Worldwide, about one-third of agricultural land has been lost to erosion over the past three decades.

Sustainable agriculture aims to create a farm landscape that resembles as closely as possible the complexity of a natural ecosystem. Plowing that brings subsoil to the surface is avoided because soil is damaged whenever it is turned over and the crop remains are buried where they cannot decay properly. The soil is instead kept covered with dead plant material or a cover crop throughout the year. The plant cover moderates soil temperatures, increases water penetration and storage, and prevents erosion by wind and water. Compost and aged manures are used to recycle nutrients on the farm rather than chemical fertilizers. Crops are rotated to break the life cycles of insect pests that tend to build up large populations when the same crops are planted year after year in the same area. Strip intercropping, in which different crops are planted in adjacent rows, is better than crop rotation because it brings a higher level of biodiversity. Corn and soybeans or cotton and alfalfa hay are examples of intercropping; the soybeans and alfalfa provide nitrogen to the soil through the nitrogen-fixing bacteria in their roots. When onions and carrots are planted together in a field, the onions mask the scent of the carrots and hide them from the carrot flies that feed on them. Planting field borders and windbreaks of shrubs and trees provides habitat for animals that can feed on insect pests. Sustainable agriculture uses biocontrol agents—beneficial insects, mites, bacteria, viruses and nematodes—to control crop pests. Many other techniques are used to maintain a healthy, sustainable farm ecosystem.

FIGURE 29.9 Solar cell array These tracking solar panels are near Barstow, California.

ESSAY 29.1 FOSSIL FUEL-FREE IN ICELAND

Iceland is an island nation located in the North Atlantic just south of the Arctic Circle. The island is part of the mid-Atlantic Ridge, and many of its mountains are active volcanoes. Iceland is blessed by an abundance of geothermal and hydro-electric power, which supplies almost all its 300,000 residents with heat and electricity. Iceland imports fossil fuel only to power cars, trucks, and its shipping fleet, which uses about half of the country's imported oil. Iceland is moving toward energy independence by converting to a hydrogen economy. The target date for conversion is 2050.

In 1997, the government of Iceland formed Icelandic New Energy, and the company bought three hydrogen-powered buses for the city bus fleet and a hydrogen fueling station to keep them running (Figure E29.1). The hydrogen is produced using electricity to split water into hydrogen and oxygen. The hydrogen

Iceland is moving toward energy independence by converting to a hydrogen economy.

is stored as a compressed gas. The buses are fitted with rooftop tanks for the hydrogen, and it takes only 6 minutes to fill a tank for a day's service. When in operation in the capital city of Reykjavík, the buses emit a trail of steam out of their exhaust pipes that is almost pure water. After the first 9 months of

operation, the buses had driven over 40,000 km, and public support for a hydrogen economy was at 93%. Icelandic New Energy is working on buying a dozen hydrogen-powered cars for government and corporate employees. The biggest problem for Iceland will be the conversion of its shipping fleet to hydrogen. Ships need a lot more fuel and run far longer than does a city bus.

E29.1 Hydrogen fuel station in Iceland

Biotechnology may be able to enhance sustainable agriculture by improving the crop plants. One strategy for increasing food production is to use plant breeding or genetic engineering techniques to improve crops. For example, "Green Revolution" crop scientist Norman Borlaug developed pest-resistant wheat varieties that greatly increased yields (from 0.75 tons per hectare to 8 tons per hectare). Beneficial genetic engineering is protecting Africa's crops from the parasitic plant *Striga* (Figure 29.10). Commonly known as witchweed, this pest attacks corn, sorghum, and millet, Africa's three most important cereals. Tiny *Striga* seeds can lie dormant in soil for as long as 20 years, germinating when they encounter a chemical called sorgolactone secreted by host plant roots. Biotechnologists have identified a gene that reduces sorgolactone production and hope that by incorporating this gene into crop plants, African crop yields will improve—at least until *Striga* adapts to lower sorgolactone levels.

In the short term, conventional agriculture can produce higher yields per acre, but in the long term, the costs of soil erosion, loss of farmland, pollution of freshwater and groundwater, and biodiversity loss outweigh the short-term gains. Even if more land is necessary for sustainable agriculture than conventional agriculture, the sustainable farmland will last.

Sustainable use of ocean resources and restoration of coastal zones are future goals

The coastal zones of the oceans are in serious trouble from pollution, and ocean fisheries are on the brink of collapse. Protecting ocean ecosystem health and mandating sustainable use of ocean resources are future policy goals. Toward these objectives, an international system of fully protected marine reserves could be established. Such reserves protect marine ecosystems and have been proven to increase the biomass of fish within their boundaries. Reserves can also replenish adjacent areas by dispersal of adult, juvenile, and larval organisms.

Scientific data indicate that the principal objective of marine fishery policy should be the protection of marine ecosystems. To this end, experts suggest that access to all fisheries should be limited and subject to licensing and regulation to assure that total catches remain below biologically safe limits. Total fishing capacity will have to be reduced in some cases. Experts recommend that a permanent fishery conservation and management fund should be established to carry out data collection, management, enforcement, and restoration work.

DNA SCIENCE

FIGURE 29.10 The parasitic plant *Striga* This problematic plant is shown here parasitizing corn (*Zea mays*). The inset is a scanning electron micrograph of a *Striga* seedling attached to a corn root.

Restoration of coastal zones will require reduction of pollution from many different sources. If the techniques of sustainable agriculture were fully implemented, the level of pollution from farm runoff would decline, and freshwater streams and rivers would deliver less waste to marine coastal zones. A halt to the destruction of wetlands and restoration of degraded wetlands throughout the watersheds that empty into coastal zones would also reduce coastal zone pollution. Conservation experts recommend that within coastal zones, critical marine habitats such as salt marshes and mangrove forests should be identified and protected from development. Government programs that subsidize harmful coastal development could usefully be redirected toward restoration activities.

Sustainability requires maintaining ecosystem services and biodiversity

Techniques developed in conservation science can be used to maintain ecosystems and their biodiversity. Restoration ecology can provide valuable methods to restore degraded ecosystems. One of the lessons that ecologists have learned by studying primary succession (Chapter 25) is that leaving restoration to natural processes takes a very long time—decades to centuries. Another lesson is that repairing the soil is often the best first step in restoring a degraded ecosystem. Adding plants with nitrogen-fixing bacterial partners increases soil nitrogen levels, which can speed recovery. Some other plants are able to remove toxic metals from soils (see Chapter 4, Essay 4.2). Once natural soil characteristics have been restored, restoration ecologists seek to add a suite of plant species that restore ecosystem function. "Nurse plants" that shade and protect more delicate species may also aid recovery. Other species are likely to recolonize on their own. Successful restoration requires a potential pool of colonists, which is why we need to protect biodiversity in nature reserves. Refer to Essay 29.2, "Sustainable Use of Neotropical Forests," for an example of sustainable use of rain forest resources.

Biodiversity "hotspots" are regions of high endemism Biodiversity hotspots are a set of 25 or so world regions especially rich in endemic species that are also threatened by human activities. Hotspots include rain forests of Hawaii, the West Indies, the Phillippines, and India. Scrublands of southern Africa, southwest Australia, and coastal California are also biodiversity hotspots. They are the last habitats for one-third of vascular plants, mammals, birds, reptiles, and amphibians. Consequently, biodiversity hotspots have a high priority for conservation efforts. If we allocate funds to protect these 25 biodiversity hotspots, we would do more to avert mass extinction than we could through any other single action. The major reason that extinction is a particular threat in these regions is that more than 1.1 billion people live within them, and population growth in the hotspots is 1.8% per year—higher than the world average of 1.3%. Large human populations fragment habitat when they plow new crop fields and build new roads, airports, and residential and commercial developments. Fragmented habitats may be capable of maintaining their species diversity if wildlife corridors are built to connect separate fragments. Such wildlife corridors may pass through residential and commercial areas by following ravines or streams and even passing under highways.

The United States, occupying 3.5 million square miles of land area (only Canada, Russia, and China are larger), harbors more than 200,000 of the Earth's 1.75 million species. These include more than 15,000 flowering plants, 34,000 fungi, 1,400 mosses, and 700 liverworts. In fact, the United States represents the richest temperate ecosystem in the world. The high biodiversity of the United States results from its great geographical diversity, with habitats ranging from tundra to deserts and tropical forests. Hotspots of species richness and

ESSAY 29.2 SUSTAINABLE USE OF NEOTROPICAL FORESTS

Neotropical forests are those occupying South and Central America and the Caribbean region. These forests provide essential ecosystem services and contain valued high biodiversity, but the people who live in or near them need to use the forests. Sustainable methods for use of the forest will help maintain forest services as well as human residents. Some forests can be sustainably used as extractive reserves. An extractive reserve is a forest from which products can be removed while the ecosystem stays largely intact (Figures E29.2A and E29.2B). The

An extractive reserve is a forest from which products can be removed while the ecosystem stays largely intact.

timber, rubber, and Brazil nut industries are important examples. Over 200,000 people harvest Brazil nuts from 20 million hectares (1 hectare = about 2.5 acres) of Amazon forest. The Brazil nut crop is worth $35 million annually and provides a large source of cash revenues. However, it is important that residents avoid overharvesting from the forest.

Neotropical forests contain billions of cubic meters of high-quality wood worth several trillion dollars as lumber. These can be managed on a sustainable basis using techniques learned from scientific research. Studies

indicate that it is good practice to (1) first conduct an inventory of valuable timber, (2) carefully plan machine movements to reduce forest damage, (3) cut vines surrounding selected trees so other trees are not dragged down during the harvest, (4) cut only the desired trees and map their locations so collectors can efficiently find them, and (5) train sawmill workers to reduce waste when sawing lumber. These efforts increase the efficiency of the harvest and thus raise profit margins, while preserving the forest. Scientific research also indicates that logging on slopes, streamsides, and other sensitive areas should be avoided; and regions with moderate to high levels of endemism and biodiversity—and all lands belonging to indigenous peoples and parks—should be off-limits to commercial wood harvest. Simple regulations that are easy to follow also help reduce impact. An example of a simple regulation is the 5/50/5 rule. The first number represents a proposed limit of 5 on the number of trees that could be removed per hectare in a single logging episode. The second number indicates a minimum of 50 years between logging episodes. The last number is a 5-m width of cleared area to be maintained around all logged stands in the first decade after logging, to reduce spread of ground fires. Following rules such as these would help prevent three factors leading to forest degradation: excessive harvest, repeated harvest at short intervals, and fires.

E29.2A Harvesting palm fruits

E29.2B Extracting latex

rarity include the southern Appalachians, Hawaii, the San Francisco Bay area, coastal and interior southern California, Death Valley (in the Mojave Desert), and the Florida panhandle (Figure 29.11).

Biomes differ in their vulnerability to extinction

Temperate and boreal forests, and arctic and alpine plant communities—which are naturally nitrogen-limited—are especially sensitive to loss of biodiversity arising from the addition of nitrogen from acid rain (Chapter 27). Because climate change effects are large at high latitudes, some species will go extinct. Other species cannot replace the extinct species in food webs because biodiversity is low. Therefore, species that were dependent on the extinct species will themselves likely go extinct in a cascade effect. Deserts and arid grasslands, where water is naturally limiting to plant growth, suffer more from increased carbon dioxide concentrations in the atmosphere. Extinction is already rampant in tropical forests because of high biodiversity and rarity, together with

rapid land-use change (Figure 29.12). Land-use changes and climate-change impacts on extinction are both expected to be high in mid-latitude ecosystems. Invasive species have their greatest effect in long-isolated southern temperate forests and islands, but they cause problems in most biomes.

Everyone can contribute to global sustainability

There are many ways that everyone can help maintain the Earth's ecosystems for present and future generations. This section describes some of those ways.

We could work toward living with a lower ecological "footprint." This is the amount of land area required for our food, water supplies, shelter, transportation, waste disposal, and government functions that support our society. On average, the world ecological footprint is 5.2 acres, but it is 24 acres for North Americans. Reducing energy consumption, eating lower on the food chain (more plants, less meat), and

FIGURE 29.11 Map of major hotspots of biodiversity in the United States (1) The Hawaiian Islands are a diversity hotspot because more than 87% of their native plants and high percentages of other organisms occur nowhere else. (2) The San Francisco Bay area retains less than 3% of its former tidal marshes, which support many imperiled species. (3) Ancient, worn-down peaks once rivaling the Himalayas, the Appalachian mountains harbor the world's richest collection of several types of aquatic organisms. (4) Death Valley in the Mojave Desert has unique "oasis species" that are specialized to live in the springs, pools, and seeps left from a once-huge lake that existed in the area more than 10,000 years ago. (5) Coastal and interior southern California's diverse geography and geology have created highly species-rich plant shrubland communities. San Diego County, with more than 1,800 native plants, has more native species and more endangered and threatened species than any other county in the continental United States, but very high human populations as well. (6) The Florida panhandle includes the largest longleaf pine forest in the world, one of the most pristine estuaries (Apalachicola Bay) in the United States, and the United States' highest diversity of woody plant species (along the Apalachicola River basin).

FIGURE 29.12 Cattle ranching in Amazonia Ranching to meet export demands for cheap beef is a major way in which tropical environments become degraded.

deemphasizing material consumption are ways to reduce one's ecological footprint.

We could volunteer to help biologists map and catalogue all of the world's organisms. We might help to find species new to science or representative species thought to have become extinct. In 2001, this very thing occurred; in Mauritius scientists found a few individuals of the flowering tree *Trochetia parviflora*, thought to be extinct since 1863.

We can help prevent mineral-rich water from infiltrating lakes and contributing to nutrient pollution—eutrophication (Chapter 28). We could plant a "rain garden" at the end of our house's downspout, but at least 10 feet from the house. The water will infiltrate the soil and favor plant growth, particularly if sedge meadow plants are chosen, because these are able to tolerate extremes of moisture.

We can help prevent ocean pollution from agricultural lands by supporting legislation that encourages farmers to use less chemical fertilizer and by voting to upgrade our local sewage treatment plant—even if we have to pay a little more for water.

We can help prevent air pollution, acid rain, and global climate change by choosing to use renewable energy—sun, wind, water, and geothermal—even if it is a bit more expensive.

We can also improve the heat retention of houses by selecting efficient windows, roofs, and appliances.

We can choose mass transportation or use a bicycle. If a car is essential, we can select a vehicle that has good gas mileage or express a preference for gasoline/electric hybrid vehicles or, when they become available, hydrogen-powered cars that do not pollute the air when operated.

We can support biodiversity conservation groups that act globally.

We can use our knowledge of plants, fungi, protists, and prokaryotes and our particular skills and expertise to foster environmental protection. Whether you are a teacher, lawyer, business executive, politician, writer, artist, or member of any other profession, you have the opportunity to make an important contribution to the sustainability of people and nature.

HIGHLIGHTS

29.1 Sustainability is the continued survival of Earth's humans while also maintaining a high-quality environment. Earth's biodiversity provides ecosystem services such as water purification, soil formation and enrichment, and many other benefits.

29.2 Humans are affecting the global environment through their increase in population numbers, which results in a transformation of much of the Earth's land surface area and a degradation of many of the Earth's ecosystems. Consumption of fossil fuels has led to significant increases in carbon dioxide in the atmosphere that have produced global warming. Global warming is affecting every ecosystem on Earth.

29.3 Human populations need to achieve stable levels in the near future so that sustainable development can provide an acceptable standard of living. Innovative energy technologies such as wind farms, photovoltaics, and hydrogen fuel offer potential solutions to energy sustainability problems. Sustainable agricultural technologies and biotechnologies such as genetic engineering of crops may help achieve food sustainability. Conservation biology and restoration ecology can provide the techniques to maintain ecosystems and biodiversity. There are many ways in which everyone can contribute to achieving global sustainability.

REVIEW QUESTIONS

1. What is sustainability? Define ecosystem services, and give several examples.

2. What is the current human population of the Earth, and what are reasonable forecasts for future population levels? What is ecologist Dr. David Pimentel's estimate of the carrying capacity of the Earth for humans? What else do his studies indicate regarding the relationship between fossil fuels and the production of food for the large global populations?

3. What are functional groups, in terms of ecosystems as well as keystone species?

4. How does the current extinction rate compare to that before modern humans appeared 40,000 years ago? How does it compare to extinction rates 65 million years ago, when an asteroid impact closed out the Cretaceous period? Are humans responsible for the extinction of some of the large mammals that roamed North America during the last ice age (12,000 years ago)? Are current plant species imperiled?

5. What are endemic species? Are they evenly distributed around the world, or do they occur in biodiversity hotspots? How does the number of endemic species in a region relate to its extinction potential? Approximately how many hotspots have been identified worldwide, and what are some examples? What are some examples of hotspot areas in the United States?

6. What are invasive species, and what are some of their characteristics?

7. List some renewable energy sources and several currently available innovative technologies to extract and store energy from them.

8. Provide several examples of past civilizations that have disappeared due to ecological decline.

9. Regarding the neotropical forests, what are extractive reserves? Describe two strategies being used today to harvest neotropical resources at a sustainable rate.

10. What are some major genetic conditions affecting small populations that have undergone habitat fragmentation?

APPLYING CONCEPTS

1. The last few chapters have focused on the major ecosystems found on Earth. Review these chapters, and then for each of the following ecosystems, discuss the prominent abiotic features (climate, geographic location, soil type, etc.), characteristic plant types, and several of the general ways in which plants are adapted to that environment: (a) polar deserts, (b) tundra, (c) boreal forests, (d) lakes, (e) freshwater marshes, (f) deserts, (g) grasslands, (h) temperate deciduous forests, and (i) tropical rain forests.

2. This chapter mentions evidence linking human hunting with the extinctions of large vertebrates living in North America some 12,000 years ago. Do you think that large vertebrates will continue to face the brunt of extinctions in the 21st century, or do you think other groups of organisms will fare even worse?

3. Use your imagination and list at least five ways in which life in the developed world is not sustainable.

4. This chapter ended by listing some ways that everyone could contribute to global sustainability. Can you add to that list by thinking of 10 other specific ways you could begin contributing to sustainability today?

Appendix I

Metric System Conversions

To Convert Metric Units:	Multiply by:	To Get English Equivalent:
Length		
Centimeters (cm)	0.3937	Inches (in.)
Meters (m)	3.2808	Feet (ft)
Meters (m)	1.0936	Yards (yd)
Kilometers (km)	0.6214	Miles (mi)
Area		
Square centimeters (cm^2)	0.155	Square inches $(in.^2)$
Square meters (m^2)	10.7639	Square feet (ft^2)
Square meters (m^2)	1.1960	Square yards (yd^2)
Square kilometers (km^2)	0.3831	Square miles (mi^2)
Hectare (ha) $(10,000\ m^2)$	2.4710	Acres (a)
Volume		
Cubic centimeters (cm^3)	0.06	Cubic inches $(in.^3)$
Cubic meters (m^3)	35.30	Cubic feet (ft^3)
Cubic meters (m^3)	1.3079	Cubic yards (yd^3)
Cubic kilometers (km^3)	0.24	Cubic miles (mi^3)
Liters (L)	1.0567	Quarts (qt), U.S.
Liters (L)	0.26	Gallons (gal), U.S.
Mass		
Grams (g)	0.03527	Ounces (oz)
Kilograms (kg)	2.2046	Pounds (lb)
Metric ton (tonne) (t)	1.10	Ton (tn), U.S.
Speed		
Meters/second (mps)	2.24	Miles/hour (mph)
Kilometers/hour (kmph)	0.62	Miles/hour (mph)

To Convert English Units:	Multiply by:	To Get Metric Equivalent:
Length		
Inches (in.)	2.54	Centimeters (cm)
Feet (ft)	0.3048	Meters (m)
Yards (yd)	0.9144	Meters (m)
Miles (mi)	1.6094	Kilometers (km)
Area		
Square inches $(in.^2)$	6.45	Square centimeters (cm^2)
Square feet (ft^2)	0.0929	Square meters (m^2)
Square yards (yd^2)	0.8361	Square meters (m^2)
Square miles (mi^2)	2.5900	Square kilometers (km^2)
Acres (a)	0.4047	Hectare (ha) $(10,000\ m^2)$
Volume		
Cubic inches $(in.^3)$	16.39	Cubic centimeters (cm^3)
Cubic feet (ft^3)	0.028	Cubic meters (m^3)
Cubic yards (yd^3)	0.765	Cubic meters (m^3)
Cubic miles (mi^3)	4.17	Cubic kilometers (km^3)
Quarts (qt), U.S.	0.9463	Liters (L)
Gallons (gal), U.S.	3.8	Liters (L)
Mass		
Ounces (oz)	28.3495	Grams (g)
Pounds (lb)	0.4536	Kilograms (kg)
Ton (tn), U.S.	0.91	Metric ton (tonne) (t)
Speed		
Miles/hour (mph)	0.448	Meters/second (mps)
Miles/hour (mph)	1.6094	Kilometers/hour (kmph)

Metric Prefixes

Prefix				Meaning
giga-	G	10^9	=	1,000,000,000
mega-	M	10^6	=	1,000,000
kilo-	k	10^3	=	1000
hecto-	h	10^2	=	100
deka-	da	10^1	=	10
		10^0	=	1
deci-	d	10^{-1}	=	0.1
centi-	c	10^{-2}	=	0.01
milli-	m	10^{-3}	=	0.001
micro-	μ	10^{-6}	=	0.000001

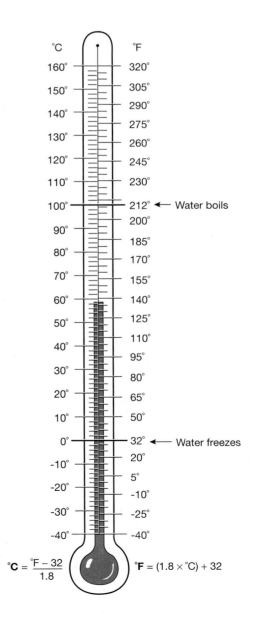

$$°C = \frac{°F - 32}{1.8} \qquad °F = (1.8 \times °C) + 32$$

Appendix II

Geological Time Scale

Eon	Era	Period	Epoch	Plant/Algal Evolutionary Events	Climate, topography, other life-forms
PHANEROZOIC (543)	CENOZOIC (65)	Quaternary (1.6)	Recent (0.01)	Multiple invasions of nonnative aquatic species resulting from human activities	Fluctuating temperatures, including several glaciation events; age of humans
			Pliocene (5)		North and South America joined by Panama land bridge; cooler climate
			Miocene (23)		Temperatures moderate; glaciation occurs in Southern Hemisphere
		Tertiary (65)	Oligocene (34)	Many modern genera of plants appear	Extensive tectonic activity; South America separated from Antarctica
			Eocene (56)	Green algae associated with formation of oil deposits	Mild to warm climate; Australia separates from Antarctica
			Paleocene (65)	Many plants become extinct; freshwater pennate diatoms appear	Extinction of dinosaurs and other organisms
	MESOZOIC (251)		Cretaceous (145)	Angiosperms diversify	Asteroid or comet impact at Cretaceous/Tertiary boundary; high sea levels; South America separates from Africa; many insect groups diversify
			Jurassic (208)	Gymnosperms dominate; early angiosperms appear; dinoflagellates undergo major radiation	Climates mild; extensive continental seas
			Triassic (251)	Forests of gymnosperms and ferns; calcium-coated marine green algae abundant	Continents form a single supercontinent; first dinosaurs and mammals appear
	PALEOZOIC (543)		Permian (290)	Several gymnosperm groups appear (conifers, cycads, Ginkgo)	Southern Hemisphere glaciation; reptiles diversify; mass extinction at the end of the period
			Carboniferous (362)	Forests of seedless vascular plants	Warm climate; extensive inland swamps; age of amphibians; origin of reptiles
			Devonian (408)	Seedless vascular plants diversify	Extensive shallow inland seas; age of fish; insects appear
			Silurian (439)	First vascular plants appear	Mild climate
			Ordovician (490)	Fossil spores suggest widespread occurrence of early bryophyte-like land plants	Mild climate until end of the period when Africa was glaciated; shallow continental seas cover much of North America; oldest crustaceans; molluscs diversify
			Cambrian (543)	Earliest fossil spore evidence for land plants; diverse red algae; diverse microfossils regarded as unicellular phytoplankton	Mild climate; continental seas; many invertebrate animal groups appear
PRECAMBRIAN (4500)	PROTEROZOIC (2500)			Molecular clock evidence for land plants at 700 mya; earliest stramenopile and dinoflagellate algae; earliest multicellular red and green algae; diverse filamentous and colonial cyanobacteria	Oldest traces of invertebrates; earliest geochemical evidence for oxygenic photosynthesis; long periods of extensive glaciation
	ARCHEAN (4500)			Oldest phytoplankton; oldest cyanobacteria/stromatolites	Arguably oldest geochemical evidence for life (> 3.7 bya); follows period of intense bombardment of Earth by debris left over from formation of the solar system; conditions incompatible with persistence of life

Note: Numbers in parentheses indicate the age in millions of years (rounded to nearest million) at which the geologic time period begins.

Answers

Chapter 1

Review Questions

1. As defined in this book, an organism must pass four criteria to qualify as a plant: (a) Plants are multicellular. There are no mature, single-celled plants. (b) Plants possess cellulose-rich walls. Many organisms, such as animals, do not possess cell walls at all; others, such as bacteria and fungi, possess walls rich in compounds other than cellulose, such as chitin. (c) Plants usually possess chlorophyll and are photosynthetic. A few plants lack chlorophyll and are not photosynthetic, but even these are descended from photosynthetic ancestors. (d) Plants are adapted to life on land, or if aquatic, then they descended from plants that were adapted to land life. Adaptations to land life are many, and will be discussed throughout this book.

2. (a) Like plants, most (or even all, depending on your definition) animals are multicellular organisms, and many are indeed adapted to land life. But animals lack cell walls and chlorophyll; they are neither photosynthetic nor do they, as a group, produce cellulose. (b) Although most fungal species are multicellular, some single-celled species (e.g., yeasts) do exist. Furthermore, although fungi do possess cell walls and many species are adapted to life on land, fungi contain neither cellulose nor chlorophyll and thus do not qualify as plants. (c) Bacteria can be easily separated from plants by their cell architecture, which is of the "prokaryotic" type. Beyond that, as a group, they consistently lack the features of plants: Although some species are multicellular, most exist as single cells. Some species do contain chlorophyll, but most do not and are not photosynthetic. Many bacteria lack a plant-type cell wall, and those with one usually do not impregnate it with cellulose. As a group, bacteria simply do not consistently fit the definition of a plant.

3. One adaptation to land life is the production of an embryo—a very young plant that arises from a fertilized egg and develops within maternal plant tissues. In effect, the mother plant nurtures her very young plantlets to give them a little help in starting out on a life of their own. Since all plants produce embryos, they are often referred to as embryophytes to separate them as a group from organisms such as algae and fungi, which do not produce embryos.

4. The major groups of land-adapted plants: (a) Bryophytes, mosses and liverworts. An example would be the *Sphagnum* moss used as peat in gardening. (b) Lycophytes, or club mosses. The most familiar example of these little-known plants is the club moss known as *Lycopodium*. (c) Pteridophytes, or ferns. Most people recognize a fern when they see one, even though fern species are not as widely known as the next two groups of plants. Plant enthusiasts may recognize ostrich fern, maidenhair fern, or staghorn fern as representative fern species. (d) Gymnosperms, best known for their coniferous, or cone-bearing, species such as pine, spruce, and redwood trees. (e) Angiosperms, or flowering plants. By far the largest group of plants, this group includes such familiar examples as apples, roses, grasses, tomatoes, and gardenias.

5. Plants help to maintain our atmospheric levels of oxygen and carbon dioxide. The oxygen in our atmosphere is mainly a by-product of photosynthesis, about half of it being produced by plants and the remainder by algae and cyanobacteria. Without these organisms to replenish the oxygen in the atmosphere, ours would be a world devoid of any significant levels of gaseous oxygen. By continually consuming carbon dioxide in the reactions of photosynthesis, plants (along with algae and cyanobacteria) help to keep the levels of this greenhouse gas from rising too high.

6. A "greenhouse" gas is an atmospheric gas that absorbs heat, or infrared, radiation. Surface features such as rock, soil, and water absorb solar radiation during the day and become warmer; typically, as this absorbed energy is radiated back to space at infrared wavelengths, the extra heat is lost and the Earth cools back down. When greenhouse gases are present in the atmosphere in significant amounts, they absorb this infrared radiation and trap the heat energy on the Earth, raising global temperatures. The result is global warming. Careful scientific measurements have documented an increase in global temperature of 0.7–1.4°C over the last century, and this correlates well with a concomitant increase in greenhouse gas levels in the atmosphere. Although the debate is often acrimonious over just how much of the increase in greenhouse gas levels is due to human activities, it is clear that if current global warming trends continue unabated, the ecological impacts will be significant and, in some cases, devastating.

7. Food webs are complex feeding relationships involving a number of different species. Typically, photosynthetic species such as plants, algae, or cyanobacteria are at the base of the web, and the web itself consists of these organisms as well as all the higher organisms that feed upon them and upon each other. When one considers the myriad species of bacteria, algae, fungi, plants, and animals that may participate in the food web of a given area, one begins to appreciate how incredibly complex such webs may be. However, the interactions between these species tend to be short-lived and not particularly intimate. Symbioses involve relationships between two or more partners, and typically these interactions are more intimate and longer term than in food webs. For example, the predator-prey relationship between a hawk and a squirrel involves a short-lived interaction; it is part of a food web. On the other hand, the single-celled animals living in the gut of a termite form an intimate association that persists for the life of the termite; it is a symbiosis.

8. Science is a process, and as such it is constantly undergoing change and revision. Although it is empirical, involving measurement, observations, experiments, and conclusions, it seeks to be as objective as possible. Out of this objectivity arises reproducibility: If physical and chemical laws govern natural events, then the same experiment repeated by multiple researchers should yield similar results. To decrease the probability that one or more observations are in error, scientists use replicates in their experiments—the more replicates showing a given result, the less likely are those observations to be erroneous. Science advances as data are summarized and scientific statements are formed. These statements may be arrived at through inductive or deductive reasoning, but all statements must be testable. Untestable statements and theories are simply outside the purview of science. Every scientist accepts that new evidence or deeper understanding may drastically change the nature of today's scientific statements, but science evolves and new ideas often supersede older ones. Another important part of science is peer review—before experimental reports and new scientific ideas can be published, peers in the field must review manuscripts. This helps to ensure that a proposed article is worthy of becoming part of the scientific literature. A similar procedure exists for obtaining grant monies to conduct research.

9. Inductive reasoning begins with particular measurements and observations, from which analyses are made and conclusions drawn. The resultant conclusions allow a scientist to formulate a more expansive generalization known as a theory. In effect, inductive reasoning begins small, with individual pieces of data (measurements, observations), and then builds up to "the big picture" (a theory). In deductive reasoning, one begins with a hypothesis, a preliminary explanation based upon accumulated knowledge and experience. One can then use that hypothesis to predict the outcome of relevant experiments, and then test the hypothesis by actually performing those experiments. Eventually the hypothesis is refined to the point where it accounts for most if not all of the experimental data.

Applying Concepts

1. A common type of mutualism occurs in our own bodies—the *E. coli* and other beneficial bacteria inhabiting our intestinal tract. We provide a nurturing environment for the bacteria while they, in turn, provide us with some vitamins and inhibit the growth of more harmful microbial species. Lichens represent a classic example of mutualism between an alga and a fungus, while belowground, fungi and bacteria form mutualistic symbioses with plant roots. A common example of commensalism is a bird building its nest high in the branches of a tree; the bird certainly benefits from its lofty home, while the tree is neither hurt nor harmed. A lichen growing on a tree trunk provides another example. Parasites abound in the world as well; examples include plant parasites such as dodder or mistletoe growing upon their plant hosts, insects such as aphids feeding off plant sap, and fleas and mange mites living on an unfortunate dog. Many bacteria or fungi parasitize their hosts, and some parasitic mites cause their plant host to form galls—round balls of plant tissue (oak galls are a common example of this). Indeed, symbioses are all around us, even in so common an area as a city park or a college campus.

2. The answers to this question will, of course, vary. Potential answers could include houseplants for aesthetics or kitchen plants (like chives) for cooking, as well as plants of all sorts for food, spices, textiles, lumber, dyes, medicines, art, rubber, fermented plant beverages, reeds for baskets, paper products, potting medium, and so forth.

Chapter 2

Review Questions

1. Scientists use both direct and indirect lines of evidence when studying the origins of plant domestication by ancient peoples. Direct evidence includes examination of actual plant material that has survived to the present day, including (1) seeds, grains, and pollen, (2) vegetative fragments such as pieces of wood, plant fibers, and leaf material, (3) inorganic plant crystals such as phytoliths, and (4) undigested human plant material in fossilized human feces (coprolites). Such material may be preserved by any number of processes, including desiccation (being dried), being charred in a fire, being buried in anoxic lake sediments, and so forth. Indirect lines of evidence include examining ancient documents, burial artifacts, art, tools, ceramics, and paint, as well as observing agricultural techniques used today by so-called primitive cultures. Last, with the advent of molecular biology, genetic analyses of both ancient crop plants and their wild relatives may yield clues to the ancestry of the crop plants.

2. Three regions involved in early plant domestication: (1) The Fertile Crescent region of the Near East, where the cereals wheat and barley, and the legumes peas, chickpeas, and lentils, were grown. (2) Several areas of the Far East, including sites in China along the Yellow and Yangtze River Valleys. People in these areas learned to grow the food crops rice and millet—both cereals, soybean—a legume, and hemp—a plant valued for its fibers. (3) Native peoples in the New World collectively domesticated more than 100 species of plants, including the cereal plant maize (corn), legumes such as peanuts and lima beans, and many other crops including sunflowers, squash, chili peppers, tomatoes, potatoes, and manioc.

3. Advantages of cereals: (1) Cereal plants can often tolerate relatively drier growing conditions than other crops, making them especially useful in drier climates. (2) Cereals have high growth rates and are quite efficient at food production. (3) Cereals are rich in carbohydrates and proteins, making them very desirable as food sources. Advantages of legumes: (1) Leguminous plants, in association with soil bacteria, are able to fix atmospheric nitrogen. Most of these nitrogen compounds eventually end up in the soil where they become available to other plants. Leguminous plants are thus a cheap source of nitrogen fertilizer. (2) Legumes are very high in protein content. One legume, the soybean, contains as much protein on a gram-for-gram basis as some animal sources. (3) Legumes can

provide the essential amino acids lysine and tryptophan, which are deficient in cereals. When legumes are consumed with cereals, a well balanced diet results. Last, both legumes and cereals are dry at maturity, facilitating easy storage for long periods.

4. The appearance of agriculture offers several crucial benefits to human society. First, since agriculture produces food so much more efficiently than hunter-gatherer practices, it also allows a given area of land to support a much larger human population. Second, the development of crop plants that store well reduced the risk of food shortages in lean years. Third, the development of reliable agricultural food sources freed a major portion of the population to pursue other interests and thus facilitated advances in all other areas of society including art, medicine, science, law, athletics, and the trades. In no small way, agriculture forms the basis of any advanced civilization.

5. Artificial selection is the selective breeding of plants (or animals) showing desirable traits so as to encourage the appearance and development of those traits in future generations. For plants, this might mean planting seed only from parents showing a certain trait, such as large fruit, thus encouraging larger fruits in the next generation of that crop. Virtually every crop plant has undergone significant modification from its wild progenitors; the domestication of corn (maize) is an example. The closest wild relative of modern corn—teosinte—produces corn kernels with a very hard, inedible coat. Further, teosinte cobs bear only a few kernels of corn, lack an enveloping husk, and shatter at maturity. In collecting seeds for the following year's crop of plants, early corn domesticators selected kernels from plants showing traits they considered desirable—larger, softer kernels, and cobs that have larger numbers of kernels, are in a protective husk, and are easier to harvest because they do not shatter at maturity. As this process of artificial selection continued over many, many years, eventually a crop plant resembling modern corn emerged.

6. The practice of monoculture involves planting large areas with a single type of crop—or in many cases, the exact same hybrid. This leads to efficient planting, maintenance, and harvesting, since the entire crop can be managed with one set of agricultural tools. The disadvantage, of course, is that the entire crop is genetically similar, or even identical. Any pest that can easily attack one plant can easily attack all its neighbors, so that the entire crop is at risk. The practice of polyculture, on the other hand, involves planting a field with different types of crop plants. This makes it more difficult for any single pest to attack the entire crop. For example, an insect or fungus that attacks tomato plants might have difficulty attacking corn and onions. The disadvantage of this technique is that each crop plant has its own special needs—different planting, watering, fertilizing, and harvesting schedules; different soil requirements; different maintenance requirements; and potentially different tools needed to plant, maintain, harvest and/or store the crop. These disadvantages increase the overall financial and labor costs of farming.

7. Many species of animals and microbes live by directly attacking plants. Plants counter these attacks by producing various defensive chemicals termed secondary compounds. These chemicals might thwart an attack by inhibiting fungal growth, by killing pathogenic bacteria, by attracting insectivorous animals, or by simply being distasteful to passing herbivores. Regions of the world such as the tropics, with their high biological diversity, harbor many organisms that are potentially harmful to plants. In dealing with the higher numbers of pathogens and pests, the plants growing in these regions typically produce a greater variety of defensive secondary compounds than plants growing elsewhere.

8. The secondary compounds of plants occur as a bewildering array of chemicals, many of which profoundly affect human physiology. Recognizing that such active compounds may have medicinal value, some companies market herbal concoctions, usually as dried herbs supplied in capsules. Such preparations typically contain dried or powdered plant material, and the active compounds are not purified. Since the concentration of secondary compounds in plants is quite dependent on growing conditions, the concentrations of these compounds in herbal preparations—which are mixtures of plants collected in different areas or

at different times—will similarly vary. Medical personnel would be concerned that casual use of such preparations that are not standardized could adversely affect the health of individuals, or that the active principles in them might interact with prescription medicines in unforeseen ways.

9. Plants produce many secondary compounds, including psychoactive compounds. Many of these chemicals help to deter herbivores that seek to eat the plants and microbes that may cause disease. Psychoactive compounds affect the nervous system of mammals and humans and may alter behavior to decrease feeding or make animals less alert and more susceptible to their own predators. Psychoactive compounds include stimulants, such as caffeine (from coffee plants), nicotine (from tobacco plants), and cocaine (from Coca plants); hallucinogens from marijuana, belladonna, peyote cactus, and *Datura*; and narcotics (heroin, morphine) from poppy plants.

Applying Concepts

1. Examining the plant remains in the surrounding vegetation as well as in the human habitation would shed much light on the types of plants and the relative proportions of the plant species in each environment. Clues that a given plant was farmed and not merely collected from the wild would include (1) rare wild plants that are found in relative abundance in human encampments; (2) plants that are not part of the local flora but nonetheless occur in relative abundance in human encampments; and (3) plants containing modifications—such as nonshattering cobs of corn—that decrease the plant's ability to survive without human intervention.

2. Of course, you could not be sure that a given plant is safe to eat unless you can unambiguously identify it as an edible plant. Lacking this ability, you can at least "stack the odds" in your favor by doing the following: (1) Eat plants that you know have many edible or at least nonpoisonous relatives, and conversely, avoid eating plants with many poisonous relatives. Examples of plant groups in the first category are the rose family—wild cherries, plums, apples, and raspberries—and the grass family, from which we obtain our cereals. Plant families having some very poisonous members include the nightshade family, which contains deadly nightshade, and the carrot family, which contains several very poisonous hemlocks. (2) Since alkaloids are bitter compounds that are toxic in high doses, it would be best to avoid any bitter-tasting plant.

3. The dark muck of wetland soils contains high concentrations of plant nutrients. This soil could be dug out and incorporated into farm soils to improve their fertility. Early Native Americans in the Amazon region transported rich dark muck from wetlands to upland farms. This transfer greatly enriched the soil and allowed production of more crops.

4. The only plants that would grow well in nitrogen-deficient soil are those capable of fixing atmospheric nitrogen for their own use. Among the crop plants, this would include the legumes; the farmer would be well advised to grow leguminous plants such as alfalfa for fodder.

Chapter 3

Review Questions

1. (a) Refer to Figure 3.1. (b) Refer to Figure 3.5. (c) A covalent bond

2. (a) An ionic bond is essentially the attractive force between oppositely charged atoms. Such atoms may be free or part of an existing molecule. Table salt is an example, where a positively charged sodium ion (Na^+) is attracted to a negatively charged chloride ion (Cl^-). A second example is seen in hydrochloric acid, where the hydrogen ion (H^+) is attracted to the chloride ion (Cl^-). (b) A covalent bond is the attractive force between atoms that are sharing electrons. Examples include molecular hydrogen (H_2), where two hydrogen atoms each share their electron; molecular nitrogen (N_2), where each nitrogen atom shares 3 electrons, for a total of 6 shared electrons; and methane (CH_4), where one carbon atom shares 4 of its 6 electrons with the four hydrogen atoms, while the hydrogen atoms each share their single electron with the carbon. (c) A hydrogen bond involves the attraction of the (+)-charged pole of one molecule with the

(−)-charged pole of another; a familiar example is the attraction of the (+)-charged hydrogen of a water molecule with the (−)-charged oxygen atom of another water molecule.

3. Because so many biologically important compounds readily dissolve in water, water is the solvent of choice for biological systems. It is difficult to imagine how the myriad reactions that normally occur in the water within a cell could, in fact, proceed so effectively in any other medium. The high cohesiveness of water gives it tensile strength, allowing a column of water to be drawn from the roots to the tops of the highest plants without a break in the flow. The high specific heat of water allows water in living organisms (as well as bodies of water in the environment) to absorb significant quantities of heat without a concomitantly large increase in temperature, thus mitigating large temperature fluctuations. Furthermore, it means that water must absorb a large amount of heat before changing from a liquid to a vapor, thereby providing a cooling mechanism (= evaporative cooling). Last, this same water vapor, upon condensing back to liquid water in the atmosphere, releases its stored heat energy, a process that helps power the weather systems blanketing our planet. Because solid water (ice) floats on liquid water, only the topmost layer of lakes and oceans freezes during the cold season. This allows the body of water to unfreeze during the following warm season instead of gradually freezing solid from the surface to the bottom.

4. Dehydration reactions are used to assemble these molecules from their simpler building blocks. The molecules listed would require the following number of dehydration reactions for assembly: (a) 1, (b) 3, (c) 3, and (d) 8.

5. The monosaccharides glucose and fructose are energy molecules, storing in their chemical bonds the radiant energy of the sun captured during photosynthesis. Virtually all cells can use their metabolic machinery to extract this chemical energy again. Humans use both of these sugars as sweetening agents (honey, for example, is a 1:1 mixture of glucose and fructose); glucose has medical and scientific uses as well. The most abundant disaccharide in plants is sucrose, which is used for short-term energy storage and transport. Humans extract this sugar from certain plants to form table sugar, which of course is used as a sweetener. The two most common plant polysaccharides are starch and cellulose. The former is used in plants for long-term energy storage, while cellulose is a structural polysaccharide used to help support the plant body. Both of these polysaccharides have dietary as well as commercial importance to humans. Starch is an important energy source to humans and livestock, and has a number of industrial and household uses (e.g., pressing clothes as well as thickening stews). Cellulose provides fiber in the diet and is exceedingly important as a major component of paper, lumber, and plant fibers (e.g., for textiles).

6. When applied to fatty acids, saturated means that the hydrocarbon tail is essentially "saturated" with hydrogen—it contains the maximum number of hydrogens possible and hence no double covalent bonds. A monounsaturated fatty acid has one double bond in its hydrocarbon tail, whereas a polyunsaturated fatty acid has two or more such double bonds. Saturated oil contains high levels of saturated fatty acids, a monounsaturated oil contains mostly monounsaturated fatty acids, and a polyunsaturated oil contains mostly polyunsaturated fatty acids. A diet emphasizing monounsaturated oils over the other two seems to be the healthiest to humans. Examples of monounsaturated oils are peanut and olive oils.

7. All amino acids possess a central carbon atom linked by covalent bonds to a hydrogen atom, an amino group (—NH_2), and a carboxylic acid group (—COOH). In addition, every central carbon atom is attached to a single side chain (the "R" group). Amino acids differ in the chemical nature of the R group they carry.

8. The four levels of protein organization are described as primary, secondary, tertiary, and quaternary. The primary structure is simply the amino acid sequence, as read from one end of the protein to the other. Secondary structure describes certain local folding motifs, such as alpha helices and pleated sheets. Some proteins possess only helices, others only sheets, and

still others both. Tertiary structure describes the three-dimensional folding of a protein into a functional molecule; when a protein loses its tertiary structure it almost invariably loses its ability to function as well. Quaternary structure describes the way some proteins assemble two or more polypeptides into a functional unit. Note, however, that some fully functional proteins consist of only a single polypeptide chain and thus lack a quaternary structure altogether.

9. The three components of a nucleotide are a sugar, one or more phosphate groups, and a nitrogenous base. Two adenine nucleotides provide examples of nucleotides that participate in a cell's metabolic activities. Adenosine diphosphate (ADP) and adenosine triphosphate (ATP) are universally involved in energy metabolism. Further examples can be seen in the B vitamins, which are complex nucleotides serving as coenzyme helpers to certain metabolic enzymes.
10. Nucleotides making up RNA and DNA are similar but differ in two ways. The sugar in an RNA nucleotide is ribose, a 5 carbon sugar, while in DNA nucleotides, the sugar is deoxyribose. Deoxyribose lacks an OH (hydroxyl) group that is present in ribose. The bases that are part of DNA nucleotides are adenine, guanine, cytosine, and thymidine. In RNA nucleotides, bases include adenine, guanine, cytosine, and the base uracil that substitutes for thymidine.
11. The three groups of major secondary compounds in plants are (1) the terpenes and terpenoids, (2) the phenolics including flavonoids, and (3) the alkaloids. The first group contains many fragrances, insecticides, and insect repellants; the second group is known for its spice flavors, antimicrobials, antioxidants that inactivate free radicals, and compounds that protect against ultraviolet radiation; the third group is known for its stimulatory, poisonous, and psychoactive compounds.

Applying Concepts

1. In the chemical analyses, the formula reported as (a) $C_4H_8O_4$ is consistent with simple 4-carbon sugars such as erythrose and threose, (b) $C_5H_{10}O_5$ is consistent with 5-carbon sugars such as ribose and xylose, (c) $C_3H_5(OH)_3$ is consistent with the formula for glycerol, and (d) $H_2NCH(CH_2SH)COOH$ is consistent with the chemical formula for the amino acid cysteine. These biological compounds are very common in living systems, suggesting that life might indeed be present on Europa. But nonbiological chemical processes may also synthesize these compounds in the absence of life. It would thus be best to proceed with optimistic caution, telling the press that the robotic lab has found an assemblage of compounds that suspiciously suggests the presence of life. At the same time, the news media should be warned not to jump to the conclusion that Europa harbors life, since these same compounds may also be created by purely chemical means in a world devoid of life. This would be a perfect opportunity to request more funding and additional future exploratory missions.
2. Starch and cellulose both consist of chains of glucose molecules linked together by covalent bonds; it is the nature of those bonds that distinguishes these two compounds. Starch is used as an energy storage molecule. Metabolic enzymes easily break the bonds linking its glucose units, so that releasing the energy of starch is a quick and easy process for a cell. Cellulose, on the other hand, is a structural carbohydrate; once laid down, it is as permanent as the girders of a skyscraper after the building is completed. This semipermanency of cellulose is reflected in the durability of the chemical bonds connecting its glucose molecules. These bonds are difficult to digest, and only a few groups of organisms can successfully degrade them. Basically, starch is easily formed and quickly degraded, while cellulose is built to endure.
3. While a substrate molecule occupies the active site of an enzyme, the amino acids forming that active site are in position to directly interact with atoms of the substrate. Thus active site amino acids are usually critical to enzyme function, participating in either substrate binding or catalysis; a mutation in one of these amino acids usually adversely affects enzyme function. Additional critical amino acids may be located far away from the active site, where they may be involved in maintaining the structural integrity of the enzyme, or in interacting with other, nonsubstrate molecules. However, these distant amino acids might simply act as noncritical spacers in the amino acid chain. For example, some amino acids are located on the periphery of the protein, their side chains pointing out into solvent and having very little interaction with neighboring amino acids. A mutation substituting one of these amino acids for another might have little or no effect on the function of that enzyme.
4. First, animal cells, like those of plants, require various mineral (i.e., inorganic) nutrients such as calcium, potassium, and iron; a healthy diet would include appropriate amounts of these substances. The human diet should also include some carbohydrates; not only do we derive energy from these molecules, but the metabolic processes that break them down form simpler molecules that in turn can serve as building blocks for other complex molecules. The diet should include some lipids, with the healthy monounsaturated oils emphasized, as well as sufficient amounts of the essential amino acids so that the body can grow and replace worn-out proteins. The body requires a variety of organic molecules known as vitamins; by definition, we cannot make these molecules and therefore must obtain them in our diet. The foods we consume should also contain sufficient fiber, or roughage (such as indigestible carbohydrates) to allow the digestive system to function efficiently.

Chapter 4

Review Questions

1. Prokaryotic cells are similar to eukaryotic cells in that both possess a phospholipid cell membrane (or plasma membrane) to delimit the cell interior from the external environment. Both types of cells use ribosomes to make proteins, although the two types of ribosomes are different from each other. Also, both types of cells contain genetic material—DNA—at least for some part of their life (some types of eukaryotic cells, such as human red blood cells, lose their nucleus and DNA as they mature). Lastly, both types of cells possess a complex metabolism for carrying on their respective vital activities. Despite these similarities, eukaryotic cells are considerably more complex and contain more unique subcellular features than their prokaryotic counterparts. For example, the eukaryotic endomembrane system consists of several organelles, including the prominent and uniquely eukaryotic endoplasmic reticulum and Golgi bodies. Also present in eukaryotic cells are a unique cytoskeleton and eukaryotic motor proteins, the latter causing various types of movements within the cytoplasm. Mitochondria are found within most (but not all) eukaryotic cells, and many others contain peroxisomes and chloroplasts. Certain cells, from primitive protozoa to some cells in the human body, bear whip-like flagella and cilia for movement, and these structures are very different from their counterparts among the prokaryotes. Finally, and inherent in the definition of eukaryotic, is the nucleus, a double-membrane-bound region containing most of the cellular DNA.
2. An integral membrane protein is a protein that is embedded in the phospholipid bilayer of a membrane; typically these proteins span the membrane from one side to the other. Receptor proteins are usually membrane proteins that bind to chemical messengers in the environment and initiate a series of intracellular events that ultimately bring about an appropriate cell response. Classic examples of receptor proteins are the hormone receptors, as found in both plant and animal cell membranes; these proteins allow chemically mediated cell-to-cell communication. Transport proteins are another type of integral membrane protein. They allow specific ions or organic compounds to cross the cell membrane. In the case of carrier and channel proteins, diffusion alone drives movement across the membrane, while active transporters or pumps use the chemical energy of ATP to move substances across the membrane against their concentration gradients. Examples of transporter proteins are the ion transporters in root cells that allow root hairs to accumulate specific minerals from very dilute soil solutions.
3. The osmotic strength of a solution can be roughly defined as the total concentration of its constituent solutes. Two solutions are isotonic if they have the same osmotic strengths. If the ionic strength of solution "A" is less than that of solution "B," then A is said to be hypotonic to B, and B is hypertonic with respect to A. If a semipermeable membrane separates two non-isotonic solutions, then water will move from the hypotonic into the hypertonic solution. Similarly, if a cell is placed into a solution that is hypertonic with respect to its cytoplasm, then water will leave the cell and move into the external solution; such a cell will shrink and become plasmolyzed. Conversely, a cell placed into a hypotonic solution is in danger of absorbing too much water and bursting. Higher animals deal with such osmotic problems by closely regulating the tonicity of their extracellular fluids—the extracellular fluids are kept isotonic with the cells. Plants cope with hypertonic extracellular fluids by engaging water conservation mechanisms, such as closing their stomata to prevent further water loss and wilting. Plants, fungi, and some bacteria and protists protect their cells from absorbing too much water from a hypotonic environment by encasing their cells in a cell wall. This prevents the cells from excessive expansion and subsequent bursting, in the same way an automobile tire prevents the inner tube from exploding when it is overinflated. Lastly, some protists, living in freshwater (very dilute) and lacking a cell wall, instead periodically pump extra water out of the cell using an organelle called a contractile vacuole.
4. Endocytosis is the bulk transport of materials into cells; exocytosis transports materials in the opposite direction. Endocytosis occurs as a pocket bulges inward from a cell membrane and eventually pinches off altogether, thus becoming a vesicle in the cytoplasm. Material once found outside the cell would then be trapped inside the vesicle. Exocytosis is the opposite process—vesicles inside the cell fuse with the plasma membrane and spew their contents to the cell exterior. Plants use exocytosis to secrete slippery mucilage around their root cells, which helps lubricate the roots as they push their way through the soil. Some plants, such as legumes, use endocytosis to bring nitrogen-fixing bacteria into the root cells. Additionally, some types of protists feed via a special type of endocytosis called phagotrophy, which they use to engulf food particles and prey cells.
5. (a) The nuclear envelope consists of two phospholipid bilayers, which delimit the nucleus of a eukaryotic cell. Inside the nuclear envelope is the ground substance of the nucleus—the nucleoplasm (b)—in which all the other nuclear components lie. The nuclear envelope is perforated by the nuclear pores (c), which allow for the passage of substances into and out of the nucleus. Bodies within the nucleus called nucleoli (d) are the sites where cytoplasmic ribosomes are produced; once assembled, the ribosomes leave the nucleus via the nuclear pores. The nucleus also contains the linear DNA molecules of eukaryotes, complexed with various proteins and referred to as chromosomes (e). All the nuclear genes are located on the chromosomes.
6. The endomembrane system is a network of interconnected phospholipid bilayers that are found within the cytoplasm of eukaryotic cells. These membranes contain embedded proteins and are specialized for certain subcellular tasks, such as producing proteins or other substances, or modifying them after they have been formed. The two major components of this system are the endoplasmic reticulum (ER) and the Golgi apparatus. There are two types of ER: smooth ER and rough ER. Smooth ER lacks ribosomes and is the site of fatty-acid and phospholipid synthesis. The cytoplasmic surface of rough ER is studded with ribosomes, and the proteins that are produced wind up inside the ER itself, where various sugars are attached to them. Proteins with attached sugars are called glycoproteins, and the glycoproteins leave the ER and travel to the Golgi apparatus. The Golgi apparatus comprises a series of flattened, membranous sacks. Glycoproteins arriving from the ER enter these sacks and undergo further chemical modifications and are then sent on their way to other parts of the cell. Additionally, in plants the Golgi apparatus also produces noncellulose polysaccharides, such as pectins and hemicelluloses.
7. (a) Any microscope that uses glass lenses and light waves to visualize a specimen is a light microscope. There are many types of light microscopes, including the following four types (b–e). (b) Compound light microscopes are typical of the microscopes found in biology classes; they are capable of magnifications of

1000× or more. Specimens must be semitransparent to be visible and are often treated with various stains to increase the contrast and/or visibility of cells or sub-cellular components. (c) Stereo dissecting microscopes are essentially two low-power microscopes placed side by side. Such systems provide true stereo views and provide typical magnification ranges of 2× to 40× although some models may magnify as much as 100×. (d) A fluorescence microscope system illuminates a specimen with high-energy light—typically ultraviolet—then images any fluoresced light emerging from the specimen (the ultraviolet light is filtered out of the imaging pathway). (e) Confocal laser scanning microscopes build their images by scanning a specimen with a precisely focused laser beam. With such a system, a very small "optical slice" of a cell can be imaged, and the portions of the cell above and below the slice plane do not appreciably degrade the image. (f and g) Any microscope that uses magnetic lenses and electron beams to image a specimen is an electron microscope. Typically electron microscopes have resolving powers several orders of magnitude higher (i.e., better) than their light microscope counterparts. (f) A transmission electron microscope images an ultrathin section of a specimen after passing an electron beam through it. Typically these sections are stained with heavy metals (such as lead and uranium salts) to increase the contrast of certain subcellular components. (g) A scanning electron microscope scans a focused electron beam across the surface of a specimen (often coated with metal) and forms an image using the electrons that bounce back. This type of microscopy is particularly suited for observing the surface features of a specimen.

8. Halophytes are plants that grow in high-salt environments, such as deserts, salt marshes, and ocean beaches. These plants have made various adaptations to life in such a saline habitat. Some plants accumulate salts in their vacuoles and organic solutes in their cytoplasm, thus counterbalancing the high salt concentrations outside the cells. Other plants, such as some mangrove trees, excrete the excess salt onto their leaf surfaces, where it can be washed or blown away. Yet other halophytes, such as the desert plant *Atriplex*, excrete the extra salt into surface structures that then break away harmlessly from the plant. Such mechanisms allow these plants to grow in areas too salty to be colonized by most plants.

9. The cytoskeleton is the structural framework of all eukaryotic cells; it consists of three types of protein fibers: microtubules, long, cylindrical aggregates of the protein tubulin; microfilaments, consisting of two strands of actin filaments twisted around each other; and intermediate filaments, which are formed from any of several other proteins. Three motor proteins associate with the cytoskeleton and produce cellular movement; they are kinesin, dynein, and myosin. Kinesin associates with microtubules and is responsible for the directed movement of many vesicles and organelles about the cell; it is also partly responsible for chromosome movement at mitosis and meiosis. Dynein also associates with microtubules and is responsible for the swimming movements of flagella and cilia. Like kinesin, dynein also plays a role in the movement of chromosomes during mitosis and meiosis. Finally, myosin interacts with actin microfilaments to produce contractile movements. Myosin is responsible for muscle contraction in animals, cytokinesis (cell division) in eukaryotes, and cytoplasmic streaming in plants.

10. (a) Cellulose is the major structural component of cell walls; it consists of long polymer fibers of the simple sugar glucose. (b) Hemicellulose, a polymer of glucose and other sugars, binds the cellulose fibers together. (c) Pectin, another polysaccharide, combines with calcium and forms a jelly-like epoxy that fills the spaces not only between the cellulose fibers but also between the cells. (d) Plasmodesmata are small, membrane-lined channels running through the walls of adjacent cells and connecting their cytoplasms. By establishing cytoplasmic continuity, they provide a direct route for the movement of small molecules and ions between cells. When a plant is wounded, the plasmodesmata of the damaged cells become plugged by another polysaccharide known as callose (e); this prevents continual loss of water and small molecules from the cellular lesion. (f) Lignin is not a polysaccharide but rather a phenolic polymer. It is not found in all cell walls, but when present, it imparts structural rigidity and waterproofing abilities.

11. Typical chloroplasts are characterized by the possession of the green pigment chlorophyll. These plastids function in photosynthesis. They often possess other pigments that mask their intrinsic green color and impart to them a blue, yellow, orange, red, or brown color. Chromoplasts are plastids that present a bright orange, yellow, or red color due to the presence of lipid-soluble carotenoid pigments. They give color to plant parts, including many flowers, fruits, and autumn leaves. Amyloplasts are plastids specialized to store starch; they typically lack pigments and appear more or less colorless.

Applying Concepts

1. (a) Although this could be done with either an LM or a TEM, it would generally be easier with the former, which allows for easier sample manipulation, easier sample staining, and easier sample sectioning. (b) Because cells are killed during preparation for electron microscopy, only LM would allow you to view living cells. Thus, only LM would be appropriate. (c) This could be most easily accomplished with a fluorescence LM; the entire cytoskeleton (= internal network of fibers) could be quickly stained with a fluorescent dye and viewed. (d) An LM does not provide sufficient resolving power and magnification to view these very small structures, so that a TEM would be required. (e) Because an SEM excels at imaging small surface details, it would be most appropriate to view these small surface scales. (f) Unfortunately, none of these microscopes can view something as small as an individual atom, and non-microscopic techniques would be required.

2. Primary cell walls are capable of stretching, so they can be laid down before and during the elongation phase of cell growth. Because primary walls are produced first, the cells must construct their secondary walls inside their primaries. Secondary walls are not capable of stretching, so they cannot be laid down until after elongation has ceased. The cell membrane plays a central role in the creation of a cell wall, and it is located within the cytoplasm, inside any wall that is present.

3. (a) Nonfunctional kinesin would be a lethal mutation to plants and animals. Chromosome separation could not occur properly, so that mitosis could not happen. Thus, zygotes could not grow. Furthermore, trafficking of cell vesicles would stop, and intracellular chaos would ensue. Cells could not grow and mature properly. (b) If dynein were nonfunctional, then microtubular sliding would stop. Flagellated sperm would not be able to swim, so fertilization would be prevented in many eukaryotes. If a zygote could be formed, mitosis could not occur properly, so they could not divide and mature. Like kinesin, nonfunctional dynein would be incompatible with normal growth and development. (c) Nonfunctional myosin would also lead to early death in both plants and animals. Because myosin is the motor operating in muscle tissue, all muscular contraction would cease, as would such vital processes as blood circulation and breathing. Another consequence to all eukaryotes would be the inability to properly complete cell division, so once again growth and development could not proceed. Clearly, motor proteins are essential to eukaryotic life.

4. Many of the nutrients entering cells do so by diffusion, and they are distributed about the cell by the same process. Similarly, many wastes leave cells by diffusion as well. Such a large cell would have a very small ratio of surface area to volume, and diffusion would probably be unable to provide nutrients and remove waste products fast enough to support life (unless some very elaborate and highly metabolic system evolved to circumvent this problem). It would be much more efficient and evolutionarily probable to construct a beach-ball-sized organism out of many smaller cells, each of which could be easily serviced by diffusion.

Chapter 5

Review Questions

1. Simply put, photosynthesis is the process by which the energy of light is used to produce carbohydrates starting from simpler precursor molecules like carbon dioxide and water, while respiration is the process whereby energy is released as more complex organic molecules like carbohydrates are broken down into simpler molecules. Both processes involve a set of enzyme-mediated chemical reactions, membrane-bound proteins that are involved in electron transport, and both result in a buildup of protons—a proton, or pH, gradient—which is tapped to produce ATP molecules. In plants, photosynthesis is associated with chloroplasts, and respiration with mitochondria and cytoplasm. Another difference is that photosynthesis uses an external energy source (light) to create high-energy compounds (ATP and NADPH) that are used to incorporate CO_2 and H_2O into organic compounds, while respiration breaks down larger organic compounds into simpler ones (sometimes all the way back down to CO_2 and H_2O), extracting energy in the process.

2. (a) Substrates, or reactants, are the initial molecules that are reacting (being altered) in a chemical reaction. (b) Products are the final molecules that result from a chemical reaction. (c) Activation energy is an initial amount of energy that must be put into a chemical reaction before it will proceed. It can be thought of as an initial energy barrier that must be overcome in order for the reaction to move forward. (d) An exergonic reaction is one that releases energy because the products possess less energy than the reactants. (e) An endergonic reaction requires an input of energy in order for the reaction to proceed, because the products have more energy than the reactants.

3. A catalyst is any substance that participates in and speeds up a chemical reaction, but is not consumed in any way during the process. (a) Enzymes are the catalysts used in biological systems. Virtually all enzymes are protein molecules, each of which is a specific catalyst in one chemical reaction. (b) The active site is the "business" end of an enzyme. It is the region where catalysis actually takes place—where substrates bind and products form. (c) Many enzymes require non-protein molecules to assist them in their normal catalytic functions; such helpers are called cofactors. Cofactors may be inorganic ions like Mg^{+2} or Fe^{+2} or they may be organic (but not proteinaceous) molecules; the latter are called coenzymes (d).

4. In freshman chemistry, many students learned the mnemonic, "Leo the lion says, 'Ger,'" to distinguish oxidation from reduction. Oxidation involves the loss of one or more electrons (Leo: losing electrons = oxidation); reduction involves the acquisition of one or more electrons (Ger: gaining electrons = reduction). Because oxygen attracts electrons so strongly, it may be the electron acceptor in an oxidation-reduction reaction, but it is not required for these reactions to occur. Most oxidation-reduction reactions in biological systems do not involve oxygen. Oxidation-reduction reactions always occur in pairs because these two processes are complementary—one molecule donates electrons while the other accepts them. In biological systems, usually a proton accompanies the electron being transferred, so that, in essence, a hydrogen atom is transferred. Thus, the most common type of oxidation-reduction reaction in biological systems involves the transfer of a hydrogen atom.

5. A pigment is any molecule that absorbs at visual wavelengths; that is, it absorbs any portion of the electromagnetic spectrum we know as visible light. The absorption spectrum of chlorophyll a shows that it absorbs strongly in the red (600–700 nm) and blue (<400 nm) regions of the spectrum. Thus, when white light falls upon this pigment, the reddish and bluish hues are selectively absorbed and mainly the green wavelengths (around 520 nm) are reflected, allowing our eyes to perceive a green color. Chlorophyll a is the crucial molecule in photosynthesis, because two special chlorophyll a molecules are at the core of every photosystem I and II reaction center. These molecules represent the final destinations of absorbed light energy, and they are the initial electron donors to the electron transfer chains of both photosystems. Without these critical chlorophyll a molecules, the light reactions of photosynthesis in plants would not occur. Accessory pigments (often called antenna pigments) increase the absorptive cross section of the reaction center chlorophyll a. Instead of just one chlorophyll a molecule absorbing light, antenna pigments permit thousands of molecules to participate in light absorption—each at

their own characteristic wavelengths—and funnel that energy to the reaction center chlorophyll. This allows both photosystems to make much more efficient use of the photons impinging upon them. Examples of accessory pigments in plants include chlorophyll b and carotenoid pigments.

6. The first detectable Calvin-cycle product in C_3 photosynthesis is a 3-carbon compound called 3-phosphoglycerate (PGA); in C_4 and CAM photosynthesis, it is a 4-carbon compound called oxaloacetate. Additional differences exist in the behavior of the stomata and of the photosynthetic cells themselves. C_3 plants normally open their stomata during the day and the mesophyll cells simultaneously perform the light and dark reactions of photosynthesis. However, these plants close their stomata on particularly hot and dry days in order to conserve water. This leads to increased oxygen tension around the mesophyll cells (because of the light reactions), with concomitant loss of fixed carbon due to photorespiration. Like C_3 plants, C_4 plants also open their stomata for gas exchange during the day. Unlike C_3 plants, the mesophyll cells of C_4 plants first fix atmospheric carbon dioxide into oxaloacetate, which is promptly converted to malate, exported to the bundle-sheath cells, and decarboxylated. The CO_2 so released enters into the Calvin cycle in the bundle-sheath cells, and PGA is produced as before. However, because of the high CO_2 tension and low O_2 tension in the sheath cells, the Calvin cycle is favored there over photorespiration. CAM plants employ a variant of the C_4 pathway, opening their stomata at night and storing the fixed CO_2 as malate in the mesophyll vacuoles. During the daytime, they close their stomata to conserve water and decarboxylate the malate; the CO_2 so released enters the Calvin cycle as before. CAM plants thus perform their C_4 reactions during the cooler, more humid nighttime hours, and perform their light reactions and Calvin cycle during the day. C_3 plants operate best in cooler temperate climates, where they do not have to close their stomata during the day to conserve water; C_4 plants do better when conditions are hot and perhaps dry as well. CAM plants are best adapted to desert conditions. Approximately 85% of plant species use the C_3 pathway alone to fix carbon.

7. In the light reactions, photosystem (PS) I and II antenna pigments absorb photons and funnel the energy to their respective reaction center chlorophyll a's. Beginning with the reaction center chlorophyll a's of photosystem II, electrons pass along to various components of an electron transport chain, through PS I reaction center chlorophyll a's, to a second electron transport chain and ultimately to $NADP^+$ which is reduced to NADPH. As the electrons are being transported from carrier to carrier between PS II and PS I, some of their energy is used to pump protons from the stroma to the thylakoid lumen. Thus a pH gradient develops there, and an ATP synthase complex harnesses the energy of this gradient to phosphorylate ADP to ATP. The electrons originally lost from the PS II reaction center chlorophyll a's are replaced with electrons snatched from water, which in the process is oxidized to $2 H^+$ and $1/2 O_2$. The net result of these reactions is that light energy is used to drive the formation of ATP and NADPH, and to split water with the concomitant release of oxygen gas. The dark, or Calvin cycle, reactions are a series of chemical reactions where CO_2 and H_2O are combined to form organic carbon. The first and best known step in this pathway is catalyzed by an enzyme called Rubisco. Along the entire cycle, unfavorable steps are driven by the energy of ATP, and reducing power is provided by NADPH. CO_2 is reduced to a 3 carbon compound called 3-phosphoglycerate, which may then be converted to the true product of photosynthesis, phosphoglyceraldehyde (PGAL). Most other organic compounds in the cell can be synthesized indirectly from PGAL.

8. In noncyclic electron flow, electrons from photosystem (PS) II reaction center chlorophyll a's are passed along an electron transport chain, through PS I, to a second electron transport chain and finally to $NADP^+$ which becomes reduced to NADPH; a pH gradient is concurrently produced, from which the ATP synthase complex synthesizes ATP molecules by phosphorylating ADP. In cyclic electron flow, electrons from PS I are passed through several of the electron transport chain carriers that normally shuttle electrons between PS II

and PS I. Passage of these electrons helps create a proton gradient to drive ATP synthesis. The electrons return to the PS I reaction center. Cyclic electron flow produces ATP; PS II is not involved, and no NADPH is synthesized.

9. Photorespiration is caused by the ability of the ribulose 1,5-bisphosphate carboxylase/oxygenase (rubisco) enzyme to combine either oxygen or carbon dioxide with ribulose 1,5-bisphosphate. When carbon dioxide is a substrate, two molecules of 3-phosphoglycerate (PGA) are eventually produced, and there has been a net fixation of carbon. When oxygen is the substrate, however, only one PGA is produced, and carbon dioxide is released. Photorespiration is considered wasteful because it is counterproductive to the Calvin cycle reactions. Through the activities of the Calvin cycle, CO_2 carbons appear in newly synthesized PGA molecules; photorespiration, by contrast, causes PGA carbon atoms to be released again as CO_2, undoing the work of photosynthesis.

10. Glycolysis is a series of 10 reactions by which a starting sugar—usually glucose—is broken down into 2 molecules of pyruvate, a 3-carbon compound. During this process, there is a net gain per glucose reaction of 2 ATP molecules and 2 NADH molecules. No carbon dioxides are released during the glycolytic reactions, so that all of the original glucose carbons are contained in the 2 pyruvate molecules. The pyruvate molecules then move into the mitochondria, where they are decarboxylated—each pyruvate becoming an acetyl group on an acetyl CoA molecule; 2 more NADH molecules are produced during this process. At this point 2 of the original 6 glucose carbons have been released as carbon dioxide, with the remaining 4 carbon atoms present as 2 acetyl groups. The acetyl groups from acetyl CoA then enter the citric acid cycle, where they first combine with a 4-carbon compound (oxaloacetate) to form a 6-carbon compound (citric acid), which is subsequently decarboxylated twice, and the resulting 4-carbon compound is used to regenerate the oxaloacetate that started the cycle. At this point all of the carbons originally held in glucose have been respired as carbon dioxide. Keeping in mind that 2 acetyl groups enter the citric acid cycle for each glucose molecule processed through glycolysis, a total of 2 more ATP molecules are made, as well as 6 NADH and 2 $FADH_2$ molecules. Both NADH and $FADH_2$ are like poker chips in that they can be "traded in" for ATP molecules. Specifically, their high-energy electrons can be indirectly used to drive the phosphorylation of ADP to ATP.

11. Electron transport begins when either NADH or $FADH_2$ donates electrons to electron carrier proteins located in the inner mitochondrial membranes of the cristae. These carrier proteins are part of an electron transport chain that shuttles electrons down an energy gradient to the ultimate electron acceptor, oxygen. As oxygen molecules accept pairs of electrons and pick up protons from solution, they are reduced to water. Concurrent with electron transport, some of the carriers in the electron transport chain also translocate protons across the inner mitochondrial membrane to the intermembrane space. This establishes a pH gradient across the membrane. ATP synthase complexes in the inner mitochondrial membrane are able to harness the proton gradient to do work; specifically, the complexes allow protons to flow through them down the proton gradient and back into the mitochondrial matrix. As this happens, the ATP synthase complexes tap into some of the kinetic energy of the moving protons to phosphorylate ADP to ATP.

Applying Concepts

1. At a gross level, photosynthesis and respiration do appear to be the opposite of each other. The goal of respiration is to oxidize organic carbon to CO_2 and water, capturing some of the released chemical energy in the bonds of several high-energy ATP molecules. The goal of photosynthesis, on the other hand, is to reduce CO_2 and H_2O to organic compounds, some of these reactions being driven by the hydrolysis of ATP. Also, both of these processes consist of two separate sets of reactions: a series of enzyme-mediated chemical reactions that occur in aqueous compartments of the plant cell, and a series of oxidation-reduction reactions occurring in the membranes of organelles

(chloroplasts or mitochondria). Nonetheless there are significant differences. For example, the chemical reactions of the two processes are quite distinct from each other, using unique enzymes and, for the most part, unique substrates. Furthermore, neither photosynthesis nor respiration can be run in reverse, so obviously neither can be the "opposite" of the other. The electron transport chains of the two processes employ different (though in some cases, similar) electron carriers, with photosynthesis actually using two different electron transport systems (photosystems I and II). Also, different adenine dinucleotide electron carriers are used in the two processes: respiration uses NADH and $FADH_2$ while photosynthesis uses NADPH. Lastly, electron transport in photosynthesis is driven by absorbed light energy, but electron transport in respiration does not produce any light in return. Other differences exist, but those described sufficiently demonstrate that, although photosynthesis and respiration are similar in some ways, they are not simply opposite reactions.

2. Some plant cells do not possess chloroplasts, while others spend some or all of their time in the dark. In either case, such cells must rely on respiration for their energy needs. Some examples are root cells, which normally do not possess chloroplasts and usually grow in darkness (in the soil), as well as the living cells in secondary xylem and phloem growing beneath a thick, opaque bark. Obviously, on a dark night respiration is the only means for any plant cell to make ATP. Often, some or all of the cells in flowers and fruits lack chloroplasts and thus use respiration for energy, even on a bright, sunny day. Thus it is clear that, like animals, respiration is a critical process for plants as well.

3. First, using many small steps produces a greater variety of small molecules to be used as precursors in the synthesis of other important biological molecules. Aside from that, the main reason to use many small steps is that they allow a much greater degree of control over the reactions than if they were done in just a few large steps. Not only would it be much harder for cells to regulate the reactions occurring in them, but it would be more difficult to capture the energy of the exergonic reactions as well as to reverse any reactions if needed.

Chapter 6

Review Questions

1. DNA is a long polymer that consists of thousands to millions of repeating chemical units called nucleotides (d) arranged to form a double helical structure similar in shape to a spiral staircase. The backbone of this molecule (the railing of the staircase) is composed of a sugar (deoxyribose) and a phosphate molecule (e), while the nitrogenous bases (f) project inward from this backbone like the steps on the spiral staircase. Each DNA polymer forms a long molecule that in prokaryotes (including mitochondria and chloroplasts) assumes the form of a closed circle, that is, the ends join together to yield a circular molecule. In eukaryotes the DNA remains linear, each molecule thus possessing two free ends. Additionally, the eukaryotic DNA associates with proteins termed histones (b), and the resulting DNA-protein aggregate condenses into small bodies termed chromosomes (a). Whether the DNA molecule is circular or linear, it carries the units of genetic information known as genes (c). Most genes contain the coded information necessary to construct a new protein molecule, although in some cases RNA is the final gene product.

2. Adenine, cytosine, and guanine are found in both DNA and RNA (a); thymine occurs only in DNA (b), and uracil only in RNA (c). All five of these bases contain nitrogen (d) and hence are called nitrogenous bases. Adenine and guanine each are double-ring compounds called purines (e); the remaining three bases, each of which form a single ring, are pyrimidines (f).

3. The pairing rules for DNA are derived from the observation by Erwin Chargaff in 1950 that, regardless of the DNA source, the ratios of adenine to thymine and of cytosine to guanine were always nearly 1.0. The Watson and Crick model of DNA structure explained these ratios perfectly: the DNA molecule is composed of two complementary strands, all of whose bases are

normally paired to a base on the other strand. In this situation, adenine always pairs with thymine and cytosine with guanine. RNA, on the other hand, is single-stranded, and thus there is no second strand for DNA-type pairing to occur. However, RNA strands can fold back on themselves in antiparallel fashion, whereupon complementary sequences may still pair. Thus RNA may contain some paired bases. If this does occur, the pairing rules in RNA are that cytosine always pairs with guanine and adenine with uracil. The DNA fragment and its complementary strand would be:

```
AATATACCG
TTATATGGC
```

4. To replicate, the two parental strands of a DNA molecule must first be pried apart from each other. Helicase enzymes help in this separation process. Once separated, binding proteins stabilize the strands so that they do not re-associate again. DNA polymerases then bind to each of the individual strands and begin synthesizing its missing complementary strand. This synthesis is accomplished by "reading" the base sequence and inserting the complementary base into the nascent strand. Polymerases do not begin their work at random but rather at very specific DNA sequences known as origins of replication. When the process is complete, there will be two DNA molecules, each consisting of one of the original parental strands and one newly synthesized strand.

5. Many genes do indeed encode the amino acid sequences of proteins. But some genes encode tRNA or rRNA sequences, which are not translated into proteins. In addition, certain sequences of DNA serve as specific recognition sequences, such as the stretches of DNA forming the origins of replication. Lastly, much eukaryotic DNA has no known function, coding or otherwise; introns are one such example.

6. Being nucleic acids, both DNA and RNA are composed of nucleotides, a sugar-phosphate backbone bearing nitrogenous bases. The bases adenine, cytosine, and guanine occur in both molecules, and the sequence of bases in either molecule can encode genetic information. Lastly, both molecules are found in all cells, from the smallest prokaryote to the most advanced eukaryote. Here are some important ways these two molecules differ: (a) DNA uses the sugar deoxyribose in its backbone while RNA uses ribose; (b) DNA uses the base thymine while RNA substitutes with the base uracil; (c) in eukaryotes, DNA resides in the nucleus while RNA is synthesized in the nucleus but mostly functions in the cytoplasm; (d) in eukaryotes, DNA is associated with proteins termed histones while RNA is not; and (e) DNA exists as a double-stranded helix while RNA is single-stranded. Additional answers are possible.

7. The three types of RNA are mRNA, rRNA, and tRNA. DNA guides synthesis of RNA in a process called transcription. The mRNA so produced essentially carries a "working copy" of one or more genes out to the ribosomes, where translation occurs. Ribosomes are the structures on which translation occurs; they are composed, in part, of rRNA, a type of structural RNA. The third RNA type, tRNA, forms the critical link between the mRNA code and the amino acids of the nascent protein. At one end of a tRNA molecule is an attached amino acid; at the other end is a three-base sequence—the anticodon—that is specific for that amino acid. By pairing of codons and anticodons during the translation process, the correct amino acids are inserted into the growing peptide chain.

8. Initiation begins with the attachment of an initiator tRNA molecule to a small ribosomal subunit. After this, the small subunit binds to an mRNA molecule in such a way that the tRNA anticodons and the mRNA codons are closely aligned. Lastly, a large subunit then binds to the complex, completing the assembly. The large subunit is oriented such that its P site is occupied by the methionine-tRNA. Elongation, the next phase, is an iterative process. An aminoacyl-tRNA binds to the A site of the large subunit. This second aminoacyl-tRNA possesses an anticodon complementary to the second codon on the mRNA. Actual elongation of the nascent polypeptide occurs as a peptide bond forms between the

two amino acids. At this point the entire peptide is found in the A site. The mRNA then indexes down one codon, so that the first tRNA—devoid of its methionine—now occupies the exit site and leaves the assembly. The second tRNA with its growing peptide chain moves with the mRNA so that it now occupies the P site. The A site is now open, but a new aminoacyl-tRNA soon occupies it. As before, the anticodon of the incoming aminoacyl-tRNA is complementary to the codon now occupying the A site. A new peptide bond is formed, as before, and the process repeats itself. In this manner, the polypeptide chain grows, one specific amino acid at a time. Eventually a stop codon is brought into the A site, and the termination phase begins. A releasing protein binds to the stop codon, causing the polypeptide chain to be released and the ribosomal assembly to fall apart. The net result is the synthesis of a new polypeptide chain whose amino acid sequence was dictated by the base sequence of the mRNA strand.

9. (a) Between prokaryotes and eukaryotes, prokaryotes have the smallest amounts of DNA. Disparities in DNA levels in excess of several thousandfold are not uncommon. (b) Among the prokaryotes, some *Mycoplasma* species contain as little as 600 kb of DNA—enough for about 760 proteins—while other bacteria contain several times more DNA. As an example, *Escherichia coli* bacteria contain 4,700 kb of DNA, enough to code for about 4,000 proteins. (c) Eukaryotic cells contain the highest levels of DNA. On the low side are "simpler" organisms such as yeast (*Saccharomyces cerevisiae*), with about 13,500 kb of DNA, while the flowering plant *Trillium* sp. contains up to 100,000,000 kb. Humans, it turns out, contain a more modest 3,000,000 kb of DNA.

10. Ribosomes are aggregates of several ribosomal RNAs (rRNAs) and ribosomal proteins. These compounds are organized into a smaller assembly, the small subunit, and a larger one, the large subunit. Ribosomes consist of one such small subunit and one large one. The small subunit contains the important binding site for the mRNA. The large subunit possesses three important sites, two of which (the A site and the P site) are filled with aminoacyl-tRNAs during the elongation phase; the third site (the exit site) serves as the point of release of a tRNA molecule after it has served its purpose during protein synthesis.

11. The newly formed mRNA in eukaryotes is a mixture of mRNA sequences that are destined to be translated and other sequences that must be removed before translation. In the processing, the noncoding sequences—called introns—are excised from the mRNA, and the remaining coding sequences—the exons—are spliced back together into a shorter mRNA molecule that is ready for translation.

12. Replication is the process by which a molecule of DNA is duplicated. Ideally, both of the daughter molecules are identical to the original parental molecule. The typical use of replication is to make a complete copy of the genetic material of a cell just prior to cell division. Transcription is the process by which the sequence information of a DNA molecule (or segment) guides the formation of a molecule of RNA. In the case of mRNA, transcription results in a working copy of the relevant genes from the DNA molecule. Translation is the ribosome-mediated process whereby a new polypeptide is assembled; the sequence information found on the mRNA strand dictates the resulting amino acid sequence of the polypeptide.

Applying Concepts

1. The original strand and the newly synthesized products are:

(complementary DNA after replication)

```
AAATGTTTGGTCAGCGAGGGTAGGTT
```

(original DNA)

```
TTTACAAACCAGTCGCTCCCATCCAA
```

(complementary mRNA)

```
AAAUGUUUGGUCAGCGAGGGUAGGUU
```

(mRNA arranged into codons)

```
AA AUG UUU GGU CAG CGA GGG UAG GUU
```

(polypeptide sequence)

```
Met·Phe·Gly·Gln·Arg·Gly·(stop)
```

During the translation process, note that bases leading up to the AUG start codon as well as bases following the UAG stop codon are ignored and untranslated. Thus translation begins with the AUG start codon and ends when one of the three stop codons (here UAG) is encountered.

2. In the following answers, mRNA sequences have been arranged into codons, where possible.
   (a) mRNA = AA AUG UUU GGU CAG CGA GGC UAG GUU. In this example, the mutation has changed a GGG codon into a GGC codon. But both of these codons code for glycine (Gly), so there would be no change in the polypeptide sequence. This is an example of a silent mutation, and the polypeptide sequence is still

   ```
   Met·Phe·Gly·Gln·Arg·Gly·(stop)
   ```

   (b) mRNA = AA AUG UUU GGU CAC CGA GGG UAG GUU. This mutation has changed a CAG codon into a CAC codon, which changes a glutamine (Gln) to a histidine (His) in the polypeptide. The new sequence would be

   ```
   Met·Phe·Gly·His·Arg·Gly·(stop)
   ```

   (c) mRNA = AAAUCUUUGGUCAGCGAGGGUAGGUU. This mutation has changed the AUG start codon into an AUC. Without the start codon, translation cannot initiate, so there will be no polypeptide product.
   (d) mRNA = AA AUG UUU GGU CAG CGA GGG UAC GUU. This mutation changed the UAG stop codon into a UAC tyrosine (Tyr) codon. Thus, translation would proceed through the UAC codon until another stop codon is encountered. Obviously the mutant polypeptide could be considerably longer than the original one, and its leading sequence would be

   ```
   Met·Phe·Gly·Gln·Arg·Gly·Tyr·Val·...
   ```

   (e) mRNA = AA AUG UUU GGU CAG CGG GGU AGG UU. This deletion is an example of a frameshift mutation; beginning with the mutated codon and proceeding downstream from it, the codons have shifted one base to the left, resulting in a completely new sequence. Starting at the first mutated codon, the protein sequence generally becomes gibberish, with translation stopping whenever the first stop codon happens to be encountered. The leading sequence of this polypeptide would be

   ```
   Met·Phe·Gly·Gln·Arg·Gly·Arg·...
   ```

   (f) mRNA = AA AUG UUU GGU CAG CGA AGG GUA GGU U. This insertion also causes a frameshift mutation, changing codons to its right by shifting them over one base. Again the stop codon disappears and a new sequence results. The leading sequence of the polypeptide would be

   ```
   Met·Phe·Gly·Gln·Arg·Arg·Val·Gly·...
   ```

3. A codon size of two would mean any of the four bases taken two at a time. This would lead to 4×4 or 16 different combinations, barely sufficient to encode 14 amino acids plus a start and a stop codon. A codon size of 3 could encode $4 \times 4 \times 4$ or 64 different combinations, so could easily encode for 50 amino acids plus punctuation. But a codon triplet would be insufficient to encode 70 amino acids; one would need to use a codon quartet, which could specify $4 \times 4 \times 4 \times 4$ or 256 different combinations—more than adequate to encode 70 amino acids plus punctuation.

4. The theory of the common origin of life on Earth states that life began as a single life-form—one cell—which, over geologic time, radiated around the planet and evolved into the many extant and extinct species that have appeared. The fact that all life-forms on the planet, with very few and minor exceptions, interpret the genetic code in the same way is powerful evidence that this theory is correct. Alternately, the existence of a multitude of groups, each interpreting the code in its own unique way, would be strong evidence that all life does not have a common ancestor but rather that the different groups of organisms have had different primordial origins. In other words, it would imply that life arose many times independently and that the descendents of those initial cells form the different groups we see

today. Such a situation would require drastic revision of our theories of the origin of life on this planet.

5. Students' answers will vary.

Chapter 7
Review Questions

1. The cell cycle is the orderly sequence of events occurring in cyclical fashion in the life of a cell. In the longer phase of this cycle—interphase—the cell components, including the DNA, are duplicated; in the shorter division phase, the duplicated components are distributed amongst the two forming daughter cells.

2. In prokaryotes, including organelles such as mitochondria and plastids, the circular DNA chromosome(s) lies supercoiled in a special region of the cell/organelle termed the nucleoid. In contrast to the nucleus of eukaryotes, the nucleoid is not membrane delimited but is rather a defined region of the bacterial cytoplasm. The linear DNA molecules of eukaryotes, on the other hand, associate with histone proteins (nucleosomes) to form highly condensed DNA-protein assemblages. Such structures are found in a double membrane-bound region of the cell called the nucleus. The DNA of mitochondria and plastids is like that in prokaryotes.

3. First, the prokaryotic chromosome replicates. The cell then elongates, and because the chromosomes are attached to the plasma membrane, this process results in the separation of the two daughter chromosomes. The cell wall and plasma membrane of this elongated cell then pinches inward until the cell separates into two, and cell division is completed.

4. A centromere is a specific region of a eukaryotic chromosome that binds the DNA to the mitotic spindle. A body that contains a single centromere, regardless of whether it carries one or two double-stranded DNA molecules, is termed a chromosome. After a eukaryotic chromosome duplicates, two identical molecules of double-stranded DNA are joined together by a single centromere; in this case, the entire structure is a chromosome and each of the two DNA molecules is a sister chromatid. During anaphase, each centromere splits, the sister chromatids separating and moving to opposite poles of the mitotic spindle. At this time each chromatid possesses its own unique centromere and is properly termed a chromosome.

5. Mitosis is division of the nucleus, whereby each of the two daughter nuclei receive one sister chromatid from every sister chromatid pair. This results in each of the daughter nuclei receiving a complete copy of every chromosome present in the parent nucleus. Cytokinesis, on the other hand, is the division of the cytoplasm and the distribution of the cytoplasmic organelles between the daughter cells. Typically, cytokinesis immediately follows mitosis, although this is not always the case. Mitosis and cytokinesis are both processes involving eukaryotic structures such as microtubules and nuclei, and thus they are properly applied only to eukaryotes.

6. Each species has a characteristic number of chromosomes. In some cells, each chromosome carries a unique assortment of genes; such cells are termed haploid. In other cells, chromosomes occur as members of pairs, meaning that each chromosome has a partner that carries the same genes; such a cell has two sets of chromosomes, two copies of each gene, and is referred to as diploid. Additionally, the chromosome pairs in such a diploid cell are called homologous chromosomes. For example, humans possess 23 unique chromosomes, so that a haploid human cell (egg or sperm) possesses only those 23 chromosomes, while a diploid human cell contains 23 pairs of chromosomes, or 46 altogether. Although the exact number of chromosomes varies from species to species, within a species, a diploid cell possesses twice as many chromosomes as a haploid one. Since every chromosome in a haploid cell carries a unique set of genes, haploid cells do not possess homologous chromosomes; the term, in fact, is not even applicable to haploid cells.

7. G$_1$ phase—Cells increase in size as they synthesize cellular components such as proteins and membranes.

S phase—Replication of DNA (duplication of chromosomes) and synthesis of associated histone proteins.

G$_2$ phase—Final preparations for cell division as cells synthesize the requisite components of the mitotic apparatus.

Prophase—Chromosomes become visible (in a light microscope) as they condense; microtubules of mitotic spindle appear; nucleolus and nuclear envelope disappear.

Metaphase—Centromeres of chromosomes align themselves in an imaginary plane along the equator of the mitotic spindle.

Anaphase—Sister chromatids separate at the centromeres and move to opposite poles of the mitotic spindle.

Telophase—Chromosomes complete separation, uncoil, and become invisible again; nucleolus and nuclear envelope reappear.

Cytokinesis—Division of the cytoplasm and distribution of the organelles to the newly formed daughter cells.

The G$_1$, S, and G$_2$ phases are part of interphase; the others are part of the division phase of the cell cycle.

8. Located at the centromere of each chromatid, kinetochores are special protein complexes that bind to the (+) end of spindle microtubules. Kinetochores play critical roles in (a) keeping sister chromatids together until all centromeres are aligned on the spindle equator (Mad2 pathway), (b) freeing chromatids so that they may separate and move to opposite poles, and (c) providing the attachment point for spindle microtubules to chromosomes, thus transmitting the motive forces necessary to accomplish (a) and (b).

9. A phragmoplast is a system of microtubules and protein filaments (actin and myosin) that occurs only in plants and some algae. It functions during cytokinesis to divide the cytoplasm. In doing this, it provides a mechanism by which plant cells may divide into two daughter cells, each producing its own complete cell wall.

Applying Concepts

1. Mitosis is the means by which eukaryotic cells distribute identical copies of the nuclear DNA (i.e., the genetic material) to two daughter cells.

2. One line of research would be to examine mitotic cells with a transmission electron microscope (TEM). Modern TEM techniques clearly show real structures—microtubules—corresponding in location to the spindle fibers. Secondly, when fluorescent antibodies directed against the protein tubulin are administered to mitotic cells, again real structures corresponding to spindle fibers are tagged and become clearly visible in a fluorescence microscope (a technique called immunomicroscopy). Also, if one examines a mitotic cell under a light microscope and uses a laser to destroy the spindle fibers to one side of a chromatid pair, the chromatids immediately move toward the opposite pole. These lines of evidence and many others strongly suggest that spindle fibers are indeed physical structures.

3. Cancer is a disease in which cells have lost their tight control over the cell cycle in general and mitosis in particular. Such frequent mitoses in cancerous tissue result in an expanding mass that forms a malignant tumor. In mitosis, the alignment and distribution of chromosomes is intimately dependent upon microtubules; indeed, without microtubules, mitosis cannot proceed properly. By inhibiting the formation of these very microtubules, compounds such as vinblastine inhibit mitosis overall and thus slow the spread of highly mitotic cancer cells.

4. The names commonly applied to these polyploidy conditions are triploid (3N), tetraploid (4N), hexaploid (6N), octaploid (8N), and decaploid (10N).

5. In properly stained tissue, each of these phases is easily identified under a light microscope. Thus, one needs only to examine prepared slides of the mitotic tissue, and then tabulate the numbers of cells seen in interphase and in each of the mitotic phases. The numbers so obtained for each phase should be in direct proportion to the amount of time spent in that phase. For example, if 5 out of 1,000 cells in mitotic tissue were in anaphase, then anaphase probably accounts for approximately 5/1,000, or 0.5% of the cell cycle.

Chapter 8
Review Questions

1. Roots, stems, and leaves compose the three basic plant organs, with stems and leaves forming the upright, aboveground organ system known as the shoot, while the roots form the buried root system. Flowers, buds, fruits, and seeds are more complex plant organs consisting of two or three of the basic plant organs.

2. These three terms are used to describe the life-cycle patterns of flowering plants. Annual plants complete their entire life cycle within a 1-year period, usually germinating early in the growing season and then flowering, setting seed, and dying before that season ends. With some annuals, such as the spring ephemerals, this may occur within a 2-month period. Biennials, on the other hand, require two years to complete their life cycle; the first year is generally a period of vegetative growth, followed by a year of flowering, fruiting, and death. Perennials, such as trees and shrubs, persist for many years; indeed, some individual bristlecone pine trees have been alive in California and Nevada for nearly 5,000 years.

3. Plasmodesmata are small, cytoplasmic connections between adjacent cells. Although plasmodesmata are essentially pores through adjacent cell walls, they are far more than simple pores. For example, each contains a cytoplasmic strand and a variety of proteins that help regulate the passage of solutes and particles from one cell to the next. Simple pores could not do that. Symplastic continuity refers to the condition in which the cytoplasm of two adjacent cells are in contact with each other through the plasmodesmata; materials such as certain ions can move from cell to cell via the plasmodesmata and never have to cross either a cell wall or a plasma membrane. Primary and secondary plasmodesmata are distinguished by how they form. Primary plasmodesmata form as new wall material is laid down between sibling cells during cytokinesis. Secondary plasmodesmata, on the other hand, form between two cells not as wall material is deposited but by selective enzymatic degradation of already formed walls; basically, small pores are enzymatically created and then strands of cytoplasm are connected through the pores. The cells involved in secondary plasmodesmata formation may or may not be sibling cells.

4. A meristem is a region where undifferentiated cells display high mitotic activity; that is, they are producing new tissue via rapid cell division. A primary meristem produces primary tissues (no wood or cork) and leads to an increase in the length of a shoot or root. A typical location for such primary meristems is at the apex of a root or shoot, in which case they are known as root or shoot apical meristems, respectively. Secondary meristems such as the vascular and cork cambia produce secondary tissues—secondary xylem (wood) and secondary phloem in the case of the vascular cambium and the cork cells of the bark in the case of the cork cambium. Despite their names, these cambia are bona fide meristems.

5. Most plant cells contain a large vacuole, a single-membrane bound organelle that occupies a central location in the cell. If the cell wall is loosened enough so that it is able to expand, then rather rapid cell growth can be effected by water uptake into the vacuole. The enlarging vacuole thus expands the cell size and causes cell growth. Animal cells, lacking a vacuole, do not possess this growth mechanism.

6. All plant life cycles alternate between a spore-producing sporophyte and a gamete-producing gametophyte. Sporophytes develop from zygotes and gametophytes from spores; they are genetically distinct individuals having different numbers of chromosomes. This type of life cycle is not seen among the animals, which do not produce spores. Plants are also capable of reproducing asexually, where an individual essentially produces clones of itself. This type of propagation may occur (a) in the wild via plantlets, sucker shoots, and rhizomes; (b) in the greenhouse via stem, leaf, or root cuttings; or (c) in the laboratory from single plant cells or plant tissue cultures via a process known as somatic embryogenesis. Higher animals lack the ability to reproduce asexually; certainly they do not do it in the wild, and even the animal cloning performed in today's research labs is not true somatic embryogenesis.

Applying Concepts

1. Plant cells joined by plasmodesmata have symplastic continuity, meaning their cytoplasms are continuous. Cytoplasmic continuity allows chemical communication, and proteins and messenger RNA (mRNA) molecules move among cells connected by plasmodesmata. On the negative side, viruses can also move through plasmodesmata, spreading disease throughout a specific interconnected plant tissue.

2. The sporophyte develops from a zygote and produces spores. The gametophyte develops from a spore and produces gametes. These two multicellular plant bodies differ in size and structure and alternate during the life cycle of a plant. In mosses the sporophyte is small and grows on its usually more conspicuous green gametophyte. The fern sporophyte is the large, leafy plant grown in gardens and homes as a houseplant. The sporophyte produces spores on the undersides of the leaves. The fern gametophyte is much smaller, about the size of a thumbnail, and a microscope is necessary to see the gamete-producing structures. In flowering plants such as daisies and oaks, the sporophyte is the large, conspicuous flowering plant. The gametophyte stages are only a few cells in size, are not green, and are located within the flowers of the sporophyte stage, on which they are dependant for food.

3. Many plants can reproduce by asexual reproduction. The common houseplant *Kalanchoë*, for example, can produce tiny plantlets along the margins of its leaves. When these plantlets drop off the leaves, they can take root in the soil and grow into adult plants that are genetically identical to their parent. In the laboratory non-reproductive cells can be taken from plants such as carrot and induced to form embryos and adult plants. This cloning process is called somatic embryogenesis, the production of embryos from somatic (body) cells rather than fusion of gametes. In both asexual reproduction and somatic embryogenesis, body cells give rise to embryos and new adult plants that are clones of the parent plant. The two processes are similar.

Chapter 9

Review Questions

1. Stems are the organs that bear the other plant organs—roots, leaves, flowers, fruits, cones, and so forth. The apical meristems of stems provide a means for plants to increase their height or length. This, in turn, further increases the available surface area and allows the production of more reproductive or photosynthetic tissue. Stems provide direct support for the aerial portions of a plant. Some stems are themselves photosynthetic, and others may sport sharp thorns for protection. The vasculature of stems allows for the transport of water, minerals, and sugars between the roots and leaves. Some stems are modified to increase their utility even further. For example, some stems store food or water, others are used for climbing and support, and yet others are purposely made hollow to house an army of protective ants! Clearly, stems play a pivotal role in the life of plants.

2. Both xylem and phloem transport materials within the plant body. Generally, xylem conducts water and minerals from the roots through the stems to the leaves, flowers, buds, and fruits. In addition, in the springtime, sugars stored in the lower parts of the plant are carried to the developing buds and leaves via the xylem. It is for this sugar solution that the xylem of maple trees is tapped for sap in the spring. Phloem, on the other hand, transports dilute solutions of organic molecules such as sugar from a "source" where they are produced or stored to a "sink" where they are utilized. This means that whereas flow in the xylem tends to be in the roots-to-leaves direction, phloem flow is not so limited and may occur in any direction, though always from source to sink.

3. Only the vascular plants possess true vascular tissue. This large assemblage of plants includes the seed plants—flowering plants and gymnosperms (pine trees and relatives)—as well as the pteridophytes (ferns) and lycophytes. Vascular tissue appeared with the early land plants hundreds of millions of years ago at a time when plants consisted of only stems, no roots or leaves. Although some bryophytes and algae do have conducting tissues, they lack lignin, a substance considered the hallmark of true vascular tissue.

4. Primary xylem and phloem are derived from the apical meristem through the activities of the procambium, while secondary xylem and phloem differentiate from a secondary meristem known as the vascular cambium. Primary xylem and phloem occur in close proximity inside elongated packets called vascular bundles. The bundles are generally wrapped in hard supporting tissue termed sclerenchyma and are separated from each other by nonvascular tissue. In monocots, vascular bundles are scattered throughout the stem, while in eudicots they are arranged in a ring near the periphery of the stem. In woody eudicot stems, vascular tissue is no longer produced by the apical meristem but instead by the vascular cambium. This cone-shaped meristem consistently produces secondary xylem on the inside and secondary phloem on the outside, an arrangement that is reflected in the anatomy of a woody stem: a large central core of wood (secondary xylem) surrounded by the peripherally located inner bark (secondary phloem).

5. (a) In cross section, the vascular cambium appears as a ring of tissue located just under the inner bark. (b) In three dimensions, the cambium is actually a cone-shaped mass of tissue broadest at the base and tapering to a point near the tip of the stem. The vascular cambium contains two types of cells: vertically elongate fusiform initials, which produce the xylem and phloem, and cuboidal ray initials, which produce the vascular rays. Longitudinal division of the fusiform initials produces two cells—one remains as an initial and the other matures into a xylem cell on the inner, or centripetal, side of the cambium. Another division of a fusiform initial will again produce two cells, one of which will again remain an initial. This other cell, this time lying to the outside of the cambium, matures into a phloem cell. Generally more xylem cells are produced than phloem cells.

6. The periderm is produced by the activities of the cork cambium, which produces cork cells. Although cork cells are dead at maturity, their cell walls are impregnated with substances that render them highly functional. Lignins, for example, confer structural strength, while suberin wards off microbes and hinders water loss. Tannins inactivate certain pathogens, while sticky gums and latexes probably serve protective functions as well.

7. In phloem, sugars and other organic compounds (or even ions) are loaded directly or indirectly into sieve elements using the chemical energy of ATP. The presence of these solutes draws water into the sieve element, since the solute-rich cytoplasm there has a lower water concentration than the surrounding tissues. This incoming water creates a hydrostatic pressure that causes water to flow along the sieve tubes and ultimately into the surrounding cells. This process continues as long as phloem loading concentrates sugars in the sieve tubes. Transpiration, on the other hand, provides the driving force for the movement of water in the xylem. As one molecule of water evaporates from the surface, attractive forces draw another molecule up to replace it. Due to the cohesive nature of water, this results in the entire water column in the xylem being lifted toward the evaporative surface. Thus, movement within the xylem is a passive process in that ATP energy is not expended.

8. Early in the growing season, water tends to be abundant and growing conditions near optimal, leading to rapid growth with the production of rather large xylem cells. Later in the season growing conditions decline, so that growth is less rapid and the resultant cells smaller. Although the change in cell size is gradual from the early to late wood within one season's growth, when observing the transition between the late wood of one season and the following season's early wood, the contrast in cell size is much more striking. This abrupt change in cell size at the late wood–early wood boundary leads to the appearance of growth rings.

9. One growth ring is usually the result of one season's growth, with good growing seasons producing thicker rings than less optimal ones. Thus, when one views the growth rings in a woody stem, the widths of the rings vary in a pattern that mirrors the suitability of the growing season that produced them. Trees growing in the same locale will experience a similar microclimate, and will hence show similar patterns in their growth rings.

By matching the growth ring patterns from the wood of both living and dead trees, one can build a record of the climate of an area dating back several thousand years, using the science known as dendrochronology.

10. Phloem contains sieve elements, either sieve cells or sieve tube members stacked end to end. The walls of the sieve elements, especially the end walls, contain sieve plates—regions having a high density of pores. Companion cells—special phloem cells developing as sister cells to the sieve elements—are especially important for phloem loading. Also present in phloem are parenchyma cells and a type of support cell with thick walls known as a fiber. Xylem tissue usually contains two types of conducting cells, tracheids and vessels. Vessels tend to be wider than tracheids and they lack end walls; they are stacked end to end and form long, continuous conduits. Both tracheids and vessels reinforce their walls with an incomplete coating of lignin; the areas without lignin are known as pits. Water enters and leaves the tracheids and vessels through their pits.

11. Horizontal stems grow parallel to the ground and usually facilitate the spread of an individual plant over a wider area. Examples are stolons, which lie on the surface of the ground; and rhizomes, which grow beneath the surface. Some stems store food reserves. For example, the reserves stored in the underground stems of crocus and gladiolus plants can be tapped for growth and development early in the growing season; such reproductive and storage stems are known as corms. Tendrils are stems (or modified leaves) that coil around objects and provide support for climbing plants. Some plants such as cacti have large, green, pleated stems that are modified to store water as well as provide a photosynthetic surface for the plant.

Applying Concepts

1. Answers to this question are many and varied. Here are a few common uses for stems: The fibers in some stems are used to make textiles like linen and rope such as hemp. Many foods either are stems or are derived from them—asparagus, broccoli, and potatoes are largely stem foods, cinnamon is derived from the bark of the cinnamon tree, and the spice ginger from a stem (rhizome) of the ginger plant; maple syrup is derived from the stem sap of maple trees; and some white sugar comes from the stem fluids of sugar cane. Bamboo for sushi mats and hanging plant rods comes from the stems of bamboo plants; wood chips in planters and particleboard is mostly derived from stems; the corks that stop wine bottles come from the bark of the cork oak; and some baskets are woven wholly or in part using stem materials.

2. The growth rings seen in many woody plants result from the contrast between the small, slow-growing cells of the late wood and the large, rapidly growing cells of the following season's early wood. In regions such as the tropics, where the growing season is more or less constant throughout the year, early wood and late wood cells are very similar in size and appearance. In such trees, growth rings may be either nonexistent or very indistinct.

3. The initial growth in length is accomplished during primary growth by the apical meristem. Plants such as the redwood then develop a secondary meristem—the vascular cambium—to produce xylem centripetally and phloem centrifugally. Since the cambium forms at the periphery of the stem, the pith and other structures present inside the cambium will remain there for the life of the tree. Thus the oldest cells, actually cell remnants at this point, will be found at the center of the stem, inside the very first growth ring.

4. A tree stem with bark has obviously already produced vascular and cork cambia and is growing via secondary growth. Secondary growth, in turn, results in increase in girth but not increase in height. Thus, as long as the nail remains in the tree trunk it will be 2 meters above the ground.

5. At the very outside, depending on the age of the stem, you might encounter the remnants of some tasty epidermis. Next you will dine on bark, including cork cells from the cork cambium and phloem cells that the vascular cambium has pushed out into the bark regions. A quick nibble and you are through the cork cambium and on to the fourth course of your meal—the scrumptious secondary phloem. Soon, however, you make your way to the vascular cambium, and

once past it you encounter the main course—xylem, perhaps sampling some ray cells along the way. The xylem, you notice, becomes less tasty—bitter even—as you eat your way into the heartwood. Eventually you find your rather unsatisfying dessert at the very center of the stem as you encounter the remnants of primary growth, namely, pith and primary xylem.

Chapter 10

Review Questions

1. The root apical meristem produces the protoderm, ground meristem, and procambium—the three primary meristems from which epidermis, cortex, and vascular tissues arise, respectively. A similar differentiation occurs from the shoot apical meristem as well.

2. For most plants, roots are essential for water and mineral uptake and for anchorage into the soil. However, many plants extend these basic functions of roots to include other duties. Some plants have fleshy roots that store carbohydrates. Roots are important in plant development, because they are the site of production of several plant hormones (cytokinins and gibberellins). Sometimes secondary compounds are synthesized in the roots and transported elsewhere, such as the poisonous alkaloid nicotine. Prop roots and buttress roots support aerial stems and help prevent them from being pushed over by wind or other forces. Pneumatophores are aerial roots that help provide oxygen to other submerged roots. A few plants produce highly specialized contractile roots, which do indeed physically contract and pull the shoot into the soil. Lastly, some parasitic plants use roots or root-like organs to directly tap into the vascular tissue of their host plant.

3. Pneumatophores, produced by some types of mangrove trees, are roots that project into the air and provide a passageway for atmospheric oxygen to reach the submerged portions of the root system. Pneumatophores contain small pores—lenticels—through which air can enter when the pneumatophores are exposed. When the tide is high and the pneumatophores are submerged, a water-repellant lining around the lenticels prevents the entry of water. Once inside, the air enters aerenchyma tissue, which has lots of air-filled spaces between its cells, and oxygen is able to diffuse down to the submerged roots. Pneumatophores essentially act as snorkels. Oxygen is required by roots in order to produce the ATP they need; because the submerged roots are not photosynthetic, they must generate this energy molecule by cellular respiration, which consumes oxygen.

4. A taproot is a single main root derived from the primary root, and it often produces many branch roots. Normally roots are produced from the primary root, the taproot, or one of its branches, so that roots arise from preexisting roots. When roots arise from other tissues—shoots or leaves, for example—they are termed adventitious. In monocots the primary root soon disintegrates and is replaced by a multitude of adventitious roots from the stem; such root systems do not contain a single dominant taproot and are known as fibrous root systems. Finally, feeder roots, produced by both taproots and fibrous roots, are simply the small, delicate roots that are most active in water and mineral uptake from the soil.

5. The branch roots closest to the tip of the main root are the youngest, most recently formed branches. As we proceed toward the root tip, we next encounter a region bearing a myriad of long, thin root hairs. At the very tip of the root is the root cap, which overlies and protects the root apical meristem. Surrounding the current, intact root cap, and dispersed among the nearby soil particles, are sloughed-off root-cap cells known as root border cells; these cells help protect the root cap from microbial and nematode worm attack. Finally, surrounding this entire region are gooey polysaccharides that were synthesized and secreted by the epidermal cells. These compounds have several functions: (a) they lubricate the roots as they push their way through the soil; (b) they aid in mineral and water absorption; (c) they help prevent desiccation of the root; and (d) they provide a more optimal growing environment for beneficial microbes.

6. At the very tip of the root—just beneath the root cap—lies the root apical meristem, which forms a zone of cell division. It is here that mitosis is actually occurring, adding more cells to the growing root tip. Immediately behind this zone lies the zone of elongation. In this region the newly formed daughter cells are taking up large amounts of water and hence increasing in size, especially length. Behind this zone is the third region, the zone of maturation. Here cells specialize and begin to take on the structure and function of mature cells. These zones are not sharply delineated and may overlap. Differentiation of phloem sieve tubes, for example, may begin in the zone of elongation.

7. One route of mineral entry into roots begins with the delicate root hairs. Transporter proteins in the cell membranes of the root hairs actively take up beneficial minerals from the soil environment and accumulate them intracellularly; water passively enters the cells as well. The mineral-laden water then moves via plasmodesmata from the epidermal cells through cortical (including endodermal) and pericycle cells until it finally enters the xylem itself. In this path, it is the root hair cell membrane that provides selectivity; certain ions are allowed to enter the root hair cell, while others are denied entry. The second route of entry involves minerals that have nonselectively entered the roots (i.e., between the root cell walls) and occupy the cortical intercellular spaces. Such intercellular minerals (and water) are denied direct entry into the vascular tissue by the Casparian strip on the endodermal cells. Such minerals must first enter the endodermal cells by crossing the cell membrane on the outer face of those cells. Once inside, the minerals can proceed to the xylem as in the first scenario. In this case, selectivity is provided by the cell membrane of the endodermal cells. Both routes are selective because, at some point, both involve passage of minerals across a cell membrane. The membranes thus act as gatekeepers.

8. (a) First of all, plants limit the uptake of aluminum into roots by not producing aluminum transporter proteins. This greatly diminishes the ability of aluminum to enter root cell cytoplasm, and hence to gain entry to the root xylem. Also, some roots release organic compounds that bind free aluminum or else they incorporate the free aluminum into their cell walls. In either case, the aluminum becomes "tied up," and is no longer available for uptake. (b) Although phosphate is a necessary mineral, it is often present at very low levels in the soil. To overcome this problem, plants enact several countermeasures aimed at either increasing the absorption of the phosphate ion or increasing the soil availability of this ion. Increasing the numbers and lengths of root hairs and of root branches is an example of the first action; the increase in root hair surface area increases ion absorption. Roots can increase soil availability of phosphate by secreting substances that dissolve bound phosphate or release it from other compounds. Either way, plants are able to concentrate phosphate to intracellular levels far exceeding those in the surrounding soil.

9. Both active transport and facilitated diffusion involve integral membrane proteins. However, active transporter proteins are able to accumulate minerals against their concentration gradients. This type of molecular work is "paid for" with the chemical energy of ATP. Cells possessing these transporters can accumulate ions even when they are present at very low levels in the environment. Facilitated diffusion does not involve a direct energy source; the driving force is the kinetic energy of the molecules themselves (in the form of diffusion). Without a direct energy source, facilitated diffusion cannot concentrate molecules; substances simply use the carrier proteins as portals to move across a membrane from high to low concentration. Roots generally use active transport to take up beneficial minerals because their soil concentrations are usually quite low.

10. The two major human uses of roots are probably as food and medicinal drug sources. Examples of the former include carrots, parsnips, sweet potatoes, turnips, rutabagas, cassava, horseradish, and sugar beet. Both mandrake (*Mandragora*) and ginseng (*Panax*) are examples of plants whose roots are used in herbal medicine, while snakeroot (*Rauwolfia serpentina*) is a source of the alkaloid reserpine, which has sedative and anti-hypertensive properties. Roots can also be a source of dye (e.g., beet roots), and they can be a source of wood for specialized art or woodworking projects.

11. To establish a partnership with nitrogen-fixing bacteria, legume roots release flavonoids into the soil. Nitrogen-fixing bacteria detect the flavonoids and respond with chemical signals known as NOD (nodulation) factors. NOD factors induce legume root hairs to curl around the nitrogen-fixing bacteria and bring them into close contact with the roots. The legume plant then constructs a passageway of cell wall material to conduct the nitrogen-fixing bacteria into the internal portions of the root. The bacteria then induce through chemical signals the formation of root nodules and the production of leghemoglobin to control oxygen levels in the nodules. Once the nodules are complete and the oxygen levels optimal, the bacteria begin fixing nitrogen as ammonia. This partnership is important because legumes require high levels of protein for their seeds, which makes them nutritious to humans, and thus have a high fixed nitrogen requirement, more than is usually present in soil.

Applying Concepts

1. You begin eating through the epidermis, perhaps—if you are at the zone of maturation—chomping on a few root hairs in the process. Next, you bore through the relatively undifferentiated cells of the cortex—things get a little tougher as you encounter the suberin on the walls of the endodermal cells. Then you break through the endodermis and sequentially encounter the pericycle, followed by a small region of primary phloem, and finally, at the very center of the root, you find the primary xylem.

2. An epidermis surrounds both roots and stems. In the zone of maturation, the root epidermis consistently produces thin-walled, transporter-laden root hairs for absorption. The stem epidermis, on the other hand, may also produce hairs of one sort or another, but these hairs are generally thick-walled and not intended for absorption; typical functions of stem hairs would be secretion of various (especially waxy) substances, deterring insect herbivores, and reflecting the intense heat and light of full sunshine. Additionally, the stem epidermis may produce stomata for photosynthesis and spines for protection, elaborations that seldom occur in roots. Of course, most stems also produce leaves and buds as well as nodes and internodes; these organs and structures are not found on roots.

3. An epidermis surrounds both primary roots and primary stems. Internal to this, roots possess a cortex, a region of relatively undifferentiated cells and significant intercellular spaces. The innermost layer of cortical cells is modified by a strip of suberin—the Casparian strip—on the four cell walls oriented perpendicular to the overlying epidermis. Because this suberin strip essentially seals off the intercellular spaces, centripetally moving substances can traverse the endodermis only by actually passing through (and not between) the endodermal cells themselves. In the root the vascular tissue lies inside the endodermis, where it occupies the center of the root. In contrast, the vascular tissue does not lie at the center of a stem; instead, it either forms a ring (eudicots) or is scattered throughout the relatively undifferentiated ground tissue (monocots). In neither case is an endodermal-type barrier present. (It should be noted that in some plants, such as corn, the root vascular tissue does not form a solid core but rather a hollow cylinder about a pith.)

4. Roots increase their length at the tips, where cells in the zone of elongation are lengthening. Essentially, elongation of the cells in this region pushes the root tip through the soil. Thus the greatest increase in distance between our 1-mm markers would occur near the root tip, in the zone of elongation. The couple of marks there, in our example, may end up being 1–2 cm apart. In the zone of maturation, and continuing up the root (away from the tip), elongation has basically ceased, so that the markers in these regions would still be 1 mm apart.

5. If root hairs were to be produced closer to the root tip than the zone of maturation, they would find themselves being pushed through the soil as the cells in the zone of elongation increased their length. This would lead to severe abrasion of the delicate hairs, and they would be sheared off as if you had dragged the root along a piece of sandpaper. Besides, the root tip itself is covered by the root cap, so that there is little opportunity for the root tip cells to send out root hairs.

6. Potatoes are indeed underground stems, modified for the storage of starch. The most convincing evidence arises from the knowledge that roots do not produce leaves, buds, nodes, or internodes. When you look closely at the "eyes" of a potato, you can recognize that the "brow" is in fact a leaf scar and the "eye" itself is a small bud. (In fact, it is these buds that sprout if you leave potatoes in your kitchen too long.) Each eye is thus a node; by drawing lines through successive nodes, you can recognize internodes. Such a structure is typical of a stem but completely foreign to a root, which is why botanists classify tubers as stems and not roots.

Chapter 11

Review Questions

1. (a) Blade—the terminal flattened, usually photosynthetic portion of a leaf.
   (b) Petiole—the stalk of a leaf, which attaches the leaf to the stem.
   (c) Stipules—small, leaflike structures, usually paired, that may occur at the base of a petiole, where the leaf joins the stem.
   (d) Axils and axillary buds—an axil is the upper angle between a leaf and the stem that bears it. Axillary buds are buds that occur in the axils.
   (e) Node—the point on a stem where one or more leaves are attached.
   (f) Internode—the region on a stem between nodes.
   (g) Sessile—refers to a leaf that lacks a petiole, its blade being borne directly on a stem.
   (h) Leaf sheath—an expansion of the leaf base that partially or totally envelops the stem; sheaths occur mainly in monocots.

2. A simple leaf is a leaf whose petiole bears only one blade; in a compound leaf, the petiole supports from two to many individual blades. In a pinnately compound leaf, the petiole is extended into a long axis from which the individual blades, or leaflets, arise, an arrangement resembling the branches, or pinnae, on a bird's feather. In a palmately compound leaf, the leaflets are all borne from a single point at the tip of the petiole.

3. Alternate leaves, the most common arrangement, are borne one leaf per node; opposite leaves occur in pairs, two per node. Generally, opposite leaves are borne 180° apart on the stem (i.e., "opposite" on the stem) and are often rotated 90° from the leaf pairs above and below them. Three or more leaves per node is an arrangement referred to as whorled.

4. Leaf epidermal cells often secrete a waxy cuticle that inhibits evaporation; some plants even possess several epidermal and cuticular layers to further retard water loss. In addition, the cuticular layer is often thicker on the upper leaf surface, which generally receives the full force of the sun's heat and light. Some leaves bear trichomes, which may act as tiny mirrors to deflect some of the excess light and heat of the sun; this would allow the leaf to remain cooler and thus reduce evaporation. In many plants, the stomata occur mainly on the lower leaf surface, away from the sun's intense light. Leaf stomata may close during the midday heat or if water loss via evaporation exceeds water uptake by the roots. CAM plants open their stomata at night, when temperatures are cooler and transpiration almost nil, storing the next day's worth of CO_2 as organic acids in the leaf; during the day, their stomata remain closed so that evaporation is minimized. Many plants shed their leaves during prolonged periods of water unavailability, such as during a hard winter or during the dry season. Plants such as cacti reduce the evaporative surface area by reducing leaves to mere spines and performing photosynthesis in their stems.

5. Hydrophytes are aquatic plants whose leaves are adapted to being wholly or partly submerged in water; xerophytes are plants adapted to grow in dry habitats; and mesophytes are plants that require a moderately moist environment. Xerophytes tend to have thick cuticles; sometimes these plants even have multiple epidermal and cuticular layers to greatly reduce water evaporation. Mesophytes typically have a thicker cuticle on the upper epidermis that, of course, is directly exposed to the desiccating effects of solar radiation. Hydrophytes grow where water is generally abundant and submerged leaves often have little or no cuticle at all; such leaves may have photosynthetic epidermal cells, in contrast to mesophytes and xerophytes. Even among hydrophytes, emergent leaves or the upper surfaces of floating leaves often have well-developed cuticles to conserve water.

6. Because most leaves are coated with a waxy cuticle, leaves have evolved pores through which gases are exchanged with the atmosphere. These pores are the stomata. Each stoma consists of two chloroplast-containing guard cells, with a pore between them. When the guard cells swell with water (i.e., become turgid), the pore opens and gas exchange occurs. Conversely, when the guard cells lose water and become flaccid, the pores close. Thus it is osmosis that opens and closes the stomata. Stomata may open during the day or night, depending on plant physiology (e.g., CAM plants open their stomata at night); but at any rate, they can be expected to be open when the plant requires gas exchange between the leaf interior and the environment.

7. Amongst the hydrophytes, floating leaves generally have stomata only on their upper (air exposed) epidermis, and submerged leaves lack stomata altogether. Xerophytes often lack photosynthetic leaves—their leaves having been reduced to spines—or they possess many stomata on their leaves but drop their leaves when water becomes scarce. Mesophytic leaves generally bear stomata mainly on their lower surface. In monocots, the stomata are arranged in rows aligned parallel to the leaf veins, whereas in eudicots, the stomata tend to be scattered about the leaf surface.

8. Generally, plants abscise their leaves either when individual leaves are worn out or when environmental changes make having leaves a liability rather than an asset. Many plants do indeed drop all of their leaves within a short period, usually when seasonal changes are about to bring a prolonged dry or cold spell. Even evergreen plants drop their leaves, although they do not do so all at once; rather, leaf abscission occurs year-round, a few leaves at a time. Some evergreens, in fact, may retain a given leaf for several decades before dropping it. Leaf abscission is preceded by leaf senescence, a period in which leaf components are disassembled and their nutrients withdrawn for recycling elsewhere in the plant. As chlorophyll molecules are degraded, the green leaf color fades and the reds and yellows of other pigments become strikingly visible. Eventually an abscission zone forms, consisting of a proximal protective layer near the stem, and a more distal separation layer. The protective layer consists of suberin-containing cork cells; forming prior to abscission, this layer seals off the wound created when the leaf breaks free and falls. The separation layer develops as a layer of thin-walled cells. When the protective layer is complete, enzymes degrade the cells of the separation layer, and eventually the leaf falls.

9. Transpiration is the process by which water moves from the roots, where it is absorbed, through the plant body until it is lost as evaporation to the atmosphere through either the stomata or the cuticle (cuticular transpiration). As a water molecule evaporates into the intercellular space from the surface of a mesophyll cell, another water molecule replaces it from the interior of that mesophyll cell. As water leaves the mesophyll cells, the intracellular solute concentration rises, and water is drawn into these cells by osmosis from surrounding mesophyll. This process continues from cell to cell until eventually replacement water is drawn out of the leaf xylem. Because of their extensive hydrogen bonding, water molecules are very cohesive—they stick together and do not break under tension. As more and more water is drawn out of the xylem, it creates a pulling force, or tension, in the xylem that is felt all the way back to the roots. This tension causes the entire column of water in the xylem to rise toward the leaf. Thus it is the transpiration-driven tension that lifts the water column in the xylem up to the leaves, and it is the cohesiveness of water itself that maintains the integrity of the column by resisting the formation of air bubbles.

10. Leaves may be adapted to help a plant climb, for example, the tendrils on the garden pea. Some leaves are modified to protect underlying soft tissues from the elements, such as bud scales that protect the young stem inside from cold, wind, and insects. Leaves may become showy, helping to attract pollinators to the flowers; an excellent example is the poinsettia plant with its red bracts. Leaves may become thick and function to store water, such as the succulent leaves of the houseplant known as the jade plant (Crassula spp.). Some epiphytes, growing high up in the canopy, do not have access to groundwater, and their leaves are arranged to form a water-holding vessel, a personal pond to meet that individual plant's water needs. Some leaves store food (e.g., onion bulbs and cabbage heads) while others are reduced to sharp, protective spines, such as in most cacti. The leaves of certain acacias are modified to house and feed a fiercely protective army of ants, while those of certain carnivorous plants have evolved mechanisms to entrap and digest small prey, usually insects. Many other leaf modifications exist, so that leaves have truly evolved a dizzying array of forms and functions.

Applying Concepts

1. Obviously, the lettuces, cabbage, and spinach are mainly leaves, as are endive, dandelion greens, and watercress. Onions and garlic are bulbs—short stems surrounded by many layers of leaves, and shallots (green onions) and leeks are primarily elongated, tubular leaves. Celery is the leaf of the celery plant, the "stalks" being the leaf petioles. Artichokes are thickened, modified leaves that surround the flowering head of the artichoke plant.

2. Generally speaking, leaves bear buds in their axils whereas stem branches do not bear buds in their "axils." A cursory examination of the celery stalks will reveal a small bud growing at the base of each, indicating that the stalks are indeed petioles and the buds at the base are axillary buds. In celery, the stem is the foreshortened cone of tissue from which the petioles arise.

3. Keep in mind the following points when examining the anatomy of these structures. Eudicot stems consist of large, parenchymatous pith cells surrounded by a ring of vascular bundles. Even in herbaceous eudicots a vascular cambium may become active and produce a small amount of secondary vascular tissue, with secondary xylem inside and secondary phloem outside. You would not expect to find buds at the bases of stems, although photosynthetic stems may certainly bear stomata on their surface. Leaves, on the other hand, contain mesophyll cells, spongy and palisade parenchyma; each has a characteristic appearance that differs markedly from pith cells. The vascular tissue is surrounded by a parenchymatous bundle sheath, and is arranged in a parallel or netted manner, but not in the ring shape characteristic of the stem. You would also expect to find axillary buds at the bases of the leaves.

Chapter 12

Review Questions

1. (a) Auxins cause cell expansion and growth. They are responsible for such plant phenomena as gravitropism, phototropism, and thigmotropism, as well as apical dominance. Some auxins (e.g., indole-3-butyric acid) are used commercially as rooting hormones, while others such as 2,4-D are broadleaf herbicides. Auxins are also important in the formation of new plants from tissue cultures.
   (b) Cytokinins stimulate cell division in plants. They are important for proper progression into mitosis, and for the establishment of bacterial and fungal partnerships with plant roots.
   (c) Ethylene influences a number of plant processes including leaf drop, cell specialization, fruit ripening, flower aging, and defense against plant pathogens. Ethylene is responsible for the hook in seedling stems as they emerge from the abrasive soil, and may be used commercially to simultaneously ripen a large batch of fruit.
   (d) Gibberellins promote general stem and leaf elongation and hence are important for proper growth of plants. In embryos, gibberellins promote the breakdown of starch and seed storage proteins into sugars and amino acids that can be used for growth.
   (e) Abscisic acid slows plant metabolism when conditions are poor and induces dormancy. It also prevents premature germination of seeds under inopportune growth conditions (e.g., when temperatures and water availability are unsuitable for growth).

(f) Brassinosteroids promote cell elongation via a different mechanism than auxins; they also protect plants from stressors such as heat, cold, salt, and so forth.

(g) Salicylic acid retards flower aging.

(h) Systemin acts as a chemical sentry to alert other tissues to the presence of tissue damage and to increase chemical defenses against microbial invaders.

(i) Jasmonic acid inhibits seed and pollen germination, slows root growth, and promotes fruit ripening and color changes.

(j) Sugars are involved in the regulation both of certain photosynthetic genes and of nitrogen balance within the plant. In addition to the effects noted earlier, salicylic acid, systemin, jasmonic acid, and sugars are also involved in the mobilization of plant defenses against microbial invaders.

2. Phytochrome is a plant pigment protein that changes in response to the quality of the light around it. Chemically, it consists of a four-ringed pigment portion bound to a protein part. Phytochrome is bluish in color because it absorbs red wavelengths of light. Specifically, when it absorbs far-red light it enters a "switched off" state, whereas absorption of red light changes it to a "switched on" state. "Switched on" or activated phytochrome moves into the nucleus, where it alters gene expression. Thus, phytochrome must be exposed to red light to become activated, and it is this form that brings about a cellular change. Examples of plant processes under the control of phytochrome are germination, flowering, seasonal dormancy, and growth patterns that minimize shading of the lower leaves.

3. These terms all describe the ways flowering in various plants responds to day length. Short-day plants will flower when the daylight hours are shorter than some critical length (which varies from species to species), while long-day plants require a day length longer than a certain critical length. These plants are actually using phytochrome to measure the length of the intervening night, so that, for example, short-day plants should more properly be referred to as long-night plants—plants that flower when the night is longer than a certain critical length. These effects are brought about as activated phytochrome slowly reverts to the inactive state during the night; the longer the night, the more complete is this reversion. Lastly, in some plants flowering is unrelated to day length; such plants are known as day-neutral plants.

4. During the night, phytochrome molecules slowly revert to the inactive, "switched off" form that can absorb red light. The longer the night, the more phytochrome molecules become inactive. With the sunrise, plenty of red light shines on the plant, and phytochrome molecules absorb the light and are "switched on." This sudden increase in the active form of phytochrome resets the plant's internal clock each day. As more active phytochrome is produced, more of the nuclear protein CONSTANS accumulates in the plant. When the day is long enough, a critical amount of CONSTANS is produced and turns on the expression of flowering genes in long day plants.

5. Plants have a number of structural features, including a waxy cuticle and some toxic secondary compounds that limit the effects of pathogens. In addition, plants are able to respond to attacks by pathogens by mounting a hypersensitive response that limits disease progression. Plants may produce hydrogen peroxide and other compounds that kill pathogens and also work to seal off the infection site. Pathogens stimulate the production of signal molecules such as NO and salicylic acid that stimulate cells to fortify their walls with lignan and callose to prevent the spread of pathogens. Whole plants may quickly respond to challenge by a familiar pathogen by producing defensive enzymes and toxic compounds. This is called systemic acquired resistance and is sometimes called "plant immunity."

Applying Concepts

1. Ethylene hastens fruit ripening in plants, and the loss of endogenous ethylene in these tomatoes suppresses fruit ripening. Thus, the tomato fruits remain green and unripe for a long period of time. The effects of the ethylene deficiency can be quickly reversed by providing ethylene gas to the fruits. The fruits will then turn the familiar red-orange as they all ripen on cue.

2. Apical dominance is the suppression of axillary bud development by auxins produced by the apical bud. By removing the apical buds, the gardener will remove this suppression and allow the lateral buds to develop, leading to a fuller, bushier shrub.

3. Recall that short-day plants do not necessarily require short growing days, but rather that the day length is shorter than some critical value. If that value happens to be, say, 16 hours, then that plant will flower when the days are still quite long. Similarly, long-day plants require that the day length be longer than some critical value; and if that value is, say, 9 hours, then you could have a long-day plant flowering when the days are, by clock, rather short. Day-neutral plants do not measure day-length (night-length) in order to flower. Thus, it would be quite possible to find a combination of short-day, long-day, and day-neutral plants that would flower together in either of the above mentioned scenarios.

4. Poinsettias are short-day plants, and require nights longer than 12 hours in order to flower. This must be complete darkness, as even a brief flash or the light of one streetlight may be sufficient to suppress flowering. To get a poinsettia to flower by mid-December, it should be placed in complete darkness from 5 p.m. to 8 a.m. every day, beginning in late September. Do not allow any light whatsoever to fall on the plant during its dark period. Poinsettia houseplants rarely re-flower because they are rarely given the necessary dark period.

5. These holes, of course, allow the produce to "breathe," although the breathing that is important here is to allow ethylene gas to escape. If this gas were not vented, the fruits would ripen much quicker, and their shelf life would be drastically shortened.

Chapter 13
Review Questions

1. Sexual reproduction has several disadvantages. In terms of energy, time, and investment of resources, it is expensive. Many scarce resources must be diverted toward all of the supporting processes in order for sexual reproduction to be successful. Mates must be found, gametes exchanged or transferred, and in many cases the young must be nurtured, at least for a while. Also, sexual reproduction is wasteful in the sense that very few gametes will successfully fuse and produce a zygote; usually the unsuccessful gametes just wither away and die. Thus there is a lot of overhead associated with sex. On the plus side, however, sexual reproduction shuffles genes (and chromosomes) around, producing individuals with new combinations of both. Via purging, it speeds the loss of deleterious genes from a population and may allow beneficial genes, once repressed, to appear. This variability is an advantage, for example, when the environment changes and a population is no longer optimally adapted to the new conditions. New traits are then needed, and sexual reproduction may supply them. However, that same variability may become a liability to a well-adapted population that is already in an optimal environment. Under such conditions asexual reproduction, which is faster, cheaper and produces genetically identical individuals, may better allow a population to quickly spread out and utilize its environment. Another advantage of asexual reproduction is frugality, in the sense that all of its progeny are potentially new individuals if the environment is suitable (in sexual reproduction, you'll recall, gametes that do not fuse with other gametes usually die, a wasteful process.) Also, in asexual reproduction a partner is not necessary, so the time and expense of finding a mate are avoided.

2. Meiosis and sexual reproduction always occur together because, at the chromosome level, they are complementary processes. Sexual reproduction—the fusion of gametes—results in a doubling of the chromosome number, as when two haploid gametes fuse to form a diploid zygote. Meiosis, on the other hand, halves the chromosome number, such as when a diploid cell undergoes meiosis and yields four haploid daughter cells. Without meiosis, successive sexual reproduction would result in continual doubling of the chromosome number in the progeny; for example, leading to successive ploidy levels of $1N$, $2N$, $4N$, $8N$, $16N$, $32N$, etc. Chromosome numbers would soon become so unwieldy that cell death would result. Conversely, meiosis alone would result in a continual diminution of the chromosome number—for example, $2N$ to $1N$ to $0.5N(?)$, which would of course

be immediately lethal. It is only when sexual reproduction is balanced by meiotic divisions that viable ploidy levels are achieved in perpetuity.

3. Meiosis and mitosis share many similarities. Both are preceded by replication of the nuclear DNA, so that each chromosome consists of two sister chromatids. Both involve a spindle apparatus, spindle microtubules, kinetochores, and chromosome movement; and both proceed through an orderly series of stages termed prophase, metaphase, anaphase, and telophase. Meiosis and mitosis are generally followed immediately by cytokinesis; and when this happens, both lead to the production of more than one daughter cell. Lastly, in most organisms, the nuclear envelope and nucleoli break down during both processes and re-form afterward. Nonetheless, there are fundamental distinctions between the two processes. Mitosis consists of one division event, but meiosis is actually two successive divisions (meiosis I and II). In meiosis I (prophase I), homologous chromosomes find each other and pair up, causing their chromatids to align with each other. At this point, aligned chromatids may exchange segments of DNA, in a process called crossing over. Next in meiosis I (metaphase I), the pairs of homologues line up along the middle of the spindle apparatus and subsequently split (anaphase I), each chromosome of a homologous pair moving to one pole or the other. These events are unique to meiosis, because homologues do not pair (and hence do not split up) in mitosis, nor does crossing over occur. Rather, in mitosis, sister chromatids split at their centromeres and are distributed to the forming daughter cells, something that does not happen in meiosis until meiosis II. The result is that in meiosis, the daughter cells are virtually always genetically distinct from the parental cell, whereas after mitosis, the daughter cells are genetically identical to their parent.

4. (a) In the gametic life cycle, individuals are diploid, and gametes represent the only haploid stage. Meiosis of cells in a diploid individual leads to the formation of haploid gametes, which then subsequently fuse during fertilization and reestablish the diploid condition. This type of life cycle is seen in all animal groups (including human) as well as in certain protists. (b) In the zygotic life cycle, individuals are haploid, they produce gametes by mitosis, and the zygotes that result from zygotic fusion are the only diploid stage. Generally zygotes begin development by dividing meiotically, thus reestablishing the haploid condition. The zygotic type of life cycle is restricted to certain protists. (c) Organisms with a sporic life cycle alternate between separate multicellular haploid and diploid individuals, a phenomenon known as alternation of generations. One individual—the gametophyte—begins life as a spore that germinates and produces a multicellular body through repeated mitoses. This individual is haploid, and specialized cells on it produce gametes directly via mitosis. After fertilization, the resulting zygote produces a second, separate individual—again through repeated mitotic divisions. This individual is diploid and is called the sporophyte. Specialized cells on the sporophyte then divide meiotically and produce the haploid spores that complete the cycle. The sporic life cycle is exhibited by all plants as well as by some protists (algae).

5. (a) Sexual reproduction is the fusion of male and female gametes to form a new, genetically distinct individual, the zygote. (b) Gametes are single cells—eggs and sperm—produced by adult organisms, which are specialized for mating (fusing). The process whereby egg and sperm fuse together to form a single cell is termed fertilization. (c) Sperm are male gametes, usually defined as the more motile of the two gamete types, whereas (d) eggs are female gametes and are usually larger and less motile than sperm. (e) Zygotes are the new, single cells that result from the fusion of egg and sperm following fertilization.

6. If sexual reproduction did indeed originate several times among the protists of long ago, the consequence would be that the modern-day descendents of those protists—the plants and animals of today—inherited the ability to perform sexual reproduction from several ancient and independent lineages. Modern DNA evidence supports this hypothesis, in that many of the genes involved in the control of sexual reproduction in plants differ considerably from their cognates in animals.

7. Haploid means that each chromosome present in a cell controls a unique assemblage of traits, so that the haploid number of an organism is essentially the number of chromosomes that comprise one complete "set." Haploidy is abbreviated as $1N$ or N. Diploidy, abbreviated $2N$, is the condition in which two complete sets of chromosomes are found in one nucleus; that is, each chromosome will have one homologue present. Whereas ploidy levels above $2N$ are generally lethal in animals, plants may be quite healthy with higher ploidy levels, so that $4N$, $6N$, and $8N$ plants are not uncommon. Note that haploidy is simply a condition in which each chromosome is unique in the combination of traits it controls; it sets no limits on the actual number of haploid chromosomes. Thus the value of N is known to vary in plant species from as few as $N = 2$ to as many as $N > 500$! In humans, $N = 23$.

8. Homologous chromosomes are chromosomes that bear the same genes (and in the same gene order), which is another way of saying that homologous chromosomes carry genes that control the same traits. In a diploid organism, one set of homologues came from the mother (the egg cell) and the other came from the father (sperm). (a) The term homologous chromosome is inapplicable to a haploid cell because, by definition, only one set of chromosomes is present. No chromosome would possess a homologue in such cells. (b) In a diploid cell, every chromosome would find one homologue present, so that each type of chromosome would be represented by a pair of homologues. (c) In a tetraploid cell, every chromosome would find three homologues present, so that each type of chromosome would be represented by four homologues.

9. Crossing over is the process whereby homologous chromosomes, paired up during meiosis I (prophase I), exchange homologous segments of DNA. This means that genes get shuffled and the resultant chromosomes are hybrids of the parental chromosomes. More properly the resulting chromosomes would be termed chimeras—consisting of discrete pieces of the parental homologues assembled together in new ways. As indicated, crossing over occurs during prophase I of meiosis I. The importance of crossing over is that progeny cells receive chromosomes that are different from either parent (because the genes have been shuffled). This increases genetic variability and ensures that the progeny are genetically unique individuals.

10. Nondisjunction is a condition wherein homologous chromosomes fail to separate at anaphase I of meiosis, leading to abnormal chromosome numbers in the daughter cells—a condition known as aneuploidy. In humans, most cases of aneuploidy are so disruptive to normal development that the fetus dies from spontaneous abortion. However, a few types of aneuploidy are known to be viable past birth. Examples include Down syndrome. In Down syndrome, if a meiotic division in a man fails to separate chromosomes 21, the result is two sperm that altogether lack chromosome 21 (a lethal condition) and two others that possess two chromosome 21's (and one copy of the other 22 chromosomes). If one of these latter sperm fertilizes a normal egg, the result is an individual with three chromosome 21's (and two of all the others). This condition is called trisomy 21, and the resulting syndrome is Down syndrome.

Applying Concepts

1. Prokaryotes do not have homologous chromosomes; in fact, the term is not even applicable to them. The prokaryotic genome typically consists of only one circular strand of DNA and not of the sets of chromosomes as is typical of eukaryotes. Furthermore, without sexual reproduction prokaryotes do not receive sets of chromosomes from each parent, and without meiosis there is no mechanism to evenly distribute homologues among the daughter cells. Thus homologous chromosomes are restricted to the eukaryotes.

2. Alternation of generations is a type of life cycle that contains two distinct, multicellular bodies, one haploid and the other diploid. Because this type of life cycle appears only in plants and certain plant-like algae, the haploid and diploid stages are referred to as gametophytes and sporophytes, respectively, with the suffix *phyte* meaning "plant." The haploid gametophytes produce gametes via mitosis, while the diploid sporophytes produce haploid spores via meiosis. Although *Obelia* does alternate between two adult stages, they are both diploid stages, and the haploid phase is represented only

by the gametes. This is not true alternation of generations. Rather, *Obelia* displays a gametic life cycle with the added twist that it can also reproduce asexually.

3. Probably the most consistent definition of the male gamete is that it is the most motile of the two. In red algae, neither gamete is self-motile, because both lack flagella. Nonetheless, one type of gamete must travel to the other, and clearly that type is more motile than the other. On this basis alone, the gametes can be distinguished as male or female.

4. Initially, when growing conditions are optimal, the best strategy would be to quickly colonize the nutrient agar while it is available. Asexual reproduction is ideal for this. As conditions in the petri dish deteriorate, sexual reproduction would be a better strategy, because its inherent variability might produce a strain of mold better able to cope with the changing, adverse conditions. Thus, you would expect to see more sexual structures at the end of this little experiment than when conditions were optimal for growth.

5. Unlike fern gametophytes, the gametophytes of flowering plants are very inconspicuous and do not lead nutritionally independent lives. They can both be found in the flower, where separate male and female gametophytes are produced. The male gametophytes are the pollen grains themselves, which at maturity consist of just three cells each, two of them sperm. The female gametophytes, on the other hand, are buried deep within the ovaries, deep inside little structures called ovules. At maturity, they are called embryo sacs and consist of just 7 cells each. Thus, although most people are aware of pollen and ovaries in a flower, it takes a trained eye to recognize the gametophytes that are found there as well.

6. During meiosis I, homologues separate and move to opposite poles. Normally each daughter cell receives a complete set of chromosomes. For example, a diploid cell entering meiosis would produce cells possessing exactly the haploid number of chromosomes, and a tetraploid cell doing the same would produce diploid daughter cells. Thus, meiosis works well when the ploidy level of the parent cell is an even number, because meiotic mechanisms normally ensure that each daughter cell receives one-half the total number of chromosomes present. Typically this means that chromosomes are distributed as "sets." In triploid individuals, three homologues are present for each chromosome type, and meiosis provides no mechanisms to ensure that the chromosomes are distributed as sets. For example, meiosis cannot ensure that one daughter cell would emerge as a complete diploid and the other as a haploid. Aneuploidy results, with each daughter cell receiving one haploid set of chromosomes and then some fraction of the remaining set. Generally such a condition is lethal to a cell, so that triploids—or any plant with an odd-numbered ploidy level—cannot produce viable gametes through meiosis. Sterility is the result of such a chaotic affair.

Chapter 14
Review Questions

1. The theory of blending inheritance that was in vogue during Mendel's time stated that traits were somehow passed directly from parents to their offspring or, more specifically, that each body part transmitted pertinent information to the growing embryo. Charles Darwin, for example, called these packets of information gemmules and suggested that they were microscopic in size and passed down from one generation to the next. Once in a developing embryo, the similar packets of information would blend together and be expressed as a mixture of the original traits. A major problem with this theory is that original traits, once mixed, should not reappear again in pure form in any future generation. This situation is similar to mixing red and white paints together—once mixed, the original pure red and white paints cannot be recovered. Genetic traits, we know, do not behave in this manner—a trait may seemingly disappear from a pedigree, only to reappear a few generations later. Clearly, neither experience nor scientific evidence supports the theory of blending inheritance.

2. A monohybrid cross is a genetic cross between two parents differing in only one trait; in a dihybrid cross, the two parents differ in two traits. An example of the former would be a cross between two pea plants,

one producing yellow seed and the other green seed. An example of a dihybrid cross would be crossing a yellow-seeded, white-flowered pea with a green-seeded, purple-flowered pea.

3. Although Mendel did choose garden peas for his experiments, it is important to realize that his experiments could have succeeded with many other species of plants—although it may have taken him a substantially longer period of time to arrive at his conclusions. However, peas did offer him several advantages. First, by Mendel's time, much was already known about growing and crossing pea plants, so that Mendel did not have to spend time discovering this knowledge and developing those skills. Neither did he have to spend time developing the different true-breeding varieties to use in his experiments, since many such peas already existed. In fact, a developing body of knowledge in rudimentary pea genetics was available to him as well, providing a base to build upon. Furthermore, peas are easy to grow and have a fast life cycle, producing several generations of plants in a single year. Also, pea plants are diploid; had peas—like many other plants—possessed higher levels of ploidy (e.g., tetraploid), their genetics could have been considerably complicated. Lastly, pea reproduction is easily controlled; not only are peas self-fertile, but they can be easily out-crossed so that the maternal and paternal parents can be selected. These advantages allowed Mendel to proceed into his experiments instead of having to spend time developing a similar system in another species.

4. In making a cross, sperm from the pollen fertilizes the eggs of the pistil. The two individuals providing the original sperm and eggs form the parental generation. All of their resulting progeny form the first filial, or F_1 generation. Crossing the individuals of the F_1 generation among themselves results in the second filial, or F_2 generation.

5. (a) When an individual has two forms of a trait, one of them expressed (and hence discernable) and the other not expressed, the former is referred to as dominant while the latter is termed recessive.

   (b) Genes are sequences of DNA (possibly RNA in certain viruses) that code for proteins and ultimately are responsible for the manifestation of a given trait; an allele is one of the alternate forms a given gene can adopt. For example, green seeds and yellow seeds are alleles for the trait of seed color in peas.

   (c) In a normal diploid individual, each cell contains two alleles for every gene, one allele being received from each parent. For any given gene, if those alleles are identical and dominant, the individual is termed homozygous dominant (for that gene or trait); if identical and recessive, homozygous recessive; and if two different alleles are present, heterozygous.

   (d) Phenotype refers to the discernable way in which a gene manifests itself (i.e., how the trait is observed); the genotype refers to the underlying genetic constitution. Thus green seed is a phenotype, and yy is a genotype.

   (e) Self-fertilization is when pollen (sperm) from the anthers of one individual fertilizes the eggs in a pistil produced by that same individual (usually within the same flower); in cross-fertilization, the pollen comes from a different individual than the eggs.

6. Mendel's model of inheritance states that (a) Information about a trait, and not the trait itself, is transmitted from parents to progeny. Mendel called this information a "factor"; today we refer to it as an allele of a gene. (b) For each trait, individuals receive two factors, one from each parent. (c) The two factors relating to a given trait need not be the same; each parent could donate a different form of a given trait. For example, regarding the trait for seed color, one parent might donate the factor for green seeds, while the other donates the factor for yellow seeds. (d) When two different factors are present in the same individual, they neither blend together nor in any other way contaminate each other. (e) When two different factors are present in the same individual, only the dominant one is actually expressed. The other, the recessive factor, is present but latent.

7. Mendel's First Law of Heredity states that the factors present in an individual remain discrete and segregate from each other during the reproductive process. This meant that traits that were latent and unexpressed could still appear unaltered in future generations. The title "Mendel's Law of Segregation" was actually

coined by biologist Hugo de Vries. Mendel's Law of Independent Assortment, so named in 1913 by geneticist T. H. Morgan, stated that different traits assorted independently of each other. For example, the assortment of the green-seed and yellow-seed factors for seed color did not in any way influence the assortment of the factors for, say, flower color. The various traits assorted independently of each other.

8. Traits that lie on different homologs (homologous chromosomes) will indeed assort independently of one another. However, a single chromosome may carry many traits, and two traits that lie on the same chromosome may not assort themselves independently at all. In fact, the closer the loci for two genes lie on the same chromosome, the more those genes will tend to assort in a dependent fashion—where one goes, the other tends to go as well. This phenomenon is called linkage, and the tighter two genes are linked, the more they will deviate from Mendel's second law.
9. In the phenomenon of complete dominance, one allele—the dominant one—is completely expressed while the other, recessive alleles are masked and not visible. In incomplete dominance, each allele is expressed, and the phenotype appears as a mixture of the underlying alleles. For example, floral color in snapdragons shows incomplete dominance. If true-breeding red-flowered and white-flowered plants are crossed, instead of finding all red or all white heterozygotes in the F_1 generation (depending on which allele was dominant), one finds all pink flowers—a mixture of the parental red and white flower types. The gene for red color codes for a functional enzyme that synthesizes a red pigment, and the gene for white codes for a non-functional enzyme that makes no red pigment. The presence of one red gene in hybrids allows synthesis of only enough red pigment for light red (pink) flowers.
10. When the loci for two genes physically lie on the same chromosome, those genes are linked. The closer the genes lie together, the more tightly they are linked and the more they will tend to be inherited together. Genes that lie far apart on the same chromosome are more loosely linked and may, in fact, assort independently if the crossover frequency between them is great enough. Crossing over is the physical exchange of DNA fragments between two homologous chromosomes, which usually results in chromosomes different from either parent. Crossing over occurs during prophase I of meiosis I, and it involves only homologous chromosomes; any given chromosome will cross over only with its homolog. Mendel did encounter linkage, although he apparently did not recognize it. Two of the seven pea traits he chose to study—pod shape and plant height—are tightly linked. Evidently Mendel did not study the assortment of these two traits from each other, because he would have seen that they did not assort independently.
11. Sometimes the expression of a trait is not dependent on a single gene but rather on the interaction of two genes. Such a phenomenon is called epistasis, and the homozygous recessive condition in either of the two genes is capable of stopping the phenotypic expression of the other. A dominant allele for each gene is required before the dominant phenotype will be expressed. In polygenic inheritance, phenotypic expression is the sum of several genes. Each dominant allele in each of the genes contributes a little toward phenotypic expression. For example, in wheat two genes control kernel color. If the *R* allele confers a red color to the kernels and r makes for white kernels, then a kernel with four *R* alleles (i.e., *RRRR*) is dark red; three *R* alleles, medium-dark red; two *R* alleles a medium red; one *R* allele, light red; and no *R* alleles (i.e., *rrrr*), white.
12. Both the X and Y chromosomes of white campion are longer than the X and Y chromosomes of humans. In addition, in humans the X chromosome is larger than the Y chromosome, but the reverse is true for the X and Y chromosomes of white campion.

Genetics Problems

Basic problems

1. (a) *Yy*. (b) *Y* and *y* gametes.
2. (a) The heterozygous blue-flowered plants are *Bb*. They produce gametes with *B* or *b*. A Punnett square gives genotypes in the ratio 1 *BB*: 2 *Bb*: 1 *bb* and phenotypic ratio 3 blue-flowered to 1 white-flowered. (b) The best way to determine the genotype of a blue-flowered plant is by a testcross to a recessive white-flowered plant. If the blue-flowered plant is *BB*, all the offspring of the testcross will be *Bb* and blue, but if the blue-flowered plant is *Bb*, half the offspring will be blue-flowered and half white-flowered.
3. (a) The data show incomplete dominance for flower color. (b) *RR* is red, *RR′* is pink, and *R′R′* is white. (c) If blending inheritance were occurring, all members of the F_2 generation should be pink. Instead, the F_2 generation shows segregation of the alleles for flower color.
4. (a) On the same chromosome because they show tight linkage. (b) The new phenotypes tall/constricted and short/round are due to crossing-over between the genes for height and pod form.

Advanced Problems

5. (a) *YyRr* produces gametes *YR*, *yR*, *Yr*, and *yr*. This is one of the dihybrid crosses carried out by Mendel. The ratio of phenotypes is 9 yellow round (*Y_R_*): 3 yellow wrinkled (*Y_rr*): 3 green round (*yyR_*): 1 green wrinkled (*yyrr*). (b) This is a testcross. *YyRr* makes gametes *YR*, *yR*, *Yr*, and *yr*. The recessive plant can only make the gamete *yr*. The phenotypes from this cross are 1 yellow round (*YyRr*:) 1 yellow wrinkled (*Yyrr*): 1 green round (*yyRr*): 1 green wrinkled (*yyrr*).
6. (a) The F_1 hybrids make the gametes *LY*, *Ly*, *L′Y*, and *L′y*.
(b)

	LY	*Ly*	*L′Y*	*L′y*
LY	*LLYY* large yellow	*LLYy* large yellow	*LL′YY* intermediate yellow	*LL′Yy* intermediate yellow
Ly	*LLYy* large yellow	*LLyy* large purple	*LL′Yy* intermediate yellow	*LL′yy* intermediate purple
L′Y	*LL′YY* intermediate yellow	*LL′Yy* intermediate yellow	*L′L′YY* small yellow	*L′L′Yy* small yellow
L′y	*LL′Yy* intermediate yellow	*LL′yy* intermediate purple	*L′L′Yy* small yellow	*L′L′yy* small purple

(c) Only 1/16 of the F_2 offspring will be true-breeding for large yellow flowers; however, such a hybrid would be valuable for commercial purposes.
7. This is an example of a modified dihybrid ratio, so it is most likely due to epistasis. (b) Disk-shaped fruit are *A_B_* with one dominant allele for each gene. Elongate fruit are *aabb* only. Plants that are *A_bb* or *aaB_* have spherical fruit. Spherical fruits are homozygous recessive for one of the two genes for fruit shape.

Chapter 15

Review Questions

1. A GM organism is a "genetically modified" organism; that is, one that has been genetically engineered, its DNA deliberately manipulated and altered. Synonyms for GM organism include genetically engineered organism (GEO), transgenic organism, or transformed organism. An actual example is the work of Ingo Potrykus and others, who transferred the genes for beta-carotene production into rice plants in such a way that the pigment actually accumulates in the grains, coloring them yellow.
2. (a) Restriction enzymes are enzymes that cut DNA at specific sequences, for example, the restriction enzyme *Eco*R I cuts at GAATTC sequences. Bacteria chemically protect these sequences in their own DNA and then use restriction enzymes to cut up any foreign DNA that has found its way into the cell. (b) Many organisms use ligases to repair damaged DNA. (c) RNA viruses use reverse transcriptase to reverse transcribe RNA into DNA, so that new viral DNA can integrate into the host-cell DNA. (d) DNA polymerases are found in all organisms and are used to create a new strand of DNA using an existing strand as a template.
3. Molecular biologists use (a) restriction enzymes as molecular scissors to cut DNA at very specific sequences. In addition, two pieces of DNA cut with the same restriction enzyme will have complementary "sticky" ends, which facilitate the process of splicing them together in a test tube. (b) Ligases are used to glue together pieces of DNA. Specifically, when two pieces of DNA are held together solely by hydrogen bonds between their bases, ligases can be used to connect the sugar-phosphate backbones together with covalent bonds. (c) Genetic engineers use reverse transcriptase to make DNA copies of purified cellular RNA. Straightforward techniques exist to extract and purify all the messenger RNA in a cell at any given moment. However, RNA is difficult to handle and analyze, in part because its single-stranded nature makes it considerably more fragile than double-stranded DNA. Reverse transcriptase allows a scientist to copy RNA into DNA, and then continue experiments with the more stable DNA. (d) DNA polymerase is used whenever a molecular biologist needs to make additional copies of an existing DNA strand. Examples include PCR reactions to amplify small quantities of DNA, as well as creating the DNA fragments necessary for analysis in a DNA sequencer.
4. A transposon, or transposable element, is a DNA segment that is able to excise itself from a chromosome and then reinsert itself at a different location. Barbara McClintock initially discovered transposons and their effects in corn plants. Because they are able to shuffle DNA pieces around the chromosome, transposons play important roles in inactivating genes, changing gene regulation, creating new genes, and adding DNA segments onto existing genes. Their ability to insert into a chromosome and then excise themselves makes transposons useful tools in the laboratory. For example, molecular biologists can allow them to insert into the chromosomes of a laboratory culture, inactivating genes in the process. This will produce characteristic phenotypes that can then be selected and catalogued. Then, the transposons can be induced to excise themselves from the chromosome, usually taking significant pieces of host DNA with them. These transposons with their chunks of host DNA can be analyzed, yielding information about the relationships between the inactivated genes and their resulting phenotypes.
5. Traditional breeding methods have been used for some ten thousand years, and have resulted in the many breeds of livestock and pets, and the many cultivars of ornamental and crop plants. Although highly developed as an art and science, these methods nonetheless suffer from several drawbacks. First, they can be very labor intensive, in some cases requiring thousands of crosses just to find a handful of progeny with desirable traits. Considerable time may be spent making those crosses and then waiting for the progeny to mature to the point where they can be assessed for desirable traits, and then selected for further breeding experiments. For example, think of how long it takes some trees to mature to the point where they produce flowers and fruit. It may be a decade or longer! Furthermore, breeding involves considerable chance. Breeders may select desirable traits, but they have little power to actually force desirable traits to appear. They must work with the genes an organism already has, and then wait for chance processes to change those genes in ways the breeder finds desirable. All in all, although breeding is effective, it can be a long, drawn-out process. Genetic engineering, on the other hand, is much more directed and expedient. Genes from very distantly related organisms can be introduced in a single afternoon, and transformed cells can be identified and isolated within days. Within a few weeks these cells can be coaxed to form young plants, which can then be assayed for the desired genetic changes. Instead of being limited to the genes and other DNA variations currently found in a species, one can introduce any appropriately sized DNA—from the most distantly related organisms, or even totally synthetic! Thus, compared to the slow process of breeding, genetic engineering gives scientists a much quicker, more powerful way to effect heritable change in an organism.
6. Yes, human antibody genes have been inserted into plants, and the new GM plants do indeed produce human antibodies that are useful in human medicine. This route offers several advantages over the use of bacterial or animal systems. For example, animals and animal cell cultures can become contaminated with viruses capable of infecting humans. This is precisely what happened with the early production of the polio

vaccine—it was produced in monkey kidneys and became contaminated with a virus known as SV-40. Plant cells, on the other hand, do not harbor viruses capable of infecting animals, including humans. Furthermore, plants can be grown quite cheaply, using minimal growth materials (media, equipment, etc.). They can produce larger amounts of antibodies than bacterial cultures in a manner that is very convenient to harvest.

7. Plasmids are small, circular pieces of DNA that often reside in bacterial cells separate from the main chromosome. Since they are usually able to replicate within the bacterial cell independent of the main chromosome, plasmids can be passed on to all the progeny of an infected cell. Plasmids are exceedingly important to bacteria in nature because they provide a route for the exchange or transfer of DNA from one cell to another. This holds true even between rather distantly related bacterial species. For example, plasmids may contain antibiotic resistance genes. A single cell containing such a plasmid may pass it on to other cells in its vicinity, thus spreading the ability to resist that particular antibiotic. Genetic engineers use plasmids in many ways, all of them centered around the ability of bacteria to undergo transformation—to internalize plasmids from the medium, and thus acquire any genes or foreign DNA those plasmids might contain. For example, if a culture of bacteria is mixed with a solution of plasmids carrying the gene for ampicillin resistance (ampicillin is an antibiotic), and those cells are then plated in a petri dish containing growth medium and ampicillin, only those cells that have actually acquired the plasmid will be able to grow; the others will lack the ampicillin resistance gene and will die. Labeling plasmids with antibiotic resistance genes thus provides the molecular biologist with a very convenient selection mechanism. Besides the antibiotic resistance, some plasmids may also be engineered to carry a foreign gene, which the bacteria will replicate and even express after transformation.

8. Bacteria are not suitable for all cloning tasks, but they are very handy at times. First, they are small, relatively simple cells that can rather easily pick up foreign DNA. Bacteria reproduce rapidly and are easily cultured on cheap media. Furthermore, properly frozen bacteria remain viable almost indefinitely. Also, a plethora of viral and plasmid vectors are known and characterized for certain bacterial species (e.g., *E. coli*), which greatly facilitates the process of cloning DNA into and out of those cells. For reasons like these, bacteria have become the standard workhorses in any modern cloning lab.

9. Four steps were described in Chapter 15. (a) The vector and the foreign DNA must be properly prepared. Generally, this means that the ends of the vector and the foreign DNA molecules must be made compatible. This can be accomplished by cutting each with the same restriction enzyme, so that they all either have compatible "sticky" ends or else they all have blunt ends (no overhanging piece of single-stranded DNA). (b) When properly prepared vector and foreign DNA are mixed, the cut ends of the DNA will stick together for a short time. This is especially true if there are sticky ends holding the DNA pieces together with hydrogen bonds. If DNA ligase is added to this mixture, it will glue together the DNA pieces with covalent bonds. The result is a perfectly re-formed piece of DNA, a plasmid that is part vector and part foreign DNA. (c) The recombinant plasmid is then used to transform *E. coli* cells. Successful transformants will internalize the plasmid with its foreign DNA. (d) Lastly, the transformed bacteria can be grown in culture vessels. Generally, appropriate antibiotics are added to the culture medium so that only cells that have actually acquired the plasmid can grow. These cultures can be used to produce large quantities of the foreign DNA for further analysis (e.g., for DNA sequencing). Or, if the foreign DNA is a gene, it can actually be expressed as a protein, which can then be purified for subsequent use or analysis.

10. Successful genetic engineering of plants is based on our current ability to grow an entirely new plant from a single cell. Because the plasmids and viruses used to clone genes into *E. coli* will not work with plant cells, another means must be found to get foreign DNA into the plant cell. One method is to use the soil bacterium

Agrobacterium tumifaciens. This bacterium is able to transfer a plasmid from itself into plant cells. If foreign DNA is cloned into the *A. tumifaciens* plasmid, then the bacterium will do the work of inserting that modified plasmid into the plant chromosomes, thereby transforming the plant cell. This gives the scientist a means of inserting foreign DNA into a plant cell and creating GM plants.

Applying Concepts

1. Transposons and viruses are both able to integrate themselves into the chromosomal DNA of their host cells. This allows the host-cell machinery to replicate the transposon/viral DNA along with that of the host. Both can also excise themselves from the chromosomal DNA. Transposons are then free to reinsert themselves at a different location on the same chromosome in the cell, or on another chromosome altogether. Viruses, on the other hand, are able to infect other cells; they are not limited to the chromosomes within one particular cell. In this sense, viruses can be thought of as transposons with the ability to move from host cell to host cell.

2. Questions posed in Chapter 15: Can genetic engineering provide sufficient food for the world, considering that farmers in poor areas are often unable to afford the GM crops? Do consumers have the right to know if the food they are purchasing is a GM food? Do GM foods pose unforeseen health dangers? Do GM foods and GM technologies pose unforeseen environmental dangers? Might the use of GM plants encourage the appearance of more vigorous, more resistant pests? Might some of the toxins used in GM plants have harmful effects on nonpest species, including the natural enemies of the pest species? Might GM plants encourage the appearance of "superweeds"—very vigorous, fast-growing weeds? What other concerns might you have regarding GM plants? For example, you might be concerned about the loss of non-GM heirloom seeds, seeds that farmers have been planting and harvesting for decades, even centuries. These cultivars might be lost if everyone switched over to purchasing the latest GM plants from a seed dealer every year. Are you concerned about the loss of genetic variability as heirloom seeds disappear? Do you feel that GM plants are of lower quality, perhaps less tasty than non-GM plants? Are you concerned that our food supply—the sine qua non of our civilization—might become a monoculture controlled by corporate giants?

3. Restriction enzymes are well known for recognizing palindromic DNA sequences, which are DNA sequences that read the same in the $5' \rightarrow 3'$ direction on each strand. Consider the *Eco*R I sequence. If you apply base complementation rules to generate the other DNA strand, you see the sequence is 5′-GAATTC-3″/3′-CTTAAG-5′. Notice that the strands have identical sequences when each is read in its $5' \rightarrow 3'$ direction. This is an example of palindromic DNA. A careful examination of the other three restriction sequences reveals a similar situation; in fact, this is a property of restriction enzymes and their sequences in general.

Chapter 16
Review Questions

1. Influenced by theology, science up to the mid-1800s maintained that the Earth was about 6,000 years old, based upon the number of generations of people described in the Old Testament. Species were believed to have had a divine origin, created by a supreme being in one instant (a concept known as "special creation"), and then remaining unchanged since. Furthermore, each species was thought to have an "ideal form," and each individual was more or less an approximation of this form. Variation within a species represented the deviation from this ideal form, not the raw material upon which natural selection could operate. A few individuals, however, did suggest that species could change. Georges Louis LeClerc, a French naturalist, suggested that the vast number of species on the planet might have evolved, via some unknown process, from a small number of founding species produced by special creation. Darwin's grandfather suggested that species might be able to change into new species. Jean Baptiste Lamarck proffered a potential mechanism (now discredited) by which evolution could occur. He believed that organisms could acquire new and heritable traits through processes such as behavior. If such changes were large enough, or if smaller changes accumulated sufficiently, then perhaps a new species could emerge. Such ideas helped pave the way for the acceptance of Darwin's theory of evolution.

2. (a) Each species has the ability to produce more offspring than can survive to maturity. If this were not the case, then the most prolific reproducers would soon overpopulate the planet. Instead, (b) over time, natural populations tend to remain rather constant in size, ultimately limited by some resource such as food, water, space, and so forth. (c) Because resources are limited, there is competition among the members of every species, and the losers do not survive to adulthood. (d) Because of variation, there exists a range of traits within a population, and these traits have a direct bearing on the chances that an individual will survive to adulthood and reproduce. In any given environment, some traits will increase the chances for survival, while others will decrease it. This leads to the inevitable conclusion that (e) the individuals most likely to survive to reproductive age and produce offspring are those possessing the most advantageous traits, a phenomenon termed "survival of the fittest." This entire process, by which the environment selects for individuals that are best adapted to it, is called natural selection. The pivotal importance of natural selection to the theory of evolution is that it provides a powerful driving force for evolution itself, causing populations to diverge and ultimately form different species.

3. Both artificial selection and natural selection lead to changes at both the molecular and the gross levels. The molecular changes are based upon heritable changes in the structure of DNA itself, while the gross changes include phenotypic changes such as the appearance of altogether new traits or a change in the frequency of existing traits. For example, white flowers may appear in a species that had previously been known to produce only yellow flowers. Or perhaps individuals will appear that consistently produce tall progeny, whereas previously tall stature was a sporadically occurring trait. However, the driving forces behind the two types of selection are different. In natural selection, natural and stochastic forces determine which individuals are "fittest" and hence will survive to reproductive age; there is no deliberate will driving the appearance of traits in some predetermined direction. In artificial selection, humans provide for the progeny and then select for the next reproductive cycle those individuals possessing desirable traits. Thus in artificial selection humans (not nature) determine which traits will be passed on to the next generation.

4. The following lines of evidence were provided in Chapter 16. (a) The process of artificial selection shows that species can be changed. No one would dispute that; for example, the ancestral dog was quite different from the many breeds existing today. It is remarkable that chihuahuas and German shepherds arose from a common ancestral dog, and that dichotomy demonstrates the extent to which artificial selection can alter a species. (b) Comparative anatomy reveals homologous structures among related organisms, and from this we may infer how natural selection can mold a common organ into different forms in order to accomplish different tasks. (c) Modern techniques in molecular biology enable scientists to track changes in proteins and DNA and to relate those changes to evolutionary history. For example, scientists may gather many specimens from one or more species and then compare the similarities of the DNA sequences of specific genes. Alternatively, they might compare the differences in sequence or structure of some critical enzyme. Such studies can provide insight on the relatedness of individuals within a species, or of several species to each other. (d) Fossils provide an incomplete record of the evolution of life on Earth. Examination of fossils and the rocks in which they occur provides information on life-forms long since extinct, including what they looked like as well as when and in what type of environment they lived. Although science has no definitive way to prove the existence of evolutionary changes of long ago, lines of evidence such as those described here provide extremely strong support in their favor.

5. Gradualism is the belief that evolution proceeds over geologic time at a steady or gradual rate, and that new species slowly and continually appear as changes accumulate. In contrast, punctuated equilibrium suggests that evolution does not occur at a constant rate, but that geologic history is interspersed with periods of high and low rates of evolution. This theory holds that relatively stable conditions allow species to thrive with little morphological change and speciation. Such periods of equilibrium are then interrupted with drastic changes in conditions, which in turn force rapid structural changes and hence rapid speciation. At present, scientists do not yet agree on whether gradualism or punctuated equilibrium best describes the evolutionary history on this planet, although the resulting controversy continues to direct research in this area.

6. (a) A population is the sum of conspecific (same species) individuals living in a certain area. The area considered may be quite small—such as the population of eastern tamarack trees (*Larix laricina*) living in a bog on a specific lake, or quite large, such as the *Larix laricina* population of the state of Wisconsin, or of the entire eastern USA. (b) The gene pool applies to a population, and is the total of all the alleles of all the genes possessed by the individuals in that population. (c) Population genetics is the branch of science concerned with gene pools and how and why those pools change over time. (d) Microevolution refers to the small-scale changes occurring in the allele frequency in a gene pool from one generation to the next. It is not evolution itself, but over time, its accumulated effects may lead to evolution. (e) Genetic drift refers to the removal of an allele from a gene pool by stochastic (random or chance) processes alone. As a hypothetical example, consider a population of plants that normally produce yellow flowers. Imagine that a mutation for white flowers occurs in one plant, and as a result five progeny plants are now alive and carrying this mutant allele for white flowers. A single flood or a ravenous cow could easily kill these five plants and thus remove the white flower allele from the entire population. (f) Gene flow is the migration of genes (alleles) between populations.

7. Both types of mutations involve changes in the DNA structure of an organism. Gene mutations tend to be rather localized and result from changes in one or, at most, several bases. Chemical agents, radiation, or a malfunctioning DNA polymerase typically induce gene mutations. An example would be a single base substitution; this could potentially lead to the substitution of one amino acid for another in the final protein. Alternatively, a single base change could obliterate a stop codon, resulting in drastic changes in the resulting protein. Chromosome mutations, on the other hand, are at a much larger scale than gene mutations. They involve rearrangements of a gene or genes, or even whole portions of chromosomes. Crossing over is one example, in which entire segments of a chromosome are exchanged. Sometimes the exchange is unequal, so that a particular chromosome will gain or lose an entire segment of DNA, resulting in gene deletions or duplications. A gene or genes may change their location on the chromosome. Yet another type of chromosome mutation involves the loss or gain of an entire chromosome or set of chromosomes, such that aneuploids or polyploids may result.

8. All eukaryotes possess a haploid number—the N number—that defines the number of chromosomes in that species, which comprise one complete set. For example, in humans, $N = 23$. In diploid organisms, meiosis typically results in the formation of haploid gametes—each gamete having received one complete set of chromosomes. Sometimes, however, things go awry and a gamete may receive one chromosome (or rarely several chromosomes) more or less than a complete set. This condition is known as aneuploidy. A prime example is Down syndrome in humans, in which an otherwise normal gamete received an extra chromosome 21, so that the individual resulting from fertilization possesses three copies of chromosome 21 (a condition known as trisomy 21). Another example is provided by Turner syndrome, in which one gamete did not receive a sex chromosome. If, during fertilization, this gamete fuses with a normal gamete, then the resulting individual will be XO in composition—a female with only one X chromosome (incidentally, the other possibility, a YO male, is lethal and such an individual would die before birth.) Polyploidy, on the other hand, results when the entire chromosome set is duplicated, and both of these sets are passed on to some of the gametes. If two such gametes fuse (as could happen with self-fertilization in plants), then the resulting zygote would be tetraploid instead of diploid. If one such gamete and a normal gamete were to fuse, then a triploid would result. Slight variations could produce pentaploids ($5N$), hexaploids ($6N$), octaploids ($8N$), and so on. In animals, polyploidy is generally lethal, but in plants polyploids may be quite healthy and even vigorous. Plants with odd-numbered ploidy levels ($3N$, $5N$, etc.) are sterile due to complications in homologue pairing at meiosis.

9. (a) In directional selection, individuals with a trait at the extreme of the character range (e.g., the tallest in stature, the most cryptically camouflaged, etc.) are favored over individuals closer to the mean or at the opposite extreme. In such situations, gene frequency will drift toward the favored trait. (b) Stabilizing selection favors individuals possessing the average trait, and (c) disruptive selection favors the individuals at either extreme of the character range (or, in other words, the mean is disfavored).

10. The morphological species concept defines species based on their structure or morphology, while the biological species concept defines species as a group of populations whose members interbreed with each other but cannot (or usually do not) interbreed with members of other populations. Thus, in the biological species concept, reproductive barriers isolate and define species.

Applying Concepts

1. Because DNA is duplicated and then distributed among the daughter cells at the time of cell division, at the cellular level mutations are at least potentially heritable. Of course a mitotic or meiotic division is required for this to occur. At the organismal level, mutations in somatic cells will not be passed on to future generations, so that only mutations in germ cells (and hence the resulting gametes) have any chance at all of being passed on to the next generation. Therefore, at the organismal level, not all mutations are heritable.

2. Bumpus's data supports stabilizing selection, because natural selection had culled out extreme characters and favored those traits near the norm. Instead of being the driving force for evolution, stabilizing selection tends to reinforce and maintain the normal phenotypes in a population and, in that sense, operates in opposition to evolution.

3. Recall that evolution is a property of populations, and natural selection operates on individuals. Recall further that only heritable changes, by definition, are passed on from one generation to the next. Thus, heritable changes are part of the gene pool and are a property of the population; as such they provide potential fodder to the evolutionary process. On the other hand, nonheritable traits are not part of the gene pool but rather are a property of the individual. Because natural selection operates on the individual, nonheritable traits (such as the accidental loss of a limb or the development of a virus-induced tumor) may play critical roles in death and survival. However, these traits are not a property of the population, so they are not important to the evolutionary process.

Chapter 17

Review Questions

1. Since many common names are local names given by people who live in the same region in which a plant or animal occurs, they have an advantage (to anyone who speaks the local language) in being easier to pronounce, understand, and remember. But this is a shortcoming to people who do not speak the local language. Further, different groups of people may assign different common names to the same plant species, or conversely, the same common name to different plant species. To add to the confusion, many organisms lack a common name altogether. Such a disparate nomenclature, lacking both convention and coordination among the various groups of people, leads to confusion, inconsistent usage, and inefficient communication. Scientific names, on the other hand, are always in Latin (or a Latinized form of another language, such as Greek). Such names may be unfamiliar to novices and hence harder for them to pronounce, understand, and remember, but they do offer several distinct advantages. Latin is a dead language—it is not the primary language of any culture today. Thus, being an unchanging language, Latin words will have the same meaning in 100 years as they do today, and the language does not contain politically offensive words or phrases. Further, strict nomenclature rules govern the naming of plants and animals. This increases consistency in applying names and decreases the incidence of inappropriate naming (such as assigning the same name to two different plant species). Last, species grouped within the same genus are related to each other and hence share certain traits; being familiar with one member of a genus imparts some knowledge about all members of that genus. Scientific names thus provide uniformity and specificity, and that is why they are preferred in professional disciplines such as science, medicine, agriculture, and commerce.

2. Scientific names consist of a noun called the generic name and an adjective called the specific epithet; for completeness, the authors of the name are often listed immediately following the specific epithet. Generic names are always listed first, are always capitalized, and are often abbreviated; specific epithets need never be capitalized and are seldom abbreviated. Generally, both parts of a scientific name are underlined, italicized, or presented in a different font; this sets the name apart from the surrounding text and indicates to the reader that the language has changed (to Latin, in this case). Only the generic name may be used by itself.

3. A species is a group of similar-appearing organisms that can freely interbreed. When a group of species seems particularly closely related, they may be placed into a loose-knit group of allied species termed a species complex. Hybrids are the offspring of two different species; the parents may or may not be members of the same genus. Hybrids are indicated by appropriate placement of an "×" in the scientific name. Intrageneric hybrids assume the name of both parents, separated by an × or are assigned an altogether new name with the × placed before the species epithet. For example, a common ornamental shrub in California is the hybrid *Abelia chinensis* × *A. uniflora*, also known as *Abelia* × *grandiflora*. Intergeneric hybrids, whose parents are members of different genera, are indicated by combining parts of the parental generic names into a new generic name, and preceding that name with an ×. An example is provided by × *Cupressocyparis leylandii* (Leyland cypress), which is a cross between Monterey cypress (*Cupressus macrocarpa*) and Alaska cedar (*Chamaecyparis nootkatensis*).

   Species may be further divided into subspecies, varieties, or cultivars. A subspecies is a subset of a species consisting of those members inhabiting a geographical region distinct from the remaining members of that species. The individuals comprising a variety possess consistent—though often subtle—differences from traits typical of the species. For example, *Carex stricta* is a grass-like plant termed a sedge that occurs throughout eastern North America. The typical variety, *C. stricta var. stricta*, produces floral spikes less than 6 cm in length, while the variant *C. stricta var. elongata* consistently bears longer spikes up to 15 cm in length. A cultivar is essentially a variant that does not occur in the wild and has been deliberately propagated for horticultural or ornamental purposes; it is usually indicated by quotation marks. Two examples of cultivars are the Leyland cypress mentioned earlier: × *Cupressocyparis leylandii* "Harlequin" and × *Cupressocyparis leylandii* "Castlewellan."

4. An herbarium is a depository for dried (and therefore dead) plant specimens and cuttings. Usually associated with major universities, the larger herbaria contain millions of plant specimens, each carefully filed away so that it is readily accessible. Each specimen acts as a reference and a voucher—proof that a given plant once grew in a given area. Each specimen is also a potential source of DNA or seeds. Botanical gardens, on the other hand, contain living specimens. While these are potentially excellent sources of DNA or seeds, there is a major drawback—growing specimens

require garden space and periodic maintenance, so the per-specimen costs are much higher in a botanical garden than in an herbarium.

5. Modern plant classification is based upon many features, including anatomical (structure of leaves, stems, roots, and flowers), biochemical (e.g., chemical characterization of secondary compounds), genetic (e.g., DNA sequence), and paleontological (e.g., fossils) lines of evidence. Flower structure—and more recently DNA sequence data—is weighted more heavily than the vegetative and biochemical characteristics since the latter seem to be much more influenced by the environment than the former. Relationships between plant groups thus are usually more evident in the flowers and DNA than in the other traits.
6. The sequence of these categories, starting with the most inclusive, is Domain—Kingdom—Phylum—Class—Order—Family—Genus—Species—Variety. Viruses are not considered as living organisms, so are not included in this classification scheme.
7. Plant taxonomy is the scientific discipline concerned with naming plants, not only assigning names to new plants but also resolving ambiguities with older names. Plant taxonomy is an integral part of the broader discipline of plant systematics, which is concerned with organizing and classifying plants into logical and related groups.
8. Carolus Linnaeus was the first scientist to use Latin binomials as scientific names, beginning sometime in the mid-1700s. He began using this format not because he felt it would catch on quickly in the scientific community, but for convenience: He preferred writing a binomial name rather than the longer, 10- to 12-word names that were typical of the time. Also typical of the time was the use of Latin as the scholarly language, which is why Linnaeus incorporated that language in his scientific names.

Applying Concepts

1. Consider first that there is no single correct answer to this question, as long as the resulting key consists of a series of two mutually exclusive choices that unambiguously distinguish the six given figures. Here is one possibility:

   A. Figure consisting of straight line segments ("sides") that join to form angular corners
      B. Figure with 3 sides and 3 corners; internal angles always sum to 180°
         C. Sides of equal length—equilateral triangle
         C. Sides of unequal length—scalene triangle
      B. Figure with more than 3 sides and 3 corners; sum of internal angles greater than 180°
         D. Figure with 4 sides and 4 corners; internal angles sum to 360°
            E. Opposite sides parallel and of equal length—rectangle
            E. One set of opposite sides parallel and of unequal length, the other set of sides not parallel but of equal length—trapezoid
         D. Figure with 5 sides and 5 corners; internal angles sum to 520°—pentagon
   A. Figure lacking straight line segments and corners, consisting of a smooth curve whose points are equidistant from an imaginary center—circle

2. The rules governing naming problems such as these are found in the International Code of Botanical Nomenclature. It describes the very specific actions that must be taken before a name can be accepted as valid. Between Prof. James and Ms. Smith, only Ms. Smith followed the rules, including a Latin description of the plant. Initially, then, Ms. Smith's name of *H. ciliaris* would have priority over Prof. James's *H. missouriensis*. However, Prof. Chang points out that even Ms. Smith's name is invalid, since the name is already in use. He corrects the error by publishing an acceptable name, which would be written as *H. smithii* (Smith) Chang. Last, since two people have used the name *H. ciliaris* to refer to two separate species, one would need to include the authority with the name (i.e., either *H. ciliaris* R. Brown or *H. ciliaris* Smith), to avoid ambiguity when referring to *H. ciliaris*.
3. Here are several benefits of each; additional answers are, of course, possible. Both botanical gardens and herbaria (1) can be sources of plant material in the form of tissue (e.g., for DNA extraction or chemical analysis) and seeds; (2) are invaluable aids for people wishing to learn how to identify various plants, since both provide named specimens; (3) are potential sources of plant material for classroom use. Botanical gardens (1) are often maintained in park-like settings, providing quiet, open areas of beauty for reflection, inspiration, and solitude as well as gathering areas for social events such as weddings; (2) provide sources of living plant material for propagation by cuttings or of pollen for plant breeding experiments; (3) provide opportunities for individuals to "browse" living examples of horticultural and landscaping plants that they might consider growing at their own homes or businesses; (4) act as living museums of horticultural plants that were once in common use but have now been largely replaced by newer cultivars; (5) provide a living laboratory for the study of various aspects of the plants growing there (for example, one could study the long-term hardiness of various imported plants, or their interactions with native insects and birds); (6) act as urban air filters, cooling and humidifying parched air, as well as replenishing its oxygen and removing airborne particulates; (7) may provide areas where new cultivars arise, either spontaneously or through plant-breeding experiments; (8) may act as conservation centers for rare plants threatened with extinction from their natural habitats. Herbaria (1) are important in scientific study since they may provide named specimens of a huge number of plants—often in the millions; (2) provide a depository for type specimens—the original specimen from which a new species was described (type specimens thus define the new species and serve as a future reference for that species); (3) provide a record of the original range of a species (as species disappear from local areas, the herbarium specimens serve as vouchers that a given species once grew in that location); (4) contain specimens that are easily shared and/or exchanged with other herbaria, thus facilitating botanical studies.
4. The International Code of Botanical Nomenclature sets forth rules that must be followed before new taxonomic names—including new species names—can be considered valid and acceptable. This body of rules thus standardizes the naming process. It ensures, for example, that every new species name is accompanied by a Latin description—usually several paragraphs—as well as a type specimen deposited in a recognized herbarium. It also provides a means to recognize invalid names, and rules to resolve naming problems such as duplicate or inappropriate names. Without the conventions of the International Code of Botanical Nomenclature, taxonomic chaos would ensue.

Chapter 18

Review Questions

1. It seems unlikely that life could have ever arisen on Venus; if it had, it would have been exterminated long ago due to the inhospitable environment. With surface temperatures over 460°C, a CO_2 atmosphere with an oppressive surface pressure of 90 atm (90 times the surface pressure on Earth), and a lack of water, Venus could not host any type of life as we know it. Mars, on the other hand, is also inhospitable, but for different reasons. Although daytime temperatures in the warmer areas of the planet may reach a balmy 22°C, nighttime temperatures may plummet to below −100°C. Furthermore, the Martian atmosphere is less than one-hundredth as dense as Earth's, most of the water is locked away as ice, and solar ultraviolet radiation bombards the surface at lethal levels. All in all, the present harsh Venusian and Martian environments would not be conducive for life.
2. Ultraviolet (UV) light is energetic enough to disrupt the chemical bonds of organic compounds; when small organisms such as single-celled algae or bacteria are exposed to high levels of UV light, the organic molecules within the cells are degraded and the cells die. In larger organisms such as humans, exposure to UV light may lead to severe skin burns, blindness, death of skin tissue, and eventually, death of the whole organism. Ozone absorbs UV radiation, and a layer of ozone in the atmosphere prevents high levels of UV light from reaching the surface. Thus, ozone protects surface life from overexposure to UV light. When life first appeared on our planet, an ozone layer was lacking, so that surface areas of the planet were bombarded with lethal levels of UV light; at that time, only the oceans—whose waters strongly absorb UV light—provided a hospitable environment where life could flourish.
3. (a) All living things possess a metabolism, defined as the sum of all the chemical processes occurring within a living cell. (b) All living things exhibit growth, defined as an increase in cell size and mass due to their metabolic activities. (c) All living things exhibit reproduction, the creation of a new individual. Reproduction provides for the continuation of the species over time. (d) Organisms respond to stimuli and show some type of movement. (e) Over time, populations of organisms adapt and evolve, which raises the overall fitness of the population and maximizes the chances of survival in any particular environment.
4. Current estimates place the birth of the Earth (in fact, the entire solar system) at about 4.6 bya. For about 800 million years the Earth was pelted with comets and asteroids. About 300 million years after this heavy bombardment ended, at 3.5 bya, life appeared on Earth and left a fossil record. By 2.2 bya, the atmospheric levels of oxygen had risen to 10% of present day levels, with a further increase to 15% by 0.6 bya. By that time, Earth's oceans were teeming with multicellular life. Approximately 450 million years ago, atmospheric oxygen concentrations had reached their current levels of 21%, and carbon dioxide levels began to drop. Had the CO_2 levels not dropped as drastically as they did, the early Earth could have experienced a runaway greenhouse effect, effectively sterilizing it. The drop in CO_2 levels coincided with the first appearance of land plants.
5. We have no direct fossil evidence for what are believed to be the very first life-forms on Earth following the bombardment period. The fossil record of that time period does contain carbonaceous graphite particles believed to have been produced by these earliest life-forms. The earliest fossils appear to be cyanobacteria and are first seen in 3.5-billion-year-old rocks from western Australia. Chemical analyses of stromatolites in Australia suggest that photosynthesis was operational at least 2.7 billion years ago. Apparently, the oxygen released by these early photosynthetic organisms did not start to accumulate in the atmosphere until about 2.2 billion years ago. Chemical analyses show that prior to this time, free oxygen was busy reacting with iron in the Earth's crust; we see the results of this today as thick layers of iron-oxide-rich rock.
6. The first stage is the formation of simple, small organic molecules from even smaller inorganic and elemental precursors. These processes would produce sugars, fatty acids, amino acids and nucleotides—the very building blocks for biological macromolecules. Next, the chemical-biological theory hypothesizes that these building blocks were assembled into such important macromolecules as proteins, carbohydrates, lipids, and nucleic acids. The third stage is the incorporation of these macromolecules into the first cell-like structures—the beginnings of life itself.
7. Most commonly, prokaryotic cells are shaped either as little spheres called cocci, as little cylinders termed rods, or as small corkscrew-shaped spirals known as spirilli. In some species, these cells exist as solitary unicells, while in others they may aggregate into a mass called a colony, or may join end to end (like pearls on a necklace) to form a linear array called a filament.
8. Thylakoids are internal arrays of membranes that provide a surface for the attachment of photosynthetic molecules such as chlorophyll. Thylakoids are found in cyanobacteria and the chloroplasts of plant cells.
9. To survive unfavorable environmental conditions, many prokaryotes produce spores—thick-walled, dormant cells that are able to germinate back to typical, actively growing cells when conditions improve again. Endospores are a type of spore produced within a parental cell and released when that cell ruptures. Akinetes are thick-walled cells full of storage materials; they are produced by many cyanobacteria. Akinetes generally settle to the bottom of lakes and ponds in the autumn, and germinate the following spring. All of these structures enable bacteria to become dormant during unfavorable environmental conditions, and then to germinate and continue growth when conditions improve.

10. Industrial nitrogen fixation is a process whereby nitrogen gas is subjected to very high temperatures and pressures in the presence of hydrogen, resulting in the formation of ammonia. It is a very energy intensive process. Biological nitrogen fixation also yields ammonia from nitrogen gas and a hydrogen source, but this feat is accomplished at normal temperatures and pressures inside certain cells using the enzyme nitrogenase. Plants, like animals, need to make a variety of nitrogen-containing compounds, proteins and DNA being two common examples. However, only nitrogen fixers are able to begin these biosynthetic processes by directly combining gaseous nitrogen with hydrogen to yield ammonia; plants can then use ammonia or one of its derivatives to form their nitrogen-containing compounds. Nitrogen fixers thus play a pivotal role in the ecosystem in that they are the only organisms that can make direct use of atmospheric nitrogen gas. Essentially, they fix nitrogen for all other life-forms to use.

11. Most forms of nitrogenase are poisoned by oxygen. Thus, in cells or tissue where nitrogenase is active, it is important that oxygen concentrations be kept low if high nitrogenase activity is to be maintained. Cyanobacteria accomplish this by producing special cells called heterocysts, wherein the oxygen-producing processes of photosynthesis have been shut off, leaving a low-oxygen environment for nitrogenase activity. Legumes go to far greater lengths to enable nitrogen fixation. These plants produce elaborate accommodations in their roots for nitrogen-fixing bacteria, and then depress oxygen levels by producing leghemoglobin, which latches onto oxygen just as the hemoglobin in your red blood cells does. The nitrogen-fixing bacteria grow happily in this environment, fixing enough nitrogen for both themselves and their leguminous host.

12. Gram staining is a two-stain laboratory procedure that allows microbiologists to discern certain differences in the cell walls of bacteria, as related to the amount of peptidoglycan present in the wall. Cell walls having an outer phospholipid envelope also tend to have lower levels of peptidoglycan; such cells stain pink with the Gram stain and are classified as Gram-negative. Cell walls with higher levels of peptidoglycan, containing little or no phospholipid envelope, generally stain blue and are said to be Gram-positive. Penicillin, you may recall, inhibits the synthesis of functional peptidoglycan. Because Gram-negative cells have little peptidoglycan in their cell walls, the penicillin class of antibiotics tend to be rather ineffectual against them. Thus, one should use some other type of antibiotic to treat Gram-negative bacterial infections.

Applying Concepts

1. Current thought is that early life on Earth evolved in what is known as a reducing atmosphere—high in methane, ammonia, hydrogen gas, nitrogen gas, carbon dioxide, and water vapor. This primordial atmosphere has since been replaced with a very oxidizing one that is rich in gaseous oxygen (21%) and low in methane, ammonia, hydrogen, and carbon dioxide. Thus, it is not possible to find a place on Earth where those primordial conditions still exist, and we are forced to simply mimic them in the laboratory. The advent of life on Earth forever changed the Earth itself.

2. By absorbing much of the incident solar UV light, the ozone layer effectively protects the ground below from the harmful effects of this harsh radiation. The destruction of large areas of the ozone layer allows more hazardous UV radiation to reach the surface. UV irradiation of living cells—be they microbial, plant, animal, or fungal—may have severe consequences in terms of damage to organic molecules and mutations in DNA. Thus, a loss of the protective ozone shield could have direct, harmful effects on all but the most sheltered species on this planet. People living in areas near the polar ozone holes are most affected by the loss of ozone now. If the ozone layer continues to be degraded, then people in many parts of the Earth may be directly harmed.

3. (a) After 1 hour, there would be 4 cells. (b) 64 cells. (c) 4,096 cells. (d) ~16.8 × 10^6 cells. (e) 2.8 × 10^14 (~300 million million!) cells. As the cells become more numerous, they compete with each other for nutrients, air, space, and water; eventually some or all of those resources become limiting, and growth rate drastically diminishes.

4. All forms of life must synthesize a plethora of proteins, nucleic acids, and other nitrogen-containing compounds in order to live. Although the most abundant nitrogen source around us is the gas in the atmosphere, surprisingly, only the nitrogen fixers can tap into this source and use it as the basis for forming their nitrogenous compounds; all other organisms must start with preexisting nitrogen-containing compounds in their environment (food, etc.) and proceed from there. Although only a small to moderate number of species may benefit directly from the activities of nitrogen fixers, virtually all species on the planet benefit from this essential service because nitrogen compounds are passed all the way up to the very top of the food chains. Without nitrogen fixers around, food chains around the world would collapse; as nitrogenous compounds decomposed, their nitrogen would be released back to the atmosphere as nitrogen gas, with no way to recover it back into the food chain. Nitrogen fixers are essential to life on this planet.

5. Books have been written on this subject, so this answer is necessarily extremely abbreviated; many other answers are possible. Genetic engineers use bacteria because they are easily grown under a wide variety of culture conditions, they grow rapidly and to high densities, and they are easily manipulated genetically. Typical uses include producing many copies of a piece of DNA of interest, providing a vehicle in which scientists can manipulate and alter DNA, preparing a fragment of DNA for DNA sequencing, and expressing a foreign gene that has been inserted into them. The latter use is extremely powerful. For example, if the gene for human insulin is properly inserted into an *E. coli* cell and that cell cultured, then the cultured cells will essentially become little protein factories, bacteria making a human protein!

6. The inhibition or blockage of any critical cellular process will most likely either stall growth or outright kill bacterial cells. To be an effective clinical antibiotic, a compound must not only do this, but it should also be selective for bacterial cells. You don't want to take an antibiotic that kills your own cells as it is killing the bacteria! Other antibiotics and their cellular sites of action include polymyxin (pokes holes in the cell membrane); mitomycin (interferes with bacterial cell division); novobiocin (inhibits DNA synthesis); and erythromycin, streptomycin, and tetracycline (interfere with protein synthesis). Many other antibiotics are known, and their modes of action are varied.

7. In the natural world, there is competition for virtually every resource. If a cell can release an antibiotic into its immediate vicinity, it may be able to inhibit the growth of nearby species and thus reserve the available resources for its own growth. Thus, antibiotic production is an attempt to gain the competitive edge over neighboring species.

Chapter 19
Review Questions

1. Similarities: Most algae and plant species contain chloroplasts for photosynthesis, store organic food, and have cell walls, usually of cellulose. Many algae and plants produce protective chemicals to ward off disease-producing microbes as well as herbivores, and many algae, like plants, possess bodies that are attached to a surface. Differences: Most algae live in aquatic or even snowy habitats, while most plants are terrestrial. The bodies of algae are also smaller and less complicated than those of their plant cousins. Indeed, some algal species exist as unicells—single cells free in the environment—while mature plant bodies are always multicellular. Having a simpler body means that algae also have fewer types of cells and tissues than plants.

2. First, not all algae are photosynthetic; some species lack chlorophyll altogether and must obtain their nutrition heterotrophically. Second, not all algae are eukaryotes, the exception being the prokaryotic cyanobacteria, or blue-green algae. Some algae do not live in aquatic habitats but rather colonize much drier environments such as in soil, on rocks, or on the surfaces of land plants and animals. Last, although algae do generally have simpler bodies than land plants, some algae such as the giant kelps have bodies that are almost as complex as higher plants. Despite these exceptions, the description of algae given here and in

the chapter text provides a good working definition of the group of organisms known as "algae."

3. An autotroph is an organism that can produce all of its organic compounds itself, usually from carbon dioxide via photosynthesis. Heterotrophs, on the other hand, are unable to conduct photosynthesis and thus must obtain some of the organic compounds they require from the environment. Examples of autotrophs are green plants such as corn, maple trees, and tulips, as well as most algae. Animals, including humans, form a large group of heterotrophs. The algae *Prototheca* and *Dinobryon* are also heterotrophic; the latter can also photosynthesize (i.e., it exhibits both autotrophic and heterotrophic nutrition).

4. The simplest algal body form is independent single cells, or unicells. Other algae assume a form that is essentially a cluster of cells, or colony, held together by some type of intercellular glue. A more complicated type of body is the filament, whose cells are joined together end-to-end to form either unbranched or branched chains. The most complicated type of algal body is tissue, a solid block of cells growing in all three dimensions; this advanced type of growth form is found in some green and brown algae. In addition to the preceding basic forms, some unicellular and colonial algae bear one or more flagella and are loosely termed flagellates. Such cells or colonies are able to swim and thus are motile in the water column. Algae may grow attached to a substrate (such as a rock) or free; such free-floating or free-swimming algae are known as phytoplankton. Last, amongst the marine algae are species large enough to be seen with the unaided eye; these are known as seaweeds.

5. Plants and many (but far from all) algae possess two distinct phases, or bodies, in their life cycles. One type of body, the gametophyte, produces egg and sperm directly; the other type of body, the sporophyte, produces sexual spores but not gametes. Some species of algae produce gametophytes and sporophytes that appear identical to the unaided eye; in the case of *Porphyra*, however, the gametophytes are macroscopic blades familiar to most people as nori seaweed, while the sporophytes are microscopic filaments that appear nothing at all like the gametophytes. So dissimilar are these two stages in *Porphyra* that they were originally thought to be two distinct organisms. It was Dr. Drew Baker's discovery that these two separate stages were in fact part of the life cycle of one organism, which we now know as *Porphyra*.

6. Chlorophyll a is a green pigment that plays a central role in the photosynthetic process. It absorbs heavily in the red and blue regions of the spectrum and is present in all photosynthetic algae and plants. Accessory pigments also absorb in various parts of the spectrum and then pass the absorbed energy to chlorophyll a. In effect they extend the absorption spectrum of chlorophyll a. Various algal groups have distinctive accessory pigments. Dinoflagellates have the brown-colored peridinin; brown algae and diatoms have the golden-brown fucoxanthin; red algae (and cyanobacteria) have red-, orange-, or blue-colored phycobilin pigments; and green algae and land plants have chlorophyll b, lutein, and beta-carotene.

7. Ecologically, algae are the primary producers in their aquatic habitats, using the sun's energy to fix carbon dioxide into organic compounds. This forms the basis of the food chain for most aquatic heterotrophs. Also, algae release at least 50% of the oxygen into the Earth's atmosphere. Some algae live in close association with animals such as corals and provide them with organic compounds. On the detrimental side, some algae produce toxic compounds that can kill fish or animals—including humans—that consume toxin-containing seafood. Economically, some algae are used directly for food, such as the red alga nori from which sushi is made. Some seaweeds provide important cell-wall polysaccharides such as carrageenan, agar and algin; diatoms produce diatomaceous earth, which has important industrial uses; and some algae contribute directly to the formation of limestone.

Applying concepts

1. Some parasitic algae may still have plastids and some enzymes and structures related to photosynthesis even though, as parasites, they no longer rely on photosynthesis. These enzymes and the metabolic pathways that

make plastids and those that make algal cell walls would be appropriate targets for antibiotics because animals do not have cell walls or plastids or carry on photosynthesis.

2. Several structural characteristics can be used to determine the relationship of the new alga to other known groups of algae. Analyses of the storage compounds (starch, non-starch polysaccharides, oil droplets, or paramylon), accessory pigments (peridinin, chlorophyll c, chlorophyll b, fucoxanthin, phycobilins, lutein, or beta-carotene), and constituents (cellulose, algin, carrageenan, agar, silica, or calcium carbonate) of the cell wall (if there is a cell wall) can help identify its relatives. Other characteristics, including the presence of flagella and their structure and stages in the reproductive cycle may also be helpful. Finally, one could compare DNA of the new alga with DNA of other algae that appeared to be related to it.

3. Single flagellated cells are very mobile and may be able to elude predators. On the other hand, they are alone in the world, so to speak. Large seaweeds are anchored to one place and their size may attract predators. But, being multicellular organisms, these seaweed cells can cooperate. For example, some cells may form gas filled vesicles that help the seaweed float and therefore be closer to sunlight for photosynthesis.

4. Dead zones are areas where oxygen levels have been depleted by the decomposition of excessive growths of algae. Few animals survive there. This algal overgrowth, in turn, is a result of high levels of nutrients, such as sewage and runoff of excess fertilizer from farms, lawns and gardens, in the water. To reduce the size and frequency of such dead zones, one should try to control large point sources of nutrient runoff—sewage treatment plants, farms that use a lot of fertilizers, farms with high concentrations of animals, and ships that may be dumping wastes directly into the water. It may also be helpful and necessary to enlist the cooperation of city residents to use low phosphate detergents, limit their use of lawn fertilizers and keep leaves and other organic material out of the streets so that they don't wash into the water and provide more nutrients for algal growth.

Chapter 20
Review Questions

1. The four divisions of fungi are (a) the chytrids; (b) Zygomycota, the zygomycetes; (c) Ascomycota, or sac fungi; and (d) Basidiomycota, or club fungi. Chytrids and some sac fungi (yeasts) are unicellular, while the other fungi possess special filaments termed hyphae. The zygomycetes include the black bread mold familiar to most people. The sac fungi include morels and yeast. The club fungi are probably most familiar to people, including many of the edible mushrooms from the grocery store as well as plant pests such as wheat rust and corn smut. During sexual reproduction in chytrids and zygomycetes, nuclear fusion occurs shortly after mating. In the sac and club fungi, pairs of parental nuclei may coexist in the same cells for a long time before fusing. After fusion, the zygote develops and eventually produces sexual spores, which in the club and sac fungi are borne on fruiting bodies more commonly known as mushrooms. Specifically, in the Ascomycota these spores are produced in sacs called asci, while in the Basidiomycota they are borne on a club-shaped pedestal termed a basidium.

2. Some fungi, such as pizza mushrooms, are important as human food, while some yeast species contribute to the formation of foods such as bread, beer, and wine. From yet other fungi we may extract hallucinogens such as LSD or medicines such as penicillin.

3. Ecologically, fungi, along with the bacteria, are extremely important as decomposers of the biological world. Without them, we would be stumbling over the bodies of plants and animals that died long ago. Growing in a mutualistic association with plant roots (i.e., mycorrhizae), some fungi greatly benefit their plant hosts. Moreover, some fungi may form mutualistic associations with algae (i.e., lichens) and thus begin the process of soil-building in such barren places as bare rock.

4. Alas, fungi play some detrimental roles as well. Fungi may cause disease, from tissue infections in humans (e.g., athlete's foot) and other animals to plant diseases

that decrease crop production and may even kill the host plant. Some species of fungi cause food spoilage, and others produce toxic compounds like aflatoxin. Some fungi, in their role as decomposers, degrade things that we would rather preserve such as stored crops and many other objects, including clothing and film, in warm humid environments.

5. Mutualistic associations of fungi and algae are known as lichens, in which the algal partner receives carbon dioxide, nutrients, water, and protection from the fungus while the fungus receives organic compounds and oxygen from the alga. Endophytes are fungi that live within the tissues of plants such as grasses; the fungus receives organic compounds from the plant and provides herbivore-repelling compounds to the plant. The roots of most plants form mycorrhizae with certain fungal species, again with the fungus receiving organic materials from the plant and providing the plant with minerals, nitrogen, or water. With ectomycorrhizae, the fungal hyphae coat the root surface of their plant partners. In endomycorrhizae, the fungi actually penetrate the cell wall and develop in contact with the plant's plasma membrane.

6. Lichens with flat, planar bodies closely adhering to a surface exhibit the crustose growth form; those with flattened, leaf-like bodies are known as foliose lichens; and those with three-dimensional, often branched forms, many of which either grow upright or hang from objects, are fruticose lichens.

Applying concepts

1. Without reproductive structures or spores to consider, you would have to examine vegetative features of the hypha. Chytrids and yeasts are unicellular, so this organism is neither of those. Zygomycete hyphae lack septae, so a lack of cross-walls would indicate that the organism in question is a zygomycete. The possession of septae would indicate the organism is either an ascomycete or a basidiomycete, and examination of the finer structure of the septa would allow you to distinguish between these two groups.

2. Since many species of fungi and bacteria are decomposers, they often compete for resources. By releasing penicillin, *Penicillium* inhibits the growth of many nearby bacteria and thus increases its own ability to compete for available resources.

3. Obviously, without fungi, fungal diseases would cease to exist. The lack of fungi would also slow decomposition of dead matter, so that bodies would linger in the environment longer after death and thus increase the incidence of certain bacterial diseases (e.g., cholera). Furthermore, animals would find it increasingly difficult to move about the planet as the surface became littered with the slowly decomposing bodies of trees, plants, and other animals. Without lichens, colonization of such inhospitable habitats as bare rock would virtually cease. Possibly the greatest impact would be on the growth of plants. Since perhaps 90% of plants depend on mycorrhizal relationships with fungi for optimal growth, plant (including crop) production would slow, and many plant species would probably disappear altogether, unable to reproduce effectively. The result would be widespread famine and the loss of many animal species. Truly, the world we know today could not exist without this important group of organisms.

4. The most likely photosynthetic organisms in such an inhospitable place would be lichens which are composed of green or blue green algae growing with a fungal partner. Lichens may grow on surfaces of rocks and have a crustose, foliose, or fruticose shape and may be a subdued gray-green color or else brightly colored—orange, yellow, or red.

Chapter 21
Review Questions

1. The following combination of traits is unique to plants: (a) an apical meristem that produces the tissues of the plant body; (b) spores with very tough walls to aid survival during dispersal in air; (c) alternation of generations, a life history in which a haploid, gamete-producing gametophyte alternates with a diploid, multicellular, spore-producing sporophyte; (d) embryos that receive nutrients and other resources from the female gametophyte that produced them; and (e) pro-

duction of spores within sporangia—multicellular enclosures. No other group of organisms possesses this collection of characteristics.

2. (a) Peat mosses are bryophytes used in gardening and for fuel; and large areas of living peat mosses in the Northern Hemisphere help dampen temperature fluctuations in the atmosphere. At one time, sterilized peat moss was used as gauze or surgical dressing in medicine. (b) Lycophytes have no economic use today, but the incompletely decomposed remains of plants that grew several hundred million years ago—many of which were lycophytes—are largely responsible for the coal deposits we use today. (c) Pteridophytes, like lycophytes, are partly responsible for today's coal deposits. Today ferns find ornamental use as landscaping plants or in gardens, and they are a common component of the tropical rain forest. In addition, the "fiddleheads" produced by certain ferns are sometimes eaten for food.

3. Phylogenies are graphic depictions—usually branching, tree-shaped diagrams—of the relationships and patterns of ancestry of a group of organisms. DNA sequences are often used in such studies, because DNA does not change with variations of the environment as easily as phenotypes can; thus DNA is thought to reflect relationships better than phenotypic characters. To use DNA in these studies, it is first extracted and purified, and the relevant gene or genes amplified by PCR. The genes are then sequenced. All the sequences so obtained are fed into a computer, where they are aligned to each other; complex algorithms then compute how related the sequences are to each other, and phylogenetic trees are constructed. Although any number of genes could theoretically be used in these studies, only a few are commonly used, partly because there is already a large and expanding database of these sequences from many other organisms. Commonly used genes include those encoding tubulin, rubisco, phytochrome, and the RNA in the small subunit of ribosomes.

4. Paleobotany is the study of plant fossils; it is very useful in estimating the geologic times when various plant groups first appeared. Plants routinely produce two very durable polymers, lignin and sporopollenin, that are decay resistant and promote fossilization. Lignin is a polyphenolic compound occurring in the walls of certain plant cells, especially tracheids and vessels, and it is largely responsible for most plant fossils. Sporopollenin is the most resistant biological material known; it is found in the walls of plant spores.

5. Four key events in the evolution of land plants: (a) The origin of land plants. The first land plants were well established by 460 million years ago. These early plants were bryophyte-like plants that possessed meristems, sporophytes, and protective sporangia. (b) The rise of vascular plants. About 400 million years ago the protracheophytes appeared—small, bryophyte-like plants still lacking lignified vascular tissue but possessing branched, independent sporophytes. These appear as intermediates between bryophytes and vascular plants. In the intervening period up to 360 million years ago, many types of vascular plants appeared, all extinct now. The earliest lacked roots and leaves but did have lignified tracheids in their stems. From 360 to 256 million years ago, seedless vascular plants included many tree species forming large forests in warm, humid swamps during the Carboniferous period. (c) The evolution of seed plants. Seed plants evolved independently several times during the Carboniferous period, and proved very effective for reproduction in the cooler, drier climate that followed in the Permian period. (d) The diversification of flowering plants or angiosperms. Some of the earliest angiosperm fossils date from about 150 million years ago; these plants possessed flowers and fruits, the hallmarks of the group.

6. Lycophylls are true leaves that contain only a single unbranched vein. Lacking an extensive conduction system, leaves with such modest venation do not grow very large and are thus sometimes termed microphylls. Lycophytes produce lycophylls. By contrast, the leaves of pteridophytes, gymnosperms, and angiosperms contain an extensively branched conduction system and regularly develop to considerable size; such leaves are called euphylls. Bryophytes produce flattened, laminar structures that look leaf-like, but are not considered true leaves because they lack vascular tissue.

7. The bryophytes comprise the liverworts, hornworts, and mosses. In all cases, the gametophyte is the dominant generation and is generally found growing on rocks, soil, or other plants. Bryophyte gametophytes are nonwoody and lack roots, leaves, and lignified vascular tissue. Typically growing close to the ground, bryophytes are anchored to the soil by rhizoids that are either single-celled tubes or multicellular filaments. Gametes are produced in specialized structures called gametangia, specifically, egg-producing archegonia and sperm-producing antheridia, both of which are surrounded by a sterile layer of cells known as the jacket. Liquid water is required for fertilization, because the sperm must swim to the eggs in the archegonia. After fertilization, the zygotes develop in situ within the archegonia, maturing into sporophytes that are nutritionally dependent upon the parental gametophyte. These sporophytes typically consist of a basal foot embedded in the gametophyte, a stalk that may raise the sporophyte in the air for better spore dispersal, and a capsule, or sporangium, in which the spores are produced.
8. The lycophytes, or club mosses, produce a dominant, branched sporophyte containing true vascular tissue (i.e., xylem tissues having lignified cells and phloem). Lycophytes produce true roots, stems, and leaves, the latter being lycophylls. Sporangia are borne on specialized leaves known as sporophylls that are often clustered together into cone-like structures called strobili. Within the strobili, meiotic divisions produce haploid spores that, after germination, develop into inconspicuous gametophytes. Most lycophytes are homosporous, and thus produce only one type of spore and only bisexual gametophytes. Today's lycophytes are mostly rather small plants and number about 100 species. However, 40-meter-tall tree lycophytes thrived in forested swamps during the Carboniferous period.
9. The 12,000 species of pteridophytes, or ferns, are widely distributed around the world. Many produce horizontal, underground stems called rhizomes, from which sprout upright stems and leaves with highly branched vascular systems (euphylls). The leaves on many—but certainly not all—ferns are highly dissected (divided) into smaller leaflets, producing very fine, lacy foliage. Clusters of sporangia called sori are borne on the undersides of photosynthetic leaves or on specialized non-photosynthetic leaves called sporophylls. Most ferns produce one type of spore (homospory) that develops into bisexual gametophytes, but some species are heterosporous and produce both microspores and megaspores. Microspores, or pollen grains, are produced in microsporangia on specialized leaves known as microsporophylls; microspores develop into male gametophytes (microgametophytes) that ultimately produce sperm. Megaspores, on the other hand, are produced in megasporangia on specialized leaves known as megasporophylls; megaspores develop into female gametophytes (megagametophytes) that ultimately yield eggs. In both homosporous and heterosporous ferns, water is required for the sperm to swim to the eggs.

Applying Concepts

1. Fossils provide evidence for the existence of plants and plant groups of long ago, many of which have no living representatives. From fossils, scientists can often reconstruct the appearance of the plant, often getting information about its internal anatomy. When studying fossils is coupled with stratigraphy and other dating methods, they can determine fairly accurately the age of a fossil and hence when a particular plant lived. Sometimes fossils are found that clearly represent intermediates between what was then an evolving plant group and its evolutionary ancestors. But of course, the fossil record is and always will be incomplete. DNA-based phylogenies, on the other hand, are excellent for showing relationships among living plants. Hundreds of thousands of plants are available for such studies, and fresh, undamaged DNA can usually be easily obtained from specimens. However, DNA studies are not feasible with long-extinct plants. Most ancient fossils do not contain DNA, and in those that do it is usually severely degraded. Further, although DNA studies are excellent at revealing relatedness of organisms, they are very weak at estimating events in geologic time, for example, exactly when two groups of plants diverged. Thus, fossil-based and DNA-based phylogenies each have their strengths and weaknesses, and the two approaches complement each other well in reconstructing the evolutionary history of plants.
2. Anatomically, the progymnosperms resembled gymnosperms in having well-developed, lignified tracheids in their secondary xylem. Recall that ferns are mostly herbaceous in habit, only one species (*Botrychium*) being known to have a vascular cambium. Progymnosperms, on the other hand, were medium-sized trees producing plenty of secondary growth from a vascular cambium. In fact, fossils of progymnosperm tree trunks were originally classified as gymnosperms based solely on this aspect of their anatomy. Eventually fossils were discovered of such trunks bearing reproductive structures. Quite unexpectedly, the reproductive structures present were not cones, as is typical of gymnosperms, but spores, as found in ferns. Thus, progymnosperm reproduction did not involve seeds, pollen, or flowers, but only spores.
3. Mosses and ferns both require a small amount of liquid water to complete their life cycles. This is because their sperm are flagellate and, to effect fertilization, they must physically swim from the antheridia where they were produced to the archegonia containing the waiting eggs. Without the liquid water, fertilization cannot take place.
4. Plants and charophytes share several important features: (a) both use chlorophyll a as their primary photosynthetic pigment, and chlorophyll b and carotenoids as accessory pigments; (b) both groups use starch as their storage carbohydrate; and (c) both form a phragmoplast during cell division. Evolving land plants met with several challenges that needed to be solved before they could fully diversify in the terrestrial environment. One problem was desiccation, which aquatic algae seldom face. A feature that developed very early to help prevent desiccation was the formation of a sterile layer of cells around the sperm-forming (antheridia), egg-forming (archegonia), and spore-forming (sporangia) tissues. A related problem was maintenance of a relatively constant internal environment, which many plants solved by the development of vascular tissue and cuticle. Body support was yet another issue. Aquatic organisms are supported and buoyed by the water itself, whereas on land gravity is relentless. Adaptations to overcome this problem include the development of lignified vascular tissue, as in lycophytes and ferns, and the development of secondary growth via a vascular cambium, as in gymnosperms and angiosperms.

Chapter 22

Review Questions

1. As a group, gymnosperms are characterized as bearing naked seeds, meaning that gymnosperm seeds are not borne in carpels and thus, at maturity, do not occur in fruits. As in the heterosporous ferns and lycophytes, microspores (pollen) are produced in microsporangia on specialized leaves called microsporophylls. The microsporophylls are then arranged into cone-like structures called male cones, or microsporangiate strobili. Ovules, on the other hand, are borne on the surfaces of ovuliferous scales, arranged to form ovulate cones or megasporangiate strobili. Pollination is effected by wind, and liquid water is not required to transport the sperm to the eggs. After fertilization, the naked seeds mature within the ovulate cones. Wood is well developed within the gymnosperms, and arises from a vascular cambium.
2. Bryophytes are non-vascular plants that do not have true xylem and phloem and therefore do not have true roots, stems, and leaves. Gymnosperms and pteridophytes are vascular plants with xylem and phloem. Both bryophytes and pteridophytes require water for the sperm to swim to the egg for fertilization. Gymnosperms do not require water for fertilization; the sperm is in the pollen grain and is transported by the wind (in most cases) to the egg. In bryophytes, the gametophyte is the dominant and most conspicuous form; in pteridophytes and gymnosperms, the sporophyte form is dominant. The fern gametophyte is often free-living although it is small. In gymnosperms, the gametophyte remains attached to the sporophyte. Gymnosperms produce seeds but neither bryophytes nor pteridophytes do.
3. Gymnosperms include four modern groups: conifers, cycads, the ginkgo, and the Gnetales. Cycads somewhat resemble palm trees but unlike palms, their stem surfaces are rough and covered with persistent leaf bases and scaly leaves. Cycads have large pinnate leaves and very large seed cones. Cycads can grow outside only in tropical and subtropical areas. People have used some cycads as food, but many cycads contain toxic compounds. Gingko is a deciduous tree with fan-shaped leaves; each tree produces pollen or ovules and seeds but not both. Fertilized ovules develop a fleshy, bad-smelling outer seed coat that looks something like a fruit but is not a true fruit. Gingko is sometimes planted as a landscape or street tree because it is beautiful and tolerates air pollution better than some other trees. Three genera of unusual plants are classified as gnetophytes. In contrast to other gymnosperms, gnetophytes contain xylem vessels in their vascular tissue. *Gnetum* plants are vines or shrubs with broad leaves and grow in tropical Asia and Africa. *Ephedra* grows in arid regions of the western United States. Its leaves are tiny brown scales and its stem carries on photosynthesis. *Welwitschia* grows in the coastal deserts of southern Africa. It has a long taproot, a short stem and two very long leaves. Conifers are the most familiar gymnosperms and include many familiar trees such as pines, fir, spruce, redwoods, and cypress. While most conifers retain their needles for several seasons there are some deciduous conifers that lose their needles before winter. Conifers produce distinct and complex ovulate cones and separate cones that produce pollen.
4. One advantage of seeds over spores is that seeds can remain dormant in the soil for many years, germinating only when conditions become favorable. The viable lifetime of spores, by contrast, is much shorter. Seeds are multicellular and larger than spores, and are thus able to store many more nutrients for the seedling. Such seed resources are available to the seedling until it is established and able to provide for itself. Also, seeds are diploid whereas spores are haploid. Diploidy confers genetic variability as well as redundancy because two copies of each gene are present. Lastly, in seed plants, sperm is transferred during pollination, a process that does not require liquid water. By contrast, many organisms reproducing by spores require that the sperm swim to the egg.
5. The evolution of seed plants occurred in two stages. The first stage was heterospory, the production of two types of spores—one producing male gametophytes (pollen) and the other, female gametophytes. Both types of gametophytes developed within spore walls, which protected them from adverse environmental conditions. The second stage in seed evolution was the development of ovules, wherein the female gametophytes developed from spores retained by the parent sporophyte. The female spore-producing sporangium (i.e., the megasporangium) was enveloped by leaf-like structures (the integuments), and the resulting female spores germinated and developed in situ, leading to an ovule with an embryo sac.
6. Gnetophytes and angiosperms share several features that are not present in other gymnosperms. Vascular tissues of other gymnosperms contain only tracheids but those of gnetophytes and most angiosperms contain xylem vessels as well. Some gnetophytes have a process of double fertilization where two sperm from one pollen tube fuse with different cells of the female gametophyte. Double fertilization also occurs in angiosperms but the process differs somewhat. Some gnetophytes, like angiosperms, do not have archegonia (female gametangia). Some experts have proposed that gnetophytes were ancestral to angiosperms but others think that the fossil record indicates that flowering plants were derived from a different group of seed plants that are now extinct.
7. Conifers are a major source of wood for building materials and pulp for paper production. They are also used as Christmas trees and as landscape plants offering shade and protection from wind. Pine nuts are used in cooking and a cancer drug, taxol, has been isolated from the conifer called yew. Some cycads have also been used as food in different parts of the world. However, cycads contain a toxic amino acid that has been associated with neurological disease. Some cycads are sold as houseplants. Gingko trees are beau-

tiful and well adapted to city life because they are not as sensitive to air pollution as some other trees. However, female (ovulate) gingko trees develop a fleshy outer seed coat with an unpleasant smell. Some people eat gingko seeds and gingko is sometimes used as a traditional medicine to improve circulation. Native Americans and early settlers to the western United States used *Ephedra*, a gnetophyte, for a medicinal tea. *Ephedra* has been promoted recently as an aid to weight loss and improved sports performance. But there have been a number of severe reactions to the drug ephedrine so that it is no longer permitted as a food supplement.

Applying Concepts

1. Gymnosperms have not evolved as many close relationships with animals as flowering plants have but there are some animals that aid in pollination and seed dispersal. Cycad cones give off an odor that attracts some beetles and these insects carry pollen from the male cones to female cones. Some gymnosperms, such as yew and juniper, produce fruitlike structures that may entice animals to help disperse seeds. Squirrels do not appear to like the smell of the fleshy seed coat of the gingko but they love the seeds. Once the rain has washed away some of the flesh, squirrels gather and hide the seeds. Some birds (crossbills) have evolved beaks to attack cones and extract seeds, and they probably help disperse some conifer seeds. Other gymnosperms must depend on the wind for pollen and seed dispersal.
2. All of these apparently negative traits are strategies to survive adverse environmental conditions. Conifers that shed their needles in the fall are reducing their surface area for water loss and freezing during cold and windy winters. Before the needles drop, the plant withdraws some important nutrients into its main body to store for growth when spring comes again. Coralline roots of cycads grow up out of the ground, and this allows their associated cyanobacteria to have the light needed for photosynthesis. These cyanobacteria can also fix nitrogen, and this helps the cycad to grow in poor soils. Dehydration occurs during seed development so that the seed can remain in a quiescent stage and not start growing until environmental conditions are right. During seed development the megaspore mother cell undergoes meiosis producing four spore products. Three of these products die, ensuring that the remaining megaspore retains all the nutrients to nourish the embryo.
3. Both gymnosperms and angiosperms produce pollen; in the former, it is produced in microsporangia located in the male cones, whereas in the latter it is produced in microsporangia (anther sacs) found in the flower. In gymnosperms, the ovules are borne nakedly on ovuliferous scales. Inside each ovule—within the megasporangium—a many-celled megagametophyte develops, eventually bearing several archegonia, each of which produces an egg. Usually, within the ovule, individual sperm fertilize the eggs, although typically only one zygote develops into an embryo. In the conifers, ovuliferous scales enlarge and become woody, the seeds maturing along with the scales. Together, scales and seeds form the familiar cones. Endosperm is not formed in gymnosperms. In flowering plants, on the other hand, the ovules are enclosed in a modified leaf called a carpel, which is part of a flower; archegonia are lacking, and the megagametophyte typically consists of only seven cells, one of which is an egg. Angiosperms use double fertilization, in which one sperm nucleus fertilizes the egg, while another separate sperm fuses with the polar nuclei to begin the endosperm. After double fertilization, seeds mature within the carpel, and carpels themselves become fruits.

Chapter 23

Review Questions

1. Simple fruits are fruits that develop from a single carpel or a small number of fused carpels, examples being tomatoes, cherries, and cucumbers. Complex fruits, on the other hand, develop from multiple pistils, either the distinct pistils of a single flower or the many pistils found in multiple flowers. Complex fruits include aggregate and multiple fruits. Aggregate fruits develop from single flowers containing multiple pistils; each pistil matures into a small fruitlet, and the fruitlets are bound together into a single "fruit" by expanded receptacular tissue. Strawberries and raspberries are familiar examples. In contrast, multiple fruits occur as many simple fruits (from individual flowers) grow together into a single fruiting structure; examples are pineapples and mulberries.
2. The most important environmental cues for germination are temperature and water availability. Some seeds require a cold period before they will germinate, which keeps them from germinating before the freezing temperatures of winter. Even if a cold period is not required, most seeds will not germinate if temperatures are below about 5°C. Other seeds need sufficient water to leach out germination-inhibiting compounds; this strategy works well in climates having a wet and a dry season, because the seeds will not germinate until the wet season has progressed sufficiently. Even if chemical inhibitors are not present, most seeds will not germinate until they have absorbed some minimum amount of water. Finally, some seeds use specialized strategies to help ensure germination under favorable conditions. For example, many plants growing in dry, fire-prone areas produce fruits that require fire to release their seeds, the expectation being that such fire-released seeds will produce the next generation of plants before fire sweeps through the area again. Another specialized strategy is a requirement that seeds be nicked before they will germinate. This nicking, or scarification, might occur when seeds pass through the gizzard of a bird, and the resulting germlings might find themselves growing in the middle of a nitrogen-rich spot of bird guano.
3. The word angiosperm comes from the two Greek words *angeion* (vessel) and *sperma* (seed). Recall that in flowering plants the ovules, after fertilization, develop into seeds whereas the carpels mature into the fruit. In effect, the "vessels" in angiosperms refer to carpels and their mature form, fruits. (a) Gymnosperms lack carpels and bear naked ovules on modified leaves (megasporophylls). Such unprotected ovules mature into naked seeds, that is, seeds without the protective tissue covering known as a fruit. Usually the seeds of gymnosperms are borne in cone-like structures; but in any case, the seeds are not covered in tissue derived from an ovary. (b) Like gymnosperms, pteridophytes do not produce carpels and fruits. Because ferns reproduce by spores, they also lack ovules and seeds. (c) Bryophytes produce neither carpels and fruits nor ovules and seeds, instead reproducing via spores as do the ferns. Also, bryophytes differ from vascular plants like angiosperms, gymnosperms, and ferns in that they lack true xylem and phloem; they also differ in the nature of their alternation of generations. In particular, bryophytes produce a conspicuous and independent gametophyte, while in vascular plants the sporophyte is the larger, dominant generation.
4. As the name suggests, monocots possess only one cotyledon in their seeds. Typically, their leaves are long and narrow with parallel venation, their flower parts are usually in threes or multiples of threes, and their pollen grains have only one pore or furrow. Dicots, by contrast, have two cotyledons in their seeds, they typically have broader leaves with netted venation, their flower parts are generally in fours or fives or multiples of four or five, and the surfaces of their pollen grains are usually marked with three pores or furrows.
5. (a) A peduncle is the stalk of an inflorescence, the inflorescence being a cluster of flowers. If more than one flower is borne on a single peduncle, then each flower will potentially have its own individual stalk, termed a pedicel. (b) In a complete flower, sepals form the outermost whorl of floral parts, and they are usually green and leaflike. Petals form the next whorl of parts, and they are often brightly colored. In some primitive angiosperms, such as magnolias, the sepals and petals are similarly large and colored; in such a case, the sepals and petals are referred to as tepals. (c) Calyx is a collective term for the entire whorl of sepals, while corolla refers to the entire whorl of petals. (d) Stamens are the whorl of floral parts that produce pollen and hence are referred to as the male reproductive parts. Carpels lie in the innermost whorl of floral parts and are the female reproductive organs. (e) Filaments and anthers are the two parts of a stamen; the filament is the stalk of the stamen, while the anther is the expanded portion containing the four pollen sacs. (f) Ovary, style, and stigma refer to the three typical parts of a carpel. The ovary is the expanded basal portion containing the ovules, the stigma is the expanded terminal portion that forms a receptive surface for pollen, and the style is the column of tissue connecting the other two.
6. (a) A complete flower possesses all four floral whorls—sepals, petals, stamens, and carpels—while an incomplete flower lacks one or more of these parts. (b) Perfect flowers possess both stamens and carpels and are therefore bisexual. Imperfect flowers are unisexual, lacking either stamens or carpels. (c) If an imperfect flower lacks carpels, it is termed staminate, and if it lacks stamens, it is called carpellate. (d) Plants that bear both staminate and carpellate flowers are monoecious, whereas plants bearing either staminate or carpellate flowers, but not both, are dioecious. (e) Ovaries are termed superior if the sepals, petals, and stamens all attach to the receptacle below the ovaries themselves. By contrast, inferior ovaries have the sepals, petals, and stamens attached above them. (f) Regular flowers exhibit radial symmetry while irregular flowers exhibit bilateral symmetry.
7. (a) Heterosporous plants—angiosperms, gymnosperms, and some ferns—all produce two types of haploid spores: megaspores and microspores. Megaspores germinate and differentiate into megagametophytes; microspores do the same and form microgametophytes. (b) Megasporangia and microsporangia are the sacs that produce megaspores and microspores, respectively. The megasporangium of an ovule is the nucellus, and inside this tissue a single megasporocyte will divide meiotically and produce four megaspores, one of which will form the embryo sac. The microsporangia are the four pollen sacs of a stamen, inside of which microsporocytes undergo meiosis and produce microspores, which eventually mature into pollen grains. (c) The megagametophyte is a small plant, typically consisting of only a few cells (e.g., 7 cells in lilies), which will eventually produce the egg. Microgametophytes are pollen grains, and they usually consist of two cells, a tube cell and a generative cell. (d) After germinating on a stigma, the generative cell divides mitotically into two sperm, and these three cells represent the mature microgametophyte. (e) A typical mature megagametophyte, as exemplified by lilies, consists of 7 cells and 8 nuclei. The entire structure is also known as an embryo sac. In the embryo sac, a large central cell contains two nuclei (the polar nuclei). Lying in the cytoplasm at the micropylar end of the central cell are three small cells, two synergids flanking a single egg. At the opposite end of the central cell lie the three antipodal cells. (f) Double fertilization refers to the process, unique to angiosperms, whereby the two sperm cells in the pollen tube each fuse with nuclei of the embryo sac: one sperm nucleus fuses with the egg nucleus to form a diploid zygote, while the other fuses with the two polar nuclei to form a triploid tissue known as endosperm.
8. Strictly speaking, a fruit is a mature ovary, usually containing seeds. If other floral parts are present besides ripened ovary, the fruit is referred to as an accessory fruit.
9. In dehiscent fruits, the ovary walls break open to release the seeds; in indehiscent fruits, the ovary wall does not break open and the seeds remain inside the fruit after it is shed from the plant.

Applying Concepts

1. Typical fruits that you might find:

   Berries: tomatoes, bananas, peppers, grapes, cranberries, dates, avocados, cucumbers, melons, watermelons, pumpkins, squash (The last five are actually special berries called pepos.)

   Pomes: apples, pears, crabapples

   Hesperidiums: lemons, oranges, limes, grapefruits

   Drupes: peaches, apricots, cherries, plums, olives, coconuts, walnuts, butternuts, pecans, cashews (In cashews, all but the seeds have been removed, because the fruit causes a poison-ivy-like dermatitis in most people.)

   Achenes: sunflower seeds

Legumes: peanuts, soybeans, green beans, peas, kidney beans, pinto beans, lentils, garbanzo beans

Nuts: hazelnuts, chestnuts, acorns

Grains: corn, rice, oats and wheat (These are all grains, although in many cases all but the endosperm has been milled away.)

Multiple fruits: pineapples, figs, mulberries

Aggregate fruits: strawberries, blackberries, raspberries

2. The described situation is one means to discourage self-pollination. Because any given flower is not producing pollen when its embryo sacs are mature, the pollen must come from other, nearby individuals.
3. Recall that flowers consist of up to four whorls of floral parts (sepals, petals, stamens, and carpels). Complete flowers possess all four of these whorls; perfect flowers possess, at minimum, both stamens and carpels. Thus, all imperfect flowers—such as staminate flowers—are by definition also incomplete because they lack at least one of the four floral parts. Some incomplete flowers, however, lack only sepals or petals. Such a flower would be incomplete (lacks sepals and petals) but perfect (has stamens and carpels), so the answer to the second question is no.
4. No. Tomatoes are the ripened ovaries of the tomato plant, and thus are true fruits (berries, actually). In common parlance, "fruit" often refers to the more or less sweet products on the produce shelves, such as apples, cherries, oranges, and bananas. Tomatoes are not typically sweet (high in sugar content), so many people think of them as vegetables. Botanically, however, there is no ambiguity—tomatoes are fruits.
5. Recall that monoecious or dioecious refers to plants that produce unisexual flowers. In such cases, the staminate and carpellate flowers are either on the same plant (monoecious) or on separate individuals (dioecious). But there is a third, very common possibility—bisexual flowers. Thus, it would be true to say that all flowering plants either produce bisexual flowers, or are monoecious or dioecious.
6. Of course, many answers are possible for this. The key given here strictly follows the descriptions given in text.

A. Simple fleshy fruits
  B. Flesh developing from ovary only, not containing accessory tissue
    C. Flesh soft throughout, not forming a pit . . . . **berry**
    CC. Inner layer of fruit forming a hard, stony pit . . . . **drupe**
  BB. Flesh arising, at least in part, from accessory tissue . . . . **pome**
AA. Simple dry fruits
  B. Fruit dehiscent
    C. Fruit arising from a single carpel
      D. Fruit splitting down one side to release the seeds . . . . **follicle**
      DD. Fruit splitting down two sides to release the seeds . . . . **legume**
    CC. Fruit arising from two or more carpels, splitting in various ways . . . . **capsule**
  BB. Fruit indehiscent
    C. Fruit arising from a single carpel; fruit wall not stony
      D. Seed attached to fruit wall at only one point
        E. Fruit without wings . . . **achene**
        EE. Fruit a winged achene . . . . **samara**
      DD. Entire seed coat fused to fruit wall . . . . **grain**
    CC. Fruit arising from multiple carpels; fruit wall stony . . . . **nut**

Chapter 24

Review Questions

1. Inbreeding is the combination of two genetically very similar genomes (e.g., self-fertilization in bisexual flowers, or matings between siblings). Outbreeding, on the other hand, is the combination of two genomes that are more diverse—still the same species, but many more allelic differences. The disadvantage of inbreeding is that it encourages the appearance of recessive traits, many of which are deleterious. Plants employ several strategies to avoid inbreeding. For instance, in some species, enzymes in the stigma prevent germination of genetically identical pollen; in other species, enzymes in the style destroy pollen tubes from genetically identical pollen. Another strategy is to ensure that, on a given individual, stamens do not produce pollen when stigmas are receptive. The pollen on such a plant can find a receptive stigma only on another individual, and similarly, the stigmas are exposed to pollen only from other individuals. Yet a third strategy is to produce unisexual flowers, with male and female flowers maturing at different times on monoecious (plants having both male and female parts) individuals, or adopting a dioecious pattern of flowering—staminate flowers on one plant, carpellate on another.
2. Some plants produce both open and closed flowers, the latter actually forcing self-fertilization. Such a pattern of floral production is advantageous in situations where cross-pollinators are not available, because these plants can self-pollinate and thus still set seed.
3. Plants provide several types of food rewards to animal pollinators. One type of reward is nectar, a solution of sugar, amino acids, and other compounds. Nectar is produced in nectaries, which are variously located in the flowers. Plants often tailor the nutritional content of nectar to the needs of their animal pollinators. Some plants provide as food reward only pollen, which is rich in protein and oils; others provide only nectar. Last, some plants use trickery to entice animal pollinators to visit their flowers and effect pollination, the animals discovering afterward that there is no food reward at all.
4. Plants use a variety of attractants to bring pollinators to them. Odor and color are the most important attractants, although flower shape and arrangement are also significant. Plants are known to use over 700 fragrances as flower attractants, the odorants falling into three major categories: (a) compounds having a floral bouquet type of scent, (b) compounds mimicking animal pheromones, and (c) fetid compounds smelling of dung or rotting flesh. Color is also important in attracting certain pollinators. Anthocyanins, anthoxanthins, betalains, and carotenoids are often used to impart bright colors to floral parts or floral bracts. In some cases insects are given runways (nectar guides) to guide them past the reproductive parts to the nectaries. Not coincidentally, many floral pigments are highly reflective (i.e., brightly colored) in the ultraviolet, which some insects can perceive. Hovering animals are attracted to flowers that hang upside down, and only they can pollinate them. Other animals are attracted to flowers having a specific shape, for example, flowers producing specific landing platforms, or even flowers that attract males of a species because they look (and smell) like females!
5. Constancy is the ability of a pollinator to recognize and repeatedly visit the flowers of a favored plant species. The advantage of this to the plant is that pollinators spend more time visiting flowers of that particular species, thus increasing the likelihood that compatible pollen will be transferred for effective pollination. Furthermore, plants are better able to tailor an appropriate reward to a constant pollinator, thus increasing the fitness of the animal species and simultaneously benefiting the plant as well. Constancy also offers an immediate benefit to the animals—by seeking rewards in familiar flowers, they minimize the amount of time spent foraging in unfamiliar ones. Because feeding is usually a very vulnerable time for an animal, by spending less time foraging, a pollinator will be less likely to be preyed upon.
6. Beetle pollination is particularly important in dry and tropical habitats, and to gymnosperms and early lineages of flowering plants. However, because beetles are covered by a hard, smooth exoskeleton, they are somewhat ineffective as pollinators. Beetle-pollinated flowers tend to be large and often flat, and they tend to offer excess pollen as a nutritional reward. Such flowers tend to be white or shades of purple, and they emit a fruity odor. Many beetle-pollinated flowers are thermogenic, which helps volatilize the floral attractants as well as provide warmth to their insect pollinators.
7. Bees make excellent pollinators. They are strong insects that can easily learn to recognize their favorite flowers; once these flowers are learned, bees exhibit constancy. Furthermore, social bees can communicate the whereabouts of desirable flowers to other workers in the hive, so that many pollinators can quickly become available to a patch of flowering plants. Bees feed on nectar and have a medium-length tongue to collect it. They are excellent conveyors of pollen, carrying it on various parts of their body such as the legs or abdomen. Bees also have good color vision and olfaction (smell); the former extends into the ultraviolet (but is not sensitive to red wavelengths), and the latter is some 10–100 times more sensitive than the human nose at detecting floral scents. Bee-pollinated flowers tend to be various shades of blue, purple, or white, often with nectar guides. Such flowers offer both nectar and pollen as food rewards.
8. Nectar-feeding flies are attracted to rather simple flowers that are often red or light in color, have little odor, and provide easy access. Such flowers provide nectar as a reward; flies recognize the sugar with sensors in their feet, and then draw up the liquid with their sucking mouthparts. Flies transfer pollen as they are foraging for nectar. Carrion flies are attracted to fetid flowers often smelling of rotten flesh, and are provided no food reward by the flower.
9. Butterflies have excellent memory and discrimination regarding flowers, and they exhibit constancy. Their vision ranges from red to ultraviolet; they use odorant sensors in their feet and antennae to smell floral fragrances. Butterfly flowers tend to be colored in shades of blue, pink, orange, or red, and they are quite fragrant. Night-flying moths are attracted to night-blooming, light-colored, fragrant flowers, and they use odor to localize the flowers. Butterflies and most moths require a landing platform, and use their long proboscis (tubular mouthpart) like a straw to suck nectar up from long, tubular flowers. As they feed, they transfer pollen that has haphazardly dusted their heads.
10. Pollinating birds have good color vision and can see red wavelengths, but their sense of smell is rather weak. Some birds require a perch, although the best-known bird pollinators—hummingbirds—do not. Bird-pollinated flowers usually lack scent, are typically brightly colored—often in shades of red—and are open during the daytime, when these birds are active. Usually such flowers offer copious amounts of dilute nectar, and they often have tubular corollas; reaching the nectaries in such flowers requires the long bills of birds.
11. Bats are active at night, are color-blind, have good olfaction (smell), and navigate via echolocation. They have excellent memory and can easily relocate their favorite plants. Bats cannot fly through foliage, so bat-pollinated flowers usually bloom in open areas—on tree trunks, or hanging below the foliage. Such flowers tend to be pale or drab in color, strongly fragrant, and open at night. The flowers tend to have wide bell shapes for easy access and are hefty enough to endure the activities of bats; many also have bat-landing platforms. Bats are especially important as pollinators in tropical areas.

Applying Concepts

1. The fig benefits by being able to flower and set seed in a closed vessel, sealed so that potentially harmful visitors cannot enter. The fig also benefits by having absolute fidelity from its pollinator. In fact, if the pollinating wasp does not carry fig pollen with it, there will be no wasp offspring. This provides powerful selective pressure on the wasp to ensure that fig flowers are successfully pollinated. The loss of a portion of its ovules is the price the fig pays for this intimate pollination service. The wasp benefits because, once in the syconium, there is no competition from other species for resources; the syconium "belongs" to the wasp. Obviously the wasp and fig are co-adapted, so that the fig provides everything the wasp needs to develop and mature, including protection from most predators. The extreme dependence each species has on the other results in an obvious danger—neither species can survive without the other, and if one is extirpated from an area, the other will die as well.
2. In *Lantana,* only the first day, yellow flowers offer nectar and pollen, and are receptive. The plants are communicating this to the pollinating butterflies by

changing the color of their flowers. It is a signal by which butterflies learn to focus on unpollinated, nectar-offering yellow flowers, and especially to ignore the third-day (and presumably pollinated) purple flowers, which offer no nectar reward. Furthermore, by maintaining the flowers several days past pollination, the entire inflorescence (containing yellow, orange, and purple flowers) is more conspicuous and attractive to butterfly pollinators at a distance, and who then discriminate between the flower types at close range.

3. Typically, bat-pollinated flowers open at night, are strongly fragrant, and are rather drab or pale in color. The flowers tend to be produced in open, accessible places instead of occurring among a lot of foliage, and they are quite robust. These characteristics allow bats to find the flowers at night and to pollinate them. Bee-pollinated flowers, on the other hand, open during the day (when bees are active), may be fragrant, and are usually some shade of blue, purple, or white; reds tend to be lacking as the dominant color, because bee eyes are not sensitive to these wavelengths.

4. As discussed in question 1, many figs absolutely require the pollination activities of the fig wasp in order to produce fruit. Such is the case with the Smyrna fig. Evidently, the wild fig stock planted on the Gates farm accidentally contained some fig wasps, and their progeny soon began pollinating the other fig trees in the orchard. This marked the beginning of the Smyrna fig industry in California. This story illustrates why knowledge of plant and animal life cycles is sometimes necessary to successfully cultivate plants outside their native range.

Chapter 25

Review Questions

1. (a) Ecology, literally "study of the house," is the study of organisms and their interactions with each other and with the chemical and physical parts of their environment. (b) A population consists of all the individuals of a given species living in a particular area, such as all the cattails in a certain marsh. (c) A community consists of all the organisms of many different species that live and interact in a particular area. (d) An ecosystem is a community and its physical environment—the living and nonliving components of a given area. (e) Biomes are ecosystems with similar climates. Species composition, however, may vary considerably. (f) The biosphere is a collective term for all the biomes on Earth; as such, it encompasses all the organisms on the planet.

2. Like individuals, populations show growth and decline, and have associated birth rates and death rates. The birth rate of a population is the number of births or germinations in that population per unit time, while the death rate is the number of deaths in that population per unit time. The net growth rate of a population is the difference between the birth and death rates, so that the growth rate will be positive in populations whose numbers are expanding (births exceed deaths) and negative in those whose numbers are shrinking (deaths exceed births). As a population expands into a new area, its growth curve usually follows the typical S-shaped curve. Early in this expansion, although reproduction proceeds unabated, the overall population increases slowly because there are so few individuals. As the number of individuals increases, the population enters the phase of growth where the number of individuals increases at some maximum rate. Eventually, one or more nutrients become limiting, and the growth rate slows down and approaches zero. The population has reached its maximum size under these conditions, which is known as the carrying capacity. If the system is closed and no new nutrients are added, then individuals will die and the population will decline. If fresh nutrients are added to the system (and old organisms and wastes removed) then the population can maintain itself, perhaps indefinitely, at some level of carrying capacity.

3. The ecological niche describes what a population does and how it fits into its community. The niche includes all the abiotic factors that delimit the area in which the population occurs, as well as interactions (such as symbioses) between populations of different species. In terrestrial systems, the two most important factors are moisture and temperature. Grasslands give way to

deserts when moisture falls below 20 cm (8 in.) per year. The temperature range for most plants lies between 0° and 40°C, but during winter dormancy fir and spruce trees of the taiga can survive winter temperatures below −40°C. The giant saguaro cactus can survive temperatures of 64°C but dies if exposed to freezing temperatures for more than 36 hours. Other factors that may assume secondary importance include the occurrence and frequency of fire, acidity or alkalinity of the soil, presence of nutrients, drainage characteristics, and length of the growing season.

4. A symbiosis is any close association between two or more species. The interactions may benefit or harm one or both species. Types of symbioses: (a) mutualism, where two populations exchange benefits (i.e., both of the interacting species receive benefits). An example from the tropics is the *Acacia* plants that harbor aggressive ants; the plant gains an army of stinging ants to protect it from herbivores as well as encroaching plants, while the ants gain a place to live and, in many cases, nutrients from the *Acacia*. (b) In commensalisms, one species benefits and does not harm the other. A common example is the crustose lichens growing on tree trunks; the lichens gain a place to grow in the sun, but the tree derives neither harm nor benefit from the interactions. (c) In parasitism, one member (the parasite) derives benefit as it grows in or on another organism (the host), which suffers harm. Typically the parasite feeds on the host, but does not cause its death in the short term. Dodder (*Cuscuta*) is an example of a plant parasite. (d) In herbivory, animals eat plant parts without killing the plant, as when cattle graze in a pasture. (e) In predation, one population (the predator) benefits while the other is killed (the prey). A familiar example is a hawk capturing and eating a squirrel.

5. Plants use a variety of chemical and physical defenses to prevent herbivores from taking full advantage of them. Some plant chemical defenses are quantitative; these are not toxic in small doses, but as an animal continues to eat the plant, the amount of plant material increases in the animal's gut. Eventually the dosage levels become high enough to prevent digestion and force the animal to stop eating. Examples of quantitative chemical defenses include tannins and resins. Qualitative chemical defenses, on the other hand, are toxic in small doses; examples include alkaloids, terpenes, and quinones. Physical defenses used by plants against herbivores include thorns, spines, and sharp leaf edges, as well as stinging hairs that break off and inject toxin. Stone plants grow with the shape and color of small rocks to avoid herbivores by resembling an inedible object. Lastly, recall that some *Acacia* plants have a symbiotic relationship with stinging ants that aggressively defend the plants against various types of attack, including those by herbivores.

6. Ecological succession is the change in community composition over time. Communities are dynamic entities, and their constituent species change over time. Usually this change is set into motion by some sort of physical disturbance, such as a flood, storm, fire, or earthquake. In a forest, succession may occur in the resulting light gap when a mature tree is felled. In any of these situations, new species may move into the disturbed area, thereby changing the species composition. Ecologists studying succession recognize two general types—primary and secondary. Primary succession begins in bare areas not previously occupied by organisms, such as on exposed granite outcroppings, the cooled lava of volcanic flows, or on land recently exposed as a glacier retreats. Secondary succession, on the other hand, occurs in an area where a community has been partially or completely removed. Examples include abandoned farmland, drained marshes, clear-cut forests, roadcuts, and areas burned by fire or cleared by bulldozer. In all these cases, succession begins as new species of plants—often pioneer species—begin colonizing the disturbed area. Pioneer species are the first to colonize bare areas, as when lichens grow on pure granite. Such species pave the way for the succession of organisms to follow.

7. The concept of a climax community still exists, although it is interpreted differently now. Originally, a climax community was thought to be the final stage of ecological succession—a stable, mature community

whose species composition was no longer in flux. More recent studies, however, have shown that climax communities do change in species composition, albeit so slowly that the changes are hardly noticed. Thus current views hold that climax communities do exhibit succession, but at a much slower rate than earlier successional stages.

8. Autotrophs are able to synthesize all the nutritive compounds they require from inorganic materials in their environment, whereas heterotrophs must obtain their organic compounds from their environment.

In many cases heterotrophs obtain their organic compounds directly from other organisms in their environment, by eating them. (a) Primary producers are autotrophs, organisms capable of using energy and inorganic molecules in their environment to synthesize all their required organic compounds. Phototrophs accomplish this using the energy of the Sun, while chemotrophs utilize the energy from inorganic compounds and geothermal sources. (b) Primary consumers are the herbivores that directly consume primary producers. Rabbits eating lettuce, and zooplankton consuming photosynthetic algae are two examples. (c) Secondary consumers eat the primary consumers, and further up the chain, tertiary consumers eat secondary consumers. (d) Detritivores are scavengers that consume detritus, the dead bodies of organisms or their remains. Two examples are earthworms, which consume detritus in the soil, and crayfish, which scavenge for detritus on the bottom of ponds. (e) Decomposers break down organic matter for a source of energy. Bacteria and fungi are two familiar examples.

9. The Earth is a sphere, whose axis of rotation is tilted 23.5° with respect to its orbital plane. Because of the Earth's shape and tilt, the equatorial areas receive far more solar energy than the polar areas, setting up uneven heat distribution that generates air cells and winds (the tilt is also responsible for seasonality). Hadley air cells rise at the equator, and as this warm, moist air ascends, rain clouds form and precipitation falls, nurturing tropical rain forests around the world. The warm air mass, now wrung of its moisture, flows poleward at altitude, and then descends at 30°N and S latitudes. Heating as it descends, the air mass draws moisture out of the land below it, favoring the formation of deserts in these areas. At the surface the air mass splits, part returning to the equator and the remainder initiating a new air cell poleward of 30°N and S. This new cell is a Ferrel cell. Ferrel cells circulate between 30° and 60°N and S latitudes, with air flowing over the surface of the Earth to 60° latitude, and then rising again. Rising air masses at 60° latitude bring moisture (and some equatorial warmth) to the forests and grasslands of the temperate zone. High in the atmosphere above 60°N and S latitudes, the air mass splits again, a part returning to 30° to complete the Ferrel cell, and the remainder continuing poleward, initiating the last cell in the series, the polar cells. Polar cells operate from 60°N or S to the respective pole. Cold air flows downward at the poles, where it picks up some moisture and then blows as surface wind back toward 60° latitude. There the air rises again, giving up its moisture to the land below it. These air cells set up the basic airflow patterns and moisture regimes for the Earth's landmasses, thus laying the foundation for our planetary biomes.

10. The air circulation patterns lay the foundation for the world's biomes by establishing the basic temperature and moisture regimes found on the landmasses. However, biomes do not strictly follow these patterns because the basic regimes are modified by three additional factors—continentality, ocean currents, and mountain ranges. Continentality refers to the situation where some landmasses are so large that their interior regions are located too far from any oceans to receive moisture-laden winds. Such inland areas are drier than would be inferred from their latitude alone. Just as air cells equilibrate heat between the warm equatorial regions and the frigid poles, so ocean currents do the same within the ocean. Rotating in vast circles, the major ocean currents turn clockwise in the Northern Hemisphere and counterclockwise in the Southern Hemisphere. These currents influence temperature and moisture levels of adjacent landmasses. The Gulf Stream, for example, carries warm water from the Caribbean northward, warming the coast of eastern

North America and Western Europe. In western South America along the coast of Chile, the Humboldt Current brings cold waters from Antarctica northward, and onshore breezes then cool the coastal areas. Because the cool air moving onshore holds significantly less moisture than it would have had the offshore waters been warmer, the Atacama Desert of Chile is considerably drier than it otherwise would be. This maritime influence thus modifies the climate in coastal areas. Mountain ranges also influence climate by wringing moisture out of the air masses flowing over them. As air rises up over a mountain range, it cools by 1°C per 100-m increase in elevation. If the mountain range is high enough, moisture condenses and falls as precipitation on the windward side. As the air continues over the mountain range, most of its moisture is removed, so that on the leeward side it descends as warm but very dry air. The areas on the leeward side of a mountain range thus lie in its rain shadow, and are considerably drier than they would have been had the mountain been absent.

11. The atmosphere, being 78% nitrogen gas, is the main reservoir for nitrogen on Earth. All organisms require nitrogen for proteins, nucleic acids, and other necessary compounds, although only a few microbial species can directly use nitrogen gas. These organisms are known as nitrogen fixers, and all other life on this planet ultimately depends on them for their nitrogen needs. These pivotal microorganisms largely control the flow of nitrogen from the atmosphere to the biosphere. The nitrogen cycle contains five main processes: (a) Nitrogen fixation, the conversion of nitrogen gas to its reduced form, ammonia, occurs via lightning discharges, industrial fixation for fertilizer production, or through biological fixation by various microorganisms. (b) Nitrification is the process by which various aerobic bacteria convert ammonia (NH_3) into nitrate (NO_3^-). (c) Assimilation is the process by which non-nitrogen-fixing organisms acquire nitrogen. Plants take up ammonia or nitrate from the soil; animals acquire nitrogen, directly or indirectly, from plants. (d) Ammonification is the process by which organic nitrogen is converted back into ammonia, which may either be taken up directly by plants or converted to nitrate. (e) Denitrification is the final process that closes the cycle. Anaerobic bacteria convert nitrate back to nitrogen gas, which is returned to the atmosphere.

12. The main reservoir for water on Earth—the hydrosphere—is all the water found in glaciers, oceans, lakes, and rivers. Water from the hydrosphere passes into the atmosphere via evaporation (or sublimation, when ice changes to water vapor) and is then carried around the world by winds. Eventually the air mass cools and water condenses out of it, falling to the ground as precipitation. On the ground, water may return to the atmosphere via evaporation, transpiration through plants, or it may join the aboveground waterways or subterranean groundwater. Eventually, all water finds its way back to the ocean.

Applying Concepts

1. (a) population; (b) ecosystem; (c) community; (d) population; (e) ecosystem; (f) biome; (g) community
2. To maintain a constant internal temperature, endotherms need to have a higher metabolic rate than ectotherms; thus endotherms need to consume more food than ectotherms. It follows, then, that a given population of prey would be able to support a larger biomass of ectothermic predators than endothermic ones because ectothermic predators burn their food at considerably higher rates than endothermic ones. Indeed, the biomass of warm-blooded predators is typically 1–3% of their prey biomass, while that of cold-blooded predators can reach 10% of their prey biomass. Following Bakker's logic, if dinosaurs were ectotherms one would expect to see considerably more predators in the dinosaur community; instead, dinosaur carnivore fossils are very rare compared to their prey.
3. crayfish, Dt; *Pediastrum*, PP; *Potomogeton*, PP; northern pike, STC; mallard duck, PC; clam, both PC and STC, since in their filter feeding they capture and eat both algae and various zooplankton; bass, STC; giant water bug, STC; water boatman, PC; and *Saprolegnia*, Dc. One possible diagram is:

STC: northern pike bass giant water bug clam

PC: water boatman clam mallard duck

PP: *Pediastrum Potomogeton*

Dt and Dc: crayfish (Dt) *Saprolegnia* (Dc)

4. If the birth rate and death rate are equal, then their difference—the growth rate—falls to zero and, in the absence of immigration, the population stops growing. When the birth rate is slightly lower than the death rate, the growth rate becomes negative, meaning that the population is shrinking.
5. Food chains are descriptions of who is eating whom. Food chains are a record of the linear flow of energy into a system, through a chain of organisms, and back out of the system. In reality, the flow of energy through a system is seldom a linear progression. Instead, the energy flows through multiple chains of organisms, all of which are interacting in various ways. The term food web more accurately describes this situation. Humans do indeed participate in a food web. Typically, we are either top-level consumers, as when we eat meat, or else primary consumers, as when we eat plant foods like lettuce or onions. Very rarely, humans are intermediate consumers, meaning that a larger predator—a shark or a crocodile, for example—kills a human.
6. The statement is based on the fact that in any food chain, only about 10% of the energy of any given level is passed on to the next trophic level. As herbivores, humans obtain only 10% of the energy available to the plants we eat; we lose the other 90% as heat and other wastes. However, as carnivores, we obtain only 10% of 10%, or 1% of the energy available to the original plants. This is because two trophic levels are involved—one from plant to herbivore, the other from herbivore to human. Therefore, it is clear that eating lower on the food chain—more plants—is indeed more eco-friendly, and that 10 times more humans could potentially be supported by consuming plant foods directly instead of feeding them to animals and then eating the animals.

Chapter 26
Review Questions

1. The two biomes of the Antarctic continent are the maritime Antarctic and the continental Antarctic, both of which are polar desert. The maritime Antarctic is found on the west coast of the Antarctic Peninsula and on some of the islands in the Southern Ocean. Summers in these regions find mean monthly air temperatures of only about 0°–2°C but the maritime influence moderates the climate so that summers are wetter and winters milder than their Arctic counterparts. The climate becomes much harsher in the continental Antarctic, however, due to the lack of the maritime influence. There, mean monthly summer temperatures barely reach above freezing (0°–1°C) with little annual precipitation and often howling, bitterly cold winds in the depths of winter. In a manner similar to the Arctic, in latitudes south of 66°33′ S, the Sun does not rise for a period in midwinter, and it does not set for a comparable period in midsummer. Because of the harsh conditions and the isolation of the continent, the vegetation of Antarctica is severely limited. In the continental Antarctic, only sparse populations of mosses, lichens, fungi, algae, protozoa (protistans), and bacteria are found, with special communities occurring in the rocks and lakes of the Dry Valleys. Vegetation is much more widespread and diverse in the maritime Antarctic, with lichens and mosses occurring frequently. In addition, Antarctica's only two flowering plants occur in this biome—the Antarctic hair grass (*Deschampsia antarctica*) and the Antarctic pearlwort (*Colobanthus quitensis*).
2. (a) Subtropical high-pressure zones—Hadley cells—predispose regions 30° N or 30° S of the equator to dry climates, due to the sinking masses of dry air. (b) Continentality may result in the formation of a desert, as in central Asia. These regions are simply too far inland to receive moisture from the oceans. (c) Ocean currents may affect the aridity of coastal deserts. Winds moving over cold ocean waters before blowing onto coastal lands pick up less moisture than if they had blown over warm waters. Thus, coastal areas will

be drier than if warm currents flowed off their shores. (d) Mountain ranges dry out air masses moving over them. As air moves up the windward side of a mountain, it cools and releases much of its moisture as precipitation. As it moves down the leeward side of the range, the air mass warms and picks up moisture from the ground. Thus an arid region forms in the so-called rain shadow of the mountain range.

3. The two main sources of water in desert lands are surface water and the deepwater table. Among plants using surface water, succulents store this water internally so that they can survive extended periods without precipitation. Trees making use of the deepwater table typically produce long, deep roots because the water table may be 9 to 60 m below the surface.
4. Desert annuals may either germinate following summer rains, when temperatures exceed 25°C or following winter rains, when temperatures are considerably cooler. Annuals spend most of their lives as dormant seeds and, once germinated, may complete their entire life cycle within 3 weeks. Such a short growing season requires that desert annuals be low growing. Indeed, some species are only 1–2 inches tall. Herbaceous perennials are similar to desert annuals in that they spend most of their lives dormant in the soil. When favorable conditions arrive, the plants shoot up from bulbs or taproots. These perennials grow quickly and store more food nutrients in their underground portions. The aboveground portions then wither and die, and the plant waits for the next rainy season. Deciduous perennials maintain significant amounts of aboveground plant tissues. When conditions turn dry and unfavorable, deciduous perennials drop their leaves and become dormant. When the rains return, aboveground buds quickly leaf out, and the plants can be actively growing and photosynthesizing almost overnight.
5. Desert succulents may have either thick leaves (leaf succulents) or thick stems (stem succulents). In either case, they generally have four adaptations to help them survive arid conditions. (a) Succulents are very good at conserving water. They generally have low surface-to-volume ratios and use CAM, both of which decrease water loss. Many of these plants employ CAM idling, which further conserves water. (b) Succulents are very good at storing water. Many contain large parenchyma cells that store water, and some succulents have pleated stems that can expand like an accordion as they swell with water. (c) Most succulents are quite tolerant of temperatures high enough to kill nonsucculents. At the high-end extreme, some cacti can withstand temperatures as high as 69°C (d) Lastly, succulents have very well developed, shallow root systems that can rapidly absorb water when it is present. Examples of leaf succulents include the agaves, the aloes, and the stone plants; stem succulents include the cacti—such as barrel cacti, cholla cacti, prickly pear cacti, and the saguaros.
6. Forbs are the non-grass herbs that inhabit the grasslands; they are mostly flowering plants, although some ferns and mosses are also included. In grasslands, the density of individual forb species is not particularly high, so that most flowering forbs, rather than depending on wind for pollination, rely on showy flowers to attract pollinators. Forbs often send roots deep into the soil in their search for nutrients and water, and some species form beneficial symbioses with nitrogen-fixing bacteria. When such plants die and decompose, they enrich the topsoil. Some forbs produce densely hairy leaves, which provide a certain amount of protection against heat and intense sunlight as well as either drying or freezing winds. Lastly, to deter grassland grazers and microbial pathogens, many forbs produce highly active secondary compounds.
7. (a) Grasslands occupy large areas of land that are able to support large populations of animals. An excellent example is the huge population of buffalo that once occupied the grasslands of the upper midwestern United States and adjacent Canada. (b) The soils in grasslands are often many feet thick and store large amounts of organic carbon. This helps makes some grassland soils exceedingly productive, so areas (c) that were once grassland are now among the most productive croplands on Earth. (d) Grasslands have had considerable impact on the course of animal and

human evolution, and have greatly affected the development of human cultures. Consider, for example, that the dominant species in grasslands worldwide are grasses, and that certain grasses (cereals) are a major part of the diet for nearly all cultures. One might even add a fifth element to this list, which is that grasslands provide habitat to a diverse number of plant species, many of which are endemic to the grassland biome. Thus grasslands help maintain the species richness and diversity on this planet.

8. Over the years, as the dead bodies of grasses and forbs accumulate and dry out during the hot, dry summers, they collectively form a highly combustible fuel for grassland fires. Whether started by natural or artificial causes, these fires may burn for many kilometers before being stopped. Soil studies have revealed that in presettlement days, a given hectare of North American prairie burned, on average, once every 5 to 30 years. Such fires killed off fire-sensitive plants—especially invading tree species—as well as reduced standing plant debris to ashes, thereby releasing nutrients to the soil for subsequent plant growth. Grazing is beneficial to low-growing plants because it helps prevent them from being shaded by taller species. Additionally, grazer excrement is an important fertilizer for the grasslands, providing nitrogen and other minerals to an already rich soil.
9. In any particular grassland, only three or four species of grass will dominate at any particular time, and those species will shift as the growing season progresses. Thus the cooler parts of the growing season will be dominated by different species than the warmer parts. During the cool season, the dominant species are typically grasses having C_3 photosynthesis; Eurasian examples of such cool-season grasses include barley (*Hordeum*), rye (*Secale*), and wheat (*Triticum*). The grasses that dominate the warmer parts of the growing season—the so-called warm-season grasses—typically show C_4 photosynthesis and are well adapted to hot, dry summers; big bluestem (*Andropogon gerardi*) is a North American example of such a plant, as is the crop plant corn (*Zea mays*).
10. Most prairie grasses are perennial monocots; their apical meristems lie underground, protected by soil from fire, grazers, drought, and winter cold. This is the reason these grasses are not permanently hurt by a prairie fire. Many growing points are underground stems—rhizomes—that can spread out and quickly colonize a large area. Rhizomes sport roots, so that the root system can become quite extensive, both vertically and horizontally; some grass roots extend as far as 4 m into the soil, which provides access to deep subsurface water and nutrients. Aboveground, grass leaves are narrow in shape, which reduces the surface area to volume ratio and thus helps conserve water. The leaves are vertically arranged and tightly packed, which collectively make a large area available for photosynthesis.
11. First, grass flowers are not showy; the density of grass plants per hectare is sufficiently high that wind pollination is quite effective for prairie grass plants. Thus, plant resources are saved by not investing in the production of showy petals or nectar. Inside the flower, grass ovaries bear two feathery stigmas to help catch wind-borne pollen. Mature grass fruits—grains—are dry, which helps them survive extended periods of drought, and grass seeds typically contain significant quantities of carbohydrates, proteins and lipids, which furnish nutrients to the young seedlings.
12. (a) Dry grasslands often border deserts and are in constant danger of being lost to desert via desertification if the plant cover is removed. Because over-grazing is the typical force behind this process, preservation of these areas may be as simple as excluding livestock from them. Another threat to dry grasslands is loss of native plant species due to invasion by foreign species. Examples are Eurasian spotted knapweeds (*Centaurea* spp.) and the European brome grass (*Bromus inermis*), plants that can quickly become noxious weeds in dry grasslands. (b) Because of their excellent soils and suitability for growing cereals, most of the world's mesic grasslands have been converted to croplands. Although remnants of mesic grasslands do exist, these areas must be actively maintained to preserve their high species diversity. If fires are suppressed in a mesic prairie remnant, for example, studies show that approximately 1% of the species are lost per year. Fires are a natural element in the mesic prairie and should be part of any maintenance program. (c) Wet grasslands support fairly luxuriant growths of grasses and forbs. Since they often border wetlands, they are effective in preventing nutrient-laden runoff from entering the waterways and polluting them. Wet grasslands are uncommon today, however, most of them having been either drained for cropland or invaded by trees and converted to forested land. The rate of decomposition is usually rather low in wet grasslands, due to the low oxygen levels, so that after a few productive years as cropland, they must then be continually supplemented with fertilizers in order to maintain their productivity. It is probably best to simply leave wet grasslands undisturbed.
13. The chaparral biomes lie within 30° to 40° N or S latitude on the western coasts of continents, and are variously known as chaparral in California, maquis around the Mediterranean Sea, matorral in Chile, fynbos in South Africa, and mallee scrub in Australia. In North America the chaparral extends from Oregon to Baja California, from sea level up to about 2,000 m elevation. Annual precipitation ranges from 20 to 100 cm, which mostly falls as rain during the winter months. Droughts, often extreme, are common. Winters are cool, with average temperatures from 0°C in the mountains to 10°C on the coast; summer temperatures may exceed 40°C. The chaparral seasons consist of a hot, sunny, dry season from late spring to early fall followed by a cool, wet season during the winter months. Another feature of the chaparral, especially in southern California, is the hot, dry Santa Ana winds that blow out of the interior deserts at speeds often exceeding 60 km/hr; such drying winds often cause major wildfires.
14. The dominant vegetation type in the chaparral is evergreen shrub. Sometimes these shrubs grow so closely together that they are impenetrable. Over 100 species of shrubs grow in California's various chaparrals, although a given region contains only 1–20 species. Because of the hot, dry conditions that prevail during summer months, many chaparral plants show adaptations similar to desert plants. These adaptations include small, tough leaves covered with a thick cuticle (sclerophyllous leaves) as well as sunken stomata; both of these adaptations reduce water loss. Typical chaparral shrub species include chamise (*Adenostoma fasciculatum*), manzanitas (*Arctostaphylos*), and ceanothus (*Ceanothus*) species. Moister areas may contain dense groves of pine (*Pinus*), cypress (*Cupressus*), or oak (*Quercus*) trees, but an herb layer is usually present only after a fire.
15. Fire is a recurrent feature of the chaparral ecosystem; on average, fire sweeps through any given chaparral area once every 20–30 years. At such times all aboveground vegetation is burned away, and subsequent new growth bursts forth from underground growing points and seeds. As the shrubs regain dominance, herbs and fire annuals decline, and within about 15 years the community has returned to its pre-fire appearance. For the most part, chaparral shrubs have adopted one of two adaptive strategies in order to survive in this environment. The seeds of obligate-seeding shrubs accumulate in the soil during the non-fire years; such seeds are quite long-lived. When a fire does sweep through the area, the heat or smoke causes these seeds to germinate. An additional 5–15 fire-free years are required for the seedlings to reach reproductive maturity and begin setting a new crop of seeds. A fire moving through the area during this maturation period could potentially remove the obligate-seeding shrubs from the area. Examples of obligate-seeding shrubs include most of the chaparral species of *Arctostaphylos* and *Ceanothus*. A second chaparral survival strategy is seen in the obligate resprouters, whose seeds are short-lived and germinate when moisture is adequate. These seedlings remain stunted as they grow under the dense canopy of the mature chaparral shrubs. After a fire has burned all the aboveground biomass, these plants resprout quickly from underground roots, rhizomes, or basal burls. Examples of obligate resprouters include some species of oak (*Quercus*) and cherry (*Prunus*).

Applying Concepts

1. The greatest similarity of these areas is that they are all extremely cold, with winter lows in excess of −50°C. At times, strong, frigid winds sweep across the Martian landscape, raising large dust storms; similar winds sweeping across continental Antarctica can cause the windchill index to plummet to unthinkable lows. Mars is very dry, since free water cannot exist on the surface in liquid form due to the thin atmosphere; the Dry Valleys of continental Antarctica are also very dry (as dry as the Arizona desert), and what little water is present is generally locked away as ice. The Dry Valleys are devoid of macroscopic life, although microbial communities are present. Although all the photographs obtained to date from the various spacecraft to Mars show that the surface of Mars is also devoid of macroscopic life, there is considerable interest that microbial life similar to that in the Dry Valleys may exist on the red planet. There are some striking differences between Mars and continental Antarctica, despite their cold similarities. For example, the atmosphere on Mars is less than 1/100th as dense as that on Earth, which severely limits the availability of oxygen. The thin Martian atmosphere allows high levels of solar UV radiation to reach the surface; much of this radiation is removed by the Earth's atmosphere (even in the presence of an ozone hole). Further, the thin Martian atmosphere would not be effective in holding heat to the planet, so that nighttime surface temperatures can become extremely cold. All in all, though, it seems that if Mars can support life, it might be very similar to lifeforms found in the depths of continental Antarctica. Certainly the climate of Antarctica presents some of the same obstacles to life as does that on Mars.
2. Obviously, additional factors besides Hadley cells are important in determining the climate and aridity of a region. In this example, maritime winds influence the climate of both of these regions. The waters of the Pacific Ocean off the coast of California range from cool to cold, so that winds coming onshore carry only moderate amounts of moisture with them. The little water they do carry is often wrung out by intervening mountain ranges, so that the westerly winds reaching Arizona and New Mexico are quite dry and conducive to desert formation. The southeastern seaboard, on the other hand, forms the northern border of the warm waters of the Gulf of Mexico. Winds blowing onshore from the Gulf are laden with water, some of which is able to penetrate eastern North America all the way up the Mississippi Valley to Canada. When this moisture falls back to the Earth, it provides significant rainfall to the entire eastern half of the United States, in particular to the southeastern seaboard. Thus the Gulf provides too much moisture for the Gulf States to be desert.
3. The main geographic feature separating these areas is the Sierra Nevada itself. As the water-laden winds rise up the western slopes, the moisture is wrung out of them and precipitated as snow (or rain). These winds then cross the tallest peaks and descend over the eastern slopes. Because most of the moisture has been extracted from the air, the eastern slopes are quite dry and the underlying land is a desert, very different from the wet coniferous forests on the western side of the mountains. Obviously, this part of the Great Basin Desert lies in the rain shadow of the Sierra Nevada.
4. An important element in the maintenance of a prairie—especially a tallgrass prairie where moisture is fairly abundant—is fire. Fire kills back the non-fire-resistant invaders and allows the fire-resistant prairie species to grow even more luxuriantly. What this landowner's restoration efforts lacked was regular burnings to maintain the prairie and keep out the invading woody species.
5. First, without the large growth of trees, grasslands were easy to colonize and grow crops. Grasslands were ready to plow or use as pasture land—tree cutting, stump pulling, and rock digging were not required. Grasslands were also incredibly productive, due to the depth of the organic and nutrient-rich soil. If not called into service as cropland, grasslands made premier grazing lands. Because of this utility, huge areas of grasslands have disappeared; in some cases, such as the Central Valley of California, nearly all the original prairie is gone. In other areas, small prairie remnants remain (such as along railroad tracks and other such small plots of land). However, without active management by fire, native species are continually lost from these remnants due to competition by

aggressive weeds and woody species. Even pasturelands in some grassland ecosystems are now dominated largely by alien species. Today, it is difficult to find examples of original grassland.

6. The frequent fires that swept across the prairies killed most of the trees and shrubs that had rooted there. With a few exceptions, most of the trees and shrubs invading from the east are not adapted to withstand fire. Their bark is rather thin, and their growing points are aboveground and thus vulnerable, as are their seeds. Fires kill such species, reestablishing and maintaining the grassland.

Chapter 27

Review Questions

1. The boreal forest, like the tundra, is circumboreal—it encircles the Northern Hemisphere across northern Eurasia and northern North America. However, its boundaries are somewhat ill-defined because there are no sharp transitions between taiga and either tundra to the north or grasslands and deciduous forest to the south. On its northern front in North America, the taiga and tundra interdigitate, with finger-like extensions of tundra running south along exposed ridges and similar extensions of taiga reaching north in sheltered valleys. Along its southern border the transition is even less well defined. To the southeast, the coniferous forest mixes with the eastern deciduous forest, giving rise in the northeastern states to a mixed conifer-broadleaf forest containing species from both biomes. In the Upper Midwest, the conifer forest gradually fades into grassland, while in the far West, the conifer forest climbs up the mountain ranges and extends far southward as montane coniferous forest. Thus, the boreal forest has no distinct boundaries.
2. Physical factors that tundra plants must cope with are strong winds, cold temperatures, and low moisture availability. Winds abrade plant parts and increase water loss by evapotranspiration. The latter is especially a problem if replacement water is still frozen in the soil. To circumvent these problems, many tundra plants avoid high winds by either growing where snows will cover and protect them, or by being very short, since wind speeds are reduced near the ground. Many exposed species retain dead leaves and stems. This accumulated dead matter serves as a windbreak and traps additional snow for protection. The short growing season of 60 to 100 days means most tundra plants are perennials. By storing at least 50% of biomass in their belowground portions, such plants are well suited to a burst of growth activity when conditions become favorable in the spring. To cope with cool temperatures, some tundra plants produce dark pigments that absorb solar radiation and thereby raise leaf and stem temperatures, or they bear an insulating layer of white hairs on their leaves and stems, which helps reduce heat loss. Water availability is also a problem; some plants are able to absorb water directly through their leaves, while others (such as certain mosses and lichens) can be dried and yet resume normal growth activity when rehydrated.
3. Because of the harsh environmental conditions in the tundra, animal pollinators are of limited availability. To overcome this, tundra plants may adopt one or more alternate means of reproduction: (a) vegetative reproduction using aboveground stems called runners; (b) apomixis, wherein seeds are produced without the sexual process; (c) autogamy, wherein self-fertilization allows an individual plant to pollinate its own ovules; and (d) wind, rather than animal vectors, being used as the pollinating agent. Some species employ several of these reproductive strategies. Lastly, in addition to pollen dissemination, many tundra plants use the high winds as a means of seed dispersal.
4. Because of the decline in temperature with altitude, tundra and taiga are able to extend southward considerable distances by occupying the tops of mountain ranges. Alpine tundra occurs above 3,500 m in the Cascade and Sierra Nevada Mountains, and extends southward to central California. Alpine tundra also occurs along the Rocky Mountains from Canada to Guatemala—where it exists at elevations above 4,500 m—and in the Appalachian Mountains of New England above 1,200 m. Montane coniferous forest extends south in the Appalachians to Tennessee and North Carolina, across the mountains of the western United States from the Rockies to the Cascades and Sierra Nevada, and from Canada south to New Mexico and southern California. Like their counterparts in the Arctic, alpine tundra and montane coniferous forest are characterized by short growing seasons, low temperatures, strong winds, and nutrient-poor soils. Beyond these similarities, considerable differences exist between the Arctic and alpine systems. Alpine systems lack the prolonged period of winter darkness, and they receive more intense solar radiation—including ultraviolet radiation—due to a combination of the thinner atmosphere and the higher overall elevation of the Sun in the sky. Large fluctuations between daytime and nighttime temperatures are common in alpine systems, and hot summer days are often followed by freezing temperatures at night. Although annual precipitation is usually higher than in the Arctic, the alpine soils are generally drier because drainage is better. Lakes and bogs are still found in the mountains, but they are not as numerous as in the Arctic. Despite some of the physical differences, alpine tundra and Arctic tundra share many of the same plant species.
5. As in the boreal forest, spruces and firs dominate the montane coniferous forests, although tree species vary considerably between the Appalachian Mountains, Rocky Mountains, and the Cascade/Sierra Nevada Mountain ranges. Generally these differences reflect history, especially glaciation, as well as speciation on isolated mountain ranges. The dominant trees of the Appalachian montane forest include the conifers balsam fir *(Abies balsamea)* and red spruce *(Picea rubens)*, as well as the broadleaf paper birch *(Betula papyrifera)* and yellow birch *(Betula lutea)*. The montane forests of the Rocky Mountains are discontinuous and vary as to their logging and fire histories, so that dominant species also vary according to location. For example, the dominant species of the Colorado Rockies include Engelmann spruce *(Picea engelmannii)* and subalpine fir *(Abies lasiocarpa)*, with disturbed areas containing lodgepole pine *(Pinus contorta)* and quaking aspen *(Populus tremuloides)*. In the Southwest, bristlecone pines (*Pinus aristata* and *P. longaeva*) dominate at timberline, with ponderosa pine *(Pinus ponderosa)* occurring at lower elevations. In the Cascades and Sierra Nevada Mountains, a suite of conifers are common: red fir *(Abies magnifica)*, white fir *(A. concolor)*, Jeffrey pine *(Pinus jeffreyi)*, lodgepole pine, western white pine *(P. monticola)*, whitebark pine *(P. albicaulis)*, mountain hemlock *(Tsuga mertensiana)*, and of course, giant sequoia *(Sequoiadendron giganteum)*.
6. In eastern North America, the temperate deciduous forest has four consistent climatic features: a marked seasonality with four distinct seasons; frosts in the wintertime; absence of droughts; and moderate to long growing seasons. Temperature and precipitation are the two factors defining the limits of this forest. Along its northern border, the eastern temperate deciduous forest changes into boreal coniferous forest where the minimum winter temperatures average −40°C, which occurs at about 48° N latitude. Precipitation is the determinant of the western border, where grassland assumes dominance as moisture levels become insufficient to support a forest. The southern limit occurs as temperatures become warm enough to support tropical vegetation. Precipitation occurs throughout the year as intermittent rain- or snowstorms pass through. Temperate deciduous forest soils are rich and deep, with considerable amounts of organics and nutrients.
7. Precipitation decreases along the east-west and the south-north gradients. The Atlantic seaboard receives some 125 cm of annual precipitation; further to the west, where grassland and forest meet, precipitation levels drop off to only 85 cm per year. Similarly, along the Gulf Coast, annual precipitation amounts to 150 cm; to the north, along the Great Lakes, it has fallen off to 75 cm. The length of the growing season also varies, from a rather modest 120 days in the north to 250 days in the south.
8. Sulfur dioxide is emitted in large quantities from smokestacks of coal-burning power plants. When this gas combines with water vapor in the air, it forms sulfuric acid. Similarly, combustion of fuels oxidizes atmospheric nitrogen to nitrates, which combine with water vapor to form nitric acid. As raindrops fall to the ground, they accumulate these acids and become quite acidic, forming what is termed "acid rain." As the protons (H^+) from these acids accumulate in the soil, they exchange with mineral cations (Na^+, K^+, Ca^{2+}) that are bound to the negatively charged soil particles. So freed, the cations wash out of the system along with anions such as sulfate and nitrate. Thus this process decreases soil fertility and makes the surrounding waterways more acidic. Acid rain has resulted in loss of plant and animal species both on land and in affected lakes; some forests have stopped accumulating biomass in the last two decades, and some lakes in New York State are so acidic they have no fish. The problem is no longer localized to the northeast; forests in the southeast are also showing decline. Leading researchers in the field state that if any recovery is to occur by 2050, sulfur emissions must be cut an additional 80% from their current levels. Nitrate levels may also need to be regulated. Before the eastern ecosystems can begin recovery, serious air pollution problems will need to be addressed.
9. An epiphyte is a plant that grows on the surfaces of other plants. Ferns cannot tolerate drying, and they produce their own moist little pot of soil high in the canopy by trapping wind- and water-borne dust and organic litter. Their light spores are easily dispersed by wind and rain. Epiphytic orchids are extremely common and diverse in the tropics. To help them in their epiphytic lifestyle, orchid roots are often photosynthetic and usually possess a velamen, an outer layer of nonliving cells with perforations so water can enter. Additionally, orchid seeds are very small and lightweight, so they are easily dispersed by wind. Bromeliads are members of the pineapple family. Their leaf bases usually overlap in such a way that a vase-like or bowl-like container is formed. Rainwater collects in this depression, forming a small pond that supports a wide variety of small animal life, from protozoa to amphibians. Bromeliads typically form bright red, sticky fruits that are attractive to birds, which distribute them to other tree branches. Lianas are woody vines that are especially common in tropical rain forests. Many lianas begin their lives on tree limbs or trunks, and only later root in the soil below.
10. Growing in a tropical climate offers several advantages to a plant. The temperatures are warm year-round, and water and sunlight are generally plentiful, so that growth and reproduction can also occur all year. The mild climate also means that plants are not subjected to the stresses of freezing temperatures and extended cold spells. Reproduction can occur in a leisurely way year-round, instead of at a frenzied rate all at once. Included among the disadvantages is the fact that high temperatures and humidity favor the growth of insect pests and disease-causing bacteria and fungi. Thus, tropical plants must be on guard against these destructive organisms all the time, especially because pest and pathogen activities are not curtailed by a cold winter. Additionally, the high nighttime temperatures promote nighttime respiration, so that carbon fixed during the day is essentially wasted away. Tropical systems exhibit high species diversity and lush growth, and the many species that are present sort themselves by height into several layers of foliage. Plants growing in this forest, especially toward the forest floor, often find themselves in fierce competition for available light. Lastly, tropical forest soils are typically very low in mineral nutrients and organic matter, and they are sometimes high in aluminum as well. Successful tropical plants must find ways to cope with all these hardships.
11. The sun passes nearly overhead every day in the tropics, and this provides intense year-round solar radiation to these areas, even when cloud cover shields the top of the rain forest from direct sunlight. On the forest floor, however, light can be extremely limited due to the heavy canopy overhead, and the forest floor is often a rather dark and gloomy place. Average annual rainfall is usually above 200 cm per year, and the humidity hovers at 60–80% during the day, often reaching 100% at night. Although rainfall is abundant and humidity high, water is often a scarce commodity high in the treetops, and plants growing there are often adapted to cope with drought. Tropical storms can generate very strong winds. These winds are generally not a problem to a mature forest because, essentially, each tree helps support its immediate neighbors.

However, at the edges of the forest or of large forest openings, trees are more vulnerable and can be more easily toppled by high winds, especially since tropical plants tend to root shallowly in the hard soil. Tropical soils are generally poor in nutrients due to heavy leaching by the abundant rainfall. Not only is the organic content of these soils also low, but to further complicate matters, aluminum levels are often rather high and the soil itself very hard and difficult to penetrate.

12. Because tropical soils are so nitrogen poor, many tropical plants rely on nitrogen-fixing microbes for their nitrogen supply. Many leguminous plants harbor nitrogen-fixing bacteria in their root nodules, and cycads produce aboveground roots containing nitrogen-fixing cyanobacteria. Nitrogen-fixing cyanobacteria are found in other plant parts as well, such as in the petioles of the tropical plant *Gunnera*. Some nitrogen-fixing microbes live freely in the tropical soils. Phosphorus can be severely limiting in tropical soils, and many tropical plants receive their phosphorus from mycorrhizal associations with soil fungi; the fungal mycelium is able to expand over a vast area of the forest floor, accumulating mineral phosphorus as it grows and transferring this nutrient to the plants it associates with.
13. A type of agriculture practiced by indigenous tropical farmers around the world is known as slash-and-burn agriculture. Generally, all the trees in a small area (under 40 acres) are cut down, and the fallen timber is left in place to dry somewhat. Eventually the entire clearing is burned, and the ash is turned into the soil as crops are planted. Within 2–3 years the soil is depleted of its meager nutrients, and the process must be repeated on a new plot of forestland. Unfortunately, the number of people this type of agriculture can support per square kilometer is quite limited, and it cannot reasonably support even a small-sized modern town. To make these lands more productive requires applications of fertilizers, insecticides, and fungicides, which most indigenous farmers cannot afford. Even before these plots are stripped of their nutrients, they quickly become overrun with aggressive weeds. Lastly, many of the fires set during the burning process have escaped and burned far more forestland than intended.
14. Because competition for light is so intense in the tropical rain forest, many species invest the resources to become a tall tree. The foliage of such trees lies in the uppermost layer of the canopy, where it is among the first to intercept incoming sunlight. Tropical tree species adopt many different growth forms to help in their quest for light in the various levels of the canopy. For example, the tallest trees are often umbrella-shaped, not branching until the top of the canopy is reached; the shorter trees often produce more spherical crowns. Tropical trees generally have smooth bark, which probably reduces the ability of epiphytes and vines to become established. Because tropical plants typically lack a taproot and root quite shallowly in the hard soil, many trees gain lateral stability by sporting large buttress roots. The great majority of tropical trees have oval, simple, entire leaves (for reasons unknown) that often end in a long, drawn-out "drip tip," presumably to drain rainwater. The leaves are evergreen, and young leaves of many tropical plants are protected from strong ultraviolet radiation by the deposition of red, purple, or bluish pigments in the central vacuole. Many tropical leaves contain bitter-tasting compounds (e.g., tannins or alkaloids) that help deter herbivores. Because there is usually little wind within the depths of the forest, most plant species use animal vectors for pollination. Canopy trees, for example, often have brightly colored flowers to attract birds and bees, while shorter trees often produce cauliflorous flowers so that their pollinators do not need to fly through the dense understory foliage. The year-round growing season in the tropics allows many plants to produce flowers and fruits year-round as well.
15. In the tropical rain forest, the year-round growing conditions allow reproduction to occur continually as well. This allows natural selection and evolution to occur at an accelerated rate, producing more species than in temperate regimes. The high plant diversity promotes the evolution of numerous animal partners for pollination and seed dispersal. In addition, the tropical forest is multitiered, with several distinct layers from the uppermost canopy to the forest floor. This provides a continuum of light, moisture, and forest resources, creating many niches in which numerous species can adapt and evolve. These factors combine to make the tropical rain forest the most species-rich biome on the planet.

Applying Concepts

1. Tundra is largely found in the Northern Hemisphere, for several reasons. First, the climate in the polar regions of the Northern Hemisphere allows the growth of tundra plants, whereas that in continental Antarctica is too extreme (dry, frigid, lack of productive soil) to support plant life. Also, northern hemispheric landmasses are closer to the poles than in the Southern Hemisphere, in the sense that plants from more temperate regions to the south were able to colonize and adapt to the Arctic tundra environment. Antarctica, however, is considerably removed from South America, Africa, and Australia; these distances made it very difficult for plants to migrate to Antarctica and form a tundra environment there. Even in maritime Antarctica, where some flowering plants can grow (*Deschampsia* and *Colobanthus*), the lack of animal pollinators would have made reproduction very difficult for many invading plants, if they could have survived the trip across large expanses of southern ocean.
2. The tundra and taiga provide resources that regularly touch our daily lives—oil and natural gas, for example, as well as metals (gold, silver) and other mining products (e.g., diamonds). The boreal forest provides huge amounts of softwood for lumber and paper pulp, and some of the bogs provide peat. The far north can have definite influence on the weather in the continental United States. For example, sometimes in the winter frigid Arctic air masses descend on the northern and eastern states, plunging the daytime temperatures below zero and generally wreaking havoc on our daily lives. How did these air masses get so cold? By stalling over the snow-covered taiga and tundra. Snow is an excellent reflector at visible-light wavelengths, and an excellent radiator at infrared (heat) wavelengths. These areas reflect sunlight during the day, and radiate heat to space at night. For every day they stall over the tundra and taiga, air masses can lose a degree of temperature. When they finally move over the United States, these air masses can be frigid indeed. The far north has sparked people's fancy for centuries. Witness a classic like Jack London's *Call of the Wild*, or consider how the howl of a wolf or the call of a loon sparks our imagination. And let us not forget the effect of the North's most famous "inhabitant," Santa Claus, on the hopes and dreams of children for centuries.
3. (a) Although both of these biomes typically have pronounced seasonality, in the boreal forests the winters are accentuated (longer, colder) and the summer growing seasons shorter (<120 days vs. 120–250 days per year) than in the deciduous forests to the south. (b) The boreal forests receive approximately 40–50 cm of annual precipitation, while the temperate deciduous forests receive 75–150 cm per year. (c) The soils of coniferous forests are rather poor, being high in undecayed organics, low in nutrients, and often acid in nature. In these cold and acid soils, activity of soil organisms such as earthworms is usually low or absent. By contrast, soils of the temperate deciduous forest are high in nutrients, neutral or even alkaline in pH, and high in soil organism activity. In these forests, earthworms continually turn the soil over, mixing the organic and mineral components into one rich soil. (d) Located further north, the boreal coniferous forests experience colder average temperatures than the deciduous forests to the south. In eastern North America, the boreal coniferous and eastern temperate deciduous forests meet at about 48° N, approximately the latitude where winter lows approach −40°C. Apparently the deciduous forests are unable to tolerate colder temperatures and hence do not occur further north than that.
4. The temperate deciduous forest also occurs in Europe and eastern Asia, and the warmer wetlands of these areas would be logical places to look for the other alligator species. Now endangered, the Chinese alligator is indeed found in eastern China along the Yangtze River, occupying habitats very similar to those of its American relative in the southeastern United States.
5. A glance at a map shows that the mountains of Europe—the Pyrenees and Alps—run in an east-west direction and, in western Europe, effectively seal the north from the south. Evidently, as the glaciers advanced from the north, the retreating species of the temperate deciduous forest became trapped between the glaciers and the mountains, with few places to act as a refuge. Thus trapped, species extinctions were higher in Europe than in North America or Asia, where southerly retreats during glacial times were not cut off by east-west mountain ranges.
6. At the edge of the forest we find a dense growth of tropical plants. Here sunlight penetrates almost to the ground, allowing the plants to form a verdant wall of tropical greenery. Wind and sunlight levels are deceptively high here, masking the gloominess of the forest interior. As we make our way into the depths of the forest, conditions change. We are at once impressed at the height of the rain forest, with the tops of some of the canopy trees towering some 80 meters (nearly 260 feet) above us. We also notice that the canopy is stratified, with trees of different heights producing layer upon layer of foliage. Epiphytes and lianas abound, and we realize that, in this world, vertical real estate—in the form of tree trunks—is as precious as horizontal real estate on the forest floor. Animal life is abuzz with activity all around us, but mostly it is concentrated above us. Regardless of what month of the year it is, the arboreal animals are surrounded by flowers and fruits. The dense canopy above prevents direct sunlight from reaching the floor, except as fleeting sun flecks, so that the forest floor is relatively clear of small plants and is, in fact, surprisingly open. Thus we realize that with all its life, the tropical rain forest is really a shady and almost gloomy place on the forest floor. Our boots sink into the soggy humus that coats the ground, but under this we find a hard, ruddy soil that is hardly nurturing to the plants around us. Even in the shade of the forest, we find little relief from the heat and humidity that permeates this environment. Truly, as Darwin remarked, "Nothing but the reality can give any idea how . . . magnificent the scene is."
7. First of all, the hardpan tropical soils are difficult to plow, requiring expensive machinery (e.g., tractors) to effectively prepare even modest-sized plots for planting. The poor nutrient status of the soils means that expensive fertilizers must be applied to make the fields productive, and much of this leaches off into the surrounding rivers and streams. The year-round high temperatures and humidity allow insect and fungal pests to reach high population levels and to spread rapidly. Likewise, rampant weed growth can quickly overrun all but the most carefully manicured fields. To abate the pests and weeds, expensive herbicides, insecticides, and fungicides must be used. Bacterial attack can also be a problem, necessitating the use of expensive antibiotics. Even if a tropical farmer has the resources to engage in the above-mentioned activities, high night-time temperatures reduce crop yield by promoting high respiration rates.
8. The high temperatures and humidity are ideal for decomposers, and the tons of litter per hectare that fall to the forest floor each year are quickly decomposed. The released nutrients are quickly absorbed by mycorrhizal fungi and plant roots, thus rapidly returning to the tropical biomass. Quick recycling is the norm here, because nutrients cannot remain in the soils for long—they would soon be leached away by the abundant rainfall. Instead, nutrients are held in the bodies of living organisms. Insoluble aluminum and iron salts may accumulate, giving the soils an orange or ruddy color, and perhaps making them mildly toxic to some plants. Little organic matter becomes incorporated into these soils, which thus become hard and difficult to penetrate. It is, in fact, surprising that such luxuriant growth can spring from such poor soils.

Chapter 28

Review Questions

1. The uppermost layers of a lake form the pelagic zone, which includes the surface and subsurface waters

receiving enough sunlight to support the growth of photosynthetic organisms such as floating plants, algae, and photosynthetic bacteria. Most lakes contain shallow, nearshore regions known as the littoral zone where rooted plants and attached protists grow. Lastly, the benthos includes the deeper waters and bottom sediments of lake. This benthic zone typically contains many decomposers, scavengers such as crawfish, and attached or bottom-dwelling animals such as the filter-feeding clams.

2. (a) Bacterioplankton are the bacteria that are found floating freely in the water column. These include both heterotrophic and photosynthetic forms. (b) The phytoplankton are all the free-floating, photosynthetic organisms. They include both the eukaryotic algae and the prokaryotic cyanobacteria. (c) Macrophytes are the aquatic vascular plants that grow in lakes and waterways, either attached or free-floating. (d) Periphyton are the algae and bacteria (and some protozoa) that live attached to the submerged portions of macrophytes. (e) The protozooplankton are essentially pelagic, or free-floating, protozoa; most of these are single-celled ciliates or flagellates. (f) The zooplankton are the small, multicellular invertebrate animals that swim through the water column; most of these are crustaceans, a group of invertebrates related to shrimp and lobsters.
3. (a) The primary producers are the photosynthetic organisms that reduce carbon dioxide to organic compounds using the energy of sunlight. Thus, starting with inorganic carbon, these organisms produce food that is used by the consumers. Examples of these organisms include cattails *(Typha)*, *Lemna*, cyanobacteria, and the green alga *Cladophora*. (b) The organisms that directly consume the primary producers are the primary consumers, including species among the protozooplankton, zooplankton, fish, insects, and waterfowl. (c) Secondary consumers, primarily fish, eat the primary consumers; and (d) tertiary consumers, such as northern pike, some predatory birds and humans, hunt and consume the secondary consumers.
4. The circulation patterns result from the fact that water is densest at 4°C and less dense above and below this temperature. In spring the entire lake is cold, so that neither the temperature nor the density of the water varies much throughout the entire water column. Wind stirs up the lake, bringing nutrient-laden waters up from the depths. In summer, the upper portion of the lake heats up while the deeper parts remain cold and dense; thus the lake becomes stratified, and the surface and deep layers cannot be mixed by wind action alone. Without the nutrients from the depths, the surface layers may become nutrient depleted. During the autumn, the surface waters again cool to the point where wind action can mix the upper waters with the deeper ones, so that nutrients are again brought up to the surface. In some climates, ice may cover the surface of the lake during the winter months, and wind-driven mixing stops until icemelt in the spring.
5. Many cyanobacteria are able to float because they produce tiny, internal, balloon-like gas vacuoles that give the cells buoyancy. Some cyanobacteria deter predators by releasing toxins or by embedding themselves in large mucilaginous coats. Many bacterioplankton are so small that they easily stay suspended in the water column. Eukaryotic phytoplankton, on the other hand, are generally larger than their bacterial cousins. They may be equipped with flagella, and they stay afloat by direct swimming. To discourage predators, some phytoplankton bear long, sharp spines or coat themselves in mucilage.
6. Oligotrophic waters are characterized by low mineral nutrient content with concomitant low productivity. Many different species are present, even though each is represented by relatively modest numbers of individuals. Such waters are typically clean and clear, and produce neither harmful algal blooms nor overgrowths of weeds. Eutrophic waters have a much higher mineral nutrient content and thus are able to support much higher levels of primary productivity. Species diversity is low in these waters, although some of the species present may at times grow excessively, leading to algal blooms and rampant weed growth. Eutrophic waters are typically cloudy, due to the presence of large numbers of cyanobacteria in the water column, and may become choked full of weeds. Eutrophic waters that produce persistent cyanobacterial blooms are described as hypereutrophic.
7. Answers: 1, E; 2, A; 3, C; 4, D; 5, B; 6, F
8. Cattails reproduce via submerged rhizomes, so that one plant often colonizes a wide area of marshland. However, the underwater portions of the plant are deprived of oxygen, and aerenchyma tissue in the leaves brings oxygen down to the submerged roots and rhizomes. In this respect the leaves act as a snorkel. Marshes are often very windy places, because they frequently border long stretches of open water and usually have few trees to act as windbreaks. Cattail leaves have several features to help them withstand these forces: (1) The leaves are light and stiff, with internal columns or chambers strengthened by numerous "cross-beams" (i.e., tissue diaphragms) running widthwise across the leaf. (2) Further strength is provided by parallel, fibrous strands running length-wise, perpendicular to the diaphragms. (3) The leaves are linear so they do not present a large surface area to the wind, and they are somewhat twisted in a manner that minimizes the lift and drag forces of the wind.
9. Many wetlands have been completely obliterated, having been filled in and reclaimed for home construction or for growing crops. Other wetlands still exist, but are not the healthy communities they once were. Drainage ditches and falling water tables have left some wetlands high and dry for parts of the year, resulting in the disappearance of many of their members. In other cases, exotic species introduced by man (e.g., inadvertently or as ornamentals) have become troublesome weeds, invading wetlands (and other communities) and outcompeting native species. Many other human activities adversely affect wetland communities—damming rivers, diverting river water for agricultural uses, using natural waterways as dumping grounds for sewage and industrial wastewater, acidification via acid rain (caused by industrial air pollutants), and nutrient runoff from farmlands, golf courses, and residential lawns. The preceding list is far from complete, and the reader should be able to think of numerous other human activities that result in wetland degradation.
10. Oceans help to moderate the Earth's temperature. Water has a high heat capacity and therefore the immense quantities of water in the oceans can absorb a great deal of heat and release it slowly. This capacity to store and release heat moderates extremes of temperature. Ocean water is also the primary source of rain for land plants and animals. As water evaporates from the ocean, it is blown over land, and some of it eventually condenses as rain and falls on terrestrial environments. The primary producers in the ocean produce much of the oxygen in the atmosphere, which is needed by most organisms for respiration, and reduce much of the carbon dioxide. Removal of carbon dioxide from the atmosphere is also important for maintaining a proper temperature on the Earth because carbon dioxide is a greenhouse gas. The ocean also provides about 90% of fish caught each year–an important source of protein for humans.
11. (a) A continental shelf is the submerged extension of a continent that slopes down from the shoreline to a depth of 100 to 200 m. (b) A continental slope is an area of rapid descent from the edge of a continental shelf down to an abyssal plain. (c) The abyssal plain is the vast, level, sediment-covered plain that extends under most of the ocean. (d) A submarine ridge is a mid-ocean mountain range that divides an ocean basin into two basins. (e) A sea mount is a submerged volcanic mountain that rises from the abyssal plain. (f) Trenches are narrow, deep depressions in the ocean floor that can reach depths of 7,000 to 11,000 m.
12. At the equator warm air masses rise and spread out at altitude to the north and south. Those air masses that move north are deflected to the right (east) by the Coriolis force. When they sink back to the surface at 30° N latitude, they form the northeast trade winds that blow from NE to SW. The trade winds impart a net flow of ocean surface waters at the equator from east to west. As a result, water piles up against the western side of the North Atlantic and is deflected by the North America landmass toward the north pole. As these northward flowing waters pass 30° N, they encounter the westerlies that blow from SW to NE. The westerlies drive the surface waters back east across the North Atlantic basin. When they reach the eastern side of the basin, Europe deflects the eastward currents back toward the equator, completing the clockwise North Atlantic gyre.
13. In temperate oceans, lying between 30° and 60° N and S latitude, seasonal variation in solar energy input means that surface temperatures also vary seasonally. In winter solar energy input is at a minimum, surface waters are cold, and the water column and nutrients mix well. As spring arrives, solar energy increases, surface waters warm, and the thermocline may begin to form. In summer solar energy is at a maximum and the thermocline is strongly developed, isolating the surface waters from deeper waters. In fall solar energy input declines, surface waters cool, the thermocline breaks down, and mixing begins again. Blooms of phytoplankton follow the periods of mixing. The main phytoplankton bloom occurs in spring following winter mixing, and a smaller phytoplankton bloom follows the fall mixing period. Zooplankton blooms follow the phytoplankton blooms. Tropical oceans do not show a seasonal pattern and have a permanent thermocline.
14. (a) The pelagic realm includes the entire volume of open ocean water. (b) The epipelagic is the uppermost pelagic zone in the euphotic zone and extends from the surface to a depth of 100–200 m. (c) The littoral zone occupies the area of ocean between the high and low tide levels. (d) The sublittoral or subtidal lies below the littoral from the low tide level to the margin of the continental shelf. (e) The benthic realm includes the area of the ocean bottom and the organisms that live there.
15. Plankton are important as the base of the oceanic food web. They may be categorized by size: femtoplankton (0.02–0.2 μm), picoplankton (0.2–2.0 μm), nanoplankton (2–20 μm), microplankton (20–200 μm), and mesoplankton 0.2–20 mm. Net plankton are all plankton larger than 20 μm. Plankton may also be categorized by kinds of organisms such as viroplankton (viruses), bacterioplankton (bacteria), protozooplankton (heterotrophic protists), phytoplankton (algal protists), and zooplankton (animals).
16. The microbial loop is the portion of the marine planktonic food web that includes organisms in the picoplankton (0.2–2 μm) and nanoplankton (2–20 μm). Picoplankton include heterotrophic bacteria, cyanobacteria, and prochlorophytes. Nanoplankton includes grazers such as ciliates and flagellates and nanophytoplankton. The classic food web includes only organisms larger than 20 μm, which were the main focus of plankton studies before the importance of smaller organisms was understood. The microbial loop joins the classic food web through the grazing of microplankton.
17. Estuaries, salt marshes, and mangrove forests are 3 littoral ecosystems that are transition zones between terrestrial and marine environments. Estuaries are coastal areas where freshwater from rivers mixes with seawater. This mixing of waters generates great variability in salinity depending on location within the estuary, tides, and the level of river flow. A number of invertebrate animals such as shrimp and shellfish live in estuaries, which are important feeding grounds for some birds and nurseries for the young of fish and other animals. Salt marshes are wetlands along ocean coasts that contain salt-tolerant herbs, grasses, and shrubs. These are important habitats for fish and wildlife, help stabilize shorelines, and filter nutrients and sediments from freshwater runoff. Mangrove forests are a transition community along tropical and subtropical coastlines. These forests provide habitat for endangered species such as the alligator and brown pelican, nurseries for important fish and shellfish species, and home for numerous crabs, shellfish and small fish.

Applying Concepts

1. Cyanobacteria like *Microcystis* produce gas vacuoles to help keep them afloat. The sharp taps on the cork of the second bottle sent shock waves through the aqueous medium, crushing many of the gas vacuoles. Many cells thus lost their buoyancy, and settled down to the bottom.

2. The remarkable preservation of this body was due to the extremely acidic pH and the cold temperatures found in the bog. These two factors virtually put a stop to the microbial decomposition that would otherwise have taken place.

3. (a) Wetlands provide habitat for many species of plants and animals. Many of these species cannot live in any other type of community. As such, wetlands are important for preserving biodiversity on this planet. (b) Many wetlands are used as rest stops and feeding stops by migrating waterfowl; loss of these wetlands would endanger the waterfowl that use them. This is one reason that sportsmen's organizations have lobbied so heavily to preserve local wetlands. (c) Wetlands provide many opportunities for leisure activities, such as fishing, bird watching, hiking, botanizing, canoeing, scuba diving, and so on. When wetlands are degraded, virtually all of these activities suffer. (d) Wetlands help absorb extra runoff during wet years, thus helping to prevent flooding downstream. (e) Wetlands help absorb nutrients in runoff so that waterways are less likely to be overrun with plant and algal blooms. (f) Wetlands are nurseries for many species of fish and birds. Many other answers are possible.

4. As mentioned earlier in Chapter 25, northern pike are tertiary consumers; they have a voracious appetite and will eat virtually anything they can. Biologists were concerned that if this aggressive predator spread to other waterways, it would pose a danger to native species of fish, such as the salmon, steelhead, and endangered species in the downstream delta regions. This is precisely what happened when pike were introduced (illegally) into the Susinta River drainage in Alaska: They have now spread to 90 lakes, destroying silver and sockeye salmon and devastating over 60% of the rainbow trout. These stories of the northern pike illustrate the dangers of introducing aggressive species into areas where they do not naturally occur and have no natural predators. Incidentally, the poisoning was unsuccessful, and some experts are now advocating a complete drainage of Lake Davis to eradicate the pike. Local examples of disruption caused by the introduction of foreign species will vary but might include such introduced species as purple loosestrife in marshes or zebra mussels in lakes and rivers.

5. Temperature would be highest at the surface of the ocean, perhaps as high as 30°C in a tropical ocean, and would decline as you descended. At a depth of 50–300 m temperature drops rapidly through a zone called the thermocline. Below the thermocline and on toward the ocean bottom, the temperature would be cold at about 4°C. Oxygen levels would be highest in the 20 m near the surface, where diffusion from the atmosphere and photosynthesis occur, and would decline with depth to a minimum at around 1,000 m. At lower depths oxygen levels rise again due to the flow of cold oxygenated waters from polar regions. Light is highest near the surface, and the upper 100–200 m of the ocean lie in the euphotic zone where light is sufficient for photosynthesis. Below the euphotic zone the ocean is permanently dark. In the euphotic zone there are numerous plankton, including bacterioplankton, phytoplankton, protozooplankton, and zooplankton. Zooplankton include tiny animals like copepods and the larvae of larger animals. Nekton such as fish, reptiles (sea turtles), and mammals (dolphins) are also present in the near-surface waters. Marine birds occur at and above the surface. Below the euphotic zone, there is a rain of detritus from above and bacterial decomposers. Fish and other animals occur because they can swim up to the surface in search of food, but their numbers decrease with depth. At the bottom the detritus accumulates and is converted into bacterial biomass that feeds a variety of consumers. Sponges, jellyfish, mollusks and starfish dominate the bottom sediments.

6. In the case of kelp forests off the coast of western North America, human hunting of sea otters along the Alaskan coast decimated their populations and at the same time kelp forests began to disappear. The otters eat sea urchins and keep the sea urchin population in check. When otter populations were decimated, the sea urchins increased and attacked and ate the kelp holdfasts. Kelp forests washed away. After otter were granted protection, their populations rebounded, sea urchins were controlled, and kelp forests returned. However, humans have been taking too many fish from coastal waters and depriving killer whales of food. The killer whales have been turning to eating sea otters, which could result in a loss of kelp forests again. Overfishing around coral reefs has also had a detrimental effect on reef ecosystems. Herbivorous fish eat attached algae that otherwise might overgrow and smother coral animals. For a while, herbivorous sea urchins kept the algae in check on Caribbean reefs, but in the early 1980s over 95% of the sea urchins succumbed to an unknown plague. This disease appeared to result from microorganisms in clouds of dust originating in the Sahara Desert that were swept across the Atlantic and reached the Caribbean Sea. After the sea urchins died, algal cover increased from 4% to 92% on the reefs around Jamaica, and coral cover declined from 52% to 3%. The condition of the reefs has not improved at present.

Chapter 29
Review Questions

1. Sustainability is a way of life wherein humans live with an acceptable standard of living while at the same time preserving the environment in a high-quality state. In real terms, this means, for example, that soils are not being depleted, metals are recycled, forests are being continually renewed, water is thoroughly scrubbed of pollutants before being returned to the waterways, and so on. Ecosystem services are processes performed by the various organisms of the Earth that help maintain the stability of the Earth's environment. Examples of such processes include enrichment of the soil by nitrogen-fixing bacteria, decomposition of dead matter by fungi and bacteria, water purification by wetland communities, the continual replenishment of oxygen in the air by photosynthetic organisms, and long-term storage of carbon (from atmospheric carbon dioxide) as limestone, peat, and petroleum.

2. The current human population of the Earth is over 6 billion people, with estimates of 7.5 billion by 2020 and 9 billion by 2050. Thus, these numbers predict that within 50 years, there will be 1.5 people for every person now living. Dr. Pimentel's research shows that we have already exceeded the carrying capacity of the United States as well as the entire Earth. His studies indicate that if the entire population of the world had a living standard one-half that of the current U.S. standard, then the carrying capacity of the Earth would be approximately 2 billion people, and that of the United States would be only 200 million. Clearly, we are already well past both of these numbers (U.S. population today approaches 300 million). Pimentel's studies further indicate that fossil fuels subsidize the production of food today. For every unit of energy contained in the food we produce, ten times that amount of fossil fuel energy is required as an input. Obviously, this very precarious house of cards could collapse when fossil fuels are depleted.

3. Functional groups are assemblages of organisms that collectively perform valuable ecosystem functions, such as nitrogen fixation, pollination, decomposition, or photosynthesis. Some species perform a much greater role in this regard than expected, given their biomass, and they are called keystone species; examples include nitrogen-fixing bacteria, mycorrhizal fungi, and pollinating bees.

4. Estimates of modern extinction rates range from 100 to 1,000 times greater than occurred before modern humans appeared some 40,000 years ago. In fact, current extinction rates are similar to those that occurred 65 million years ago, when an asteroid collision caused massive extinctions and brought the Cretaceous period to a close. There is now archeological evidence suggesting that extensive human hunting was linked to the disappearance of many large mammals in North America some 12,000 years ago, including the sabertooth cats and giant sloths. As to plants, today more than 10% of plant species worldwide face imminent extinction, with some 34,000 species at immediate risk. Thirty percent of the plant species in the United States are in danger of extinction.

5. An organism that possesses low dispersal ability and therefore occurs only in a particular area is known as an endemic species. The number of endemic species in a region is correlated with the risk for loss of biodiversity in that region, so that the degree of endemism provides an estimate of the potential for extinction. Globally, some 25 biodiversity hotspots have been identified (i.e., regions around the world especially rich in endemic species), all of which are endangered by human activities. These areas include the West Indies, southwestern Australia, the Philippines, India, and the rain forests of Hawaii. These areas are the last natural habitat left for approximately one-third of the vascular plant and terrestrial vertebrate species. In the United States, biological hotspots include Death Valley (part of the Mojave Desert), the Florida panhandle, coastal and interior parts of southern California, Hawaii, the southern Appalachians, and the San Francisco Bay area. Approximately 12% of the world's species occur in the United States, the richest temperate ecosystem in the world.

6. Invasive species are usually foreign species showing aggressive growth when introduced into new areas that lack the native control mechanisms, such as diseases (pathogens) or predators (herbivores). When displaced into new areas, such species often compete aggressively for resources, so that they grow rapidly and overrun native species. Here are some characteristics of invasive species: (a) they often develop rapidly from seedling to adult stages, (b) they usually produce many progeny, (c) they typically already have well-developed dispersal mechanisms for their seeds or for vegetative growth, (d) they are usually very tolerant of a wide variety of ecological conditions, (e) they generally vary their appearance and/or behavior, and (f) many species are capable of reproducing asexually.

7. Renewable energy sources include wind, sunlight, flowing waters, geothermal heat, and biomass derived from plants and animals. The energy of wind is generally harvested using electrical windmills on large wind farms; several are currently in use in the United States. Solar energy is generally harvested using photovoltaic technology—using solar cells to directly generate electricity from sunlight. This technology has advanced rapidly over the past decade, and now large rolls of photovoltaic material are available that completely replace the shingles on the roof of a house or other building. The most common way to generate electricity from flowing waters is via hydroelectricity produced by damming rivers. Other methods exist, however, such as extracting energy from moving seawater as it shifts between high and low tide. The energy of the Earth's heat—geothermal energy—is generally extracted by using geothermal heat to convert water into steam, and then driving an electrical turbine with the steam. There are currently a number of innovative technologies using plant and animal biomass, including the production of combustible methane from manure as well as the conversion of spent deep-frying fat from restaurants into automobile fuel. Examples of innovative energy storage technologies include storage of hydrogen, batteries, flywheels, superconducting storage rings, ultracapacitors, and compressed gas. The storage of energy as hydrogen is particularly exciting nowadays. Hydrogen is easily generated from water by simply passing an electrical current through it (electrolysis), and state-of-the-art fuel cells convert hydrogen back to electricity with only water as a by-product. An added benefit is that hydrogen is a rather safe fuel. Unlike methane and other hydrocarbon gases—which are heavier than air and thus accumulate over the floor or ground as an explosive mixture—or liquid fuels, which soak into porous substances they come into contact with, escaped hydrogen dissipates very rapidly because it is so light.

8. Around a thousand years ago, the Anasazi had a thriving civilization in the southwestern United States, where they were well known for their multistoried cliff dwellings. A thousand miles to the east, at a southwestern Illinois site known as Cahokia, a city had been planned and erected, and it contained more people at the time than did London, England. Both of these cultures were skilled in farming, and they raised squash and corn, among other crops. By 1300 Cahokia and all the Anasazi sites were deserted, victims of ecological decline due to environmental changes and population pressures. Neither culture could adapt to the

altered conditions, and the cultures declined as the peoples either died or dispersed. Similar fates were met by the Mayas on the Yucatán peninsula of Mexico, as well as by the Aztecs and Toltecs in the highland valleys of Mexico. In each case, ecological decline had so degraded the environments of these cultures that the disappearances of the cultures themselves were actually hastened.

9. An extractive reserve is a forest whose resources are removed in such a manner and at such a rate that the ecosystem itself remains largely intact. One way that neotropical forests are being harvested is by following established procedures favoring sustainability. For example, scientific studies have indicated that it is good practice to (a) first inventory the valuable timber in an area, (b) plan movements of heavy machinery so as to reduce damage to the forest, (c) physically isolate selected trees (e.g., by cutting the vines that surround them) so that neighboring trees will not be dragged down during the harvest, (d) harvest only the marked trees, (e) carefully map the selected trees so that they can be easily found with a minimum of searching, and (f) reduce waste and maximize lumber output at the sawmill. In addition, logging should be avoided in sensitive areas, such as along slopes and streams, as well as in parks, lands of indigenous peoples, and regions of moderate to high biodiversity. A proposed simple rule for loggers to follow is the 5/50/5 rule: Remove no more than 5 trees per hectare per logging session; allow an area to lie fallow for 50 years before it is logged again; and to minimize the spread of fire, maintain a 5-meter-wide cleared border around all logged stands for at least one decade after logging. Another strategy being used in the Peruvian Amazon is known as the strip-shelter belt system, wherein timber crops are harvested in long, narrow strips that alternate with strips of natural forest. Periodically, the timber strips are clear-cut, and then allowed to be recolonized by natural species migrating in from the neighboring natural strips.

10. In habitat fragmentation, a large continuous population is broken down into small, isolated ones. When plants grow in small, isolated or sparse groups, they receive fewer visits from pollinators and thus set fewer seeds. Furthermore, the number of genetically unique mates is lower in small populations, leading to an increase in the amount of self-fertilization and inbreeding. This, in turn, can result in inbreeding depression—a condition of inbred populations whereby vitality decreases as the number of deleterious genes increases. A second genetic condition affecting small populations and making them more vulnerable to extinction is genetic drift, where chance events result in the loss of certain alleles from the population. Evidently, when breeding populations contain fewer than 250,000 individuals, genetic drift has such a severe impact that these populations may be doomed for extinction.

Applying Concepts

1. (a) Polar deserts occur in Antarctica as both the maritime and continental Antarctic ecosystems and in the High Arctic of Canada, Alaska, and Greenland where they are called herb barrens. In the Antarctic, mean monthly summer air temperatures are less than 2°C while in the Arctic mean monthly summer temperatures are 3°–7°C. Moisture is limiting; winters are extremely harsh, and the growing season is very short. Mosses, lichens, algae, fungi, and bacteria are the predominant organisms in both the Arctic and Antarctic. What little soil is present in Antarctica is generally frozen and many organisms survive in rocks and cracks between rocks. The maritime Antarctic contains Antarctica's two flowering plants, *Deschampsia antarctica* and *Colobanthus quitensis*. Flowering plants in the herb barrens are located in areas with higher levels of moisture near snowfields. Plants are adapted to survive drying conditions, are low growing to avoid wind, and in some cases have dark pigments to absorb more heat from the Sun. *Deschampsia* can produce seeds without any sexual process, which allows it to reproduce in an environment with few if any pollinators.

(b) Tundra is widespread and circumboreal in the Northern Hemisphere. Tundra soils are low in nutrients and high in undecomposed organic matter. Generally the soils are frozen (permafrost), moisture is limited, and the growing season is short. High temperatures during the warm months average less than 12°C. Trees are lacking on the tundra, and shrubs and herbs dominate the vegetation. In the Low Arctic, plant communities are more diverse, and grasses, sedges, and shrubs rooted in mosses and lichens dominate large areas of wetlands and bogs. Tundra plants are short and often retain dead leaves and stems to protect them from high winds. Most plants are perennials and store food in underground stems or bulbs to allow for rapid growth in the spring. As in polar deserts, many plants can produce seeds without pollinators.

(c) Boreal forests are also circumboreal; they are sandwiched between tundra to the north and the grasslands and temperate forests to the south. Temperatures and precipitation levels are both low but higher than in the tundra. Soils are acidic and poor both in nutrients and drainage characteristics. Conifers such as black spruce and balsam fir dominate boreal forests; species diversity is low. Below the conifers there is a layer of shrubs, an herb layer, and a surface layer of mosses or lichens. Plants growing here must adapt to short growing seasons (less than 120 days), snowfall, low temperatures, and limited moisture. Conifers have thick cuticles on leaves, are dormant in winter, have shallow roots and associated mycorrhizal fungi to absorb moisture and minerals, and are cone-shaped to allow snow to slide off.

(d) Lakes are depressions in which freshwater has accumulated. Many lakes have a seasonal turnover of water and nutrients. Water is not limiting, but plants and other lake dwellers must cope with other problems: winter covering of snow and ice, buoyancy problems for free-floating species, light availability deep in the water column, high water pressures in deeper waters, and low oxygen levels in the water and especially in the sediments. Lake inhabitants include bacterioplankton and phytoplankton, floating ferns and plants, and rooted plants; the latter cannot grow deeper than about 10 meters due to the water pressure. Rooted plants have long stems or petioles to allow leaves to reach the surface and may have aerenchyma tissue to conduct oxygen to the roots.

(e) Freshwater marshes are extremely productive areas occurring where the water is not acidic and calcium levels are relatively high. Marshes are found growing throughout temperate and tropical areas where there is a transition between terrestrial habitats and rivers or lakes. Plants growing in these marshes must contend with intense sunlight, frequent high winds, and poor oxygenation of the submerged stems and roots. The hallmark plant of the freshwater marsh is the cattail *(Typha),* although certain sedges and grasses are also common. Cattails have aerenchyma tissue to conduct oxygen to the roots, and their leaves are narrow and flexible to withstand wind.

(f) Deserts tend to occur 30° N or S of the equator, where air circulation patterns (Hadley cells) produce dry, descending air masses. Local factors, such as rain shadows and maritime influences, may also be important. Soils are very dry and sandy, with little organic matter. Plant survival strategies include a rapid growth phase with dormant periods during the dry spells (e.g., ephemerals and deciduous perennials) or succulency—conserving water and storing large amounts within plant tissues. Many plants carry on CAM photosynthesis to conserve water. Cacti, agave, and paloverde are characteristic desert plants.

(g) Grasslands are temperate areas dominated by grasses; typical grasslands have fewer than one tree or shrub per acre. Soils are deep, rich, and highly productive, with much organic matter. Grasslands typically have low precipitation, frequent fires, extremes of temperature, and large herds of grazing mammals. Plants adapt to these conditions with a long, deep root system and growing points located at or below ground level where they are shielded from grazers and fires.

(h) Temperate deciduous forests typically have abundant year-round precipitation, an absence of droughts, marked seasonality, winters that include frost and usually snow, and a relatively long growing season. The trees growing in this forest are adapted to the cold winters by losing their leaves. Plants growing on the forest floor survive competition for light by leafing out and producing flowers quickly in spring before the trees leaf out. The soils are deep and rich in organics and nutrients. Trees typical of the eastern temperate deciduous forest include sugar maple, American beech, oaks, hickories, and basswoods.

(i) Tropical rain forests occur in the equatorial regions and have continual high temperatures and humidity, producing year-round growing conditions. The soils are very poor, the nutrients being held in the biomass. Species diversity is extremely high, with numerous species occupying the many niches. Trees occupy various levels of the canopy from top to bottom, and support many epiphytes and lianas. Flowering and fruiting occur year-round. Many flowering plants have dedicated animal pollinators to carry pollen between widely dispersed plants, and many plants produce secondary compounds that are toxic to the abundant insects and other herbivores.

2. Today, the damage to the world's ecosystems has been as extensive as it has been severe; some systems, in fact, have been completely obliterated (for example, the prairie that once occupied the Central Valley in California). It is rather easy to imagine the demise of large vertebrate populations—for example, the shrinking African elephant populations or the extremely endangered black rhinoceros population. Worldwide conservation groups are protecting many of these animals, and many are also represented in the world's zoos. However, we do not even know all of the small, inconspicuous species of animals in the world—for example, all the species of beetles or other insects—so we do not know how many of these have already gone extinct. Today, 34,000 plant species are at risk of extinction, and more than a few have already been lost forever. As important as protection and conservation of large vertebrates is, it seems clear that as extinctions continue to increase, plants and small animals will be hit the hardest.

3. Of course, answers to this question will be varied and numerous. Thus, the following list is merely a sampler. (a) The developed world's consumption of petroleum is not sustainable at current levels. Petroleum is not a renewable resource, and estimates are that the world's oil reserves will run dry before the middle of the 21st century (before 2050). Clearly, current levels of oil consumption cannot be sustained much longer. (b) Much of the world's electrical generation capacity is not sustainable using natural gas, petroleum products, or coal. Although coal and natural gas supplies are predicted to outlast oil supplies, none are renewable. This dilemma is currently sparking renewed interest in alternative energy technologies such as solar and wind power. (c) Our consumption of uncontaminated freshwater is not sustainable at current rates if there is also to be sufficient water to maintain wetland ecosystems. Underground aquifers are being tapped dry, and surface freshwaters are being increasingly diverted or, in some cases, rendered virtually useless via pollution. For example, the waters of the Colorado River no longer reach the Pacific Ocean (they have been diverted). Fresh, clean water is becoming an increasingly rare commodity. (d) Our increasing appetite for most plastics is not sustainable, since the petroleum they are based on is not itself renewable. (e) Food production using current agricultural practices is not sustainable, in part because of the high dependence on petroleum-based energy sources. Nitrogen fertilizer, for example, requires large amounts of energy to manufacture.

4. Again, answers will be numerous and varied; the following list barely scratches the surface, and readers should have no difficulty devising similar lists. (a) Recycle newspapers, to lessen impact on forest reserves and save the energy of paper production. (b) Shower instead of bathe, since typical showers use considerably less water than bathing. (c) If possible, turn off electrical appliances in unoccupied rooms. (d) If possible, reduce heating/cooling in seldom occupied

or unoccupied rooms. Make occupied rooms comfortable, but reduce the heating/cooling to the others. (e) In the cooler months of the year, reduce the thermostat a few degrees and wear a sweater inside. (f) Plan automobile trips so that multiple errands can be accomplished on a single trip. (g) Reduce the amount of packaging entering your home. For example, buy food in bulk (less packaging) and subdivide it at home. (h)

Visit local farmer's markets and buy locally grown produce, if possible. This saves the large transportation costs associated with food that is not produced locally. (i) If there is a recycling center near you, recycle glass, metals, paper, and plastics. (j) Find new uses for what you would otherwise discard. For example, some of the square plastic containers from the grocery store, when trimmed properly, make excellent drawer

organizers. (k) Eat lower on the food chain. Increase the amount of fruits and vegetables in your diet, and decrease the dairy and meat. (l) When painting, choose paints with low levels of volatile organic compounds, because these contribute to atmospheric pollution when they evaporate (and they are simply bad to breathe). (m) Perhaps most important, teach others—especially children—how to be more eco-friendly!

Glossary

ABC model An explanation for the genetic control of the development of different flower parts.

Abscise To drop leaves, flowers, or fruits.

Abscisic acid (ABA) A plant hormone that is involved in dormancy and closure of stomata.

Abscission zone A special region of cells that develops at the base of leaves, flowers, or fruits, and plays a role in leaf, flower, and fruit drop.

Absorption spectrum The pattern of light absorbed by a pigment.

Abyssal plain The bottom of the deep ocean.

Accessory fruit A type of fruit in which the fleshy portion is derived from tissues other than those of the ovary. In apples and pears the core is the ovary, and floral parts and the receptacle form the fleshy outer portions. In the strawberry the receptacle forms the large red "fruit" and the "seeds" are actually fruits called achenes.

Accessory pigments Pigments that capture light energy and pass it along to chlorophyll *a*.

Achene A type of simple, dry indehiscent fruit containing a single seed attached to the fruit wall at only a single point. Sunflower fruits and the fruits embedded on the surfaces of strawberries are examples.

Acid A substance that releases hydrogen ions (protons) in water.

Acid bog A type of freshwater wetland in which the littoral vegetation is dominated by *Sphagnum* moss and the waters are acidic and low in mineral nutrients.

Actinomorphic A type of flower symmetry that is radial (also known as regular flower symmetry).

Activation energy The amount of energy that must be put into a chemical reaction to cause that reaction to proceed and to break and re-form the chemical bonds.

Active site A region of an enzyme that binds chemical substrates, and where chemical modifications of substrates occur.

Active transport Movement of a dissolved substance such as a mineral ion across a membrane that requires the expenditure of energy as ATP.

Adaptive radiation The rapid evolution of a single species into many new species with different forms and habitats. Well-documented examples occur on island chains.

Adenosine triphosphate (ATP) A nucleotide that consists of the base adenine, ribose sugar, and three phosphate groups and functions as the main energy carrier in cellular metabolism.

Adventitious roots Roots developing from an unusual location such as stems or even leaves. From the Latin *adventicius* meaning "not belonging to."

Aerenchyma Spongy tissue having large air spaces within the parts of aquatic plants that facilitates gas exchange and buoyancy.

Aerial roots Roots that extend down from branches to anchor in the soil. In the tropical banyan tree these aerial roots form massive columns that support the heavy branches.

Aerobic Requiring or occurring in the presence of oxygen, such as a chemical reaction.

Aflatoxins Cancer-causing compounds produced by some fungi that may accumulate in foods, especially stored grains, fruits, and spices.

Agar Water-binding polysaccharides derived from red seaweeds that are useful in increasing the viscosity of food products and in preparing nutritive media for laboratory cultivation of microbes.

Aggregate fruits Fruits formed from a single flower with multiple pistils in which each pistil develops into a fruitlet. Raspberries and blackberries are examples.

Akinete A thick-walled resting spore formed from a vegetative cell by many species of cyanobacteria. The akinete allows the cyanobacteria to survive through adverse conditions.

Alga (singular), algae (plural) Cyanobacteria, plus protists that usually (though not always) possess chloroplasts, photosynthetic storage materials, and cell walls. Algae primarily live in aquatic habitats and are typically simpler in structure and reproduction than land plants.

Algin Water-binding, slimy polysaccharides derived from brown seaweeds that have many applications in food and other industries.

Alginates Polysaccharides (polymers of mannuronic and guluronic acids and their mineral salts) occurring within the intercellular matrix of brown algae that confer flexibility, desiccation resistance, and mineral harvesting capacity, and are also useful to humans as industrial materials.

Alkaloids Nitrogen-containing chemical compounds that have physiological effects on animals, protect plants against herbivory, and are widely used as stimulants (such as the caffeine in coffee or nicotine in tobacco products) or medicines (such as the cancer-treatment podophyllotoxin).

Allele One of two or more different forms of a gene.

Allelopathy A type of plant competition in which one plant species produces chemicals that inhibit growth of individuals of the same or different species growing near it.

Allopatric speciation The process of speciation in which populations of a single species become geographically isolated from each other and are subjected to differential evolutionary forces. Allopatric comes from the Greek *allos* meaning "other" and *patra* meaning "fatherland."

Allopolyploidy A common mechanism of sympatric speciation in plants where two distinct species produce a fertile hybrid through a doubling of the chromosome number.

Alternate A type of leaf arrangement in which there is just one leaf or bud at each node.

Alternation of generations A life cycle in which a haploid (*N*) gametophyte produces gametes by mitosis that fuse into a diploid (2*N*) zygote that germinates and grows into a sporophyte. The sporophyte produces spores by meiosis that form new gametophytes.

Alveolates A group of related protists (dinoflagellates, ciliates, and apicomplexans) that characteristically have vesicles (alveoli) at their cell periphery.

Amino acids The nitrogen-containing organic acids that can be linked together to form proteins.

Amyloplasts Non-green, starch-rich plastids with few thylakoids that occur in roots and other starch storage tissues.

Anaerobic Occurring under or requiring the absence of oxygen, as in anaerobic chemical reactions or anaerobic microorganisms.

Analogous A term applied to structures that perform a similar function but have different evolutionary origins. The tendrils of peas and those of grapevines both provide climbing support, but those of peas derive from leaves while the tendrils of grapevines are modified stems.

Anaphase A stage of mitosis in which chromatids separate and travel to different spindle poles; a stage of meiosis in which homologous chromosomes (meiosis I) or chromatids (meiosis II) separate and travel to different poles.

Anaphase-promoting complex (APC) A cluster of proteins that induces the breakdown of cohesin proteins, thus facilitating the separation of chromatids during mitosis.

Angiosperms The flowering plants, whose ovules are produced within an ovary and whose seeds occur within a fruit that develops from the ovary.

Antenna complex The part of a photosystem whose pigments gather light energy and channel it to reaction center chlorophyll.

Anther The part of the stamen that contains the pollen sacs.

Antheridium (singular), antheridia (plural) The type of gametangium in seedless plants that produces sperm and is enclosed in a sterile jacket layer of tissue.

Anthocyanins Water-soluble red or blue pigments that may color vacuoles of flowers, fruits, or leaves.

Antibiotic An organic substance that retards the growth of or kills microorganisms. Examples are penicillin or streptomycin.

Antibiotic resistance genes Genes present in microorganisms that confer resistance to antibiotics.

Anticodon A series of three nucleotide bases that occurs on the basal loop of tRNA molecules, and binds to codons of mRNA that is associated with ribosomes.

Apical dominance Suppression of lateral bud growth by the apical bud or meristem.

Apical meristems Tissues that undergo mitosis and thereby generate new primary tissues at the shoot and root apices of plants.

Apical-basal polarity   Having a distinct top and bottom.

Apomixis   A form of asexual reproduction in which certain flowering plants produce seeds without fertilization.

Apomorphy   A derived character that has evolved from a more primitive character (plesiomorphy) occurring in an ancestor.

Apoplastic   A type of intercellular transport associated with cell membrane transport proteins; contrasts with symplastic transport.

Apoplastic loading   Movement of materials into phloem cells via cell membrane carrier proteins, through cell walls.

Aquaculture   The cultivation of aquatic organisms such as fish, shellfish, and seaweeds for human food and other products.

Arbuscular mycorrhizae   Associations between particular types of fungi whose hyphae form bushy structures between the cell walls and cell membranes of certain plants' root cells.

Arbuscules   Bushy fungal hyphal structures produced between the walls and cell membranes of root cells of particular plants.

Archaea   One of the three great domains of living organisms. The Archaea contains prokaryotic microorganisms often found in extreme environments.

Archegonium (singular), archegonia (plural)   The vase-shaped gametangium in seedless vascular plants in which an egg is produced within a sterile jacket layer of cells.

Arroyo   A small steep-sided watercourse with a flat floor that is usually dry except after rains.

Artificial selection   Breeding organisms so as to obtain desirable characteristics in succeeding generations.

Ascomycetes   A group of fungi characterized by sexual spores produce within sacs (asci).

Ascus (singular), asci (plural)   In ascomycete fungi, a cell in which two nuclei fuse to form a zygote that immediately undergoes meiosis to produce ascospores.

Asexual reproduction   Reproduction that does not involve fusion of gametes or meiosis, and which generates progeny that are genetically identical to the parent.

Assimilation   The process of nitrogen uptake carried out by all organisms.

Atom   The smallest possible unit of an element that has all of the properties of that element; composed of a nucleus of subatomic particles (protons and neutrons) surrounded by a cloud of orbiting electrons.

Atomic mass number   The sum of the number of protons and neutrons in an atomic nucleus.

Atomic number   The number of protons in an atomic nucleus.

ATP   Abbreviation for adenosine triphosphate

ATP synthase complex   The complex of enzymes that forms ATP from ADP and phosphate with the energy derived from the passage of protons down an electrochemical gradient. ATP synthase complexes are located in the thylakoid membranes of chloroplasts and in the inner mitochondrial membranes.

Autopolyploidy   The process by which a single species can give rise to another species by the doubling of its own chromosome number in some of its progeny.

Autosomes   Non-sex chromosomes.

Autotroph   An organism that is capable of producing its own organic food from inorganic materials.

Autotrophy   The capability of producing organic food from inorganic materials; photosynthesis is a form of autotrophy (but not all autotrophs are photosynthetic; chemosynthetic bacteria are autotrophic, but not photosynthetic).

Auxins   A group of plant hormones that are involved in cell enlargement and other aspects of plant development.

Auxotrophy   Having a nutritional requirement for one or more vitamins.

Axil   The upper angle between a stem and a branch or the petiole of an attached leaf.

Axillary buds   Buds that occur in the axil of a leaf.

Bacterioplankton   Bacteria that swim or float freely in aquatic environments.

Bacterium (singular), bacteria (plural)   A member of one of the two domains of prokaryotes (the domain Bacteria).

Basidiomycetes   A group of fungi whose sexual spores are produced at the surfaces of club-shaped structures (basidia).

Basidium (singular), basidia (plural)   A club-shaped cell of basidiomycete fungi in which nuclear fusion and meiosis occur during production of basidiospores at the cell surface.

Benthic   Relating to the bottom of a body of water.

Benthos   The lake habitat consisting of the deeper, dark waters and bottom sediments and their associated organisms.

Berry   A simple fleshy fruit with one to several carpels in which the flesh is soft throughout the fruit.

Beta-carotene   A photosynthetic accessory (carotenoid) pigment that helps plants harvest light during photosynthesis; confers yellow, orange, or red colors to plant parts; and is also an important nutrient in the diets of animals.

Biennial plant   A plant that requires two growing seasons to complete its life cycle. Biennial plants store food reserves from the first growing season and flower and set seed in the second season.

Bilaterally symmetrical   A type of flower symmetry in which the flower can be divided into two equal halves by only one plane of division. Orchids and snapdragons are examples.

Binary fission   The division of single-celled protists or prokaryotes into two progeny cells of equal size.

Binomial scientific name   The two-part scientific name, consisting of generic name and specific epithet, which is given to each unique type of organism.

Biodegradable   Capable of being broken down (degraded) by microorganisms such as fungi and bacteria.

Biodiversity   Term for the biological diversity, the richness of numbers and individuals of different species in an environment.

Biofilms   An organic film consisting of attached bacteria (and/or eukaryotic microbes) and their adhesive mucilage.

Biological nitrogen fixation   The process by which atmospheric nitrogen is reduced to ammonia by certain free-living and symbiotic bacteria.

Biological species concept   The definition of species based on reproductive and genetic isolation. The species is defined as a group of populations whose members interbreed with each other but cannot or do not interbreed with members of other populations.

Biome   Ecosystems with similar climate such as the tropical rain forest biome. Ecosystems in the same biome may contain different communities of species.

Biosphere   The sum of all the biomes with all their organisms on the Earth.

Biotechnology   The genetic manipulation of biological organisms by humans to better suit them for human needs, or the industrial applications of biological organisms.

Bisexual flowers   Flowers that are perfect in the sense that they possess both male (stamens) and female (carpels) reproductive organs.

Blade   The wide, flat portion of a typical leaf (also known as the lamina) or similar structures in kelps (certain brown algae).

Boreal coniferous forest (or boreal forest)   The forest that occurs at high latitude in the Northern Hemisphere and encircles the Earth across Eurasia, Alaska, and Canada. The forest is dominated by coniferous trees such as spruce and fir.

Botany   The scientific study of plants, together with associated microorganisms (prokaryotes and fungi) and photosynthetic protist groups (algae).

Bracts   Modified leaves, some of which are colorful and serve in attraction of pollinators to flowers.

Branch roots   A root that arises from an older root. The youngest branch roots occur closest to the root tip. Also called a lateral root.

Brassinosteroids   Steroid plant hormones that promote cell elongation and help protect plants from heat, cold, salt, and herbicide injury.

Brown algae   A group of mostly marine seaweeds whose chloroplasts are colored brown by the accessory pigment fucoxanthin.

Bryophytes   Mosses, liverworts, and hornworts; the non-vascular embryophytes.

Bud scales   The leaf-like structures that protect developing buds (young shoots).

Bud   A young shoot that will produce new stems with leaves and/or flowers.

Buffer   A substance that is capable of neutralizing acids or bases and thus able to stabilize the original acidity or basic pH of a solution.

Bulb   A short, conical underground stem surrounded by fleshy food-storage leaves.

Bulliform cell   Large epidermal cells that occur in rows in grass leaves, and which are thought to play a role in leaf folding and unfolding.

Bundle sheath One or more layers of parenchyma or sclerenchyma cells that surround the vascular bundles (veins) of some plants.

Bundle-sheath extensions Cells that connect the bundle sheath to the upper and lower leaf epidermis.

Buttress roots Roots produced by the trunks of many tropical trees that act as projecting supports to stabilize the trunks in thin tropical soils and high winds.

C_3 pathway The series of photosynthetic reactions in which 3 carbon dioxide molecules become reduced, giving rise to glyceraldehyde 3-phosphate; also known as the Calvin cycle.

C_3 plants Plants that utilize the Calvin cycle ($= C_3$ pathway).

C_4 pathway The carbon-fixation pathway in which carbon dioxide is first fixed to phosphoenolpyruvate (PEP) to yield the 4-carbon compound oxaloacetate.

C_4 plants Plants that use the C_4 pathway for initial carbon dioxide fixation and the C_3 pathway or Calvin cycle for production of simple sugars.

Calcium-binding proteins Proteins that occur in the cell membrane, cytoplasm, or nucleus of eukaryotic cells and that bind calcium ions and then interact with other cell proteins, in the process transmitting signals from the outside to the inside of cells.

Callose A complex branched polysaccharide that blocks up plasmodesmata at the site of wounds in plants.

Callus A mass of undifferentiated plant cells.

Calyx The whorl of sepals in a flower.

CAM plants Plants that employ crassulacean acid metabolism and use the C_4 pathway at night to fix carbon dioxide and the C_3 pathway in the daytime to produce simple sugars.

Capsule A simple, dry dehiscent fruit consisting of two or more carpels that may split open in a variety of ways. Capsules may split open along the carpels, form a cap that pops off, or form rows of pores to shed seeds like a salt shaker.

Carbohydrates Organic molecules composed of carbon, hydrogen, and oxygen in the ratio of 1 : 2 : 1; sugars, starch, and cellulose are examples of carbohydrates.

Carbon dioxide An atmospheric gas that is used by plants in photosynthesis to generate organic food; also a so-called "greenhouse gas" whose atmospheric increase has been related to global warming.

Carbon sink Any site of long-term carbon storage, such as peatlands and the woody trunks of tropical trees.

Carpel A folded and fused leaf-like structure that contains ovules.

Carpellate Flowers that lack stamens.

Carrageenan A polysaccharide component of the extracellular matrix of red algae that is used in the food industry.

Carrying capacity The maximum size of a population of organisms that the environment can support.

Caryopsis The simple indehiscent dry fruit found in cereals such as corn, wheat, and barley, and also called a grain. The fruit contains a single seed whose outer seed coat is fused to the fruit wall.

Casparian strip A ribbon-like layer of water-repellent suberin that is wrapped around the top, bottom, and side walls of endodermal cells in the innermost cortex layer.

Catalyst A substance that speeds up the rate of a chemical reaction without itself being consumed; enzymes function as catalysts.

Cell A unit of cytoplasm that is delimited by the cell membrane.

Cell biology The branch of biology that specializes in the study of cells.

Cell cycle The sequence of events leading to, and including, cell division.

Cell division The formation of two progeny cells from one parental cell.

Cell membrane The outer membrane composed of phospholipid molecules arranged in a bilayer (together with associated proteins, and often sterols) that surrounds all eukaryotic and prokaryotic cells. Also called the plasmalemma or plasma membrane.

Cell plate A structure that forms at the equator of a spindle during early telophase of mitosis (or meiosis) and which, by the end of cytokinesis, has developed into the middle lamella.

Cell wall The rigid portion of the extracellular matrix of plants, fungi, bacteria, and many protists.

Cellulose A carbohydrate formed of long chains of glucose molecules that is a major component of plant cell walls (and those of some protists and some bacteria). Cellulose is also produced by various bacteria for use as an attachment material.

Cellulose-rich cell wall The outermost rigid layer of cellulose fibers that in plant cells lies outside the cell membrane.

Centromere The often constricted region of a chromosome where chromatids are held most closely together.

Cereals Plants whose fruits are grains; examples include corn (maize), wheat, barley, and rice.

Characters Structural, biochemical, or molecular features of organisms that are used to deduce evolutionary relationships.

Checkpoint Processes that occur during the cell cycle that allow the cycle to continue, unless defects are present, in which case the cell cycle stops.

Chemical bonds Ionic, covalent, or electrostatic (hydrogen bond) interactions that hold atoms, ions, molecules, and compounds together.

Chemical-biological theory The theory that treats the origin of life on Earth as a process consisting of a series of stages with increasing levels of organization.

Chihuahuan Desert North American desert in northern Mexico and parts of southern Texas and New Mexico. Big Bend National Park lies in the Chihuahuan Desert.

Chinampas agriculture A type of cultivation used by past and present inhabitants of Mexico to grow crops in rows of raised ground in swampy regions.

Chitin The nitrogen-containing polysaccharide that forms much of the cell walls of fungi, and is also produced by some animals.

Chlorarachniophytes A group of protists included among the filose amoebae (Rhizaria) having spider-shaped cells that contain secondary green plastids.

Chlorophyll The green photosynthetic pigment of plants, many protists, and some bacteria.

Chlorophyll a The form of chlorophyll that is most directly associated with conversion of light energy into chemical energy.

Chlorophyll b An accessory pigment that transfers absorbed light energy to chlorophyll a; found in green algae, a few other protists, and land plants.

Chloroplast The organelle or plastid that contains chlorophyll in plants and algae and is the site of photosynthesis.

Chromatids The two daughter strands of a chromosome that are produced by DNA replication and remain attached at the centromere.

Chromatin The darkly staining colored material composed of DNA and proteins in eukaryotic chromosomes.

Chromoplasts The plastids that contain bright yellow, orange, or red carotenoid pigments.

Chromosomes Threadlike arrays of DNA with associated histone proteins; found in the nuclei of eukaryotes.

Chytrids A group of early-diverging fungi whose hyphae lack cross-walls.

Cilia (plural), cilium (singular) Short, hair-like extensions from some eukaryotic cells that can bend and flex, allowing the cells to move through water.

Ciliates A group of protists whose surfaces are completely or partially covered with cilia (short extensions of the cell that function in motility).

Cisternae The pancake-like membrane sacs that are arranged in stacks to form Golgi bodies.

Citric acid cycle The alternate name for the Krebs cycle in which the first product formed is citric acid.

Clade A group of related organisms that is defined by at least one shared, derived character.

Cladistic analysis A technique of phylogenetic analysis whereby organisms' relationships are inferred by examining the distribution of their characteristics.

Cladistics A formal method of organizing organisms on the basis of relationships established by the presence or absence of derived characters.

Classes A category of classification in which related orders of organisms are grouped together.

Climax community A mature community formed toward the end of a process of ecological succession. Climax communities show very slow changes in species composition.

Closed nutrient system The system in tropical rain forests by which nutrients are passed from decaying organisms and detritus to living through mycorrhizal fungi without leakage into the soil.

Cocci The term for the spherical shape of many prokaryotic cells.

Coccolithophores Single-celled or colonial planktonic organisms that are covered with scales composed of calcium carbonate.

Coccoliths Calcium carbonate scales produced by many haptophyte algae.

Codons Sequences of three nucleotide bases, known as triplets, which occur in linear arrays on messenger RNA.

Coenzyme A (CoA) The coenzyme to which the 2-carbon acetyl group is attached to yield acetyl CoA.

Coenzymes Organic molecules that assist enzymes in performing their functions.

Coevolution Reciprocal evolutionary influences exerted by two or more organisms on each other.

Coevolutionary plant–animal associations Associations between plants and animals that have influenced the evolution of both participants.

Cofactors Nonprotein components such as metal ions that are required by certain enzymes for their function.

Cohesin A type of protein that links chromatids together prior to their separation in the final stages of mitosis and meiosis.

Cohesion-tension theory The concept that water movement in xylem tissues can be ascribed to two related phenomena, the cohesion of polar water molecules that allows water to exist as thin streams that are resistant to breakage, and the tension placed on a stream of water within a plant when water molecules at the plant's surface are removed by transpiration.

Collenchyma A type of flexible supporting tissue composed of collenchyma cells.

Collenchyma cells Elongated living cells whose primary walls are unevenly thickened with the carbohydrate pectin.

Colonies A type of microbe body in which few to many cells are held together by mucilage or cell-wall material.

Common names The informal names given to organisms by people living in the same area.

Community All the organisms of different species in a particular area.

Companion cell A type of specialized cell that is produced by the same division that gives rise to a sieve tube member in angiosperm phloem; provides materials to sieve tube elements.

Competition A negative interaction among organisms that utilize the same environmental resources.

Complementary The relationship between the two strands of nucleotides in DNA whereby the base A is linked to the base T, and the base G is linked to the base C, via hydrogen bonding. Complementary DNA (cDNA) is a single-stranded molecule of DNA that has been synthesized by using messenger RNA (mRNA) as a template and the enzyme reverse transcriptase.

Complete Flowers that possess all four floral whorls: sepals, petals, stamens, and carpels.

Compound Any chemical substance that contains two or more elements, joined by covalent bonds, in definite proportions.

Compound leaves Leaves having blades that are divided into two or more leaflets.

Compound light microscope A light microscope that has an objective lens and one or two eyepieces (ocular lenses) mounted on the same tube.

Cones Club-shaped clusters of sporophylls containing sporangia. Also called strobili.

Conidia Asexual spores that are produced at the tips of fungal hyphae.

Constancy The ability of pollinators to recognize and repeatedly visit specific flower species. Also called fidelity.

Continental shelf A shallow nearshore plain of varying width that borders a continent and a slope to the ocean bottom.

Continental slope The often steep slope from the continental shelf to the ocean floor.

Contractile roots Roots on certain herbaceous plants, such as dandelions and hyacinths, which are able to shorten by collapsing their cells.

Contractile vacuole Membranous sacs in some freshwater protists that accumulate excess water in the cell and periodically expel it.

Controls The unmanipulated replicates in an experimental design used for comparison to the experimental replicates to decrease the probability that observations are in error.

Convergent evolution The process by which unrelated organisms in a similar environment evolve similar adaptive structures and physiology.

Coralline algae Red algae whose extracellular matrix is rich in calcium carbonate.

Coralloid roots Thickened, branched roots that somewhat resemble corals that are produced above-ground by cycads, and which contain nitrogen-fixing cyanobacterial endosymbionts.

Cork A tissue, composed of dead cells with thick suberized cell walls (also known as phellem) that is produced by a cork cambium and lies at the outer part of a plant stem's periderm.

Cork cambium A meristematic tissue that generates cork cells toward the periphery of a woody plant's stem or roots, and may also produce some parenchyma cells toward the inside; also known as phellogen.

Corm A type of underground stem with food storage.

Corolla The whorl of petals in a flower.

Cortex The tissue in roots and stems lying between the epidermis and the vascular tissues.

Cotyledon The embryonic seed leaves in flowering plants that contain the nutritive tissue derived from the endosperm.

Covalent bond The strongest type of chemical bond, in which electrons are shared between elements.

Crassulacean acid metabolism (CAM) The metabolic pathway in which carbon dioxide is fixed at night by the C_4 pathway and simple sugars produced in the day by the C_3 pathway. The name derives from the stonecrop family of flowering plants (Crassulaceae) where this pathway was first recognized.

Cristae The folds of the inner membrane of the mitochondrion where the enzymes for electron transport and oxidative phosphorylation are located.

Cross-fertilization In plants, the fusion of gametes derived from flowers on different individuals.

Crossing over The exchange of corresponding segments of chromatids of homologous chromosomes during meiosis that results in new gene combinations.

Cross-pollination The transfer of pollen between different individuals of the same species.

Cryptochrome A blue- or ultraviolet-absorbing pigment that helps plants to discern their light environments.

Cryptomonads A group of unicellular, flagellate algae that have brown, red, or blue-green-colored chloroplasts (or in a few cases, are colorless).

Cultivar A variety of plant that occurs only in cultivation.

Cuticle Material found on the plant epidermal surface; usually includes both a microbe-resistant polyester (cutin) and waxes that help plants retain water.

Cutin A polyester material deposited by plant epidermal cells onto their outer surfaces, which helps prevent attack by disease microorganisms; a portion of plant's waxy cuticle.

Cyanobacteria Prokaryotic cells containing chlorophyll a and capable of photosynthesis that produces oxygen.

Cyclic electron flow The light-driven flow of electrons through photosystem I that produces ATP.

Cysts Resistant stages produced by many protists that enable survival in stressful periods or environments.

Cytokinins A class of compounds that function as plant hormones.

Cytoplasm The watery living material of the cell inside the cell membrane and outside the nucleus.

Cytoplasmic streaming The rapid movement of cytoplasm within cells due to the action of various motor proteins.

Cytoskeleton The skeleton of a cell composed of protein fibers.

Day-neutral plants Angiosperms that lack a specific daylength requirement for flowering.

Dead zone A region of near shore ocean water that is depleted of oxygen to the point that animal life cannot exist, as the result of excess mineral pollution from sewage and agricultural effluents supplied by rivers.

Decomposers Bacteria, fungi, and other organisms that break down dead organisms and organic matter, releasing smaller components; release mineral nutrients so that they are available for uptake by living plants or algae.

Deductive reasoning The scientific process that begins with the formulation of a hypothesis, a preliminary explanation based on some prior knowledge, and the design of an experiment to test the validity of that hypothesis.

Deep-water swamps Swamps found in warm regions that typically contain large trees such as cypress. The Okefenokee Swamp in southern Georgia is an example.

Dehydration Loss of water; a type of chemical reaction characterized by loss of water molecules.

Dehydration synthesis The synthesis of organic polymers from monomers by the removal of water during the formation of each new bond.

Dermal tissues The outermost tissues of a plant; the epidermis or the periderm (of woody plants).

Desertification The conversion of grasslands and semiarid lands into desert-like regions through human overexploitation, usually by overgrazing, the extension of farming into marginal lands, and removal of plant cover for fuel.

Desiccation tolerance The ability to undergo significant tissue water loss without dying.

Determinate growth Growth that occurs only for a limited time period.

Development The process whereby a single cell is transformed into a multicellular adult body by cell division, cell growth, and cell specialization.

Diatoms A group of the strameopile (heterokont) protists that have porous silica enclosures (walls).

Dichotomous key A tool used for organism identification, whereby the user makes a series of choices between two alternative character descriptions.

Differentiation A developmental process during which a relatively unspecialized cell or tissue becomes specialized in form or function.

Diffusion The net movement of particles such as ions or molecules from a more concentrated area to a less concentrated area by the random motion of the particles.

Dihybrid cross A cross between two plants or varieties that differ in two traits.

Dinoflagellates A group of mostly unicellular, flagellate algae that may have golden (or less often green or blue-green) chloroplasts, or lack chloroplasts.

Dioecious Plants in which male (staminate) and female (carpellate) flowers are found on separate plants, such as holly, date palm, and willow.

Dioxin A highly potent carcinogen (cancer-causing agent) that is inadvertently produced during the chemical manufacture of certain herbicides and during wood pulp processing.

Diploid Having two sets of chromosomes.

Directional selection A process of selection that favors individuals with a trait at the extreme of its character range over individuals with the average or opposite extreme for that trait.

Disaccharides Carbohydrates consisting of two sugars.

Discicristates A group of protists (including euglenoid flagellates) that have disk-shaped mitochondrial cristae.

Disruptive selection A process of selection that favors individuals with the extremes of the range of a trait over those with the average for that trait.

Dissecting microscope Light microscopes used for the study of specimens that are too large or thick to observe under a compound microscope.

DNA Deoxyribonucleic acid; the genetic information storage material.

DNA fingerprint The product of a process whereby DNA is digested into smaller fragments that are separated to form banding patterns on a gel material, or a series of peaks representing the size distributions of the DNA fragments; can be used to link DNA evidence with its source, as in criminal investigations.

DNA ligase The enzyme that catalyzes the covalent bonding of the $3'$ end of a DNA fragment to the $5'$ end of another DNA entity.

DNA polymerases Enzymes that help link nucleotides together to form the nucleic acid DNA.

DNA repair nucleases Enzymes found in the cell nucleus that help repair damaged DNA.

DNA science Methods and concepts associated with the study of DNA and other nucleic acids.

Domestication The transformation of wild plants or animals into agricultural crops or livestock.

Dominant The form of a trait that is expressed in the next generation when two true-breeding varieties with contrasting forms of that trait are crossed.

Dormancy Condition in which seeds, spores, or tissues within plants do not grow until specific environmental signals or cues occur.

Double fertilization A unique characteristic of flowering plants in which two sperm nuclei fuse with separate nuclei to produce a diploid zygote and a triploid endosperm.

Double helix Two polymer strands twisted about each other, as in DNA.

Drupe A simple fleshy fruit in which the inner layer of the fruit forms a hard stony pit tightly bound to the seed. Cherries, peaches, and olives are examples.

Dynein A motor protein that transforms the chemical energy of ATP into cell motion.

Ecology The study of organisms' interactions with their environments and with each other.

Economic botany The study of the uses of plants by modern industrialized societies.

Ecosystem The community of organisms in a defined area plus its physical environment.

Ecosystem services Services provided by the operation of ecosystems such as soil formation and enrichment, water purification, oxygen production and carbon dioxide storage, nitrogen fixation, temperature control, and crop pollination.

Ectomycorrhizae Beneficial symbioses between temperate forest tree roots and soil fungi whose hyphae coat tree-root surfaces. Trees provide organic food to the fungi, and receive mineral nutrients harvested from soil by the fungi.

Egg A female gamete, usually non-motile and larger than the corresponding male gamete (sperm).

Electron microscope A microscope that has about 100 times the magnifying power of a light microscope because it forms images with electrons and magnetic lenses.

Electron transport chain A linked series of electron acceptor molecules that pass electrons down them in one direction and lower the energy of the electrons in the process.

Element A substance composed of only one kind of atom.

Embryo A young stage of an organism that develops from a single-celled zygote by cell division and differentiation.

Embryo sac The mature female gametophyte or megagametophyte in angiosperms.

Embryonic root The first plant organ to emerge when seeds germinate. Also called the radicle.

Embryophytes Photosynthetic organisms whose embryos develop within the protective, nutritive confines of maternal tissues; the land-adapted plants (and aquatic plants descended from embryophyte ancestors).

Empirical Based upon experiment and observation.

Endemic species Species that occur only in a particular area and are usually rare.

Endergonic A type of chemical reaction that requires an input of energy to proceed. In endergonic reactions, the potential energy of the reactants is less than that of the products.

Endocytosis The formation of a pocket in the cell membrane that pinches around an external particle or dissolved material and forms a vesicle around it. The vesicle then detaches from the membrane and floats free in the cytoplasm. Endocytosis brings material into a cell.

Endodermis The cells of the innermost cortex layer that form a barrier between the cortex and the vascular tissue.

Endomembrane system The internal membrane system of eukaryotic cells that consists of the endoplasmic reticulum (ER), Golgi bodies, and vesicles and vacuoles.

Endomembranes The phospholipid membranes with their embedded proteins that occur within the cytoplasm.

Endomycorrhizae Mutually beneficial partnerships between plants and fungi that grow into plant roots; the plants provide organic food and the fungi contribute mineral nutrients obtained from soil.

Endophytes Microbes (especially fungi) that live compatibly within the tissues of plants.

Endoplasmic reticulum (ER) A network of flattened membrane sacs or tubes and vesicles that arise from them. Smooth ER makes fatty acids and phospholipids; rough ER makes proteins.

Endosperm The triploid nutritive tissue formed as one result of double fertilization in angiosperms.

Endospores Spores produced inside certain bacterial cells as an adaptation to survive environmental stress. The spores are released upon the death of the parental cell.

Endosporic In modern seed (and some non-seed) vascular plants, the gametophytes are formed endosporically, literally "inside the spore." In flowering plants, for example, the male gametophyte lies inside the microspore wall.

Endosymbiont An organism that lives stably within the cells or tissues of another (host) cell or organism.

Endosymbiosis The incorporation of entire living organisms into a host cell by phagotrophy.

Endosymbiotic theory The theory that the chloroplasts and mitochondria of eukaryotic organisms arose by endosymbiosis of prokaryotic organisms. A wide range of observations and experiments support this theory.

Environmental remediation The use of organisms such as bacteria to process wastewater, clean up toxic wastes and spills, and break down various substances that endanger human health and the environment.

Enzymes Proteins that act as biochemical catalysts and regulate the rates at which biological processes occur within cells.

Epicotyl The upper portion of the shoot of an embryo or seedling that lies above the seed leaves (cotyledons) and below the next leaf.

Epidermis The outermost, primary tissues of plants; located at the surfaces of leaves and young roots and stems.

Epipelagic Referring to the upper layer of ocean water from the surface to the depth allowing sufficient light to penetrate to allow photosynthesis.

Epiphytes Plants that grow on the surfaces of other plants.

Epistasis A genetic condition in which two genes combine to produce a trait. A defect in either gene prevents expression of the trait.

ER See endoplasmic reticulum.

Ethnobotany The study of past human uses of plants or present-day plant uses by traditional societies.

Ethylene A gaseous plant hormone that is involved in fruit ripening, leaf abscission, and other plant processes.

Eudicots The largest and most diverse group of flowering plants, distinguished by pollen with three pores or furrows.

Euglenoids A group of discicristate protist flagellates characterized by protein strips lying just beneath the cell membrane, and carbohydrate food storage granules known as paramylon in the cytoplasm; some contain green plastids.

Eukarya One of the three domains, along with Archaea and Bacteria, which include all the organisms on Earth. The Eukarya is the eukaryotes.

Eukaryotes Organisms whose cells have their DNA contained within a membrane-bound nucleus.

Eukaryotic cells Cells possessing a nucleus, cytoskeleton, and other features lacking from prokaryotic cells.

Euphylls Leaves having a branched vascular system that leaves a leaf gap in the stem's vascular system at the point where the leaf's vascular system connects to that of the stem.

Eutrophic waters Waters having a high mineral content and primary productivity but low species diversity.

Eutrophication The process by which clean waters are converted into polluted or eutrophic waters.

Evenness The distribution of individual organisms among the species in a community.

Evolution The process by which organisms originate and change through time as a result of natural selection and adaptation to their environments.

Evolutionary relationships Relationships among organisms based on ancestry.

Excited state The condition of an electron when it absorbs energy from visible light and jumps into a higher orbital position around the nucleus of its atom.

Exergonic A type of chemical reaction in which the potential energy of the products is less than that of the reactants and energy is released.

Exocytosis The process in eukaryotes by which particles or dissolved substances may be delivered outside the cell membrane by enclosing them in vesicles. The vesicle fuses with the cell membrane and dumps the contents to the outside.

Exons Regions of DNA that encode proteins or other nucleic acids.

Experiment A test or trial; a procedure or operation undertaken to test a principle or hypothesis.

Extinct No longer living, as in dinosaurs are extinct.

Extinction The process by which a species becomes extinct. Human activities are driving many species to extinction through habitat fragmentation, global climate change, and spread of invasive species.

F_2 The second filial generation obtained by crossing members of the F_1 or first filial generation.

Facilitated diffusion Passive transport by carrier proteins that does not require energy input.

Families Aggregates of related genera of organisms.

Fatty acids The monomeric constituents of fats, oils, and other lipids.

Feeder roots Fine peripheral roots active in uptake of water and mineral ions in the soil and often very near the soil surface.

Female gametophyte The megagametophyte formed from the megaspore. In most flowering plants, the female gametophyte consists of seven cells.

Fermentation The process that follows glycolysis under anaerobic conditions and leads to a build-up of lactate or ethanol.

Fertile Crescent A region of the modern-day Near East that is associated with early evidence for the origin of agriculture.

Fibers A type of specialized plant cell that is elongate, and usually has a tough cell wall that may or may not include lignin; these cells may be alive or dead at maturity.

Fibrous system The system of roots common in monocots where there are many highly branched roots rather than a single prominent root.

Filament (1) A type of microbial body in which cells occur in a linear array; some filaments are branched, and some are unbranched. (2) The stalk-like structure that bears the anthers in flowering plants.

First filial generation (F_1) The hybrid offspring obtained by crossing two true-breeding parental varieties that differ in one or more traits.

Fixed nitrogen Ammonia and the nitrogen-containing compounds derived from it that can be used by plants and microorganisms.

Flagellates Organisms that can swim because their cells bear flagella; may occur as single cells or motile colonies.

Flagellum (singular), flagella (plural) Elongate extensions present on eukaryotic cells that can bend and flex to move the organism through water.

Flavonoid A type of phenolic compound that may function in plants in UV protection or production of some types of colors in flower petals or other plant parts.

Floating plants Any of various bryophytes, ferns, or flowering plants that form surface growths in sheltered areas of lakes, wetlands, and canals.

Flower The reproductive structure of angiosperms that contains at least stamens or one or more carpels, and usually both of these plus petals and sepals.

Follicle A dry, dehiscent fruit formed from a single carpel that splits along one side to release seeds. Examples are milkweeds and columbines.

Food chain A chain of organisms in a natural community in which each link in the chain consumes the organisms below it and is consumed by the organisms above it.

Food web The network of interrelationships between producers and various consumers and decomposers in an ecosystem.

Fossil (Latin *fossils* meaning "dug up") The remains, impressions, or traces of organisms that have been preserved in sedimentary rocks.

Fossil fuels Coal, oil, and natural gas (methane); the altered remains of ancient plants and protists that are burned for use as energy sources by humans.

Founder effect The loss of genetic diversity due to genetic drift when a small number of organisms found a new population.

Freshwater marshes Freshwater wetlands characterized by the presence of cattails, sedges, and certain grasses.

Fruit The structure that develops from the ovary (or several ovaries), and sometimes adjacent flower parts, after pollination and fertilization have been achieved; a ripened ovary bearing seeds.

Fruiting bodies Reproductive structures of fungi that produce the sexual spores, and usually emerge from substrata (e.g., mushrooms).

Functional groups Groups of organisms that play the same ecological roles in an ecosystem.

Fungi The kingdom of heterotrophic, eukaryotic organisms characterized by presence of chitin-containing cell walls, absorptive nutrition, and reproduction by means of spores.

Fusiform initials Vertically elongate cells of the vascular cambium that give rise to vertically-arranged secondary xylem and secondary phloem of woody plants.

G_1 phase The portion of the cell cycle between a previous mitosis and the S phase (in which DNA is replicated).

Gametangium (singular), gametangia (plural) A structure in which gametes are formed.

Gametes The haploid reproductive cells (produced by adult individuals) that are specialized for mating.

Gametophore In bryophytes, the portion of the gametophyte body that bears the gametangia.

Gametophyte The gamete-producing body of a plant or an alga; usually assumed to be haploid.

Gemma (singular), gemmae (plural) In bryophytes, a small piece of the gametophyte body that can develop into a new plant. Production of gemmae is a form of asexual reproduction.

Gene cloning The process of inserting a selected gene into a vector to transform an organism such as a bacterium for the purpose of producing multiple copies of that gene.

Gene flow The movement of genes between populations caused by the migration of individual organisms (or their reproductive structures, such as pollen, seeds, or gemmae) between those populations.

Genera Groups of related species of organisms.

Generic name The noun that forms the first component of an organism's scientific name.

Genes The units of genetic information that contain the coded information for synthesis of proteins and nucleic acids.

Genetic code The series of nucleotide triplets in DNA and codons in RNA that determine the sequence of amino acids in proteins; each codon specifies one of 20 amino acids or three "stop" signals.

Genetic drift The loss of an allele from a small population by random or chance events.

Genetic recombination The occurrence of new gene combinations in the progeny of parents. Crossing over is one mechanism that gives rise to new combinations of alleles.

Genomic library A library of DNA fragments that results from cloning all the genes of a particular organism into suitable bacterial hosts.

Genophore Term for the circular DNA molecule that contains most of a prokaryotic cell's genes.

Genotype The genetic makeup of an organism.

Giant herbs Plants found in tropical forests that are as tall as trees but lack wood produced by a vascular cambium.

Gibberellins A group of chemically related plant hormones that control cell elongation and other plant growth and development processes.

Glaucophytes A group of protists that have blue-green primary plastids that are similar to cyanobacterial cells in some ways.

Global warming Increase in global atmospheric, ocean, and surface temperatures that has been linked to increase in atmospheric "greenhouse gases" resulting from human activities since the beginning of the Industrial Age.

Glycerol A 3-carbon compound to which three fatty acids can be attached to form lipids, or to which two fatty acids plus a phosphate can be attached to form phospholipids.

Glycogen A carbohydrate that serves as an energy-storage molecule in fungi and various organisms other than plants.

Glycoproteins Proteins to which chains of sugars are added in preparation for their transport outside the cell.

Glyoxysomes Organelles that transform storage lipids into sugar for use as energy in seed germination.

GM organisms Genetically modified organisms whose DNA has been altered so that they are more useful to people.

Golden algae The informal name for a large and diverse group of algae that usually contain chloroplasts colored golden or brown by large amounts of the accessory pigment fucoxanthin; formal name Ochrophyta.

Golgi apparatus The apparatus consisting of one or more Golgi bodies, each of which is a stacked array of membrane sacs. The Golgi apparatus produces, modifies, and distributes cell materials to their correct locations.

Golgi bodies The stacked array of membrane sacs that make up the Golgi apparatus.

Gradualism The Darwinian idea that evolution proceeds at a steady or gradual rate as genetic changes slowly accumulate.

Grains The simple indehiscent dry fruit found in cereals such as barley and wheat. See also caryopsis.

Grass flowers Small, often greenish, and inconspicuous flowers that lack petals and are wind pollinated.

Grass fruits See grains and caryopsis.

Gravitropism The growth responses of shoots or roots to the force of gravity.

Great Basin Desert One of the four main North American deserts, the Great Basin Desert lies at high altitude between the Sierra Nevada and Cascade Ranges in California and the Rocky Mountains of Colorado.

Green algae The informal name for a large and diverse group of algae that usually contain green chloroplasts; formal name Chlorophyta.

Greenhouse gases Gases that absorb heat, thereby warming Earth's atmosphere; water vapor, methane, and carbon dioxide are the three most important.

Ground meristem One of the three primary meristematic tissues that give rise to the primary tissues. Ground meristem gives rise to the cortex and pith tissues.

Ground tissues All primary tissues of shoots and roots other than the epidermis and vascular tissues.

Growth To increase in size by the addition of material.

Guard cells The specialized epidermal cells that form the boundaries of stomata, the pores in plant surfaces; changes in the water pressure in guard cells cause the pores to open (allowing gas exchange) or close (to retain water within plants).

Guttation The production of water droplets at the surfaces of leaves, as the result of root pressure when soil moisture level is high.

Gymnosperms The group of embryophytes that have seeds not enclosed in fruits (ovules are not enclosed in ovaries); includes conifers, *Ginkgo*, cycads, and Gnetales.

Gyres Giant surface ocean currents having a circular motion.

Habitable zone The zone around a particular star where water can be stable as a liquid.

Halophytes Plants able to grow in coastal salt marshes or saline deserts. From the Greek *hals* "salt" and *phytos* "plant."

Haploid Having only one set of chromosomes; contrast to diploid.

Haptophytes A group of protists that have a haptonema, a coiled flagellum-like structure that aids in feeding or obstacle avoidance.

Heat shock proteins Proteins produced in response to heat stress and which help to prevent other cell proteins from losing function by coagulating.

Helicases Enzymes that untwist DNA.

Hemicellulose A polysaccharide found in cell walls that is similar to cellulose but more soluble.

Herb Nonwoody flowering plants with a short-lived stem or stems. Herbaceous (adjective): With the characteristics of an herb.

Herbaceous Referring to non-woody plants.

Herbaria Collections of identified, dried or otherwise preserved plant specimens.

Herbivores Animals that consume plants or algae.

Herbivory A type of nutrition in which parts of plants, fungi, or algae are consumed without killing them.

Hesperidium A type of berry with a leathery rind containing oil packets.

Heterocysts Specialized cells, which are produced by certain cyanobacteria, in which nitrogen fixation occurs.

Heterokonts A group of protists that produce cells having two structurally-distinct flagella (=stramenopiles).

Heterosis Alternate term for hybrid vigor—the phenomenon where hybrids are often larger, more vigorous, and produce larger seed crops than any pure-breeding single variety.

Heterospory The formation of two different types of spores, microspores and megaspores.

Heterotrophic Requiring organic food from the environment.

Heterotrophs Organisms that are not autotrophic, and thus must obtain organic food from their environments.

Heterozygous An individual that possesses two different alleles for a single gene.

Histones Proteins that coat the nuclear DNA to form chromosomes of eukaryotes.

Homologous Structures that have a common evolutionary origin but may have different present functions and structures.

Homologous chromosomes Members of pairs of chromosomes (one derived from each parent) that are similar in size and gene loci.

Homospory The formation of spores that are all of about the same size.

Homozygous An individual that possesses two copies of the same allele of a gene.

Hormones Chemical substances produced in one part of a plant or animal body that move to other locations where they influence development and function.

Hornworts One of the three groups of living bryophytes. Hornworts are small herbaceous plants named for the horn-like shape of their sporophytes.

Horticulture The cultivation of flowers, fruits, and ornamental plants.

Host The organism on or in which a parasite lives and feeds without killing it.

Hybrid vigor See heterosis.

Hybrids The result of crossbreeding between two varieties, species, or genera.

Hydrogen bond A relatively weak electrostatic attraction between portions of two or more molecules.

Hydrophilic A term used to describe substances that readily dissolve in water.

Hydrophobic A term used to describe substances that do not readily dissolve in water, but which do readily dissolve in lipids.

Hydrophytes Plants that grow in very wet environments where they are wholly or partly submerged in water.

Hydroponics Growth of plants in nutrient-rich water rather than in soil.

Hypereutrophic A lake having very high levels of mineral nutrients and with persistent cyanobacterial blooms.

Hypersensitive response A response of plants to attack by microbial pathogens that involves localized tissue death as a way of containing the disease microbes.

Hypertonic Having a concentration of dissolved solutes great enough to draw water across a cell membrane. If a cell is immersed in a hypertonic solution, the concentration of dissolved substances is greater in this solution than the concentration inside the cell, and water will be drawn out of the cell.

Hyphae The tubular filaments that compose the body (mycelium) of a fungus.

Hypocotyl The portion of an embryo or seedling that lies between the seed leaves (cotyledons) and the radical (embryonic root).

Hypothesis A tentative explanation for some observations or phenomena that serves as a guide in designing an experiment.

Hypotonic If a cell is immersed in a hypotonic solution, the concentration of dissolved substances in that solution is less than the concentration of those substances inside the cell, and the cell will tend to take on water from the solution.

Identification keys Book or electronic resources that can be used to identify organisms.

Imperfect A flower is imperfect (and unisexual) if it lacks either stamens or carpels.

Imperfect fungi Ascomycete or basidiomycete fungi that have lost the ability to produce sexual spores.

Inbreeding The combination of two genomes from organisms that are genetically similar, as in self-pollination in plants.

Incomplete A flower is incomplete if it lacks one or more of the four basic whorls of flower parts.

Incomplete dominance The genetic condition in which neither of two alleles of a gene is dominant over the other, and the heterozygous individual has a phenotype intermediate between the two homozygous phenotypes.

Indeterminate growth Relatively unrestricted or unlimited growth; for example, production of many nodes, leaves, and other organs by a single plant apical meristem.

Inductive reasoning The scientific process that begins with the collection and analysis of data and observations and leads to the formulation of a generalization called a theory.

Industrial nitrogen fixation The industrial process by which atmospheric nitrogen is fixed or reduced to ammonia under high temperatures and pressures.

Inferior When the sepals, petals, and stamens are attached above the ovary or ovaries of a flower, the ovary or ovaries are said to be inferior.

Inflorescence A cluster of flowers.

Inorganic acid An acid that does not contain a carbon-hydrogen bond.

Inorganic compound A compound that contains no carbon, only a single carbon atom, or carbon atoms that are not covalently bonded to other elements; using this definition, water, carbon dioxide, and diamonds are inorganic compounds.

Integral membrane proteins Proteins embedded in the phospholipid bilayer of cell membranes; integral membrane proteins act as transporters of nutrients, wastes, or chemical signals.

Intermediate filaments One of three types of long, thin protein fibers that comprise the cytoskeleton. Intermediate filaments help maintain nuclear shape.

Internode The space along a stem between nodes (locations of leaf and bud emergence).

Interphase The portion of the cell cycle between successive cell divisions.

Introns Regions of genes that do not encode proteins or nucleic acids; occur between exon segments.

Ion A charged atom that has either gained or lost electrons.

Ionic bonds Chemical linkages based on the electrical attraction of positively charged regions of one molecule to negatively charged regions of another.

Irregular Flowers such as orchids that show bilateral symmetry.

Isotonic The condition when the solution surrounding a cell has the same concentrations of dissolved substances as occur inside the cell. There will be no net movement of water into or out of the cell in isotonic solution.

Isotopes Alternate forms of a chemical element that differ in the number of neutrons in the atom's nucleus, but not in chemical properties.

Jasmonic acid One of a group of plant hormones that are involved in protective responses to environmental stress.

Keystone species Species that are critical to the survival of many other life-forms.

Kinesin One of the three types of small motor proteins in cells. Kinesin is associated with microtubules.

Kinetochore A specialized protein complex at the centromere of each pair of chromatids, forming a chromosome at nuclear division.

Kranz anatomy A wreath-like arrangement of mesophyll cells that occurs around a layer of specialized bundle sheath cells; both layers surround vascular bundles in leaves of plants.

Law of independent assortment Mendel's second law that states that genes for traits located on different non-homologous chromosomes assort independently of each other during meiosis.

Law of segregation Mendel's first law of heredity that states that the alleles for a gene segregate independently of each other in individuals that possess both alleles.

Leaching The removal of mineral nutrients from soils by the action of abundant rainfall.

Leaflets The components of a leaf blade that is divided into two or more parts.

Leaves The dorsiventral (flattened, with distinct upper and lower surfaces) organs of vascular plants that are specialized for photosynthesis; true leaves contain at least one vein (vascular bundle).

Legumes Plants that are members of the pea (bean) family; also their distinctive fruits, which develop from one carpel and open along both sides.

Lenticels Interruptions in the cork layer of bark on woody plants; provide gas exchange for inner stem tissues.

Lianas Woody vines common in tropical rain forests.

Lichen A mutualistic partnership of a green alga and/or cyanobacteria with a fungus to form a body distinct from that exhibited by either partner alone.

Ligase See DNA ligase

Lignin The complex phenolic cell wall that endows the xylem and other tissues of plants with compression and decay resistance. It is largely responsible for the preservation of plants as fossils.

Linked The genetic condition that arises when two or more genes are located close together on the same chromosome. Linked genes tend to be inherited together.

Lipids Hydrophobic organic molecules that serve as components of cell membranes or as energy storages in cells; fats, oils, steroids, phospholipids, and carotenoids.

Littoral The shoreline area of the ocean or a lake that is submerged all or part of the time.

Liverworts One of the three main groups of bryophytes. Liverworts are small herbaceous plants some of which are shaped much like a lobed mammalian liver.

Locus The site on homologous chromosomes where the alleles for a gene are located.

Long-day plants Angiosperms that flower in spring or summer because they require a daylength longer than a critical period in order for flowering to commence.

Lycophylls Leaves having a single, unbranched vein that does not leave a gap in the stem's vascular system at the point where the two vascular systems attach.

Lycophytes Members of a group of vascular plants that have lycophylls (= microphylls), leaves typically having only a single, unbranched vein.

M phase A phase of the cell cycle during which mitosis and cytokinesis occur.

Macroevolution Evolution that involves the origins of new species in a lineage over long periods of time.

Macronutrients Minerals that are required by plants in relatively large amounts; carbon, hydrogen, oxygen, nitrogen, phosphorus, sulfur, calcium, potassium, and magnesium.

Macrophytes Rooted or floating flowering plants found in lakes.

Male gametophyte The haploid male plant that lies inside the microspore; gives rise to the sperm.

Mass flow Flow of a mass of material; transport of organic materials from their sources to site of consumption along a turgor pressure gradient that develops from differences in osmotic properties of adjacent cells; also known as pressure flow.

Matrix The fluid-filled area inside the inner membrane of the mitochondrion; the site of the Krebs cycle.

Megaphyll A leaf having a branched system of veins (= euphyll).

Megasporangium The sporangium in which megaspores are produced.

Megaspore The large haploid spore that gives rise to the female gametophyte.

Megaspore mother cell A cell that undergoes meiosis to produce megaspores in heterosporous plants (= megasporocyte).

Megasporocyte See megaspore mother cell.

Megasporophyll The leaf-like structure that bears the megasporangium.

Meiosis The two successive nuclear divisions that reduce the number of chromosomes from diploid to haploid, producing either haploid gametes (in animals and certain protists) or haploid spores (in plants and certain protists).

Mendel's first law of heredity See law of segregation.

Mendel's second law of heredity See law of independent assortment.

Mesophyll The main photosynthetic tissue of a leaf, located between upper and lower epidermis.

Mesophyll cells The parenchyma cells located between upper and lower epidermis of a leaf.

Mesophytes Plants that require a moderately moist environment.

Messenger molecules Intracellular molecules that transmit signals from one part of a cell to another.

Messenger RNA (mRNA) The form of RNA that transmits genetic information from genes (DNA) to ribosomes, where the information is used to synthesize proteins.

Metabolism The sum of the chemical reactions and processes occurring within a cell.

Metaboly A form of squirming movement that does not involve flagella or extrusion of mucilage.

Metaphase The stage of mitosis or meiosis during which chromosomes (mitosis) or homologous chromosome pairs (meiosis) are aligned along the spindle center.

Methane (CH_4), A gas that contributes to global warming.

Methanogens Prokaryotes that produce methane.

Methanotrophs Prokaryotes that consume methane.

Microbes Organisms so small that a microscope is normally needed to observe them: prokaryotes, many stages of fungi, and most species of algae as well as other protists are included.

Microbial loop The sum of complex nutritional interactions among microorganisms, microscopic animals and algae.

Microevolution Changes in the frequencies of alleles in a population over a number of generations.

Microfilaments One of three types of long, thin protein fibers that make up the cytoskeleton. Microfilaments are made of two intertwined strands of the protein actin.

Micronutrients Minerals required for plant growth in relatively small amounts.

Microphylls Small leaves with only one vein found in the lycophytes (= lycophylls).

Microsporangium The sporangium that gives rise to microspores.

Microspores The tiny spores whose walls enclose male gametophytes.

Microsporocytes The diploid cells in the pollen sacs that divide by meiosis to produce four haploid microspores.

Microsporophyll The leaf-like structure that bears one or more microsporangia.

Microtubules One of three types of long, thin protein fibers that make up the cytoskeleton. Microtubules are constructed from many molecules of the protein tubulin.

Mimicry A form of convergent evolution in which one organism evolves to resemble another organism or an inanimate object.

Mitochondrion (singular), mitochondria (plural) Organelles present in eukaryotic cells that are the location for the generation of energy in the form of ATP by the process of respiration.

Mitosis A form of nuclear division in which a set of chromosome copies resulting from DNA replication are distributed to progeny cells.

Mitotic spindle A structure composed of microtubules that aids in separation of chromatids (mitosis) or homologous chromosomes (meiosis) to progeny cells.

Mixotrophy A type of protist nutrition in which cells are capable of photosynthesis as well as heterotrophy (consumption of particulate or dissolved organic food).

Mojave Desert One of the four North American deserts; the Mojave Desert lies in southern California, Nevada, and part of Arizona.

Molecule A substance that is composed of two or more atoms of the same or different elements linked by chemical bonds.

Monoculture The planting of fields with a single type of crop.

Monoecious A plant with both staminate and carpellate flowers on the same individual.

Monohybrid cross A cross between two true-breeding varieties that differ in only one trait, such as flower color in garden peas.

Monophyletic group A group of organisms whose members have descended from a single common ancestor.

Monosaccharides Simple sugars such as glucose or fructose; simple sugars can be linked to form more complex polysaccharides.

Monospores The non-flagellate, asexual reproductive cells of red algae.

Monosymmetric See bilaterally symmetrical

Morphological species concept The definition of species based on morphological criteria alone, that is, based on visible structures or anatomy.

Mosses One of the three main groups of bryophytes. Mosses often form short mats or turfs.

Motor protein A type of protein that can use the chemical energy of ATP to accomplish movement of a cell or its internal components.

Mucigel A gluey polysaccharide that coats the tips of roots and the root hair region. Mucigel aids the root in passage through the soil and prevents root drying.

Multicellular Body composed of more than a few cells, particularly when those cells can be demonstrated to communicate with each other by means of intercellular connections or diffusible substances.

Multicellular bodies Body types that are composed of more than one cell, and in which intercellular communication occurs.

Multicellularity Having a body composed of more than one cell and in which cells communicate with each other.

Multiple fruits The fruits formed when the fruitlets made by individual flowers in an inflorescence fuse into a single large fruit, as in pineapples.

Mutation An inheritable change in a gene.

Mutualism A symbiotic relationship between two or more organisms in which each derives some benefit.

Mutualists Organisms that form part of a mutually beneficial symbiotic relationship.

Mycelium A mass of interconnected fungal hyphae that forms the body of a fungus.

Mycobionts Fungal components of lichens or other symbiotic associations.

Mycoheterotrophy A form of plant nutrition in which organic food materials are channeled from autotrophic plants to heterotrophic plants by way of mycorrhizal fungi.

Mycorrhizae Mutualistic symbioses between particular fungi and the roots of plants; the fungi obtain organic food from plants, and plants obtain soil minerals from fungal hyphae.

Mycorrhizal fungi Fungi that form nutritional associations with plant roots.

Myosin One of the three types of small motor proteins and the major component of animal skeletal muscle.

NAD$^+$ Nicotinamide adenine dinucleotide, an electron carrier in cell respiration.

NADP$^+$ Nicotinamine adenine dinucleotide phosphate, an electron carrier in photosynthesis.

Nanoplankton Floating or swimming organisms whose diameter lies between 2 micrometers and 20 micrometers (a micrometer is one millionth of a meter).

Natural products Materials that are derived from living things; secondary compounds of plants (those not involved in basic metabolism, but rather used in specific ways such as protection or reproduction).

Natural selection The differential reproductive success of organisms differing in phenotypic traits that results from the interactions of organisms with their environments.

Nectar The sugary solution that attracts pollinators to certain flowers.

Nectaries The glands in some flowers that produce nectar.

Netted veins A type of leaf venation in which the veins are connected to form a net pattern.

Nicotinamide adenine dinucleotide (NAD$^+$) An electron acceptor molecule that functions in the oxidation–reduction reactions of respiration.

Nicotinamide adenine dinucleotide phosphate (NADP$^+$) An electron acceptor molecule that operates in oxidation–reduction reactions in photosynthesis.

Nitrification The process by which certain aerobic bacteria convert ammonia into nitrate.

Nitrogen fixation The process of reducing nitrogen gas to ammonia.

Nitrogenase The oxygen-sensitive enzyme that plays a major role in the fixation of atmospheric nitrogen into ammonia.

Nitrogen-fixing bacteria Bacteria capable of reducing atmospheric nitrogen gas (N_2) to ammonia (NH_3).

Node The point on a stem where one or more branches, leaves, or buds are attached.

Noncyclic electron flow The one-way passage of electrons from water through photosystem II to photosystem I and finally to NADP$^+$ in the light reactions of photosynthesis.

Nondisjunction The failure of homologous chromosomes to separate at anaphase of meiosis I, leading to the formation of abnormal gametes or spores.

Nucellus The tissue within the ovule from which the megasporocyte develops, equivalent to a megasporangium.

Nuclear division The formation of two progeny nuclei from one parental nucleus.

Nuclear envelope The structure composed of two membranes that encloses the interior of the nucleus of a cell.

Nuclear pores Pores in the membranous envelope of the nucleus, formed by proteins that control materials' entry to and exit from the nucleus.

Nucleic acids Acidic organic molecules that are composed of chains of nucleotides; DNA and RNA.

Nucleoid Aggregations of DNA in the cells of prokaryotes.

Nucleolus (singular), nucleoli (plural) A small, spherical body in the nucleus of eukaryotes where ribosomal RNA is assembled into the ribosomal subunits.

Nucleoplasm The substance of the eukaryotic nucleus inside the nuclear envelope.

Nucleotide An organic molecule that is composed of a sugar (typically ribose or deoxyribose), phosphate, and a base (adenine, thymine, cytosine, and guanine for DNA; adenine, uracil, cytosine, and guanine for RNA).

Nucleus (1) The double-membrane-bound structure within eukaryotic cells that contains the chromosomes. (2) The center of an atom that contains the protons and neutrons.

Nutation A dancelike motion enabling plant shoots to locate and attach to supporting structures.

Nut A type of simple indehiscent dry fruit, such as acorns and chestnuts, in which the fruit wall is stony and the nut derived from multiple carpels.

Oligotrophic Lakes and streams that are clean, clear, and low in mineral nutrients. Oligotrophic waters usually have a high species diversity.

Oomycetes Heterotrophic protists that resemble fungi in some ways, but which are not closely related to fungi; the agents of some important plant diseases.

Opposite A type of leaf arrangement in which there are pairs of leaves or buds at each node.

Organ system Plant structures, such as shoots, buds, and flowers, which are composed of more than one type of organ.

Organelles Membrane-enclosed structures, such as mitochondria and plastids, in eukaryotic cells.

Organic acid An acid that includes carbon atoms; compare to an inorganic acid, which lacks carbon atoms.

Organic compounds Compounds that contain two or more carbon atoms that are chemically linked to other types of atoms (typically hydrogen, oxygen, and nitrogen).

Organs A structure composed of two or more tissues; e.g., leaves, roots, stem, flower parts.

Origin of replication The site on a DNA molecule where replication begins.

Osmosis The movement of water into or out of cells, across a differentially permeable membrane, in response to differences in the concentrations of solutes (dissolved substances) such as sugars or ions.

Osmotrophy A form of nutrition in which organic food is consumed as dissolved molecules.

Outbreeding Mating that occurs between genetically different organisms of the same species. Also known as outcrossing.

Outer bark All of the tissues outside of the innermost periderm (cork, cork cambium, and associated parenchyma cells) in a woody plant.

Ovary The broad, round lower portion of the carpel in flowering plants, where the ovules are located. Fruits develop from ovaries.

Ovule The structure contained within the ovary that gives rise to the female gametophyte and, when fertilized and mature, a seed.

Ovuliferous scales The scale-shaped structure upon which ovules are borne in ovulate (female) cones of conifers.

Oxidation The loss of an electron from an atom or molecule.

Oxidation-reduction reactions The coupled loss of an electron from one atom or molecule and gain of an electron by another atom or molecule.

Oxidative phosphorylation The process by which ATP is generated from ADP and inorganic phosphate in the cell's mitochondria.

Oxygen An element that is common in living things; also, the diatomic molecule O_2 that makes up about 21% of Earth's atmosphere.

Ozone Atmospheric molecular gas composed of three oxygen atoms (O_3). Ozone is very important in screening the surface of the Earth from UV radiation.

Paleobotany The study of fossil plants.

Paleontology From the Greek *palaios* meaning "old." Paleontology is the study of fossil organisms from past geological eras.

Palisade layer An internal layer of column-shaped leaf photosynthetic cells arranged with their longest dimensions perpendicular to the leaf surface.

Palmately compound leaves Leaves whose leaflets all arise from a common point at the end of the petiole; palm-shaped.

Palmately netted veins The arrangement of plant leaf veins in which several major veins branch from a common point; sugar maple leaves, for example.

Parallel veins A type of leaf venation in which the principle veins lie parallel to each other.

Parasite An organism that lives on or in another organism (the host) and derives its nutrition from that organism. In the population interaction called parasitism, one population (the parasite) benefits and the other (the host) is harmed by the relationship.

Parasitic plants Plants such as dodder and mistletoe that obtain water, minerals, and in some cases organic food from plant hosts.

Parasitism A type of nutrition in which nutrients are obtained from a living organism that is thereby harmed.

Parenchyma A tissue of plants consisting of thin-walled living cells that are capable of division when mature, and which compose much of plant bodies.

Parenchyma cells Thin-walled cells that function in photosynthesis or storage, and which are capable of division when mature.

Pathogens Heterotrophic organisms that obtain their food from living cells or tissues, thereby causing disease.

PCR Polymerase chain reaction; a process for making many copies of specific genes (DNA sequences).

Pectin A hydrophilic polysaccharide found in the cell walls of plants.

Peduncle The stalk that bears a flower or cluster of flowers.

Peer review The review of a scientific paper or grant proposal by other scientists who are peers in the sense of possessing equal qualifications and abilities.

Pelagic Related to the open waters of the ocean or lakes.

Pelagic zone The surface and subsurface waters of lakes that are well illuminated and away from the nearshore waters.

Pepo A type of fruit that is a fleshy berry with a hard rind.

Peptide bond The type of covalent chemical bond that links amino acids together in proteins; loss of a water molecule occurs during formation of each peptide bond.

Peptidoglycan Substance composed of protein and carbohydrate and found in the cell walls of most bacteria.

Perfect Term for flowers that have both stamens and carpels and are therefore bisexual.

Perianth The calyx (the whorl of sepals) and the corolla (the whorl of petals) considered together.

Pericycle The cylinder of tissue that surrounds the core of xylem and phloem.

Periderm The outer protective layer of a woody stem or root; generated during secondary growth; consists of cork tissue, cork cambium, and associated parenchyma cells; replaces the primary epidermal tissues generated during primary growth.

Periphyton Algae and bacteria that grow attached to the surfaces of aquatic plants.

Peristome The upper part of the moss capsule that is specialized to discharge spores.

Permafrost The part of the soil in polar regions that is permanently frozen. Permafrost lies below the surface, which may undergo freeze-thaw cycles.

Peroxisome An organelle in eukaryotic cells that is bounded by a single membrane and contains the enzyme catalase to break down hydrogen peroxide.

Petals The whorl of nonreproductive flower parts lying just above the sepals. Petals are usually brightly colored to attract pollinators.

Petiole The stalk that connects a leaf blade to the stem.

pH scale A continuous set of values from 0 to 14 that indicate the negative logarithm of the hydrogen ion concentration of solutions; values lower than 7 indicate acidic solutions; values higher than 7 indicate alkaline (basic) solutions.

Phagotrophy The form of endocytosis associated with particle feeding. Many protists are phagotrophic on bacteria.

Phenolics The simple monomer phenol (which includes an aromatic ring and at least one hydroxyl group) and polymers made of similar monomers; flavonoids and lignin are examples.

Phenotype The physical characteristic of an organism. In garden peas, for example, purple flower color is a phenotype, but the genotype could be *PP* or *Pp*.

Phloem The vascular plant tissue in which organic compounds, including sugars, are conducted in a watery solution.

Photoautotrophs Photosynthetic organisms.

Photobionts Green algae or cyanobacteria that occur in lichens.

Photoperiodism Response to the timing and periodicity of day and night, as a way of detecting seasonal change.

Photophosphorylation The formation of ATP from ADP and inorganic phosphate during photosynthesis in the chloroplast.

Photorespiration The metabolic process in C_3 plants that results when the enzyme rubisco combines with oxygen instead of carbon dioxide and produces only one molecule of PGA instead of two PGA. The efficiency of photosynthesis is considerably reduced.

Photosynthesis Production of high-energy organic compounds (sugar and derivatives) from low-energy inorganic compounds (carbon dioxide and water); the conversion of light energy into chemical energy by most plants and algae and some bacteria.

Phototropin A blue light-sensitive molecule involved in phototropism, growth in response to the direction of light.

Phototropism The growth of plant shoot tips toward a light source.

Phragmoplast A system of microtubules and other protein components that functions in plant cytokinesis (division of the cytoplasm) and primary cell wall formation.

Phragmosome A layer of cytoplasm that forms across the region of cytoplasm where cell division will later occur, and into which the nucleus moves in preparation for nuclear division.

Phylogenetic systematics A field of biology in which biologists classify organisms according to their ancestry (evolutionary relationships).

Phylogenetic trees Branching diagrams that illustrate concepts of evolutionary relationships among organisms. Also called phylogenies.

Phylum (pl. phyla) A grouping of related classes of organisms; equivalent to a division, a term used in plant biology.

Phytoalexins Compounds produced by plants as a defensive response to attack by bacteria or fungi.

Phytochrome A type of molecule found in bacteria, some algae, and plants that allows these organisms to perceive red and far-red light.

Phytoliths Plant-derived crystals used to identify plant remains.

Phytoplankton Algae and cyanobacteria that float or actively move through water.

Picoplankton A size class of floating or swimming organisms (plankton) that range from 0.2 micrometers to 2 micrometers (a micrometer is one millionth of a meter) in diameter.

Pilus (singular), pili (plural) Thin tubes that link conjugating bacterial cells and through which DNA can be transmitted.

Pinnately compound leaves Leaves that have leaflets arranged along a central axis that is an extension of the petiole.

Pinnately netted veins A type of leaf vein pattern in which major veins branch off along a central vein; elm leaves are examples.

Pioneer species Species that are the first to enter an area that is undergoing succession. Lichens and cyanobacteria are pioneer species on bare rock.

Pistil The entire carpel or several fused carpels consisting of stigma, style, and ovary.

Pith The tissue that occupies the central region of a stem or root, within a cylinder of vascular tissue; typically composed of unspecialized parenchyma cells.

Pits Recessed cavities in plant cell walls where secondary wall materials are absent.

Plankton Floating or swimming organisms living in aquatic habitats.

Plant growth substances Organic compounds produced in one part of the plant that have effects elsewhere in the plant; plant hormones.

Plants Organisms that are composed of many cells, have cellulose-rich cell walls, have chlorophyll in chloroplasts, and are photosynthetic (or, if non-photosynthetic, originated from photosynthetic ancestors), and are adapted in many ways to life on land (or, if aquatic, are descended from land-adapted plants).

Plasmids Small circular pieces of DNA that occur in bacterial cells in addition to the single main circle of DNA, the genophore. Plasmids can be used as vectors to move pieces of DNA between organisms.

Plasmodesmata Narrow cytoplasmic extensions through cell walls; link the cytoplasm of adjacent cells; characteristic of cells arranged in tissues; reflect the capacity for intercellular communication.

Plasmolysis The shrinkage of the cytoplasm that occurs when water leaves the cell by osmosis. The water leaves because the concentration of dissolved substances outside the cell is greater than the concentration inside the cell.

Plastid Organelle bound by two membranes; may contain chlorophyll, carotenoids, or starch.

Plumule The first bud of an embryo shoot, located above the seed leaves (cotyledons).

Pneumatophores Aerial roots on mangrove trees that provide oxygen to the submerged portions of the roots.

Polar compound A molecule that has distinct electrically charged poles.

Polar covalent bond   A covalent chemical bond in which the electrons are unequally shared.

Polar molecules   Molecules having distinct positively-charged and negatively-charged ends.

Polar nuclei   The two nuclei that lie in the center of the female gametophyte or embryo sac. After fertilization with a sperm nucleus, the three nuclei give rise to the triploid endosperm tissue.

Pollination   The transfer of pollen from anther to stigma in angiosperms or from microsporangia to ovules in gymnosperms.

Pollination syndromes   The pattern of coevolved traits between pollinators and their flowers.

Polyculture   An agricultural method in which several types of crops are grown together.

Polygenes   Genetic traits that are controlled by several genes, such as height in humans and kernel color in wheat.

Polymerase chain reaction (PCR)   A procedure for amplifying a specific sequence of DNA by subjecting it to numerous cycles of polymerization followed by separation of complementary strands. Two DNA primers, one complementary to each end of the selected DNA sequence, are used to amplify the selected sequence.

Polyploidy   The condition in which the cells of an organism contain more than two complete sets of chromosomes.

Polysaccharides   Carbohydrates that are composed of many sugars linked together.

Polysymmetric   See actinomorphic or regular symmetry.

Pomes   A type of fleshy fruit where the flesh derives from an enlarged receptacle. Apples and pears are examples.

Population   All of the individuals of a single species occupying a given area.

Population genetics   The branch of biology that combines the study of evolution with genetics by examining the gene pools of populations and how and why they change over time.

Potential energy   Energy of position. In an atom, electrons possess a certain potential energy by virtue of their position relative to the atomic nucleus. The further an electron is from the nucleus, the greater the potential energy.

Practical salinity units   The mass in grams of dissolved compounds in 1000 g of seawater. An average sample of seawater has 35 g of dissolved compounds in 1000 g of seawater or 35 practical salinity units, abbreviated as 35 psu.

Prairies   Grasslands with few or no trees.

Predation   A type of nutrition in which the consumer kills the consumed.

Predator   An organism that feeds by consuming other organisms (the prey). In the population interaction called predation, one population benefits and the other suffers mortality.

Preprophase band   A ringlike band of cytoplasmic microtubules that forms near the cell membrane at the site where a new wall will join following mitosis and cytokinesis.

Pressure flow   See mass flow.

Primary compounds   The macromolecules that make up most of the bodies of all life-forms; include carbohydrates, lipids, proteins, and nucleic acids.

Primary endosymbiosis   Having prokaryotic endosymbionts or organelles derived from them.

Primary phloem   Phloem tissues that are formed by primary growth (the activity of apical meristems).

Primary plastids   Plastids that are derived from endosymbiotic cyanobacteria.

Primary succession   The ecological process of changing communities that begins on new or barren ground where a community has not previously existed.

Primary tissues   The first tissues produced by root apical meristem and shoot apical meristem.

Primary xylem   Xylem tissues that are formed by primary growth (the activity of apical meristems).

Procambium   One of the three primary meristems produced by the apical meristems of roots and shoots. Procambium gives rise to the vascular tissues.

Products   General term for the final molecules after a chemical reaction.

Progymnosperms   A group of extinct plants known only from fossils that possessed wood similar to that of modern gymnosperms, but which lacked seeds.

Prokaryotes   Organisms belonging to the domains Bacteria and Archaea that lack a membrane-bound nucleus in their cells.

Prokaryotic   Adjective form of prokaryote.

Prop roots   A type of adventitious root that arises from a stem above the soil surface and helps support the plant.

Prophase   An early stage in the process of nuclear division, during which the chromosomes become shorter and thicker and move toward the center of the developing spindle apparatus.

Proplastids   Young plastids that have not yet differentiated into chloroplasts, amyloplasts, or chromoplasts; present in cells of the growing tissues of root and shoot tips.

Protective coloration   Any color or color pattern that hides an organism from potential harm. Some plants, for example, have the color and form of small stones, and this characteristic conceals them from herbivores.

Proteins   Organic molecules composed of carbon, hydrogen, oxygen, nitrogen, and sulfur; macromolecules that are composed of amino acids.

Protists   Eukaryotic organisms that are not classified in the Kingdoms Fungi, Animalia (animals; metazoa), or Plantae (land plants, embryophytes); algae, protozoa, slime molds and other organisms.

Protocells   Structures produced in the laboratory that resemble simple cells in some ways.

Protoderm   One of the three primary meristems produced by the apical meristem. Protoderm gives rise to epidermis.

Protonema   A filamentous or platelike mass of green cells one-cell thick that arises from the germination of a moss spore. The protonema is part of the gametophyte generation that develops into the gametophore.

Protozooplankton   Swimming or floating (planktonic) non-photosynthetic protists.

Pteridophytes   Non-seed vascular plants; when leaves are present, they have branching vascular systems. Ferns, horsetails, and whisk ferns are examples.

Punctuated equilibrium   A model that proposes that the process of evolution is marked by long intervals of little change interrupted or punctuated by brief intervals of rapid change and speciation.

Punnett square   A box diagram developed by British geneticist Reginald Punnett to represent the genetic combinations in a cross between parental varieties.

Radial symmetry   Being divisible into two equal parts by many longitudinal planes; having parts arranged like rays.

Radicle   The embryonic root.

Rain roots   Thin roots quickly formed by the main roots of desert succulents in response to rainfall.

Rain shadow   An area of arid land that forms in the leeward side of a mountain range.

Range   The area occupied by all the separate local populations of a species.

Ray initials   Cube-shaped cells of the vascular cambium that contribute to the horizontal growth of woody tissues; form the secondary xylem and phloem of rays in wood.

Reactants   The initial molecules present before a chemical reaction occurs. Also called the substrates of that reaction.

Reaction center   Within a photosystem, the complex of chlorophylls and proteins that transform light energy into chemical energy.

Receptacle   The enlarged area at the end of a peduncle to which the flower parts are attached.

Receptor molecules   Molecules that typically trigger cascades of chemical reactions that precipitate cellular responses.

Receptor proteins   Special proteins in the cell membrane that bind to chemical messengers from the environment and transmit signals to the cytoplasm to initiate a response.

Recessive   The trait in the generation that is masked or not expressed when a cross is made between two true-breeding parental varieties that possess contrasting forms of that trait.

Recombinant DNA   Term used to describe the incorporation of foreign DNA into the normal DNA of an organism.

Red algae   A group of mostly marine eukaryotic algae that includes unicells and more complex multicellular forms, often characterized by red-colored accessory pigments (including phycoerythrin) that obscure the presence of green chlorophyll.

Reduction   Gain of an electron by an atom; from the Latin word "reductio" meaning "to bring back," which references the metallurgical process by which metal elements are released–brought back–from their oxides.

Regular   Flowers such as tulips and poppies that show radial symmetry.

Replicates The duplicate runs that form the controls or experiments in an experimental design. Replicates are intended to reduce the probability of error.

Resistance gene A plant gene that confers resistance to a microbial pathogen.

Respiration The chemical process by which simple sugars such as glucose are oxidized to carbon dioxide and water with the liberation of energy in the form of ATP.

Restoration ecology The branch of ecology that aims to restore communities that have been severely disturbed by human activities.

Restriction enzymes Enzymes, derived from bacteria, which cut the DNA strand at specific nucleotide sequences.

Restriction fragment length polymorphisms (RFLPs) Variations in the numbers of DNA fragments of different lengths that result when DNA is cut with restriction enzymes.

Reverse transcriptase An enzyme produced by certain RNA viruses, such as that of AIDS, to translate its own RNA-encoded genes into DNA for insertion into a host cell's chromosomes.

RFLP Abbreviation for restriction fragment length polymorphism.

Rhizaria A group of protists that often have delicate processes extending from the cells (includes chlorarachniophytes, radiolarians, and foraminiferans).

Rhizoids In bryophytes, rhizoids are long, tubular single cells (liverworts and hornworts) or multicellular filaments (mosses) that anchor the plant body to the substrate.

Rhizomorphs Bundles of fungal hyphae that are thick enough to be seen more easily than individual hyphae; elongated and often hardened strands that resemble roots and can be found between the wood and bark of fallen trees or stumps; store and transmit food.

Ribosomal RNA (rRNA) Ribonucleic acid (RNA) that is found in ribosomes.

Ribosomes Small particles, composed of RNA and protein, which are the sites of protein synthesis in cells.

Riparian wetlands Wetlands that border streams or rivers.

RNA polymerase A type of enzyme that catalyzes the synthesis of RNA, using the sequence of bases in DNA as a template.

Rods One of the three basic forms of bacterial cells.

Root apical meristem (RAM) The tissue in the last few millimeters of the growing root tip that gives rise to the primary tissues.

Root border cells Cells shed from the root cap and dispersed into the surrounding soil where they help prevent attack by microbes and nematodes.

Root cap The thimble-shaped cap of tissue at the tip of the root that is produced by the root apical meristem.

Root hairs Tubular extensions from epidermal cells in a zone near the root tip.

Root system One or more roots that comprise the anchoring and mineral- and water-absorption organs of plants.

Rooted macrophytes Aquatic vascular plants that are rooted in the sediments of the littoral zones of lakes.

Rough ER The rough endoplasmic reticulum that is covered in ribosomes; produces glycoproteins that the cell exports and other cell materials.

Rubisco The abbreviation for the key enzyme in the Calvin cycle, ribulose 1,5-bisphosphate carboxylase/oxygenase, which attaches a molecule of carbon dioxide to the 5-carbon sugar ribulose 1,5-bisphosphate.

Salicylic acid A compound similar to that of common aspirin that acts as a hormone (chemical messenger) in plants.

Salinity The concentration of salt in water.

Samara Simple indehiscent dry fruits containing a single seed and wings. Samaras are winged achenes, as in elms.

Saprobes Heterotrophic organisms that obtain their food from non-living organic materials (contrast with pathogens).

Savanna Grasslands having more trees than is typical for prairies.

Scanning electron microscopes (SEMs) Microscopes that generate an image by scanning a beam of electrons across a specimen that has been coated in a very thin layer of gold, palladium, or platinum.

Schizocarp A type of fruit that at maturity splits into two or more one-seeded parts (maple fruits are examples).

Scientific name A two-word name, derived from Latin or Greek, given to each individual species at the time of its formal description by a taxonomist (scientist who specializes in the naming of organisms).

Sclereids Star- or stone-shaped cells having tough cell walls thickened by addition of lignin.

Sclerenchyma Tissue composed of sclereids and fibers, both having tough cell walls that resist damage by mechanical stress or attack; occurs in seed coats, nut shells, and the pits of stone fruits such as peaches.

Sclerophyllous Leaves that are small, hard, and covered in a thick cuticle, frequently with sunken stomata. Sclerophyllous leaves are common on chaparral shrubs.

Seamounts Submarine mountains rising from the ocean floor.

Seaweeds Algae that live in the ocean and are large enough to see with the unaided eye; usually attached to the ocean bottom or shoreline rocks, but sometimes are free-floating.

Second filial generation (F_2) The generation obtained by crossing members of the first filial generation (F_1). The reappearance of individuals with the recessive trait indicate that inherited characters do not blend in hybrids.

Secondary compounds Molecules that are produced by plants, algae, or fungi that are restricted in their distribution; alkaloids, terpenoids, and phenolics.

Secondary endosymbiosis Having eukaryotic endosymbionts or plastids derived from them.

Secondary growth Production of new plant tissues by secondary meristems (vascular and cork cambia); increases girth of woody plants (compare to primary growth, which increases plant length).

Secondary meristems Regions of cell division that increase girth, rather than height or length of a plant body; the vascular cambium and cork cambium.

Secondary phloem Phloem tissues that arise by the activity of the vascular cambium, a type of secondary meristem.

Secondary plastids Plastids that are derived from endosymbiotic eukaryotes whose plastids were of primary origin (red or green algae).

Secondary succession Succession occurring on an area where a prior community has been removed, such as abandoned farmland or an area cleared by fire.

Secondary tissues Tissues that arise by the activity of secondary meristems.

Secondary xylem Xylem tissues that arise by the activity of the vascular cambium, a type of secondary meristem.

Secretory cells Specialized cells that produce secondary compounds such as tannins or resins.

Securins Proteins that control the separation of sister chromatids during nuclear division.

Sedge meadow A wetland located near lakes and dominated by sedges (*Carex*) with characteristic triangular stems.

Seed A structure that develops from a fertilized ovule in seed plants (gymnosperms and angiosperms).

Segregating Term applied to the reappearance of recessive traits in offspring of hybrids. The recessive traits reappear because the chromosomes (and their genes) from each parent separate in meiosis.

Self-fertilization Process that occurs in some plants in which a single flower produces both male and female gametes that can fuse and produce healthy offspring.

Self-pollination The capability of an individual plant to pollinate its own flowers by means of bisexual flowers or separate staminate and carpellate flowers on the same individual plant.

Semiconservative replication DNA replication in which the replicated double helix consists of one old strand and one newly synthesized strand.

Sepals The outermost whorl of floral parts, which are leaflike and usually green.

Separase An enzyme complex that in metaphase breaks down molecules (cohesins) that in previous stages held chromatid arms together.

Septa (singular, septum) The cross walls that divide fungal hyphae into cells.

Septum A partition or cross wall in fungi or prokaryotes.

Sessile Attached at the base (for example, leaves that lack petioles and thus emerge directly from stems).

Sex chromosomes Chromosomes that bear gender-determination gene loci.

Sexual reproduction Production of single-celled gametes that undergo fusion to form a zygote.

Sexual spores Spores that are produced as the result of sexual reproduction.

Shared, derived characters Features that are present in some, but not all of the members of a group because of inheritance from a common ancestor.

Sheath A leaf base that wraps around the stem.

Shoot The aboveground organ system of a plant, usually upright, but sometimes growing horizontally along the ground.

Short-day plants Plants that usually flower in autumn because they require a daylength shorter than some critical value in order for flowering to commence.

Shrub A woody plant with several main stems.

Sieve elements Elongate tubular structures in phloem of vascular plants; composed of sieve cells or sieve tube members arranged end to end.

Signal transduction The transmission of information from one part of a cell to another by means of chemical interactions (see messenger proteins).

Silique A simple dehiscent dry fruit that splits open on two sides. The fruits of broccoli and cabbage are examples.

Simple leaves Leaves that have a single, entire blade.

Simple sugars Monosaccharides.

Sister chromatids The two DNA molecules that result from DNA replication.

Slash and burn In tropical rain forests, the agricultural practice of cutting down all the trees in a small area, burning them, and planting crops for two to three years before moving on to a new site. Also called swidden, shifting, or ladang agriculture.

Smooth ER Smooth endoplasmic reticulum, the part of the internal network of flattened membranous sacs, which constructs fatty acids and phospholipids.

Solutes Materials that dissolve in water.

Somatic embryogenesis Development of an embryo from non-reproductive parts of the plant body.

Sonoran Desert One of the four North American deserts, the Sonoran Desert extends across southern Arizona and California into Mexico.

Soredia Asexual reproductive structures formed by lichens and dispersed by wind; small clumps of fungal hyphae that surround a few photobiont cells.

Sorus (singular), sori (plural) Clusters of sporangia on the leaves of ferns.

Specialized cells Cells having particular structure and function.

Species A group of organisms that are similar because they are descended from a common ancestor, and can interbreed.

Species complexes A group of closely related species.

Species diversity The number of species in a community combined with the distribution of numbers of individual organisms among those species (the evenness).

Species richness The number of species in a community.

Specific epithet The second part of a scientific name of an organism.

Specific heat The amount of heat that must be absorbed or lost for one gram of a substance to change temperature by one degree Celsius.

Sperm Male gametes.

Spindle assembly checkpoint The mechanism by which anaphase of cell division is held up until all chromosomes are attached to the spindle and at the equator.

Spirillum (singular), spirilli (plural) The corkscrew-shaped form typical of many kinds of bacteria.

Spirochaetes Spiral-shaped prokaryotes.

Spongy layer A leaf tissue composed of loosely assembled cells having chloroplasts.

Sporangium (plural, sporangia) A structure in which spores are produced.

Spore (1) A reproductive cell that is capable of growing into a new organism without fusing with another cell. (2) A thick-walled cell produced by certain prokaryotes that are under stress.

Sporophyte A multicellular, spore-producing life phase of an organism.

Sporopollenin The tough, resistant substance that forms the outer coat of pollen grains and other spores of embryophytes.

Stabilizing selection Selection that favors individuals with the average condition for a particular trait.

Stamen The structure in flowers, consisting of the filament and anther, in which pollen is produced.

Staminate A flower that lacks carpels.

Start codon The sequence AUG in messenger RNA; also encodes the amino acid methionine.

Stem The shoot of vascular plants; contains vascular tissue and typically produces leaves and reproductive organs; occurs above-ground in most cases, but can include rhizomes and other forms of underground stems.

Stereo dissecting microscope A microscope having a separate set of optics for each eye allowing three-dimensional views of objects.

Steroids A type of lipid that has a structure with four rings.

Stigma The area of the pistil that receives the pollen grains.

Stipules Small leaflike structures that occur at the junction of leaf petioles and stems.

Stoma (plural, stomata) Minute pores in leaf and other plant epidermal surfaces, defined by two guard cells that enable pores to open and close in response to changes in environmental conditions.

Stomatal closure The swelling of stomatal guard cells so that the pore between them becomes closed.

Stop codons One of three sets of messenger RNA triplets (UAG, UAA, or UGA) that signal termination of protein synthesis.

Storage roots Fleshy roots on biennial plants where carbohydrates are stored during the first year of growth.

Stramenopiles A group of protists characterized by particular types of hairs on flagella (=heterokonts).

Stratification The layering in tropical rain forests that provides a wide range of conditions and therefore many niches for a high diversity of species.

Streptophytes A group combining the embryophytes and charophyceans.

Stroma The viscous fluid inside the chloroplast in which the membranous thylakoids are embedded.

Stromatolites Large rocky mounds formed of calcium carbonate by colonies of cyanobacteria or purple bacteria in shallow marine waters. Stromatolites have a geological history dating back more than 3 billion years.

Style In the flower's pistil, the style is the column of tissue between the stigma and the ovary through which the pollen tubes grow.

Suberin A waxy waterproof substance found in the Casparian strip and used to seal off the stem after leaf abscission.

Sublittoral The region of a lake below the deepest growing rooted vegetation; the region of an ocean shoreline between the depth of the lowest of low tides and the edge of the continental shelf.

Submarine ridges Underwater mountain chains that bisect the ocean basins and mark the edges of the Earth's crustal plates.

Subspecies A group of organisms that inhabit a particular geographical area or type of habitat different from other members of the same species.

Substrate The material to which an organism is attached; the material upon which an enzyme acts.

Subtropical high pressure zones The zones of high pressure created at 30° N and S by descending air masses from the two Hadley cells.

Sugar Molecules that have the general formula CH_2O with a carbonyl group and several hydroxyl groups.

Sugoshin A type of protein that helps stabilize the connection between daughter chromatids in mitosis and meiosis.

Supercoiled domains Regions of coiling of DNA in bacterial chromosomes.

Superior An ovary is superior when the sepals, petals, and stamens are attached below it on the receptacle.

Sustainability The concept of preserving a high-quality environment while maintaining a minimally acceptable standard of living for human beings.

Symbiosis Organisms that live in intimate ecological association, for example, the fungi and green algae or cyanobacteria that together form lichens.

Sympatric speciation Reproductive isolation and speciation that occurs without geographic isolation of a population. Sympatric speciation may occur by hybridization and polyploidy or the sudden appearance of a few genes with major impact on reproductive biology.

Symplastic A type of transport that involves movement of substances from cell to cell via plasmodesmata.

Symplastic continuity The unity of cytoplasmic units (cells) that are linked by plasmodesmata.

Symplastic loading Movement of materials into phloem cells via plasmodesmata.

Synapomorphy A shared, derived character that defines a clade.

Syngamy The fusion of two gametes to form a zygote.

Synthesis reaction Formation of polymers from monomers by removal of water molecules; also known as dehydration synthesis.

Synthetic theory of evolution The modern theory of evolution that combines Darwin's theory of evolution with population genetics and molecular biology.

Systematics The science of organizing and classifying organisms.

Systemic acquired resistance (SAR) The ability of plant tissues removed in time or distance from a previous pathogen attack to resist future or continued attack.

Systemin A peptide that functions as a plant hormone, serving particularly in defense reactions.

Taiga The boreal coniferous forest that encircles the Northern Hemisphere of Earth.

Taproot The single main root that develops from the embryonic radical.

Taproot system A root system with a single main root from which many branch roots develop.

Taxonomy The science of discovering new organisms and naming them.

Telophase The final phase of mitosis, during which the two complete sets of chromosomes are fully separated.

Tendrils A leaf or portion of a stem modified into a thin, coiled structure that aids in stem support in some angiosperms.

Tepals Sepals and petals of similar shape and color.

Terpenes and terpenoids Major types of secondary chemical compounds produced by plants.

Tertiary plastids Plastids derived from endosymbiotic eukaryotes whose plastids were of secondary origin.

Testcross A cross in which an organism showing the phenotype of the dominant trait is crossed with one showing the recessive trait to determine if the dominant phenotype is homozygous or heterozygous for the dominant trait.

Theory In the scientific process of inductive reasoning, a theory is a generalization derived from a large body of observations and experiments that best explains that body of data.

Thermocline The region in a water body in which temperature declines abruptly, distinguishing warmer, upper, oxygen richer waters from colder, lower, oxygen-depleted waters.

Thigmotropism Response to contact with a solid object; for example, the coiling of plant tendrils around a support.

Thylakoids Thin, plate-like, membranous vesicles found in cyanobacteria and the chloroplasts of photosynthetic protists and plants; the site of the light reactions of photosynthesis.

Timberline The boundary area where trees disappear in mountains or at high latitudes. Also referred to as treeline.

Tissue culture The cultivation of individual plant cells in the laboratory so that they form undifferentiated masses of tissue or young plants when placed under the appropriate hormonal conditions.

Tissue systems The three major groups of primary tissues; outer dermal tissues, vascular tissues, and ground tissues located between dermal and vascular tissues.

Tissues Groups of similar cells organized into units having particular function and intercellular communication.

Trace elements Elements that are required by plants only in small amounts.

Tracheids The major type of water-conducting cells found in non-flowering plants, and one of the two major types of water-conducting cells abundant in angiosperms (vessel elements are the other). The elongate cells are dead at maturity, have tapered ends, are arranged end to end, and have walls that are partially coated with lignin.

Transcription The process by which information is copied from DNA to mRNA.

Transfer RNA (tRNA) The form of RNA that carries amino acids to the site of protein synthesis on ribosomes.

Transformed Term applied to an organism that has taken up foreign DNA, as in bacterial cells that have taken up plasmids with the gene for human insulin.

Translation The process by which information encoded in messenger RNA is used to synthesize specific proteins.

Transmission electron microscope (TEM) An electron microscope that forms images by transmitting an electron beam through extremely thin sections of specimens stained with heavy metals.

Transpiration The loss of water vapor from plants at their surfaces, primarily through stomata.

Transport proteins Proteins bound in the membranes of cells that transfer solutes across the membranes; also known as carrier, channel, or pump proteins.

Transposons A length of DNA containing one or more genes that is capable of moving from one location in a chromosome to another location in the same or a different chromosome; a "jumping gene"; first discovered in corn.

Tree A woody plant having a single main stem with few branches on the lower part.

Trenches Narrow, deep troughs that break the abyssal plains, especially along the islands and continents around the Pacific Ocean.

Trichomes Spiky or hairlike projections from the epidermal surfaces of plants, especially those of leaves, that offer protection from excessive light, ultraviolet radiation, extreme air temperature, or attack.

Triplets Series of three nucleotide bases.

Trophic cascade An ecosystem model that predicts that organisms at the tops of food chains or food webs will control populations below.

Tumors Cancerous growths that result from uncontrolled cell division.

Tundra The treeless polar biome dominated by shrubs and herbs in the Northern Hemisphere.

Turgid The condition of cells when they are firm and swollen due to high water content.

Turgor pressure Pressure within a cell resulting from the uptake of water.

Ultraviolet radiation (UV) The short-wavelength radiation that lies just below visible violet light in the electromagnetic spectrum. UV radiation can break the chemical bonds of organic molecules and cause mutations in DNA.

Unicells The body type of organisms that consist of only one cell. The adjective form is unicellular.

Unicellular bodies Bodies that are composed of only a single cell.

Uniformitarianism The principle that the rock formations of the past were produced by normal geological processes observable today such as erosion, sedimentation, glaciation, and volcanism operating over long periods of time.

Unisexual Flowers that have only one set of reproductive structures and are either staminate or carpellate.

Vacuole A space within the cytoplasm that is usually large, centrally located, and filled with a watery sap. One of the distinguishing characteristics of plant cells.

Varieties Members of the same species that have small but consistent differences from each other.

Vascular bundles Primary conducting tissues that occur in a group; a vein in a plant.

Vascular cambium The meristematic tissue that generates secondary xylem and phloem (wood and inner bark); occurs in woody plants.

Vascular plants Plants that contain vascular tissues (xylem and phloem).

Vascular rays Ray parenchyma cells and ray tracheids, both produced by ray initials of the vascular cambium.

Vascular tissues The tissues in a plant body that transport water, minerals, and organic compounds.

Vectors Viruses or plasmids capable of carrying a segment of DNA from one organism into another, which becomes transformed.

Veins Vascular bundles, consisting of groups of primary xylem, primary phloem, and associated tissues.

Vernalization The induction of flowering by cold.

Vesicles Small spherical bodies surrounded by a membrane and containing materials to be excreted from cells by exocytosis.

Vessel elements One of the cells that together comprise a vessel.

Vessels   Tubes in the vascular bundles of most angiosperms (and some other plants) that are composed of linear series of elongate vessel elements having perforations in their walls, and which conduct water, minerals, and other materials.

Vibrios   Comma-shaped prokaryotes.

Viroplankton   Floating viruses in water bodies.

Watershed   A region where all the water drains toward a common exit such as a stream.

Whorled   A type of leaf arrangement in which there are three or more leaves or buds at a node; the arrangement of floral parts in concentric circles.

Wood   Lignin-rich secondary xylem tissue.

Woody plants   Plants that produce large amounts of wood (secondary xylem).

Xeromorphic   Dry-form leaves that are small and often light colored to reflect heat. Xeromorphic leaves are common in desert and chaparral ecosystems.

Xerophytes   Plants having adaptations to dry habitats.

Xylem   Specialized tissue that conducts water and minerals in plant vascular systems; dead at maturity.

Zeaxanthin   A carotenoid produced by plants in response to high light stress and that provides protection from light damage to cells.

Zooplankton   Swimming or floating animals of water bodies.

Zoopores   Flagellate asexual spores produced by many protists (and certain early-diverging fungi).

Zygomorphic   See bilaterally symmetrical.

Zygomycetes   A group of early-diverging fungi characterized by thick-walled zygotes (zygospores) that also function as resistant spores.

Zygote   The cell that results from sexual fusion of two gametes, commonly egg and sperm.

Photo and Illustration Credits

FRONTMATTER
vii Claudia Lipke; x–xii Lee Wilcox; xiii top Photo Researchers, Inc., bottom Inga Spence/Visuals Unlimited; xiv–xv Lee Wilcox; xvi top D. Cavagnaro/DRK Photo, bottom Lee Wilcox; xvii–xx Lee Wilcox; xxi top Michael Quinton/Minden Pictures, bottom Lee Wilcox; xxii top Lee Wilcox, bottom NASA/John F. Kennedy Space Center; xxiii top Fred Bruemmer/DRK Photo, bottom Jim Brandenburg/Minden Pictures; xxiv Lee Wilcox; xxv top Constantino Petrinos/Z. Legacy Images Resource Centers, bottom Tim Fitzharris/Minden Pictures; xxvi top Dr. Zeyaur Khan/Striga, bottom, James Graham

PART I
CHAPTER 1 CO1 Lee Wilcox; CO1.1 Lee Wilcox; 1.1–1.4 Lee Wilcox; 1.5b insets Lee Wilcox (blackbird: Dr. Edgar Spalding); 1.6a Dr. Morley Read/Science Photo Library/Photo Researchers, Inc., 1.6b Linda Graham, 1.6c Charles O. Cecil/Visuals Unlimited, 1.6d,g Lee Wilcox, 1.6e Inga Spence/Visuals Unlimited, 1.6f Walter H. Hodge/Peter Arnold, Inc.; 1.7 re-drawn from Harwood, W. S. (2004) A new model for inquiry, Is the scientific method dead? *Journal of College Science Teaching* The National Science Teachers Association; 1.8 (Earth: NASA Headquarters); 1.10 Lee Wilcox; 1.11 Linda Graham; 1.12 Lee Wilcox; 1.13 Marie Read/Animals Animals/Earth Scenes; 1.14 Lee Wilcox; 1.15 Linda Graham; 1.16 Lee Wilcox; 1.17 Mary Bauschelt Florilegia; E1.1 Lee Wilcox **CHAPTER 2** CO2 Lee Wilcox; CO2.1 Lee Wilcox; 2.1 Lee Wilcox;2.2 Reprinted with permission from Smith, B.D. (1997) May 4 276:932–934 The initial domestication of *Cucurbita pepo* in the Americas 10,000 years ago. © 1997 American Association for the Advancement of Science; 2.3 Lee Wilcox; 2.4 re-drawn and used with permission from Diamond, J. (1997) Nov 14 278:1243. Location, location, location: the first farmers. © 1997 American Association for the Advancement of Science; 2.5 Lee Wilcox; 2.6 modified and used with permission from Iltis, H. (1983). Nov. 25 222:887. From teosinte to maize: the catastrophic sexual transmutation. © 1983 American Association for the Advancement of Science; 2.7–2.8 Lee Wilcox; 2.9 re-drawn from illustration by Eric Mose; 2.10 Lee Wilcox; 2.11 Claudia Lipke; E2.1A Frostburg State University Biology Website, E2.1B Linda Lyons; E2.2 Inga Spence/Visuals Unlimited

PART II
CHAPTER 3 CO3 Lee Wilcox; CO3.1 Lee Wilcox; 3.7 Lee Wilcox; 3.10a,b Lee Wilcox, 3.10c Dr. Martha Cook; 3.11a Lee Wilcox, 3.11b Photo Researchers, Inc.; 3.13a Jim Strawser/Grant Heilman Photography, Inc., 3.13b–f Lee Wilcox; 3.15 Lee Wilcox; 3.24–3.27 Lee Wilcox; E3.1A,B NASA **CHAPTER 4** CO4 Lee Wilcox; CO4.1 Hilda Canter-Lund, Freshwater Biological Association, Ambleside, UK; 4.1–4.2 Lee Wilcox; 4.3a Lee Wilcox, 4.3b Dr. Martha Cook; 4.4a–c Linda Graham, 4.4d Dr. Roy Brown; 4.5a Lee Wilcox, 4.5b Sonia Cook; 4.6a Lee Wilcox, 4.6c Linda Graham; 4.8 Linda Graham; 4.12a,b Lee Wilcox; 4.13 Professor Edward Hasselkus; 4.14 Patti Murray/Animals Animals/Earth Scenes; 4.16 Linda Graham; 4.17–4.18 Lee Wilcox; 4.21 Lee Wilcox; 4.22b Dr. Martha Cook; 4.23 Linda Graham; 4.25a,b Dr. James Busse; 4.26a Dr. Martha Cook; 4.27b,c W.P. Wergin, Courtesy Dr. Eldon H. Newcomb; 4.28a Linda Graham, 4.28b Lee Wilcox; 4.29 Lee Wilcox; 4.30 Dr. Eldon H. Newcomb; 4.31 Lee Wilcox; E4.2 Dr. David E. Salt **CHAPTER 5** CO5 Lee Wilcox; CO5.1 D. Nunuk/Photo Researchers, Inc.; 5.18 Lee Wilcox; 5.20 Lee Wilcox; 5.24a Dr. Martha Cook, 5.24b Lee Wilcox; E5.1A TH FOTO-WERBUNG/Phototake NYC, E5.1B Nigel Cattlin/Holt Studios Int'l./Photo Researchers, Inc. **CHAPTER 6** CO-6 Dr. Gary Wedemayer; CO6.1 Lee Wilcox; 6.12 Lee Wilcox; E6.1 Dr. David Spooner **CHAPTER 7** CO7 Inga Spence/Visuals Unlimited; CO7.1 Lee Wilcox; 7.1 Linda Graham; 7.2–7.3 Photo Researchers, Inc.; 7.6 Linda Graham; 7.10 bottom Lee Wilcox; 7.11 Photo Researchers, Inc.; 7.12b from Van Hooser, A., Brinkley, B.R. (1999). Methods for in situ localization of proteins and DNA in the centromere-kinetochore complex. *Methods in Cell Biology* 61:57–80 Academic Press ©1999; 7.14 re-drawn and used by permission from Buchanan, et al. (eds.) © 2000 by the American Society of Plant Biologists **CHAPTER 8** CO8 Photo Researchers, Inc.; CO8.1 Lee Wilcox; 8.3 Lee Wilcox; 8.4a Dr. Martha Cook; 8.6–8.7 Lee Wilcox; 8.10 George Thompson; 8.11–8.15 Lee Wilcox; 8.16b re-drawn and used with permission from Sack, F. (2004) Jun 4 304:1461–1462. Yoda would be proud:Valves for land plants. © 2004 American Association for the Advancement of Science; 8.18–8.21 Lee Wilcox; 8.22a Lee Wilcox; E8.1A–C Lee Wilcox, E8.1D Peter Arnold, Inc., E8.1E,F Lee Wilcox **CHAPTER 9** CO9 Lee Wilcox; CO9.1 Photo Researchers, Inc.; 9.1 Lee Wilcox; 9.2 modified and reprinted with permission from Gensel, P. and Andrews, H. (1987) The evolution of early land plants. *American Scientist* 75:478–789; 9.3 Lee Wilcox; 9.5–9.7 Lee Wilcox; 9.11 Lee Wilcox; 9.16 Lee Wilcox; 9.17 Steve Kaufman/Peter Arnold, Inc.; 9.18 Lee Wilcox; 9.20a Lee Wilcox; 9.21–9.26 Lee Wilcox; 9.27a Tom Till/DRK Photo, 9.27b,c Dr. Gary Wedemayer; 9.28 Lee Wilcox; 9.29b Lee Wilcox; 9.30 C. C. Lockwood/DRK Photo (Inset: Dr. Gary Wedemayer); 9.31 Linda Graham; E9.1A–D Lee Wilcox; E9.2A Lee Wilcox, E9.2B John Gerlach/DRK Photo **CHAPTER 10** CO10 Lee Wilcox; CO10.1 Linda Graham; 10.1 Lee Wilcox; 10.2 Norma Wilcox; 10.3–10.4 Lee Wilcox; 10.5 Linda Graham; 10.6 Walter Hodge/Peter Arnold, Inc.; 10.7a Nancy Sefton/Photo Researchers, Inc.; 10.8–10.12 Lee Wilcox; 10.15–10.16 Lee Wilcox;10.17b Dr. James Busse; 10.18–10.19 Lee Wilcox; 10.24 Visuals Unlimited; E10.1B Buttner Batyrbukd./OKAPIA/Photo Researchers, Inc.; E10.1C http://www.alaskanative.net/339.asp; E10.2 Catherine Ursillo/Photo Researchers, Inc. **CHAPTER 11** All photos by Lee Wilcox **CHAPTER 12** CO12 Jason Lindsey/Perceptive Visions; CO12.1 Tom Bean/DRK Photo; 12.6 Lee Wilcox; 12.7a A. Riedmiller/Peter Arnold, Inc., 12.7b George & Judy Manna/Photo Researchers, Inc.; 12.10 insets Lee Wilcox; 12.11 re-drawn and used with permission from KKlejnot, J. and Lin, C. (2004) Feb 13 303:965–966. A *CONSTANS* experience brought to light. © 2004 American Association for the Advancement of Science; 12.12–12.17 Lee Wilcox; E12.1 Lee Wilcox

PART III
CHAPTER 13 CO13 Lee Wilcox; CO13.1 Lee Wilcox/Dr. Martha Cook; 13.1–13.2 Lee Wilcox; 13.5a,c Linda Graham; 13.6–13.7 Lee Wilcox; 13.8 re-drawn and used with permission from Alexopoulos, C. J., Mims, C. W., and Blackwell, M. (1996) *Introductory Mycology.* John Wiley and Sons, Inc. and the McGraw-Hill Companies; 13.9–13.10 Lee Wilcox; 13.15 Lee Wilcox; 13.18 Lee Wilcox; E13.1A Lee Wilcox, E13.1B Ross M. Horowitz/Getty Images Inc., Image Bank; E13.2A re-drawn and used with permission from Page, S. L. and Hawley, R. S. (2003) Aug 8 301:785–798.Chromosome choreography: The meiotic ballet. © 2003 American Association for the Advancement of Science; E13.2B reprinted with permission from *Nature.* Bradshaw, H. D. et al. Genetic mapping of floral traits associated with reproductive isolation in monkeyflowers *(Mimulus).* Vol. 376:762–765 © 1995 Macmillan Magazines, Ltd. **CHAPTER 14** CO14 D. Cavagnaro/DRK Photo; CO14.1 Nigel Cattlin/Holt Studios Int'l/Photo Researchers, Inc.; 14.1 Mendel/Heredity/Visuals Unlimited; 14.2a Lee Wilcox; 14.11 Lee Wilcox; 14.17c Lee Wilcox; 14.19 Lee Wilcox; E14.1 SPL/London/ Photo Researchers, Inc., E14.2B–D Lee Wilcox **CHAPTER 15** CO15 Lee Wilcox; CO15.1 Andrew Syred/SPL/Photo Researchers, Inc.; 15.1 Lee Wilcox; 15.2 Hans Gelderblom/Getty Images Inc.-Stone Allstock; 15.3 Lee Wilcox; 15.4 Doug Sokell/Visuals Unlimited; 15.5 Lee Wilcox; 15.11 Matt Meadows/Peter Arnold, Inc.; 15.12 Pallava Bagla/Corbis/Sygma; 15.13 Inga Spence/Visuals Unlimited **CHAPTER 16** CO16 Frans Lanting; CO16.1 Minden Pictures; 16.3a Christian Grzimek Okapia/Photo Researchers, Inc., 16.3b Galen Rowell/Corbis/Bettmann, 16.3c Ralph Reinhold/Animals Animals/Earth Scenes, 16.3d Tui De Roy/The Roving Tortoise Nature Photography; 16.4a–d Petren, K, Grant, B.R., Grant, P.R. (1999) A phylogeny of Darwin's finches based on microsatellite DNA length variation. *Proceedings of the Royal Society of London,* Series B 266:321–329; 16.5–16.8 Lee Wilcox; 16.9a M. Klindword/Visuals Unlimited, 16.9b Lee Wilcox; 16.10a Jim Pojar, 16.10b–d Robert Gustafson; 16.11 Lee Wilcox; 16.19b–d Dr. Gerald C. Carr CBC University of Hawaii-Manoa; 16.21 modifed from Ownbey, M. (1950) Natural hybridization and amphiploidy in the genus *Tragopogon. American Journal of Botany* 37:494–495; 16.22 Reprinted with permission from Appenzeller, T. (1999) Jun 25 284:2108–2110 Test tube evolution catches time in a bottle. © 1999 American Association for the Advancement of Science; E16.1 reprinted with permission from *Nature.* Bradshaw, H. D. et al. Genetic mapping of floral traits associated with reproductive isolation in monkeyflowers *(Mimulus).* Vol. 376:762–765 © 1995 Macmillan Magazines, Ltd.

PART IV
CHAPTER 17 CO17 Lee Wilcox; CO17.1 Lee Wilcox; 17.1–17.7 Lee Wilcox; 17.8 Jaime Plaza Van Roon/Auscape International Pty. Ltd.; 17.9–17.10 Lee Wilcox; 17.11 insets Lee Wilcox; 17.12–17.14 Lee Wilcox; E17.1A Dr. Ray F. Evert, E17.1B Nigel Cattlin/Holt Studios Int'l/Photo Researchers, Inc., E17.1C Lee Wilcox; E17.2A,D Linda Graham, E17.2B Missouri Botanical Garden, E17.2C James Graham **CHAPTER 18** CO18 Lee Wilcox; CO18.1 Lee Wilcox; 18.1–18.2 Lee Wilcox; 18.4 Dr. William Schopf; 18.5 Dr. Andrew H. Knoll, Harvard University; 5.7 from Fox, J. L. Proteinoids: Clues to cellular origins? *BioScience* (1985) 35:74–75 © American Institute of Biological Sciences (photo by Robert Syren); 18.9 Lee Wilcox; 18.10a Dr. Ralph Lewin, 18.10b–d Lee Wilcox; 18.11 Dr. Norma Lang (Inset: Lee Wilcox); 18.12 Linda Graham; 18.13–18.14 Lee Wilcox; 18.15 Linda Graham; 18.16 Lee Wilcox; 18.17a Lee Wilcox, 18.17b Linda Graham; 18.19 Re-drawn with permission from Pennisi, E. (2004) Jul 16 305:334–335 Researchers trade insights about gene swapping. © 2004 American Association for the Advancement of Science;18.20 Lee Wilcox; E18.1A,B NASA/John F. Kennedy Space Center; E18.1C,D Kathie L. Thomas-Keprta, Lunar and Planetary Institute **CHAPTER 19** CO19 Lee Wilcox; CO19.1 A. Winter & P. Friedinger; 19.1a Lee Wilcox, 19.1b James Graham; 19.2a Lee Wilcox, 19.2b Dr. Kalle Oli; 19.3a,b Lee Wilcox, 19.3c,d Linda Graham; 19.4a Lee Wilcox (Inset: Linda Graham), 19.4b Claudia Lipke; 19.5 Dr. Eldon H. Newcomb; 19.6a Lee Wilcox, 19.6b Linda Graham, 19.6c Dr. Gary. Wedemayer and Lee Wilcox; 19.7 Linda Graham; 19.8 Re-drawn with permission from Baldauf, S. L. (2003) Jun 13 300:1703–1706 The deep roots of eukaryotes. © 2003 American Association for the Advancement of Science;19.9 photo by A. Kleijne, in Winter, A. and Siesser, W. G. (eds.) (1999) *Coccolithophores.* Cambridge University Press, Cambridge. UK.; 19.10a Andrew Syred/Photo Researchers Inc., 19.10b (left) Lee Wilcox (right) Linda Graham, 19.10c Linda Graham; 19.11a Dr. Ken-Ichiro Ishida/Mr. Shuhei Ota, 19.11b,c James Graham; 19.12 Lee Wilcox; 19.16 Lee Wilcox; 19.17a Lee Wilcox, 19.17b Lee Wilcox and Dr. Gary Wedemayer; 19.18 Dr. Evan Lau; 19.19 Lee Wilcox; 19.20 insets Lee Wilcox (vegetative filament: Dr. Martha Cook); 19.21–19.22 Lee Wilcox; 19.23 insets Lee Wilcox (carposporophyte: Linda Graham); 19.24a Dr. Mariette Cole, 19.24b Lee Wilcox; 19.25 Richard L. (Larry) Blanton, Ph.D.; E19.1A Lee Wilcox, E19.1B BIOS (F. Denhz)/Peter Arnold, Inc., E19.1C Springer/Verlag GmbH & Co KG, E19.1D Jürgen Berger and Dr. Peter Overath, Planck Institue

Index

A

ABC model of flower development, 424–25, 502
Abies balsamea (balsam fir), 517, 518
Abies lasiocarpa, 519
Abiotic signals, 212, **212,** 465
Abscisic acid (ABA), 142, 199, 213, **214,** 217, 223
Abscission, 202–3, 217
Abscission zone, 202–3, **203**
Absorption spectrum, 87, **88**
Abyssal plain, 558
Acacia, 204, 503
Accessory fruit, 433
Accessory pigments, 87, 349
Acer rubrum (red maple), 522
Acer platanoides, **194**
Acer saccharinum (silver maple), **457, 524**
Acer saccharum (sugar maple), 77, 151, 155, 157, **162,** 168, 521
Acetic acid, 36
Acetobacter, 325
Acetyl coenzyme A (acetyl CoA)
    formation in mitochondrial matrix, 97
    formation of, 99, **101**
Acetyl group, 98
    in Krebs cycle, 99–100
Achene, 434, **435**
Achillea (yarrow), 419
Acid bog, **553**
Acid rain, 577
    aluminum in soil and, 184
    effect on eastern temperate deciduous forest, 528–29, **529**
    effect on freshwaters, 549–50
Acids, 36–37
Actin filaments, 125
Actinidia deliciosa (kiwifruit), 446
Actinobacteria, 329
Actinomorphic, 451
Activation energy, 47, **49**
Active site, 48
Active transport, 184
Adansonia (baobab), **455,** 503
Adaptation, sexual reproduction and, 228–29
Adaptive radiation, 292–93, 295
Adenine, 50, **50,** 51, 106, **106,** 108, 110
    cytokinins derived from, 215
Adenosine diphosphate (ADP), 51, **51,** 52, 83, **83**
Adenosine monophosphate (AMP), 83, **83**
Adenosine triphosphate (ATP), **51,** 51–52, 83, **83**
Adenostoma fasciculatum (chamise), **507,** 507–8
Adiantum (maidenhair fern), **394**

ADP (adenosine diphosphate), 51, **51,** 52, 83, **83**
Adventitious roots, 176–77, **177,** 500
Aerenchyma, 175, 223, 549, 553
Aerial roots, 173, **173**
Aerobic respiration, 98–102
    fungi and, 360–61
Aesculus (buckeye), **139**
Aflatoxins, 371
Agar, 349, 354
Agaricus (portabella mushroom), 369, **371**
Agarose, 349
Agassiz, Louis, 280
Agathis, 530, **530**
Agave, 96, 203, **204,** 208, 493, 495
Age distribution, 462–63, **463**
Agent Orange, 213
Aggregate fruits, 435
Agriculture
    domestication of plants, 267–69
    environmental degradation from, 575
    genetically engineered crops, 273–74
    on grasslands, 496–97
    importance of plant-animal coevolutionary interactions in, 442–43
    mesic grasslands and, 504–6
    monoculture, 536–37
    origins of, 20–22
        reasons for, 22
        site of first domestication, 22–23
    polyculture, 536, 537
    productivity of, 580
    prokaryotes and, 330–31
    slash-and-burn, 536, **536**
    spread of, **22**
    sustainable, 574, 581–82
    traditional methods, **27,** 27–29, 536, **536**
    tropical forest, 535–36
Agrobacterium, 325
Agrobacterium tumefaciens, 122, 273
Air cells, 475, **475**
Air quality, lichen as monitors of, 373
Akinetes, 326, **327**
Alanine, **48**
Alaskan spruce forests, 11
Alaskan violets, 286, **287**
Alcohol fermentation, 98, 99
Alder (*Alnus sinuata*), 469, **469**
Aleurone, 502, **502**
Alfalfa (*Medicago sativa*), **24, 158,** 187
Algae, 5, **7,** 335, 336. *See also* Green algae; Red algae
    biotechnical applications, 354–55
    cells, **59**
    coralline, 566
    cryptoendolithic, 489–90
    desert, 489–90
    distinguishing characteristics, **379**

diversity of reflecting evolutionary events, 345–48
evolutionary relationships of, **342**
features distinguishing groups of, **342**
feeding methods, 348–49
flagellate, **340**
as food, 11
freshwater, 548
gametic life cycles, 237–38
lichens and, 13
ocean, 559–60
parasitic, 340
phosphorus availability and growth of, 550–51
photosynthetic, 87–89
plant association with, 5
in polar desert, 485
reproduction in, 240, 350–54
sporic life cycle, 240
structural adaptations, 349–50
    cell coverings, 349
    food storage, 350
toxins produced by, 343
Algal blooms, **314,** 315, 339, 343, 545, 550, 562, 577
Alginates, 354, 565
Alkaloids, 52, 54–55, 121, 466, 539
Alleles, 238, 249, 250
    change in frequency in gene pool, 287–91
    independent assortment of, 256–57, **257**
    segregation of, **251,** 251–52
Allelopathy, 467, 492
Allergens, 227
    in GM crops, 275
    pollen grains as, 422
Alliaria petiolata (garlic mustard), 526, **526**
Allium cepa (onion), 138, 156, **179,** 204, **204, 433,** 436, **437**
Allopatric speciation, 292, 296
    geographic isolation and, **292,** 292–93
Allopolyploidy, 293, **294**
Alnus sinuata (alder), 469, **469**
Aloe, 19, **19,** 203, 495
Aloe barbadensis, 29
Aloe vera, 495
Alpha helix, 45, **48**
Alpine ecosystems
    human effects on, 519–20
    retreat of glaciation and, 514
Alpine pennycress (*Thlaspi caerulescens*), 78, **78**
Alternaria, **369**
Alternate leaf arrangement, 192, **194**
Alternation of generations, 240–42, 378, **430**
Althea (hollyhock), 419
Aluminum
    effect on roots, 184
    in tropical soil, 535